The Encyclopedia of
the Solid Earth Sciences

This work is dedicated to Richard S. Thorpe,
Subject Editor for Geochemistry,
who died on 22 August 1991

The Encyclopedia of
the Solid Earth Sciences

EDITOR-IN-CHIEF
PHILIP KEAREY
Department of Geology
University of Bristol

SUBJECT EDITORS
P.A.ALLEN, A.M.EVANS, A.S.GOUDIE,
A.HALLAM, P.KEAREY, R.G.PARK,
R.S.THORPE AND D.J.VAUGHAN

OXFORD
BLACKWELL SCIENTIFIC PUBLICATIONS
LONDON EDINBURGH BOSTON
MELBOURNE PARIS BERLIN VIENNA

© 1993 by
Blackwell Scientific Publications
Editorial Offices:
Osney Mead, Oxford OX2 0EL
25 John Street, London WC1N 2BL
23 Ainslie Place, Edinburgh EH3 6AJ
238 Main Street, Cambridge
   Massachusetts 02142, USA
54 University Street, Carlton
   Victoria 3053, Australia

Other Editorial Offices:
Librairie Arnette SA
2, rue Casimir-Delavigne
75006 Paris
France

Blackwell Wissenschafts-Verlag
Meinekestrasse 4
D-1000 Berlin 15
Germany

Blackwell MZV
Feldgasse 13
A-1238 Wien
Austria

First published 1993

Set by Excel Typesetters Company, Hong Kong
Printed in Great Britain at The Alden Press, Oxford and
bound by Hartnolls Ltd, Bodmin, Cornwall

DISTRIBUTORS

Marston Book Services Ltd
PO Box 87
Oxford OX2 0DT
(*Orders*: Tel: 0865 791155
       Fax: 0865 791927
       Telex: 837515)

USA
Blackwell Scientific Publications, Inc.
238 Main Street
Cambridge, MA 02142
(*Orders*: Tel: 800 759-6102
          617 876-7000)

Canada
Oxford University Press
70 Wynford Drive
Don Mills
Ontario M3C 1J9
(*Orders*: Tel: 416 441-2941)

Australia
Blackwell Scientific Publications Pty Ltd
54 University Street
Carlton, Victoria 3053
(*Orders*: Tel: 03 347-5552)

A catalogue record for this title
is available from the British Library

ISBN 0-632-03699-0 (pbk)

Library of Congress
Cataloging-in-Publication Data

The Encyclopedia of the Solid Earth Sciences/
Editor-in-chief, Philip Kearey:
   Subject editors. P.A. Allen . . . [*et al.*].
        p.   cm.
   Includes bibliographical references
   and index.
   ISBN 0-632-03699-0 (pbk)
   1. Earth sciences—Dictionaries.
   I. Kearey, P.
QE5.E517   1993
550'.3—dc20

Contents

List of Contributors

EDITOR-IN-CHIEF

P.KEAREY

SUBJECT EDITORS

P.A.ALLEN [Sedimentology]

A.M.EVANS [Economic Geology]

A.S.GOUDIE [Geomorphology and Environmental Geology]

A.HALLAM [Paleontology and Stratigraphy]

P.KEAREY [Geophysics and Miscellaneous Topics]

R.G.PARK [Structural Geology and Geodynamics]

R.S.THORPE [Geochemistry, Volcanology, Magmatism, Igneous and Metamorphic Rocks]

D.J.VAUGHAN [Mineralogy]

CONTRIBUTORS AND ADDRESSES

JA   J.ALEXANDER *Department of Geology, University of Wales College of Cardiff, Cardiff CF1 3YE*

PAA   P.A.ALLEN *Department of Earth Sciences, University of Oxford, Oxford OX1 3PR*

RJA   R.J.ALDRIDGE *Department of Geology, University of Leicester, Leicester LE1 7RH*

DEGB D.E.G.BRIGGS *Department of Geology, University of Bristol, Bristol BS8 1RJ*

DJB   D.J.BARBER *Department of Physics, University of Essex, Colchester CO2 3SQ*

DQB   D.Q.BOWEN *Institute of Earth Studies, University of Wales College of Aberystwyth, Aberystwyth SY23 3DB*

JFWB J.F.W.BOWLES *Mineral Science Ltd, 109 Asheridge Road, Chesham HP5 2PZ*

JLB   J.L.BEST *Department of Earth Sciences, University of Leeds, Leeds LS2 9JT*

MDB   M.D.BRASIER *Department of Earth Sciences, University of Oxford, Oxford OX1 3PR*

PJB   P.J.BRENCHLEY *Department of Earth Sciences, University of Liverpool, Liverpool L69 3BX*

SB   S.BOWLER *Department of Earth Sciences, University of Leeds, Leeds LS2 9JT*

SDB   S.D.BURLEY *Department of Geology, University of Manchester, Manchester M13 9PL*

TGB   T.G.BLENKINSOP *Department of Geology, University of Zimbabwe, P.O. Box 167, Harare, Zimbabwe*

TPB   T.P.BURT *School of Geography, University of Oxford, Oxford OX1 3TB*

ARIC A.R.I.CRUIKSHANK *Leicestershire Museums, Arts and Records Service, 96 New Walk, Leicester LE1 6TD*

GRC   G.R.COOPE *School of Earth Sciences, University of Birmingham, Birmingham B15 2TT*

JCC   J.C.CRIPPS *Department of Geology, University of Sheffield, Sheffield S3 7HF*

JEC   J.E.CHISHOLM *Department of Mineralogy, British Museum (Natural History), London SW7 5BD*

RAC   R.A.CLARK *Department of Earth Sciences, University of Leeds, Leeds LS2 9JT*

DLD   D.L.DINELEY *Department of Geology, University of Bristol, Bristol BS8 1RJ*

AME   A.M.EVANS *Department of Geology, University of Leicester, Leicester LE1 7RH*

IJF   I.J.FAIRCHILD *School of Earth Sciences, University of Birmingham, Birmingham BL5 2TT*

ASG   A.S.GOUDIE *School of Geography, University of Oxford, Oxford OX1 3TB*

AH   A.HALLAM *School of Earth Sciences, University of Birmingham, Birmingham B15 2TT*

GMH   G.M.HARWOOD *School of Environmental Sciences, University of East Anglia, Norwich NR4 7TJ*

JDH   J.D.HANSOM *Department of Geography and Topographic Science, University of Glasgow, Glasgow G12 8QQ*

JMH   J.M.HANCOCK *Department of Geology, Imperial College, London SW7 2BP*

RAH   R.A.HOWIE *Department of Geology, Royal Holloway and Bedford New College, Egham TW20 0EX*

ALAJ A.L.A.JOHNSON *Department of Earth Sciences, University of Derby, Derby DE3 1GB*

DGJ   D.G.JENKINS *Jenkins Science Foundation, Pitton, Rhossili, Swansea SA3 1PH*

GALJ G.A.L.JOHNSON *Department of Geological Sciences, University of Durham, Durham DH1 3LE*

AESK **A.E.S.KEMP** *Department of Oceanography, University of Southampton, Southampton SO9 5NH*

PK **P.KEAREY** *Department of Geology, University of Bristol, Bristol BS8 1RJ*

RK **R.KAY** *Environmental Risk Assessment Unit, School of Environmental Sciences, University of East Anglia, Norwich NR4 7TJ*

ADL **A.D.LAW** *BP, Brittanic House, Moor Lane, London EC2Y 9BU*

PTL **P.T.LEAT** *British Antarctic Survey, High Cross, Madingley Road, Cambridge CB3 0ET*

JDM **J.D.MARSHALL** *Department of Earth Sciences, University of Liverpool, Liverpool L69 3BX*

NJM **N.J.MIDDLETON** *School of Geography, University of Oxford, Oxford OX1 3TB*

RMW **R.MUIR-WOOD** *Newbridge House, Clapton on the Hill, Gloucester GL54 2LG*

DMP **D.M.PRESCOTT** *Shell International Petroleum Company, The Hague, The Netherlands*

IP **I.PARSONS** *Department of Geology and Geophysics, University of Edinburgh, Edinburgh EH9 3JW*

JEP **J.E.POLLARD** *Department of Geology, University of Manchester, Manchester M13 9PL*

KTP **K.T.PICKERING** *Department of Geology, University of Leicester, Leicester LE1 7RH*

RGP **R.G.PARK** *Department of Geology, University of Keele, Keele ST5 5BG*

DAR **D.A.ROTHERY** *Department of Earth Sciences, Open University, Milton Keynes MK7 6AA*

DAR **D.A.ROGERS** *Lomond Associates, De Quincey House, 48 West Regent Street, Glasgow G2 2RA*

DER **D.E.ROBERTS** *Department of Geology, North Staffordshire Polytechnic, Stoke-on-Trent ST4 2DE*

RR **R.RIDING** *Department of Geology, University of Wales College of Cardiff, Cardiff CF1 3YE*

RBR **R.B.RICKARDS** *Department of Earth Sciences, University of Cambridge, Cambridge CB2 3EQ*

GS **G.STUART** *Department of Earth Sciences, University of Leeds, Leeds LS2 9JT*

MJS **M.J.SIMMS** *Department of Geology, University of Bristol, Bristol BS8 1RJ*

NJS **N.J.SNELLING** *British Geological Survey, Keyworth, Nottingham NG12 5GG*

RAS **R.A.SPICER** *Department of Earth Sciences, University of Oxford, Oxford OX1 3PR*

RDS **R.D.SARRE** *North Devon College, Barnstaple, Devon EX16 7RX*

ATB **A.THOMAS-BETTS** *Department of Geology, Imperial College, London SW7 2BP*

DHT **D.H.TARLING** *Department of Geology, Polytechnic Southwest, Plymouth PL4 8AA*

RST **R.S.THORPE** (Deceased) *Department of Earth Sciences, Open University, Milton Keynes MK7 6AA*

DSU **D.S.URCH** *Department of Chemistry, Queen Mary and Westfield College, London E1 4NS*

DJV **D.J.VAUGHAN** *Department of Geology, University of Manchester, Manchester M13 9PL*

HAV **H.A.VILES** *St. Catherines College, University of Oxford, Oxford OX1 3UJ*

AW **A.WATSON** *Environmental Geosciences, 23 Prospect Street, Holliston, MA 01746, USA*

AEW **A.E.WRIGHT** *School of Earth Sciences, University of Birmingham, Birmingham B15 2TT*

CW **C.WOODWARD** *Geological Museum, Exhibition Road, London SW7 2DE*

OWT **O.WILLIAMS-THORPE** *Department of Earth Sciences, Open University, Milton Keynes MK7 6AA*

VW **V.WINCHESTER** *23 Warnborough Road, Oxford OX2 6JA*

VPW **V.P.WRIGHT** *Postgraduate Research Institute for Sedimentology, University of Reading, Reading RG6 2AB*

Preface

The preparation of such a comprehensive work has involved a large number of contributors. To them we offer our thanks. The idea of this encyclopedia was first mooted in 1985 by Peter J. Smith, who did much of the early administration, organization and commissioning. I took over from Peter in 1988, and offer him my thanks for all his initial efforts. The encyclopedia has thus been some seven years in preparation, and I hope that the result of this lengthy period of gestation will be useful to all those who have occasion to refer to it.

PK

Introduction

The aim of *The Encyclopedia of the Solid Earth Sciences* is to provide a comprehensive modern reference text for all the subdisciplines of the Earth Sciences, supplying definitions of all appropriate terms and longer items which provide a review of the present state of knowledge. Well-established terms and those of modern usage are all included.

The Encyclopedia is primarily intended for professional Earth scientists and those specializing in related subjects. It should also provide an important reference for students of the Earth Sciences and those needing information on terms in current usage.

The Encyclopedia contains over 2700 entries of varying length. These are arranged in alphabetical order according to the exact wording of the entry title (hereafter called the keyword). Both the keywords and their derivatives (e.g. gravity anomaly and anomaly, gravity) appear in the index. Where keywords occur within any entry, other than their own, they appear in bold type. Synonyms of keywords appear in brackets after them. Cross-reference is made to other particularly relevant keywords at the end of the entry. Additionally, there are more than 2750 other indexed entries including synonyms. These are shown in the text in italic type. These indexed items and their derivatives also appear in the index. In the index the principal page reference to any particular item is marked in bold type. This principal reference is the actual entry of a keyword, and for other indexed items usually refers to the page(s) on which the item is defined. The whole index contains over 8000 entries.

The longer entries are followed by a reading list, which gives the principal references to be followed if more information is sought on the subject(s) covered by the keyword.

PK

SI, c.g.s. and Imperial (Customary US) Units and Conversion Factors

Quantity	SI name	SI symbol	c.g.s. equivalent	Imperial (US) equivalent
Mass	kilogram	kg	10^3 g	2.205 lb
Time	second	s	s	s
Length	metre	m	10^2 cm	39.37 in
				3.281 ft
Acceleration	metre s^{-2}	m s^{-2}	10^2 cm s^{-2} = 10^2 gal	39.37 in s^{-2}
Gravity	gravity unit	g.u. = μm s^{-2}	10^{-1} milligal (mgal)	3.937×10^{-5} in s^{-2}
Density	megagram m^{-3}	Mg m^{-3}	g cm^{-3}	3.613×10^{-2} lb in^{-3}
				62.421 lb ft^{-3}
Force	newton	N	10^5 dyne	0.2248 lb (force)
Pressure	pascal	Pa = N m^{-2}	10 dyne cm^{-2} = 10^{-5} bar	1.45×10^{-4} lb in^{-2}
Viscosity	pascal sec	Pa s	10 poise	
Energy	joule	J	10^7 erg	0.7375 ft lb
Power	watt	W = J s^{-1}	10^7 erg s^{-1}	0.7375 ft lb s^{-1}
				1.341×10^{-3} hp
Heat flow	watt m^{-2}	W m^{-2}	23.9 μcal cm^{-2} s^{-1}	
Thermal conductivity	watt m^{-1}°C	W m^{-1}°C	0.00239 cal cm^{-1} s^{-1}°C^{-1}	
Heat production	watt m^{-3}	W m^{-3}	2.39×10^{-7} cal cm^{-3} s^{-1}	
Temperature	T	°C*	°C	(1.8T + 32)°F
Current	ampere	A	A	A
Potential	volt	V	V	V
Electric field	volt m^{-1}	V m^{-1}	10^6 e.m.u.	
Electric charge	coulomb	C = A s	10^{-1} e.m.u.	
Capacitance	farad	F = C V^{-1}	10^{-9} e.m.u.	
Resistance	ohm	Ω = V A^{-1}	Ω	Ω
Resistivity	ohm m	Ω m	10^2 Ω cm	3.281 ohm ft
Conductance	siemen	S = Ω^{-1}	mho	mho
Conductivity	siemen m^{-1}	S m^{-1}	10^{-2} mho cm^{-1}	0.3048 mho ft^{-1}
Dielectric constant	dimensionless			
Magnetic flux	weber	Wb = V s	10^8 maxwell	
Magnetic flux density (B)	tesla	T = Wb m^{-2}	10^4 gauss (G)	
Magnetic anomaly	nanotesla	nT = 10^{-9} T	gamma (γ) = 10^{-5} G	
Magnetizing field (H)	ampere m^{-1}	A m^{-1}	$4\pi 10^{-3}$ oersted (Oe)	
Inductance	henry	H = Wb A^{-1}	10^9 e.m.u.	
Permeability of vacuum (μ_0)	henry m^{-1}	$4\pi 10^{-7}$ H m^{-1}	1	
Susceptibility	dimensionless	k	4π e.m.u.	
Magnetic pole strength	ampere m	A m	10 e.m.u.	
Magnetic moment	ampere m^2	A m^2	10^3 e.m.u.	
Magnetization (J)	ampere m^{-1}	A m^{-1}	10^{-3} e.m.u. cm^{-3}	

*Strictly, SI temperatures should be stated in degrees Kelvin (K = 273.15+°C). In this work, however, temperatures are given on the more familiar Centigrade (Celsius) scale.
Table after Markowitz, W. (1973) SI, the international system of units. *Geophysical Survey* **1**: 217–41.
c.g.s., centimetre, gram, second; e.m.u., electromagnetic unit.

The Encyclopedia of
the Solid Earth Sciences

A

abnormal pressure A formation pressure that exceeds the hydrostatic pressure exerted by a column of water containing 80 000 p.p.m. total solids (1114 kg m^{-3}) is termed abnormal. Pressures greater than hydrostatic are important in the processes of generation, **migration**, accumulation and preservation of fluid *hydrocarbons*. They are often referred to as **overpressures** or *geopressures*. [AME]

North, F.K. (1985) *Petroleum Geology*. Allen & Unwin, Boston.

abrasion The process by which rocks are mechanically worn down by the frictional effect of debris-charged wind, water or ice. The transporting medium must move the abrading agent, e.g. **sand**, pebble or boulder across the rock surface in order to wear it down by scratching, grinding and polishing. The rate of abrasion is thus principally related to the nature and velocity of the transporting medium and the relative hardness of the **abrasives** and abraded surfaces. Examples include the lowering of intertidal rock surfaces by abrading sediments under waves and the wearing down of desert surfaces by the impact of aeolian **sand**. [JDH]

Boulton, G.S. (1974) Processes and patterns of glacial erosion. In: Coates, D.R. (ed.) *Glacial Geology*, pp. 41–87. State University of New York, Binghampton.
Robinson, L.A. (1977) Marine erosive processes at the cliff foot. *Marine Geology* **23**: 257–71.

abrasive Any natural or artificial substance suitable for cutting, grinding, polishing or scouring. Upwards of 25 minerals and rocks of diverse composition are used as natural abrasives. They include **diamond, corundum,** *emery,* **garnet, quartz sand, chalk, bauxite, diatomite** and **pumice.** Some natural abrasives are being supplanted by artificial products, e.g. natural **diamonds** by synthetic **diamonds, corundum** by *carborundum*.

Natural abrasives are divided into three groups:
1 High grade, e.g. **diamond, corundum,** *emery,* **garnet**.
2 Siliceous abrasives, i.e. various forms of **silica**, such as **sandstone, quartzite, flint, chert, sand, pumice, diatomite** and others.
3 Miscellaneous, including buffing and polishing powders, e.g. **bauxite, magnesite, chalk, kaolinite, talc**.

Natural abrasives may be used:
1 In their natural form (sometimes after grinding and grading), e.g. **sand, pumice**.
2 After shaping, e.g. grindstones.
3 After being ground into grains or powders and made up into wheels or papers, e.g. **diamond** cutting wheels, drill bits, *emery* and sandpaper. [AME]

Jenson, M.L. & Bateman, A.M. (1979) *Economic Mineral Deposits*. John Wiley & Sons, New York.

absolute age The age of a rock or formation with respect to the present. The absolute age is determined by using one of the *radiometric dating* methods or paleontological techniques. This compares with methods to determine relative age using stratigraphic relationships, which can only determine the order in which geological events took place. [PK]

absolute gravimeter A **gravimeter** capable of measuring the absolute value of **gravity**. Such instruments are based on the accurate timing of the period of oscillation of a pendulum or the time taken for a body to fall a known distance. Numerous repeated observations are necessary to achieve an accuracy of about 10 **gravity units** (g.u.). Classically, measurements were restricted to the laboratory, but recently a portable instrument based on the falling body technique has been devised. [PK]

absolute permeability The ability of a rock to allow a particular fluid, e.g. oil, at 100% saturation to pass through it. If other fluids (gas and/or water) are present the absolute permeability is of little practical significance as the fluids create complex mutual interference and there is an **effective permeability** for each fluid in the presence of the others. [AME]

North, F.K. (1985) *Petroleum Geology*. Allen & Unwin, Boston.

absolute plate motion The motions of the tectonic *plates* can only be determined relative to each other during geological time (see **relative plate motion**). **Paleomagnetic** and *paleoclimatic* evidence provides some limits on the absolute movement of a *plate* between different latitudes and *plate* orientation within such latitudinal constraints, but do not define absolute changes in longitude. Geodetic measurements have now become sufficiently precise to determine present-day motions relative to astronomical coordinates (see **very long baseline interferometry**), but cannot be applied in the past. For such earlier times, it is generally assumed that **hotspots** remain approximately fixed relative to the **mantle** and motions relative to this framework are often termed absolute motions. However,

hotspot localities are known to change slowly relative to each other so that movements within this are not strictly absolute. Absolute motions are commonly estimated assuming minimum total motion of all *plates*. At the moment, this leads to a net motion towards the northern Pacific, but this is mainly a reflection of the distribution of **subduction zones** rather than being true absolute motion relative to the Earth's rotational coordinates. [DHT]

Chase, C.G. (1978) Plate kinematics: the Americas, East Africa, and the rest of the world. *Earth and Planetary Science Letters* **37**: 355–68.

abyss A very deep chasm or ocean trough, particularly used for ocean floors 3000 m below sea-level. The term is also used for a terrestrial ravine, deep gorge or chasm. [NJM]

acanthite (Ag_2S) An **ore** mineral of **silver**. [DJV]

accelerated erosion The increased rate of **erosion** resulting from human activity. Generally, accelerated erosion is caused by deliberate modification of the natural vegetation cover for the purposes of agriculture. However, increased *soil erosion* may be caused by changes in natural **runoff** processes as a result of urbanization or artificial drainage of wetlands. Also, the changes induced by human activity may not be deliberate; destruction of the vegetation cover may be caused by fires, or by footpath **erosion**, or by off-road vehicles — such as in some areas of coastal and desert **dunes** in North America.

The most dramatic forms of accelerated erosion are often caused by the removal or modification of natural vegetation. Rapid **erosion** may occur following forest clearance because the bare *soil* surface is exposed to **runoff** or the action of wind. Today, the problems that result from deforestation are seen most clearly in areas of rain forest which have been recently cleared. However, similar accelerated erosion probably occurred in Europe when woodland was cleared some 1000 to 2000 years ago. The same effects have been felt more recently in New Zealand and North America. In these areas, processes of **erosion** may still be responding to the changes in vegetation cover resulting from agricultural expansion over the past one or two centuries.

Accelerated erosion often continues long after the initial response to forest or bush clearance. Even in areas which have been in agricultural use for many years, the amount of *soil* lost may be greater than the amount produced by natural processes. Often this imbalance will result in almost imperceptible changes, the net effects of which may not be appreciated until the problem is beyond remedy. In many parts of Africa, the long-term effects of **erosion**, which is gradually accelerating in response to bush clearance and increased over-grazing, have only recently been recognized.

On cultivated hillslopes, *soil creep* and *slope-wash* are greater than under forest cover. It has been estimated that in the contiguous United States about 4 000 000 000 tonnes of *soil* are washed into streams and rivers each year; 75% of this is from agricultural land. Another 1 000 000 000 tonnes are lost annually through **erosion** by wind. Today, in parts of the mid-western United States, dust storms generated by aeolian **erosion** of *soil* are becoming more frequent owing to modern farming practices such as the removal of hedgerows to increase the size of individual fields, and the ploughing of fields in spring rather than during the wetter winter months (Chagnon, 1983).

The problems caused by accelerated erosion often extend far beyond the loss of agricultural productivity — though this in itself can have an enormous impact as was the case during the 1930s in the American Dust Bowl. Of equal significance, is the effect that rapid *soil erosion* has on sedimentation rates in lakes, rivers, and estuaries. The increased sediment budgets of streams draining agricultural land have a detrimental effect on water quality and can cause excessive sedimentation in reservoirs. Remedial procedures must tackle the source of the problem; this often involves the implementation of **soil conservation** measures and the modification of agricultural practices. [AW]

Carter, L.J. (1977) Soil erosion: the problem persists despite the billions spent on it. *Science* **196**: 409–11.
Chagnon, S.A. (1983) Record dust storms in Illinois: causes and implications. *Journal of Soil and Water Conservation* **38**: 58–63.
Chepil, W.S. & Woodruff, N.P. (1963) The physics of wind erosion and its control. *Advances in Agronomy* **15**: 211–302.
Phillips, J.D. (1986) The utility of the sediment budget concept in sediment pollution control. *The Professional Geographer* **38**: 246–52.
Pimentel, D., Terhune, E.C., Dyson-Hudson, R. *et al.* (1976) Land degradation: effects on food and energy. *Science* **194**: 149–55.

accelerator radiocarbon dating A method of **radiocarbon dating** in which a **mass spectrometer** is used for the detection of the carbon-14 atoms (as opposed to beta counters used in conventional **radiocarbon dating**). The method is also referred to as *Accelerator Mass Spectrometry (AMS) dating*. The senstivity of the **mass spectrometer** allows dating of small amounts of archeological organic material such as bone, ivory and cloth, which contain insufficient carbon-14 for conventional **radiocarbon dating**. Thus, for example, dates can be obtained from 0.2 to 5 g samples of bone, with errors at least as good as those from conventional radiocarbon (typically ±700 years on a date of *c.* 26 000 BP) which requires 100 times the sample mass for these materials (100–500 g for bone). Accelerator radiocarbon dating is therefore particularly useful for sites with few or poorly preserved remains, where well-stratified charcoal may not be available. [OWT]

accessory mineral A mineral present in small quantities (generally less than *c.* 10%) within a rock, which is not significant in terms of use of the mineral composition for purposes of nomenclature or classification. (See **essential mineral**.) [RST]

accretion The process by which inorganic objects increase in size through the attachment of additional material to their surface, as with the growth of hailstones. [ASG]

accretion of continents Continents are characterized by areas of some $1000\,km^2$ of rocks of broadly similar age (*c.* $\pm200-300\,Ma$), commonly comprising either an ancient *orogenic belt*, or a continental block that originated elsewhere and subsequently became welded onto the continent. In the **Archean** (older than $2.5\,Ga$), the individual units involved were small, usually with dimensions of *c.* $1000\,km$ by $50-100\,km$. In the **Proterozoic** and later (post $2.5\,Ga$), the units involved have been mostly on much larger scales, comprising both *orogenic belts* and pre-existing continental blocks. This situation is most clearly defined in North America where the oldest parts of the Canadian Shield are bordered by younger orogenic belts. Generally, these younger structures are not cross-cutting, which suggests that accretion has largely been by the addition of new **continental crust**, with little splitting of the older crustal block. However, **continental splitting** appears to have become increasingly common during the last Ga. [DHT]

Condie, K.C. (1982) *Plate Tectonics and Crustal Evolution*. Pergamon, Oxford.
Taylor, S.R. & McClennan, S.M. (1985) *The Continental Crust: its Composition and Evolution*. Blackwell Scientific Publications, Oxford.

accretion vein A type of mineral **vein** in which there has been a repetition of opening by **fracturing** and infilling of the newly created open spaces with more minerals. [AME]

accretionary prism A wedge of sediments. Normally referring to sediments accumulating, or having accumulated, immediately adjacent to an *ocean trench* which are also known as *subduction complexes*. They form as the sedimentary cover of the downgoing **lithosphere** is progressively scraped off by the leading edge of the overriding *plate*, forming a sequence of thrust-bound packets of sediments, which generally young towards the *island arc*. This sedimentary prism is generally transparent to *seismic waves* and its low density largely accounts for the strong negative **gravity anomaly** close to and including the *oceanic trenches*. [DHT]

acid clay A clay that releases hydrogen ions in a water suspension, e.g. certain varieties of **Fuller's earth**. [AME]

acmite ($NaFe^{3+}Si_2O_6$) A sodic-pyroxene mineral, also known as *aegirine*. (See **pyroxene minerals**.) [DJV]

acoustic basement In *hydrocarbon exploration*, the deepest structure of interest. This is likely to be the boundary between sedimentary rocks and the igneous and/or metamorphic rocks beneath. When imaged using **seismic reflection** methods, the latter rocks are termed 'acoustic basement'. The usual seismic signature of crystalline basement rocks is a 'reflection-free' zone containing only incoherent weak reflections and **diffractions** with little lateral continuity. [RAC]

acoustic impedance The product of seismic velocity and bulk density. An analogy can be made with the electrical impedance of a material when it conducts electrical energy — a material which propagates acoustic energy has an acoustic impedance such that for a given acoustic pressure a large particle velocity is produced in a material of low acoustic impedance and vice versa. **Reflection** and **transmission coefficients** of an interface or boundary between two media are dependent on the contrast between their acoustic impedances. [GS]

Anstey, N.A. (1977) *Seismic Interpretation: The Physical Aspects*, pp. 43–7. IHRDC Publications, Boston.

acoustic log A general term describing geophysical measurements taken down a borehole which measure various properties of acoustic wave propagation. The **sonic log** measures the travel time of the compressional wave over a fixed distance and thus the velocity of the formation. The *cement bond log* is used to test the quality of the cement bond between the casing of a borehole and the surrounding rock by measuring the **attenuation** of the amplitude of the signal. Long spacing sonic logs and variable density logs investigate the whole wave train from which both compressional and shear wave velocities can be derived as well as frequency, amplitude and **attenuation** information. (See **geophysical borehole logging**.) [GS]

Serra, O. (1984) *Fundamentals of Well Log Interpretation* Vols 1 and 2. Developments in Petroleum Science 15A and 15B. Elsevier, Amsterdam.

acritarchs Microscopic, hollow, organic walled vesicles ($5-500\,\mu m$ in diameter), useful for the biostratigraphy of late *Precambrian* to **Paleozoic** argillaceous rocks of marine origin. The term 'acritarch' is informal since it applies to resting cysts or spore capsules of uncertain origin but which are likely to have been produced by several different groups of protists. It was erected to contain those problematical 'hystrichospheres' that could not be transferred to the **dinoflagellates**. Acritarchs are differentiated on the nature of the pylome (the escape hole), shape and symmetry of the vesicle, presence of spine-like processes, surface sculpture, and nature of the wall (single or double layered). 'Sphaeromorph' is a loose term used for spherical forms without spines or crests (e.g. *Leiosphaeridia*); 'acanthomorph' is a term applied to spinose forms (e.g.

Baltisphaeridium). Like other *palynomorphs*, acritarchs are usually extracted from marlstones and **shales** with hydrofluoric acid, concentrated, bleached and mounted on glass slides for reflected light microscopy using oil immersion objectives. Rocks that are too altered for this method may still yield acritarchs in thin section.

The earliest acritarchs are simple spherical vesicles without regular pylomes, extracted from **Proterozoic shales** about 1400 Ma old. Some of these may be the dis-aggregated remains of benthic *cyanobacteria*. Such simple vesicles were joined by more complex polygonal and frilled forms during the *Riphean* (*c.* 700 Ma) but these suffered decline during the latest *Precambrian glaciation* so that terminal *Precambrian* assemblages are mostly of simple sphaeromorphs. A second radiation of acritarchs, includ-ing many acanthomorphs, took place over the *Precambrian–Cambrian boundary* interval at about the same time as the explosive evolution of invertebrate groups. Their main **biostratigraphic** potential lies in this interval and in **Cambrian** to **Devonian** strata. They suffered a gradual decline from **Carboniferous** to **Triassic** times, a minor resurgence in the **Jurassic–Cretaceous**, and continued decline into the *Tertiary*. In comparison with **dinoflagellate** cysts, acritarchs were of minor importance through most of the **Mesozoic** and **Cenozoic**. The geographical distri-bution of acritarchs generally supports the view that many were planktonic and some forms originally placed here have since been removed to the planktonic prasinophycean *algae* or to the **dinoflagellates**. [MDB]

Loeblich, A.R., Jr. (1970) Morphology, ultrastructure and distribution of Paleozoic acritarchs. *Proceedings of the North American Paleontologists Convention, Chicago*, Part G, 705–88.
Tappan, H. (1982) *The Paleobiology of Plant Protists*. W.H. Freeman, San Francisco.

actinolite ($Ca_2(Mg, Fe)_5Si_8O_{22}(OH)_2$) A green **amphibole mineral**. [DJV]

active margin A continental margin that is also a **plate boundary**, typically a **subduction zone**. A **passive con-tinental margin**, in contrast, does not correspond to a **plate boundary**. (See **plate tectonics**.) [RGP]

adhesion structure A sedimentary structure produced by 'adhesion', a process whereby dry **sand** grains blown onto a damp surface are held there by the surface tension of water which rises by capillary action between the grains.

'*Adhesion ripples*', also termed '*aeolian micro-ridges*' and '*anti-ripplets*' are the bedforms which form by adhesion under a unidirectional wind. They are straight or sinuous parallel ridges orientated transverse to the wind, with spacings of several millimetres to a few centimetres and heights of up to a few millimetres. They have super-imposed button-like protruberances, and are steeper, often overhanging, on their upwind sides, whereas their downwind **lee slopes** are shallow (Fig. A1). As deposition proceeds, they migrate to windward. *Adhesion ripples* are very similar in appearance to *rain-impact ripples*, which have the opposite orientation with respect to the wind.

'*Climbing adhesion ripple structures*' form when net de-position occurs by adhesion to a surface kept damp by the capillary rise of moisture (Hunter, 1973; Kocurek & Fielder, 1982). As shown in Fig. A1, the windward climbing of *adhesion ripples* deposits well-sorted **sand** as cross-laminae in tabular sets up to a few centimetres thick, which have also been termed '*adhesion pseudo cross-laminations*' because their mode of formation is quite unlike that of more common *cross-laminations* (Hunter, 1973). The laminae are several millimetres thick, have irregular, crinkled boundaries with amplitudes of up to a few mil-limetres, and consist of a couplet of coarser over finer **sand**, reflecting the tendency of coarser **sand**, moving by **creep**, to accumulate in the *adhesion ripple* troughs, whilst the finer **saltation** population adheres to the crests. The *cross-laminae* have inclinations between 20° and 60°. Com-monly, they are convex-up, as a result, it is thought, of declining rates of climb as the depositional surface dries and adhesion becomes less efficient.

'*Quasi-planar adhesion stratification*', or '*adhesion lamina-tion*' is a flat, or low angle, lamination produced in well-sorted **sand** by adhesion under strong winds, on surfaces which are only slightly damp, or beneath vertical grain fall. The corresponding bedform is the '*adhesion plane bed*'. The lamination is distinguishable from dry **sand** aeolian stratification by its crinkly appearance and from subaqueously-deposited flat lamination by its very good sorting and lack of mud *intraclasts* and **primary current lineation**.

Adhesion under frequently shifting strong winds can produce irregular protruberances with submillimetre to several millimetre amplitudes and spacings, scattered

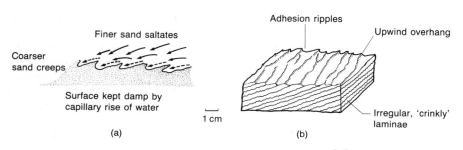

Finer sand saltates
Coarser sand creeps
Surface kept damp by capillary rise of water
1 cm
(a)

Adhesion ripples
Upwind overhang
Irregular, 'crinkly' laminae
(b)

Fig. A1 Adhesion structure. (a) Formation and (b) characteristics of climbing adhesion ripple structure. (After Hunter, 1973 and Kocurek & Fielder, 1982.)

over a sand surface. These bedforms have been termed *'adhesion warts'*. Net deposition by adhesion on such surfaces ought theoretically to produce an irregular, undulatory type of lamination. However, whereas the formation of the other adhesion structures has been observed widely both experimentally and in natural environments, there is some doubt as to whether *adhesion warts* are a genuine bedform. Kocurek & Fielder (1982) were unable to reproduce them experimentally, producing instead frequently realigning *adhesion ripples*. However, they note that wartlike features may form when adhesion occurs onto, and modifies, pre-existing sedimentary structures on a **sand** surface. [DAR]

Hunter, R.E. (1973) Pseudo-cross lamination formed by adhesion ripples. *Journal of Sedimentary Petrology* **43**: 1125–7.
Kocurek, G. & Fielder, G. (1982) Adhesion structures. *Journal of Sedimentary Petrology* **52**: 1229–41.

adiabatic The relationship between pressure and volume when a substance is expanded or compressed without giving out or taking in heat. [DHT]

adit A subhorizontal tunnel driven into a hillside to give access to mine workings or for exploratory purposes. Such tunnels normally have a slight upward gradient for drainage purposes. [AME]

adularia ($KAlSi_3O_8$) A colourless, translucent variety of potash feldspar. (See **feldspar minerals**.) [DJV]

advanced argillic alteration A form of **wall rock alteration** characterized by **dickite**, **kaolinite**, **pyrophyllite** and **quartz**. **Sericite** is usually present and frequently **alunite**, **pyrite**, **tourmaline**, **topaz**, *zunyite* and amorphous **clay minerals**. This is one of the more intense forms of **alteration**, often present as an inner zone adjoining many **base metal vein** or **pipe** deposits associated with acid **plutonic** stocks as at Butte, Montana and Cerro de Pasco in Peru. It is also found in *hot spring* environments and in telescoped shallow, **precious metal** deposits. [AME]

advection (**1**) The rate of change of turbulent (kinetic) energy per unit volume along a mean **streamline**.
(**2**) The lateral transport of mass. Usually referring to lateral motions of **mantle convective** currents. [JLB/DHT]

aenigmatite ($Na_2Fe_5^{2+}TiO_2(Si_2O_6)_3$) A rare **titanium**-bearing silicate mineral. [DJV]

aeolian bounding surface An **erosion** surface which truncates strata within wind-deposited sediments, usually **dune** and **interdune sand**. Higher order bounding surfaces truncate lower order ones. The lowest (usually third) order of bounding surface usually considered is a **reactivation**

surface within the set of cross strata deposited by a single bedform. Next come set boundaries, produced by scour in the lee of a bedform as it migrates over the deposits downwind. A climbing **draa** produces two orders of these, by scour in front of the bedform itself and in front of its individual superimposed **dunes**. Isolated **dunes** produce only a single order. As well as the climbing of bedforms, other mechanisms which may produce higher order bounding surfaces include: (i) **deflation** over all or part of an **erg**, resulting in scour to the cohesive or cemented **sand** at the **water table** — the hypothesis of Stokes (1968), which produces a subhorizontal *'Stokes surface'*; (ii) a rise of the **water table** consequent on lacustrine or marine transgression, or climatic change; (iii) stabilization of an **erg** by vegetation during episodes of more humid or less windy climate (see **aeolian sand sheet**).

Identification and interpretation of lower order bounding surfaces are readily done at outcrop, but for the higher orders they require correlation over many kilometres to detect if they climb relative to time marker horizons, coupled with detailed studies of the sediments immediately above and below to assess whether they were deposited close to the **water table**. (See **interdune**, **adhesion structure**.) [DAR]

Brookfield, M.E. (1977) The origin of bounding surfaces in ancient aeolian sandstones. *Sedimentology* **24**: 303–32.
Stokes, W.L. (1968) Multiple parallel truncation bedding planes — a feature of wind-deposited sandstone formations. *Journal of Sedimentary Petrology* **38**: 510–15.

aeolian sand sheet An extensive area of wind-deposited **sand** which has only rare **slip-faced** bedforms and therefore contrasts with **dune–interdune** systems. Bedforms include **wind ripples** and *aeolian plane beds* (see also **aeolian stratification**) either on subhorizontal surfaces, or on regular or irregular relief created by sporadic **nebkhas** and shadow **dunes** associated with vegetation, by isolated **barchan dunes**, or by coarse-grained, low relief slipfaceless bedforms known as *'zibars'* (see **dune**). **Sand sheet** deposits are dominated by *wind ripple lamination* inclined at low angles and punctuated by subhorizontal **erosion** surfaces, but also include minor volumes of the other main types of dry **sand aeolian stratification**, along with **adhesion structures**, *dikaka* (see **nebkha**), soil horizons including **duricrusts**, **deflation** *lags, granule ripples* (see **wind ripple**) and even deposits and/or **erosion** surfaces of fluvial, marsh, pond, **playa lake** and, in coastal settings, washover origin.

The typical occurrence of **sand sheets** in modern environments is around the margins of **dune** fields, but they also occur within **ergs**, or completely separate from **dunes**. From modern and ancient deposits, it is apparent that they range from thin **sand** veneers to deposits several metres thick, so they do not form just where **sand** supply is limited. Kocurek & Nielson (1986) conclude that they occur where **dune** formation is inhibited by such factors as, in warm arid environments, (i) high **water tables**, (ii)

Table A1 Aeolian stratification. Characteristics of basic types of aeolian stratification. (After Hunter, 1977.)

Depositional process	Character of depositional surface	Type of stratification	Dip angle	Thickness of strata Sharpness of contacts	Segregation of grain types Size grading	Packing	Form of strata
Tractional deposition	Rippled	Subcritically climbing translatent stratification	Stratification: low (typically 0–20°, max. *c.* 30°) Depositional surface: similarly low	Thin (typically 1–10 mm, max. *c.* 5 cm) Sharp, erosional	Distinct Inverse	Close	Tabular, planar
		Supercritically climbing translatent stratification	Stratification: variable (0–90°) Depositional surface: intermediate (10–25°)	Intermediate (typically 5–15 mm) Gradational	Distinct Inverse except in contact zones	Close	Tabular, commonly curved
		Ripple-foreset cross-lamination	Relative to translatent stratification: intermediate (5–20°)	Individual laminae: Thin (typically 1–3 mm)		Close	Tabular, concave-up or sigmoidal
		Rippleform lamination	Generalized: intermediate (typically 10–25°)	Sharp or gradational, non-erosional		Close	Very tabular, wavy
	Smooth	Plane-bed lamination	Low (typically 0–15°, max.?)		Individual laminae and sets of laminae: Indistinct Normal and inverse, neither greatly predominating	Close	Very tabular, planar
Largely grain fall deposition	Smooth	Grain fall lamination	Intermediate (typically 20–30°, min. 0°, max. *c.* 40°)	Sets of laminae: Intermediate (typically 1–10 cm) Sharp or gradational, non-erosional		Intermediate	Very tabular, follows pre-existent topography
Grain flow deposition	Marked by avalanches	Sand flow cross-stratification	High (angle of repose) (typically 28–34°)	Thick (typically 2–5 cm) Sharp, erosional or non-erosional	Distinct to indistinct Inverse except near toe	Open	Cone-shaped, tongue-shaped, or roughly tabular

surface binding or **cementation**, (iii) periodic marine or fluvial flooding, (iv) predominance of a coarser grain size than that typical of **dunes** (fine to medium **sand**) and (v) vegetation. Factors (iv) and (v) are thought to be the most important in modern environments. The relative importance of these and other factors in controlling ancient **sand sheet** occurrence can be inferred from the context of the deposit and from the nature of the subsidiary stratification types punctuating the *wind ripple lamination* of the deposit itself. All five factors are also important in causing **sand** to accumulate where it would otherwise be dropped only temporarily during transport. [DAR]

Fryberger, S.G., Ahlbrandt, T.S. & Andrews, S. (1979) Origin, sedimentary features and significance of low angle eolian 'sand sheet' deposits, Great Sand Dunes National Monument and vicinity, Colorado. *Journal of Sedimentary Petrology* **49**: 733–96.
Kocurek, G. & Nielson, J. (1986) Conditions favourable for the formation of warm-climate aeolian sand sheets. *Sedimentology* **33**: 795–816.

aeolian stratification The **bedding** and lamination produced when wind-blown sediment accumulates on a dry surface (see **adhesion structure**). Characteristic aeolian stratification types are best documented for sand-grade sediment.

Hunter (1977) has described the stratification types produced by each of the main processes which deposit dry **sand** on aeolian **dunes**, **ripples** and flat surfaces (see Table A1). *Wind ripple lamination* is produced by climbing **wind ripples**. Each lamina, like the bedform which produced it, is inversely graded and well packed. The most common form is *subcritical translatent stratification* produced, mostly in fine to medium **sand**, by **ripples** climbing at an angle below the critical angle necessary for the preservation of **stoss side** deposits (see **climbing ripple**). Each **ripple** deposits a planar tabular millimetre-scale inverse graded lamina above an erosion surface, so the lamination is very well defined. *Supercritical translatent stratification* is much rarer and forms when rates of deposition are high enough to preserve **stoss** deposits. Lamina contacts are gradational in consequence. In both types of translatent stratification, laminae which demark the position of the surface of a supercritically climbing rippled **sand** bed at an instant (i.e. over a time interval much shorter than a ripple's lifetime) are only rarely visible. They are termed '*ripple-foreset cross-laminae*' and '*rippleform laminae*', for sub- and supercritical translatent strata respectively, and are probably caused by erratic variations in wind transporting power. Their orientation represents the migration direction of **ripples**. **Wind ripples** migrate up, down and along **dune lee** and **stoss slopes** as well as on the subhorizontal surfaces of **interdunes** and **aeolian sand sheets**, so translatent laminae occur both as subhorizontal and cross-strata. A related type of stratification, confined to flat lying deposits and **dune** toesets, consists of translatent laminae of fine to medium **sand** alternating rhythmically with trains of inverse graded lenses, several millimetres to centimetres thick, of granules and/or coarse to very coarse **sand**. The lenses represent coarse **sand** or *granule ripples* (see **wind ripple**). They are usually symmetrical, but cross laminae may be present to indicate migration directions.

Sand flow cross strata are restricted to the deposits of slip (avalanche) faces of **dunes** where they are produced by *grain flows* of dry **sand** and so are characteristically inverse graded and poorly packed (see **sediment gravity flow**). Avalanching of rain- or dew-dampened **sand** produces brittle *soft-sediment deformation* (see **wet sediment deformation**). On small **dunes** (up to a few metres high) *sand flow* strata take the form of tongue-shaped lenses or lobes, but *sand flows* on larger **slip faces** tend to merge laterally, so towards the base of the **slip face**, at least, their deposits are tabular. In vertical sections parallel to cross stratal dip, *sand flow* strata appear either as angular wedges thinning downslope if the flows did not reach the base of the **slip face**, or as less regular masses with sigmoidal boundaries where they have piled up at its base. Towards the top of **slip faces**, where they overlie slump scars, they have **erosional** lower boundaries. Lower down, their bases are sharp, but non-erosive.

Grain fall lamination is produced by fall-out of **sand** from suspension in zones of **flow separation** and so occurs both as foreset and toeset stratification (see **dune**). It has intermediate packing and is much less distinct than the other stratification types, being defined by sharp or gradational grain size variations on a scale from centimetres to that of a single grain, which probably result from wind fluctuations. Its **sand** is generally finer than any intercalating

sand flow strata, which consist of material which was previously deposited much higher up a **slip face**.

Plane bed lamination is deposited, in **interdunes** or on **aeolian sand sheets**, on the flat **sand** beds produced by winds too rapid for **ripples** to form. It is analogous to the upper **flow regime** flat bedding of aqueous currents, though it lacks **primary current lineation**. Except for its confinement to low angle strata, its characteristics are similar to those of *grain fall lamination*. The grain size fluctuations which define it probably result both from fluctuations of wind velocity and by a like-seeks-like sorting mechanism where grains fit best between grains of the same size. Clemmensen & Abrahamsen (1983) describe a stratification type, which probably results from similar processes in a similar setting, acting on coarser sediment, as *granule-rich horizons*: well-sorted flat laminated **sand** with modes in fine and coarse to very coarse sand and with granules scattered over closely spaced laminae.

The recognition of **aeolian stratification** types is of use both in the identification of wind deposits and in their interpretation; since the types occur in different proportions on various parts of **dunes**, **interdunes** and **aeolian sand sheets**, '*stratification maps*' showing their distribution in a section through an aeolian deposit can be used in combination with **dip** measurements to infer the nature of the bedform(s) which deposited the sediment. Caution is necessary, however, at both levels of interpretation, for subaqueous *sand flow* stratification, in particular, is superficially similar to its aeolian equivalent, and even in modern aeolian deposits 20% or more of the stratification may be difficult to classify. It is particularly difficult to distinguish between certain low angle strata which may be grain fall, **plane bed**, or translatent strata deposited by **ripples** with particularly low angles of climb. Difficulties are increased by very good sorting of **sand** and, in ancient sediments, by the obscuring effects of **compaction** and other **diagenetic** processes. [DAR]

Clemmensen, L.B. & Abrahamsen, K. (1983) Aeolian stratification and facies association in desert sediments, Arran Basin (Permian), Scotland. *Sedimentology* **30**: 311–39.
Hunter, R.E. (1977) Basic types of stratification in small aeolian dunes. *Sedimentology* **24**: 361–87.

aeolianite Cemented **dune sand**. The **sand** may be made up of grains of **quartz**, calcium carbonate, **gypsum**, or other minerals, or mixtures of these. The cement is usually calcium carbonate but other water-soluble minerals, such as **gypsum**, may consolidate the sand-size particles. The term aeolianite, in its broadest definition, may be applied to all consolidated wind-blown sediments, regardless of when or where **cementation** or lithification occurred. However, the term usually refers to **dune sands** which have been cemented at or near the surface. The cement may originate as grains within the **dune sand**; these are dissolved by infiltrating rainwater or by **groundwater** which subsequently evaporates, reprecipitating the minerals around the **sand** grains and cementing them together.

Alternatively, the mineral cements may be introduced in **groundwater** or through leaching of saline dust or spray deposited on the **dune** surface. As little as 8.0% calcium carbonate as cement within the bulk material is sufficient to consolidate the **sand** (Yaalon, 1967). However, most aeolianites contain between 30 and 60% calcium carbonate as both cement and sand-sized fragments of sea-shells and **Foraminifera**.

Most aeolianites are found along coasts in regions within 40° of the Equator where the climate has a marked dry season. The deposits are widespread in those regions with Mediterranean climates, and also in warm deserts where **dune sands** contain **evaporite** minerals. In wet, temperate regions and in the tropics, leaching of the **sands** precludes the accumulation of water-soluble cements. Though **cementation** can produce hard, dense aeolianites, few aeolian **sandstones** occurring in the geological record have been interpreted as subaerially lithified aeolianites. [AW]

Gardner, R.A.M. (1983) Aeolianite. In: Goudie, A.S. & Pye, K. (eds) *Chemical Sediments and Geomorphology*, pp. 265–300. Academic Press, London.
Yaalon, D.H. (1967) Factors affecting the lithification of eolianite and interpretation of its environmental significance in the coastal plain of Israel. *Journal of Sedimentary Petrology* 37: 1189–99.

AFMAG An **electromagnetic induction method** for ground and **airborne surveys**. Natural variations of the **geomagnetic field** in the audiofrequency range (**sferics**) serve as the source. Near a lateral **electrical conductivity** anomaly the plane of polarization of the field is tilted from the horizontal, and two receiver coils are used to detect the dip angle and azimuth of the resultant **magnetic field** at two frequencies, along profiles perpendicular to the expected strike of the anomaly investigated. The projections on to the horizontal plane of the polarization field vectors appear to radiate from the location of the conductivity anomaly, providing a quick method of identifying the target zones.

In the airborne system, two fixed orthogonal receiver coils, at 45° to the horizontal and with their axes pointing towards the flight direction, are towed behind the aircraft.

AFMAG is particularly suited for exploring large, deep-seated conductive targets. [ATB]

Keller, G.V. & Frischknecht, F.C. (1966) *Electrical Methods in Geophysical Prospecting*. Pergamon, Oxford.

aftershock Small **earthquake** following, and in the same region as, a large, 'main', **earthquake**. The main shock does not release all the accumulated **stress** at its *focus*, but also transfers **stress** to nearby areas on the same or other **faults**; these residual or transferred **stresses** cause the aftershocks. The locations of aftershocks can show the shape and extent of movement on the **fault** plane. Hundreds of aftershocks or more can occur over weeks to months following the main shock, gradually weakening in strength but still capable of causing further damage, such as to already weakened structures. An empirically derived relationship by Omori in 1894 is still accepted:

$$N(t) = N_o . (t + c)^{-K}$$

where $N(t)$ is the number of aftershocks per unit time at time t after the mainshock, and N_o, c ($c.$ 0) and K ($c.$ 1.0–1.4) are constant for any particular region. (See **foreshock, seismicity**.) [RAC]

Page, R. (1968) Aftershocks and microaftershocks of the great Alaskan earthquake of 1964. *Seismological Society of America Bulletin* 58: 1131–68.

agalmatolite A varietal name for the compact form of the mineral **pyrophyllite**. [DJV]

agate Concentric layers of **chalcedony** (microcrystalline **quartz**) having different colours and **porosity**. [DJV]

age of the Earth In the 19th century, age estimates were based on either biblical evidence ($c.$ 3003 BC), the estimated rate of cooling for the Earth ($c.$ 20–80 Ma), and estimates based on the rates of **erosion** and the thickness of sediments (10^{8-9} years). The discovery of radiogenic heat production not only caused revision of the cooling history of the Earth and planets, but the *radioactive* isotopes provided the main modern *dating* methods using the ratio between parent and daughter isotopes to determine when the minerals became a closed chemical system. The oldest continental rocks so far dated, have an age of slightly younger than 4 Ga (3.89 Ga in west Greenland, although similar ages are also known from Siberia and **cratonic** North America). In contrast, the oldest ages for meteorites and lunar rocks suggest an age of 4.66 Ga for their formation, a few million years after the primordial formation of the elements of the solar system by means of fast nuclear processes during a supernova explosion. Studies of uranium–lead isotopes, in particular, also suggest a similar age for the time that the Earth's **mantle** originally formed a closed chemical (isotopic) system. The Earth therefore formed at the same time as the other planets of the solar system, although it is still not clear if the Sun is of the same age or older. [DHT]

Brown, G.C. & Mussett, A.E. (1980) *The Inaccessible Earth*. Allen & Unwin, London.
Press, F. & Sevier, R. (1982) *Earth*. W.H. Freeman, San Francisco.

aggradation The process by which a land surface is built up by material deposited by geomorphological processes, such as rivers, wind and **waves**. Material is deposited by a river, for example, when the **load** becomes too great to be transported. This may occur for a number of reasons, including an increase in material to be carried by the river, a loss of discharge or velocity of flow, or a rise in base level. Base level changes occurred for most world streams at the end of the last *glaciation*, so that many rivers filled their pre-glacial channels particularly in the lower reaches as *Pleistocene ice sheets* melted and world sea levels rose. The construction of dams on river courses also changes the base level of a river upstream of the dam and leads to aggradation of any reservoir behind the dam. Wind-blown sediments may aggrade very large regions, as evidenced by the world's **loess** deposits. The term is also used to describe the building of marine **beach deposits**. [NJM]

aggregate (1) A mass of rock fragments or mineral grains or a mixture of both.
(2) Any solid material, e.g. gravel, slag, crushed stone, used alone as in railway **ballast** or road foundations or mixed with cement or bitumen to form concrete or tar macadam. [AME]

aggressivity A measure of the capacity of water to dissolve calcium carbonate. The term is used in the context of **limestone solution** studies. The aggressivity of a **karst** water depends upon the amount of carbon dioxide and any organic acids in the water. As water passes through a **limestone** area its aggressivity decreases as more calcium carbonate is dissolved, until a state of saturation equilibrium is reached. The amount of calcium carbonate present in a saturated water depends upon the water temperature and partial pressure of carbon dioxide of the air with which the water is in contact. Aggressivity can be renewed by mixing, leading to **mixing corrosion**. [HAV]

airborne geophysical survey A number of geophysical exploration methods can be undertaken from the air, and are very widely used as, once the initial high capital cost of the equipment has been recouped, they represent a highly cost-effective means of gaining information. Vast areas can be covered very rapidly and there is rarely any necessity for prior ground access. Consequently data can be obtained over regions hostile to land surveying, such as ice caps, **swamps**, deserts and seas. A major problem in such surveys is accurate position fixing and navigation. If no electronic fixing system is available, recourse is commonly made to terrain photographs taken simultaneously with the geophysical measurements.

Gravity surveys can be performed from aircraft, but are of low accuracy because of large errors in applying corrections. **Eötvös corrections** may be as great as 16 000 g.u. at a speed of 200 knots, and vertical accelerations with a periodicity of greater than the instrumental averaging time cannot easily be corrected. In the future, with the development of autopilots and height stabilizers, it should be possible to improve on the present accuracy of about 100 g.u. Airborne gravity is thus a technique used only for coarse regional surveys, and is not used in *mineral exploration*.

Magnetic surveys are usually performed from fixed-wing aircraft. The sensor of a **fluxgate** or **proton magnet-** **ometer** is either towed at a sufficient distance to remove it from the **magnetic field** of the aircraft, or mounted as a 'stinger' in the tail, in which case the field of the aircraft must be compensated. The survey pattern is arranged so that there are numerous cross-overs of flight lines to check navigation and to allow the assessment of **diurnal corrections**. A further advantage of the aeromagnetic method is that the height of the survey causes the attenuation of effects originating at shallow depths so that the anomalies of deep-seated structures are defined.

Airborne **electromagnetic** surveys are commonly of either quadrature or fixed-separation type. Quadrature systems use a transmitter in the form of a loop of wire strung between the wingtips and tail of a fixed-wing aircraft and a receiver mounted in a 'bird' is towed behind the plane. Such systems can only measure the phase difference between the primary and secondary *electromagnetic radiation*. They thus only provide an indication of the presence of an anomalous body, although taking readings at two or more frequencies can resolve the response of good and bad conductors. Fixed separation systems provide more information about the anomalous body and measure the **real** and **imaginary components** of the secondary electromagnetic field. To do this without large error, it is necessary to maintain the transmitter and receiver at an accurately fixed distance, and this is achieved by mounting them on the wing-tips of a fixed-wing aircraft, on the nose and tail of a helicopter or on a beam towed behind a helicopter. An alternative, providing greater penetration but more difficult to accomplish, is to use two planes flying in tandem some 300 m apart, one carrying the receiver and the other the transmitter. This method uses a rotating electromagnetic field to compensate for relative rotation of transmitter and receiver about an horizontal axis. A more sophisticated system is **INPUT**®, which employs a transient electromagnetic field and only measures the secondary field when the primary is turned off. The tilt-angle methods **VLF** and **AFMAG** can also be used with an airborne receiver. These **electromagnetic** methods suffer from the drawback of relatively low penetration of the order of 50 m. Also not all electromagnetic anomalies have economically important sources, such as **graphite** and bodies of water.

Airborne **radiometric** methods are all based on measurements of gamma radiation, which is the only type to propagate over the distances relevant to airborne work. Surveying is performed with a *scintillation meter* or *gamma-ray spectrometer*. The low speed of a helicopter is often an advantage in such work for the discrimination of multiple anomalies.

Commonly magnetic, electromagnetic and radiometric methods are used simultaneously from the same aircraft, thus providing multiple datasets in a very time- and cost-effective fashion. [PK]

Airy hypothesis An Earth model to explain **isostasy** by means of deep 'roots' of **continental crustal** density extending into the **mantle** to account for high surface topography and 'antiroots' of **mantle** material extending into the crustal rocks to account for low lying areas, as in the ocean basins. [DHT]

aklé Complex **dune** form with interlocking crescentic **dune** ridges. The ridges are generally parallel. Aklé is found in areas of high **sand** supply forming transverse to the wind direction. Also known as '*fishscale dune pattern*'. [RDS]

alabandite (MnS) A relatively rare **ore** mineral of **manganese**. [DJV]

alabaster Mineral used extensively in antiquity for statues, tombs, paving stones and smaller items such as bowls. Amongst those who valued alabaster were the Sumerians, Egyptians, Minoans and Mycenaeans. Sources of alabaster in Greece and Crete have been characterized by their **strontium isotope** ratios, thus enabling artefacts of alabaster used by the Minoan and Mycenaean cultures to be assigned to a Cretan source. Schools of alabaster carving in England were famous in the medieval period, and carvers in Nottingham, the main production centre, obtained alabaster from quarries in Derbyshire and Staffordshire, making altars, tombs and shrines which were often exported to Europe. [OWT]

alas A steep-sided, flat-floored depression, sometimes containing a lake, found in areas where local melting of *permafrost* has taken place. It is one manifestation of **thermokarst**. [ASG]

albite (NaAlSi$_3$O$_8$) Sodic end-member of the **plagioclase feldspar** *solid-solution* series. (See **feldspar minerals**.) [DJV]

alcove An arcuate, steep-sided cavity on the side of a rock outcrop which has been produced by water **erosion**, especially *spring sapping* or **solutional** processes. [ASG]

alcrete An aluminium-rich **duricrust** resulting from the accumulation of aluminium sesquioxides within the zone of **weathering**. It is an alternative name for indurated **bauxite**. [ASG]

alexandrite Gem variety of the mineral **chrysoberyl**. [DJV]

Algoma-type iron formation A type of **banded iron formation** developed as oxide, carbonate and sulphide facies; **iron** silicates are often present in the carbonate facies. This BIF type generally ranges from a few centimetres to a hundred or so metres in thickness and is rarely more than a few kilometres in strike length; exceptions to this observation occur in Western Australia where late

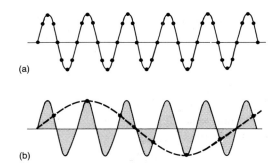

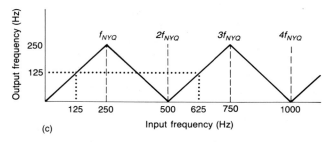

Fig. A2 Aliasing. (a) Sine wave frequency less than Nyquist frequency. (b) Sine wave frequency showing the fictitous frequency that is generated by aliasing. (c) Relationship between input and output frequencies for a sampling frequency of 500 Hz (Nyquist frequency f_{NYQ} = 250 Hz). (After Kearey & Brooks, 1991.)

Archean deposits of economic importance are found. **Oolitic** and granular textures are absent or inconspicuous and the typical texture is a streaky **lamination**. It is characteristic of the **Archean greenstone belts** and shows a *greywacke*–volcanic association suggesting a **geosynclinal**-type environment and a volcanic source for the **iron**. [AME]

Goodwin, A.M. (1973) Archean iron-formations and tectonic basins of the Canadian Shield. *Economic Geology* **68**: 915–33.

aliasing The phenomenon in sampling a waveform where frequencies higher than one half the sampling interval (called the **Nyquist frequency**, f_{NYQ}) — i.e. wavelengths smaller than twice the sampling interval, in a sampled signal produce false low frequency distortion of the true lower signal frequencies. For example, frequencies higher than the **Nyquist** by 50 Hz, f_{NYQ} + 50 Hz will appear in the spectrum of the sampled signal at the lower frequency f_{NYQ} − 50 Hz. The latter frequency is said to be an alias of the first. To avoid aliasing frequencies when sampling a signal, frequencies above the **Nyquist** must be removed by an *anti-alias filter* before sampling. (See Fig. A2.) [GS]

Kearey, P. & Brooks, M. (1991) *An Introduction to Geophysical Exploration* (2nd edn). Blackwell Scientific Publications, Oxford.

alkali feldspar General term for feldspars of the NaAlSi$_3$O$_8$–KAlSi$_3$O$_8$ series. (See **feldspar minerals**.) [DJV]

alkali lake A salt lake, usually found in arid regions. **Dissolved** salts include large amounts of sodium and potassium carbonate in addition to sodium chloride and other alkaline salts, e.g. Lakes Magadi, Natron and other lakes in the East African Rift Valleys. [AME]

alkaline igneous rock Rock with a high concentration of alkalis (Na_2O, K_2O) relative to **silica** (SiO_2), so that it is commonly silica-undersaturated, containing **normative nepheline** (or **leucite**), and characterized by modal **alkali feldspar**, **feldspathoids**, alkali-rich **pyroxenes** and **amphiboles** and **phlogopite**. Such rocks are also enriched in certain elements (Zr, Nb, Rb, Ba, Ti, P) which occur in trace or minor amounts in non-alkaline rocks. The textures of such rocks encompass all of those described under **'igneous rocks'** and include crystalline aphanitic and phaneritic types, glassy varieties and **pyroclastic rocks**. Although alkaline rocks are relatively rare (forming less than 1% of all igneous rocks), the chemical and mineralogical variability means that they are described by more names than all other igneous rocks.

Aphanitic alkaline rocks commonly form a spectrum between alkali **basalt** and **basanite**, through **hawaiite**, *mugearite* and **benmoreite** to **trachyte** and **phonolite**. Other aphanitic alkaline rocks include *shoshonite* (associated with silica-oversaturated lavas), **nephelinite**, **kimberlite** and **lamprophyre**. These rocks have phaneritic equivalents which include alkali **gabbro** (alkali **basalt**), *theralite* (**basanite**), **syenite** (**trachyte**), **feldspathoidal syenite** (**phonolite**) and **ijolite** (**nephelinite**). Although most of these rocks are relatively soda-rich, certain alkaline rocks are potash-rich (K_2O/Na_2O over 3), and these include **kimberlite**. Alkaline **lavas** and **pyroclastic rocks** commonly contain **mantle**-derived ultramafic **xenoliths**.

Alkaline igneous rocks are characteristic of intraplate (or 'within-plate') regions that are tectonically stable or experiencing **lithospheric rifting** and are extremely rare in tectonically active regions, such as *convergent plate margins*. They hence occur on intraplate *oceanic islands* (e.g. Hawaii) and within continental **rifts** (e.g. East African rift system).

The occurrence of alkaline rocks in both oceanic and continental regions and the presence of **mantle**-derived **xenoliths** indicate formation within **mantle** source regions. The small volume of alkaline **magmas** and their high concentrations of incompatible elements and varied mineralogy therefore implies derivation by a small degree of partial melting at great depth of heterogeneous alkali-rich **mantle peridotite**. Distinctive associations of alkaline rocks (e.g. alkali **basalt**–**hawaiite**–*mugearite*–**benmoreite**–**trachyte**) may reflect fractional crystallization processes. Alkaline igneous rocks are associated with certain mineral deposits, and **kimberlite diatremes** are the major source of **diamond**. [RST]

Fitton, J.G. & Upton, B.G.J. (eds) (1987) *Alkaline Igneous Rocks.* Geological Society Special Publication No. 30. Blackwell Scientific Publications, Oxford.

Wimmaenauer, W. (ed.) (1974) *The Alkaline Rocks.* John Wiley & Sons, New York.

allanite $((Ca, Ce)_3(Fe^{2+}, Fe^{3+})Al_2O(SiO_4)(Si_2O_7))(OH)$ A double-island silicate mineral (*sorosilicate*) that occurs as a minor accessory constituent in many igneous rocks. [DJV]

allemontite (AsSb) A rare mineral occurring in **vein** deposits. [DJV]

allochem (an abbreviation of *allochemical constituent*) A term introduced by Folk to cover organized aggregates of **calcite**. The word is synonymous with grain or particle and is now not widely used. Folk recognized four main types of allochems in **limestones**: *intraclasts* (reworked fragments of lithified carbonate sediment), **ooids**, fossils (skeletal material) and pellets (in Folk's usage these are sand-sized grains composed of carbonate mud, now referred to by the more general term *peloid*). [VPW]

Folk, R.L. (1962) Spectral subdivision of limestone types. *American Association of Petroleum Geologists Memoir* 1: 62–84.

allochthon Piece of **crust** that has originated at a distance from the rocks that now lie adjacent to it; adj. 'allochthonous'. The term is used for relatively large **nappes** or *thrust sheets*, or for displaced or exotic **terranes**. (cf. **autochthon**). [RGP]

allogenic stream A stream flowing through an area where it does not gain any discharge. Allogenic streams are found in arid regions where discharge is derived from rainfall in the upper **catchment**. They are also found in areas of permeable rock — such as **limestone** or **chalk** — where streams originating on impermeable rocks in the upper part of the **catchment** have sufficient discharge to maintain their flow through the area which does not contribute **runoff**. In some cases, however, **stream flow** may be maintained when **groundwater** supplements the discharge. [AW]

allometric growth A biological term which has been used in **geomorphological** research which states that a change in size of the whole is accompanied by scale-related changes in its parts. The term is used in opposition to the term *isometric growth*. [HAV]

alluvial aquifer An **aquifer** composed of unconsolidated sediment. In many parts of the world such **aquifers** provide the main or only source of **groundwater** and they form the bulk of the world's developed **aquifers**. They are generally of recent origin, little compacted or cemented and thus highly permeable. **Sands** and **gravels** in large, wide valleys of major rivers can provide a very important **groundwater** resource, e.g. the Ganges **delta**. [AME]

alluvial architecture The three-dimensional distribution of sedimentary facies in an **alluvial deposit**. Alluvial

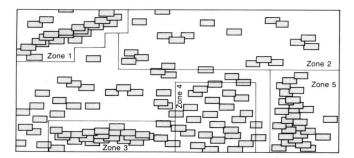

Fig. A3 Alluvial architecture. Zone 1: well developed offset stacking. Zone 2: random stacking — body grouping not obvious or poorly developed. Random packing with no pronounced stacking trend. Zone 3: lateral stacking — coalescence of bodies gives rise to a body group pattern of extensive 'sheet' geometry — multilateral sandstone body. Zone 4: slight offset stacking — younger bodies preferentially lying lateral to older ones. Zone 5: vertical stacking.

architecture includes the lateral and vertical facies pattern, the geometry and dimensions of component facies elements, their **interconnectedness** and the distribution of **bedding** and sedimentary structures that result from the temporal evolution of **aggrading** alluvial systems.

The controls of alluvial architecture are of two types, *autocyclic* and *allocyclic*. *Autocyclic* controls are inherent to the sedimentary system and include: channel type, sediment and water discharge, early **diagenesis**, **compaction** and some mechanisms of **avulsion**. *Allocyclic* controls are those factors external to the basin that affect sedimentation, such as climate, source area geology, tectonic and volcanic activity and sea-level change.

The deposition of alluvial sediments is primarily controlled by the dimensions and thus the tectonic setting of the **basin** and the rate of sediment and water supplied to that **basin**. *Allocyclic* controls are, therefore, of major importance to the overall architectural style but the detailed architecture of **alluvial deposits** depends mainly on *autocylic* processes. Galloway (1981) considered that because 'alluvially deposited stratigraphic units consist of a mosaic of individual depositional systems and components, accurate descriptions and interpretations necessitate a hierarchical, three-dimensional approach to the analysis of alluvial stratigraphy'. Alluvial architectural models and descriptions should, therefore, be considered on all scales from individual **sandstone** body to sedimentary **basin**.

On a **basin** scale, alluvial architecture may be described by the shape of the **alluvial deposits** relative to the basin geometry and the **sandstone** percentage distribution. These factors are related primarily to the tectonic setting and **basin** dynamics. Large scale geometrical differences are seen between alluvial deposits of **foreland**, **strike-slip**, extensional and sag **basins**. The percentage of **basin** fill that is alluvial also affects the overall architecture. The

river-lain deposits may interfinger with marine, lacustrine or aeolian deposits and are frequently seen to wedge out in a particular direction. Within this framework channel **sandstone** body stacking patterns distinguish a number of different alluvial architectural styles. Channel **sandstone** bodies may have random packing, spaced regular packing, lateral stacking, vertical stacking and offset stacking, as seen in Fig. A3.

The interpretation of **sandstone** body stacking patterns depends on the scale of observations. For example, within a succession with an overall random pattern, small outcrops may indicate a vertical, lateral or offset stacking pattern. In many areas of **alluvial deposits**, **sandstone** will be preferentially exposed because of its greater resistance to **weathering** and this may lead to an incorrect interpretation of the stacking pattern and overall **sandstone** percentage. Multistorey stacking patterns result from persistent lateral restriction of channel position that may be due to long-lived patterns of differential subsidence or may follow major incision events. Vertically offset stacking commonly results from the influence of buried **sandstone** bodies on the topography and thus river channel position on the depositional surface. This topographic effect results from differential compaction of the channel belt and overbank deposits. The amount of offset in a multistorey pattern decreases with increasing topographic restriction. Multilateral channel sandstone body patterns may result from deposition in large **braided rivers** with laterally shifting channels or from superimposition of channels during a period of low net **aggradation** (Bridge & Leeder, 1979).

On the scale of individual **sandstone** bodies, alluvial architecture describes the distribution of facies within and around the **sandstone** body and allows consideration of the pattern of grain size and bedform distribution related to the processes of deposition.

The architecture of a channel deposit depends on the size and nature of the channel, its residence period at one site between **avulsions**, its lateral stability, **aggradation**, load characteristics, process of channel abandonment and post-**avulsion** alterations that may result from subsequent deposition, **erosion**, **diagenesis**, **compaction** and **deformation**.

Channel **sandstone** bodies may be classified on the basis of their external geometry and internal morphology. The primary division is between ribbons and sheets and subdivision is based on dimensions and internal complexity (Friend, 1983). [JA]

Bridge, J.S. & Leeder, M.R. (1979) A simulation model of alluvial stratigraphy. *Sedimentology* **26**: 617–44.

Friend, P.F. (1983) Towards the field classification of alluvial architecture or sequence. In: Collinson, J.D. & Lewin, J. (eds) *Modern and Ancient Fluvial Systems*, pp. 345–54. Special Publications of the International Association of Sedimentologists No. 6. Blackwell Scientific Publications, Oxford.

Galloway, W.E. (1981) Depositional architecture of Cenozoic Gulf Coastal Plain fluvial systems. In: Ethridge, F.G. & Flores, R.M. *Recent and Ancient Nonmarine Depositional Environments: Models for Exploration*. Special Publication of the Society of Economic Paleontologists and Mineralogists **31**: 127–55.

alluvial deposit Accumulation of material deposited from rivers, streams or non-marine flood water within a channel or on an alluvial, coastal or deltaic plain. A stratigraphic unit consisting of alluvial deposits may be made up of **gravel**, **sand**, **mud**, **coal** and chemical precipitates. The nature and distribution of alluvial facies (**alluvial architecture**) depend upon a large number of factors, including channel type, vegetation cover, channel density, source area geology, climate, tectonic setting and surface **deformation**.

In general, alluvial deposits may be divided between **channel-belt deposits** and **floodplain** deposits; this division often corresponds with grain size variation. Channel-belt sediments are deposited in the immediate vicinity of the channel and within the strip of land over which the channel migrates between **avulsions**. The **floodplain** sediments are those that are deposited from flood water outside the area of the channel-belt but still under the influence of the channel or its tributaries and distributaries. **Channel-belt deposits** include major channel **sand** or *conglomerate* bodies, channel abandonment facies such as clay plugs and proximal **levée** deposits. **Floodplain** deposits may consist of sheet **sands**, muds and silts deposited from moving or stagnating flood water, autocthonous **coal**, shallow lacustrine deposits, all of which may be subsequently affected by **pedogenic** activity.

Attempts have been made to classify alluvial deposits on the basis of river plan geometry into meandering, braided and anastomosing river deposits, but many of the criteria used for this classification are unsound: the processes occurring in rivers of different types are similar, leading to architectural similarities. In addition, the plan form of any river may change downstream or through time. More useful classifications of alluvial deposits are based on the grade of channel sediment or on sediment body geometry.

The division of alluvial deposits on the basis of channel load characteristics is between two end members of the continuum **bedload** dominated to **suspension**-load dominated rivers. The use of this classification in the rock record assumes a representative preservation of channel and floodplain sediments which may not always be the case.

Bedload river deposits may be subdivided into those that are **gravel** or **sand** dominated. In general the **bedload** dominated deposits tend to be low **sinuosity** river deposits and the coarse members may be laterally extensive, often approximating to a sheet geometry.

Classification of alluvial deposits on the basis of sediment body geometry is particularly difficult in areas of poor outcrop and in the subsurface. The main divisions are based on the geometry of the coarse members into pod, shoestring (or ribbon), tabular and sheet-like. It is also possible in exceptional areas to recognize divergent (distributary) or parallel **sandstone** body geometries. This approach is of major benefit to **alluvial architectural** studies but as yet a uniform terminology of geometries has not emerged.

In most considerations of **alluvial architecture** a twofold division between fine and coarse material is made which is generally considered to be the division between channel-belt and overbank deposits. The proportional division of alluvium between coarse **channel-belt deposits** and overbank fines is controlled by channel type and magnitude, rate of lateral channel migration, **avulsion** and **floodplain** vertical accretion. In practice, distinguishing the extent of these two facies associations is difficult as channel-belt deposits include major channel **sand** bodies, channel plugs and proximal **levée** deposits. Overbank deposits consist of *lagoon*, lake, marsh, **crevasse splay** and **sheetflood** deposits, all of which may be altered by **pedogenic** activity.

The nature of **paleosols** in **overbank deposits** and their distribution is of major importance in understanding alluvial sedimentation. Most work in this area has concentrated on **calcretes** with only a few studies of other types of **paleosol**. **Pedogenic** processes are ubiquitous in the terrestrial environment and the nature of those processes reflect climate, vegetation, depth and fluctuations of the **watertable**, local topography and sedimentation rate. Mature *soils* indicate a prolonged period of **pedogenic** activity resulting from infrequent flooding and reflect distance from active channels or periods of channel entrenchment.

Floodplain lakes may also cause the preservation of distinctive facies patterns within alluvial deposits. They may either form organic-rich mud deposits, frequently containing well preserved plant and animal fossils or they may contain coarse clastic material with a **delta** morphology due to fluvial progradation into the lake.

In many areas, persistent facies distributions reflect long-lived topographic features, often resulting from continued surface **deformation**. This differential subsidence may be the result of tectonic activity, compaction, salt solution or volcanic activity. Tectonically subsiding areas may be occupied by lakes or relatively wet lands, while rising areas support better drained, mature *soils*. Within-basin tectonism, such as active syndepositional upwarps, can cause downstream fluctuations in slope and hence fluvial style.

In the lower reaches of the alluvial environment, on a coastal or deltaic plain, there is increasing marine influence on the sediments. This marine influence may be in the form of a saline wedge intrusion near the base of the channel, tidal back-up causing fluctuations in flow velocity and water level, current reversals or the introduction of marine micro-organisms. The gradation between alluvial, tidal and fully marine conditions frequently causes controversy in the interpretation of coastal deposits. Notably, in coastal plain areas with low gradients, major *storm* events associated with high **tides** may cause flooding of alluvial channels and the introduction of brackish or marine micro-organisms to **floodplain** deposits in an area that is dominantly fluvial in its characteristics. (See **alluvial architecture**.) [JA]

Collinson, J.D. (1986) Alluvial sediments. In: Reading, H.G. (ed.) *Sedimentary Environments and Facies*, pp. 20–62. Blackwell Scientific Publications, Oxford.

Rust, B.R. & Koster, E.H. (1984) Coarse alluvial deposits. In: Walker, R.G. (ed.) *Facies Models*, pp. 53–69. Geoscience Canada Reprint Series 1.

Walker, R.G. & Cant, D.J. (1984) Sandy fluvial systems. In: Walker, R.G. (ed.) *Facies Models*, pp. 71–89. Geoscience Canada Reprint Series 1.

alluvial fan A semi-conical, downstream fanning, sediment accumulation predominantly of alluvial origin, resulting from loss of transporting capacity due to horizontal flow expansion at a site of gradient reduction. Alluvial fans form along mountain fronts, **fault** scarps, sides of major valleys and at the margins of **glacier** ice. The size and slope of a fan is controlled by the topographic amplitude, source area geology, climate, differential subsidence and surface titling. Two end-member fan types have been distinguished: humid or stream-dominated and semi-arid or dry alluvial fans between which a gradation exists. Semi-arid alluvial fans are constructed by deposition from ephemeral water flow and **debris flows**. Wet fans are dominated by channelized perennial flow, are generally more laterally extensive and have a lower gradient than semi-arid fans. Alluvial fans may consist of conglomerate, **sand** or mud grade material and show a proximal to distal reduction in grain size with a related change in dominant sedimentary structures. Adjacent fans may coalesce to form an alluvial plain termed a **bajada**. [JA]

Heward, A.P. (1978) Alluvial fan sequence and megasequence models: with examples from Westphalian D – Stephanian B coalfields, northern Spain. In: Miall, A.D. (ed.) *Fluvial Sedimentology*. Canadian Society of Petroleum Geologists Memoir **5**: 669–702.

Hooke, R. Le B. (1967) Processes on arid-region alluvial fans. *Journal of Geology* **75**: 438–60.

alluvial ore deposit A **placer deposit** formed by the action of running water in a stream or river channel. [AME]

almandine ($Fe_2Al_2Si_3O_{12}$) A garnet group silicate mineral. (See **garnet minerals**.) [DJV]

altaite (PbTe) A very rare **ore** mineral of lead. [DJV]

alteration Any change, chemical or mineralogical, brought about in a rock by chemical or physical action. **Hydrothermal solutions** frequently give rise to **wall rock** alteration around any mineral or **ore** deposit formed from them. Surface waters give rise to **weathering** and some **groundwater** movement to **supergene enrichment**. [AME]

alum Potash alum is $KAl(SO_4)_2 . 12H_2O$. Natural alum minerals only approximate to this formula and are a series of double sulphate isomorphs with potash alum.

Manufactured alums are prepared by treating various **aluminium**-rich rocks, e.g. **bauxite**, clay, with acid. Alum dehydrated at a dull red heat yields a porous, friable material known as *burnt alum*. Alum has many industrial uses, e.g. the dyeing, leather and paper industries, waterproofing fabrics, water purification.

Alum was used in early tanning, dyeing and glassmaking industries and alum quarries were exploited by the ancient Egyptians. Assyrian texts record medicinal use of alum. [AME/OWT]

alum shale Alum-bearing **shale** in which the **alum** has usually formed during the **weathering** of pyritous **shale**. The **pyrite** is oxidized to sulphuric acid and **limonite** and the acid reacts with the **shale** to produce **alum**. [AME]

aluminium (Al) At. no. 13, at. wt. 26.98154, m.p. 660.37°C, d. 2.702. This metal does not occur native but is a constituent of numerous silicates and also occurs as hydrated oxides particularly in the rock **bauxite** from which it is recovered commercially. It is the third most common element in the Earth's **crust** of which it forms 8.1%. [AME]

alunite ($KAl_3(SO_4)_2(OH)_6$) A mineral used in the production of **alum**. [DJV]

alunitization The development of new **alunite** in rocks undergoing **hydrothermal alteration**. It may be the dominant form of **alteration** at low temperatures; in moderate and high temperature **alteration** it may be present as an additional mineral, sometimes in large quantities. It may also form in silicate host rocks during **supergene enrichment**. [AME]

alveoles Small **weathering** depressions forming a closely spaced network of holes. They are often found in coastal areas and are also known as *honeycombs* or *stone lattice*. Alveoles are similar features to **tafoni**, and are regarded by many authors as being small **tafoni**, up to a few centimetres in diameter and depth. Alveoles are best developed on horizontal or gently sloping rocks, and are especially common on **sandstones**. There is some debate over the mechanisms responsible for their formation. Salt crystallization and other physical processes are commonly invoked, although chemical and biological processes may be important in some cases. Rock structure helps control the pattern of alveoles. [HAV]

Mustoe, G.E. (1982) The origin of honeycomb weathering. *Geological Society of America Bulletin* **93**: 108–15.

Trenhaile, A.S. (1987) *The Geomorphology of Rock Coasts*, pp. 48–51. Clarendon Press, Oxford.

amalgam (Ag, Hg) A naturally occurring *solid solution*. [DJV]

amazonite A green variety of the mineral **microcline**. (See **feldspar minerals**.) [DJV]

amazonstone A green variety of the mineral **microcline** valued as a semi-precious stone. (See **feldspar minerals**.) [DJV]

amber A mineral formed from fossilized resin valued in antiquity for the production of small decorative objects, e.g. beads and spacer-plates for necklaces. It may also have been prized because of its apparently magical property of attracting small particles when rubbed (as a result of static electricity). Areas on the south-east Baltic coast where amber may be found have been exploited since the Neolithic and Bronze Age, and records exist of both Greek and Roman amber collecting expeditions to the Baltic area. Smaller sources exist elsewhere in Europe (France, England, Rumania, Hungary and Sicily) and analyses of archeological amber by chemical means, *infrared spectroscopy* and *gas chromatography* have attempted to relate artefacts to a particular source area. [OWT]

amber ice Ice containing a dispersion of fine-grained sediments which impart an amber-like appearance. It is usually located in the uppermost sections of basal ice. [ASG]

amblygonite ($LiAlFPO_4$) A rare mineral found in **granite pegmatites** and exploited as a source of **lithium**. [DJV]

amethyst A purple variety of the mineral **quartz**. [DJV]

amosite An asbestiform variety of the **amphibole mineral cummingtonite**, *'brown asbestos.'* [DJV]

Amphibia/amphibians Class of vertebrate animals capable of free life on dry land, but requiring water in which to lay eggs and in which larval forms live until they metamorphose into the adult state.

The Amphibia can be most easily divided into two groups, the extinct forms known by the term '*Labyrinthodontia*' (**Devonian–Jurassic**), and the living forms comprising the *Urodela* (salamanders), the *Anura* (frogs and toads) and the *Apoda* (legless, snake-like animals). These latter are brigaded together as the *Lissamphibia*. There is in addition a further group of extinct forms whose relations to the other two are uncertain, the *Lepospondyli*. Lissamphibians are first seen in the **Triassic**, but the *lepospondyls* do not extend beyond the Lower **Permian**. The *lissamphibians* are supposed to show *labyrinthodont* affinities, although the *apodans* may be related to an Order within the *Lepospondyli*, the *Microsauria*.

The earliest body fossils of amphibians are from the Upper **Devonian** of Greenland, which were recovered in the 1930s and of which only tantalizing, brief descriptions exist to date. Two genera have been recorded from Greenland, *Acanthostega* and *Ichthyostega*, both of which seem very close to their *rhipidistian* **fish** (*Crossopterygii*) ancestors, even to the extent of having ossified fin rays in their tails. This is a fish-like character unknown in all other amphibians. However, tetrapod trackways are now known from the Lower **Devonian** of Victoria, Australia, so that the commonly known *rhipidistian* species are probably too late in time to be the actual ancestor of the known fossil amphibians. Current research will be able to clarify much of the mystery surrounding these animals.

Classification of the extinct forms has traditionally been based on the pattern of their vertebrae, as well as on skull morphology. The 'rhachitomous' (= *Temnospondyli*) vertebra is close to that of the *rhipidistian* **fish** and seems to be designed to allow maximum torsion in the body while 'walking'. This would be an advantage in an animal with the limbs splayed laterally from a relatively stubby body and possessing a relatively large head as did these *labyrinthodonts*. Any animal with this combination of features would have a problem with its centre of gravity when walking. A reverse analogy with the geodetic type of construction seen in some early (20th century) airships and aeroplanes has been made for these rhachitomous vertebrae, which allows maximum torsion, whereas the geodetic structure resists it.

The contrasting pattern of vertebra is the 'embolomerous' type, where each segment has a doubling up of the central elements. This adaptation was apparently to allow the maximum of lateral flexibility in a long eel-like body. At one time it was thought that the rhachitomous and embolomerous (= *anthracosaur*) vertebrae were mutually exclusive in any one individual, but in some cases in *rhipidistians*, the body may have rhachitomous vertebrae and the tail region embolomerous.

The latter pattern of vertebra could be more easily adapted to become that type seen in the **Reptilia**, the gastrocentrous vertebra. This has only one main central element. However there is a range of other extinct amphibians with a single centrum to each vertebra, but these are not homologous with the reptilian type. These vertebrae are found in a range of **Permo-Carboniferous** forms, e.g. the *Nectridea* and *Aistopoda* (= *Lepospondyli*) of small size and sometimes of bizarre morphology. The paradox of this type of centrum is that it conforms closely to those of the *Lissamphibia*, but there are few, if any, other characters in the *Nectridea* or *Aistopoda* which might link them specifically with the modern Orders. The *Microsauria*, from the **Carboniferous**, seem to come closest of the amphibian orders to the **Reptilia**. The other contestants for this role of reptilian ancestor are the *Anthracosauria*, also of **Carboniferous** age.

Whereas the early amphibians were adapted to living their adult lives on land, by the **Permian** they were becoming secondarily aquatic and modifying their shapes

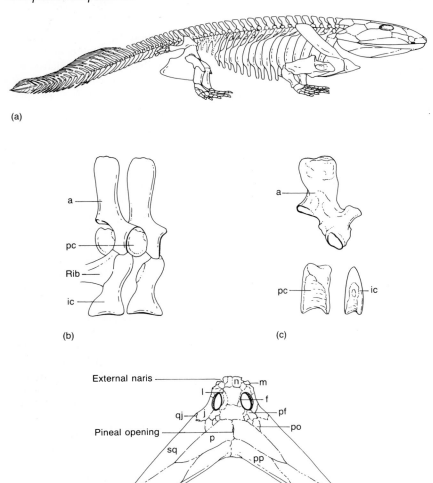

(a)

(b)

(c)

External naris

Pineal opening

(d)

Fig. A4 Amphibia/amphibians. (a) *Ichthyostega*, a primitive labyrinthodont from the Upper Devonian of East Greenland. Original about 1 m long. (After Jarvik, 1955.) (b) *Ichthyostega* from the Upper Devonian. The intercentrum (ic) is a crescent and the pleurocentra (pc) separate elements. a, neural arch. (After Jarvik, 1955.) (c) The embolomere *Eogyrinus*; both intercentra and pleurocentra are complete cylinders. (After Panchen, 1966.) (d) The skull of *Diploceraspis*, a horned nectridean of the early Permian. f, frontal; j, jugal; l, lacrimal; m, maxilla, n, nasal; p, parietal; pf, postfrontal; po, postorbital; pp, postparietal; qj, quadratojugal; sq, squamosal. (After Beerbower, 1963.)

to be dorso-ventrally flattened. This has been regarded as a degenerate feature, but one family within the *Nectridea* (the Keraterpetontidae) can show how this feature can be regarded as helping towards being a more efficient aquatic predator. In this family the skull in the genera *Diploceraspis* and *Diplocaulus* is boomerang-shaped and has been shown to have the properties of a swept-back wing of a high-speed aircraft. These two genera could rise steeply through the water to catch small prey swimming past and return quickly to their resting places. The two forms were adapted to contrasting environments; the former to quiet water and the latter to seasonally running rivers of a *monsoonal* region in the south-western United States.

Close analogy between the later temnospondyls and skates and rays (flattened *Chondrichthyes*) is not to be found. Rather these temnospondyls may have lived a life close to that of the modern *Crocodilia*. The truth lies close

to knowing that deep-bodied forms such as *Eryops*, living before the emergence of the **Reptilia** were efficient terrestrial predators and the flattened forms equally efficient aquatic predators.

The use of modern amphibians as models of forms intermediate between the **fish** and the **reptiles**, combined with their traditional use to teach dissection techniques, has lead to an erroneous view of the capabilities of their extinct relatives. Apart from the fact that the *Anura* (frogs and toads) are highly specialized in their bony skeleton and have an exceedingly poor fossil record prior to the **Jurassic**, their physiology is also undoubtedly totally at variance with that of the *Labyrinthodontia*.

Although the salamanders (*Urodela*) have a superficial resemblance to the large extinct forms, they share with the frogs a soft mucus-covered skin which is highly permeable to gases. Frog lungs provide only about one-third of

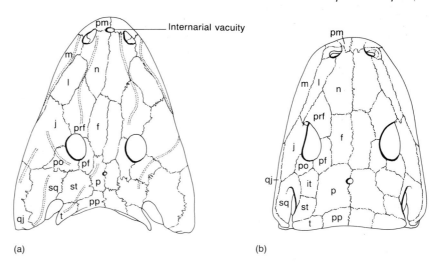

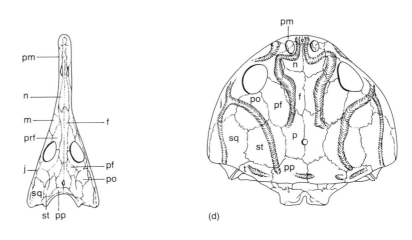

Fig. A5 Amphibia/amphibians. Skulls of various extinct amphibians to show the range of outline shapes. (a) *Rhinesuchus*, an Upper Permian temnospondyl from South Africa. (After Watson, 1962.) (b) *Seymouria*, a Lower Permian anthracosaur from the USA. (After White, 1939.) (c) *Wantosaurus*, a Lower Triassic trematosaur. (After Lehman, 1961.) (d) *Batrachosuchus*, a Lower Triassic brachyopoid from South Africa. (After Watson, 1956.) Drawings are not to scale. a, neural arch; f, frontal; ic, intercentrum; it, intertemporal; j, jugal; l, lacrimal; m, maxilla; n, nasal; p, parietal; pc, pleurocentrum; pf, postfrontal; pm, premaxilla; po, postorbital; pp, postparietal; prf, prefrontal; qj, quadratojugal; sq, squamosal; st, supratemporal; t, tubular.

the oxygen requirements of an active animal and many salamanders have lost their lungs altogether. Therefore the skin in these forms assumes a very important role in gas exchange. To assist this, the blood circulation in these forms allows the systemic flow to circulate to the body with a high $P\text{CO}_2$, bypassing the lungs. The ventricle of the heart is undivided, so permitting partially oxygenated blood from the lungs to mix with oxygen-deficient blood from the returning body circulation and so be cycled to the skin. It may be that using the skin as a gas-exchange organ allows respiration to continue at low levels of metabolism, such as during hibernation or estivation.

In contrast to the modern Orders, the extinct amphibians almost always had an extensive armour of dermal bony plates all over the body. Even if there was a vascular supply to the epidermis overlying the dermal armour, it is unlikely that such large forms as *Paracyclotosaurus* (**Triassic**) or *Eryops* (**Permian**) could have had an efficient gas-exchange system in their skin because of their relatively large bulk compared with the total body surface area. The assumption which also has to be made is that

oxygen levels in the atmosphere have remained constant over the last 300 Ma. If the large *labyrinthodonts* had been able to make use of skin respiration, atmospheric oxygen levels would have been much higher than compared to the present.

The modern amphibians, in common with their extinct relatives, always have to lay their eggs in water and the hatched larval forms (tadpoles) undergo a process of metamorphosis before reaching adulthood. Evidence for this in the fossil record lies in the preservation of external gill structures in small temnospondyl-like fossils known as branchiosaurs. Associated with an aquatic life-style is the presence, in **fish** and amphibians, of lateral line organs (pressure and vibration-sensing devices running in definite patterns on the surface of the head and body). Therefore the presence of either or both of these in a fossil tetrapod indicates that it must be an amphibian.

The status of *Seymouria*, a very reptilian-looking fossil tetrapod from the Lower **Permian** of Texas changed dramatically when its young stages were found to possess lateral-line organs. Up to that point it had been regarded as the earliest representative of the class **Reptilia**, possessing all the characters to be expected in such an animal.

It is not known what process was called into action that allowed a tetrapod laying a soft, jelly-covered egg in water to change to one laying a shelled egg with its associated desiccation-resisting membranes, on land. However, it is only at this level of organization that the real differences between an amphibian and a **reptile** can be made, and the fossil record cannot help. (See Figs A4, A5 and Table A2.)
[ARIC]

Alexander, R.McN. (1981) *The Chordates* (2nd edn). Cambridge University Press, Cambridge.

Beerbower, J.R. (1963) Morphology, paleoecology and phylogeny of the Permo-Pennsylvanian amphibian *Diploceraspis. Bulletin of the Museum of Comparative Zoology* **130**: 31–108.

Carroll, R.L. (1987) *Vertebrate Paleontology and Evolution.* W.H. Freeman, Chicago.

Jarvik, E. (1955) The oldest tetrapods and their forerunners. *Science Monthly* **80**: 141–54.

Lehman, J.-P. (1961) Les stégocéphales du Trias de Madagascar. *Annales de Paléontologie* **47**: 111–54.

Panchen, A.L. (1966) The axial skeleton of the labyrinthodont *Eogyrinus attheyi. Journal of Zoology, London* **150**: 199–222.

Romer, A.S. (1966) *Vertebrate Paleontology* (3rd edn). Chicago University Press, Chicago.

Watson, D.M.S. (1956) The brachyopid labyrinthodonts. *Bulletin of the British Museum (Natural History), Geology* **3**: 233–63.

Watson, D.M.S. (1962) The evolution of the labyrinthodonts. *Philosophical Transactions of the Royal Society, London* **245B**: 219–65.

White, T.E. (1939) Osteology of *Seymouria baylorensis* Broili. *Bulletin of the Museum of Comparative Zoology* **85**: 325–409.

amphibole minerals Silicate minerals whose internal structure consists of a double chain of linked silicate tetrahedra, with cations occupying sites formed between oxygen ions at the edges of the chains.

Table A2 Amphibia/amphibians. Classes, subclasses and orders

Class Amphibia
Subclass Labyrinthodontia
Order Ichthyostegalia
Temnospondyli
Anthracosauria
Subclass Lepospondyli
Order Nectridea
Aistopoda
Microsauria
Subclass Lissamphibia
Order Proanura
Anura
Urodela
Apoda

STRUCTURE. The amphibole structure (Fig. A6) has the appearance of a chain of linked hexagonal rings of SiO_4 tetrahedra. The 'bases' of the tetrahedra lie roughly in a plane parallel to the chain length. The apices of the tetrahedra, together with the oxygen of a hydroxyl group (OH) near the centre of the ring, lie approximately in a second, parallel, plane. The line of tetrahedra is staggered, or kinked, as in the **pyroxene** structure: the amphibole chain can be imagined as a pair of laterally linked **pyroxene** chains. The repeat along the chain length comprises two pairs of tetrahedra and is approximately 0.53 nm in length.

The chains are held together by metal ions at sites formed mainly by the non-bridging oxygens (those linked only to one Si). In particular, strong bonds form between apical oxygens and cations, so that chains associate in pairs with cations sandwiched between them. This gives

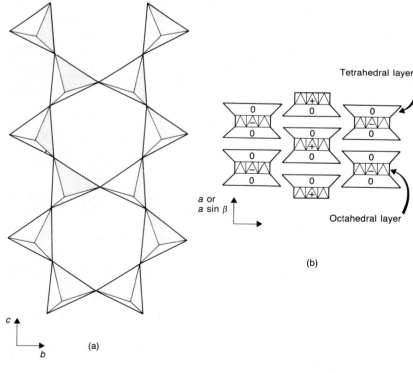

Tetrahedral layer

Octahedral layer

a or
a sin β

(b)

(a)

Fig. A6 Structure of **amphiboles**. (a) Chain structure viewed along the length of the chain. (b) Chain structure viewed end-on, showing the stacking of I-beams and the location of cations within the structure. (After Papike, 1969 and Papike & Ross, 1970.)

the 'I-beam' building block which is characteristic of amphiboles as of **pyroxenes** (Fig. A6). I-beams are held together in the overall structure by cations at sites formed by oxygens at the edges of the I-beam, and to some extent by cations which may occupy a large site formed by cavities in the backs of adjacent I-beams.

The direction of polymerization of the double chain unit defines the Z- or c-axis of the **crystal structure** according to the normal convention. Most naturally occurring amphiboles are monoclinic. The exceptions are orthorhombic: **anthophyllite** and **gedrite**, the lithium-containing amphibole **holmquistite**, and *joesmithite* which has **beryllium** in the tetrahedral chain. The structure gives rise to four crystallographically distinct sites for metal ions, two within the I-beam and two between the edges of adjacent I-beams, and to one large site, not always occupied, formed between their backs; and to two or (in orthorhombic species) four distinct tetrahedral sites, within the chain structure. The available ionic species are ordered to varying degrees among these sites.

CHEMISTRY AND NOMENCLATURE. The overall chemistry of the amphiboles can be represented as

$$A_{0-1}B_2C_5T_8O_{22}X_2$$

where in common amphiboles A is vacant or is Na or K; B is Na, Ca, Mg or Fe^{2+}; C is Mg, Fe^{2+}, Fe^{3+} or Al; T is Si or Al; and X is hydroxyl (OH) with some replacement by F or Cl, or by O^{2-}.

Many cations which are not readily accommodated in higher temperature minerals such as **olivine** (and which accumulate in the residual liquid of the later stages of igneous processes) can be taken up into the amphibole structure. In addition to the major ions listed above, the amphibole structure can accommodate the smallest cation, **lithium**, and much larger ions such as **potassium**. The wide range of amphibole composition is reflected in the number of different varieties which have been characterized and named.

Amphibole nomenclature was rationalized in 1978 on the basis of crystal chemistry. There are four principal groups: the iron−magnesium−manganese amphibole group; the calcic amphibole group; the sodio-calcic amphibole group: and the alkali amphibole group. Classification details are given in Table A3. The interested reader is referred to Leake (1978) for fuller details, including the names now recognized and those which have been formally abandoned.

Some chemical variations in amphiboles are continuous. For example, there is continuous *solid solution* between **tremolite** $Ca_2Mg_5Si_8O_{22}(OH)_2$ and **actinolite** $Ca_2Fe_5^{2+}Si_8O_{22}(OH)_2$. However, simple replacement relationships of this kind are rare. More commonly, linked substitution takes place involving several cations, sometimes of different valency. For example, the transition from **anthophyllite** to **gedrite** in the orthorhombic structure involves not only a change in the Mg/Fe^{2+} ratio but also the replacement of Si by trivalent Al in the tetrahedral sites; the charge balance is maintained by introduction of Na (at A in the formula) and of significant trivalent ions (Al and Fe^{3+}) replacing divalent Mg and Fe^{2+}.

In general, where the chemistry of the different species is very different — specifically, where divalent Mg or Fe are replaced either by trivalent Al or Fe^{3+}, or (separately or at the same time) by Ca and/or Na — the adjustment of the structure to accommodate the different charge and size pattern is reflected in incompatible structures and in consequence there is no *solid solution*. For example, there is a miscibility gap between **anthophyllite** and **gedrite** except at elevated temperatures and, similarly, only limited miscibility between **tremolite−actinolite** and the Al-containing **hornblende**. The **tremolite−actinolite** series is continuous: however, simple replacement of more than about 30% of the Mg in the orthorhombic structure of **anthophyllite** $(Mg_7Si_8O_{22}(OH)_2)$ by Fe^{2+} causes a change to the monoclinic structure of the **cummingtonite−grunerite** series.

Frequently a mineral which forms as a single phase may, as it cools, exsolve lamellae of one phase within the other. Assemblages of three amphiboles formed in equilibrium (e.g. **hornblende** + **anthophyllite** + **cummingtonite**) and of four (including both **anthophyllite** and **gedrite**) are known.

OCCURRENCE. Amphiboles occur in both igneous and metamorphic rocks. From the earliest days they have played a central role in studies of metamorphic rocks where they are formed from other, usually more hydrous, assemblages or by hydration of igneous rocks. At higher metamorphic grade they become unstable by reactions involving their dehydration, e.g.

hornblende + **quartz** → *orthopyroxene* + **augite** + **plagioclase** + water

Conversely, at the surface they weather to more hydrous sheet silicates (e.g. **anthophyllite** to **talc**). Thus, amphiboles are stable at intermediate conditions of temperature, pressure and H_2O content. For example, the stability field of the compositionally simple **anthophyllites** at a water pressure of 100 MPa extends over the approximate range 500 to 800°C.

Table A3 **Amphibole minerals.** The recognized amphibole groups. (After Leake, 1978.)

General formula: $A_{0-1}B_2C_5T_8O_{22}X_2$

(a) When $(Ca + Na)_B < 1.34$, then the amphibole is a member of the iron−magnesium−manganese amphibole group

(b) When $(Ca + Na)_B > 1.34$ and $Na_B < 0.67$, then the amphibole is a member of the calcic amphibole group. Nearly all such natural amphiboles have $Ca_B > 1.34$

(c) When $(Ca + Na)_B > 1.34$ and $0.67 < Na_B < 1.34$, then the amphibole is a member of the sodio-calcic amphibole group. Such natural amphiboles usually contain $0.67 < Ca_B < 1.34$

(d) When $Na_B > 1.34$, then the amphibole is a member of the alkali amphibole group

Amphiboles occupy a central position in the range of bulk chemistries of rocks. They can occur, for example, in SiO_2-rich rocks with **quartz**, and in SiO_2-poor rocks with **olivine** and **nepheline**. Similarly, they crystallize or break down by reactions between minerals which lie on either side of them in composition, e.g. **pyroxenes** and sheet silicates. It is this central position in terms of thermodynamic stability, chemical composition and structure which enables them to play a wide role in metamorphism of a wide variety of rocks, so long as H_2O, or equivalent F or Cl, is available. A meteoritic amphibole with fluorine and O^{2-} in place of OH has been reported.

Oxidation of amphiboles proceeds most readily for iron-containing species and probably involves simultaneous oxidation of the hydroxyl at the centre of the chain (OH^- to O^{2-}) and of nearby Fe^{2+} to Fe^{3+}, thus maintaining charge balance. This may happen initially without disruption of the **crystal structure**, but further oxidation results in its breakdown.

In the igneous environment, the characteristic amphibole is **hornblende** which may form at a late stage in the crystallization of almost any except **olivine**-containing compositions (**olivine gabbro, peridotite, dunite**). It is commonly the dominant mafic phase in rocks of intermediate SiO_2 content (**diorites** and **quartz diorites**), competing with **biotite** as the bulk composition becomes more silicic and potassic. There is a great deal of overlap between the stability fields of amphibole and of **mica**.

Pyroxene and amphibole are the important chain silicate structures in terms of rock-forming minerals. However, in recent years several non-classical chain structures have been discovered. These phases were first found as domains within *orthoamphiboles*, using techniques of **electron micrography**. In some cases, the domains of multi-chain structures are large enough to be identified as distinct **pyriboles**. Small scale dislocations to the chain structure are now widely known in amphiboles and their presence adds further complexity to the already complex character of the group.

RECOGNITION. The wide chemical variation of the amphiboles is reflected in their range of colours: for example, white (**anthophyllite**), dark green (**hornblende**), blue (**riebeckite**). Increasing darkness of colour generally correlates with increasing iron content. Varieties containing significant Mn may be pinkish. A **streak** test normally shows white.

The chain structure of the amphiboles gives rise to their characteristic prismatic habit. In some varieties (**anthophyllite, tremolite**, and the fibrous variety of **riebeckite** formerly known as **crocidolite**) this can take the extreme, fibrous form of **asbestos**. In others, the typical occurrence is as radiating 'bundles' of needle-like crystals.

Amphiboles normally have vitreous **lustre**, and perfect **cleavages** which can be seen in hand specimen or thin section. Hardness is normally 5 or 6 on Mohs' scale. Density is normally between 3.0 and $3.5\,Mg\,m^{-3}$ increasing with iron content; higher densities are known at extreme compositions, e.g. a density of over $3.8\,Mg\,m^{-3}$ for a ferro-**gedrite** in composition close to $Fe_5Al_2 . Si_6Al_2O_{22}(OH)_2$ from Kitakami, north-east Japan.

In thin section, amphiboles appear as crystals elongated parallel to (010). They are readily recognized by the commonly occurring diamond pattern of **cleavage** traces, at angles of about 56° and 124°, seen when viewed normal to the c-axis. For monoclinic amphiboles, the angle between c- and the Z-axis varies between about 10° and 30°. **Refractive indices** vary between about 1.6 (**anthophyllite** high in Mg) and 1.7. Coloured species are pleochroic, the green to brown pleochroism of **hornblende** being particularly characteristic. Most optical properties such as pleochroism, **refractive indices** and birefringence are composition dependent and are determined to a first approximation by **iron** content. Where **exsolution** has taken place, it may be possible to see **exsolution** lamellae (narrow domains of one amphibole within the other) under the microscope: however, in many cases **exsolution** takes place on a sub-microscopic scale. [ADL]

Deer, W.A., Howie, R.A. & Zussman, J. (1978) *Rock Forming Minerals* Vol. 2B: *Double Chain Silicates* (2nd edn). Longman, Harlow.

Ernst, W.G. (1968) *Amphiboles*. Springer-Verlag, New York.

Leake, B.E. (1978) Nomenclature of amphiboles. *American Mineralogist* **63**: 1023–52.

Papike, J.J. (ed.) (1969) *Pyroxenes and Amphiboles: Crystal Chemistry and Phase Petrology*. Special Paper of the Mineralogical Society of America **2**.

Papike, J.J. & Ross, M. (1970) Gedrites: crystal structures and intracrystalline cation distributions. *American Mineralogist* **55**: 1945–72.

Veblen, D.R. (ed.) (1983) *Amphiboles and Other Hydrous Pyriboles — Mineralogy*. Reviews in Mineralogy Vol. 9A. Mineralogical Society of America, Washington, DC. [This and the following reference comprise course notes from a summer course arranged by the MSA and contain authoritative material covering a wide field of amphibole mineralogy.]

Veblen, D.R. & Ribbe, P.H. (eds) (1983) *Amphiboles — Petrology and Phase Relations*. Reviews in Mineralogy Vol. 9B. Mineralogical Society of America, Washington, DC.

amphidromic point A nodal point in the ocean or sea at which there is no vertical rise or fall of **tide**. A simplified tidal oscillation may be viewed as a **standing wave** with a nodal line of no vertical rise and fall of water level. However, the Earth's oceans and seas are subject to the effects of the Earth's rotation. These **Coriolis effects** act such that there are nodal points (amphidromic points) rather than nodal lines around which **tidal currents** swing in an anti-clockwise direction in the Northern hemisphere. Tidal range increases away from amphidromic points. On tidal charts an amphidromic system comprises co-tidal lines or lines of equal phase which radiate from amphidromic points, and co-range lines of equal amplitude or tidal range which encircle amphidromic points. Although there is no rise or fall of water at amphidromic points, **tidal currents** tend to be fastest at them. [AESK]

Howarth, M.J. (1982) Tidal currents of the continental shelf. In: Stride, A.H. (ed.) *Offshore Tidal Sands: Processes and Deposits*, pp. 10–26. Chapman & Hall, London.

Bowden, K.F. (1983) *Physical Oceanography of Coastal Waters*. Ellis Horwood, Chichester.

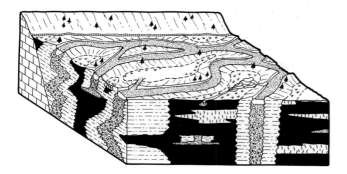

Fig. A7 Block diagram of an **anastomosing river**. In this typical reach, channel sediments (gravel symbol) are bounded by sandy silt (dash, dot symbol) of the levées, which turn into muds and silty muds of the wetlands (black). Peats are shown by small vertical wiggles. Note channel aggradation without significant lateral accretion, the channel pattern being stabilized by the muds and organic material of the wetlands, which are hard to erode. (After Smith & Smith, 1980.)

amygdale A **vesicle** in a **lava flow** or a **pyroclastic rock** that has been filled in by low-temperature minerals such as **calcite, zeolites** or **quartz**. [PTL]

anabranch A river channel pattern where the width of the islands in a channel is relatively wide (generally more than about three times the river width at average discharge). [ASG]

analcime ($NaAlSi_2O_6H_2O$) A **zeolite mineral**. [DJV]

anastomosing river A river whose plan form is characterized by branching and rejoining sinuous channels that divide and rejoin on a length scale many times the individual channel widths. The channels tend to be laterally stable with localized meander migration. Deposition is dominantly by vertical accretion of the **floodplain**. Anastomosing rivers occur in low gradient swampy areas often resulting from upstream tilting of the **floodplain** or the development of a downstream barrier. They may also develop during periods of low precipitation where a river flows over an older **braidplain**. (See Fig. A7.) [JA]

Rust, B.R. (1981) Sedimentation in an arid-zone anastomosing fluvial system: Cooper's Creek, Central Australia. *Journal of Sedimentary Petrology* 51: 745–56.
Smith, D.G. & Smith, N.D. (1980) Sedimentation in anastomosed river systems. Examples from alluvial valleys near Banff, Alberta. *Journal of Sedimentary Petrology* 50: 157–64.

anatase (TiO_2) A relatively rare mineral form of TiO_2. (See **oxide minerals**.) [DJV]

anatexis A process of partial melting and fusion that produces a melt of smaller proportion than that of the original solid rock, so leaving a refractory, unmelted residuum. In such cases the composition of the melt depends upon the phase relationships within the solid at the T–P conditions of melting. Examples of the process include **mantle** anatexis in which **mantle peridotite** is partially melted to form **basalt magmas** and **crustal** anatexis in which **crustal** rocks are partially melted to form **granite** and **granodiorite magmas**. [RST]

andalusite (Al_2SiO_5) An important mineral in metamorphic rocks, an island silicate (*nesosilicate*). [DJV]

Andean mountain belt This South American *orogenic belt* is generally considered as typical of mountain formation as a consequence of **subduction** bordering a continent, without the presence of an intervening *marginal basin*. The number of *exotic terranes*, i.e. (continental blocks derived from elsewhere and carried by **seafloor spreading** to collide and become incorporated with the *orogenic belt*) is far fewer than for the Western Cordillera of North America, and differs drastically from the continent–continent collisions within the **Himalayan mountain belt**. However, there are little understood motions along major **fault** systems that complicate the situation in Venezuela and Colombia, while the existence of the area as an intermittently active orogenic area through most of the **Phanerozoic** suggests that a simple **subduction** model is inadequate. [DHT]

andesine A **plagioclase feldspar** mineral. (See **feldspar minerals**.) [DJV]

andesite A volcanic rock of intermediate chemical composition, commonly **porphyritic**, containing **phenocrysts** of zoned sodic **plagioclase** (e.g. **andesine**) together with mafic minerals such as **biotite, hornblende** and **pyroxene** in a groundmass composed of the same minerals commonly with more sodic **plagioclase** and **quartz**. Andesite is the extrusive equivalent of **diorite**, and grades into **latite** with increasing **alkali feldspar** content and into **dacite** with increasing **alkali feldspar** and **quartz**.

Andesite is a characteristic of **calc-alkaline basalt–andesite–dacite–rhyolite** associations erupted at **island arcs** and **active margins**. The petrogenesis of such associations is related to the **subduction** of oceanic **lithosphere** below oceanic **island arcs** or continental **lithosphere**. This leads to dehydration of **oceanic crust** (which has experienced seafloor **alteration**) releasing fluids and trace elements into the overlying **mantle** wedge. Partial melting of such subduction-enriched **mantle**-generated **basaltic** magmas which experienced *fractional crystallization* and other processes of **magmatic diversification** to yield **magmas** ranging from dominantly **basaltic** (in volcanic

island arcs) to progressively more evolved **basalt**−andesite and andesite−**dacite** associations at **active margins**. [RST]

Gill, J.B. (1981) *Orogenic Andesites and Plate Tectonics*. Springer-Verlag, Berlin.

Thorpe, R.S. (1982) *Andesites: Orogenic Andesites and Related Rocks*. John Wiley & Sons, Chichester.

andradite ($Ca_3Fe_2Si_3O_{12}$) A garnet group silicate mineral. (See **garnet minerals**.) [DJV]

anelasticity Time-dependent elasticity; i.e. basically elastic (recoverable) **strain** that deviates from the ideal elastic property of instantaneous **recovery**. [RGP]

angle of contact The angle between a liquid and a solid. If a liquid has a high tendency to 'wet' the solid surface, it will spread out over it forming a thin film and the angle of contact between the liquid and solid surfaces will be substantially zero. If, on the other hand, it has less tendency to wet the surface, it will form discrete globules rather than films and there will be a finite angle of contact with the solid. [AME]

Price, M. (1985) *Introducing Groundwater*. Allen & Unwin, London.

angle of internal friction The theoretical angle relating **shear stress** τ required for sliding to the *normal stress* (σ) acting on a plane. The angle is defined as the inverse tangent of the *coefficient of friction* (μ) in either *Amonton's Law* for frictional sliding:

$$\tau = \mu\sigma$$

or the *Coulomb criterion* for failure:

$$\tau = S + \mu\sigma$$

where S is the *cohesive strength*. The angle of internal friction is

$$\phi = \tan^{-1}\mu$$

For a *Coulomb failure criterion*, ϕ can be used to predict the angle between the **failure** surface and the maximum *principal stress*, γ:

$$\gamma = 45° - \phi/2$$

However, there is no satisfactory physical interpretation for the concept of internal friction, which was created by analogy with the angle of slope for sliding under **gravity**. [TGB]

Jaeger, J.C. & Cook, N.G. (1979) *Fundamentals of Rock Mechanics* (3rd edn), pp. 95–8. Chapman & Hall, London.

anglesite ($PbSO_4$) A mineral formed from **weathering** of primary **lead ores**. [DJV]

angular velocity Velocity measured as the rate of change of angle between two lines; used specifically in *plate* motion to refer to the rate of angular rotation of one *plate* with respect to another, about their common axis of rotation. Unlike the *tangential velocity*, which varies from zero at the *pole of rotation* to a maximum at 90° from the pole, the angular velocity is constant for a specific *plate* rotation. (See **plate tectonics**.) [RGP]

anhydrite ($CaSO_4$) An **evaporite** mineral. [DJV]

anisotropic Not isotropic. An isotropic substance exhibits no directional variation in physical properties. The term 'anisotropic' in structural geology is borrowed from crystallography, where its meaning is more precise, and signifies a lack of **preferred orientation** of **fabric** or structural arrangement, i.e. a random or near-random structural state. [RGP]

ankaramite A **porphyritic**, **phenocryst**-rich alkali *olivine basalt* (**pyroxene** > **olivine**) with a groundmass of **augite**−*titanaugite* **microlites**, Ca-rich **plagioclase** (**labradorite**) and **biotite**. Ankaramites occur in association with **basanite** and alkali **basalt**. [RST]

ankerite ((Ca, Fe)CO_3) A **carbonate mineral** of the **dolomite** group. [DJV]

annabergite ($Ni_3(AsO_4)_28H_2O$) *Nickel bloom*, a green secondary mineral of **nickel**. [DJV]

annealing Movement of grain boundaries and grain growth in a polycrystalline rock. This can be driven by elevated temperatures or can occur slowly over long time periods. Annealing is a **recovery** process, as **deformation** features such as **fractures**, lattice defects and non-equant grain shapes can be obliterated. In a monomineralic rock such as **marble**, annealing produces a characteristic texture, of equant grains and a predominance of 120° grain boundary intersections. The structure in three dimensions resembles froth in a bottle, for example. Annealing usually produces an increase in grain size, with a drop in surface energy resulting from the decrease in proportion of grain boundary volume to total rock volume. (See **grain size coarsening**, **recovery**, **normal grain growth**, **dislocation**.) [SB]

annular drainage A curved or ring-like drainage pattern in which the *subsequent streams* follow curved courses as discontinuous portions of what appear to be near concentric circles. This form of structural drainage guidance occurs particularly around dissected **domes** or **basins**. [NJM]

anomalous lead Lead having an isotopic ratio that produces a demonstrably incorrect model lead age. [AME]

anorthite ($CaAl_2Si_2O_8$) Calcic end-member of **plagioclase feldspar** *solid-solution* series. (See **feldspar minerals**.) [DJV]

anorthoclase An **alkali feldspar** mineral. (See **feldspar minerals**.) [DJV]

anoxic Of, or pertaining to, an environment without or with only very low levels of dissolved oxygen. 'Anoxic' has been used to describe conditions both in the water column and within sediment **pore-waters** where the concentration of dissolved oxygen is insufficient to support metazoan benthos, resulting in the preservation of primary structures such as lamination in the sediment. This limiting level of oxygen concentration has been cited as $0.5\,ml\,l^{-1}$ (Demaison & Moore, 1980) although more recent studies of benthic activity within the **oxygen minimum zone** suggest that this lower limit is $0.1\,ml\,l^{-1}$. Environments with less than this limiting level are also termed *azoic* or *anerobic*. After available oxygen has been utilized, *anerobic* organisms use other oxidants to break down organic matter in the order: nitrate, manganese oxide, **iron oxide**, sulphate. The term 'anoxic' has been applied to systems in which oxygen and other oxidants have been exhausted and in which sulphate is being reduced to oxidize organic matter (Calvert in Brooks & Fleet, 1987). Such environments are characterized by the presence of free sulphide and are also termed '**euxinic**' (from the ancient Latin name for the Black Sea, Pontus Euxinus). Thus, anoxic conditions are the result of a high flux of organic **carbon** and/or a lack of oxygen replenishment to bottom waters. Ancient sediments deposited under anoxic conditions may be identified by their preserved lamination and by the presence of **pyrite**, although a lack or scarcity of **pyrite** may reflect lack of free iron rather than lack of sulphide. Environments which promote the development of anoxic conditions include silled **basins** such as the Black Sea in which there is little renewal of bottom waters; the upwelling zones of continental margins where high fluxes of organic **carbon** utilize large amounts of available oxygen; large tropical lakes in which a permanent **thermocline** and stratification of the water column prevent replenishment of bottom waters. It is important to note that sediments deposited in anoxic bottom-waters need not be **black shales** as they may contain only a small amount of organic **carbon**. The term *'anoxic event'* has been used to refer to geographically widespread but stratigraphically limited intervals of organic-rich deposition, e.g. the Turonian *anoxic event*. Anoxia is the state of being anoxic. (See **black shale**.) [AESK]

Brooks, J. & Fleet, A.J. (eds) (1987) *Marine Petroleum Source Rocks*. Blackwell Scientific Publications, Oxford.

Demaison, G.U. & Moore, G.T. (1980) Anoxic environments and oil source bed genesis. *American Association of Petroleum Geologists Bulletin* **64**(8): 1179–209.

antecedent drainage A drainage system that has maintained its general direction across an area of localized **uplift** by cutting down at a faster rate than the rate of **uplift**. It is, with **superimposed drainage**, one of the main ways in which river systems become discordant to geological **structures**, and it is an explanation for the development of many gorges. [ASG]

antecedent platform theory The theory that **coral** *reefs* and **atolls** formed by upward growth following colonization of submarine banks built up to a suitable depth by **uplift** or **accretion**. The central problem of *reef* development concerns the upward growth of *reefs* from beyond depths at which **corals** can survive. Alternative views include the subsidence theory which regards *reefs* as the result of upward growth of ancient *reefs* fringing land that has since **tectonically** subsided. The *glacial control theory* suggests that marine platforms abraded during *glacial* low sea-levels were the sites of post-glacial *reef* growth. Current opinion favours a combination of subsidence and *glacial control theory* to explain *reef* landforms. [JDH]

Bird, E.C.F. (1984) *Coasts*. Basil Blackwell, Oxford.

anthophyllite (($Mg, Fe)_7Si_8O_{22}(OH)_2$) An **amphibole mineral**. [DJV]

anthracite A high rank **coal** with a very low volatile content and a high **carbon** content. (See **coal**, **coal classification**.) [AME]

anthracitization The conversion during metamorphism of **bituminous coal** into **anthracite**. [AME]

anthraxolite A hard, black **asphalt** material that occurs in **veins** and masses in sedimentary rocks, especially **oil shale**. [AME]

anthropogeomorphology The study of the role of humans as geomorphological agents. There are few spheres of human activity that do not create landforms, either directly or indirectly. Landforms produced by direct anthropogenic processes are relatively obvious in form and origin, and are frequently created deliberately and knowingly. On the other hand, landforms produced by indirect anthropogenic processes are often less easy to recognize: often they do not so much involve the operation of a new process or processes as the acceleration or deceleration of natural processes. They are the result of environmental changes brought about inadvertently by human technology. [ASG]

Brown, E.H. (1970) Man shapes the earth. *Geographical Journal* 136: 74–85.

Goudie, A.S. (1986) *The Human Impact on the Environment*. Basil Blackwell, Oxford.

anticlinal trap An **oil trap** consisting of a geometrical arrangement of strata that dip away from a relatively high central area. It is not necessarily an anticlinal fold *per se* and may not have been generated by the normal processes of **folding**. In this wider sense the majority of gas- and **oilfields** may be said to occur in anticlinal traps. [AME]

North, F.K. (1985) *Petroleum Geology*. Allen & Unwin, Boston.

anticline **Fold** containing older rocks in its *core*. The term is also used synonymously, but incorrectly, for **antiform**. An anticlinal **fold** may close in any direction, downwards and sideways as well as upwards, and its attitude must be further defined by the use of terms such as **antiform**, or by a description of *limb* orientation. [RGP]

antidune Bedform whose crest is transverse to the flow and which is essentially symmetrical in streamwise section and associated with in-phase water surface **waves** which mirror the bed surface profile. These bedwaves are commonly of low amplitude and occur in trains associated with supercritical flows (**Froude number** is greater than 1). As such they are largely found in fast, shallow flows. Antidunes display a cycle of growth whereby their height increases until the water surface wave breaks upstream. Consequently, although preservation is rare, upstream dipping stratification, separated by internal **erosion surfaces**, may be produced due to the breaking of successive **waves**. (See **flow regime**.) [JLB]

antiferromagnetism The form of **ferromagnetism** (*s.l.*) in which the magnetic lattices within a grain are exactly equal and oppositely **spontaneously magnetized** by quantum-mechanical super-exchange forces. Antiferromagnetic materials have zero external spontaneous **magnetic field**. The common **iron oxide**, **hematite**, has a fundamental antiferromagnetic structure, but its lattices are distorted, giving rise to a weak parasitic magnetization that means that it behaves as a weak antiferromagnetic material. [DHT]

antiform **Fold** that *closes* upwards, i.e. in which the *limbs* diverge downwards. [RGP]

antigorite ($Mg_3Si_2O_5(OH)_4$) A platy layer silicate (*phyllosilicate*) mineral of the **serpentine** group. [DJV]

antimony (Sb) A rare mineral in **vein** deposits.

Antimony was added to early glass, particularly from the 6th century BC onwards, as a decolorant. (See **native element**.) [OWT/DJV]

antithetic Opposed to the prevailing sense of **vergence** or asymmetry. The term is applied most commonly to *thrusts* and **thrust belts**, but may be used for any kind of structure or **fabric**. (See **synthetic**.) [RGP]

antlerite ($Cu_3SO_4(OH)_4$) A secondary mineral formed from **alteration** of **copper ores**. [DJV]

apatite ($Ca_5(PO_4)_3(F, Cl, OH)$) A mineral important as a source of **phosphate** for fertilizers. [DJV]

apex Term used in USA mining law to denote the outcrop of a **vein** or *lode* that reaches the surface or the highest point of a **vein** or *lode* that does not extend to the surface. [AME]

apex law A peculiar USA mining law, abandoned by mutual agreement in some US orefields, whereby the holder of the mining rights to the outcrop or **apex** of a **vein** may mine it down dip, even though it passes outside the vertical projection of his own claim boundaries. This law has caused endless confusion and litigation in orefields where the **veins** are complex or **faulted**, or where the **ore** is not in **veins** at all. [AME]

API gravity The Americal Petroleum Institute standard for expressing the specific gravity of oils. In relation to the density (ρ), API value = $141.5/\rho - 131.5$. This gives water, under standard temperature and pressure (STP) conditions, a value of 10° API. Oils with API values higher than 30° are considered light, 30°–22° medium and less than 22° heavy. [AME]

aplite A fine- to medium-grained **igneous rock** occurring as thin (generally less than 200 mm) **veins** within coarser-grained hypabyssal or **plutonic** rocks. Aplites may range from **gabbro** to **granite** in composition but the term used without qualification is taken to indicate a rock of **granitic** composition and such aplites may occur within a wide range of silica-oversaturated igneous rocks. Aplites have a distinctive subhedral–anhedral fabric with a sugary or saccharoidal texture. Aplites are commonly associated with **pegmatites**, often showing complex intrusive relationships. While **pegmatites** are interpreted as forming from volatile-rich residual **magmas**, the uniform fine grain size for aplites indicates formation by squeezing out of residual **granite magma** (associated with loss of volatiles causing crystallization) into **fractures** or **joints** formed during a late stage of cooling of the intrusion. [RST]

apophyllite ($KCa_4(Si_4O_{10})_2F, 8H_2O$) A layer silicate (*phyllosilicate*) mineral. [DJV]

apophysis A **vein** or protruberance, typically with a closed end, that is connected either visibly, or by inference, to a larger body of intrusive rock; often used to describe small (centimetre-scale) branches of **dykes**, **sills** or mineral **veins**. [RGP]

apparent dip The angle of inclination of a given plane (usually **bedding**) with the horizontal, measured in a plane that is not orthogonal to the **strike** of the given plane (see **true dip**). The angle of apparent dip measured in a series of vertical planes varies from zero parallel to the **strike**, to a maximum in the direction of **true dip**.

If the angle of apparent dip in two different directions is measured, the **true dip** can be calculated (e.g. using a *stereogram*). [RGP]

apparent polar wander If **paleomagnetic** poles are plotted for different ages for any given tectonic block, they form a pattern, usually curved, that shows a change with time averaging some $0.3° \, \text{Ma}^{-1}$ during the last 300 Ma, although each block has a different shape and rate of change. This pattern is termed the apparent polar wander curve or path for that tectonic unit as it arises from the movement of each block relative to the mean geomagnetic pole and not to an actual motion of the pole. The paleomagnetic pole, for any given time, appears to correspond with the Earth's axis of rotation and therefore can be used to determine the *paleolatitude* of the tectonic block and its rotation relative to the pole. [DHT]

Khramov, A.N. (1987) *Paleomagnetology*. Springer-Verlag, Berlin.
Tarling, D.H. (1983) *Palaeomagnetism*. Chapman & Hall, London.

apparent resistivity The **electrical resistivity** measured over an inhomogeneous medium. In the case of the **resistivity method**, the measured resistivity over such a medium varies with the location and separation of the electrodes. This apparent resistivity is a complex weighted mean of the resistivities present in the subsurface. [PK]

appinite A term applied to a heterogeneous group of medium- to coarse-grained **hornblende**- and **biotite**-rich **igneous rocks** varying from subalkaline (**diorite**) to alkaline (**syenite**) in composition and containing prominent prismatic **hornblende phenocrysts** (and groundmass **hornblende**). Appinites form small intrusions (e.g. **dykes**), or occur as a marginal facies of related **plutonic** rock types. [RST]

applied geomorphology The field of **geomorphology** which applies geomorphological knowledge and research techniques to solving environmental problems and providing support for **environmental engineering** projects. While it may be argued that geomorphological research should always be applied to answering pertinent questions, the term applied geomorphology is usually reserved for those areas of research which pertain directly

to utilitarian issues. Often the role of the applied geomorphologist is to evaluate environmental hazards and recommend courses of action which minimize disruption of the environment (see **environmental engineering geomorphology** and **environmental impact statement**). As such, the geomorphologist's brief is to provide a service for planners, engineers and environmentalists who sponsor multidisciplinary studies of interactions between human beings and the environment. As **geomorphology** has become more concerned with processes of landform evolution, and with the human impact on natural landscapes (Goudie, 1981), so geomorphologists have become increasingly valuable as scientists capable of tackling complex environmental questions.

There are four broad areas of applied geomorphological research: mapping of landforms and surficial materials; process studies; investigations of the causes of environmental change; and evaluations of human modification of the natural environment. While the methodologies involved in these areas of research are often different, many studies — notably those for **environmental impact statements** — can involve aspects of all four.

The first stage of many geomorphological investigations is the production of a map of the topography and surface features of the area of interest. Geomorphological mapping may involve the use of *aerial photography* or *satellite imagery* (see **remote sensing**) to identify landforms or different types of terrain. In all cases, however, geomorphic maps require field checking. This may involve simple verification that features identified on the imagery have been correctly interpreted. In other cases, the field study may require more detailed topographic surveys or sampling of surficial materials for laboratory analysis. Some of these investigations may themselves provide the information required; resource surveys, for example, may be undertaken to identify surface sediments suitable for building materials, or mineral deposits suitable for exploitation. In other cases, the geomorphological survey may provide the base map for a more specific study of, for example, the distribution of *soils* or water resources.

Studies of geomorphic processes have become an increasingly important area of **geomorphology** over the past two or three decades. They are especially significant in applied geomorphology because they provide an understanding of how the natural environment is evolving. Moreover, a knowledge of how landforms have developed enables the geomorphologist to evaluate what natural processes are at work shaping the present landscape. In order to achieve this level of understanding, geomorphic phenomena must be monitored in the field, or their evolution must be ascertained from historical sources. Field monitoring of gullying or coastal **erosion**, the movement of **dunes** or **glaciers**, and sedimentation in rivers or lakes, can provide valuable information on the rates of change. However, many natural processes are slow or episodic. Hence, they are difficult or impossible to monitor in the field over short periods. In such cases, geomorphologists may employ archeological data, archival records such as early maps and historical accounts, to reconstruct former environmental conditions. Today, in many parts of the world, *aerial photograph* coverage is available for the past few decades. Even in some of the remotest parts of Africa, for example, *aerial photographs* dating back 40 years or more are available. By comparing relatively long-term data from such records with more detailed, short-term measurements from field monitoring, geomorphologists can begin to quantify the rate of environmental change. This becomes particularly important when it is necessary to estimate the likely frequency of episodic phenomena such as mass movements or floods. Similarly, the evaluation of other natural hazards — such as desert **dune** encroachment, *soil erosion*, or gullying — must be based upon accurate data obtained through rigorous empirical studies.

Having mapped the geomorphic features of an area, and perhaps monitored the processes which currently shape the landscape, the next phase of many geomorphological studies is to identify the causes of any changes which have occurred in the environmental processes. This is especially important when human agencies have influenced geomorphic processes and events. While the detrimental effects that **accelerated erosion** can have on agricultural productivity or sedimentation in rivers may be obvious, it is essential to determine the extent to which human activities have altered or exacerbated natural phenomena (see **anthropogeomorphology**). This can prove particularly difficult in the developed world where human alteration of the natural environment has been both pronounced and prolonged. Elsewhere, the absence of historical records may again make it difficult to assess the impact of human beings on the environment. In these cases, the geomorphologist can employ sedimentological, archeological, and **palynological** data — as well as a wide range of additional information — to facilitate *paleoenvironmental reconstructions*. Having achieved this, however, it must be acknowledged that subsequent environmental changes may be caused by natural forces — such as climatic change — as well as human influences.

Accidental or incidental human disruption of the environment can have a marked effect on geomorphic processes. Though accepting that natural responses to human interference may be complex and unpredictable, the fourth broad area of applied geomorphology focuses on environmental modification, and ways in which environmental hazards may be avoided or tolerated in order to facilitate economic development. This is the area of **environmental engineering geomorphology**. The applied geomorphologist may be called upon to recommend or evaluate engineering procedures intended to remedy environmental hazards — for example, the construction of groynes along eroding **beaches**, or sand fences in sandy deserts. In other cases, geomorphologists are involved at the planning stage of engineering projects.

A proposed facility, such as a road, airport, or industrial development, can be designed and located to minimize potential environmental hazards. Applied geomorphologists have become increasingly involved in such decisions involving engineering projects in arid regions where salt **weathering** and wind-blown **sand** pose an environmental threat, and where poorly managed irrigation schemes have increased the likelihood of *soil* salinization. Also in Arctic environments, where *permafrost* can create engineering problems, and in developing countries where **accelerated erosion** has resulted from the wholesale removal of natural vegetation as population pressure has increased the demand for agricultural land, there is a growing need for applied geomorphological research.

This broad range of topics which falls within the scope of applied geomorphology extends from the identification of hazardous waste sites in industrialized, urban centres to evaluation of potential natural resources in developing countries. In the industrialized world, growing concern about environmental pollution has led to an increase in the funding of applied geomorphological research. Research projects are being sponsored not only by government agencies but also by engineering corporations. There is an increasing awareness that geomorphological studies are an essential part of many **environmental impact statements**. In the developing world, geomorphological research has long been funded by foreign government organizations such as the US Agency for International Development, the Canadian International Development Agency, the French Office de la Recherche Scientifique et Technique d'Outre-Mer, and the British Overseas Development Ministry. Today, through the efforts of these and other international bodies such as the United Nations Development Program and the EEC, many government departments in developing countries have trained geomorphologists on staff. The application of geomorphological expertise to such a wide range of environmental issues, holds great promise for the future of **geomorphology**. [AW]

Coates, D.R. (ed.) (1976) *Geomorphology and Engineering. Proceedings of the 7th Binghamton Symposium in Geomorphology*. Dowden, Hutchinson & Ross, Stroudsburg, PA.
Coates, D.R. (1981) *Environmental Geology*. John Wiley & Sons, New York.
Cooke, R.U. & Doornkamp, J.C. (1974) *Geomorphology and Environmental Management*. Clarendon Press, Oxford.
Goudie, A.S. (1981) *The Human Impact*. Basil Blackwell, Oxford.
Hails, J.R. (ed.) (1977) *Applied Geomorphology*. Elsevier, Amsterdam.
Jones, D.K.C. (1980) British applied geomorphology: an appraisal. *Zeitschrift für Geomorphologie* Supp. **36**: 48–73.
Verstappen, H.T. (1983) *Applied Geomorphology*. Elsevier, Amsterdam.

applied geophysics (*geophysical exploration*) Methods to determine information about the subsurface from measurements made at or near the Earth's surface which reflect changes in the physical properties at depth. The methods can be applied to all scales from localized **site investigations** for civil engineering applications to deducing the major internal subdivisions of the Earth. The principal surveying techniques employed are **gravity**, **magnetic**, **seismic** (**refraction** and **reflection**), electrical (**resistivity**, **induced polarization** and **self potential**) **electromagnetic** and **radiometric**. **Gravity**, **magnetic** and **seismic** methods can be used at sea, and **gravity**, **magnetic**, **electromagnetic** and **radiometric** methods can be used from the air. **Airborne geophysical surveys** are particularly time- and cost-effective. (See **airborne geophysical surveys** and **shipborne geophysical surveys**.) [PK]

Kearey, P. & Brooks, M. (1991) *An Introduction to Geophysical Exploration* (2nd edn). Blackwell Scientific Publications, Oxford.

aquamarine A greenish-blue **gem** variety of the mineral **beryl**. [DJV]

aquiclude A formation which may be **porous** and contain water but which is not geologically speaking significantly **permeable**. Aquicludes are important in controlling the movement of water in adjacent **permeable** formations above or below them. [AME]

aquifer A geological formation, group of formations or part of a formation that contains sufficient saturated **permeable** material to yield significant quantities of water to **wells** and **springs**. [AME]

aquifer test A test in which known quantities of water are withdrawn from or added to a **well** and the resulting changes in **head** during or after the time of discharge or addition are measured. [AME]

aquitard A formation that allows water to move through it, but at a much lower rate than through adjacent **aquifers**. [AME]

aragonite **Polymorph** of calcium carbonate ($CaCO_3$) — orthorhombic system, hardness = 3.5, specific gravity = 2.9. Aragonite is a major constituent of present-day carbonate accumulations, being the skeletal component of many invertebrate groups. Non-skeletal aragonite (in the form of **ooids** and marine cements) is also common in the Recent, while aragonite mud covers large areas in modern carbonate provinces. Although much of this mud can be attributed to breakdown of the tests of organisms, a certain proportion may be derived from '*whitings*'. These are suspensions of aragonite needles that form milky white patches some tens or hundreds of metres long in the blue waters, and have been observed in the Bahamas, the **Persian Gulf** and the **Dead Sea**. The cause of '*whitings*' is unclear. Stirring of bottom sediments may be responsible, but spontaneous **precipitation** from super-saturated waters in response to sudden chemical changes in the sea has also been suggested.

Aragonite is more soluble than **calcite** due to the lower symmetry of the aragonite crystal lattice. Consequently, with increasing age the proportion of aragonite in **limestones** decreases, through progressive **dissolution** or **neomorphism** to **calcite**. Textural and chemical criteria (e.g. elevated residual Sr^{2+} levels) can be used to determine the former presence of aragonite. Based on this information, Sandberg (1983) suggested that for certain periods of the **Phanerozoic**, non-skeletal aragonite formation was suppressed (during high stands of sea-level) or enhanced (during low stands of sea-level). Changes in the ratio of dissolved Mg and Ca in the oceans, and variations in the partial pressure of CO_2 in the atmosphere, linked to **plate tectonics** and **eustasy**, have been suggested as the controlling factors in the **precipitation** of non-skeletal aragonite (see **Bahama Banks**, **carbonate buildups**, **trace elements in carbonate rocks**.) [DMP]

Sandberg, P.A. (1983) An oscillating trend in non-skeletal carbonate mineralogy. *Nature* **305**: 19–22.

Archean The older eon of the *Precambrian*, that part of the Earth's history before the appearance of life forms with hard parts that gave rise to body fossils at the beginning of the **Cambrian** period. The younger eon of the *Precambrian* is the **Proterozoic**. The Archean extends from the formation of the Earth *c.* 4600 Ma until 2500 Ma, although outcropping rocks at the Earth's surface are not found much older than 3800 Ma. These rocks have been divided in Upper, Middle and Lower Archean Eras with boundaries at approximately 3000 Ma and 3500 Ma, although these have not been formally agreed. In view of the lack of any evidence for early events, and the universal reference to the 3800 Ma rocks as 'Earliest Archean', there have been suggestions that rocks older than, say, 4000 Ma should be assigned to a *Hadean* or *Priscoan* Eon.

During this very early period of time, the Earth presumably endured a massive bombardment by *meteorites*, for which a detailed stratigraphy has been worked out for the Moon and for many of the terrestrial planets, and it may be significant that the massive bombardment of the Moon ceased about 4000 Ma. Any evidence of this early phase of bombardment on the Earth has been destroyed by later events in the Archean during which the major part of the **continental crust** formed. The only evidence of events earlier than 4000 Ma so far discovered are a few **zircon** grains in some Archean **quartzites** from Western Australia which have given ages of *c.* 4100 Ma. This probably indicates that granitic **crust** had started to form by this time and was being actively eroded to give mature arenaceous sediments in early Archean time.

Rocks which formed during the rest of Archean time are recognized to have formed in two major tectonic environments, **greenstone–granite terrains** and shelf-volcanic **granulite–gneiss terrains**. Linear *orogenic belts* are atypical of Archean crustal structure, although linearity is evident in some **greenstone** terrains, so although the two major terrain types are generally termed 'belts' they are not, in general, similar to later *orogenic belts*. Many features of Archean crustal evolution are very different from those of more recent times and in addition the atmosphere and oceans were of very different chemistry. All these features led to a distinctive Archean phase of crustal evolution.

Greenstone belts are predominantly composed of mafic and ultramafic lavas and overlying sediments, which in general are not strongly deformed and whose maximum metamorphic grade is generally only of *greenschist* facies. They are typically not linear but have an irregular basinal shape, usually steep at the margins. Although several periods of **deformation** can normally be recognized, these rarely produce large areas of penetratively deformed **schists** and the volcanic and sedimentary structures are surprisingly well preserved. Great structural thicknesses of lavas are found and these were regarded as stratigraphic thicknesses. This has recently been questioned and repetition by thrusting is thought to have duplicated the succession in many cases.

The most characteristic feature of the volcanicity is the presence of ultramafic lavas and of highly magnesian basic lavas called **komatiites**. The ultramafic lavas often display spinifex-textured **olivines**, large radiating clumps of needle-like crystals ascribed to quenching of the lava. The lavas were so hot that **erosion** of their channels by melting has been demonstrated and this led, in some cases, to melting of sulphide-rich sediment which then segregated as an immiscible (Ni, Co, Cu) sulphide-rich layer at the base of the flow. These ultramafic **komatiites** are attributed to very high degrees of partial melting of the **mantle** at temperatures in excess of 1700°C.

A general sequence is seen in many **greenstone belts** from mafic and ultramafic **komatiites**, through **tholeiites** to more acid **calc-alkaline** volcanics followed by sediments which include **cherts**, **banded iron formation**, **sandstones** and *turbidites*. Deposition or emplacement onto nearly contemporaneous **tonalitic gneiss** basement is followed by the intrusion of late **granites** which sometimes pierce the **greenstone belt** itself. The classic area where many of the features were first described is the Barberton Mountain Land of South Africa where a sequence of ultramafic and mafic lavas, the Onverwacht Group (3500 Ma), was regarded as a volcanic pile 15 km thick. More recent studies suggest that *thrust* stacking of the sequence is present and that it represents only 3 km of **oceanic crust obducted** onto the 3500 Ma Ancient Gneiss Complex of bimodal meta-volcanics and sediments and **tonalitic batholiths**. The succeeding sedimentary groups, the Fig-Tree Group and Moodies Group, represent foredeep basin deepening with the Fig-Tree composed of iron formation, **chert** and mudstone (2 km thick) with a transition to shallow-marine and braided alluvial **sandstones** in the 3300 Ma Moodies Group (3 km thick). The whole was folded and interleaved before and during the emplacement of late **trondhjemites** between 3200 and 3000 Ma, probably related to **subduction** at a *continental margin*.

Similar sequences have been recognized in many other **shields**, notably the Abiti belt of Canada and the Pilbara and Yilgarn blocks of Western Australia. The Abiti is the largest known **greenstone belt** 800 km long and 200 km wide and has many large bimodal volcanic centres overlying the **komatiitic** and **tholeiitic plateaux**. These are regarded as **rift basins** dissecting volcanic arcs with subsequent **tonalites** completing the cycle, related again to a **subduction zone** environment.

The Pilbara also has later **calc-alkaline** volcanics associated with **rift basins** but these overly extensive **tholeiitic basalts** (3600 Ma) extruded over an area of 450 km by 200 km onto sialic **crust**. **Chert**, carbonate and even **evaporite** sediments followed the bimodal volcanics and, after phases of **deformation**, thick **sandstones**, conglomerates and *turbidites* in several separate **basins** point again to an active tectonic environment for the late sediments (c. 3000 Ma) culminating in the intrusion of *adamellitic* **plutons**.

Although **greenstone belt** formation is widespread in time throughout the Archean (from c. 3600–2600 Ma for the initial volcanism) each belt seems to have had a very extended history and goes through a complete cycle of crustal activity lasting sometimes 500 Ma. Most recent studies suggest some form of **plate-tectonic** *continental margin* but there is as yet no known relationship with the other predominant terrain — the shelf-volcanic terrains.

These terrains are generally termed *granulite-gneiss belts* and contrast with the **greenstone belts** in their very high grade of metamorphism, mostly reaching high **amphibolite** or **pyroxene granulite** facies. The characteristic sediments are **quartzite**, **limestone** and pelite with some huge **banded iron formation** deposits. The oldest sedimentary sequence known, the Isua Group of western Greenland, contains a **banded iron formation** and this passes laterally into **marbles** indicating a chemical origin for the iron formation and **limestone**. Most Archean **marbles** are *dolostones* but some major **limestones** do exist. The association of these mature sediments with *amphibolites* (representing basic volcanism) is very common in this environment, and the whole sequence is quite unlike that of **greenstone belts**. The other major rock units, quite widespread in these terrains but which are unique to the Archean, are the *layered-anorthosite complexes*. These are very extensive intrusions which often retain cumulate textures and have much more abundant *anorthosite* and *leucogabbro* than is usual for basic intrusions. They are also often very rich in **chromite**.

The sequence of events worked out in western Greenland from 3800 to 2800 Ma is typical of these terrains. After the deposition of the Isua Supracrustals (3800 Ma), the area was intruded by a complex of **tonalitic** to **dioritic** intrusions mainly as subhorizontal sheets, which were then converted to **gneisses** (the Amitsoq gneiss, 3700 Ma). Much later (c. 3300 Ma) intrusion of the Ameralik **tholeiitic dykes** throughout the whole terrain provides a time marker enabling the distinction from the later sediments and intrusions not cut by the Ameralik **dykes**. These sediments, the Malene Supracrustals, are of the typical **quartzite–limestone**–pelite type with basic volcanics. A huge *anorthosite*, the Fiskenaesset complex, intruded the area after this sedimentation. All these units were interleaved tectonically and intruded by the tonalitic Nûk **gneisses**, much of the intrusion taking advantage of the layering produced by the interleaving. Subsequently the whole area was multiply deformed to give the **dome** and **basin** interference structure typical of this type of terrain, and raised to very high temperatures and pressures at *c.* 2800 Ma. This late Archean **granulite** facies metamorphism is another feature of global occurrence leading to the production of stabilized crustal blocks with a thick, refractory, **lower crust** which controlled the pattern of geodynamic evolution throughout the rest of Earth history. This sequence of events in the *granulite–gneiss terrains* clearly represents a very active tectonic environment in the Archean and has been compared to modern Andean environments. The mature shelf sedimentary environment with its implications of a stable continental landmass seems to be at variance with the much more energetic orogenic environment of the later stages of the cycle. This paradox has still to be resolved.

The Archean is also remarkable for the presence of the earliest life forms. The Fig-Tree and Onverwacht sediments of South Africa have yielded spheroidal and filamentous *bacterium*-like micro-organisms and in the Witwatersrand Group (2800 Ma), a **sandstone** sequence in a **cratonic basin** on top of the **greenstones**, *algal* mats were widespread in pools in the **braided rivers** and acted as traps for detrital **gold** particles. **Stromatolites** are known from a number of late (3000–2600 Ma) **greenstone belts** and have also been discovered in the mid-Archean Warrawoona Group in the Pilbara Block. The presence of abundant **graphite** in the shelf-type sediments of the **granulite–gneiss** *terrains* may also indicate the presence of organic material but they are all at much too high a metamorphic grade to preserve the fossils. [AEW]

Glover, J.E. & Groves, D.I. (eds) (1981) *Archaean Geology*. Special Publication of the Geological Society of Australia 7.
Kroner, A. & Greiling, R. (eds) (1984) *Precambrian Tectonics Illustrated*. Schweizerbart, Stuttgart.
Windley, B.F. (1977) *The Evolving Continents* (2nd edn). John Wiley & Sons, Chichester.

archeomagnetic dating Archeological artefacts which contain **iron oxides** and which have been heated to temperatures above *c.* 650°C (for example, hearths, pottery, kilns) acquired a **thermoremanent magnetization**. This thermoremanent magnetism reflects both the direction (declination and inclination) and the intensity of the *Earth's magnetic field* at the time of heating and re-cooling. These magnetic properties will remain the same until the sample is re-fired, except for a small proportion of grains with low **blocking temperature** in which the magnetization may be realigned at normal ambient temperatures.

Since both direction and intensity of the **geomagnetic field** has altered gradually (though non-linearly) throughout the Earth's history, measurement of **thermoremanent magnetization** can provide a method of dating fired objects.

Dating based on magnetic direction requires that the sample be found *in situ* (i.e. not moved since firing). It is therefore suitable for pottery kilns and hearths, but not for pottery removed from the kiln in antiquity. The direction of the thermoremanent magnetism is measured by an astatic or a **spinner magnetometer** and compared with a calibration curve of known past magnetic directions (which will only be valid for an area up to *c.* 800 km across, as direction is dependent on position on the Earth's surface).

Dating based on magnetic intensity does not require an *in situ* sample; this involves reheating the sample in the laboratory, but error may result from uncontrolled alterations in the mineralogy (and therefore magnetism) in the sample during heating.

In practice most archeomagnetic dating is still done to add to calibration curves rather than to determine independent dates, and work has been done on this particularly in south-eastern Europe, the Ukraine and Japan.

The reversal of polarity of the *Earth's magnetic field* at intervals in the past (see **reversed magnetization**) also provides a potential method for dating of paleolithic hearths. (See **archeomagnetism**.) [OWT]

Tarling, D.H. (1983) *Palaeomagnetism*. Chapman & Hall, London.

archeomagnetism Paleomagnetism undertaken over archeological time-scales and usually using archeological materials. It has mainly been used for **archeomagnetic dating** based on geomagnetic **secular variation**. As these variations are only locally coherent, i.e. over an area of some 100 000 km^2, such patterns must be constructed for different parts of a continent. Furthermore, as geomagnetic observations only commenced in the late 16th century, well-dated archeological materials must be used to construct the local reference **secular variation** curve for earlier times and are hence subject to the errors in the original archeological dating. Conveniently, however, such errors reduce as more data become available as all observations from the 100 000 km^2 region can be used to develop the regional reference curve, although a geomagnetic model, usually of an inclined **geocentric dipole**, is required for correction of the observations to a central point. When such curves are not available, relative dating is still possible as materials fired at the same time should have been magnetized in similar directions within that area and also the field strength would be similar, although both require a geographic correction based on the inclined **geocentric dipole** model. In addition to dating, the fact that any object or structure that was heated in its entirety will have become magnetized uniformly can be used to assist in the reconstruction of the object by placing all fragments in the original magnetic alignment. The magnetic properties of some archeological artefacts, such as **obsidians**, can also be attributed to specific sources by matching their magnetic properties with those of the source areas. (See **archeomagnetic dating**.) [DHT]

Aitken, M.J. (1974) *Physics and Archaeology*. Clarendon Press, Oxford.
Tarling, D.H. (1983) *Palaeomagnetism*. Chapman & Hall, London.

Archie's formula An empirical law expressing the **electrical resistivity** of a porous rock:

$$\rho = a\rho_w\phi^{-m}$$

where ϕ is the **porosity** per unit volume of the rock, ρ is its bulk resistivity and ρ_w is the resistivity of the **pore fluid**. a and m are empirical parameters, but their values are approximately 1 and 2 for the normal range of **porosities**. The more significant errors in Archie's law arise from uncertainties in ρ_w and, in fact, the ratio of ρ to ρ_w is not always a constant for a given **porosity** as suggested by the equation. [ATB]

Keller, G.V. & Frischknecht, F.C. (1966). *Electrical Methods in Geophysical Prospecting*. Pergamon, Oxford.

arfvedsonite ($Na_3Fe_4^{2+}Fe^{3+}Si_8O_{22}(OH)_2$) A deep green **amphibole mineral**. [DJV]

argentite (Ag_2S) High temperature (>179°C) form of the **silver** sulphide **ore** mineral **acanthite**. [DJV]

argillization The general process of formation of **clay minerals** in rocks hosting mineral deposits, i.e. a type of **wall rock alteration**. The most important subtypes are **advanced argillic alteration** and **intermediate argillic alteration**. [AME]

armoured mud ball (*mud ball, mud pebble, pudding ball* and *clay ball*) Roughly spherical lump of cohesive sediment, which generally has a diameter of a few centimetres. Many examples are lumps of clay of cohesive mud that have been gouged from stream beds or banks by vigorous currents. They often occur in **badlands** and along ephemeral streams, but can also be found in tidal channels and on **beaches**. [ASG]

Bell, H.S. (1940) Armored mud balls: their origin, properties and role in sedimentation. *Journal of Geology* 48: 1–31.

armouring The term used to describe heterogeneous river bed material in which coarse grains are concentrated in a thin layer, typically 1–2 grains thick, at the bed surface in sufficient concentration to inhibit transportation of the underlying fine material. The armoured layer is coarser and better sorted than the underlying bed material. A genetic distinction can be made between an *armoured surface* and a *paved bed*. An *armoured surface* is characteristic of an equilibrium channel and is periodically disrupted and mobilized during floods but reforms. A *paved bed*, however, is more stable and is markedly coarser than the substrate and often occurs in a channel experiencing degradation. It forms as a *lag* deposit of the coarsest component of scoured bed material too heavy to be transported away. [NJM]

Gomez, B. (1984) Typology of segregated (armoured/paved) surfaces: some comments. *Earth Surface Processes and Landforms* 9: 19–24.

aromatics Benzene-based compounds having the general formula C_nH_{2n-6} that occur in many **crude oils**, commonly in amounts less than 1% and only occasionally exceeding 10%. [AME]

arrival time (*onset time*) The exact time at which some particular **seismic wave** arrives at a sensor. In exploration **seismology**, it is measured to 0.001s accuracy and is relative to the time of source detonation. In global **seismology**, accuracies of some 0.1–0.05s are typical because other signals or background noise obscure the exact arrival. For waves such as *P* or *S waves*, a precise instant of arrival exists; where it is clearly visible, the onset is termed 'impulsive', otherwise it is 'emergent'. Arrival times of other types of waves (e.g. the maximum amplitude of a *surface wave*) might also be specified. (See **hypocentre**.) [RAC]

Simon, Ruth B. (1981) *Earthquake Interpretation*: A Manual for Reading Seismograms. Kauffmann, Los Altos.

arroyo (*wash, dry wash, coulée*) A steep-sided, ephemeral stream channel of desert regions. The term arroyo is of Spanish origin and is usually applied to dry water courses in the south-western United States and northern Mexico. Arroyos are the same type of feature as **wadis** in North Africa and the Middle East. These landforms differ slightly from *gullies* in that they are flat-floored and often contain fluviatile sediments. *Gullies*, in contrast, are generally narrow, V-shaped **erosional** landforms.

Several metres of poorly-sorted sediments may accumulate on the floors of arroyos, deposited there by occasional **runoff** from the surrounding desert landscape. This *flash-flood* **runoff** is laden with sediment and may turn into a **mudflow** as the water is absorbed by the older fluvial deposits in the bed of the arroyo. However, these accumulations of sediment may be **eroded** during major floods every few decades. Phases of arroyo entrenchment may be related to significant changes in climate or land use rather than being responses to occasional *storms* with a low periodicity. If this is so, depositional and **erosional** structures in arroyo sediments may be valuable *paleoenvironmental indicators*. [AW]

Cooke, R.U. & Reeves, R.W. (1976) *Arroyos and Environmental Change in the American South-West*. Clarendon Press, Oxford.
Dodge, R.E. (1902) Arroyo formation. *Science* **15**: 746.

arsenic (As) A relatively rare mineral found in certain **veins**. Arsenic is found in concentrations of up to 3% in European early **copper** implements; the addition of arsenic enhanced the metallurgical properties of the **copper** and may have been intentional, either through selection of high-arsenic copper **ores** (found particularly in Germany) or through controlled smelting. Arsenical **copper** declined in importance with the advent of **tin** and later **lead bronze** and this is exemplified in the Copper and Early Bronze Ages of England (later 3rd millenium to mid-2nd millenium BC) where arsenic contents of **copper** implements decline from 1–3% in the Beaker period (earliest metal-using culture in Britain) to mainly less than 1% in the second phase of the Early Bronze Age Wessex Culture. Apart from the recognition of **tin** as an alternative to arsenic, it may have been noticed that arsenic produced toxic fumes during smelting. (See **native element**.) [OWT/DJV]

arsenopyrite (FeAsS) A mineral found in many metal sulphide **ores**. [DJV]

artesian aquifer An **aquifer** whose water is under sufficient pressure to drive it to the surface when the **aquifer** is penetrated by a **well** or borehole. Today such **aquifers** are frequently called *confined aquifers*. The classic case is illustrated in Fig. A8. A dipping **aquifer** lies between two **aquicludes** or confining beds and crops out to form a **recharge** area. Clearly the water-saturated volume below the **water table** will be at a pressure greater than atmospheric. Water will rise up a **well** through the upper **aquiclude** until the water column balances the **aquifer** pressure. With a **well** sited some way down dip, as at A, water will flow out of the **well** which is then termed artesian. If there were many **wells** into the **aquifer** with their water levels joined by an imaginary surface, that surface would indicate the static head of water in the aquifer. This is called the **potentiometric** or *piezometric* **surface**. If a large number of **wells** are sunk into the **aquifer** then discharge may outstrip replenishment and this will lower the **potentiometric surface** to a level like that of B. At the same time **springs** and **seepages** at C may cease to flow. If discharge becomes too great the **wells** may lose their artesian nature and water will then have to be pumped out which will in turn lower the **potentiometric surface** still further. A good example of a former **artesian basin** where this has happened is the London Basin. [AME]

Price, M. (1985) *Introducing Groundwater*. Allen & Unwin, London.

artesian basin An area of **artesian wells** fed from an **artesian aquifer** which has a basin-shaped structure. [AME]

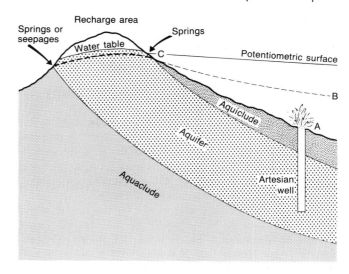

Fig. A8 Artesian aquifer. A simple explanation of an artesian well. (After Price, 1985.)

artesian discharge Discharge of water from a **well** or borehole that has penetrated an **artesian aquifer**. [AME]

artesian head The hydrostatic head of an **artesian aquifer** or of the water in the **aquifer**. [AME]

artesian pressure The hydrostatic pressure of artesian water at the land surface. [AME]

artesian spring A **spring** from which water flows under **artesian pressure**, usually through an opening in the **aquiclude** that overlies the **aquifer**. [AME]

artesian well A **well** fed from an **artesian aquifer**. [AME]

Arthropoda/arthropods A diverse group of invertebrates characterized by an exoskeleton and jointed appendages. The body is bilaterally symmetrical, the trunk formed of a series of similar somites which are usually united in specialized divisions or tagmata (e.g. thorax, abdomen). The exoskeleton is made up of chitin and proteins and may be further strengthened by mineralization, usually with calcium carbonate ($CaCO_3$), rarely with calcium phosphate ($CaPO_4$). Growth is facilitated by moulting and as each moult is a potential fossil, groups with mineralized exoskeletons are overrepresented in the fossil record. Arthropods were the dominant invertebrates in most communities throughout geological time, and are still spectacularly successful today, representing over 80% of all living animal species and occupying every conceivable ecological niche. They are the only invertebrates to have become adapted for flight.

The living arthropods fall into three major groups, the *crustaceans*, *chelicerates* and *uniramians*. The *trilobites*, which became extinct at the end of the **Paleozoic**, constitute a fourth. The living representatives of the major groups are the result of over 400 Ma of independent evolution. They are differentiated by the arrangement of the segments and corresponding limbs of the head, by the nature of their embryological development and by the mode of operation of the jaw (mandibular mechanisms). Some authorities consider that these differences indicate an independent origin for each of the major arthropod groups, implying separate phylum status for the *crustaceans*, *chelicerates*, *uniramians* and *trilobites* (e.g. Schram, 1978).

Although this polyphyletic model for the origin of the arthropods is based on a wealth of data on the living groups, it continues to be controversial. Some paleontologists (e.g. Hessler & Newman, 1975) have drawn attention to similarities between the limbs and feeding mechanism in *trilobites* and those of the *crustaceans* and *chelicerates*. On this basis they argue that these three aquatic groups should be united in a single phylum, distinct from the terrestrial *Uniramia*. In addition, a number of the arthropods known from exceptionally preserved Cambrian faunas such as the *Burgess Shale* cannot be assigned to any of the four major groups, but suggest the possibility of intermediate morphologies. The classification below acknowledges the possibility of a common origin for

Table A4 Stratigraphic ranges of the main groups of **arthropods** mentioned in the text

```
Phylum Arthropoda
    Subphylum Trilobita (L. Cambrian–U. Permian)
        Order Agnostida (L. Cambrian–Ordovician)

    Subphylum Crustacea (Cambrian–Recent)
        Class Branchiopoda (L. Devonian–Recent)
            Order Conchostraca (L. Devonian–Recent)
        Class Ostracoda (Cambrian–Recent)
            Order Phosphatocopida (U. Cambrian)
        Class Maxillopoda (U. Silurian–Recent)
            Subclass Cirripedia (U. Silurian–Recent)
        Class Malacostraca (Cambrian–Recent)
            Subclass Phyllocarida (Cambrian–Recent)
                Order Leptostraca (Recent)
            Subclass Eumalacostraca (Devonian–Recent)
                Order Decapoda (U. Devonian–Recent)

    Subphylum Chelicerata (Cambrian–Recent)
        Class Xiphosura (Silurian–Recent)
        Class Eurypterida (Ordovician–Permian)
        Class Arachnida (Silurian–Recent)
            Order Scorpionida (Silurian–Recent)
            Order Araneida (Carboniferous–Recent)

Phylum Uniramia (Silurian–Recent)
    Superclass Myriapoda
        Class Diplopoda (Silurian–Recent)
        Class Chilopoda (Cretaceous–Recent)
        Class Arthropleurida (Devonian–U. Carboniferous)
    Superclass Hexapoda (Devonian–Recent)
```

the *trilobites*, *crustaceans* and *chelicerates*, and separates them (Phylum Arthropoda) from the Phylum *Uniramia*. Many entomologists, however, ally the *crustaceans* with the myriapods-hexapods (*uniramians*) in a group *Mandibulata*. The debate on arthropod interrelationships will continue for some time. Stratigraphic ranges of the main groups of arthropods are given in Table A4.

Trilobites are particularly important as fossils due to their heavily calcified dorsal exoskeleton. They are characterized by trilobation (the exoskeleton comprises an axis flanked by lateral lobes) and by a transverse division into three tagmata (cephalon, thorax and pygidium). In the majority of *trilobites* the exoskeleton of the cephalon is sutured to facilitate moulting. The *trilobites* reached a peak of diversity in the late **Cambrian** and early **Ordovician**; only a small number persisted through the **Carboniferous**, and they disappeared by the end of the **Permian**. There are about 150 families but there is, as yet, no consensus on their grouping into higher taxa. *Trilobites* were exclusively marine. They display a range of variations on a basic body plan and included nektonic/pelagic representatives as well as the more common epifaunal and infaunal forms. Only a small number of examples preserve evidence of *trilobite* appendages, the most completely known being *Olenoides* from the **Cambrian** *Burgess Shale*. The cephalon usually bears a pair of uniramous antennae and three pairs of biramous limbs, with an inner segmented and outer filamentous branch. The limbs of the trunk are similarly biramous and essentially undifferentiated. The *trilobites* are important in **biostratigraphy**, particularly in deeper water facies in the **Cambrian** (agnostids) and in shallower shelly facies thereafter.

The *crustaceans* also appear in the **Cambrian**. The body is normally divided into a head (which may bear a carapace), a thorax and abdomen. The head bears five pairs of limbs: two pairs of antennae followed by a mandible and two pairs of maxillae which are used in feeding. The *crustaceans* display greater morphological variability than any other major arthropod group. They are mainly marine, but some live in brackish and fresh water and a few are terrestrial. Only those forms with calcified cuticle have a significant fossil record; they include representatives of four classes: *Branchiopoda*, *Ostracoda*, *Maxillopoda* and *Malacostraca*.

The most important fossil *branchiopods* are the *Conchostraca*. The body and limbs of these small bivalved arthropods are enveloped by the carapace. The valves are usually marked by growth lines which reflect changes in size as a result of successive moults. Fossil conchostracans are useful biostratigraphically in some non-marine sections.

The *ostracodes* have a bivalved carapace which is usually calcareous and encloses the body. The importance of these microfossils in **biostratigraphy** is a reflection of the enormous diversity of carapace morphology. Examples of fossil *ostracodes* with preserved appendages are rare. The earliest examples are the late **Cambrian** phosphatocopidans which retain some primitive features. The antennae are

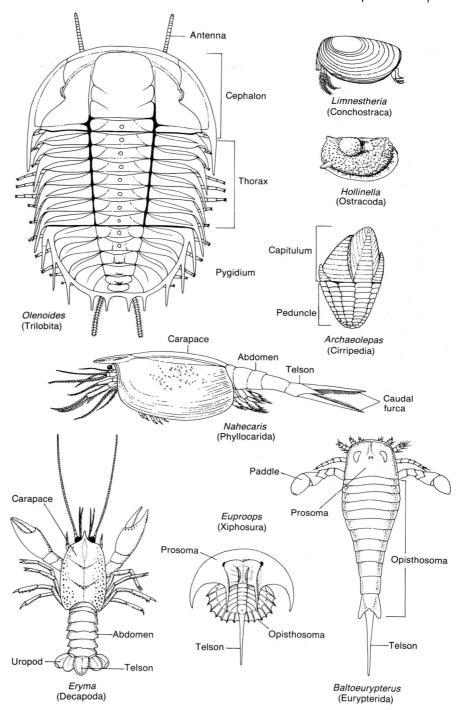

Fig. A9 Arthropoda. Fossil arthropods. (After Moore, 1955, 1969; Whittington, 1975; Brasier, 1980 and Clarkson, 1986.)

equipped with endites and may have assisted with feeding, and the number of trunk limbs in some species exceeds that in living *ostracodes*. The Order *Phosphatocopida* is also unusual in having a phosphatic carapace.

The only *maxillopodans* with a significant fossil record are the *barnacles* (*Cirripedia*). *Barnacles* are attached to the substrate by the first antenna and enclosed within a carapace which may consist of calcareous plates. They filter feed using thoracic limbs modified as cirri. In spite of their early origins, *barnacles* only become abundant in the fossil record in the **Mesozoic** and *Tertiary* with the radiation of forms with calcareous exoskeletons. They illustrate the degree to which the basic *crustacean* body plan may be modified.

[33]

The *malacostracans* are the most abundant fossil *crustaceans* apart from the *ostracodes*. A carapace covers the head and part of the trunk, which is composed of a clearly distinct thorax and abdomen, and terminates in a telson. Two subclasses have an extensive fossil record, the *Phyllocarida* and *Eumalacostraca*.

The *phyllocarids* have a large bivalved carapace. The abdomen consists of seven somites as opposed to six in other *malacostracans*, and it terminates in a telson with a simple paired appendage (caudal furca). Evidence of the appendages of fossil *phyllocarids* is rare but the most completely known, *Nahecaris* from the **Devonian** *Hunsrück Slate* of Germany, shows clear similarities to the living *phyllocarids*, the *leptostracans*.

The *eumalacostracan* carapace is not bivalved. There are only six abdominal somites and the telson forms part of a tail fan with the appendages of the last abdominal somite (uropods). The *eumalacostracans* include a diversity of 'shrimps' as well as the more advanced lobsters and crabs (*decapods*). The *decapod* carapace is fused with the somites of the thorax, and the first three thoracic limbs (maxillipeds) are modified for feeding. The **Paleozoic** record of *decapods* is fragmentary. They radiate in the **Mesozoic** and more markedly in the *Tertiary* with a pronounced increase in the diversity of crabs.

The *chelicerates* also appear in the **Cambrian**. The body is divided into a prosoma (cephalothorax) and opisthosoma (abdomen). There are usually six prosomal appendages. There are no antennae (in contrast to the *trilobites* and *crustaceans*); the first pair are chelicerae used in feeding. The other prosomal appendages are used in both feeding and locomotion. The opisthosomal appendages either function in respiration or are reduced.

The most spectacular **Paleozoic** *chelicerates* are the *eurypterids*. These large predatory arthropods (some approaching 2 m in length) have a relatively small prosoma with the last appendage usually modified as a swimming paddle. The long opisthosoma has 12 somites, the anterior seven bearing gills; the telson may be a spine or paddle. The *eurypterids* originated and radiated in the marine environment but shifted to predominantly brackish–fresh water during their geological history, a few of the later examples becoming amphibious. The horse-shoe crabs are the only living *chelicerates* which are primarily aquatic, but they are the remnant of a diverse subclass, the *Xiphosura*. The *xiphosures* have a large prosoma, a variable number of opisthosomal somites and a long pointed telson. The living representatives and some of the fossils are marine. In the past some occupied brackish–fresh water and rare examples were amphibious; it has been suggested that the **Carboniferous** *Euproops* may have climbed **coal swamp** vegetation.

The majority of living *chelicerates* are terrestrial and consequently these groups have a poor fossil record. The *scorpions*, however, are closely related to the extinct *eurypterids*, and most of the **Paleozoic** examples were aquatic gill-breathers. Some fossil *scorpions* reached lengths of 1 m. Scorpion cuticle contains a unique hyaline layer which may account for its prevalence in **Carboniferous coals**. Other terrestrial groups, such as the *Araneida* (*spiders*) are only well represented in exceptional preservations, such as the **Eocene** Baltic **amber**.

The *uniramians* are essentially terrestrial and respire with tracheae or through the body wall (the uniramous limbs lack a gill branch). The head bears a pair of antennae followed by mandibles and two pairs of maxillae used in feeding. The *myriapods*, which have many trunk somites, include the herbivorous *Arthropleura* which inhabited the **Carboniferous** coal swamps where it reached lengths of up to 2 m. The *hexapods* (*insects*) have a distinct head, thorax with three pairs of limbs, and abdomen. They require exceptional circumstances for preservation, but their abundance and diversity ensure a substantial record. Most of the major steps in insect evolution took place before the end of the **Carboniferous**. Modern families predominate from the **Cretaceous** when they evolved in association with the flowering plants. Beetle elytra are common in **Quaternary** sediments and are used as climatic and environmental indicators. (See Fig. A9.) [DEGB]

Brasier, M.D. (1980) *Microfossils*. Allen & Unwin, London.

Carpenter, F.M. & Burnham, L. (1985) The geological record of insects. *Annual Review of Earth and Planetary Sciences* **13**: 297–314.

Clarkson, E.N.K. (1986) *Invertebrate Palaeontology and Evolution* (2nd edn). Allen & Unwin, London.

Hessler, R.R. & Newman, W.A. (1975) A trilobitomorph origin for the Crustacea. *Fossils and Strata* **4**: 437–59.

Kjellesvig-Waering, E.N. (1986) A restudy of the fossil Scorpionida of the world. *Palaeontographica Americana* **55**.

Moore, R.C. (ed.) (1955, 1969) *Treatise on Invertebrate Paleontology, part P, Arthropoda 2, part R, Arthropoda 4 (1)*. Geological Society of America and University of Kansas Press, Boulder, CO & Lawrence, KS

Murray, J.W. (ed.) (1985) *Atlas of Invertebrate Macrofossils*. Longman, Harlow.

Schram, F.R. (1978) Arthropods: a convergent phenomenon. *Fieldiana, Geology* **39**: 61–108.

Schram, F.R. (1986) *Crustacea*. Oxford University Press, Oxford.

Whittington, H.B. (1975) Trilobites with appendages from the Middle Cambrian, Burgess Shale, British Columbia. *Fossils and Strata* **4**: 97–136.

artificial recharge Replenishment of an **aquifer** by pumping suitably treated water into it. This water may be taken from surface supplies, e.g. rivers, or from other **aquifers**, when it can be conveyed by pipeline or by making use of a suitable river. [AME]

asbestos The name given to a group of **industrial minerals** characterized by an extreme fibrous ('asbestiform') habit and including the **serpentine** mineral **chrysotile** and the amphiboles **crocidolite** and **amosite**. (See **amphibole minerals**.) [DJV]

aseismic ridge A ridge, within an oceanic basin, that is not characterized by regular seismic activity, although occasional **earthquakes** of low **magnitude** may occur, such as the Walvis–Rio Grande Ridges (oceanic) in the South Atlantic Ocean and the Lomonosov Ridge (continental) in the Arctic. Such ridges may have formed by the passage of the **lithosphere** over a near-stationary **hotspot** in the **mantle**, so that their locus defines the direction of *plate* movement. [DHT/PK]

Kearey, P. & Vine, F.J. (1990) *Global Tectonics*. Blackwell Scientific Publications, Oxford.

aseismic slip Slip occurring without **earthquakes**. Aseismic slip rates are generally less than $0.1\,ms^{-1}$ and aseismic **strain rates** less than $10\,s^{-1}$. [TGB]

ash (1) The non-combustible inorganic residue remaining when **coal** is burnt. It represents the bulk of the mineral matter in the **coal**. High ash **coals** are generally less valuable than low ash **coals** as they have a lower calorific value and produce greater amounts of waste for removal after combustion.
(2) Fine-grained magmatic material mostly formed from the glass walls between adjacent bubbles in *vesiculating* **magma**. [AME/PTL]

ash flow (*pyroclastic flow*) A concentrated dispersion of hot, solid, juvenile volcanic fragments in gas which moves as a *gravity-controlled flow*. The deposits such ash flows generate are poorly sorted, and non-stratified, and consist of **ash**, **pumice** and/or **scoria**, blocks of non-vesicular material, and liberated crystals (formerly *phenocrysts*). Commonly, dense lithic fragments are normally-graded, and **pumice** (or **scoria**) fragments are reversely-graded. This indicates that the active flows have densities between those of dense lithic clasts and **pumice** (Sparks, 1976; Wilson and Head in Lipman & Mullineaux, 1981). Ash flows are thought to move by **laminar flow**, or as a plug riding on a basal laminar layer (Sparks, 1976; Wright & Walker, 1981), although some are thought to experience local turbulence (e.g. Walker *et al.*, 1980). The normal and reversed grading of lithic and pumic clasts, respectively, is used to distinguish flow units — the products of individual ash flows — in ash flow deposits produced in a single **volcanic eruption**. Most ash flow deposits are strongly ponded in topographic depressions but, because of their high velocity of movement, many can surmount hills over $500\,m$ high which lie in their path.

Ash flow deposits range in volume from around $0.01\,km^3$ to over $1000\,km^3$ (Smith, 1960). The smaller volume deposits ($<1\,km^3$) are usually **scoria** and **ash**, or block and **ash** deposits. These ash flows may be generated; (i) by disruption and fragmentation of **domes** or **lava flows**, (ii) by gravitational collapse of dense parts of *plinian-style eruption* columns (see **volcanic eruption**), (iii) from dense clouds of **tephra** which do not form high eruption columns, but which appear to 'boil-over' from **craters**, (iv) from sideways 'directed blasts' from **volcanic vents**

(Fisher & Heiken, 1982). Such small-volume ash flows are usually accompanied by a high cloud rising turbulently above the active flow. These clouds are known as **nuées ardentes**, and may deposit a layer of **ash** overlying the ash flow deposits.

Ash flow deposits which have volumes exceeding $1\,km^3$ are usually pumiceous and associated with **caldera** collapse. Such deposits are called **ignimbrites**. [PTL]

Chapin, C.E. & Elston, W.E. (eds) (1979) *Ash Flow Tuffs*. Special Paper of the Geological Society of America **180**.
Fisher, R.V. & Heiken, G. (1982) Mt Pelée, Martinique: May 8 and 20, 1902, pyroclastic flows and surges. *Journal of Volcanology and Geothermal Research* **13**: 339–71.
Lipman, P.W. & Mullineaux, D.R. (eds) (1981) *The 1980 Eruptions of Mount St Helens, Washington*. Professional Paper of the United States Geological Survey **1250**.
Smith, R.L. (1960) Ash flows. *Geological Society of America Bulletin* **71**: 795–842.
Sparks, R.S.J. (1976) Grain size variations in ignimbrites and implications for the transport of pyroclastic flows. *Sedimentology* **13**: 147–88.
Walker, G.P.L., Wilson, C.J.N. & Froggatt, P.C. (1980) Fines-depleted ignimbrite in New Zealand — the product of a turbulent pyroclastic flow. *Geology* **8**: 245–9.
Wright, J.V. & Walker, G.P.L. (1981) Eruption, transport and deposition of ignimbrite: a case study from Mexico. *Journal of Volcanology and Geothermal Research* **9**: 111–31.

ashlar A block of **building stone** with straight edges. [AME]

asphalt The solid members of the **petroleum** group of substances are known as asphalt, *bitumen* or *tar*. They occur in productive petroliferous regions, in **oil shale** and occasionally in other rocks. These solids may be **brittle** or friable, or plastic and malleable. The second type is really a fluid *hydrocarbon* of high **viscosity** and can occur as lakes, sheets or *tar mats* at or near the surface or at former surfaces that are now **unconformities**. [AME]

North, F.K. (1985) *Petroleum Geology*. Allen & Unwin, Boston.

asphalt seal An **asphalt** accumulation acting as a **seal** to oil accumulations. The **asphalt** has formed by a drying up process (*inspissation*) of oil that reaches the surface. When the resulting **seal** covers dipping beds an oil accumulation often forms beneath it. [AME]

asphaltene High molecular weight aggregate that occurs in solid **petroleum** (see **asphalt**). Asphaltenes are low in hydrogen and contribute to the high **viscosity** of heavy oils. They are also the main carriers of the metals in heavy oils. The metals, with much of the **sulphur** and nitrogen, are mainly bound to the asphaltene molecules. [AME]

asphaltic pyrobitumen A black, structureless **asphalt** that is infusible and insoluble in carbon disulphide. It generally contains less than 5% oxygen. [AME]

asphalt-based crude oil (*asphalt-base oil*) **Crude oil** composed dominantly of naphthenic components. Only about 15% of world **crude oil** supplies are of this type. They comprise the so-called *black oils* of Venezuela, Mexico, parts of the Gulf Coast and many Russian crude oils. [AME]

assay The determination of the amount (assay value) of a metal or metals in an **ore**. The process of determining these amounts is called assaying. In order to delineate the boundaries of an **orebody** in which the level of mineralization gradually decreases to a background value many samples will have to be collected and assayed. The boundaries thus established are called *assay limits* or *assay boundaries*. [AME]

assimilation The physical and/or chemical incorporation of rock material into a foreign or unrelated **magma**. The process is most commonly (but not exclusively) applied to assimilation of **crustal** materials into **mantle**-derived **magmas** such as **basalts**. Such **magmas** may be referred to as 'contaminated'. Evidence for **crustal** assimilation may be the presence of crustal **xenoliths** and **xenocrysts**. For example disaggregated **xenoliths** and **xenocrysts** of **quartz** and **alkali feldspar** (possibly showing some evidence of chemical reaction) in **basalt** is commonly taken as evidence of **crustal** contamination of the host **basalt**. In addition, chemical evidence may be taken as possible evidence of **assimilation** of **crustal** material by **basalt magma**. [RST]

associated gas **Petroleum** gas formed as a by-product of the generation of **petroleum**. It may accompany oil in a **reservoir** or form a separate gas accumulation within an **oilfield**. **Petroleum** gas not accompanying oil is called *non-associated gas*. [AME]

asterism A property of some mineral crystals that arises from the way in which light is reflected back from oriented inclusions. In 'star' **rubies** and **sapphires**, inclusions arranged in three crystallographic directions at 120° to each other scatter light in such a way as to produce a six-pointed star. [DJV]

asthenosphere The 'soft' part of the **mantle** immediately underlying the **lithosphere** within which it is conventionally considered that the application of natural **stress** fields, $\geq 100\,\text{MPa}$, results in **plastic** rather than **brittle deformation**. In the oceans, the top of the asthenosphere corresponds to the top of the seismic **low velocity zone** which is interpreted as indicating the presence of fluids, usually attributed to partial melting where the **mantle** melting temperature is most closely approached. Such partially molten material would deform plastically when stressed, rather than accumulate strains to be eventually released as **earthquakes** following **brittle failure**. However, the **viscosity** of the **low velocity zone** is the same as that of the underlying **mantle**, $10^{21-22}\,\text{Pa s}$, and the fluids present may not necessarily indicate partial melting but could be, for example, water released from *amphibolites* at temperatures of *c*. 800°C (or, less likely, from **serpentinites** at *c*. 600°C). This boundary is therefore isothermal, on all models, but the actual temperature could be 1200°C, 800°C or 600°C. It is very shallow near the crest of an *oceanic ridge*, *c*. 20 km, and increases in thickness according to the square root of the age of the oceanic **lithosphere** so that it is mostly 120 to 180 km deep in the ocean basins older than about 100 Ma. The presence of the asthenosphere beneath the continents is not well established as the **low velocity zone** is absent beneath the **cratonic** blocks where **carbonatites** and **kimberlitic** intrusions, with associated **xenoliths**, indicate that the top of the asthenosphere is at least 250 km and probably 350–450 km deep. [DHT]

Bott, M.H.P. (1982) *The Interior of the Earth.* Arnold, London.
Tarling, D.H. (1980) Lithosphere evolution and changing tectonic regimes. *Quarterly Journal of the Geological Society, London* **137**: 459–66.

astrophyllite $((K, Na)_3(Fe, Mn)_7(Ti, Zr)_2Si_8(O, OH)_{31})$ A rare **titanium**-bearing silicate mineral. [DJV]

asymmetric valley A valley in which the slope of one side is steeper than the other. Such valleys may be the result of geological structure, as when streams experience **uniclinal shifting** and so undercut one valley side. Such features are common in *periglacial* regions where they are the result of differences in slope aspect, and thus amounts of solar radiation received, which produces great variation in the strength of **frost weathering** and *solifluction* processes. In Britain and Europe many asymmetric valleys are thought to be relict landforms, developed under *periglacial* conditions during the *Pleistocene*. The 'active' slopes, on which the *periglacial* processes were most effective, were either steepened or reduced in angle more rapidly than the 'inactive' slopes, thus producing the asymmetrical shape. Examples from southern Britain indicate that no general rules can be applied as in the Chilterns the steeper slopes face towards the south-west whereas in the Marlborough Downs the asymmetry is reversed. [NJM]

Churchill, R.R. (1982) Aspect-induced differences in hillslope processes. *Earth Surface Processes and Landforms* 7: 171–82.
Long, D. & Stoker, M.S. (1986) Valley asymmetry: evidence for periglacial activity in the central North Sea. *Earth Surface Processes and Landforms* **11**: 525–32.

atacamite $(Cu_2Cl(OH)_3)$ A secondary **copper** mineral. [DJV]

atoll An annular form of **coral**–*algal reef* found in oceanic waters. Generally, atolls consist of an irregular elliptical rim of **coral**–*algal reef* around a central *lagoon*. Charles Darwin put forward an early theory to explain the development of atolls (which are largely made up of organisms such as *scleractinian* **corals** which only grow in shallow waters) in deep oceanic waters. Darwin suggested that atolls were the last form in a genetic sequence of *reefs* which developed around a subsiding, volcanic island. *Fringing reefs* and **barrier reefs** were postulated to be the first stages in the evolutionary sequence. The subsidence theory of the origin of atolls has been vindicated by more recent deep drilling of *reefs* which has discovered volcanic basement rocks at various depths below the reef surface. So, for example, boreholes on Eniwetok, Marshall Islands, western Pacific in 1951–2 revealed a sequence of *reef* **limestones** on top of **Eocene limestone** above **basalt** basement at a depth of 1411 m.

An alternative theory of atoll development was put forward by R.A. Daly, who envisaged that lowered sea-levels and sea temperatures during *Pleistocene glaciations* would have killed off **coral** *reefs*. Marine planation and subaerial **erosion** of these relict *reefs* would have created platforms upon which new **coral** grew once conditions ameliorated. Daly proposed that growth would take place on the outside edges of such platforms, producing the annular shape of contemporary atolls. Recent work, however, suggests that during *glacial* periods **coral** *reef* growth was not dramatically curtailed and that the marine planation envisaged by Daly (at least on current process rate evidence) could not have been achieved during the *glacial* period timespan.

Atolls are primarily a feature of tropical seas, as their distribution is controlled by the growth requirements of the *reef* forming organisms. There are approximately 400 known atolls in the world today, most of which are located in the Indo-Pacific. Only about 10 lie outside this zone in the Atlantic. Atolls vary greatly in size and shape as well as in the depth of the central *lagoon*. *Lagoon* depths average about 40 m, but vary in depth between near sea-level and approximately 150 m. In the Caribbean, atoll *lagoons* are 5–15 m deep, regardless of the size of the atoll, whereas in many atolls *lagoon* depth is related to atoll diameter. Atolls often occur in groups or chains, as for example in the Maldive Islands, Indian Ocean; the Marshall Islands, western Pacific; and the Seychelles, Indian Ocean. Not all atolls are at sea-level; there are many elevated atolls, and many drowned atolls. Sunken atolls occur when the rate of **coral** growth has not equalled sea-level rise, while elevated atolls occur where there has been local **uplift** or other events which have led to emergence of the *reef* surface. Aldabra Atoll, in the Seychelles archipelago, is an example of an elevated atoll where the land surface is elevated by up to 8 m above present sea-level. Elevated atolls are characterized by a **karst** topography developed upon exposed surfaces, due to subaerial solutional **erosion**.

There are several different morphological zones of atolls, which are characterized by varying hydrological and ecological characteristics. On the seaward side there are very steep slopes, and there are often spur-and-groove features on the *reef* front itself. The *reef* flat sometimes has an *algal* ridge on the seaward side (common in the Indo-Pacific, but rare on Caribbean atolls). Towards the *lagoon* side of the *reef* flat a low island or **cay** may form. Such islands are usually composed of *sand*-size material, which may be indurated in places to form **beachrock**. The *lagoon* itself often has a pronounced topography, and may contain small *patch reefs*, and *micro-atolls* which are individual massive **coral** colonies. [HAV]

Menard, H.W. (1982) Influence of rainfall upon the morphology and distribution of coral reefs. In: Scrutton, R.A. & Talwani, M. (eds) *The Ocean Floor*, pp. 305–11. John Wiley & Sons, Chichester.

Montaggioni, L.F., Richard, G., Bourouilh-Le Jan, F., *et al.* (1985) Geology and marine biology of Makatea, an uplifted atoll, Tuamotu Archipelago, central Pacific Ocean. *Journal of Coastal Research* **1**: 165–71.

Rougerie, F. & Wauthy, B. (1986) Le concept d'endo-upwelling dans le fonctionnement des atolls-oasis. *Oceanological Acta* **9**: 133–48.

atollon A small **atoll** which lies on the flank of a larger one. [HAV]

atomic absorption spectroscopy (AAS) Method of **chemical analysis** in which a solution of the sample is injected into a flame, typically of air/acetylene, where the sample is atomized (outer electrons excited). Light of the wavelength of an element to be determined is provided by a hollow cathode lamp and focused onto the flame. The amount of light from the lamp absorbed by the sample atoms in the flame may be used to give a measure of the concentration of that element. The procedure is repeated for each element of interest. AAS is used in archeological studies to characterize sources and artefacts of, for example, **obsidian** and **flint** in order to determine the extent of trade in these implements. Its disadvantages for archeological work are that it is partially destructive of the artefact under examination and also is of limited sensitivity. Recent developments of *inductively-coupled plasma emission spectrometry* makes AAS less important. [OWT]

atomic orbital A descriptor of the spatial distribution of an electron in an atom. The application of *quantum mechanics* to the system of an electron in close proximity to a proton, i.e. a hydrogen atom, gives a set of equations (wave functions) which are differentiated from each other by the specific values that can be given to constants in the functions. These constants are *quantum numbers* and can only have integral values. Each set of values for these *quantum numbers* — n (principal), l (angular), m_l (magnetic) — defines a different wave function. From each wave function the energy and spatial distribution of the electron can be calculated. n can have the values 1, 2, 3, . . . etc., which correspond to different energy states of the hydrogen atom of decreasing stability.

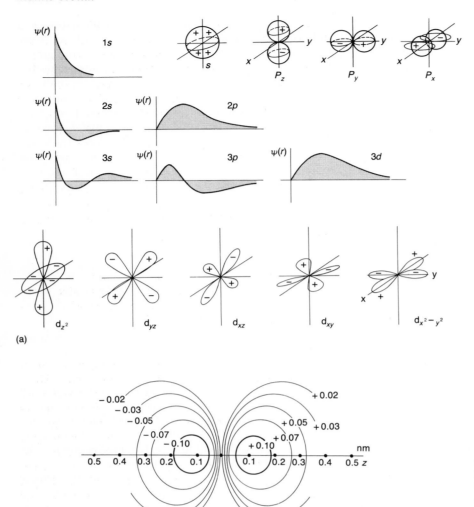

(a)

(b)

Fig. A10 Atomic orbital. (a) Radial functions (Ψ)(r)) for n = 1, 2 and 3: polar coordinate representations of angular functions (l = 0, s; l = 1, p and l = 2, d). (b) Cross-section through the wave function for a hydrogen $2p_z$ orbital in a plane containing the Z-axis.

It is found that as the principal quantum number increases so does the number of substates associated with different values of the angular quantum number (l). l can take integral values from 0 to $n - 1$. Thus for n = 1, l can only equal 0 — for n = 2, l = 0 and 1, etc. Different values of l are usually distinguished by letters as follows: l = 0, s; l = 1, p; l = 2, d; l = 3, f. Wave functions are conveniently referred to by giving the numerical value of n first and then the letter corresponding to l. A further subdivision is possible, which depends upon m_l. For a specific value of l, m_l can have all possible integral values in the range m_l, $m_l - 1$, . . . , 0, -1, . . . $-(m_l - 1)$, $-m_l$. This gives rise to a hierarchy of functions which increases in number and complexity with n, i.e. 1s (one), 2s (one) 2p (three), 3s (one), 3p (three), 3d (five), etc.

Whilst the actual behaviour of an electron in an atom can never be known, it is convenient to regard the electron as moving in the space around the nucleus; the value of the wave function multiplied by its complex conjugate at any point determines the probability of the electron's

being at that point. Because electrons are thought of as moving around a nucleus, atomic wave functions are also called atomic orbitals. The total wave function is the product of 'radial' and 'angular' functions, the former depending on the distance of the electron from the nucleus, r, and the *quantum numbers*, n and l, the latter being a function of the polar coordinate angles θ and ϕ controlled by *quantum numbers l* and m_l. Thus the 'angular' functions determine orbital shape. Examples of both radial and angular functions are shown in Fig. A10(a) together with a cross-section through an 'atomic orbital' derived from their product.

The above description of wave functions for the hydrogen atom can be adapted for all atoms. The nuclear charge in the basic equations must be modified to take account of the increase in the number of protons and also the 'screening' effect which electrons in inner shells (characterized by small values of n) will have on this increased nuclear charge for the electrons in outer orbitals. The electron can be thought of as a very small spherical

[38]

spin clockwise or anticlockwise and each orbital can hold a maximum of two electrons provided that they have opposite spins. But the introduction of more than one electron brings electron–electron repulsion and profound computational (and conceptual) complications. It also breaks the degeneracy of states with the same *n* but different *l*. The electronic structures of the elements may be understood by the 'aufbau' principle in which electrons are added successively to the most stable available orbital, $1s-2s-2p-3s-3p-(4s^2, 4p^1, 3d)-4p-5s$, etc. (See Fig. A10.) [DSU]

Eyring, H., Walter, J. & Kimball, G.E. (1944) *Quantum Chemistry*. John Wiley & Sons, New York.
Pauling, L. & Wilson, E.B. (1935) *Introduction to Quantum Mechanics*. McGraw-Hill, New York.
Woodgate, G.K. (1980) *Elementary Atomic Structure*. Oxford University Press, Oxford.

attached groundwater Groundwater retained on the walls of pores in rocks above the **water table**. [AME]

attenuation The progressive reduction in amplitude of waves as they propagate. For **seismic waves** it has several causes:

1 Scattering. Sub-wavelength scale inhomogeneities in the path of **seismic waves** deflect and scatter the wave energy.
2 Intrinsic or **anelastic** attenuation. Even in homogeneous rocks, transmission of wave energy is not perfect: mechanisms such as grain boundary friction absorb a fraction $(2\pi/Q)$ of the wave energy per cycle. Q is the '*Quality Factor*'. Typical *P wave* Q is 100–500 in the **crust**, 1000–5000 in the **mantle**. *S wave* Q is about one-quarter of *P wave* Q. t^* (travel time/average Q over the wave path) is also used, especially for **teleseismic** *P waves*.
3 Mode conversion and partitioning. When wave energy meets an interface in the Earth, it is reflected, transmitted and converted into other forms of **seismic wave**; for example, an incident *P wave* produces reflected and transmitted *P* and *S*. At curved boundaries, diffraction can also occur.
4 Geometric spreading. As it travels further from its source, wave energy occupies a greater area and so amplitudes are reduced (just as the more distant a light is, the duller is seems). *Body waves* spread out in three dimensions, *surface waves* in only two and so are attenuated more slowly with distance.

(1)–(3) are frequency-dependent: i.e. the degree of attenuation differs for waves of different frequency. Scattering and **anelastic** attenuation are hard to separate experimentally. [RAC]

Aki, K. & Richards, P.G. (1980) *Quantitative Seismology* (2 Vols). W.H. Freeman, San Francisco.

attrital coal A term of the US Geological Survey system for the field description (including **logging boreholes**) of coals. Three constituents are described: *fusain*, *vitrain* and attrital coal. The first two are considered as being larger clastic units within a matrix of finely divided attrital coal. The thickness and concentration of the *fusain* and *vitrain* are recorded quantitatively and the attrital coal is described according to five categories of **lustre:** bright, moderately bright, midlustrous, moderately dull and dull. [AME]

Ward, C.R. (ed.) (1984) *Coal Geology and Coal Technology*. Blackwell, Scientific Publications, Melbourne.

attrital–anthraxylous coal A lustrous **coal** in which the ratio of *anthraxylon* (vitreous **coal** components derived from woody tissues) to **attritus** varies from 1:1 to 1:3. [AME]

attritus A microcomponent of **coal** used in the Thiessen–Bureau of Mines (US) system of **coal classification** based on thin section study. It includes spores, cuticles, resins and granular opaque matter. [AME]

augen gneiss A type of **gneiss** exhibiting a planar or linear **fabric** consisting of aligned lensoid shapes. These are the 'augen' (German for 'eyes'). This *shape fabric* defines the **gneissosity** of the **gneiss**. Many augen gneisses result from the **deformation** of coarse-grained igneous rocks such as **granites** or **gabbros**, often containing a primary igneous **porphyritic** or *megacrystic* **texture**. The augen represent deformed megacrysts or crystal aggregates typically composed of more felsic minerals (**feldspars**, **quartz**, **epidote**, etc.) in a matrix of more mafic and finer-grained felsic material. Other types of augen gneiss are essentially **porphyroblastic** in origin, derived from crystalline, **schistose**, metamorphic rocks by progressive increase in crystal size. In such **gneisses**, the augen may consist of other aluminosilicate minerals such as **kyanite**, **staurolite** or **garnet**, for example, as well as **feldspar**. Augen gneisses provide useful *strain markers*, albeit of a rather crude nature, in many **gneissose** terrains where alternative indicators are uncommon or lacking. [RGP]

augite ((Ca, Na)(Mg, Fe, Al) (Si, Al)$_2$O$_6$) A dark green to black **pyroxene mineral**, the most common pyroxene and an important rock-forming mineral. [DJV]

aulacogen A non-active *rift valley*, often now partially or completely filled with sediments and containing mineralization associated with its original activity. Often a result of a failed rifting event along one of the arms of a **triple junction** that originally formed during **continental splitting**. [DHT]

Dewey, J.F. & Burke, K. (1974) Hot spots and continental break-up: implications for collisional orogeny. *Geology* **2**: 57–60.

Table A5 Authigenesis. Some of the major authigenic minerals in sediments

Mudrocks			Sandstones			Limestones		
Authigenic mineral	Common occurrence	Abundance	Authigenic mineral	Common occurrence	Abundance	Authigenic mineral	Common occurrence	Abundance
Pyrite	Replacive	Very common	Quartz Feldspar	Overgrowths	Extremely common	Aragonite	Cements	Common in Recent sediments
Siderite Calcite Dolomite Ankerite	Pore-filling	Very common, particular species present varies	Hematite Anatase Pyrite	Replacements, grain coats	Common to specific environments	Calcite Dolomite	Cements and replacements	Extremely common
Glauconite Zeolite Celestite	Replacive	Rare, restricted development	Calcite Dolomite Ankerite Siderite	Cements, often replacive	Generally extremely common, particular species present varies	Anhydrite Gypsum	Cements	Rare, restricted development
Illite Kaolinite Chlorite Smectite	Pore-filling	Extremely common, particular species present varies	Anhydrite Gypsym Halite	Pore-filling cements	Rare, restricted development	Chalcedony	Cements and replacements	Common
			Illite Kaolinite Chlorite Smectite	Grain coats, pore linings and pore-fillings	Extremely common, particular species present varies	Kaolinite Smectite	Pore-fillings	Variable; generally rare but abundent in specific environments
						Glauconite Apatite	Replacements	Rare, restricted to specific environments

aurichalcite ($(Zn, Cu)_3(CO_3)_2(OH)_3$) A rare secondary mineral formed from **alteration** of **base metal ores**. [DJV]

authigenesis (adj. authigenic) Literally 'generation *in situ*'; usually applied to diagenetic mineral formation in sediments subsequent to their deposition (see **diagenesis**). Authigenic minerals are thus distinct from detrital (transported) minerals and formed in place within the host sediment in which they now occur.

A wide diversity of authigenic minerals are known from most types of sediments (Table A5). They can often be distinguished from detrital sedimentary components on the basis of their textural occurrence and morphology (Fig. A11). Many authigenic minerals form overgrowths to detrital grains (e.g. **quartz**, **feldspar**) whilst others may occur as pore-filling cements (e.g. **calcite**, **chalcedony**), replacements (e.g. **dolomite**, **pyrite**) or pore-linings (**clay minerals**). Processes of authigenic mineral formation may thus include **precipitation** direct from a **pore-fluid**, **replacement** of one mineral by another, recrystallization of a pre-existing mineral or by structural transformation of one mineral **polymorph** to another **polymorph** by **neomorphism**. (See also **cementation**, **neoformation of clays**.) [SDB]

autochthon Refers to the local derivation of a large structural unit, e.g. a **nappe** or **terrane**, in relation to adjoining rocks. In contrast, the terms *parautochthonous*

and **allochthonous** are used for units that have travelled short or long distances, respectively, from their site of origin. [RGP]

autosuspension The fluid condition where the hydraulic forces due to motion provide sufficient energy input to maintain the flow with its suspended load. Autosuspension is only approached for thick **turbidity currents** composed of low-density particles moving down a high-gradient slope. Where the particle fall velocity is large, autosuspension is impossible. [KTP]

Pantin, H.M. (1979) Interaction between velocity and effective density in turbidity flow: phase plane analysis, with criteria for autosuspension. *Marine Geology* **31**: 59–99.

autunite ($Ca(UO_2)_2(PO_4)_2 . 10-12H_2O$) An **ore** mineral of **uranium**. [DJV]

aven A French term for a vertical or steeply sloping shaft connecting a **cave** to the surface in **limestone** terrain. The term is synonymous with **pothole** or **swallow hole** in English. [HAV]

aventurine The name given to a **gem** variety of **quartz** (or sometimes **oligoclase**) resulting from inclusions of coloured minerals such as **hematite** (red) or **chromium mica** (green). The phenomenon is sometimes called *aventurescence*. [DJV]

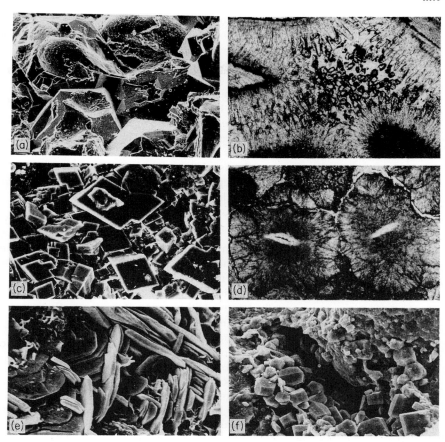

Fig. A11 Authigenesis. Authigenic minerals in sediments. (a) Quartz overgrowths (SEM). (b) Radial aragonite (thin section). (c) Euhedral dolomite crystals (SEM). (d) Spherulitic siderite replacements (thin section). (e) Hematite overgrowths (SEM). (f) Pore-lining, euhedral anatase crystals (SEM).

average velocity (1) For a river or stream, discharge divided by cross-sectional area normal to the flow.
(2) For **groundwater**, the volume of water passing through a given cross-section, divided by the **porosity** of the material through which it moves. [AME]

avulsion The process by which a channel changes its course by a sudden diversion of the channelled water. The period between avulsions is controlled by the rate of topographic change, discharge fluctuations and the occurrence of triggering events such as major **earthquakes**. Between avulsions the channel position may change gradually by bank migration, **anabranch** switching and meander abandonment. The processes of avulsion are poorly understood but are generally considered to result from erosional or deformational breaching of channel **levées** and diversion of the current to a more energetically efficient course. When the process takes place persistently from one point, such as at marked breaks in **floodplain** slope or in width of alluvial valley, it is known as *nodal avulsion*. [JA]

Smith, N.D., Cross, T.A. Dufficy, J.P. & Clough, S.R. (1989) Anatomy of an avulsion. *Sedimentology* **36**: 1–23.

axe Flat tool designed mainly for tree-cutting, but also probably used for general woodwork, hoeing, even butchering. Similar to the adze, which differs from the axe in having the plane of the cutting edge at right angles to the haft. Axes in antiquity were made largely of stone before the advent of metallurgy, but some were of bone or shell. Stone axes may be flat or perforated (battleaxe), and later metal types include palstaves and socketed axes. Stone axes were initially roughed out of the rock by flaking and many were then polished very finely, especially along the cutting edge.

A variety of rocks was used for axe manufacture and hard rocks of fine-grained texture were preferred. **Flint** and **chert** were commonly used in Europe, though many of these axes were unpolished or edge-polished only, presumably because of the difficulty of polishing these hard materials using the most readily available **abrasive, quartz**. Other rocks used for axe manufacture include *greywackes* (commonly in Australia and New Zealand and rarely in England), micaceous **sandstone**, occasional other sedimentary rocks such as **limestone** and **shales** and volcanic rocks. The last group includes many **basalts**, used in Europe, Hawaii, Australia, New Zealand and the Middle and Near East. **Andesites** were important in South

America, **tuffs** second in importance to **flint** in England and Wales and **dolerites** were also used in the UK and France. A group of rocks which may be termed **greenstones** include metamorphosed basic igneous rocks, such as **schists** and *amphibolites*, and were used for axes in northern, western and central Europe, in the Mediterranean, Australia, central America and Ghana. **Hornfels** was also used in the UK and in New South Wales. Rocks of coarse texture were less commonly used becase they were less easy to rough out by flaking and less readily polished. They occur occasionally, for example, as rare **granite** axes in England. Beautifully polished **jade** (**nephrite** and **jadeite**) axes are found in Europe and were almost certainly for ceremonial rather than functional purposes. [OWT]

axe factory A locality where large numbers of waste flakes and **axe** rough-outs indicate stone **axe** manufacturing or at least roughing-out. The products need not necessarily have been finished (polished) at the factory site. Such factories are normally located within areas of fine-grained rock suitable for **axe** manufacture. Axe factories are known in Norway, Brittanny, Australia (New South Wales, Victoria), New Guinea and at five localities in Britain. The British axe factories and the distribution of their products have been the object of a long-standing study begun in 1936. Thirty-two major **axe** groups (not including **axes** of **flint** and **chert**) have been identified on the basis of the mineralogy of the rock types, but for only five of these groups have the axe factories been located. These five areas, seemingly operative during the Neolithic period (mid-4th millenium to later-3rd millenium BC) are: (i) Great Langdale and Scafell Pike (Lake District); (ii) Penmaenmawr (north Wales); (iii) Tievebulliagh and Rathlin Island (Northern Ireland); (iv) Mynnydd Rhiw (south Wales); and (v) Killin (Perthshire, Scotland). The products of the more important of the axe factories are found in significant concentrations up to 550 km from the source areas, indicating widespread trade in these objects during the prehistoric period (though some would argue that the role of *glaciation* in disseminating the raw materials of the **axe** groups has been insufficiently acknowledged in British **axe** provenance studies). [OWT]

axial dipole A single magnet aligned along an axis. This term usually refers to the model of the time-averaged **geomagnetic field** in which the north and south **magnetic** poles correspond exactly with the geograpic (rotational) poles. [DHT]

axial flattening strain Type of **strain** characterized by one **principal strain** direction being much shorter than the other two, which are similar. (See **strain ellipsoid**.) [RGP]

axial plane cleavage/foliation Set of **cleavage** or **foliation** planes that are parallel or near-parallel to the *axial plane* of a **fold** or set of **folds**, and are by inference genetically related to the formation of the **folds**. [RGP]

axial rift A valley, usually some 2 to 3 km deep and 30 to 50 km wide, that characterizes the mid-Atlantic Ridge and most of the mid-Indian Ocean Ridge. It is cut by **transform faults** which offset it by tens to hundreds of kilometres. The median line of the rift is characterized by moderate tensional seismicity and volcanicity, and also by a strong **magnetic anomaly**. [DHT]

axially-symmetric extension **Extension** in one **principal strain** direction and equal **shortening** in all directions at right angles to it. (See **strain ellipsoid**.) [RGP]

axially-symmetric shortening **Shortening** in one **principal strain** direction and equal **extension** in all directions at right angles to it. (See **strain ellipsoid**.) [RGP]

axinite $((Ca, Fe, Mn)_3Al_2(BO_3)_3(Si_4O_{12})(OH))$ A complex ring silicate (*cyclosilicate*) found in cavities in **granites** or **contact metamorphic** zones surrounding granitic intrusions. [DJV]

axis of rotation Imaginary axis through the centre of the Earth about which the relative movement of any two *plates* on the surface of the Earth takes place. The axis penetrates the Earth's surface at the *poles of rotation*, also known as *Eulerian poles*. (See **plate tectonics**, **Euler's theorem**.) [RGP]

azurite $(Cu_3(CO_3)_2(OH)_2)$ A secondary **copper** mineral. Azurite was used for jewellery and small ornaments in antiquity. [DJV/OWT]

B

B-type lead An isotopically **anomalous lead** whose **lead** isotopic ratios yield a model **lead** age older than the host rocks within which the **lead** occurs, whether in an **orebody** or a rock. This is held in some interpretations to indicate that the **lead** has been derived by remobilization of **lead** deposited in the **crust** during an older event. [AME]

back The roof of a **drift**, **adit**, gallery, crosscut, drive, tunnel, **stope**, etc., i.e. the ceiling of any underground working. [AME]

backfill The refill introduced into mine workings either to support the worked areas and prevent or ameliorate subsidence effects or to dispose of waste materials, or for a combination of both purposes. [AME]

backfold A **fold** in a mountain belt with **vergence** opposed to that of the belt in general. These **folds** are the equivalent of **backthrusts** in **thrust belts**, and can occur over kilometre-wide zones, as in the Vanoise region of the French Alps. The term '*retrocharriage*' is also applied to these **folds**. [SB]

backshore The area of a **beach** which lies between the normal high **tide** level and the point reached by the highest *storm* waves or **tides**. The upper limit may include the cliff foot and rocks affected by severe *storm* waves or the **foredune** area of a **sand dune** system if similarly affected. The backshore area of a low sandy coast is often dominated by a **beach berm** with a steep seaward face to above high water mark, a prominent crest and gentle landward slope. On pebble **beaches** the backshore **beach ridge** can be built to considerable altitude, e.g. 13 m above high tide on Chesil Beach, southern England, and several **beach ridges** may occur associated with *storms* of varying magnitude. Often the landward face of the **beach ridge** is affected by overwash processes and large *overwash fans* are deposited in the **backshore** zones of **barrier islands** and some low-lying **spits**. In comparison with the **beach** face, the **backshore** suffers infrequent **wave** modification and inundation and so is a favourite habitat of the pioneer plants that are able to tolerate sediment movement and high salinity. [JDH]

backswamp An area of waterlogged land adjacent to a river. Backswamps are common in areas where **aggradation** of fluvial deposits along the river channel has formed **levées**. During peak discharge, overtopping of the **levées** results in inundation of the backswamp portion of the **floodplain** which may extend to the sides of the river valley. The fluvial sediments which accumulate in the backswamp are characteristically fine-grained and rich in organic material derived from the decay of **swamp** and marsh vegetation. Artificial drainage of backswamp areas is often undertaken in order to exploit these fertile deposits for agriculture. However, agricultural development and human settlement in such areas may be subject to catastrophic disruption when there is overbank discharge or **levées** are breached. [AW]

backthrust Low angle *reverse fault* in a **thrust belt** which shows an overthrusting direction opposite to the overall sense of the belt, i.e. the **faults** have **vergence** opposed to the **vergence** of the belt overall. For example, in the Caledonian Moine **thrust belt** of north-west Scotland, the majority of *thrust faults* show movement of the **hanging wall**, i.e. **vergence**, to the west-north-west; backthrusts have similar **strike**, but show displacement of the **hanging wall** to the east-south-east. (See **backfold**.) [SB]

backwash The gravity-fed return flow of water formed after a **wave** breaks on a **beach**. Together with sediment size, it is an important factor influencing the **beach** gradient. If the **beach** is in equilibrium with **wave** conditions then the amount of sediment moved up by the high velocity **swash** exactly equals the amount moved down by the lower velocity, but longer duration, backwash. The volume of this backwash is reduced by percolation into the **beach** face so that on a coarse pebble beach a steeper slope is necessary to maintain equilibrium between **swash** and backwash than on a **sand beach** with little percolation and a proportionally higher backwash. Disruption of this equilibrium by, for example, an increase in **wave** height during *storms* may result in an increase in the volume of backwash which leads to enhanced downbeach sediment movement and a reduction in **beach** gradient.

On **beaches** where a range of sediment sizes exist, the low backwash velocities may not be sufficient to move coarse sediment carried upbeach by the higher velocity **swash**. The result is a progressive upbeach increase in sediment size and **beach** gradient. [JDH]

Pethick, J.S. (1984) *An Introduction to Coastal Geomorphology*. Arnold, London.

backwearing The retreat of a hillslope through **erosion** without change in the morphology of the slope; also referred to as **parallel retreat**. Backwearing occurs when the rate of **erosion** on different segments of the hillslope is proportional to the inclination. Hence, the steepest slopes experience most **erosion** and the gentlest slopes the least; slope units with the same gradient are eroded the same amount. If the slopes at the top of the hill or **escarpment** erode more rapidly than the lower slopes, slope morphology will change — under these conditions, *downwearing* rather than backwearing will occur. [AW]

back-arc spreading The **oceanic crust** between a **subduction zone** dipping towards a continent and the **continental crust** is thought to originate by similar processes to those at an *oceanic ridge*, but the *oceanic magnetic anomaly* patterns, **heat flow**, and topography are not well defined in most instances and it is assumed that the creation of new **oceanic crust** in a *marginal sea (back-arc basin)*, as a result of **convective** upwelling over the **Benioff–Wadati zone** is more diffuse that at *oceanic ridge* systems. It is also possible that such anomalies are less obvious than in the open ocean due to the masking effects of greater thicknesses of sediments. [DHT]

Karig, D.E. (1971) Origin and development of marginal basins in the western Pacific. *Journal of Geophysical Research* **38**: 435–48.

Kearey, P. & Vine, F.J. (1990) *Global Tectonics*. Blackwell Scientific Publications, Oxford.

badland A landscape of numerous deep gullies and ravines separated by sawtoothed ridges, small **mesas** and **buttes**. The areas are usually devoid of vegetation because *soils* have been stripped by **erosion**. The landscapes usually develop through incision and **erosion** of weakly cohesive materials. Examples of badlands are developed on bedrock such as **shale** or clay, and on sediments such as volcanic **ash**, **loess**, and **colluvium**. The materials are readily eroded by water often because they loose **cohesion** when wet. Clays may provide the **cohesion**, and upon wetting — especially by water rich in sodium ions — the clays are dispersed and the residual material erodes by **slaking**. Rapid **erosion** can occur on clayey materials with high *exchangeable sodium percentages*. Rainwater falling on these materials is quickly enriched in sodium aiding dispersion of clays; in addition, **infiltration** rates are low, so surface **runoff** is readily generated. These physical and chemical properties of the surficial materials may promote the development of characteristic erosional features such as deeply fluted walls, jagged ridges and isolated pinnacles.

Badlands often occur in semi-arid regions where rainfall is seasonal or sporadic. In addition to dissection of the terrain by surface **runoff** and **stream flow**, other processes such as *mass-wasting*, **piping** and **tunnel erosion** can be important.

The term badland was first coined by French-Canadian trappers who described the landscape of parts of south-western South Dakota as *les mauvaises terres pour traverser*, 'the bad lands to cross'. This area of eroding clay bedrock covers more than $5000 \, km^2$ wherein lies the Badlands National Monument. [AW]

Bryan, R. & Yair, A. (eds) (1982) *Badland Geomorphology and Piping*. Geo Books, Norwich.

Bahama Banks A series of *carbonate* blocks (*platforms*) developed on the subsiding eastern margin of North America. These blocks are composed of a thick sequence (over 5 km) of shallow-water **limestones** of *Pleistocene* to **Cretaceous** age. The banks extend for 1400 km from off the Florida shelf to north-east of Cuba. The north-west section is particularly well known because of extensive studies of the **Quaternary** carbonate sediments, and consists of two large platforms, the Great Bahama Bank and Little Bahama Bank. These are separated from the Florida shelf by a deep channel, the Florida Straits, and are dissected by a number of deep channels. The Great Bahama Bank covers a huge area yet it is only submerged to depths of less than 10 m and is the site of extensive carbonate sedimentation. Around the platforms the margins slope steeply to depths of several hundred metres in only a few kilometres on the continental side and to depths of over 3000 m on the ocean-ward side. [VPW]

Schlager, W. & Ginsburg, R.N. (1981) Bahama carbonate platforms — the deep and the past. *Marine Geology* **44**: 1–24.

bajada Confluent **pediment** slopes and **alluvial fans** forming a low-lying area at the foot of the mountain front encircling desert **basins**. [ASG]

balanced section A **cross-section** in **folded** or **faulted** regions constructed so that the **bed** lengths or areas in the deformed state are equivalent to those on a *restored section*, which represents the arrangement of beds, etc., before **deformation**. The two sections are known as 'deformed' (after **strain**) and '*restored*' (inferred pre-**deformation**) sections. The aim of producing such sections is to produce a model compatible with all observations (field, borehole, seismic, gravity, etc.) that can be related or restored to a feasible pre-**deformation** state. Certain qualities are needed to ensure this. The structures used to continue the deformed state section below surface, and away from field and borehole data, should maintain the observed structural style, making the deformed section *admissible*. For a **cross-section** to represent accurately the displacement of, for example, *thrust sheets*, the tectonic transport direction must be in the section plane, and **deformation** must be **plane strain**. This means that all the rock represented in the deformed section must also be represented in the *restored section*. Movement on **faults** in and out of the section plane means that the rock represented in the two sections differs, so that the sections cannot balance. Distributed non-**plane strain** and volume change (**dilation**) can be taken into account in section construction (Hossack, 1978), in cases where they can be quantified. In folded regions, restorable **cross-sections** can only be constructed when folds are *cylindrical*. These methods have been widely used in **foreland thrust belts** where there is a single transport direction, such as the Rocky Mountains Foothills (Dahlstrom, 1970).

A valid balanced **cross-section** represents the observed structures and an inferred structure at depth that can be systematically related to the original undeformed state. Balanced sections form testable hypotheses for the structure of **fold** and **thrust belts**, and are successfully used in *hydrocarbon exploration*. These techniques can be applied to folded, *thrust-faulted* and extensionally **faulted** regions. One of the most important aspects of their use is the necessity to consider the incremental development of the final **fold/fault** structure. A section constructed in order to be restorable requires the constructor to deduce the order of imposition of the different **strains**, i.e. **fault** movements, **folding**, distributed **strain**. As the superposition of **strain** is non-commutative, the order is important. The latest **strain** increment must be removed first, to remove the effects of that strain from the previously existing distribution. So in an **imbricate structure** for example, the order of **fault** movement must be deduced in order to produce balanced sections. The systematic 'unstraining' which balancing a section requires allows the total **shortening** or **elongation** to be quantified. If sections are constructed with the additional constraint that shortening be minimized, for example by choosing **fault** trajectories with minimum offsets, then a balanced section produces a model for the structure of an area compatible with available data and giving a minimum displacement estimate (Butler, 1987). [SB]

Butler, R.W.H. (1987) Thrust sequences. *Journal of the Geological Society* **144**: 619–34

Dahlstrom, C.D.H. (1969) Balanced cross sections. *Canadian Journal of Earth Sciences* **6**: 743–57.

Elliott, D. (1983) The construction of balanced cross sections. *Journal of Structural Geology* **5**: 101.

Hossack, J.G. (1978) The correction of stratigraphic sections for tectonic strain in the Bygdin area, Norway. *Journal of the Geological Society* **135**: 229–41.

balas ruby A red **gem** variety of the mineral **spinel**. (See **oxide minerals**.) [DJV]

ball clay A clay typically made up of 70% **kaolinite** plus **illite**, **quartz**, **montmorillonite**, **chlorite** and small amounts (2–3%) of carbonaceous material. Ball clays are fine-grained and this together with the presence of carbonaceous material renders them highly plastic and imparts a green strength superior to that of **kaolin (china clay)**. Ball clay also has a low **mica** content and is extremely refractory. These properties and its long vitrification range explain its importance to the ceramics industry. A further bonus is that the carbonaceous material burns on firing to produce a white or near-white product. [AME]

ballas A variety of **industrial diamond** having the form of dense, globular aggregates that are extremely hard and tough. [AME]

ballast Crushed or broken rock or slag used for road foundations or railway tracks. [AME]

band theory A group of **molecular orbitals** (MO) which extend throughout a solid and which have similar energies comprise a band. When atoms (or ions) are arranged in a lattice so that each atom has many neighbours, the **atomic orbitals** interact to produce such a band of MOs. The number of MOs whose energies lie in a given energy range measures the *density of states*. This band theory approach to calculating the electronic stucture of solids is widely used for metals and alloys; it is also useful for semiconductors but is employed less frequently for insulators.

In real materials, some bands are quite narrow so that one band is separated from another by a gap on the energy scale, the *band gap*. If the least tightly bound electrons are in orbitals within a band, then only a negligible amount of energy is required to excite an electron into an empty orbital where it could be highly mobile, leading to good conductivity of heat and electricity and a shiny 'metallic' appearance. By contrast, if the least tightly bound electrons lie at the top of a band, easy excitation is not possible — the energy barrier of the band gap must be surmounted — and an insulator results. Narrow *band gaps* characterize semiconductors. Bands can suffer local distortions due to the presence of impurity atoms. Such atoms are added to semiconductors to modify their electrical properties. [DSU]

Cox, P.A. (1987) *The Electronic Structure and Chemistry of Solids*. Oxford University Press, Oxford.

Kittel, C. (1976) *Introduction to Solid State Physics*. John Wiley & Sons, New York.

banded coal Megascopically, **coal** can be classified into two broad groups, banded or *humic coals* and non-banded or **sapropelic coals**. Banded coals are stratified with layers or bands of organic material of varying appearance. Individual layers are usually no more than a few centimetres in thickness. These **coals** are a heterogeneous mixture of a wide range of plant materials which were deposited as **peats** composed of small to large fragments of plant debris.

The non-banded coals on the other hand are homogeneous, tough and often have a marked conchoidal fracture. They commonly consist of masses of *spores* or *algal* material. [AME]

Ward, C.R. (1984) *Coal Geology and Coal Techology*. Blackwell Scientific Publications, Melbourne.

banded iron formation (BIF) (*cherty iron formation, itabirite, jaspillite*) A finely layered **iron**-rich sediment. The layers are generally 5–30 mm thick and in turn they are commonly **laminated** on a scale of millimetres or fractions of a millimetre. The layering consists of **silica** layers (in the form of **chert** or better crystallized **silica**) alternating with layers of **iron** minerals. The simplest and commonest BIF consists of alternating **hematite** and **chert** layers. The content of alumina is less than 1%, contrasting with **Phanerozoic** *ironstones* which normally carry several per cent of this oxide. There are four important facies of BIF.

1 *Oxide facies*. This is the most important facies and it is divided into **hematite** and **magnetite** subfacies according to which **iron oxide** is dominant. There is a complete gradation between the two subfacies. An **oolitic** texture is common in some **hematite**, suggesting a shallow water origin, but in others the **hematite** may have the form of structureless granules. The 'chert' varies from fine-grained cryptocrystalline material to mosaics of intergrown **quartz** grains. In the much less common **magnetite** subfacies layers of **magnetite** alternate with iron silicate or carbonate and cherty layers. Oxide facies BIF typically averages 30–35% Fe and these rocks are mineable provided they are amenable to **beneficiation** by magnetic or gravity separation of the iron minerals.

2 *Carbonate facies*. This commonly consists of interbanded **chert** and **siderite** in about equal proportions. It may grade through **magnetite–siderite–quartz** rock into the oxide facies, or, by the addition of **pyrite**, into the sulphide facies.

3 *Silicate facies*. Iron silicate minerals are generally associated with **magnetite**, **siderite** and **chert** which form layers alternating with each other. Most of the iron in this facies is in the ferrous state, suggesting deposition in a reducing environment. Carbonate and silicate facies BIF typically run 25–30% Fe, which is too low to be of economic interest. They also present **beneficiation** problems.

4 *Sulphide facies*. This consists of **pyritic** carbonaceous argillites — thinly banded rocks with organic matter plus **carbon** making up 7–8%. This facies clearly formed under *anerobic* conditions. Its high **sulphur** content precludes its exploitation as an iron **ore**.

Precambrian BIF can be divided into two principal types, **Algoma** and **Superior**. Goodwin, in a study of **Algoma-type** BIF in the Canadian Shield, showed that facies analysis was a powerful tool in elucidating the **paleogeography** and could be used to outline a large number of **Archean basins**. His section across the Michipicotin Basin is shown in Fig. B1.

BIF forms one of the Earth's great mineral treasures. It occurs in stratigraphical units hundreds of metres thick and hundreds or even thousands of kilometres in lateral extent. Substantial parts of these **iron** formations are directly usable as a low grade iron **ore** and other parts have been the **protores** for higher **grade** deposits. The great bulk of **iron** formations of the world was laid down in the very short time interval of 2600–1800 Ma, but BIF is not restricted to this period, older and younger examples being known. [AME]

Evans, A.M. (1993) *Ore Geology and Industrial Minerals: An Introduction* (3rd edn). Blackwell Scientific Publications, Oxford.
Gross, G.A. (1970) Nature and occurrence of iron ore deposits. In: *Survey of World Iron Ore Resources*, pp. 13–31. United Nations, New York.

bank erosion Bank erosion plays an important part in the maintenance of equilibrium channel form. A variety of erosional processes may be involved, including detachment of individual particles and mass failure of the bank by rotational **failure** or undercutting. Much bank erosion is achieved by high discharges close to **bankfull**. The timing of a flood may be important in that less **stress** is required for **erosion** of wet river banks in winter, particularly if they have been weakened by frost (Lawler, 1986). In addition, **failure** after river levels have fallen is also more likely at that time. Bank material characteristics exert an important control on **erosion**, both in terms of the type

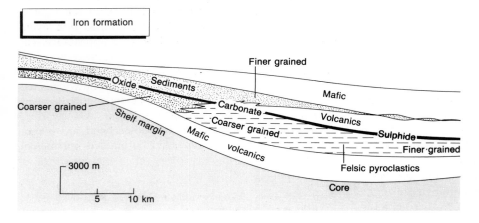

Fig. B1 Reconstructed stratigraphic section of the Michipicotin Basin showing the relationship of the oxide, carbonate and sulphide facies of **banded iron formation** to the configuration of the basin and associated rock types. (After Evans, 1993.)

of process operating and the rate of **erosion**. Bank retreat is particularly affected by the percentage of silt and clay in the bank material since these provide **cohesive strength** at times when frictional **strength** is reduced by the effect of high **pore fluid pressure**. Bank vegetation may also exert a strong control over bank stability and so may have some influence on channel form. Bank erosion is particularly important in the development and migration of meanders. [TPB]

Lawler, D.M. (1986) River bank erosion and the influence of frost: a statistical examination. *Transactions of the Institute of British Geographers* **NS 11**: 227–42.

banket A name given to the **gold–uranium**-bearing **quartz**-pebble conglomerates of the early **Proterozoic** Witwatersrand Goldfield of South Africa. The name is not applied to similar **Proterozoic** deposits in other parts of the world. These conglomerates have a matrix rich in **pyrite**, **sericite** and **quartz**. The **gold** and **uranium** minerals (principally **uraninite**) occur in the matrix together with a host of other detrital minerals. These are ancient **placer deposits** that mark an important metallogenic event in the early **Proterozoic**, as this type of metal concentration is rare in the **Archean** and essentially absent from younger rocks. [AME]

Pretorius D.A. (1981) Gold and uranium in quartz-pebble conglomerates. *Economic Geology 75th Anniversary Volume*, 117–38.

bankfull discharge The maximum discharge which can be contained within a river channel without overtopping the banks. Though this may be difficult to define precisely for some channel cross-sections, the concept is straightforward, allowing channel flow to be distinguished from overbank flow. Because the channel is completely full, many authors have argued that bankfull discharge best represents the range of morphologically significant discharges and is therefore conceptually equivalent to the 'dominant discharge'. Bankfull discharge has been successfully correlated with measures of channel form such as width, depth and meander wavelength. The recurrence interval for such flows may vary widely; for some humid temperate rivers, an average recurrence of about 1.5 years has been identified. [TPB]

bar A non-SI unit of **pressure** or **stress**, now superseded by the **pascal**. 1 bar = 10 Pa. [RGP]

bar finger sand Deposit resulting from the progradation of distributary channel-mouth systems where the distributary channels are fixed by the cohesive muds of the **delta** front and interdistributary bays. Bar finger sands are therefore markedly elongate, at high angles to the mean coastline orientation. In the Mississippi Delta they range up to 30 km long, 5–8 km wide and have an average thickness of 70 m (Fig. B2). (See **delta**.) [PAA]

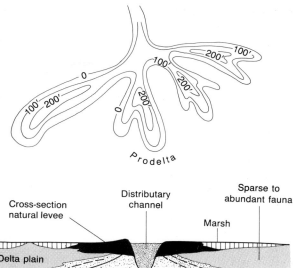

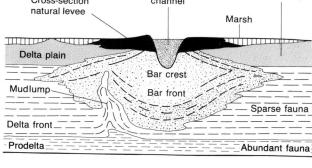

Fig. B2 Bar finger sands of the Mississippi delta. (After Fisk, 1961.) The influence of the diapiric mudlumps on sand-body shape is now regarded as more significant than originally depicted and the bar fingers are thought to be composed of thick pods of sand between diapirs and thin connecting sand intervals above the diapirs.

Fisk, H.N. (1961) Bar finger sands of the Mississippi Delta. In: Peterson, J.A. & Osmond, J.C. (eds) *Geometry of Sandstone Bodies — A Symposium*, pp. 29–52. American Association of Petroleum Geologists, Tulsa, OK.

barchan Isolated crescent-shaped **dune** with a gentle windward or **stoss slope** and a **lee slope** with a **slip-face** bounded by two arms which point downwind. Windward slopes typically have slope angles of 8° to 15°. A barchan is a form of **transverse dune**. Because of its shape it can move over long distances with little change in form. **Sand** eroded from the windward slope becomes trapped in the still air beyond the crest and is deposited on the **lee slope**. Barchans form in areas where there is a uni-modal wind direction and a moderate **sand** supply moving over an area of restricted **sand** supply, commonly a **gravel** pavement or **sand sheet**. Sizes vary; heights recorded range from 0.3 m, this being the minimum height for a **slip-face** to form, to 30 m. Migration rates of 5 to 10 m a^{-1} are usual. Widespread distribution in many desert and coastal environments. [RDS]

Howard, A.D., Morton, J.B., Gad-El-Hak, M. & Pierce, D.B. (1978) Sand transport model of barchan dune equilibrium. *Sedimentology* **25**: 307–38.

barite (BaSO$_4$) An **industrial mineral** valued for its high density. [DJV]

barium (Ba) At. no. 56, at. wt. 137.33. An alkaline earth element occurring principally as **barite** and **witherite**. [AME]

barranca Steep-sided erosional gully or ravine. Comparable to a **donga**. [ASG]

barrel The traditional unit for expressing quantities of oil. One barrel contains 42 US gallons or 35 Imperial gallons. The relation between tonnes and barrels varies of course with the density of the oil. One tonne of average world **crude oil** is equivalent to 7.33 barrels. In the future oil reserves will probably be expressed in cubic metres. [AME]

barrier boundary A hydraulic boundary to an **aquifer**. Particularly those that prevent a **cone of depression** from expanding beyond them, for example, where an **aquifer** abuts against impermeable material at a **fault**. [AME]

barrier island A largely sandy, elongated shore-parallel island separated from the shore by a *lagoon* or marsh. Barrier islands are relatively common coastal features around the world, with barrier coasts making up 10–14% of the global coastline. Barrier islands are characteristically formed on depositional coasts with low gradients and abundant supplies of loose sediment. Unlike barrier **spits**, barrier islands are not necessarily attached to the shore-line. Good examples of barrier islands are found on the coast of North and South Carolina in the USA, where they occur along a 300 km stretch of the coastline.

There are two major barrier island morphologies distinguished by tidal range; microtidal barriers (tidal range <2 m) and mesotidal barriers (tidal range 2–4 m). No barriers form where the tidal range is over 4 m. Their morphology is controlled by, and varies systematically with, the incident wave–tidal regime. Microtidal barriers are long linear features dominated by storm washover processes and are separated by minor **tidal inlets**. Mesotidal barriers are short and stunted in comparison, with a characteristic drumstick shape. The tidal inlets and tidal **deltas** of mesotidal barrier systems are well developed, reflecting the greater influence of tidal currents (Fig. B3).

Barrier islands are composed of well-sorted and cross-bedded **sands** and pebbles, with wind-blown **sand** on top. At the seaward side of these islands the sandy facies grade into fine **sands** silts and then marine muds. At the landward side there are lagoonal muds and silts, marsh **peats** and organic clays and sometimes **floodplain** deposits. Barrier islands occur particularly in low to middle latitudes with low to medium tidal range and in swell wave environments. There have been many suggestions put forward as to how barrier islands develop and the sea-level conditions involved. In some cases, barrier islands are formed under conditions of sea-level rise. The rising sea-level causes a *storm beach* or *dune ridge* to be separated from the shore by flooding the back beach to produce a *lagoon*. The newly formed barrier island will only grow if there is a suitable supply of sediment. In some cases, barrier islands may be formed from **spits** which are built up by *storm* **wave** action and then breached by **wave** attack to form islands. On coasts where sea-level is rising, barrier island systems tend

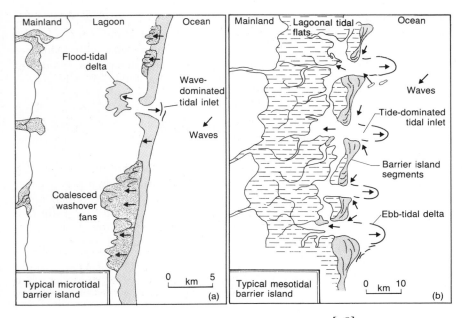

Fig. B3 Morphologic models of **barrier island** systems. (a) Microtidal and (b) mesotidal. (After Hayes, 1979.)

to move landwards due to washover of sediment under storm conditions.

Ancient barrier island deposits are enormously varied, as a result of the complex interaction of oceanographic processes, sediment supply and sea-level variation that lead to their evolution. They are often important *hydrocarbon* **reservoirs**. [RK/HAV]

Davies, R.A. (ed.) (1985) *Coastal Sedimentary Environments*. Springer-Verlag, Berlin.

Hayes, M.O. (1979) Barrier Island morphology as a function of tidal and wave regime. In: Leatherman, S.P. (ed.) *Barrier Islands from the Gulf of Mexico*, pp. 1–27. Academic Press, New York.

Oertel, G.F. & Leatherman, S.P. (eds) (1985) *Barrier Islands*. Marine Geology Special Issue No. 63. Elsevier, Amsterdam.

barrier reef A coastal, mainly organic *reef*, separated from the shore by a *lagoon*. Barrier reefs are one of the four main types of **coral–algal reef** as recognized today, the others being *fringing reefs*, **atolls** and *table reefs*. Charles Darwin proposed that barrier reefs were part of a genetic sequence of *reef* development around islands in tropical seas. He suggested that **coral** growth around a subsiding volcanic island would lead to *fringing reef* development, followed by barrier reef development, and finally **atoll** formation as the central island subsided. Examples of barrier reefs around such islands include those of Mayotte, Comoros Islands in the Indian Ocean and Santa Cruz in the western Pacific. Barrier reefs are also found on continental coastlines, as illustrated by the Belize barrier reef. Most barrier reefs are about 300–1000 m wide. A notable exception is the **Great Barrier Reef** on the Queensland coast of Australia which is a much larger feature. The **Great Barrier Reef** stretches for 1950 km along the coast and is actually composed of over 2500 separate *reefs*. It is backed by a *lagoonal* sea which is tens of kilometres wide. [HAV]

basal sapping The process whereby a slope is undercut at the base. It can describe the retreat of a hillslope or **escarpment** by **erosion** along a **spring**-line (mid-latitudes), by salt weathering (arid environments) or by glacial **erosion** (*cirque*). Sapping may also form some of the landforms on Mars (Higgins, 1984). [ASG]

Higgins, C.G. (1984) Piping and sapping: development of landforms by groundwater outflow. In LaFleur, R.G. (ed.) *Groundwater as a Geomorphic Agent*, pp. 18–58. Allen & Unwin, Boston.

basal tar mat A development, as a band or zone, of heavy oil and tar at, or immediately below, the oil–water interface of a productive **oilfield**. Such mats may plug the **reservoir** pore space with *bitumen* and prevent natural **water drive** from maintaining the **reservoir** pressure necessary for production. Thick tar mats do, however, mark the bases of some of the world's largest **oilfields** including the supergiant Sarir field in Libya. These mats may result from *bacterial* attack or the action of excess gas causing **precipitation** of the **asphalt** content of the oil. [AME]

basalt An aphanitic mafic **igneous rock** composed of **plagioclase** (more calcic than An_{50}) and **pyroxene**, with or without **nepheline**, **olivine** or **quartz** and with accessory Fe–Ti oxide. Basalts may be classified on the basis of modal **phenocryst**/groundmass mineralogy, but due to the fine grain size are commonly classified on the basis of chemical and **normative composition** as shown below (Table B1).

In terms of the normative classification alkali basalt is termed **silica**-undersaturated, **olivine tholeiite** is termed **silica**-saturated and **quartz tholeiite** (with normative **quartz**) is termed **silica**-oversaturated.

The occurrence in oceanic and continental areas, the presence of **mantle**-derived ultramafic **xenoliths** and the results of experimental petrology of basaltic rocks indicate an origin by partial melting of **mantle peridotite**. Experimental studies indicate that the different basalt types originate by *partial melting* and/or *fractional crystallization* of **mantle**-derived **magmas** at different depths. Alkali basalts originate by lower degrees of *partial melting* at greater depth, possibly of relatively alkali-rich **peridotite**, in comparison with **tholeiitic** basalts which have formed as a result of a greater degree of *partial melting* and/or equilibration at shallower **mantle** depths. Mid-*ocean ridge* basalts (MORB) are characterized by high calcium (CaO) and low alkali ($Na_2O + K_2O$) concentrations and are believed to result from high degrees of partial melting of alkali-poor **peridotite mantle** below *ocean ridges*.

Basalts were widely used in antiquity as **building stone** and for monumental sculpture, for example the Olmec statues of Mexico, Roman, Greek and Egyptian sarcophagi and Buddhist temples of India. Basalt was also used for smaller objects such as lamps and querns. Basalt querns and **millstones** mainly of carefully selected young **vesicular** rock, were common in Europe from the Bronze Age until recent times. [RST/OWT]

Morse, S.A. (1980) *Basalts and Phase Diagrams*. Springer-Verlag, Berlin.

Basaltic Volcanism Study Project (1981) *Basaltic Volcanism on the Terrestrial Planets*. Pergamon, New York.

basanite A silica-undersaturated alkali **olivine basalt** containing **olivine**, *clinopyroxene* and **plagioclase** together with >5–10% (depending on definition) of normative and modal **feldspathoid** (**nepheline** or **leucite**). Basanites occur in association with other **alkaline igneous rocks**. [RST]

base metal A term employed in the mining industry to include **copper**, **lead**, **zinc** and **tin**. [AME]

Table B1 Basalt classifications

Basalt type	Diagnostic normative minerals
Alkali (olivine) basalt	Nepheline + olivine
Olivine tholeiite	Olivine + hypersthene
Quartz tholeiite	Quartz + hypersthene

base station A reference location used during a geophysical survey. In a gravity survey it is a site where the absolute value of **gravity** is known, and is also used to monitor instrumental **drift**. In a **magnetic survey** it may be used to monitor **diurnal variation**. [PK]

base surge A cloud consisting of a turbulent, dilute mixture of solid material, and possibly condensed water, in gas which travels rapidly along the ground away from the site of a large explosion or **volcanic eruption**. Base surges are common in some kinds of **volcanic eruption**, notably **phreatomagmatic eruptions**. Base surges dump suspended **tephra** to form deposits that mantle topography, but which also tend to thicken into depressions. The deposits are usually laminated and cross-bedded. Normal **dunes** and **antidunes** are both commonly present. Individual bedsets are about 1–10 cm thick and interpreted to represent deposition from a single surge (Walker, 1984). Bedsets are internally laminated and cross-bedded, and are commonly separated from other surge bedsets by air-fall layers. Base surges are regarded as being a different volcanic phenomenon from the high-concentration **ash flows**. Many authors (e.g. Cas & Wright, 1987) regard base surges as a subgroup of **pyroclastic** surges, the latter group also including ground surges. Ground surges are laminated and cross-bedded deposits which usually underlie, or are interbedded with, **ash flow tuffs** (Fisher, 1979), and may be generated directly from the eruption at the vent, or from the advancing front of the active **ash flow**. (See **tephra**, **tuff ring**.) [PTL]

Cas, R.A.F. & Wright, J.V. (1987) *Volcanic Successions Modern and Ancient*. Allen & Unwin, London.
Fisher, R.V. (1979) Models for pyroclastic surges and pyroclastic flows. *Journal of Volcanology and Geothermal Research* **6**: 305–18.
Walker, G.P.L. (1984). Characteristics of dune-bedded pyroclastic surge bedsets. *Journal of Volcanology and Geothermal Research* **20**: 281–96.

basin This term is used geologically in two different senses. In a strictly structural sense, it describes a **synformal** structure with a circular or near-circular outcrop, i.e. where the *limb* **dips** are radial. Such a structure is defined by circular *stratum contours*. The term is also used, however, in a topographic or stratigraphical sense to describe a depression that either contains sediments, or is capable of receiving sediments. Basins in this sense may be either marine or continental and exhibit a wide variation in size and shape. Most basins are **tectonic** in origin, produced by active depression of the **crust**, due, for example, to **extensional** or gravitational forces. Numerous types of basins have been distinguished: e.g. **foreland basins**, *intermontane basins*, etc. [RGP]

basin plain Flat or near-flat, generally smooth, floor to a **basin**, mainly covered by sheet-like sediments. The term is often used synonymously with *abyssal plain* or *basin floor*.

Typically, basin plains are situated in deep water, occurring in small slope-basins, large deep lakes and flooring the major oceans as *abyssal plains* of up to about $1.5 \times 10^6 \, \text{km}^2$. Their smooth, essentially flat surfaces result from terrigenous and **pelagic** sediments covering any pre-existing uneven topography. Basin plains are the ultimate terrigenous sediment trap, commonly as the distal part of continental rises and **submarine fans**. They also form the optimum site for the accumulation of **pelagic** biogenic and chemogenic sediments such as **oozes**, **red clays** and **manganese nodules**. Basin plain sediments, up to hundreds of metres thick, provide an important geochemical contaminant of primary **mantle** when **subducted** in **Benioff zones**. [KTP]

basin-and-range A type of terrain, as found in Utah and Nevada, where there are **fault** block mountains interspersed with **basins**. The **basins** often contain lakes, some of which may only fill up in **pluvials**, and are flanked by **pediments** and **alluvial fans**. [ASG]

bastard ganister A siliceous rock resembling a **ganister** but containing more interstitial matter and often with incomplete secondary **silicification**. [AME]

bastard rock An impure **sandstone** with thin lenticular layers of **shale** or **coal**. [AME]

Bath stone A soft, creamy yellow, **oolitic limestone** quarried near Bath, England and much used as a **building stone**. [AME]

batholith A large-volume intrusion (with no observed floor), having a surface area exceeding $100 \, \text{km}^2$ and comprising individual **plutons** ranging from **gabbro** to **granite** in composition, with an overall **diorite–granodiorite** composition. Batholiths are **composite intrusions** and comprise many individual **plutons**, ring intrusions and other minor intrusions. [RST]

batt (1) A British term for any hardened clay other than **fireclay**.
(2) A British term for a compact black carbonaceous **shale** that splits into fine laminae and is often intercalated with thin layers of **coal** or *ironstone*.
(3) Another term for *bastard coal* — thin layers of impure **coal** found in the lower part of **shales** immediately above a **coal seam**; or any **coal** with a high **ash** content. [AME]

battery ore **Manganese ore** of a **grade** suitable for the production of manganese dioxide for use in dry cell batteries. Battery grade ore contains a minimum of 80% MnO_2, less than 0.05% of metals and no nitrates. [AME]

[50]

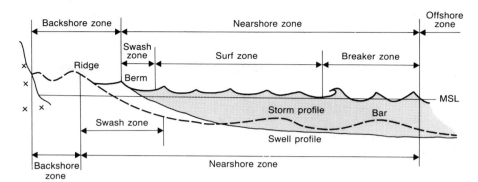

Fig. B4 Beach. Schematic diagram of wave zones on a beach and of storm and swell wave profile changes. Note the extension of the swash zone under storm waves and the development of an erosional low gradient storm profile with bar development in the breaker zone.

bauxite An earthy rock composed almost entirely of aluminium hydroxide, the chief **ore** of **aluminium**. The mineralogy of bauxites depends on their age. Young bauxites are **gibbsitic** but, with age, **gibbsite** gives way to **boehmite** and **diaspore**. **Residual deposits** of bauxite have formed as a result of intense chemical **weathering** of pre-existing rocks such as in tropical climates having a high rainfall. All soluble material has been dissolved away and even **iron oxides** have been leached out. This process is known as *bauxitization*. Bauxite will form on any rock with a low **iron** content or one from which the **iron** is removed during **weathering**. Bauxites are vulnerable to **erosion** and most deposits are therefore post-**Mesozoic**. Some eroded bauxite has been redeposited to form what are called transported or sedimentary bauxites.

Bauxite deposits are extremely variable in their nature and geological situations and have been classified in many different ways. A simple threefold classification by Grubb is based on the topographical levels at which they were formed. High level bauxites generally occur on volcanic or igneous source rocks and form thick blankets up to 30 m, which cap **plateaux** in tropical to subtropical climates; they are dominantly **gibbsitic**. Low level bauxites occur along tropical coastlines. They are distinguished by the development of *pisolitic textures* and are often **boehmitic**. Grubb's third class is that of sedimentary bauxites. (See **karst bauxite**.) [AME]

Hutchison, C.S. (1983) *Economic Deposits and Their Tectonic Setting*. Macmillan, London.

Maynard, J.B. (1983). *Geochemistry of Sedimentary Ore Deposits*. Springer-Verlag, New York.

bayou A swampy creek or backwater. The name originates from the southern United States, and notable examples of bayous are found along the lower Mississippi and its **delta**. [HAV]

beach An accumulation of sediment deposited by **waves** and currents in the shore zone. They are common features along sea and lake coasts. Beach sediments are derived mainly from river and **glacier** sources with more limited contributions from biogenic processes and **cliff erosion**. Much of this sediment is first deposited in the offshore zone and then moved into the nearshore zone by waves and currents. All beaches adjust their form to accommodate changes in waves and sediment supply. Two extremes of profile type can be recognized and between these lies a continuum of potential configurations. High frequency storm waves move larger amounts of water per unit time than do low frequency swell waves and this produces high beach **water tables**. Independent of sediment size considerations, storm waves create not only strong **swash** velocities but also enhanced **backwash** via limited percolation due to these high beach water tables. Net offshore sediment movement and a lower nearshore gradient results, much of the sediment being deposited in the nearshore zone as **bars**. The erosional beach state produced is called a *storm* or **dissipative beach**. The accretional *swell* or *reflective beach* state is related to low frequency swell waves, low beach water tables, higher percolation and thus less backwash. This results in net onshore sediment movement, building of beach ridges or berms and profile steepening.

Variations in sediment size modifies this pattern. Coarser sediments allow greater percolation and backwash reduction leading to steeper beach slopes, whereas finer sediments are linked to lower gradients. On beaches where a range of sediments is found, backwash velocities may not be sufficient to remove coarse sediment carried upbeach by the high velocity swash, and a progressive upbeach increase in sediment size and beach gradient results.

In addition to profile adjustment, beaches also change their plan form. Again, two extremes can be recognized. Where the net sediment movement alongshore is minimal, the beach is built parallel to the crests of the incoming **waves**. Such **swash** alignment is common on irregular coastlines. The alternative is where sediment drift alongshore is substantial and the beach is built parallel to the direction of this drift (see **longshore drift**). Drift alignment is common on smooth coasts where sediment movement is unimpeded.

Most beaches lie somewhere between these extremes of profile and plan. A major influence on beach development through time is change in sea-level. **Raised beaches** are commonly observed features of old shorelines that have

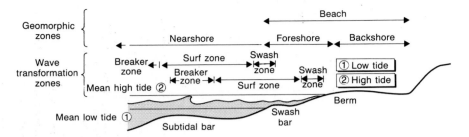

Fig. B5 Beach deposits. Generalized beach profile and adjacent environments showing geomorphic zones and subdivisions based on wave transformations.

since become elevated, but submerged shorelines also occur. Where the rate of sea-level rise has been slow, the rate of coast sedimentation may have kept pace creating a coastal *strandplain* composed of multiple **beach ridges**. (See Fig. B4.) [JDH]

Davies, J.L. (1980) *Geographical Variation in Coastal Development.* Longman, Harlow.
Komar, P.D. (1976) *Beach Processes and Sedimentation.* Prentice-Hall, Englewood Cliffs, NJ.
Pethick, J.S. (1984) *An Introduction to Coastal Geomorphology.* Arnold, London.

beach cusp Rhythmic lunate feature along the upper **swash** zone of **sand** or **gravel beaches**, with coarser material on horns pointing towards the shore. Beach cusps of 1–60 m in width have been recorded, their formation is considered be due to a regularity of **wave** action, and their size related to **wave** energy. Best examples are usually to be seen on **beaches** with poorly sorted sediment sizes. [RDS]

Kuenen, P.H. (1948) The formation of beach cusps. *Journal of Geology* **56**: 34–40.

beach deposits A **beach** is a zone of unconsolidated sediment that extends from the uppermost limit of **wave** action to the mean low tide mark. Its landward limit is commonly associated with an abrupt change in slope and/or composition. This definition is purely geomorphic, with physical process boundaries (defined by **wave** transformation zones) varying spatially as the **tide** rises and falls (Fig. B5).

The physical processes that operate on the beach face are associated with the dissipation of wave energy firstly through breaking (*breaker zone*), then through wave transformation to surf (*surf zone*) and finally into **swash** and **backwash** (*swash zone*). Surface sedimentary bedforms indicate the high flow power operating on the **beach** face, with a predominance of current lineated **plane beds**, *rhomboid ripples* and **antidunes**.

Studies of modern and sub-recent (*Holocene*) **barred** beach deposits indicate that a range of sedimentary structures can be preserved, but with an overall dominance of gently dipping, parallel lamination (see Fig. B6). Intertidal bars, of varying genesis, are common on many modern **beaches**, and hence a wide variety of sedimentary structures would intuitively be expected to be preserved within beach deposits. The dominance of **swash**-produced seaward-dipping parallel lamination, in all but very rapidly prograding or aggrading coastlines, occurs because of the depth to which *storm* waves scour during *storms*.

Gently-dipping parallel-laminated sands, occurring in erosive or sharply bounded sets, are the general form of paleobeach deposits. The angle and direction of **dip** represents the primary slope of the beach face. A speculative model for the explanation of such sequences is shown in Fig. B7 from Allen (1982). This assumes the preservation of a number of **swash**-produced laminae every **tide** as a set. The laminae consist of up to 20 mm thick couplets of a fine and/or heavy mineral-rich layer which grades upwards into a coarser and/or heavy mineral-poor layer. A number of such tidally produced sets are preserved during the fairweather period before the next *storm* scour occurs. The set **dip** is likely to increase upwards representing the steepening of the **swash** slope during fairweather conditions. This model uses first order cyclicity, or minimum time period, that can account for set preservation in beach deposits. The major problem in the interpretation of beach deposits is that higher order cycles can also be envisaged. A third order cycle, for instance, replacing the laminae scale with the *storm* cycle; the tidal scale becomes the major *storm* return period and so on. This problem has yet to be resolved.

Finally, with the lack of diagnostic criteria in the recognition of ancient beach sediments, interpretation is made mainly on their stratigraphic position within a nearshore sequence, and the dominance of parallel lamination. Recognition of bona fide beach deposits is always problematic. [RK]

Allen, J.R.L. (1982) *Sedimentary Structures, Their Character and Physical Basis* Vol. 2. Developments in Sedimentology 30B. Elsevier, New York.
Roep, Th.B. (1986) Sea level markers in coastal barrier sands: examples from the North Sea Coast. In: Van de Plassche, O. (ed.) *Sea-level Research: a Manual for the Collection and Evaluation of Data.* Geo Books, Norwich.

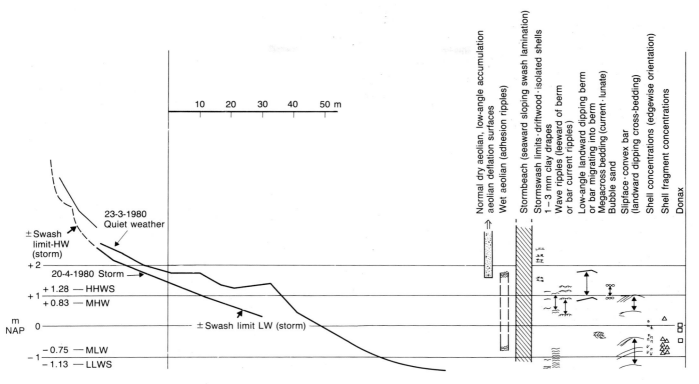

Fig. B6 Beach deposits. Beach profiles from a modern fetch-limited beach on the coast of The Netherlands. The left-hand part of the figure shows the typical quiet-weather profile and the storm profile is emphasized by heavy lines. The right-hand part of the figure gives the type and depth distribution of sedimentary structures which may also be recognized from fossil barrier sediments from a variety of settings. (After Roep, 1986.)

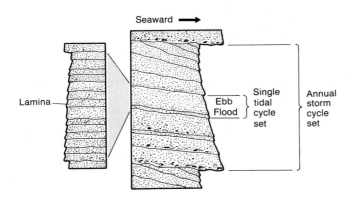

Fig. B7 Beach deposits. Speculative model depicting cycles of lamination (storm, tide, swash−backwash) in beach sands. (After Allen, 1982.)

beach placer deposit **Placer deposit** present on modern or ancient **beaches**. The economically important minerals of these deposits are **cassiterite**, **diamond**, **gold**, **ilmenite**, **magnetite**, **monazite**, **rutile**, *xenotime* and **zircon**. Heavy mineral accumulations have sharp bases and form discrete laminae within the normal **beach sand**. They are especially well developed during *storm* wave action. The **beaches** on which such accumulations are forming today include many upon which trade winds impinge obliquely and ocean currents parallel the coast, these two factors favouring **longshore drift**. For their preservation either the shoreline must prograde or the sea-level must fall to remove the heavies from **wave** activity. [AME]

beach ridge A prominent linear accumulation of wave-deposited sediment found on many **beaches**. Pebble **beaches** usually have several **swash**-formed ridges developed at high tide and exposed as the **tide** falls from spring to neap level. **Backshore** ridges built above high tide level are formed by pebbles thrown or carried to altitude by the plunging or surging breakers associated with successive *storms*. On sandy **beaches**, the flatter ridges at the top of **beaches** are called **berms**, whereas near low tide on low gradient **beaches**, several low ridges may be separated by runnels. [JDH]

King, C.A.M. (1972) *Beaches and Coasts*. Arnold, London.

beachrock A lithified deposit found on **beaches** between high and low water marks. It is a common feature of many tropical **beaches**, but is also found occasionally outside the tropics, e.g. around the Mediterranean and Red Sea coastlines. Beachrock is commonly found on the **beaches** of tropical **cays**. Characteristically, beachrock consists of planar laminations, with individual units 10–100 mm thick, and with a seaward **dip** of 5 to 10°. Beachrock is usually formed of **beach** material **cemented** *in situ*. The cements are formed as seawater evaporates as it drains through the beach towards low tide and are normally **aragonite**. Beachrock usually forms in sheltered **beaches** and may develop very quickly, as evidenced by the presence of human artefacts cemented into beachrock in some areas. Micro-organisms may play a role in the **cementation** process, by aiding the **precipitation** of calcium carbonate. Once formed, beach rock is a relatively persistent feature upon which a suite of **erosional** features are developed, such as pits and runnels. [HAV]

Scoffin, T.P. & Stoddart, D.R. (1983) Beachrock and intertidal sediments. In: Goudie, A.S. & Pye, K. (eds) *Chemical Sediments and Geomorphology*, pp. 401–25. Academic Press, London.

bed/bedding Layer of rock deposited at the Earth's surface and bounded above and below by distinct surfaces (*bedding planes*); these usually mark a break in the continuity of sedimentation, i.e. a cessation of sedimentation, or a period of **erosion**, or a change in type or source of sediment. Beds are normally sedimentary, but may also consist of volcanogenic material. A thickness in the range of centimentres to metres is normally implied. 'Bed' is more or less synonymous with *stratum*, but the latter term is normally used only in the plural (e.g. 'Silurian strata'). Beds may be relatively homogeneous in composition and internal structure and represent more or less continuous deposition. However the term is also used for a sedimentary unit composed of numerous thin distinct layers. The term 'bedded' means composed of beds; thus 'bedded rocks', 'thin-bedded' 'cross-bedded', etc; 'bedding' is used as a collective noun for the beds in a particular outcrop or area: thus 'the bedding **dips** to the west'. The term is also used to describe various characteristics of the beds, thus 'cross-bedding', 'graded bedding'. (See **unconformity**.) [RGP]

bed roughness In uniform flow, the force driving the flow is exactly balanced by the *total bed resistance*. In **gravel-** bed rivers, friction is the important source of bed roughness. In mobile sandy sediments, a sequence of bedforms caused by increasing flow intensity are moulded to introduce form resistance. These forms include **ripples**, **dunes** and **antidunes**. At the channel scale, roughness forms an integral part of the overall consideration of hydraulic geometry. *Coefficients of roughness*, such as those of Manning and Chézy, are commonly used as the basis for estimations of flood discharge. (See **Chézy equation**, **Manning equation**.) [TPB]

bedding plane schistosity Preferred alignment of planar minerals parallel to **bedding** in metamorphic rocks. The term is usually applied to a **fabric** produced by **deformation**, but is also used for a structure of unknown origin that could be primary. Bedding schistosity has been attributed to **load pressure**, but most examples are probably due to coincidence of *strain axes* with bedding due to **transposition** at high **strains**. (See **schistosity**.) [RGP]

bedding plane slip The process of displacement of a **bed** relative to an adjacent **bed** during *flexural slip folding*. This process may produce **slickenside striations** on the bedding plane (Fig. B8). [RGP]

bedding plane thrust *Thrust fault* parallel to *bedding*. In thrust zones, such faults are termed *flats*. These alternate with **ramps**, which cut up through the strata. (See **thrust belt**.) [RGP]

bedding–cleavage intersection The mutual intersection of sets of planar **bedding** and **cleavage**. Such intersections may produce an obvious linear structure at outcrop, particularly where *slaty cleavage* cuts regularly-bedded or laminated sediments. The *intersection lineation* is often parallel or near-parallel to related *fold axes*. (See **lineation**.) [RGP]

bedform theory A grouping of hypotheses which attempt to explain the physical reasons for the initiation, development, stability and characteristics of bedforms produced in the natural environment, whether by air, water or higher **viscosity** flows. Many of the processes associated with each flow type have common links, and bedform theory, although in its infancy, should seek to apply knowledge of fluid mechanical processes with the movement of sediment, with the generation of bedforms and sedimentary structures and importantly with explanations for their variability within different environments.

A developing example of bedform theory concerns the generation of bedforms in **sand** grade sediment under unidirectional flows. Work up until the 1970s linked bedform stability to the major controls of grain size and flow power (Fig. B9) as well as flow depth. Recent research

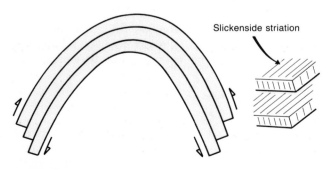

Slickenside striation

Fig. B8 Bedding plane slip.

has begun to concentrate upon the physical processes responsible for the generation and stability of these bedform types: the initiation of **ripple** development and growth; the influence of **viscous sublayer** structure upon bedform characteristics such as the generation of **lower stage plane beds** and **bars**; the influence of vertical turbulence damping upon the modification of **dune** bedforms to **upper stage plane beds**; the role of macroturbulence within the boundary layer upon the generation of bedforms and the influence of mutual particle interference in coarser **sands** upon bedform type. The development of these fluid mechanical explanations for the generation and stability of bedforms will provide the required basis for development of predictive and numerical models for the growth and characteristics of bedforms. [JLB]

Allen, J.R.L. (1985) *Principles of Physical Sedimentology*. Allen & Unwin, London.
Jackson, R.G. (1976) Sedimentological and fluid dynamic implications of the turbulent bursting phenomenon in geophysical flows. *Journal of Fluid Mechanics* **77**: 531–60.
Leeder, M.R. (1983) On the interactions between turbulent flow, sediment transport and bedform mechanics in channelized flows. In: Collinson, J.D. & Lewin, J. (eds) *Modern and Ancient Fluvial Systems*, pp. 5–18. Special Publications of the International Association of Sedimentologists No. 6. Blackwell Scientific Publications, Oxford.

bedload Material transported along the river channel floor by rolling, sliding or **saltation**. Initiation of bedload movement requires a critical flow velocity to be reached which will provide the minimum **shear stress** capable of moving the bed sediments. *Flow competence* is the maximum particle size capable of being transported. Bedload

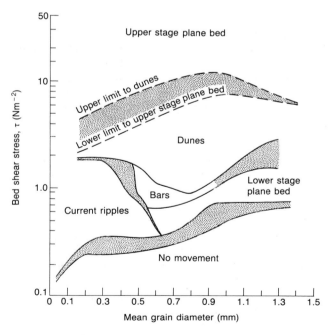

Fig. B9 Bedform theory. Bedform phase diagram of bed shear stress plotted against mean grain size illustrating the stability fields of bedforms generated in sand-grade sediment under unidirectional flows. The stippled areas denote regions of overlap between bedform fields. (After Leeder, 1982.)

is derived from the channel bed, but this source is replenished by **bank erosion** and by supplies from hillslope sources such as gullies and **landslides**. [TPB]

bedload transport Once a critical **shear stress** is exceeded, sediment is moved either as **bedload** or suspended load. Bedload transport involves the sediment that remains essentially in contact with the bed. This includes rolling, where grains are in continuous contact with the bed and **saltation** where contact is intermittent. Additionally, **creep** may occur due to the displacement of grains downstream through collision with incoming saltating grains. **Saltation** and **creep** are generally of greater significance in aeolian environments where the effects of fluid **viscosity** are minimal. (See **suspension**.) [JLB]

Yalin, M.S. (1977) *Mechanics of Sediment Transport*. Pergamon, Oxford.
Middleton, G.V. & Southard, J.B. (1984) *Mechanics of Sediment Transport* (2nd edn). Short Course No. 3 Lecture Notes. Society of Economic Paleontologists and Mineralogists, Tulsa, OK.

beidellite ((Ca, Na)$_{0.3}$Al$_2$(OH)$_2$(Al, Si)$_4$O$_{10}$. (H$_2$O)$_4$) A member of the **montmorillonite** group of **clay minerals**. [DJV]

bell/bell hole A cone-shaped **nodule** or *concretion* in the roof of a **coal seam** which may fall without warning to leave a cavity called a bell hole. [AME]

belt of no erosion This zone was defined by R.E. Horton (1945) as that area of hillslope, extending away from the **interfluve**, where surface **runoff** was not capable of **eroding** the *soil* surface. In practice, this zone can be equated with the unrilled part of the slope. Horton determined the distance from the drainage divide at which **erosion** began by balancing the tractive force of the flow against the resistance of the *soil* surface. Such belts of no erosion are usually confined to bare *soils*, often in **badland** areas, although even there it may be shallow *throughflow* rather than true surface **runoff** which initiates the rilling. In well-vegetated areas, surface **runoff** may be rare, so that the concept of a belt of no erosion may have little value. [TPB]

Horton, R.E. (1945) Erosional development of streams and their drainage basins: hydrophysical approach to quantitative morphology. *Geological Society of America Bulletin* **56**: 275–370.

benefication The process of separating wanted minerals from unwanted materials (**gangue**) during the exploitation of a mineral deposit. In general it is a concentrating process, preparing the product for the refining stage. In most cases this involves crushing the mined material to a fine size and then employing various processes of sorting and separation to prepare a concentrate for direct marketing, or conveying to a smelter. [AME]

Wills, B.A. (1981) *Mineral Processing Technology*. Pergamon, Oxford.

Benioff–Wadati/Benioff zone The active seismic zone that dips down below **island arc** systems at **subduction zones**. This was originally mapped as a zone of **earthquake** activity on a single dipping plane but, in areas where seismic stations are adequate to define *earthquake foci* with sufficient precision, the individual *earthquake foci* lie on two planes, some 10 to 20 km apart. In most areas, the dip of these zones is initially shallow, becoming steeper below some 150 km. However, there are many exceptions, and the **earthquake** activity is often discontinuous, particularly around 200 km depth, sometimes referred to as the '**seismic gap**', and various configurations for the planes are known to exist, many including sections in which the **earthquake** activity remains at a constant depth for several tens of kilometres, such as along several sections perpendicular to the Peru–Chile Trench. It is probable that most **subduction zones** are characterized by the double plane and that similar variety in the precise configuration of the planes exists, although they all eventually become steeper with increased distance from the *oceanic trench*. It is probable that many of the **earthquakes** within the Benioff–Wadati zones are actually confined within the subducting **oceanic crust**, but their distribution is complicated. As the subducting *plate* begins to bend beneath the *oceanic trench*, tensional **earthquakes** predominate close to the upper surface of the **lithosphere** and are underlain by mostly compressional **earthquakes** where the **mantle** rocks immediately below the **oceanic crust** are compressed by the lithospheric bending. After the **lithosphere** has descended to about 100 km, the pattern of **earthquake** activity within the Benioff–Wadati zone(s) is believed to occur mainly within the subducting plate in response to internal **deformation**, while the upper of the two **earthquake** planes originates by the interaction of the top of the downgoing **lithosphere** with the **mantle**. The deepest recorded **earthquake** was at 680–710 km, but no **earthquakes** have been detected by modern instruments at depths greater than 650 km, which is the depth at which the Benioff–Wadati zone is considered to terminate. It is not established whether this is because the rigidity of the subducting lithospheric *plate* then becomes unable to accumulate sufficient **strain** to result in **brittle fracture** or whether the **mantle** beneath is too rigid to allow further penetration of the subducting *plate*. **Seismic tomography** suggests the thermal anomaly associated with the descending **lithosphere** extends deeper into the **mantle**, but these cooler zones are not accompanied by seismic activity. (See Fig. B10.) [DHT]

Hasegawa, A., Umino, N. & Takagi, A. (1978) Double-planed deep seismic zone and upper-mantle structure in the northeastern Japan arc. *Geophysical Journal of the Royal Astronomical Society* **54**: 281–96.

benitoite (BaTiSi$_3$O$_9$) A rare ring silicate (*cyclosilicate*) mineral. [DJV]

benmoreite An alkali **lava** intermediate in composition between *mugearite* and **trachyte**. Benmoreite has **normative**

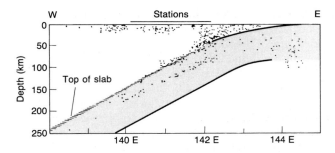

Fig. B10 Benioff–Wadati zone. Section showing the focal depth distribution of small earthquakes along an east–west line across the north-eastern part of the Japan arc, showing the double planed deep seismic zone detected by the Tohoku University seismic network. The hatched zone shows the position of the boundary between the descending slab and the overlying mantle as inferred from *ScS* and *ScSp* arrivals, assuming that the descending slab has seismic velocities 6% above those of the adjacent mantle. (After Hasegawa *et al.*, 1978.)

plagioclase of composition *c.* An$_{10}$–An$_{30}$ (i.e. **oligoclase**), a *Differentiation Index* of *c.* 60–75 (these values depending on definition) and may be **silica**-saturated or under-saturated. [RST]

bentonite A clay rock composed mainly of **montmorillonite** from **alteration** of volcanic **ash**. (See **clay minerals**.) [DJV]

berm (1) In cross-section (shore perpendicular), a triangular feature with a horizontal to slightly landward dipping top surface (berm top) and a more steeply dipping seaward facing surface found on certain **beaches**. The junction between the two slopes (berm crest) marks the morphological boundary between the foreshore, which is the area affected by normal astronomical **tides**, and the *storm* influenced section of the **beach** (backshore).

Only static and constructional **beaches** contain berms, whereas **beaches** undergoing **erosion** have a continuous foreshore-backshore profile.
(2) A term used in **opencast mining** or quarrying to describe a relatively narrow, man-made shelf or bench, often carrying a roadway, that breaks the continuity of a slope. [AME/RK]

Thomas, K.V. & Baba, M. (1986) Berm development on a monsoon-influenced microtidal beach. *Sedimentology* **33**: 537–46.

beryl (Be$_3$Al$_2$(Si$_6$O$_{18}$)) A ring (*cyclosilicate*) mineral. [DJV]

beryllium (Be) At. no. 4, at. wt. 9.01218, d. 1.85. The lightest alkaline earth. The principal sources are **hydrothermal deposits** of *bertrandite* and **beryl** from **pegmatites**, although beryllium is present in many other minerals. It is used in alloys and in nuclear reactors. BeO is used in ceramics and also in reactors. [AME]

beryllonite (NaBePO$_4$) Rare **gem** mineral. [DJV]

bessemer ore **Iron ore** in which the **phosphorus** content is less than 0.045%. A premium price is paid for this **ore**. [AME]

Besshi-type deposit (*Kieslager deposit*) A type of volcanic-associated **massive sulphide deposit** which is usually an **ore** of **copper** and **zinc** with, or without, **gold** and **silver** values. These deposits appear to be associated with intra-plate **basalts** but may be situated within deep water sediments, *greywackes* and other *turbidites*. The depositional environment appears to have been a back-arc, deep marine one often accompanied by **basaltic** volcanism and lying within a zone of **rifting**. These deposits are known so far from the early **Proterozoic** and the **Paleozoic**. The type occurrence is in late **Paleozoic** terrain of Honshu, Japan. [AME]

Fox, J.S. (1984) Besshi-type volcanogenic sulphide deposits — a review. *CIM Bulletin* **77** (April): 57–68.

beta (β) pole diagram On a *stereogram*, the β pole or axis is the point of intersection of the **great-circle** traces representing successive positions of a folded surface. The β diagram is the *stereogram* of a **fold** represented in this way. For an alternative method, see **pi (π) diagram**. [RGP]

Turner, F.J. & Weiss, L.E. (1963) *Structural Analysis of Metamorphic Tectonites*. McGraw-Hill, New York.

bifurcation ratio The ratio of the number of streams of a particular order to the number of streams of the next highest order. It gives a quantitative expression of the rate at which a stream network bifurcates. [ASG]

billabong An aboriginal word from Australia which is used to describe a class of permanent and semi-permanent waterbodies on riverine **floodplains**, formed by the geomorphological action of the mainstream. Most common is the cut-off meander or **oxbow** lake. However, other forms occur, e.g. backwater remnants of migrating channels probably isolated by concave bank bench deposits. [ASG]

Hillman, T.J. (1986) Billabongs. In: de Deckker, P. & Williams, W.D. (eds) *Limnology in Australia*, pp. 457–70. Junk, Dordrecht.

biodegradation The decomposition by *bacteria* of **crude oil**. It is brought about by aerobic (oxidizing) *bacteria* that have been carried into the oil **reservoir** by **meteoric waters**. [AME]

bioherm A discrete lens-like body of organic origin, e.g. an aggregation of *in situ* colonial organisms such as **corals**, or a shell bank. The term is non-genetic and refers to the shape only. The lens is recognized by being surrounded by a different lithology.

A *biostrome* is a bedded accumulation of skeletal material such as a shell bed. [VPW]

Heckel, P.H. (1974) Carbonate build-ups in the geologic record: a review. In: Laporte, L.F. (ed.) *Reefs in Time and Space*. Special Publication of the Society of Economic Paleontologists and Mineralogists **18**: 90–155.

biokarst A **karst** landform, usually small scale, produced mainly by organic action. The term **phytokarst** is also used, although strictly speaking this should refer to features produced by plants alone. Biokarst features are especially prominent in the coastal zone where a suite of organisms bore into and abrade rock surfaces, producing pits and runnels and contributing to the formation of a coastal **notch**. Constructional, as well as **erosional**, landforms may also be classed as biokarst where organisms control the **precipitation** of calcium carbonate as, for example, in certain *tufas* and *reef* forms. [HAV]

Viles, H.A. (1984) Biokarst: review and prospect. *Progress in Physical Geography* **8**: 523–43.

biostasy The state of environmental stability when climate and geomorphic conditions favour **weathering** and *soil formation* rather than **erosion** and **denudation** of the landscape. The term was used by Erhart (1956) to refer to periods of stability when neither climatic change nor **tectonic** disturbances disrupt the normal vegetation cover, and when surficial materials undergo chemical decomposition. Periods of instability (when the environment is in a state of **rhexistasy** caused by climatic change or **tectonic** activity) disrupt the vegetation cover and **erosion** of the land surface prevails. [AW]

Erhart, H. (1956) *La Genèse des Sols en tant que Phénomène Géologique*. Masson, Paris.

biostratigraphy The subdivision and correlation of stratal sequences by means of fossils. The pioneer work in this field was done at the beginning of the 19th century by William Smith in England and Cuvier and Brongniart in northern France, respectively on **Jurassic** and **Tertiary** strata. Later in the century the concepts of *stages* and *zones* (or *biozones*) were introduced, the former consisting of several of the latter, by A. d'Orbigny and A. Oppel as *chronostratigraphic units* independent of facies, and it is solely upon the fossil sequence, now known to be the result of evolution and *extinction*, that the **Phanerozoic** systems are based.

The *zone* is the basic unit of biostratigraphic correlation and several different types have been distinguished. An *assemblage zone* is based on the temporal range of a group of species, while a *concurrent range zone* is based on the co-occurrence of two or more species. The best type of *zone*, where it can be established, is the *lineage zone*, based on an evolutionary succession of species of a given genus, because only in this way can stratigraphic gaps or hiatuses be ruled out in given instances, as can migration of species from one region to another. The establishment of a good zonal stratigraphy is hampered by facies variations, because particular organisms only occur in certain types of facies that reflect their environmental tolerances. Correlation of marine and non-marine strata, for example, is difficult unless parallel zonal schemes are set up for corresponding fossil groups and the strata are found to interleave in some localities.

The best *zonal index fossils* are those with a wide geographic but short temporal range, which ideally occur in a wide range of facies. *Ammonites* are particularly good for the **Mesozoic** and **graptolites** for the Lower **Paleozoic**. Much modern stratigraphic research is conducted in the subsurface, by study of material in borehole cores, and hence it is more practical to utilize microfossils. **Conodonts** have proved especially valuable biostratigraphic indices for the **Paleozoic**, **dinoflagellates** for the **Mesozoic** and planktonic **Foraminifera** for the Cenozoic. [AH]

Donovan, D.T. (1966) *Stratigraphy — An Introduction to Principles*. John Wiley & Sons, New York.
Eicher, D.L. (1976) *Geologic Time* (2nd edn). Prentice-Hall, Englewood Cliffs, NJ.
Kauffman, E.G. & Hazel, J.E. (eds) (1977) *Concepts and Methods of Biostratigraphy*. Dowden, Hutchinson & Ross, Stroudsburg, PA.

biotite $(K(Mg, Fe)_3)(AlSi_3O_{10})(OH)_2)$ Black layer silicate (*phyllosilicate*) mineral of the mica group. (See **mica minerals**.) [DJV]

biotitization A form of **wall rock alteration** in which new, normally fine-grained, **biotite** is developed in rocks permeated by **hydrothermal solutions**. It is sometimes the main feature of **potassic alteration**. The secondary **biotite** often replaces and *pseudomorphs* primary **biotite**, but also grows in other parts of the altered *host rock*. [AME]

birds/Aves Feathered vertebrate animals which share a cleidoic egg with the **reptiles** and **mammals**. Modern birds are part of one of the most successful radiations known in the animal kingdom, but their ancestry and early history is still very poorly documented. The single undisputed belief is that the ancestral bird was an *archosaur* of one kind or another, but which is still very much in doubt.

The most famous fossils of all time are the skeletons found in the *Solnhoffen Lithographic Limestone* of Bavaria and termed *Archaeopteryx lithographica*. These represent bird-like creatures preserved with their feather impressions. The general proportions of these skeletons are those of one of the small bipedal (crow-sized) *theropod dinosaurs*, the *Coelurosauria* (Upper **Trias**–**Cretaceous**) also known from these deposits but with much longer forelimbs. The term *Archaeopteryx* strictly speaking refers to a single feather found in the lithographic **limestones**, and the skel-

etons were originally ascribed to the genus *Griphosaurus* when first described in 1861 only a few years after the publication of Darwin's *Origin of Species*. These bird-like specimens therefore represented one of the first apparent occurrences of a 'missing link', so eagerly sought after by Darwin's followers. Considerable controversy has followed them ever since.

Notwithstanding the similarity of the *Archaeopteryx* skeleton to that of a *coelurosaur* such as *Compsognathus*, a *dinosaur* origin of birds is not universally accepted. Very detailed work on advanced *thecodontian* **reptiles** and primitive *Crocodilia* of the Upper **Triassic** has demonstrated several resemblances between them and the suite of primitive characters assembled for *Archaeopteryx* itself.

In many ways *Archaeopteryx* makes a very poor bird. Although possessing a furcula (wish-bone) it does not have an ossified sternum, nor is there the closely built box-like thorax of the Recent bird. It would also appear that there was no '*foramen triosseum*' in the shoulder girdle which is so important in modern birds as a route for the wing elevator muscles. *Archaeopteryx* thus would seem to have been a poor flyer, capable of a powered downstroke of the wing, but with a very poor recovery action. The long feathered tail was present to enhance stability during passive flight (gliding) much in the same way as the early *Pterosauria* also had long tails. The hind limbs are long, with digitigrade feet, but, strangely for a presumed primary biped, the forelimbs are also long (*c.* 80% of the hind limb). There are three clawed digits on the forelimbs, whose homologies are with *coelurosaur* 'fingers', and not with modern birds. In the skull the jaws are still lined by teeth, which may be more like *crocodilian* teeth than those of *thecodont* **reptiles**. Some of these characters are plainly primitive for birds as a whole, some are still enigmatic and some on the other hand, such as the lack of hollow bones, would seem to preclude a close relationship with the *Coelurosauria*.

There are two theories in general currency concerning the way in which flight may have arisen in birds, or their ancestors. There is the 'running flapper', up from the ground theory contrasted with the 'arboreal glider' or down from the trees theory. The first was elaborated in the 1920s and envisaged a three-phase evolutionary process. In the beginning there would be a 'wingless', feathered running insectivore, followed by a predator using newly evolved wings to help its running by 'rowing' with them, i.e. running and flapping. Once established with functioning wings the now partially flighted *Proavis* could take to the trees and evolve into something like *Archaeopteryx*. Unfortunately this theory ignores the fact that flapping reduces the ground friction of the feet in a bird and so would not help in improving locomotion. Variations of this theory envisage that a feathered 'wing' could be used to catch insects, either by knocking them down, or by being used as a net to trap them.

In the 'tree-downward' theory, it is proposed that the pre-bird *archosaur* was an arboreal insectivore which used

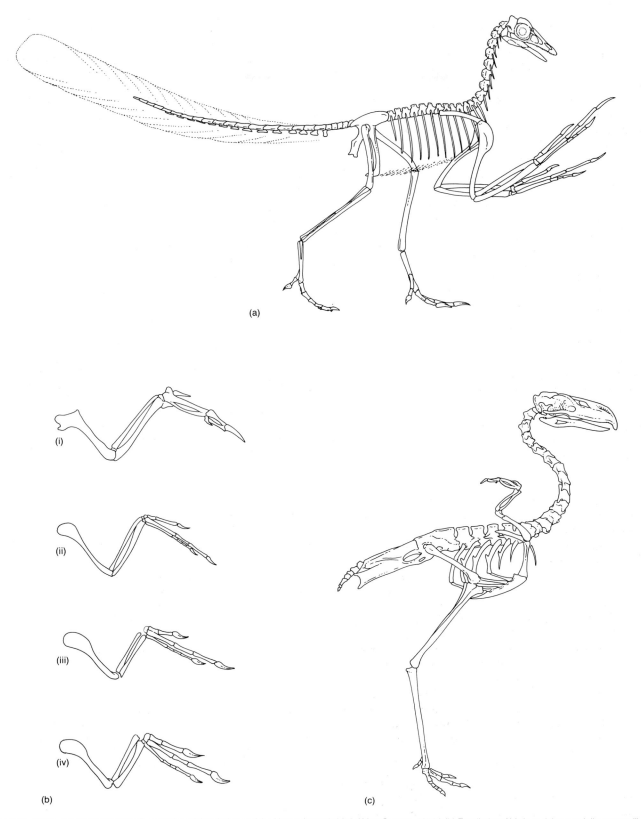

Fig. B11 Birds. (a) *Archaeopteryx*. Restoration of the skeleton of the Upper Jurassic bird. (After Ostrom, 1976.) (b) Forelimbs of birds and theropod dinosaurs; (i) modern pigeon, (ii) *Archaeopteryx*, (iii) *Ornitholestes*, (iv) *Deinonychus*. (After Ostrom, 1975.) (c) Phorusrhacus (*Phororhacos*). A gigantic non-flying predaceous bird from the Tertiary of South America, approximately 1.5 m tall. (After Andrews, 1901.)

Table B2 Birds/Aves

Class Aves
  Subclass Archaeornithes
    Order Archaeopterygiformes

  Subclass Neornithes
    Superorder Odotognathidae
      Order Hesperornithifomes
    Superorder Paleognathae
      Order Tinamiformes
        Struthioniformes
        Rheiformes
        Casuariformes
        Aepyornithiformes
        Dinornithiformes
        Apterygiformes

    Superorder Neognathae
      Order Gaviiformes
        Podicipediformes
        Procellariiformes
        Sphenisciformes
        Pelecaniformes
        Ciconiiformes
        Anseriformes
        Falconiformes
        Galliformes
        Ralliformes
        Diatrymiformes
        Ichthyornithiformes
        Charadriiformes
        Columbiformes
        Psittaciformes
        Cuculiformes
        Strigiformes
        Caprimuligiformes
        Apodiformes
        Coliiformes
        Trogoniformes
        Coraciiformes
        Piciformes
        Passeriformes

its clawed front limbs to help climb on branches and trunks. Progress on the level would be by a bounding, greyhound-like gait and quadrupedal. This would give the forelimbs the powerful pectoralis muscles as a pre-adaptation for flight. An implication of this is that such an animal would be closer to a *thecodontian* than a bipedal *dinosaur*. This mode of life would be linked with the ability to glide between trees, as in modern flying squirrels. The clawed forelimbs would help the pre-bird to climb tree trunks so as to exploit another food reservoir. However, both of these theories beg the question of the origin of feathers themselves, the fundamental character which defines the whole taxon.

Much of the land surface of **Pangea** during **Triassic** times was covered by deserts, with the implication that those in low latitudes would have marked diurnal temperature variations. All the postulated bird ancestors were 'small' (pigeon–crow-sized) and as small animals they would be subjected to relatively greater thermal stress

than larger animals were. Sediments deposited at this time are a rich source of both *coelurosaur* and *thecodont* **reptile** fossils, and so they must have been able to combat such thermal stress when alive, in part by evolving an insulating layer of one sort or another, to protect from the very cold nights or to act as a heat shield by day; they were almost certainly not endothermic, i.e. warm-blooded, or capable of maintaining a constant body temperature.

Feathers have a histological similarity to scales of **reptiles** and the probability exists that the one did evolve into the other in the line leading to birds. Apart from *Archaeopteryx* there are no feathers known in the **Jurassic** or earlier, but as feathers are very delicate structures it is not surprising that it took something like the *Solnhoffen Limestone* to preserve an impression of one. But if the physiological stress theory holds good then perhaps feathers may have been present in a wide range of small *Archosaurs* by the Upper **Triassic**. Flight could have evolved in more than one lineage of small insectivorous *archosaur* exploiting the upper canopy, using the tree-downward path. If that was the case, then the definition of a 'bird' may have to be reconsidered.

Archaeopteryx is the sole representative of the *Archaeornithes*, but recently in the lower **Cretaceous** several relatively advanced true birds have been reported from Mongolia, China and Patagonia. These discoveries are still being assessed, but clearly represent a distinct lineage closely related to modern birds and not to *Archaeopteryx* (Sereno & Rao, 1992). There were also a number of toothed birds which had adopted what appears to be a secondarily flightless way of life, and were similar in construction to modern divers, i.e. fish-eaters. These are the *Odontognathidae*.

Living birds are divided into the *Paleognathae* and *Neognathae*. The former are typified by the ostrich, cassowary and kiwi, and are all flightless. However they possess all the hallmarks of having been descended from flighted ancestors and seem to represent a peculiarity of the relics of **Gondwanaland**, where during the early *Tertiary* some very large flightless birds adopted the role of dominant carnivore pending the arrival of the *Eutherian* **mammals**.

The *Neognathae* comprise all the rest of modern birds and more properly are the subject matter of behavioural zoologists than Earth scientists.

Birds are relatively rare as fossils because of their delicate bones, but under some circumstances they can form a large proportion of any bone accumulation such as in **cave** deposits or fissure fillings. (See Fig. B11 and Table B2.) [ARIC]

Andrews, C.W. (1901) On the extinct birds of Patagonia. 1. The skull and skeleton of *Phororachos inflatus* Ameghino. *Transactions of the Zoological Society, London* **15**: 55–86.

Carroll, R.L. (1987) *Vertebrate Paleontology and Evolution*. W.H. Freeman, Chicago.

Ostrom, J.H. (1975) The origin of birds. *Annual Review of Earth and Planetary Science* **3**: 55–77.

Ostrom, J.H. (1976) Archaeopteryx and the origin of birds. *Biological Journal of the Linnean Society, London* **8**: 91–182.

Romer, A.S. (1966) *Vertebrate Paleontology* (3rd edn). University Press, Chicago.

Sereno, P.C. & Rao, C. (1992) Early evolution of avian flight and perching: new evidence from the Lower Cretaceous of China. *Science* **255**: 845–8.

bismuth (Bi) A rare **ore** mineral found in certain **vein** deposits. (See **native element**.) [DJV]

bismuthinite (Bi_2S_3) An **ore** mineral of **bismuth**. [DJV]

bituminous coal The general name for those **coals** of intermediate **rank** that lie between low **rank brown coals** and **lignites** and the high **rank anthracites**. In other words it is the normal **coal** type and generally consists of mixtures of both **banded** and **sapropelic coals**. It is in general soft, rich in volatile *hydrocarbons* and burns with a smoky yellow flame. The fixed **carbon** content is 46–86% and the calorific value varies from 19.3 to 36.3 MJ kg^{-1}. [AME]

black gold **Placer gold** coated with a black or dark brown material (often **manganese** oxides). The yellow colour of the **gold** may not be seen unless the coating is removed. [AME]

black sand An alluvial or **beach placer** with a considerable content of heavy dark minerals, generally **magnetite** or **ilmenite**. It may be economically important as a source of these minerals or others such as **rutile**, **monazite** and **zircon**, and, occasionally, **gold**. [AME]

black shale A black or very dark grey mudstone rich in organic **carbon**. There are many more or less rigorous definitions although a very broad usage is common. The lower limit of weight per cent content of organic **carbon** (C (org)) for a sediment to be considered a black shale varies with different definitions from 0.5% to 5%. (0.5% can hardly be regarded as rich, since marine muds contain an average of 0.35 wt.% C (org)). The black colour need not correspond to a significant C (org) content but may rather be due to the presence of fine-grained sulphides. The current breadth of usage results, in part, from the variety of **Cretaceous** 'black shales' encountered in **deep sea drilling** in the Atlantic and the considerable research effort expended on the study of **petroleum** *source rocks*. Black shales are generally marine muds although the organic matter they contain may be of terrigenous, marine or mixed origins. Their **fissility** is due to the compaction of their organic and clay portions. The lamination which enhances this **fissility** is variable in origin from a primary *varve*-type alternation of organics and clay/silt to secondary compaction of an originally unlaminated sediment. Models proposed for the origins of **Cretaceous** Atlantic black

shales during the 1970s generally appealed to the presence of **anoxic** bottom waters generated by basin-wide stagnation or thermohaline-related expansion of the **oxygen minimum zone**. However, analysis of the organic content using organic geochemical and palynofacies techniques documented substantial amounts of terrigenous organic matter (Habib in Schlanger & Cita, 1982). These results, together with modern oceanographic research on organic matter preservation, suggest that a high sedimentation rate is the most important factor in the genesis of black shales in oceanic environments. The term *'sapropel'* has been applied to organic-rich deposits of *Neogene* age occurring in the Mediterranean and the Black Sea.

Black shales often contain unusual concentrations of **trace elements** such as U, V, Cu, Ni. Some are sufficiently rich in **uranium** to form **protores** of this element. Others may be important *host rocks* for **vein** mineralization for which they may have been the source of **sulphur**. (See **anoxic, upwelling**.) [AME/AESK]

Journal of the Geological Society, London (1980) Thematic issue on black shales **137**: 123–94.

Schlanger, S.O. & Cita, M.B. (eds) (1982) *Nature and Origins of Cretaceous Carbon-rich Facies*. Academic Press, New York.

black smoker Plume of *hydrothermal water* that issues from vents in the ocean floor. The black colour is due to tiny particles of sulphide minerals. The **precipitated** sulphides build up chimney-like edifices and other structures. Water temperatures up to 400°C have been measured. The minerals deposited by the vent solutions include various forms of iron sulphide, **pyrrhotite**, **pyrite**, **marcasite**, **chalcopyrite**, **sphalerite**, **barite**, **anhydrite** and **gypsum**. **Manganese** minerals are deposited as a coating on the rocks around the vents. An exotic fauna lives in the warmer oceanic water around the vents. Remains of vents have been found in Cretaceous **massive sulphide deposits** of Cyprus and of the peculiar fauna associated with these vents in similar deposits in Oman, giving strong support to the **exhalative** hypothesis for the formation of **massive sulphide deposits**. [AME]

blanket bog A type of **bog**, often composed of **peat**, which drapes upland terrain and infills hollows in areas of high precipitation and low evapotranspiration. [ASG]

bleaching A minor and weak form of **wall rock alteration** in which rock colour is rendered much less intense and may become light grey or white. [AME]

blind layer A layer or zone which cannot be detected by first arrival **seismic refraction** methods since waves sampling the region are refracted downwards and not recorded at the surface as **head waves**. The layer may either have a velocity lower than that above it or may have a velocity intermediate between that above and below but not of sufficient velocity difference or thickness to produce

first arrivals. The presence of a blind layer causes depths calculated from a refraction interpretation of deeper layers to be erroneous. [GS]

Soske, J.L. (1959) Discussion on the blind zone problem in engineering geophysics. *Geophysics* **24**: 359–416.

blind valley A river-cut valley which terminates in a precipitous cliff and which is itself steep-sided. Blind valleys are common in **limestone** areas where subterranean flow and **sapping** contributes to their formation. [ASG]

block caving A large scale method of unsupported underground mining. In caving the **ore** is encouraged to collapse into the excavated areas. In so doing it is normally sufficiently broken to be handled. Block caving is carried out on a massive scale and is thus a very cheap underground mining method. Overlying rocks are of course disrupted and the method cannot be used where surface subsidence is untenable. [AME]

blocking temperature The temperature at which a **thermal magnetic remanence** is acquired and becomes fixed within a magnetic grain for at least 20 min after the removal of any external magnetizing field. Also used for the temperature at which gases (usually radiogenic, e.g. ^{40}Ar) become locked within a rock for geologically meaningful lengths of time. [DHT]

bloodstone Green and red **chalcedony** also known as **heliotrope**. [DJV]

blue asbestos (*crocidolite*) $Na_2Fe_5Si_8O_{22}(OH)_2$ So called because of the colour of its fibres which may be short to long with good flexibility. The fusion point is only 1193°C, but resistance to acids and alkalis is good. It is the most dangerous **asbestos** to health and its use is banned in several European countries. [AME]

blue ground At the surface **kimberlite** may be **weathered** and **oxidized** to '*yellow ground*' which gives way at depth to fresher 'blue ground' and '*hardebank*' — resistant **kimberlite** that often crops out and does not disintegrate easily upon exposure. [AME]

Blue John Various forms of mauve to purple **fluorite** found in Derbyshire, England. It is frequently banded and was in great demand in the 19th century for the manufacture of vases and other ornaments. [AME]

bluehole Like a sapphire set in turquoise, a circular, steep-sided hole which occurs in **coral** *reefs*. The classic examples come from the Bahamas (Dill, 1977), but other examples are known from Belize and the **Great Barrier Reef** of Australia (Backshall *et al.*, 1979). Although vol-

canicity and *meteorite* impact have both been proposed as mechanisms of formation, the most favoured view is that they are the product of **karstic** processes (e.g. collapse **dolines**) which acted at times of low sea-level when the *reefs* were exposed to subaerial processes. [ASG]

Backshall, D.G., Barnett, J., Davies, P.J., *et al.* (1979) Drowned dolines — the blue holes of the Pompey Reefs, Great Barrier Reef. *BMR Journal of Australian Geology and Geophysics* **4**: 99–109.
Dill, R.F. (1977) The Blue Holes — geologically significant sink holes and caves off British Honduras and Andros, Bahama Islands. *Proceedings of the 3rd International Coral Reef Symposium, Miami* **2**: 238–42.

bluestone Name commonly given to the imported (non-local) stones of the Stonehenge monument, UK. The bluestones include **dolerites** (mainly with a distinctive **ophitic texture** and a spotted appearance), rhyolitic **lava** and **tuffs** and **sandstones**, and comprise both standing stones of the monument and fragments excavated at and near Stonehenge. Some of these stones have a bluish tinge when wet, hence presumably the name. The source of these stones is assumed to be the Preseli Hills of south Wales, mainly because of the occurrence of similar 'spotted' **dolerites** in this area. There is also an **axe** group of this spotted **dolerite** recognized in Britain. [OWT]

bodden An irregularly shaped coastal inlet found on the south Baltic coast. Boddens are produced by a rise in sea-level over an uneven lowland surface. Some good examples are found on the island of Rügen. [HAV]

boehmite (AlO . OH) An **ore** mineral of **aluminium** (found in **bauxite**). (See **oxide minerals**.) [DJV]

bog A type of wetland, also known as *peatland* or *mire*, composed largely of accumulations of semi-decomposed plant material. Bogs form either due to accumulation of drainage water over impermeable rocks, glacial **drift** or *permafrost*, or due to an accumulation of water caused by high precipitation to evaporation ratios. The term bog is generally restricted to mires where precipitation is the main cause of wetness, rather than **groundwater** mires, which are known as *fens*. Bogs occur in all continents and peatlands have been estimated to cover over 500 million hectares world-wide. Characteristically, bogs are acid, nutrient deficient environments with species-poor vegetation including sedges, rushes and mosses. The bog mosses (*Sphagnum spp*) play an important role in the formation and maintenance of bogs and are capable of holding over 10 times their dry weight of water. There are several types of bog including blanket, raised, string and palsa forms. The acid, waterlogged environment of bogs favours the preservation of archeological and other remains, as evidenced by the discovery in a Cheshire bog in 1984 of 'Pete Marsh', a preserved medieval man. Bogs, like many other wetlands, are often damaged by human activity, such as **peat** digging. [HAV]

Maltby, E. (1986) *Waterlogged Wealth*. IIED Earthscan, London.

Gore, A.J.P. (ed.) (1983) *Mires: Swamp, Bog, Fen and Moor*. Ecosystems of the World 4A and B. Elsevier, Amsterdam.

bog burst The sudden disruption of a **bog** (an area of waterlogged ground characterized by thick accumulations of organic material) so that there is a release of water and **peat**, which may then flow over a considerable distance. [ASG]

bog iron ore A soft, spongy, porous **limonite** deposit found in **bogs**, marshes, **swamps**, shallow lakes, etc. It often contains plant debris, clay and other clastic material and may be tubular, pisolitic, nodular, concretionary or in massive, irregular aggregates. It is formed by **precipitation** from iron-bearing waters and is of no economic importance at the present day. [AME]

bogaz An elongated, deep ravine in **karst** terrain. Bogazes are formed by **solutional** widening of large **joints** and resemble giant **grikes**. [HAV]

Bohm lamellae Planar arrays of microscopic *fluid inclusions*, considered to form by **annealing** of **deformation lamellae**. [TGB]

bolson A type of trough or basin found in arid and semi-arid areas. The term is from the Spanish *bolsón*, 'large purse', and is most commonly applied to desert **basins** in the south-west United States and northern Mexico. Typically, the flat-floored depression is centred around a salt pan (**playa**) which is the focus of local **runoff** from the surrounding hills or mountains. The closed depressions are often of **tectonic** origin and contain thick accumulations of clastic sediments and **evaporites**. [AW]

bonanza A mining term for a rich body of **ore** or part of an **orebody**. Normally used of **precious metal** deposits, particularly **gold**. [AME]

bond clay A highly plastic clay used to bond relatively non-plastic materials in the fabrication of ceramics or other moulded products. The bonding may be in the green state (cold); upon firing to vitrification temperature adjacent ceramic materials that have higher vitrification temperatures are bonded. [AME]

boninite Olivine- and **pyroxene**-bearing **plagioclase**-poor **pillow lavas** commonly containing quench-textured **pyroxene** and **accessory magnesiochromite** in a glassy groundmass. The **lavas** are characterized by relatively high MgO (over 9%), Cr and Ni in comparison with **calc-alkaline andesite** of similar SiO_2 concentration (over 55% SiO_2). Boninites are also characterized by low TiO_2, Ta, Zr, Hf and REE in comparison with other lavas of intermediate chemical composition. Boninites were first described from the Bonin **island arc** but have been identified in many other volcanic arcs and in older volcanic successions such as **ophiolite** complexes.

Models for the petrogenesis of boninites involve (i) depletion of **peridotite mantle** in Ti, Zr and REE by *partial melting* to form **ocean ridge basalt**, leaving a **harzburgite–wehrlite** residue, (ii) enrichment of the refractory residue in elements released by **subduction** and dehydration of oceanic **lithosphere** and (iii) subsequent melting to yield boninitic **magmas** as defined above. Boninites therefore provide valuable evidence for the evolution of ocean floor and volcanic arc **magmas** in modern and ancient oceanic settings. [RST]

Crawford, A.J. (ed.) (1989) *Boninites*. Unwin Hyman, London.

boomer A marine **seismic source** in which a high voltage, discharged through a coil embedded in an epoxy-resin block, produces **eddy currents** in a spring-loaded aluminium plate which rapidly repels the plate. An implosion is produced by water rushing into the low pressure region generated between the plate and the block. This type of source is used for high resolution marine **seismic reflection** surveys, mainly for engineering **site investigations**. [GS]

boracite ($Mg_3ClB_7O_{13}$) A mineral that is a source of **borax**. [DJV]

borax ($Na_2B_4O_5(OH)_4 . 8H_2O$) An important **industrial mineral** used in the chemical industry. [DJV]

borehole gravimeter A specialized **gravimeter** for use in boreholes. Its main use is to determine average formation densities over given vertical intervals. The instrument is so expensive that it is only used in the most stable of holes. (See **geophysical borehole logging**.) [PK]

bornite (Cu_5FeS_4) An **ore** mineral of **copper**. [DJV]

bort A term used to refer to badly coloured or flawed **diamonds** without **gem** value. (See **native element**.) [DJV]

bottoming The downward termination of an **orebody**, usually an **epigenetic** one, due to structural or economic causes. [AME]

boudin/boudinage Segmentation of a **competent** layer (such as a **quartz**-rich sediment or igneous intrusion) into parallel, elongate, structures (boudins), which may have rectangular to elliptical cross-sections. The term derives from the French for 'sausage'. Boudinage (*sensu strictu*) results from layer-parallel **extension** in a single direction perpendicular to the axes of the boudins, and may be preceded by the development of **pinch-and-swell structure**

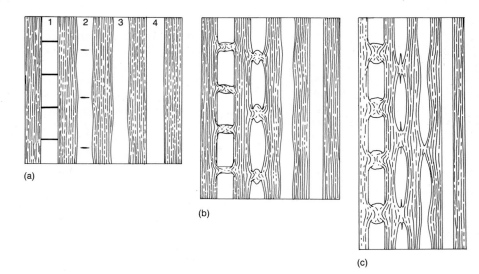

Fig. B12 Boudin. The progressive development of boudinage. The competent bands 1, 2, 3 and 4 are arranged in decreasing order of competence and band 4 has the same properties as the matrix.

which forms boudins when the thinner areas (necks or pinches) are cut by discrete **fractures** (Fig. B12). The layer is segmented due to loading by surrounding, less competent material. *Chocolate tablet (chocolate block) boudinage* is the segmentation of a competent layer into equidimensional structures resulting from layer parallel **extension** in two directions. Gaps between boudins are commonly filled by either flow of the surrounding material or new mineral growth. *Internal boudinage* is the development of *sinusoidal* thickening and thinning (similar to **pinch-and-swell structure**) within a homogeneous **anisotropic** material, also resulting from layer-parallel **extension** but due to development of an instability. (See **necking**.) [TGB]

Hobbs, B.E., Means, W.D. & Williams, P.F. (1976) *An Outline of Structural Geology*, Chapter 6. John Wiley & Sons, New York.

Bouguer anomaly **Gravity** measurement which has been corrected for the effects of **latitude**, **elevation** (including the **Bouguer correction**), *Earth tides*, **topography** and, where appropriate, **Eötvös** effects. Variations in the Bouguer anomaly reflect density variations in the subsurface. The Bouguer anomaly is usually the most suitable for interpretation. [PK]

Bouguer correction (BC) A correction for the gravitational effect of the rock between the observation point and datum level, usually applied by assuming the rock has the form of an infinite slab.

$$BC = 2\pi G\rho t$$

where G is the **gravitational constant**, ρ the density of the surface rock and t its thickness.

At sea the correction is equivalent to replacing the water by rock of a specified density. (See **reduction**.) [PK]

boulangerite ($Pb_5Sb_4S_{11}$) A sulphosalt mineral. [DJV]

boulder tracing An old, but still highly effective, method of *mineral exploration* in which mineralized boulders are traced back to their source. Most use of this method has been in glaciated terrains, particularly in Scandinavia where it has led to the discovery of some of Europe's largest **base metal** mineral deposits such as Outokumpu (**copper**) in Finland and Laisvall (**lead**) in Sweden. [AME]

Bouma sequence Vertical fivefold division of a bed produced by deposition from a waning **turbidity current**, named after Arnold Bouma who first described the sequence in 1962. From base to top, the complete Bouma sequence comprises: lower structureless, coarsest-grained division, A; upper flow regime parallel-lamination, or **plane bed**, usually in coarse- to fine-grained **sand**, B; *current ripple lamination*, in fine **sand** to silt, C; lower flow regime parallel-lamination, or **plane bed**, in silt, D; structureless to very fine-scale laminated mud-rich division, E. **Sole marks**, including large-scale scours, may occur at the base of a bed. The sequence may show incomplete development, e.g. base- or top-absent Bouma sequence. Horizontal to vertical *bioturbation* (*ichnofacies*) may affect any part of a bed. **Wet sediment deformation** can be present, especially **convolute-lamination** in the Tc division. (See Fig. B13.) [KTP]

boundary layer The region in any flow which is affected by the frictional resistance of the boundary over which it is moving. This resistance causes a retardation of velocity close to the boundary and creates **shear stresses** within the fluid. Boundary layers may either be **laminar** or **turbulent**.

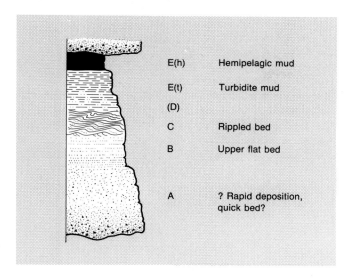

Fig. B13 Five divisions of the **Bouma sequence**. A = massive or graded; B = sandy parallel laminations; C = rippled and/or convoluted; (D) = delicate parallel interlaminations of silt and mud; E(t) = mud introduced by the turbidity current; E(h) = the hemiplegic background mud of the basin.

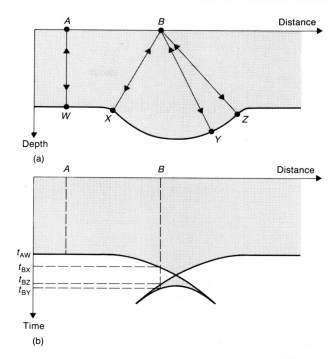

Fig. B15 Bow-tie effect. (a) A sharp synclinal feature in a reflecting interface. (b) The resultant 'bow-tie' shape of the reflection event on the non-migrated seismic section. (After Kearey & Brooks, 1984.)

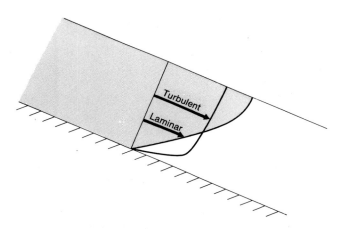

Fig. B14 Boundary layer. Comparison of velocity profiles in laminar and turbulent flow down an inclined plane. The profiles are drawn so that they represent about the same total discharge, or volume rate of flow, per unit width of the plane (i.e. the areas under the two curves are about the same).

Flow velocity tends to zero at the boundary, where adsorbed molecules are stationary, and then increases away from the boundary as frictional resistance declines. The edge of the boundary layer, where no velocity gradient and hence no **shear stresses** exist, is defined where the velocity reaches 0.99 of the freestream velocity. (See Fig. B14.) [JLB]

Leeder, M.R. (1982) *Sedimentology: Process and Product*. Allen & Unwin, London.
Middleton, G.V. & Southard, J.B. (1984) *Mechanics of Sediment Transport* (2nd edn). Short Course No. 3 Lecture Notes. Society of Economic Paleontologists and Mineralogists, Tulsa, OK.

bournonite ($PbCuSbS_3$) A sulphosalt mineral. [DJV]

bow-tie effect The appearance on a **seismic reflection stacked** section of a buried structure which focuses reflected energy on un**migrated seismic reflection** records. This effect is seen where the curvature of an upwardly concave reflector is such that the reflected seismic energy is focused before it is recorded at the surface. The curvature of the reflector is such that a single point on the surface records three separate reflections from different parts of the reflector. The set of reflections that pass through the focus is called the reverse branch. Tight **synclinal** folds produce a bow-tie effect and care must be taken in interpreting un**migrated** sections that the reverse branch of the reflection is not interpreted as an **anticline**. (See Fig. B15.) [GS]

Kearey, P. & Brooks, M. (1991) *An Introduction to Geophysical Exploration* (2nd edn). Blackwell Scientific Publications, Oxford.

Bowen's reaction series A series of minerals which crystallize from molten rock of a specific chemical composition, wherein any mineral formed early in the chain will later react with the melt, forming a new mineral further down the series. [ASG]

bowenite

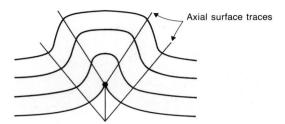

Fig. B16 Box fold.

bowenite A yellow-green variety of **serpentine**, used as a substitute for **jade**. [DJV]

box fold A composite **fold** with two **antiformal** *hinges* situated between two **synformal** hinges (or vice versa). The *axial planes* (or axial surfaces) of these individual **folds** are arranged in **conjugate** pairs. (See Fig. B16.) [RGP]

boxwork A honeycomb-like pattern that is commonly found in **gossans** and that develops when residual **limonite** remains in the cavity formed when a sulphide grain is **oxidized**. The colour and structure of the boxwork is frequently diagnostic of the predecessor mineral and so can be of great importance to the prospector. [AME]

Jensen, M.L. & Bateman, A.M. (1979) *Economic Mineral Deposits*. John Wiley & Sons, New York.

brachiopods/Brachiopoda Solitary, bilaterally symmetrical, unsegmented marine animals with a bivalved shell and a complex food-gathering apparatus, the lophophore. Attachment to the seafloor is commonly effected by a flexible fleshy stalk, or pedicle, although some are cemented or free-lying. Brachiopods are all assigned to the phylum Brachiopoda, and there are two classes: the *Inarticulata*, in which the valves are held together solely by muscles, and the *Articulata*, in which the valves are hinged by teeth and sockets. Fossil brachiopods are common and diverse, with a stratigraphical range from Lower **Cambrian** to *Holocene*, although the group diminished in diversity globally during the late **Mesozoic** and has subsequently been relatively unimportant.

CLASS INARTICULATA. Brachiopods with chitinophosphatic or, occasionally, calcareous valves, no hingement, and a lophophore unsupported by skeletal material. Inarticulates are more abundant and varied than articulates in the **Cambrian** and showed diversification in the **Ordovician**, but several subgroups had become **extinct** by the end of the **Devonian**. They are often most abundant in nearshore environments and, unlike the articulates, may have been able to tolerate variable or reduced salinity. The group is of little value in **biostratigraphy**, with many genera long-ranging; the 'living fossil' *Lingula*, a burrowing inarticulate which has survived from the **Ordovician** to the present day, is a classic example.

CLASS ARTICULATA. Brachiopods with posteriorly hinged **calcite** valves and a complex lophophore which normally has a skeletal support. The pedicle, when present, emerges from a notch or foramen at the posterior end of the pedicle (ventral) valve; the opening, or delthyrium, is triangular and may be partially closed by a pair of deltidial plates. A corresponding notch, or notothyrium, may be present on the opposing smaller brachial (dorsal) valve. The internal skeletal support for the lophophore, the brachidium, is attached at the posterior end of the brachial valve. The valves are operated by two sets of muscles, the adductors to close them and the diductors to open them. When open, the anterior end of the shell gapes and water is drawn across the lophophore for respiration and filtering of nutrient particles. Most fossil and living articulates are accommodated in seven orders.

Order Orthida: Biconvex articulates, characteristically with broad, straight hinge lines. The shells are impunctate and there is no brachidium. Lower **Cambrian** to Upper **Permian**.

Order Strophomenida: Diverse articulates with convex pedicle valves and, usually, planar or concave brachial valves. The hinge line is broad and straight and the pedicle foramen normally closed. Lower **Ordovician** to Lower **Jurassic**.

Order Pentamerida: Biconvex impunctate shells, commonly smooth, often with two diverging plates in the brachial valve and a median septum in the pedicle valve for muscle attachment. Middle **Ordovician** to Upper **Devonian**.

Order Rhynchonellida: Biconvex impunctate shells, usually with coarse ribs on each valve meeting at a zigzag commissure. A pedicle foramen is characteristically developed in the beak-like posterior tip (umbo) of the pedicle valve. The lophophore is spirally coiled, but supported only by short spikes (crura) in the brachial valve. At the anterior end a concave axial depression, or sulcus, is often present on the pedicle valve, with a corresponding fold on the brachial valve. Middle **Ordovician** to *Holocene*.

Order Atrypida: Biconvex impunctate shells, with short, rounded hinge and pedicle foramen or notch. The brachidium is spiral, with the apex directed dorsally or laterally. Middle **Ordovician** to Upper **Devonian**.

Order Spiriferida: Biconvex shells, usually impunctate, with broad hinge lines. Often a well-developed triangular interarea is present between the hinge line and the umbo on the pedicle valve. The spiral brachidium is laterally directed. Upper **Ordovician** to **Jurassic**.

Order Terebratulida: Biconvex articulates with short, rounded hinge and pedicle foramen. Valves have perforations (punctae) on the inner surface. The brachidium is looped. Lower **Devonian** to *Holocene*.

Living brachiopods occupy environments from intertidal rock pools to ocean depths, but are most numerous on the off-shore shelf. Early **Cambrian** forms occurred in shallow habitats, but spread and diversified in the **Ordovician** so that by the end of the period all parts of the shelf had been

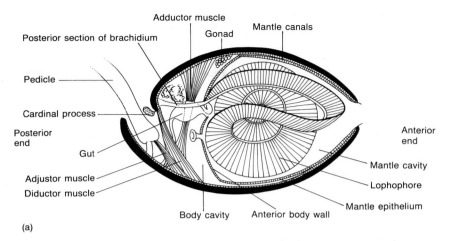

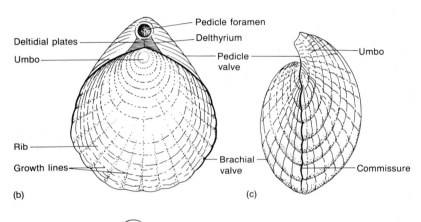

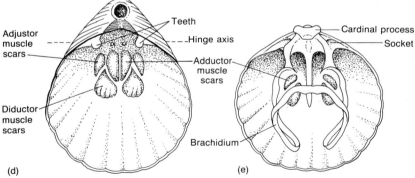

Fig. B17 Brachiopods. The extant brachiopod *Magellania flavescens*. (a) Stylized median section. (b) Dorsal view of the shell. (c) Lateral view of the shell. (d) Internal view of the pedicle valve. (e) Internal view of the brachial valve. (After Clarkson, 1986.)

colonized. In general, the articulate orders with the less complex lophophores, the orthides, strophomenides and rhynchonellides, are dominant in sediments deposited in shallower waters. Deeper water environments, where suspended food may be less readily available, are characterized by pentamerides, atrypides and spiriferides, which have complex brachidia. The development of a sulcus, seen in the rhynchonellides and some atrypides, spiriferides and pentamerides, apparently allowed separation of inhalant and exhalant water currents. Adaptations to living on soft muddy sediments include large resting

areas, as in the flat pedicle valves of some strophomenides or the broad interareas of some spiriferides. Pedicle attachment is particularly important in higher-energy, nearshore habitats, where ribbing is also commonly developed.

Associations of brachiopod species with similar ecological tolerances often occurred in bands subparallel to the shoreline. In the **Paleozoic**, when brachiopods were particularly abundant, they often dominated benthic marine communities and changes in brachiopod assemblages are important guides to changing environments and water depths. Hence, detailed patterns of transgression and regression can be recognized by mapping the distribution of recurrent associations of brachiopods; the classic example relates to the early **Silurian** transgression of Wales and the Welsh Borderland (Ziegler *et al.*, 1968).

Although controlled in their distribution by facies and, at times, by provincialism, brachiopods have proved of some **biostratigraphical** utility, especially in the **Paleozoic**. Some correlations have been effected using concurrent brachiopod ranges, while more refined schemes based on particular evolving lineages are of importance in the **Ordovician** and **Silurian**. (See Fig. B17.) [RJA]

Clarkson, E.N.K. (1986) *Invertebrate Palaeontology and Evolution* (2nd edn), pp. 130–65. Allen & Unwin, London.

Cocks, L.R.M. (1985) Brachiopoda. In: Murray, J.W. (ed.) *Atlas of Invertebrate Macrofossils*, pp. 53–78. Longman, Harlow.

Rudwick, M.J.S. (1970) *Living and Fossil Brachiopods*. Hutchinson, London.

Ziegler, A.M., Cocks, L.R.M. & McKerrow, W.S. (1968) The Llandovery transgression of the Welsh Borderland. *Palaeontology* **11**: 736–82.

braided river Multi-thread channels usually found in high-energy fluvial environments with steep valley gradients, large and variable discharges, non-cohesive channel banks, and dominantly **bedload transport**. The distributary channels of a braided river are less efficient than a single channel, with increased width, reduced depth and steeper gradient maintaining comparable velocities; this implies a higher rate of expenditure of stream power in braided than in single-thread streams. Braids have been shown to replace meanders at higher slopes and discharges (Leopold & Wolman, 1957). Braids are typically found in proglacial environments, but wherever located, they are associated with **gravel** beds of variable sediment size with highly mobile channel forms. [TPB]

Leopold, L.B. & Wolman, M.G. (1957) *River Channel Patterns*. Professional Paper of the United States Geological Survey, **282B**.

braided stream A stream that in planform divides and rejoins around channel **bars** that are of a size comparable to the channel width. Braided streams may be differentiated from **anastomosing streams** on the basis of braid intensity (two times the total bar length divided by the channel reach length) and from **meandering streams** by the **sinuosity** (channel length divided by channel reach length). The choice of class boundaries is arbitrarily taken as sinuosity less than 1.3 and a braid intensity greater than 1 but these do not represent natural threshold values.

Alluvial braided streams may be subdivided on the basis of load characteristics, channel stability and discharge characteristics. *Braid bars* may be emergent only at low flow stage or they may be semipermanent features with plant colonization. A large number of facies models have arisen as a result of the wide range of channel types that are classified as braided streams. [JA]

Carsen, M.A. (1984) The meandering-braided river threshold: reappraisal. *Journal of Hydrology* **74**: 315–34.

Rust, B.R. (1978) A classification of alluvial channel systems. In: Miall, A.D. (ed.) *Fluvial Sedimentology*. Canadian Society of Petroleum Geologists Memoir **5**.

Schumm, S.A. (1972) *The Fluvial System*. John Wiley & Sons, New York.

braidplain An extensive region of gently sloping terrain composed of interlapping **alluvial fans** resulting from the sediment discharge of *glacial* meltwater streams. [ASG]

bravoite ((Ni, Fe)S_2) A rare **ore** mineral. [DJV]

brazilian emerald A variety name for green **gem tourmaline**. [DJV]

brazilianite (NaAl$_3$(PO$_4$)$_2$(OH)$_4$) A rare yellow-green **gem** mineral found in **pegmatite deposits**. [DJV]

breaching thrust A *thrust fault* that offsets a pre-existing, structurally higher **fault** or **fold** structure. This is in contrast to the smooth branching of a lower *thrust* onto a higher one, typical of structures such as *duplexes* (see **imbricate structure**). Breaching thrusts are good evidence of the order of *thrust* movement, and so are useful in the construction of **balanced sections**. [SB]

break-back thrust A *reverse fault* in an overall *piggyback thrust system* that forms within existing thrust structures, rather than closer to the **foreland**. It thus forms out-of-sequence in this thrust system. Such a **fault** post-dates and truncates existing thrust structures (**faults**, **folds**) in both **hanging wall** and **footwall**. Identification of such **faults** affects estimates of **thrust belt shortening** obtained from **balanced sections**. (See **imbricate structure**.) [SB]

breast Mining term for the **face** of a mine working. [AME]

breithauptite (NiSb) A **nickel ore** mineral. [DJV]

brick clay Clay suitable for making bricks. The materials used for brickmaking vary from unlithified clays to **mudrocks** such as mudstone, **shales** and even some soft **slates**. After being ground and moistened with water, the materials must be capable of taking a good shape by moulding or pressure and of retaining that shape without detrimental shrinkage, warping or cracking when the bricks are dried and then fired. A high clay content will make the material too plastic and 20% is about right. Calcium carbonate should run 5–15%; it helps to prevent shrinking during firing. A high proportion of finely disseminated carbonaceous material will act as a built-in fuel and lower heating costs. [AME]

bright coal Coals are sometimes described with reference only to their relative brightness according to the scale: bright coal, banded bright coal, **banded coal**, banded dull coal, *dull coal*. [AME]

bright spot A local increase in **seismic reflection** strength which does not owe its origin to focusing effects. Usually bright spots are associated with very strong **seismic reflections** from the top of shallow **sandstone reservoirs** filled with gas. Thus they are often thought of as direct indicators of *hydrocarbon* on a **seismic reflection** section. Unfortunately a **sandstone** with a few per cent gas saturation yields a bright spot similar in strength to one containing commercial quantities of gas. [GS]

brine Concentrated NaCl in water. Brines in nature are generally far more complex than this and are found in most parts of the **crust** and in all rock types, either as interpore water or in channel ways such as **faults**. **Hydrothermal solutions**, whether responsible for mineralization or present in *hot spring* activity, are all complex brines containing cations such as Na^+, Ca^+, K^+, Mg^+, in many thousand p.p.m. Cl^- is usually the most important anion. These solutions can leach **base** and other **metals** from the rocks through which they pass and may carry important amounts of these. Natural brines can be important sources of metals, e.g. **lithium** which is recovered commercially from brines in the USA and Chile. [AME]

brittle Property of material that deforms by **fracture** without appreciable **viscous** or **plastic** behaviour. The brittleness of a material depends on the amount of *crystal plasticity* accompanying the **fracture** process: this can be quantified by the ratio of *fracture toughness* to *surface energy*, which varies from approximately 1 for highly brittle materials in which no **plasticity** occurs during **fracture**, to very large values in non-brittle materials which fracture by coalescence of plastically formed voids. A brittle **stress–strain** relationship consists of perfectly elastic behaviour up to a *failure strength*; non-brittle (brittle–ductile or ductile) relationships involve increasing amounts of non-elastic, *permanent strain* before **failure**. Brittle materials can therefore also be defined as those with less than 3% permanent **strain** before **failure**, those with 3–5% permanent pre-**failure strain** as transitional, and those with greater than 5% as ductile. A slightly different definition classes brittle materials as those showing a **stress** drop after **failure** compared to ductile materials which have no **stress** drop after **failure**. 'Brittle' is also used as a field term to describe discontinuous **strain** distribution, in contrast to continuous **strain**, which is referred to as 'ductile'. (See **ductile deformation**.) [TGB]

Lawn, B.R. & Wilshaw, T.R. (1975) *Fracture of Brittle Solids*. Cambridge University Press, Cambridge.

Logan, J.M. (1979) Brittle phenomena. *Reviews of Geophysics and Space Physics* **17**: 1121–32.

Paterson, M.S. (1978) *Experimental Rock Deformation — The Brittle Field*. Springer-Verlag, Berlin.

Rutter, E.H. (1986) On the nomenclature of mode of failure transitions in rocks. *Tectonophysics* **122**: 381–87 (letter).

brittle–ductile transition The transition from **deformation** by **fracture** (cataclasis) to **deformation** by *crystal plasticity*. The transition is favoured by increasing **confining pressure** and temperature, and suppressed by increasing **pore-fluid pressure**. It is often taken to occur at a depth in the Earth's **crust** where the **stress** supported by a **brittle failure** criterion (such as the *Coulomb* or *Griffith criterion*) is equal to that supported by a crystal-plastic flow law. The transition has also been defined in terms of **stress–strain** behaviour as the pressure at which the **stress** required to form a **fault** is equal to the **stress** required for sliding along the **fault**; this implies that the transition occurs when there is no **stress** drop at **failure**. Another definition uses the criterion of non-elastic pre-**failure strain**: the transition occurs when this reaches a value of 3–5%.

It has been suggested that the transition should be divided into two stages: a lower transition from **failure** with a **stress** drop to frictional sliding with no **stress** drop (*cataclastic flow*), and an upper transition to **deformation** which is completely pressure insensitive.

Brittle–ductile transitions occur between the **upper** and **lower continental crust** and between the **lithosphere** and **asthenosphere**. (See also **brittle** and **ductile deformation**.) [TGB/DHT]

Byerlee, J.D. (1968) Brittle–ductile transition in rocks. *Journal of Geophysical Research* **73**: 4741–50.

Heard, H.C. (1960) Transition from brittle fracture to ductile flow in Solenhofen limestone as a function of temperature, confining pressure, and interstitial fluid pressure. In: Griggs, D. & Handin, J. (eds) *Rock Deformation*. Geological Society of America Memoir **79**: 193–226.

Tullis, J. & Yund, R.A. (1977) Experimental deformation of Dry Westerly Granite. *Journal of Geophysical Research* **82**: 5705–18.

brochantite ($Cu_4SO_4(OH)_6$) A secondary **copper** mineral. [DJV]

bromargyrite (AgBr) A rare supergene mineral. [DJV]

bromine (Br) At. no. 35, at. wt. 79.904. Bromine occurs naturally as bromides in seawater and in some natural **brines** and salt deposits. Over half the world production is from **brines** such as the Smackover Brines recovered from wells in Arkansas that run to over 6000 p.p.m. bromine. Bromine is extracted commercially from seawater although the average concentration is only about 60 p.p.m. It is produced from the **Dead Sea**, from the water of Searles Lake, California and is also a by-product from the processing of potash deposits in a number of countries. [AME]

bronze Alloy of **copper** and **tin**, normally in a ratio of *c.* 9:1, used extensively for tools, weapons and vessels in antiquity in the Old World. The first bronzes were produced in Mesopotamia around the middle of the 3rd millenium BC, and bronze was in general use in almost all areas of the world except the Americas between *c.* 2000 and 1000 BC. The archeological term 'Bronze Age' refers to the period of time during which bronze metallurgy was invented, adopted and improved. Bronze was gradually superseded by **iron** from about 1000 BC onwards, though the former continued to be used for decorative objects. The addition of **tin** to **copper** improved the hardness of the resulting alloy and also made casting easier to achieve without flaws. Chemical analyses of bronzes from the UK show that some early products contained high concentrations of **arsenic** (2–3%) while later in the Bronze Age **lead** became a common addition, perhaps in order to economize on the amount of **tin** needed. Attempts to chemically characterize metal objects and assign them to **ore** sources are made difficult by inhomogeneity of **ore bodies** and by re-melting, mixing and re-use of metal artefacts. [OWT]

bronzite ((Mg, Fe)SiO_3) An *orthopyroxene* mineral. (See **pyroxene minerals**.) [DJV]

brookite (TiO_2) A relatively rare form of **titanium** oxide. (See **oxide minerals**.) [DJV]

brousse tigrée Vegetation banding, which may include grassland patterns but which generally consists of bands of more closely spaced trees alternating with bands of sparser vegetation. Its nature and origin have been well described thus by Mabbutt & Fanning (1987, p. 41): 'All are developed in arid or semi-arid areas, in open low woodlands or tall shrublands, with average annual rainfalls of between 100 and 450mm; they occur on slopes of the order of 0.25%, too gentle for the development of drainage channels, but steep enough to maintain organized patterns of **sheetflow**; these slopes are mantled with *alluvium* or **colluvium** and the patterns are independent of bedrock. The associated *soils* are earths, and sandier crests or clay flats in the same areas do not have tree bands. The bands of denser vegetation, termed *"vegetation arcs"* run close enough to the contour to serve as *form lines*; hence they tend to be convex downslope on **interfluves** and convex upslope in shallow drainage ways. In drier areas the banding may be restricted to the better-watered depressions, but it is commonly best-developed on low **interfluves**, with the intervening depressions marked by uniformly dense tree cover. Such tracts of more concentrated **sheetflow** have been named *"water lanes"*.

The bands commonly occur in fairly regular sequences or ladder-like "tiers" downslope, the tiers being bounded by *water lanes*. Tree bands may extend up to a kilometre or more along the contour, but in detail they are commonly slightly irregular, "burgeoning here and becoming attenuated there; dying out and succeeding one another *en echelon*".

The downslope distance between bands ranges from 70–500m, although it is mainly between 100 and 250m, and the interband intervals are commonly between two and four times as wide as the bands.' [ASG]

Mabbutt, J.A. & Fanning, P.C. (1987) Vegetation banding in western Australia. *Journal of Arid Environments* **12**: 41–59.

brown coal A soft, low **rank coal** which includes **lignite** and *sub-bituminous coal*. The distinction between these and **bituminous coal** is based on the calorific value and other chemical properties, i.e. a calorific value of less than $19.3\,MJ\,kg^{-1}$ and a fixed **carbon** content of 46–60%. In Europe the term brown coal is more commonly used, but in the USA and other countries brown coals are usually classed as either **lignite** or *sub-bituminous coal*, depending principally on their chemical characteristics. [AME]

brucite (Mg(OH)$_2$) An **industrial mineral** found in **limestones** and as an **alteration** product of magnesium silicates. [DJV]

Bryozoa Moss animals, all colonial, largely marine, in which, most commonly, a calcareous skeleton (zooecia) houses the zooid and builds up the colony (zooaria). Some groups are soft bodied. Colonies have variable lifestyle: encrusting, creeping, erect, chain-like and may be polymorphic.

Calcareous zooecia are often exquisitely preserved tubes or tightly packed containers, each with an orifice distally, possibly an operculum, hinged proximally, and a conspicuous frontal wall. Muscular contraction, coupled with swelling of a water-filled sac proximal to the orifice, allows the operculum to open and extrudes the zooid. Specialized zooids occur, at least partially to protect the colony, with mandibles and deterrent bristles. Calcareous ovicells are associated with the distal ends of the zooecia, often overlapping slightly the proximal part of the frontal wall of the next zooecium in the chain.

The zooid itself has a retractible ring of tentacles around a mouth, a U-shaped gut and an anus outside the tentacles. The whole is extruded or retracted as above or, in the case of soft bodied bryozoans, by transverse (compressional) and retractor muscles respectively. Ciliated currents pass phytoplankton down the funnel formed by the tentacles, to the mouth.

Colonies grow by asexual budding from a connecting stolon which has a funicular tube along its length; a branch of this leads to a minute thread which connects to the zooid's gut, near its base. New colonies develop after sexual reproduction and, after settling, by growth from an original zooid, the ancestrula.

There are three classes, two ranging from the **Ordovician** to the Recent, whilst one, of freshwater forms, is *Tertiary* to Recent. The taxonomy below class level reflects the major events undergone by bryozoans: thus of the two main fossil classes two orders range from the **Ordovician** to Recent; three from the **Ordovician** to **Permian** or **Triassic**; and one is **Jurassic** to Recent. They were strongly affected by events at the end of the **Paleozoic** although only one order actually became extinct in the **Permian**. Great diversity was effected in the **Ordovician**. The cyclostome bryozoans, the only calcareous order to survive into the **Mesozoic**, actually reached a peak of development there but have declined to the Recent.

Bryozoans occur in all oceans, have a great range of depth tolerance, but are more constrained by needing a hard substratum for settling. The intertidal zone is too disturbed an environment for real success, hence the sub-littoral zone and deeper is preferred. Their food being phytoplankton, a majority of bryozoans are relatively shallow and they may play an important role at times as *reef* frame builders or sediment binders. [RBR]

Nielsen, C. & Larwood, G.P. (eds) (1985) *Bryozoa: Ordovician to Recent*, i–viii, 1–364. Olsen & Olsen, Fredensborg.

Bubnoff unit Unit providing a useful means to quantify the rate of operation of diverse **geomorphological** processes as a rate of ground loss (perpendicular to the surface) or slope retreat. A unit equals 1 mm per 1000 years, equivalent to $1\,m^3\,km^{-2}\,a^{-1}$ (Fischer, 1969). [ASG]

Fisher, A.G. (1969) Geological time–distance rates: the Bubnoff unit. *Geological Society of America Bulletin* **80**: 594–652.

buchite A fine-grained or glassy metamorphic rock (**hornfels**) produced by sintering or fusion of a clay-rich sedimentary rock as a result of intense thermal/**contact metamorphism**. Buchites are commonly devoid of metamorphic fabric and commonly occur at the margin, and as inclusions within **basalt/dolerite** intrusions. [RST]

buckling Folding resulting from compression of a layered rock along the layering, i.e. *layer parallel shortening*. This sort of **folding** is seen in a beam shortened solely by compression at its ends. In rocks, the folded layers generally maintain their original thicknesses and form Class 1, *parallel* or *concentric folds* (see **fold**). **Ptygmatic folds** are common in isolated single layers such as **veins**.

Buckle folds are seen in single layers and multilayer sequences, in both natural and experimental/analogue examples; modelling of *layer parallel shortening* in layered media has greatly increased understanding of *buckle folds*. Analogue models, consisting of an isolated more **competent** layer within a less **competent** matrix, show *buckle folds* developing from irregularities in, for example, the layer thickness, or in the **rheology**. Theoretical modelling in terms of layers of differing **viscosity** (Biot, 1961; Ramberg,

1964) shows that material instabilities will amplify to form *buckle folds*. **Folds** develop along the layer, but they grow at different rates. For a layer of a given thickness, **folds** of a particular wavelength amplify fastest and so become the dominant wavelength as **deformation** continues. Biot derived a relationship between layer thickness, dominant wavelength and **viscosity** contrast:

$$W_d = 2\pi t \sqrt[3]{\eta_1/6\eta_2}$$

where W_d = dominant wavelength, t = layer thickness, η_1 = viscosity of isolated layer, η_2 = viscosity of matrix.

This holds for model **deformation** with analogue materials where **viscosities** are known and appears to hold for natural examples, where the properties can be estimated.

Multilayered sequences, where there are many subparallel layers of differing **viscosity**, can also respond to *layer parallel shortening* by buckling. The relative thicknesses and competences of the layers affect the behaviour

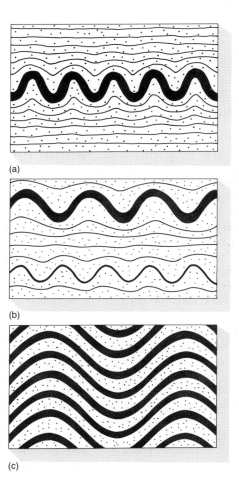

(a)

(b)

(c)

Fig. B18 Buckling. The drawings show buckle folds which have developed in isolated, more competent layers (black) in a less competent matrix, and in a sequence of more and less competent layers. (a) Single competent layer. (b) Disharmonic folding. (c) Harmonic folding. (After Ramsay & Huber, 1987.)

of multilayered models. Where the less **competent** layers are relatively thin, **kink bands** and **chevron folds** form. Where the less **competent** layers have thicknesses close to those of the more **competent** ones, more rounded **folds** of varying class develop, as described under **fold**. The **fold** class tends to vary with the competence contrast between the layers. More **competent** layers tend to form Class 1 **folds**, and the adjacent less **competent** layers, Class 2 (similar) or Class 3 **folds**. Thus an alternating sequence of more and less **competent** layers can produce a sequence of **folds** of different class. Some combinations of **fold** classes can continue indefinitely, without **decollement** between layers. It is more common in natural multilayer sequences to find buckling of individual widely-spaced layers, each with dominant wavelengths depending on their thicknesses. As with description of **fold** geometries, the extent to which **fold** models can be applied to natural examples provides further valuable information on their origins. (See Fig. B18.) [SB]

Biot, M.A. (1961) Theory of folding of stratified viscoelastic media and its implications in tectonics and orogenesis. *Geological Society of America Bulletin* **72**: 1595–620.
Ramsay, J.G. (1967) *Folding and Fracturing of Rocks*. McGraw-Hill, New York.
Ramsay, J.G. & Huber, M.I. (1987) *The Techniques of Modern Structural Geology* Vol. 2: *Folds and Fractures*. Academic Press, London.
Ramberg, H. (1964) Selective buckling of composite layers with contrasted rheological properties: a theory for simultaneous formation of several orders of folds. *Tectonophysics* **1**: 307–41.

building stone Any rock suitable for use in construction. Building stone that is quarried, cut and dressed into regularly shaped blocks is termed *dimension stone*. For use as a building stone a rock should possess mineral purity, that is be free from decomposition products produced by secondary **alteration** of the rock. It should have the necessary mechanical strength to sustain loads and **stresses** in service (i.e. a high *crushing strength*). The *durability* (resistance to **weathering**) should be high if the stone is to be used in the polluted atmospheres of cities, sulphur dioxide being particularly deleterious. The hardness, workability and directional properties are likewise important. The hardness ranges from that of soft **coquina** or **limestone** to that of **granite**. Workability depends partially on hardness. **Limestone** is easy and cheap to dress but **granite** is expensive. Workability also depends on the presence or absence of planes or incipient planes of splitting (**joints**). **Porosity** and **permeability** determine the water content and therefore the susceptibility to frost action. The texture affects the workability; fine-grained rocks split and dress more readily than coarse ones. Lastly the colour should be permanent and not affected by **weathering**, nor should there be **accessory minerals** such as **pyrite**, that will **oxidize** and produce unsightly stains. [AME]

Jensen, M.L. & Bateman, A.M. (1979) *Economic Mineral Deposits*. John Wiley & Sons, New York.

bulk finite strain The total **finite strain** of a volume of material calculated on the assumption that the **strain** is homogeneous. [RGP]

bulk material A general term for material used in the constructional industry such as clays for brick and tile making; **limestone** and **shale** for cement, which is added to either **sands** and **gravels** or crushed rock to prepare concrete; and hard rock for **building stone**, *road stone*, **aggregate** and **ballast**. [AME]

bulk modulus Also known as the *incompressibility*, K; a physical property of a material, defined as the ratio of the hydrostatic pressure to the **dilation** in an elastic material. (See **elastic moduli**.) [RGP]

buoyancy Term used in the oil industry when discussing the upward **migration of oil** (or gas) through the heavier water during secondary **migration**. All **crude oils** float on saline water (**brine**) and nearly all on fresh water. Thus *hydrocarbons* are physically pushed ahead of water moving through water-wet rocks, i.e. they are subjected to a buoyant force. [AME]

burial diagenesis (*mesogenesis* of Schmidt & McDonald, 1979) Includes all the physical, chemical and biological processes that act upon a sediment during burial as it is gradually removed from the influence of the depositional environment into the mesogenetic realm (see **regimes of diagenesis**) and persists until the onset of metamorphism or structural **inversion** and exposure to **meteoric water diagenesis** (see also **diagenesis**). During burial a multitude of physical and chemical reactions are initiated in response to changes in the ambient geochemical environment. Conceptually, subsequent to *eogenesis* (see **depositional environment related diagenesis**) the sediment will comprise a mixture of those detrital components that were initially stable in the *eogenetic* **pore-waters** and new authigenic minerals (see **authigenesis**) which have reached stability (or metastability) through reaction with the **pore-waters**. To effect any further change in the diagenetic assemblage will require a change in the ambient physico-chemical conditions of pressure, temperature or **pore-fluid** chemistry. The equilibrium composition of the detrital assemblage should move in response to changes in these variables. As a result, at elevated pressure and temperature or in different **pore-fluids**, many minerals stable under *eogenetic* conditions become unstable and will react to produce new, stable authigenic minerals. These reactions may be driven by pressure, temperature, chemistry or a combination of any of these parameters.

The physical processes of **compaction** all combine to gradually reduce **porosity** with increasing burial. Initially, **porosity** is reduced through the effects of mechanical compaction, principally as a result of grain reorientation

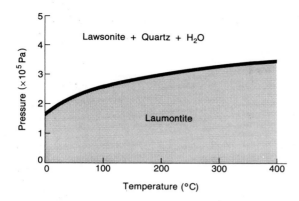

Fig. B19 Burial diagenesis. The laumontite to lawsonite reaction as an example of a pressure sensitive reaction involving a large volume change.

and **brittle fracturing**. Realignment of clay particles in **mudrocks** is particularly effective in reducing initial **porosities**, but the process also takes place in **sandstones**. At greater depths of burial, **quartz** becomes soluble at grain-to-grain contacts and compaction proceeds via the process of **pressure dissolution**. Whilst the effects of **pressure dissolution** are well known in **sandstones**, the relative importance of this process has not yet been fully documented in **mudrocks**.

As a direct result of compaction the volume of intergranular pore space is reduced (see **porosity**) and the interstitial **pore-waters** originally trapped within the gradually subsiding sediment will be expelled. The amount of physical compaction is most intense during the initial period of burial (the first 500 m is sufficient to compact clays with an initial **porosity** of 70–90% to **mudrocks** with only 30% **porosity**). Subsequently, **pore-water** expulsion falls off drastically and major additions of waters to the *mesogenetic* regime rely on dehydration reactions. It follows that the amount of water available from compaction in a given subsiding **basin** is finite.

In parallel with the physical processes of **compaction** that result from burial, both inorganic and organic constituents of **clastic** sedimentary **rocks** additionally undergo a variety of chemical depth-related changes in response to increases in pressure and temperature. Increasing pressure with burial will generally favour the **precipitation** of more dense minerals in pressure-sensitive mineral reactions. Pressure sensitive reactions involve net changes in molar volume:

volume change $(\Delta V_R) = \Sigma MV$ products $- \Sigma MV$ reactants

If ΔV_R is a negative value — that is, the reaction products are more dense than the reactants — then the reaction should proceed at higher pressures. For example, the reaction of **laumontite** to **lawsonite** is strongly pressure dependent (Fig. B19) and involves a large volume change:

laumontite $=$ lawsonite $+ 2$ quartz $+ 2H_2O$

$MV_R = 404.4 \quad MV_1 = 101.3 \quad MV_{2q} = 22.7 \quad MV_{2w} = 18.1$

$\Delta V_R = (182.9 - 404.4) = -221.3$

There are, however, relatively few diagenetic reactions that are purely pressure sensitive. Most diagenetic reactions are also strongly temperature dependent. Increasing temperature during burial promotes many reactions for several reasons. In the first instance, and most fundamentally, increasing temperature adds energy to the reacting system. Reaction rates will, therefore, increase and reaction barriers are more readily overcome (Fig. B20). Secondly, increasing temperature increases the *entropy* (S) of the diagenetic system. *Entropy* is a measure of disorder in a chemical system. Except at absolute zero, all substances have entropy, although the degree of *entropy* depends largely upon the physical state of a given component. Anhydrous solids are characterized by low entropy, hydrous solids and liquids intermediate *entropy* whilst gases are characterized by high *entropy*. With increasing temperature, therefore, there is a tendency for liquids and gases to become the stable phases. During burial diagenesis, if the change in *entropy* of a reaction (ΔS_R) is positive (that is, S increases), then the reaction should proceed with increasing temperature.

$\Delta S_R = \Sigma S$ products $- \Sigma S$ reactants

Fig. B20 Burial diagenesis. Schematic presentation of the energy changes associated with a typical diagenetic reaction demonstrating the necessity for energy input to overcome reaction barriers. Hydrolysis of feldspar according to the reaction

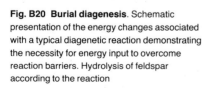

$KAlSi_3O_{8(s)} + 4H_3O^+$
$= K^+ + 3SiO_{2(aq)} + Al^{3+} + 6H_2O$

is illustrated as an example.

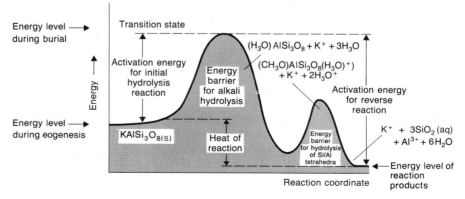

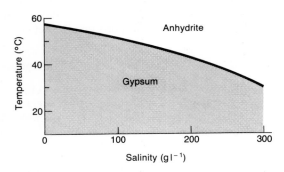

$$CaSO_4 + 2H_2O \rightleftharpoons CaSO_4 + 2H_2O$$
Hydrous mineral   Anhydrous mineral

Fig. B21 Burial diagenesis. The gypsum to anhydrite dehydration as an example of a temperature-sensitive reaction.

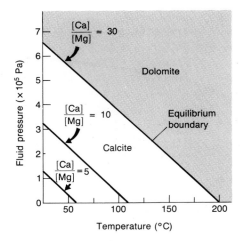

Fig. B22 Burial diagenesis. Stability relationships between calcite and dolomite expressed as a function of pressure and temperature. Equilibrium boundary at various [Ca:Mg] ratios (5, 10, 30) shown for the reaction

$$Mg^{2+} + 2CaCO_3 = CaMg(CO_3)_2 + Ca^{2+}$$

Note that 'dolomitization' takes place at progressively lower Mg^{2+} concentrations with increasing pressure and temperature.

If ΔS_R is negative the reactants are stable,
If ΔS_R is positive the products are stable,
If ΔS_R is zero the reaction is independent of temperature.

The concept of *entropy* explains why many burial diagenetic reactions are dehydration reactions. Bonding H_2O to a crystal decreases the *entropy* of the water molecule relative to the **pore-water**:

$$N \cdot H_2O \rightleftharpoons N + H_2O$$
hydrous mineral

e.g. $SiO_2 + NaAlSiO_6 \cdot H_2O \rightleftharpoons NaAlSi_{13}O_8 + H_2O \; \Delta S_R$
   quartz   analcime   albite
$$= +3.6 \, cal/mole \, K$$

Dehydration reactions thus increase the *entropy* of reaction products and will be thermodynamically favourable at higher temperatures. With increasing depth of burial less hydrous minerals will tend to be stable (see Fig. B21, e.g. **gypsum → anhydrite, smectite → illite, zeolites → feldspars**). Increasing temperature also generally increases the solubility of minerals in aqueous **pore-fluids**. Minerals which are more soluble at higher temperatures are referred to as prograde (e.g. **silica**; see **silica solubility**) and their total solubility will increase with increasing burial. Conversely, a cooling, upwardly migrating compactional flow will tend to result in **precipitation** of prograde minerals due to the temperature decrease. By contrast, the solubility of minerals involving reactions which generate a gas phase decreases with increasing temperature as the solubility of gases decreases with increasing temperature. Such minerals exhibit retrograde solubility, as exemplified by the behaviour of **carbonate minerals**:

$$CaCO_3 + H_2CO_3 \rightleftharpoons Ca^{2+} + 2HCO_3^+$$
$$\Updownarrow$$
$$CO_2 + H_2O$$

Temperature also directly influences the ability of aqueous ionic species to be accepted into mineral lattices. Most cations in aqueous solutions are hydrated and are enclosed in a hydration sphere of water molecules bound to the cations by dipolar bonds. With increasing energy (i.e. temperature) in the reacting system the hydration effect is reduced and ions are more readily accommodated into mineral lattices. The ions Fe^{2+} and Mg^{2+}, for example, are strongly hydrated at surface temperatures and require considerable amounts of energy to enter the carbonate lattice. The reaction

$$Mg^{2+} (or \, Fe^{2+}) + 2CaCO_3 \rightleftharpoons CaMg(Fe)(CO_3)_2 + Ca^{2+}$$
         calcite   dolomite/ankerite

will thus proceed spontaneously at lower Mg^{2+}/Ca^{2+} ratios at progressively higher temperatures due to decreased hydration effects. Progressively less Mg^{2+} is required to **dolomitize calcite** with increasing temperature. This reaction is also favoured by increasing pressure as **dolomite** is more dense than **calcite**. During burial diagenesis iron-magnesian carbonates will thus tend to be stable over **calcite** (Fig. B22).

Thus, during burial diagenesis, the detrital or *eogenetic* mineral components of the sediment will react with the ambient pressure–temperature environment in an attempt to shift towards equilibrium. The equilibrium conditions will, however, also change with burial and it is likely that true thermodynamic equilibrium is rarely attained during diagenesis. Additionally, changes in the interstitial **pore-water** composition may take place rapidly through the migration of depth-related reaction products from adjacent sediments. At elevated temperatures and pressures the sediment framework assemblage is generally much more reactive and changes in the **pore-water** chemistry should

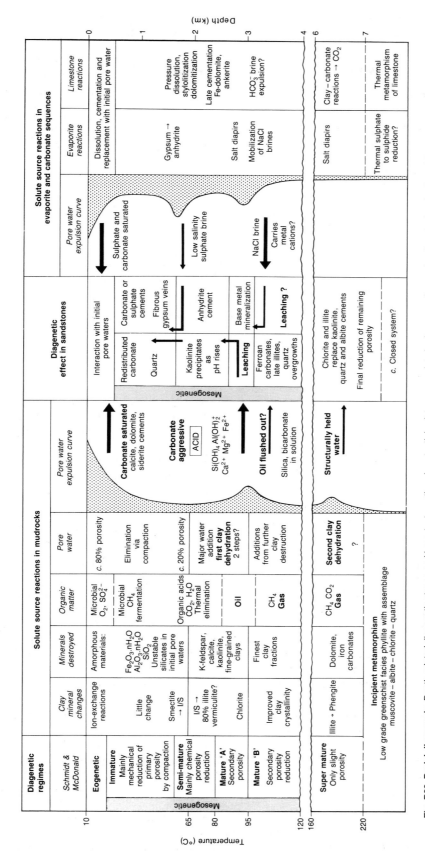

Fig. B23 Burial diagenesis. Depth-related diagenetic stages and important reactions in potential solute sources (mudrocks, evaporites and carbonates) and solute sinks (porous sandstones) with associated trends in pore-water evolution.

effect rapid and widespread changes in the detrital and authigenic mineralogy.

It is this tendency for large scale sediment–water interaction that characterizes burial diagenesis. All the various lithological components of a gradually subsiding sedimentary **basin** will respond to the changes in the controlling physico-chemical conditions. Minerals may **dissolve, precipitate** or be **replaced; pore-fluids** will thus be modified (see **pore-waters/fluids and their evolution**) and, as a result of pressure, will be squeezed out of compacting sediments into high **permeability** carrier beds. The constituents of a subsiding **basin** are thus likely to interact in a complex manner and should not be considered in isolation. An attempt to integrate many recent ideas on burial diagenetic trends and possible interactions between solute sources and solute sinks as a result of **pore-water** expulsion is summarized in Fig. B23. The sparsity of data for **evaporites** and carbonates reflects the general lack of studies that address burial diagenetic reactions in such sediments. The inorganic and organic constituents of mudrocks are much more comprehensively known. **Clay mineral** reactions and the **smectite** to **illite** transformation:

$$K^+ + smectite \rightarrow illite + (FeMg) + SiO_2 + H_2O$$

may make copious quantities of water and a variety of solutes available for either immediate precipitation as cements or migration during **pore-water** expulsion. Thermal **decarboxylation** of organic matter liberates additional water, organic acids, CO_2 and, ultimately, *hydrocarbons* Reactions between clay and carbonate at temperatures around 180°C also generate additional CO_2, e.g.

$$kaolinite + dolomite + quartz + H_2O \rightarrow Mg\text{-chlorite} + calcite + CO_2$$

which may be available for transport. Water expelled from **mudrocks** during compaction or from dehydration reactions may dissolve chloride and sulphate from **evaporite** sequences and the resulting **brines** contain complex various metal cations for transport.

Within this general framework of diagenetic response to burial, **pore-waters** within the subsurface will inevitably tend to migrate from low **permeability**, high pressure regimes to high **permeability**, lower pressure regimes (see **subsurface fluid migration**). Although exact flow paths may be complex and tortuous, the net movements of **pore-fluids** will be upwards and out towards **basin** margins and structural highs along permeable carrier beds.

Mass solute transfer will, therefore, tend to take place from the potential *source sediment* (**mudrocks, evaporites, carbonates,** organic-rich sediments) to the porous **reservoir** sediments (**sandstones, carbonates**) via permeable conduits (carrier beds, **faults**). If, for any reason, this subsurface fluid movement is prevented, **overpressuring** develops as the fluid pressure exceeds the normal hydrostatic pressure. (See also **neoformation of clays, organic matter diagenesis**.) [SDB]

Land, L.S., Milliken, K.L. & McBride, E.F. (1987) Diagenetic evolution of Cenozoic sandstones, Gulf of Mexico Sedimentary Basin. *Sedimentary Geology* **50**: 195–225.

Magara, K. (1976) Water expulsion from clastic sediments during compaction — directions and volumes. *American Association of Petroleum Geologist Bulletin* **60**: 543–53.

Schmidt, V. & McDonald, D.A. (1979) The role of secondary porosity in the course of sandstone diagenesis. In: Scholle, P.A. & Schluger, P.R. (eds) *Aspects of Diagenesis*. Special Publication of the Society of Economic Palaeontologists and Mineralogists **26**: 175–207.

burial metamorphism A form of metamorphism characteristic of thick (*c.* 5–10 km) sedimentary/volcanic sequences which have experienced relatively rapid burial within sedimentary **basins** and/or *oceanic trenches*. The products of burial metamorphism do not show, or have relatively weak development of, metamorphic structures such as **penetrative deformation** (see **foliation**) and often preserve relict sedimentary/volcanic **fabrics** obscured by static metamorphic recrystallization. Such metamorphism is characterized by low T–P conditions (i.e. low **geothermal** gradient) and encompasses the transition from **diagenesis** through **zeolite** facies to higher grades of *blueschist/ glaucophane schist* metamorphism. [RST]

burrstone (*buhrstone*) Siliceous rock quarried at La Ferté-sous-Jouarre (Seine-et-Marne, France) for **millstones**, from at least the 16th century AD and perhaps earlier. [OWT]

burst Upward movement of low velocity fluid away from the lower regions of the turbulent **boundary layer**. These turbulent motions are characterized by a downstream velocity component less than the time-average and upward components of vertical velocity. These ejections of low momentum fluid are associated with cyclic burst–**sweep** events in **turbulent boundary layers** and increase in size and frequency with boundary roughness. Large burst events are associated with large scale form roughness and the onset of macroturbulence. Burst events have been speculated to provide the local vertical anisotropy in turbulence which enables the maintenance of sediment in **suspension**. (See **hydraulically smooth and rough**.) [JLB]

Jackson, R.G. (1976) Sedimentological and fluid dynamic implications of the turbulent bursting phenomenon in geophysical flows. *Journal of Fluid Mechanics* **77**: 531–60.

Bushveld Complex An enormous differentiated igneous complex in South Africa (Fig. B24), which is generally considered to have resulted from the repeated intrusion of two main **magma** types into partly overlapping, conical intrusions. It is one of the world's greatest storehouses of economic minerals and contains enormous reserves of **chromium, platinum,** other **platinum** group metals and **iron**. Additional products include **copper, nickel, gold, fluorite, tin** and **vanadium**. [AME]

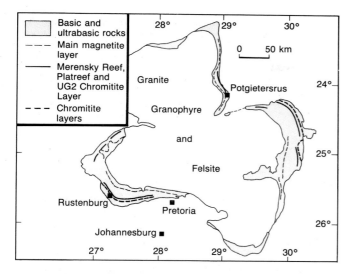

Fig. B24 Sketch map of the **Bushveld Complex**. (After van Gruenewaldt, 1977.)

van Gruenewaldt, G. (1977) The mineral resources of the Bushveld Complex. *Minerals Science and Engineering* **9**: 83–95.

bustamite ((Mn, Ca, Fe)SiO$_3$) A pink to brownish **pyroxenoid mineral**. [DJV]

butane (C$_4$H$_{10}$) A member of the paraffin series. A colourless gas with a faint odour. There are two isomers, *n*-butane and iso-butane. Both are present in **natural gas**. [AME]

butte A small, flat-topped hill isolated on an otherwise flat plain. The hill often has steep sides and it is usually capped with a resistant layer of rock which is commonly thought to represent a remnant of a former land surface. [NJM]

butte témoin A flat-topped **outlier**, situated on the scarp side of a **cuesta**, of which it was once a part. Most true **buttes** are butte témoins as témoin means 'witness' or 'evidence', in other words of the former extent of a land surface. [NJM]

bysmalith A large igneous intrusion or **plutonic** plug, roughly cylindrical in shape, which has been forced up into the overlying rocks. It arches up and **fractures** the overlying rock or becomes exposed at the surface. [NJM]

bytownite A **plagioclase feldspar** mineral. (See **feldspar minerals**.) [DJV]

C

cadmium (Cd) At. no. 48, at. wt. 112.41. This metal is used extensively in electroplating and in alloys. It is produced as a by-product from the smelting of **zinc**-bearing **ores**. [AME]

Caen stone A light, cream-coloured to yellow Jurassic **limestone** much used as a **building stone** in northern France and England. Type locality is Caen, Normandy. [AME]

Cairngorm stone A semi-precious form of smoky **quartz**. (See **gem**.) [DJV]

caking coal (*binding coal*) Any **coal** that softens, melts and agglomerates on heating and on quenching produces a hard cellular **coke**. Not all caking coals are good *coking coals*. [AME]

calamine The old name for secondary **zinc** minerals, notably **smithsonite**. [DJV]

calanque A coastal inlet which tends to be of a gorge-like form. They are widespread around the Mediterranean and may be **karstic dry valleys** which have been partially drowned, as a result of the **Flandrian transgression**. Their positions may be **fault** controlled. In Mallorca they are called *calas*. [ASG]

Nicod, J. (1951) Le problème de la classification des 'calanques' parmi les formes de côtes de submersion. *Revue de Geomorphologie Dynamique* **2**: 120–7.
Paskoff, R. & Sanlaville, P. (1978) Observations géomorphologiques sur les côtes de l'archipel Maltais. *Zeitschrift für Geomorphologie* NF **22**: 310–28.

calaverite ($AuTe_2$) A **gold ore** mineral. [DJV]

calc-alkaline Calc-alkaline rocks are characterized by higher concentrations of calcium (CaO) in relation to alkalis (Na_2O, K_2O) in comparison with **alkaline igneous rocks**. The term therefore describes **igneous rocks** in which the dominant **feldspar** is **plagioclase** rather than **alkali feldspar**. The term was originally defined in terms of the *alkali-lime index* (the SiO_2 value at which the CaO = $Na_2O + K_2O$ in a graph of these quantities against SiO_2 for a suite of related rocks) which enables rock associations to be classified as shown in Table C1.

Calc-alkaline and calcic associations define the **basalt**–basaltic **andesite**–**andesite**–**dacite**–**rhyolite** association, of which (calc-alkaline) **andesite** is the characteristic component. The association may be subdivided on the basis of SiO_2 and K_2O concentrations. This association occurs above destructive **plate margins** as in **island arcs** and **active continental margins** and may hence be termed the *orogenic **andesite** association*. This association shows characteristic variations in the relative abundances and compositions of the different components within these different settings. Certain **island arcs** are dominanted by **basaltic andesite** associations with minor **andesite**. The rocks have low K_2O, show trends of Fe-enrichment and are termed the **island arc tholeiite** association. Other **island arcs** and **active continental margins** are dominated by **andesite** and in general the proportion of more acid rocks (**andesite**, **dacite**, **rhyolite**) increases as the crustal thickness increases such that in continental margins with the thickest **crust** (central Andes) **basalts** are rare or absent. Prior to the development of **plate tectonic** *theory*, a large number of hypotheses was proposed to account for the petrogenesis of the orogenic **andesite** associations. The petrogenesis of such calc-alkaline rocks is now explained in terms of the descent of oceanic **lithosphere** at a destructive **plate margin**. This process is accompanied by metamorphism and dehydration of the descending **oceanic crust**. The hydrous fluids released on dehydration rise into the overlying **mantle** wedge causing **metasomatism** and *partial melting of the* **mantle peridotite** to yield **basaltic magmas**. During uprise these **magmas** experience **magma diversification** to yield **basalt**–**andesite**–**dacite**–**rhyolite** associations. [RST]

Gill, J.B. (1981) *Orogenic Andesites and Plate Tectonics*. Springer-Verlag, Berlin.
Thorpe, R.S. (ed.) (1982) *Andesites: Orogenic Andesites and Related Rocks*. John Wiley & Sons, Chichester.

calcareous algae Calcareous algae, including in common usage *cyanobacteria* (*blue-green algae*), are those in which $CaCO_3$ deposited as a result of life-processes provides a skeleton for the plant or for a discrete part of it. Calcareous algae are diverse and common in shallow marine and freshwater carbonate environments throughout the **Phanerozoic**. Their sedimentological importance in both reefal and bioclastic **limestones** is matched by their sensitivity as environmental indicators, and several groups also have **biostratigraphic** value.

Among modern marine algae, 11% of chlorophyte,

Table C1 Calc-alkaline rock associations

Alkali-lime index	Term for rock suite
<51	alkalic
51–56	alkali-calcic
56–61	calc-alkaline
>61	calcic

6% of rhodophyte and 2% of phaeophyte genera have calcified species. In freshwater, calcified species occur in all charophyte genera, 12% of *cyanobacteria* and 1% of chlorophyte genera. Reasons for calcification presumably include protection and support, but the fact that the majority of algae are uncalcified makes the adaptive significance of calcification problematic. In both algae and *cyanobacteria* calcification is closely associated with organic matter, mainly polysaccharides, and is favoured by photosynthetic uptake of CO_2 and HCO_3^-. Sites of calcification range from inside the cell wall (charophyte oosporangia), within the wall (coralline red algae), to outside the wall (udoteacean green algae and charophyte internodes) and within external mucilaginous envelopes (dasycladalean green algae and *cyanobacteria*). Modern freshwater algae and *cyanobacteria* **precipitate calcite**. Dasycladaleans and udoteaceans, together with nemalialean and squamariacean red algae, are **aragonitic**. Corallinacean red algae **precipitate** magnesian **calcite**.

There are six major groups of fossil calcareous red and green algae, plus *cyanobacteria* and the problematic *receptaculitids* (Fig. C1), and at least two additional heterogeneous groupings in the phylloid algae and *Solenoporaceae*.

CYANOBACTERIA. *Cyanobacteria* differ from algae in being prokaryotic, but they are nevertheless commonly known as *blue-green algae*. The mucilaginous sheath in *cyanobacteria*, which acts as the principal sediment-trapping agent in **stromatolite** formation, is also the site for crystal nucleation in calcification. Calcified filaments result in small tubes, which may be tangled, as in *Girvanella*, or form erect radiating masses, as in *Ortonella* and *Cayeuxia*. Comparison with modern analogues, usually in freshwater, allows precise interpretation of affinity. For example *Hedstroemia*, which is common in the Lower **Paleozoic**, is very similar to some Recent *Rivularia* species. These and other filamentous forms occur as layers and dendritic masses in *reefs*, and also construct **stromatolites** and **oncoids** (spherical **stromatolites**) with a distinctive tubiform skeletal microstructure which distinguishes them from the clotted fabric typical of **stromatolites** formed by sediment trapping.

Calcified coccoid *cyanobacteria* are less common, but may be represented by the problematic **Paleozoic** fossil *Renalcis* and its relatives. In the **Cambrian** *Renalcis* is associated with the dendritic *Epiphyton*, which may be a *cyanobacterium*, and the shrublike *Angulocellularia*, which is related to Recent *Schizothrix*. All three genera are individually very small, but locally are so abundant that they make massive *reef* **limestones** in association with *Archeocyatha*.

Calcified *cyanobacteria* are common fossils from the early **Cambrian** until the **Cretaceous**. **Cenozoic** calcified marine *cyanobacteria* are very rare. This may reflect a long-term change in seawater chemistry since cyanobacterial calcification appears to be environmentally dependent.

RHODOPHYTA. *Corallinaceae* is the most important group of calcareous red algae (*Rhodophyta*) and is exclusively marine. Modern **scleractinian coral** *reefs* commonly have a significant component of crustose corallines such as *Lithophyllum* and *Porolithon* particularly in the wave-swept *reef* crest. *Rhodoliths* (red algal nodules) are common from tropics to high latitudes. They may be up to 20 cm in diameter. Articulated corallines, such as Recent *Corallina* and *Jania*, have dendritic jointed thalli with calcified segments linked by uncalcified geniculae. Thus, these plants disarticulate after death.

Cellular tissue in both crustose and articulated corallines is generally divided into a basal or inner hypothallus and an external perithallus. Reproductive bodies are characteristically internal and occupy conceptacles in the perithallus. Variation in tissue, pit connections and sporangial conceptacle morphology provide the basis for classification into 40 crustose and 15 articulated genera, including both fossil and modern forms.

One of the earliest corallinaceans is *Archaeolithothamnium* (**Jurassic** to Recent), but the origins of the group have

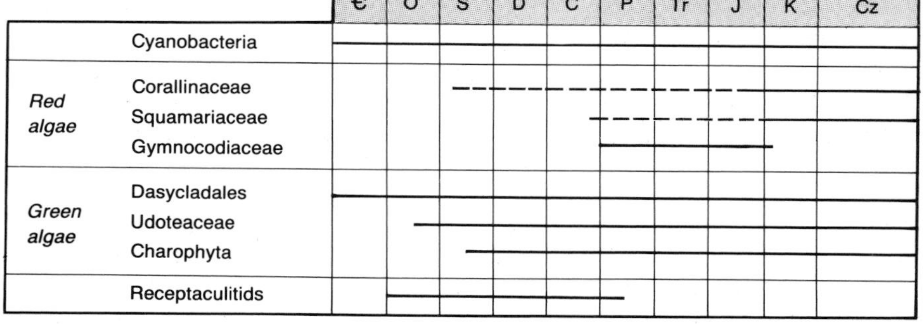

		€	O	S	D	C	P	Tr	J	K	Cz
	Cyanobacteria										
Red algae	Corallinaceae					–	–	–	–		
	Squamariaceae							–	–	–	–
	Gymnocodiaceae										
Green algae	Dasycladales										
	Udoteaceae										
	Charophyta										
	Receptaculitids										

Fig. C1 Calcareous algae. Geological distribution of major groups of calcified benthic algae and cyanobacteria. Solenoporaceae and phylloid algae are heterogeneous groupings of several of these. The affinities of the receptaculitids are uncertain.

been sought in so-called *ancestral corallines* of the **Carboniferous** and **Permian** such as *Archaeolithophyllum*. Some **Silurian** solenoporaceans have tissue closely comparable with coralline perithallus, but they lack hypothallus and conceptacles.

Squamariaceae are more weakly calcified and form thinner layers than crustose corallines, but this is a related family ranging from **Cretaceous** to Recent which may have representatives among late **Paleozoic** phylloid algae.

GYMNOCODIACEAE. *Gymnocodiaceae* range from **Permian** to **Cretaceous**. *Gymnocodium* and *Permocalculus* are the most common genera and are most conspicuous in the **Permian**. They pose a problem of affinity because their vegetative structure is broadly similar to that of green algae but they possess sporangia resembling those of red algae. Originally regarded as dasycladalean, and then udoteacean, green algae the *Gymnocodiaceae* are now generally compared with red algae such as *Galaxaura*, which is a weakly calcified member of the Nemaliales (*Corallinaceae* and *Squamariaceae* belong to the separate order Cryptonemiales).

DASYCLADALES. The *Dasycladales* typically are erect marine plants with a central stem cell from which branches radiate. Calcification is external and produces a crust on the surface of the stem cell through which lateral branches protrude. The resulting skeleton is thus a tube (stem cell) with pores (branches). It is readily distinguished from that of udoteacean green algae which possesses internal filaments. Classification of fossil forms relies mainly on details of branching style and the gametangia which may be preserved if the calcified layer is sufficiently thick. These fossils are usually fragmentary and it can be difficult to reconstruct the overall form of the original thallus.

Calcified dasycladaleans range from **Cambrian** to Recent and are most abundant and diverse in the **Mesozoic**. In most **Paleozoic** genera the lateral branches are arranged irregularly (aspondyl). Sporangia are not preserved and may have been borne in the stem cell (endospory). In the **Mesozoic** and **Cenozoic** the laterals are in whorls (euspondyl) and the sporangia are external. There are only some eight modern genera, such as *Acetabularia*, *Cymopolia* and *Neomeris*, and these occur in subtropical and tropical seas. Another Recent genus, *Bornetella*, is only weakly calcified, but it is comparable with **Ordovician** cyclocrinitid algae in having radial branches terminating in hexagonal plates which form a globular head on the central axis. The rather superficial similarity between cyclocrinitids and *receptaculitids* has been the main reason for supposing that the latter may be related to dasycladaleans.

UDOTEACEAE. The calcified *Udoteaceae*, previously often included in the Codiaceae, are characterized by internal medullary filaments which branch and curve outwards to form an exterior cortical layer. Typically they are segmented (*Halimeda*) or fanlike (*Udotea*). Some calcified forms, such as *Penicillus*, make a major contribution to

modern carbonate sedimentation but have a poor fossil record due to near total post-mortem disintegration. In addition to bushy or fanlike examples, phylloid forms such as *Anchicodium* have been described from the late **Paleozoic**. A further complication in the fossil record is that a variety of irregular filamentous forms, including *Cayeuxia*, *Garwoodia* and *Ortonella*, some of which are certainly *cyanobacteria*, have mistakenly been placed in the Codiaceae/*Udoteaceae*. Nevertheless, *Halimeda*-like forms (*Arabicodium* and *Boueina* from the **Mesozoic**) and the not very dissimilar **Paleozoic** *Dimorphosiphon* and *Litanaia* provide a line of development from the Ordovician onwards.

CHAROPHYTA. The *Charophyta* are relatively large, bushy plants with some features intermediate between those of algae and bryophytes. They typically occur in calcareous freshwater lakes and ponds, although they are also present in brackish water and have been reported from hypersaline environments too. Their specialized reproductive organs, which include heavily calcified oosporangia (although not all species calcify), together with cytological features, distinguish them from chlorophytes and they are now usually placed in a separate algal division, the *Charophyta*. They range from **Silurian** to Recent, but the details of this fossil record are based almost entirely only on the calcified oosporangia, also termed gyrogonites, which form millimetre-sized spherical to ovoid calcitic fossils with distinctive spirally-ribbed exteriors. The vegetative plant is relatively weakly calcified and only poorly preserved.

There are more than 100 genera of fossil and modern charophytes, with greatest diversity from the **Cretaceous** to **Oligocene** when they are biostratigraphically useful in lacustrine deposits. Recent examples include *Chara* and *Nitella*.

PHYLLOID ALGAE. *Phylloid*, or leaflike, *algae* are common in the **Carboniferous** and **Permian** where they are locally important in *reef* building. They are a heterogeneous group, probably including green algae and red algae, united by their flattened, often curved and crustose, morphology. Udoteacean-like phylloids include *Eugonophyllum*, *Anchicodium* and *Ivanovia*. Generally they are only a few centimetres across and less than 1 mm thick. They possess a filamentous interior with medulla and cortex. Phylloids which may be red algae include both *Archaeolithophyllum*, which has been regarded as an ancestral **coralline**, and squamariacean-like forms. These have cellular internal structure. Recrystallization of these fossils commonly destroys their distinguishing features and emphasizes their external morphological similarity.

SOLENOPORACEAE. The *Solenoporaceae* have recently been shown to be a heterogeneous group, not all of which are algae. Traditionally they have been regarded as **Cambrian** to early *Tertiary* red algae resembling corallinaceans but generally lacking tissue differentiation (into hypothallus and perithallus) and internal reproductive organs. However, it is probable that the group has included calcified *cyanobacteria* as well as red algae, and confusion is heightened by the likelihood that *Solenopora* itself is a *metazoan*. It was first described as a chaetetid and appears to be restricted to the **Ordovician**. Some **Silurian** solenoporaceans, in contrast, possess delicate cellular structure similar to the perithallus of corallinaceans and it is possible that these may be the ancestors of coralline red algae. In the **Mesozoic**, similar, but larger-celled, forms such as *S. jurassica* may also be related to corallines. Locally these have a pink colour suggested to be original pigmentation, but analyses have not yet confirmed this. From the **Cambrian** onwards another group of solenoporaceans occurs which lacks cross-partitions in their filaments. These more closely resemble *cyanobacteria* than red algae. Hence the *Solenoporaceae* as it has traditionally been recognized is a very varied group of superficially similar, but fundamentally different, organisms. Nevertheless it remains probable that some of these are ancestors of the *Corallineaceae*.

RECEPTACULITIDS. Receptaculitales/Receptaculitida are variously regarded as a group of green algae, **sponges**, or a distinct phylum. They range from **Ordovician** to **Permian**, but are most common in the middle **Paleozoic**. They have a globular or pear-shaped form with internal radiating rods (rays) terminating distally in plate-like facets, which are united in an outer wall lacking pores but with an opening at the top. At its broadest the group also includes genera allied with *Amphispongia*, *Calathium*, and *Cyclocrinites*, the latter generally being regarded as a dasycladalean. However, the *receptaculitids* may better be regarded as a problematic group, allied with calathiids and radiocyathids, which is morphologically approximately intermediate between the sponge-like *Archaeocyatha* and the algal Cyclocrineae, without necessarily being closely related to either.

PROBLEMATICA. Numerous problematic fossils, apart from *receptaculitids*, have been placed with questionable justification in the calcareous algae. These are usually morphologically simple forms which have a skeletal organization broadly similar not only to algae but also to some protozoans and lower invertebrates, such as **foraminifers** and **sponges**. Others, such as solenopores and *phylloid algae*, are heterogeneous groupings of a variety of algae.

Much remains to be learned about these organisms and about ancient and modern calcareous algae and *cyanobacteria* in general. They have considerable potential as indicators of depth, salinity and temperature in paleo-environmental studies and it is likely that variations in their diversity through time, when better understood, will help elucidate long-term changes in global conditions. [RR]

Tappan, H. (1980) *The Paleobiology of Plant Protists*. W.H. Freeman, San Francisco.
Wray, J.L. (1977) *Calcareous Algae*. Developments in Paleontology and Stratigraphy 4. Elsevier, Amsterdam.

calcareous ooze (*carbonate ooze*) A **pelagic** biogenic sediment composed of the skeletal remains of calcareous microfossils such as *coccolithoporids*, **foraminifers** or *pteropods*, e.g. **coccolith** ooze. Calcareous ooze covers about half of the ocean floor (one quarter of the Earth's surface) above the **calcite/carbonate compensation depth**. [AESK]

calcite ($CaCO_3$) The most common **carbonate mineral**, the main component of **limestone**. [DJV]

calcite/carbonate compensation depth (CCD) The level in the ocean at which the rate of **dissolution** of calcium carbonate equals the rate of supply. Above this depth **calcareous ooze** is the predominant sediment, largely composed of **foraminiferal** and **coccolith** remains, but is replaced by **red clay** or **siliceous ooze** below it. The presence of the CCD is due to the increasing **dissolution** of **calcite** with increasing pressure and decreasing temperature (i.e. depth). However, the factors governing the depth of the CCD in the oceans are complex and depend on an interaction between productivity, (calcium carbonate) flux rates, bottom water temperature and chemistry (mainly CO_2 content), sedimentation rates, benthic activity and **pore-water** chemistry. The most fundamental control is probably the mass balance between the utilization of calcium carbonate in surface productivity and its **dissolution** at depth.

The CCD varies widely in time and space, but tends to parallel depth contours in any given location and possesses a variation in relief of over 2 km, from less than 3 km in depth in high latitudes to over 5 km in low latitudes at the centre of the ocean basins. The relatively shallow depth of the CCD in high latitudes is due to the colder and more **corrosive** bottom waters of these areas. Elevation of the CCD near continental margins is related to the higher productivity giving rise to more CO_2-rich bottom waters. The most marked regional variation in the depth of the CCD is between the Pacific (4.2–4.5 km) and the Atlantic (5 km) and is related to circulation of waters between the two oceans. Bottom waters formed in the Atlantic (**thermohaline circulation**) are exported to the Pacific which in turn exports surface waters to the Atlantic (Berger, 1976). Pacific bottom waters are therefore generally older, more enriched in CO_2 and hence more corrosive to calcareous tests resulting in a higher CCD. Thus the Atlantic is dominated by **calcareous ooze** while the Pacific has abundant **red clay** and **siliceous ooze**.

An analogous level to the CCD is the *aragonite compensation depth* (ACD) which marks the level of disappearance of *pteropods* which have **aragonite** tests. The ACD is generally substantially shallower than the CCD. Another commonly used level concerning calcium carbonate preservation on the ocean floor is the **lysocline**, which is a measure of the preservation of microfossil assemblages and lies above the CCD. Variation in depth of the CCD through time has been extensively documented by **deep sea drilling**. [AESK/DMP]

Fig. C2 Calcrete. (a) The indurated horizon above the hammer is a 'hard-pan' calcrete or petrocalcic horizon. It is a very mature Pleistocene calcrete, which is overlain by a rubble of reworked calcrete. Coastal area south of Adelaide, South Australia. (b) Jurassic nodular calcrete overlain by fluvial channel sandstone (base shows loading features). The nodules are of calcite and represent paleosol nodules. Upper Jurassic, Porto Novo, near Torres Vedras, Portugal.

Berger, W.H. (1976) Biogenous deep sea sediments. In: Riley, J.P. & Chester, R. (eds) *Chemical Oceanography* Vol. 5 (2nd edn), pp. 260–338. Academic Press, New York.

Berger, W.H. & Winterer, E.L. (1974) Plate stratigraphy and the carbonate line. In: Hsü, K.J. & Jenkyns, H.C. (eds) *Pelagic Sediments: on Land and Under the Sea*. Special Publications of the International Association of Sedimentologists No. 1. Blackwell Scientific Publications, Oxford.

Kennett, J.P. (1982) *Marine Geology*. Prentice-Hall, Englewood Cliffs, NJ.

Goudie, A.S. (1983) Calcrete. In: Goudie, A.S. & Pye, K. (eds) *Chemical Sediments and Geomorphology*. Academic Press, London.

Netterberg, F. (1980) Geology of southern African calcretes — 1: Terminology, description, macrofeatures and classification. *Transactions of the Geological Society of South Africa* **83**: 255–83.

calcrete A term used for near-surface terrestrial material composed dominantly but not exclusively of calcium carbonate which occurs in a variety of forms from powdery and nodular to highly indurated. It results from **cementation** and the displacive and replacive introduction of **calcite** into *soils*, indurated rock and sediment mainly in arid and semi-arid regions where vadose and shallow **groundwaters** become saturated with respect to calcium carbonate. Calcretes are not restricted to *soils* (*pedogenic calcrete*) where much of the **calcite precipitation** is biologically induced, but extensive calcretization also occurs in the capillary fringe just above or below the **water-table** to form **groundwater** or phreatic calcretes. The term is not used to describe **cave** deposits (*speleothem*), *tufas*, **travertines** or **beachrock**. It is effectively synonomous with the term **caliche**. (See **duricrusts**.) (See Fig. C2.) [VPW]

caldera A depression in the Earth's surface formed by collapse into an underlying *magma chamber*. Collapse occurs during, and in response to, removal of **magma** from the *magma chamber*, either by its eruption or by its migration to another part of the volcano's plumbing system. Calderas are usually approximately circular in plan and have diameters greater than about 1.5 km. That is, they have diameters at least three times the average diameter of **craters** formed by volcanic explosions (see **volcanic eruption**), although some **maars** have diameters up to 3 km (Wood in Lipman *et al.*, 1984). The structure underlying a caldera — the subsided block, the associated **ring dykes** (if any) and *magma chamber* — is called a *cauldron*.

Two basic types of caldera can be distinguished. The first type, called shield calderas, occur on the summits of basaltic **shield volcanoes**. These calderas are up to 20 km in diameter and the collapse takes place in response to draining of **magma** from the underlying *magma chambers* into a fissure. The drained **magma** may erupt as a **fissure eruption** on the flanks of the **shield volcano**. Shield calderas are exemplified by the large **shield volcanoes** of Hawaii which experienced frequent caldera collapse during their growth.

In the second type of caldera (**ash flow** calderas), collapse occurs in response to eruption of **magma** from the underlying *magma chamber*, from **volcanic vents** situated within, or very close to, the caldera. The erupted **magma** is usually of intermediate or silicic composition and is always erupted as **tephra**, usually to form **ash flow tuffs**. **Ash flow** calderas occur in a range of volcanic settings, from the summits of 'stratovolcanoes' (i.e. volcanoes of dominantly **andesitic** composition, which consist of layers of **lava** and **tephra** increasing in thickness towards a central vent), to within initially essentially flat lying volcanic fields consisting of **lava flows** and **domes** of silicic, intermediate and mafic composition. The size of **ash flow** calderas ranges from about 3 km to about 45 km, although two very large **ash flow** calderas, one at Yellowstone, Wyoming and the other at Toba, Sumatra, have diameters of 70–100 km; related **pyroclastic** deposits have volumes exceeding 1000 km^3. Caldera collapse is thought to occur either as coherent subsidence of a piston-like block, by chaotic piecemeal foundering of many blocks, by funnel-shaped collapse into a central *cauldron* much smaller than the caldera depression, or by sagging. Classification of **ash flow** calderas has been based largely on the nature of the subsided cauldron blocks (e.g. Smith & Bailey, 1968; Macdonald, 1972; Williams & McBirney, 1979) but these are never visible in young calderas. Some models for catastrophic caldera collapse and **ash flow** eruption require outward-dipping ring **faults**, but most observed ring **faults** are vertical or inward-dipping (Lipman in Lipman *et al.*, 1984). Many large calderas experienced post-caldera resurgence — up-doming or up-faulting of a central part of the caldera (Smith & Bailey, 1968). Post-caldera eruptions of **magma** broadly similar in composition to the caldera-forming **ash flows** are common both outside and within many calderas.

Ash flows erupted during caldera formation can be of **batholithic** dimensions (Smith, 1979) and represent an instantaneous 'quench' sample of the underlying *magma chamber*. Many, but not all, caldera-forming **ash flows** are chemically zoned. This potentially provides useful information on *magma chamber* processes, especially when the magma chemistry of pre- and post-caldera lavas and **pyroclastic** deposits are also studied. (See **ignimbrite, plateau lava, shield volcano**). [PTL]

Lipman, P.W., Self, S. & Heiken, G. (eds) (1984) Calderas and associated igneous rocks. *Journal of Geophysical Research* **89**: 8219–841.
Macdonald, G.A. (1972) *Volcanoes*. Prentice-Hall, Englewood Cliffs:, NJ.
Smith, R.L. (1979) Ash-flow magmatism. In: Chapin, C.E. & Elston, W.E. (eds) *Ash Flow Tuffs*. Special Paper of the Geological Society of America **180**: 5–27.
Smith, R.L. & Bailey, R.A. (1968) Resurgent cauldrons. *Geological Society of America Memoir* **116**: 613–62.
Williams, H. & McBirney, A.R. (1979) *Volcanology*. Freeman, Cooper & Co, San Francisco.

caliche (**1**) A term used to describe the nitrate deposits of the hyper-arid coastal desert of the Atacama in South America.
(**2**) The calcium carbonate **duricrusts** (**calcretes**) of many other desert areas. [ASG]

californite A **gem** substitute for **jade** comprised of compact green **vesuvianite**. [DJV]

calläis Term used by archeologists for stone used to make small blue or green beads found in Neolithic and Copper Age contexts in western Europe. Chemical analysis shows that 'calläis' beads are mostly of **variscite** and **turquoise** and occasionally of **malachite**. [OWT]

cambering The result of warping and sagging of relatively **competent** rock strata which overlie beds of less **competent** strata, notably clay. The plastic nature of the clays causes the overlying rocks to flow towards adjacent valleys, producing a convex top to valley-side slopes. Cambering is often associated with the development of **valley bulges** and seems to have been accelerated under extreme *Pleistocene periglacial* conditions. [ASG]

Cambrian The oldest group of rocks which contain relatively common shelled fossils and as such is the first system in the **Phanerozoic**. The age range of the system is still very uncertain. Estimates of the base range from 580 to 530 Ma and those for the top from 520 to 505 Ma (Fig. C3).

The Cambrian System takes its name from Cambria, the old name for Wales. 'Cambrian' was first used in a joint communication to the British Association in 1835 by Adam Sedgwick and R.I. Murchison. Sedgwick referred particularly to the rocks between Anglesey and Berwyn Hills, but also to rocks in Cumberland. The base of the Cambrian was never clearly defined and the top was a matter of dispute until Lapworth defined the **Ordovician** System in 1879. The definition of the upper and lower boundaries of the Cambrian is currently under review by the International Commission on Stratigraphy. It is likely that the base of the Cambrian will be defined at or near the base of the *Tommotian Stage*, where a distinctive suite of small shelly fossils make their first appearance. Suitable candidates for a type section occur at Meishucun, Yunnan, China, and the Aldan River, east Siberia. Other important sections through the *Precambrian–Cambrian boundary* occur in Newfoundland, Iran and India.

Traditionally, in Britain, the top of the Cambrian has been taken as the top of the *Tremadoc*, but international opinion generally favours the placing of the *Cambro-*

Cambrian System

Period	Series — Epoch	British Stages — Age	British Biozones		Ma	Siberia	China	Australia	N. America
O	Tremadoc		Dictyonema flabelliforme		505	Olentian (Kazakhstan)		Datsonian	Trempealeauan
Cambrian (Є)	Merioneth (Late Cambrian)	Dolgellian	Acerocare			Shidertinian	Fengshan	Payntonian	Franconian (Croixian)
			Peltura scarabaeoides / Peltura minor / Protopeltura praecursor				Changshan	'Pre-Payntonian'	
			Leptoplastus						
			Parabolina spinulosa					'Post-Idamean'	
		Maentwrogian	Olenus & Agnostus obesus			Tuorian	Gushan	Idamean	Dresbachian
			Agnostus pisiformis		525			Mindyallan	
	St David's (Middle Cambrian)	Menevian	Lejopyge laevigata	*Paradoxides forchhammeri*		Mayan	Zhangxia	Albertian	
			Solenopleura brachymetopa						
			Ptychagnostus lundgreni & P.(G) nathorsti						
			Ptychagnostus punctuosus	*Paradoxides paradoxissimus*			Xuzhuang		
			Hypagnostus parvifrons						
			Tomagnostus fissus & Ptychagnostus atavus						
		Solvan	Ptychagnostus gibbus			Amgan			
			Eccaparadoxides oelandicus pinus	*E. oelandicus*			Maozhuang	Templetonian	
			Eccaparadoxides insularis		540				
	Comley (Early Cambrian)	Lenian	Anabaraspis / Lermontovia dzevanovskii & Paramicmacca / Bergeroniellus expansus / Bergeroniellus micmacciformis			Lenan	Longwangmiao	Ordian	
		Atdabanian	Judomia & Dipharus attleborensis			Petrotsvet	Canglangpu	'Lower Cambrian'	Waucoban
			Judomia (& Fallotaspis)						
		Tommotian	Dokidocyathus lenicus						
			L. bella — Dokidocyathus regularis				Qiongzhusi		
			L. tortuosa						
			Ajacicyathus sunnaginicus		530				
PЄ	Ediacaran	Poundian	Anabarites trisulcatus		590?	Yudoma			

Fig. C3 Stratigraphic table showing the division of the **Cambrian** System into Series, Stages and Biozones in Britain and a selection of stratigraphic sequences of stages elsewhere.

Ordovician boundary at the base of the *Tremadoc*, and it seems likely that this stratigraphic level will be given formal recognition.

In many countries the Cambrian is divided into three series based on sections in Britain; in ascending order the *Comley Series*, the *St Davids Series* and the *Merioneth Series*. The lower part of the *Comley Series* is partly represented in the Welsh Borderlands but is better known from sections in Siberia where the *Tommotian Stage* and overlying *Atdabanian Stage* are defined. Although the Welsh Series names are internationally recognized it is not always easy to apply them in practice. In most regions of the world the Cambrian is divided into a Lower, Middle and Upper Cambrian but usages differ in different countries and local names are applied in some areas.

The **biostratigraphic** subdivision of the Cambrian, above the *Tommotian Stage*, is effected using the *trilobite* faunas. Correlation of *trilobite* zones is commonly possible on a regional scale and sometimes on an intercontinental scale. The provincialism of *trilobite* faunas usually prevents world-wide correlation.

Although *Metazoa* appeared probably about 150 Ma before the start of the Cambrian, there was a major *metazoan* radiation at the start of the period. The *Tommotian Stage* is characterized by a large variety of small conical and tubular fossils, many of them of uncertain taxonomic status,

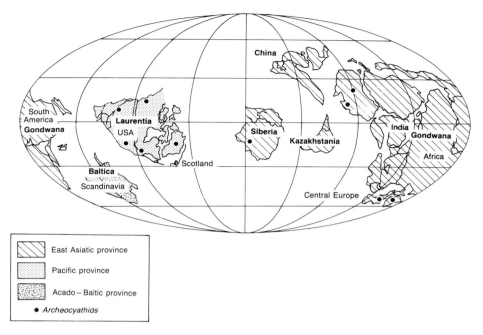

Fig. C4 A possible reconstruction of continental distribution in the late **Cambrian** and the main biogeographic provinces determined principally on the distribution of trilobites. (After Scotese *et al.*, 1979.)

composed of phosphate (e.g. *hyolithelminthids*, *gastropods* and *tommotiids*), together with calcareous *hyolithids*, *archeocyathids*, **sponges** and a variety of **trace fossils**. *Trilobites*, **echinoderms** and small **brachiopods** appear in the succeeding *Atdabanian Stage*. The acquisition of a mineralized skeleton, which could be formed of phosphate, **calcite** or **aragonite**, or an agglutinated skeleton (composed of detrital particles) evolved independently in several different groups in this early Cambrian evolutionary phase.

The fossils most commonly found in the main part of the Cambrian are distinctly different from many of those found later in the **Paleozoic** and commonly include *trilobites*, inarticulate **brachiopods**, *archeocyathids* (mainly lower Cambrian), **sponges**, *monoplacophorans* and *hyolithids*. The diversity of the faunas was generally low. However the Middle Cambrian *Burgess Shale* of British Columbia (Canada) preserves a mainly soft-bodied fauna of approximately 100 genera belonging to 12 major groups, suggesting Cambrian faunas were probably more taxonomically diverse and ecologically varied than they usually appear. Most Cambrian fossils come from shelf deposits and it appears that deep water and **pelagic** habitats were only colonized by degrees through the Cambrian. Many elements of the Cambrian fauna became extinct during the period, to be succeeded by a new and distinct fauna in the **Ordovician**.

The distribution of continents in the Cambrian was determined by the break-up of one *supercontinent* in the very late *Precambrian*. The newly formed blocks consisted of a large **Gondwanaland** continent, **Laurentia**, *Baltica* and three continents, Siberia, *Kazakhstania* and China, which probably separated progressively during the Cambrian (Fig. C4). All the continents apart from *Baltica* lay partly in equatorial latitudes.

The paleobiogeography of the Cambrian faunas partly reflect the continental distribution of the time but were partly influenced by climate and depositional environment. There is no agreement about the number of Cambrian faunal provinces but many authors recognize an Acado–Baltic province and a Pacific province which together comprise an Olenellid Realm in the Lower Cambrian, and an East Asiatic province or Redlichiid Realm (Fig. C4). There are also some distinctive features of the faunas of Siberia, China and *Kazakhstania* which might be sufficient to distinguish them as separate provinces. The faunas of the Pacific province and Acado–Baltic province are so different that they were probably geographically isolated from one another. In contrast there seems to be a gradual transition from the East Asiatic influence in Morocco to the pure Acado–Baltic faunas of Scandinavia, suggesting there was complete biological interconnection. Likewise there was some faunal mixing between the Siberian and Pacific provinces and China and the Australian and Antarctic part of **Gondwanaland**.

The broad character of Cambrian sediments was determined partly by sea-level, partly by continental distribution and partly by regional tectonics. Sea-level started to rise in very late *Precambrian*, probably as a result of **continental splitting** and the formation of new *ocean ridge* systems.

The sea transgressed across many continental platforms at the start of the Cambrian period so that in many places Lower Cambrian shallow marine sediments, often **quartzites**, are found to lie **unconformably** on *Precambrian* rocks. It is mainly in the **basins** that there are continuous sequences through from the *Precambrian* into the Cambrian, though some platforms, such as that in Siberia, were continuously covered by sea from the late *Precambrian* onwards. Sea-level rose throughout most of the Cambrian period, though there may have been some brief **eustatic** falls.

The large continental regions and their **passive margins** which lay within the tropics during the Cambrian were either emergent or mainly covered by shallow *epeiric seas* in which carbonates accumulated. Consequently, much of the Cambrian of the USA, Arctic Canada, Greenland, and Siberia consists of **limestones** and **dolomites**. Dominantly **sandstone** and **shale** sequences are mainly confined to the marginal basins of these continents. In contrast the Baltic Platform, which lay outside the tropics, has a dominantly clastic cover, with a distinctive **black shale** unit, the Alum Shales, in the upper Cambrian.

On the active edges of continental regions thick sequences of **shales** and *turbidites* accumulated in *trenches*, **forearc**, **back-arc** and other marginal **basins** in a **subductive** regime. Thick *turbidite* sequences of Cambrian age are known, amongst other places, from the Welsh Basin, the Leinster Basin in Ireland, parts of Quebec Province and Newfoundland. Cambrian sequences of 'geosynclinal' aspect are also reported from the Cordillera Oriental of Columbia, southern Bolivia and northern Argentina.

Most Cambrian sequences pass up conformably into the **Ordovician**. However there was a major episode of **deformation** of the Cambrian in Scotland (the *Grampian event*) and less pronounced **deformation** in parts of Iberia in late Cambrian or early **Ordovician** times. [PJB]

Burrett, C. & Richardson, R. (1980) Trilobite biogeography and Cambrian tectonic models. *Tectonophysics* **63**: 155–92.
Conway Morris, S. (1986) The community structure of the Middle Cambrian Phyllopod Bed (Burgess Shale). *Palaeontology* **29**: 423–67.
Glaessner, M.F. (1984) *The Dawn of Animal Life — A Biohistorical Study*. Cambridge University Press, Cambridge.
Holland, C.H. (ed.) (1971) *Cambrian of the New World*. Wiley Interscience, New York.
Holland, C.H. (ed.) (1974) *Cambrian of the British Isles, Norden, and Spitsbergen*. John Wiley & Sons, Chichester.
Palmer, A.R. (1977) Biostratigraphy of the Cambrian System — a progress report. *Annual Review of Earth and Planetary Science* **5**: 13–33.
Rodgers, J. (ed.) (1956) *El Sistema Cambrico, su paleogeografia y el Problema de su Base-Symposium. 20th International Geology Congress, Mexico* (2 parts).
Scotese, C.R., Bambach, R.K., Barton, C., Van der Voo, R. & Ziegler, A.M. (1979) Paleozoic base maps. *Journal of Geology* **87**: 217–77.

camptonite An alkali **lamprophyre** composed of *phenocrysts* of **amphibole** (*barkevikite* and/or *kaersutite*), clinopyroxene (**augite**), **olivine** and/or **biotite/phlogopite** in a groundmass of Ca-rich **plagioclase**, **amphibole** and pyroxene with subordinate **alkali feldspar (plagioclase > alkali feldspar)**, **feldspathoids**, **apatite**, Fe–Ti oxides and **carbonate minerals**. [RST]

cancrinite ($Na_6Ca(CO_3)(AlSiO_4)_6 . 2H_2O$) A rare **feldspathoid mineral**. [DJV]

cap rock (1) A virtually impermeable rock normally lying above a gas or oil **reservoir** and sealing in the *hydrocarbons*. Often called the *roof rock* or **seal**. Clay, **shale** and **evaporite** are the most common cap rocks.
(2) The secondary sheath around and over the tops of **salt domes**. It commonly consists of gypsiferous or **anhydrite-bearing limestone**. [AME]

North, F.K. (1985) *Petroleum Geology*. Allen & Unwin, Boston.

carat (c) The unit of weight of **diamonds**. This was formerly defined as 3.17 grains (avoirdupoids or troy) but the international (metric) carat is now standardized as 0.2 g. This carat should not be confused with that used to state the number of parts of **gold** in 24 parts of an alloy (usual American spelling *karat*). [AME]

carbon (C) At. no. 6, at. wt. 12.011. Occurs in both the free (**diamond**, **graphite**) and combined states (**carbonate minerals**). [AME]

carbona A Cornish term for large masses of rich **tin ore**. A typical carbona was about 30 m long by 10 m high and wide. [AME]

carbonado A black, cryptocrystalline mixture of **diamond** with **graphite** or amorphous **carbon**. This is one of the varieties of **industrial diamond**. [AME/DJV]

carbonate buildup Commonly used to describe a *reef* complex in the geological record. However, it can be used, in the broader sense, to refer to any large stratigraphic accumulations of **limestones** and **dolomites**.

In recent years there has been much interest in the study of the large scale aspects of carbonate buildups, such as their geometries and internal architectures, and on their major controlling factors such as **tectonics** and sea-level changes. Three main types of large-scale geotectonic settings are recognized for carbonate buildups, each with distinctive sedimentary associations allowing their recognition in ancient **limestones**. These three settings are *platforms*, *shelves* and *ramps* (Fig. C5).

Carbonate platforms are extensive, flat-topped areas from hundreds to thousands of kilometres wide. They are covered by shallow water and correspond to the term *epeiric sea*. Such settings developed during the drowning of **cratonic** areas and best developed during phases of high

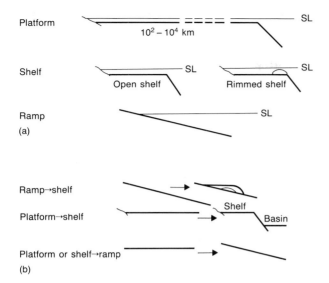

Fig. C5 Carbonate buildup. (a) The three major depositional settings of shallow marine carbonates: platform, shelf and ramp. (b) The common transitions from one to the other. (After Tucker, 1985.)

absolute sea-level such as the **Cambrian** and **Ordovician**, although other major platforms developed during the **Devonian** and the **Triassic**. No large platforms occur at present, but the isolated platforms of the **Bahama Banks** provide analogues. The transition of platforms into adjacent **basins** may be via gentle (ramp-like) or steep margins. Ancient platform deposits are recognized because of the occurrence of areally very extensive, shallow-water deposits.

A *carbonate shelf* resembles a platform but is less extensive, being from a few to hundreds of kilometres wide. They are flat-topped with steep slopes into the **basin** and the shelf edge may be either bordered by a rim of *reefs* or **sand** shoals (rimmed shelf) or open to oceanic influences (open shelf). The east Florida Shelf is an example of a rimmed shelf. A problem arises in the use of the term shelf in relation to its other wider usage in the sedimentology of siliciclastic deposits. On non-carbonate shelves rims are absent and there is usually sufficient slope from the shoreline to the shelf edge for a variety of depth (and energy) related zones to be recognized. The classical carbonate shelf has virtually no gradient and is protected by a shoal or *reef* rim with a shallow, low energy, shelf *lagoon* in its lee. Thus carbonate shelf environments differ radically from those of non-carbonate shelves. Shelf **limestone** sequences are recognized in the geological record by the occurrence of relatively extensive, low energy, shallow water deposits associated with *reef* or **sand**-grade **limestones**, and by the association of slope or slope-related deposits on the basinward side.

A *carbonate ramp* is a gently sloping surface (slope of less than 1°) that passes into progressively deeper water basinward. If the slope to the **basin** is continuous and even, it is said to be a homoclinal ramp (e.g. the southern

coast of the **Persian Gulf**). Distally steepened ramps also show a low gradient topography but steepen basinward with a significant slope into the **basin**. This change, or break, in slope occurs in deeper water and so is not defined by a rim of shallow-water *reefs* or **sand** shoals. The Yucatan 'shelf' has been interpreted as a modern example. Ramp sequences are recognized in the geological record by having shoal deposits developed in a nearshore setting where wave base impinges on the seafloor, and not offshore at a shelf-break. They also exhibit a distinctive depth-related facies distribution. Essentially, ramps resemble non-carbonate shelves and it is possible to recognize **wave**, **tide** or *storm* dominated ramp sequences in the geological record.

These settings are subject to influences such as sediment **accretion**, sea-level changes and **tectonics**. Transformations between types are commonly seen with ramps evolving to shelves being a common trend. Tilting in extensional regimes can cause platforms to break up and shelves and ramps to form. [VPW]

Read, J.F. (1982) Carbonate platforms of passive (extensional) continental margins: types, characteristics and evolution. *Tectonophysics* **81**: 195–212.
Read, J.F. (1985) Carbonate platform facies models. *American Association of Petroleum Geologists Bulletin* **69**: 1–20.
Tucker, M.E. (1985) Shallow-marine carbonate facies and facies models. In: Brenchley, P.J. & Williams, B.P.J. (eds) *Sedimentology: Recent Developments and Applied Aspects*, pp. 147–69. Geological Society Special Publication No. 18. Blackwell Scientific Publications, Oxford.

carbonate minerals A common group of minerals in which the essential structural unit is the $(CO_3)^{2-}$ ion. Although there are some 60 known carbonate minerals, many of them are comparatively rare and some of the less common species are hydrated, contain hydroxyl or halogen ions, or are compounds with silicate, sulphate or phosphate radicals. The more common rock-forming carbonate minerals which are dealt with here are **calcite**, **magnesite**, **rhodochrosite**, **siderite**, **dolomite** and **aragonite**.

Calcium carbonate is **polymorphous** and exists in at least five modifications. The two **polymorphs** commonly found in nature are **calcite** and **aragonite**; *vaterite* is a metastable form.

Calcite ($CaCO_3$) is one of the most ubiquitous minerals; in addition to being an important rock-forming mineral in sedimentary environments, it also occurs in metamorphic and igneous rocks and is a common mineral of hydrothermal and secondary mineralization. The structure of **calcite** can be described by analogy with the cubic structure of **halite**: Na and Cl ions are replaced by Ca and $(CO_3)^{2-}$ ions respectively and the unit cell is distorted by compression along a triad axis to give a face-centred rhombohedral cell. The distortion of the cube is necessary to accommodate the large planar CO_3 groups which contain a carbon atom at the centre of an equilateral triangle of oxygens (Fig. C6). **Calcite** is commonly colourless or white but sometimes grey, yellow, or light shades of pink,

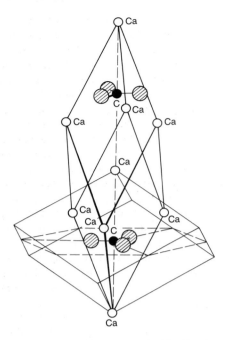

Fig. C6 Carbonate minerals. The structure of calcite. The elongated cell is the true rhombohedral unit cell (Z = 2); the cleavage rhombohedron cell corresponding with a face-centred rhombohedron is also shown. (After Ewald & Hermann, 1931.)

green or blue. It has a perfect rhombohedral **cleavage**, Mohs' hardness 3, specific gravity 2.72 and shows brisk effervescence in cold, dilute HCl. Optically, its refractive indices are ε 1.486, ω 1.658, giving it a birefringence of 0.172 and allowing the transparent rhombs sometimes found (**Iceland spar**) to show double images. Chemically it is generally fairly pure $CaCO_3$ but magnesian **calcites** are important constituents in biological materials.

The **calcite–aragonite polymorphism** is brought about by the fact that the radius of the calcium ion is very close to the limiting value for the transition from the rhombohedral carbonate structure type to that of the orthorhombic carbonates. **Calcite** is the low-pressure **polymorph**; it can be transformed to **aragonite** by grinding at room temperature for a few hours. Experimental work shows that **calcite** melts incongruently to liquid and vapour at 1310 ± 10°C at 100 MPa pressure, though in the presence of water vapour at this pressure, **calcite** begins to melt at 740°C.

In sedimentary rocks **calcite** is the principal constituent of most **limestones**. It occurs both as a primary precipitate and in the form of fossil shells. **Calcite** is the stable form of $CaCO_3$ and although approximately equal numbers of organisms make their shells of **calcite** or **aragonite** (or, as for some of the Mollusca, of both), the **aragonite** eventually undergoes recrystallization to **calcite**. Low-temperature magnesian **calcites** are particularly common in calcareous **sponges**, **echinoids**, *crinoids*, **brachiopods** and **calcareous algae**; they also occur in marine **calcite**

cements and in some freshwater *tufas*. Precipitated **calcite** in sediments occurs in the **calcrete** of surface **limestones**, and in **travertine** deposits where, in **limestone** regions, underground streams may carry considerable quantities of calcium in solution as calcium bicarbonate. On reaching the open air, the rise in temperature brings about the release of CO_2 and the growth of mosses and reeds at the exit may extract further CO_2 from solution, leading to the **precipitation** of $CaCO_3$ as **calcite**. The **precipitation** of $CaCO_3$ in freshwater lake marls and in marine conditions is generally in the form of **aragonite** (see later); in time, however, the **aragonite** mudstones or **aragonite** *ooliths* generally recrystallize to **calcite** mudstones or to an *oolitic* rock with calcite *ooliths*. **Calcite** also occurs as a cementing medium in sedimentary rocks, both in **limestones** and **sandstones**. When sedimentary **calcite** undergoes metamorphism it recrystallizes to form a **marble**. If the original **limestone** contained other materials in addition to $CaCO_3$, the **calcite** may react to give various mineral assemblages; e.g. in the presence of **quartz** the reaction $CaCO_3 + SiO_2 \rightleftharpoons CaSiO_3$ (**wollastonite**) + CO_2 may take place. **Calcite** frequently crystallizes in the later stages of hydrothermal mineralization, where it occurs in **veins** and cavities. Well-formed crystals may be found in **amygdales** in basic igneous rocks, where they are often associated with **zeolites** or with **quartz**. **Calcite** of primary igneous origin is found in **carbonatites**.

Magnesite ($MgCO_3$) is a trigonal carbonate, which generally occurs as white, finely-crystalline masses but may be yellow with the substitution of minor amounts of iron for magnesium. It dissolves with effervescence in warm dilute HCl. The most common occurrence of **magnesite** is as an **alteration** product of various **magnesium**-rich igneous and metamorphic rocks, e.g. **serpentinites**. It also occurs in **evaporite deposits** and as the result of Mg **metasomatism** of sediments.

Rhodochrosite ($MnCO_3$) is trigonal and typically forms rose-pink or pink crystals. It is often locally altered to brown or black oxides of **manganese**. It is typical of high-temperature **metasomatic** deposits, associated with other **manganese** minerals such as **rhodonite**, **spessartine**, etc. It is also known from hydrothermal **vein** deposits, and in the Lower **Cambrian manganese ore** of North Wales it is found in metamorphosed sediments.

Siderite ($FeCO_3$) is a trigonal carbonate which frequently occurs as yellowish brown to dark brown crystals with perfect rhombohedral **cleavage**. It commonly alters to a hydrous ferric oxide, typically **goethite**, but **hematite** and **magnetite** may also be associated. Its specific gravity of 3.96 is considerably higher than that of other common carbonates. **Siderite** most commonly occurs in bedded sedimentary rocks; it is the chief **iron**-bearing mineral in clay *ironstones*. In the **oolitic Jurassic** *ironstones* of the English Midlands it is one of the principal **ore** minerals, together with **chamosite** and hydrated **iron oxides**. **Siderite** is also found as a hydrothermal mineral in metallic **veins**, where it may often be a manganoan variety.

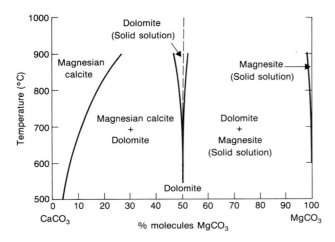

Fig. C7 Carbonate minerals. The CaCO₃–MgCO₃ system at CO₂ pressures sufficient to prevent decomposition of the carbonates. (After Goldsmith & Heard, 1961.)

Dolomite, $CaMg(CO_3)_2$, is an important carbonate, crystallizing in the trigonal system but with a structure with a slightly lower symmetry than that of **calcite**. It is typically colourless or white but the common replacement of some Mg by Fe^{2+} produces a yellow or brown colour in a dolomite rock. It has Mohs' hardness 3.5–4 and specific gravity 2.86, both values being higher than for **calcite**; it is poorly soluble in cold dilute HCl unless freshly powdered. Characteristically, the rhombohedral crystal faces are often curved or saddle-shaped.

Artificial **dolomite** has never been directly **precipitated** in the laboratory from solutions at ordinary pressures and temperatures. Its field of stability for moderate temperatures and high CO_2 pressures has been determined (Fig. C7). The thermal dissociation of **dolomite** takes place in two steps: firstly, $CaMg(CO_3)_2 \rightarrow CaCO_3 + MgO + CO_2$, at around 800°C, followed by the breakdown of the **calcite** component, $CaCO_3 \rightarrow CaO + CO_2$, at just over 900°C. Commercially **dolomite** is important as a refractory, for which purpose it is calcined at about 1500°C resulting in a sintered mixture of MgO (**periclase**) and CaO. It is also used in the extraction of magnesia from seawater.

Dolomite typically is a mineral of sedimentary environments, though there are important occurrences in metamorphic and hydrothermal **metasomatic** deposits. A few sedimentary deposits are known to have contained **dolomite** initially, such **dolomite** being termed primary. Secondary **dolomite** is formed where the primary **aragonite** and **calcite** of **limestones** have reacted with magnesium compounds to form **dolomite**. Primary **dolomite** may thus include **dolomite** associated with **evaporite deposits**, as in the **Permian evaporites** of north-east England. Secondary **dolomite** formed from **limestone** by **metasomatic alteration** can be divided into two classes. One occurs over relatively wide areas at one horizon and was formed very soon after deposition of the **limestone**, the **dolomitization** probably having occurred while the sediment was in an uncon-

solidated condition on the seafloor, this being referred to as *penecontemporaneous dolomitization*. In the other type of occurrence, the **dolomite** formed much later than the lithification of the **limestone**, the magnesian solutions having entered along **faults** and **joints** in the rock; this is known as *subsequent dolomitization*. Well-crystallized **dolomite** also occurs in hydrothermal **veins**, often associated with **ores** of **lead**, **zinc** and **copper**, or with **fluorite**, **barite**, **calcite**, **siderite** and **quartz**. Dolomitic or dolomitic-sideritic **carbonatites** (*beforsites*) are known from many **carbonatite** complexes. **Ankeritic dolomite**, in which appreciable **iron** and **manganese** occur partially replacing **magnesium**, is a typical mineral of both hydrothermal and low-temperature **metasomatism**. The low-temperature **ankeritic veins** occur in clay *ironstones* and along **cleats** and **joints** in coal.

Aragonite, the orthorhombic CaCO₃ **polymorph**, is typically colourless or white, but may show repeated **twinning** leading to pseudohexagonal groups. Most **aragonites** are relatively pure, but some contain appreciable **strontium**, the ionic radii of Sr and Ca being fairly similar. The equilibrium curve defining the stability fields of **calcite** and **aragonite** (Fig. C8) confirms that **aragonite** is metastable at room temperature and pressure. **Aragonite** has a higher specific gravity (2.94) than **calcite**, lacks the perfect rhombohedral **cleavage** of **calcite** and **dolomite** and effervesces readily in cold dilute HCl. Many organisms with calcareous skeletons build their shells of **aragonite**, though this is gradually converted to **calcite**. Primary **precipitation** of CaCO₃ from seawater under tropical conditions also gives **aragonite**, typically producing **aragonite** muds or, under suitable conditions, **aragonite ooliths**. It is also found in *pisolites* and *sinter* deposits from **geysers** and *hot springs* and may occur as a secondary mineral in cavities in volcanic rocks. It is a metamorphic mineral of widespread occurrence in the Franciscan *blueschist* terrain of California. [RAH]

Carlson, W.D. (1980) The calcite–aragonite equilibrium: effects of Sr substitution and anion orientational disorder. *American Mineralogist* **65**: 1252–62.

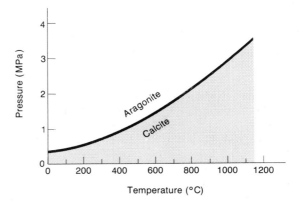

Fig. C8 Carbonate minerals. Phase relations in the system CaCO₃. (After Carlson, 1980.)

Ewald, P.P. & Hermann, C. (1931) Zeitschrift für Kristallographie, Kristallgeometrie, Kristallphysik, Kristallchemie. *Strukturbericht* 1913–1928.

Goldsmith, J.R. & Heard, H.C. (1961) Subsolidus phase relations in the system CaCO₃–MgCO₃. *Journal of Geology* **69**: 45–74.

Reeder, R.J. (ed.) (1983) *Carbonates: Mineralogy and Chemistry.* Reviews in Mineralogy No. 11. Mineralogical Society of America, Washington, DC.

carbonate-apatite (Ca₅F(PO₄CO₃, OH)₃) A variety of **apatite**. [DJV]

carbonate-hosted base metal deposit Strata-bound mineral **deposit** of **lead** and **zinc**, ±**copper** and also, sometimes principally, of **fluorite** and **barite** that occur in thick sequences of **limestone** or **dolomite**. Most **lead** and **zinc** production in Europe and the USA comes from this type of deposit. In Europe there are important fields in Central Ireland, the Alps, southern Poland and the Pennines of England. In the United States there are the famous Appalachian, Tri-State, south-east Missouri and Upper Mississippi districts. There are also important fields in north Africa (Tunisia and Algeria) and Canada. There are very few deposits of this type in the *Precambrian*.

The **orebodies** are variable in type. In the British Pennines, **vein orebodies with ribbon** *ore shoots* occupying *normal faults* are the main deposit type in the northern field (Fig. C9). In the southern Pennines the **veins** occupy *tear (wrench) faults*. The **orebodies** in the Tri-State Field, USA, are in **solution** and collapse structures, **caves** and underground channelways connected with **karst** topography. In Ireland, the **orebodies** vary from **stockwork** brecciation zones to **stratiform mineral deposits**. At Pine Point, Canada, the **ores** are in interconnected small-scale **solution** cavities which it has been suggested are the result of the **dissolution** of carbonate rocks by corrosive fluids generated by a reaction between **petroleum** and sulphate ions.

Average **ore grades** range mainly from 3 to 10% combined Pb + Zn with individual **orebodies** running up to 50%. Tonnages generally range from a few tens of thousands up to 20 Mt. The characteristic minerals of this **ore** association are **galena**, **sphalerite**, **fluorite** and **barite** in different ratios to one another varying from field to field. **Pyrite** and especially **marcasite** may be common and **chalcopyrite** is important in a few deposits. **Calcite**, **dolomite**, other carbonates and various forms of **silica** usually constitute the main **gangue** material.

With regard to the geological environment, the most important feature is the presence of a thick carbonate sequence, because thin carbonate layers in **shales** seldom contain important deposits of this type. The fauna and lithologies of the host rocks show that they were mostly formed in shallow water, near-shore environments of warm seas, and a plot of major carbonate-hosted deposits on *paleolatitude* maps shows a grouping of these deposits in low latitudes. The warmer climates of low latitudes encourage the development of *reefs* and so the frequent association of these deposits with *reefs* and carbonate mudbanks is not surprising.

Sangster has divided carbonate-hosted base metal deposits into two major *types*: (i) *Mississippi Valley*, and (ii) *Alpine*. Other workers do not make this distinction and

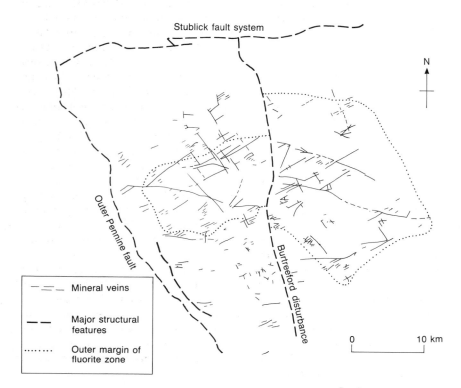

Stublick fault system

N

Outer Pennine fault

Buttreeford disturbance

– ·=· – Mineral veins

— — Major structural features

· · · · · · Outer margin of fluorite zone

0   10 km

Fig. C9 Carbonate-hosted base metal deposit. Vein system of the Alston block of the Northern Pennine Orefield, England. Note the three dominant vein directions. (After Evans, 1993.)

refer to all low temperature carbonate-hosted deposits as being of *Mississippi Valley type*. There is little doubt that the majority of deposits of this class have been formed from epigenetic **hydrothermal solutions**. A few show **syngenetic** features and the occasional deposit shows evidence of both **epigenetic** and **syngenetic deposition**, e.g. the Mogul Mine at Silvermines, Ireland, where the lower **ore** along a **fault** is epigenetic and yields meteoritic values for the sulphur isotopic ratios but the upper **orebody** appears to be syngenetic and to mark the time of **ore** formation. [AME]

Edwards, R. & Atkinson, K. (1986) *Ore Deposit Geology*. Chapman & Hall, London.
Evans, A.M. (1993) *Ore Geology and Industrial Minerals: An Introduction* (3rd edn). Blackwell Scientific Publications, Oxford.
Sangster, D.F. (1983) Mississippi Valley-type deposits: a geological mélange. In: Kisvarsanyi, G., Grant, S.K., Pratt, W.P. & Koenig, J.W. (eds) *International Conference on Mississippi Valley-type Lead–Zinc Deposits*, pp. 7–19. University of Missouri-Rolla, Rolla.

carbonation (1) The process of introducing carbon dioxide into water.
(2) A process of chemical **weathering** through the **replacement** of minerals by carbonates. This commonly involves the reaction of minerals with carbonated water and includes the **dissolution** of calcium carbonates by **meteoric water** and **groundwater** containing dissolved atmospheric carbon dioxide. Rainwater contains about 0.03% by weight carbon dioxide, making it weakly acidic. When it comes in contact with **carbonate minerals**, such as those making up **limestones**, **dolomites**, and **chalk**, the carbonated water reacts, forming water-soluble bicarbonate. The reaction is conventionally expressed as:

$$CO_2 + H_2O + CaCO_3 \rightleftharpoons Ca(HCO_3)_2$$

This is somewhat of an oversimplification of a complex chemical process which is not fully understood. [AW]

Trudgill, S. (1985) *Limestone Geomorphology*. Longman, London.
Trudgill, S.T. (ed.) (1986) *Solute Processes*. John Wiley & Sons, Chichester.

carbonatite Igneous rock that contains over 50% of **carbonate minerals**. They may be subdivided by the *dominant* carbonate mineral; **calcite** ($CaCO_3$) carbonatites are termed **sövites**, **dolomite** ($Ca, Mg_2(CO_3)_2$) carbonatites are termed *beforsites*, **siderite** ($FeCO_3$) carbonatites are termed *ferrocarbonatites*, and *natrocarbonatites* are composed of Na, Ca, K carbonates. They occur as extrusive **lava flows** of **pyroclastic rocks** and as minor intrusions which include **dykes** and **sills**, and are commonly associated with **alkaline igneous rocks** which include **ijolite** and **nephelinite** *melanephelinite*. Intrusive carbonatites may be associated with alkali-**metasomatized** country rocks termed **fenites**. Rocks of the carbonatite–ijolite–nepheline association may contain economic deposits enriched in Al, P, Ti, F, Fe, Cu, Zr, Nb, Ba, REE, Th and U.
Carbonatites occur within alkaline rock provinces associated with continental **rift** systems (e.g. the East African rift system) and rifted continental margins (e.g. around the South Atlantic Ocean). In common with other **alkaline igneous rocks**, the rocks of the carbonatite–ijolite–nephelinite association are derived from **magmas** formed by small degrees of *partial melting* of **mantle peridotite**.
During ascent through the **lithosphere** such **magmas** may yield volatile-rich carbonatite fluids as a result of fractional crystalization and/or **exsolution**. Depending upon the ambient T–P conditions and the composition of coexisting silicate **magma** these **magmas**, correspond to the carbonatite types defined above. [RST]

Le Bas, M.J. (1977) *Carbonatite–Nephelinite Volcanism*. John Wiley & Sons, New York.

carbonatization A form of **wall rock alteration** in which **dolomite**, **ankerite** and other carbonates may be developed in the **wall rocks** of **epigenetic mineral deposits**. **Dolomitization** is most commonly associated with low temperature **carbonate-hosted base metal deposits**.
Other carbonates may be developed in silicate rocks, especially where **iron** is available, and **ankerite** may then be common, particularly in the calcium–iron environment of carbonatized basic **igneous rocks** and volcaniclastics. This is particularly the case with many **vein gold** deposits, for example the Mother Lode in California where **ankerite**, **sericite**, **albite**, **quartz**, **pyrite** and **arsenopyrite** are well developed in the altered **wall rocks**. [AME]

Carboniferous The first system to be established formally early in the 19th century and was based on the widely occurring **coal**-bearing sequence found in the British Isles. The succession was traditionally divided into three lithological groups, from the bottom the marine Carboniferous Limestone or Lower Carboniferous, the **deltaic** Millstone Grit and the **delta-swamp** and terrestrial Coal Measures. Biostratigraphical divisions were set up by the beginning of the 20th century with the three lithological groups needing separate schemes based on basal **coral–brachiopod** zones, central *goniatite–bivalve* zones and upper nonmarine *bivalve* zones. Later refinements have introduced a classification of the Carboniferous System which is internationally accepted (see Fig. C10). It is a compromise bringing together accepted *period*, epoch and age names from Western Europe, the former USSR and the USA to form a comprehensive set of stratal divisions for the whole Carboniferous System.
Three great continents and associated smaller land masses divided the wide oceans of the Carboniferous World (Fig. C11). The continent of **Laurasia**, made up of Europe, Greenland and North America, held a dominant position with its long southern shore within the equatorial belt. This southern shore passed over the equator during the period, taking Western Europe into the northern hemisphere. To the south a second large continent called

Carboniferous period (International divisions in part provisional)			Carboniferous system (Regional Divisions)		
Periods and subperiods	Epoch	Age	Western Europe	USSR	USA
Permian		Asselian	Autunian	Asselian	Wolfcampian
Carboniferous — Pennsylvanian	Gzelian	Noginskian	Silesian — Stephanian C (Coal Measures / Upper Carboniferous)	Gzelian	Virgilian
		Klazminskian	Stephanian B		
	Kasimovian	Dorogomilovskian	Stephanian B	Kasimovian	
		Chamovnicheskian			
		Krevyakinskian	Stephanian A		Missourian
	Moscovian	Myachkovskian	Cantabrian	Moscovian (Middle Carboniferous)	Desmoinesian
		Podolskian	Westphalian C		
		Kashirskian			Atokan
		Vereiskian	Westphalian B		
	Bashkirian	Melekesskian	Westphalian A	Bashkirian	Morrowan
		Cheremshanskian	G$_2$		
		Yeadonian	G$_1$		
		Marsdenian	R$_2$ (Millstone Grit)		
		Kinderscoutian	R$_1$		
Carboniferous — Mississippian	Serpukhovian	Alportian	Namurian H$_2$	Serpukhovian (Lower Carboniferous)	Chesterian
		Chokierian	H$_1$		
		Arnsbergian	E$_2$		
		Pendleian	E$_1$		
	Visean	Brigantian	Dinantian — Visean D$_2$ D$_1$ (Carboniferous Limestone)	Visean	Meramecian
		Asbian			
		Holkerian	S$_2$		
		Arundian	C$_2$S$_1$		
		Chadian			
	Tournaisian	Ivorian	Tournaisian C$_1$	Tournaisian	Osagean
		Hastarian	Z K		Kinderhookian
Devonian		Famennian	Strunian		Louisiana

Fig. C10 Stratigraphic table showing the regional and international divisions of the **Carboniferous**. (After Harland *et al.*, 1989.)

Gondwanaland lay in the southern temperate zone and partly over the southern pole. This continent was composed of the assembly of Africa, South America, Australia, India, Madagascar and Antarctica and it also drifted northwards. The third large continent, called *Angaraland*, was composed mainly of northern Asia (the former USSR east of the Urals) and it lay in the northern hemisphere.

The Carboniferous Period was a time of closing seas and coalescing continents. Thus the ocean separating the southern shore of **Laurasia** from **Gondwanaland** gradually narrowed leading to collision **orogeny**. *Collisional mountain belts* began to form in the middle *Pennsylvanian* of the Ouachita–Marathon belt in south-east USA and affected the Appalachian belt (eastern USA) and western Europe by the end of the *period*. Similarly the ocean between **Laurasia** on the west and *Angaraland* on the east gradually narrowed. **Subduction** of **oceanic crust** led to **active** continental **margins** and **Andean-type mountain belts** such as the andesitic volcanic pile that formed mountain ranges across central Europe during the lower and middle *Pennsylvanian*. **Continental accretion** is the world-wide setting for the Carboniferous and it is not restricted to

Laurasia but affected all the major continents. **Subduction** and *continental collision* are self-destructive processes that all but removed the evidence from which Carboniferous geography could be deduced and in some regions, such as Tibet, China and south-east Asia, even less is known because little research has been conducted.

In this geographical setting a world-wide lowering of sea-level would be anticipated owing to the steady elimination of intercontinental seaways and the increase in volume of the ocean **basins**. This is confirmed in the general Carboniferous stratigraphical succession that is found on the continental margins. The initial marine carbonate phase, the *Mississippian*, is followed by *Pennsylvanian* regression and progradation of **deltas**, initially in a mainly lower **delta** plain environment, and continued with upper **delta** plain terrestrial **swamps** and dry desert interior hinterlands. At the end of the *Mississippian* regressive changes became more rapid with the onset of continental *glaciation* on **Gondwanaland** which lay partly over the Antarctic. Increase in land ice over what is now South America, Africa, Antarctica, India and Australia caused lowering sea-level because water was abstracted from the

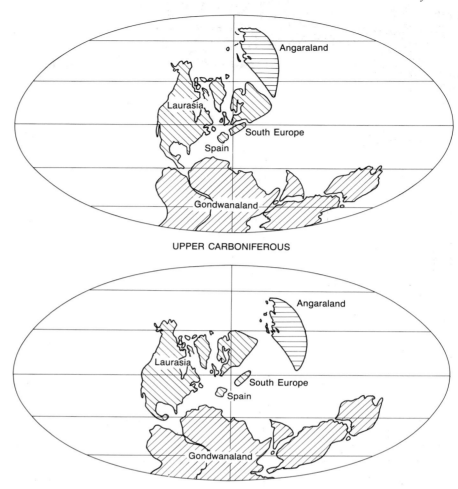

Fig. C11 A possible reconstruction of continental distribution in the **Carboniferous**.

seas. Coalescing land masses in a setting of **eustatic** lowering of sea-level finally led to land connections between all the continents, but it was late **Permian** before this process was complete.

The best documented region of Carboniferous deposition is the equatorial southern margin of **Laurasia** stretching from the Russian Donetz, through western Europe, Britain, Ireland and to western USA. Along this *paralic belt*, *Mississippian* carbonates gave way to *Pennsylvanian* **deltaic** and **delta swamp** deposits. A full succession of strata is present over much of this region owing to continual sinking of sedimentary **basins**. A widespread facies across this region is **deltaic** and fluvial deposition in pronounced cycles of the type: **limestone**, **shale**, **sandstone** and **coal** (coalbelt **cyclothems**) repeated many times. In Britain and Ireland these sedimentary cycles, called *Yoredale facies*, are particularly well developed near the top of the *Mississippian* and in eastern and central USA and the Donetz Basin, former USSR similar cycles extend to the top of the Carboniferous. This cyclic deposition is an expression of widespread marine transgression over the

lower **delta** plain followed by regression caused by **delta** progradation. Another widespread facies is the upper **delta** plain, fluvial and **delta swamp** or **coal measures**. It is found from low in the *Mississippian* near to the shorelines, but it is especially associated with emergence and falling world-wide sea-levels of the *Pennsylvanian*. Along the tropical southern shoreline of **Laurasia**, thick **peat** deposits accumulated in extensive **delta swamps** which, after burial and compaction, produced the celebrated Carboniferous **coal seams**. Thick seams of high quality **coal** occur in this region from the former USSR westwards to the USA and constitute some 16% of the **coal** resources of the world; total Carboniferous **coal** amounts to over 23% of world resources. **Delta swamp** facies are also associated with inland lakes, thick **coal seams** are developed in a *limnic* setting in central Europe where this facies straddles the *Carboniferous–Permian boundary*.

Carboniferous floras and faunas were diversified and luxuriant. The land flora, based on spore-bearing plants such as the *lycopods*, spread over most of the climatic zones of the world (see **land plants**). Abundant marine

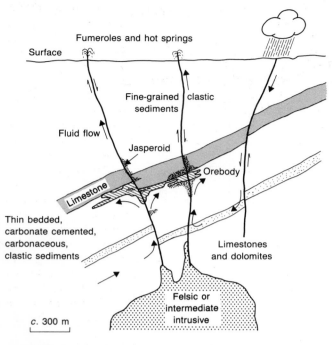

Fig. C12 Schema for the formation of **Carlin-type deposits**. (After Evans, 1993.)

life, particularly characterized by *algae*, **foraminiferans**, rugose **corals**, **brachiopods** and *crinoids*, inhabited the shallow seas where thick and widespread bioclastic **limestones** and **oolites** were laid down. **Fish** and **molluscs** had become established in freshwater, a variety of **amphibians** dwelt in the tropical **swamps** and the first **reptiles** appeared. [GALJ]

Harland, W.B., Armstrong, R.L. & Cox, A.V. (1989) *A Geological Time Scale*. Cambridge University Press, Cambridge.
Dias, C.M., Granados, L.F., Wagner, R.F. & Winkler Prins, C.F. (eds) (1983, 1985) *The Carboniferous of the World* Vols I and II. IUGS Publications No. 16, and No. 20. Instituto Geologico y Minero de España and Empresa Nacional Adaro de Investigaciones Mineras, S.A., Madrid, Spain.

Carlin-type deposit A subtype of **disseminated gold deposits**. Carlin deposits occur in carbonate-bearing carbonaceous sediments and volcanics. In the **ores calcite** has been replaced by **pyrite** and fine-grained **silica**, together with **gold, arsenic, thallium, antimony** and **mercury**. Whatever the host rock, **permeability**, allowing the establishment of circulating **geothermal systems**, was very important and **stable isotope** studies suggest that all the introduced materials were probably leached from the underlying rocks. A schema for the formation of these deposits is given in Fig. C12. **Wall rock alteration** in the form of **silicification**, **jasperoid** formation, **argillization**, **propylitic alteration** and **alunitization** is common. [AME]

Bagby, W.C. & Berger, B.R. (1986) Geological characteristics of sediment-hosted, disseminated precious-metal deposits in the western United States. In: Berger, B.R. & Bethke, P.M. (eds) *Geology and Geochemistry of Epithermal Systems*, pp. 169–202. Society of Economic Geologists, El Paso.
Evans, A.M. (1993) *Ore Geology and Industrial Minerals: An Introduction* (3rd edn). Blackwell Scientific Publications, Oxford.
Tooker, E.W. (ed.) (1985) *Geological Characteristics of Sediment- and Volcanic-hosted Disseminated Gold Deposits — Search for an Occurrence Model*. United States Geological Survey Bulletin **1646**.

carnallite $(KMgCl_3 . 6H_2O)$ An **evaporite** mineral. [DJV]

carnelian (*cornelian, carnellian*) A semi-precious stone, red **chalcedony**. This variety of **chalcedony** was used for beads and seals in antiquity, for example by the people of the Indus Valley civilization. (See **gem**.) [OWT/DJV]

carnotite $(K_2UO_2)_2(VO_4)_2 . 3H_2O)$ An **ore** mineral of **uranium**. [DJV]

carse An alluvial **floodplain** beside a river or estuary. The term is perhaps derived from 'carr', an area of **swamp**, marsh, or **bog** which has been reclaimed by artificial drainage. [AW]

cascade fold Type of fold produced by gravity collapse. (See **gravity collapse structure**.) [RGP]

case-hardening The process of surficial induration of porous rocks, or the crust formed by this phenomenon. Case-hardening occurs on exposed **limestones** and **sandstones** in tropical and subtropical regions. It is the result of infilling of pore spaces by mineral cements. These cements, which may be calcium carbonate, **silica**, or iron minerals, are **precipitated** from evaporating **meteoric water**, *soil* moisture, or **groundwater**. Case-hardening of crystalline rocks such as **granites** may result from **precipitation** of dissolved **iron oxides** or **silica** near the surface of exposed bedrock or loose boulders when moisture held within the rock evaporates. The breaching of such case-hardened crusts or veneers may expose more **erodible** rock beneath. This can lead to cavernous **weathering** and the formation of **tafoni** and '*rock doughnuts*' (Blank, 1951). [AW]

Blank, H.R. (1951) 'Rock doughnuts', a product of granite weathering. *American Journal of Science* **249**: 822–29.
Ireland, P. (1979) Geomorphological variations of 'case-hardening' in Puerto Rico. *Zeitschrift für Geomorphologie* Supp **32**: 9–20.
White, W.A. (1944) Geomorphic effects of indurated veneers on granites in the southeastern states. *Journal of Geology* **52**: 333–41.

cassiterite (SnO_2) The major **ore** mineral of **tin**. (See **oxide minerals**.) [DJV]

cat's eye A chatoyant variety of the mineral **chrysoberyl**. (See **chatoyancy**.) [DJV]

cataclasis (*cataclastic flow*) **Deformation** by **fracture** and sliding of **rigid** particles without internal **strain**. Rotation of particles and *grain boundary sliding* are often important elements of cataclasis, which may be recognized principally by the presence of **fractures** and by evidence for sliding or rotation of rigid particles that are internally undeformed. The fracturing process itself may be dominant or subordinate as in the case of *particulate flow*. Cataclastic deformation is characterized by increasing **strength** with **confining pressure**, and **dilatancy**. **Strain rate** and temperature have very little effect on **strength**. [TGB]

catastrophism A mode of thought that ascribes important change in the physical environment to the action of major events of high magnitude, low frequency and normally short duration. It is often placed in direct opposition to *uniformitarianism*. In the late 18th century and early 19th century, when the Earth was seen to have been founded in 4004 BC and to have suffered the deluge (Noah's flood), catastrophist ideas were widespread among geomorphologists and paleontologists. In recent years catastrophist ideas have returned in the shape of *neocatastrophism*, and it has been recognized that many landforms result from catastrophic processes and that sudden changes may contribute to species evolution (the concept of *punctuated equilibria*). [ASG]

Berggren, W.A. & Van Couvering, J.A. (eds) (1984) *Catastrophes and Earth History*. Princeton University Press, Princeton, NJ.

catchment A term synonymous with the total area of a drainage **basin**. It refers implicitly to the land area of the **basin** in that it emphasizes the total area draining to the point in question. [TPB]

catena A term introduced by Milne (1935) to describe a sequence of *soils* found in succession along a single hillslope. The *soils* are related to each other in terms of their proximity and (usually) by having a similar parent material. However, the *soil* profile may change radically along the slope in response to variations in site characteristics, such as drainage and slope steepness, and because of the position of the site with respect to the whole catena: for example, **erosion** by *soil wash* is distance dependent. In upland Britain a typical catena will consist, from the top of the slope to the bottom, of a *podsol*, often peaty, then a brown earth deepening downslope, with a *gley* in the valley floor, and possibly a **basin peat** beyond this. The *soil* catena is thus an expression of the interrelationship between **geomorphological** and **pedogenetic** processes. [TPB]

Milne, G. (1935) Some suggested units of mapping and classification particularly for East African soils. *Soil Research* **4**: 3.

cathodoluminescence The luminescence generated in minerals under electron bombardment. The technique is used extensively in petrographic studies of sedimentary rocks, particularly in elucidating **diagenetic** trends, and can also be used to detect compositional zonation of crystals in igneous and metamorphic rocks.

To produce cathodoluminescence (CL), an electron beam generated by a hot- or cold-cathode electron gun is directed onto the uncovered, polished surface of a thin section or rock specimen. The entire operation takes place in a vacuum chamber which is mounted on a microscope stage to allow direct observation of luminescence.

The exact cause of the CL remains a matter for some debate. In **calcite** and **dolomite**, luminescing zones correspond to regions of inhomogeneity in the crystal lattices. These can be due to lattice defects (see **defects in crystals**), or to the presence of certain trace elements — **manganese** is commonly regarded as an activator of CL, whereas ferrous **iron** in the lattice will quench or inhibit luminescence. Many studies have been made to relate the onset and quenching of CL to the level of impurities in **calcite** and **dolomite**, but widely accepted conclusions have yet to be reached.

The use of CL in **diagenetic** studies of **limestones** is now well established. The data from CL complements that available from simple staining techniques (using Alizarin Red S and Potassium Ferricyanide), allowing the recognition of complex zonations in coarse-grained **carbonate minerals**, particularly sparry **calcite**. Variations in CL intensity reflect changing trace element concentrations in the crystal lattice, which were themselves controlled by minute changes in the chemistry (redox potential, salinity, etc.) of the **pore fluids** from which **precipitation** took place. Regional correlation of these variations in certain formations has led to the concept of *'cement stratigraphy'*. Using this method, highly detailed models for the burial history and **pore-fluid evolution** of a formation can be constructed. (See **calcite**, **cementation**, **diagenesis**, **dolomitization**, **staining for carbonate minerals**, **trace elements in sedimentary carbonates**.) [DMP]

causse A French term for a **limestone plateau** characterized by closed depressions, **caves** and **avens**. Les Grandes Causses of central France are formed by a number of such plateaux around the Tarn river. Cvijic regarded causse landscapes as being halfway between **holokarst** and **merokarst** landscapes. [HAV]

cave A hole or fissure in a rock, usually defined as being large enough for a person to enter. Caves are found in many types of rock in coastal and inland situations, and may be formed by a range of processes including **solution**, **wave** attack or tectonic activity. Large cave systems exist in **lava flows**, for example, as found in Hawaii. However, the most spectacular cave systems, as well as the largest number of caves, are found in **limestones**. In carbonate rocks, caves develop through solutional widening of **joints**. There is a great range of caves found within **limestone** areas, in terms of size and nature. One of the largest cave chambers is the Sarawak Chamber in Good Luck Cave, Sarawak, which is 700 m long, 400 m wide and 280 m high. Cave types range from single rooms, short passages and open shafts (**potholes**) to complex, three-dimensional systems of passages, shafts and chambers. Some caves are completely filled with water, others are totally dry. There has been much debate over cave formation in **limestone** areas. The earliest ideas focused on catastrophic causes, especially Earth movements. Now it is accepted that most **limestone** caves form through the action of water but there is still some debate over whether this action occurs in the **vadose** or **phreatic zones**. In many cases, caves are initiated in the **phreatic** (saturated) **zone** leading to circular cave tubes. A lowering of the **water table** leads to **vadose** conditions which produce canyons cut in the floors of formerly phreatic tubes. The overall form of any cave system is highly influenced by the structural geology of the area, which influences jointing frequency and directions. The diversity of caves in **limestone** areas means that there are many different processes and combinations of processes involved in cave development. Caves often contain deposits such as **speleothems** and cave sediments, which can be used to provide *paleoenvironmental* information which gives clues as to the nature and timing of cave formation. [HAV]

Jennings, J.N. (1985) *Karst Geomorphology*. Basil Blackwell, Oxford.

cavitation erosion **Erosion** occurring during the **turbulent flow** of meltwater at high velocity over rough bedrock beneath **glaciers**. Local areas of low pressure may be created in the water. 'If the pressure falls as low as the vapour pressure of the water at bulk temperature, macroscopic bubbles of vapour (cavities) will form. The cavitation bubbles grow and are moved along in the fluid flow until they reach a region of slightly higher local pressure where they will suddenly collapse. If cavity collapse is adjacent to the channel wall localized but very high impact forces are produced against the rock. This action may give rise to mechanical **failure** of the channel . . .' (Drewry, 1986, p. 68). [ASG]

Drewry, D. (1986) *Glacial Geologic Processes*. Arnold, London.

cay A carbonate island formed of unconsolidated sediments and found on **coral–algal reefs**. Cays are characteristically made up of **sand** and **gravel beaches** in the intertidal and supratidal zones, with shingle ramparts on the landward side. Some cays are vegetated and therefore relatively stable features, but all cays are vulnerable to high energy events such as hurricanes or typhoons. [HAV]

celestite ($SrSO_4$) An **industrial mineral**. [DJV]

celsian ($BaAl_2Si_2O_8$) The **barium feldspar mineral**. [DJV]

cementation The **diagenetic** process by which **authigenic** minerals are precipitated in the pore space of sediments which thereby become lithified or consolidated (see **authigenesis**, **precipitation**, cf. **porosity**). Most **authigenic** minerals may form cements although the habit of these cements varies with mineral type and type of host sediment undergoing cementation (Fig. C13). If the cement is the same mineral as that which constitutes the host sediment lithology, then the cement may form overgrowths; such overgrowths are commonly in optical continuity with their host detrital grain and are described as **syntaxial**. Mineral cements of differing composition from the host framework, by contrast, typically form pore-filling cements which may be either peripherally replacive to the detrital grains or may replace whole detrital grains. Minerals with a fibrous or platy habit (such as **clay minerals**) more characteristically form grain-coating or pore-lining cements (see **neoformation of clays**).

In coarse clastic sediments (**sandstones**, conglomerates), **quartz** overgrowths around detrital **quartz** grains, **feldspar** overgrowths on **feldspar** grains and pore-filling carbonate or sulphate minerals are common cementing agents (Fig. C13). **Authigenic clay minerals** (**chlorite**, **illite**, **smectite** and **kaolinite**) are common grain coating and pore-lining cements. In **limestones calcite** often forms overgrowths around detrital skeletal grains (particularly **echinoid** plates). **Calcite** is a major pore-filling cement in many **limestones** (Fig. C13), and may also cement **limestones** with *micrite* through secondary grain enlargement (see **neomorphism**). **Dolomite** is also widespread in **limestones** but often occurs as a replacement of pre-existing **carbonate minerals**. Cementation in fine-grained clastic sediments (**mudrocks**) is less well-documented but carbonates are known to occur extensively as **nodules** and intergranular cements. **Pyrite** is a common cementing mineral and in the more silty **mudrock** lithologies authigenic **quartz** overgrowths may be an important cement.

Cementation may occur at any time in the diagenetic history of the sediment, ranging from very early during *eogenesis* (see **depositional environment related diagenesis**) as in **calcrete**, **beachrock**, *concretions* or **silcrete**, or very late during *mesogenesis* (see **burial diagenesis**) after **secondary porosity** generation or, for example, in the water zone of *hydrocarbon reservoirs*. Particularly in carbonate sediments, extensive cementation may also take place after **uplift** and exposure to ingress of **meteoric waters** (see **meteoric water diagenesis**). [SDB]

Cenozoic (*Kainozoic*) The collective term for the **Paleocene** to **Quaternary**. It consists of the *Tertiary* (**Paleocene** to **Pliocene**) and the **Quaternary**. The older Tertiary (**Paleocene** to **Oligocene**) is known as *Paleogene*, the younger Tertiary (**Miocene** and **Pliocene**) as *Neogene*. [AH]

Sandstones

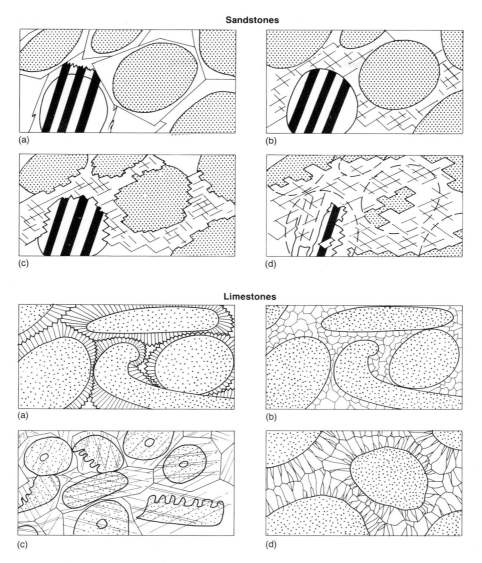

Limestones

Fig. C13 Cementation. Types and habit of the more common pore-filling cements in sandstones and limestones. Sandstones: (a) syntaxial quartz and feldspar overgrowths; (b) passive pore-filling carbonate cement; (c) peripherally grain replacive carbonate; (d) extensive grain replacement by carbonate cement. Limestones: (a) acicular aragonite cement; (b) equant blocky pore-filling calcite cement; (c) syntaxial calcite overgrowths to detrital crinoid fragments; (d) bladed pore-filling calcite cement.

cerargyrite (AgCl) A secondary mineral forming in the oxidized zone of **silver veins**. [DJV]

cerianite ((Ce, Th)O_2) A very rare **ore** mineral of **thorium**. (See **oxide minerals**.) [DJV]

cerium (Ce) At. no. 58, at. wt. 140.12. The most common of the lanthanides produced after the extraction of **thorium** from **monazite**. Used in various alloys, as CeO_2 in glass polishing, ceramic coatings and in incandescent gas mantles. [AME]

cerussite ($PbCO_3$) An important supergene **ore** of **lead**. [DJV]

cesium (Cs) At. no. 55, at. wt. 132.9054. An alkali metal. Because of its large ionic radius it is concentrated into late stage magmatic differentiates, especially **granite pegmatites**, in a few of which it occurs as **pollucite**, its only **ore** mineral. [AME]

cesium-vapour magnetometer A portable **magnetometer**, often used on satellites for measurement of the total **magnetic field**, based on the gyromagnetic constant of cesium. [DHT]

chabazite ($Ca_2Al_2Si_4O_{12} . 6H_2O$) A **zeolite mineral**. [DJV]

chalcanthite ($CuSO_4 . 5H_2O$) A secondary **copper** mineral. [DJV]

chalcedony (SiO_2) A microcrystalline form of **silica** found lining cavities in rocks. Chalcedony was used in antiquity mainly for the production of beads and seals (including **carnelian** objects). Occasionally it was used as a substitute for **flint** in tool prodoction where the latter was scarce. Chalcedony objects are recorded from North American Indian cultures and from the Sudan. (See **gemstone**.) [OWT]

chalcocite (Cu_2S) An **ore** mineral of **copper**. [DJV]

chalcopyrite ($CuFeS_2$) The major **ore** mineral of **copper**. [DJV]

chalcosiderite ($(CuFe_6(PO_4)_4(OH)_8 . 4H_2O)$) A relatively rare secondary mineral. [DJV]

chalk Chalk was used in prehistory for small sculpted objects, presumably because it is soft and easily carved. Chalk figurines and phalli and a symbolic **axe** have been found in Neolithic sites in southern Britain. Despite its softness, chalk was occasionally used as a **building stone** and for milestones by the Romans during their occupation of Britain. [OWT]

chalybite An old name for **siderite**. [DJV]

chamosite ($(Fe^{2+}, Mg, Fe^{3+})_5Al(Si_3Al)O_{10}(OH, O)_8$) An **ore** mineral of **iron** exploited in so-called **minette** or **Clinton ironstones**. (See **clay minerals**.) [DJV]

Chandler wobble Studies of the location of the Earth's rotational axis, mostly by monitoring the astronomically determined latitudes at different observatories, show that it has a free period of oscillation with a periodicity of 435 days. This wobble has an amplitude of some 100 m arcsec and appears to have a decay time of some 40 years, but the cause of its excitation is still uncertain as atmospheric effects operate on too short a time-scale. It seems to be mainly due to changes in the rotation rate of the Earth as modified by changes in the magnetic coupling between the Earth's **core** and **mantle**. It has been postulated that there are relationships between the Chandler wobble and **earthquake** activity but the actual relationship is not well established. [DHT]

Lambeck, K. (1980) *The Earth's Variable Rotation*. Cambridge University Press, Cambridge.

channel capacity The maximum **discharge** capable of being contained within a river channel. This is usually calculated as the product of channel width and mean depth. Capacity–basin-area relationships are often used to identify changes in channel size produced by impacts such as urban expansion or the building of a dam. (See also **bankfull discharge**.) [TPB]

channel resistance Four scales of flow resistance are considered to exist: particle or skin resistance; the resistance imparted by the development of bedforms such as **ripples** and **dunes**; additional resistance provided by flow around bends in sinuous channels; and spill resistance which is an additional source of energy loss caused by changes in the flow pattern at high **Froude numbers**. Estimation of average resistance coefficients for channels is an important part of flood discharge reconstruction. (See also **bed roughness**, **Chézy equation**, **Manning equation**.) [TPB]

channel sample A sample obtained by cutting channels across the face of exposed **ore** and collecting the chips, fragments and dust from each channel to make up a sample for **assaying** or mineralogical examination. [AME]

channel storage The capacity of the channel network to contain a flood discharge. This volume forms an important control on the change in the shape of a flood **hydrograph** as it passes through the channel system. **Flood routing** algorithms such as the *Muskingum method* take this effect into account in order to predict the attenuation of the **hydrograph** as it moves downstream. [TPB]

channel types These range from single thread channels (which may be straight, sinuous or **meandering**) to multi-thread (or **braided**) forms. This variety of channel patterns reflects varying energy conditions. (See also **meandering stream**, and **braided river**.) [TPB]

channel-belt deposit The sediments deposited by a river between successive **avulsions** in the area of **floodplain** directly affected by the channel activity. These deposits will include channel **bar** and *lag* deposits as well as clay plugs and proximal **levée** sediments. Channel-belt deposits gradually elevate themselves, through **aggradation**, above the adjacent **floodplain** producing an alluvial ridge as in the present-day Mississippi (Fig. C14). This leads to a metastable condition eventually resulting in **avulsion**. [JA]

Fisk, H.N. (1944) *Geological Investigations of the Alluvial Valley of the Lower Mississippi River*. Mississippi River Commission, Vicksburg, Miss.

chargeability The parameter in time-domain **induced polarization** surveys indicating the presence of polarizable mineralization, mainly sulphides. When a direct electric current flows through the ground these minerals acquire a polarization due to the accumulation of charges which decays once the current flow ceases. The area under the decay curve of the polarization potential, after normalization for the initial (saturation) polarization, is the

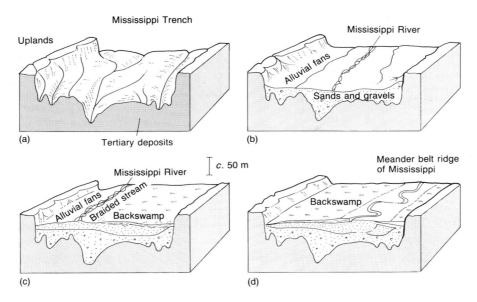

Fig. C14 Channel belt deposit. Block diagram of the Mississippi Valley. (a) Late Pleistocene entrenchment at sea-level minus 130 m. Braided river carries gravels to the Gulf of Mexico down a high slope. (b) Holocene aggradation at sea-level minus 30 m. (c) Holocene aggradation at sea-level minus 6 m. (d) The modern Mississippi meandering channel deposits sand, silts and clay. Note the final upward-fining sequence. (After Fisk, 1944.)

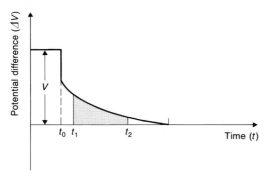

Fig. C15 Chargeability. Idealized induced polarization decay curve. When the current is switched off at $t = t_0$, the voltage decreases immediately, but then decays to zero gradually. The hatched area, after normalization for V and the time interval $t_2 - t_1$, is the chargeability.

chargeability of the medium, usually expressed in units of milliseconds (Fig. C15). The chargeabilities vary from a few thousand milliseconds for **massive sulphides** to about 10 ms in barren rocks such as **granite** and **limestone**. [ATB]

Keller, G.V. & Frischknecht, F.C. (1966) *Electrical Methods in Geophysical Prospecting.* Pergamon, Oxford.
Telford, W.M., Geldart, L.P. & Sheriff, R.E. (1990) *Applied Geophysics* (2nd edn). Cambridge University Press, Cambridge.

charnockite An *orthopyroxene*-bearing (anhydrous) **granite** of igneous or metamorphic origin commonly found in **granulite**-facies metamorphic terrains. The petrogenesis reflects **magma** genesis and/or metamorphism at high temperature and pressure. The name derives from that of Job Charnock (d. 1693), the founder of Calcutta (India), from whose tombstone the rock was first described. [RST]

chatoyancy A property exhibited by **minerals** arising from the way in which light is reflected back from the surface so as to give a silky and banded appearance. It results from the scattering of light by closely packed parallel fibres, or by very fine inclusions or cavities that are arranged in parallel lines. Examples of chatoyancy are shown by 'satin spar' gypsum, a **gem** variety of **chrysoberyl** called 'cat's eye' and by 'tiger's eye', another semi-precious stone formed when fibrous **crocidolite** is replaced by **silica**. [DJV]

cheirographic coast A coastline characterized by a succession of deep bays and long promontories produced by **tectonic uplift**, subsidence and associated complex **faulting** and **folding**. Although the term appears to have fallen from common usage, a cheirographic coast appears to be a specifically **tectonic** version of an inundated *discordant coast* where the structural trend runs transverse to the coastline. Examples of this may be the indented Yampi coast of the Kimberley region in Western Australia and the Kerry coast of south-western Ireland. [JDH]

chelation (origin Greek *chele*, claw) The complexing of metal atoms or cations in organic ring compounds where the metal is usually attached to the **carbon** atoms of the ring through intermediate nitrogen, oxygen or **sulphur** atoms. The term *chelate* refers to the very strong, multifunctional, claw-like bond the organic molecule (termed the *ligand*) forms with the metal. *Ligands* with more than one point of attachment to the central metal atom are known as polydentate and form very stable complexes. It is the multifunctional nature of the metal–*ligand* bond that gives the complexes their great stability by drastically reducing the statistical probability of random thermal energy simultaneously breaking all the points of attachment. Very powerful chelating agents with up to six points of attachment even form stable complexes with non-complexing simple ions such as Li^+. Metal ions complexed with chelating agents are literally 'wrapped up' in the *ligand* and are effectively removed from the solution just as if they had been **precipitated**.

Organic *ligands* with the property of chelating, including peptides, porphyrins and organic acids, are a normal by-product of the biological degradation of organic matter. Chelation plays an important role in **diagenesis** by enhancing the solubility of many metal ions in solution (including As, Ag, Al, Ba, Co, Pb, Zn, etc.) leading to the concentration of economically important metalliferous deposits associated with **soil weathering** profiles, marine *eogenesis* and **migration of oil and gas** (see **organic matter diagenesis**). Chelation of **aluminium** with organic acids may be an important subsurface **diagenetic** mechanism for transporting large quantities of **aluminium** in solution from **mudrocks** to **sandstones** (see **mass solute transfer, secondary porosity**). [SDB]

Stumm, W. & Morgan, J.J. (1981) *Aquatic Chemistry.* John Wiley & Sons, New York.

Surdam, R.C. & Crossey, L.J. (1985) Organic–inorganic reactions during progressive burial; key to porosity and permeability enhancement and preservation. *Philosophical Transactions of the Royal Society, London* **A315**: 135–56.

chemical and instrumental analysis of minerals

Techniques which provide unequivocal identification of minerals; the means to quantify their chemistry and structure; and illustrate the processes which occur in the development of minerals and mineral textures.

Classical methods of *wet chemical analysis* require that clean mineral separates can be obtained and taken into solution. Each element is determined by a distinct and appropriate method, generally involving either *colorimetry* or *titration*. Given a suitable specimen which is unzoned and free from surface impurities or inclusions, these classical methods are accepted as providing the best quality of analysis.

X-ray diffraction indicates the mineral structure and provides definitive identification of separated minerals as single phases. If more than one mineral is present the identification becomes more difficult or impossible.

The *electron microprobe* examines the chemistry of small phases and intergrowths by use of a focused electron beam in a vacuum. The beam acts upon a polished rock or mineral section and the technique examines the resultant X-ray spectrum. Two methods of detecting and measuring the X-ray spectrum are in use. Crystal *spectrometers*, following *Bragg's Law*, with gas-filled proportional counters are used to provide low detection limits and high resolution between elements. Energy dispersive detectors admit a wide band from the X-ray spectrum into a single, lithium drifted, silicon detector and the elements are separated by

the relative energy of their characteristic X-rays. *Energy dispersive analysis* provides the means to look at the whole spectrum at a single measurement, but it has a higher detection limit compared with the former method, which is known as *wavelength dispersive analysis*. The technique may be quantified and an analysis obtained of a small volume, measured in a few micrometers, at the section surface. Major, minor and **trace elements** above about 100 p.p.m. may be determined for all elements other than the light elements (H to Ne). Unlike the classical methods, this provides the means to examine mineral compositions in the context of their association with other minerals in a rock and the means to look at variations in composition of a mineral due to zoning, development of lamellae or the development of different generations. The ability to provide maps of the relative concentration of individual elements is often useful in this type of study. Since the invention of this method in the 1950s it has become, especially since the 1970s, the most commonly used technique of mineral analysis. Because of its ability to examine a wide range of elements with the same technique quickly and in relation to their appearance under the microscope, this technique has largely replaced classical chemical methods of analysis. Since the technique is non-destructive it is also possible to make other measurements such as *reflectance* and *microhardness* on exactly the same spot as the analysis. This permits an integrated study with an unambiguous relationship between the results by different methods. Quantitative analysis is obtained by relating X-ray count rates from each element to the count rate from a standard of known composition. The relationship is complicated by three categories of phenomenon which affect the X-ray production. These are the mean atomic number (Z) of the specimen, the X-ray absorption (A) by the specimen and the **X-ray fluorescence** (F) of the specimen. The correction procedure to accommodate these effects, known as the *ZAF* correction, includes several lengthy formulae and the size of the correction depends upon the

Table C2 Chemical and instrumental analysis of minerals. A typical electron microprobe analysis of a clinopyroxene

Element (as oxide)	Wt%
SiO_2	49.99
TiO_2	0.51
Al_2O_3	3.12
Cr_2O_3	0.13
V_2O_3	0.07
FeO	8.86
MnO	0.09
MgO	14.52
CaO	22.12
NiO	0.01
Na_2O	0.33
K_2O	0.01
Total	99.76

unknown concentration of the element in the specimen. For this reason an iterative procedure is followed in the calculation and computers are used to enable this to be done. (See Table C2.)

The *ion microprobe* uses a beam of ions, usually O^-, Ar^+ or Cs^+, to bombard a polished section in vacuum. An area $5-20\,\mu m$ across is damaged and ionic species ejected from the specimen surface. These are attracted into a **mass spectrometer** where they may be identified and quantified. Used in a similar way to the *electron microprobe*, the measured ions may be used for mapping the distribution of elements as well as analysis of their concentrations with the added advantage of being able to distinguish isotopes. Other advantages are the ability to detect **trace elements** down to levels below 1 p.p.m. and the ability to detect the light elements including hydrogen. The ability to make isotopic measurements has led to its use in U–Pb **zircon** dating and the study of the distribution of stable isotopes. The ionic species ejected from the specimen surface do not consist solely of single elements or isotopes but also consist of groupings of ions of different elements. The mass of such a group often corresponds to the mass of a single element. Since these ionic groups constitute a significant proportion of the ejected ions, interferences in the mass spectra are a major problem. No quantitative correction procedure has yet been developed and quantitative measurements depend on comparison with standards very close in chemical and isotopic composition to the specimen. Because of these problems, the *ion microprobe* is not yet routinely used in geological studies.

Attempts to overcome these problems have led to the development of instruments which use an ion beam to eject ionic groups from a located area. This is followed by acceleration, stripping and further acceleration stages before input to a **mass spectrometer**. The object of the acceleration and the stripping is to reduce the ionic groups to individual ions. A few such instruments are under development world-wide and have found particular applications such as determination of **platinum**-group elements in rocks at the p.p.b. level or studies of **carbon** compounds. [JFWB]

Reed, S.J.B. (1989) Ion microprobe analysis — a review of geological applications. *Mineralogical Magazine* **53**: 3–24.
Sweatman, T.R. & Long, J.V.P. (1969) Electron-probe analysis of minerals. *Journal of Petrology* **10**: 332–79.

chemical remanent magnetization (*crystalline remanent magnetization*) (CRM) The magnetization acquired as **ferromagnetic** grains grow, in an external **magnetic field**, through their **blocking** volume. [DHT]

chemostratigraphy The use of chemical signatures in stratal sequences for the purpose of correlation. Fluctuations in *carbon* and oxygen isotopes have proved of particular value in *Tertiary* and **Quaternary** strata and increasing use is being made of *carbon isotopes* for times as old as the late *Precambrian*. **Strontium isotopes** are increasingly being explored as a chemostratigraphic tool and amino acid residues have proved their worth in the **Quaternary**. [AH]

chenier plain Progradational coastal plain consisting of alternate coarse clastic, roughly coast-parallel, linear ridges and intervening muddy sediments. They occur in many localities where intertidal *marsh* sedimentation occurs, the common factor being that they undergo periodic phases of **erosion**. Each isolated **sand** ridge represents the coarser fraction of the coastal mudplain that is *winnowed* and deposited by a minor transgression caused either by a relative sea-level change or, more usually, by variations in mud supply. When the transgression declines, *marsh* sedimentation restarts seaward of the chenier ridge. These transgressive/regressive cycles result in the production of multiple chenier **sand** ridges and intervening *marsh* deposits to form a chenier plain. [RK]

chert (SiO_2) A granular microcrystalline variety of **quartz**. Chert was commonly used for artefacts in both the Old World and the New in antiquity and in more recent stone-using cultures. It was widely used in the Paleolithic and large numbers of chert artefacts were found at Olduvai Gorge in east Africa and Chou Koutien in China. It was also used in Europe, Ceylon, the Sudan, Australia and in north and south America by pre-Hispanic peoples. Artefact types include **axes** and blades (cutting tools). Some studies have concentrated on relating chert artefacts to their geological sources. North American chert sources exploited by the Woodland (early- and pre-Indian) Cultures are visually distinguishable, whilst those used in Neolithic Europe have been studied by **chemical analysis**. Microfossils in cherts of Western Australia provide a basis for identification of the source of these artefacts, and **oxygen isotopes** distinguish between certain chert **nodules** used in the Paleolithic of east Africa. **Flint** is a type of chert very commonly found on Stone Age sites. [OWT/DJV]

chesterite ((Mg, Fe)$_{17}Si_{20}O_{54}(OH)_6$) A *biopyribole* mineral produced as a microscopic **alteration** of **anthophyllite**. [DJV]

chevron construction Method of constructing the geometry of a *listric* or non-planar **fault** from the shape of its folded **hangingwall**, in either *extensional* or *thrust* **fault** *systems*. The method assumes constant *heave* (horizontal component of displacement) and is illustrated in Fig. C16. For each uniform component of *heave* AB, BC, CD, etc., a vertical line is drawn cutting the fold surface at points B, C, D, etc., and the diagonals BB', CC', DD', etc., drawn. These diagonals represent the orientation of the **fault** in each of the successive segments BC, CD, etc., and thus, starting from the known position AA', the **fault** can be

chevron fold

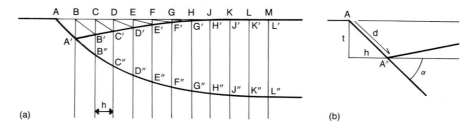

Fig. C16 (a) The **Chevron construction** for fault profile using hanging wall rollover geometry. The vertical grid has a horizontal spacing of 1 heave unit. Diagonals drawn from regional to rollover, e.g. BB′, parallel the fault in that heave segment. (b) Detail of fault to show displacement resolved into vertical throw and horizontal heave components. α is the dip of the fault.

reconstructed by adding successive portions A′B′, B′C′, C′D′ and so on.

Although the method offers a rapid and convenient method of reconstructing **fault** systems, it is only an approximation, as shown by Williams & Vann (1987). The construction assumes that *heave* is conserved; consequently there must be a reduction of both displacement and *throw* as the *listric fault* flattens out. There is no theoretical justification for this, nor for the implication of vertical **simple shear** as the deforming mechanism in the **hanging wall**. Williams & Vann suggest more realistic modifications of the technique to overcome these limitations. [RGP]

Williams, G.D. & Vann, I. (1987) The geometry of listric normal faults and deformation in their hangingwalls. *Journal of Structural Geology* **9**: 789–95.

chevron fold Type of **fold** characterized by straight *limbs* and sharp *hinges*. *Accordion fold* is an outdated synonym. (See **fold, kink**.) [RGP]

Chézy equation A formula derived by Chézy (1775) which defines mean channel velocity on the basis of the **hydraulic radius**, channel slope and a *roughness coefficient*:

$$v = C/(R \cdot s)$$

where v = mean flow velocity, s = channel slope, R = **hydraulic radius** (cross-section area divided by wetted perimeter), and C = the 'Chézy' *roughness coefficient*. [TPB]

Chézy, A. de (1775) Mémoire sur la vitesse de l'eau conduite dans un regole. Reprinted in 1921 *Annals des Ponts et Chaussées* **60**.

chiastolite A variety of **andalusite** with dark coloured carbonaceous inclusions arranged to form a cruciform design. [DJV]

chimney A term applied to an **orebody** of this shape that lies within an **ore**-locating structure such as a **vein**. [AME]

china clay A white, soft, plastic clay composed of **kaolinite** with a low **iron** content. Commercial deposits carry 10–95% **kaolinite**. Both quality and quantity can be improved by **benefication**. Non-clay minerals must be removed, usually by mixing with water and employing drag boxes, classifiers or cyclones. **Bleaching** decolours **iron**-bearing clays, **flotation** removes **iron oxides**. Major uses are in the ceramic and paper industries. Besides in paper, kaolin is used as a filler in plastics, rubber and paint. It also has applications in cosmetics, catalysts, inks, insecticides, food additives and filters.

It is formed by the **alteration** of **feldspar**-rich rocks: **granite**, **gneiss**, *arkose*, etc., by **weathering** and/or **hydrothermal alteration**. Good examples of deposits formed from **granite** occur in the south-west of England where there are funnel- or trough-shaped deposits that narrow downwards and extend as deeply as 230 m. [AME]

Constructional and Other Bulk Materials. The Earth's Physical Resources Block 4. Open University Press, Milton Keynes.

chine A narrow ravine or canyon which runs down to the sea, particularly in southern England. Well-developed chines can be found on the coasts of Hampshire and the Isle of Wight. [NJM]

chip sample A collection of rock chips taken either in a continuous line across a **face** or at random intervals over a **face**. In some cases, where samplers are familiar with a particular mine, it can be as accurate as **channel sampling**. [AME]

chloanthite ($(Ni, Co)As_{3-x}$) An **ore** mineral of **nickel** and **cobalt**. [DJV]

chlorapatite ($Ca_5(PO_4)_3Cl$) A variety of **apatite**. [DJV]

chlorargyrite (AgCl) An alternative name for **cerargyrite**. [DJV]

[102]

chlorite ((Mg, Fe)$_3$(SiAl)$_4$O$_{10}$(OH)$_2$. (Mg, Fe)$_3$(OH)$_6$) A hydrous layer silicate mineral (*phyllosilicate*) found in certain metamorphic sedimentary and **altered igneous rocks**. [DJV]

chloritization A form of **wall rock alteration** in which significant amounts of new **chlorite** are developed in rocks affected by **hydrothermal solutions**. The **chlorite** may be present on its own or with **quartz** or **tourmaline** in simple assemblages, but other **propylitic** minerals are usually present. The development of secondary **chlorite** may result from the **alteration** of mafic minerals already present in the country rocks, or, less commonly, from the introduction of **magnesium** and **iron** by the **hydrothermal solution**. [AME]

chloritoid ((Fe, Mg)$_2$Al$_4$O$_2$(SiO$_4$)$_2$(OH)$_4$) A mineral that appears superficially similar to **chlorite** but is actually an island silicate (*nesosilicate*). [DJV]

chondrodite (Mg$_5$(SiO$_4$)$_2$(F, OH)$_2$) An island silicate mineral (*nesosilicate*) of the **humite** group. [DJV]

chott A large, seasonally-flooded desert **basin**. The term is applied especially to the large, **tectonic basins** of North Africa — for example, Chott Djerid and Chott el Fedjedj, in southern Tunisia; and Chott Melrhir and Chott ech Chergui, in Algeria. These are continental **sabkhas**, several of which lie below sea-level. They are flooded during the cool, wet winter months by surface **runoff** and rising **groundwater**. During the summer, the flood waters evaporate, **precipitating gypsum**, **halite** and other water-soluble minerals. Some of the North African chotts were probably perennial lakes during the late *Pleistocene* when mean annual temperatures were lower and, perhaps, there was greater rainfall. [AW]

Mensching, H. (1964) Zur Geomorphologie Südtunesiens. *Zeitschrift für Geomorphologie* NF **8**: 424–39.

chromite (FeCr$_2$O$_4$) A **spinel** mineral, the major **ore** mineral of **chromium**. (See **oxide minerals**.) [DJV]

chromium (Cr) At. no. 24, at. wt. 51.996. A hard silver-white metal resistant to **oxidation**, widely used in electro-plating and as a ferroalloy metal. Cr compounds are used as pigments in glass, etc., leather tanning, textiles, as catalysts, oxidizing agents and in refractories. The only **ore** mineral is **chromite** which is won from **orthomagmatic deposits**. (See Figs O17 and O18, pp. 444–5.) [AME]

chrysoberyl (BeAl$_2$O$_4$) A rare mineral found in **granitic** rocks, **pegmatites**, mica **schists** and river **sands** and **gravels** and valued as a **gemstone**, particularly the varieties **alexandrite** and **cat's eye**. [DJV]

chrysocolla (Cu$_4$H$_4$Si$_4$O$_{10}$(OH)$_8$) A secondary **copper** mineral. [DJV]

chrysolite An old name for **olivine**. [DJV]

chrysoprase An apple green variety of **chalcedony**. [DJV]

chrysotile (Mg$_3$Si$_2$O$_5$(OH)$_4$) A fibrous variety of the mineral **serpentine**; asbestiform varieties are exploited as a 'white' **asbestos**. (See **clay minerals**.) [DJV]

chute bar A sedimentary deposit formed at the downstream end of a high stage bar-top channel. Sediment transported through the chute channel at high stage avalanches into the main channel to form unusually thick tabular *cross-sets*. The slip faces of the chute bar are convex downstream in planform with a morphology and construction similar to a small *Gilbert delta*. [JA]

McGowen, J.H. & Garner, L.E. (1970) Physiographic features and stratification types of coarse-grained point bars: modern and ancient examples. *Sedimentology* **14**: 77–111.
Levey, R.A. (1978) Bed-form distribution and internal stratification of coarse-grained point bars, Upper Congaree River, S.C. In: Miall, A.D. (ed.) *Fluvial Sedimentology*. Canadian Society of Petroleum Geologists Memoir **5**: 105–27.

cinnabar (HgS) The major **ore** mineral of **mercury**. Cinnabar was occasionally used by prehistoric and early historic cultures as a colorant in the same way as red **ochre**. It was used in particular by the Romans to colour parts of inscriptions. It may have been used as a body paint by the Amerindians of Arizona, perhaps with some detrimental effects on the health of these people. Cinnabar from eastern Europe was exploited by the prehistoric peoples of this area. [OWT/DJV]

cinnamon stone A brown variety of the **garnet mineral grossular**. [DJV]

citrine A semi-precious stone, a clear yellow variety of **quartz**. (See **gem**.) [DJV]

claim A piece of land staked out by a geologist, miner, etc., in which he has certain rights to explore for and exploit minerals. These rights (and the obligations that go with them) vary from country to country. [AME]

classification of ore deposits Many classifications have been put forward over the centuries. In older classifications of ore deposits much emphasis was placed on the mode of origins of deposits, with the result that as ideas concerning these changed, many classifications became obsolete. A good example of this occurs with the volcanic-associated **massive sulphide deposits** which, 30 years ago, were generally held to be formed by **replacement** at considerable depths within the **crust**, but are now thought to be the product of deposition in open spaces at the volcanic– or sediment–sea water interface. They have, therefore, moved from the class of **hydrothermal-replacement deposits** to that of *volcanic*-exhalative *deposits*.

Later classifications have been based upon commodity (**copper** deposits, **iron** deposits, etc.) morphology, environment and origin. Commodity and morphological classifications may be of value to economists and mining engineers, but they lump too many fundamentally different deposit types together to be of much use to geologists. In the past, **ore** geologists have been inclined to favour genetic classifications but in recent years there has been a swing away from such ideas towards environmental–rock-association classifications. Good examples of this trend are to be found in the recommended reading. [AME]

Dixon, C.J. (1979) *Atlas of Economic Mineral Deposits*. Chapman & Hall, London.

Stanton, R.L. (1972) *Ore Petrology*. McGraw-Hill, New York.

Wolf, K.H. (ed.) *Handbook of Strata-bound and Stratiform Ore Deposits*. Elsevier, Amsterdam.

clastic rock A sedimentary rock made up of particles ('*clasts*'), including single crystals, of pre-existing rocks, which have been **weathered**, **eroded** and, usually, transported by physical and biological processes. The terms '*siliciclastic*' and '*terrigenous*' refer to rocks and sediments where the *clasts* are predominantly made of **silicate minerals**.

The basic subdivision of *clastic sediments* and rocks is according to *clast* size. The most widely accepted classification is that of the *Udden–Wentworth scale* (Table C3) which has a ratio of two between successive class boundaries when expressed in millimetres. Clast sizes are often expressed in '*phi units*', where phi (ϕ) $= -\log_2 x$, x being the grain size in millimetres. Table C3 shows the terms used for *clasts* and unconsolidated sediment, as well as those for rocks made up of indurated sediment. Note that, in this context, '*clay*' includes all particles finer than 3.9 microns, not just **clay minerals**. For full definitions of the three categories of clastic rocks, see **rudite**, **sandstone** and **mudrock**. Although all clastic rocks and sediments do belong to one of these categories, many of course have intermediate features, so prefixes are useful, as in the case of 'pebbly sandstone' for a **sandstone** with a scattering of pebbles, but not enough (30%) to make it a **rudite**. Additionally a term, '*mictite*', '*mixtite*', or '*diamictite*' is useful for a clastic rock with a very wide range of grain sizes, such as a pebbly sandy mudstone. '*Diamicton*' is the unconsolidated equivalent.

As well as its clasts, a clastic rock is made up of diagenetic *cement* (see **cementation**, **diagenesis**) and a

Table C3 Clastic rock. Grain size scale for sediments and sedimentary rocks. (After J.A. Udden & C.K. Wentworth and modified after Tucker, 1981.)

mm	phi	Class terms		Sediment classification	Rock classification
256	−8	Boulders			
128	−7				
64	−6	Cobbles			
32	−5			Gravel,	
16	−4	Pebbles		rudaceous	Conglomerates
8	−3			sediments	Breccias } Rudites
4	−2				
2	−1	Granules			
1	0		v. coarse		
0.5	1		coarse	Sand,	
0.25	2	Sand	medium	arenaceous	Sandstones
0.125	3		fine	sediments	Arenites
0.0625	4		v. fine		
0.0312	5		coarse		
0.0156	6		medium		
0.0078	7	Silt	fine	Silt	Siltstones
0.0039	8		v. fine		
		Clay		Clay	Claystones

variable proportion of primary and secondary '*matrix*': this matrix is particulate material, of clastic and **authigenic** (see **authigenesis**) origin respectively, which is finer than the grains which define the rock, though it is not always easy to distinguish between the three. According to the definition given above, the matrix of a **rudite** can be of **sand** and/or mud, whereas that of a **sandstone** can only be of mud. Some authors, however, would consider a **sandstone's** matrix as material only of clay grade, whereas others would set the maximum grain size at medium silt.

The classification of a clastic sediment is usually based upon estimates of average grain size made, for progressively finer deposits, by eye, with a hand lens, or with the teeth (silt feels gritty whereas clay does not). However, it is often desirable to measure *grain size distribution* more accurately since, along with **fabric**, *grain shape* and *surface texture* (see below), it defines the '*texture*' of a clastic sediment, description of which is fundamental to interpretations of depositional and post-depositional history (see **facies**, **diagenesis**).

The most detailed studies of *clast size distribution* use data measured in the field for coarse **gravels** and **rudites**, with sieves for unconsolidated silts, sands and fine **gravels**, by point counting with a microscope for their indurated equivalents and, for indurated and unconsolidated muds respectively, with the **electron microscope** and devices, such as the '*sedigraph*', which record the relative settling velocities of particles in water. The *mean, modal and median grain size, sorting, skewness* and *kurtosis* of the distribution can then be calculated either directly from the data, by the method of moments, or from histograms, smoothed frequency curves and cumulative frequency curves. Table C4 gives relevant formulae. The graphic representations are themselves of value in giving a visual impression of the

Table C4 Clastic rock. Formulae for the calculation of grain size parameters

Parameter	Formulae of Folk & Ward (1957) for use with cumulative frequency plots. ϕ_n is the grain size in phi units at the nth percentage frequency.	Method of moments. f is the percentage fraction in each class interval. $M\phi$ is the mid-point of each class interval in phi units.
Mean	$M = \dfrac{\phi_{16} + \phi_{50} + \phi_{84}}{3}$	$\bar{X} = \dfrac{\Sigma f M\phi}{100}$
Sorting (standard deviation)	$\sigma = \dfrac{\phi_{84} - \phi_{16}}{4} + \dfrac{\phi_{95} - \phi_{5}}{6.6}$	$\sigma = \sqrt{\dfrac{\Sigma f (M\phi - \bar{X})^2}{100}}$
Skewness	$Sk = \dfrac{\phi_{16} + \phi_{84} - 2\phi_{50}}{2(\phi_{84} - \phi_{16})} + \dfrac{\phi_{5} + \phi_{95} - 2\phi_{50}}{2(\phi_{95} - \phi_{5})}$ (Sk is the inclusive graphic skewness)	$\alpha_3 = \dfrac{\Sigma f (M\phi - \bar{X})^3}{100\sigma^3}$ (α_3 is the moment coefficient of skewness)
Kurtosis	$K_G = \dfrac{\phi_{95} - \phi_{5}}{2.44(\phi_{75} - \phi_{25})}$ (K_G is the graphic kurtosis)	$\alpha_4 = \dfrac{\Sigma f (M\phi - \bar{X})^4}{100\sigma^4}$

distribution. In this respect, the most useful graphs are smoothed frequency curves with arithmetic scales and log probability plots of cumulative frequency which give straight lines for normal distributions.

The mean is of more significance than the mid-point value of a grain size distribution, the median. Modes, the mid-points of the most frequent size classes, are also significant. Sediments can have more than one mode and are described as 'unimodal', 'bimodal' or 'polymodal'.

The *sorting* of a sediment, the standard deviation of the *grain size distribution*, is diagnostically significant in being controlled in part by the processes of transport and deposition which have affected the sediment and by the duration of their activity. Wind and water currents, for example, are far more effective than viscous **debris flows** at separating grains of different diameter and so their deposits are generally better sorted. Environments where sediment is repeatedly reworked, such as **beaches**, tend to produce much better sorted sediments than, for example, **submarine fans**. However, there are other controls on *sorting* including source rock crystal or *clast* size distribution and grain size; **gravel** and mud are less easily transported by wind and water than **sand**, so they tend to have poorer *sorting*. For these reasons, and because of the possibility of **reworking** of deposits of one environment or process by another, sorting should never be used in isolation as **paleogeographic** evidence.

Verbal terms used to describe *sorting* are as follows:

standard deviation < 0.35	very well sorted
$0.35-0.5$	well sorted
$0.5-0.71$	moderately well sorted
$0.71-1.00$	moderately sorted
$1.0-2.0$	poorly sorted
> 2.0	very poorly sorted

For many studies, sufficient knowledge of *sorting* can be obtained by using visual comparison charts, such as those of Pettijohn *et al.* (1973) in conjunction with a thin section, or, in the field, a hand lens.

The *skewness* of a *grain size distribution* is a measure of its symmetry and is controlled by similar factors to sorting, and similar caveats apply to its use in *paleoenvironmental* interpretation. An example of control by depositional environment is the tendency of **beach sands** to have a negative skew, i.e. an excess of coarse grains compared to a normal distribution, because of the effectiveness of repeated wave action in *winnowing* away fine particles. Verbal terms for degrees of *skewness* are:

Sk greater than $+0.30$	strongly fine-skewed	} 'fine-
$+0.30$ to $+0.10$	fine-skewed	tailed'
$+0.10$ to -0.10	near-symmetrical	
-0.10 to -0.30	coarse-skewed	} 'coarse-
less than -0.30	strongly coarse-skewed	tailed'

Kurtosis measures the peakedness of a distribution curve. The relevant descriptive terms are as follows:

K greater than 1	(relatively peaked curve)	*leptokurtic*
$K = 1$	(normal distribution)	*mesokurtic*
K less than 1	(relatively flattened curve)	*platykurtic*

As a measurement in isolation, *kurtosis* is now thought to have little geological significance, but it has been used in conjunction with the other parameters above in attempts to characterize sedimentary environments by means of scatter diagrams where one parameter is plotted against another. It has been shown that it is possible to distinguish modern aeolian, **beach** and fluvial **sands** on skewness-sorting plots. However, the same limitations apply to the interpretative value of such plots as to the individual parameters. The approach has, in any case, been superseded in *paleoenvironmental* analysis of clastic rocks by the detailed interpretation of sedimentary structures.

The texture of a clastic rock also encompasses the several aspects of *grain morphology*. The first of these is *grain shape*, which is measured by the ratios of the long, medium and short axes of a grain. Figure C17 shows how the ratios define the four classes of shape, 'equant' (cubes or spheres), 'oblate' (discs), 'prolate' (rods) and 'bladed'. The *sphericity* of a grain measures the degree to which its shape approaches that of a sphere. Its theoretical definition is the ratio between the surface area of the grain and that

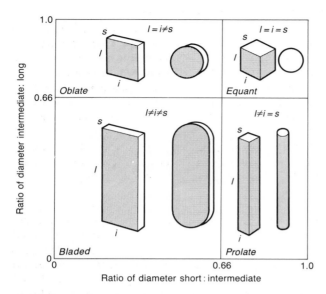

Fig. C17 Clastic rock. Zingg's four classes of grain shape based on the ratios of the long (*l*), intermediate (*i*) and short (*s*) diameters. (After Tucker, 1981.)

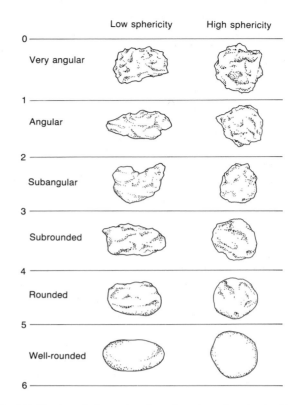

Fig. C18 Clastic rock. Categories of roundness for sediment grains. For each category a grain of low and high sphericity is shown. (After Pettijohn *et al.*, 1973.)

of a sphere with the same volume, but since measurement of the area of an irregular particle is next to impossible, a practical definition is $(V_p/V_{cs})^{1/3}$, where V_p is the volume of the grain (measured by its displacement of a liquid) and V_{cs} is that of the smallest sphere which could enclose the grain. Alternatively, the ratio of maximum and minimum cross-sectional areas may be used to define the '*maximum projection sphericity*', $(s^2/li)^{1/3}$, where *l*, *i*, and *s* are the long, intermediate and short axes of the grain respectively.

Both the shape and the sphericity of *clasts* have been shown to change due to **abrasion** during transport by fluid media, but by far the most important control upon them is the isotropy or otherwise of the rock or crystal of which they are made, i.e. the existence or not of planes of weakness such as **joints**, **cleavage** and **bedding**. Certain sedimentary processes and environments affect *clasts* of different shapes in different ways; **wave swash** and **back-wash**, for example, concentrate discs at the top of a pebbly beach and spheres at its base.

More generally significant geologically is the *roundness* (or *angularity*) of a *clast*, which is a measure of the sharpness or otherwise of its corners. Its theoretical definition is as the ratio between the average radius of circles drawn within the corners and edges of the clast and the radius of the largest circle which can be drawn within the clast. Practical assessment depends on the use of visual estimation by comparison with pictures of two sets (with very different sphericities) of six standard *clasts*, each of which represents one of the classes in the *visual roundness scale* (Fig. C18).

In **sands** and **gravels**, *roundness*, like *sorting*, tends to increase with the degree of transport, reworking and

therefore **abrasion** undergone by a clast. Indeed, *roundness* and **sorting** are the main aspects of the concept of '*textural maturity*' of a sediment (see **sandstone** for a discussion of *compositional maturity*), a mature sediment being one with high values of both, and inferred therefore to have a history of protracted **reworking**. Very intense **abrasion**, however, may cause angular **fractures** to form, and similar limitations apply to the *paleoenvironmental* significance of *roundness* values as to grain size parameters. The trend of increasing *roundness* with transport does not apply to *clasts* finer than **sand**, which tend to be more cushioned by transporting media and whose population may be augmented, during the transport of mixed sediment, by angular fragments abraded from larger grains.

Fabric is the aspect of clastic rock texture concerned with *clast* orientation and *packing*. Rocks have an *isotropic fabric* if composed either of equant *clasts*, or of non-equant *clasts* with no **preferred orientation**. An example of an *anisotropic fabric* is the common **preferred orientation** of clay flakes parallel to **bedding** in **mudrocks**, which results both from **compaction** and from the orientation taken up by the flakes as they settle from **suspension**. Examples in coarser sediments include the **preferred orientations** of wave-rolled prolate *clasts* normal to the **wave** propagation direction, or of unidirectional current-deposited rods both parallel and normal to flow, the latter being the orientation of rolling *clasts*, the former that of *clasts* carried in suspension and that which offers most resistance to re-entrainment by a flow once a *clast* is deposited. Another **fabric** resulting from the *clast* orientation which most resists re-entrainment is the *imbrication* of oblate *clasts*, where the plane containing their long (a) and intermediate (b) axes dips at a small angle (<20°) upstream. There are two types: one with a majority of long axes horizontal, i.e. normal to flow, results from the imbrication of *clasts* which have been rolled as **bedload**, e.g. in aqueous currents. The other type, with most (a) axes sub-parallel to the flow, results from the transport of *clasts* within a flowing medium, such as a **debris flow** (see **sediment-gravity flow**).

The *packing* of *clasts* is important as a control of **porosity** and **permeability**. *Packing* increases during compaction, but in unconsolidated sediment it depends mainly on *sorting*; in poorly sorted sediments, small *clasts* are available to fit into the spaces between larger *clasts* and **porosity** may be 25% less than in well-sorted sediments with the same mean grain size. Even perfectly sorted spheres can be arranged in different ways, however. Two end members are *cubic packing* (where orthogonal lines join sphere centres), which gives 48% **porosity**, and *rhombohedral packing* (26% **porosity**). The reticulate arrangement of flocculated clay flakes can give muds initial **porosities** of as much as 90%. In **sands** and **gravels**, poor *packing*, with local *cubic coordination*, is favoured by high rates of deposition. Vibration by current fluctuations, animals, or **earthquakes** can shake such sediments into a better *packing*. This results in fluid expulsion and can cause fluidized flow (see **sediment-gravity flow, wet sediment deformation**).

Grain surface texture is a textural property which reflects the **abrasive** and **corrosive** processes which have affected a *clast* during **erosion**, transport, deposition and **diagenesis** and therefore, like other textural attributes, should be used with great caution only as corroborative evidence for the final mode of deposition of a sediment. Examples of distinctive textures are *glacial* striations on pebbles, the frosted, dull appearance of **sand** subjected to protracted aeolian transport and the microscopic v-shaped depressions produced on **quartz sand** grains by subaqueous collisions. The scanning **electron microscope** is the best tool for the study of surface textures. Krinsley & Doornkamp (1973) show many examples. [DAR]

Blatt, H., Middleton, G.V. & Murray, R.C. (1980) *Origin of Sedimentary Rocks*. Prentice-Hall, Englewood Cliffs, NJ.

Folk, R.L. & Ward, W. (1957) Brazos River bar: a study in the significance of grain size parameters. *Journal of Sedimentary Petrology* **41**: 1045–58.

Krinsley, D.H. & Doornkamp, J.C. (1973) *Atlas of Sand Surface Textures*. Cambridge University Press, Cambridge.

Pettijohn, F.J., Potter, P.E. & Siever, R. (1973) *Sand and Sandstone*. Springer-Verlag, Berlin.

Tucker, M.E. (1981) *Sedimentary Petrology*. Blackwell Scientific Publications, Oxford.

clathrates The *hydrocarbon*-bearing ice molecules which constitute *gas hydrates*. Frozen water in the subsurface may contain *hydrocarbons* (*methane*, ethane, etc.) bound into the ice lattice. Such formations are restricted to regions of low temperature and high pressure — *gas hydrates* have been discovered in the deep ocean basins and extensively in the *permafrost* of Alaska and Siberia. These *permafrost* deposits, extending up to 3 km in depth, are potentially huge sources of *hydrocarbons*, but economic production techniques have yet to be developed. The origin of clathrates is unclear. Both shallow biogenic and inorganic crustal mechanisms have been invoked. [DMP]

clay dune Aeolian **dune** formed of clay-dominated aggregates rather than the more normal **quartz sand**-sized grains. They generally consist of about 30% of clay-sized material (though occasionally the figure may exceed 70%), and may also contain some **quartz** grains, silt and **evaporite** crystals. They form as the result of the **deflation** of coastal *lagoonal* and inland dried lake sediments, which then accumulate as **lunettes** on the **lee sides** of the depressions. They are widespread in southern Australia, parts of the Argentinian pampas, in the Kalahari and in Texas, especially in areas where the mean annual precipitation is between 200 and 500 mm (Bowler, 1973). [ASG]

Bowler, J.M. (1973) Clay dunes: their occurrence, formation, and environmental significance. *Earth Science Reviews* **9**: 315–38.

clay ironstone A sediment consisting largely of **siderite** that occurs in thin bands or **nodules** in some argillaceous rocks. The clay ironstones of the British Coal Measures were an important source of **iron** during the industrial revolution. [AME]

clay minerals The constituents of clays can be assigned to two groups: those called clay minerals, which by their nature give to the clay its plastic properties; and the others which are **accessory** or 'non-clay minerals' such as **feldspars**, **micas**, ferromagnesian minerals and **quartz**. In general terms the occurrence of as little as 10% of clay minerals gives a sediment a degree of the typical properties of a clay. The clay minerals have a number of characteristics in common. Their structures are those of *phyllosilicates*, based on composite layers built from components with tetrahedrally and octahedrally co-ordinated cations (see **mica minerals**). Most of them occur as microscopic platy particles in fine-grained aggregates which when mixed with water yield materials which have varying degrees of **plasticity**. This is due to the large ratio of surface area to weight in these particles (ranging from about $10\,m^2\,g^{-1}$ to several hundred $m^2\,g^{-1}$), so that the water adsorbed in a monomolecular layer on the platy surfaces causes the particles to adhere and to slip over one another when a clay mass is deformed.

Chemically, the clay minerals are hydrous silicates (principally of **aluminium**, **magnesium**, **iron** and **potassium**) which, on heating, lose adsorbed and constitutional water, and at high temperatures yield refractory materials. Important differences among the clay minerals, however, lead to their subdivision into several main groups. The four most important layered clay mineral groups are *kandites*, **illites**, **smectites** and *vermiculites*, which have characteristic basal spacings of approximately 7, 10, 15 and 14.5 Å respectively, although for some (the *kandite* mineral **halloysite**, **smectites** and *vermiculites*) the layer separation is variable since swelling may occur through the intercalation of water or organic liquids, and shrinkage may result from dehydration. The clay minerals *palygorskite* and **sepiolite** have chain-like structures and are less common than the layered clay minerals.

The particles of clay minerals may be crystalline or amorphous, platy or fibrous and, though nearly always small, may vary from colloidal dimensions to those above the limit of resolution of an ordinary microscope. Because of their typical small size they are generally studied by **X-ray diffraction** and *electron diffraction* methods. Important thermal effects occur when clay minerals are heated and these can be studied by *thermogravimetric* and *differential thermal analysis* (Mackenzie, 1957). *Infrared spectrometry* is used to study the hydroxyl groups in clay minerals. **X-ray diffraction** is the principal method used for the identification of clay minerals, together with relevant thermal and chemical tests (Brindley & Brown, 1984).

Chemical composition may vary according to the replacement of Si, Al and Mg by other cations, the nature and quantity of inter-layer cations and the water content. The clay minerals vary in their dehydration and breakdown characteristics and in their decomposition products, but they also differ in their cation exchange properties according to the nature of their inter-layer cations and residual surface charges. Their uses are many, some for example being particularly suitable for **drilling muds**, some for catalysts in **petroleum** processing, some as fillers or for surface coating in paper manufacture, as well as for clarification of wines, emulsification, cosmetics, etc.

The clay minerals are the main constituents of the argillaceous sediments, which on accumulation and **compaction** yield **shales** or **mudrocks**. The clays are generally the products of **weathering** or **hydrothermal** alteration, different clay minerals resulting according to physicochemical conditions and the nature of parent materials, e.g. **feldspars**, volcanic glasses, **micas**, **chlorites**, etc.

The *kandite* group includes **kaolinite**, **dickite**, **nacrite**, **halloysite** and meta-halloysite. **Kaolinite**, $Al_4[Si_4O_{10}](OH)_6$, is generally white and is the most important member of the *kandite* group; **dickite** and **nacrite** are rarer **polymorphs**. **Halloysite** with formula $Al_4Si_4(OH)_8O_{10} \cdot 8H_2O$ has a single layer of water molecules between its structural sheets, giving an inter-layer distance (d_{001}) of approximately 10 Å rather than 7.2 Å; the inter-layer water may be

replaced by glycol which causes an increase in d_{001} from 10 Å to about 11 Å. **Kaolinite**-group minerals (the *kandites*) are formed principally by the **hydrothermal alteration** or **weathering** of **feldspars**. The rocks so altered are usually the more acid types (**granites**, **granodiorites**, etc.). The **kaolinite** produced by **alteration** sometimes occurs *in situ* (e.g. in the **hydrothermal deposits** associated with the St Austell **Granite** in Cornwall). The name is from the Chinese *kao lin* (white hill), although the type material has been shown recently to be **halloysite**, hence the name China clay. **Kaolinite** is the most important clay mineral for the manufacture of fine china, stoneware, tiles, etc., the properties of the clay–water system being fundamental in the initial forming process.

The **illite** group represents the clay-grade micas and includes **illite** (named after the State of Illinois), *hydromicas*, *phengite* and **glauconite**. **Illite** has the general formula $K_yAl_4[Si_{8-y}Al_yO_{20}](OH)_4$ with $y < 2$ and usually 1–1.5; it thus differs from **muscovite** in being richer in Si and poorer in K. In *phengite*, K is not deficient and excess Si is compensated by replacement of $[Al]^6$ by $[Mg, Fe^{2+}]^6$. The more general term **sericite**, used for a fine-grained white mica, covers **muscovite**, *phengite*, *hydromuscovite* and mixed-layer aggregates. **Illites** are the dominant clay minerals in **shales** and **mudrocks** and may also occur in other sediments such as impure **limestones**. They may have been produced after deposition by the degradation of **feldspars** or sometimes by the **alteration** of other clay minerals. **Illites** may also have a **hydrothermal** origin and they are often found in **alteration** zones around *hot springs* and metalliferous **veins**. There is some experimental evidence that for both **hydrothermal** and sedimentary occurrences the formation of **illite** is generally favoured by alkaline conditions and by high concentrations of **aluminium** and **potassium**.

The **smectite** group includes **montmorillonite**, **beidellite**, **nontronite**, **hectorite**, **saponite** and *sauconite*. All are 'swelling' clay minerals in that they can take up water or organic liquids between their structural layers, and all show marked cation exchange properties. They are useful in the preparation of ceramics, **drilling muds**, paper, rubber, paints and moulding sands and are important in agriculture. **Montmorillonite** (named for Montmorillon, France) has the formula $(Na, Ca)_{0.3}(Al, Mg, Fe)_4[Si_4O_{10}](OH)_2 \cdot nH_2O$ whereas **saponite** has Mg > Al. **Beidellite** is, by comparison, poorer in Si and richer in Al. **Nontronite** contains appreciable Fe^{3+} and **hectorite** contains some Li substituting for Mg. The cation exchange capacity (c.e.c.) of clays in the **smectite** group is high and values are commonly in the range 80–120 meq./100 g clay, but vary with particle size and the nature of the cation, usually being greater for Ca than Na. The principal cause of cation exchange in **smectites** is the imbalance of charge in the structural layers and not the presence of unsatisfied surface valencies. The amount of inter-layer water adsorbed varies according to the type of **smectite**, Ca **smectites** usually taking up two layers of water molecules in each space, while in Na **smectites** the amount is variable and in general they show a greater swelling capacity. Various organic molecules can also be accommodated in inter-layer spaces, and because of this smectites (the principal constituent of **Fuller's earth**) are extensively used as decolorizing agents, for purifying fats and oils and in the refining of **petroleum**. **Montmorillonite** and **beidellite** are the main components in **bentonite** clay deposits formed by the **alteration** of eruptive **igneous rocks**, usually **tuffs** and volcanic ash, which also contain **cristobalite, zeolites, quartz, feldspar, zircon**, etc. **Smectites** also occur as **hydrothermal alteration** products around metalliferous **veins** and near *hot springs* and **geysers**. An alkaline environment, the availability of calcium and paucity of potassium and the weathering of basic rocks in conditions of poor drainage are all factors which appear to favour the formation of **smectites**. **Saponite** mainly occurs associated with mineral **veins** but is also found in **amygdaloidal** cavities in **basalt**. **Nontronite** is found both in mineral **veins** (often with **opal** and **quartz**) and as an **alteration** product of volcanic glass.

Vermiculite's name is derived from the Latin *vermiculare* (to breed worms), alluding to the peculiar **exfoliation** phenomenon exhibited when specimens are rapidly heated. In its natural state the mineral has little useful application, but when exfoliated it provides a low density material with excellent thermal and acoustic insulation properties. Its general formula is (Mg, Fe^{2+}, Al)$_3$[(Al, Si)$_4$O$_{10}$](OH)$_4$. 4H$_2$O but the charge deficiency caused by tetrahedral substitution of Al for Si is compensated by about 0.6 divalent cations, or their equivalent, between the layers. In natural specimens these are most commonly Mg^{2+}; accordingly the c.e.c. of *vermiculites* (100–260 meq./100 g) is the highest of the clay minerals. The **exfoliation** which occurs when *vermiculite* is heated suddenly to about 300°C or more is due to the rapid generation of steam which cannot escape without buckling and separating the structural layers. One of the two main types of occurrence of *vermiculite* is as an **alteration** product of **biotite** either by **weathering** or by **hydrothermal alteration**. The second major occurrence is in the contact region between acid intrusive and basic or ultrabasic rocks, when it may be associated with **serpentine**, **talc**, **corundum** and **apatite**. *Vermiculites* are also found associated with **carbonatites** and in metamorphosed **limestones**. [RAH]

Bailey, S.W. (ed.) (1988) *Hydrous Phyllosilicates Exclusive of Micas.* Reviews in Mineralogy No. 19. Mineralogical Society of America, Washington, DC.

Brindley, G.W. & Brown, G. (eds) (1984) *Crystal Structures of Clay Minerals and their X-ray Identification* (3rd edn). Mineralogical Society Monograph **5**.

Deer, W.A., Howie, R.A. & Zussman, J. (1966) *An Introduction to the Rock-Forming Minerals.* Longman, London.

Farmer, V.C. (ed.) (1974) *The Infrared Spectra of Minerals.* Mineralogical Society Monograph **4**.

Gard, J.A. (ed.) (1971) *The Electron-Optical Investigation of Clays.* Mineralogical Society Monograph **3**.

Grim, R.F. (1968) *Clay Mineralogy* (2nd edn). McGraw-Hill, New York.

Mackenzie, R.E. (ed.) (1957) *The Differential Thermal Investigation of Clays.* Mineralogical Society Monograph **2**.

Newman, A.C.D. (ed.) (1987) *Chemistry of Clays and Clay Minerals.* Mineralogical Society Monograph **6**.

clear water erosion The channel **erosion** caused by rivers whose sediment load has been removed by the construction of a dam and a reservoir. With a reduced sediment load, the size of which will depend on the trap-efficiency of the reservoir, incision occurs rather than **aggradation**. This can create problems for bridges and other man-made structures downstream. [ASG]

cleat The closely spaced **jointing** found in **coal seams**. There are usually two approximately orthogonal sets of **joints** roughly normal to the **bedding**. The **fracture** planes are much more closely spaced than the **joints** in the adjacent non-coal strata. Cleat planes are important in the exploitation of **coal** both in regard to the cutting of the **coal** and because they affect the size of the fragments produced. The latter point is important in the design of **coal** processing plants and in **coal** utilization and marketing. [AME]

cleavage (1) The property of a mineral whereby it breaks along regular crystallographic planes (related to the internal **crystal structure**). Cleavage is labelled by combining the Miller index of the crystallographic plane with a term describing the perfection of the cleavage (in descending order of perfection: ϵminent, perfect, distinct, imperfect). (2) A **foliation** produced by **deformation** at low metamorphic grade, along which a rock preferentially splits. Cleavage is tectonically produced **fabric** in distinction to **bedding**-parallel **fissility** which is of sedimentary or compactional origin. The degree of localization of cleavage planes may vary from a *spaced cleavage*, in which cleavage planes are separated by uncleaved slices of rock (**microlithons**), to thoroughly **penetrative**. Cleavage can be classified by mechanism of formation into four end-member types of cleavage, shown in the *cleavage tetrahedron* (Fig. C19), but all gradations may occur between them and several types and orientations may be present in a rock. The three types of cleavage on the base of the tetrahedron are all *spaced cleavages*.

1 *Slaty cleavage*. Defined by parallel alignment of inequant *phyllosilicate* grains too small to be visible to the naked eye, e.g. **illite, chlorite, muscovite, biotite**. This is the most

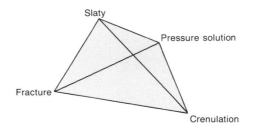

Fig. C19 End-member types of **cleavage**. (After Ramsay & Huber, 1983.)

common type of cleavage responsible for the ease of splitting of **slates**. The **foliation** may be defined by distinct bands (**domains**) of well-orientated *phyllosilicates* separating domains of other minerals (e.g. **quartz**) in which any *phyllosilicates* present may be equant and show no **preferred orientation**. **Domains** may have lenticular shapes surrounded by *phyllosilicate* films, or the cleavage minerals may form 'beards' at ends of inclusions of other materials. There is a complete spectrum of *slaty cleavage* from the highly domainal to thoroughly **penetrative** in which the **fabric** is completely homogeneous but **anisotropic**. The origin of *slaty cleavage* remains a matter of controversy: two basic mechanisms have been suggested. The mechanical model proposes that **preferred orientation** is produced entirely by either rigid rotation or **buckling** of pre-existing *phyllosilicate* grains. Extensive **buckling** of an earlier fabric would produce a *crenulation cleavage* (see below). A metamorphic origin by recrystallization and growth of *phyllosilicates* in a new direction has also been suggested. When material enters an intermediate fluid phase, this results in a *pressure-solution cleavage* (see below). If the *phyllosilicate* grain size is coarse enough to be visible, the foliation is referred to as a **schistosity**. (See also **gneissosity**.)

2 *Fracture cleavage*. A cleavage defined by closely spaced **fractures**. The tabular rock bodies between **fractures** are called **microlithons**. *Fracture cleavage* is most common in **competent** units, e.g. *psammites*, where it may coexist with and grade into *slaty cleavage* in interlayered *pelites*. The **fracture** spacing may be quite variable from microscopic to metre scale. The origin of *fracture cleavage* is not clear; it has been suggested that **fractures** form by parting on earlier cleavage surfaces to allow more competent units to accommodate **folding**.

3 *Crenulation cleavage*. Defined by folding of a pre-existing **fabric** on a microscopic scale. *Crenulation cleavage* generally forms by buckling of a former *slaty cleavage*. The **foliation** results from parallel alignment of grains in the limbs of the tight to isoclinal buckles, which may be faulted. When the micro-folds have an angular, **chevron** geometry, the *crenulation cleavage* is defined by **kink bands** in the earlier **foliation**. The localization of *preferred orientation* on limbs of buckles means that *crenulation cleavage* is always **domainal**. Well-developed *crenulation cleavage* leads to a mechanical differentiation in the rock components; *phyllosilicates* are concentrated in the **crenulation domains** leaving the remaining components in the hinge areas, ultimately resulting in **transposition**. When the primary **foliation** is deflected in a sense indicating **extension**, the secondary cleavage is known as an *extensional crenulation cleavage*. Two *crenulation cleavages* may develop in some cases, intersecting in an angle of 60–90° and symmetrical about the earlier **foliation**: these are *conjugate crenulation cleavages*. The mechanical origin of *crenulation cleavage* by **buckling** of an earlier **fabric** is not in doubt, but the process may be assisted by any amount of solid-state or solution transfer of material, or by **fracture**. This type of cleavage is also referred to as *strain-slip cleavage*.

4 *Pressure solution cleavage*. Defined by closely spaced solution surfaces (**stylolites**). The tabular bodies between solution surfaces may also be referred to as **microlithons**. Solution surfaces defining the **foliation** are commonly planar and do not have the characteristic interlocking aspect of **stylolites**. Their spacing is as variable as *fracture cleavage* surfaces, to which they may be similar; they may also have similarities to *crenulation cleavage* surfaces. Preferential solution and removal of non-phyllosilicates (e.g. **quartz**) from the solution surfaces causes differentiation of the rock components and assists the development of **transposed foliations**; these may cause a marked division of the rock into light and dark stripes. The origin of *pressure solution cleavage* is primarily due to solution transfer; mechanical **buckling** and **fracture** may also play a part, as suggested by the similarity of **pressure-solution** to other types of cleavage.

All types of cleavage are commonly associated with **folds**, to which they are often approximately axial-planar. In this case, the cleavage-**bedding** intersection defines a **lineation** which is parallel to the *fold hinge* line (Fig. C20).

A *cleavage fan* is a radiating pattern of cleavage surfaces. *Convergent fans* radiate from a focus on the concave side of a folded layer (also known as normal fans). *Divergent (reverse) fans* radiate from a focus on the convex side of the layer (Fig. C20). *Convergent fans* are similar to the pattern of the *XY* plane of *principal strain* produced during folding by *tangential longitudinal strain*, while divergent fans are comparable to the pattern due to *flexural slip*. This similarity suggests that cleavage forms in the *XY* plane of **principal strain** (perpendicular to the axis of maximum **shortening**). Both convergent and divergent fans may coexist in a folded multilayer in which the *convergent fans* are found in more **competent** layers (perhaps deforming by tangential longitudinal **strain**) (stippled in Fig. C20) and divergent fans in the less competent layers (deforming by *flexural slip*) (blank in Fig. C20). The change of orientation between adjacent layers is an example of *cleavage refraction*.

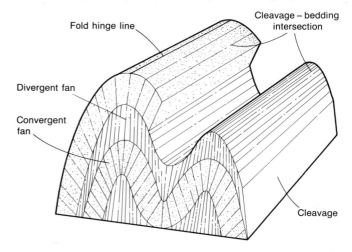

Fig. C20 Geometric relationships between **cleavage** and folded bedding. (After Ramsay & Huber, 1983.)

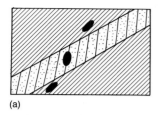 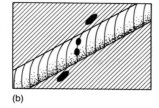

(a)   (b)

Fig. C21 Cleavage refraction in a more competent layer within a less competent layer.

Cleavage refraction is the change of orientation of cleavage usually across layer boundaries. The change in orientation may be gradual (producing a curved cleavage surface (Fig. C21b), or sharp (Fig. C21a)), depending on the nature of the layers, but the cleavage is recognizably the same feature in both cases and may be traced from one layer into the next. It is often accompanied by a change in the nature of cleavage, for example from *slaty cleavage* in *pelitic* units to *fracture cleavage* in psammitic units. In multilayers of contrasting **competence**, cleavage often changes orientation between the less and more **competent** layers. Characteristically the more competent layers (e.g. psammitic) exhibit a *convergent cleavage fan* with respect to a **fold** while less **competent** (e.g. pelitic) layers have an *axial planar* or *divergent cleavage fan* (Fig. C20). This pattern mimics the *XY* planes of **principal strain** in layers deformed by *tangential longitudinal strain* and *flexural slip* respectively. The origin of *cleavage refraction* is therefore considered to reflect a change in orientation of the *XY* plane of **principal strain** in different layers.

Cleavage transection is the oblique intersection between cleavage and a *fold axis*. In a perfectly **axial-plane cleavage** the *fold axis* lies within the cleavage plane. However, small angles may exist between the *fold axis* and the cleavage plane, described by the amount and sense of rotation of the acute angle from the *fold axis to* the cleavage, either clockwise or anticlockwise (Fig. C22). *Cleavage transection* is generally taken to imply formation of cleavage and **fold** at separate times, possibly relating to a rotating **strain** field. If **folds** and cleavage are both produced in a **shear zone**, and the order of folding and cleavage can be established, the sense of transection may be used as a **shear** sense indicator.

The intersection of the cleavage and *fold limbs* can therefore define position on the **fold** relative to adjacent *fold axes* (Fig. C23), which is an important aid to mapping where *fold hinges* are not visible. This relationship between cleavage and *fold limbs* is described more accurately by the term *cleavage vergence*. This is the horizontal direction of rotation, through the acute angle, from the cleavage to an earlier **fabric** within the plane normal to the cleavage/**fabric** intersection. The acute intersection of cleavage with a *fold limb* in the plane perpendicular to the axis of intersection can be used to define a unique direction of *cleavage vergence* (Fig. C23). For an **axial plane cleavage**, this changes by 180° across the *fold axial surface*. The *cleavage*

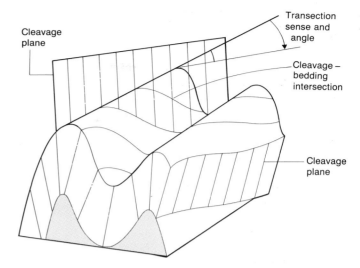

Fig. C22 Cleavage transection.

vergence is therefore very useful to define position with respect to a *fold axis* when the *hinge* area of the **fold** is not visible.

It is widely assumed that cleavage marks the *XY* plane of *principal finite strain*, normal to the axis of maximum **shortening**. This has been demonstrated to be true in some cases to within the limits of measurement accuracy by **strain** analysis of deformed objects, and can be justified by comparing observed and modelled cleavage patterns. However, **shear** displacements of a consistent sense are clearly visible along some cleavage planes, in which case they cannot represent a plane of **principal strain**. These cases (common for *crenulation* and *fracture cleavages*) may either be cleavages formed initially parallel to the *XY* plane of **principal strain** and subsequently used as **shear** surfaces, or they may form in planes of **shear strain** at the outset. Cleavages may therefore have a variety of rela-

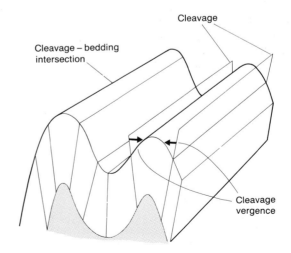

Fig. C23 Cleavage vergence.

tionships to the orientation of **finite strain**. Since many of the cleavage-forming mechanisms are **strain**- and temperature-dependent, there is also no simple relationship between magnitude of **strain** and cleavage formation.

Cleavage is characteristic of **deformation** under low grades of metamorphism. It will not form in **diagenetic** conditions but may occur at any stage of the *anchizone* or *epizone*, and is common throughout the *greenschist* facies. At higher grades, *slaty cleavage* becomes **schistosity**. [TGB]

Bell, A.M. (1981) Vergence: an evaluation. *Journal of Structural Geology* **3**: 197–202.
Means, W.D. (1977) Experimental contributions to the study of foliations in rocks: a review of research since 1860. *Tectonophysics* **39**: 329–54.
Ramsay, J.G. & Huber, M.I. (1983) *The Techniques of Modern Structural Geology*. Academic Press, London.
Treagus, S.H. (1983) A theory of finite strain variation through contrasting layers, and its bearing on cleavage refraction. *Journal of Structural Geology* **5**: 351–68.
Williams, P.F. (1977) Foliation: a review and discussion. *Tectonophysics* **39**: 305–28

cleavelandite A white, platy variety of the **plagioclase feldspar** mineral **albite**. (See **feldspar minerals**.) [DJV]

climatic geomorphology This branch of **geomorphology** is the study of landforms determined by climate. The subject is generally regarded as originating towards the end of the 19th century when attempts were made to explain distinctive landforms encountered by scientific explorers during expeditions into remote areas. The concept of climatic geomorphology is that a specific climate will produce a characteristic suite of landforms.

On the basis of work in south-western USA, W.M. Davis, regarded by some as the founder of the subject, proposed a **cycle of erosion** in which all landscapes are a product of three variables: structure, process and time. His initial treatise, based on the landscapes observed in temperate climates, which Davis considered developed under 'normal' **erosional** processes, was later expanded to incorporate arid and *glacial* regions, which he regarded as deviants from his normal cycle. The geographical cycle proposed by Davis was by implication related to climate, though European workers criticized him for the lack of emphasis on the importance of climatic control on geomorphic processes.

At the beginning of the 20th century, theoretical approaches to the classification of morphoclimatic regions were also being carried out in Europe. However, attempts, such as that by Penck in 1910, have been criticized both due to over-concern with climate and the climatic factors considered. The approach taken by Penck was essentially **hydrological**. He recognized three climatic provinces based on whether evaporation was greater, less or equal to precipitation. In a particularly perceptive article, de Martonne (reprinted in Derbyshire, 1973) used humidity and temperature as climatic factors in his classification of six morphoclimatic regions. Although he only presented qualitative evidence, he related climatic variations to relief forms with particular regard to the morphology of his regions. He also introduced the importance of scale in the recognition of structurally controlled features in small-scale studies compared to large-scale climatic control observed between regions. This was pursued in later years by Büdel, Tricart, Cailleux and Birot. Others have used mean annual temperature and precipitation as climatic factors on which they base their morphoclimatic regionalization, for example Peltier in 1948, who recognized nine climatic regimes which he regarded as geomorphologically significant. Following the cycle approach of Davis, Peltier considered there to be a cycle of landform evolution within each of his climatic regimes.

In the early 1960s more quantitative approaches were attempted. Tanner (1961), for example, developed a model which combined climatic data, based on precipitation and potential evaporation data collected from various parts of the world, with geological controls on landform evolution. He suggested that four main climatic regions could be recognized and he incorporated these into his model, together with three geological controls: rocks with horizontal and dipping **beds**, where structure and/or stratigraphy control the **weathering**, and isotropic rocks. His model gives examples of the general landforms and topography that result within these broad categories of climate and structure. Peltier (1962), through a terrain analysis in various climatic regions, quantified mean slope angles within sample areas. By considering the number of drainageways per mile under a variety of climatic regimes it was found that different topographic slope and topographic texture were obtained in different climatic regions.

The most significant contributions to the subject have been from Büdel, Tricart and Cailleux. Büdel proposed seven large-scale climatic-morphological zones based not only on the **denudational** and **erosional** processes predominating in different climatic regions but also on the product of those processes. In his regionalization on a global scale Büdel recognizes three main templates of continental relief-forming: extreme valley cutting in the unglaciated polar zone, extreme planation in subtropical zones and *glacial* action in the subglacial relief zone (Fig. C24). The two former he regarded as the key templates in terms of process activity, and that they were of far greater significance in *paleoclimates* than the present day. Tricart and Cailleux classify morphoclimatic regions based on principle and subdivisions of climatic and biogeographic zones together with *paleoclimatic* factors. The morphoclimatic regions proposed by Tricart & Cailleux (1965) attach much importance to the role of vegetation in the complex interrelationship between climate and landforms.

Classification of continental surfaces into climatic zones with characteristic landform suites has been an important aspect of climatic geomorphology. The work has been mainly undertaken by the French and German Schools, and is useful in that it demonstrates the complications that

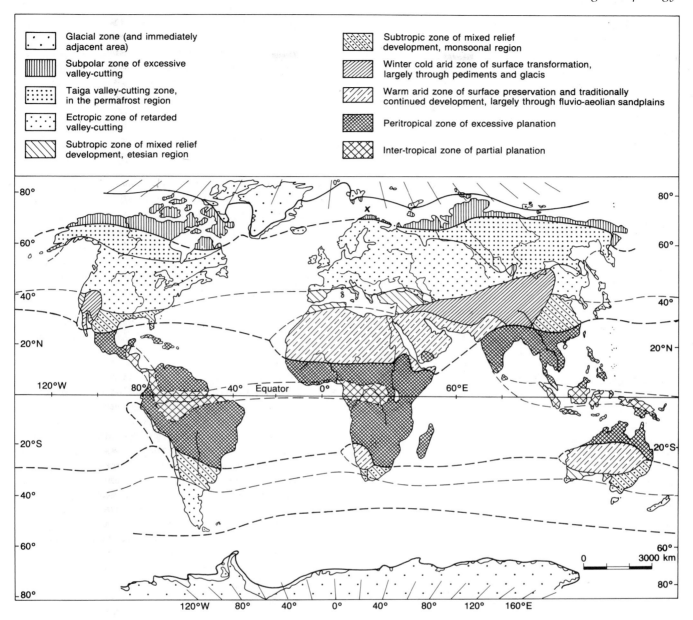

Fig. C24 Climatic geomorphology. Present-day morphoclimatic zones. (After Büdel, 1982.)

arise when largely theoretical concepts are applied to the real world. Two of the main criticisms of such work have been that the parameters used are too crude to provide meaningful morphoclimatic regions, and that such large-scale regionalization is of little practical value.

Throughout the development of the subject many limitations have been raised. One frequent criticism has been the lack of work into the processes determining landform evolution. As early as 1926, Passarge recognized that climate/geomorphological regions were untenable as, in many parts of the world, climatic influence is tempered by vegetation and **weathered** rock material. On the other hand, Büdel (1982) maintains that since 95% of mid-latitude features are relic in terms of present processes, interpretation of relief in order to determine the relief-forming mechanisms is a useful approach. A second major criticism, and one of the most fundamental, has been the premise, now widely accepted as erroneous, that present-day features can be regarded as being solely a function of present-day processes (climate). This problem is addressed in the development of **climato-genetic geomorphology**. [RDS]

Büdel, J. (1982) *Climatic Geomorphology.* Princeton University Press, Princeton, NJ.

Davis, W.M. (1899) The geographical cycle. *Geographical Journal* **14**: 481–504.

Derbyshire, E. (ed.) (1973) *Climatic Geomorphology*. Macmillan, London.

Holzner, L. & Weaver, G.D. (1965) Geographic evaluation of climatic and climato-genetic geomorphology. *Annals of the Association of American Geographers* **55**: 592–602.

Peltier, L.C. (1962) Area sampling for terrain analysis. *The Professional Geographer* **14**: 24–8.

Stoddart, D.R. (1969) Climatic geomorphology. In: Chorley, R.J. (ed.) *Water, Earth and Man*, pp. 473–85. Methuen, London.

Tanner, W.F. (1961) An alternative approach to morphogenetic climates. *Southeastern Geology* **2**: 251–7.

Tricart, J. & Cailleux, A. (1972) *Introduction to Climatic Geomorphology*. Longman, London.

climato-genetic geomorphology The study of landforms in terms of paleo- as well as current climatic influence. The term was proposed by Büdel in 1963, when he saw the task of climato-genetic geomorphology as distinguishing between relief generations made by previous climates in the present-day relief. It is founded on the as yet unsubstantiated premise that the influence of particular climates is expressed systematically in a suite of characteristic landforms. The importance of fossil climates in attempts to explain landforms was recognized as early as the time of Louis Agassiz in the 19th century and by others early in the 20th century (e.g. de Martonne (1913) in Derbyshire (1973)).

Büdel recognizes three major relief generations in the mid-latitudes. The earliest are the *Tertiary* **etchplain** systems, that span from as early as the Late **Cretaceous** to the Upper **Pliocene**. This period was one of stable conditions with landform development associated with processes of current peritropical, seasonally wet zones. The surfaces generated during this period are not followed by present-day valley networks. The second period, which spans the Late **Pliocene** to the Lower *Pleistocene*, is regarded as a period of transition, with no present-day morphoclimatological equivalent. The third period is the *Pleistocene* **ice age**, which completely transformed the previous relief through the action of continental ice sheets, ice fields in mountainous areas and the development of deep *permafrost* in *periglacial* areas. The period was one of extreme valley cutting as characterized in the present-day polar and sub-polar morphoclimatologic regions.

The warm post-glacial *Holocene* Büdel regards as having had little impact on the landforms of the mid-latitudes, other than to modify the relief with the development of a deep *soil* cover. [RDS]

Büdel, J. (1982) *Climatic Geomorphology*. Princeton University Press, Princeton, NJ.

Derbyshire, E.D. (ed.) (1973) *Climatic Geomorphology*, pp. 61–75. Macmillan, London.

climbing dune Depositional **dune** which develops against an obstacle such as a hill or cliff. Climbing dunes are dependent on a topographical feature for their development and are therefore essentially stationary **dune** forms. Often found on the more gentle slope angles along an **escarpment**, sometimes in association with *echo dunes*, which form on steeper slope angles. Wind tunnel modelling by Tsoar (1983) showed that climbing dunes can develop on slope angles of up to 50°. [RDS]

Tsoar, H. (1983) Wind tunnel modelling of echo and climbing dunes. In: Brookfield, M.E. & Ahlbrandt, T.S. (eds) *Aeolian Sediments and Processes*. Developments in Sedimentology No. 38, pp. 247–59. Elsevier, Amsterdam.

climbing ripples The preservation of **ripple cross-stratification** in which the internal set boundaries are inclined at some angle to the horizontal giving the impression that the **ripples** are climbing over one another (Fig. C25). The angle of ripple climb is determined by the ratio of horizontal bedload movement to vertical **aggradation**. As this ratio decreases the ripple climb angle increases, generating in turn thin laminae, various angles of ripple climb and eventually draped laminae where only vertical **aggradation** occurs. Where the angle of climb exceeds the angle of the bedform **stoss side** the entire bedform profile may be preserved. [JLB]

Allen, J.R.L. (1968) *Current Ripples*. North Holland, Amsterdam.

Allen, J.R.L. (1984) *Sedimentary Structures: Their Character and Physical Basis*. Elsevier, Amsterdam.

Ashley, G.M., Southard, J.B. & Boothroyd, J.C. (1982) Deposition of climbing-ripple beds: a flume simulation. *Sedimentology* **29**: 67–79.

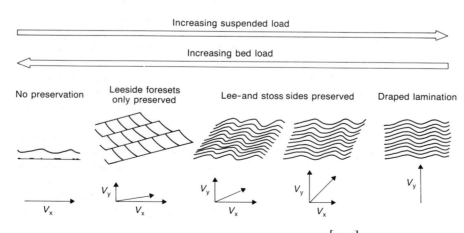

Fig. C25 Climbing ripples. The dependence of ripple climb angle and the stratification produced upon the ratio of horizontal bedload movement (V_x) to vertical aggradation through suspension fallout (V_y). (After Ashley *et al.*, 1982.)

clinochlore A variety of the layer silicate mineral **chlorite**. [DJV]

clinoenstatite (MgSiO₃) A *clinopyroxene* mineral. (See **pyroxene minerals**.) [DJV]

clinoferrosilite (FeSiO₃) A *clinopyroxene* mineral, the **iron** end-member. (See **pyroxene minerals**.) [DJV]

clinohumite (Mg₉(SiO₄)(F,OH)₂ An island silicate mineral (*nesosilicate*) of the **humite** group. [DJV]

clinohypersthene ((Mg, Fe)SiO₃) A *clinopyroxene* mineral. (See **pyroxene minerals**.) [DJV]

clinometer Instrument used in the field to measure the inclination of a planar surface (**dip**) or a **lineation** (**plunge**), and often combined with a compass in order to measure the orientation of planes or lines with reference to geographic co-ordinates. [RGP]

clinozoisite (Ca₂Al₃O(SiO₄)Si₂O₇(OH)) A double island silicate mineral (*sorosilicate*) of the **epidote** group. [DJV]

clint Term originating from Northern England used to describe tabular blocks of **limestone** within **limestone pavements**. Clint blocks are divided by **grikes**, or runnels, which form along **joints** and other planes of weakness. The German term for clints is *Flachkarren*. Clint surfaces are often flat, although **kamenitzas** and other **karren** features may develop upon them. The size and shape of clints depends upon rock **structure** and the intensity of **erosional** processes. [HAV]

Goldie, H.S. (1978) Morphometry of limestone pavements of Farleton Knott (Cumbria, England). *Transactions, British Cave Research Association* 8: 207–24.

Clinton ironstone Oolitic hematite–chamosite–siderite rock. The **iron** content is about 40–50% and it is higher in Al and P than **banded iron formation**. It also differs from BIF in the absence of **chert** bands, the **silica** being mainly present in iron silicate minerals and as clastic **quartz** grains. Clinton ironstones form lenticular beds usually 2–3 m thick and appear to have formed in shallow water along the margins of continents. It is common in rocks of **Cambrian** to **Devonian** age. One of the best examples is the **Ordovician** Wabana Ore of Newfoundland. This rock once formed an important source of **iron ore** in North America. [AME]

Gross, G.A. (1970) Nature and occurrence of iron ore deposits. In: *Survey of World Iron Ore Resources*, pp. 13–31. United Nations, New York.

clintonite (Ca(Mg, Al)₃₋₂Al₂Si₂O₁₀(OH)₂) A layer silicate (*phyllosilicate*) and a *brittle mica*. (See **mica minerals**.) [DJV]

clitter A scatter of large, **granite** boulders on the land surface — especially such as those found on Dartmoor in south-west England. The blocks forming clitter are not piled or stacked such as the core boulders creating a **tor**. Rather, they are a type of block-stream or *blockfield* which perhaps developed during the formation of the **tor**. Clitter may comprise a surface *lag* of isolated **corestones** exhumed during **erosion** of a **deep weathering** profile on **granite**. [AW]

cluse A steep-sided valley cutting through a mountain ridge. Originally, such a valley in the Jura Mountains of eastern France, but now used as a more general term. [AW]

coal A readily combustible organoclastic sedimentary rock composed mainly of lithified plant remains and containing more than 50% by weight of carbonaceous material and inherent moisture. *Coalification* progresses from moist, spongy partially decomposed vegetation such as **peat**, through **brown coal** (**lignite** and *sub-bituminous coal*), **bituminous coal**, semi-anthracite to **anthracite**. This transition is a response to **diagenesis** associated with burial and tectonic activity. These are the **banded** or *humic* **coals**. The other main group contains the non-banded or **sapropelic coals** derived from *algae*, spores and finely divided plant material. During *coalification* the percentage of **carbon** increases, volatiles and moisture are gradually eliminated, the calorific value increases and so does the *reflectance* of the *vitrinite* content.

Microscopic examination indicates that coal consists of particles and bands of different kinds of carbonaceous material. These are the coalified remains of plant material that existed at the time of **peat** formation. They differ from each other in their morphology, hardness, optical properties and chemical characteristics. They are analogous to the minerals that form other rocks, but they are organic materials characterized by botanical structure rather than crystallographic properties and are called **macerals**.

In addition to the assemblage of **macerals**, the general appearance, chemical composition and petrographical properties of coal are affected by the effects of post-depositional increases in pressure and temperature. A coal little affected by burial or tectonism such as **brown coal** is called a soft or low **rank** coal, whilst one much modified by these processes is called a hard or high **rank** coal.

Coal may be described in various ways, e.g. as an organic sediment with certain chemical and physical properties which determine its economic value and uses. It may also be considered from the petrological point of view as a mildly metamorphosed sediment with properties dependent on the nature of the original materials or **maceral** assemblage (coal type) and the degree of diagenesis or metamorphism which it has suffered (**coal rank**).

The accumulations of **peat** that were altered to **bituminous coals** originated in ancient **swamps** and marshes in **deltas**, *lagoons* and **estuaries**. Close comparisons can be made between the repetitive sedimentary cycles (**cyclothems**) of **coal measures** and Recent **deltaic** deposits. Extensive and thick **peat** deposits are forming today from vegetation of fresh to brackish water marshes and **levée**-flank **swamps** in coastal interdistributary **basins** and from cypress-gum **swamps** in broad inland flood **basins** of the Mississippi **deltaic** plain. Older **peat** accumulations in this **delta** have been buried beneath layers of clastic sediment and are on their way to becoming coals.

Coals occur almost entirely in **Phanerozoic** strata. The best-known are the high **rank Carboniferous** coals of Europe, Asia and North America and those of **Permian** age present throughout the former continent of **Gondwanaland**. The **Mesozoic** is also important for high **rank** coals in North America and Australia. The *Tertiary* strata of Europe, North America, Australia and parts of south-eastern Asia contain important deposits of *sub-bituminous coal* and **lignite**. Reserves of coal recoverable under present-day technical and economic conditions are estimated to be 0.899×10^{12} tonnes. (See **lithotypes in banded coal, macerals** and **microlithotypes of coal**.)

The first use of coal as fuel by man is recorded from Czechoslovakia during the later part of the Paleolithic period (*c.* 35 000–8000 years BP). [AME/OWT]

Ward, C.R. (ed.) (1984) *Coal Geology and Coal Technology.* Blackwell Scientific Publications, Melbourne.
Wilson, C.L. (1980) *Coal — Bridge to the Future.* Ballinger, Cambridge, MA.

coal ball Spheroidal to irregularly shaped mass of minerals in a **coal seam**. **Calcite**, **dolomite**, **siderite** and **pyrite** are the common constituents. [AME]

coal basin A sedimentary **basin** within which economically important **coal seams** have been deposited. **Deltas** that formed on stable, **rifted** continental margins are the most likely to contain extensive **coal** deposits. These tend to be developed close to the **basin** margins and are termed *paralic*. The **Carboniferous coals** of northwestern Europe were deposited in shallow, near-coastal areas and are good examples of *paralic* coal basins. River **deltas** in freshwater lakes may also develop in intracratonic settings in continental interior **basins** and **graben**, or in **strike-slip basins**. These are known as *limnic basins* and some young examples are important for their **lignite** deposits. [AME]

coal classification Various properties are used for the classification of **coal**. These include the percentages of **carbon**, hydrogen and volatiles, the specific energy or *calorific value* and the coking and agglomerating properties. These properties may be measured and reported on in various ways. Most analyses are performed on air-dried samples. This method excludes the surface moisture always present in **coal** that is mined, shipped and delivered to the customer. Analyses on this material are reported as 'as received' or 'as sampled'. Other analyses may be presented as 'dry' or 'moisture free', representing **coal** after removal of both surface and inherent moisture. For data described as 'dry, **ash**-free' (d.a.f.) the analysis is recalculated after subtraction of the **ash** and moisture content whereas 'dry mineral matter-free' (d.m.m.f.) excludes volatile mineral matter (e.g. CO_2 from carbonates, SO_2 from sulphides and H_2O from clays) as well as **ash**.

Many different schemes have been drawn up for the classification of coals and most require reference to tables that are too extensive to reproduce here. Seyler's classification dating from 1933 is still much used (Fig. C26). It is based on plots of **hydrogen** versus **carbon** or, alternatively, of volatile matter versus calorific value. Most **coals** plot within the curved band. Low **rank coals** have lower **carbon** and higher **hydrogen** contents than high **rank coals** and fall on the right of the diagram. Carbonaceous coals (**anthracites**) with very low **hydrogen** contents plot at the other end of the band. The meaning of other terms can be read off the diagram.

The American Society for Testing and Materials (A.S.T.M.) classification is widely used in North America and other parts of the world. It is based on variations in the **fixed carbon** content and the *calorific value*. The International Coal Classification drawn up by the United Nations Economic Commission for Europe uses a series of numbers to represent the chemical and physical properties of the **coal**. In the United Kingdom the scheme devised by

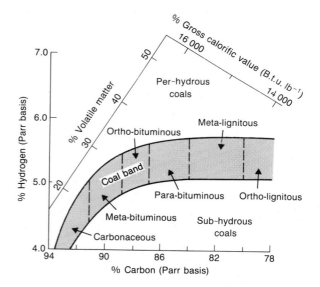

Fig. C26 Seyler's classification of **coal**. All results are given on a dry, mineral free basis (Parr formula). (After Ward, 1984.)

the former National Coal Board (now British Coal) is used. This is again a series of numbers expressed on a d.m.m.f. basis to represent the chemical and physical properties that represent the usefulness of the **coal** for **coke** production and steam-raising applications. The Australian Standard Coal Classification is similar to the International Coal Classification. [AME]

Ward, C.R. (ed.) (1984) *Coal Geology and Coal Technology*. Blackwell Scientific Publications, Melbourne.
Williamson, A. (1967) *Coal Mining Geology*. Oxford University Press, Oxford.

coal gas (*town gas*) The general name for gas used for heating and lighting which is produced by the distillation of **bituminous coal** and has a *calorific value* of about $18\,MJ\,m^{-3}$. Various processes have been used but all produce a gas containing more than 50% **hydrogen** and 10–30% *methane*. The other main components are higher *hydrocarbons*, CO, CO_2 and N_2. [AME]

coal measures Any series of strata containing economically workable **coal seams**. Derived from the old British name for the uppermost of the three lithological divisions of the **Carboniferous** System. [AME]

coal petrology The study of the constituent parts and textures of **coal**. For macroscopic description **coals** are divided into **banded coals** and **sapropelic coals** (see also **lithotypes in banded coal**). Although some petrographic studies of **coal** employ thin sections, most modern examinations use polished sections of crushed **coal** particles mounted in epoxy resin. The microscopic technique is similar to that used in studies of opaque minerals in rocks and **ores**. The nature and shape of the particles (the **macerals**) is noted and their *reflectance* measured. [AME]

coal quality maps Various types of contour maps that portray changes in the nature of a seam throughout a mine area or **coalfield**. They may delineate areas of **coal** most suitable for mining and the degree of quality variation that may be expected as exploitation proceeds. The most frequently used are *iso-ash* and *iso-sulph* maps indicating trends in the **ash** and total **sulphur** contents. [AME]

coal seam A bed of **coal**. Coal seams may vary in their lithotype profiles (see **lithotypes in banded coal**) to a considerable extent, but some have characteristic profiles that remain fairly constant over wide areas. Others may contain marker **beds** of distinctive character which can be recognized in many parts of a **coalfield** and used for correlation purposes. A log of *durains*, *fusains* and non-coal materials in the seam is particularly useful. [AME]

coal tar Tar produced by the distillation of **bituminous coal**. Further distillation produces benzene, toluene, xylene, anthracene, phenol, etc. [AME]

coalfield A region rich in deposits of **coal**. [AME]

coastal aquifer An **aquifer** that reaches out from the coast, so that part lies under the sea and is accessible to seawater. The denser saline **groundwater** penetrates beneath the freshwater in the **aquifer** to lie below it in the coastal areas. Deep **wells** close to the coast may penetrate the saline water layer. On the other hand, pumping from a **well** in the overlying freshwater may cause the saline water–freshwater interface to rise and salt water may be drawn into the **well**. [AME]

coastal bar A linear accumulation of mainly sandy coastal sediment generally lying submerged in the nearshore zone, more or less parallel to the coastline. Insufficient knowledge of bar types and processes has resulted in confusion in the literature over the use of the term. The description **bar** has been used for features intermittently exposed by **tides** (*tidal bars*), permanently submerged (*submerged* or *offshore bars*), located across a river mouth or harbour (*harbour bar*) and even **bars** permanently exposed above water level (*bay bar*), although this last category is now called a *barrier-beach* to avoid confusion. Amongst the **wave**-formed bars there appears to be a simple classification between those permanently submerged and those exposed intertidally. The most common of the *permanently submerged bars* occurs in the nearshore as a response to the breaking of steep, high energy **waves**. These **waves** move sediment onshore seaward of the break-point and offshore landward of it and so sediment accumulates in a *break-point bar*. Such bars do not build above sea-level since **wave** breaking intensifies as the water depth over the bar decreases thus increasing the rate of seaward sediment transport. As with all bars, *break-point bars* may occur singly or as multiple features. *Crescentic bars* lie just seawards of low water and are regularly spaced arcs of between 100 m and 2000 m in length. The horns of these crescents point landwards and may be zones of deposition associated with currents produced by edge **wave** interaction with incident **waves**.

Some bars are found both in the nearshore and intertidal zones of low energy environments. *Transverse bars* form at right angles to the shore, whereas up to 10 *multiple bars* are often found parallel to the shore. Both types require restricted **fetch** lengths and low **wave** energy conditions for their formation. The remaining bar types are those found only in the intertidal zone and are related to the action of **wave swash** on mainly low-angled foreshores. One type, called **ridge and runnel**, is low in amplitude, appears to be very stable in position and forms and develops under conditions of plentiful sediment supply and high **tidal** range. The more common type of *intertidal bar* is the **sandwave** or **swash bar** which migrates rapidly onshore under **swash** action and is destroyed by storm action. They eventually weld themselves onto the upper foreshore to produce a **berm** and represent the most common mechanism of **beach** adjustment to **wave** conditions on sandy **beaches**. [JDH]

Greenwood, B. & Davidson-Arnott, R.G.D. (1979) Sedimentation and equilibrium in wave formed bars: a review and case study. *Canadian Journal of Earth Sciences* **16**: 312–32.
King, C.A.M. (1972) *Beaches and Coasts*. Arnold, London.

coastal floods The inundation by a temporary rise in sea-level of low-lying coastal areas not normally submerged. Areas most at risk are the flat coastal plains, **estuarine** and **deltaic** regions such as the Sundarbans of the Bay of Bengal, The Netherlands, south-east England and the Gulf of Mexico coast of the United States. There are three main ways in which coastal floods occur. In descending order of magnitude these are **tsunami**, **storm surges** and high **tides**. The Japanese word **tsunami** is used for the **waves** produced by **earthquakes** and **volcanic eruptions** which create **waves** of hundreds of kilometres in length, **wave** periods of up to 30 minutes and **wave** velocities of up to $800 \, km \, h^{-1}$. The magnitude of such **waves** at the coast depends on the offshore topography and the 30 m **waves** which may be produced cause catastrophic flooding. For example, off the Chilean coast, in May 1960, a high-**intensity earthquake** caused large **waves** on the adjacent coast, soon followed by **waves** up to 11 m high which devastated the town of Hilo in Hawaii. A few hours later Hokkaido and Honshu in Japan were extensively flooded and many people drowned by **waves** which were now up to 40 m high. Severe coastal flooding may also be caused by **storm surges** related to deep depressions, hurricanes or tropical cyclones. Low pressure systems allow the sea surface to elevate beyond normal levels by about 10 mm for every 1 mbar pressure fall. Strong onshore winds associated with such events further elevate sea-level leading to extensive flooding. Sea-levels of 3 m in excess of normal occurred in eastern England and The Netherlands in 1953, *monsoonal* 'king tides' caused widespread coastal flooding of the northern coast of Australia and Hurricane Eloise raised water levels at Panama City by 6 m in 1975.

High **tides** may also cause coastal flooding especially where they are augmented by high **river discharges**, or by long-term sea-level rise. High **tides** serve to pond back **river discharge**, leading to flooding of the lower **floodplain**. In other coastal areas, flooding may be the result of a combination of processes. For example, the cities of Venice and London are both subject to subsidence, the effects of which are exacerbated by the gradual rise in global sea-level as well as **storm surges**. Sea-level in Venice has increased by 300 mm in 100 years whereas high tides at London Bridge have increased by 830 mm over the last 100 years. Coastal flooding in such areas is potentially disastrous yet the main flood threat for the future lies in the present rise of 20–30 mm each decade of mean global sea-level. Further, because of the projected increase in global temperatures and the consequent accelerated polar melting and thermal expansion of the oceans, sea-levels are predicted to rise faster than before (Titus *et al.*, 1985). Such scenarios have profound implications for coastal flood control.

Responses to coastal flooding have traditionally focused on the building of protective walls, e.g. the dykes of The Netherlands and the murazzi of Venice. Rebuilding and strengthening of the sea walls of eastern England following the 1953 storm surge did much to alleviate the impact of subsequent events. In **estuaries** where valuable land or urban areas are threatened, movable flood barriers have been constructed, e.g. at Hull on the Humber Estuary, at Barking and London on the Thames **estuary** and in the **delta** region of The Netherlands. Alternatives to civil engineering projects include land-use zoning, flood-proofing of buildings and emergency flood procedures. [JDH]

Bird, E.C.F. (1984) *Coasts* (3rd edn). Basil Blackwell, Oxford.
Hopley, D. (1974) Coastal changes produced by tropical cyclone Althea in Queensland, December 1971. *Australian Geographer* **12**: 445–56.
Titus, J.G., Leatherman, P., Everts, C.H. & Kriebel, D.L. (1985) *Potential Impacts of Sea-level Rise on the Beach at Ocean City, Maryland*. US Environmental Protection Agency, Washington, DC.

coastal notches Horizontal, slot-like recesses **eroded** at the base of a **sea cliff**, usually at high-water mark. Their formation is the principal way in which cliffs are undermined. There are three main groups of processes responsible for notch development. On most coasts the pressures exerted by breaking **waves** leads to quarrying of the cliff foot and the availability of rock fragments further enhances notch development by **abrasion**. Notches also form as a result of both the **solution** of rocks by seawater and the **erosion** of rocks by browsing and burrowing marine organisms, especially in carbonate rocks and on tropical coasts. [JDH]

Bird E.C.F. (1984) *Coasts* (3rd edn). Basil Blackwell, Oxford.

coated gold Native **gold** with a rusty or tarnished appearance due to the development of a surface film of **iron oxide**. Such **gold** is difficult to amalgamate or cyanide. (See **black gold**.) [AME]

coaxial This describes any progressive **deformation** during which the **principal strain** axes do not rotate with respect to reference lines within the rock. It is also used to refer to groups of **folds** which have a common axial direction. When referring to **strain** histories, *non-coaxial deformation* refers to **strain** histories in which lines of maximum and minimum **elongation** do rotate. In this sense, these terms refer to changes occurring as **deformation** proceeds, and so are distinct from the term *rotational*, which considers only the initial and final positions of the **strain** axes with respect to an external reference frame. Coaxial and *non-coaxial deformation* histories may result in *rotational* or *non-rotational strain*. Rock **fabrics** can be used to identify coaxial **strain** histories (Law *et al.*, 1984). [SB]

Law, R.D., Knipe, R.J. & Dayan, H. (1984) Strain path partitioning within thrust sheets: microstructural and petrological evidence from the Moine thrust zone at Loch Eriboll. *Journal of Structural Geology* **6**: 477–97.

cobalt (Co) At. no. 27, at. wt. 58.9332, d. 8.92. Obtained commercially from **copper** and **silver ores**. The main use is in alloys. Cobalt compounds are used in paints, varnishes and catalysts. [AME]

cobalt bloom An old name for the mineral **erythrite**, a secondary **cobalt** mineral. [DJV]

cobaltite ((Co, Fe)AsS) A relatively rare **ore** mineral of **cobalt**. [DJV]

coccoliths Minute calcareous plates of major importance in the formation of **chalks** and deep sea oozes; these microfossils provide a fine-scale **biostratigraphy** for many pelagic deposits of **Jurassic** to *Pleistocene* age. Coccolith plates (1–25 μm in diameter) collectively form a protective armour around the coccolithophorid, a unicellular planktonic *alga*. The term 'calcareous nannoplankton' is a wider term, used to cover coccoliths and other remains (e.g. discoasters, nannoliths) commonly associated with them but made by other groups of less than 60 μm overall diameter.

Heterococcoliths are built of varying submicroscopic elements, with discs of circular or elliptical outline (shields), constructed of radially arranged plates, enclosing a central area which may be empty, crossed by bars, filled with a lattice or produced into a spine. Holococcoliths are usually smaller, built of submicroscopic **calcite** crystals that invariably disintegrate after they are shed. *Discoasters* are stellate shields, usually built of six radiating rays. These remains can be obtained from a few pellets of marl (preferably) or **chalk** by gentle crushing with pestle and mortar, smearing the residue onto a glass slide with distilled water and observing with transmitted light under a powerful petrological microscope. The interference figure formed under polarized light is often studied. Smears may also be studied on scanning *electron microscope* stubs.

Living coccolithophores are photosynthetic and live in the top 200 m of the water column, where they play an important role in the food chain of the oceans. After death, coccoliths fall away from the parent organism and sink gradually downwards, suffering progressive **dissolution** or disaggregation as they proceed towards the calcium carbonate compensation depth (*c.* 3000–5000 m). Coccolithic remains are therefore best preserved in shelf or slope deposits, especially in marls where recrystallization has not obscured their diagnostic features.

The first generally accepted fossil coccoliths are of late **Triassic** age but their major diversification took place during early **Jurassic** times. About 12 calcareous nannoplankton zones can be recognized through the **Jurassic**. A climax of abundance and diversity was reached in the late **Cretaceous chalks** (e.g. *Deflandrius*) but extinguished by the *Cretaceous–Tertiary boundary* event. Some 26 zones have been erected for the **Cretaceous** and a similar number for the *Tertiary*. Coccoliths of the latter period are often much larger in size than those of the **Cretaceous**, while warm water assemblages often contain rosette-shaped *discoasters*. [MDB]

Bolli, H.M., Saunders, J.B. & Perch-Nielsen, K. (eds) (1985) *Plankton Stratigraphy*. Cambridge University Press, Cambridge.
Hamilton, G.B. & Lord, A.R. (eds) (1982) *A Stratigraphical Index of Calcareous Nannofossils*. Ellis Horwood, Chichester.
Tappan, H. (1980) *The Paleobiology of Plant Protists*. W.H. Freeman, San Francisco.

cockpit karst A **karst** landscape characteristic of humid tropical areas consisting of star-shaped closed depressions separated by steep residual hills. The German term **kegelkarst**, or *cone karst*, is also applied to such landscapes. The use of the word 'cockpit' to describe the closed depressions originates from Jamaica, where this type of landscape is found. Cockpit karst is also found in Puerto Rico, Papua New Guinea and Java. Star-shaped cockpits are more complex morphologically than some other closed depressions found in **karst** areas such as **dolines**, although they play a similar role in the landscape. Lehmann (1936) put forward an early theory to explain the development of cockpit karst from his observations at Gunung Sewu, Java which was the starting point for modern investigations of tropical **karst** evolution. The cockpits themselves are largely formed through surface **solution**, and centripetal drainage systems develop within them. [HAV]

Aub, C.F. (1969) The nature of cockpits and other depressions in the karst of Jamaica. *Proceedings of the 5th International Speleological Congress* **M15**: 1–7.
Lehmann, H. (1936) *Morphologische Studien auf Java*. Engelhorn, Stuttgart
Williams, P.W. (1972) Morphometric analysis of polygonal karst in New Zealand. *Geological Society of America Bulletin* **83**: 761–96.

coesite (SiO_2) A high-pressure form of **silica**. [DJV]

cohesion The **strength** of bonding between particles or surfaces. Cohesion or *cohesive strength* is specifically used in rock mechanics to mean the inherent **shear strength** of a plane across which there is no *normal* **stress**. The concept is also commonly used to distinguish cohesive **fault rocks** (e.g. **mylonites**) from incohesive **fault rocks** (*gouges*). [TGB]

coke The dense but porous product of the carbonization of **coal** in an oven. It is used as a fuel, as a reducing agent in blast furnaces and for the preparation of producer gas. Coke for the iron and steel industry is usually dense and strong. When the residue after carbonization is a non-porous powder of granular mass it is known as *char* and is briquetted for industrial use.

Only **caking coals** with a specific range of **rank** and properties are suitable for coke making. Low **rank coals** (**brown coal**) and high **rank coals** (**anthracite**) do not cake. Much of the coal's **ash** content is retained in the coke and, as it is an unwanted component since it increases slag volume in the blast furnace and therefore fuel costs as well as consuming **lime**, coking coals should have a low **ash** content. **Sulphur** and **phosphorus** should also be low as a large proportion is passed on to the coke and these elements have a deleterious effect in steel manufacture. In most cases the volatile content of the **coal** should also be low.

Coke oven feeds usually contain a blend of different **coals** and it is therefore important to select the correct mass ratios of each component to produce a coke of the required specification. Various physical and petrographic properties must be measured and the *cokability index* of a **coal** can then be determined using the appropriate equations or graphs. [AME]

Ward, C.R. (ed.) (1984) *Coal Geology and Coal Technology*. Blackwell Scientific Publications, Melbourne.

colemanite ($CaB_3O_4(OH)_3 . H_2O$) A mineral valued as a source of **borax**. [DJV]

collophane A name given to massive crypto-crystalline varieties of **apatite** that constitute the bulk of phosphate rock and fossil bone. [DJV]

colluvium Sediment which has been transported by weakly selective processes such as *mass-wasting* and *slope-wash*. Colluvium accumulates at the base of cliffs — where it may be referred to as **talus** — or on lower hillslopes, where it is typically poorly sorted and poorly stratified. These characteristics differentiate colluvium from **alluvium** which is laid down by fluvial processes which sort the particles according to their size (and perhaps shape), producing distinct sedimentary structures. Colluvium is often derived from weathered bedrock (**eluvium**, **saprolite**) on upper hillslopes. Because the processes of wash and creep which erode the **weathered** rock are not highly selective of the size of particles which are transported, the colluvium may have a strong physical resemblance to **eluvium**. However, especially further downslope, colluvium may interfinger with **alluvium**, and **paleosols** and **stonelines** may interrupt the colluvial sequence revealing its sedimentary origin.

In parts of the south-western United States, colluvium has been accumulating in bedrock hollows on hillsides since the late *Pleistocene* or early *Holocene* (Dietrich & Dorn, 1984). Earlier, wetter climatic conditions promoted periodic **landsliding** of the colluvial deposits thereby emptying the depressions. In contrast, throughout much of southern Africa, more arid conditions during the late *Pleistocene* promoted devegetation of hillslopes and **erosion** of destabilized **eluvium**. This resulted in massive *colluviation* on lower **pediment** slopes and in broad valleys where river-flow had become sporadic. Sheets of this colluvium reach thicknesses of 10 m or more. They contain an assortment of **paleosols** and stone artefacts which have provided a wealth of *paleoenvironmental* information (Price Williams *et al.*, 1982; Watson *et al.*, 1984).

In Africa and parts of South America, many colluvial deposits are currently undergoing **erosion** (see **badland**, **donga**, **gully erosion**). While the causes of this are uncertain, perhaps being related to changing climate or patterns of land use, it seems that the physical and chemical characteristics of the sediments make them especially susceptible to *gullying*. The colluvium is made up of about 60% **sand**-sized material, 10–30% silt, and 10–30% clay. It is the clay component that provides **cohesion** when the material is dry. Upon wetting, however, exchangeable sodium is liberated from the surface of the clay particles and the sodic water causes dispersion of the clayey materials. This results in slaking of the sediment and rapid incision by surface **runoff** forming **dongas** and **badlands**. [AW]

Dietrich, W.E. & Dorn, R. (1984) Significance of thick deposits of colluvium on hillslopes: a case study involving the use of pollen analysis in the coastal mountains of Northern California. *Journal of Geology* **92**: 147–58.

Price Williams, D., Watson, A. & Goudie, A. (1982) Quaternary colluvial stratigraphy, archaeological sequences and palaeoenvironment in Swaziland, southern Africa. *Geographical Journal* **148**: 50–67.

Watson, A., Price Williams, D. & Goudie, A. (1984) The palaeo-environmental interpretation of colluvial sediments and palaeosols of the late Pleistocene hypothermal in southern Africa. *Palaeogeography, Palaeoclimatology, Palaeoecology* **45**: 225–49.

colour index The total percentage of mafic (Mg- and Fe-rich) minerals (e.g. **olivine**, **pyroxene** and **amphibole**) present within the modal analysis of an **igneous rock**. The colour index may be used to classify **igneous rocks** as follows: *leucocratic* (colour index = 0–33%), *mesocratic* (colour index = 34–66%) and *melanocratic* (colour index = 67–100%). [RST]

columbite (($Fe, Mn)Nb_2O_6$) The **ore** mineral of **niobium** found in **granitic** rocks and **pegmatites**. (See **oxide minerals**.) [DJV]

columnar jointing A **joint** pattern, related to cooling of large bodies of volcanic rock. Thick **lava flows**, especially **basalts**, develop vertical cooling **joints** which neatly divide the rock into many polygonal columns. The result is that the flows are columnar jointed into a stack of vertical, polygonal posts. Columnar jointed flows commonly have a multi-tiered structure, in which layers of the flows with regular, vertical **joints** (the *colonnade*) alternate with layers with more irregular, commonly fan-like **joints** (the *entablature*). Columnar jointing also occurs in **dykes** and **sills** (normal to the walls of the intrusion), and in welded **ash flow tuffs**. [PTL]

Cas, R.A.F. & Wright, J.V. (1987) *Volcanic Successions Modern and Ancient*. Allen & Unwin, London.
Macdonald, G.A. (1972) *Volcanoes*. Prentice-Hall, Englewood Cliffs, NJ.

combe (*coombe, comb, coomb*) A small, steep-sided valley. Commonly, in **chalk** country, a deep and narrow stream valley which is dry for most of the year. [AW]

combination trap A combination of structural and strati-graphic **oil traps**. [AME]

combined gold Gold combined with **tellurium** to form some of the telluride minerals. This **gold** can be over-looked in the assessment of **gold** prospects. [AME]

common depth point (*CDP*) A place on a reflecting inter-face which produces **seismic reflections** for a number of different combinations of source to receiver locations on the surface. **Seismic reflection** data are normally collected using the common depth point (CDP) technique, which involves the configuration of source and receivers such that, with horizontally-layered strata, a group of seismic records with the same CDPs, i.e. which sample the same reflecting point, known as a CDP gather, can be created and **stack**ed together after appropriate processing to produce a single CDP **stack**ed trace with an enhanced **signal to noise** ratio plotted to overlie the common depth points. (See Fig. C27.) [GS]

Mayne, W.H. (1962) Common reflection point horizontal stacking techniques. *Geophysics* **27**: 927–8.

common lead A 'mixed lead' formed by the addition of radiogenic **lead** to primeval **lead**. Such **leads** have been considered by some workers to have had a very simple history, e.g. primeval **lead** in a uniform, deep source rock (the **mantle**) has had its lead isotopic ratios continually changed by the addition of radiogenic **lead** from the associated **uranium** and **thorium**. If the **lead** is removed at some point in time by, say, *hydrothermal activity*, and **precipitated** in the upper **crust** as **galena** it will provide a

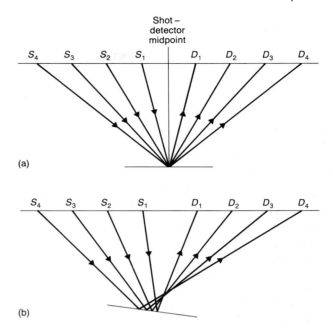

Fig. C27 Common depth point (CDP) reflection profiling. (a) A set of rays from different shots (*S*) to detectors (*D*) all reflected off a common point on a horizontal reflector. (b) The common depth point is not achieved in the case of a dipping reflector.

record of the isotopic ratios present at the time of its separation from the associated **uranium** and **thorium** in the source rock. [AME]

Cumming, G.L. & Richards, J.R. (1975) Ore lead isotope ratios in a continuously changing Earth. *Earth and Planetary Science Letters* **28**: 155–71.

compaction The process of volume reduction and con-sequential **pore-water** expulsion within sediments in re-sponse to normal **shear**-compressional **stresses** due to the increasing weight of overburden load. Compaction can be expressed as a percentage of the original **porosity** of the sediment, as a percentage of the thickness of the original sediment or by specific compressibility values based on **strength** or *rigidity*. All these parameters are strongly influenced by the lithology of the sediment, and relative **porosity** values are a convenient method of express-ing compaction during burial commonly employed by sedimentologists.

Siliciclastic and carbonate muds have very high initial **porosities** (>80%) compared with moderately sorted coarse-grained clastic sediments (<40%). Compaction takes place as the sediment package is gradually buried by younger sediments and the overburden load increases. The resulting degree of compaction is largely dependent upon the ratio of fine to coarse material in the sediment and the lithology of the sediment framework components. In argillaceous **mudrocks**, overburden pressure results in a marked **preferred orientation** of the platy component

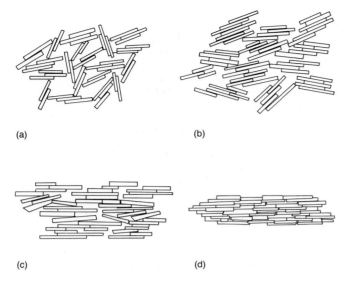

(a)　　　　　　　　　　　　(b)

(c)　　　　　　　　　　　　(d)

Fig. C28 Orientation fabrics in mudrocks as a result of overburden **compaction**. (a) At deposition — randomly linked groups or chains of clay particles. (b) Initial compaction — collapse of links and formation of orientated domains. (c) Shallow burial — high degree of orientation as a result of shearing. (d) Deep burial — highly fissile compacted mudrock with all grains showing preferred orientation. Loss of most porosity.

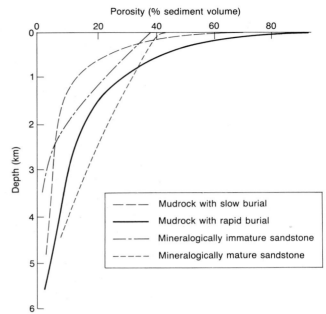

Fig. C29 Compaction. Idealized depth–porosity plots for mudrocks and sandstones. The range of porosity values at a given depth reflects variation in burial rate (or geothermal gradient) and detrital composition.

particles (Fig. C28). When initially deposited, the microstructure of clays comprises linked groups or chains of small particles in and between denser flocs, aggregates or larger particles. During compaction the links between particles collapse and form domains of orientated particles. Shearing develops with increasing overburden and a high degree of **preferred orientation** results. Ultimately, all platy grains assume an orientation with their elongate axes perpendicular to the pressure direction, giving the **mudrock** a distinct **fissility**. The extent of compaction in argillaceous **mudrocks** is dependent upon many factors, including type of **clay minerals** present, particle size, presence (or absence of) microfossils and the type and concentration of interstitial **pore solutions**. However, there is a gradual and systematic loss of **porosity** generally recorded with increasing overburden (Fig. C29). Initially, flocculated clays undergo the greatest degree of compaction within the first metre of burial, and on reaching depths of around 300–500 m, argillaceous muds initially having 70–90% **porosity** have remaining **porosities** of less than 30%. After deeper burial, fissile **shales** invariably have **porosities** of less than 10%. In **sandstones** composed predominantly of competent detrital grains (**quartz**, **feldspar**, etc.) the potential for **porosity** loss through grain re-orientation is much less pronounced. Nevertheless, freshly deposited **sands** are characterized by low particle concentrations with an open packing grain framework (Fig. C30). This open system tends to evolve towards a close packing grain framework during the initial stages of burial compaction and is accomplished by the processes of grain slippage, grain rotation, bending of elongate grains

and **brittle fracturing**. Such reorientation of grain **fabrics** is collectively referred to as mechanical compaction and generally takes place in the first 1 km of burial, reducing intergranular **porosities** by around 25–30% of the starting **porosity**. By contrast, **sandstones** with a significant proportion of less competent detrital grains (mudflakes, volcanic rock fragments, etc.) undergo intense **plastic deformation** whereby ductile grains are squeezed around the more competent grains. Much greater **porosity** losses are accommodated in these **sandstones** than in **quartz**-rich lithologies under comparable overburden **stresses** with the result that virtually all intergranular **porosity** may be destroyed at less than 2 km of burial (Fig. C29).

After initial **porosity** losses have been taken up through the processes of mechanical compaction, further **porosity** destruction in coarse-grained clastics is accomplished by the processes of chemical compaction. According to *Riecke's Principle*, the solubility of minerals increases with increasing effective **stress** at grain contacts. **Dissolution** of detrital grains may thus take place at grain contacts given suitable overburden **stress** values. The grain **stress** which causes such **pressure dissolution** is the **effective stress** (S_e) per surface area of grain contact given by

$$S_e = (S_g - S_h)\frac{100}{K}$$

where S_g = geostatic pressure, S_h = hydrostatic pressure, and K = the grain contact area expressed as a percentage of the total horizontal cross-section of the **sandstone**.

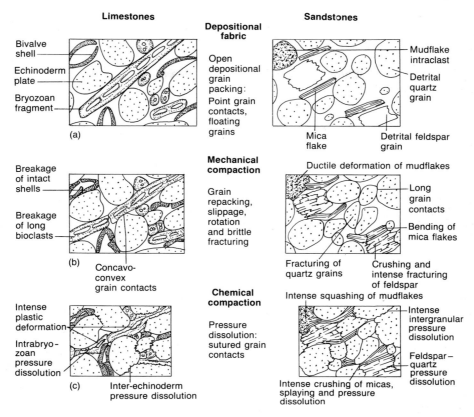

Limestones

Bivalve shell
Echinoderm plate
Bryozoan fragment

(a)

Depositional fabric

Open depositional grain packing: Point grain contacts, floating grains

Sandstones

Mudflake intraclast
Detrital quartz grain
Mica flake
Detrital feldspar grain

Breakage of intact shells
Breakage of long bioclasts

(b)

Concavo-convex grain contacts

Mechanical compaction

Grain repacking, slippage, rotation and brittle fracturing

Ductile deformation of mudflakes
Long grain contacts
Bending of mica flakes
Fracturing of quartz grains
Crushing and intense fracturing of feldspar

Intense plastic deformation
Intrabryo-zoan pressure dissolution

(c)

Inter-echinoderm pressure dissolution

Chemical compaction

Pressure dissolution: sutured grain contacts

Intense squashing of mudflakes
Intense intergranular pressure dissolution
Feldspar–quartz pressure dissolution
Intense crushing of micas, splaying and pressure dissolution

Fig. C30 Schematic diagram illustrating the progressive fabric changes in limestones and sandstones resultant of **compaction** during gradual burial. (a) Surficial sediment. (b) Shallow burial. (c) Deep burial.

Pressure dissolution is a compactional response of the sediment during burial in an attempt to increase the percentage of grain contact area so as to distribute the **effective stress** over a larger surface. As a result grain contacts can be observed to increase from point, through straight-elongate to concavo-convex and ultimately inter-penetrative or sutured (Fig. C30). Although an **effective stress** is a prerequisite to drive **pressure dissolution**, the mechanics and chemistry of the process are poorly understood.

Similar compactional processes can be documented for carbonate sediments (Fig. C30). The loss of **porosity** with increasing overburden pressure in **lime** muds is directly comparable to that recorded from argillaceous muds. Both mechanical and chemical compaction are operative in **oolitic** and bioclastic *grainstones*, having been demonstrated to reduce intergranular **porosities** by as much as 75%. Depending upon the extent of early **cementation**, **ooliths** may be squashed, spalled or fractured. Intact skeletal material is initially buckled or broken across fulcrum points whilst hollow grains (articulated *bivalves*, *gastropods*, **Foraminifera**, etc.) may be squashed. At greater overburden pressures **oolitic** and bioclastic material may experience **plastic deformation** and, ultimately, suffer **pressure dissolution** at grain contacts. As in **sandstones**,

large volumes of framework sediment may be dissolved by this process. Due to the inherent heterogeneity typical of carbonate sediments, **stylolites**, which are planes of preferential **pressure dissolution**, commonly develop during deep burial. **Stylolites** are usually recognized by the seam of insoluble clayey and organic residue left behind after carbonate **dissolution**. [SDB]

Chilingarian, G.V. (1983) Compactional diagenesis. In: Parker, A. & Sellwood, B. (eds) *Sediment Diagenesis*, pp. 57–167. NATA ASI Series C, Vol. 115, Reidel, Dordrecht.

compensator EM method An **electromagnetic induction method** which uses a constant transmitter–receiver separation over a survey area. With such a local transmitter, the primary field at the receiver would be far stronger than the secondary field and the accurate detection of the secondary field becomes difficult. In compensator methods such as the *Slingram* system, sensitivity is enhanced by compensating for the primary field, and the pre-setting of receiver controls to read zero secondary field, over non-anomalous ground at the start of a survey.

It is usual, though not necessary, to use horizontal coils separated by tens of metres for the transmission and reception of signals, and normally measurements are made at two frequencies in the audio-frequency range, differing by a factor of about four, in each survey. A reference signal is supplied from the transmitter directly to the receiver via a direct cable link so that the **real** and **imaginary components** of the secondary field may be directly calibrated as percentages. Interpretation is by comparison with standard curves theoretically computed or obtained from laboratory model studies.

The method is used in **airborne geophysical surveys** although only the gross features of conductivity anomalies are obtainable. [ATB]

Keller, G.V. & Frischknecht, F.C. (1966) *Electrical Methods in Geophysical Prospecting.* Pergamon, Oxford.

Strangway, D.W. (1969) *Electromagnetic Parameters of Some Sulphide Ore Bodies. SEG Mining Geophysics Vol. I, Case Histories*, pp. 227–42. Society of Exploration Geophysicists, Tulsa, OK.

competent Term referring to the relative **strength** of a rock. Stronger rocks (e.g. **sandstones**, **limestones**) are said to be more competent than weaker (e.g. **shales**), especially when **deformed** together as interlayered sequences. Competent layers may exhibit **boudinage** in **extension**. [RGP]

composite intrusion A type of **multiple intrusion** which is composed of **magmas** of contrasted composition. Such intrusions include composite **dykes** and **sills**. A common example is an intrusion in which an early intrusion of basic **magma** was followed by an intrusion of intermediate or acid **magma**. [RST]

composite seam A **coal seam** made up of two or more distinct **coal** beds that have come into contact when intervening strata have wedged out. [AME]

compositional layering Set of layers distinguished from each other by differences in composition. The term is usually applied to metamorphic rocks (e.g. **gneisses**) where the origin of the layering is in doubt. In addition to sedimentary **bedding**, compositional layers may be produced by metamorphic or igneous processes aided or modified by **deformation**. (See **gneissosity**.) [RGP]

concealed coalfield A **coalfield** hidden beneath deposits of younger or overthrust strata. [AME]

concentration factor For the formation of an **orebody** the element or elements concerned must be enriched to a considerably higher level than their normal crustal abundance, and the degree of enrichment is termed the concentration factor. Some example figures are **aluminium** — 3.75, **iron** — 5, **copper** — 80, **gold** — 250, **manganese** — 389, **lead** — 4000. [AME]

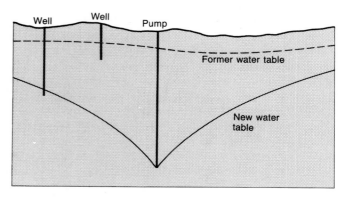

Fig. C31 Cone of depression of water table formed by pumping and its effect upon nearby wells.

condensate A very light **crude oil** with an **API gravity** greater than 50°. **Butane** is the lightest *hydrocarbon* that can occur in condensates. Commonly produced with *wet gas*. [AME]

cone of depression When water is removed from an **aquifer** surrounding a **well** by pumping then the **water table** (or the **potentiometric surface** in an **artesian aquifer**) is lowered around the **well** and a **hydraulic gradient** from all directions is established towards the **well**. The **water table** in homogeneous **aquifers** then takes the form of an inverted cone called the cone of depression (Fig. C31). If pumping is continuous then the cone of depression deepens and spreads out laterally. Continued **drawdown** lowers the **water table** still further and nearby shallow **wells** may go dry. [AME]

cone sheet A type of **ring intrusion** characterized by inward-dipping margins presumed to be towards a centre corresponding to the upper part of a *magma chamber*. Cone sheets are small intrusions, with outcrops separated by curved 'screens' of *country rock*, and the geometrical form indicates emplacement associated with **uplift** of a central conical block above the pressured *magma chamber*. [RST]

confining pressure The radial **stress** applied to a cylindrical specimen deformed by axially symmetric loading. In a conventional triaxial rock mechanics test, radially uniform **stress** is applied by a fluid to a cylindrical specimen, usually across a jacket, at the same time as an axial **stress**. In these conditions, two of the *principal stresses* are equal: they define the 'confining pressure'. [TGB]

conical fold Special type of **fold** that approximates to the shape of a conical surface. Such folds may be produced at *culminations* and *depressions* in a non-cylindroidal folded

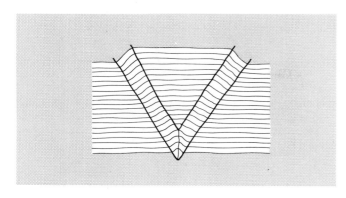

Fig. C32 Conjugate kink bands.

surface (i.e. with variably plunging *fold axes*) or at the ends of *cylindroidal folds* as the *amplitude* decreases to zero. [RGP]

conjugate Term applied to a pair, or two sets, of **faults**, **shear zones**, asymmetric **folds**, **kink bands**, etc., characterized by opposing **dips** (in the case of **faults**) or opposing **vergence** (in the case of **folds**). (See Fig. C32.) [RGP]

conjunctive use The combined use of **groundwater** and surface water. **Artificial recharge** is one example of this augmentation of water supplies. [AME]

connate water Water deposited with and included in a sediment. As the sediment is buried and indurated much of its connate water is altered in composition (often becoming a **brine**), diluted or displaced by other waters or largely expelled. Besides being the source of the water in most **oilfield brines**, saline connate water expelled from deeply buried **shale basins** may move up **dip** as **hydrothermal solutions** carrying **base metals** that they have leached from the **shales**. These metals may be deposited in **basin** margin **limestones** to form **lead–zinc** and other deposits. [AME]

conodonts Enigmatic, extinct, marine animals represented in the fossil record primarily by the scattered elements of their mineralized feeding apparatus (Fig. C33). Conodont elements are normally between 0.2 and 2 mm in size (the largest known is 14 mm long) and composed of **apatite**. They occur, often abundantly, in marine strata of the Upper **Cambrian** to uppermost **Triassic**. The **apatite** is in the form of thin growth lamellae which in the true conodonts ('euconodonts') were added sequentially to the outside of the element; there may be a division into an upper crown and a basal body. The lamellae of the crown are continuous around the upper surface, but open basally to form a basal cavity; the basal body, when present, fills and extends beyond this cavity. In primitive elements the crown is entirely hyaline, but in advanced forms the lamellar structure is interrupted by patches of opaque white matter, which is finely crystalline and porous.

Element morphology is highly varied within three major shape categories: coniform, ramiform and pectiniform. Coniform elements vary in curvature, cross-sectional shape and the development and distribution of costae or other ornament. Ramiform elements show variation in the number and distribution of processes, the degree and style of denticulation and the extent of basal cavity development. Pectiniform elements may be straight or arched blades, or may be laterally thickened and extended to form complex, ornate, platform-like units.

The conodont feeding apparatus comprised a set of elements, normally of several different morphological types. These elements were mostly symmetrically paired across the apparatus midline and appear to have acted in lateral opposition. Rare occurrences of undisturbed apparatuses preserved on **bedding** planes, particularly in **Carboniferous shales**, have given direct evidence of the apparatus structures of some taxa. The most common type comprises a pair of platform-like pectiniform elements, a pair of arched blades and a set of seven to eleven ramiform elements of varied morphology, of which the axial member is bilaterally symmetrical and unpaired. A feature of the ramiform set is the inclusion of a symmetry transition series, in which elements increase in asymmetry with distance from the apparatus axis. Other apparatuses may have been composed entirely of ramiform elements, or solely of various coniform elements, within which symmetry transition series may also be identified. Apparatuses in which coniform elements occurred together with ramiform elements are rare in post-**Ordovician** strata.

The conodonts possessed no mineralized skeleton apart from their elements, and for many decades after the discovery of isolated elements in 1856 their biological affinities were widely debated. Publications variously suggested relationships with *algae*, plants, conulariids, aschelminthes, gnathostomulids, **molluscs**, *annelids*, **arthropods**, tentaculates, chaetognaths, and *chordates*. Several of these assignments were based on superficial morphological similarities between structures in extant organisms and selected conodont elements. The soft parts of the conodont body were unknown until 1982, when the first of several fossil specimens was discovered in the **Carboniferous** *Granton Shrimp Bed* of Edinburgh, Scotland. These specimens, preserved through replacement of the soft tissues by calcium phosphate, are of elongate animals, 40–60 mm long, with a bilobed head behind which a segmented trunk terminates in an asymmetrical tail (Briggs *et al.*, 1983; Aldridge *et al.*, 1986). The conodont apparatus lies within and behind the head, with a group of ramiform elements to the anterior separated from pairs of arched blade and platform elements positioned in sequence behind. The affinities of the animals appear to be with the jawless craniates, perhaps closest to the myxinoids (hagfish), although some authorities prefer other assignments. The apparatus may have functioned

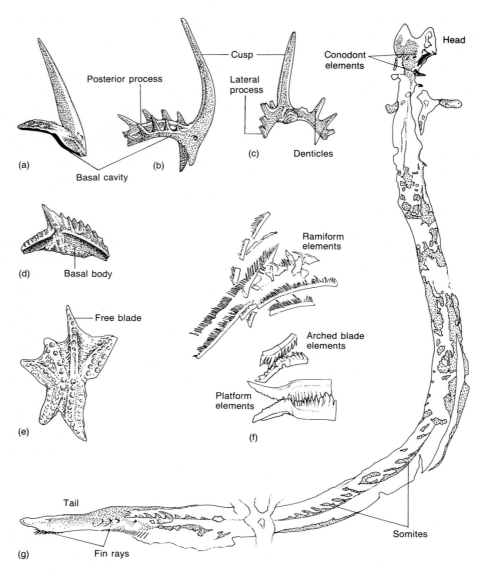

Fig. C33 Conodont. (a) Lateral view of coniform element of *Phragmodus*, ×40. (b) Lateral view of ramiform element of *Paraprioniodus*, ×27. (c) Posterior view of ramiform element of *Ctenognathodus*, ×27. (d) Lateral view of pectiniform (arched blade) element of *Pterospathodus*, ×27. (e) Upper view of pectiniform (platform) element of *Aulacognathus*, ×27. (f) Lateral view of a Carboniferous apparatus preserved undisturbed on a bedding plane, ×15. (g) Lateral view of *Clydagnathus*?, the first known complete fossil conodont.

as a food gathering and processing structure, with the anterior ramiform set grasping prey and passing it back to the pectiniform pairs for shearing and grinding.

Not all conodont animals may have been elongate. A single poorly-preserved specimen from the **Silurian** of Wisconsin, USA, displays an apparently broader, flatter body (Smith *et al.* in Aldridge, 1987). The apparatus of coniform elements in this animal is also arranged in accordance with a bilateral grasping mode of operation.

Late **Cambrian** conodont elements were dominantly smooth or finely striated cones with little morphological variation. A major radiation in the early **Ordovician** was accompanied by the development of denticulated processes and the morphological differentiation of elements within apparatuses. This diversification reached a peak in the *Arenig*, when the variety of apparatus types was greater than at any other time. Conodonts were relatively unaffected by the *mass extinction* at the end of the **Permian**, but disappeared in the latest **Triassic**. Conodont elements are important in **biostratigraphy** throughout their stratigraphical range. Conodont biozonal schemes for all the **Paleozoic** systems and the **Triassic** are in wide use, with particular resolution available in the Lower **Ordovician** and Upper **Devonian**. Elements are abundant in many marine facies, and their easy extraction from calcareous rocks by treatment with dilute organic acids (e.g. acetic acid) has led to a particular utility in the **biostratigraphy** of carbonate sequences.

Although conodonts as a group display wide environmental tolerance, several species were ecologically specialized. Biofacies belts commonly lie parallel to ancient shorelines and changes in conodont faunas may reflect transgressive and regressive events. Faunal provincialism has been recorded at various times through conodont history, and was especially marked in the **Ordovician** when water temperature may have been the dominant factor controlling distributional patterns.

Conodont elements are important indicators of the thermal history of strata and of oceanic geochemistry. During heating, organic material within the lamellar structure of the elements breaks down, and **carbon** fixing produces a continuous colour change from yellow, through brown to black. Element colour can be calibrated with temperature, assuming heating lasted for periods of 1–500 Ma, and a *colour alteration index* (CAI) has been developed for the assessment of *thermal maturation*. CAI 1 (pale yellow) corresponds to <50–80°C and CAI 5 (black) to 300–350°C; at higher temperatures (CAI 6–8) elements become grey, then opaque white, then clear. The index has applications in assessing local and regional heating during tectonic or igneous events and in investigating the thermal history of sedimentary **basins**; it also provides thermal cutoffs for **hydrocarbon generation**.

The **apatite** of conodont elements retains an enriched but unfractionated geochemical signature of the seawater in which it was deposited. Analyses of **strontium** concentrations, **strontium isotope** ratios, neodymium isotope ratios and trace-element concentrations in conodont elements have provided the basis for chemostratigraphic studies of several **Paleozoic** sequences. **Oxygen isotope analysis** provides evidence of changing marine *paleotemperatures* and some correlation is apparent between periods of high water temperature and high conodont diversity. [RJA]

Aldridge, R.J. (ed.) (1987) *Palaeobiology of Conodonts*. Ellis Horwood, Chichester.

Aldridge, R.J., Briggs, D.E.G., Clarkson, E.N.K. & Smith, M.P. (1986) The affinities of conodonts — new evidence from the Carboniferous of Edinburgh, Scotland. *Lethaia* **19**: 279–91.

Austin, R.L. (ed.) (1987) *Conodonts — Investigative Techniques and Applications*. Ellis Horwood, Chichester.

Briggs, D.E.G., Clarkson, E.N.K. & Aldridge, R.J. (1983) The conodont animal. *Lethaia* **16**: 1–14.

Robison, R.A. (ed.) (1981) *Treatise on Invertebrate Paleontology, Part W, Supplement 2, Conodonta*. Geological Society of America and University of Kansas Press, Boulder, CO and Lawrence, KS.

Conrad discontinuity A seismic discontinuity in the **continental crust** that is commonly seen at a depth of 10–12 km in Europe and has been observed, irregularly, elsewhere. It is generally taken to mark the boundary between the **upper** and **lower continental crust**, i.e. a difference in the concentration of basic rocks combined with a change in the metamorphic grade, but in Europe its occurrence seems to correlate with the presence of *Tertiary*

volcanism and it may therefore be related to the remnants of *magma chambers* at these depths. [DHT]

constant separation traversing A **resistivity method** in which the source–receiver separation is maintained constant. The **depth of penetration** then remains constant to a first approximation and so any variations in the parameter measured will be due to lateral changes in the subsurface electrical properties within that depth. This type of surveying, which is also known as *electric profiling* or *trenching*, contrasts with **vertical electrical sounding** where the objective is to obtain resistivity variations with depth. [ATB]

contact metamorphism Thermal metamorphism developed within rocks adjacent to the contact of an igneous body. The metamorphism reflects the high temperature of the intrusion and may also be influenced by expulsion of **hydrothermal solutions** from the intrusion and/or mobilization of **groundwater** within the surrounding *country rocks*. Contact metamorphism varies from being inconspicuous in relatively dry **igneous rocks** or **sandstones** but may be more pronounced in sedimentary rocks such as **shale/mudrock** and impure **limestones**. As a result of the factors noted, zones of contact metamorphism or *contact aureoles* may have dimensions ranging from centimetres to kilometres in width and may show complex zonation of **fabric** and mineralogy reflecting the thermal and chemical variations within the metamorphosed *country rocks*. [RST]

continental crust The base of the continental crust is the **Mohorovičić discontinuity** which marks the boundary between rocks of an average density of less than *c.* $3.0 \, \text{Mg m}^{-3}$ from those of the mantle, with densities of some $3.3 \, \text{Mg m}^{-3}$. It is divided into an upper **granitic** layer with characteristic densities of about $2.7 \, \text{Mg m}^{-3}$ within which most **earthquakes** occur, and a lower crust, with densities around $3.0 \, \text{Mg m}^{-3}$ and a composition which is somewhat more basic than the overlying **granodioritic** crust, but is in the **granulite** grade of metamorphism. The boundary between the upper and lower crust is not always clear, but is usually between 10 and 12 km, and sometimes marked by the **Conrad discontinuity**. The continental crust varies in thickness, being mostly between 30 and 40 km, but is very variable in *orogenic belts* where it may change from only a few kilometres thick to 70–80 km. [DHT]

Tarling, D.H. (1978) (ed.) *Evolution of the Earth's Crust*. Academic Press, London.

Taylor, S.R. & McClennan, S.M. (1985) *The Continental Crust: its Composition and Evolution*. Blackwell Scientific Publications, Oxford.

continental drift The hypothesis, largely attributed to A. Wegener, that the continent blocks are, on geological time-scales, moving relative to each other. It is broadly similar to **plate tectonics**, but involves the motions of only the **continental crust** and was promulgated at a time when the main features of the ocean floors were unknown and it was assumed that they formed by continental thinning combined with widespread igneous intrusions, i.e. the movements were thought to be largely confined to the continental blocks and that these could pass through or override all oceanic materials. However, some continental drift models, particularly those of Holmes, required the generation of new **oceanic crust** predominantly within the *ocean ridge* systems as then known. However, the term **plate tectonics** now subsumes that of continental drift as *plates* may comprise both **continental** and **oceanic crust**, together with substantial parts of the upper **mantle**, i.e. the **lithosphere**, and is therefore on a larger scale than continental drift itself. [DHT]

Holmes, A. (1964) *Principles of Physical Geology*. Nelson, London.
Le Grand, H.E. (1988) *Drifting Continents and Shifting Theories*. Cambridge University Press, Cambridge.
Wegener, A. (1966) *The Origin of Continents and Oceans* (Translation of 4th edition, 1929). Methuen, London.

continental reconstruction As the present and past continents have had different configurations in the geological past, an important study is to reconstruct **paleogeographies** (ancient geography) for different areas and times. Various methods can be used for this purpose, although these should be considered as complementary rather than individually conclusive.

1 The simplest is that continental fragments, separated during the last 200 Ma, should mostly still be capable of being assembled on the basis of their geometric shapes, i.e. jig-saw fitting. This cannot be done by simply using two-dimensional maps, because of their necessary distortion of the Earth's surface, but it can be done on a globe and by computer fitting. Computer fitting has the advantage that the statistical properties of different reassemblies can be tested. However, the geometric fitting is always hindered by defining the edges to be fitted, as the **passive continental margins** are distorted by faulting and also masked by sediments that build out and extend beyond the original continental edge. Nonetheless, remarkably good reassemblies have been made of both **Laurentia** (North America, Greenland and Europe) and of **Gondwanaland**, i.e. reconstructing the continents prior to the formation of the Atlantic and Indian Oceans.

2 Such geometric fits cannot be used to determine intermediate continental reconstructions as the matching of any two continental edges is undertaken by the rotation of just one of the two continents about a single *Euler pole*, while the actual motion will normally have been about different poles for different times and rotated by different amounts and rates. Such intermediate positions can, however, be determined relatively simply by use of linear *oceanic magnetic anomalies* and **fracture zones**. The **frac-**ture zones, for any given period, form **small circles** that are concentric about the *Euler pole* and the magnetic anomalies, for the same period, originally radiated away from the same *Euler pole*. (In other words, the **fracture zones** are equivalent to latitude lines and the *oceanic magnetic anomalies* equate to lines of longitude if the *Euler pole* is imagined as the Earth's rotational pole.) Each of the **finite rotations** involved for different *Euler pole* locations and for different rotations can be summed to yield the same pole for reconstructing the continental edge fits, but these must be undertaken in the correct sequence and retain one continent fixed relative to its present-day coordinates.

3 The use of *oceanic magnetic anomalies*, constrained by **fracture zones**, allows the previous relative positions of the two continents to be determined, but does not allow for any motion common to both continental *plates*, i.e. they could both be moving northwards at the same time, but this would not be shown in reconstructions based only on their relative position to each other. **Paleomagnetic** definition of the average geomagnetic pole position and the coincidence between this and the Earth's axis of rotation enables continental reconstructions to be undertaken in which any latitudinal movement relative to the rotational pole can be determined, as can the orientation of each continent relative to the pole position. However, such **paleomagnetic** reconstructions cannot provide any control on the absolute longitude, i.e. the matching of **polar wander** paths defines the relative longitudinal shift, but does not allow for any common longitudinal motion affecting both plates equally. Nonetheless, **paleomagnetic** data still provide quantitative measures of past continental configurations and should be applicable throughout geological time if the **remanent magnetization** of suitable rocks can be determined. (See Fig. C34.)

4 The test of all such reconstructions is that they must be consistent with the geological evidence — structural, paleontological, *paleoclimatic*, etc. Such observations can also form the basis to determine past continental reconstructions as, for example, the distribution of fossil plants and animals, *paleobiogeography*, led to the original concept of connected land masses in the geological past. Similarly, the existence of ice-sheet *glaciation* in now disparate continents, but at the same age, can be used to determine probable continental configurations. Nonetheless, such geological data are largely qualitative rather than quantitative and are mainly useful for testing the validity of reconstructions determined by the foregoing methods. Such reconstructions then provide a basis on which to assess what other factors may have operated to account for the observed distribution of fossil types, etc., e.g. the distribution of predators, mountain barriers, etc. [DHT]

Bullard, E.W.C., Everett, J.E. & Smith, A.G. (1965) The fit of the continents around the Atlantic. *Philosophical Transactions of the Royal Society, London* **A258**: 41–51.
Smith, A.G. & Hallam, A. (1970) The fit of the southern continents. *Nature* **225**: 139–44.

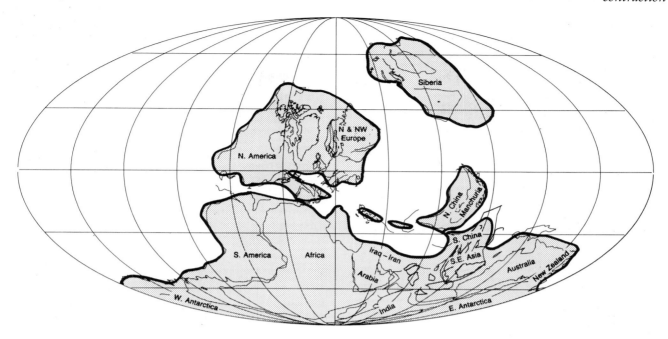

Fig. C34 Continental reconstruction. The Upper Carboniferous *c.* 300 million years ago. The reconstruction is based primarily on paleomagnetic data.

continental splitting The initiation of a new ocean within a continental area clearly requires the original continent to be split. It is commonly supposed that the East African Rift system is a current example of such activity. The sequence of events is usually considered to be a doming of the continent, followed by a splitting at the crest of the **dome** with associated volcanism. Further **rifting** and more concentrated volcanic activity eventually leads to the formation of **oceanic crust** within the central *rift valley*. This is a somewhat oversimplified model as the evolution of the margins of the Atlantic suggests that this model only applies in specific areas, e.g. bordering the Walvis and Rio Grande ridges, but that initial doming is not a necessary precursor to continental splitting. Such splitting is normally attributed to **mantle convective** motions, but while **mantle convective** currents must have always operated in order to transfer heat to the Earth's surface, there are times in the Earth's history, such as during the early **Proterozoic**, when continental splitting appears to have been rare. [DHT]

Tarling, D.H. (1980) Lithosphere evolution and changing tectonic regimes. *Quarterly Journal of the Geological Society, London* **137**: 459–66.
Bott, M.H.P., (1981) Crustal doming and the mechanism of continental rifting. *Tectonophysics* **73**: 1–8.

contourite Muddy or sandy sediment reworked and deposited by semi-permanent or permanent **thermohaline** ocean currents flowing parallel/subparallel to the strike of continental slopes and rises — hence *'contour currents'*. *Contour currents* may winnow finer grained sediments to leave a discontinuous coarse-grained **sand**/**gravel** *lag*, or transport fine-grained **sands** and muds to construct elongate lobate mounds or *'sediment drifts'*. Deposits of *contour currents*, contourites, show traction-produced sedimentary structures or appear homogeneous and structureless. Sediments may be *bioturbated*. Contourites may show normal and **inverse grading**, reflecting fluctuating current velocities. Long-term fluctuations in mean current velocity may be on the order of 2000–10 000 years for a sequence 500 mm thick. [KTP]

contracting Earth Prior to the discovery of *radioactive* heating in the Earth early this century, it was assumed that an originally hot Earth was cooling and hence contracting. Such contraction was often considered as the explanation for the compressional forms of mountain belts, analogous to wrinkles on a drying apple. *Radioactive* heating means that such contraction is not a necessary requirement and even a small degree of expansion is conceivable. It has been proposed that changes in the **gravitational constant** could cause either contraction or **expansion** of the **Earth** as a response of phase changes in the **mantle** to changes in the hydrostatic pressure resulting from a change in the **gravitational constant**. Evidence from Mars and the Moon suggests that any contraction of the Earth, since its formation, is unlikely to have exceed 5% and may be close to zero. [DHT]

contraction Decrease in length of a line during **deformation**. A contractional **strain** is measured using the expressions for **elongation**, giving negative values, or **stretch**, giving values less than 1. (See **strain**.) [SB]

contributing area The area of a **catchment** contributing to storm **runoff**. The term was originally used by Betson (1964) to refer to those restricted parts of the **catchment** ('partial areas', or 'partial contributing areas') which produced **infiltration**-excess **overland flow**. The term has also been applied to those parts of the **basin** which act as variable source areas producing saturation-excess **overland flow**. [TPB]

Betson, R.P. (1964) What is watershed runoff? *Journal of Geophysical Research* **69**: 1541–52.

convection Motion within a fluid by means of which heat is transferred from one area to another, maintaining the material at approximately constant **viscosity** and low **thermal gradient**. Within the Earth, convection is probably driven mostly by radiogenic heat production within the **mantle** itself, possibly with additional heat provided from processes within the **core**, such as the latent heat released as the inner **core** continues to solidify and the **core**'s internal radiogenic heat production. The differences in temperature between up-going and down-going limbs within the **mantle** are likely to be very small, of the order of 1–2°C, which accounts for the uniform properties of the **lower mantle** in particular, but even the phase changes in the **transition zone** do not appear to either inhibit or enhance convective flow. This convection, throughout the **mantle**, maintains it at a **viscosity** of 10^{21-22} Pa s. Some authorities consider that there are two or more convective systems within the **mantle** — one in the upper and one in the lower — but the constant **viscosity** and the evidence of **seismic tomography** indicate that only one system is involved. [DHT]

Davies, P.A. & Runcorn, S.K. (eds) (1980) *Mechanisms of Continental Drift and Plate Tectonics*. Academic Press, London.
Tritton, D.J. (1988) *Physical Fluid Dynamics*. Clarendon Press, Oxford.

convolute lamination Internal structure typically found in fine **sand**-to-silt-grade sediment comprising disturbed or contorted lamination, varying from regular to irregular wavelength and amplitude **folds**. *Axial surfaces* of **folds** are usually perpendicular to recumbent with respect to **bedding**. No scale implied, but generally seen in layers or **beds** up to decimetres thick and common in *turbidite* successions. The structure is produced mainly by **liquefaction** with subordinate **fluidization** processes. [KTP]

convolution The mathematical operation defining the change in shape of a signal resulting from its being **filtered**. This operation of convolving *a* with *b* is usually denoted as *a* ∗ *b*. A **filter** can be defined by its impulse response, which is the output of the **filter** when the input is a very sharp impulse of very short duration — a 'spike'. For digitally sampled signals, which essentially consist of a series of impulses of varying amplitude, convolution is equivalent to the superposition of impulse responses, weighted according to their input amplitude. Convolution of two waveforms can be shown to be equivalent to the multiplication of their amplitude spectra and the addition of their phase spectra in the frequency domain. Convolution is applied widely in geophysical data processing, where digital **filtering** of data is required. [GS]

Kanasewich, E.R. (1981) *Time Sequence Analysis in Geophysics*. University of Alberta Press, Edmonton.

Cooley–Tukey method Also known as the Fast *Fourier Transform* (FFT), the Cooley–Tukey method was first formulated in the early 1960s. It is an algorithm which performs the *Fourier transformation* of digital signals *N* samples long in a time proportional to $N\log_2(N)$, rather than N^2 if the transform formulae are programmed 'as they are written'. This substantial increase in speed makes the *Fourier transformation* of long signals feasible, and so facilitates manipulation of digital signals in the frequency domain, where many signal-processing methods (e.g. **convolution**) are much simpler. (See **power spectrum analysis**.) [RAC]

Press, W.H., Flannery, B.P., Teukolsky, S.A. & Vetterling, W.T. (1986) *Numerical Recipes: the Art of Scientific Computing*. Cambridge University Press, Cambridge.

Coorong Region Part of the South Australia carbonate province south-east of Adelaide in which stepwise coastal progradation along a coastline several hundred kilometres in length has occurred over the last 700 000 years yielding a coastal plain about 100 km wide. A series of NNW-SSE-trending *barrier beach*–**dune** systems are separated by topographic lows representing coastal *lagoons* that evolved to ephemeral *lagoons* and ephemeral lakes with time as they were filled with sediment. The modern Coorong *lagoon* is ephemeral at its south-eastern end and passes into a chain of **groundwater**-fed lakes which are notable for their variety of **carbonate mineral** precipitates including **dolomite** (see **dolomitization**). Non-carbonate **evaporites** occur, but readily redissolve, and the salts tend to be transported out of the system. The Coorong system is a useful analogue for ancient dolomitic carbonate sediments with marine-derived sediment and evidence both for freshwater and hypersalinity. [IJF]

copper (Cu) A native metal mineral (see **native element**). Copper was exploited by man from *c.* 6000 BC, when native copper was used in the Near East to make pins, awls and beads. Native copper was used in the North American Great Lakes area at *c.* 4000 BC. During the 5th millenium BC copper smelting was developed in the Near East and copper **ores** (carbonate, oxides and sulphides) were exploited to produce **axes** and daggers and, still later, coins. The period of time during which copper metallurgy was discovered and developed is known as the Copper Age. During the succeeding Bronze Age, copper retained its importance as a constituent in the alloy **bronze**. Copper was also used in antiquity as a colorant in glass, while copper carbonates produced paint pigments and were occasionally used to glaze **steatite** beads. [OWT/DJV]

copper pyrites An old name for **chalcopyrite**. [DJV]

coquina A type of carbonate rock consisting of mechanically sorted debris, and especially shells. It is often very porous. [HAV]

coral–algal reef A marine, largely biogenically controlled, accumulation of calcium carbonate, which can also be viewed as a complex and productive ecosystem. Skeletal carbonates provide much of the reef framework and organisms also assist in the **erosion** of skeletal material and the production of reefal sediments. The distribution of coral–algal reefs is controlled by environmental factors. Reef-forming **corals** grow best in warm, clear waters. Preferred conditions include an average water temperature of 21°C or more, salinities of 30–40 p.p.t. and a lack of suspended sediment. **Corals** also grow only in shallow waters, due to a lack of light at depth, so a suitable shallow platform is required. Most **coral** growth occurs in waters shallower than 25 m, although some species grow down to 165 m. These factors mean that coral reefs are characteristic of mud-free tropical coastlines between latitudes of about 30° N and S.

There are four main types of oceanic coral–algal reef recognized today, i.e. *fringing reefs*, **barrier reefs**, **atolls** and *platform reefs*. Reefs formed on continental coastlines are often multiple, complex features which are more difficult to categorize as indicated by Maxwell (1968). Several theories have been put forward to explain the development and distribution of different coral–algal reefs (see **atoll**).

Coral–algal reefs are complex ecosystems with very different environmental conditions from those in the surrounding seas. An important component in modern reefs is the zooxanthellate *scleractinian* **corals** which form most of the primary reef framework. These **corals** have only been reef-builders since the **Jurassic**. They are colonial organisms and can reproduce both sexually and asexually. There are over 80 genera of reef **corals** in the Indo-Pacific, but only approximately 20 genera in the Caribbean reef province. The growth form of **corals** varies according to taxonomy and environment. In the wave zone **corals** need to be more resistant than in quieter waters. Massive **corals**, such as the species *Montastrea annularis*, are more resistant to mechanical damage than branching forms such as *Acropora palmata*. Calcareous, encrusting organisms attach themselves to cavities within the primary reef framework, forming the secondary reef framework. Major modern encrusting organisms within reefs include crustose

coralline *algae*, **corals**, **bryozoans**, *gastropods*, and serpulid worms.

The reef framework is continuously being eroded by physical and biological processes, such as **wave erosion**, grazing and boring. There are many bioeroding organisms found within the reef ecosystems which provide an important source of reefal sediments. The reef framework and sediments become cemented by **diagenetic** processes as active growth declines. Reef growth and **cementation** lead to the development of three main sedimentary facies, i.e. *reef-flat facies*, *forereef facies* and the *backreef facies*. Fossil coral–algal reefs occur in the geological record and are often important **petroleum reservoirs**.

Contemporary reef ecosystems are vulnerable to catastrophic events, such as hurricanes and **tsunamis** which may cause devastation to the reef flat and mass mortality of **corals**. The **geomorphology** of coral–algal reefs reflects both features of coral growth and **erosion**. Characteristic geomorphological features on reefs are *spur-and-groove topography* on the reef front, and the *algal ridge* which is found on some *reef flats*. Low islands (**cays**) may form upon coral–algal reefs under suitable conditions. Due to the changing relative levels of land and sea, coral–algal reefs may become submerged or elevated. Many **guyots** are capped with **coral**, for example, and there are well-known elevated flights of *coral terraces* (such as those on the coast of the Huon Peninsula, New Guinea). [HAV]

Maxwell, W.G.H. (1968) *Atlas of the Great Barrier Reef.* Elsevier, Amsterdam.
Stoddart, D.R. (1969) Ecology and morphology of recent coral reefs. *Biological Reviews* **44**: 433–98.

corals Marine, solitary or colonial, polypoid animals, represented in the fossil record by their calcareous skeletons which show radial or near-radial symmetry (Fig. C35). All corals belong to the Class *Anthozoa* of the Phylum *Cnidaria* (= *Coelenterata*). Most fossils are accommodated in three subclasses, *Zoantharia*, *Rugosa* and *Tabulata*. The subclass *Heterocorallia* comprises a very small group of late **Devonian** to late **Carboniferous** corals which may be close to the *Rugosa* or may be a separate group that acquired a skeleton independently. The subclass *Octocorallia* (= *Alcyonaria*) is represented by rare fossils, but may include the earliest *anthozoans* in the supposed pennatulaceans (sea-pens) of the late *Precambrian*. Calcareous spicules found in Lower **Silurian** strata have been considered to be from the spicular skeletons of octocorals, and possible complete specimens have been described from the Lower **Ordovician**.

The *anthozoan* polyp has a body wall of two layers, the ectoderm and endoderm, between which is a thin layer of structureless jelly, the mesoglea. The body cavity (enteron) is lined by endoderm, which is infolded to create compartments separated by radial partitions (mesenteries). The body is basically cylindrical with an upward-facing mouth surrounded by tentacles which catch food, primarily living animals. A tubular gullet leads from the

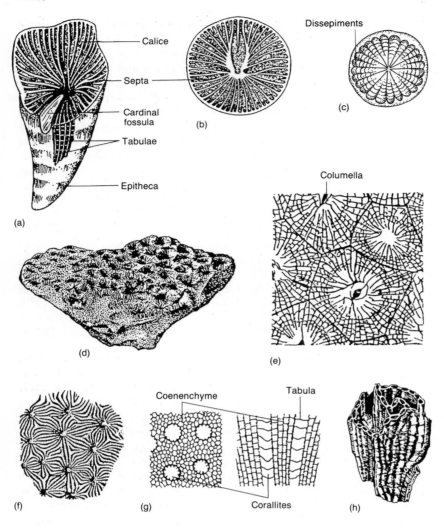

Fig. C35 Corals. (a) A Carboniferous solitary rugose coral, *Amplexizaphrentis*, with part of the epitheca removed to show internal structure of corallum. (After Clarkson, 1986.) (b) Transverse section of *Amplexizaphrentis*. (After Clarkson, 1986.) (c) Transverse section of a Cretaceous scleractinian coral, *Blothrocyathus*, ×0.4. (After Moore, 1956.) (d) A Jurassic colonial scleractinian coral, *Isastrea*, ×1.6. (After British Museum, 1962.) (e) Transverse section of a Carboniferous cerioid rugose coral, *Lithostrotion*, ×2.4. (f) A Jurassic thamnasterioid scleractinian coral, *Thamnasteria*, ×1.6. (After British Museum, 1962.) (g) Transverse and vertical sections of a Silurian coenenchymal tabulate coral, *Heliolites*. (After Clarkson, 1986.) (h) A Silurian catenoid tabulate coral, *Halysites*, ×0.6. (After Clarkson, 1986.)

mouth to the enteron. There is no separate anus, waste being ejected through the mouth.

SUBCLASS ZOANTHARIA. Solitary or colonial *anthozoans* with polyps bearing six (or multiples of six) tentacles. *Zoantharians* which secrete a calcareous, commonly **aragonitic**, exoskeleton (the corallum) are assigned to the Order *Scleractinia*, which has a fossil record from the Middle **Triassic** to the *Holocene*. The scleractinian skeleton consists primarily of an outer wall (epitheca) and numerous radial septa. The polyp occupies a cup (calice) on the upper surface of the skeleton and overlaps at the margin to form an edge zone. The number of septa increases during growth; initially six prosepta are secreted between the mesenteries, followed by essentially cyclic insertion of additional septa in the septal spaces. The septa are generally plate-like and are commonly differentiated into size orders reflecting the sequence of insertion. A peripheral area occupied by small, sub-horizontal, domed skeletal elements (dissepiments) may be developed between the septa, and an axial structure, or columella, is often present. The columella may be in the form of a simple styliform rod, or may be constructed from a complex of numerous elements. Classification of the *Scleractinia* is based principally on septal microstructure.

In *colonial corals*, the corallum is built by several polyps, each contributing a corallite to the skeleton. New corallites are added by asexual division during growth, either within or outside the wall of the parent corallite; incomplete division is common, leading to development of radially confluent centres or meandroid valleys. Individual corallites may be polygonal and separated by walls (a cerioid corallum; in *scleractinian* corals these walls are of dissepimental or septal origin), lack walls and can be joined by confluent septa (thamnasterioid) or, by dissepiments alone (aphroid), or have walls separated by areas of intercorallite tissue (plocoid). The colony may be massive in form or branch in a variety of ways; for example, in dendroid types the branching is irregular and divergent, while in phaceloid forms the branches are subparallel.

Major *scleractinian* coral *reefs* are known from the late **Triassic** to the present. Most *reef* corals are hermatypic, with the endoderm containing large numbers of symbiotic *algae* (zooxanthellae). The ecological requirements of the corals and *algae* restrict *reef* development to warm (>18°C), shallow waters. *Reefs* are rigid, wave-resistant structures and, as well as corals, harbour a diverse biota of other organisms, some of which, such as **calcareous algae**, may contribute to frame-building. The earliest *scleractinians* were hermatypic, but constructed only small patch *reefs*. Ahermatypic forms, which lack *zooxanthellae*, appeared in the **Jurassic**. Most are solitary, but ahermatypic corals are environmentally widespread, occurring at depths of up to 6000 m and tolerating temperatures close to freezing point.

SUBCLASS RUGOSA. Solitary or colonial **Paleozoic** *anthozoans* with bilaterally symmetrical, calcareous corallites in which septa are inserted in four loci. Primary septa comprise a cardinal and a counter septum on the plane of bilateral symmetry, a pair of alar septa adjacent to the cardinal septum and a pair of counterlateral septa adjacent to the counter septum. Subsequent metasepta are inserted on the cardinal side of the alar and counterlateral septa; a cardinal fossula may develop around the cardinal septum, where no septa are inserted. Short minor septa are normally emplaced between the metasepta. A marginal zone of dissepiments may be developed and an axial structure may be present. Transverse partitions (tabulae) occur in the axial area or across the entire corallite interior if a zone of dissepiments is absent. The region occupied by tabulae is termed the tabularium.

Corallum form is highly variable. Solitary *rugosans* may be discoidal, conical or cylindrical, with different degrees of curvature and, occasionally, twisting. Colonial forms may be massive or branching in habit, and cerioid, thamnasterioid, aphroid, or astraeoid (lacking corallite walls and with intermingling septa) in style. In cerioid rugosans the corallite walls are epithecal in origin.

The relationship of the *Rugosa* to the *Scleractinia* is uncertain, but it is now considered unlikely that the *scleractinians* evolved from the *rugosans*. Subdivision of the *Rugosa* is based on patterns of morphological similarity and on microstructural features, especially of the septa, where the crystalline components may be of fibro-normal, trabecular or combined types. Trabecular tissue comprises pillars of radiating calcareous fibres (trabeculae) lying in the plane of the septum.

The stratigraphical range is Middle **Ordovician** to Upper **Permian**. Early *rugosans* are mostly small and solitary, with the *reef* environment first being exploited in the mid-**Silurian** by both solitary and colonial forms. *Rugosans* never became major *reef*-builders, but associated with *tabulates* and **Stromatoporoids** in **biohermal** build-ups, or occurred in isolated patches around *reef* flanks. Evolutionary bursts occurred in the late **Devonian** and early **Carboniferous**, and to a lesser degree in the **Permian**. The stratigraphical ranges of *rugosan* species are generally too long or too uncertain for them to be of primary importance in **biostratigraphy**, but broad biozonal schemes have been devised, especially for parts of the **Devonian** and **Carboniferous**.

Fine growth ridges on the epithecal surface of some well-preserved corals have been identified as daily growth increments, with broader monthly banding and annual annulations also recognizable. Counts of these ridges on **Devonian** rugosans have led to the suggestion that the **Devonian** year averaged 400 days, a conclusion consistent with the astronomical calculation of the decreasing rate of the Earth's rotation through *tidal friction*.

SUBCLASS TABULATA. Colonial **Paleozoic** *anthozoans* with calcareous tubular corallites. Tabulae are prominent, while septa are inconspicuous or absent. When present the septa are spinose and commonly number 12. A marginal zone of annular lamellae or, when corallite walls are absent, of tubular or vesicular connecting tissue (coenenchyme) may be developed. Where walls are present, corallites are often connected through mural pores or, where the corallites are separated (fasciculate), via connecting tubules. Some genera have trabecular skeletal tissue in the septa and walls.

Tabulates may be massive, foliaceous, dendroid, phaceloid or creeping in habit. Form may be cerioid (e.g. the Favositida), catenoid (with corallites linked laterally like the posts in a palisade fence, e.g. the Halysitida), or coenenchymal (e.g. the Heliolitida).

Tabulates are known from the Lower **Ordovician** to the Upper **Permian**, and have been used as biostratigraphical indices in the **Ordovician** to Devonian, particularly in the former USSR. The origins of the *Tabulata* are obscure, and the nature of any relationship with the *Rugosa* uncertain. They are best represented in carbonate environments and are generally unimportant in argillaceous or arenaceous sediments. They are especially abundant in *reef* facies, and are particularly associated with **Stromatoporoids** in **bioherms** and *biostromes* of late **Ordovician** to **Devonian** age. Numbers were greatly reduced in the **Carboniferous** and **Permian**. [RJA]

British Museum (Natural History) (1962) *British Mesozoic Fossils*. p. 53.
Clarkson, E.N.K. (1986) *Invertebrate Palaeontology and Evolution*, pp. 80–116. Allen & Unwin, London.
Hill, D. (1981) Coelenterata: Anthozoa, Subclasses Rugosa, Tabulata. In: Teichert, C. (ed.) *Treatise on Invertebrate Paleontology, Part F, Supplement 1* (2 Vols). Geological Society of America and University of Kansas Press, Boulder, CO and Lawrence, KS.
Moore, R.C. (ed.) (1956) *Treatise on Invertebrate Paleontology, Part F, Coelenterata*. Geological Society of America, Boulder, CO.
Scrutton, C.T. & Rosen, B.R. (1985) Cnidaria. In: Murray, J.W. (ed.) *Atlas of Invertebrate Macrofossils*, pp. 11–46. Longman, Harlow.

cordierite $((Mg, Fe)_2Al_4Si_5O_{18} \cdot nH_2O)$ A hydrated aluminosilicate mineral (a ring or *cyclosilicate*) found in metamorphic rocks. [DJV]

core (1) It has been known for centuries that the density of interior of the Earth was greater than that of its surface rocks as the Earth's moment of inertia is approximately 0.3

and would be about 0.4 if the Earth were homogeneous. The first distinct evidence for the core, which comprises some 16% of the Earth's volume and 31% of its total mass, was determined from the passage of **earthquake** waves which where strongly refracted by it, giving rise to a '*shadow zone*' in localities at *epicentral angles* of more than 110°. More sophisticated studies of seismic **arrival times** and amplitudes by I. Lehmann in 1936 also indicated the presence of an inner core. Subsequent studies of refracted and reflected *body waves* and of *surface waves* have since allowed three core zones to be defined. The outer core occurs at a depth of 2900 km and extends down to approximately 4980 km, while the inner core extends from the Earth's centre, at 6370 km, to about 5120 km, leaving a possible **transition zone** between 4980 and 5120 km. The outer core does not transmit *shear waves*, i.e. it has zero *shear modulus* and is therefore liquid, while the inner core does transmit them poorly, indicating that it is solid, but essentially 'soft', for the transmission of **seismic waves**. Travel time studies indicate an increase in *P wave* velocity from about $7.91 \, km \, s^{-1}$ close to the **mantle** to $10.44 \, km \, s^{-1}$ close to the **transition zone**, corresponding to a density of about $9.94 \, Mg \, m^{-3}$, while the inner core has *P wave* velocities of about 11.2 to $11.3 \, km \, s^{-1}$, suggesting a poorly defined density of about $13.0–14.0 \, Mg \, m^{-3}$. The fluidity of the outer core is also shown by *solid-earth tides* and induced **eddy currents** indicate that it is strongly **electrically conductive**. The structure of the outermost 200 km of the core is complex, as is the **transition zone**, with some evidence for abrupt velocity transitions, i.e. layering, in both localities. The core/**mantle** boundary may also be irregular (i.e. have a topography) as there are slight differences (*c*. 10–20 km) between its depth calculated using *surface waves* and that based on reflected *body waves*. The composition of the outer core is predominantly **iron**, being the only common element of appropriate density and electrical conductivity, but contains some 10–15 wt% of lighter elements, probably **sulphur** in the form $Fe_{0.9}S$, but could be oxygen. Such a composition would indicate a temperature above *c*. 3700°C, the melting point of such a mixture at these pressures (as indicated by shock wave studies). The inner core has a density which probably corresponds to a fairly pure **iron** composition which would, at such pressures, have a solidus slightly below 6000°C and the temperature is thus considered to be only slightly lower, accounting for its solid but soft seismic transmission properties. The **geomagnetic field** is thought to be driven by **convective** motions in the **electrically conducting** outer core which could be driven by *radioactive* elements within the core, heat released by the solidification of the inner core, stirring by differential motions between the core and **mantle**, etc. (See Fig. M2, p. 382.) [DHT]

(2) Cylindrical pieces of rock obtained using a diamond drill. The drilling is normally undertaken for exploratory purposes in engineering works, *mineral exploration*, scientific research and so on, frequently to obtain core samples. [AME]

Bott, M.H.P. (1982) *The Interior of the Earth*. Arnold, London.
Jacobs, J.A. (1987) *The Earth's Core* (2nd edn). Academic Press, London.
Lambeck, K. (1980) *The Earth's Variable Rotation*. Cambridge University Press, Cambridge.
Melchior, P. (1986) *The Physics of the Earth's Core*. Pergamon, Oxford.

corestone A large cobble or boulder of relatively unweathered rock which is found within a **deep weathering** profile. Corestones of crystalline rocks such as **basalt**, **dolerite** or **granite** are spherical. They develop as **weathering**, induced by percolating water, proceeds along **joints** which formed as the **igneous rocks** cooled and contracted. Gradually, the angular blocks become rounded boulders set within an unconsolidated, clayey matrix (see **spheroidal weathering**). **Erosion** of the weathered material may exhume the corestones which then form **clitter** and **tors**. [AW]

Coriolis effect/force An effect caused by the Earth's rotation which deflects a body of fluid or gas moving relative to the Earth's surface from a straight line. This deflection is to the right in the northern hemisphere and to the left in the southern hemisphere. The effect is due to a Coriolis force F_C which produces an acceleration whose magnitude varies with the speed of the particle and increases towards higher latitudes, being zero at the equator.

$$F_C = 2\omega v \sin \phi$$

where ω is the Earth's angular velocity, v the velocity of the moving material and ϕ is the latitude. [AESK/DT]

Pond, S. & Pickard, G.L. (1983) *Introductory Dynamical Oceanography* (2nd edn). Pergamon, Oxford.

corniche An organic protrusion growing out from steep rock surfaces at about sea-level, and providing a narrow pavement or sidewalk-like path at the foot of **sea cliffs**. Comparable rock ledges caused by **erosional** processes and coated with organic material are termed **trottoirs**. Corniches are often formed of **calcareous algae**. They are largely intertidal, being best developed in the inlets of exposed coasts and generally protrude about 0.2–2.0 m. Vermetids and serpulids may contribute to their development. [ASG]

Cornish stone Crushed, partially **weathered granite** with appreciable **feldspar** and **kaolinite** contents; exploited in Cornwall for many centuries and used in the manufacture of bone china. [AME]

corrasion The process of mechanical **erosion** of a rock surface by material transported across the surface in water, wind, ice or by mass movement. The resulting effect on the rock surface is known as **abrasion**. [NJM]

correlation The process of equating a set of information from a particular formation or stratigraphical level at one location as representing a similar set of information at another location. The set of information being correlated can be a signal from a **seismic record** or a **geophysical borehole log**, or a physical, chemical or biological characteristic of the rock type. Mathematically a correlation function can be computed which provides a measure of the degree of similarity between a pair of traces at various relative offsets. The *auto-correlation function* describes the correlation of a trace with itself and measures similarity of shape and periodicity along the trace, while the **cross-correlation** function measures the degree of similarity between two different traces. [GS]

Anstey, N.A. (1964) Correlation techniques — a review. *Geophysical Prospecting* **12**: 355–82.

corrosion A general term for chemical **weathering** of rock. The term is used in opposition to **corrasion** which is the downwearing of rock through mechanical **abrasion**. [HAV]

corundum (Al₂O₃) A mineral exploited industrially for *refractories*. (See **oxide minerals**.) [DJV]

cottonballs Colloquial term for the fine fibrous crystals of the borate mineral **ulexite**. [DJV]

couloir A deep gorge or ravine on the side of a mountain, especially in the Alps. Some of them are more or less parallel and may be the result of *avalanche erosion*. The term is also applied to the parallel depressions that occur between wind-sculpted bedrock outcrops (**yardangs**) in desert areas. [ASG]

counter lode/vein (*caunter lode/vein*) A *lode* running in a different direction from the usual direction for the district. [AME]

counterpoint bar A river bar deposited on the concave side of a channel bend due to a change in meander migration pattern or flow conditions. Counterpoint bars have been recorded from modern **gravel** bed streams and from suspended load streams. [JA]

covellite (CuS) An **ore** mineral of **copper**. [DJV]

crabhole A small abrupt depression in the ground surface which can vary in diameter from a few centimetres to more than a metre, and in depth from around 50 to 600 mm. They occur in sediments which are prone to vertical cracking and horizontal packing. [ASG]

Upton, G. (1983) Genesis of crabhole microrelief at Fowlers Gap, Western New South Wales. *Catena* **10**: 383–92.

crack-seal mechanism Filling of a **vein** by repeated cycles of **extensional fracture** followed by **cementation**. Many **veins** have evidence that **vein**-fills grew incrementally, in the form of inclusion bands of *host rock* parallel to the **vein** wall, and changes in the direction of growth of crystals in the **vein** may be observed (see Ramsay & Huber, 1987). [TGB]

Ramsay, J.G. & Huber, M. I. (1987) *The Techniques of Modern Structural Geology* Vol. 2: *Folds and Fractures*. Academic Press, London.

crackle brecciation The name given to the **fractures** which are usually healed with veinlets of **quartz** and economic minerals that form the **stockwork** mineralization characteristic of many **porphyry copper**, **disseminated molybdenum** and **tin stockwork** deposits. It is named after its resemblance to the fine network of cracks in the glaze of some porcelain and pottery. Zones of crackle brecciation in **porphyry copper deposits** are circular to irregular in outline, always larger than the **orebody** and fade out in the surrounding, low grade, outer **propylitic alteration**. This brecciation is thought to be due to the expansion resulting from the release of volatiles from the host **magma** during **retrograde boiling**. [AME]

Philips, W.J. (1973) Mechanical effects of retrograde boiling and its probable importance in the formation of some porphyry ore deposits. *Transactions of the Institution of Mining and Metallurgy (Section B: Applied Earth Sciences)* **82**: B90–8.

crater A large, bowl-shaped depression in the Earth's surface. It may be caused by volcanic activity, being situated at the summit or on the flank of a volcano. Alternatively, the circular depression may have been caused by the impact of a *meteorite*. [NJM]

craton The relatively stable interior of a continental *plate*. Prior to **plate tectonic** theory, **continental crust** was regarded as divided, for a given geological period, into cratons and **mobile** or *orogenic belts*. In **plate tectonic** terms, the craton represents that part of a continental *plate* not affected by contemporaneous **plate boundary** activity. The term is more frequently used to subdivide regions of **Precambrian** shield, e.g. the 'Tanzanian craton', the 'Superior craton', etc. A piece of **crust** that forms part of a **mobile belt** in one period may become part of a craton in a subsequent period, after *stabilization* or 'cratonization' has occurred, i.e. when the tectono-thermal activity characteristic of the **mobile belt** has ceased and the area has become stable. [RGP]

creep Slow, but more or less continuous, **deformation** process occurring below the *elastic limit* producing time-dependent **strain** in response to prolonged application of

stress. The term covers all **deformation** that is not purely **elastic** (i.e. **viscoelastic**, **viscous** and **plastic**), and is used particularly in the context of laboratory testing of materials and in **engineering geology**.

The main forms of creep are (i) *cataclastic flow*, in which individual grains or fragments physically move relative to each other, (ii) *dislocation creep*, in which parts of crystals glide past each other along crystalline dislocations, (iii) *grain boundary gliding*, (iv) *Coble creep*, which is by the diffusion of individual atoms along grain boundaries, (v) *Nabarro–Herring creep*, which involves the diffusion of atoms within the crystalline lattice, and (vi) *recrystallization*. Each process is dependent on the magnitude of the applied **stress**, the duration of its application, and the ambient temperature. Creep processes can also be divided into linear and power-law creep, according to the response to a particular **stress** applied under the same conditions to different materials. Creep is conventionally divided into three stages (see Fig. C36), representing characteristic changes that take place in the **strain**–time curve at constant **stress** during the **deformation**. *Primary* or *transient creep* is the initial stage of **viscoelastic strain** characterized by a concave-downwards **strain**–time curve. This stage is followed by *secondary creep* characterized by steady-state *viscous flow* with a constant **strain**–time slope. The final stage is known as *tertiary creep*, marked by accelerating strain leading to **failure**.

Laboratory experiments on natural rock materials carried out over periods of several months under low load **stresses** at **strain rates** of around $10^{-7}\,\mathrm{s}^{-1}$ are thought to simulate natural creep processes, although the latter take place at **strain rates** several orders of magnitude lower.

Creep strain may be represented by an empirical expression of the type:

$$\varepsilon = \varepsilon_e + \varepsilon_1(t) + v_t + \varepsilon_3(t)$$

where ε_e is the instantaneous **elastic** strain, $\varepsilon_1(t)$ the transient creep, v_t the steady-state creep and $\varepsilon_3(t)$ the accelerating creep; t is time. The transient creep $\varepsilon_1(t)$ may be expressed in terms of the modified *Lomnitz Law*:

$$\varepsilon_1(t) = A[(1 + at)^\alpha - 1]$$

where A and α are constants for a particular **rheology**, but vary with temperature and confining pressure. The steady state creep, V_t, may be expressed in terms of a **viscosity** η such that:

$$\dot{\varepsilon} = \sigma/3\eta \quad (\textit{Maxwell substance})$$

or

$$\dot{\varepsilon} = (\sigma - \sigma_0)/3\eta \quad (\textit{Bingham substance})$$

A more general law may be derived from linear **rheological** theory. (See **deformation mechanism**.) [RGP/DHT]

Ashby, M.F. & Verrall, R.H. (1977) Micromechanisms of flow and fracture, and their relations to the rheology of the upper mantle. *Philosophical Transactions of the Royal Society, London* **288A**: 59–95.

Jaeger, J.C. & Cook, N.G.W. (1976) *Fundamentals of Rock Mechanics*. Chapman & Hall, London.

crenulation Small-scale folding or kinking. (See *crenulation cleavage*.) [TGB]

crest line The line joining the topographically highest points of a folded surface. The crest line is normally parallel to the *hinge line* but is only coincident with it in *upright folds* (Fig. C37). (See **fold**.) [RGP]

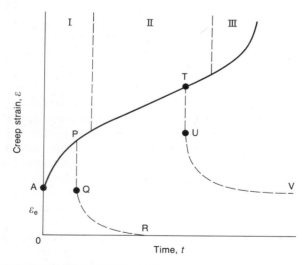

Fig. C36 Typical **creep** strain–time curve. If constant stress is applied to a material, an instantaneous elastic strain is followed by a region I of primary or transient creep, followed by a region II of secondary or steady-state creep, followed by region III of tertiary or accelerating creep. If the applied stress is suddenly removed at P in region I, the strain–time curve takes the form PQR in which PQ is instantaneous and QR tends asymptotically to zero; there is no permanent strain. If the stress is removed at T in region II, the curve takes the form TUV which leads asymptotically to a value of permanent strain.

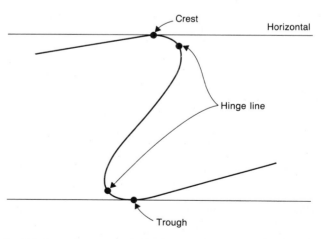

Fig. C37 Crest line. Fold profile showing how the crest and trough generally differ from the hinge lines when the axial surface is dipping. (After Hobbs *et al.*, 1976.)

Hobbs, B.E., Means, W.D. & Williams, P.F. (1976) *An Outline of Structural Geology*. John Wiley & Sons, New York.

Cretaceous System named after *creta*, the latin for **chalk**. This distinctive type of sediment represents the upper half of the system in northern Europe. The Cretaceous period is one of the longest in **Phanerozoic** time, occupying some 80 Ma from approximately 145 Ma to almost exactly 65 Ma (Fig. C38).

The break-up of the **Pangean** landmass, which had started earlier, accelerated during the Cretaceous. Africa finally split from South America, the last land connection being between Brazil and Nigeria. A seaway between the North and South Atlantic may have been initiated in the *Aptian* and was certainly continuous during the Late *Albian* onwards; the Benue Trough in Nigeria grew as an **aulacogen** from the **triple junction**. At the same time there was an acceleration in the widening of the North Atlantic; from the beginning of the *Aptian* to the end of the Santonian (approximately 30 Ma) the lateral spread was no more than 400 km short of the widening during the whole of the **Cenozoic**. During this time, primarily *Barremian* to *Cenomanian*, black and grey-green carbonaceous **shales** of the Hatteras Formation were deposited in the main basin of the North Atlantic instead of the pelagic carbonate oozes that dominated the centre of this ocean in much of Cretaceous–**Cenozoic** time. The Bay of Biscay trough opened during the Late *Aptian* to *Santonian*; this may have coincided with the main initiation of the Rockall Trough. Madagascar separated from the African Plate and India from the Antarctic, probably before the end of the Early Cretaceous. By middle Late Cretaceous the northern margin of the Indian Plate had started to impact with the Lhasa Block in south China along the southern margin of Tibet; in Kashmir–northern Afghanistan the **Tethys Ocean** closed. In the far western region of North America the Kula Plate impinged against the North American **craton** at speeds of up to $100 \, mm \, a^{-1}$, with a northward latitudinal displacement of 20° and rotations of up to 60°.

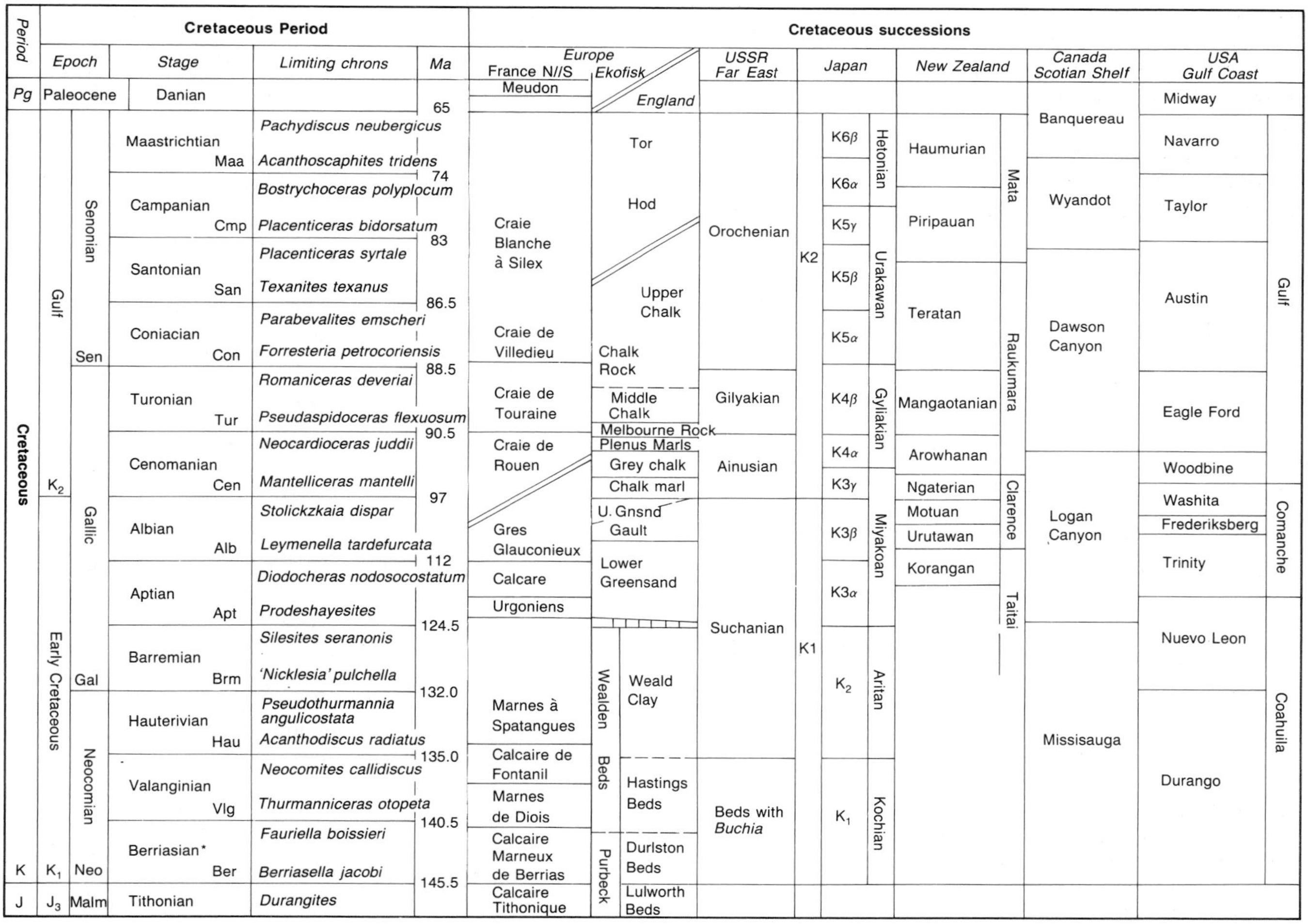

Period	Cretaceous Period				Cretaceous successions							
Period	Epoch	Stage	Limiting chrons	Ma	Europe France NII/S / Ekofisk	USSR Far East	Japan		New Zealand		Canada Scotian Shelf	USA Gulf Coast
Pg	Paleocene	Danian		65	Meudon / England						Banquereau	Midway
Cretaceous	Senonian / Gulf / Sen / K2	Maastrichtian Maa	Pachydiscus neubergicus / Acanthoscaphites tridens	74	Tor		K6β	Hetonian	Haumurian	Mata	Wyandot	Navarro
		Campanian Cmp	Bostrychoceras polyplocum / Placenticeras bidorsatum	83	Craie Blanche à Silex	Hod	K6α	Urakawan	Piripauan	Raukumara		Taylor
		Santonian San	Placenticeras syrtale / Texanites texanus	86.5		Upper Chalk	K5γ / K5β		Teratan		Dawson Canyon	Austin
		Coniacian Con	Parabevalites emscheri / Forresteria petrocoriensis	88.5	Craie de Villedieu	Chalk Rock	K5α					
	Gallic / Sen	Turonian Tur	Romaniceras deveriai / Pseudaspidoceras flexuosum	90.5	Craie de Touraine	Middle Chalk / Melbourne Rock	Gilyakian	K4β	Gyliakian	Mangaotanian		Eagle Ford
	K2 / Gallic / Cen	Cenomanian Cen	Neocardioceras juddii / Mantelliceras mantelli	97	Craie de Rouen	Plenus Marls / Grey chalk / Chalk marl	Ainusian	K4α / K3γ		Arowhanan / Ngaterian	Clarence	Woodbine
	Gallic / Alb	Albian Alb	Stolickzkaia dispar / Leymenella tardefurcata	112	Gres Glauconieux	U. Gnsnd / Gault		K3β	Miyakoan	Motuan / Urutawan	Logan Canyon	Washita / Frederiksberg
	Early Cretaceous / Gal	Aptian Apt	Diodocheras nodosocostatum / Prodeshayesites	124.5	Calcare / Urgoniens	Lower Greensand		K3α		Korangan	Taitai	Trinity
	Gal / Brm	Barremian Brm	Silesites seranonis / 'Nicklesia' pulchella	132.0	Marnes à Spatangues	Weald Clay	Suchanian	K1 / K2	Aritan			Nuevo Leon
	Neocomian / Hau	Hauterivian Hau	Pseudothurmannia angulicostata / Acanthodiscus radiatus	135.0	Calcaire de Fontanil	Hastings Beds					Missisauga	Coahuila
	Neocomian / Vlg	Valanginian Vlg	Neocomites callidiscus / Thurmanniceras otopeta	140.5	Marnes de Diois	Beds with Buchia		K1	Kochian			Durango
K / K1	Neo / Ber	Berriasian* Ber	Fauriella boissieri / Berriasella jacobi	145.5	Calcaire Marneux de Berrias	Durlston Beds						
J	J3 / Malm	Tithonian	Durangites		Calcaire Tithonique	Lulworth Beds						

Fig. C38 The **Cretaceous** period and selected successions. (The numerical ages are rounded to 0.5 Ma intervals.) In parts of the Boreal Realm, Ryazanian is often substituted for Berriasian. (After Harland *et al.*, 1989.)

With regard to *igneous activity*, apart from local developments such as **ophiolites** in Oman and the north-west Himalaya, the Cretaceous is renowned for: (i) major complex intrusions in the Cordillera of the western USA and Canada starting in the mid-Cretaceous and culminating in the Boulder Batholith (*Campanian–Maastrichtian*). Associated lavas started as **andesites, rhyolites** and **trachytes. Ash**-falls from the Elkhorn Volcanics in Idaho provided **bentonites** over 2000 km thick in the seaway that yield the most accurate *radiometric dates* for the Late Cretaceous; (ii) in west-central India the *Deccan traps* are **tholeiitic basalts** more than 1200 m thick erupted during the Late Cretaceous to **Eocene** over an area of some 500 000 km^2.

Sea-levels at the start of the period were below the average of today but there was an overall **eustatic** rise during the Early Cretaceous with an acceleration from Late *Albian* to Early *Turonian*. Apart from a strong fall in the late *Turonian*, sea-levels stayed high during the Late Cretaceous, only showing a sharp fall again near the end of the *Maastrichtian*. The abnormally high sea-levels of the Late Cretaceous have been variously estimated as 150 to 600 m above present-day levels; only 18% of the Earth's surface was left as land, compared with 28% today.

Quaternary style *glacial* temperatures are shown by the *Albian–Cenomanian* flora in Alaska, whilst the Sverdrup Basin in Canada contains true *glendonites* (stellar **pseudomorphs** of **calcite** after *ikaite*) and **dropstones** (ice transported pebbles). Re-analysis of some high latitude floras confirms that warm intervals must have been much warmer than today. Variations between these extremes occur in major cycles in which the *Valanginian* and Late *Aptian* were particularly cold, the *Hauterivian, Barremian,* Early *Aptian, Albian, Turonian* and much of the *Santonian* to mid-*Campanian* seem to have been particularly warm, with **coals** developed in several regions. Small-scale *Milankovitch cycles*, with variations in clay to **limestone** ratios, are found in all stages and are prominent in the *Barremian, Cenomanian* and *Turonian*. Floral and faunal differentiation between high and low latitudes does not support an equable distribution of climates.

Chalk and **coccolith**-rich **pelagic** *micritic* **limestones** are not limited to this period, but it was the high sea-level of the Late Cretaceous that carried these **chalks** on to the continents from north-west Europe eastwards to the Ukraine and east of the Caspian, and through the western interior seaway in the USA. The **coccolithic** component (sometimes more than 90% of the rock) was secreted as the stable, low-Mg **calcite** by haptophycean *algae*. The *Urgonian* is a massive landscape-dominating Upper *Barremian* to Lower *Aptian* **limestone** facies in southern Europe whose name is commonly associated with *rudists* but is rarely formed of a *rudist* framework.

Carbon-rich shales that mark a severe deficiency of oxygen in the bottom waters of the oceans (*anoxic event*), extended briefly into shelf seas several times during the Cretaceous; the most widespread occurred high in the *Cenomanian*.

The most important event for life on Earth was the appearance of the *angiosperms* (flowering **land plants**) in the *Barremian*; by the mid-Late Cretaceous they were a major component of the world's floras. Three major floral provinces are recognizable on land through much of the period.

Boreal and **tethyan** faunas became less marked after the Early *Turonian* but are still recognizable by *ammonites, belemnites, gastropods,* **corals** and *rudists*; the last group were aberrant *bivalves* specially characteristic of the **tethyan** Cretaceous. Late in the period many of the distinctive animal groups of the Cretaceous became extinct; *ammonites, belemnites, rudists,* inoceramid *bivalves,* and on land the *dinosaurs*. In most regions these **extinctions** occurred several million years before the end of the period, and were preceded by a contraction in variety and geographical distribution. The most important terminal Cretaceous **extinctions** were in the calcareous *nannoplankton* and planktic **Foraminifera**, both of which almost disappeared, having been very prolific indeed. **Plants** on **land** were hardly affected.

More than half the world's oil and gas comes from Cretaceous **reservoirs**. [JMH]

Caldwell, W.G.E. & Kauffman, E.G. (eds) *The Evolution of the Cretaceous Western Interior Basin.* Special Paper of the Geological Association of Canada. (In press.)

Hancock, J.M. (1986) Cretaceous. In: Glennie, K.W. (ed.) *Introduction to the Petroleum Geology of the North Sea* (2nd edn) pp. 161–78. Blackwell Scientific Publications, Oxford.

Harland, W.B., Armsbrong, R.L. & Cox, A.V. *et al.* (1989) *A Geologic Time Scale.* Cambridge University Press, Cambridge.

Kemper, E. (1987) Das Klima der Kreide-Zeit. *Geologisches Jahrbuch, Reihe A* **96**.

Krasilov, V.A. (1985) *Cretaceous Period: Evolution of the Terrestrial Crust and Biosphere.* Academya Nauk USSR, Moscow. [Russian]

Matsumoto, T. (1978) Japan and adjoining areas. In: Moullade, M. & Nairn, A.E.M. (eds) *Phanerozoic Geology of the World II, The Mesozoic* pp. 79–144. Elsevier, Amsterdam.

Plus many papers in *Cretaceous Research*, Academic Press, London.

crevasse splay A flood event or repeated flood events that result from breaching of a channel bank or **levée** depositing relatively coarse sediment in the overbank environment. Crevasse splays are distinct from overbank **sheetfloods** in that they are point sourced and always associated with **erosion** near the trunk channel. Crevasse splay **sandstones** may be distinguished from **overbank** flood sheet **sandstones** in the rock record only where channelized features and **erosional** bases may be identified near the source channel sediment body.

Crevasse splay deposits thin and fine away from the channel bank breach as they spread into the overbank environment due to reductions in flow velocity. Progressive diversion of water through the channel bank breach may result in crevasse splay progradation and deposition of a coarsening upward sequence. If water continues to be diverted through the crevasse, **avulsion** may result at the point of channel breach. [JA]

cristobalite (SiO_2) A high temperature form of **silica** found in volcanic rocks. [DJV]

critical angle (i_c) The angle between the propagation path of a ray and the normal to an interface between two media along which the **refracted** wave travels as a **head wave**. The velocity of the first medium needs to be less than that of the second. If the velocities of the media in which the ray is propagating from and to are V_i and V_r respectively, *Snell's Law* states that

$$\frac{\sin i_c}{V_r} = \frac{\sin 90°}{V_r} \quad \text{hence } \sin i_c = \frac{V_i}{V_r} \quad \text{providing } V_i < V_r$$

(See Fig. D12, p. 160.) [GS]

critical distance The distance from a **seismic source** at which the travel time of the reflection and refraction from an interface are equal. This is also the distance from the source at which the reflection, incident on a horizontal interface at the **critical angle**, emerges, has its highest amplitude and is the closest to a source at which a refracted **head wave** exists. [GS]

crocidolite ($NaFe_3^{2+}Fe_2^{3+}Si_8O_{22}(OH)_2$) A sodic **amphibole mineral** that occurs with an **asbestiform** habit, the so-called *'blue asbestos'*. [DJV]

crocite ($PbCuO_4$) A rare mineral found in the oxidized zones of **lead** deposits. [DJV]

cross fold **Fold** whose *axis* trends transversely to the general **fold** trend in a region. The term is purely descriptive and carries no implication about relative age or origin. Thus cross folds may be earlier, later or coeval with the 'main' set of **folds** in an area. [RGP]

cross lode/vein A *lode* or **vein** that intersects a larger or more important one. [AME]

cross-correlation A measure of the **correlation** or similarity between two signals. [GS]

cross-section Diagrammatic representation in the vertical plane of the geology of an area, usually constructed from a **geological map**. Reasonable assumptions are made about the way in which structures visible at the surface continue downwards, and may be supplemented by data from boreholes, **wells**, etc. Cross-sections may be drawn along a particular line or lines on the map, chosen to illustrate the vertical structure effectively. The combination of map and cross-section should ideally give a good three-dimensional picture of the geological **structure** of an area.

In complex areas, several lines of section may be used to give a better coverage of the structural variation. [RGP]

cross-stratification The characteristic **bedding** structure produced by the migration of bedforms with inclined depositional surfaces. Cross-stratification occurs on a wide variety of scales, from millimetre scale (when it is more correctly called *cross-lamination*) up to tens of metres in vertical height in the case of large aeolian **dunes** (**draas**). The geometry of the stratification resulting from the migration of bedforms is nearly always highly three-dimensional, since the bedforms themselves rarely have perfectly linear **crest lines** and rarely migrate in a regular fashion. The description of the geometry of cross-stratification can therefore yield information on the nature of the bedforms responsible and thereby on the *paleoflow* conditions.

The fundamental unit of cross-stratification is the *cross-set*, a unit of sediment bounded by **erosional** or non-erosional, horizontal or inclined surfaces. Within the *cross-set*, cross-stratal surfaces are inclined with respect to the lower bounding surface. For a *cross-set* to be preserved, the bedforms must climb over these lower bounding surfaces, causing net **aggradation** of the bed (Fig. C39). Failure to climb results in the destruction of the down-flow deposit by the up-flow bedform during the migration process.

Classification schemes for modern bedforms cannot be directly applied to the preserved cross-stratification because the morphological properties and the behavioural properties of modern bedforms are significantly different from the observations that can be made from cross-stratification. For example, the height, wavelength, asymmetry and sinuosity of a modern bedform can be directly measured, but the time–behaviour is generally not well understood. In contrast, it is difficult to interpret bedform shapes from cross-stratification, but the history of sediment transport is easier to ascertain.

A modern approach is to group cross-stratification and ancient bedforms into four main classes based on two basic criteria:
1 The plan-form shape of the bedform, i.e. two-dimensional (linear **crest lines**) or three-dimensional (irregular, sinuous, curved **crest lines**).
2 Changes in morphology or path of climb, i.e. invariable (no change in morphology or path of climb) and variable (change in morphology or path of climb).

Bedforms can be further subdivided into transverse, oblique or longitudinal varieties according to the **crest line** orientation with respect to the sediment transport vector (Rubin, 1987).

Cross-stratification can also be described according to the geometry of the *cross-sets*. Where the lower bounding surface is sub-horizontal in three dimensions, the resulting *cross-set* is termed tabular, and is thought to be produced by the migration of linear-crested **dunes** (Fig. C40). Where the lower bounding surface is highly curved so that cross-sets are scoop- or spoon-shaped, the term *trough cross-*

cross-stratification

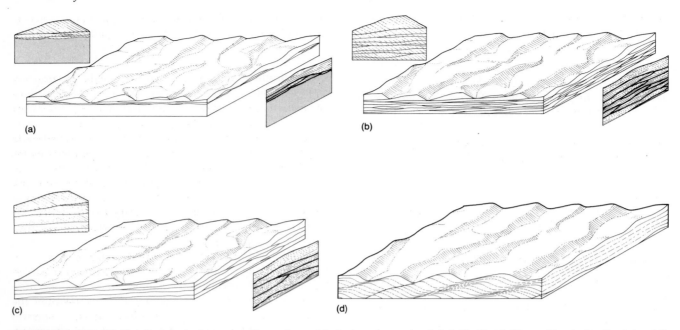

Fig. C39 Cross-stratification. Cross-lamination produced by small current ripples for various angles of climb. The block is 50 cm × 50 cm in size. Flow is from left to right. (a) Zero angle of climb. (b) Angle of climb very small. (c) Angle of climb still small, but greater than in (b). (d) Angle of climb large. In (a), (b) and (c) the middle segments of both the flow-parallel and flow-transverse sections are shown enlarged by a factor of 1.75.

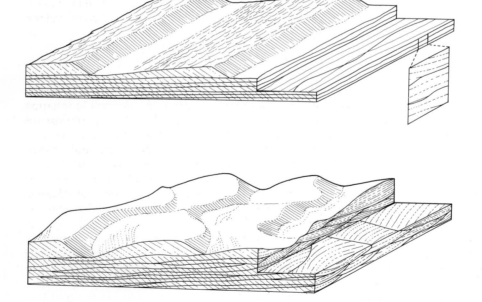

Fig. C40 Block diagram showing large-scale tabular **cross-stratification** formed by migration of two-dimensional large ripples. Flow is from left to right. The length of the sides of the block could range from 10 m to as much as a few hundred metres.

Fig. C41 Block diagram showing large-scale trough **cross-stratification** formed by migration of large three-dimensional ripples. Flow is from left to right. The length of the sides of the block could range from a few metres to a few tens of metres.

stratification is used. *Trough cross-stratification* is thought to result from the migration of irregular, sinuous or linguoid **dunes** (Fig. C41). The internal cross-stratification may be planar with an angular contact with the lower bounding surface, or tangential to asymptotic with repect to the lower bounding surface. The nature of the cross-stratification planes depends not only on the plan-form shape of the

bedform, but also on the method of sediment transport. High amounts of material in suspension cause long asymptotic toesets to be developed, whilst bedforms experiencing predominantly bedload transport produce planar angular foresets as a result of the avalanching process on **slip-faces**.

Increasingly, styles of stratification are being described which result from the migration of bedforms lacking well-developed **slip-faces**, or result almost entirely from fall-out of suspended material (see **hummocky cross-stratification**). [PAA]

Rubin, D.M. (1987) *Cross-bedding, Bedforms and Paleocurrents.* Concepts in Sedimentology **1**. Society of Economic Paleontologists and Mineralogists, Tulsa, OK.

crossite An **amphibole mineral** between **glaucophane** and **riebeckite** in composition. [DJV]

crossover distance The distance from the **seismic source** at which the refracted **head wave** from an interface becomes a **first arrival**. [GS]

crucible swelling number A simple test to evaluate the amenability of a **coal** for **coke** manufacture. A small sample of crushed **coal** is heated to a temperature of 800°C in a standard crucible. After a specified time, or when all the volatiles have been driven off, a small **coke** 'button' remains. The profile (not diameter) of this is compared with a number of standard profiles to read off the swelling number of the **coal**. The scale is from 1 to 9. For blast furnace purposes swelling numbers of 4–6 are best. [AME]

crude oil The liquid members of the spectrum of materials that constitute **petroleum**. Crude oil is a mixture of *hydrocarbons*. The principal components belong to the *paraffin (methane)* and *naphthene series*. The *paraffin series* consists of straight-chain *hydrocarbons* of the general formula C_nH_{2n+2}. These compounds range from *methane* (CH_4) upwards, with each heavier molecule having a further CH_2. The first four are gases at S.T.P. The first liquid is *n*-pentane (C_5H_{12}) and the first solid is *n*-hexadecane ($C_{16}H_{34}$); *n*-hexadecane and higher members are solid paraffin waxes.

The *naphthene series* has the general formula C_nH_{2n} and consists of saturated **carbon** ring compounds. The lowest two members are gases, cyclopentane (C_5H_{10}) is the first liquid and it and cyclohexane are the commonest in most crude oils. **Aromatics** are generally unimportant.

Paraffin-base crudes, the most valued of all oils, only form about 2% of modern world supplies. Naphthenic dominated crudes (**asphalt-based crude oils**) provide about 15%. Most modern crude oils are of mixed base including nearly all Middle East, Mid-Continent (of the US) and North Sea oils.

The non-hydrocarbon constituents include **sulphur**, nitrogen and oxygen together with organic compounds of some heavy metals, mainly **nickel** and **vanadium**. **Sulphur** is present in nearly all crudes in quantities up to about 3%. It is higher in heavier than in light oils and occurs mainly in polyaromatics. A small amount occurs as elemental **sulphur** or H_2S. Oils with detectable H_2S are called *sour crudes*. Oil refineries prefer low sulphur (*sweet*

crudes) and will pay a premium price for them. Nitrogen and oxygen are present in much smaller quantities. **Vanadium** runs 30–300 p.p.m. in crudes and **nickel** 20–85 p.p.m.

The relative densities of oils are expressed on the API scale (see **API gravity**). The **viscosities** of oils vary with the density and the amount of dissolved gas. Light oils have **viscosities** below 30 mPa s whereas heavy asphaltic oils have **viscosities** in the thousands of millipascal-seconds range. The *pour point* is also used as a measure of **viscosity**. This is the lowest temperature at which the oil will flow under specified conditions. High contents of paraffin wax produce pour points above 40°C in some crudes, whilst other crudes may flow at temperatures as low as −36°C. Refractive indices vary from 1.42 to 1.48 in passing from lighter to heavier oils.

The majority of scientists consider that crude oil is of biogenic origin but, because of evidence such as the association of *bitumens, hydrocarbons*, gases and oils with some volcanic activity, a few suggest a deep-seated inorganic origin by degassing of the lower **crust** or **mantle**. Still others, citing the existence of *hydrocarbons* in the atmospheres of some celestial bodies, invoke an extraterrestrial origin.

The close association between oil and sediments, as with **coal**, has long convinced most geologists that there is a genetic connection. Various lines of evidence suggest that crude oil is derived from source sediments and migrates into **reservoir** rocks to form accumulations beneath **oil traps**. [AME]

North, F.K. (1985) *Petroleum Geology.* Allen & Unwin, Boston.

crust On the Earth and other planets, the crust is formed by the observable (and submarine, in the case of the Earth) surface rocks and commonly equates with the **lithosphere** on other planets. On the Earth, the crustal rocks are distinguished chemically from the underlying **mantle** rocks, being of lower density, and comprise either **oceanic** or **continental crust** which are separated by a sharp **seismic discontinuity**, the **Mohorovičić discontinuity** (*Moho*), from the underlying **mantle** rocks with densities of $3.3\,Mg\,m^{-3}$. On the Earth, the crust does not equate with the **lithoshere**, but is simply its uppermost part. (See Fig. M2, p. 382.) [DHT]

Bott, M.H.P. (1982) *The Interior of the Earth.* Arnold, London.

crustiform banding Layers (crusts) of different mineral composition formed during infilling of open spaces. This structure is most commonly developed when dilatant zones along **faults**, solution channels in **karst** topography, etc., are permeated by mineralizing solutions. The walls act as nucleating surfaces and the first-formed minerals grow on them. If the solutions change in composition, there may be a change in mineralogy and later deposited crusts of different composition may give the **vein** so

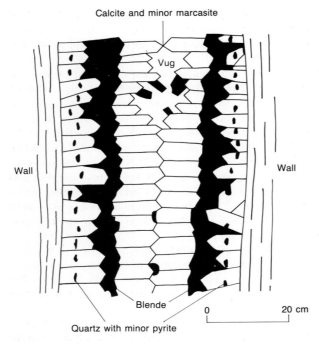

Calcite and minor marcasite

Vug

Wall

Wall

Blende

0   20 cm

Quartz with minor pyrite

Fig. C42 Section across a vein showing **crustiform banding**. (After Evans, 1993.)

formed a layered appearance (Fig. C42) called crustiform banding. With repeated opening of a fissure by **faulting** the banding may become asymmetrical. [AME]

Evans, A.M. (1993) *Ore Geology and Industrial Minerals: An Introduction* (3rd edn). Blackwell Scientific Publications, Oxford.

cryogenic magnetometer A **magnetometer** that uses liquid helium to maintain the detecting system, usually a *squid*, at superconducting temperatures. The instruments are faster and are somewhat more sensitive than most other forms of **magnetometer**. [DHT]

cryolite (Na_3AlF_6) A fluoride mineral valued industrially as a *flux*. [DJV]

cryptomelane (KMn_8O_{16}) A mineral occurring in **manganese ores**. [DJV]

cryptovolcano A circular structure of highly deformed strata believed to have been caused by volcanic activity but lacking any direct evidence thereof. The type example is Steinheim Basin, Germany. [NJM]

crystal field theory A theory which describes how the energy levels of atoms, and in particular the *d* levels of transition metal ions, are perturbed by a regular array of nearby negative charges. If an isolated atom is surrounded by a field of uniform negative charge then the electrons of the atom will all become less tightly bound due to electrostatic repulsion. If the surrounding charge is not uniform, but is concentrated at specific points in space — other atoms for example — then a situation similar to that in a crystal results. The field generated by such an array of point charges is the '*crystal field*'.

Simple cases correspond to frequently encountered *ligand* environments, e.g. octahedral and tetrahedral co-ordination. In the octahedral case (point group O_h) the charges at the six *ligand* sites transform as a_{1g}, t_{1u} and e_g and the valence shell **atomic orbitals** (AO) of the central atom as: s, a_{1g}; p, t_{1u}; $d(x^2 - y^2)$ and $d(z^2)$, e_g and $d(xy)$, $d(xz)$ and $d(yz)$, t_{2g}. The crystal field will therefore perturb the energies of the s, p, $d(x^2 - y^2)$ and $d(z^2)$ making them less stable but $d(xz)$, $d(yz)$ and $d(xz)$ will be unaffected. This can easily be seen by reference to the AO angular functions (see **atomic orbital**) where *ligands* approaching along Cartesian axes will interact with s, $p(x)$, $p(y)$, $p(z)$, $d(z^2)$, $d(x^2 - y^2)$ and avoid $d(xy)$, $d(xz)$ and $d(yz)$. The energy difference induced between the two sets of d orbitals is the '*Crystal Field Splitting*' and is sometimes given the symbol Δ, sometimes 10 Dq.

Transition metal ions are those in which the d shell is being filled and the crystal field affects this filling. If the field is weak then the ground electronic state is achieved by allocating electrons singly, first to the (d) t_{2g} orbitals and then to the (d) e_g orbitals before placing two electrons into one orbital where they would have to face greater electron–electron repulsion. The converse is true when the crystal field is strong so that (d) t_{2g} is much more stable than (d) e_g giving rise to the following possible configurations:

Number of d electrons	1	2	3	4	5	6	7	8	9	10
Weak field case										
(d) t_{2g}	1	2	3	3	3	4	5	6	6	6
(d) e_g	0	0	0	1	2	2	2	2	3	4
No. unpaired electrons	1	2	3	4	5	4	3	2	1	0
Strong field case										
(d) t_{2g}	1	2	3	4	5	6	6	6	6	6
(d) e_g	0	0	0	0	0	0	1	2	3	4
No. unpaired electrons	1	2	3	2	1	0	1	2	1	0

The colours of transition metal ions are associated with electronic transitions between the t_{2g} and e_g orbitals, and can be expected at longer wavelengths for weak fields and at shorter wavelengths for strong fields. The position is greatly complicated by electron–electron interactions so that one configuration can give rise to many possible electronic states. As the magnetic properties of atoms and ions are directly dependent on the number of unpaired electrons it is clear that the strength of the crystal field will determine directly the **paramagnetism** of a transition metal ion.

Ion size, and therefore hydration energies and ion–lattice stabilization energies, are also affected by the strength of the crystal field. It is also found that filled and half-filled t_{2g} and e_g shells are especially stable (e.g. weak field Mn^{2+}, Fe^{3+}; strong field Fe^{2+}). Sub-shells of degenerate orbitals that are unevenly filled with electrons (e.g. $e_g{}^3$ in Cu^{2+}) are subject to forces of distortion (the Jahn–Teller effect) which seek to remove the high symmetry that gives rise to the degeneracy. Thus the environment of many transition metal ions is somewhat distorted.

The strong field case is rarely encountered in mineralogical situations where the field is usually generated by **oxygen** or **sulphur** *ligands*. Also, minerals often include a wide range of transition metal ions in trace amounts. In such situations the environment of the ion will be dictated by the overall energy requirements of the bulk lattice and the crystal field will be different from the optimum for that ion. This can lead to unexpected colour and magnetic effects (e.g. traces of Cr^{3+} in **corundum** impart a red not green colour to **ruby**).

The crystal field effects due to a tetrahedral (point group T_d) arrangement of *ligands* causes the octahedral d shell splitting described above to be inverted, two (e) orbitals being more tightly bound than the t_2 trio. If the *ligand* groups are no longer regarded as point charges but as atoms or ions capable of bonding interactions with the transition metal ion to form molecular orbitals, then crystal field theory is transformed into '*ligand field theory*' which is capable of giving a much more detailed and accurate description of the electronic structure of transition metal compounds and complexes. [DSU]

Ballhausen, C.J. (1962) *Introduction to Ligand Field Theory*. McGraw-Hill, New York.
Van Vleck, J.H. (1935) Valence strength and the magnetism of complex salts. *Journal of Chemistry and Physics* **3**: 807–13.

crystal morphology The detailed study of the shape and appearance of crystals, more especially the regularity and symmetry of their faces and their varied facial development, which are spectacularly shown in many fine mineral specimens. The scope of crystal morphology may now be said to extend to the regular internal arrangement of the atoms, which underlies the regular external appearance, and to the precise geometrical description of all kinds of features found in and shown by crystals.

The regularity and symmetry of crystal faces requires precise mathematical description. Steno was the first to observe the exact nature of the regularity, that in a given compound, the angles between corresponding faces were always the same (*Law of Constancy of Angle*). The measurement of *interfacial angles* was once widely used as a means to identify minerals (and other crystals) and the *Barker Index of Crystals* was prepared for this purpose.

Goniometers were developed to measure the angles between crystal faces accurately. The crystal was mounted at the centre of a series of circles each, with an accurate angular scale, which allowed it to be turned into any orientation. An optical collimator and telescope were used to locate each crystal face by bringing it into the position where it reflected a light beam from the collimator into the telescope. In this way the relative angular positions of all the faces could be found and represented on a **stereographic projection** which correctly displays angular relationships.

CRYSTAL SYMMETRY. The symmetry of a crystal may be described in terms of symmetry elements and operations. An object possesses a symmetry element if the operation of that element brings the object into coincidence with itself in the starting orientation. An alternative description is that a set of identical faces, each bearing the same relationship to all the others, may be generated from one starting face by carrying out the symmetry operation.

A symmetry operation may be:

1 reflection in a plane (mirror plane),

2 rotation about an axis, in steps of $(360/n)°$ so that the object comes into self-coincidence n times per complete revolution,

3 inversion at a point, usually called a centre of symmetry, i.e. transfer it to a diametrically opposite position on the other side of the point so that for every feature with coordinates x, y, z relative to the centre of symmetry there is another inverted at $-x$, $-y$, $-z$,

4 roto-inversion at a point, usually called an inversion axis, in which rotation about an axis is combined with inversion at a point.

The regular arrangement of atoms in a crystalline solid requires regular repetition of a structural unit and only symmetry elements consistent with this occur in crystals. These *crystallographic symmetry elements* are (i) mirror planes; (ii) two-, three-, four- and sixfold rotation axes (diads, triads, tetrads and hexads); (iii) centres of symmetry; (iv) three-, four- and sixfold inversion axes.

The symmetry of a crystal consists of one or more symmetry elements acting at a point in the centre of the crystal and such a group of symmetry elements is called a *point group*. As an example, the *point group* symmetry of a crystal of **zircon** is illustrated in Fig. C43. For the crystal-lographic symmetry operators, there are 32 possible *point groups* which correspond to the 32 *crystal classes* into which a crystal may be placed according to the symmetry it possesses. These *crystal classes* are grouped into seven *crystal systems*, each of which is characterized by particular types of symmetry element (Table C5). A shorthand notation, the *Hermann–Maugin* symbol, has been devised to describe the symmetry of any crystal; it is defined in *International Tables for Crystallography* (Volume A) and in most texts on crystal morphology (see bibliography).

A set of crystal faces which are equivalent to each other by the *point group* symmetry or which can be generated by operating the symmetry elements of the *point group* on one starting face is called a *form*. The faces of a *special form* bear a specific relationship to the symmetry operators: they may be parallel or perpendicular to a rotation axis or

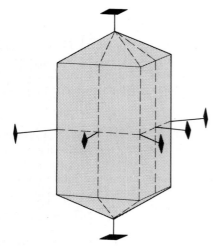

Fig. C43 Crystal morphology. A zircon crystal. The point group symmetry elements consist of (a) one tetrad axis, (b) four diad axes perpendicular to the tetrad axis and at 45° to each other, (c) four mirror planes each parallel to the tetrad axis and one of the diad axes, (d) a fifth mirror plane perpendicular to the tetrad axis and containing the four diad axes. (After Bishop, 1967.)

Table C5 Crystal morphology. The seven crystal systems

System	Characteristic symmetry	Conventional unit cell
Triclinic	None or centre	$a \neq b \neq c$; $\alpha \neq \beta \neq \gamma$
Monoclinic	Diad and/or mirror plane	$a \neq b \neq c$; $\alpha = \gamma = 90°$; $\beta > 90°$
Orthorhombic	Three mutually perpendicular diads and/or mirror planes	$a \neq b \neq c$; $\alpha = \beta = \gamma = 90°$
Trigonal	One triad axis	$a = b \neq c$; $\alpha = \beta = 90°$; $\gamma = 120°$ or $a = b = c$; $\alpha = \beta = \gamma \neq 90°$
Tetragonal	One tetrad axis	$a = b \neq c$; $\alpha = \beta = \gamma = 90°$
Hexagonal	One hexad axis	$a = b \neq c$; $\alpha = \beta = 90°$; $\gamma = 120°$
Cubic	Four triad axes	$a = b = c$; $\alpha = \beta = \gamma = 90°$

a mirror plane. The faces of a *general form* have no such particular relationship to the symmetry operators. The maximum number of equivalent faces occurs in the *general form*: *special forms* usually have fewer faces.

THE SPECIFICATION OF CRYSTAL FACES: THE LATTICE, UNIT CELL AND MILLER INDICES. The mathematical description of crystal faces depends on the underlying regularity of the atomic arrangement in a crystalline solid. In thinking of this, we may envisage a three-dimensional array of points, regularly spaced, each of which has around it the same arrangement of atoms in the same orientation. This is a *lattice* and is described geometrically in terms of a *unit cell* which will in general be a parallelepiped with edge lengths a, b, c and angles α (= angle between b and c), β (= $c \wedge a$), γ (= $a \wedge b$).

Since the crystal owes its symmetry to the regularity of the internal arrangement, the *lattice* cannot have lower symmetry than the crystal and the *unit cell* edge lengths and angles are constrained by the symmetry of the *crystal system* as noted in Table C5.

The faces of a crystal lie parallel to lattice planes, i.e. to sets of parallel planes which pass through *lattice* points. These are specified by taking the first *lattice* plane out from an origin and looking at its intercepts on the unit cell edges a, b, c. In general these will be a/h, b/k, c/l. h, k, l are integers and are called the *Miller indices* (hkl): they specify the set of *lattice* planes and the crystal face parallel to them (Fig. C44). The same notation is used to specify the orientation of planar features in a crystal, such as twin boundaries, **cleavage** planes, interfaces and the **slip-plane** under **stress**.

In measuring the angles between crystal faces, it is convenient to measure a set of faces called a *zone* which are all parallel to the same direction (the *zone axis*). Such sets of faces with parallel edges are a feature of many crystals (Fig. C45). Such a direction is parallel to a row of lattice points defined by the coordinates U_a, V_b, W_c of the first lattice point out from the origin (Fig. C45c). The symbol [UVW] is used to denote the *zone axis* or any other direction in a crystal, e.g. the direction of surface striations or the slip direction under **stress**. A useful relationship is the *Weiss zone law* which defines the condition that a plane (hkl) is parallel to a direction [UVW] or that a face (hkl) lies in the zone [UVW], when

$$hU = kV + lW = 0$$

The angles between crystal faces specified by their indices (hkl) depend on, and can be calculated from, the ratios of the *unit cell* edge lengths, $a/b : 1 : c/b$ (known as the axial ratios) and the angles α, β, γ of the unit cell. From measured *interfacial angles*, it is possible to deduce the axial ratios and angles which define the shape of the unit cell but its actual size remained unknown until the advent of **X-ray diffraction** as a method of studying the atomic arrangement in crystals.

VARIATIONS IN CRYSTAL HABIT. Despite their regularity and symmetry, crystals of a particular mineral will often show variations in *habit*, i.e. in the relative development of faces. Faces which are large on some crystals may be small or non-existent on others.

The largest crystal faces would be expected to be those of lowest surface energy, not just because this minimizes the total energy of the crystal but also because growth occurs most rapidly normal to faces of high surface energy, thereby extending those of lower energy. Simplistically, the favoured low-energy faces will be those parallel to lattice planes with a high density of lattice points (*Law of Bravais*) for those planes will cut the smallest area of each *unit cell* and so break the smallest number of bonds in the structure. These will be the lattice planes with the widest spacings and the simplest *Miller indices*. A more rigorous treatment developed by Hartmann and Perdok looks for sequences of strong bonds in the struc-

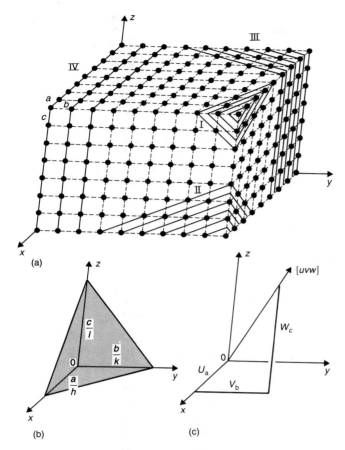

(a)

(b)

(c)

Fig. C44 Crystal morphology. (a) An array of lattice points with dashed lines parallel to the axes, *x, y, z*, along which the unit cell edge lengths are *a, b, c*.

Four sets of parallel lattice planes, I–IV, are shown by solid lines. Taking set I, the first plane out from the origin at the unseen corner will cut the axes at *a*, *b*/2, *c*/2 and the planes have Miller indices (122). Similarly, the other sets of lattice planes have indices II: (212), III: (210), IV: (010). The planes III are parallel to the *z*-axis and may be said to cut it at infinity, i.e. at *c*/0: thus the *l* index is 0.

(b) Definition of Miller indices. The plane (*hkl*) cuts the axes at *a*/*h*, *b*/*k*, *c*/*l* where *h, k, l* are integers.

(c) Definition of zone symbol. The direction [*UVW*] starts from the origin and passes through the lattice point U_a, V_b, W_c with *U, V, W* integers. (After McKie & McKie, 1986.)

ture, the preferred growth faces lying parallel to planes containing such periodic bond chains. The morphology of many crystals can be satisfactorily accounted for in this way, but the theoretical models necessarily assume unconstrained isotropic growth conditions. Such is by no means always the case in nature, which provides a rich variety of crystal morphologies reflecting varied environments and changing growth conditions. [JEC]

Berry, L.G. & Mason, B. (1983) *Mineralogy: Concepts, Descriptions, Determinations*, Chapter 2. W.H. Freeman, San Francisco.
Bishop, A.C. (1967) *An Outline of Crystal Morphology*. Hutchinson, London.
McKie, D. & McKie, C. (1986) *Essentials of Crystallography*. Blackwell Scientific Publications, Oxford.

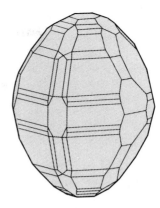

Fig. C45 Crystal morphology. A crystal of sulphur showing several prominent zones, i.e. sets of faces all parallel to a particular direction. These are most easily recognized by their parallel edges. (After Phillips, 1963.)

Phillips, F.C. (1963) *An Introduction to Crystallography*. Longman, Harlow.

crystal structure The regular arrangement of atoms which makes up every crystalline solid. This regular arrangement can be described in terms of parallelepipeds (*unit cells*), each containing the same atoms in the same positions that are repeated indefinitely in three dimensions.

A full description of a crystal structure requires the size and shape of the parallelepiped (i.e. the edge lengths *a, b, c* and angles α, β, γ of the *unit cell*) and the coordinates *x, y, z* which define the position of each atom in that unit.

HOW A CRYSTAL STRUCTURE CAN BE DETERMINED. **X-ray diffraction** is now established as the standard technique for crystal structure determination. It depends on the fact that the wavelength of X-rays is the same order of magnitude as the separation of atoms and the size of the unit cell. The regular spacing of the *unit cells* makes them a three-dimensional analogue of a simple optical **diffraction** grating. The direction of the diffracted beams from a single crystal depends on the X-ray wavelength and the size and shape of the *unit cell*. The **diffraction** pattern from an arrangement regularly repeated in three dimensions is itself three-dimensional and various types of camera and counter diffractometer have been designed to examine **X-ray diffraction** patterns from single crystals (see bibliography). The dimensions of the unit cell can be determined directly from the directions of the diffracted beams and the symmetry of the **X-ray diffraction** patterns gives some information of the symmetry of the arrangement of atoms.

In order to get information on the atoms present and their position in the *unit cell*, we have to look at the intensities of the diffracted beams which depend on the scattering power of the atoms (proportional to atomic number) and their position in the *unit cell*. Given the atomic positions, it is possible to calculate the amplitude of the diffracted beams and their relative phases. However, all that can be measured is the intensity of the diffracted beams, proportional to the square of the amplitude. The information contained in the phases of the diffracted beams is lost and one cannot deduce the atomic positions from the diffracted intensities alone. This *phase problem* represents the greatest obstacle to crystal structure determination.

A variety of methods have been devised to get round the *phase problem* but essentially they amount to different kinds of trial and error. The first stage is to obtain a reasonable trial structure or partial structure, perhaps based on a known structure of a chemically similar or related compound or one with a similar or related *unit cell*. If the calculated diffracted intensites for the trial structure fit the observed ones reasonably well, one can try to deduce a structure using the observed intensities with the phases for the trial structure. This will often yield a better or more complete trial structure. In the final stages of this process, least-squares refinement minimizing the sum over all the observed diffracted beams of $(|F_o| - |F_c|)^2$ (F_o = observed structure amplitude = square root of observed intensity; F_c = calculated structure amplitude for the trial structure) can be used to obtain the atomic coordinates which fit the observed structure amplitudes best. The quality of the structure finally determined is judged by the *R*-factor, which with modern methods of data collection should be below about 0.05.

HOW CRYSTAL STRUCTURE DATA ARE PRESENTED. The results of a crystal structure determination normally include the *unit cell* dimensions, space group symbol, atomic coordinates and thermal vibration parameters for the atoms. The atomic coordinates are given as fractions of the *unit cell* edges and for the asymmetric unit only, i.e. coordinates are not given for positions that are equivalent by symmetry. The space group symbol describes the symmetry elements of the *unit cell* which define the set of equivalent atomic positions arising from the symmetry. The coordinates of these equivalent positions are tabulated for every space group in *International Tables for Crystallography* where the notation is fully explained. With the coordinates of these equivalent positions, one has sufficient information to draw projections of the crystal structure.

Crystal structure determinations are normally accompanied by a table of bond distances and inter-bond angles for nearest neighbour atoms.

HOW CRYSTAL STRUCTURES ARE DESCRIBED. Crystal structures are usually described in terms of *coordination number* and *coordination pholyhedra*, these concepts being most useful for structures with predominantly ionic bonding such as the oxides and the silicates.

The *coordination number* of a cation is the number of anions most closely surrounding it. These anions may be thought of as lying at the corners of a polyhedron, the *coordination polyhedron*. Examples of the different

ways of representing coordination in diagrams of crystal structures are shown in Fig. C46 for coordination numbers 3, 4, 6, 8 and 12. In oxygen compounds, Si is usually linked to four oxygen atoms which lie at the corners of a tetrahedron (Fig. C46b); Mg is usually linked to six oxygen atoms at the corners of an octahedron (Fig. C46c); B is usually linked to three oxygen atoms at the corners of a triangle (Fig. C46a); large cations such as K, Na, Ca can have higher *coordination numbers* (Fig. C46d and e).

In ionic compounds, the *coordination number* is determined by how many of the larger anions can pack around a smaller cation without ceasing to 'touch' the cation. This depends on the ratio of their ionic radii: the larger the anion and the smaller the cation, the fewer anions can pack around the cation and the lower the *coordination number*.

Having established the coordination from each atom or ion to its neighbours, the next stage in describing the crystal structure is to look at the way in which the *coordination polyhedra* link together. The *coordination polyhedra* may be fitted together sharing corners, edges or faces and the number of those elements which are shared may also vary. The structural classification of the rock-forming silicates, for example, is based on how many corners each Si tetrahedron shares with other Si tetrahedra. So there are *nesosilicates* with separate tetrahedra which share no corners with other tetrahedra (but do, of course share elements with other polyhedra), *sorosilicates* with pairs of tetrahedra sharing one corner, *cyclosilicates* and *inosilicates* with rings and chains respectively formed when each Si tetrahedron shares two corners with adjacent tetrahedra, *phyllosilicates* with layers formed when each Si tetrahedron shares three corners with others, and *tectosilicates* in which each tetrahedron shares all four corners with others forming a three-dimensional framework. Many other structural classifications are based on types of *coordination polyhedra* and the way they are linked together.

An alternative description of crystal structures sometimes adopted is based on the packing of the largest atoms, which in ionic structures are the anions. Usually the largest atoms lie in one of the ideal close-packed arrangements which have both tetrahedrally and octahedrally coordinated interstices able to accommodate the smaller atoms (cations in ionic structures) depending on their size. [JEC]

Bragg, W.L. & Claringbull, G.F. (1965) *Crystal Structures of Minerals.* Bell, London.

Dent Glasser, L.S. (1977) *Crystallography and its Applications.* Van Nostrand Reinhold, New York.

Evans, R.C. (1966) *An Introduction to Crystal Chemistry* (2nd edn). Cambridge University Press, Cambridge.

Ladd, M.F.C. & Palmer, R.A. (1985) *Structure Determination by X-ray Crystallography* (2nd edn). Plenum, New York.

Megaw, H.D. (1973) *Crystal Structures: A Working Approach.* W.B. Saunders, Philadelphia.

Woolfson, M.M. (1970) *An Introduction to X-ray Crystallography.* Cambridge University Press, Cambridge.

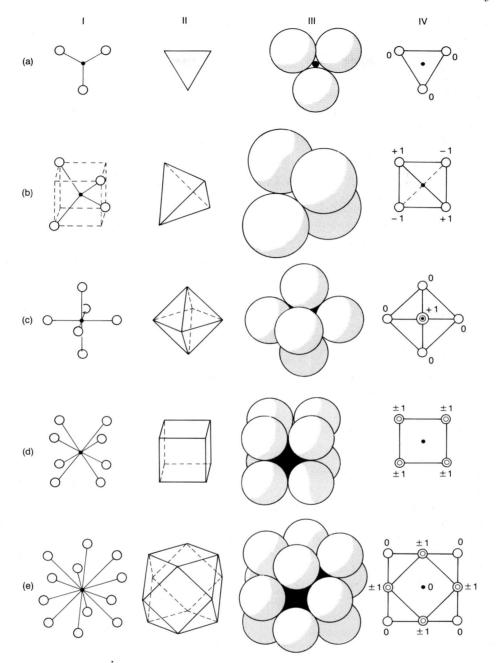

I II III IV

(a)

(b)

(c)

(d)

(e)

Fig. C46 Crystal structure. Coordination polyhedra shown from left to right for each coordination number as (I) a ball and spoke diagram of atoms and bonds, (II) an outline of the polyhedron showing the edges, (III) a packing diagram showing the relative sizes of the atoms and (IV) in plan with atom heights marked.

The polyhedra illustrated are for (a) 3-coordination (triangle), (b) 4-coordination (tetrahedron), (c) 6-coordination (octahedron), (d) 8-coordination (cube) and (e) 12-coordination (cuboctahedron). (After Megaw, 1973.)

crystalline massif A piece of **continental crust** formed of metamorphic or igneous rocks, which is regarded as relatively stable compared with surrounding areas. Some crystalline massifs are formed from basement rocks and are contrasted with more deformed sedimentary cover. The term may be used in the context of an *orogenic belt*.

Thus the Aar, Mount Blanc and Monte Rosa massifs in the Alps are composed of *Hercynian* crystalline basement and exhibit relatively weak *Alpine* **deformation** in contrast to the **Mesozoic** cover in the **Alpine** fold-thrust belt. Another type of crystalline massif represents a region where both basement and cover have remained relatively stable compared with surrounding areas. The Bohemian massif is an example of this type in the context of the *Hercynian Orogeny* of western Europe. [RGP]

cuesta An asymmetrical ridge produced by differential **erosion** of gently dipping strata. It consists of a steep scarp face, a well-defined crest and a gentle back slope which is generally accordant with the **dip** of the resistant strata which form the cuesta. [NJM]

cuirass Any well-cemented **duricrust** which mantles the landscape, protecting underlying, unconsolidated material from **erosion**. **Calcrete, silcrete,** and **ferricrete** — as well as other types of **weathering** and *soil* crusts — commonly form cuirasses in the tropics and subtropics. Because these crusts resist **erosion,** particularly in arid and semi-arid regions, they are often relict features which preserve the landscapes on which they formed. In doing so, cuirasses can provide important information on an area's geomorphic and *paleoenvironmental* history. [AW]

culm The **Carboniferous** rocks of south-western England were formerly known as the Culm Measures on account of the occurrence at a few localities of soft, sooty **coal** known in Devon as culm. [AME]

cummingtonite $((Mg, Fe)_7Si_8O_{22}(OH)_2)$ An **amphibole mineral**. [DJV]

cumulate igneous rock An **igneous rock** characterized by a framework of touching or interlocking mineral crystals and grains that are interpreted as having been concentrated by crystal–**magma** concentration processes from a parental **magma** (see **magma, magmatic diversification**). Cumulates include a variety of mafic/felsic intrusive rocks which represent crystalline residua from which **magma** has been removed (to a further *magma chamber* or for surface eruption).

The concept of igneous cumulates was originally developed from the study of ultramafic/mafic layered intrusions characterized by distinctive mineral layering. Such layering may reflect variation in the grain-size, modal-proportions, textural character and mineral composition of the minerals. The concept was originally formulated from the concept that such rocks were formed by concentration of minerals by *crystal settling*. Since it is now recognized that a variety of processes contribute to the formation of such layered igneous rocks, the term 'cumulate' is now redefined so that *crystal settling* is a possible but not essential process in the origins of the described rocks.

Such cumulate ultramafic/mafic rocks occur within continental anorogenic igneous associations and within **ophiolite** complexes. [RST]

Wager, L.R. & Brown, G.M. (1968) *Layered Igneous Rocks.* Oliver and Boyd, Edinburgh.
Irvine, R.N. (1982) Terminology for layered intrusions. *Journal of Petrology* **23**: 127–62.

cuprite (Cu_2O) A secondary **copper** mineral, ruby-red in transparent crystals. (See **oxide minerals**.) [DJV]

Curie point/temperature The temperature at which magnetic alignment by internal quantum mechanical forces, which give rise to **ferromagnetism** (*s.l.*), is overcome by the randomizing effects of thermal vibrations. Above this temperature, **ferromagnetic** minerals only have **paramagnetic** properties. [DHT]

Curie–Weiss Law The **susceptibility** of a **ferromagnetic** material is inversely proportional to the absolute temperature. [DHT]

cuspate fold profile Characteristic shape of a folded surface separating layers of differing **competence**. **Folds** closing (see **fold closure**) in one direction (e.g. upwards) have a broad arcuate shape, whereas those closing in the opposite direction have a cusp shape. The cusps point towards the more **competent** layer, i.e. the material of higher **strength** or **viscosity**. (See Fig. C47.) [RGP]

Ramsay, J.G. & Huber, M.I. (1987) *The Techniques of Modern Structural Geology* Vol. 2: *Folds and Fractures.* Academic Press, London.

cut-off grade The lowest **grade** of **ore** that can be economically produced from an **orebody**. [AME]

cycle of erosion (*geographical cycle*) The sequence of **denudational** forms and processes which exist between the initial **tectonic uplift** of an area of ground and its reduction to a gently undulating planation or **erosion** surface (**peneplain**) close to *base level*. Under this theory (Davis, 1899), landforms are seen to evolve through time in a sequential manner, from youth to maturity to old age. However, climatic and **tectonic** changes mean that any simple progression of the type envisaged by Davis is unlikely. The concept was initially applied to humid landscapes, but later, supposedly distinctive, sequences were added by Davis and other workers to account for landform evolution in arid, **karstic**, coastal, *glacial* and *periglacial* environments. [ASG]

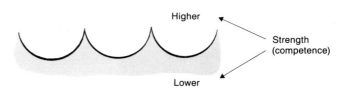

Fig. C47 Cuspate fold profile.

Chorley, R.J. (1965) A re-evaluation of the geomorphic system of W.M. Davis. In: Chorley, R.J. & Haggett, P. (eds) *Frontiers in Geographical Teaching*. Methuen, London.

Davis, W.M. (1899) The geographical cycle. *Geographical Journal* **14**: 481–504.

cyclothem A sequence of sedimentary **beds** deposited as a single cycle of sedimentation. The concept was widely applied to **coal**-bearing strata, where the sequence was held to commence with a **sandstone** layer followed by **shale** and sometimes freshwater **limestone**. This in turn was succeeded by clay (**fireclay**) and then a **coal seam**. Following the **coal** a marine incursion deposited more **shale** and then the cycle would be repeated. A variety of tectonic, climatic or sedimentological controls was invoked to explain this and other idealized cycles. Nowadays facies sequences are studied rather than idealized cycles and various depositional models of **coal**-bearing sequences have been devised. [AME]

Reading, H.G. (1986) *Sedimentary Environments and Facies*. Blackwell Scientific Publications, Oxford.

Ward, C.R. (1984) *Coal Geology and Coal Technology*. Blackwell Scientific Publications, Melbourne.

cymatogeny A term introduced by L.C. King to describe large-scale warping of the Earth's **crust** over distances from tens to hundreds of kilometres with little **deforma**-tion of the crustal rocks (see **diastrophism**). Cymatogenic movement occurs at a smaller scale than **epeirogeny** and is less violent than **orogeny**. Nevertheless, cymatogeny can result in sufficient vertical warping to produce mountain ranges, though broad, domal **uplifts** are more usual. The forces which drive cymatogeny are assumed to originate in the Earth's **mantle**. [AW]

King, L.C. (1967) *The Morphology of the Earth*. Oliver & Boyd, Edinburgh.

cymophane A chatoyant variety of **chrysoberyl**. (See **chatoyancy**.) [DJV]

Cyprus-type deposit A type of volcanic-associated **massive sulphide deposit**, which is usually an **ore** of **copper**, often with some **gold** values and occasionally carrying economic amounts of **zinc**. There is a strong spatial association with **basaltic pillow lavas**, usually in the upper part of **ophiolite** complexes. Clastic sediments are minor or absent and the depositional environment is deep marine with **tholeiitic** volcanism. Like **ophiolite** complexes, they are largely confined to the **Phanerozoic**, although a few deposits have been found recently in the **Proterozoic**. Good examples are the **Cretaceous** deposits in Cyprus and those in the **Ordovician** Bay of Islands Complex, Newfoundland. [AME]

D

D (deformation phase) In analysing the **structure** and **deformation** history of a region, it is common practice to use the letter D followed by a number or subscript to denote each separate phase of **deformation** that is distinguished. Thus D1, D2 and D3 would denote three successive **deformations**. It is also usual to label individual **folds**, and planar and linear structures in the same way. Thus for a given **deformation**, D2 say, **folds** are designated F2, **foliations** S2, and **lineations** L2. [RGP]

dacite A silica-oversaturated intermediate-acid **calcalkaline lava**, which lies between **andesite** and **rhyolite** in composition. Dacites occur as **lavas** and **pyroclastic rocks** and commonly comprise *phenocrysts* of **plagioclase**, minor **olivine, pyroxene, amphibole, biotite** and Fe–Ti oxide in a fine-grained siliceous groundmass. [RST]

Barker, F. (ed.) (1979) *Trondhjemites, Dacites and Related Rocks*. Elsevier, Amsterdam.

dalmatian coast A coastline characterized by island chains and long inlets formed by marine inundation of a coast-parallel system of ridges and valleys. The inundation may be caused by **tectonic** subsidence or sea-level rise or both. The name derives from the Dalmatian coast of the Yugoslavian Adriatic Sea. A dalmatian coast is a flooded version of a *concordant coast* where the coast-parallel trend of relief produces a straight and regular outline, unless inundation isolates the ridge crests, producing chains of islands. [JDH]

dambo A linear depression without a well-marked stream channel which is characteristic of old, relatively gently sloping land surfaces, especially in the tropics. Frequently forming reticulate networks, they are often marked by the presence of *termite mounds*, small **pan**-like depressions, and vegetation assemblages different from the surrounding woodland and savanna. [ASG]

Thomas, M.F. & Goudie, A.S. (eds) (1985) Dambos: small channel-less valleys in the tropics. *Zeitschrift für Geomorphologie* Supp. **52**.

danburite ($Ca(B_2Si_2O_8)$) A framework silicate (*tectosilicate*) mineral. [DJV]

dating of ore deposits When **orebodies** form part of a stratigraphical succession, e.g. certain **iron** and **manganese** deposits, their age is not disputed. The ages of **ortho-** **magmatic deposits** may be almost as certainly fixed. By contrast, **epigenetic deposits** may be very difficult to date, especially as there is now abundant evidence that many of them may have resulted from polyphase mineralization. There are three methods that can be used: the field data, *radiometric* and **paleomagnetic** *age determinations*. Unfortunately, the field evidence may be vague and *radioactive* elements and magnetic minerals are uncommon in most **ore** deposits. This leads to much uncertaintly about the age of many **epigenetic deposits**. [AME]

Evans, A.M. (1993) *Ore Geology and Industrial Minerals: An Introduction* (3rd edn). Blackwell Scientific Publications, Oxford.

datolite ($CaB(SiO_4)(OH)$) An island silicate (*nesosilicate*) mineral found as a secondary mineral in cavities in **basaltic lavas** and similar rocks. [DJV]

daya A small, silt-filled, solutional depression found on **limestone** surfaces in some arid areas of the Middle East and North Africa. [ASG]

Mitchell, C.W. & Willimott, S.G. (1974) Dayas of the Moroccan Sahara and other arid regions. *Geographical Journal* **140**: 441–53.

de-asphalting A **degradation** process involving the **precipitation** of the **asphalt** content of **crude oils**. This may result from *bacterial* or aqueous attack but is generally attributed to the introduction into **crude oils** of excess **natural gas**. The gas may **migrate** into the oil from below or deeper burial may increase the pressure on an oil pool having a **gas cap**. The excess gas precipitates **asphalt** improving the quality of the remaining oil, but problems may result if the **reservoir** pore space becomes plugged with *bitumen*. (See **basal tar mat**.) [AME]

dead line (1) Mineralization; the depth in a well-explored **orefield** below which no economic mineralization is known and therefore none expected.
(2) Oil; a level (determined by the grade of metamorphism) below which no significant amount of oil is likely to be found. [AME]

Dead Sea A perennial saline lake, 78 km long by 15 km wide, which lies in a desert **pull-apart basin** between Israel and Jordan; lake depths vary from an average 185 m in the north to about 6 m in the south. The altitude of the lake is 396 m below sea-level, the lowest lake surface in

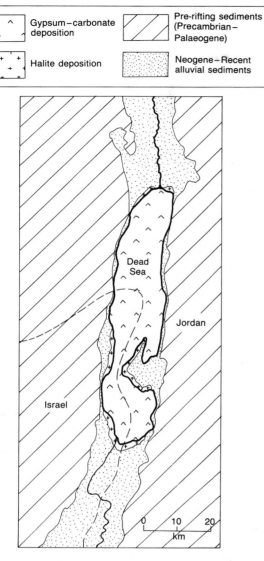

Fig. D1 Map of the **Dead Sea**, showing inverted 'bull's eye' pattern of evaporite deposition.

Legend:
- ^ ^ Gypsum–carbonate deposition
- + + Halite deposition
- /// Pre-rifting sediments (Precambrian–Palaeogene)
- :::: Neogene–Recent alluvial sediments

Dead Sea

Jordan

Israel

0 10 20 km

Today the Dead Sea is **precipitating gypsum** and some **aragonite**, commonly in the form of *whitings*, although the water is approaching **halite** saturation. Moreover, **halite** is **precipitated** around the lake margins, giving rise to an 'inverted bull's eye' pattern of **evaporite** formation (see **evaporite basin**), with a discontinuous ring of saline deposits around a central deposit of less soluble salt. Stratification of **brine** layers within the Dead Sea in the first part of this century resulted in **anoxic** bottom conditions and little **gypsum precipitation** on the lake bottom, as **gypsum**, formed at or near the lake surface, was largely destroyed by subsequent **reduction** as it settled through the lower stratified **anoxic brine** masses. Recent overturn of the lake **brine** layers has destroyed the stratification and **gypsum precipitation** on the lake bottom has greatly increased since overturn. Changes in climate have resulted in wide fluctuations in water input in the past; lake bottom cores show a thin late *Holocene* carbonate package (30 cm), evidencing a **pluvial** period, overlying a more arid early *Holocene*–late *Pleistocene* **halite** package.

The Dead Sea is one of the world's greatest sources of chemical raw materials, with salt deposits estimated at 43×10^9 tonnes; exploitation of these salt reserves commenced in 1931, with **bromine** production, and has continued until the present day. Evaporating **pans** progressively precipitate **halite** then **carnallite**; the **carnallite** is then treated by filtration and flotation processes for chemical separation and purification. **Bromine** and potash salts are the main outputs, but common salt and magnesium chloride also are economic by-products. [GMH]

debris flow Generally laminar (rarely turbulent) **sediment gravity flow** process in which large particles, up to boulder size, derive support mainly from the buoyancy and cohesive strength of a sediment–water slurry or 'matrix'. Some additional *clast* support is provided by a combination of **dispersive pressure** and excess **pore fluid pressures**. In some debris flows, larger *clasts* may not actually be suspended in the matrix but remain in contact through rolling, sliding and intermittent bouncing downslope — these latter debris flows tend to produce *clast-supported conglomerates* with as little as 5% clay–water content during flow. The general lack of grading, stratification and recognizable **fabric** suggests that sediment was deposited en masse with little or no inertial interactions between *clasts*. Sediment deposited from a debris flow is called a '*debrite*'. *Debrites* range from mud matrix-supported to *clast-supported conglomerates* with a mainly silt/**sand**-grade matrix. Beds tend to be sheet-like in geometry, possibly with pebbles, cobbles and boulders sunk into underlying sediments and projecting above the main surface of the deposit. [KTP]

decarboxylation Literally, the loss of carbon dioxide (CO_2) from organic matter. The term is applied in **burial diagenesis** to describe the thermocatalytic decomposition

the world. The lake gains its name from the apparent lack of life therein, although research from the 1940s onwards has demonstrated different genera of *bacteria* and *algae* to live within the lake waters. Climatic conditions around the Dead Sea produce near-continuous high air pressure, high temperatures (maximum 40°C) and relatively low humidity combined with long summers, resulting in a high rate of evaporation (annually more than $2 \times 10^9 \, m^3$); average rainfall is only 51 mm. Resultant salinities within the lake are 10 times that of open ocean water. The lake is mainly fed by the Jordan River (Fig. D1), although this is supplemented by smaller streams and **springs**. Fluctuations in water input have led to recorded variations in lake water levels in excess of 9 m over the last 60 years.

of oxygen-containing *kerogen* to liberate CO_2 (see **organic matter diagenesis**):

$$RCO_2H \rightarrow RH + CO_2$$

where R denotes a high molecular weight organic polymer.

Continental-derived humic types of *kerogen* provide the most CO_2 largely due to the decomposition of carbonyl (C=0), methoxyl ($-OCH_3$) and phenolic hydroxyl ($-OH$) groups. The CO_2 released by decarboxylation dissolves in the enclosing **pore-waters** to produce carbonic acid:

$$CO_2 + H_2O = H_2CO_3 = H^+ + HCO_3^-$$

and may, therefore, be responsible for the generation of subsurface **secondary porosity**.

Decarboxylation of oliphatic (fatty) acids also takes place during **burial diagenesis** through the decomposition of $-COOH$ groups to produce *methane* (CH_4):

$$CH_3COOH + H_2O \rightarrow CH_4 + H_2CO_3$$

Decarboxylation of organic matter is associated with de-hydration (loss of H_2O); oxygen and hydrogen are lost most readily from the organic matter, whilst **carbon** is lost the least readily. As a result the **carbon** content of the *kerogen* residue increases and the H:C ratio decreases. [SDB]

decibel scale A logarithmic scale used to measure power or amplitude ratios. Conversion to a decibel scale, symbolized by dB, is undertaken by calculating $20 \log_{10}$ of the amplitude ratio or $10 \log_{10}$ of the power ratio. Thus an amplitude increase of 2 is equal to 6 dB and a decrease of 100 is equal to -40 dB. [GS]

décollement Refers to a surface in the Earth across which there is a discontinuity in **strain**, **folding** or **fold** style. The term may be used as an adjective or as a noun to describe either the structure or the process. The term *detachment* is used almost synonymously with décollement, and both terms can be used to refer to the undeformed surface at the boundary of **strained** or **faulted** areas, as shown in Fig. D2. These terms are also applied to the boundary between **allochthonous** sheets of rock and the underlying **autochthon**, whether or not there is everywhere a contrast in **deformation** intensity or style. They can also be used to describe the process of *detachment*.

A décollement surface, particularly in folded rocks, is

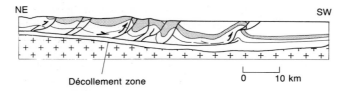

Décollement zone

0   10 km

Fig. D2 The **décollement** below a fold and thrust belt, Papua. (After Smith, 1965.)

often associated with considerable tectonically-produced thickness variations in an adjacent unit. These **strains** accommodate the changes in **deformation** style across the décollement surface; the position of décollement is often lithologically controlled, especially where there is strong contrast in **rheology** in a deformed sedimentary sequence. For example a layer of **evaporite** minerals acts as a décollement to the overlying Jura **folds**. In **thrust belts** such as this, the *sole thrust* acts as a décollement, and **faulting** and **strain** are generally restricted to the overlying rocks. **Extensional fault systems** also typically show décollement surfaces, separating regions where **strain** is accommodated by different mechanisms, e.g. faulting above, **creep** below. Identification of décollement surfaces is of importance in assessment of the **strain** continuity in **lithospheric deformation**. [SB]

Smith, J.G. (1965) Orogenesis in western Papua and New Guinea. *Tectonophysics* **2**: 1–27.

deconvolution The process of undoing the effect of the application of a previous **convolution** or **filtering** operation. Deconvolution is usually undertaken by designing an inverse filter and then **convolving** it with the signal. Deconvolution is an important part of seismic data processing and is used to improve **seismic records** by removing the detrimental effects on the waveform due to the passage of the waves through the Earth, such as reverberant seismic pulses, the receiver response and the **attenuation** of the higher frequencies in a signal. [GS]

Ziolkowski, A. (1983) *Deconvolution*. IHRDC Publications, Boston.

deep-focus earthquake Most **earthquakes** occur at shallow depth (<70 km); that they occur at any greater depths was first recognized by H.H. Turner in 1922.

Earthquakes at depths of over 300 km are termed 'deep-focus' earthquakes, and those from 70 km to 300 km '*intermediate focus*'. Deep- and *intermediate-focus earthquakes* occur only in planar zones dipping down at between 30° and 90° but typically about 45° from convergent **plate boundaries**, and this pattern (the '**Benioff–Wadati zone**') has allowed us to recognize the slab-like shape of tectonic *plates* as they descend into the Earth's **mantle**. The deepest known **earthquake** was at 720 km.

These **earthquakes** are caused by the **stresses** within the relatively cold and **brittle slabs**; either extensional as the slabs are pulled down by their own weight (*intermediate-focus* events) or compressional as they meet resistance (deep-focus events), and NOT by friction at their outer surfaces.

The numbers and **magnitudes** of *intermediate-* and deep-focus **earthquakes** are relatively small; only about 35% and 5% respectively of total numbers and 12% and 3% respectively of energy release. (See **hypocentre**; **seismicity**.) [RAC]

Isacks, B., Oliver, J. & Sykes, L.R. (1968) Seismology and the new global tectonics. *Journal of Geophysical Research* **73**: 5855–900.

deep lead A buried alluvial **placer**. (See **alluvial deposit**.) [AME]

Deep Sea Drilling Project (*DSDP*) An international programme to drill the surface of the Earth in deep water. It began with coring operations on 11 August 1968 on Leg 1, terminating with Leg 96 on 8 November 1983, both scientific cruises being in the Gulf of Mexico. In the intervening 15 years the drilling ship, D/V *Glomar Challenger*, gathered data in all the world's major oceans with participating scientists from the USA, Germany, Japan, United Kingdom, France and the former USSR. Scientific planning was conducted through the *Joint Oceanographic Institutions for Deep Earth Sampling* (**JOIDES**) panel of over 250 members, constituting up to 24 committees, panels and working groups. The deepest hole drilled was to 1740 m sub-bottom in water depths of 3910 m. In total, 624 sites were drilled world-wide. Amongst the successes of the DSDP programme was the demonstration of the validity of the hypothesis of **seafloor spreading**, the age of the ocean basins and their processes of development, the nature of **oceanic crust**, the sedimentary history of **passive continental margins**, and *ocean ridge* processes. Technological advances resulted in improved coring, core-recovery and *down-hole logging* methods. The DSDP programme with *Glomar Challenger* has been superseded by the *Ocean Drilling Program* (*ODP*), using a new advanced drill ship, the **JOIDES** *Resolution*, registered as SEDCO/BP 471, that began operations in 1984. [KTP]

deep weathering The disintegration and chemical decomposition of bedrock to depths of several tens of metres beneath the landsurface. Deep weathering proceeds as **meteoric water** percolates through pores, **joints**, and fissures in the bedrock dissolving and chemically altering the constituent minerals. Crystalline rocks such as **basalts**, **dolerites**, **granites**, and **gneisses** weather through the breakdown of minerals such as **feldspars**, **pyroxenes**, and **epidote**. Often only **quartz** crystals will remain relatively unaffected by chemical decomposition. Initially, **weathering** will be along the lines of water movement. Gradually, however, moisture penetrates the surface of blocks of solid rock along the boundaries between crystals. As the process proceeds, **corestones** of sound rock shrink in size as their surfaces are **weathered** (see **spheroidal weathering**). This eventually leads to the accumulation of thoroughly rotted rock as **saprolite**. Provided that the **saprolite** is not eroded at the surface, or alternatively if the **weathering front** advances at least as fast as the land surface is lowered, a deep weathering profile develops. This may extend to depths of well over 100 m. Such profiles are most common on flat landscapes in the humid tropics. Here, moisture availability promotes chemical **weathering** of the bedrock and maintains a thick vegetation cover which retards **erosion**. On ancient landscapes, even relatively resistant rocks such as **rhyolites** which have a high **silica** content may exhibit deep weathering profiles extending to depths of 20 m or more beneath the land surface.

The movement of minerals in solution through deep weathering profiles may result in the development of **laterites**. **Bauxites**, for example, are accumulations of aluminium hydroxides which form as a result of the chemical leaching of **silica** and iron minerals at the top of weathering profiles in the tropics. Some **ferricretes** also form in leached horizons where there is a relative accumulation of iron minerals. In southern Africa, characteristic deep weathering profiles have been used to differentiate ancient *planation surfaces*.

The **erosion** of unconsolidated material produced by deep weathering leads to exposure of the base of the weathering profile — the **weathering front**. In places where the bedrock's susceptibility to chemical **weathering** was not uniform, **erosion** may lead to the exposure of the more resistant zones of rock as *bornhardts* and **inselbergs**. [AW]

Goudie, A.S. (1981) Weathering. In: Goudie, A.S. (ed.) *Geomorphological Techniques*, pp. 139–55. Allen & Unwin, London.
Ollier, C.D. (1969) *Weathering*. Oliver & Boyd, Edinburgh.
Pye, K. (1985) Granular disintegration of gneiss and migmatites. *Catena* **12**: 191–9.
Pye, K. (1986) Mineralogical and textural controls on the weathering of granitoid rocks. *Catena* **13**: 47–57.

defects in crystals Crystalline solids are normally said to have a regular arrangement of atoms extending indefinitely in three dimensions. However, complete perfection is not always achieved and defects may occur in which the otherwise regular arrangement is altered locally.

Three types of defect are possible according to whether the departure from perfect regularity occurs at a point, along a line or extends over a plane in the **crystal structure**.

POINT DEFECTS. The simplest type of *point defect* is a *vacancy* or *Schottky defect*: an atom is missing from its expected position in the regular arrangement (Fig. D3a) and is completely absent from the crystal, with a small departure from stoichiometry.

Another type of *point defect* is an *interstitial*: an atom, perhaps an impurity or in excess of the stoichiometric formula, occupies one of the small gaps which exist between the atoms in the regular arrangement (Fig. D3b).

A *Frenkel defect* is formed when an atom is transferred from its usual position into an immediately adjacent interstitial site not normally occupied by an atom, producing what is effectively a vacancy-interstitial pair.

Point defects are responsible for the colour of some materials because they perturb the electronic structure of the solid. But the main importance of *point defects* is in providing a mechanism for self-diffusion by enabling atoms to interchange position with vacancies. The temperature dependence of the *diffusion coefficent* leads to an activation energy which can often be related to the

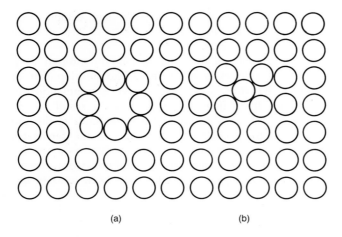

(a)

(b)

Fig. D3 Defects in crystals. (a) Vacancy and (b) interstitial in a simple regular two-dimensional arrangement of atoms. The atoms around the point defects undergo adjustments to their positions. (After Hull, 1975.)

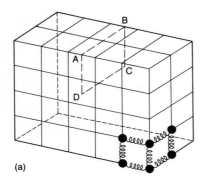

(a)

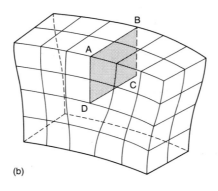

(b)

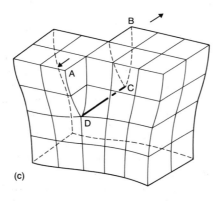

(c)

Fig. D4 Defects in crystals. (a) Simple cubic arrangement of atoms thought of as hard spheres with bonds like springs able to accommodate some bending and stretching. (b) Edge dislocation CD formed by inserting an extra half plane of atoms ABCD. The atoms on either side of this half plane are displaced in directions normal to the dislocation line. (c) Screw dislocation CD formed by displacing atoms on either side of the plane ABCD relative to each other in the direction AB, parallel to the dislocation line. (After Hull, 1975.)

energies of formation of *Schottky* or *Frenkel defects*. The presence of *point defects* will make diffusion easier and may assist reactions which proceed by diffusion mechanisms such as those involved in **metasomatism**.

LINE DEFECTS: DISLOCATIONS. The idea of *line defects* called **dislocations** originated in metallurgy to account for the **strength** and **deformation** behaviour of metals, but such faults may occur in all kinds of materials.

Figure D4(a) shows a model of a simple cubic arrangement of atoms, the bonds being thought of as springs able to accommodate some bending and stretching. An *edge dislocation* CD (Fig. D4b) can be imagined as being formed when an additional half plane of atoms is inserted at

ABCD and the surrounding atoms adjust their positions to accommodate it. The displacement of the structure at the **dislocation** is normal to the *dislocation line*.

At a *screw dislocation* (Fig. D4c), the atoms on either side of ABCD are shifted relative to each other in the direction AB, the structural displacement being parallel to the *dislocation line*.

In general, the *dislocation line* need not be straight and zigzags, tangles and loops have all been observed. Nor does the displacement at the **dislocation** have to be a whole *unit cell: partial dislocations* whose displacement is a fraction of a *unit cell* (or lattice vector) may occur, for example, when a *stacking fault* (see below) terminates within a crystal.

[154]

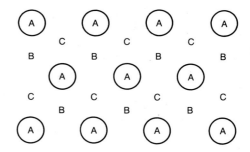

Fig. D5 Defects in crystals. Close-packing of atoms viewed normal to the layers. The atoms A lie in one close-packed layer. Atoms in the next layer must occupy one of the two sets of hollows B or C between atoms in the A layer. (After Evans, 1964.)

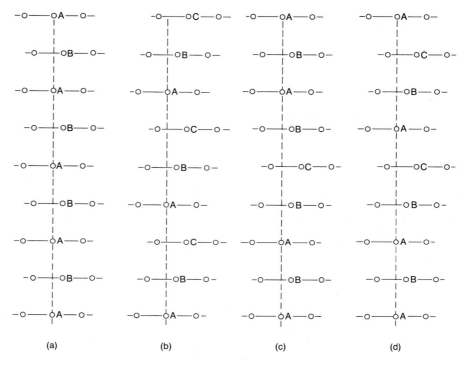

Fig. D6 Defects in crystals. Sequences of close-packed layers viewed along a direction in the plane of the layers which lies left-right in Fig. D5. (a) Regular hexagonal close-packed sequence ABABAB. . . . (b) Regular cubic close-packed sequence ABCABC. . . . (c) Error in the layer stacking with one isolated layer in the wrong position C in an otherwise regular ABAB . . . sequence. (d) Going upwards, there is initially a regular ABAB . . . sequence, then an error C as in (c) but the stacking then changes to the new regular sequence CABCABC. . . .

(a)   (b)   (c)   (d)

Dislocations act as **stress** concentrators in the solid and their presence explains why most materials have much less than the theoretical **strength** predicted from bond energies. But **dislocations** can also help to accommodate misfit at an interface, for example, and thereby relieve **strain**.

Dislocations can move through the solids as a result of:
1 atomic displacements under applied **stress** (*glide*), in which the atoms move in a direction parallel to the structural displacement at the **dislocation**, or
2 diffusion of atoms (*climb*) in a direction normal to the structural displacement at the **dislocation**.

The density and distribution of **dislocations** in a solid depends on the **stresses** and heat-treatment it has undergone. **Dislocations** in minerals are therefore studied as an aid to understanding the **deformation** and thermal history of rocks.

PLANAR DEFECTS. Some **crystal structures** may be described in terms of a sequence of layer modules along a particular direction. In the simplest case, all the layer modules are the same but may be displaced in different directions relative to the previous layer. Normally the sequence is regular but errors may occur: such errors are known as *stacking faults* and lie parallel to the plane of the layer. Like **dislocations**, they can be described in terms of a structural displacement in this case parallel to the *fault plane*.

The simplest case arises for structures based on the close-packing of atoms. Starting with a close-packed layer of atoms A (Fig. D5), the next layer must lie in one of the sets of hollows B or C. Let us suppose it occupies position B (position C is in fact equivalent with a different choice of reference axes). The third layer may now either occupy position A, giving the hexagonal close-packed sequence A B A B . . . , or position C, giving the cubic close-packed sequence A B C A B C . . . These sequences are shown in Fig. D6(a and b) projected on the *a* plane normal to the layer and lying vertical in Fig. D5.

Whether the third layer occupies position A or C will make very little difference to the overall energy, and accordingly errors in the stacking sequence can easily occur. Examples are shown in Fig. D6(c and d): the first has an isolated layer in the wrong position in an otherwise regular sequence while the second fault amounts to a changeover from one regular sequence to another. *Stacking faults* of these and other kinds are often found in metals and simple compounds with structures based on close-packing.

In minerals, the layers usually consist of a complex grouping of several different atoms, but as in the simple case, successive layers can have different relative displacements. The **clay minerals** have well-defined structural layers which are frequently stacked in disordered sequence, but in addition some clays contain mixtures of different kinds of layers which may or may not be in ordered sequence.

Stacking faults may also be related to **polytypism**, the existence of chemically similar phases with different regular layer stacking sequences. For example, transformations between the **pyroxene polytypes** (they have different stacking sequences of similar layers) can give rise to *stacking faults* which are often found in partially transformed specimens which may have been either cooled too rapidly or heated insufficiently for equilibrium to be reached. However, *stacking faults* can also arise when growth from the melt or solution is rapid or under applied **stress**.

It is possible to describe structures in terms of more than one type of layer module and these may even have different composition and atomic arrangement so long as they are able to fit together. The relationship among **pyroxenes**, **amphiboles** and **micas** (*biopyriboles*) may be described in this way (as a *polysomatic series*) with two types of (010) layer module, **pyroxene**-like (P) and **mica**-like (M). **Pyroxenes** have a sequence of separate silicate chains PPPP . . . , **micas** a sequence of silicate chains linked to each other forming sheets MMM . . . **Amphiboles** have chains linked in pairs with the sequence MPMP . . . and triple chain structures with the sequence MMPMMP . . . have been reported. Since the proportions of the two-layer modules P and M vary, so do the compositions of the minerals. But irregular sequences can also occur with, for example an isolated triple chain in **amphibole**, MPMPMPMMPMPMP. The *chain width error* constitutes a *planar defect* but the displacement vector at the **fault** now has a component normal to the **fault** plane unlike a simple *stacking fault* for which the displacement vector is parallel to the **fault** plane. Because the two modules have different compositions, a **fault** of this type will lead to a departure from the ideal composition, i.e. to *non-stoichiometry*. Planar **faults** like this provide a way in which the crystal structure can adapt to changing chemistry and can play an important part in chemical reactions such as those which occur in the **alteration** of **amphibole** and **talc**. [JEC]

Brindley, G.W. & Brown, G. (eds) (1984) *Crystal Structures of Clay Minerals and their X-ray Identification*, Chapters 2 and 3. Mineralogical Society, London.
Evans, R.C. (1966) *An Introduction to Crystal Chemistry* (2nd edn). Cambridge University Press, Cambridge.

Fyfe, W.S. (1964) *Geochemistry of Solids: An Introduction*, Chapters 12 and 13. McGraw-Hill, New York.
Hull, D. (1975) *Introduction to Dislocations*. Pergamon, Oxford.
Thompson, J.B. (1978) Biopyriboles and polysomatic series. *American Mineralogist* **63**: 239–49.

deflation The process by which material below a certain size is removed from a ground surface by wind. Larger particles are left behind as a '*deflation lag*'. Wind action is most effective on dry, unconsolidated, fine-grained sediments that are sparsely protected by vegetation thus; deflation is particularly common in desert and semi-desert regions. The grain size and mass of material moved by the wind determines the exact type of movement. Large **sand** grains, shingle and pebbles may creep along the surface, while **sand**-sized particles bounce along by **saltation**. The smallest, **dust**-sized, particles may be carried to great heights and transported over thousands of kilometres from their source. Where large quantities of material are deflated the event may be called a *sandstorm* or *dust storm*.

In the rock record, *deflation lags* are most readily discernible where **sand** with scattered larger clasts deposited, for instance, by rivers, has been deflated over a wide area to leave an approximately planar layer of material of granule grade and above. They may be the only indication of aeolian activity in a sedimentary succession dominated by fluvial deposits. [NJM/DAR]

deformation The geological process in which the application of a **force** results in a change in shape or form. Deformation produces geological **structures** such as **folds**, **faults**, and **fabrics**. The reaction of a rock body to an applied **force** is defined as a **stress**, and the geometrical rearrangements produced within the rock body by **stress** are defined as the **strain**. In addition to purely geometrical effects, deformation has thermal implications: work is done by the force in producing displacements, and energy may be converted from thermal to mechanical or vice versa. Deformation is often accompanied by metamorphic reactions that produce changes in the crystalline **fabric** of a rock. Such changes may be endothermic or exothermic. The *deformation history* is the chronological sequence of deformational events that occurred, or can be recognized, in a rock or a region. The elucidation of the *deformation history* (*polyphase deformation*) depends on the recognition of the relative age of different **structures**, such as **folds**, **faults** and **fabrics**, using cross-cutting or **overprinting** relationships. Progressive changes in movement direction or in **strain** *axes* can often be preserved as orientation changes in linear structures, in particular **growth fibres**. (See **structure**, **stress**, **strain**.) [RGP]

deformation band A tabular zone of a crystal lattice differing in crystallographic orientation from the surrounding mineral grain. Deformation bands are seen within minerals such as **quartz**, **olivine** and **feldspars**, and indicate **deformation** involving **dislocation** movement with **recovery**,

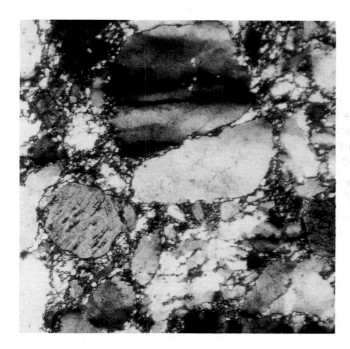

Fig. D7 Photomicrograph of **deformation bands**, developed across the page in the quartz grain at the top of the picture. The grain is about 1 mm in diameter.

i.e. *dislocation climb*. Slight distortion of a crystal lattice such as that of **quartz** results in undulatory extinction; localization of lattice misorientation into deformation bands, as shown in Fig. D7, results from **recovery**. *Transmission electron microscopy* of **quartz** shows that they consists of many prismatic **subgrains**. Deformation band boundaries are thus **subgrain** boundaries, comprising ordered walls of **dislocations**, and undulatory extinction is found within the band. Increased **recovery** changes the lattice structure to make single larger **subgrains**, without internal lattice **deformation**. (See **deformation lamellae, crystal defect**.) [SB]

White, S.H. (1973) The dislocation structures responsible for the optical effects in some naturally deformed quartzites. *Journal of Materials Science* **8**: 490–9.

deformation lamellae Narrow planar zones within mineral grains such as **quartz**, **feldspar** and **olivine**, related to **deformation** of the crystal lattice. Many origins are cited for these features; White (1973) discusses the interpretation of optical features in **quartz** using *transmission electron microscopy*. Many lamellae seen optically in **quartz**, as shown in the photomicrograph, are narrow **subgrains** with boundaries parallel to the basal plane, and hence are formed by lattice **deformation** and **recovery** (Fig. D8). Narrow **subgrains** along healed **fractures** are also seen as deformation lamellae optically, as are narrow **deformation twins** in minerals such as **calcite**. In some cases the lamellae seen represent **subgrains** formed along many closely-spaced **slip** surfaces within the lattice. In these

Fig. D8 Photomicrograph showing **deformation lamellae** in a quartz grain. In the grain in the bottom right corner of the picture, the deformation lamellae are inclined to the right at about 20°, and are slightly curved. The lines inclined to the left are trails of bubbles and inclusions marking healed fractures.

cases, and in these cases only, deformation lamellae represent an episode of **deformation** at a high **strain rate**. (See **deformation band, crystal defect**.) [SB]

White, S.H. (1973) The dislocation structures responsible for the optical effects in some naturally deformed quartzites. *Journal of Materials Science* **8**: 490–9.

deformation mechanism The mechanism by which **deformation** is accommodated at a **microscopic** scale. Deformation mechanisms can be divided into three classes: **cataclasis**, *intracrystalline plasticity* and *diffusive mass transfer*. **Cataclasis** is deformation by **fracture** and sliding of rigid particles. This class therefore includes *particulate flow, cataclastic flow* and *grain-boundary sliding*.

Grain boundary sliding is the displacement of grains relative to one another along grain boundaries. This often accompanies other processes, to contribute to composite deformation mechanisms, particularly with **fracture** in **cataclasis** to allow *cataclastic flow*. Sliding along grain boundaries accommodated by diffusion is known as *super-*

plasticity; it may occur also with solution transfer or *crystal plasticity*. Diagnostic textures to identify grain boundary sliding are parallel alignments of grain boundaries and rectangular grain profiles. There is no internal distortion of the rigid particles during **deformation**. **Cataclasis** is always accompanied by volume change and is pressure dependent; however, it is independent of temperature and **strain rate**.

Intracrystalline plasticity is **deformation** by the movement of **dislocations** through a *crystal lattice*. It is possible to distinguish pure *dislocation glide* from *dislocation creep*, in which both *glide* and *climb dislocations* occur (see **dislocation**). Shape changes occur within single crystals and a variety of crystal-plastic features form, including *undulose extinction*, **deformation bands**, **kink bands**, **deformation lamellae**, **subgrains** and both shape- and crystallographic **preferred orientations**. New grains may form by recrystallization due to rotation of **subgrains** or grain boundary migration. *Intracrystalline plasticity* conserves volume independently of pressure, but is sensitive to temperature and **strain rate**.

Diffusive mass transfer is the movement of material by diffusion of vacancies, atoms, molecules or ions. When the diffusion pathway is through the crystal lattice, this mechanism is known as *Coble creep*; diffusion at grain boundaries is referred to as *Nabarro–Herring creep*; these are both solid-state diffusion processes, in contrast to diffusion in the dissolved state, which is known as *solution transfer*, and commonly referred to as **pressure solution** when driven by **stress**. *Diffusive mass transfer* is dependent on temperature and **strain rate**.

Most rock **deformation** occurs by a combination of two or three of the above classes. Thus, for example, *diffusive mass transfer* may be accompanied by grain boundary sliding (**cataclasis**): this combination is known as *superplasticity* and may operate in very fine-grained **mylonites**. As well as this combination of mechanisms within a single phase, different phases are likely to combine different mechanisms during the same **deformation**: for example, **feldspar** may deform cataclastically adjacent to **quartz** which is flowing by *intracrystalline plasticity* in *greenschist* conditions.

The operative deformation mechanism of a phase depends on intrinsic factors (i.e. its material properties such as *fracture toughness*, *bond strength* and *diffusion coefficient*), and extrinsic factors, of which the most important are the *deviatoric* and **hydrostatic** components of **stress**, temperature and grain size. If the intrinsic properties are known, the deformation mechanism can be predicted for a given set of extrinsic conditions. This is usually done by using a *failure criterion* to determine whether **cataclastic** deformation will occur, and a flow law to determine which crystal plastic or diffusive mass transfer mechanism will provide the greatest **strain rate**. The results can be plotted on a *deformation mechanism map*, which shows the active deformation mechanism in coordinates of **stress**, temperature and grain size.

All deformation mechanisms may be enhanced by chemical reactions or *phase* changes: this is known as *transformation-enhanced deformation*. Reaction may in fact be an integral part of **deformation**. **Cataclasis** may be favoured by reactions involving volume change due to **strain** incompatibilities. *Subcritical crack growth* may be highly dependent on the chemical environment due to reactions responsible for weakening bonds at the crack tip. Dislocation mobility and hence *intracrystalline plasticity* may also be increased by reactions that stretch atomic bonds. Higher **potential** gradients introduced by chemical reactions will increase *diffusion coefficients* and hence enhance *diffusive mass transfer*. [TGB]

Lawn, B.R. & Wilshaw, T.R. (1975) *Fracture of Brittle Solids*. Cambridge University Press, Cambridge.
Nicolas, R. & Poirier, J.P. (1976) *Crystalline Plasticity and Solid-state Flow in Metamorphic Rocks*. Wiley Interscience, London.

deformation of orebodies **Folding** and/or metamorphism of **orebodies** may change radically the original form of **orebodies** and such changes can directly influence the exploration for, and the development of, these **orebodies**. Intense **folding** can result in the effective thickening of **orebodies** or in their attenuation and shredding into small sections so that their original form and their genetic type may be very much in doubt. [AME]

Evans, A.M. (1993) *Ore Geology and Industrial Minerals: An Introduction* (3rd edn). Blackwell Scientific Publications, Oxford.

deformation path Line showing successive states in a history of **deformation**. Since it is impossible to show all three aspects of **deformation** (**strain**, rotation and translation) for all but the simplest **deformation**, deformation paths are usually simplified to show only one or two aspects, e.g. **strain** and rotation. These can be plotted in three-dimensional coordinates of λ_1, λ_2 and ϕ where λ_1, λ_2 are the principal **quadratic elongations** and ϕ is the angle of rotation with respect to a reference direction, for a two-dimensional analysis. In this framework (Fig. D9), irrotational **deformations** plot in the $\phi = 0$ or basal plane of the diagram. Deformation paths are thus to be dis-

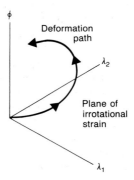

Fig. D9 A **deformation path** showing the progressive state of deformation by the principal quadratic elongations λ_1, λ_2 and the rotation ϕ.

tinguished from *strain paths* because they include extra information on, for example, rotation. [TGB]

Ramsay, J.G. & Huber, M.I. (1987) *The Techniques of Modern Structural Geology.* Academic Press, London.

deformation twinning Systematic reorientation of part of a **crystal structure** during **deformation** such that the deformed zone is geometrically related to the rest of the mineral. Deformation twins can have similar relationships with the host crystal lattics as simple growth twins. Many deformation twins form **kinks** in the lattice, achieved by bond breakage and re-formation, allowing **bulk strain** of the mineral grain.

Twin formation is a common type of **deformation** in **calcite** and other carbonates. **Twins** form sharp-sided straight *domains*, with a systematic relationship to the host **crystal structure**. With increasing **strain**, the **twinned** volumes of the crystals increase, either by **twin** broadening, or by an increase in the number of **twins**. The percentage of the crystal **twinned** can be used as a measure of **strain**. Twin orientations are used to deduce **stress** orientations at the time of **deformation**. (See Fig. D10.) [SB]

Fig. D10 Deformation twinning.

degradation of oil Modification of **crude oil** by differential **solution**, excess **natural gas** (see **de-asphalting**) **oxidation** or bacterial attack (see **biodegradation**). Oil accumulations in **reservoir** rocks must have displaced water that previously occupied the pore space. Contact with the water may give rise to modifications in the nature of the *hydrocarbons* as their various components have very different relative solubilities, although all their absolute solubilities are low. The non-hydrocarbons or S–N–O compounds are the most soluble and the most easily removed. The **aromatics** are more soluble than the *paraffin* and *naphthene series* compounds (see **crude oil**). As well as acting as solvents, surface **recharge** waters will bring oxygen into oil **reservoirs**, but many workers are dubious about the oxidation of *hydrocarbons* by oxygenated water. [AME]

North, F.K. (1985) *Petroleum Geology.* Allen & Unwin, Boston.

delamination The process whereby the **crust** detaches from the **mantle** during collisional **orogeny**, or an **upper crustal** layer detaches from a lower. Oxburgh (1972) introduced the term *'flake tectonics'* to describe the detachment or delamination of an **upper crustal** layer from the African *plate* which became thrust over the European *plate* as the high-level Austro-Alpine nappes. The **lower crustal** layer, together with the underlying **mantle** part of the **lithosphere** were considered to be **underthrust** below the European *plate* (Fig. D11a). Other authors (e.g. Bird, 1978) have suggested crustal delamination to explain the doubling of *crustal* thickness beneath the Himalaya. **Crust** of the main Indian *plate* was thought to underthrust the **crust** of the

northern part of the Indian *plate* for which the **mantle** part of the **lithosphere** had become detached (delaminated), allowing the **underthrust** crustal slab to move along the **crust–mantle** boundary (Fig. D11b). Delamination is thought to be favoured by the presence of relatively weak zones in the crust provided by major changes in composition and thus **rheology** (see Kusznir & Park, 1987).

Thin-skinned behaviour during delamination tectonics can also be related to areas where crustal rocks have been extended, commonly by sliding between subhorizontal layers, as occurs in the deeper parts of *listric faults*. [RGP/DHT]

Bird, J.M. (1978) Initiation of intracontinental subduction in the Himalaya. *Journal of Geophysical Research* **83**: B10, 4975–87.
Kusznir, N.J. & Park, R.G. (1987) The extensional strength of the continental lithosphere: its dependence on geothermal gradient, and crustal composition and thickness. In: Coward, M.P., Dewey, J.F. & Hancock, P.L. (eds) *Continental Extensional Tectonics*, pp. 35–52. Geological Society Special Publication No. 28. Blackwell Scientific Publications, Oxford.
Oxburgh, E.R. (1972) Flake tectonics and continental collision. *Nature* **239**: 202–15.

delay time In **seismic refaction** surveying, 'delay time' is the difference between the actual **arrival time** of a **head wave** at a surface **geophone** and the (shorter) time that it would have taken to reach a point on the refracting boundary beneath the **geophone** (see Fig. D12). There are various means of determining the delay time, and from it the depth of the refracting boundary can be estimated.

In global seismology, delay time or 'residual' is the difference between the actual **arrival time** of a **seismic**

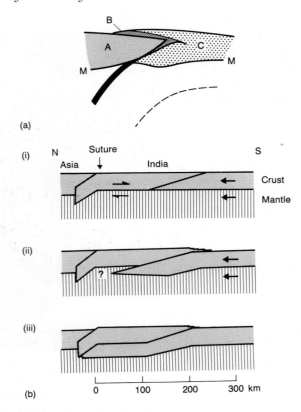

(a)

(i)

(ii)

(iii)

(b)

Fig. D11 Delamination. (a) The flake tectonic mechanism. The diagram shows overthrusting of the upper part of the crust of continent C over that of continent A, and the underthrusting of the lower part of C following the subducted oceanic crust (black). B represents marine sedimentary cover from between the two continents. M = Moho. (After Oxburgh, 1972.) (b) Diagram suggesting delamination of the crust during emplacement of the High Himalayan nappe. For crustal overlap to occur, the mantle part of the lithosphere must detach in the region of overlap shown by ? in (ii). (After Bird, 1978.)

wave and that which is predicted by some standard model of the velocities within the Earth. The standard model is usually one with only radial variations in velocity, such as the *Jeffreys–Bullen model*. The residual reflects local velocity differences relative to the 'standard' model. [RAC]

delayed recovery The removal (**recovery**) of **temporary strain** a measurable time after removal of the deforming **stress**. This is characteristic of rocks showing **visco-elastic** behaviour, rather than **elastic** behaviour, in which **recovery** is instantaneous. As an example, the rate of **recovery** of crystal **plastic deformation** is controlled by factors such as temperature (See **elastic strain**, **deformation mechanisms**.) [SB]

dell Small headwater valley which is characteristically sediment-choked and **swampy**. Dells frequently occur at the head of deep gorges on **plateaux** surfaces and may be analogous to **dambos**. Notable dells have developed on **sandstone** on the Woronora Plateau of New South Wales, Australia (Young, 1986). [ASG]

Young, A.R.M. (1986) The geomorphic development of dells (upland swamps) on the Woronora Plateau, N.S.W., Australia. *Zeitschrift für Geomorphologie* NF **30**: 317–27.

delta Term first applied by Herodotus (490 BC) to the triangular subaerial surface at the mouth of the River Nile, because of the similarity in shape with the fourth letter of the Greek alphabet Δ. A delta is a constructional sedimentary body, up to thousands of square kilometres in areal extent, formed where fluvial systems interact with fresh to fully marine water to deposit sediments as a result of flow expansion and deceleration as channellized flows become unconfined. Provision of sediment from fluvial outflows occurs at a faster rate than basinal processes can redistribute the sediment.

The processes at river mouths are varied, depending on the buoyancy of the incoming water compared with sea-water, outflow velocity and the nearshore bathymetry. Three situations are possible with regard to relative buoyancy. Where river waters and basinal waters are equally dense, immediate three-dimensional mixing of the water bodies takes place causing considerable sediment deposition; this is a situation known as *homopycnal flow*. Where the incoming water is denser than the basinal water the former flows as bottom-hugging density currents, causing sediment to travel a long way from the point of

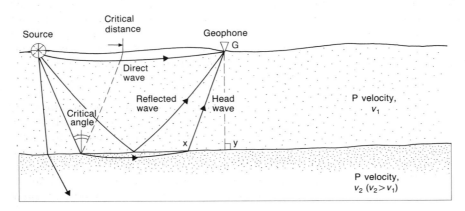

Fig. D12 Schematic illustration of some terms used in seismic refraction exploration. The **delay time** at G is the time taken by the head wave to travel from X to G minus the time the head wave would take to travel from X to Y. The critical distance is the closest distance at which a head wave can exist.

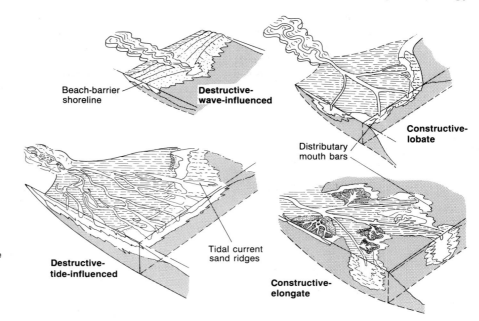

Fig. D13 Delta types according to the relative importance of fluvial processes (constructive types) and basinal processes (destructive types). (After Fisher *et al.*, 1969.)

river water entry; this situation is termed *hyperpycnal flow*. Where river waters are less dense than basinal waters, they take the form of a buoyant surface plume or jet, the condition of *hypopycnal flow*.

Outflow velocity and nearshore bathymetry combine to influence the type of outflow dispersion. For example, where nearshore slopes are steep and inflow velocities are large, outflows are dominated by inertia and steep-fronted *Gilbert-type deltas* or *mouth bars* build out into the **basin**. However, if nearshore slopes are shallow, the river outflow may be dominated by friction, causing a rapid deceleration and wide spreading of the jet. This often causes a bifurcation of the river channel, producing an intervening *middle ground bar*.

Deltas are shaped by fluvial, **wave** or **tidal** processes. Where **wave** power is large, sediment is transported along the coast in the direction of **longshore drift**, constantly eroding and redistributing the delta front. Where tidal range is large, deltas may take the form of deep marine embayments with elongate tidal **sand** ridges and subtidal channels (Fig. D13).

Environments associated with deltas include: the finer-grained subaqueous prodelta and delta front with distributary *mouth bars* and abundant **wet sediment deformation**, **slumps** and **slides**; the emergent delta top that includes *distributary channels* (see **bar finger sands**), *inter-distributary bays* with intruding **crevasse splays** and channels and vegetated *swamp* areas. The delta top comprises a lower delta plain, influenced by marine processes, and an entirely freshwater upper delta plain free from marine influences. Deltas may be mainly mud- or **sand**-rich, be wholly contained on a continental shelf or build via steep slopes into deep-water **basins**. [PAA]

Fisher, W.L., Brown, L.F., Scott, A.J. & McGowen, J.H. (1969) *Delta Systems in the Exploration for Oil and Gas.* Bureau of Economic Geology, University of Texas, Austin, TX.

Wright, L.D. (1977) Sediment transport and deposition at river mouths: a synthesis. *Geological Society of America Bulletin* **88**: 857–68.

demagnetization The **natural remanent magnetization** of rocks is usually dissected into its constituent components by partial demagnetization, i.e. using incrementally increased heating or applied **magnetic fields** to remove components of differing **blocking temperature** or *coercivity*. (See also **paleomagnetism**.) [DHT]

demantoid A green variety of the mineral **andradite** garnet. (See **garnet minerals**.) [DJV]

dendrochronology Dating of trees and wood from archeological sites by measuring and counting the annual rings present. Tree-ring thicknesses vary according to environmental factors such as rainfall and temperature, and a distinctive sequence in annual rainfall or temperature will result in a distinctive sequence of tree-rings. Patterns of tree-ring spacing and thicknesses can be compared, sometimes using statistical methods, to correlate wood from different buildings or excavations. Dendrochronology has been used for relative and absolute dating in the south-western USA, Alaska and Europe for material from the prehistoric and Medieval periods. It has also provided important checks on **radiocarbon** dates, using particularly the long-lived bristle-cone pine trees of California, USA. [OWT]

dendrogeomorphology A branch of **dendrochronology** (the general science of dating annual growth layers in woody plants and the exploitation of associated environmental information). It seeks to use **dendrochronological** information to understand certain geomorphological features. For example, the growth rings of trees are affected by such environmental impacts as inclination, corrosion of bark, **shear**, burial, exposure, inundation and nudation. These impacts are in turn sometimes the result of specific geomorphological processes, including **faulting**, shoreline warping, volcanism, floods, mass wasting, *avalanches*, glacial fluctuations, etc. Tree-ring dating can thus permit the establishment of the frequency or last occurrence of such events (Shroder, 1980). Another major influence on the growth of tree rings in suitable species is climate, so that *dendroclimatology* is the branch of **dendrochronology** which seeks to extract climatic information from tree-rings. [ASG]

Shroder, J.F. (1980) Dendrogeomorphology: review and new techniques of tree-ring dating. *Progress in Physical Geography* **4**: 161–88.

density measurement Knowledge of rock densities is necessary for both the **reduction** and interpretation of **gravity** measurements. Densities can be determined in the laboratory or *in situ*. Laboratory methods involve weighing rock samples in air and then suspended in water to determine the volume, and hence the density. If the sample is porous, a further pair of measurements made after soaking the rock overnight in water provide the wet density.

In situ methods include density measurements down a borehole using *gamma-gamma logging* (see **geophysical borehole logging**) or a downhole gravimeter (see **borehole gravimeter**). An analogous method to the latter is the measurement of **gravity** down mineshafts with a standard

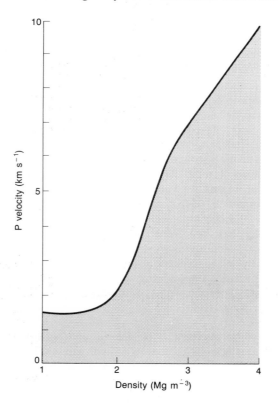

Fig. D14 Density measurement. The relationship between P wave velocity and density. (After Nafe & Drake, 1963.)

gravimeter. Nettleton's method involves measuring **gravity** over a small topographic irregularity. **Bouguer anomalies** (including **topographic corrections**) are then computed for a range of reduction densities, and the profile with the least correlation with the topography provides the most reasonable density. A final method relies on the well-developed curvilinear relationship between seismic *P*-wave velocity and density (Fig. D14) demonstrated by Nafe & Drake (1963). Consequently, if a **seismic refraction** experiment has been performed in the area of interest, knowledge of the *P* wave velocity allows estimation of the density to an accuracy of about $0.01 \, \text{Mg m}^{-3}$, even if the layer involved never appears at the surface. [PK]

Kearey, P. & Brooks, M. (1991) *An Introduction to Geophysical Exploration* (2nd edn). Blackwell Scientific Publications, Oxford.
Nafe, J.E. & Drake, C.L. (1963) Physical properties of marine sediments. In: Hill, M.N. (ed.) *The Sea* Vol. 3, pp. 794–815. Interscience Publishers, New York.

denudation This incorporates both **weathering** of material at the Earth's surface and its removal by processes of **erosion**. Estimates of denudation rates are important both for the interpretation of landform evolution over long periods of time and for more short-term concerns such as the **erosion** of topsoil. Effectively, denudation rates equate with **erosion** rates. [TPB]

denudation chronology The attempt by geomorphologists to reconstruct the **erosional** history of the Earth's surface using **erosional** remnants to reconstruct the history of the Earth where the stratigraphic record was interrupted or unclear. Techniques were developed to help in the identification of such **erosional** remnants (e.g. superimposed contours, altimetric frequency curves, etc.) and much effort was expended in trying to fathom out whether the surfaces were the product of subaerial or marine **denudation**. However, much of the evidence upon which schemes of denudation chronology were based relied upon morphological criteria of limited reliability (Rich, 1938), though some of the more successful schemes were able to supplement the morphological evidence with information gained from deposits resting on the *planation surfaces* (e.g. Wooldridge & Linton, 1955). Once the dominant approach in British and American geomorphology, denudation chronology is now much less central to the discipline, though notable studies are still undertaken (e.g. Jones, 1980). [ASG]

Jones, D.K.C. (ed.) (1980) *The Shaping of Southern England*. Academic Press, London.
Rich, J.L. (1938) Recognition and significance of multiple erosion surfaces. *Geological Society of America Bulletin* **49**: 1695–772.
Wooldridge, S.W. & Linton, D.L. (1955) *Structure, Surface and Drainage in South-East England* (2nd edn). Philip, London.

depletion The loss of water from surface or **groundwater reservoirs** at a greater rate than that of their **recharge**. [AME]

depletion allowance A proportion of the income from mining or oil production that is not subject to tax. It is a method of recognizing that once an **orebody** or **oilfield** is being exploited it has become a wasting asset and that production will ultimately exhaust. [AME]

depletion drive (*solution gas drive/dissolved-gas drive*) This mechanism, which drives **reservoir** oil towards a **well** bore, uses the energy of the gas in solution in the oil. As oil is withdrawn the **reservoir** pressure drops and the gas expands driving the oil in the direction of the pressure gradient (towards the well bore). [AME]

depositional-environment-related diagenesis (*eogenesis*, see **regimes of diagenesis**) **Diagenetic** processes which fall into the regime where the composition of the interstitial **pore-waters** is largely controlled by the gross physical, biological and chemical characteristics of the depositional system.

Many diagenetic reactions take place in association with the depositional environment. The driving mechanism for many *eogenetic* reactions is well explained by the '*law of mineral stability*' which simply states that 'minerals are in thermodynamic equilibrium only in the environment in which they form' (Keller, 1969). This concept has widespread applications in *eogenetic* systems. On deposition, an original detrital mineral assemblage comprises a mixture of minerals formed under widely differing temperature–pressure regimes, typically higher than those prevailing at or near the Earth's surface. Therefore, in the sedimentary environment, the detrital mineral assemblage is inherently unstable (at best metastable) and during *eogenesis* will tend to react with the ambient environment through *eogenetic* processes towards an equilibrium state.

The inherent relative instability of the detrital assemblage during *eogenesis* results in sediments being 'finger-printed' by the interstitial **pore-waters** prevailing immediately following deposition. These **pore-waters** are directly influenced by the depositional environment and, as a corollary, provided the effects of *eogenesis* are preserved and can be recognized within a given sediment, then the *eogenetic* assemblage can be used to place constraints on the depositional environment. Thus the depositional environment can be considered to create a broad division of diagenetic processes in most sedimentary systems. That is not to say, however, that any one particular **authigenic** mineral is indicative of any particular depositional environment. Rather the *whole eogenetic* assemblage can be considered to reflect three extremes of depositional environment:

1 Marine environment, characterized by slightly alkaline **pore-waters** initially with around 350 p.p.t. dissolved solids.

2 Warm, wet, often subtropical to temperate, non-marine environments, where intense **weathering** is typically developed in the source area, have initially acidic or **anoxic** **pore-waters** with low concentrations of dissolved species.

3 Hot, dry, semi-arid to arid, desert environments characterized by low precipitation and little source area chemical **weathering**, in which surface interstitial **pore-waters** are typically highly oxidative and alkaline.

In marine environments, the interaction of reducing organic matter and oxidizing inorganic detrital components with marine **pore-waters**, modified by the reaction products of *bacterial* metabolism (see **organic matter diagenesis**), causes rapid *eogenetic* **alteration** of the shallow sediments (Fig. D15).

Marine **sandstones** are characterized by an early **authigenic** suite of minerals which includes **illite**, interstratified **illite–smectite**, **potash feldspar** overgrowths, **quartz** (or **silica** polymorphs), **pyrite**, various carbonate cements, and, in suitable settings, a variety of green sheet silicates (*glaucony*, *bertherine*, *celadonite* and **chlorite**; Fig. D16).

In warm, wet, vegetated, non-marine environments, organic matter is usually abundant and not completely oxidized at the sediment surface so that, as with marine sediments, *bacterial* processes dominate. Mineral **authigenesis**, however, reflects the relative paucity of potassium, magnesium and sulphate ions in solution

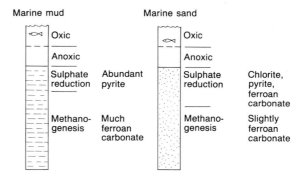

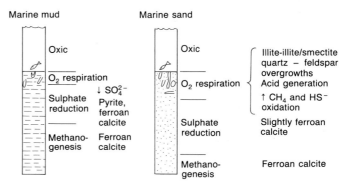

Fig. D15 Depositional-environment-related diagenesis. Schematic diagenetic pathways during eogenesis of marine sediments.

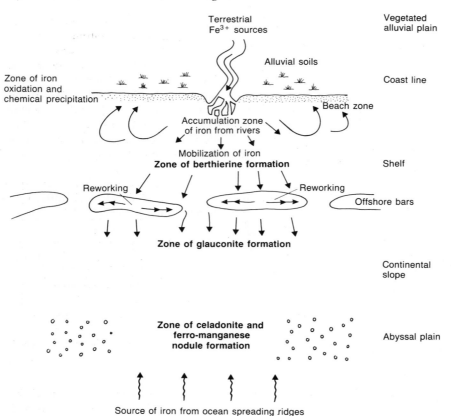

Fig. D16 Depositional-environment-related diagenesis. Distribution of eogenetic green iron silicates in the marine depositional environment.

in freshwater. **Quartz** overgrowths, dissolved **feldspar** grains, **pore**-filling **kaolinite**, altered **muscovites** and iron-rich carbonates (ferroan **calcite** or **siderite**) are typical **authigenic** minerals reported from quartz *arenites* and *arkoses* (Fig. D17). With the introduction of more labile grains, **chlorite** may be developed in preference to **kaolinite**.

In hot, dry, continental desert sediments, intensive oxidative conditions, low **water tables** and stable geomorphic surfaces result in any organic matter undergoing complete **oxidation** at the sediment surface. All iron

remains in the ferric state and cannot enter the carbonate lattice. *Eogenetic* carbonates of either **calcrete** or **evaporite** origin remain non-ferroan. **Hematite** is the stable **iron oxide** and imparts a distinct red colour to the sediments. In the alkaline **groundwater**, **smectites**, **quartz** and **feldspar** overgrowths, **calcite** and **dolomite** and, where **evaporitic** conditions develop, **gypsum** or **anhydrite** are the characteristic **authigenic** minerals (Fig. D18). [SDB]

Burley, S.D., Kantorowicz, J.D. & Waugh, B. (1985) Clastic diagenesis. In: Brenchley, P.J. & Williams, B.P.J. (eds) *Sedimentology; Recent Developments and Applied Aspects*, pp. 189–226. Geological Society

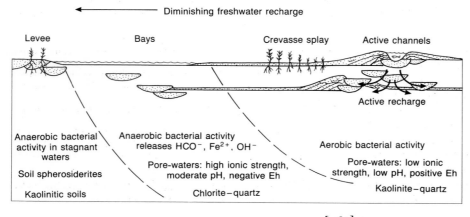

Fig. D17 Depositional-environment-related diagenesis. Facies related eogenetic modifications in warm, wet, vegetated, non-marine environments.

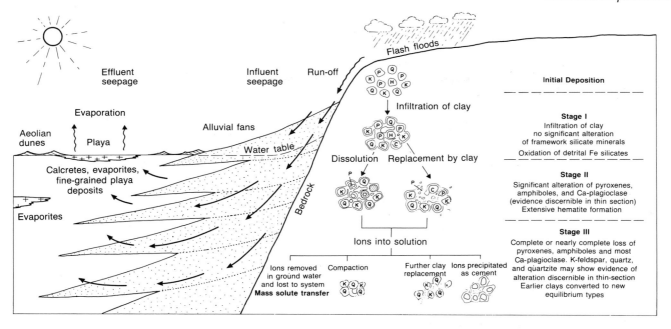

Fig. D18 Depositional-environment-related diagenesis. Eogenetic modifications in their basin setting for hot, dry, desert environments. Q, quartz; K, K-feldspar; C, Ca-plagioclase; P, pyroxene; H, hornblende.

Special Publication No. 18. Blackwell Scientific Publications, Oxford.

Keller, W.D. (1969) *Chemistry in Introductory Geology* (9th edn). Lucas, Columbia, Missouri.

Walker, T.R., Waugh, B. & Crone, A.J. (1978) Diagenesis in first cycle desert alluvium of Cenozoic age, southwestern United States and northwestern Mexico. *Geological Society of America Bulletin* 89: 19–32.

depositional remanent magnetization The **natural remanent magnetization** carried mainly by unconsolidated sediments that was acquired during the deposition of already magnetized grains. Broadly equivalent to **detrital remanent magnetization**. [DHT]

depression spring A **spring** that forms when the land surface reaches down to the **water table**. [AME]

depth of compensation The depth at which the pressure of all overlying rocks is the same throughout the Earth. (See **Pratt hypothesis, isostasy**.) [DHT]

depth of penetration The approximate depth at which (in electrical methods) the current flow becomes insignificant, and hence the **limiting depth** for conductivity investigations. The concept applies to both galvanic **resistivity methods** and **electromagnetic induction methods**, with somewhat different significance. In the former, the current penetration depends upon the current electrode separation (and to some extent on the resistivity configuration), and is of the order of the electrode separation. In the latter, the relevant parameter is *skin depth*, proportional to the product of the inverse square root of the frequency of the electromagnetic wave and the **electrical conductivity** and **permeability** of the medium. [ATB]

desert varnish A dark, shiny coating found on rocks and pebbles most commonly in arid regions. The orange (**manganese**-poor) to black (**manganese**-rich) varnish or patina is generally composed of **iron** and **manganese oxides** and clays. Rock varnish is usually less than 1.0 mm thick — much thinner than most **weathering rinds**. The oxides and clays which form it are probably derived from aeolian inputs and accrete most effectively under non-acid environmental conditions. Some evidence suggests that micro-organisms play an important role in the formation of varnish and that variations in the **iron/manganese** ratios through a coating reflect changes in environmental conditions. Since some coatings of desert varnish are tens of thousands — perhaps hundreds of thousands — of years old, they may prove extremely valuable in *paleoenvironmental* research. [AW]

Dorn, R.I. (1984) Cause and implications of rock varnish microchemical lamination. *Nature* **310**: 767–70.

Dorn, R.I. & DeNiro, M.J. (1985) Stable carbon isotope ratios of rock varnish organic matter: a new paleoenvironmental indicator. *Science* **227**: 1472–4.

Dorn, R.I. & Oberlander, T.M. (1982) Rock varnish. *Progress in Physical Geography* **6**: 317–67.

Whalley, W.B. (1983) Desert varnish. In: Goudie, A.S. & Pye, K. (eds) *Chemical Sediments and Geomorphology*, pp. 197–226. Academic Press, London.

desert/stone pavement A surficial layer of rock pebbles and cobbles which covers unconsolidated, finer-grained materials in an arid region. They are most common in warm deserts, though similar features occur in the *permafrost* zone. The surficial **gravel** may accumulate as a residual *lag* when finer materials are eroded by *sheetwash* on shallow slopes or are **deflated** by wind — a process which can create desert pavements over extensive, flat landscapes. Once it has developed, the pavement protects the underlying **sands** and silts from further **erosion**. Some desert pavements may form when vertical sorting mechanisms move larger clasts upwards through fine-grained materials. This process can occur in deserts when periodic wetting and drying causes volumetric changes in salts or **clay minerals** — especially **smectites** — resulting in upward displacement of large particles. In cold regions, the same upward migration of pebbles and cobbles occurs as a result of cyclic freezing and thawing of the *soil* zone. [AW]

Cooke, R.U. (1970) Stone pavements in deserts. *Annals of the Association of American Geographers* **60**: 560–77.
Ollier, C.D. (1966) Desert gilgai. *Nature* **212**: 581–3.

detachment fault **Fault** marking the displacement along a **detachment horizon** or *décollement* plane. In **thrust belts**, such a **fault** is known as the *sole thrust*, or in **extensional fault systems**, *the sole fault*. (See **thrust belt**.) [RGP]

detachment horizon A surface, normally parallel to **bedding**, along which an upper layer has become detached from the rock below and moved independently during **deformation**. The term is synonymous with **décollement** plane. (See also **delamination**.) [RGP]

detrital remanent magnetization (DRM) The **natural remanent magnetization** carried by detrital grains within a sediment. (See also **depositional remanent magnetization**.) [DHT]

deuteric alteration The **alteration** resulting from reactions between primary magmatic minerals and **hydrothermal solutions** that separated from the same **magma** at a late stage in its consolidation. Thus the deuteric alteration of **granites** may produce various amounts of minerals such as **tourmaline**, **topaz**, **fluorite**, **muscovite**, lithium **mica**, **cassiterite** and **wolframite** that replace the primary **feldspars** and **micas**. The development of **greisen** is often considered to be the result of deuteric alteration. [AME]

development well **Well** used to develop and produce oil rather than an exploratory well many of which may be classified as 'wild cat'. [AME]

deviation from axial linearity (*DEVAL*) Small bathymetric offset of an *ocean ridge* crest. The origin of DEVALs is obscure as yet; they may represent the last stage of melt

Table D1 Devonian stage names, 1987

Upper Devonian	Famennian
	Frasnian
Middle Devonian	Givetian
	Eifelian
Lower Devonian	Emsian
	Pragian
	Lochkovian

See Ziegler, W. & Klapper, G. 1985; also Ziegler, W. in House *et al.*, 1979.

segregation where small **magma** pockets ascend beneath the **crust**, individual DEVALs having a lifetime of $\leqslant 10^4$–10^5 years; they may mark the edge of an event of volcanic construction, and so provide a convenient site for the next pulse of **magma**. [PK]

Langmuir, C.H., Bender, J.F. & Batiza, R. (1986) Petrological and tectonic segmentation of the East Pacific Rise, 5°30′N. *Nature* **322**: 422–9.

devolatilization The loss of volatiles (and concomitant increase in **carbon**) during *coalification*. (See **coal**.) [AME]

Devonian A stratigraphic system of **Paleozoic** rocks lying above the **Silurian** System and below the **Carboniferous** (or *Mississippian*) System and named in 1839 by A. Sedgwick and R.I. Murchison in consequence of their studies in south-west England (Rudwick, 1985). Three conventional series comprise the system; seven stages, in ascending order *Lochkovian, Pragian, Emsian, Eifelian, Givetian, Frasnian* and *Famennian* are based upon distinctive faunas (Table D1). The system outcrops in all continents to a global total of some 7 750 000 km²; the Period lasted from about 410 Ma to 355 Ma during which time continental *plate* movement and **orogenic** activity was vigorous.

Recent formal definitions: The International Commission on Stratigraphy (of IUGS) accepts the following definition of the system and its divisions based upon **conodont** zones:

Base of the Upper Devonian Series and base of the *Frasnian* Stage = base of the lower zone of *Polygnathus asymmetricus*, the first appearance of *Ancyrodella rotundiloba*.

Base of the Middle Devonian Series and base of the *Eifelian* Stage = base of the zone of *Polygnathus costatus partitus*.

Base of the Lower Devonian Series (base of the system) and base of the *Lochkovian* Stage = base of the zone of the **graptolite** *Monograptus uniformis uniformis* and of the **conodont** *Icriodus woschmidti*.

Other stages based upon marine **biostratigraphy** have also hitherto been used in USA, the former USSR and China. Modern correlation practice is toward the global standards listed above (Dineley, 1984). *Radiometric age dating* within acceptable limits is increasingly consistent.

Amongst the wide variety of sedimentary facies, the continental (post-orogenic) *molasse 'Old Red Sandstone'* is characteristic. Marine sediments include a high proportion of clastics (both mature and immature) and carbonates; **black shales** are widespread and conspicuous in the Middle and Upper Series. Carbonates, widely distributed and including many *reefal* (*algal*, **stromatoporoidal** and other) bodies with distinctive faunas in the Middle and Upper Series, are very rare in the Lower. **Evaporites** include widespread sulphates and chlorides and also minor alkali salts. Such formations are especially widespread in the former USSR. The earliest **coal seams** are of Late Devonian age, and **petroleum** occurs in Canadian Devonian rocks. Both marine and terrestrial volcanics were produced in similar ratios to those in other **Paleozoic** systems but in total rose from about 16% of contemporary products in the early Devonian to about 25% in the late Devonian. In *orogenic belts* large **plutons** of acidic rocks were emplaced.

Devonian paleontology is distinctive in the rise of vascular **land plants** from primitive *psilophytes* to massive and complex *lepidophytes*, in the great diversification of neritic marine invertebrates and in the advance of vertebrates from faunas dominantly of *agnatha* to those of *gnathostomes* (see **fish**) with the earliest **amphibian** tetrapods (House *et al.*, 1979). Marine *protozoan* and other microfossils are abundant. *Algae*, **Stromatoporoidea**, *rugosa* and *tabulata* are distinctive, the latter at their acme. **Brachiopod** faunas were dominated by *spiriferoids* but *strophomenids* and *orthoids* were numerous and *rhynchonellids* increasing. Of the **Mollusca** the most important were the *cephalopoda*; *dacryoconarida* were also abundant in the early part of the period. **Arthropod** species included a diminishing number of *trilobites* and *ostracods*, *merostomata* and *insects*. **Echinoid** diversity diminished but *crinoids* were especially abundant and diverse. **Conodonts** achieved their acme but the **graptolites** (monograptids) became extinct by the end of the early Devonian. **Biostratigraphic** correlation on regional and intercontinental scales is most successful by means of **conodonts**, *ammonoids* and *dacryoconarids*; *trilobites*, **corals** and **brachiopods**; *spores* and some vertebrates are also amenable. The early Devonian Rhenish magnafacies and the Bohemian (or Hercynian) magnafacies were first differentiated in Europe as representing nearshore and oceanic sediments and biotas respectively. More recently, the Old World Realm and the Eastern Americas Realm have been distinguished as major warm- to temperate-water marine biological entities while the Malvinokaffric Realm is thought to be a southern hemisphere cold- or cool-water marine realm. Much remains to be done to correlate the marine with the non-marine facies: **palynology** seems to offer the best means for this. The well-established paleobiogeographic realms and provinces early in the period gave way to a progressively more cosmopolitan biosphere. At the end of *Frasnian* time occurred the global **extinction** of many marine forms, possilby as a result of *bolide impact* (McLaren, 1982).

The continents were globally concentrated in southern and equatorial latitudes, with a **'Gondwanaland'** *supercontinent* close to the north of which lay the separate blocs of **Laurentia**-Baltica, *Kazakhstania*, Siberia, North China, South China and four or more minor blocks (see Fig. D19a and b). The equatorial seas had very high organic productivity; **swamp** forests in adjacent land masses also thrived towards the end of the period. *Glacial* deposits are known in South America and southern Africa.

Orogenies during Devonian time were prolonged and extensive and included the *Caledonian-Acadian* (north-west Europe–north-east North America), the *Mauretanian* (north-west Africa), the *Ellesmerian* (Canadian Arctic), the *Antler* (north-west America) and others in China, Australia and possibly Antarctica.

Epeiric (**eustatic**) movements of the sea seem to have been widespread and significant, several cycles of transgression–regression being well known in North America, Europe and China (Johnson *et al.*, 1985). Rise of sea-level was apparently progressive until almost the end of the period. Both the **eustatic** and the orogenic geographical changes have been attributed to *ocean ridge* growth and to *plate-margin* **cratonic** collisions. Paleontological evidence suggests a Devonian year of about 400 days and a Devonian day of some 21 hours: **tidal** behaviour would have been correspondingly different from today's. [DLD]

Dineley, D.L. (1984) *Aspects of a Stratigraphic System: the Devonian.* Macmillan, London.

House, M.R., Scrutton, C.T. & Bassett, M.G. (eds) (1979) *The Devonian System.* Special Paper in Palaeontology No. 23. The Palaeontological Association, London.

Johnson, J.G., Klapper, G. & Sandberg, C.M. (1985) Devonian eustatic fluctuations in Euramerica. *Geological Society of America Bulletin* **96**: 567–87.

McLaren, D.J. (1982) *Frasnian–Famennian Extinctions.* Special Paper of the Geological Society of America **190**: 477–84.

Rudwick, M.J. (1985) *The Great Devonian Controversy: The Shaping of Scientific Knowledge among Gentlemanly Specialists.* University of Chicago Press, Chicago.

Scotese, C.R., Bambach, R.K., Barton, C., Van der Voo, R. & Ziegler, A.M. (1979) Paleozoic base maps. *Journal of Geology* **87**: 217–77.

Ziegler, W. & Klapper, G. (1985) Stages of the Devonian System. *Episodes* **8**: No. 2, 104–9.

Ziegler, W. & Werner, R. (eds) (1985) Devonian series boundaries: results of world-wide studies. *Courier Forschungsinstitut Senckenberg* **75**.

dewatering The removal of **groundwater** in order to lower the **water table** for excavation or other work to take place within the dewatered area. The dewatering can be temporary, e.g. for construction purposes, or permanent, e.g. when it is desired to keep **groundwater** out of the basements of buildings. Various methods are employed for dewatering. One is to use rows of small diameter boreholes to pump out the water. During mining operations in areas of topographical relief, drainage of rocks above valley bottom level may be accomplished by driving **adits** beneath the working areas. [AME]

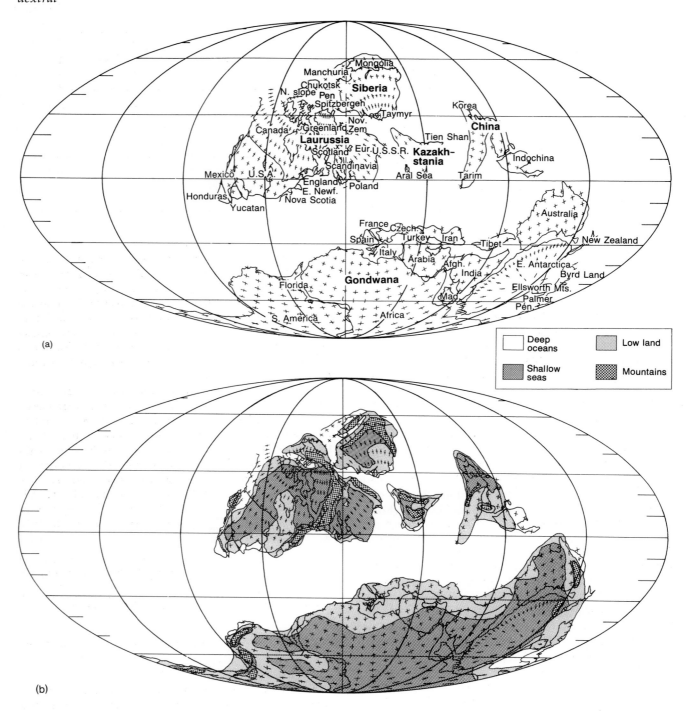

Fig. D19 Early **Devonian** world geography, showing (a) positions of the major land elements and (b) paleogeographic features. (Deep oceans, unshaded; shallow seas, light-shading; low land, intermediate shading; mountains, dense shading.) (After Scotese *et al.*, 1979.)

dextral Sense of relative movement across a boundary or zone at which the opposite side to the observer moves to the right. The term is used for the sense of **strike-slip** displacement on **faults** and **shear zones**, thus *'dextral fault'*. The converse is **sinistral**. (See **fault**.) [RGP]

diachronism Refers to lithostratigraphic units that are time-transgressive; in other words, they do not correspond to chronostratigraphic units. A good example is the 'ideal' sequence of strata deposited by a shallow sea that gradually transgresses over a land surface of irregular relief. In

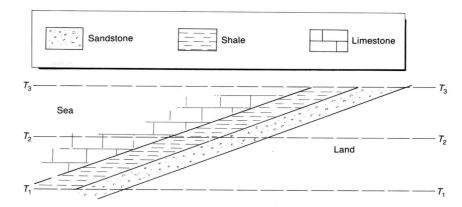

Fig. D20 Schematic vertical section illustrating **diachronism** in a marine transgressive succession. T$_1$, T$_2$ and T$_3$ represent time planes.

Fig. D20 T$_1$, T$_2$ and T$_3$ represent successive time planes. [AH]

diagenesis The sum of all those physical, chemical and biological post-depositional processes prior to the onset of metamorphism by which originally sedimentary assemblages and their interstitial **pore-waters** react, and attempt to reach equilibrium, with their evolving geochemical environment.

Two important points of this definition are worth stressing. Firstly, diagenesis encompasses a broad spectrum of post-depositional modifications to sediments. Defining the beginning of diagenesis is not easy to accomplish. In deep marine clastic sediments, organic matter is altered in the water column as it settles to accumulate on the seafloor; such **alteration** can be considered diagenetic. **Alteration** to sediments in the **weathering** zone prior to burial beneath an **unconformity** can also be clearly considered as part of the diagenetic cycle; so therefore can be the complete destruction of surficial rocks and **supergene enrichment** during intense **weathering**. At the other extreme, defining the end of diagenesis is also difficult although Heroux *et al.* (1979) provided a comprehensive appraisal of the indices which can be used to separate diagenesis from metamorphism. However, during deep sedimentary burial, diagenesis and metamorphism reflect a geological continuum that will vary with local burial rate, **geothermal** gradient, sediment lithology and subsurface fluid flow, and no single limit can be considered definitive. In the loosest sense, therefore, diagenesis can be considered as everything that contributes to making up a sedimentary rock, from its **weathering** in the depositional **basin** to its metamorphism during deep burial.

The second point of interest relates to the fundamental driving mechanism behind diagenesis. Each time the mineral–**pore-water** assemblage (be it the original sedimentary assemblage or a diagenetically modified assemblage) is exposed to instability, either through changes in temperature, pressure or **pore-fluid** chemistry, then the sedimentary assemblage reacts via the interstitial **pore-fluids** in an attempt to equilibrate with the newly estab-

lished conditions. Diagenesis is, therefore, dynamic; as the depositional or burial history of the sedimentary **basin** evolves, so does the diagenetic assemblage and fabric. True thermodynamic equilibrium, however, is probably rarely attained; **basins** are rarely static, the kinetics of diagenetic reactions are slow and metastability is extremely common at diagenetic temperatures and pressures. In this sense, then, diagenesis can be considered simply as low temperature geochemistry.

Despite this discussion it is emphasized that at present there is no universally accepted definition of diagenesis, the term having been employed in variously more and less restricted senses by many authors. Similarly, there is no current definitive delimitation of diagenesis either with respect to the processes of **weathering** and metamorphism or in absolute temperature–pressure terms. A useful review of the concepts of diagenesis and the usage of the term is provided by Larsen & Chilingar (1967). (See **regimes of diagenesis**, **diagenetic processes**, **pre-diagenetic controls on diagenesis**.) [SDB]

Heroux, Y., Chagnon, A. & Bertrand, C. (1979) Compilation and correlation of major thermal maturation indicators. *American Association of Petroleum Geologists Bulletin* **63**: 2128–44.
Larson, G. & Chilingar, G.V. (eds) (1967) *Diagenesis in Sediments.* Developments in Sedimentology No. 8. Elsevier, Amsterdam.

diallage A variety of the **pyroxene mineral diopside**. [DJV]

diamagnetism The fundamental magnetization of all substances resulting from electrons orbiting a nucleus. If a **magnetic field** is applied, the spin magnetizations precess to produce a field in opposition to the applied field. The **susceptibility** is usually $\leq 10^{-5}$ (SI) and may be superimposed by stronger **paramagnetic** or **ferromagnetic** properties. [DHT]

diamond (C) The naturally-occurring high pressure form of **carbon** valued as an **industrial mineral** because of its great hardness and as a **gemstone**. (See **native element**.) [DJV]

Table D2 Diamond deposit. World production of natural diamonds (Mc) in 1986 (bedrock and placer deposits)

Australia	29.2
Zaire	20.5
Botswana	13.0
Former USSR	12.0
Republic of South Africa	10.2
Namibia	1.0
Central African Republic	0.6
Others	3.1
Total	89.6

diamond deposit The ultimate bedrock source of diamonds are the igneous rocks **kimberlite** and **lamproite**. This source is described in this entry. Important amounts of diamond are also recovered from **beach** and **alluvial placer deposits**.

Not all **kimberlites** and **lamproites** contain **diamond** and, in those that do, it is present only in minute concentrations. For example, in the famous Kimberley Mine (RSA), 24 Mtonnes of **kimberlite** yielded only 3 tonnes of **diamond**, or one part in eight million. From the revenue point of view it is not necessarily the highest **grade** mines that provide the greatest return — what matters is the percentage of **gem** quality **diamonds**. Each mining region and even each mine has a different percentage, and some may be 90% or more while others may rely on only an occasional **gemstone** to boost their income. Total world production of natural **diamonds** is about 90 Mc a^{-1}, of which about 45 Mc are of industrial **grade**. In addition about 100 Mc a^{-1} of synthetic **industrial diamonds** are manufactured. The leading world producers are shown

in Table D2 and the general occurrence of **kimberlites** and world bedrock **diamond** fields in Fig. D21.

Many near-surface, diamond-bearing **kimberlites** and **lamproites** occur in pipelike **diatremes** (often just called **pipes**), which are small, generally less than 1 km^2 in horizontal area and often grouped in clusters. Some **diatremes** are known to coalesce at depth with **dykes** of non-fragmental **kimberlite**. These **dykes** are thin, usually less than 10 m in diameter but may be as much as 14 km long.

Some recently formed **diatremes** terminate at the surface in **maars**; these volcanic **craters** may be filled with diamondiferous lacustrine sediments to a depth of over 300 m. The **kimberlite** beneath the sediments may be relatively barren. Below the flared **crater** area is the vertical or near vertical **pipe** itself which typically has walls dipping inwards at about 82° and a fairly regular outline producing the classic, carrot-shaped **diatreme** (Fig. D22) that may exceed 2 km in depth. At the surface **kimberlite** may be weathered and oxidized to '*yellow ground*', '**blue ground**' and '*hardebank*'. In the upper levels of **pipes** the **kimberlite** is usually in the form of so-called *agglomerate* (really a tuffisitic breccia with many rounded and embayed fragments in a finer grained matrix) and **tuff**. The rounded fragments are often **xenoliths** of metamorphic rocks from deeper **crust**, or **garnet-peridotite** or **eclogite** from the **upper mantle**. Their rounded nature is attributed to a gas-fluidized origin.

Kimberlite has traditionally been considered to be the only important primary source of **diamond**; however, **lamproites** in Arkansas and Western Australia carry significant amounts of **diamond**. **Lamproitic craters** are

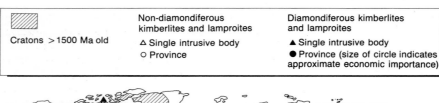

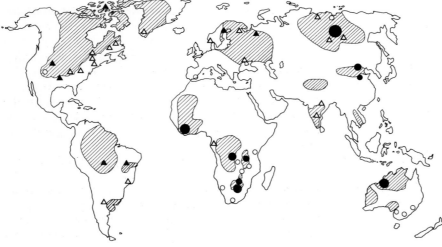

Fig. D21 Diamond deposits. Distribution of diamondiferous and non-diamondiferous kimberlites and lamproites. (After Evans, 1993.)

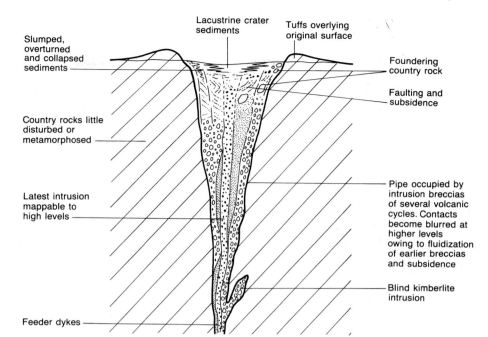

Slumped, overturned and collapsed sediments

Lacustrine crater sediments

Tuffs overlying original surface

Foundering country rock

Faulting and subsidence

Country rocks little disturbed or metamorphosed

Latest intrusion mappable to high levels

Pipe occupied by intrusion breccias of several volcanic cycles. Contacts become blurred at higher levels owing to fluidization of earlier breccias and subsidence

Blind kimberlite intrusion

Feeder dykes

Fig. D22 Diamond deposits. Schematic diagram of a kimberlite diatreme (pipe) and maar (volcanic crater below ground level and surrounded by a low tuff rim). The maar can be up to 2 km across. (After Evans, 1993.)

generally wider and shallower than those of **kimberlites**, and fertile **lamproites** appear to be the **silica**-saturated *orendites* and *madupites* which carry **sanidine** rather than **leucite**.

Kimberlites and **lamproites** are generally regarded as having been intruded upwards through a series of deep-seated tension **fractures**, often in areas of regional **doming** and **rifting**, in which the **magmas** started to consolidate as **dykes**. Then highly gas-charged **magma** broke through explosively to the surface at points of weakness, such as cross-cutting **fractures**, to form the explosion **vent** which was filled with fluidized fragmented **kimberlite** or **lamproite** and **xenoliths** of country rock.

Extremely high temperature and pressure are required to form **diamond** rather than **graphite** from pure **carbon** — of the order of 1000 K and 3.5 GPa, equivalent in areas of 60 km thick **continental crust** to a depth of about 117 km. As **coesite** rather than **stishovite** is found as inclusions in **diamonds**, and the inversion curve for these two **silica polymorphs** is equivalent to a depth of about 300 km, the approximate range for **diamond** genesis is 100–300 km.

For many decades there has been a very active debate as to whether **diamonds** crystallized from the **magmas** which cooled to form the igneous rocks in which they are now found (**phenocrysts**), or whether they were picked up by these **magmas** as exotic fragments derived from the **diamond** stability field within the **upper mantle** (**xenocrysts**). The much greater abundance of **diamonds** in **eclogite xenoliths** than in the surrounding **kimberlite** suggests that they have been derived from disaggregated **eclogite**. The former hypothesis infers that successful exploration consists of finding **kimberlites** and **lamproites** but the latter implies a more sophisticated approach. Furthermore,

in southern Africa, **kimberlites** erupted within the confines of the **Archean craton** are diamondiferous, while those in adjacent younger *orogenic belts* are barren; a general point which might be deduced on a global scale from Fig. D.21.

Most **kimberlites** and **lamproites** are younger than 200 Ma, but dates on **garnet** inclusions in **diamonds** from two southern African kimberlites (Kimberley and Finsch, 160 km apart and about 90 Ma old) are of greater age than 3000 Ma. It seems highly probable that the **diamond** hosts are of a similar great age.

This and other evidence is leading workers to the opinion that **diamonds** grew stably within the **upper mantle**, probably either in **eclogite** or **ultramafic rocks** or both. The estimated conditions of equilibration for ultramafic suite minerals co-existing with **diamonds** suggest that this growth takes, or took, place in a layer between about 132 and 208 km in depth beneath continents and 121–197 km beneath oceans, at temperatures of 1200–1600 K, provided that **carbon** is (or was) present. Thus, any **magma** that samples a diamondiferous zone of this layer may bring **diamonds** to the surface if it moves swiftly enough. The speed of ascent of such **magmas** has been calculated to be around 70 km h^{-1}. Slow ascent could allow time for the absorption of **diamonds** by transporting **magmas** as the pressure decreases, but there is some uncertainty over this.

The diamondiferous layer in the **upper mantle** is probably discontinuous because its formation requires the existence (and preservation) of a thick 'cold' **crust** above, otherwise **diamond** generating (and destroying) temperatures may be present at too shallow a depth. Absence of the layer may account for the absence of **diamonds** from

[171]

many **kimberlites**, but another important reason is that many **kimberlitic magmas** were developed at too shallow a level to be able to sample it. [AME]

Evans, A.M. (1993) *Ore Geology and Industrial Minerals: An Introduction* (3rd edn). Blackwell Scientific Publications, Oxford.

Nixon, P.H. (1980) The morphology and mineralogy of diamond pipes. In: Glover, J.E. & Groves, D.I. (eds) *Kimberlites and Diamonds*, pp. 32–47. Extension Service, University of Western Australia, Nedlands.

Rogers, N. & Hawkesworth, C. (1984) New date for diamonds. *Nature* **310**: 187–8.

diapir/diapirism Diapirism refers to the buoyant upwelling of volumes of rock surrounded by rock of higher density. Diapirs are the bodies of low density rock resulting from this upward movement and **deformation**. The low density rocks involved are most commonly **halite** and other **evaporites**, and **granitic igneous rocks**.

Salt (**halite**) diapirism, and that of other **evaporite** minerals such as **gypsum**, has been extensively studied because the **deformation** in the surrounding rocks often results in *hydrocarbon reservoir* formation. Density contrast is the controlling factor in diapiric activity; during **compaction** the overall density of the sediment pile increases from typically less than that of salt to greater, so that salt in a sedimentary sequence can become buoyant during burial. The morphology of **salt diapirs**, shown in Fig. D23, varies from undulations in the surface of sedimentary salt layers, through antiformal **salt domes** and pillars, to isolated pillows and balloon shapes. **Salt diapirs** are found in sedimentary **basins** such as the North Sea, and at

passive continental margins. **Evaporite** bodies can reach the surface and flow out as 'glaciers'. In experiments with analogue materials (summarized by Dixon, 1975), these forms make up a sequence as **flow** into the upper parts of the body increases. This **flow** involves high **strain**, which can be seen in exposed diapirs as contorted layering, **boudinage** and **flow folds**. Diapiric activity can follow the development of a **decollement** along an **evaporite** or **halite** layer, when **deformation** causes irregularity in the layer boundary. Salt behaviour is discussed extensively by Lerche & O'Brien (1986).

Sediments surrounding diapirs are also deformed, when the **evaporitic** rock is highly strained. **Anticlines** and **periclines** are found above diapirs, with **synclines** around the rim. Upward movement of the diapir can detach these **folds**, so that they can be truncated by salt. Local **unconformities** can also result. **Extensional faulting** is common above **salt domes** and above parts of a salt layer depleted by **flow** into an adjacent structure.

Diapiric activity is also important in the emplacement of **granitic** bodies in the **crust**, to form **batholiths** (see Atherton & Tarney, 1981) and **mantled gneiss domes**. In these cases the density contrast can occur both between different rock types and between **magma** and its surroundings. Diapiric emplacement of **igneous bodies** can result in concentric **foliation** development around the margins of the bodies, with **flattening strains** above the body. Diapiric emplacement associated with **faulting** may produce complex **strain** patterns in and around igneous intrusions. (See **granite greenstone terrane, salt diapir/salt dome**.) [SB]

Atherton, M.P. & Tarney, J. (1981) *The Origin of Granite Batholiths*. Geochemistry Group of the Mineralogical Society, Shiva, Kent.

Dixon, J.M. (1975) Finite strain and progressive deformation in models of diapiric structures. *Tectonophysics* **28**: 89–124.

Lerche, I. & O'Brien, J.J. (1986) *The Dynamical Geology of Salt and Related Structures*. Academic Press, London.

Taylor, J.C.M. (1990) Upper Permian — Zechstein. In: Glennie, K.W. (ed.) *Introduction to the Petroleum Geology of the North Sea* (3rd edn). Blackwell Scientific Publications, Oxford.

diaspore (AlO.OH) A component mineral of **bauxite**, the **ore** of **aluminium**. [DJV]

diastrophism Large-scale **tectonic deformation** of the Earth's **crust**; also the landforms and terrain created by these processes. Diastrophic processes include large-scale warping of the **crust** (**epeirogeny** and **cymatogeny**), mountain building (**orogeny**), and **faulting** and **folding** of crustal rocks. Linear and horizontal movements such as those involved in **continental drift**, and also vertical movements such as **uplift** and subsidence of the Earth's **crust**, are all considered diastrophic processes. An understanding of the mechanisms influencing diastrophism is, therefore, integral to the study of a wide range of **tectonic** phenomena. These range from the effects that the weight of large bodies of water or ice have on localized crustal movements, to geophysical forces in the Earth's **mantle**

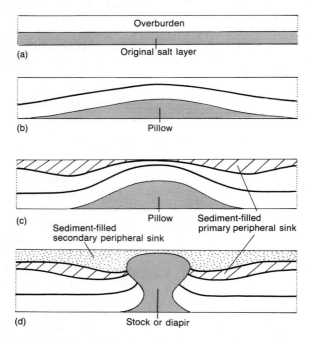

Fig. D23 (a)–(d) Sequential development of a salt **diapir** or stock and salt pillows from a sedimentary salt layer, with associated distortion of the surrounding sedimentary layers. (After Taylor, in Glennie, 1990.)

and **asthenosphere** which drive *plate* movements on a global scale. [AW]

diatomaceous earth An alternative name for **diatomite**. [DJV]

diatomite Fine-grained hydrated **silica** formed by the sinking from near-surface and accumulation on the sea floor of the siliceous tests of *diatoms*. [DJV]

diatreme A vertical **pipe** or upward flaring funnel-shaped volcanic intrusion composed of a chaotic breccia of blocks of country rock, a **magmatic** component and possibly **mantle**-derived **xenoliths** and **xenocrysts**. Most diatremes are 200 m–2 km wide and up to about 2 km deep. They pass downward into **dykes**. Diatremes were intruded forcefully as a mixture of mafic **magma**, **volcanic gases** and *accidental lithic* blocks and clasts. In the best-known examples, the mafic **magma** is **kimberlite**, and some **kimberlite** diatremes contain **xenocrysts** of **mantle**-derived **diamond** (Cox, 1978). The **volcanic gases** in **kimberlite** diatremes were **magmatic**, derived from the **mantle**. In other diatremes, the mafic **magma** is **basalt**, and the gases in these were probably mostly steam derived from **groundwater** (Lorenz, 1985).

Some diatremes preserve a stratigraphy of the country rocks they intrude, and some shallow diatremes contain bedded **pyroclastic** deposits (Francis, 1970). In others, blocks of country rock are chaotically arranged, some situated structurally above, others below, the horizon in the country rock from which they were derived. These relationships suggest that the interior of the diatremes experienced down-sagging or down-faulting during emplacement. The original volcanic surface expression of diatremes are therefore thought to have been collapse structures, possibly **maars**. (See **volcanic vent**.) [PTL]

Cox, K.G. (1978) Kimberlite pipes. *Scientific American* **238** (4): 120–32.
Francis, E.H. (1970) Bedding in Scottish (Fifeshire) tuff-pipes and its relevance to maars and calderas. *Bulletin Volcanologique* **34**: 697–712.
Lorenz, V. (1985) Maars and diatremes of phreatomagmatic origin: a review. *Transactions of the Geological Society of South Africa* **88**: 459–70.

dichroite Name given to the **gem** quality variety of the mineral **cordierite**. [DJV]

dickite ($Al_2Si_2O_3(OH)_4$) A **clay mineral**. [DJV]

differential stress The **stress** difference, in two dimensions, between the maximum and minimum **stress**. The term is used, for example, in experimental **deformation** where two of the *principal stresses* are equal, and greater or less than the third. [RGP]

diffraction The phenomenon of radial scattering of seismic energy at abrupt discontinuities in reflecting surfaces or structures whose radius of curvature is shorter than the wavelength of the incident waves. Diffraction is particularly common when **faults** cut reflecting horizons. Diffractions should be collapsed back to their point of diffraction if the correct **migration** has been applied. On a global scale direct *P* waves are diffracted around the **core** at **epicentral distances** greater than 104°. [GS]

diffusion in sediments The tendency of random (Brownian) movement of molecules and ions in **pore-waters** to act to reduce chemical gradients. The gradients are often expressed in terms of concentration, but strictly activity gradients or, if temperature changes occur, chemical potential gradients induce the diffusion. Studies of **pore-waters** of sediments undergoing **early diagenesis** commonly reveal strong concentration gradients reflecting relatively rapid chemical reactions. *Fick's first law of diffusion* is:

$$J_i = -D_i\, dC/dx$$

where J_i is the flux of component i (mass transported per unit area per unit time), dC/dx is the concentration gradient and D_i is the *diffusion coefficient* in an unrestricted environment. In sediments, diffusion is slowed because of the obstructions formed by sediment grains and the effective *diffusion coefficient* is approximately $\phi^2 D_i$ where ϕ is the fractional **porosity**. At near-surface temperatures diffusion is a more active agent of chemical transport than **pore-water** flow on the millimetre–centimetre scale and can act effectively over distances of up to tens of metres in the absence of lateral flow. *Diffusion coefficients* increase with temperature whereas **porosities** decrease on sediment burial. The net effect is to increase the relative importance of diffusion as a transporting agent during **burial diagenesis**. Diffusion is important along intercrystalline boundaries, especially during chemical compaction, but is difficult to model quantitatively. (See also **pore-waters and their evolution, pressure dissolution, Peclet number**.) [IJF]

Lerman, A. (1979) *Geochemical Processes. Water and Sediment Environments.* John Wiley & Sons, New York.

digenite (Cu_9S_5) A **copper ore** mineral. [DJV]

dilatancy The volume increase of a granular material when it is subjected to **stress** and approaches **failure**. It is caused by changes in the crack and pore space distribution in the rock. In seismological terms, the relevant **stress** is the long-term regional **stress** associated with tectonic forces; the **failure** is the sudden **stress** release which is an **earthquake**. Dilatancy is thus of importance to **earthquake prediction**. Dilatancy influences various measurable bulk physical properties of the rock:
— the ratio of *P wave* velocity to *S wave* velocity (because *S* velocity is so sensitive to the fluid content of the cracks)
— **electrical resistivity** (which too depends on fluid content) and associated magnetic properties

— ground level and water levels in **wells**

— acoustic emissions from *microearthquakes* as cracks propagate.

S wave splitting has been proposed as a means of observing dilatancy. Cracks with a common alignment will cause *S waves* to be 'split' into a fast wave polarized parallel to cracks and a slow wave polarized perpendicular to them. Recognizing and understanding systematic changes in one or more of these parameters could allow an incipient **earthquake** to be recognized. (See **earthquake prediction, seismic waves**.) [RAC]

Bolt, B.A. (1978) *Earthquakes — A Primer*. W.H. Freeman, San Francisco.

dilation (*dilatation*) A shape change or **strain** which involves a change in volume. The dilation is defined as:

$$\frac{\text{change in volume}}{\text{initial volume}}$$

This definition includes cases of volume decrease (negative dilation) as well as volume increase (positive dilation), which is the common English usage of the term. Changes in volume are included in many progressive **strains** and may be caused by **compaction**, **cleavage** development or opening and infilling of **fractures**. These processes modify the **finite strain** and must be taken into account in **strain** analysis (Ramsay & Wood, 1973). [SB]

Ramsay, J.G. & Wood, D.S. (1973) The geometrical effect of volume change during deformation processes. *Tectonophysics* **16**: 263–77.

dilation vein A **vein** formed by the infilling of a pre-existing open space developed in a **fault** or fissure. [AME]

diluvialism A form of **catastrophism** in which it is believed that the landscape was shaped by Noah's Flood, as reported in the book of Genesis. Before the true origin of *glacial* deposits was recognized, such materials, called 'drift' were ascribed to a great deluge, when 'waves of translation' covered the Earth. By the 1830s the recognition of the complex stratigraphy of the drift and the discovery of the importance of the **Ice Age** greatly weakened the diluvial viewpoint. [ASG]

Davies, G.L. (1969) *The Earth in Decay*. Macdonald, London.

dinoflagellates Microscopic, unicellular protists that produce organic-walled vesicles (often called '*dinocysts*'), generally from 30 to 60 µm across, of widespread importance in **biostratigraphy** of **Jurassic** to *Pleistocene* marine sediments. Dinoflagellates' motile cells are characterized by two flagellae, by distinctive photosynthetic pigments and nuclear structures and are placed by botanists in the Class Dinophyceae within the Division Pyrrophyta of the *Algae*. They are major elements of the phytoplankton and blooms

may lead to toxic 'red tides'; they may also act as endosymbionts in **corals** and **Foraminifera**.

Although the motile cells are not themselves preserved, they may produce resting cysts of a resistant sporopollenin-like material. These produce a fossil record which is highly selective because only certain cyst-forming taxa are represented. In the order Peridiniales, the architecture of each cyst reflects the highly ordered armour of plates (tabulation) around the living cell. As this tabulation varies distinctly between taxa, dinocysts can provide a useful **biostratigraphy**, especially in marlstones and **shales**. Their use has been further encouraged in recent exploration for *hydrocarbons* because they are little affected by modern drilling techniques, they are not susceptible to **dissolution** at depth and they provide a *colour index* of *thermal maturation* of the sediment. Like other *palynomorphs*, dinoflagellates are usually extracted from marlstones and **shales** with hydrofluoric acid, concentrated, bleached and mounted on glass slides for reflected light microscopy using oil immersion objectives.

Several different categories of cyst are recognized: those that developed in contact with the inner wall of the enclosing motile cell, with clearly reflected tabulation (proximate cysts); those that developed further within the cell, connected to the cell wall by spines indirectly reflecting tabulation (chorate cysts; proximochorate cysts are intermediate between these); non-spinose forms that developed close to the cell wall but contain large cavities between the two-layered cyst wall (cavate cysts). Escapement of the young motile stage takes place through an opening (archeopyle).

Although *dinocysts* with tabulation are suspected in **Silurian** rocks, the main radiation began with proximate forms in the late **Triassic** and early **Jurassic**, while proximochorate, chorate and cavate types appeared by middle **Jurassic** times. **Cretaceous** *dinocysts* provided a diverse assemblage of mainly chorate and proximochorate types, and cavate cysts appeared in the late **Cretaceous**. Dinoflagellate diversity has declined gradually since the *Albian*. [MDB]

Bolli, H.M., Saunders, J.B. & Perch-Nielsen, K. (eds) (1985) *Plankton Stratigraphy*. Cambridge University Press, Cambridge.
Sarjeant, W.S. (1974) *Fossil and Living Dinoflagellates*. Academic Press, London.
Tappan, H. (1982) *The Paleobiology of Plant Protists*. W.H. Freeman, San Francisco.

diopside ($CaMgSi_2O_6$) A white to light green *clinopyroxene* mineral. (See **pyroxene minerals**.) [DJV]

dioptase ($Cu_6(Si_6O_{18}) \cdot 6H_2O$) A ring silicate (*cyclosilicate*) mineral. [DJV]

diorite A medium to coarse-grained intrusive **igneous rock** composed of **plagioclase feldspar** (more Ab-rich than An_{50}) together with **quartz** (below 20%), **amphibole** and/

or **pyroxene**. Diorite therefore grades into **tonalite** with increase of **quartz**, and into *monzonite* with increase in **alkali feldspar** (over 10%). Diorite is thus equivalent to **andesite** in composition, and diorite intrusions may represent subvolcanic/plutonic intrusions formed in a similar **island arc/active** continental **margin** to those produced for **andesite lavas**.

Diorite was used in antiquity for the carving of Sumerian and Egyptian statues, bowls and vases, and is recorded as a building material, particularly in South Africa, Australia and Canada. [RST/OWT]

dip The inclination or tilt of a planar surface, e.g. *bedding* or **foliation**. The dip of a plane is measured as an angle from the horizontal to the plane, measured in the vertical plane perpendicular to the horizontal line (**strike**) in the plane. The direction of dip (i.e. the direction in which the plane dips downwards from the surface) is measured either directly as a compass bearing (azimuth) or in relation to the **strike** direction, which is 90° from the dip direction. Thus a **bed** may be said to **dip** at 30° SE, if the **strike** direction is not specified, or at 30°S to 110° if it is. A dip arrow is the conventional representation of dip direction on a **geological map**. The position of the observation is indicated by the arrowhead and the amount of dip in degrees is placed alongside the arrow. On many maps, this symbol is replaced by a line parallel to the **strike**, with a short tick indicating the dip direction. [RGP]

dip fault Relatively steep **fault** whose **strike** is parallel to the **dip** of the **bedding**. [RGP]

dip isogon Line joining points of equal **dip** in a folded multilayer system. Dip isogons were used by Ramsay (1967) to define the shape of a multilayer **fold** profile in a systematic way (see Fig. F28, p. 262). The arrangement of the dip isogons for different **dips** may be used to distinguish different *fold styles*. (See **fold**.) [RGP]

Ramsay, J.G. (1967) *Folding and Fracturing of Rocks*. McGraw-Hill, New York.

dip slope Topographic slope parallel to the **dip** of bedding. The amount of **dip** is usually less than the **bedding dip**. In areas of gently to moderately dipping strata, the topography may reflect the structure by alternations of dip slopes and **escarpments** or scarp slopes, parallel to the **strike**. [RGP]

dip-slip Movement parallel to the direction of **dip** of an inclined **fault** or other planar surface. Thus a dip-slip **fault** is that class of **fault** categorized by displacement parallel to the **dip** rather than parallel to the **strike**, as in a **strike-slip fault**. (See **fault**.) [RGP]

dipole field The **magnetic field** due to two oppositely polarized **magnetic poles** of the same strength. In most instances this refers to the magnetization of individual grains, an **orebody**, or the **geomagnetic field** itself which predominantly consists of a single inclined **geocentric dipole**. [DHT]

direct seismic wave The wave that travels the most direct route from a **seismic source** to a detector. The direct wave travels through the body of the material and not along its surface. [GS]

direct-shipping ore (*lump ore*) An **ore**, e.g. **iron ore**, that can be transported from a mine direct to a smelter without the necessary for **benefication**. A minor amount of crushing or grinding may be required. [AME]

dirt band A thin bed of **shale** or other inorganic rock in a **coal seam**. [AME]

discharge of a river The volume of water passing a given point on the river course in a specified time interval. It is the cross-sectional area normal to the flow multiplied by the water velocity. [AME]

discharge of groundwater The natural outflow of water from **aquifers**. The natural flow of **groundwater** is from **recharge areas** on high ground to low-lying discharge areas where it usually finds its way into river systems. Water from **coastal aquifers** may, however, discharge direct into the sea. Water will flow from an **unconfined aquifer** wherever the ground surface intersects the **water table** to form a **seepage** or **spring**. Water will flow from a *confined aquifer* (**artesian aquifer**) when the **potentiometric surface** is above the ground surface and where there is some form of permeable channelway through the overlying **aquiclude**. [AME]

Price, M. (1985) *Introducing Groundwater*. Allen & Unwin, London.

dish structure Slightly concave-upwards structure highlighted by a thin clay-rich layer (0.2 to 2 mm thick) in a background of **sandstone**, produced by the upward escape of **pore fluid**. The three-dimensional shape is of shallow dishes a few centimetres across. Dish structures are often associated with *pillar structures*. These structures are also due to water escape, but characteristically penetrate essentially flat laminae lacking well-developed dishes. (See **wet sediment deformation**.) [PAA]

disharmonic folds Folds in a layered sequence in which the amplitude, wavelength or class of **fold**, i.e. the **fold**

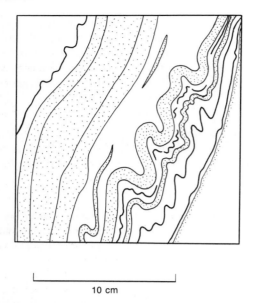

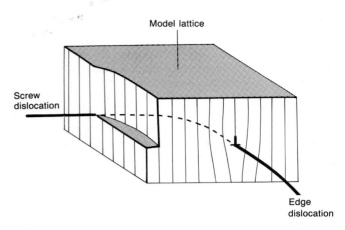

Fig. D25 Sketch of an edge, screw, and mixed **dislocation** formed by the presence of a partial half-plane within a crystal lattice.

Fig. D24 Line drawing of **disharmonic folds** in galena layers from an orebody. Notice the variation in fold style in layers of different thickness. (After Hobbs *et al.* 1976.)

style, changes from layer to layer. Thus **decollement** surfaces are a part of any such folded sequence, running parallel to layering. Disharmonic folds appear to develop because of differing **rheology** in the different layers. *Buckle folds*, in which the fold wavelength depends on individual layer thicknesses, are frequently disharmonic. (See Fig. D24.) [SB]

Hobbs, B.E., Means, W.D. & Williams, P.F. (1976) *An Outline of Structural Geology*. John Wiley & Sons, New York.

dislocation A linear or curvilinear discontinuity such as the termination of a half-plane within a crystal lattice. Some dislocations originate as *stacking faults* (see **crystal defects**) i.e. imperfections in the arrangement of atoms during crystal growth. The term dislocation is also used in a more general sense to refer to a surface across which there is a loss of continuity, and often displacement within rock. **Fractures** and **faults** can be considered as planes of dislocation.

In many minerals such as **olivine** or **quartz**, higher densities of dislocations are associated with crystal **plastic deformation**. The distortion achieved by a dislocation depends on its geometry; most are a combination of the *edge* and *screw dislocations* shown in Fig. D25.

Two important **deformation mechanisms** involving dislocations are *dislocation glide* and *dislocation climb*. *Dislocation glide* is a process of **deformation** occurring by formation and movement of dislocations within a crystal lattice, such that each dislocation remains within its own lattice plane. It is seen in metals and alloys in low temperature deformation, giving rise to the name *'cold working'*. This **deformation mechanism** occurs by continual breaking and

re-formation of bonds, and is common in low temperature **deformation** (e.g. at less than half of the melting temperature) in **quartz** and **calcite**. Other dislocations and **crystal defects** impede the movement of a dislocation in its own lattice plane ('tangling') so that **deformation** becomes increasingly more difficult, and the material is said to *strain harden*.

Dislocation climb is a process of **deformation** occurring by the formation and movement of dislocations in a crystal lattice such that dislocations can move out of their original lattice planes. The process occurs in the **deformation** of metals and alloys at elevated temperatures, when it is known as *'hot-working'*. The movement of dislocations out of their planes (by *'cross-slip'*) distinguishes this mechanism from *dislocation glide*. In *dislocation climb* **deformation**, dislocations avoid interaction with crystal lattice defects, and do not tangle to produce *strain hardening*. **Recovery**, by reordering and annihilation of dislocations, can lead to recrystallization during **deformation**, and allows the accumulation of high **strains**. This **deformation mechanism** can occur during *dynamic recrystallization* in the formation of **mylonites**. It is seen in minerals such as **quartz**, **calcite**, **olivine** and **plagioclase**, at temperatures greater than about half the mineral melting temperature. [SB]

Friedel, J. (1964) *Dislocations*. Pergamon, London.

dispersion The dependence of propagation velocity on **seismic wave** frequency. *Body waves* travel with a specific velocity regardless of their frequency, but *surface waves* exhibit dispersion. They sample to a depth proportional to their wavelength and, since velocity increases with depth, lower-frequency (longer-wavelength) surface waves travel faster than higher-frequency ones. An example of a dispersed wave is shown with **earthquake magnitude**. Dispersion is measured by two velocities: the *'phase velocity'* (that of a specific crest or trough) and the *'group velocity'* (that of the energy in the signal). Their variation with frequency can be used to determine the velocity–depth distribution that causes it. (See **seismic waves** and Fig. E4, p. 196) [RAC]

Bullen, K.E. & Bolt, B.A. (1985) *An Introduction to the Theory of Seismology* (4th edn). Cambridge University Press, Cambridge.

dispersive pressure Term introduced by Bagnold (1954) for the fluid condition where grains are supported above a bed in a dispersed state due to collisions between grains and/or grain interactions through near misses, resulting in a viscous force with a strong vertical component. To maintain such a 'dispersive pressure', the vertical component of *turbulence* (root-mean-square of the instantaneous velocity) must equal or exceed the **settling velocity** of the grains. The effects of dispersive pressure, or the effective normal **stress**, in a fluid are particularly significant at high sediment concentrations although relevant down to as little as 9% by volume. Grain support and transport is considerably enhanced by dispersive pressure in the traction-carpet or inertial regime at the base of a flow. Dispersive pressure equals **shear stress**/tan ϕ where ϕ is the angle of internal friction. (See **sediment gravity flow**.) [PAA]

Bagnold, R.A. (1954) Experiments on a gravity-free dispersion of large solid spheres in a Newtonian fluid under shear. *Proceedings of the Royal Society, London* **A225**: 49–63.

displacement The distance over which one body of rock has moved relative to another across a line or plane, e.g. of a **fault**. [RGP]

displacement pressure In oil geology the replacement of one fluid by another. For example, most **reservoir** rocks contained water before *hydrocarbons* entered them. The *hydrocarbons* preferentially occupy the coarser pore spaces driving the water into finer spaces where it is held by capillary forces and from which it may not be displaced. To move from the *source-rock* pore spaces into the **reservoir** pore spaces the capillary pressure of the oil must exceed the displacement pressure of the water in the **reservoir**-rock pore space. The displacement pressure is defined as the smallest capillary pressure necessary to force *hydrocarbons* into the largest interconnecting pores of a water-wet rock. [AME]

North, F.K. (1985) *Petroleum Geology*. Allen & Unwin, Boston.

disseminated deposit Deposit in which **ore** minerals are peppered throughout the body of the host rock in the same way as **accessory minerals** are disseminated through an **igneous rock**; in fact, they often *are* **accessory minerals**. A good example is that of **diamonds** in **kimberlites**; another is that of some **orthomagmatic nickel–copper** deposits such as the La Perouse Layered Gabbro, Alaska, which contains disseminated sulphide mineralization throughout its entire thickness of about 6 km. This deposit

has over 100 Mt grading about 0.5% **nickel** and 0.3% **copper**. [AME]

disseminated gold deposit A form of **disseminated deposit** in which **gold** is finely disseminated and usually invisible as the grain size is normally less than 1 μm. The range of deposit subtypes is considerable. The four most important are: (1) **Carlin-type** (p. 94), (2) disseminated and **stock-work gold–silver** deposits in intrusive igneous rocks (sometimes called *porphyry gold deposits*). (3) disseminated **gold–silver** occurrences in volcanic flows and associated volcaniclastic rocks and (4) disseminated deposits in tuffaceous rocks and *iron formations*.

Subtype 2 occurs in highly fractured zones of irregular outline which have been healed by veinlets, **veins** and stringers of auriferous **quartz**. These zones are marked by considerable **hydrothermal alteration** of the host rock. Deposits of this subtype occur in *orogenic belts* in all the continents and range from **Archean** to **Phanerozoic** in age.

Orebodies of subtype 3 are not very widely exploited as they have only become important with the recent rise in the **gold** price. The **orebodies** occur in large diffuse volumes of **alteration** in **rhyolites**, **andesites** or **basalts**.

Subtype 4 is common in the **greenstone belts** of the *Precambrian* as disseminations in sheared and highly altered **tuffs** and various facies of **banded iron formation** (BIF) of which sulphide-bearing carbonate and silicate facies are the most important. The Homestake Mine, the biggest **gold** mine in the USA, is a good example of a BIF-hosted deposit. [AME]

Boyle, R.W. (1979) The geochemistry of gold and its deposits. *Geological Survey of Canada Bulletin* **280**.

Evans, A.M. (1993) *Ore Geology and Industrial Minerals: An Introduction* (3rd edn). Blackwell Scientific Publications, Oxford.

disseminated or porphyry molybdenum deposit Deposit with many features in common with a **porphyry copper deposit**. Average **grades** are 0.1–0.45 MoS₂. Host intrusions vary from *quartz monzodiorite* through **granodiorite** to **granite**. **Stockwork** mineralization is more important than disseminated mineralization and the **orebodies** are associated with simple, multiple or composite intrusions or with **dykes** or breccia **pipes**. There are three general **orebody** morphologies (Fig. D26) and tonnages range from 50 to 1500 Mt. The **molybdenite** occurs in (i) **quartz** veinlets carrying minor amounts of other sulphides, oxides and **gangue**, (ii) fissure **veins**, (iii) fine **fractures** containing molybdenite paint, (iv) breccia matrices and, more rarely, (v) disseminated grains. **Supergene enrichment**, which can be very important in **porphyry copper deposits** is generally absent or minor.

Disseminated molybdenums have been divided into Climax and *quartz monzonite* types. The Climax types generally have high trace or accessory contents of **tin** and **tungsten**, intense **silicification** associated with their **wall rock alteration**, and are low in **copper** compared with

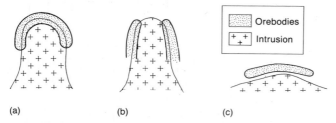

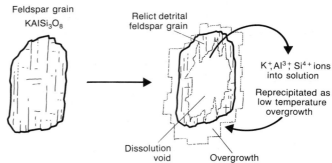

Fig. D26 Disseminated molybdenum deposit morphologies. (a) Inverted cup. The ore zone is in both host intrusion and country rocks, e.g. Climax, Colorado and Henderson, Colorado. (b) Hollow cylinder. The ore zone is in both host intrusion and country rocks, e.g. Pine Grove, Utah; Mount Emmons, Colorado and Kitsault, British Columbia. (c) Tabular or inverted bowl. The ore zone is generally in country rocks only, e.g. Mt Tolman, Washington and Endako, British Columbia. (After Evans, 1993.)

Fig. D27 Dissolution may be congruent or incongruent but ions enter into solution. Ions may be lost to the system (open) or reprecipitated (closed), e.g. dissolution of detrital potassium feldspar.

the other type which lies in the transition zone to **molybdenum**-bearing **porphyry coppers**. The **alteration** patterns are very similar to those found in **porphyry copper deposits**, with potassic **alteration** and **silicification** being predominant. The most detailed study is on the Urad and Henderson deposits (Colorado) where, associated with the Henderson **orebody**, there is a central *potassic zone* carrying secondary **potash feldspar** and **biotite**. Succeeding this are **quartz-topaz**, phyllic, argillic and **propylitic** zones. There is a silicified zone which lies largely within the *potassic zone*. The Henderson **orebody** is roughly coincident with the *potassic* and silicified *zones*.

The close spatial association of the **orebodies** of most deposits with the *potassic zone* of **alteration** suggests a **magmatic** source for the mineralization. [AME]

Edwards, R. & Atkinson, K. (1986) *Ore Deposit Geology*. Chapman & Hall, London.

Evans, A.M. (1993) *Ore Geology and Industrial Minerals: An Introduction* (3rd edn). Blackwell Scientific Publications, Oxford.

dissipative beach A **beach** of low gradient with a **bar** or **bars** which absorb and dissipate much of the energy of incoming **waves**. The dissipative beach is usually characteristic of a fully **erosional** stage of **beach** development achieved under high-energy **waves** which reduce the **beach** gradient and maximize **bar** development. The other extreme is the *reflective beach* which has a steep profile with a well-developed **berm** or **beach ridge**. This results from constructive **waves** building up a steep foreshore from which a large proportion of the incoming **wave** energy is reflected. Intermediate stages are characterized by rapid **beach** changes. [JDH]

Davies, J.L. (1980) *Geographical Variation in Coastal Development* (2nd edn). Longman, London.

dissolution The **diagenetic** process by which a solid phase is dissolved by an aqueous **pore solution** leaving behind a space or cavity within the host sediment (see **porosity**). If all the solid phase is gradually dissolved bit by bit (atom by atom, ion by ion), leaving behind a fresh surface of undissolved solid of the same composition as the original solid phase, then the dissolution is congruent. Such dissolution is typical of ionic salts (e.g. NaCl) and most silicates (**quartz**, **feldspar**) and is, therefore, probably dominant in natural clastic systems (Figs D27 and D28). Alternatively, if the dissolution is selective to individual ionic or molecular components of the solid phase, whereby the solid phase left undissolved is different in composition from the original solid phase (either because of selective leaching or because the ratio of dissolved components is different from that of the original solid), then the dissolution is referred to as incongruent. Such dissolution is typical of many **carbonate minerals**.

Dissolution reactions are either controlled by the rate of transport of ions to and away from the reacting surface, or by the reaction rate at the solid–solution interface. In the former case, transport controlled dissolution is characterized by rapid, non-specific **corrosion** and is typical of fast dissolution by strongly concentrated solutions or of dissolution of highly soluble minerals. In the second case, by contrast, surface reaction controlled dissolution is relatively slow and more specific, being characteristic of slow

Fig. D28 Dissolution. Partially dissolved detrital potassium feldspar grain with authigenic overgrowths and relict feldspar lamellae as revealed by scanning electron microscopy.

dissolution of relatively insoluble minerals in dilute solutions of low chemical reactivity. Surface reaction controlled dissolution is probably typical of many **diagenetic** dissolution reactions and tends preferentially to corrode sites of excess free-surface energy (surface **dislocations**, **abraded** grain surfaces, **fracture** surfaces) and produces distinct crystallographically controlled etch features such as well-defined pits or notches (see **precipitation**).

Dissolution occurs in **aquifers** below **recharge areas** by the dissolving of rock salts by **groundwater**. **Precipitation** infiltrating *soil* dissolves CO_2 from the *soil* atmosphere forming weak carbonic acid which is able to dissolve calcium carbonate from underlying rocks. Dissolution of nitrates may occur in the *soil* and of any sulphates present in the bedrock. If the **aquifer** dips beneath an **aquiclude** these conditions may continue for some way down **dip** so increasing the content of dissolved salts. Further down **dip** dissolution gives way to ion exchange with dissolved ions being *adsorbed* on to **clay minerals**. [SDB/AME]

Berner, R.A. (1978) Rate control of mineral dissolution under earth surface conditions. *American Journal of Science* **278**: 1235–52.
Price, M. (1985) *Introducing Groundwater*. Allen & Unwin, London.

distribution/service reservoir A large tank situated at a high point in the area it is to serve, which provides a sufficient **head of water** for flow by **gravity** through the distribution mains to the consumers. [AME]

diurnal variation Daily variation in the **geomagnetic field** affecting all the **geomagnetic elements**. Quiet variation of up to a few tens of nT originates from the varying pattern in the movement of charged particles in the ionosphere. Extreme variation of high amplitude and frequency (known as a *magnetic storm*) results from the arrival of charged particles from sunspot activity. A magnetic survey should be curtailed during a *magnetic storm* as the diurnal correction cannot be accurately computed. [PK]

divergent erosion The difference between **erosion** in low latitudes where, according to Bremer, chemical **weathering** is most effective on substantial *planation surfaces* and weakest on steeper slopes, and mid-latitudes, where **erosion** is weakest on horizontal surfaces and most effective on slopes. [ASG]

Bremer, H. (1971) Flüsse, Flachen- und Stufenbildung in den feuchten Tropen. *Würzburger Geographische Arbeiten* **35**.

diving seismic wave A seismic wave that has been progressively refracted back to the surface by a strong velocity gradient, rather than a **discontinuity** as for a **head wave**. It has a curved ray path and, provided that the velocity in the Earth increases with depth, variation of **arrival time** with distance can be used to derive the velocity versus depth distribution using the *Weichert-Herglotz inversion* technique. [GS]

dolerite A fine- or medium-grained mafic **igneous rock** which typically forms a minor intrusion such as a **dyke** or **sill**. In terms of mineralogy and chemical composition, dolerite is equivalent to **basalt**.

Dolerite is sufficiently fine-grained for stone tool manufacture and was used for this purpose as early as the Lower Paleolithic in Africa. Neolithic polished **axes** were made of dolerite in Europe, commonly in Finland and in Britanny where a dolerite **axe factory** exists, and less commonly in Britain, where the Whin Sill was exploited for **axe** manufacture. Later, the Romans incorporated parts of the quartz dolerite Whin Sill into Hadrian's Wall, and also made milestones of North Wales dolerite. Perhaps the most famous archeological use of dolerites is in the **'Bluestones'** of Stonehenge, some of which are of distinctive 'spotted' dolerite believed to be from the Preseli Hills. The same type of rock was used for less famous prehistoric circles in South Wales. [RST/OWT]

doline A simple closed depression, usually circular or oval in plan, found in **karst** terrain. The English terms **swallet** and **swallow hole**, and the American term **sinkhole**, are also applied to such depressions. Although dolines are relatively simple features, they vary greatly in size and origin. Jennings (1985) recognizes five types of doline, i.e. collapse doline solution doline, subsidence doline, subjacent **karst** collapse doline, and alluvial streamsink doline. Processes leading to the formation of dolines include solution at or near the surface, **cave** collapse, **piping** and subsidence. In most cases a combination of processes is responsible. Dolines may form in bare or *soil*-covered **limestone** or in other rocks overlying **limestone** formations, such as those found in South Wales forming in conglomeratic **sandstone**. Dolines play an important role in **karst hydrology**, acting as foci for the transmission of water from the surface underground. [HAV]

Gunn, J. (1981) Hydrological processes in karst depressions. *Zeitschrift für Geomorphologie* **25**: 313–31.
Jennings, J.N. (1985) *Karst Geomorphology*, pp. 106–20. Basil Blackwell, Oxford.

dolomite ($CaMg(CO_3)_2$) The **carbonate mineral** found in magnesian **limestones** and in the rock 'dolomite'. [DJV]

dolomitization Formation of the mineral **dolomite** or rock *dolostone*, usually considered to be by replacement of a calcium carbonate precursor. **Dolomite** is a complex mineral species displaying much variability in cation composition and ordering. Ideal **dolomite** has alternating layers of calcium and magnesium ions, but most ancient examples and nearly all modern ones have excess calcium. *Holocene* near-surface **dolomite** is also poorly-ordered (magnesium appears in calcium layers, weakening the intensity of superstructure reflections seen by **X-ray diffraction**) with irregular microstructures seen by *Transmission Electron Microscopy* (*TEM*). Most **dolomites** display in *TEM* a 50–200 Å modulation whose origin is unknown. *Saddle dolomite* is coarsely-crystalline and calcium-rich with pronounced lattice curvature and is attributed to elevated temperatures (>50–100°C).

The usual reaction proposed for dolomitization is an exchange of calcium for magnesium ions with conservation of carbonate. However, study of **pore-water** chemistry in **Deep Sea Drilling Project** sites, where dolomitization is related to *bacterial* reaction occurring in **early diagenesis**, and coexistence of **dolomite** and unaltered **aragonite** in **sabkhas** and tidal flats indicates that partially or wholly cementing origins may be applicable for dolomite in these circumstances. In the **Coorong Region**, primary **dolomite precipitation** as yoghurt-textured mud occurs in ephemeral lakes. Nevertheless, many *dolostones* contain evidence of replacement of **allochems**, including mimetic types like the micron-sized crystals replacing coralline *algae* in *Tertiary* carbonates of The Netherlands Antilles. The chemical reaction responsible for most ancient *dolomicrites* remains obscure.

The paradox of dolomitization is that **dolomite** is thermodynamically favoured in most natural waters, yet it only forms exceptionally in *Holocene* sediments and is also uncommon in *Tertiary* sediments. This relates to kinetic factors: solutions which are highly supersaturated for **dolomite** are generally also supersaturated for **aragonite** and magnesian **calcite**, but only the latter minerals can precipitate rapidly. Solutions weakly supersaturated for **dolomite** require long time periods for complete dolomitization, but rapidly fluctuating *Neogene* sea-levels have hampered such dolomitization in shallow-marine settings.

Many thermodynamic and kinetic factors are known to retard or accelerate **dolomite** formation, but it is unclear whether there are only a few chemically distinct models of dolomitization, or a complete spectrum of dolomitizing solutions influenced to some extent by several of these factors. Higher magnesium:calcium ratios increase the thermodynamic drive for dolomitization, as do increased temperatures. The temperature effect is vital for epigenetic dolomitization. Primary **dolomite** and other magnesium-bearing carbonates **precipitating** in ephemeral lakes in the **Coorong Region** form from solutions of high magnesium: calcium and high **pH**. Magnesium ions are less hydrated at high **pH** which aids magnesium-bearing carbonate **precipitation**; another factor may be a relatively high anion: cation ratio. Production of carbonate alkalinity related to *bacterially*-secreted enzymes has also been suggested as a kinetic factor. Experiments at elevated temperatures indicate that dissolved sulphate retards dolomitization of $CaCO_3$, but field observations do not support this as an overriding control. The common association of **dolomite** with reducing conditions could be attributed to decomposition of amino acids which have also been shown experi-

mentally to retard dolomitization. Whereas some authors regard **dolomite** as an **evaporite** mineral and hence link its formation to high salinities, others postulate that low-salinity conditions (with no lowering of magnesium–calcium) promote dolomitization. Salinity has often mistakenly been used instead of degree of supersaturation, and salinity *per se* is probably largely irrelevant. Assessing the *paleosalinities* of ancient **dolomites** is made difficult by the observation that many, but not all of them, have undergone oxygen isotope exchange, presumably by re-crystallization (see **oxygen isotopes in sedimentology**).

A number of different spatial settings for dolomitization have been proposed or demonstrated (Fig. D29). One general constraint for any such model is that the rate of supply of magnesium by **diffusion** or **advection** must be adequate for the time-scale of dolomitization: usually sea-water is directly or indirectly the source of this magnesium. Elements of the Coorong model may apply to ancient coastal *dolostones* and one close analogue from the Middle **Proterozoic** of northern Australia has been described. **Dolomite** formation in peritidal carbonates by normal seawater aided by tidal pumping is now known from Florida. Although the **evaporitic** association of **sabkha dolomite** from the **Persian Gulf** and elsewhere is obvious, and ancient **sabkha** cycles have frequently been recognized, the controls on **dolomite** distribution are still controversial. Dolomitization by **brines** refluxing from *salinas* or *lagoons* has had a chequered history as a hypothesis; it works well on a small scale, but large-scale modern analogues are lacking. The same criticism applies to dolomitization from modern meteoric-marine mixing zones; also many mixing zones are **corrosive** to all **carbonate minerals**. The model is feasible, however, given that mixed waters like cold seawater (see below) may allow simultaneous $CaCO_3$ **dissolution** and **dolomite precipitation**. **Dolomite** formation in organic-rich marine sediments during **early diagenesis** can be extensive; a link with *bacterial* reactions and magnesium supply by **diffusion** is established, but it is still unclear why **calcite** forms in some organic-rich sediments and **dolomite** in others. Dolomitization of *reef* **limestones** beneath **atolls** and dolomitization of peri-platform **ooze** on carbonate slopes by cold seawater has recently been demonstrated with an inferred **convective** flow mechanism. In the burial setting, slow **advective** flow especially if focused, or thermal **convection**, provide the mechanism for movement of magnesium (see **subsurface fluid migration**).

The assessment of the effectiveness of the various mechanisms capable of affecting shallow marine carbonates is very difficult even in modern settings because of the lack of sufficiently stable conditions. A possible positive covariation of **dolomite** abundance and sea-level also has to be considered. The question remains as to whether ordinary seawater (albeit slightly diluted or evaporated) is a sufficiently potent dolomitizing agent given sufficient time, or whether ancient *dolostones* on the whole require special chemical conditions. [IJF]

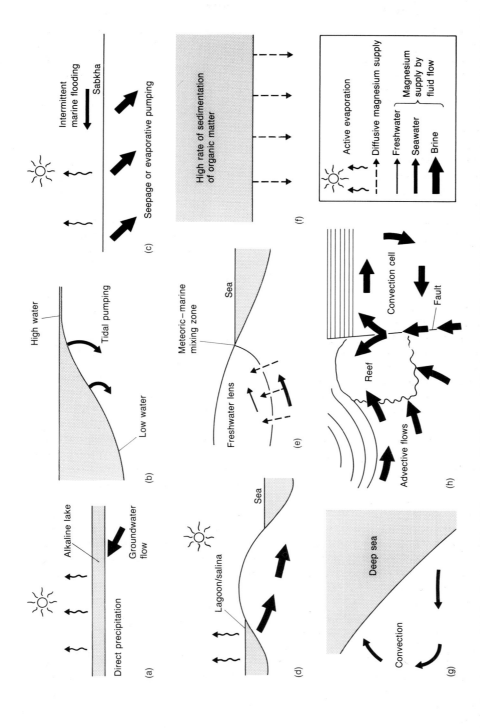

Fig. D29 Comparison of some spatial **dolomitization** models. (a) Coorong model of direct precipitation in ephemeral lakes. (b) Peritidal model, Florida Keys, shown at high tide. (c) Sabkha. (d) Seepage reflux. (e) Mixing-zone. (f) Dolomite in deep-sea/lagoonal organic-rich sediments. (g) Dolomitization of atolls or carbonate slope sediments. (h) Subsurface with advective flow induced by gravity (compactional or hydrostatic head), or rising of deep basinal fluids along faults, or convection.

Given, R.K. & Wilkinson, B.H. (1987) Dolomite abundance and stratigraphic age: constraints on rates and mechanisms of Phanerozoic dolostone formation. *Journal of Sedimentary Petrology* **57**: 1068–78.

Hardie, L.A. (1987) Dolomitization: a critical view of some current views. *Journal of Sedimentary Petrology* **57**: 166–83.

Machel, H. & Mountjoy, E.W. (1986) Chemistry and environments of dolomitization — a reappraisal. *Earth Science Reviews* **23**: 175–222.

Zenger, D.H., Dunham, J.B. & Ethington, R.L. (1980) *Concepts and Models of Dolomitization*. Special Publication of the Society of Economic Palaeontologists and Mineralogists **28**.

domain Part of a larger area or volume that is considered to be more homogeneous structurally than the whole. In dealing with complex areas, it is often convenient to subdivide them into domains or subareas whose structure can be understood more easily. The term is used by Turner & Weiss (1963) for statistically homogeneous **fabric** subareas in a heterogeneous whole. Domains may be at scales varying from the large map area (**macroscopic**) to the **microscopic**. In crystallographic study, domains with differing structural characteristics are recognized, both in a crystalline aggregate on the scale of a thin section, and within individual crystals. [RGP]

Turner, F.J. & Weiss, L.E. (1963) *Structural Analysis of Metamorphic Tectonites*. McGraw-Hill, New York.

dome (1) A volcanic feature, formed by the accumulation of extruded **magma** above a **volcanic vent**. Domes are steep-sided and normally approximately circular in plan. They consist of viscous **magma**, unable to move away from the vent as **lava flows**. They consist typically of a ragged, essentially flat-topped, central portion, surrounded by **talus** aprons. They are commonly surmounted by small **craters**, formed either by explosions or by sagging of **magma** back down the vent. Domes range in size up to about 1.5 km across and 800 m high. Collapse of unstable parts of growing domes, under the influence of **gravity**, may generate **ash flows**. Domes may be formed in single **volcanic eruptions**, or from repeated eruptions from the same vent. Domes have been called endogenous when new **magma** is intruded into the core of the dome, or exogenous when new **magma** is mostly extruded on top of the older parts of the dome. Most domes are **rhyolitic**, **dacitic** or, rarely, **trachytic** in composition, i.e. **magmas** which have a high **viscosity**.
(2) An **antiformal fold** structure characterized by a circular or near-circular outcrop pattern. (See **pericline**.) [PTL/RGP]

Macdonald, G.A. (1972) *Volcanoes*. Prentice-Hall, Englewood Cliffs, NJ.

dome dune Circular or elliptical **dune** with a poorly developed or absent *slip-face*. Dome dunes are commonly found in desert areas that experience particularly strong winds, which restricts the normal upward growth of the **dune** crests. Sizes range from 0.1 to 1 km across. [RDS]

donga A *gully* or **arroyo** — especially those found in southern Africa. The term derives from the Zulu *udonga*, 'a wall', which describes the nearly vertical sides of the small ravines. Southern African dongas reach depths of 30 m or more and may be up to 3.0 km long. Individual dongas may have a linear or dendritic morphology and, in some cases, their walls are characteristically fluted. They are formed by the **erosion** of unconsolidated or weakly **cohesive** surficial materials by surface **runoff** and subsurface **piping**. In some areas, dongas are recent **erosional** features; some large dongas have formed in a few decades as a result of bush clearance and overgrazing which have disturbed the protective vegetation cover. Elsewhere, the examination of maps dating from the 18th and 19th centuries suggests that some dongas predate the arrival of agriculturalists and are now essentially stable landforms. [AW]

down-plunge projection/view A reconstructed profile of a **fold** structure drawn perpendicular to the **plunge** of the *fold axis*. [RGP]

downthrow In a **dip-slip fault**, one side has been displaced downwards relative to the other, and is termed the downthrown side or block. [RGP]

draa Large-scale or mega-**dune** formed by a regional wind pattern. The name is the arabic word for 'arm'. Draa, like **dunes**, can be divided into longitudinal and transverse forms. Also known as *complex dunes* and *'uruq*. Often regularly spaced with **dunes** migrating over the top of them. [RDS]

drag fold Term used in the past to describe **minor** or *parasitic folds* thought to have formed by **shear** in an incompetent layer sandwiched between two **competent** layers folded by flexural **slip**. Whitten (1966) points out the disadvantage of this genetic use of the term when the origin of the **folds** so described is not known. The term is, however, also used for **folds** produced in association with **fault** movement. (See **fault drag**.) [RGP]

Whitten, E.H.T. (1966) *Structural Geology of Folded Rocks*. Rand McNally, Chicago.

drag force The force exerted by a fluid upon a surface that acts in a direction parallel to the flow itself. This force is instrumental in determining the settling behaviour of particles and together with the **lift force** is critical in controlling the initiation of sediment movement. If the mean bed **shear stress** is given as τ, then the mean drag force per grain on a bed, F_D, is given by

$$F_D = \tau/n$$

where n is the number of grains per unit area. A common expression of the drag force is as a dimensionless parameter termed the *drag coefficient* which is a proportionality coefficient relating the **shear stress** to the kinetic energy of a fluid. The *drag coefficient*, C_D, may be given in its general form by

$$C_D = F_D / \tfrac{1}{2} \rho U^2 . d$$

where ρ is the fluid density, U the mean flow velocity and d a measure of grain size (e.g. cross-sectional area). The dimensionless *drag coefficient* represents a parameter which can be used in comparing the drag force between objects of differing size and in different velocity flows. An example of the application of the *drag coefficient* is given when interpreting the settling behaviour of grains. If C_D is plotted against particle **Reynolds number** for **quartz** spheres settling in water a distinctive curve is produced (Fig. D30). At low **Reynolds numbers** the *drag coefficient* is directly proportional to the **Reynolds number** and the behaviour of the settling particle may be given by *Stokes' Law*. However, at **Reynolds numbers** of approximately 24, **flow separation** begins to occur in the lee of the settling grain and produces higher values of drag and resistance than would be expected from *Stokes' Law*. This trend continues for a wide range of **Reynolds numbers** until approximately 10^6, where a dramatic drop in the *drag coefficient* corresponds to the onset of **turbulence** in the boundary layer before flow separation occurs. [JLB]

Middleton, G.V. & Southard, J.B. (1984) *Mechanics of Sediment Transport* (2nd edn). Short Course No. 3 Lecture Notes. Society of Economic Paleontologists and Mineralogists, Tulsa, OK.

drainage coefficient The amount of **runoff** drained from a given area in 24 h. [AME]

drainage density An essential morphometric characteristic of a drainage **basin** which describes the average length of stream channel per unit area. Drainage density represents the degree of fluvial dissection present within the **basin**. Values vary widely from permeable **limestones** where the density may be as low as 4 miles mile^{-2} (equivalent to 2 km km^{-2}) to more than a thousand miles mile^{-2} of **basin** in **badland** areas (above 600 km km^{-2}). Drainage density may be difficult to calculate from map data because of uncertainty over the consistency of the 'blue lines' included on a map of given scale. Use of contour crenulations may describe the total fluvial dissection of the **basin** but may be unrepresentative of current drainage conditions. Also, the network is temporally variable depending upon how wet the **catchment** *soils* are. Lithology and climate provide the main controls of drainage density. Network density may also increase with time as headwater tributaries gradually extend into undrained areas. [TPB]

drainage network The hierarchical system of channel links existing within the drainage **basin**. Two aspects have been studied in particular: the topological properties of the network and the **drainage density**. The topological hierarchy consists of links, both external (headwater tributaries) and internal. Various numerical methods can be used to determine the magnitude and order of each link (see **stream order**). The density of channels within the **basin** represents a balance between channel head stream **erosion** and depositional infill by slope processes, and thus a compromise between energy losses of flow over slopes and in channels. Identification of the network is complicated by the fact that the network is temporally dynamic, since the channel network expands into its '*variable source areas*' seasonally and during *storm* events.

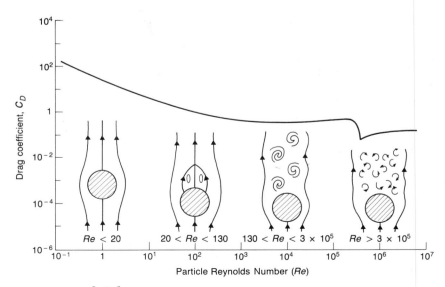

Fig. D30 Drag force. A plot of drag coefficient, C_D, against particle Reynolds number for quartz spheres settling through water and the associated flow patterns at different Reynolds numbers. (After Allen, 1985.)

Because of this variability, the length of stream networks shown on maps is inconsistent, and tributary links tend to be omitted as map scale decreases — thus **stream order** becomes a function of map scale. The existence of **dry valleys** provides a further source of uncertainty; some authors have preferred to use contour crenulations as a measure of total fluvial dissection rather than rely on the uncertain evidence of 'blue lines' on maps. [TPB]

drainage ratio The ratio between **runoff** and precipitation for a given area in a given time period. [AME]

drainage well A 'well' used to drain excess *soil* or surface water into an **aquifer**. [AME]

dravite A brown **tourmaline** mineral containing **manganese**. [DJV]

drawdown The loss of **head of water** around a **well** from which pumping is taking place. It is the difference in height between the rest water level (position of the **potentiometric surface** before pumping) and the pumping water level (see Fig. D31). (See **cone of depression**.) [AME]

dreikanter A pyramidal-shaped *clast* or pebble which has been worked by wind **abrasion** on three sides or 'facets'. **Abrasion** on several sides is thought to arise from the wind winnowing out finer grains from beneath the pebble until it falls over and a fresh facet is exposed. (See **ventifact**.) [RDS]

drift A horizontal underground tunnel that follows a **vein** or is parallel to the **strike** of an **orebody**, as distinct from a *crosscut* which cuts a **vein** or **orebody** at a high angle. [AME]

drill hole The term more commonly used in metalliferous mining circles for borehole. [AME]

drill string The sum total of equipment extending down a **drill hole** during the drilling operation, principally consisting of the drill pipes with the drill bit at the bottom. [AME]

drilling mud A mixture of water, clay and **barite** which is forced down the inside of the **drill string** of an oil drilling rig to cool and lubricate the drill bit. The mud returns to the surface bringing up with it the drill cuttings which are subjected to detailed examination. The **barite** is added to increase the density of the mud which serves to create a pressure head to minimize gas escape from porous formations. [AME]

dropstone A *clast* which falls through a water column into soft or partly-unconsolidated sediment disrupting **bedding** and other structures. *Clasts* are typically released from melting *icebergs* and sea ice. [ASG]

dry placer A **placer** that cannot be exploited for lack of the necessary water supply. [AME]

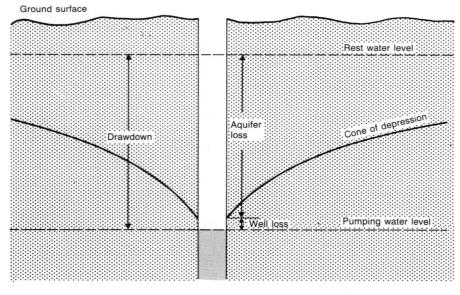

Fig. D31 Draw down.

dry valley A valley which is seldom, if ever at the present time, occupied by an active stream channel. These features are common, especially on sedimentary rocks such as the **Cretaceous chalks** of southern England. They display a considerable range of shapes and sizes, from mere indentations in the line of an **escarpment**, to great, long, winding chasms. An enormous range of hypotheses has been put forward to explain why all the different types are generally dry (Table D3). These have been summarized in Goudie (1984, Section 4.15). [ASG]

Goudie, A.S. (1984) *The Nature of the Environment.* Basil Blackwell, Oxford.

ductile deformation **Deformation** resulting in macroscopically continuous **strain**. A ductile **stress–strain** relationship is one in which large non-elastic, *permanent strains* (e.g. 75%) are achieved before **failure**. Alternative definitions are a **stress–strain** relationship which has no **stress** drop after **failure**, or a *flow strength* which is independent of **confining pressure**. These three characteristics of **stress–strain** behaviour are commonly associated.

The term ductile is commonly used to describe **deformation** in the field, where **stress–strain** relations cannot be known; in these circumstances it must refer to observed **strain** distribution visible on a **macroscopic** scale. Ductile **strain** distributions may be either localized (for example in a **shear zone**) or homogeneous (in uniformly deformed *tectonite*); their common feature is that the **strain** is macroscopically continuous. According to this definition therefore, it is possible to have **deformation** which is simultaneously ductile and **brittle**, for example in the case

of a **fold** (macroscopically continuous **strain**, therefore ductile) formed by small displacements on individual **fractures** (**brittle deformation**). [TGB]

Heard, H.C. (1960) Transition from brittle fracture to ductile flow in Solenhofen limestone as a function of temperature, confining pressure, and interstitial fluid pressure. In: Griggs, D. & Handin, J. (eds) *Rock Deformation.* Geological Society of America Memoir **79**: 193–226.
Rutter, E.H. (1986) On the nomenclature of mode of failure transitions in rocks. *Tectonophysics* **122**: 381–7 (letter).

dumortiorite $(Al_2O_3(BO_3)(SiO_4)_3)$ An island silicate (*nesosilicate*) mineral. [DJV]

dune An accumulation of unconsolidated material which is shaped by wind into a landform distinguishable from the surrounding topography. Dunes are usually composed of **quartz sand** though they can also be composed of clay, **gypsum** or carbonate grains. Dunes vary in shape and size depending on the availability of **sand**, the local topography, the wind regime and other climatic factors (Fig. D32). Several classifications of dune types have been proposed. Most are based on whether the dune has a transverse or linear form (Table D4). (See **aklé, barchan, climbing dune, dome dune, draa, falling dune, foredune, fulje, nebkha, parabolic dune, reversing dune, seif, star dune, transverse dune.**) [RDS]

Bagnold, R.A. (1941) *The Physics of Blown Sand and Desert Dunes.* Methuen, London.
Brookfield, M.E. & Ahlbrandt, T.S. (eds) *Eolian Sediments and Processes.* Developments in Sedimentology No. 38. Elsevier, Amsterdam.
Greeley, R. & Iversen, J.D. (1985) *Wind as a Geological Process.* Cambridge University Press, Cambridge.
McKee, E.D. (1979) *A Study of Global Sand Seas.* Professional Paper of the United States Geological Survey **1052**.

dune stabilization The prevention of **erosion** of a **sand dune** or the immobilization of a moving **dune**. Dune stabilization is undertaken in coastal areas where **dunes** on **barrier islands** provide protection from **storm surges** or where **sand deflated** from **dunes** encroaches upon roads or residential areas. Stabilization is often required to remedy dune **erosion** resulting from human interference. This may be caused by the construction of houses or the destruction of vegetation where the **dunes** are in recreational areas. Along the eastern seaboard of the United States and in The Netherlands, stabilization is achieved by limiting access to the **dunes**, by constructing fences which trap wind-blown **sand** and by re-establishing a protective vegetation cover. In temperate environments where rainfall is adequate, the planting of indigenous *dune grasses* such as marram (*Ammophila arenaria*), American beachgrass (*A. brevigulata*) and panicgrass (*Panicum amarum*) has proven very effective. These species are fairly tolerant of drought and

Table D3 Hypotheses of **dry valley** formation

Uniformitarian
1 Superimposition from a cover of impermeable rocks or sediments
2 Joint enlargement by solution through time
3 Cutting down of major through-flowing streams
4 Reduction in catchment area and groundwater lowering through scarp retreat
5 Cavern collapse
6 River capture
7 Rare events of extreme magnitude

Marine
1 Non-adjustment of streams to a falling *Pleistocene* sea-level and associated fall of groundwater levels
2 Tidal scour in association with former estuarine conditions

Paleoclimatic
1 Overflow from proglacial lakes
2 Glacial scour
3 Erosion by glacial meltwater
4 Reduced evaporation caused by lower temperatures
5 Spring snow-melt under periglacial conditions
6 Runoff from impermeable *permafrost*

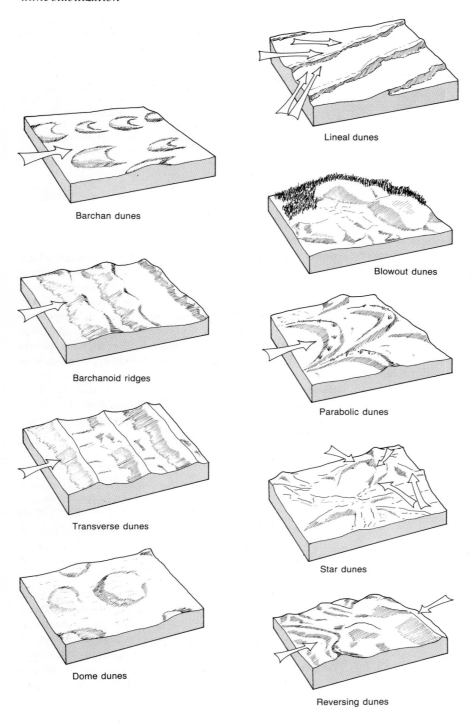

Lineal dunes

Barchan dunes

Blowout dunes

Barchanoid ridges

Parabolic dunes

Transverse dunes

Star dunes

Dome dunes

Fig. D32 Dune. The morphology of selected desert dune types after the classification of McKee (1979).

Reversing dunes

high salinity, and as well as preventing **erosion** by wind, actually promote **sand** accumulation and **dune** growth.

In arid regions, **dune** encroachment may threaten desert communities, the communications infrastructure and industrial facilities. Here, aridity often precludes the use of vegetation as a **sand** control measure. Some success has been achieved by planting belts of indigenous, drought-

resistant trees such as *Tamarix aphylla* in Saudi Arabia. Once established, the tree belts disrupt the airflow and promote deposition of windborne **sand**. However, the trees are quickly overwhelmed by the advance of large **dunes** which may be more than 10 m high and move 20 or 30 m each year (Watson, 1985). In such areas, the **dunes** must either be physically removed, forced to dissipate

Table D4 Classification of **dune** types. (After McKee, 1979 and Greeley & Iversen, 1985.)

Type	Name	Form	Slipfaces	Geomorphic features
Transverse	Barchan	Crescent-shaped	1	Uni-modal wind, all grains mobile, no vegetation
Transverse	Barchanoid or aklé	Interlocking crescentic ridges	1	Uni-modal wind, all grains mobile, no vegetation
Transverse	Transverse ridge	Asymmetric ridges	1	Uni-modal wind, all grains mobile, no vegetation
Transverse	Zibar	Low-angle ridges	0	Coarse and medium grain size fractions
Transverse	Parabolic	U-shaped	0/1	Partial vegetation cover
Longitudinal	Seif/linear	Symmetrical ridges	1/2	Bi-modal wind, all grains mobile
Longitudinal	Reversing	Asymmetrical ridges	2	Opposing wind directions, all grains mobile
Others	Star	Arms radiating from central peak	3+	Multidirectional winds
Others	Dome	Elliptical or circular mound	0	Strong winds

naturally through **deflation**, or they must be stabilized.

Removal of **dunes** is the most effective way of dealing with the problems they pose. However, the volume of **sand** may be enormous and transporting it may not be economically viable. Desert **dune sand** is often unsuitable for use in cement because it is saline and because it is difficult to compact; it is also unsuitable for use as fill or **ballast**. The dissipation of **sand dunes** can be achieved by altering their aerodynamic profiles by excavating a trench through the centre of the **dune** parallel to the direction of movement, or by stabilizing a portion of the **dune**'s surface. This results in destruction of the **dune** as it advances. Alternatively, by starving the **dune** of its **sand** supply from upwind (by stabilizing the source areas or erecting sand fences), the dune gradually ablates as it advances and eventually it disappears. Dissipation of **dunes** using these techniques is appropriate only when the installation that is threatened can tolerate increased amounts of **sand** drift as the **dune** ablates. The technique may be used upwind of aerial power-lines to reduce the height of the **dunes**, allowing them to pass beneath the cables.

The stabilization of **sand dunes** is aimed at halting their advance toward sensitive installations such as roads, railways and industrial or residential facilities. It is usually undertaken using **sand** fences or surface stabilization techniques. The construction of fences on the **dunes**, perpendicular to the direction of sandflow, immobilizes the **dune** and causes it to grow through **accretion**. While such procedures can greatly reduce the amount of wind-blown **sand** further downwind, huge artificial **dunes** may develop. These require careful management and maintenance of fences to prevent **deflation** or mobilization.

Surface stabilization of **dunes** with **gravel** coatings, artificial salt crusts (produced by spraying the **dune** with saline water), or oils and chemicals has been widely employed in the Middle East and Central Asia. The surface stabilizers either mantle the **dune** or penetrate the sand and bind the grains preventing **deflation**. Either the whole **dune** surface may be treated or strip stabilization may be adopted — this involves treatment of strips aligned perpendicular to the direction of **dune** movement and leaving untreated areas between. In most cases, the stabilized surface allows unimpeded movement of wind-blown **sand** across the **dune**. One exception to this is the use of coatings of heavy, waxy **crude oils** which trap some of the airborne **sand** (Mirakhmedov, 1983). While these procedures immobilize encroaching **dunes**, they do not halt wind-blown **sand**; this must be trapped using fences or vegetation belts, or it must be tolerated. [AW]

Adriani, M.J. & Terwindt, J.H.J. (1974) Sand stabilization and dune building. *Rijkswaterstaat Communications* **19**: 1–68.

Kerr, R.C. & Nigra, J.O. (1952) Eolian sand control. *American Association of Petroleum Geologists Bulletin* **36**: 1541–73.

Mirakhmedov, M. (1983) Stabilization of mobile sands with heavy crude. *Problems of Desert Development* **1983–1**: 88–9.

Savage, R.P. & Woodhouse, W.W. (1968) Creation and stabilization of coastal barrier dunes. *Proceedings of the Coastal Engineering Conference, London* **1**: 671–700.

Watson, A. (1985) The control of wind blown sand and moving dunes: a review of the methods of sand control in deserts, with observations from Saudi Arabia. *Quarterly Journal of Engineering Geology* **18**: 237–52.

Willetts, B.B. & Phillips, C.J. (1978) Using fences to create and stabilize sand dunes. *Proceedings of the Coastal Engineering Conference, USA* **2**: 2040–50.

dunite A medium to coarse-grained ultramafic rock (*peridotite*) composed of more than 90% **olivine**. Dunites may form **olivine**-rich layers within intrusions composed of **cumulate igneous rocks**, may occur as **xenoliths** within **alkali basalt** or may form more homogeneous masses that are regarded as parts of complete or fragmented **ophiolite** complexes. [RST]

duricrust Hard (Latin *durus*, 'hard'), mineral-cemented crust of variable thickness occurring within **weathering** profiles (**saprolite**) or the *soil* zone. Most commonly, duricrusts associated with **deep weathering** profiles are composed mainly of **silica** (**silcrete**), aluminium oxides (**alcrete** or **bauxite**), or **iron oxides** (**ferricrete**). However, in addition to these **lateritic** forms, certain desert *soil* crusts are also classed as duricrusts. These include crusts cemented by calcium carbonate (**calcrete**), **dolomite** (*dolocrete*), calcium sulphate (**gypsum** crusts or *gypcrete*), **aluminium** (*alcrete*) and sodium chloride (**halite** crusts or **salcrete**). In some cases, the crusts may contain more than one principal mineral cement. For example, **silcretes** commonly contain significant amounts of **aluminium** and **iron oxides**, and some are rich in **manganese** and **titanium** minerals. Similarly, some **calcretes** may be heavily silicified and some **gypsum** crusts exhibit features indicative of calcretization. The term **laterite** is commonly used for duricrusts containing **iron** and **aluminium**, and **bauxite** for deposits sufficiently rich in **aluminium** to be economically exploitable.

The formation of duricrusts associated with **deep weathering** profiles is commonly attributed to the chemical mobilization of rock **weathering** products in humid tropical and subtropical environments. The relative enrichment in **iron** and **aluminium oxides** at the top of **lateritic** profiles probably results from the mobilization of **silica** which is leached to greater depths. Today, this process is occurring in areas where rainfall is between about 500 and 1200 mm annually and where mean annual temperatures are between about 20° and 25°C. It is likely that the development of the chemical conditions required for **silica eluviation** is related to vegetational and micro-organic environments. **Silica** enrichment leading to **silcrete** formation may result from **illuvial** accumulation of **silica** deep in **weathering** profiles. However, **silcretes** are common today in semi-arid regions, notably in southern Africa and parts of Australia. This may indicate that some **silcretes** are formed under different environmental conditions from **ferricretes** and **alcretes**.

Silcretes are also found in arid areas where **calcretes** and **gypsum** crusts are often the most widespread types of duricrust. In these environments the crusts are usually no more than 4.0 m or so thick — whereas the crusts in **deep weathering** profiles can reach thicknesses of well over 10 m. The desert crusts form in areas where mean annual rainfall is less than about 300 to 400 mm — thick **halite** crusts may be preserved only where rainfall is less than about 50 mm. The crusts' development is dependent upon aridity and the maintenance of specific chemical environments in the *soil* zone or at the **water table**. In deserts, **silcretes** probably form only in the highly alkaline environments of some saline **basins**. Many **calcretes**, **gypsum** crusts and **halite** crusts form as **illuvial accretions** in areas where atmospheric salt deposition occurs. However, these crusts also develop as **groundwater** and lacustrine **evaporites** in many arid regions. The different duricrusts have specific different climatic requirements and their occurrences in the geological record can be used to assess *paleoclimates*.

Duricrusts are an important component of many landscapes — though in themselves they are not landforms. **Ferricretes** and **silcretes** often cap ancient **pediment** surfaces and **plateaux**. Once exhumed through **erosion** of the overlying, unconsolidated materials, the hard crusts resist further **erosion** and preserve the now relict landscapes upon which they formed. Relict **ferricretes** in the humid tropics may attest to former periods of lower rainfall because induration does not occur under wet conditions. In contrast, the presence of **ferricretes** and **weathering**-profile **silcretes** in areas which today are arid, argues that climatic conditions were formerly wetter. In parts of arid Australia and southern Africa, some of these duricrusts may date from the early *Tertiary* because they not only mantle ancient landscapes but also required long periods of time to accrete. A **ferricrete** between about 0.50 and 1.0 m thick may require 300 000 to 750 000 years to form.

The duricrusts of deserts are more ephemeral. Most are prone to degradation through **dissolution** if the climate becomes wetter. However, examples of very ancient **calcretes** and **gypsum** crusts occur in the geological record. Moreover, the presence of thick duricrusts of this type in certain areas provides a strong argument for the great antiquity of some deserts.

Duricrusts have economic significance and provide materials for road construction and **iron** and **aluminium ores**. [AW/VPW]

Goudie, A.S. (1973) *Duricrusts in Tropical and Subtropical Landcapes.* Clarendon Press, Oxford.

Goudie, A.S. (1983) Calcrete. In: Goudie, A.S. & Pye, K. (eds) *Chemical Sediments and Geomorphology*, pp. 93–131. Academic Press, London.

Goudie, A.S. (1984) Duricrusts and landforms. In: Richards, K.S., Arnett, R.R. & Ellis, S. (eds) *Geomorphology and Soils*, pp. 141–58. Allen & Unwin, London.

McFarlane, M.J. (1983) Laterites. In: Goudie, A.S. & Pye, K. (eds) *Chemical Sediments and Geomorphology*, pp. 7–58. Academic Press, London.

Summerfield, M.A. (1983) Silcrete. In: Goudie, A.S. & Pye, K. (eds) *Chemical Sediments and Geomorphology*, pp. 59–91. Academic Press, London.

Watson, A. (1983) Gypsum crusts. In: Goudie, A.S. & Pye, K. (eds) *Chemical Sediments and Geomorphology*, pp. 133–61. Academic Press, London.

dust Solid particles of varying size, but all less than 0.08 mm in diameter, that are suspended in the air. The material may originate from *soil*, blown from a dry and sparsely vegetated surface, from volcanoes, ejected into the atmosphere during eruption, from industrial and domestic pollution, or from interstellar matter. The period of time dust remains in the atmosphere depends upon the mass of the grains, their composition and the height to which they are carried. Volcanic particles may reach the stratosphere and remain there for several years, encircling the Earth. *Soil* dust may be carried up to 8 km above the Earth's surface and be transported over large distances in the upper winds, material from the Gobi Desert, for example, has been traced over 10 000 km to Alaska.

Appreciable quantities of atmospheric dust cause a haze and if blowing *soil* dust reduces visibility to 1000 m or below the condition is known as a *dust storm*. *Dust storms* are common in the world's dry lands and in many areas occur on more than 15 days each year, in extreme cases more than 80 days (Middleton *et al.*, 1986). *Dust storms* have a variety of environmental implications and represent a significant hazard to human occupation of arid areas in particular (Péwé, 1981). They are composed of valuable topsoil and nutrients when blown from agricultural lands, and atmospheric dust can adversely affect human health, radio communications, transport facilities and meteorological processes.

One of the classic features of dust deposition is **loess** and parts of the world's ocean floor are mantled with dust deposited over geological time. Analysis of seabed material from the North Atlantic coast of West Africa indicates that dust deposition during the cold dry *interpluvials* of the *Pleistocene* was more intense than today (Goudie, 1983). Dust storms on the planet Mars are more frequent than on Earth, and some individual Martian events obscure the atmosphere of the entire planet (Surek, 1982). [ASG]

Goudie, A.S. (1983) Dust storms in space and time. *Progress in Physical Geography* **7**: 502–30.

Middleton, N.J., Goudie, A.S. & Wells, G.L. (1986) The frequency and source areas of dust storms. In: Nickling, W.G. (ed.) *Aeolian Geomorphology*, pp. 237–96. Unwin Hyman, Boston.

Péwé, T.L. (ed.) (1981) *Desert Dust Origin, Characteristics and Effect on Man*. Special Paper of the Geological Society of America **186**.

Zurek, R.W. (1982) Martian great dust storms: an update. *Icarus* **50**: 288–310.

dyke A tabular minor intrusion of near-vertical attitude that cuts across horizontal or gently dipping planar structures (e.g. stratification or **foliation**) within the surrounding rocks. Dykes are hence analogous to **veins** and vary from around a centimetre to many metres (and less commonly to hundreds of metres) in width. Dykes may be of basic, intermediate or acid chemical composition. However, the commonest dykes are of basic/mafic composition and are termed **dolerite** (*diabase*). Mafic dykes occur as parallel intrusions on a regional scale (regional **dyke swarm**), as radially disposed intrusions around an igneous centre (radial **dyke swarm**) and as circular intrusions (**ring-dykes**) by subsidence above a **basaltic** *magma chamber*. [RST]

dyke swarm Set of **dykes**, usually subparallel and with a common origin. Dyke swarms are often associated with major plutonic/volcanic centres, and their frequency diminishes as they are traced away from such centres. Good British examples of dyke swarms are the *Tertiary* Mull swarm, which extends for 400 km from the Mull centre to Yorkshire, and the early **Devonian** Etive swarm in the Glencoe area, where the **dykes** account for nearly 30% of the outcrop area. The geometry of dyke swarms has been used to reconstruct *paleostress fields* on the assumption that the **dykes** represent a series of **extensional** fissures. (See **stress**.) [RGP]

dynamic correction A **moveout time correction** applied to seismic data, whose magnitude depends on the travel time as well as the source and detector location and velocity of the material above the reflector. This is in contrast to a **static correction** which is independent of travel time and only dependent on location. The normal **moveout** correction is a dynamic correction which is applied to **seismic records** to remove the **moveout** of reflections in a **common depth point** gather such that they can be **stacked**. [GS]

dynamic equilibrium The equilibrium of a system is measured in terms of the degree to which its internal state or its output is adjusted to its inputs. Open systems are not in static equilibrium because they are continuously importing energy. However, the interconnecting structure of most physical systems means that they have the ability to self-regulate. Negative feedback operates such that, when change is introduced into the system via change in one of the system variables, its transmission through the system directs the effect of the change back to the initial variable. This means that any change in the energy status of the system will result in a change in the system variables which will, in turn, lead to a new system equilibrium; the negative feedback has the effect of stabilizing or damping down the original disturbance. This tendency for self-regulation is sometimes called *dynamic homeostasis*; its effect is usually indicated by the existence of high **correlation** between system variables, and through the tendency for observed values of individual variables to assume characteristic values. Where a system oscillates around a stable average value, this type of equilibrium is termed a steady state. However, many physical systems also undergo progressive change in their condition through time. Thus the term dynamic equilibrium is applied to a condition of oscillation about a mean value which is itself trending continuously through time. Such a condition might be recognized in a hillslope whose convexo-concave form is controlled by the operation of creep and *wash* processes, but whose maximum slope angle is gradually declining over time in response to progressive **erosion**. In that oscillation about the mean value may often be much greater than changes in the trend itself, since a system can often only be viewed over a relatively short time period, an impression of steady-state equilibrium, sometimes called *quasi-equilibrium*, is achieved. Equilibrium is manifest in the tendency towards a mean condition of the form of the system, recognizable statistically. Thus, in the study of river channels, empirical work on hydraulic geometry has provided the basis to identify an equilibrium state in terms of the maintenance *dynamic equilibrium*

of an average channel form over time. Self-regulation of natural systems is complicated by the existence of *thresholds*. Where these occur, the system may undergo drastic alternation between equilibrium states, a condition known as *dynamic metastable equilibrium*. In **geomorphology**, dynamic equilibrium is analogous to the concept of *grade*. [TPB]

Chorley, R.J. & Kennedy, B.A. (1971) *Physical Geography: A Systems Approach*. Prentice-Hall, Englewood Cliffs, NJ.

E

early diagenesis of subaqueous sediments Diagenesis near the interface of freshly deposited sediment and overlying water. The major processes are: *bacterial* decomposition of organic matter, **compaction** (particularly of muds), molecular diffusion (see **diffusion in sediments**), *bioturbation*, *adsorption* and *ion-exchange*, mineral **dissolution** and **precipitation**. By utilizing profiles of **pore-water** chemistry with depth, the rates of these processes can be quantified by solving the *general diagenetic equation* (Berner, 1980). (See also **pore-waters/fluids and their evolution**.)

In most cases, the mineralogical changes that occur are triggered by *bacterial* processes. A sequence of *bacterial* reactions occurs as a sediment is progressively buried (Fig. E1; Table E1), successive reactions decreasing in free energy yield. **Oxidation** by molecular oxygen can be thought of as the reverse of photosynthesis. It occurs in the water column prior to sedimentation and will persist in **diagenesis** until oxygen or organic matter-depletion occurs. **Pore-waters** remain oxygenic if the sediment was originally poor in organic matter (e.g. most continental and some marine **sandstones**) or if the sedimentation rate is so slow that organic decomposition is completed very close to the sediment–water interface as in the abyssal **red clays**. *Nitrate-reduction* occurs next, but is quantitatively minor. *Manganese-reduction* is an important process in certain **pelagic** sediments allowing the formation of manganoan carbonates. In some cases, divalent **manganese** diffuses upwards and is reoxidized to MnO_2 leading to the development of a **manganese**-rich layer in the sediment, or contributing to **manganese nodules** on the sediment surface. *Iron-reduction* is placed next in the sequence, but also occurs within the succeeding *sulphate-reduction* and methanogenic zones because of the varying physical condition of oxidized **iron** in the sediment. This reaction is vital for the generation of **pyrite** and ferroan carbonates. It also generates considerable alkalinity (Table E1) facilitating **carbonate mineral precipitation**. *Sulphate-reduction* is the best understood of the *bacterial* reactions and is known to be important in nearly all fine-grained marine sediments. Sulphate availability for reduction is the major control on the fate of reactive detrital **iron** since highly insoluble **iron** sulphides precipitate very rapidly where sulphate and **iron** reduction are both occurring. *Sulphate-reduction* also occurs in other diagenetic environments, for example around **salt domes** and in **groundwaters**. *Methanogenesis* or **fermentation** covers a complex of processes which have wide application in biotechnology as well as **diagenesis**. The product *methane* may diffuse out of the sediment, especially in freshwater settings where *methanogenesis*

commences at shallow depths, or be reoxidized within the sediment.

The thickness of these bacterial zones is highly variable. For example, reactions preceding *sulphate-reduction* are compressed to within the uppermost centimetre in most fine marine sediments, but are expanded in organic-poor *hemipelagic* sediments. The top of the methanogenic zone varies between a few centimetres depth in some freshwater settings to several hundred metres depth in some **pelagic** sediments. Early diagenesis ceases when a significant rise of temperature (to around 50°C) is reached and organic matter is decomposed by abiotic **decarboxylation**.

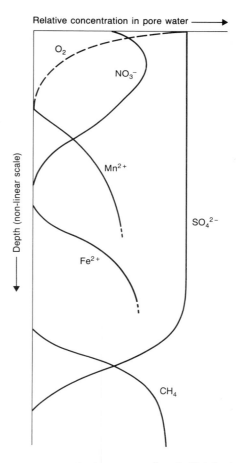

Fig. E1 Early diagenesis of subaqueous sediments. Variation in pore-water chemistry with depth providing evidence for the sequence of bacterial reactions. Decreases in manganese, iron and methane concentrations are due to upward diffusion and reoxidation. (After Froelich *et al.*, 1979 and Matsumoto & Matsuhisa, 1985.)

An extensive suite of minerals forms in response to bacterially-induced changes in **pore-water** chemistry. Various **carbonate minerals** occur, bearing calcium, magnesium, **iron** and **manganese** in different proportions. The carbonates often have characteristic *carbon isotope* signatures reflecting the origin of carbonate by organic matter decomposition. Particularly common occurrences are non-ferroan **calcite** accompanying **pyrite** in the *sulphate-reduction* zone of organic-rich sediments; variably ferroan **dolomite** in marine *hemipelagic* sediments; **siderite** in **deltaic** sediments and marine *ironstones*; ferroan carbonates in the methanogenic zone of marine organic-rich sediments. The carbonates are usually largely cementing and commonly form **nodules** or *concretions*. Other non-silicates include **silica** (**chert nodule** distribution often follows that of organic matter; see **silicification**); and *phosphates*, **iron**-phosphates in freshwater and calcium-phosphates in marine water. Isotopic studies of **carbon** and **sulphur** in substituents in the phosphate lattice help characterize the relationship to **organic matter diagenesis**. Among silicates, the distribution of **iron**-rich **smectites** which age to **glauconite** indicates their origin at the interface of oxidizing and reducing conditions created by *bacterial* activity. *Berthierine* requires *iron-reduction* for its formation, among other factors.

Other changes in early diagenesis are related to the characteristics of depositional waters. **Corrosion** and **alteration** of unstable minerals (e.g. siliceous tests; volcanogenic grains; **aragonite** in deep-water sediments) may be active. Corresponding **precipitation** of diagenetic **silica**, **zeolites** and **calcite** may result. In shallow seas active **cementation** by **aragonite** and magnesian **calcite** may occur. Fixing of potassium on detrital clays and other *ion-exchange* processes are also important. In restricted marine or lacustrine settings saline waters induce **precipitation** of **evaporite** mineral cements. [IJF]

Berner, R.A. (1980) *Early Diagenesis: A Theoretical Approach.* Princeton University Press, Princeton, NJ.

Coleman, M.L. (1985) Geochemistry of diagenetic non-silicate minerals: kinetic considerations. *Philosophical Transactions of the Royal Society, London* **A315**: 39–56.

Froelich, P.N., Klinkhammer, G.P., Bender, M.L. *et al.* (1979) Early oxidation of organic matter in pelagic sediments of the eastern equatorial Atlantic: suboxic diagenesis. *Geochimica Cosmochimica Acta* **43**: 1075–90.

Matsumoto, R. & Matsuhisa, Y. (1985) Chemistry, carbon and oxygen isotope ratios, and origin of deep-sea carbonates at sites 438, 439, and 584: inner slope of the Japan trench. *Initial Reports of the Deep Sea Drilling Project* **87**: 669–78.

earth hummocks A form of patterned ground on which rounded hummocks produce an irregular net pattern. The hummocks are produced by *frost heaving* in the *soil* and are, therefore, particularly characteristic of areas experiencing *periglacial* conditions. [NJM]

Earth movement A general term for **crustal deformation**, more used in older literature. Holmes (1944), in his textbook on physical geology, discusses vertical movements of the Earth's surface together with **orogenesis** as 'Earth movements'. [RGP]

Holmes, A. (1944) *Principles of Physical Geology* (1st edn) (see also 3rd edn, 1978). Nelson, Sunbury-on-Thames.

earth pillar A column of soft earthy material, often clay, capped by a boulder which protects the pillar from the rapid subaerial **denudation** which the surrounds have suffered. It is a feature typical of a semi-arid **badland** area. It is also commonly developed in glacial *morainic* deposits. In France the feature is known as a *'demoiselle'*. [NJM]

earthquake The sudden release of accumulated stress by **shear failure** along a planar subsurface zone (the '**fault plane**'). This causes vibrations (**seismic waves**) to be radiated out into the surrounding rock and up to the surface. If enough energy is released in the earthquake, close to the source the waves may be large enough to be felt as a shaking of the ground. They may shake structures such as buildings and bridges, etc., and even damage or destroy them. Both the cause (the rock **failure**) and the effect (perceived ground shaking) are termed an earthquake. There are three types of actual earthquake: tectonic, volcanic, and man-made; but similar ground shaking can be caused by mining explosions and under-

Table E1 Early diagenesis of subaqueous sediments. Mass balance for early diagenetic bacterial reactions expressed to highlight acid or alkali generation (developed from Coleman, 1985). Two alternatives are given for sulphate reduction: if sulphur escapes the system as H_2S gas, no acid is generated. This is beneficial to carbonate precipitation

Zone	Reactants		Oxidant			Products	Acid	Alkali
Oxidation	CH_2O		O_2	$\rightarrow$		HCO_3^-	H^+	
Mn-reduction	CH_2O	H_2O	$2MnO_2$	$\rightarrow$	$2Mn^{2+}$	HCO_3^-		$3OH^-$
Nitrate-reduction	CH_2O		NO_3^-	$\rightarrow$	$\frac{1}{2}N_2$	HCO_3^-		OH^-
Fe-reduction	CH_2O	$3H_2O$	$2Fe_2O_3$	$\rightarrow$	$4Fe^{2+}$	HCO_3^-		$7OH^-$
Sulphate-reduction	CH_2O		$\frac{1}{2}SO_4^{2-}$	$\rightarrow$	$\frac{1}{2}S^{2-}$	HCO_3^-	H^+	
Sulphate-reduction	CH_2O		$\frac{1}{2}SO_4^{2-}$	$\rightarrow$	$\frac{1}{2}H_2S$	$H\overset{.}{C}O_3^-$		
Methanogenesis	CH_2O	$\frac{1}{2}H_2O$		$\rightarrow$	$\frac{1}{2}CH_4$	$\frac{1}{2}HCO_3^-$	$\frac{1}{2}H^+$	

ground nuclear tests. (See **man-made earthquake, nuclear explosion seismology, tectonic earthquake, volcanic earthquake.**) [RAC]

Bolt, B.A. (1978) *Earthquakes — A Primer.* W.H. Freeman, San Francisco.

earthquake engineering The study of the hazards posed by **earthquakes** to buildings and other structures such as dams, power stations and bridges. The seismologist must assess, for some specified region, time-scale and level of probability, the likely *focal mechanism* and maximum *magnitude* of **earthquakes** liable to occur. For intraplate regions, this requires a detailed study of historical, as well as instrumentally-detected, **seismicity**. From a knowledge of near-surface characteristics (such as types and extent of superficial deposits and bedrock) estimates can be made of what horizontal and vertical ground accelerations and velocities could be produced. These data can then be passed on to engineers, who design structures so that they do not fail under these conditions. [RAC]

Dowrick, D.J. (1987) *Earthquake-resistant Design for Engineers and Architects* (2nd edn). John Wiley & Sons, New York.
Institute of Civil Engineers (1985) *Earthquake Engineering in Britain.* Thomas Telford, London.

earthquake intensity A subjective measure of **earthquake** 'size' — it attempts to quantify how the **earthquake** is experienced at the surface, at and around the *epicentre*. Intensity is quoted on a scale from I to X or XII, and written in Roman numerals to avoid confusion with **magnitude**. Very close to the *epicentre* these effects are caused directly by the rock movement in the **earthquake**; further away, it is the *surface waves* (especially *Love waves*) which are perceived.

The intensity at a given location is determined by cataloguing the 'felt effects' or damage caused. The information needed is usually obtained from field surveys and questionnaires. Most agencies have a standard form of questionnaire, which is distributed directly to individuals or displayed in newspapers with responses invited (the example given, Fig. E2, is from the British Geological Survey). The reports are compared to descriptions on a standard scale, and assigned an intensity. For example, 'felt indoors by only few: felt indoors by several: felt indoors by many and outdoors by a few . . .' are all different intensities. Intensity is harder to assess in sparsely populated areas. The intensities assigned to every felt report are posted on a map and the zones of equal intensity defined. The boundaries between these zones are termed **isoseismals**.

Over 40 intensity scales have been developed in the last 200 years or so, reflecting regional differences and evolution of building design and cultural conditions. The most notable and widely used are:

Rossi–Forel 1878, by de Rossi of Italy and Forel of Switzerland; a 10-point scale.

Table E2 Earthquake intensity. The table gives an approximate relationship between various earthquake intensity scales; ground velocity and acceleration; body- and surface-wave magnitudes; and energy release. Velocity and acceleration are 'average maximum values', the latter given as fractions of the acceleration due to gravity (about 9.81 m s^{-2}). Magnitudes and energy release are appropriate for shallow earthquakes. (Data from Bath (1973), Bolt (1978) and US Dept. of Interior/Geological Survey.)

Modified Mercalli	Rossi–Forel	JMA	Velocity (mm s^{-1})	Acceleration (g)	m_b	M_s	$\log_{10}$ energy (ergs)
I	I	0					
II		I					
	II						
III	III	II					
IV	IV		10–20	0.015–0.02			
	V						
V		III	20–50	0.03–0.04			
	VI						
VI		IV	50–80	0.06–0.07			
	VII				5.9	5.4	20
VII			80–120	0.10–0.15			
	VIII				6.3	6.1	21
VIII		V	200–300	0.2–0.3			
	IX				6.7	6.8	22
IX			450–550	0.50–0.55			
					7.1	7.5	23
X	X	VI	>600	>0.6			
					7.5	8.2	24
XI	—	VII	—	—			
					7.9	8.9	25
XII	—	—	—	—	—	—	—

For earthquakes in the UK, the following relationship between MM intensity, epicentral distance R (in km) and surface-wave magnitude M_s has been found:

$$I = 1.36 + 1.80\,M_s - 1.18 \ln R - 0.0019\,R$$

Thus the M_s 5.1 Welshpool earthquake of 2 April 1990 was felt I = III to IV at a range of 200 km.

Mercalli 1917, more detailed 12-point scale by the Italian volcanologist Mercalli. Twice revised.

Modified Mercalli 1931, revision of *Mercalli scale* by Wood & Neumann for building styles in USA. Further refined by Richter in 1956. The most common scale, abbreviated to MM.

JMA 1950, by the Japanese Meteorological Agency; scale of 0–VII suitable for local building styles.

MSK 1964, by Medvedev–Sponheuer–Karnik East European 12-point scale in use in the former USSR, similar to the MM scale.

Many factors determine the pattern of intensity generated by a given **earthquake**; not just its energy release (**magnitude**), but also its depth and focal mechanism (i.e. direction of rock movement). Local variations are caused by topography (focusing and de-focusing of **seismic waves**) and near-surface geology; wave amplitudes are generally much greater in materials with slow wave velocities such as soils and drift cover. Much effort has been expended in setting up empirical relations between

---

SURVEY OF SHEFFIELD EARTH TREMOR

8 FEBRUARY 1990

You are invited to participate in a study of earth tremors which recently occurred in your area.
Would you please complete the questionnaire below and return to

GLOBAL SEISMOLOGY UNIT
FREEPOST, EDINBURGH, EH9 OLX

In answering these questions you may consult people in the immediate vicinity, but it is most important to disregard reports of effects in more distant places. If you felt or heard nothing, please answer questions 1 to 7, as negative replies are valuable.

1 At the time of the observations where were you? (Address and post code) ....

2 What time was the tremor? ....

3 What did you feel? ....

4 What did you hear? ....

5 What did others nearby feel or hear? ....

6 Were you indoors or out? ....

7 Were you sitting, standing, lying down, sleeping, active, listening to radio or TV? ....

8 Were you alarmed or frightened? ....

9 Was anyone nearby alarmed or frightened? ....

10 Did windows or doors rattle? (Please give details) ....

11 Did anything else rattle? (Please give details) ....

12 Did any hanging objects swing? (Please give details) ....

13 Did anything fall or upset? (Please give details) ....

14 Was there any damage? (Please give details) ....

15 Have you any other observations or further details on above questions? ....

BRTISH GEOLOGICAL SURVEY
Global Seismology Unit, Murchison House, Edinburgh

Fig. E2 Example of the form of questionnaire used to determine **earthquake intensity**.

points on intensity scales and actual variables such as ground velocity or acceleration; much is dependent on the frequency of the vibration as well as other cultural circumstances. Table E2 gives a rough relation between some of the major intensity scales in use and equivalent ground velocities and accelerations.

A notable example of this last effect is Mexico City, which is built on a 'bowl' of dried-up, low-velocity, lake sediments whose shape and size leads them to resonate at frequencies of around 0.5 Hz — typical of **seismic waves**, and also the natural frequency of 15–25 storey buildings. Consequently, serious damage can be caused by even

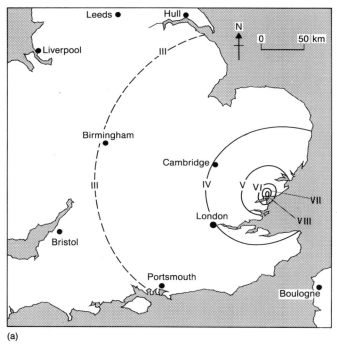

(a)

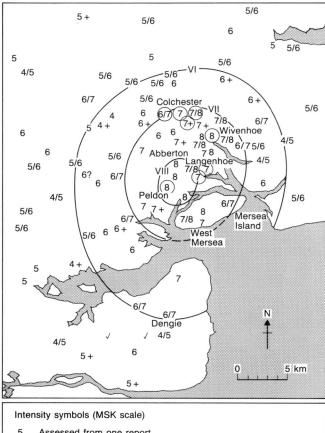

Intensity symbols (MSK scale)

5	Assessed from one report
⑤	Assessed from a number of reports
5 +	Equal to or greater than value shown
✓	Earthquake felt, intensity not assessed
VI	Isoseismal

(b)

Fig. E3 Earthquake intensity. Isoseismal maps of the 22 April 1884 Colchester earthquake. (a) Compilation for the whole of southern England. (b) Detail of the epicentral region. (After Skipp *et al.*, 1985.)

distant **earthquakes** such as that of September 1985, 300 km away. In the October 1989 Santa Cruz, California, **earthquake**, the worst damage was tens of kilometres away in parts of San Francisco built on reclaimed shorelands, while closer structures on bedrock were unharmed.

Evaluating intensities has a clear role in **earthquake** hazard assessment; by mapping felt effects and zoning them, suitable building design standards can be specified. For historical **earthquakes**, where no instrumental records are available, intensities can often be estimated from contemporary accounts. The pattern of **isoseismals** allows an approximate *epicentre* and **magnitude** to be calculated. The example shown, Fig. E3, is for the Colchester, UK, earthquake of 22 April 1884, one of the most damaging known in Britain in historic times. (See **earthquake engineering, isoseismal, Modified Mercalli scale, Rossi–Forel scale**.) [RAC]

Bath, M. (1973) *Introduction to Seismology*: Birkhauser Verlag, Zurich.
Bolt, B.A. (1978) *Earthquakes — A Primer*. W.H. Freeman, San Francisco.

Skipp, B.O., Shilston, D.T., Gutmanis, J.C. & Adams, R.D. (1985) Colchester revisited. In: Institute of Civil Engineers. *Earthquake Engineering in Britain*, pp. 189–206. Thomas Telford, London.
US Dept. of Interior/Geological Survey, National Earthquake Information Center. *Preliminary Determination of Epicentres*. (Monthly publication: technical review included every January.)

earthquake magnitude A measure of the amount of energy released by an **earthquake**, estimated from the amplitude of the **seismic waves** it produces. It was introduced by Charles Richter in 1935 as a means of ranking the 'size' of shallow **earthquakes** in California. It has come to have very widespread usage, and has evolved to have several definitions using different types of **seismic waves**. Magnitude formulae have one term based on the logarithm of ground displacement or velocity, added to further distance and/or depth correction term(s) which account for geometric spreading and **attenuation**, and also contain the (arbitrary) definition of a magnitude 0 **earthquake**. Magnitude scales in common use are:

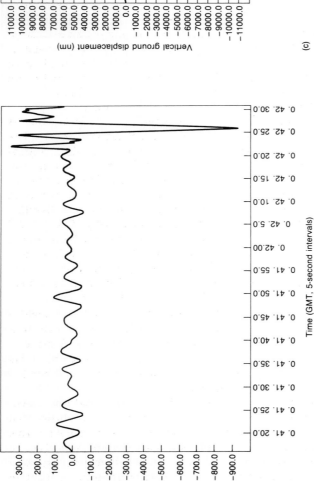

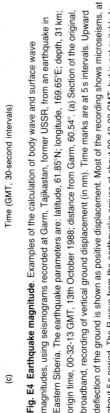

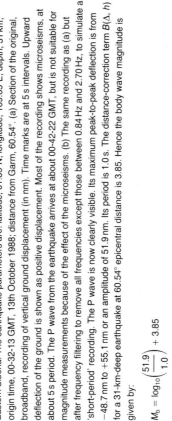

(c)

Fig. E4 Earthquake magnitude. *Examples of the calculation of body wave and surface wave magnitudes, using seismograms recorded at Garm, Tajikastan, former USSR, from an earthquake in Eastern Siberia. The earthquake parameters are: latitude, 61.85°N; longitude, 169.65°E; depth, 31 km; origin time, 00-32-13 GMT, 13th October 1988; distance from Garm, 60.54°. (a) Section of the original, broadband, recording of vertical ground displacement (in nm). Most of the recording shows microseisms, at about 5 s period. The P wave from the earthquake arrives at about 00-42-22 GMT, but is not suitable for magnitude measurements because of the effect of the microseisms. (b) The same recording as (a) but after frequency filtering to remove all frequencies except those between 0.84 Hz and 2.70 Hz, to simulate a 'short-period' recording. The P wave is now clearly visible. Its maximum peak-to-peak deflection is from −48.7 nm to +55.1 nm or an amplitude of 51.9 nm. Its period is 1.0 s. The distance-correction term $B(\Delta, h)$ for a 31-km-deep earthquake at 60.54° epicentral distance is 3.85. Hence the body wave magnitude is given by:*

$$M_b = \log_{10}\left(\frac{51.9}{1.0}\right) + 3.85$$

Hence M_b is 5.57. (c) A later section of the broadband recording, showing the Rayleigh wave and its dispersion: the lower frequency components have travelled more quickly and so arrive earlier than the higher-frequency waves. (Note the change in time-scale from (a) and (b); time-marks are at 30 s intervals). The Rayleigh wave has a peak-to-peak deflection of 22 728 nm or amplitude 11.36 microns, at a period of 21 s, hence its surface wave magnitude is given by:

$$M_s = \log_{10}\left(\frac{11.36}{21}\right) + 1.66 \log_{10}(60.54) + 3.3$$

which gives M_s = 5.99. The average values of M_b and M_s from all stations world-wide were 5.4 and 5.7 respectively.

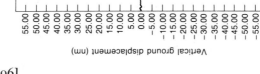

Local magnitude M_L: Richter's original definition of magnitude, for crustal-depth **earthquakes** within 600 km of the *seismograph*, based on wave amplitudes and a simple distance-dependent correction.

Body wave magnitude m_b: Gutenberg & Richter's extension of M_L to provide magnitudes for **earthquakes** recorded world-wide, based on amplitude and period of the first few cycles of the *P wave* corrected for both distance and depth of source. (Fig. E4: also see **Richter earthquake scale**.)

Surface wave magnitude M_s: Uses *Rayleigh* or *Love surface waves* at about 0.05 Hz frequency (20 s period) with a distance-dependent correction:

$$M_s = \log_{10}(A/T) + 1.66 \log_{10}D + 3.3$$

where amplitude A is in microns, period T is in seconds, and distance D is in degrees. M_s is of less use for deep **earthquakes** since they are less efficient in generating **surface waves**. An example of M_s calculation is shown.

Numerous refinements of all magnitude scales have been offered, as have rough relationships between them and with other parameters such as **intensity**, **fault** rupture length, etc. For example,

$$\log E = 12.28 + 1.44 M_s$$

where E is the energy released as **seismic waves** in ergs. Note a change of 1 magnitude unit is equivalent to a factor of 10 in ground displacement or a factor of 30 in energy. Many factors make magnitude awkward to use:

1 Earthquakes concentrate their seismic energy at frequencies below a limit called the *corner frequency*, which decreases as the size of the **earthquake** increases. For very large **earthquakes** the *corner frequency* can fall below 1 Hz and even 0.05 Hz; m_b and even M_s 'saturate' and become unrepresentative — all the energy is in lower-frequency waves. This happens at m_b 7–7.5 and M_s 7.5–8. Only the 'moment magnitude' scale (see **seismic moment**) is independent of these effects.

2 The local structure beneath a given station or **earthquake** will bias the magnitudes it records. Thus a simple mean of all reported magnitudes is incorrect: for small **earthquakes**, only stations biased to large magnitude will detect them, so the mean is too great. The reverse happens for great **earthquakes** when many *seismographs* are over-loaded. Complex statistical methods are needed to allow for this. Precisions of better than ±0.2–0.3 are hard to acheive.

3 Distance-correction terms for amplitudes are dependent on velocity (because of focusing and defocusing) and **attenuation**, and so vary regionally. The definitions of m_b and M_s given here are global-average values as used by the International Seismological Centre, Newbury, UK. Many *local magnitude* scales have been offered, as have attempts to determine magnitude from other parameters such as the duration of a seismic signal. (See **earthquake intensity**, **Richter scale**, **seismic waves**.) [RAC]

earthquake prediction The aims of earthquake prediction are: specific, short-term 'prediction' (i.e. time, place, **magnitude** of a future **earthquake**); and long-term 'forecasting' (i.e. general probability of **earthquakes** of given **magnitude** in a given time period and region).

The human cost of **earthquakes** is huge: about 10 000 lives annually and rising as the population increases and concentrates in cities. The economic consequences are also very great indeed. Prediction of earthquakes is one of the most important research aims of **seismology**. Considerable optimism in the 1960s and 1970s (when **plate tectonic** theory provided a general framework for understanding **earthquake** activity) has yet to be substantiated. While forecasting is fairly well established, specific prediction has yet to become reliable or routine.

Three approaches to prediction can be identified:
1 Theoretical. Theoretical models of how rocks and **faults** behave as **strains** build up prior to an **earthquake** have both evolved from, and led to, observational studies. There are two alternative models describing the progressive development of **microfractures** and how they in turn influence physical properties such as water flow, **electrical conductivity** and **seismic wave** velocities. The *'dilatancy-diffusion'* and the *'dilatancy-instability'* models both start with growing numbers of *microcracks* in the **hypocentre** region, but differ as to how these lead to a catastrophic **failure** (see Fig. E10, p. 205). In the 'instability' model, the cracks 'avalanche' to relax the build-up of **stress** slightly; in the alternative model, the cracks lock up, and an influx of water causes the main **failure**.

2 Statistical. On scales of thousands of years, rates of *plate* movement and hence the frequency of **earthquakes** is believed to be essentially constant along a given **plate boundary** or **seismic belt**. By combining modern **seismogram** records, inferences from sedimentary deposits and evidence in historical archives, a long-term general record of **seismicity** stretching back hundreds or even thousands of years can be constructed. This allows recognition of '**seismic gaps**' (areas with abnormally little current activity), 'migration' of events along **seismic belts**, and of 'recurrence rates' or 'return times' of **earthquakes**, from which the probability of an **earthquake** of a given **magnitude**, happening within a given period can be estimated. Such data form the basis of earthquake forecasting.

3 Observational. This relies on precursory phenomena which, if recognized and correctly interpreted, may constitute warning of an imminent **earthquake**. Some precursors have been recognized entirely empirically; others derive from an understanding of the physical and chemical behaviour of stressed rock.

(a) Seismic velocities and **anisotropy**. **Seismic waves** travel more quickly parallel to dilatant **fractures** than across them (because the crack fill is slow material). This effect may be small — only a few percent — but is more marked for *S waves*. Thus two effects are expected when the **microfractures** in the subsurface grow in number: first, the *P wave/S wave* velocity ratio will change, and second, if the cracks are generally aligned parallel, *S waves* will be polarized or 'split' into a fast and slow *S wave*, oscillating parallel and obliquely respectively to the cracks.

(b) Seismic activity. Many **earthquakes** are preceded by abnormal seismic activity. This may be an increase in small **earthquakes** in an otherwise aseismic area, or the cessation of the normal 'background' activity in a **seismic belt**. The smaller **earthquakes** leading to a main **earthquake** are called **foreshocks**.

(c) Surface deformation. The build-up of **strain** across a 'locked' **fault** can sometimes lead to a measurable bulge of the Earth's surface, detectable by accurate surveys of elevation and/or *tiltmeters*.

(d) Geomagnetic/geoelectric. There is empirical evidence of small changes in the strength of the **geomagnetic field** in the area of an imminent **earthquake**, and of the large-scale **electrical resistivity** of the subsurface rocks, due probably to its changing water content and perhaps piezoelectric effects.

(e) Geochemical/hydrogeological. The level of water in wells, and the amount of the radioactive gas *radon* dissolved in the **groundwater**, have both been found to change before and during an **earthquake** (once again, because of the changing numbers of **microfractures** and their effect on water content).

(f) Triggering correlations. Attempts have been made to correlate the occurrence of **earthquakes** with non-geological factors, such as climate and tidal attraction of the Moon and planets, to identify any triggering forces which may add to tectonic forces and cause the **earthquake**. Only a few cases have shown a link, such as **reservoir-induced earthquakes** linked to rainfall and the timing of some **earthquakes** near the Equator being related to *Earth tidal* forces.

Another widely-quoted precursor, little understood but still less studied, is abnormal animal behaviour. Reports of unusually quiet or unusually disturbed activity are common but do not yet reveal any systematic features. A possible mechanism may be very high frequency sound emitted by the subsurface when large-scale crack growth is taking place. Some observatories in the former USSR and China routinely monitor small collections of animals.

To date, only some five or six specific predictions have been accurate. Forecasting has been more successful (15 or more), usually based on identifying **seismic gaps**. Whether or not to announce a prediction in advance is a complex sociological and political as well as scientific question; the gains from a successful prediction must be weighed against losses from a false alarm (possible casualties in panic, economic disruption, etc.) given the level of confidence in the prediction. False alarms that had significant social impact include August 1976 Canton–Hong Kong and 1981 Peru. A particularly successful specific prediction was the M_s 7.3 Haicheng–Yingkow, N.E. China, **earthquake**, when up to 90% of buildings were damaged or destroyed but almost no casualties were incurred among the 3 000 000 population as they had been warned and moved out-of-doors. Nevertheless, the unpredicted occurrence of a **magnitude** 7.6 **earthquake** 18 months later at nearby Tangshan (estimated 650 000 deaths and 780 000 injuries) highlights the current shortcomings of specific prediction ability. At present, the most reliable course remains to evaluate **seismicity** on a statistical basis, then to design and build all structures to withstand the largest likely **earthquake**. (See **earthquake engineering, foreshock, aftershock, seismic creep, seismic gap, seismicity**.) [RAC]

Olson, R.S. (1989) *The Politics of Earthquake Prediction*. Princeton University Press, Princeton, NJ.
Press, F. (1975) Earthquake prediction. *Scientific American* **232**, 14–23.
Rikitake, T. (1976) *Earthquake Prediction*. Elsevier, Amsterdam.
Simpson, D.W. & Richards, P.G. (eds) (1981) *Earthquake Prediction: An International Review*. American Geophysical Union Maurice Ewing Series 4, Washington, DC.

earthquake swarm A prolonged series of numerous small to moderate events without any single clear 'main' event. **Volcanic earthquakes** frequently occur in swarms. In a major swarm in the Matsushiro region of Japan, hundreds of thousands of **earthquakes** up to **magnitude** 5 took place over two years, the most in one day being 6780. Recognizing the difference between swarms and **foreshocks** or precursors to **volcanic eruptions** is vital in hazard assessment. [RAC]

earthworms in geomorphology Earthworms, segmented *annelid* worms, are among the most ancient groups of terrestrial animals and some 3000 species are known. They play a major role in modifying the action of surface processes. They modify *soil* profiles by burrowing and construction, moving material within and between *soil* horizons, with accompanying mixing of organic and inorganic *soil* constituents, disintegration and reformation of aggregates, and changes in **porosity**, aeration and water **infiltration** (Lee, 1983). As Darwin stated (1881, p. 316):

> The plough is one of the most ancient and valuable of man's inventions; but long before he existed the land was in fact regularly ploughed, and still continues to be thus ploughed by earthworms. It may be doubted whether there are many other animals which have played so important a part in the history of the world, as have these lowly organised creatures.

[ASG]

Darwin, C. (1881) *Vegetable Mould and Earthworms*. Murray, London.
Lee, K.E. (1983) Earthworms of tropical regions — some aspects of their ecology and relationships with soils. In: Satchell, J.E. (ed.) *Earthworm Ecology from Darwin to Vermiculture*, pp. 179–93. Chapman & Hall, London.

echinoderms/Echinodermata A group of coelomate deuterostomes possessing a multi-element **calcite** endoskeleton and a coelomic hydrostatic system of canals and tube feet, called the water vascular system. The skeletal plating and the water vascular system typically are arranged with pentaradial symmetry.

The name echinoderm is in allusion to the spiny exterior of certain representatives (Greek: *echinos* = sea-urchin, hedgehog, *derma* = skin). The three diagnostic characters, the **calcite** endoskeleton, water vascular system and pentaradial symmetry, render them amongst the best characterized of all invertebrate phyla, though in some forms the radial symmetry may be lost or the **calcite** endoskeleton greatly reduced.

The **calcite** endoskeleton confers considerable geological importance on the group. It is formed of distinct ossicles, each deposited as a single crystal of high **magnesium calcite** in an organic mesenchymatous network to form a reticulate stereom microstructure. This stereomic construction has a number of advantages. It is lightweight, highly economical in calcium carbonate, ideal for the attachment of muscles and ligaments which can be looped round the trabeculae, and inhibits **fracturing** and crack propagation. Usually the skeleton constitutes a very considerable proportion of the animal's mass and forms a

framework for the body. Reconstruction of the skeleton of a fossil echinoderm often represents a reasonable approximation to its original form and is an important source of information concerning its phylogenetic relationships and mode of life.

CLASSIFICATION. The phylum Echinodermata can be divided into two subphyla: *Pelmatozoa* and *Eleutherozoa*. Pelmatozoans differ from eleutherozoans in having the aboral surface expanded to form a functional stalk and the water vascular system extended into elongate appendages. Five extant classes are recognized; *crinoids*, in the subphylum *Pelmatozoa*, and *asteroids, ophiuroids, echinoids* and *holothuroids* within the *Eleutherozoa*. More than 20 other classes have been proposed for various **Paleozoic** groups (Sprinkle, 1983) but in the most recent classification scheme, using *cladistics* (Paul & Smith, 1984; Smith, 1984a), these have been assigned to the stem groups of extant classes or are regarded as unnatural, paraphyletic groups.

Two early **Paleozoic** groups cannot be assigned to either subphylum. *Carpoids* (L. **Cambrian–Carboniferous**) are a

Table E3 Classification of the **Echinodermata**

Pelmatozoa			
Crinoids	Echmatocrinus	M. Cambrian	
	Camerata	Ordovician–Permian	
	Flexibilia	Ordovician–Permian	
	Inadunata	Ordovician–Permian	Paraphyletic. Includes ancestors of flexibles and articulates
	Articulata	Triassic–Recent	
Lepidocystoids		L. Cambrian	Close to the common ancestry of crinoids and cystoids
Cystoids	Rhombifera	Ordovician–Devonian	Both of these groups are paraphyletic or even polyphyletic
	Diploporita	Ordovician–Devonian	
	Blastoidea	Ordovician–Permian	
	Paracrinoidea	Ordovician	
	Eocrinoidea	Cambrian–Silurian	A paraphyletic group defined on the basis of primitive characters
Eleutherozoa			
Asteroids		Ordovician–Recent	Paleozoic asteroids represent the stem group of the class and can be divided into a number of plesions according to the order of appearance of synapomorphies. Post-Paleozoic asteroids are a monophyletic group and have been independently classified by Blake (1987) and Gale (1987)
Echinoids	Perischoechinoidea	Ordovician–Carboniferous	A convenient, though paraphyletic, grouping of most Paleozoic echinoids
	Cidaracea	Permian–Recent	
	Euechinoidea	Triassic–Recent	
Edrioasteroids	Stromatocystitidae	Cambrian	Primitive sister group to other edrioasteroids
	Edrioasterida	Cambrian–Devonian	
	Cyathocystida	Cambrian–Devonian	
	Isorophida	Cambrian–Carboniferous	
Ophiuroids	Stenurida	Ordovician–Devonian	Close to the common ancestry of ophiuroids and echinoids with holothuroids
	Oegophiurida	Ordovician–Carboniferous	
	Ophiurida	Silurian–Recent	
	Phryonophiurida	Devonian–Recent	
Holothuroids	Dendrochirotacea	Silurian–Recent	
	Aspidochirotacea		
	Apodacea		
Ophiocistioids		Ordovician–Devonian	Close to the common ancestry of echinoids and holothuroids

bizarre group of **calcite**-plated organisms which differ from true echinoderms in their total lack of radial symmetry. They have been considered to have closer affinities with *chordates* (Jefferies, 1986) and clearly diverged from true echinoderms very early in the history of the phylum. *Helicoplacoids* (L. **Cambrian**), are the most primitive true echinoderms and are placed in the stem group of both pelmatozoans and eleutherozoans. A third group, the genus *Camptostroma*, is the earliest pentaradiate echinoderm and lies close to the common ancestry of both pelmatozoans and eleutherozoans.

The classification of the various groups within the *Pelmatozoa* and *Eleutherozoa* is summarized in Table E3 and in the *cladogram* of Fig. E5. Two groups of pelmatozoans are recognized, *crinoids* and *cystoids*. Crinoids (M. **Cambrian**–Recent) are grouped into four subclasses. The camerates, flexibles and articulates are monophyletic groups, but the inadunates are paraphyletic and include the ancestors of both flexibles and articulates. The *cystoids* (L. **Cambrian**–**Permian**) comprise several important groups. *Blastoids*, and the relatively minor paracrinoids, are both monophyletic groups but the rhombiferans, diploporites and eocrinoids are all unnatural paraphyletic, or even polyphyletic, taxa. Another minor group given class status by Sprinkle, the lepidocystoids (L. **Cambrian**), lie close to the common ancestry of *crinoids* and *cystoids*.

There are five main groups of eleutherozoans; *asteroids* (L. **Ordovician**–**Recent**), *ophiuroids* (L. **Ordovician**–

Recent), *echinoids* (**Ordovician**–Recent), *holothuroids* (**Silurian**–Recent) and *edrioasteroids* (L. **Cambrian**–U. **Carboniferous**). The phylogeny and classification of post-**Paleozoic** *asteroids* has been revised, using *cladistics*, by Blake (1987) and Gale (1987). Both are agreed that the post-**Paleozoic** *asteroids* are a monophyletic group but the two proposed classifications are otherwise quite different. Blake recognizes eight orders within the group whilst Gale has only four. **Paleozoic** *asteroids* represent the stem group of the class and can be divided into a number of plesions. Crown Group *echinoids* comprise the relatively conservative Cidaracea and the much more diverse Euechinoidea which includes a wide range of regular and irregular forms (Smith, 1984b). The various **Paleozoic** *echinoids* are grouped together in the Perischoechinoidea. The *Edrioasteroids* are divided into four orders (Smith, 1985), though all are comparatively minor groups. The present *ophiuroid* classification is based on grades of organization rather than on phylogenetic relationships and hence is rather unsatisfactory. *Holothuroid* classification is similarly in need of revision, though their fossil record is poor. *Ophoicistioids* (L. **Ordovician**–**Devonian**) appear to represent an intermediate between *echinoids* and *holothuroids*.

MODE OF LIFE/PALEOECOLOGY. Echinoderms are exclusively marine, usually stenohaline, though a few may be tolerant of brackish conditions. They range from intertidal

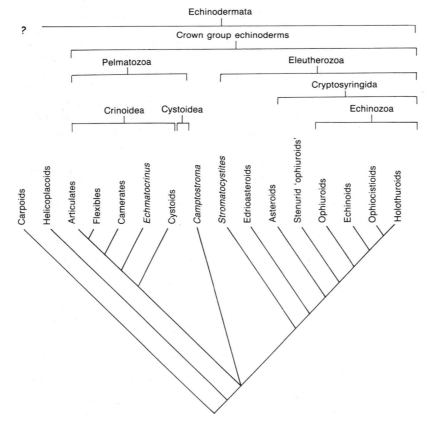

Fig. E5 Echinodermata. Cladogram to show how the major echinoderm groups fit into the classification scheme. (After Smith, 1984a.)

[200]

to abyssal habitats. With the exception of a few *holothuroids* and a small number of extinct *crinoids*, adult echinoderms are exclusively benthic.

Pelmatozoans are a largely sessile group although many *crinoids* are not permanently attached to one site. A few extant *crinoids* have limited crawling or swimming abilities whilst several groups of fossil *crinoids* were pseudopelagic, attaching to floating objects such as driftwood, or pelagic. *Crinoids* show probably the maximum size range among echinoderms; the largest, the extinct *Seirocrinus*, had a stem up to 20 m long and arms 0.5 m long, whilst the smallest microcrinoids had a cup less than 2 mm across and lacked arms entirely. Pelmatozoans are entirely filter feeders, either passively extracting detritus and plankton from currents or actively generating their own feeding currents.

Eleutherozoans show a much greater diversity of habitat within the benthic realm, although only a few swimming *holothurians* have managed to escape it. Infaunal as well as epifaunal forms are known among all four subclasses, especially *echinoids* and *ophiuroids*. Filter-feeding is encountered in many *ophiuroids* and *holothuroids* and a few *asteroids* and *echinoids*, but otherwise most eleutherozoans are detritus feeders, scavengers or grazers. In most cases it is possible to deduce the mode of feeding of a fossil echinoderm from its morphology although a few *echinoids* may leave direct evidence of their feeding strategies preserved in the form of **trace fossils**. *Gnathichnus* is a pentaradiate pattern of scratches sometimes found on shells and produced by the teeth of regular *echinoids* grazing on *algae*, whilst *Scolicia* is a burrow or surface furrow produced by deposit-feeding irregular *echinoids*.

Most echinoderms pass through a planktic larval stage prior to settlement which, since most adults are benthic, represents the major dispersal phase. The length of this larval stage varies considerably both within and between the five classes, though generally it is shorter among *crinoids* than among the others.

PRESERVATION. Since the ossicles of the multi-element echinoderm skeleton are composed of **calcite** they have a high preservation potential. However, generally they are held together only by soft tissues and hence become rapidly disarticulated following death and the subsequent decay of the soft parts. Stereomic interlocking of the plates in some *echinoid* tests may enable them to remain intact for some time but most other echinoderms are usually reduced to their constituent ossicles within a matter of days. Dissociated *crinoid* ossicles were sometimes sufficiently abundant in the mid to late **Paleozoic** to form a major part of some carbonate sequences, though rarely has this occurred among post-**Paleozoic** forms.

The preservation of intact echinoderm skeletons generally requires unusual conditions of deposition, either *obrution* or *stagnation* (see **fossil preservation**). In *obrution* deposits they are smothered by a rapid sediment influx which kills and buries them more or less simultaneously.

Pelmatozoans are more susceptible to death in this way but even the most active *Eleutherozoa* may succumb if the enveloping sediment layer is too deep for them to escape or is sufficiently fine to clog their water vascular system. Most instances of echinoderm *Lagerstätten* are preserved in this manner but some also occur in *stagnation* deposits, where *post mortem* disarticulation is inhibited by **anoxic** conditions which exclude disruptive benthos and slow down the rate of decay of soft tissues. However, typically the echinoderm fauna of such *stagnation* deposits is restricted to pelagic and pseudopelagic *crinoids*. The largest known echinoderm, the **Jurassic** pseudopelagic *crinoid Seirocrinus*, is typically found in such situations.

EVOLUTION. Echinoderms have experienced considerable fluctuations in diversity during the **Phanerozoic**, though it is unlikely that they had a long *Precambrian* history (Paul & Smith, 1984). The earliest dichotomy was that between *carpoids* and true echinoderms, closely followed by the divergence of *helicoplacoids* and other echinoderms (Figs E5 and E6). By the end of the Lower **Cambrian** the basic body plans for both pelmatozoans and eleutherozoans had been established. All of the extant classes, with the possible exception of the *holothuroids*, were established by the **Ordovician**. Several minor groups were extinct or in decline by the end of the Lower **Paleozoic** but the others, particularly the *crinoids*, show an overall increase in diversity. There was a decrease in diversity from the **Carboniferous** into the **Permian** followed by a major **extinction** at the end of the **Permian**. Among the *echinoids*, *asteroids* and *crinoids* it is likely that only one or two genera of each crossed the *Permian–Triassic boundary* (Simms, 1990). The main post-**Paleozoic** diversification of echinoderms lasted from the mid-**Triassic** into the early **Jurassic** with a more gradual expansion thereafter. Pelmatozoans, dominant throughout the **Paleozoic**, declined subsequently by comparison with the *eleutherozoans*. No echinoderm group appears to have been significantly affected by events at the *Cretaceous–Tertiary boundary*. [MJS]

Blake, D.B. (1987) A classification and phylogeny of post-Palaeozoic sea stars. *Journal of Natural History* 21: 481–528.

Clark, A.M. (1968) *Starfishes and Their Relations*. British Museum of Natural History, London.

Gale, A.S. (1987) Phylogeny and classification of the Asteroidea (Echinodermata). *Zoological Journal of the Linnean Society* 89: 107–32.

Hyman, L.H. (1955) *The Invertebrates* Vol. 4: *Echinodermata*. McGraw-Hill, New York.

Jefferies, R.P.S. (1986) *The Ancestry of Vertebrates*. British Museum of Natural History, London.

Paul, C.R.C. & Smith, A.B. (1984) The early radiation and phylogeny of echinoderms. *Biological Review* 59: 443–81.

Simms, M.J. (1990) The radiation of post-Palaeozoic echinoderms. In: Taylor, P.D. & Larwood, G.P. (eds) *Major Evolutionary Radiations*, pp. 287–304. Systematics Association Special Volume No. 42. Clarendon Press, Oxford.

Smith, A.B. (1984a) Classification of the Echinodermata. *Palaeontology* 27: 431–59.

Smith, A.B. (1984b) *Echinoid Palaeobiology*. Allen & Unwin, Hemel Hempstead.

Smith, A.B. (1985) Cambrian eleutherozoan echinoderms and the early diversification of edrioasteroids. *Palaeontology* 28: 715–56.

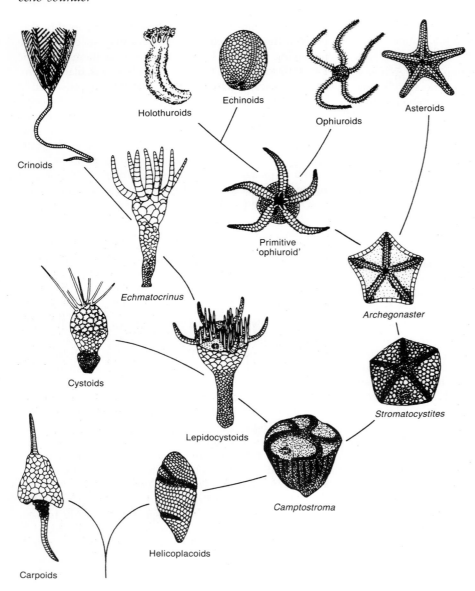

Crinoids

Holothuroids

Echinoids

Ophiuroids

Asteroids

Primitive 'ophiuroid'

Echmatocrinus

Archegonaster

Cystoids

Stromatocystites

Lepidocystoids

Camptostroma

Carpoids

Helicoplacoids

Fig. E6 Echinodermata. Evolutionary tree for the more important fossil and extant echinoderm groups. (After Paul & Smith, 1984.)

Sprinkle, J.S. (1983) Patterns and problems in echinoderm evolution. *Echinoderm Studies* **1**: 1–18.

echo sounder A device to calculate water depth by measuring, close to the sea surface, the travel time of an acoustic wave reflected from the sea bed. The velocity of propagation of the wave in seawater is used to convert the travel time to depth. An echo sounder is sometimes known as a *fathometer*. A piezoelectric transducer is used to send and receive the acoustic energy usually at frequencies of 30–80 kHz. [GS]

economic geology The sum total of geological studies that can be profitably utilized by man for the exploration for and exploitation of materials in the Earth's **crust**. [AME]

economic yield The maximum rate at which water can be withdrawn from an **aquifer** without creating a deficiency or damaging the water quality. [AME]

eddy current Electric current loops induced within a conductive body by a time-varying **magnetic field**. Eddy currents, whose magnitude will depend upon the size and **electrical conductivity** of the body, will in turn generate (secondary) **magnetic fields**. It is the measurement and interpretation of these secondary **magnetic fields** which forms the basis of the **electromagnetic induction methods**. [ATB]

eddy viscosity The component of resistance to **deformation** in a fluid produced by the generation of eddies within a **turbulent flow**. This component of **viscosity** is in addition to the **kinematic viscosity** of a fluid and is not constant for a given fluid or temperature. It can be considered as a coefficient of momentum transfer. If a particle of fluid of density ρ is displaced a distance of *l* before its momentum is changed by the new environment, it is found that

$$\eta = \rho l^2 \left(\frac{\delta u}{\delta y}\right)$$

where η is the eddy viscosity and $\delta u/\delta y$ is the velocity gradient in the **boundary layer** (in the direction normal to the boundary). [JLB/PAA]

Francis, J.R.D. (1975) *Fluid Mechanics for Engineering Students*. Arnold, London.

edenite ($NaCa_2Mg_5AlSi_7O_{22}(OH)_2$) A variety of the **amphibole mineral hornblende**. [DJV]

edge water If an oil pool in a **reservoir** rock is surrounded laterally by water-saturated rock, this water is termed edge water. [AME]

effective permeability The ability of a rock to allow the passage of one fluid such as gas in the presence of other fluids, e.g. oil and water. The presence of more than one fluid creates complex mutual interference that retards the movement of fluids. [AME]

effective porosity The fraction, expressed as a percentage, of a given mass of rock or *soil* that consists of interconnecting interstices. [AME]

effective stress The difference between applied *normal stress* and **pore-fluid pressure**.

In tensor notation, the effective stress, usually denoted by primed values, e.g. σ', is given by:

$$\begin{bmatrix} \sigma'_{11} & \sigma'_{12} & \sigma'_{13} \\ \sigma'_{21} & \sigma'_{22} & \sigma'_{23} \\ \sigma'_{31} & \sigma'_{32} & \sigma'_{33} \end{bmatrix} = \begin{bmatrix} \sigma_{11} - \rho & \sigma_{12} & \sigma_{13} \\ \sigma_{21} & \sigma_{22} - \rho & \sigma_{23} \\ \sigma_{31} & \sigma_{32} & \sigma_{33} - \rho \end{bmatrix}$$

The *law of effective stress* states that in the presence of **pore-fluid pressure**, the *normal stress* σ across a plane will be reduced by the **pore-fluid pressure** *p*. Therefore the **shear resistance** (τ) will also be reduced in the *Coulomb failure criterion* or *Amonton's law of frictional sliding*:

$$\tau = S + \mu(\sigma - p)$$

where *S* is the **cohesion**. Reduction of *normal stress* and hence **shear resistance** by **pore-fluids** can therefore cause **failure** in rocks at lower values of applied **stress** than in the dry state (Fig. E7). The presence of low **shear** resistance at the base of large *thrust sheets* due to **pore-fluid**

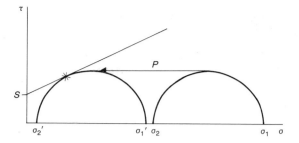

Fig. E7 Effective stress. The reduction of normal stress (σ_1, σ_2) by pore-fluid pressure (*P*) to create effective stress (σ_1', σ_2'). This causes failure as indicated by the intersection of the semi-circle with the failure envelope at the asterisk.

pressures has been invoked to explain how large thin *thrust sheets* can be transported for many kilometres without internal **deformation**. The converse effect, in which reduction of **pore-fluid pressures** due to **dilatancy** causes strengthening, is known as *dilatancy-hardening*, and considered to accompany several precursory **earthquake** phenomena (see **earthquake prediction**). The existance of high **pore-fluid pressures**, causing low effective stress, has been proved in many **wells** drilled into sedimentary basins. [TGB]

Hubbert, M.K. & Rubey, W.W. (1959) Role of fluid pressure in mechanics of overthrust faulting. *Geological Society of America Bulletin* **70**: 115–206.
Jaeger, J.C. & Cook, N.G.W. (1979) *Fundamentals of Rock Mechanics* (3rd edn), pp. 219–27. Chapman & Hall, London.

efficiency of a water well When **drawdown** occurs (see Fig. D31, p. 184) part of the drawdown results from water flowing through the **aquifer** to the **well** and this is called the *aquifer loss*. The other part, the *well loss*, results from water flowing into the **well** across the **well** face. The ratio of *aquifer loss* to total drawdown is a measure of the efficiency of the well. [AME]

Egyptian blue Fused mixture of **quartz**, **lime** and **copper ore** ground to fine powder for use as a blue pigment in antiquity. The Romans employed it in wall painting. [OWT]

Ekman spiral A graphical representation of the current velocity distribution with depth resulting from applied wind **stress** on the surface of a water body (Fig. E8). Due to the **Coriolis effect** and the frictional resistance of underlying water, the current generated on the surface is at 45° to the right of the wind (in the northern hemisphere). In successively lower layers of water the direction becomes systematically deflected and the current progressively decreases in magnitude such that the net volume transport of water is perpendicular to the wind direction (to the right in the northern hemisphere and to the left in the southern hemisphere). The thickness of the layer (*Ekman layer*) which undergoes this net displacement

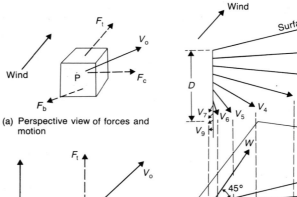

(c) 'Ekman flow' – water velocities decreasing and rotating with increasing depth

(d) Ekman spiral

Resultant volume transport at right angles to wind

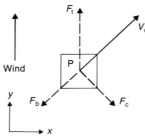

(a) Perspective view of forces and motion

Fig. E8 Wind-driven currents — northern hemisphere. F_t is a frictional stress caused by the wind on the water surface and acts in the direction of the wind. The Coriolis force F_c acts at right angles to the direction of motion, tending to cause the flow at the surface V_0 to be deflected. A retarding frictional stress F_b develops between the moving upper layer of water and the layer below, and acts opposite to V_0. (a) Forces and motion of a surface parcel of fluid P. (b) Plan view of forces and motion. (c) Water velocity as a function of depth. (d) **Ekman spiral**.

(b) Plan view, forces and motion (steady state)

(*Ekman transport*) depends on the wind speed and the latitude (magnitude of the **Coriolis effect**). A typical value for the thickness of the *Ekman layer* in mid-latitudes is 100 m. The Ekman spiral also depicts the effects of bottom friction on current pattern within the bottom *Ekman layer* but with the sense of deflection reversed to that of the surface *Ekman layer*. [AESK]

Pickard, G.L. & Emery, W.J. (1982) *Descriptive Physical Oceanography: An Introduction* (4th edn). Pergamon, Oxford.
Pond, S. & Pickard, G.L. (1983) *Introductory Dynamical Oceanography* (2nd edn). Pergamon, Oxford.

elastic constants/moduli Constants that define the elastic properties of an isotropic medium. *Elasticity* is the property of a medium whereby **deformations** disappear completely when the **stress** which causes them is removed. For small **deformations** Hooke's law states that the **stress** set up in an elastic medium is proportional to the **strain** applied and the elastic moduli relate these two quantities for various types of **deformation**.

Bulk modulus or *incompressibility*, k: the ratio of **stress** to **strain** under simple *hydrostatic pressure*.

$$k = \frac{\text{change in pressure}}{\text{dilation}}, \text{ where}$$

$$\text{dilation} = \frac{\text{change in volume}}{\text{original volume}}$$

Shear modulus, rigidity modulus or **Lamé's constant**, μ: the ratio of **shear stress** to **shear strain** under lateral **deformation**.

$$\mu = \frac{\text{shearing force per unit area}}{(\text{shear deformation/distance between shear planes})}$$

Lamé's constant, $\lambda = (\textbf{bulk modulus} - 2/3 \text{ *shear modulus*})$: for *Poisson solids* **Lamé's constant** and *shear modulus* are equal.

Poisson's ratio, σ: the ratio of transverse **strain** to longitudinal **strain** when a rod is extended.

$$\sigma = \frac{\text{transverse deformation per unit width}}{\text{longitudinal deformation per unit length}}$$

Poisson's ratio has a value of 0.25 for a *Poisson solid* — a homogeneous elastic solid for which the *shear modulus* and **Lamé's constant** are equal.

Young's modulus, E: the ratio of **stress** to **strain** when a rod is extended or compressed.

$$E = \frac{\text{deformation force per unit area}}{\text{change in length per unit length}}$$

Axial modulus, ψ: the ratio of the longitudinal **stress** to longitudinal **strain** in the case where there is no lateral **strain**.

The velocities of **seismic waves** can be expressed in terms of the elastic moduli and density ρ

$$\text{Velocity of compressional wave} = \sqrt{\frac{k + \frac{4}{3}\mu}{\rho}}$$

$$\text{Velocity of shear wave} = \sqrt{\frac{\mu}{\rho}}$$

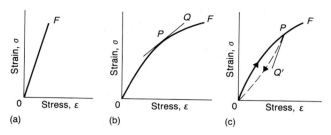

Fig. E9 Elastic deformation. (a) Linearly elastic material. (b) Perfectly elastic material showing tangent modulus *PQ*. (c) Elastic material with hysteresis, showing loading and unloading cycle. (After Jaeger & Cook, 1976.)

Ratio of compressional to shear wave velocities

$$= \sqrt{\frac{(0.5 - \sigma)}{(1 - \sigma)}} \quad \text{[GS]}$$

elastic deformation (*elasticity*) **Deformation** that is instantaneously and totally recoverable. Linear elastic deformation obeys the law of linear elasticity (**Hooke's law**), which states that **strain** is directly proportional to **stress** and the constant of proportionality is known as **Young's modulus**, the **elastic constant**, or *stiffness* of the material (Fig. E9). Perfectly elastic deformation implies a unique but not necessarily linear relation between **stress** and **strain**, such that the same **stress–strain** path is followed on both *loading* and unloading; in this case, the variable parameter relating **stress** to **strain** is the *tangent modulus* (Fig. E9b). More generally, elastic deformation may occur by different paths during loading and unloading; the difference is known as *hysteresis* (Fig. E9). **Anisotropic** materials may have different **elastic constants** in different directions; the relation between **stress** and **strain** is fully given by 36 components of the **elastic constant** or *stiffness* tensor, or their inverse, the **elastic moduli** or *compliance tensor*. Other important quantities in the theory of linear elasticity are: the **bulk modulus** or *incompressibility* (the ratio of *hydrostatic stress* to volumetric strain) and its reciprocal, the **compressibility**; the *modulus of rigidity* or *shear modulus* (the ratio of **shear stress** to **shear strain**), and **Poisson's ratio** (the ratio of lateral expansion to vertical contraction in uniaxial loading), or its reciprocal, *Poisson's number*. The *elastic limit* or *yield stress* is the **stress** above which elastic behaviour is no longer observed.

A material that fails within its elastic range by **fracture** is said to exhibit **brittle** behaviour. Many materials exhibit a combination of elastic and **viscous** behaviour of which one or the other may predominate, depending on temperature, *hydrostatic pressure*, **differential stress** and **strain rate**. (See also **strength, elastoviscous deformation, viscoelasticity**.)

Elastic behaviour is limited to **strains** of 5–10% in rocks, is transient, and generally not seen where finite **strain** is measured. The major exceptions arise in areas of active **faulting**, such as the San Andreas *fault zone*, or other

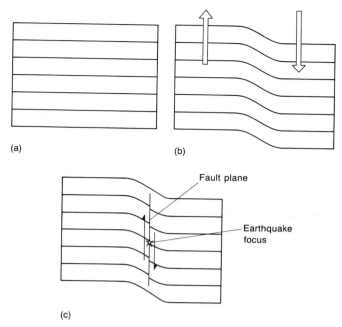

Fig. E10 Schematic representation of the **elastic rebound theory**, shown as the distortion of linear features. For an earthquake with almost entirely horizontal movement on a vertical plane, such as the San Andreas system in the western US, the diagram could represent a plan view of features such as roads, fences, etc.

active **deformation**. In the San Andreas **strike-slip fault** system, **deformation** at the Earth's surface occurs over a wide zone of **shear** before an **earthquake**. Afterwards, when displacement occurs on a **fault** segment, little or no distributed **strain** remains around the **fault**. [TGB/RGP/SB]

Jaeger, J.C. & Cook, N.G.W. (1976) *Fundamentals of Rock Mechanics*. Chapman & Hall, London.

elastic rebound theory The generally accepted model of an *earthquake's mechanism*, which arose from the 1906 San Francisco **earthquake**. H.F. Reid compared **geodetic** data spanning over 50 years. He showed how the build-up of **strain** in rocks is expressed in distortion of the Earth's surface. The **earthquake** happens where the **strain** exceeds the **strength** of the rocks and causes a break which diminishes in size away from the *focus*. For the simple near-horizontal movements of Californian **earthquakes**, ground distortion is easily visible as bending of once-straight features such as fences and rows of trees (Fig. E10). (See **earthquake prediction, focal mechanism solution**.) [RAC]

elastic strain A shape change (**strain**) resulting from **elastic deformation**; the process of **elastic deformation**, i.e. transient compression and **extension** such that the rock returns to its original shape after removal of the imposed **stress** (i.e. **temporary strain**). This **elastic behaviour**, in accordance with **Hooke's law** (see Jaeger & Cook, 1976) is limited to **strains** of 5–10% in rocks, is

elastica

Fig. E11 Elastica.

transient, and generally not seen where **finite strain** is measured. The major exceptions arise in areas of active **faulting**, such as the San Andreas *fault zone*, or other active **deformation** (see **neotectonics**). In the San Andreas **strike-slip fault** system, **deformation** at the Earth's surface occurs over a wide zone of **shear** before an **earthquake**. Afterwards, when displacement occurs on a **fault** segment, little or no distributed **strain** remains around the **fault**. (See **elastic strain**, **elastic rebound theory**, **seismic slip**, **aseismic slip**.) [SB]

Jaegar, J.C. & Cook, N.G.W. (1976) *Fundamentals of Rock Mechanics*. Chapman & Hall, London.

elastica Special type of **fold** profile in which the *fold angle* has a negative value. In such a **fold**, the *limbs* of an **anticline**, for example, converge downwards due to the excessive curvature of the *hinge* (see Fig. E11). Folds of this type have been termed **ptygmatic**. [RGP]

Ramsay, J.G. & Huber, M.I. (1987) *The Techniques of Modern Structural Geology* Vol. 2: *Folds and Fractures*. Academic Press, London.

elasticity of aquifer In an **artesian aquifer** the water is under considerable pressure and it is compressed so that more water is present than would be if the pressure throughout the **aquifer** were atmospheric. Although this effect on a small scale is not very significant, the sum total in a regional **aquifer** is enormous. [AME]

elastoviscous deformation **Deformation** consisting of **elastic** and **viscous** behaviour in series. This is also known as the *Maxwell model*, where the application of a **stress** results in an instantaneous **elastic strain** followed by **flow** at a constant **strain rate**.

On release of the applied **stress**, **strain** and internal **stress** are relaxed exponentially. Elastoviscous behaviour has been invoked to explain some experimental results, and also as a component of more complex models of rock behaviour. [TGB]

Ramsay, J.G. (1967) *Folding and Fracturing of Rocks*. McGraw-Hill, New York.

elbaite The **lithium**-rich variety of the mineral **tourmaline**. [DJV]

electric log Electrical **resistivity** vs depth plot obtained from surveys down a borehole, sometimes obtained from induction methods. The region of investigation, limited to a small radius around the hole and a small vertical thickness about the tool, is dependent on the electrode configuration and the resistivities of the formation and the **drilling mud**. Resolution of horizontal layers also depends on the electrode spacings, typically tens to hundreds of centimetres. The measured **apparent resistivity** will be strongly affected by the mud **resistivity** and elaborate sets of standard curves have been produced by commercial loggers to interpret logs from oil wells. (See **geophysical borehole log**.) [ATB]

Keller, G.V. & Frischknecht, F.C. (1966) *Electrical Methods in Geophysical Prospecting*. Pergamon, Oxford.

electrical conductivity Reciprocal of **electrical resistivity**. [ATB]

electrical resistivity Property of a material representing the resistance it offers to current flow. Resistivity is defined by **Ohm's law** for a continuous medium:

$$E = \rho j$$

or,

$$j = \sigma E$$

where E is the electric field, j the current density, ρ the resistivity and σ the conductivity. σ is a tensor, but in isotropic media resistivity may be represented as the resistance between opposite sides of a unit cube, and has units of ohm-metre.

Resistivity of near-surface rocks is dominated by **pore fluids** (see **Archie's law**). **Minerals** have highly variable resistivities, with many sulphides having low resistivities and most oxides high resistivities. However, few **minerals** occur in high enough concentrations to modify the bulk resistivities significantly, **magnetite**, iron sulphides and **graphite** being notable exceptions. [ATB]

Keller, G.V. & Frischknecht, F.C. (1966) *Electrical Methods in Geophysical Prospecting*. Pergamon, Oxford.

electrolytic polarization (1) Build-up of charges on metal electrodes because of their interaction with electrolytic solutions within the *soil* pores. This is to be avoided on potential electrodes as it tends to distort current flow in their vicinity and produce erroneous potential readings. **Non-polarizable electrodes** are necessary to eliminate this effect in polarization studies (see **self potential** and **induced polarization**). In galvanic **resistivity methods** slowly alternating currents (a few Hz) are employed to prevent the build-up of charges.
(2) Term used synonymously with **membrane polarization**. [ATB]

Kearey, P. & Brooks, M. (1991) *An Introduction to Geophysical Exploration* (2nd end). Blackwell Scientific Publications, Oxford.
Keller, G.V. & Frischknecht, F.C. (1966) *Electrical Methods in Geophysical Prospecting*. Pergamon, Oxford.

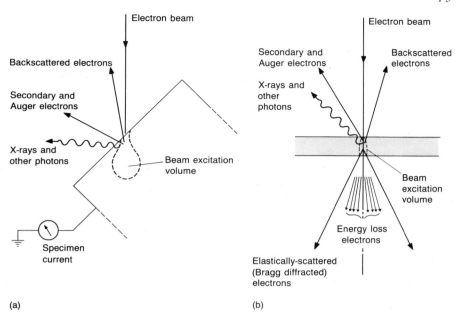

Fig. E12 Electron microscopy. (a) Scanning electron microscopy. (b) Transmission electron microscopy.

(a)

(b)

electromagnetic induction methods (*electromagnetic methods*) Exploration methods based on the **magnetic fields** of induced currents within the conductive Earth by time-varying electromagnetic fields. The primary fields may be man-made (e.g. **VLF**, **INPUT**®, COMPARA-TOR . . . etc.) or natural (e.g. **AFMAG**, **magnetotelluric** method). The frequency of the primary field ($<c$. 50 kHz) is such that no significant secondary field is produced in an 'average' Earth. A conductive body will produce a secondary field due to **eddy currents**, dependent on the **electrical conductivity**, dimensions and location of the conductor. Of the two main categories of electromagnetic exploration systems, *dip-angle methods* merely measure the direction of the combined primary and secondary fields, whereas *phase component systems* measure the relative magnitudes of the secondary field, both in and out of phase with the primary. [ATB]

Grant, F.S. & West, G.F. (1965) *Interpretation Theory in Applied Geophysics*. McGraw-Hill, New York.
Keller, G.V. & Frischknecht, F.C. (1966) *Electrical Methods in Geophysical Prospecting*. Pergamon, Oxford.

electron microscopy The use of beams of high energy electrons to form images of the surfaces or internal structures of materials. This approach, heralded by the construction of the first electron microscope in the late 1930s, has expanded to incorporate not only imaging and *electron diffraction*, but also the chemical *microanalysis* of materials by the detection and processing of the various signals generated by the action of an electron beam on a specimen.

The basis for the use of electrons in imaging is the *de Broglie relationship* between wavelength λ and momentum p

$$\lambda = h/p$$

where h is *Planck's constant*. The velocity of an electron, and hence its momentum, is determined by the accelerating voltage. High energy electrons have wavelengths ($\simeq 3.7$ pm at 100 keV) much less than those of visible light ($\simeq 500$ nm for green light). Using radiation of shorter wavelengths in microscopy improves resolution. Electrons are easily scattered by collisions with gas molecules, so that an electron microscope must have a good internal vacuum ($\simeq 10^{-5}$ torr). An electron beam can be focused by means of electromagnetic lenses, thus facilitating the design of a transmission instrument analogous to the optical microscope. Developments since 1950 have led to the design of several other types of electron microscope, each best suited to particular purposes.

As indicated in Fig. E12(a and b), electrons are scattered when they interact with a solid, but both forward-scattered electrons and the transmitted beam only exit the specimen if it is thin ($\simeq 300$ nm for 100 keV electrons). The incident electrons also generate secondary electrons and photons of varying energies, which exit from the upper specimen surface. The latter, and inelastically forward-scattered electrons, are characteristic of the atoms comprising the target (specimen) and therefore can be utilized in compositional analysis.

SCANNING ELECTRON MICROSCOPY (SEM). This technique is primarily used for the investigation of surface structure and composition. Bulk samples are irradiated with electrons (typically $\simeq$ 10 keV in energy), causing several effects, as indicated in Fig. E12(a). With **minerals**, the mostly widely used imaging methods employ secondary electrons and backscattered electrons. No special specimen preparation methods are usually employed, but electrical charging of insulating specimens is suppressed by coating with a thin conducting layer (usually **gold**).

A focused beam of electrons is continuously scanned over a small area of the specimen. Some of the emitted secondary or backscattered electrons are collected and the resulting signals are used to vary the voltage on the grid of a cathode ray tube (CRT), whose screen is scanned in synchronism with the specimen. The resulting CRT image shows mainly topographic detail when secondary electrons are detected, and a resolution of 25 nm in this imaging mode is readily attainable. Backscattering of electrons increases with the atomic charge on the target atoms, so that the contrast in a backscattered electron image indicates changes in average atomic number, and is useful in the qualitative identification of minerals. Different magnifications are produced by altering the area of specimen scanned by the primary beam and typically range up to 100 000 times. Images on the CRT are recorded photographically. Signals like specimen current and Auger electrons are mostly used outside the earth sciences, e.g. for semiconductors, but X-rays excited by electrons striking the specimen are of universal value for *microanalysis* (EDX), as described below. The great value of SEM lies in the broad similarities between the properties of secondary electron images and those formed by the unaided eye, an adequate depth of focus at all magnifications and easy preparation of samples.

TRANSMISSION ELECTRON MICROSCOPY (TEM). Electrons transmitted through a crystalline specimen provide information about its internal **microstructure**. High energy ($\simeq$100 keV) electrons are focused by lenses to irradiate a chosen area of specimen and post-specimen lenses enable high magnification images (up to 1 000 000 times) to be formed and focused on a phosphor-coated plate, giving resolutions of $\simeq$0.3 nm. The regular atomic arrangments in crystals and mineral grains cause the diffraction of electrons into discrete beams, according to the *Bragg law* $n_\lambda = 2d_{hkl} \sin \theta$, θ being very small ($\lambda \ll d_{hkl}$), unlike in **X-ray diffraction**. Imaging is achieved by using a small aperture to select either the undiffracted beam emerging from the specimen (bright field) or one of the diffracted beams (dark field). **Microstructural defects in crystals** are imaged because they cause local distortions of the lattice, leading to variations in diffraction contrast. Comparisons of this contrast under different diffraction conditions enable the **strain** fields of defects, Burger vectors of **dislocations**, etc. to be analysed. Bringing all the diffracted beams to a focus produces a pattern that is characteristic of the mineral and its crystallographic orientation with respect to the incident electron beam. Such diffraction patterns are sections through the reciprocal lattice and are thus easy to interpret. Magnified images and the corresponding diffraction patterns are recorded photographically by directly exposing film to electrons.

The combination of high magnification imaging of **microstructure** with crystallographic and microchemical data (by means of *X-ray microanalysis*, see below) makes *TEM* a very effective tool in studying **deformation** processes, *phase transformations*, **exsolution**, etc. *Convergent Beam Electron Diffraction* (CBED) is also able to determine the symmetry (*point groups* and space group) of small crystals and second phases. Because *TEM* offers such powerful crystallographic techniques, *TEM* microscopes have specimen stages that enable grains to be precisely oriented during study. Other types of stage enable specimens to be heated, cooled, or **strained** during observation.

The preparation of a specimen for *TEM* is seldom simple, because the electron-optics require specimens to be located within the objective (focus) lens (which limits their diameter to $\simeq$ 3 mm), while for transmission of electrons the specimen must be $\simeq$ 200 nm thick. Crushed fragments can be used but rock and mineral samples are more often prepared by a process called *ion-milling*. Beams of argon ions (or atoms), fired at a piece of demounted, polished thin section, gradually erode its surfaces until small holes form, between which there are electron-transparent regions.

High voltage microscopes (>200 kV, which gives both greater specimen penetration and better resolution) and microscopes dedicated to *scanning transmission* (STEM) operation are also manufactured (STEM also improves penetration). Some *TEM* instruments are fitted with scanning coils, enabling *SEM*-type imaging to be obtained, while other possible attachments include a *magnetic spectrometer*, enabling the energy lost by inelastically scattered electrons to be measured. The energy loss is characteristic of the elements in the specimen and this technique (EELS) is useful for chemical *microanalysis* (light elements, in particular).

Imaging of the fundamental **structures of crystals** is now within the capabilities of many transmission microscopes. It is known as lattice imaging. This high resolution technique requires very thin specimens and the electron beam to be parallel to a low-index crystal direction. Many **minerals** have rather large interplanar spacings ($\simeq$1 nm), making it relatively easy to image lattice fringes which correspond to close-packed planes of atoms. Obtaining the ultimate resolution in order to reveal more details of the atomic arrangements requires purpose-designed instruments. The resulting images must be interpreted with care, usually by comparison with computer-simulated images of the structure.

MICROANALYSIS BY ENERGY DISPERSIVE X-RAY ANALYSIS (EDX). Both *SEM* and *TEM* commonly employ a semi-conductor detector-based system to measure the energies of the X-rays generated from the specimen by electrons. The X-ray photons are sorted by energy and displayed as a spectrum of X-ray lines, characteristic of the elements in the volume of the specimen with which the beam has interacted. The interaction volume for *SEM* is large (Fig. E12a), so that to obtain quantitative *micro-*

analysis requires corrections for absorption, fluorescence and energy-dependent detector sensitivity. The thin specimen used in *TEM* gives a small interaction volume (Fig. E12b), so that absorption and fluorescence are small effects. The intensity ratio between two X-ray lines is thus taken to be directly proportional to the concentration ratio of the corresponding elements, through an instrument factor that corrects for detector sensitivity and can be determined from standards. [DJB]

Goldstein, J.I., Newbury, D.E., Echlin, P., *et al.* (1981) *Scanning Electron Micoscopy and X-ray Microanalysis*. Plenum, London.
Hirsch, P.B., Howie, A., Nicholson, R.B. Pashley, D.W. & Whelan, M.J. (1965) *Electron Microscopy of Thin Crystals*. Butterworths, London.
Loretto, M.H. (1984) *Electron Beam Analysis of Materials*. Chapman & Hall, London.

electron spin resonance (ESR) Method of dating some geological and archeological inorganic materials. The technique involves the detection of unpaired electrons which have resulted from the action of ionizing radiation and/or heating. The sample is placed in a strong **magnetic field** and unpaired electrons are detected by their absorption of microwaves. The method has been developed for dating of **cave** *stalagmites* and archeological ceramics. It can also be used to identify **flints** which have been subjected to heat treatment in antiquity (in order to improve their working quality). [OWT]

electrum A naturally occurring **gold–silver** alloy. (See **native element**.) [DJV]

eleolite A massive variety of the mineral **nepheline**. (See **feldspathoid minerals**.) [DJV]

elevation correction The correction applied to a **gravity** or magnetic measurement to compensate for the decrease in amplitude of the field with height. In **gravity reduction** it also includes correction for the attraction of the rocks between observation point and datum and for topography. It is very small for **magnetic fields** and is generally ignored. [PK]

elevation energy The potential energy of a given mass of water with respect to its elevation above a prescribed datum level. [AME]

elongation The relative change in length of a line during **deformation** with respect to its underformed length. It is expressed as a ratio:

$$e = \frac{\text{deformed length} - \text{undeformed length}}{\text{undeformed length}}$$

A *contraction* is a decrease in length of a line during **deformation**. A contractional **strain** is measured using the expressions for elongation, giving negative values, or *stretch*, giving values less than 1.

The elongation is positive for **extension**, and negative for **contraction**.

An *elongation lineation* is a type of **lineation** formed by a set of parallel elongate objects in deformed rock. Such a **lineation** may mark the maximum *principal finite strain* direction and is useful in indicating the movement direction in high-**strain shear zones**, **mylonites**, etc. [SB/RGP]

Means, W.D. (1978) *Stress and Strain*. John Wiley & Sons, New York.

eluviation The movement of *soil* material through the *soil* zone. The transport of minerals in solution or material in suspension may be lateral as a result of **throughflow** down slopes or it may be vertical as a result of *leaching*. Soil horizons which have lost material such as clay or have been leached of water-soluble minerals, are described as eluvial; the upper horizons of podsolic *soils*, for example, are eluvial. Horizons which gain material through eluviation are described as **illuvial**. Some types of desert **duricrust** — notably **calcretes** and many **gypsum** crusts — are **pedogenic illuvial accretions**. [AW]

eluvium *In situ* **weathered** bedrock. Eluvium is material produced by **weathering** of bedrock but which has not been transported by geomorphic processes. The transport of eluvium by slope processes such as *mass-wasting* results in the deposition of **colluvium**. In the tropics and subtropics, **deep weathering** of crystalline rocks produces thick accumulations of eluvium which are referred to as **saprolite**. [AW]

emerald A precious stone, the deep green **gem** variety of the mineral **beryl**. [DJV]

en echelon Arrangement of parallel lines or planes (e.g. **dykes**, fissures, *fold axes*) in which each unit is of finite length and is displaced laterally from those on either side in a consistent sense (Fig. E13). [RGP]

enargite (Cu_3AsS_4) A sulphosalt **ore** mineral of **copper**. [DJV]

endlichite A variety of secondary **lead** mineral intermediate between **vanadinite** and **mimetite**. [DJV]

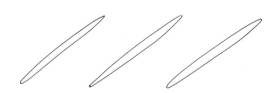

Fig. E13 En echelon arrangement.

endrumpf A **peneplain** or land surface that has been reduced to a flat plain or gently undulating landscape by **erosive** processes. [ASG]

engineering geological mapping Conventional geological maps do not fulfil the needs of the civil engineer or the engineering geologist in several respects. On conventional maps rocks are grouped together because they are of the same age or origin irrespective of their markedly different geotechnical properties. There is a lack of quantitative information on the physical properties of rocks, the presence of discontinuities, the extent of **weathering** and **groundwater** conditions, all of which are vital in engineering planning and operations.

Regional Engineering Geology Maps (Scale 1:10 000 or smaller), which are generalized and intended for use during the preliminary planning stage, should show units defined by their engineering properties (e.g. **weathering** with its distinctive properties may cross rock units) and should be in terms of a descriptive engineering rock or *soil* classification.

Engineering Geology Plans are more detailed and also display **cross-sections** plus the walls and floors of trial pits. These are of much wider use than the regional map and are drawn at a larger scale. The general-purpose scale is 1:5000 and is used for **site investigation** for reservoirs, dams, tunnels, road systems, etc. A 1:1250 scale is used for more detailed presentations of the sites of, for example, bridges or buildings. The sections of inspection trenches or trial pits would normally be presented at scales between 1:500 and 1:100.

In order to prepare an engineering geology map a large-scale published topographic map may be used as a base for completely new mapping, or existing geological maps would be enhanced with relevant data. For detailed studies, however, it is normal practice to resurvey the area and produce a completely new purpose-drawn map. These maps should contain such geological information as a description of the rocks and *soils*, the degree of **weathering** and **alteration**, geological boundaries, **shear** surfaces and zones, **joints**, **faults**, nature and type of mass movements such as **landslides**, **cambering**, **seismicity**, caverns, **hydrological** conditions and properties, **springs** and **seepages**. In addition, the localities of boreholes, trial pits and other **site investigations** should be marked together with old mine shafts or **adits**, limits of old workings of mines and quarries (open and filled), tips and other made-up ground, areas of subsidence and disturbed ground. [JCC]

Geological Society of London (1972) *The Preparation of Maps and Plans in Terms of Engineering Geology.* Report by the Geological Society Engineering Group Working Party. *Quarterly Journal of Engineering Geology* **5**: 293–382.
International Association of Engineering Geology (1976) *Engineering Geological Maps: A Guide to their Preparation.* UNESCO Press, Paris.

engineering geology The application of geological information, techniques and principles to the design, construction and maintenance of engineering works. *Soils* or rocks involved in engineering works may be used either *in situ* or as construction materials. *In situ* uses include the ground beneath a foundation, in a slope or the medium involved in a tunnelling, mining or other excavation process. Examples of constructional materials include decorative and structural **building stones**, **aggregates** for concrete or roadstone manufacture, and fills. The consequences of not fully taking into account the behaviour of geological materials under the conditions imposed by engineering works can be severe and lead to increased maintenance costs, loss of serviceability, structural failure or even loss of life. Hence the collection and interpretation of geological and geotechnical data are an essential prerequisite for rational engineering design and construction work. It is also desirable that geological expertise is applied in land use planning, the disposal of solid and liquid wastes, particularly hazardous ones (see **sanitary landfill**), as well as the exploitation of **groundwater**, **hydrothermal** and *hydrocarbon* resources. Increased awareness of the need for the care of the environment by the use of engineering solutions which harmonize with their surroundings has heightened the significance of the subject.

Definitions of engineering geology, which vary in respect of the emphasis they give to different aspects of the subject, exist in the literature. One of the most useful definitions is provided by the Association of Engineering Geologists (Anon., 1969). It advocates that the purpose of engineering geology is to ensure that geological factors affecting the design, construction, operation and maintenance of engineering works and the development of **groundwater** resources are recognized, adequately interpreted, utilized and presented for use in engineering practice. A historical view of engineering geology is provided by Anderson & Trigg (1976) who point out that, although some attention was paid to geological factors in mining engineering, not until well into the 20th century was geological information universally or uniformly used in the civil engineering industry. This is in spite of the fact that the civil engineer William Smith (1769–1839) made important advances in early geological studies and in 1815 published a geological map of southern England. Increasing labour costs, the requirement for large structures, particularly dams, and a greater need for construction on difficult ground gradually brought about the practice of commissioning geological reports for engineering works. Early geological studies often provided little more than an inventory of the materials present with little or no regard to factors of significance to their engineering behaviour. Furthermore, in many cases, these geological investigations were carried out without reference to the engineering proposals.

In practice, the ground conditions relevant to particular engineering proposals are determined by carrying out a **site investigation**. A knowledge of the composition and mode of formation of rocks and *soils* coupled with an awareness of the implications of these factors to engineering design or construction operations equips the engineering geologist to make informed predictions regarding the spatial distribution and behaviour, both long and short term, of geological materials. Engineering designs are then based on numerical analysis of the behaviour of the materials involved. Predicting the distribution of rocks and *soils* between points at which they can be determined, for instance in boreholes and temporary or natural exposures, requires an appreciation of the geological **structure**. Engineering geologists may advise on the locations of boreholes, the samples to be tested and the type of tests to be performed as well as providing an interpretation of **strength**, compressibility and other test results. There are a number of reasons why, in carrying out the interpretation of geotechnical tests, a knowledge of the mode of formation and composition of the materials may be important. For example, removing samples from the ground and testing them in a laboratory inevitably causes some disturbance to the material. Even carrying out *in situ* testing does not entirely overcome this difficulty since changes in the value of *in situ* **stress** during the preparation of the test site may modify the properties of the material. Certain rocks display **anisotropy**, metastability or non-linear **stress–strain** characteristics while others retain high levels of residual **stress**. In some cases the presence of relatively minor amounts of certain **minerals** can radically modify the engineering properties, particularly in respect of resistance to **weathering** of the material. Finally, owing to the presence of **joints**, fissures and other discontinuities within rock masses, laboratory tests on small samples of rocks and *soils* may not provide an adequate indication of the behaviour of the material *en masse*.

The aspects of geological science liable to be applied most frequently in engineering geological work are *structural geology*, petrology, sedimentology, **hydrogeology** and **geophysics**. Disciplines of less obvious direct relevance to engineering needs, including paleontology, crystallography and **geotectonics**, may also be utilized.

Relevant associated sciences include glaciology, pedology, and **geomorphology**. The engineering geologist needs also to possess a basic understanding of many aspects of engineering, in particular *soil mechanics*, rock mechanics, foundation engineering, the techniques of ground property improvement, tunnelling, **open cast** and deep mining, quarrying, **hydrology**, excavation design and slope stability analysis. Although many engineering geologists undertake design work in relation to earth materials in engineering, these operations may also be carried out by civil or geotechnical engineers.

Unfortunately, an inadequate appreciation of the geological conditions in engineering design can stem from difficulties with incorporating mainly qualitative geological information into designs. The selection of relevant data and their transmission in a suitable form is a vital aspect of the effective communication of geoscience data between geologists and engineers. This has been greatly assisted by the standardization and quantification of geological descriptions, an appropriate approach having been fostered by the widespread use of the recommendations of working parties set up by the Engineering Group of the Geological Society of London (see Anon., 1977) and other bodies with similar aims. In these, particular attention is paid to geological factors of greatest potential significance

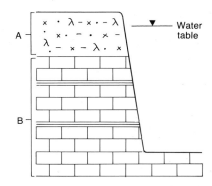

Fig. E14 Engineering geology. Engineering assessment for an excavation.

Table E4 Engineering geology. Engineering assessment for an excavation

Stratum	Description	Engineering assessment		
		Excavation method	Short-term stability	Long-term stability
A	Firm brown and grey-brown mottled medium dense to dense, in part laminated, sandy silt with a little clay. Occasional roots at top and some limestone gravel towards base	Face shovel or backacter excavator	Excavations deeper than 1.0 m will require support. Danger of slumping and flow of material where silt laminae occur	Slope cut back to shallow angle to be determined by stability analysis. Requires drainage works and protection from erosion
B	Light brown, thinly, medium and thickly bedded, slightly weathered and fresh, moderately strong and strong limestone with some thin clayey bands and medium spaced subvertical intersecting silty clay infilled joints in two sets	Will be rippable where moderately weathered. Thick units of fresh, strong limestone will require blasting prior to excavation	Will stand vertically during construction period	Excavation liable to break back to vertical discontinuities as clay infills soften. Rock bolting and masonry infill required to secure slope

in engineering work. Accordingly, in addition to colour, grain size and lithology, the guidelines also provide definitions for the description of **strength**, **weathering** grade and the spacing and other features of discontinuities.

Engineering geological investigations usually relate specifically to the site of proposed engineering works. Table E4 and Fig. E14 illustrate an interpretation for an excavation in terms of the excavation method as well as the short- and long-term stability of the side slopes. Studies may also be carried out to determine the causes of engineering failures involving the ground and to provide data for the design of remedial measures. On occasions investigations consist of a broadly based exploration of an area to provide general data about the ground conditions. Figure E14 is an example of an environmental geological map in which a consideration of the geological conditions, surface topography and **groundwater** conditions has given rise to a zonation in terms of land instability. An important further function of many engineering geologists is to anticipate the location, effects and frequency of a variety of geological hazards. These include volcanoes, seismic activity, natural and man-induced ground movements and slope instability. Advice concerning remedial or avoidance measures may also be required.

An increased awareness of the importance of engineering geology has been reflected in the UK by the establishment, respectively in *c.* 1964 and 1947, of the Engineering Group of the Geological Society of London and the British Geotechnical Society, a section of the Institution of Civil Engineers. Bodies with similar aims in the USA and Europe respectively include the Association of Engineering Geologists and the International Association of Engineering Geology. Several technical journals that cover advances in the subject share a similar vintage. Of these the *Quarterly Journal of Engineering Geology* (Geological Society of London, 1967), *Engineering Geology* (Elsevier, 1967), the *Bulletin of the International Association of Engineering Geology* (1970) and *Engineering Geology* (Association of Engineering Geologists, 1964) are probably the most widely read. Relevant papers are also published in *Géotechnique*, the *International Journal of Rock Mechanics Mining Sciences and Geomechanical Abstracts, Rock Mechanics,* the *Journal of Geotechnical Engineering,* the *Canadian Geotechnical Journal* and *Ground Engineering*. [JCC]

Anderson, J.G.C. & Trigg, C.F. (1976) *Case Histories in Engineering Geology*. Elek Sciences, London.

Anon. (1969) *Newsletter of the Association of Engineering Geologists* **12**: No. 4, 3.

Anon. (1977) The logging of rock cores for engineering purposes. *Quarterly Journal of Engineering Geology* **10**: 45–52.

enhanced/tertiary oil recovery The techniques employed when primary and secondary recovery techniques have declined beyond economic acceptibility. These techniques are commonly thermal, e.g. the heating of viscous oil by injecting steam into the **reservoir** rook. [AME]

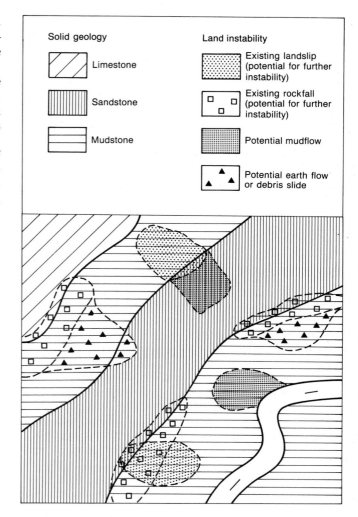

Fig. E15 Engineering geology. Part of an environmental geology map showing land instability hazard.

enstatite ($MgSiO_3$) The magnesian end-member of the *orthopyroxene solid solution* series of minerals. (See **pyroxene minerals**.) [DJV]

entrainment equivalence A term which expresses the equivalence between different grains on a sediment bed in relation to the fluid forces acting upon that bed at the onset of sediment movement: two grains have entrainment equivalence when they begin to move at the identical fluid **shear stress**. This term is dependent not only upon grain size, shape and density but also upon the interrelationships between grains in a mobile bed and the feedback these have upon the initiation of sediment transport. Entrainment equivalence is increasingly being recognized as important in the formation of **placer deposits** rather than just **hydraulic equivalence** alone. [JLB]

enveloping surface Imaginary surface joining the **crest** or **trough lines** of a set of **folds**. Successive enveloping

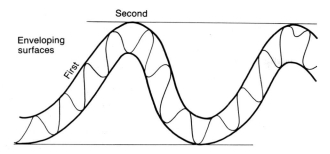

Fig. E16 A complex profile may be simplified by drawing one or more **enveloping surfaces** joining the parasitic fold hinges.

surfaces may be employed in order to understand and simplify complex **fold systems** (Fig. E16). [RGP]

environmental engineering geomorphology A branch of **applied geomorphology** which deals with the study of geomorphic features and processes in relation to environmental management and engineering. Environmental engineering geomorphology usually comprises one aspect of a multidisciplinary investigation of a site or region where planners require an assessment of the likely environmental impact of an engineering project. In other cases, environmental engineering geomorphologists may be called upon to undertake surveys of the natural resources of an area or to assess potential environmental hazards at a site where development is proposed.

The analytical techniques employed in environmental engineering geomorphology are the same as might be used in any geomorphological study. Preliminary geomorphological mapping of the area may be facilitated using **remote sensing** imagery including *aerial photographs*, but group surveys are the fundamental research tool of the geomorphologist. Having mapped the terrain, monitoring of geomorphic processes and analysis of surficial materials provide data which will quantitatively characterize the local geomorphic phenomena. This information may then be used to develop laboratory simulations of natural processes or to formulate computer models. In both cases, the techniques attempt to replicate geomorphic processes under controlled, artificial conditions. In this way, natural phenomena which may be difficult to monitor in the field owing to technical limitations or the constraints of time can be studied in detail.

With the increasing awareness of potential human impact on the environment, the demand for professional environmental engineering geomorphologists has dramatically expanded. It is today one of the most rapidly growing fields of **geomorphology**. In North America and western Europe, concern about the effects of urban and industrial growth on the natural environment are prompting numerous government-sponsored studies of environmental issues. In North America, for example, the United States Environmental Protection Agency and

Environment Canada are responsible for identifying hazardous waste sites and for directing clean-up programmes. Such projects require the expertise of soil scientists, hydrogeologists and ecologists as well as geomorphologists who, ideally, are in a position to synthesize the findings of scientists and engineers working in diverse fields. Within an urban or industrial landscape, the environmental engineering geomorphologist must attempt to gauge the influence that human beings have had on the geomorphic process. This may involve studies of pollution of *soil*, water and atmosphere, but human influence on rates of **erosion** may be equally detrimental to the natural environment.

The effects that river channelization and storm drains have on river discharge and rates of fluvial **erosion** are critical in areas where river **flood plains** have been used for agriculture and urban development. Similarly, the development of coastal areas — particularly for recreational purposes — may result in potentially catastrophic modification of the environment. Along the eastern seaboard of the United States, extensive **barrier island** systems have been developed with little regard for the preservation of natural sediment budgets. The contruction of groynes and sea walls reduces the supply of **sand** to the coastal **dunes** which are then susceptible to **ablation** — especially if the protective vegetation cover is disturbed. Eventually, this degradation of the **barrier island** leads to more frequent breaching by *storm surges*, which destabilizes the system still further. At an advanced stage of deterioration, occupation of the **barrier island** is seemingly foolhardy. Nevertheless, some large urban centres — such as Atlantic City, New Jersey — continue to develop despite their tentative position. At this point, the role of the environmental engineering geomorphologist may be to offer short-term solutions to problems caused by environmental forces which can neither be modified nor accommodated in the long term.

Such dilemmas are not unusual. While the identification of natural or man-made environmental hazards is an important aspect of environmental engineering geomorphology, it does not necessarily follow that the perceived hazard can be overcome. In the case of the populated **barrier island** which is prone to catastrophic flooding, the only effective solution to the problem may be to relocate the inhabitants. Even on this scale, however, this may be impractical. On a regional scale, where large urban areas are threatened by desertification, **sand** encroachment, or **earthquakes**, such a solution is all but impossible. In effect, the role of the environmental engineering geomorphologist is to propose ways in which the environmental threat can be tolerated or ways of modifying the environment in order to alleviate the hazard — at least temporarily.

Studies of **sand** and **dust** control in arid regions provide examples of ways in which environmental hazards may be tolerated or alleviated. In many such areas, the flow of wind-blown **sand** and airborne **dust** is a regional phenomenon. Huge volumes of material are mobilized by atmospheric forces which cannot be controlled. However, on a local scale, roads and other installations can be designed to tolerate the flow of **sand** and prevent the accumulation of drifts. If **sand** flow cannot be accommodated, as is the case around airports, techniques of trapping **sand** may be employed; barriers, fences, and tree-belts can accomplish this. Similarly, the encroachment of moving **sand dunes** can be halted temporarily by mechanically altering the **dunes'** aerodynamic profiles, or by **dune stabilization** using gravel, salt crusts, oil, and chemicals.

Alleviation of an environmental threat by interfering with natural geomorphic processes can in itself exacerbate the hazard. Hence, in the arid zone, the stabilization of **dunes** and trapping of wind-blown **sand** results in the accumulation of sandbanks which must be carefully monitored to prevent their mobilization. **Sand** fences must be replaced as they become buried by accumulating **sand**, and tree belts must be monitored to ensure that they are not depleted by drought or disease.

It may appear that by seeking to minimize the environmentally disruptive effects of engineering projects, while at the same time minimizing environmental hazards, the environmental engineering geomorphologist is attempting to achieve an impossible compromise. However, often such compromises are necessary only when there has been a lack of planning or insufficient geomorphological information during the planning process. Hence, there are increasing restrictions on the residential and commercial development of **barrier islands** along the Atlantic Coast of the United States. Also, in the Middle East, there is a growing awareness that **sand** control schemes are far more efficient when knowledge of the geomorphic processes is applied at the design stage rather than only when remedial measures are required. By undertaking geomorphological surveys of a site at the outset of an engineering or planning project, it should be possible to avoid or minimize potential environmental hazards and environmental disruption. Hence, today, in parts of the Middle East, environmental engineering geomorphologists are involved in the planning of new roads, airports, and industrial and residential facilities. They evaluate potential problems which may be caused by blowing **sand** and **dust**, or by salt **weathering**; their recommendations may call for the selection of alternative sites, the implementation of preventative engineering work, or the design of schemes to protect the installations from environmental hazards.

The scope of environmental engineering geomorphology is broad. While some aspects fall within the areas of expertise of geologists and sedimentologists, or hydrologists and hydrogeologists, many are the exclusive domain of the geomorphologist. Resource surveys of surficial materials, for example, usually require a thorough knowledge of regional geology, sedimentology, *soils* and **hydrology** — all of which are encompassed by **geomorphology**. Similarly, studies of *soil* **erosion**, *gullying*, mass

movements, aeolian processes, the **weathering** of rock and building materials and many more environmental processes require an understanding of the complex environmental interactions which geomorphology strives to explain. [AW]

Coates, D.R. (ed.) (1976) *Geomorphology and Engineering. Proceedings of the 7th Binghamton Symposium in Geomorphology.* Dowden, Hutchinson & Ross, Stroudsburg, PA.

Coates, D.R. (1981) *Environmental Geology.* John Wiley & Sons, New York.

Cooke, R.U. & Doornkamp, J.C. (1974) *Geomorphology and Environmental Management.* Clarendon Press, Oxford.

Jones, D.K.C., Cooke, R.U. & Warren, A. (1986) Geomorphological investigation, for engineering purposes, of blowing sand and dust. *Quarterly Journal of Engineering Geology* **19**: 251–70.

Nordstrom, K.F. & Psuty, N.P. (1980) Dune district management: a framework for shorefront protection and land use control. *Coastal Zone Management Journal* **7**: 1–23.

Watson, A. (1985) The control of wind-blown sand and moving dunes: a review of the methods of sand control in deserts, with observations from Saudi Arabia. *Quarterly Journal of Engineering Geology* **18**: 237–52.

environmental impact statement (*EIS*) A summary of the information obtained from a study of the impact that an action will have or has had on the environment. Generally referring to the effect of a proposed action — such as the implementation of an engineering project or land use scheme — the impacts may be direct and planned, or they may be indirect, unplanned, and undesirable. In either case, they may be long-term and irreversible, so it may be impossible to remedy problems caused by unforeseen impacts.

It is the purpose of the EIS to provide a thorough and succinct assessment of the likely effects that a planned course of action will have on the environment. It will typically comprise the following:
1 a description of the proposed action;
2 the likely environmental impact;
3 the likely positive and negative effects of the action;
4 the likely duration of the impacts;
5 the range of direct and indirect impacts;
6 an assessment of whether the impacts are reversible or irreversible;
7 the geographical scale of the impacts.
The EIS may also incorporate an evaluation of the likely implications if the proposed course of action is not implemented. Generally, an EIS will be confined to an assessment of likely changes in the physical environment. In some instances, however, broader social and economic impacts may also be evaluated. This evaluation of cause and effect must balance available technical data against assessments which are necessarily speculative or subjective. Hence, an EIS is not necessarily a definitive, objective assessment of cause and effect. [AW]

Heer, J.E. & Hagerty, D.J. (1977) *Environmental Assessments and Statements.* Van Nostrand Reinhold, New York.

eolian See **aeolian**

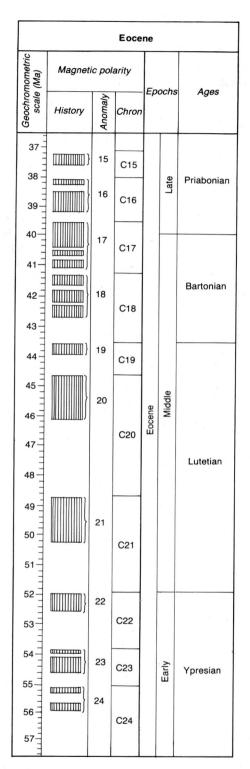

Eocene				
Geochromometric scale (Ma)	Magnetic polarity		Epochs	Ages
	History	Anomaly	Chron	

Fig. E17 Eocene.

Australia was still attached to Antarctica at the Tasman Rise but continued to drift northwards, with its marsupial fauna, towards warmer latitudes. At *c.* 45 Ma India collided with Eurasia. The **seafloor spreading** between Greenland and Scandinavia formed the Reykjanes–Iceland–Jan Meyen–Spitzbergen ridge, the Faeroes and Rockall Plateau thus splitting off the European continent from North America (Pomerol, 1982). The major Alpine tectonic event occurred prior to the end of the Eocene.

The warm climate continued through to the Late Eocene. Diversity of calcareous nannofossils, particularly *discoasters*, increased during the Early Eocene; during the Middle Eocene many species disappeared and only a few new ones appeared in the remainder of the Eocene. The last appearance datum of all rosette-shaped *discoasters* is in the Late Eocene.

Rapid evolution of the larger **Foraminifera** *nummulites* took place in the Eocene while many larger benthic **Foraminifera** also became extinct. The planktonic **Foraminfera** underwent a further burst of evolution through the Eocene, but a large number became extinct in the Late Eocene.

More advanced **mammals** displaced many older forms in Europe and North America. New arrivals included true rodents and squirrel-like species. More advanced whales evolved (many-toothed and whale-bone varieties) at *c.* 40 Ma, together with insectivores, shrews, and hedgehogs.

In the Eocene, vegetation took on a modern appearance with **land plants** having Recent affinities becoming forest-dominant (Thomas & Spicer, 1987). The warmer global climate had tropical and subtropical floras extending to latitudes 50° (N and S) with broad-leaved deciduous and coniferous forests in the higher latitudes (Wolfe, 1980).

In the Eocene the following extant floral orders appeared: Theales, Thymeliales, Nepenthales, Polemoniales, Dipsacales, Scrophulariales, Cyperales, Eupteliales, Saxifragales and Liliales (Thomas & Spicer, 1987). The first megafossils of grass are recorded from the Eocene and these plants became increasingly abundant. [DGJ]

Berggren, W.A., Kent, D.V. & Flynn, J. (1985) Jurassic to Paleogene: Part 2 Paleogene geochronology and chronostratigraphy. In: Snelling, N.J. (ed.) *The Chronology of the Geological Record*, pp. 141–95. Memoirs of the Geological Society No. 10. Blackwell Scientific Publications, Oxford.

Pomerol, C. (1982) *The Cenozoic Era*. Ellis Horwood, Chichester.

Thomas, B. & Spicer, R. (1987) *Evolution and Palaeobiology of Land Plants*, pp. 1–304. Croom Helm, London.

Wolfe, L.A. (1980) Tertiary climates and floristic relationships at high latitudes in the Northern Hemisphere. *Palaeogeography, Palaeoclimatology, Palaeoecology* 30: 313–25.

Eocene The Eocene Epoch has a duration of *c.* 15 Ma from the end of the **Paleocene** at *c.* 52 Ma to the base of the **Oligocene** at *c.* 37 Ma and has been divided into four stages (Berggren *et al.*, 1985) (Fig. E17).

Eötvös correction (EC) A correction applied to **gravity** measurements taken from a moving vehicle (ship or aircraft). Depending on the heading, the vehicle's motion generates a centripetal acceleration which either reinforces or opposes **gravity**.

$$EC = 75.03\,V \sin \alpha \cos \phi + 0.04154\,V^2 \text{ g.u.}$$

where V is the speed in knots, α the heading and ϕ the latitude. [PK]

epeirogeny/epeirogenesis Very large-scale tectonic movements which result in **uplift** and subsidence of areas of the **continental crust** or oceanic **basins** over thousands of kilometres (see **diastrophism**). Epeirogenic upwarping has formed the stable, **cratonic** portions of some continental interiors. The crustal rocks are not significantly deformed during this process, unlike during **orogeny**; nor are regional metamorphism or igneous intrusion commonly associated with epeirogeny. It has been suggested, however, that during the intervals between epeirogenic episodes the uplifted landscape is eroded. The sedimentary strata which accumulated during these cycles have been termed sequences in North America.

The forces which drive epeirogeny are not well understood. They may be generated by changes in the volumetric expansion of the **asthenosphere** caused by **olivine** hydration, or *phase transitions* in the **mantle**. Alternatively, **convection** currents within the **mantle** may cause **crustal** warping — doming above rising *plumes*, and subsidence above heat sinks. Similar topographic features, though with greater relief, have been identified on the surface of the Earth's **core**; these have been tentatively attributed to **convection** within the **mantle**. [AW]

Holmes, A. (1944) *Principles of Physical Geology* (1st edn; see also 3rd edn, 1978). Nelson, Sunbury-on-Thames.

epicentral distance The distance between an **earthquake epicentre** and *seismometer*, defined in terms of the angle in degrees that they subtend at the Earth's centre. Thus from either pole to any point on the Equator is 90°; from London to San Francisco is about 77°. Distances in degrees should be multiplied by 111.1 to obtain the equivalent distance in kilometres along the **great-circle** route at surface. (See **shadow zone**, Fig. S21, p. 559; **teleseism**.) [RAC]

epidote ($Ca_2(Al,Fe)Al_2(SiO_4)(Si_2O_7)(O,OH)_2$) A group of apple-green hydrated calcium aluminosilicate minerals with double *island silicate* (*sorosilicate*) structures. [DJV]

epigenetic deposit A deposit formed after the host rocks in which it occurs. A good igneous analogy is a **dyke**; an example among mineral deposits is a **vein**. [AME]

epilimnion The upper, oxygenated and circulating layer of a stratified lake. This layer is warmed by solar radiation and its base is marked by a rapid temperature gradient into a cooler lower layer (**hypolimnion**). The plane where the temperature gradient is greatest is the **thermocline**. The epilimnion varies from less than 10 m to over 50 m in depth. It is the zone of maximum productivity of photic organisms. [PAA]

epitaxy Refers to cement overgrowths which are of a different mineralogy from the substrate grain. **Calcite** overgrowths on **quartz** grains are an example of epitaxial cements.

Some dispute exists over the terms 'epitaxial' and '**syntaxial**'. To most workers in the field of siliciclastic **diagenesis**, the two words are quite distinct in their usage, as is suggested by the entries in this encyclopaedia. However, in carbonate **diagenesis**, the two are synonymous, and refer to cement growths that are in optical continuity with the substrate grain. (See **diagenesis, syntaxy**.) [DMP]

Pettijohn, F.J., Potter, P.E. & Siever, R. (1987) *Sand and Sandstone* (2nd edn). Springer-Verlag, New York.
Bathurst, R.G.C. (1975) *Carbonate Sediments and their Diagenesis* (2nd enlarged edn). Elsevier, Amsterdam.

epithermal deposit **Epigenetic** mineral **deposit** formed at low temperatures near the Earth's surface. The probable depth of formation ranges from near-surface to about 1500 m and the temperature is about 50–200°C. These deposits often occur as **veins** in *normal fault* systems, **joints** or other fissures and they have usually been formed by the infilling of open spaces and not by **replacement**; they therefore frequently display **crustiform banding**. They are important sources of **lead**, **zinc**, **gold**, **silver**, **mercury**, **antimony**, **uranium**, **fluorite**, **baryte** and other elements and minerals. This type of mineralization may vary abruptly with depth and often has only a small vertical range. [AME]

epsilon cross-stratification A type of lateral **accretion** structure with gently inclined curved **bedding** surfaces with **dip**-direction approximately perpendicular to *paleocurrent* direction measured from small sedimentary structures within the epsilon cross-stratified unit. **Sandstone beds** are often seen to fine and thin up-**dip** with an associated change in small-scale sedimentary structures related to a reduction in **flow** velocity. Down **dip**, **beds** may be truncated by steep **erosion** surfaces. Epsilon cross-stratification results from lateral migration of the convex (inner) bank of a channel bend and may make up the entire thickness or upper part of a coarse member in an **alluvial deposit** (Fig. E18). [JA]

Allen, J.R.L. (1963) The classification of cross-stratification units, with notes on their origin. *Sedimentology* 2: 93–114.
Allen, J.R.L. (1965) The sedimentation and palaeogeography of the Old Red Sandstone of Anglesey, North Wales. *Proceedings of the Yorkshire Geological Society* 35: 139–85.
Jackson, R.G. II. (1978) Preliminary evaluation of lithofacies models for meandering alluvial streams. In: Miall, A.D. *Fluvial Sedimentology*. Canadian Society of Petroleum Geologists Memoir 5: 543–76.

epsomite ($MgSO_4 . 7H_2O$) An **evaporite** mineral. [DJV]

equipotential surface A surface on which the **potential** has the same value everywhere and on which the deriva-

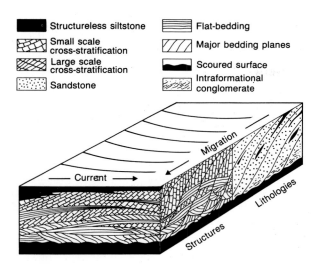

■	Structureless siltstone	▤	Flat-bedding
▨	Small scale cross-stratification	⧄	Major bedding planes
▨	Large scale cross-stratification	▬	Scoured surface
⋮	Sandstone	⧄	Intraformational conglomerate

Fig. E18 Epsilon cross-stratification. Idealized model to show the main features of lateral accretion (epsilon) cross-bedding, based on examples from the Old Red Sandstone. Major bedding surfaces (epsilon cross-stratification planes) dip at between 4° and 14°. Note the large vertical exaggeration. (After Allen, 1965.)

tive of the **potential** in any tangential direction, and therefore the tangential field intensity, is zero. If follows that the **force** at any point on an equipotential surface is normal to it. For example, the equipotential surfaces of a point charge are spherical, with radial field lines. The theoretical gravitational equipotential surface is the 'reference ellipsoid' (**spheroid**), the actual equipotential surface being the **geoid**. [ATB]

erg A large **sand**-covered region with **dunes** or **sand sheets**, also known as *sand seas*. Globally, there are 58 ergs with areas greater than 12 000 km², mostly in regions with annual precipitation of less than 150 mm. The largest is the Rub Al Khali in Saudi Arabia with an area of over 560 000 km². Ergs normally form in desert **basin** areas where **sand** collects after moving from areas of higher exposure. The form of the surface **sand** reflects the regional wind patterns and other climatic parameters, which control the growth of vegetation. [RDS]

Wilson, I.G. (1971) Desert sandflow basins and a model for the development of ergs. *Geographical Journal* **137**: 180–99.

erodibility The resistance of a *soil* to the entrainment and transport of its particles by an agent of **erosion**. The most important factors influencing a *soil*'s erodibility are its mechanical and chemical properties, including texture, aggregate stability, **shear strength**, **infiltration capacity** and organic and chemical content. Erodibility is also affected by topographical position, angle of slope and disturbances created by human action such as tillage on agricultural lands. [NJM]

erosion The process whereby *soil* particles are detached from the *soil* mass and transported away. The main agents of erosion are water, wind, mass movement and ice, all of which entrain particles and transport them. When sufficient energy is no longer available for transport the material is deposited. Thus, erosion removes the products of **weathering** and is an important part of the process of **denudation**. The most important detaching agent is the impact of raindrops striking the Earth sending *soil* particles flying into the air (see **raindrop impact erosion**). In areas where wind erosion dominates, the action of bouncing **sand**-sized particles is important in dislodging other grains. In both cases the collision involves a transfer of energy, enabling new particles to move. *Soil* particles are also loosened from the parent mass by **weathering** processes such as alternate wetting and drying, frost action and biochemical action. *Soil* may also be disturbed by tillage operations and the trampling of people and livestock.

The amount and rate of *soil* loss from a particular area depends upon the **erosivity** of the eroding agent, the **erodibility** of the *soil*, and the degree to which the *soil* is protected by such factors as vegetation, rocks and boulders. To these factors should be added the effects of slope, topography and human and animal disturbance. The relative importance of these influences may vary with the scale of analysis. Many of these factors are affected by climate, and on a world scale it has been shown that erosion reaches a maximum in areas with an effective mean annual precipitation of 300 mm, producing a sediment yield of over 300 tonne $km^{-2}a^{-1}$ (Langbein & Schumm, 1958). At precipitation totals below 300 mm erosion decreases as precipitation decreases. Above 300 mm, however, as precipitation increases so does vegetation cover, which offers increasing protection to the soil. Other evidence suggests that *soil loss* may increase again at higher precipitation and therefore **runoff** levels (Douglas, 1967). Such estimates are based on erosion by water; it has also been demonstrated that wind erosion rates are at a maximum in semi-arid regions and indeed that on a world scale rates of aeolian **deflation** may be of a similar order of magnitude to fluvial erosion rates (Goudie, 1978).

Erosion rates also vary through time; many erosion regimes display seasonal variations, and there is a gradation between catastrophic *soil erosion* events and many events of small magnitude. Superimposed on the natural frequency–magnitude patterns of erosion are changes brought about by human interference in the ecosystem. Changes in land use that may result in **accelerated erosion** include many poor agricultural practices, clearance of vegetation, construction activities, mining and recreational activities.

Soil erosion has been a problem to human activity since land was first cultivated. Although the erosion hazard is traditionally associated with agriculture in tropical and semi-arid areas, it is now being recognized in regions devoted to forestry, transport and recreation and in countries with temperate climates such as Britain (Morgan, 1980). Conservation of *soil* and the development of effective strategies for erosion control are dependent upon measurement and modelling of erosion processes. One of the most widely used predictive equations for modelling erosion due to rainfall is the **universal soil loss equation** (Wischmeier & Smith, 1962). [NJM]

Douglas, I. (1967) Man, vegetation and sediment yield of rivers. *Nature* **215**: 925–8.
Goudie, A.S. (1978) Dust storms and their geomorphological implications. *Journal of Arid Environments* **1**: 291–310.
Langbein, W.B. & Schumm, S.A. (1958) Yield of sediment in relation to mean annual precipitation. *Transactions of the American Geophysical Union* **39**: 1076–84.
Morgan, R.P.C. (1980) Soil erosion and conservation in Britain. *Progress in Physical Geography* **4**: 24–47.
Morgan, R.P.C (1986) *Soil Erosion and Conservation*. Longman, Harlow.
Richter, G. (1980) On the soil erosion in the temperate humid area of central Europe. *GeoJournal* **4**: 279–87.
Wischmeier, W.H. & Smith, D.D. (1962) Soil loss estimation as a tool in soil and water management planning. *International Association of Scientific Hydrology Publications* **59**: 148–59.

erosion surface A flattish plain resulting from **erosion**. The identification of the height, distribution and origin of erosion surfaces of different ages was the basis of a major approach to **geomorphology**, **denudation chronology**. They are generally regarded as the end point of an **erosion** cycle, for example an **endrumpf**, **peneplain**, *pediplain*, **etchplain**, or *wave-cut platform*, but the precise origin of any particular example is always difficult to determine. Erosion surfaces, truncating complex strata, are an important feature of highland areas like central Wales (Brown, 1960), but have also been identified in lowland England (Wooldridge & Linton, 1955). [ASG]

Brown, E.H. (1960) *The Relief and Drainage of Wales*. University of Wales Press, Cardiff.
Wooldridge, S.W. & Linton, D.L. (1955) *Structure, Surface and Drainage in South-East England* (2nd edn). Philip, London.

erosivity The potential ability of an eroding agent, such as rainfall, **runoff** or wind, to cause **erosion**. The erosivity of a rain storm, for example, will depend upon such factors as raindrop size, mass and velocity, the rainfall intensity and duration. Measures of erosivity are based on the kinetic energy of the eroding agent, but the erosivity does not necessarily produce actual data for *soil loss* as this will depend on the **erodibility** of the *soil* and the protection to the *soil* afforded by vegetation and other surface coverings. [NJM]

erythrite ($Co_3(AsO_4)_2 . 8H_2O$) 'Cobalt bloom' a pink secondary **cobalt** mineral. [DJV]

escarpment The steeper slope of a **cuesta**. Often used as a synonym for a **cuesta**. The escarpment or *scarpslope* cuts

across the **bedding** plane of the underlying sedimentary or volcanic rocks, while the gentler *dipslope* parallels the bedrock's **bedding**. Escarpments also occur where **erosion** or **uplift** has juxtaposed a gently sloping or flat-topped upland area and a low-lying trough. The Great Escarpment which occurs around the margin of much of southern Africa marks the outer edge of the ancient African Planation Surface, where it meets the block-**faulted** and heavily eroded coastal margin of the continent. [AW]

essexite An alkaline **gabbro** composed primarily of **plagioclase**, **hornblende**, **biotite**, *clinopyroxene* (*titanaugite*) with minor **alkali feldspar** and **nepheline**. Essexite grades into *theralite* by decrease in **alkali feldspar** and increase in **nepheline**. [RST]

essonite A variety of **grossular** garnet. (See **garnet minerals**.) [DJV]

estuary A semi-enclosed body of water that is open to the sea and contains water from both land drainage and seawater. The upstream limit of an estuary is usually regarded as being the point at which there are measurable differences in water chemistry due to seawater inputs. The interaction of the saline and fresh water, which fluctuate both tidally and seasonally, result in complex biological, **geomorphological** and chemical processes. The salinity difference between the inputs of ocean and river water often leads to a marked vertical gradient of water density, with the denser seawater penetrating upstream along the floor of the estuary below the fresh water moving downstream. Estuaries provide unique environmental systems of great economic and scientific value. [RDS]

Nichols, M.N. & Biggs, R.B. (1985) Estuaries. In: Davis, R. (ed.) *Coastal Sedimentary Environments*, pp. 77–186. Springer-Verlag, New York.

etchplain A broad, **erosional** land surface common on the continental **shields** of the humid tropics. The plains are developed in areas of deeply **weathered** crystalline rocks where **erosion** does not deeply incise the landscape; rather, surface lowering produces a flat topography. Differential rates of **weathering** of the bedrock — probably resulting from small differences in the chemistry of the rock — may lead to the exhumation of **inselbergs** which provide relief to the monotony of the landscape. [AW]

Ollier, C.D. (1969) *Weathering*. Oliver & Boyd, Edinburgh.

euclase ($BeAl(SiO_4)(OH)$) A rare **beryllium** silicate mineral with a ring silicate (*cyclosilicate*) structure. [DJV]

eucryptite ($LiAlSiO_4$) A mineral formed by **alteration** of **spodumene**, a chain silicate (*inosilicate*) used as a source of **lithium**. [DJV]

Euler's theorem This theorem states that any motion, confined to the surface of a sphere, can be defined as a rotation about a specific axis, the points where this axis intersects the surface being termed the *Euler poles*. This theorem enables the motion of a lithospheric plate across the Earth's surface to be defined in terms of the location of its *Euler pole* and the angle of rotation to provide a **finite rotation**. [DHT]

Le Pichon, X., Francheteau, J. & Bonnin, J. (1973) *Plate Tectonics*. Elsevier, Amsterdam.

eustasy The noun corresponding to the adjective eustatic, which refers to global changes of sea-level. The term was introduced at the end of the 19th century by the Austrian geologist Eduard Suess, following an analysis of the stratigraphic record which suggested that at least some of the more important marine transgressions and regressions in the past, such as the late **Cretaceous** transgression, were world-wide in extent. Eustatic changes of sea-level can be inferred from several different types of study. The variable spread of epicontinental seas with time can be inferred from analysis of the distribution of marine deposits in successive stratigraphic units. Marine transgressive and regressive deposits on former land margins can be correlated with inferred sea-deepening and shallowing events within the former marine **basins**. Clearly, to establish global as opposed to merely regional events due to local **tectonics**, it is necessary to correlate such changes extensively across and between continents. In recent years, the technique of **seismic stratigraphy** has been used to recognize eustatic changes in the past from the regressions or transgressions of sedimentary sequences recognized from analyses of **seismic reflection** data.

Two quite different mechanisms can be proposed to account for global sea-level changes. Melting and freezing of polar ice caps leads respectively to rises and falls of sea-level, and has clearly been of great significance during the **Quaternary**. For long periods in the **Phanerozoic**, however, there is no evidence of polar ice caps and much evidence of climatic equability. Changes of sea-level during these times must have been due to changes in the cubic capacity of the ocean basins as a consequence, for example, of changes in **seafloor spreading** rates affecting *ocean ridge* volumes, changes in the length of *ocean ridge* systems, development of *ocean trenches* and *collision of continents*. Such changes would have caused the sea-level to rise or fall at a rate about three orders of magnitude less than those induced by polar *glaciation* and deglaciation, and both the speed and extent of pre-**Quaternary** sea-level changes remain somewhat controversial. However, it is difficult to see why the **convective** forces feeding the ridges should be spasmodic, as studies of the *oceanic magnetic anomalies* suggest that **seafloor spreading** has been essentially continuous throughout the history of the present ocean basins. [AH/DHT]

Hallam, A. (1984) Pre-Quaternary sea-level changes. *Annual Review of Earth and Planetary Science* **12**: 205–43.

Haq, B.U., Hardenbol, J. & Vail, P.R. (1987) Chronology of fluctuating sea levels since the Triassic. *Science* **235**: 1156–67.

Vail, P.R., Mitchum, R.M. & Thompson, S. (1977) Seismic stratigraphy and global changes of sea level. *American Association of Petroleum Geologists Memoir* **26**: 51–97.

eutaxitic texture A texture observed in most **welded tuffs**, defined by flattened glassy discs in an **ashy** matrix. The discs are irregular and flame-shaped in cross-section, and known as *fiamme*. They represent devesiculated **pumice** or **scoria** *clasts* (in some cases possibly non-*vesiculated* **magma** blebs) which were flattened during the welding process. Eutaxitic texture is most common in welded **ignimbrites**, but also occurs in other **welded tuffs**, including welded air-fall **tuffs**. Eutaxitic texture can be the result of loss of **porosity** of the **tuff** alone, during compaction and welding, and is not necessarily a result of secondary (rheomorphic) **flow**. *Parataxitic texture* is similar to eutaxitic texture, but with greatly streaked-out *fiamme*, a result of secondary **flow**. [PTL]

euxinic Describes an environment of restricted circulation where stagnant or *anerobic* conditions exist. The name 'euxinic' is derived from the Roman name for the Black Sea and is typified by conditions at the base of the Black Sea today where organic-rich **black shales** are deposited. Euxinic conditions develop in any water body that remains stratified for a long period of time; such conditions are therefore common in deep **evaporite basins** where the rapid accumulation of organic debris may form a *hydrocarbon source rock*. [GMH]

evaporative drawdown Drawdown in **evaporite basins** comprises the loss in **brine** volume and the consequent lowering in **brine** level which takes place as evaporation proceeds and, as such, is a feature which characterizes all isolated **evaporite basins**. Within an **evaporite basin**, the relatively rapid rates of **evaporite** deposition, together with evidence of exposure in the centre of the **basins** in at least some of the **evaporite** facies, demonstrate falls in **brine** level of tens of metres (Fig. E19) and within a time interval that is commonly so short that it is not resolvable stratigraphically. As the majority of **evaporites** appear to have formed in shallow-water or brine-pan environments, frequent past evaporative drawdown is interpreted to have taken place. Even those **evaporites** inferred to have formed on the floors of pre-existing deep **basins** may have only formed in a few tens of metres of **brine** and preceding and overlying **evaporites** commonly exhibit some evidence of desiccation. Drawdown results in a desiccated **basin** within a topographic low, thus providing a focus for surrounding **groundwater** migration, for any surface **runoff** and for seepage through the barrier separating the **basin** from any open ocean (Fig. E19). [GMH]

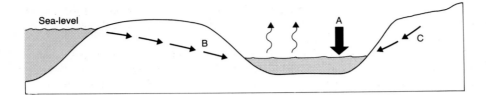

Fig. E19 Evaporative drawdown. Evaporation exceeds fluid input from precipitation (A), seawater reflux (B) and groundwater reflux plus spring input (C).

evaporative pumping The upwards movement of **groundwater** towards a **deflation** surface as a result of intense evaporation at the sediment–air interface. It can only take place on an exposure surface where evaporation greatly exceeds precipitation and is a common process on **sabkha** surfaces, both in marginal marine and continental environments. The volume of **groundwater** constantly lost at the sediment–air interface is replaced by more **groundwater**, a process which may well increase **groundwater** flow beneath a **deflation** surface, and thus catalyse penecontemporaneous **diagenesis** within these sediments. [GMH]

evaporite Rock composed of mineral(s) which form by **precipitation** from concentrated **brines**. Evaporites may form in both subaerial and subaqueous depositional environments and, by **diagenetic** replacement of precursor sediments, within the subsurface. Within any depositional environment, an excess of evaporation over water input is necessary for evaporites to form, and concentration of the parent **brine** is commonly achieved by evaporation at the air–fluid interface. **Brine** formation may also occur through subsurface concentration processes, including ion filtration, and by **brine** freezing. The greatest masses of evaporites are formed with seawater as an initial parent fluid; thus their mineralogy (Table E5) reflects seawater composition (Table E6), although as the system evolves the **brines** display progressively less marine characteristics. Non-marine evaporites, with a more restricted distribution (Table E5), have significantly different compositions and may contain hydrated carbonate minerals such as **trona**.

Within any body of concentrated **brine**, primary **precipitation** of evaporite minerals may take place at the fluid–air interface, within the **brine** column, at the sediment–**brine** interface or between grains in the underlying soft sediments. Complete evaporation of a shallow **brine** body leads to formation of evaporites both as a crust and within the surficial sediments; capillary fluid movement and subsequent evaporation may enhance the amount of evaporite minerals thus precipitated (see **evaporative pumping**). However, many evaporites are not deposited as direct precipitates but form at varied burial depths either by replacement of non-evaporitic sediments, or by **diagenetic** modification of existing primary evaporites.

Table E5 Common non-marine and marine **evaporite** minerals. (After Hardie, 1984.)

Non-marine evaporite minerals	
Halite	NaCl
Gypsum	$CaSO_4 . 2H_2O$
Anhydrite	$CaSO_4$
Epsomite	$MgSO_4 . 7H_2O$
Mirabilite	$Na_2SO_4 . 10H_2O$
Trona	$NaHCO_3 . Na_2CO_3 . 2H_2O$
Marine evaporite minerals	
Gypsum	$CaSO_4 . 2H_2O$
Anhydrite	$CaSO_4$
Halite	NaCl
Polyhalite	$K_2MgCa_2(SO_4)_4 . 2H_2O$
Sylvite	KCl
Carnallite	$MgCl_2 . KCl . 6H_2O$
Bischovite	$MgCl_2 . 6H_2O$
Kieserite	$MgSO_4 . H_2O$

Table E6 Evaporites. Composition of modern ocean water. (After Krauskopf, 1979.)

Dissolved species	p.p.m.	% of total
Cl^-	18 000	55.05
Na^+	10 770	30.61
SO_4^{2-}	2 715	7.68
Mg^{2+}	1 290	3.69
Ca^{2+}	412	1.16
K^+	380	1.10
HCO_3^-	140	0.41
Br^-	67	0.19
H_3BO_3	26	0.07
Sr^{2+}	8	0.03
F^-	1.3	0.005

Although evaporites have been found from the *Precambrian* (as old as 3.4 Ga) to modern times, there are almost no modern large evaporite depositional environments comparable to ancient evaporite-forming basins, Lake Macleod (see **evaporite basins**) in Western Australia being one major exception. Moreover, the abundances of evaporites have oscillated significantly during the geological past, with peaks of evaporite formation during the Early **Cambrian**, **Permian**, and **Jurassic–Cretaceous** plus lesser concentrations during the **Triassic** and *Tertiary*. This is partially a result of continental distribution and low sea-level stands during those periods, with major continental masses and enclosed, or, initially, semi-enclosed, **basins** situated beneath climatic belts of descending warm, dry air. Today, these subtropical high-pressure atmospheric circulation zones are centred about 30° north and south of the Equator, and range between 10° and 40°. Evaporites form in a variety of tectonic settings, occurring in interior, intracontinental sag **basins**, where an incoming marine transgression becomes trapped on continental shelves, where marine waters are isolated by the formation of an offshore barrier, and during the marine incursions associated with early phases of crustal downwarp as a result of the extensional processes which lead to **continental splitting**. With continued **extension** and subsidence, evaporites which form in these extensional terrains come to underlie thick sedimentary sequences and, where involved in later *plate collision* processes, control the resultant styles of upper **crustal deformation**, as in the Alps and the Himalaya.

The extreme solubility of evaporites is a further relevant factor when considering preservation of ancient evaporites as, in addition to **dissolution** during, or shortly following, deposition in non-arid areas of the world, evaporite **dissolution** in the shallow subsurface has removed major quantities of former evaporites (see **evaporite-related collapse breccias**).

The solubility of evaporites results in their rare occurrence at outcrop, apart from in arid areas. The least soluble evaporite, **gypsum**, is the most common both at outcrop and in the subsurface, remaining where more soluble **halite** and other salts are dissolved. However, even where **gypsum** *is* present at outcrop, it rarely occurs in an unaltered form, and commonly results from the hydration of **anhydrite** on **uplift** (see **gypsum–anhydrite equilibrium**). Data limitations are also present in the subsurface as, apart from in restricted areas of **halite** or potash mines, evaporites are rarely cored. Together with the absence of comparable modern environments, their susceptibility to rapid **diagenetic alteration**, the dominance in the literature of chemical models of evaporite formation, and the absence of many unaltered sections available for study, explains why, until recently, evaporites were considered purely as chemical precipitates rather than as sediments. This has led to an inadequate knowledge both of evaporite facies models and of their relevance in **basin** stratigraphy. There are many controversies, therefore, concerning evaporite genesis and depositional models.

Evaporites have formed in a variety of depositional environments. Primary sedimentary structures described include *ripple lamination*, **cross-stratification**, **graded bedding**, **mass flowage**, and *desiccation cracks*, structures similar to other clastic sequences. In addition, evaporites can mimic other sediment types, with **gypsum** and **halite** 'ooids', evaporite *turbidites* and subaqueous 'reefs' of upstanding huge **gypsum** crystals (in places building up to

several metres high). Evaporites are formed in three major physical environments; continental, coastal **sabkha** and subaqueous marine.

Continental evaporites are rare in the geological record, probably reflecting their ephemeral nature and constant upwards recycling by **groundwater** flowage; they are more susceptible to early **dissolution** than evaporites of other environments. Continental evaporites are deposited both subaqueously and within soft sediments in **playa lakes** and their margins (or continental **sabkhas**). They form from evaporation of **groundwater** and **evaporative pumping** in addition to surface **runoff** and may, in places, be **pedogenic**. Continental evaporites are usually interbedded with *red beds*, particularly clay-rich **playa lake** sediments and those of the surrounding mud flats. Minerals found include **halite**, *bittern salts* and **gypsum**, in addition to **trona**, *mirabilite* and other minerals reflecting the dominance of **groundwater** ions (Ca, Mg, Na, CO_3, HCO_3, SO_4, Cl).

Coastal **sabkha** evaporites are deposited in a supratidal environment as **early diagenetic** replacements within arid coastal plain sediments which form an equilibrium geomorphic surface, just above the intertidal area in both siliciclastic and carbonate environments. Thin, commonly ephemeral, evaporites may also form at the **sabkha** surface. On the **sabkha** surfaces high rates of evaporation at the sediment–air interface produce **brines** which deposit **gypsum** and, in places, **anhydrite** and **halite**, within the **sabkha** sediments; **halite** may also form as an ephemeral crust. Both **gypsum** and **anhydrite** form in **nodules** which, with continued growth, may develop a mosaic, or 'chicken-wire', form, a texture many times confused with diagenetic nodular **anhydrites**.

Subaqueous marine evaporites form in both shallow and relatively deep water (see **evaporite basins**). In shallow (0–10 m) marine environments sedimentary structures common to other clastic environments are commonly developed, although in protected environments *algal* activity is significant. Coarse palisades of upright palmate **gypsum** crystals, well-developed in the *Messinian* (**Miocene**) evaporites of the Mediterranean, grew in relatively quiet environments, and formed structures analogous to carbonate *reefs*. Many shallow marine evaporite sediments contain evidence of periodic exposure, with **karstic dissolution** of both sulphates and chlorides. Other such sediments contain evidence of influxes of less saline waters, with **dissolution** and etching of grains and palmate crystals. Deep water evaporites accumulate below wave base; the exact depth of water in which 'deep water' evaporites form is difficult to determine, and may be little in excess of 10 m, particularly where **brines** have exceeded **halite** solubility and are thus of considerable density compared with modern ocean water. These sediments include gravity deposits, with *turbidites*, **slumps** and **debris flows**. The most characteristic 'deep water' evaporite sediments are, however, thinly (1–10 mm) laminated sulphates and/or **halite**, individual laminae of which have been traced for up to several hundred kilometres within one **evaporite basin**.

Evaporites have a considerable economic importance, both in their own right and in association with *hydrocarbons* and **base metal** sulphide deposits. In addition to exploitation of evaporites for the minerals included (**gypsum** for plasterboards, **halite** for domestic use and for frost prevention, **potassium** salts for fertilizers), they supply a large proportion of world chemical industry requirements of *sodium*, **potassium**, *chlorine*, *bromine* and **sulphur**.

Evaporites are important in the sourcing of *hydrocarbons*, the formation of potential **reservoir** structures and as seals to *hydrocarbon* reservoirs. Since many evaporite-forming environments are **anoxic** or **euxinic**, organic matter is commonly preserved in marine evaporites and associated sediments, thus providing an oil-prone *hydrocarbon source rock*. Loading of evaporites by the overlying sedimentary pile leads to flowage and the formation of **salt domes**, thus generating positive subsurface structures in sedimentary **basins** devoid of tectonic activity. Potential *hydrocarbon* traps associated with **salt domes** form both on the flanks, generally faulted, and in **anticlines** overlying the dome crests. Evaporites also form excellent seals to *hydrocarbon reservoirs*, as they are both plastic and relatively impermeable. Evaporites are the seal to many of the North Sea gas fields and of **oilfields** in the Michigan and Williston Basins of the United States.

Economic **base metal** sulphide deposits commonly occur in ancient marine **evaporite basins** and associated with evaporitic **brines**. Deposits in carbonates fringing ancient **evaporite basins** include **galena** (PbS) and **sphalerite** (ZnS) together with **baryte** ($BaSO_4$) and **fluorite** (CaF_2). *Fluid inclusion studies* indicate that these minerals were formed at relatively low temperatures ($100–200°C$) where metal-rich **brines**, probably originating from the evaporites, encounter hydrogen sulphide within the flanking carbonate facies.

Since the late 1980s evaporites have also become important as possible depositories for *radioactive* waste. Although some evaporites have been investigated, to become later rejected, for *nuclear waste disposal* (such as the **Permian** evaporites of north-east England) research is continuing in other evaporite formations, including the **Permian** evaporites of Texas/New Mexico, USA, where an isolation disposal site is currently under construction. [GMH]

Dean, W.E. & Schreiber, B.C. (eds) (1978) *Marine Evaporites*. Short Course No. 4. Society of Economic Paleontologists and Mineralogists, Tulsa, OK.
Hardie, L.A. (1984) Evaporites, marine or non-marine? *American Journal of Science* **284**: 193–240.
Kirkland, D.W. & Evans, R. (eds) (1973) *Marine Evaporites: Origin, Diagenesis and Geochemistry*. Dowden, Hutchinson & Ross, Stroudsburg, PA.
Krauskopf, K.B. (1979) *Introduction to Geochemistry*. McGraw-Hill, New York.
Sonnenfeld, P. (1984) *Brines and Evaporites*. Academic Press, London.

evaporite basin Topographic low where evaporation is in excess of fluid input and the contained **brines** are sufficiently saturated for a range of **evaporite** minerals to be deposited; most major ancient evaporite basins contain at least some evaporitic sediments of a marine origin (see **evaporites**). Evaporite basins can be intracontinental, or develop during extensional processes as a prelude to active **rifting** and ocean growth. Thus the active **rifting** of the **Permian** and **Triassic** has led to many ancient **evaporite** sequences in either *failed rift* basins, such as the North Sea (*Zechstein* **evaporites**) and the US Gulf Coast (East Texas–Louisiana–Alabama Werner–Louann **evaporites**) basins, or in **evaporite** sequences underlying **passive continental margins**; for example, the North Atlantic margins of the US East Coast and West Africa–Portugal. In many such settings later sediment loading has resulted in *halokinesis* (see **salt dome**) and subsequent destruction of depositional patterns. Furthermore, information on continental margin **evaporites** is poor and commonly restricted to confidential seismic lines, in large part due to the great depth of burial.

As with **evaporites** themselves, the study of ancient evaporite basins is hampered by the lack of large modern analogues and much research has used **evaporite precipitation** in artificial *salt ponds* (or *solar ponds*, or *salinas*) as a guide to evaporite basin evolution. Only Lake Macleod, Western Australia and the Gulf of Karabogaz-Gol, east of the Caspian Sea, in the Turkmen Republic, are sufficiently large and sufficiently well documented to be used as analogues and even these examples are considerably smaller than many ancient evaporite basins, such as the Mediterranean during the *Messinian*, the **Permian** *Zechstein Sea*, the Michigan Basin or the Western Canadian **Devonian** Elk Point Basin. In addition, today Lake Macleod only naturally deposits **gypsum** and the Gulf of Karabogaz-Gol only **gypsum** and **halite**, whereas most ancient examples are known to contain potassium salts in addition to **gypsum** and **halite**. Furthermore, the intense **diagenetic alteration** of most ancient **evaporite** sequences obscures the evolution of the evaporite basin itself; for example, in the Zechstein Basin the transition between **gypsum** and **halite alteration** is veiled both by the **alteration** of **gypsum** to **anhydrite** (see **gypsum–anhydrite equilibrium**), thus destroying many of the original sedimentary textures, and by the later **diagenetic** addition of considerable amounts of **polyhalite**, further obliterating sedimentary plus early **diagenetic fabrics** (see **evaporite**).

Within an evaporite basin, the rates of sediment deposition may be extremely high (Table E7), much greater than for any other sediment type. Such rates demonstrate that evaporite basins can be filled extremely rapidly; for example, more than 2 km of salt in the *Messinian* of the Mediterranean accumulated in less than 2 Ma. Rates of deposition are comparable in more ancient evaporite basins; deposition of over 300 m of **evaporites** occurred within one **conodont** zone (perhaps some 500 000 years) in the Western Canadian Basin, with some, if not much,

Table E7 Evaporite basin. Rates of deposition of evaporites from marine water and from marine-fed marginal deposits, mostly artificial salt (or solar) ponds. (After Schreiber & Hsü, 1980.)

Sediment type	Area of formation	Observed rates of deposition
Sulphates and carbonate	Sabkha	Thickness of 1 m 1000 a^{-1} with 1 km progradation 1000 a^{-1}
Sulphates	Salt pond (subaqueous)	1–40 m 1000 a^{-1} over entire basin
Halite	Salt pond (subaqueous)	10–100 m 1000 a^{-1} over entire basin

of this time interval also representing episodes of non-deposition. Such rates require large-scale **evaporite** deposition to have taken place within pre-existing deep basins. The lack of fauna and flora, other than specialized organisms, within evaporite basins leads to a lack of **biostratigraphic** control, except from formations above or below the **evaporites** themselves. However, the rapidity of **evaporite** deposition together with **evaporative drawdown** may be so short (in geological terms) that **biostratigraphic** breaks would not be recognizable, even were any fauna present.

The rapid rate of **evaporite** accumulation also indicates that **evaporites** may have very thin, or no, laterally equiv-

alent sediments. Drawdown, with **evaporite** sedimentation in the basin centre, means that no basin margin sediments are deposited, i.e. that the **evaporite** equivalents are *disconformities*. In addition, a change in basin **brine** salinity, say from **gypsum** to **halite** saturation, will occur throughout the entire basin at any one time, so that there may be no lateral transition of subaqueous **halite** to subaqueous **gypsum** sediments. Lateral and vertical contacts in evaporite basins are, therefore, commonly very abrupt.

In spite of these difficulties, most evaporite researchers utilize three simplistic patterns of zonation within evaporite basins (Fig. E20a–c). The 'bull's eye' pattern (Fig. E20a) is based on a simple salt pond, where continued desiccation results in a concentric zonation of **evaporite** minerals, the least soluble in the outer ring to the most soluble in the centre. With such a distribution there is minimal influx or reflux once desiccation has commenced, and zone boundaries are controlled by topographic elevation. Hsü (1972) believed that this pattern explained the distribution of the *Messinian* **evaporites** (Fig. E21). The 'tear-drop' pattern (Fig. E20b) is based on the Gulf of Karabogaz-Gol, where there is a measurable influx of water over a sill at the basin entrance (Fig. E22). Here, the most soluble **evaporite** mineral is deposited away from the influx point, with the least soluble adjacent to the influx. Unlike the 'bull's eye' pattern, this model necessitates lateral concentration gradients within the **brines** of the basin. This, in turn, means that the interface between the inflowing

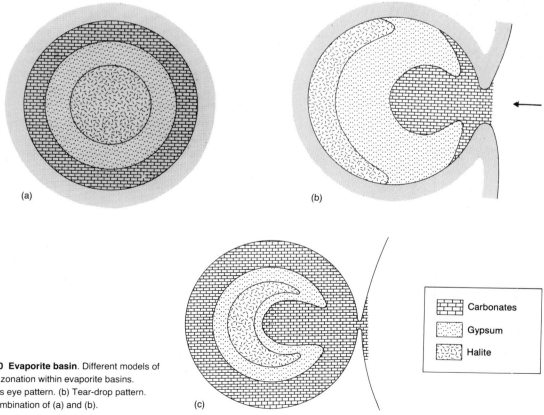

(a)

(b)

(c)

Carbonates

Gypsum

Halite

Fig. E20 Evaporite basin. Different models of mineral zonation within evaporite basins.
(a) Bull's eye pattern. (b) Tear-drop pattern.
(c) A combination of (a) and (b).

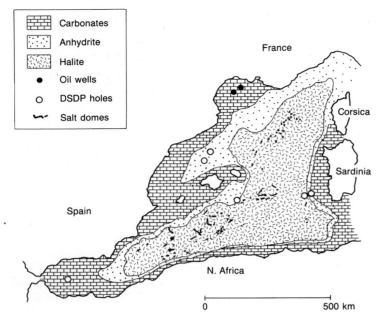

Fig. E21 **Evaporite basin**. The distribution of evaporites during the Messinian in the Mediterranean Balearic Basin. (After Schreiber & Hsü, 1980.)

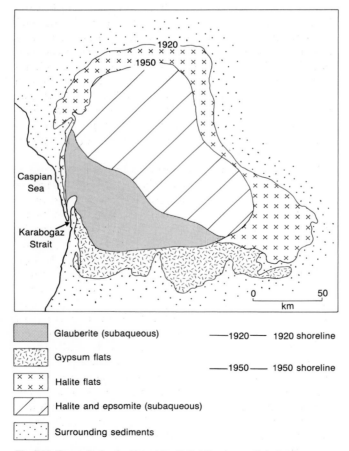

Fig. E22 **Evaporite basin**. Map of the Gulf of Karabogaz-Gol, showing evaporite sedimentation and changes in lake shore configuration. (After Schreiber & Hsü, 1980.)

surface water and the outflowing denser **brines** must lie below wave base, or mixing of the waters will occur and the depositional pattern become disrupted. The Elk Point **evaporites** of Western Canada were originally cited as an ancient example of the 'tear-drop' pattern (Fig. E23), but, more recently, anomalous thickening of these **evaporites** has been attributed to large-scale **alteration** and removal of the more soluble salts within the subsurface. The third pattern of **evaporite** zonation (Fig. E20c) is a composite of the two other models, combining drawdown with periodic marine input, or reflux of marine fluids through a permeable barrier.

A simple comparison of many ancient evaporite basins demonstrates that the majority follow the 'bull's eye' pattern indicating that most ancient evaporite basins are the result of isolation from any neighbouring ocean, and had negligible input over, or through, any intervening barrier. This hypothesis is supported by calculations imposing strict limits on the size of any basin inlet. In any evaporite basin, however, there is additional input from precipitation and subsequent **runoff**, perennial streams and rivers, **groundwater**, **springs** and **seepage** inflow through a permeable barrier. These additional inputs reinforce the hypothesis of isolation of ancient evaporite basins.

This isolation contrasts with the Gulf of Karabogaz-Gol (Fig. E22) which covers some $18\,000\,km^2$ east of the Caspian Sea. **Evaporites** have been forming intermittently within the area during the *Pleistocene* and *Holocene*. Today, inflow into the Gulf of Karabogaz-Gol is solely through the inlet, the Karabogaz Strait, to the Caspian Sea. Although this inflow amounts to some $9.3\,km^3\,a^{-1}$, the level of

[224]

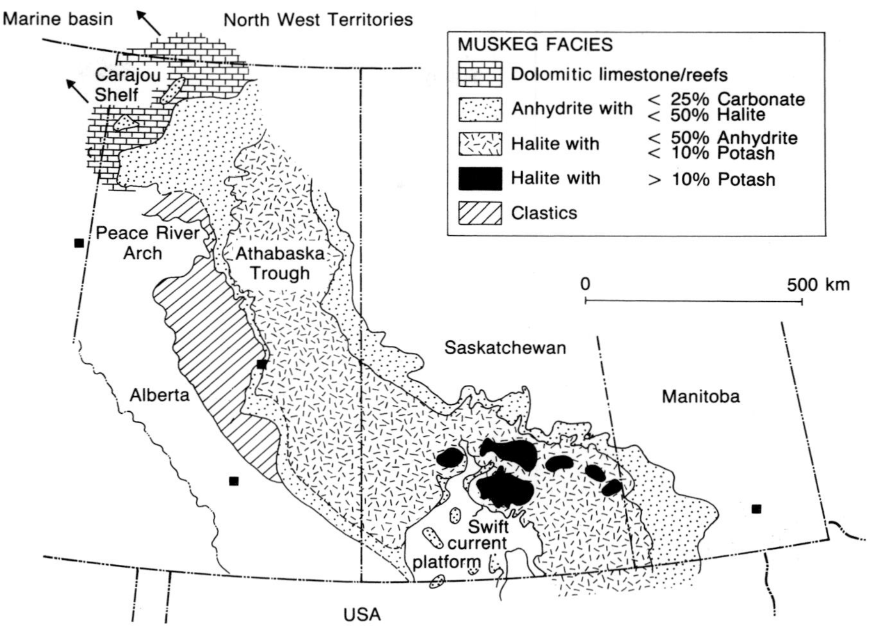

Fig. E23 Evaporite basin. The distribution of evaporites within the Western Canadian Elk Point Basin, initially considered a 'tear-drop' pattern, but now thought the result of subsurface dissolution of the more soluble evaporite facies. (After Schreiber & Hsü, 1980.)

the Gulf of Karabogaz-Gol remains several metres below that of the Caspian Sea and, contrary to original postulations of the 'tear-drop' model, there is no bottom outflow. In addition, in the Gulf of Karabogaz-Gol the amount of input is very dependent on the water level of the Caspian Sea, which itself has fallen since the 1950s, resulting in higher **brine** concentrations in the Gulf of Karabogaz-Gol and decreasing lake area. Although the waters of the Caspian Sea have a salinity only about one third that of normal seawater, they are richer in calcium and sulphate, and poorer in sodium and chloride, than average seawater. Thus, although evaporation within the Gulf of Karabogaz-Gol has produced more (non-artificial) **brine** concentrations than anywhere else on Earth, the mineralogy deposited is not typical of true marine **evaporites** (Kirkland & Evans, 1973). Unfortunately, there are no modern published sedimentological data on the evolution of this gulf, although there are published data on **brine** geochemistry and ancient configurations of the gulf shores (reprinted in Kirkland & Evans, 1973).

One other modern evaporite basin, Lake Macleod, Western Australia, is a marine basin which precipitates **gypsum** and ephemeral **halite**, and is approximately 110 km long by 40 km wide (Fig. E24); its surface lies 4 m below mean sea-level and it is isolated from the open ocean by a 15 km wide barrier of *Pleistocene* and older carbonate sediments. During most of the time, about 95% of the surface of Lake Macleod is dry, with small pools which precipitate **gypsum** and carbonates only in the northern area. The carbonates form *pisolitic* mounds and contain **tepees**; they develop where ocean water seeps through the permeable barrier into the topographically lower lake area. **Carbonate minerals** present are **aragonite**, (proto)**dolomite** and *huntite*. Periodically, during times of high **runoff**, the river system to the south of Lake Macleod breaches its **levees** and, as water flows downslope and over the broad exposed surface, the lake surface becomes inundated. This water dissolves both **gypsum** and ephemeral **halite**, resulting, on subsequent evaporation, in a basin-wide thin layer of **gypsum**, similar to the laminae which characterize 'deep water' evaporite basins. However, such laminae are commonly monominerallic and contain little or no organic matter, unlike their counterparts in ancient evaporite basins.

Within many ancient evaporite basins, thin millimetre-scale organic-rich laminae of sulphate, sulphate–carbonate or sulphate–**halite** can be traced laterally over many thousands of kilometres. The lateral continuity of these **laminites**, together with their abundance in **evaporite** formations, has supported the idea that these are cumulates and have a deep water origin; thus any vertical changes in **evaporite** mineralogy within the **laminites** must therefore reflect a change in salinity of the basin **brines**, or a mixing between previously stratified **brines**. Individual laminae may represent seasonal precipitates, annual events or aperiodic deposits, but must have been deposited in the same time interval over the entire basin, representing isochronous events. Stratification of the **brine** body within the evaporite basin is suggested by the organic-rich nature of these **laminites**, with the basal **brine** layers being both denser and **anoxic**, compared with less dense, more oxic surface waters. The extent and thickness of such **laminites** in ancient **evaporite** sequences also supports the hypothesis that ancient giant **evaporites** developed in pre-existing topographic basins.

Although there are possible modern analogues for

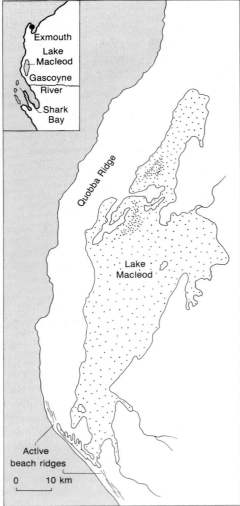

Evaporite flats, exposed throughout most of year

Pond areas, with active marine seepage

Exmouth
Lake Macleod
Gascoyne River
Shark Bay

Quobba Ridge

Lake Macleod

Active beach ridges

0   10 km

Fig. E24 Evaporite basin. Map of the Lake Macleod Basin, Western Australia.

Kinsman, D.J.J. (1974) Evaporites: relative humidity control on primary mineral facies. *Journal of Sedimentary Petrology* **46**: 273–9.

Kirkland, D.W. & Evans, R. (1973) *Marine Evaporites: Origin, Diagenesis and Geochemistry*. Benchmark Papers in Geology. Hutchinson & Ross, Stroudsburg, PA.

Schreiber, B.C. & Hsü, K.J. (1980) Evaporites. In: Hobson, G.D. (ed.) *Developments in Petroleum Geology 2*, pp. 87–138. Applied Science Publishers, London.

evaporite deposit Deposit from which salts of various types may be economically recovered either for use as the purified salt or to obtain one or more of the elements they contain. The principal **evaporite** minerals of economic importance are: borates (e.g. **borax**, **kernite**, **colemanite**), **celestine**, **gypsum** and **anhydrite**, **halite**, nitrates, **potassium** salts (e.g. **sylvite**, **carnallite**), sodium carbonates (e.g. **trona**) and sodium sulphate minerals.

Borates have been used by mankind since the earliest civilizations, but the big demand came early in this century and led to the large-scale exploitation of deposits in Turkey and the USA. These two countries produce nearly 90% of the world's requirements. The principal uses are in glass and glass fibres, fire retardants, soaps and detergents, vitreous enamels and agrochemicals. Borate deposits normally occur in lacustrine sediments and are not accompanied by important accumulations of other salts. The boron salts appear to have been derived from nearby volcanic rocks and as they are very soluble their preservation requires rapid burial by muds or other rocks of low **permeability**.

Celestine is the sole source of **strontium** and it occurs in bedded deposits within sediments and also in the **cap rock** of **salt domes**. The principal producers are Mexico, Turkey, Spain and the UK.

Gypsum and **anhydrite** deposits occur in both basinal and **sabkha evaporites**. When evaporation exceeds inflow to lakes or nearly isolated arms of the sea, $CaSO_4$ follows $CaCO_3$ in the theoretical sequence of deposition of salts from seawater. This explains its very common occurrence in *red bed* sequences such as the **Permo–Triassic** of northwestern Europe and its common occurrence throughout the world. World production of **gypsum** and **anhydrite** is around $79\,Mt\,a^{-1}$.

Over half the world's **halite** production of about $150\,Mt\,a^{-1}$ is consumed by the chemical industry to produce numerous sodium and chlorine chemicals that in turn are used for the preparation of thousands of chemical products. **Halite** beds occur with other sedimentary rocks, especially **anhydrite**, **shale** and **dolomite**. **Halite** is also found in **salt diapirs**. It is exploited either by underground or *solution mining*. **Halite** is being deposited at the present day in some **sabkha** deposits, e.g. in Baja California, but the enormous deposits of the past appear to be largely of basinal origin.

Next in the sequence of **precipitation** of seawater salts are the **potassium** salts which show a complicated mineral succession due to fractional **dissolution** and reprecipitation. Their **precipitation** requires extreme evaporation of

ancient giant evaporite basins, many characteristics of ancient **evaporite** sequences cannot be found in modern examples. It is therefore invalid to apply the theory of *uniformitarianism* to ancient evaporite basins. However, large evaporite basins have been described from almost all periods of the **Phanerozoic**, and recent research is documenting their existence through the *Precambrian*. The main conclusions of modern research on ancient evaporite basins are that almost all large **evaporite** deposits were formed within basins which were isolated from the open ocean, and that these basins formed from the flooding of pre-existing topographic lows. [GMH]

Hsü, K.J. (1972) Origin of saline giants: a critical review after the discovery of the Mediterranean evaporite. *Earth Science Reviews* **8**: 371–96.

seawater and they are therefore less common than **halite** deposits. Nevertheless, many countries have potash deposits and world production is around $30\,Mt\,a^{-1}$. Their mineralogy can be complex and there are many double salts with **magnesium**, in addition **magnesium** salts such as **epsomite** may be intermixed. The principal producers are the former USSR, Canada, Germany, the USA and France.

Sodium carbonates and sulphates are recovered either by extraction from the **brines** of **alkali lakes** or marshes or by mining beds of these salts. However, about 60% of the world's supply of sodium sulphate is by-product material from chemical factories.

Most nitrogen for ammonia and nitrate production comes from the atmosphere. Natural production is almost entirely from sodium nitrate-bearing **calcretes** in the Atacama Desert of northern Chile which run 8–15% $NaNO_3$. [AME]

Harben, P.W. & Bates, R.L. (1984) *Geology of the Nonmetallics*. Metal Bulletin Inc., New York.
Reading H.G. (ed.) (1986) *Sedimentary Environments and Facies*. Blackwell Scientific Publications, Oxford.

evaporite-related collapse breccia A *collapse breccia* comprises chaotic, angular fragments of lithified rock and forms when an underlying **evaporite** formation is dissolved wholesale. **Dissolution** may be due to **uplift** to near-surface environments or may take place within the deep subsurface. Evaporite-related collapse breccias commonly develop within dolomitized carbonate sequences which have undergone early lithification. **Dissolution** of **evaporites** and subsequent collapse of the overlying formation therefore results in the production of rigid carbonate *clasts*. The degree of brecciation is dependent on the thickness of the former underlying **evaporites**; thin beds of former **evaporites** produce thin, laminar collapse breccias, whereas **dissolution** of thick **evaporite** formations may demonstrate several generations of collapse, with initial collapse commonly in pipe form and later collapse more pervasive with large-scale foundering. Recognition of near-surface **evaporite dissolution** therefore has important consequences for **engineering geology**. In the subsurface, **dissolution** fronts of **halite** and more soluble salts exist which, if not recognized, may result in false interpretations of ancient **evaporite** distributions (see **evaporite basins**).

Associated with carbonate collapse-breccias, **dissolution** of **gypsum** produces calcium-rich sulphatic waters which may react with both the brecciated and underlying formations, causing partial or complete calcitization of **dolomite** (*dedolomitization*). [GMH]

exceptional fossil deposits Deposits, often termed '*fossil Lagerstätten*', that fall into two major groups, *concentration* and *conservation deposits*. *Concentration deposits* are exceptionally rich accumulations. They include condensation deposits, formed by winnowing or slow rates of sedimentation (e.g. *ammonite* beds), **placer** deposits, which are concentrated by transport (e.g. bonebeds), and concentration traps (e.g. fissure fills). Exceptional preservations which yield soft tissues are termed *conservation deposits*. Three main processes may play a part in the inhibition of decay and promotion of **diagenesis** which lead to their formation. These are *stagnation*, *obrution* (rapid burial) and the role of *cyanobacterial* films in sealing the sediment. The type of mineral preserving the soft tissue depends on a number of factors, particularly the rate of burial, concentration and distribution of organic material in the sediment, salinity, and the composition of the organism itself. The most common minerals are **pyrite**, phosphate (**apatite** and **vivianite**) and carbonates such as **calcite** and **siderite**.

Conservation *Lagerstätten* occur in the fossil record at a frequency of about one per 15 Ma. Some of the most significant are the following:

Messel Oil Shales (**Eocene**), Darmstadt, Germany. Lacustrine: articulated vertebrates, some with body out-lines preserved in **siderite**, *insects* and plants.

Solnhoffen Limestone (U. **Jurassic**), Bavaria, Germany. Marine: articulated vertebrates, including *pterosaurs* and *Archaeopteryx*, shrimps, limulines and *insects*.

Mazon Creek, Francis Creek Shale (U. **Carboniferous**), Illinois. Terrestrial–marine: over 350 species of plant and 320 species of animal preserved in **sideritic** nodules in the marine Essex and non-marine Braidwood assemblages.

Granton Shrimp Bed (L. **Carboniferous**), Edinburgh, Scotland. Restricted marine: soft parts of **conodonts**, shrimps, worms and hydroids, preserved in phosphate.

Hunsrück Slate (L. **Devonian**), Rhineland, Germany. Marine: **arthropods**, **molluscs**, **echinoderms**, **fishes** and plants, preserved in **pyrite**.

Rhynie Chert (L. **Devonian**), Aberdeenshire, Scotland. Early **land plants** and **arthropods**, preserved in **silica**.

Waukesha Dolomite, Brandon Bridge (L. **Silurian**), Wisconsin. Marine: unusual assemblage including **arthropods**, worms, **conodont** with soft parts, preserved in phosphate.

Burgess Shale, Stephen Formation, (M. **Cambrian**), British Columbia. Marine: most diverse **Cambrian** biota with about 120 genera, preserved in alumino-silicates, including a significant proportion of soft-bodied groups which cannot be assigned to living phyla.

Pound Quartzite (*Vendian*), Flinders Ranges, Australia. Shallow marine: diverse fauna, mainly of medusans and other *cnidarians*, with some segmented organisms, preserved as casts and impressions.

Exceptionally preserved fossils are important for three major reasons. Firstly, they provide evidence of the history of soft-bodied taxa. The normal fossil record consists only of hard parts. Analyses of living communities show that about 60% of taxa lack mineralized parts, and have no prospect of becoming fossilized in normal circumstances.

Similarly, comparison of exceptionally preserved biota with shelly faunas of similar age shows that as few as 20% of genera may be normally represented in the fossil record. Thus the fossil record of leeches (Class Hirundinea), for example, consists of two species from the **Jurassic** *Solnhofen Limestone* and a possible example from the **Silurian** *Waukesha Dolomite*. About 20% of marine classes in the **Phanerozoic** are known only from three conservation *Lagerstätten*, the *Burgess Shale*, *Hunsrück Slate* and *Mazon Creek*.

Secondly, a number of the soft-bodied organisms in exceptionally preserved biotas represent groups which are extinct. The **Cambrian** *Burgess Shale* includes over 20 genera which cannot be assigned to living phyla, and there is no clear evidence of interrelationships between them. Small numbers of such problematica occur in later **Paleozoic** *Lagerstätten*; it appears that their diversity is greatest in the **Cambrian** and declines thereafter. The ecospace vacated when problematica became extinct was filled by radiation from the ranks of surviving phyla. Without exceptional preservations like the *Burgess Shale*, which provide a true indication of the former diversity of life, we would know little of this major radiation of body plans.

Normally the soft parts of fossils can be restored by comparison with their living relatives. It is clearly impossible, however, to treat extinct groups in this way and here exceptionally preserved fossils play their third important role. The nature of **conodonts** (**Cambrian** to **Triassic**), for example, whose microscopic tooth-like elements are used extensively in **biostratigraphy**, remained a puzzle for some 120 years after their discovery in 1850. The soft parts were finally discovered in the **Carboniferous** *Granton Shrimp Bed*, revealing the **conodonts** as primitive *chordates* with an elongate trunk and fins.

Some conservation *Lagerstätten* are not representative of typical assemblages because the conditions which inhibit scavengers and decay are usually inimical to life. Certain sedimentary settings, however, host exceptional preservations at several localities or geological horizons. Environments transitional between marine and non-marine provide one example. Low salinity **pore-water** inhibits decay and promotes the **diagenetic** formation of a **siderite** *concretion* around the organism. The *Mazon Creek* biota is the most important example, and similar deltaic–estuarine biotas occur in the late **Carboniferous** elsewhere in North America and in France (at Montceau-les-Mines). **Coal balls** are widespread in extensive **Carboniferous coal seams** and provide information on **swamp** plant ecosystems. Their formation, which involves permineralizing **peat** with **calcite**, may also involve a marine influence. An understanding of the context of exceptional fossil deposits allows the development of models to explain their preservation and thereby predict additional occurrences. [DEGB]

Whittington, H.B. & Conway Morris, S. (eds) (1985) Extraordinary fossil biotas: their ecological and evolutionary significance. *Philosophical Transactions of the Royal Society, London* **B311**: 1–192.

Whittington, H.B. (1985) *The Burgess Shale*. Yale University Press, New Haven, CT.

excess mass The difference in mass between an anomalous body of rock and the country rock which would fill its volume if it were not there. Excess mass can be calculated unambiguously from **gravity anomalies** by a surface integration. The actual mass of the anomalous body can only be determined if the densities of body and country rock are known. The technique finds application in mining exploration for the estimation of **ore reserves**. (See **Gauss's theorem**.) [PK]

Kearey, P. & Brooks, M. (1991) *An Introduction to Geophysical Exploration* (2nd edn). Blackwell Scientific Publications, Oxford.

exfoliation (**desquammation**, *onion-skin weathering*, **spheroidal weathering**) The **weathering** of rocks and boulders by spalling of surficial layers from a few millimetres to several metres thick. Exfoliation on the surface of bedrock exposures — particularly of plutonic rocks — may result from the release of **lithostatic pressure** as the rock is exhumed by **erosion** of the overlying rocks. The surface layers exfoliate as the rock expands when compressional forces decrease. However, exfoliation also occurs on blocks of rock and on boulders as a result of physical and chemical **weathering** phenomena (see **desquammation** and **spheroidal weathering**). Another mechanism may be the detachment of surficial rinds owing to salt crystallization just beneath the surface of the rock. Exfoliation in deserts was formerly widely believed to result from repeated heating and cooling of rock surfaces (**insolation weathering**). However, laboratory experimentation under carefully simulated conditions has failed to replicate this process. [AW]

Bradley, W.C. (1963) Large-scale exfoliation in massive sandstones of the Colorado Plateau. *Geological Society of America Bulletin* **74**: 519–27.

Ollier, C.D. (1967) Spheroidal weathering, exfoliation and constant volume alteration. *Zeitschrift für Geomorphologie* NF **11**: 103–8.

Twidale, C.R. (1973) On the origin of sheet jointing. *Rock Mechanics* **5**: 163–87.

exhalite and exhalative processes Chemical deposit formed from exhalations, dominantly **hydrothermal** and most commonly debouching on to the seafloor. These terms are most generally used in discussing economic mineral deposits, but exhalites may be of no economic importance, such as the sulphide and **manganese** oxide deposits around the **black smokers** at oceanic depths and the ferruginous **cherts** capping **Kuroko-type deposits**. Many of the world's great **stratiform mineral deposits**, whether present in volcanic or sedimentary environments, are considered by most geologists to have been formed by volcanic- or sedimentary-exhalative processes, for example the huge volcanic-associated **massive sulphide deposits**

of the Pyrite Belt of southern Spain and Portugal and the sediment-hosted **lead–zinc** deposits of Broken Hill, Australia or Sullivan, Canada. Other deposits, particularly those associated in space and time with volcanic activity, are also considered by many workers to be exhalites, e.g. the **banded iron formations** of **Algoma-type** and certain **disseminated gold deposits** in *iron formation*.

Somewhat different exhalites from those forming around **black smokers** develop during the early stages of opening of an ocean basin. Mineralization of this type is found along the median zone of the Red Sea which has hot **brines** and metal-rich muds in some of the deep **basins**, many of which are defined by the 2000 m contour. The largest of these deeps is the Atlantis II which is 14 × 5 km and 179 m deep. At the base, multicoloured sediments usually 20 m thick, but up to 100 m, rest on **basalt**, and piston coring has demonstrated the presence of stratification and various sedimentary facies high in **manganites**, **iron oxides** and **iron**, **copper** and **zinc** sulphides. These sediments can run up to 20% **zinc** but the top 10 m in the Atlantis II deep run 1.3% **copper** and 3.4% **zinc**, and feasibility studies are under way for their recovery. Above the sediments, the deeps contain **brine** solutions; these and the sediments run up to about 60°C and 30% sodium chloride. It has been suggested that the sediments and **brines** have originated either from the ascent of juvenile solutions, or from the recirculation of seawater which has leached salt from the **evaporites** flanking the Red Sea and passed through hot **basalt** from which it has leached metals. The second hypothesis meets with more favour at the moment to explain the geneses of these and **massive sulphide deposits**, with circulation of seawater as depicted in Fig. M5, p. 387. [AME]

Rona, P.A. (1985) Hydrothermal mineralization at slow-spreading centers: Red Sea, Atlantic Ocean and Indian Ocean. *Marine Mining* **5**: 117–45.
Scott, S.D. (1985) Seafloor polymetallic sulfide deposits: modern and ancient. *Marine Mining* **5**: 191–212.

expanding Earth The discovery of radiogenic heat production meant that the Earth was probably not contracting at a significant rate and that it may even be expanding. Minor expansion cannot be ruled out on the basis of **paleomagnetic** observations, but major changes with increased age of the Earth can be excluded, while a slight contraction, probably less than 5%, is indicated by studies of the surface of the Moon and planets. A more extreme view is that there has been an expansion of the Earth's radius by some 33% in the last 250 Ma and that this has caused an expansion in the Earth's surface by 66%, which is evidenced by the presence of the oceans. Other evidence in support of this view has been suggested as claimed misfits in **continental reconstructions**, although other explanations usually seem more plausible. The mechanism proposed for such major expansion is a change in the universal **gravitational 'constant'** resulting in an expansion of the Earth's interior as a result of phase

changes consequent upon the reduction in hydrostatic pressure. [DHT]

Carey, S.W. (1976) *The Expanding Earth*. Elsevier, Amsterdam.
Owen, H.G. (1976) Continental displacement and expansion of the Earth during the Mesozoic and Cenozoic. *Philosophical Transactions of the Royal Society, London* **A281**: 233–91.

expansive soil *Soil* which shrinks and swells according to its moisture content. *Soils* and rocks containing sodium **montmorillonite** clays are especially susceptible, though alkali *soils* and **pyrites** also exhibit volume change properties. The process is especially serious in areas with extreme seasonal climates, or during periods of extreme drought. It can cause major structural problems for buildings. Holtz (1983) has estimated that in the USA the damage caused by expansive soils is $6 billion per annum (at 1982 prices), and that the losses exceed those caused by **earthquakes**, tornadoes, hurricanes and floods combined. [ASG]

Holtz, W.G. (1983) The influence of vegetation on the swelling and sinking of clays in the United States of America. *Géotechnique* **33**: 159–63.

exploratory well A **well** drilled in the hope of finding oil or gas where none has been found before. This compares with **wells** drilled to permit the exploitation of known deposits, which are known as **development** or production **wells**. There is now a nomenclature for oil and gas **wells** that is used throughout most of the non-communist world and which was drawn up in 1948 by the Committee of Statistics on Exploratory Drilling of the American Association of Petroleum Geologists. [AME]

exposed coalfield A **coalfield** whose **coal**-bearing strata crop out at the surface. (See **concealed coalfield**.) [AME]

exsolution The development of two or more compositionally different phases from a *solid solution*, usually by cooling from a high temperature at which the *solid solution* is the stable state. Another term for exsolution, used outside the earth sciences, is **precipitation**. Exsolution occurs widely in mineral systems, e.g. **augite** exsolves from **pigeonite** in **pyroxenes**. The second phase starts to exsolve within a cooling *solid solution* (i.e. a mineral) of particular composition at a temperature given by the solvus in the phase diagram. Exsolution may also occur when a solution that has been maintained as a single phase outside its normal temperature–composition stability field (e.g. by rapid cooling) is raised to a higher temperature and held there.

The formation of the second phase depends upon the clustering of atoms to form stable nuclei, which requires the attainment of nuclei of a critical size. Both *nucleation* and subsequent growth of nuclei depend upon diffusion. Within grains and in the absence of **crystal defects**, the nuclei are randomly and uniformly distributed, a characteristic recognized in the term then applied — *homogeneous nucleation*. However, stable clusters of solute atoms are most easily formed where there are irregularities in the **crystal structure** (since the **strain** energy of a nucleus is thus reduced), so that exsolved phases tend to appear first at internal defects, e.g. grain boundaries, dislocations. This is called *heterogeneous nucleation*. Nuclei at such defects tend to grow preferentially (because defects enhance diffusion rates). This causes local depletion of solute, as witnessed by zones around the anomalously large particles that are denuded of small precipitates.

Exsolution can also occur because of a change in bulk composition of a mineral, as when minerals containing Fe^{2+} are oxidized, e.g. **spinels** in the **magnetite–ulvöspinel** system. Exsolution often leads to **microstructures** that are characteristic of the system, the exsolution temperature, and the mineral composition. Exsolution may also be accompanied by *phase transformation* (e.g. the exsolution of **perthites** in **alkali feldspar** *solid solutions* of intermediate composition). Exsolution without a *nucleation* step is also possible through the transformation mechanism that occurs in *solid solutions* with unstable compositions, known as *spinoidal decomposition*.

The degree of lattice continuity at the interface between an exsolved phase and the host mineral is described in terms of coherency. Both full and partial coherency between the lattices imply crystallographic relationships between the two (**epitaxy** or *topotaxy*), which is why exsolution **microstructures** are frequently characteristic of systems. The second phase may be coherent with a host phase when nucleated, but become less coherent on growth, because of **elastic strains** generated at the interfaces. Since many minerals are of low symmetry and elastically anisotropic, exsolved phases often grow as lamellae parallel to low index planes. In recent years studies of the relationships between exsolved phases and their hosts have been greatly advanced by *transmission electron microscopy*. [DJB]

Putnis, A. & McConnell, J.D.C. (1980) *Principles of Mineral Behaviour.* Blackwell Scientific Publications, Oxford.

extension An increase in length of a line or lines during **deformation**; a **strain** involving extension. As a measure of **strain**, extension is defined in the same way as **elongation**. Extension can be used to describe part of a larger-scale **deformation** distribution, or can represent the bulk effect of a **strain** episode.

In regional **deformation**, extension is associated with the maximum **principal strain** direction of the **strain ellipsoid**. Extension often results in **boudinage** of layers of rock. In rocks with less pronounced **rheological** layering, **shear bands** and local layer thinning may indicate the extension direction. In non-**coaxial deformation** sequences, such as **simple shear**, the orientation of the extension direction changes with time, and so some layers

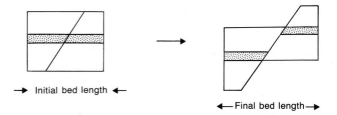

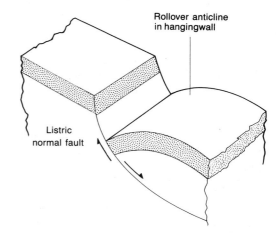

Fig. E25 Extension of a horizontal bed by a normal fault.

Fig. E26 Extension. 'Rollover anticline' in the hangingwall of a listric normal fault. The exact shape of the fold depends on the deformation mechanisms operating, as well as fault shape.

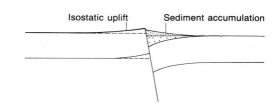

Fig. E27 Extension. Isostatic uplift and accumulation of sediment around an emergent normal fault. Thickening of sedimentary layers towards the fault surface results from repeated faulting of this kind, which is therefore known as 'growth faulting'.

may experience a sequence of extension and **contraction** with time.

Localized **strain** and **fault** zones can also affect bulk extension. Such **faults** are initially oriented in space so that surfaces which they cut move apart subhorizontally, to form *normal faults*, as in Fig. E25. Seismic sections through **extensional fault systems** (see Bally *et al.*, 1981) show *listric fault* profiles. **Displacement** on such non-planar **faults** results in **deformation** around the **fault**. In general, this is seen as a *'rollover'* anticline in the **hangingwall**, as shown in Fig. E26. The shapes of these **anticlines**, and the pattern of **deformation** are used to construct the shape of the **fault** profile at depth, using **balanced section**

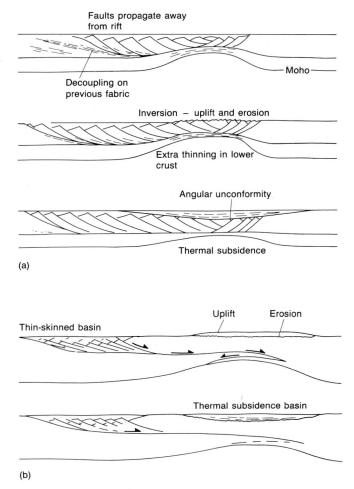

(a)

(b)

Fig. E28 Extension. Models for the structure of extensional sedimentary basins, which show two differing 'end-member' models for the link between fault-dominated strain in the upper crust and distributed strain below. The position of greatest thermal subsidence can be beneath or offset from the maximum fault subsidence. (After Coward, 1986.)

techniques. In a region of closely spaced *listric normal faults*, existing **fault** surfaces can be rotated by displacement on **faults** in their **footwall**. Thus rotated extensional **faults** can have *reverse* or even *thrust* geometry. **Deformation** by displacement on arrays of closely spaced planar **faults** can also rotate domino-like blocks of rock between the **faults**.

Crustal extension results in **isostatic** depression of the **hangingwall** and **uplift** of the **footwall** (Jackson & McKenzie, 1983) around an **extensional fault**, so that variable thicknesses of sediment build up in extensional tectonic settings, as shown in Fig. E27. This *synsedimentary faulting* is useful for dating **fault** activity where sediment ages are established.

Faults accommodate large-scale **crustal** and **lithospheric** extension in the *formation of sedimentary basins* such as the North Sea, in continental **rifting** and in areas such as the

Basin-and-Range Province of western North America. In these areas dominated at the surface by many closely spaced **faults**, there is debate about the relationship between the **faults** at depth, and how they relate to the more distributed **strain** characteristic of lower levels in the **lithosphere**. Two models are proposed for the deep structure of such sedimentary **basins**: that of McKenzie (1978), and that of Wernicke (1985), discussed by Coward (1986). These models involve an initial 'tectonic' phase of extension of the continental **lithosphere**, with associated subsidence. This leads to upwelling of the **asthenosphere**, and an increase in **heat flow**. With time, the **heat flow** decays to its normal values, and there is a phase of 'thermal' subsidence. In the McKenzie model, based on geophysical data from the North Sea, the thinning of the lower **lithosphere** takes place immediately below the **faulting**, separated by a **decollement**. The Wernicke model, based on **basin-and-range** structures, considers that the **faults** in the upper **lithosphere** branch onto a gently-dipping crustal **shear zone**, which maintains continuity between upper and lower **lithosphere** extension. However, in this case lower **lithosphere** extension, leading to the thermal phase of subsidence, can take place away from the site of tectonic subsidence. It is likely that most natural examples fall between these two models, as shown in Fig. E28. Detailed stratigraphic analysis of **basin** fill thicknesses and facies allows modelling of the development and extent of the subsidence phases. This is important in *hydrocarbon exploration*, when subtle changes in sediment patterns affect **reservoir** potential. [SB]

Bally, A.W., Bernouilli, D., Davis, G.A. & Montadert, L. (1981) Listric normal faults. *Oceanalogica Acta: Actes 26ième Congrès International Géologique, Colloque des marges continentales* 87–101.
Coward, M.P. (1986) Heterogeneous stretching, simple shear and basin development. *Earth and Planetary Science Letters* **80**: 325–36.
Jackson, J.A. & McKenzie, D.P. (1983) The geometrical evolution of normal fault systems. *Journal of Structural Geology* **5**: 471–82.
McKenzie, D.P. (1978) Some remarks on the development of sedimentary basins. *Earth and Planetary Science Letters* **40**: 25–32.
Wernicke, B. (1985) Uniform sense simple shear of the continental lithosphere. *Canadian Journal of Earth Sciences* **22**: 108–25.

extensional cleavage Set of planes or narrow zones, oblique to a pre-existing planar structure, such that the displacements on the **cleavage** result in net **extension** parallel to the pre-existing planes. Used particularly for **crenulation cleavages** affecting previously **foliated** rocks (Fig. E29). (See also **shear band**.) [RGP]

Platt, J.P. & Vissers, R.L.M. (1980) Extensional structures in anisotropic rocks. *Journal of Structural Geology* **2**: 397–410.

extensional fault **Fault** across which **extension** has taken place. The term is usually applied to a *normal fault* but may also be used for an **extensional fracture** (**fissure** or **joint**) that has opened to allow infilling of extraneous material (e.g. mineral **veins**). [RGP]

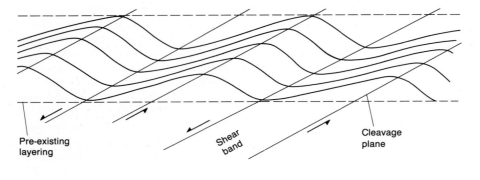

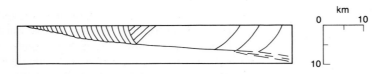

Pre-existing layering

Shear band

Cleavage plane

Fig. E29 Extensional cleavage.

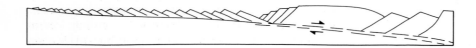

Fig. E30 Extensional fault system. Method of accommodating extension by block rotation above a low-angle extensional fault-shear zone. (After Wernicke, 1981.)

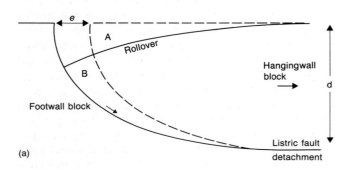

Fig. E31 Extensional fault system.

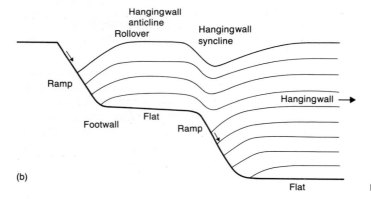

extensional fault system Set of related **faults** on which the individual displacements produce a net **extension** in the system as a whole. Such fault systems are characteristic of *extensional* or *divergent tectonic regimes*, which are associated (i) with *constructive plate boundaries*, at oceanic and continental **rift** zones; (ii) with *destructive plate boundaries*, in regions of **back-arc spreading** such as the **Basin-and-Range** Province of the western USA, and in oceanic *marginal basins*; and (iii) with certain continental intraplate regions marked by **rifts** and extensional **basins**.

Classical views on faulting and on the formation of **graben** and **rifts** visualized **extension** being accommodated by **dip-slip** movements on steep *normal faults* or by the filling of *extensional* **fissures** by **magma**. However more recently the importance of low-angle *normal faults* has been emphasized by Wernicke (1981) and others as a method of achieving the much larger extensions (of 100% to 200%) required, for example, in the **Basin-and-Range** Province. Wernicke & Burchfiel (1982) demonstrate three principal geometric elements in extensional fault systems: (i) low-angle *normal faults* acting as *detachment planes*; (ii) rotated *fault blocks*; and (iii) curved or *listric fault* surfaces (Fig. E30). The widespread occurrence of such geometries has been confirmed by **seismic records** obtained in **petroleum** exploration of marine **basins**, and is summarized by Gibbs (1984). The *listric fault* achieves a rotation in the **hangingwall** as a geometric consequence of the displacement, and may be accompanied by an accommodation anticline, known as a *rollover anticline* (Fig. E31). The detachment or *sole fault* in extensional systems has the same function as that in *thrust fault systems*, and may possess a similar *ramp-flat* geometry (Fig. E31b) necessitating complex accommodation **strains** in the **hangingwall**. The accommodation may take place by the formation of a set of **antithetic faults** that have the effect of extending and thinning the **hangingwall** in order to relieve the accommodation **strain** (Fig. E32). As **extension** proceeds, the *sole fault* may migrate into the **footwall**, producing a set of **synthetic** *'horsetail' listric faults*, known as a *listric fan* (Fig. E32b), in an analogous way to the propagation of *thrusts* into the **foreland** in *piggyback thrust sequences*. The *fault blocks* formed in this process are termed *riders*. The migration of the *sole fault* into the **footwall** may create an extensional *duplex* bounded by a *roof fault*. Figure E33 shows the complex type of geometry that may be pro-

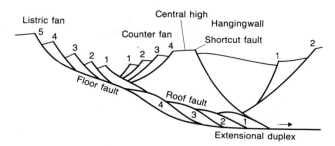

Fig. E33 Extensional fault system. Synthetic faults in footwall.

duced by a combination of **hangingwall** and **footwall** collapse resulting from progressive **extension**. The rotated blocks formed by these processes form **half graben** at the surface; these may become filled with sediment.

Upper-**crustal extension** on **fault** systems of this type may be transferred to mid-**crustal** levels and below by low-angle **shear zones** such as those found in *Precambrian* basement complexes. Deep **seismic profiling** across the **Mesozoic basins** off northern Scotland and in the North Sea indicate that these **shear zones** probably cut through the whole thickness of the **crust** and may continue into the **mantle** part of the **lithosphere**. It is probable that intra-plate extensional displacements may be transferred by this means to distant **plate boundaries**.

Extensional fault systems frequently contain steep **faults** with **strike-slip** displacements. Such **faults** are integral to the system and transfer displacement from one **dip-slip** plane to another. They are therefore termed *transfer faults*, and are smaller-scale counterparts of oceanic **transform faults**. Figure E34 shows two **dip-slip faults** linked by a *transfer fault*, each detaching on the same *sole fault*. *Transfer faults* may separate **imbricate fault–fold** packages that are geometrically distinct and uncorrelatable across the **fault**. [RGP]

Gibbs, A.D. (1984) Structural evolution of extensional basin margins. *Journal of the Geological Society, London* **141**: 609–20.
Park, R.G. (1988) *Geological Structures and Moving Plates*. Blackie, Glasgow.

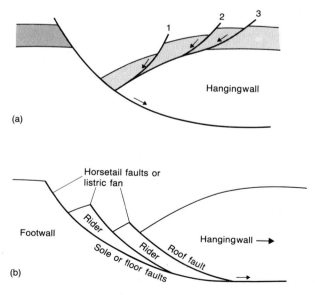

(a)

(b)

Fig. E32 Extensional fault system. Antithetic faults in hangingwall.

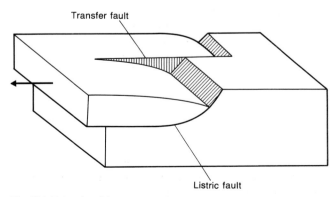

Fig. E34 Extensional fault system. Two dip-slip faults linked by a transfer fault.

Wernicke, B. (1981) Low angle normal faults in the Basin and Range — nappe tectonics in a extending orogen. *Nature* **291**: 645–8.

Wernicke, B. & Burchfiel, B.C. (1982) Modes of extensional tectonism. *Journal of Structural Geology* **4**: 105–15.

extinction Most species that have ever existed on this planet have gone extinct and it has been increasingly recognized in recent years that the extinctions were not uniformly spaced in time. Instead they tend to be concentrated in a limited number of *mass extinction* episodes, affecting a large percentage of the world's biota. The five most important of these are: end-**Ordovician**, late **Devonian** (*Frasnian–Famennian* boundary), end-**Permian**, end-**Triassic** and end-**Cretaceous**. All the events involve marine invertebrates and the three youngest involve terrestrial vertebrates also; significant correlative changes in **land plants** have not been recognized. The fact that so many organisms of different biology and habitat have gone extinct at the same time suggests strongly that the underlying cause or causes must have involved a drastic change in the physical environment rather than something specific to any one particular group. Thus nearly a hundred hypotheses of *dinosaur* extinction at the end of the **Cretaceous** have been proposed, but most of these are highly fanciful and unsupported by evidence, and ignore the fact that *dinosaurs* are only one of many groups that went extinct at this time.

Traditionally paleontologists have sought an explanation for *mass extinctions* in terms of events intrinsic to this planet, such as global regressions of the sea, climatic changes or episodes of exceptionally intense volcanism and/or tectonism. There is indeed a correlation between falls of sea-level inferred from study of the stratigraphic record and episodes of *mass extinction* on both major and minor scales, but there is as yet no general agreement on the precise causal relationship, nor on the closeness of correlation. For several decades there has been speculation about possible extraterrestrial influences, such as periodic increases in cosmic ray or solar proton flux, that would damage the ozone layer of the outer atmosphere and permit increased incidence of dangerous ultraviolet light at the Earth's surface. Such speculation was not taken seriously in the absence of supporting evidence until a group of Berkeley scientists (Alvarez *et al.*, 1980) discovered an abnormal enrichment of the **platinum**-group element **iridium** precisely at the *Cretaceous–Tertiary boundary* as determined by microfossils. The exact coincidence of the **iridium**-enriched clay with *mass extinction*

of calcareous plankton was held to be highly significant, and has since been found at many localities across the world. It was held that the degree of **iridium** enrichment cannot be explained by purely terrestrial processes and instead Alvarez and co-workers invoked the *impact* of a 10-km-diameter *asteroid*, which would have caused the expulsion of an immense dust cloud that would have inhibited photosynthesis for several weeks to months, and hence caused the *mass extinction* of plankton, with chain reactions through the rest of the biosphere. Although this *impact hypothesis* has received independent support from the discovery at a number of localities of **quartz** grains with laminar structures indicative of shock induced by immense pressures, of the sort characteristic of *meteorite* impact, others are sceptical. Many paleontologists argue that a large proportion of the end-**Cretaceous** extinctions are too selective and too gradual to be plausibly accounted for by the Alvarez hypothesis, and a number of scientists have invoked a compound scenario involving sea-level and temperature fall in conjunction with an episode of intense volcanism, involving phenomena intrinsic to the Earth, pointing out that such a scenario can account for the facts more plausibly than one involving impact.

Despite the controversial nature of the *Cretaceous–Tertiary boundary impact hypothesis*, many have been encouraged to seek an extraterrestrial explanation for other *mass extinction* horizons. The University of Chicago paleontologists, for example (Raup & Sepkoski, 1984), have claimed, on the basis of statistical analysis, a 26 Ma extinction periodicity within the last 250 Ma, which has led to speculation of periodic impact by *bolides* (the general term for extraterrestrial bodies such as *asteroids* or *comets*). The reliability of both the statistics and the dating have been challenged, however, and the supporting evidence from geochemistry has proved disappointing to impact supporters, for all major extinction events apart from the *Cretaceous–Tertiary boundary*. As noted above, even for this intensively studied event many of the key data can be interpreted in radically different ways. [AH]

Alvarez, L.W., Alvarez, W., Asaro, F. & Michel, H.V. (1980) Extraterrestrial causes for the Cretaceous–Tertiary boundary extinction. *Science* **208**: 1095–108.

Nitecki, M.H. (ed.) (1984) *Extinctions.* University of Chicago Press, Chicago.

Raup, D.M. & Jablonski, D. (eds) (1986) *Patterns and Processes in the History of Life.* Springer-Verlag, Berlin.

Raup, D.M. & Sepkoski, J.J. (1984) Periodicity of extinctions in the geologic past. *Proceedings of the National Academy of Sciences* **81**: 801–5.

Stanley, S.M. (1987) *Extinction.* Scientific American Books, New York.

F

fabric The pervasive features of a rock body resulting from **deformation**. The features that contribute to the rock fabric are known as *fabric elements*, such as for example, **cleavage**, **fracture**, **lineation**, grain shapes, grain boundaries and crystallographic orientations. They also include geometrical abstractions such as **fold axial trace**. Bodies of rock that are homogeneous with respect to the orientation of a *fabric element* define *fabric domains*; for example, areas with constant *fold axis* orientation, or areas of constant **quartz** *c*-axis orientation. Although *fabric elements* must be pervasive, the concept of pervasiveness is scale-dependent, so that a feature which is not pervasive at one scale (for example, a **fracture**) may become part of the fabric at a larger scale (if the **fracture** belongs to part of a regional **fracture** system). *Microfabrics* are pervasive features revealed at a microscopic scale.

The *fabric symmetry* is the class of symmetry shown by a *fabric element*. This concept classifies **anisotropic** fabrics into symmetry classes of which the most common are orthorhombic (three mutually perpendicular planes of symmetry), monoclinic (one plane of symmetry) and triclinic (no symmetry). In the case of monoclinic symmetry, three *fabric symmetry* axes *a*, *b*, and *c* can be defined by the *ac* plane (the symmetry plane), the *b* axis (normal to the symmetry plane) and the *a* axis (perpendicular to *b* in the principal fabric plane). Figure F1 shows, for example, the *fabric symmetry* axes for a monoclinic fabric produced by an asymmetrically folded **bedding** plane: the *b* axis is parallel to the *fold axis*, and is sometimes referred to as the principal fabric axis. The orthogonal fabric axes of a monoclinic symmetry have been called the *fabric cross* (Turner & Weiss, 1963), and can be used to define the orientation of other *fabric elements* by indices in a similar way to Miller indices in crystallography (e.g. see Hancock (1985) in which orientations of **fractures** with respect to **folds** are described by such a system).

Kinematic symmetry axes a, b and *c* can be defined by a similar convention as the *fabric symmetry* axes: *a* is the direction of **flow** and *ab* the plane of **flow**, assuming **plane strain**, **simple shear**. *Sander's symmetry principle* states that the symmetry of the fabric is the same or less than the kinetic symmetry producing the fabrics; this principle has been used to infer the kinematics of **deformation** from observed fabrics. However, since much **deformation** is not **plane strain** and **simple shear**, monoclinic symmetry axes are often inappropriate. Furthermore, the final symmetry of the fabric may depend on the original *fabric symmetry*, in which case the *fabric symmetry* does not relate directly to the kinematic symmetry. These are two severe limita-

tions to Sander's approach, which is therefore no longer widely used, although it is historically important. [TGB]

Hancock, P.L. (1985) Brittle microtectonics: principles and practice. *Journal of Structural Geology* **7**: 437–57.
Turner, F.J. & Weiss, L.E. (1963) *Structural Analysis of Metamorphic Tectonics*. McGraw-Hill, New York.
Sander, B. (1950) *Einführung in die Gefugekunde der Geologischen Körper*. Springer-Verlag OMG, Vienna.

face A mining and quarrying term for an exposed surface (apart from the **back** or floor) from which rocks or minerals are being removed. [AME]

facet Any of the surfaces of a cut **gemstone**. Not to be confused with natural crystal faces. [AME]

facing **Folds** of **bedding** are said to face along their *axial planes* in the direction in which the **beds** become younger. **Folds** are thus said to face upwards or downwards, or in a particular geographic orientation (e.g. NW). [RGP]

fahlband A band of metamorphic rock carrying disseminated sulphides. The sulphides are more abundant than **accessory minerals** but too few to form an **orebody**. These bands usually weather to a rusty brown colour. [AME]

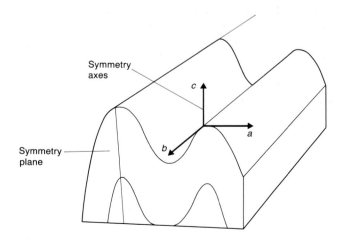

Fig. F1 The **fabric** symmetry axes.

[235]

failure Loss of **strength**. The *failure strength* or *stress* is the **stress** at which failure occurs. The *failure criterion* is the relationship between *principal stresses* which gives the condition for failure. The form of the *failure criterion* may be theoretically predicted or empirically determined. The most common of the former are as follows.

1 *Tresca's criterion*: failure occurs when the maximum **shear stress** reaches a constant value, which is a material property.

2 *Maximum octahedral shear criterion*: failure occurs when the *octahedral* **shear stress** reaches a constant value, which is a material property.

3 *Maximum strain energy of distortion criterion*: failure occurs when the **strain** energy of distortion reaches a constant value which is a material property.

4 *Coulomb criterion* (also known as *Coulomb–Mohr* or *Coulomb–Navier criterion*): failure occurs when the **shear stress** across a plane reaches a constant value (the **cohesion**) plus a constant (the *coefficient of friction*) times the *normal stress* across the plane.

$$\tau = S + \mu\sigma$$

where τ is the **shear stress**, S is the **cohesion**, μ is the *coefficient of friction* and σ is the *normal stress*. This can also be expressed in terms of the *principal stresses* σ_1 and σ_3:

$$\sigma_1[(\mu^2 + 1)^{1/2} - \mu] - \sigma_3[(\mu^2 + 1)^{1/2} + \mu] = 2S$$

The *Coulomb failure criterion* is shown on a **Mohr diagram** of **shear stress** against normal **stress** as a straight line.

5 The *Griffith failure criterion* is the relationship between *principal stresses* which gives the condition for failure, based on the postulate that failure occurs when the potential energy of a pre-existing crack (**Griffith crack** or **flaw**) system remains constant or decreases with an increase in crack length. The fundamental criterion is that the **tensile stress** σ, is

$$\sigma \geqslant (2E\lambda/\pi C)^{1/2}$$

where E is the **Young's modulus**, λ the surface energy of the crack, and C the crack half-length. In biaxial compression, this gives

$$(\sigma_1 - \sigma_3)^2 - 8T_0(\sigma_1 + \sigma_3) = 0 \quad \text{if} \quad (3\sigma_3 + \sigma_1) > 0$$

or

$$\sigma_3 = -T_0 \quad \text{if} \quad (3\sigma_3 + \sigma_1) < 0$$

Where σ_1, σ_3 are the principal **stresses**, T_0 is the *tensile strength* of the material, and **stress** is taken as positive for compression. This *failure criterion* therefore predicts that the failure strength is independent of the intermediate *principal stress* σ_2, that the ratio of uniaxial compressive to *tensile strengths* is 8; and makes no allowance for the effects of friction on a closed **Griffith flaw**. None of these is realistic for most rocks: attempts to overcome these difficulties have led to various theories building on the

pioneering work of Griffith, and including most importantly: the *modified Griffith theory*, which introduces sliding between the crack surfaces with a *coefficient of friction*, μ. The *failure criterion* then becomes:

$$\sigma_1[(1 + \mu^2)^{1/2} - \mu] - \sigma_3[(1 + \mu^2)^{1/2} + \mu]$$
$$= 4T_0(1 + \sigma_m T_o)^{1/2} - 2\mu\sigma_m$$

where σ_m is the **stress** necessary to close the crack.

6 The *extended Griffith criterion* (or *Griffith–Murrell criterion*) (Murrell, 1963) extends the two-dimensional Griffith analysis to a triaxial **stress** state, yielding the criterion:

$$(\sigma_1 - \sigma_2)^2 + (\sigma_2 - \sigma_3)^2 + (\sigma_3 - \sigma_1)^2 = 24T_0(\sigma_1 + \sigma_2 + \sigma_3)$$

and predicts a better ratio of 12 between the uniaxial compressive and tensile **stresses**.

However, all types of *Griffith failure criterion* are strictly only criteria for the propagation of cracks rather than failure, which is a complex interaction of cracks. The importance of the theories is the thermodynamic approach, which is the basis of much of *fracture mechanics* theory. [TGB]

Jaeger, J.C. & Cook, N.G.W. (1979) *Fundamentals of Rock Mechanics* (3rd edn). Chapman & Hall, London.
Murrell, S.A.F. (1963) A criterion for brittle fracture of rocks and concrete under triaxial stress and the effect of pore pressure on the criterion. Proceedings 5th Rock Mechanics Symposium, University of Minnesota. In: Fairhurst, C. (ed.) *Rock Mechanics*, pp. 563–77. Pergamon, Oxford.
Price, N.J. (1966) *Fault and Joint Development in Brittle and Semi-brittle Rock*. Pergamon, Oxford.

falling dune Depositional **dune** which forms on the **lee slope** of an obstacle such as a hill or cliff where **sand** collects after having been moved from an area of greater exposure. Falling dunes have steep slope angles often at or near the *angle of repose* of **sand**. As they are dependent on a topographical feature for their development they are essentially a *stationary dune* form. In appearance often similar to **climbing dunes**. [RDS]

famatinite (Cu_3SbS_4) A sulphosalt mineral of **copper**. [DJV]

fan shooting A form of **seismic refraction** surveying where the detectors are located at a near constant distance from the **shot** in a radial fan-like **seismic array**. A characteristic travel-time curve is established in an area of uniform geology. **Delay times** computed from the observed minus characteristic travel-times are then mapped out from intersecting fan-arrays to delineate anomalous structures. This form of refraction shooting was used early in **seismic** oil **exploration** to find **salt domes**, where fast arrivals travelling through the high velocity salt helped locate the domes. [GS]

faro A small, elongate *reef* with a *lagoon* up to 30 m deep,

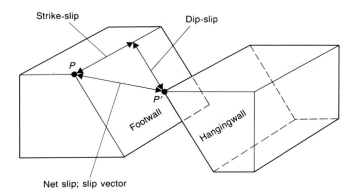

Fig. F2 Fault. Before faulting, points *P* and *P'* were coincident. The fault is an oblique slip fault, with normal and left lateral components of dip-slip and strike-slip respectively. (After Hobbs *et al.*, 1976.)

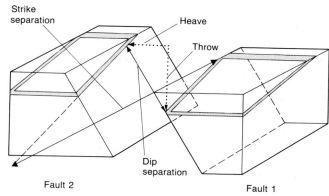

Fig. F3 Fault. Same pair of blocks as in Fig. F2 showing an offset layer (shaded) and terminology for fault offsets. Offset of a planar feature across a fault can be measured parallel to the dip, giving the dip separation or parallel to the strike, giving the strike separation. Vertical and horizontal components of the dip separation are respectively throw and heave (After Hobbs *et al.*, 1976.)

that forms part of the rim of a **barrier reef** or of an **atoll**. [NJM]

fault A surface of discontinuity across which there has been **shear displacement**. The vector connecting a point to its original position before **displacement** is known as the *slip* or *displacement vector* and its length defines the *net slip*. This can be resolved into a horizontal component parallel to the fault, the **strike-slip** component, and a vertical component parallel to the **dip** of the fault, the **dip-slip** component (Fig. F2).

The offset of a planar feature measured in a horizontal section parallel to the fault is the *strike separation* and in a vertical section normal to the fault, the *dip separation*. In this vertical section, the *dip separation* can be factorized into horizontal and vertical components, which are termed the *heave* and *throw* respectively (Fig. F3). Because faulting of planar features, e.g. **bedding**, is more commonly observed than the displacement of linear features or points, *strike* or *dip separation* can be measured more often than **strike-** or **dip-slip**. The difference between the separation and the slip can be seen by comparing Figs F2 and F3.

The rock bodies above and below an inclined fault are known as the **hangingwall** and **footwall** respectively. The intersection of a feature with the fault surface is a *cut-off point* or *cut-off line*. Faults with a dominant component of **strike-slip** are known as *strike-slip faults*; when vertical, they are referred to as *transcurrent* or *wrench faults*. **Transform faults** are large *transcurrent faults* which terminate in other **plate boundaries** such as ridges or trenches, and are therefore discontinuities across which **strike-slip displacements** are occurring between *plates*. **Strike-slip** is defined as dextral (right-lateral or right-handed) or sinistral (left-lateral or left-handed) by the direction of movement of the block on the opposite side of the fault to an observer (Fig. F2).

Dip-slip faults, with a dominant component of **dip-slip displacement**, are termed *normal* when the **hangingwall** is

displaced downwards relative to the **footwall**. *Normal faults* with a **dip** of less than 45° are sometimes known as *lag faults* or *low angle (extension) faults*. The opposite type of **dip-slip** fault, in which the **hangingwall** moves up relative to the **footwall**, is a *reverse fault*; a *reverse fault* with a low angle of **dip** is known as a *thrust* or *overthrust fault*. *Thrusts* are commonly linked together in a set or *thrust system* (see **thrust belt**, **imbricate structure**). *Oblique slip faults* have similar magnitudes of both **strike-** and **dip-slip** components of **displacement**.

A *rotational fault* is a fault along which movement causes rotation of one or both adjacent blocks. *Listric faults* are spoon-shaped *rotational faults* in which the **hangingwall** is rotated towards the fault in the same sense as the movement along the fault (see Fig. E31, p. 232). These are found in **extensional fault systems** (see also **extension**).

A *scissor fault* exhibits a reversal of **displacement** sense along the *fault plane*. There is an axis on the *fault plane* where no movement occurs; on either side of the axis, the **displacement** is in the opposite sense so that the action may be compared to opening a pair of scissors. This is most commonly seen along a subvertical fault, where the downthrown block changes side across the scissor axis.

Faults often occur together in a number of geometrical associations. A *fault set* is a group of faults with similar orientations and *displacement vectors*.

Conjugate faults are two equally developed faults with opposite senses of slip. The intersection of the fault is generally normal to the *displacement vectors* of both faults (Fig. F4), and the two faults are generally inferred to have originated at the same time in response to the same **stress**.

En echelon faults are parallel or subparallel faults slightly offset from each other in the direction perpendicular to the fault surface (Fig. F5). *En echelon faults* are usually inferred to have a common origin.

Second order or *splay faults* are minor faults or *fault sets* at

fault

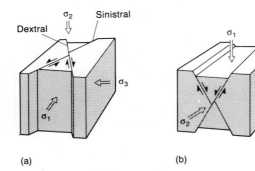

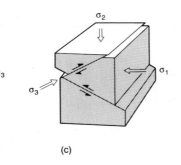

Fig. F4 Three classes of **fault**. (a) Strike-slip fault; (b) normal fault; (c) thrust fault, showing their hypothetical relationship to the maximum, intermediate and minimum principal stresses, σ_1, σ_2 and σ_3. For each class of faults, two conjugate faults are shown.

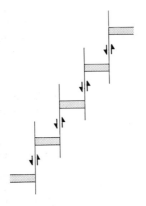

Fig. F5 An en echelon array of dextral **faults** in plan view.

acute angles to a major fault with the same sense of slip. Further sets of minor faults with a similar relationship to *second order faults* are termed *third order faults*, and so on. *Splay faults* have been interpreted as forming in a **stress** field reorientated by the main fault, or in the **stress** field ahead of a propagating fault.

A *fault block* is a body of rock partially or completely defined by faults.

Multiple faults are a set of equally developed faults with a common geometrical relationship. Like *conjugate faults*, *multiple faults* or *fault sets* may form simultaneously in a common **stress** field; they may have an orthorhombic symmetry.

A *fault zone* is a tabular volume containing many faults and **fault rocks**. Within *fault zones* there may be subsidiary faults known as *Reidel* or *R1 faults*, *conjugate Reidel* (or *R2*) *faults*, *P shears* and *Y shears*. **Fault rocks** are rocks found within *fault zones* and produced by fault processes. They include *fault gouges*, *breccias*, *cataclasites*, **mylonites** and *cements* (see **fault rock**).

Fault surfaces may be smooth and polished, in which case they are known as **slickensides**. They may also show linear features of several different types, including *slickenside striations* or *slickenlines*, *wear grooves* and *growth fibres*. These are often taken to indicate the direction of the *displacement vector* and their length may give the net slip of the movement which created them. They may originate in

several ways: some are due to scratching of the surface by hard particles (*wear grooves*), and some are due to growth of fibrous cements across the fault surface. Others, however, are clearly longer than the net slip, and cannot be due to relative motion of the fault; these may be original features of the shape of the fault surface. A second group of fault surface features are roughly perpendicular to the **lineations**; these are *steps* or *risers*. The *steps* in the fault surface give the linear features an asymmetry. If the short faces defined by the *steps* point towards the sense of movement of the block sliding over them, they are *congruous* (e.g. *wear grooves*); otherwise they oppose the sense of movement and are *incongruous* (e.g. *growth fibres*). Fault surfaces may show layers of **lineations**, which are not always parallel. The **lineations** can only be used to define the sense of movement if the process forming them is known.

The orientations of fault surfaces and *slip vectors* have been used in a number of ways to deduce the **stress** responsible for them. The pioneer work by Anderson in this field forms the basis of the approach. Faults represent **failure** of rock in response to imposed **stress**. By analogy with triaxial tests, faults are assumed to form at an angle less than 45° to the maximum *principal stress* (σ_1). The minimum *principal stress* (σ_3) is perpendicular to σ_1 in the plane normal to the fault surface; since faults may form equally either side of σ_1; **conjugate faults**, typically enclosing an angle of 60°, may develop in a *triaxial stress system* such that the intermediate *principal stress* (σ_2) is parallel to the line of intersection between them (Fig. F4). Anderson assumed that one of the *principal stresses* was due to **gravity** and therefore always vertical. If it is σ_1, then *normal faults* form; if σ_2 is vertical, *transcurrent faults* form, and the third category of faults (*reverse*), are predicted from a vertical minimum *principal stress* σ_3. This categorization does indeed explain the commonest types of fault observed, but exceptions are provided by high angle *reverse faults*.

A considerable spread of fault orientations and *slip vectors* may result from a single **deformation**. This spread can be used in more sophisticated analyses to deduce *principal stress* orientations, making the simple assumptions that slip occurs in the direction of maximum **shear**

[238]

stress and that a fault forms at an acute angle to maximum *principal stress*.

Both these methods and the approach of Anderson are limited by the assumptions necessary to relate the fault to the **stress** orientation; this may be complicated, for example, by **anisotropy** in rock **strength**. Strictly speaking, the methods using fault orientations and *slip vectors* deduce the orientation of the **strain** field. This forms the basis of more recent alternative views of faulting, in which the orientation of faults is a response to the bulk **strain** rather than **stress**. One of these views suggests that generally at least four sets (multiple *fault sets*) rather than two *conjugate sets* will form with orthorhombic symmetry about the **principal strain** axes; another suggests that faults form in the directions of no *infinitesimal* longitudinal *strain* (see **strain ellipsoid**).

Earthquakes occur on faults; slip accumulated during these events is known as **seismic slip** in distinction to that during intervening periods when the fault may be inactive, or moving by **creep** (*aseismic slip*). Monitoring of the pattern of first ground movement following the **earthquake** can allow the orientation of the fault to be deduced by the method of **focal mechanism solution**, which can also provide information on the **stress** responsible for the **earthquake** faulting. [TGB]

Hobbs, B.E., Means, W.D. & Williams, P.F. (1976) *An Outline of Structural Geology* (2nd edn). John Wiley & Sons, New York.
Park, R.G. (1983) *Foundations of Structural Geology*. Blackie, Glasgow.
Sibson, R.H. (1977) Fault rocks and fault mechanisms. *Journal of the Geological Society, London* **133**: 191–213.
Price, N.J. (1966) *Fault and Joint Development in Brittle and Semi-brittle Rock*. Pergamon, Oxford.

fault drag Deflection of a marker adjacent to a **fault**. Curvature of the marker consistent with the sense of **displacement** on the **fault** is *normal drag* in contrast to *reverse drag*, which is contrary to the sense of displacement (Fig. F6). The existence of both types of drag, sometimes even on the same **fault**, cautions against the use of drag to deduce sense of displacement. Neither type of drag necessarily follows or is due to **fault** movement; the deflection may precede faulting. (See **drag fold**.) [TGB]

fault propagation Process whereby a **fault** extends itself along its length. (See **fault**.) [RGP]

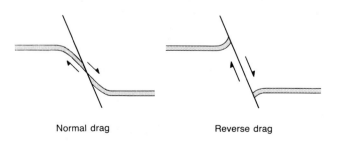

Normal drag   Reverse drag

Fig. F6 Fault drag.

fault rock Rock produced by the faulting process (see **fault**). Table F1 shows a classification of such rocks, based on Sibson (1977). The following types of fault rock are identified: *fault breccia*, *fault gouge*, **pseudotachylite**, *crush breccia*, *cataclasite* (including *protocataclasite* and *ultracataclasite*), **mylonite** (including *protomylonite*, *ultramylonite* and *blastomylonite*).

A *fault breccia* is a non-foliated, incohesive, fault rock with more than 30% visible fragments surrounded by matrix.

Fault gouge is an incohesive fault rock with less than 30% visible fragments surrounded by matrix. A number of typical *fault gouge* features are well known from both experimentally produced and natural gouges (Fig. F7). These include a *P-foliation*, consisting of a *phyllosilicate* **preferred orientation** or layering at an acute angle to the **shear** direction, *Reidel (or R1) shears* at a high angle to the **shear** direction, *conjugate Reidel or R2 shears* in conjugate orientations to the *Reidel shears*, *Y shears* parallel to the **shear** direction, and *ductile stringers* from hard inclusions drawn out in the **shear** direction (Fig. F7). All these features are useful for deducing the sense of **shear** in the gouge.

Crush breccia is a cohesive fault rock, with less than 10% of matrix, produced by tectonic reduction in grain size, in which the fragments are larger than 5 mm. Fragments between 1 and 5 mm define a *fine crush breccia*, and below 1 mm, a *crush microbreccia*, which is a non-foliated, cohesive fault rock containing fragments less than 1 mm large in up to 10% of a finer grained matrix.

Cataclasite is a non-foliated, cohesive rock containing 10–50% of fragments in a finer grained matrix occurring in localized zones (*fault zones*). *Cataclasite* is the intermediate member of the *cataclasite* series *protocataclasite–cataclasite–ultracataclasite*, divided at proportions of 50% and 90% matrix. Although originally described as 'random-fabric', *cataclasites* may have a crude **foliation** defined by **preferred orientation** of inequant fragments, **fractures**, or *phyllosilicates* within the matrix. There is therefore a transition between *cataclasites* and **mylonites** via *cataclasites* that have some degree of **foliation**. This definition is non-

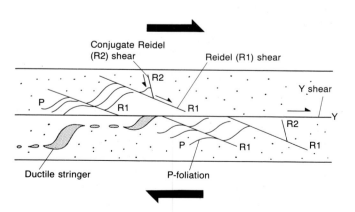

Conjugate Reidel (R2) shear
Reidel (R1) shear
R2
P   R1   R1   Y shear
R1
Y
R2
P   R1   R1
Ductile stringer   P-foliation

Fig. F7 Fault rock. Principal gouge features for a right lateral gouge zone.

Table F1 Classification of **fault rocks**. (After Sibson, 1977.)

Incohesive		Random fabric		Foliated	Proportion of matrix
		Fault breccia (visible fragments > 30% of rock mass)		?	
		Fault gouge (visible fragments < 30% of rock mass)		?	
Cohesive	Glass/devitrified glass	Pseudotachylite		?	
	Nature of matrix — Tectonic reduction in grain size dominates grain growth by recrystallization and neomineralization	Crush breccia (fragments > 0.5 cm) / Fine crush breccia (0.1 cm < fragments < 0.5 cm) / Crush microbreccia (fragments < 0.1 cm)			0–10%
		Protocataclasite	Cataclasite Series	Protomylonite — Mylonite Series	10–50%
		Cataclasite		Mylonite	50–90%
		Ultracataclasite		Ultramylonite	90–100%
	Grain growth pronounced	?	Phyllonite varieties	Blastomylonite	

genetic, but other definitions stipulate that *cataclasites* are products of **cataclasis**. [TGB]

Chester, F.M., Friedman, M. & Logan, J.M. (1985) Foliated cataclasites. *Tectonophysics* **3**: 139–46.

Rutter, E.H., Maddock, R.H., Hall, S.H. & White, S.H. (1986) Comparative microstructures of natural and experimentally produced clay-bearing fault gouges. *Pure and Applied Geophysics* **124**: 1–30.

Sibson, R.H. (1977) Fault rocks and fault mechanisms. *Journal of the Geological Society, London* **133**: 191–213.

fault termination The end of a **fault**, where the displacement is reduced to zero. Geometric problems exist in the material surrounding a fault termination; the displacement must be accommodated by **strain** in the walls of the **fault**. Alternatively, a **fault** may terminate against another **fault** or a **fold**, in which case the displacement may be transferred. (See **fault**.) [RGP]

fayalite (Fe_2SiO_4) The **iron** end-member of the **olivine** *solid solution* series of minerals. (See **olivine minerals**.) [DJV]

feather edge The line of intersection on a map of a stratigraphic boundary with a higher stratigraphic boundary such as an **unconformity**, e.g. 'the feather edge of the base of the Coal Measures on the base of the **Triassic**'. The feather edge marks the zero **isopachyte** of the rock body between the two boundaries in question. [RGP]

feldspar minerals The most abundant mineral group in the **crust** of the Earth, making up about 60% by volume of the **igneous rocks**. They are framework aluminosilicates of sodium, **potassium** and calcium and can form under conditions ranging from those in the **upper mantle**, through crustal igneous and metamorphic conditions, to growth in near-surface **pore-waters** during sedimentary **diagenesis**. Feldspar is the main constituent of the crust of the Moon. Although of minor economic value (for glazes and as **abrasives**) their abundance gives them a central role in many rock-forming processes and they form one of the main components of classification schemes for igneous and some metamorphic rocks. **Dissolution** of feldspars

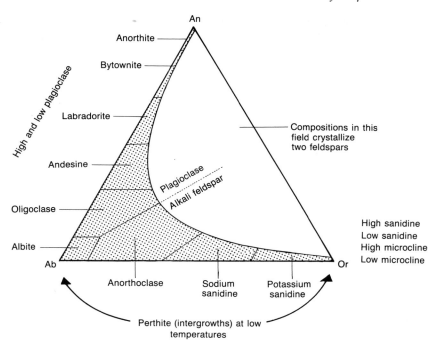

Fig. F8 Feldspar minerals — nomenclature. (After Smith & Brown, 1988.)

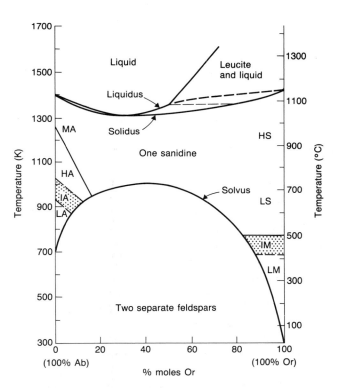

Fig. F9 Feldspar minerals. Phase diagram for the alkali feldspars. The liquidus and solidus are shown for atmospheric pressure; they move rapidly to lower temperature with increasing water vapour pressure. The solvus is for ordered feldspars that are structurally discontinuous; in perthites with a continuous or coherent structure, compositions are given by a coherent solvus inside the curve given. (After Brown & Parsons 1989. Melting curves after Tuttle & Bowen 1958.)

in **clastic** sedimentary **rocks** leads to the development of **porosity**, which is of importance in **petroleum reservoirs**, and feldspars are the main source of **clay minerals** during **diagenesis** and *soil* formation.

Common feldspars are *solid solutions* of three end-member components, **anorthite** $CaAl_2Si_2O_8$(An), **albite** $NaAlSi_3O_8$(Ab) and **orthoclase** $KAlSi_3O_8$(Or) (Fig. F8). Although all three components are usually present, in most cases feldspars have compositions close to either the Ab–An edge of the diagram, when they are called **plagioclase**, or to the Ab–Or edge, called **alkali feldspar**. **Alkali feldspars** are therefore a relatively simple *solid solution* series, in which Na^+ and K^+ substitute for each other. **Plagioclase** *solid solution* is more complex, substitution of Na^+ by Ca^{2+} necessarily being accompanied by substitution of Si^{4+} by Al^{3+}, to achieve charge balance, an example of a coupled substitution. Phase diagrams for these *solid solutions* are shown as Figs F9 and F10. Various **polymorphs** occur, indicated on Figs F8, F9 and F10. The reason for the **polymorphism** is discussed later. There are three other, rare, naturally occurring feldspars: *buddingtonite*, $NH_4AlSi_3O_8$; *reedmergnerite*, $NaBSi_3O_8$; and *celsian* $BaAl_2Si_2O_8$. About forty synthetic compounds with the feldspar structure have been prepared, including analogues in which **gallium** substitutes for silicon.

Feldspar is the most abundant mineral of **granitic** rocks and for this reason is a familiar constituent of the polished slabs used in shop facades. It is usually pale in colour, ranging from white, through cream, pale or medium brown, to a strong pinkish red. Less commonly feldspars may be pale green. The opacity and translucency of many feldspars results from numerous tiny *micropores*, on a

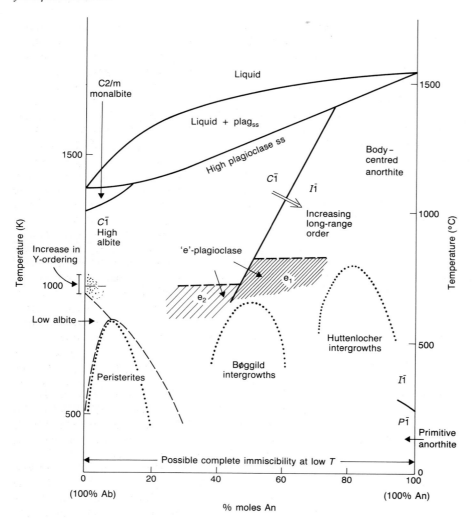

Fig. F10 Feldspar minerals. Phase diagram for the plagioclase feldspars. The liquidus and solidus are shown for atmospheric pressure. Only continuous boundaries are known with any certainty. Dotted curves are probably coherent spinodals.

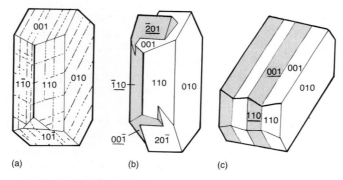

Fig. F11 Feldspar minerals. Morphology and twinning in feldspars.
(a) Orthoclase, showing common habit and prominent cleavages.
(b) Interpenetrant Carlsbad twin. (c) Albite twins.

scale of about 1 μm and below, sometimes accompanied by microscopic grains of **micas**, **clay minerals** and **iron oxides**. When such *micropores* or **alteration** are absent, feldspars may be transparent and glassy in appearance, particularly when occurring as *phenocrysts* in volcanic

rocks, or dark green or black when seen in the mass in plutonic rocks. A familiar example of this type is the variety of **syenite**, *larvikite*, a common architectural stone in which the feldspar shows a striking blue *iridescence*. Such feldspars, and also those showing a silvery *iridescence*, known as **moonstone**, are semi-precious and used in jewellery (see **gem**). Some relatively rare feldspars are intrinsically black in colour. The density of feldspars is in the range 2.56 (**potassium feldspar**) to 2.61 (**albite**) and 2.76 Mg m^{-3} (**anorthite**). Important crystallographic features of feldspar are shown in Fig. F11. All **plagioclases** have triclinic symmetry, but **alkali feldspars** are either monoclinic or triclinic. Feldspars usually form blocky tablets, slightly flattened on (010), although in many **granitic** rocks they are shapeless because of interference during growth. In some **alkaline rocks** the **alkali feldspars** are extremely flattened on (010). **Plagioclase** in **basaltic** rocks is usually in the form of flat tablets, either aligned by **flow** or sometimes growing as stellate aggregates.

Most feldspars have two marked **cleavages**, parallel to (010) and (001). In monoclinic feldspars these are exactly at right angles, while in triclinic feldspars they are nearly at right angles. Some **alkali feldspars** in volcanic rocks may have poorly developed **cleavages** and a glassy appearance, but **quartz**, which is the next most abundant mineral in **granite**, has no **cleavage** or colour and can be readily distinguished. The **feldspathoids**, **nepheline** and **leucite**, are much like feldspar in appearance and most easily recognized by their crystal shape. Some feldspars with a habit made up of {110} faces, and called **adularia**, may have crystal shapes not unlike **calcite**, but feldspars are harder (6 on Mohs' scale) and have only two, not three, good **cleavages**. It is not easy to reliably distinguish the different sorts of feldspar in hand specimens; in **granites**, in which both **alkali feldspars** and **plagioclase feldspar** often occur together, the former is often pink and the latter white.

Feldspars are frequently **twinned** and formed of two or more individuals growing 'back-to-back' in a characteristic crystallographic relationship. Feldspars in **granite** can often be seen to be composed of two individuals in a relationship known as *'Carlsbad' twinning* (Fig. F11). **Plagioclase** often forms as a sandwich of numerous twinned individuals each usually less than 1 mm thick, called *albite twins*, parallel to (010), and these *'polysynthetic' twins* can sometimes been seen on the (001) **cleavage** surface, looking rather like grooves in a gramophone record (Fig. F11). Such twins cannot form in monoclinic feldspars, and **twinning** in triclinic **alkali feldspars** is usually on too fine a scale to be visible in hand specimen, so the feature is distinctive of **plagioclase**. Another type of *polysynthetic twin*, called *pericline twinning*, at right angles to the **albite twin plane**, is possible in all triclinic feldspars, but is most common in **microcline** and **anorthoclase**.

In thin section, feldspars are colourless and transparent, with low relief. Refractive indices (β) range from 1.522 (**microcline**) to 1.533 (**albite**) and 1.583 (**anorthite**). Most show at least slight turbidity (caused by the pores noted above) and this serves to distinguish them readily from **quartz**. The turbidity, which is caused by interaction between the feldspars and percolating aqueous fluids, can be developed to the extent that the feldspars become nearly opaque. This is usually described as '**alteration**' by petrologists. Under crossed polarizers feldspars show low interference colours, up to pale first order yellow in **anorthite**. The **twinning** is distinctive: **plagioclase** showing *albite twinning* has a zebra-striped appearance, and **anorthoclase** (Fig. F8) and **microcline** have distinctive intersecting *albite* and *pericline twins*, looking like tartan. **Microcline** is characteristic of **granites**, but **anorthoclase** occurs mainly in volcanic rocks, so the two are unlikely to be confused. The relationship to the **cleavages** is diagnostic, both (010) and (001) **cleavages** being visible together with the tartan in **anorthoclase**, only (010) in **microcline**.

Phase relationships of **alkali feldspar** are shown in Fig. F9. The temperatures of the beginning and end of freezing curves (liquidus and solidus) are strongly sensitive to the applied water vapour pressure (P_{H_2O}), 0.2 GPa

pressure lowering the liquidus and solidus temperatures by *c.* 270°C compared with melting at atmospheric pressure. This effect is of profound importance in the generation and movement of silicate **magmas** in the **crust**. **Alkali feldspars** crystallized at low P_{H_2O} can have any composition at all in the *solid solution* series (although compositions near the minimum, at about $Ab_{60}Or_{40}$ are common in **syenites**) but at $P_{H_2O} \geqslant 0.4$ GPa the solidus intersects the solvus and two feldspars crystallize simultaneously, their compositions being depicted by the solvus curve with respect to temperature. Feldspars which crystallize at very low temperatures, in sedimentary **diagenesis**, for example, have nearly pure end-member compositions, because of the presence of the solvus.

Alkali feldspars that crystallize at high temperatures, above the solvus, usually **exsolve** as they cool to give complex intergrowths of end-member feldspars, called **perthites**. Only very rapid cooling can lead to the preservation of homogeneous **alkali feldspars**. Cooling under dry conditions gives rise to regular fine-scale lamellar intergrowths (up to about 1 μm in width) called **microperthites** (or *cryptoperthites*, when sub-optical). **Diffraction** of light by fine-scale lamellar intergrowths of this type is the cause of the *iridescence* of the feldspars in *larvikite*. The dimensions of the lamellae depend on cooling rate and their orientation on the elastic properties of the **crystal structure**. Cooling in the presence of an aqueous fluid produces irregular 'deuteric', or patch **perthites**, on a much coarser scale, up to 1 mm, and their formation is accompanied by the development of turbidity.

Feldspars are framework silicates in which Si–O and Al–O tetrahedra are linked together to form an infinite three-dimensional framework with the alkali and alkaline earth cations in the large interstices. The aluminium and silicon ions may be arranged in a regular or ordered way over the possible structural sites, or in a random or disordered arrangement. **Sanidine** is a disordered variety of **potassium feldspar**, with monoclinic symmetry, stable at high temperatures; **microcline** is the ordered form, stable at low temperatures, and is triclinic. The ordering process is rather slow, and the disordered arrangements are easily 'frozen in' to low temperature. A continuum of ordering states exists leading to the sequence high **sanidine**–low **sanidine**–intermediate **microcline**–low **microcline** in **potassium feldspar**, and to high and low **albite** (Fig. F9). The ordering mechanism leads to the distinct 'tartan' **twinning** of microcline. **Orthoclase** itself is a common form of monoclinic **potassium feldspar** found in slowly cooled rocks. It has a structure composed of **domains**, similar to the twins of **microcline**, but on the scale of a few unit cells and hence invisible in the optical microscope. In nature, the difficult transformation from **orthoclase** to **microcline** very probably requires interactions between the feldspar and aqueous fluids, or external **stress**.

All feldspars also exhibit rapid 'displacive' changes in lattice angles with changing temperature which cannot be 'frozen in'. In **alkali feldspar** this leads to a symmetry change from monoclinic sodium **sanidine** to triclinic **anorthoclase** (Fig. F9). Phase relationships in **plagioclase** (Fig. F10) are much less well understood than **alkali feldspar**. The liquidus and solidus form a simple loop, **anorthite** melting at a much higher temperature (1550°C) than **albite** (1100°C). **Plagioclases** with (Si, Al)–O disorder (high **plagioclases**) form a comparatively simple *solid solution*. The amount of order at solidus temperatures increases towards **anorthite** and natural **anorthites** are always highly ordered. The phase relationships at low temperatures, between ordered **albite** and **anorthite**, are very complex. There are many different types of intergrowth, almost all on a sub-optical scale. Three regions of **exsolution** have been recognized, leading to the *peristerite, Bøggild* and *Huttenlocher intergrowths* (Fig. F10). All may cause *iridescence*, the best known being that exhibited by **labradorite**. In addition, many intermediate **plagioclases** have a fine-scale slab-like microtexture, known as '*e-plagioclase*', detectable only by **X-ray** or *electron* **diffraction**. The regions of the phase diagram (Fig. F10) in which these intergrowth textures are found are very tentative with respect to temperature. It is probable that, strictly speaking, intermediate low-**plagioclase** is not stable and that a mixture of ordered low **albite** and **anorthite** is the stable arrangement. In practice, the breakdown is so slow that it is only partly approached, even on a geological time-scale, leading to the various fine-scale intergrowths. The low temperature relationships of **plagioclase** have important implications for mineral reactions in a variety of metamorphic rocks. [IP]

Brown, W.L. (ed.) (1984) *Feldspars and Feldspathoids. Structures, Properties and Occurrences.* NATO ASI Series, Series C, Vol. 137. Reidel, Dordrecht.

Brown, W.L. & Parsons, I. (1989) Alkali feldspars: ordering rates, phase transformations and behaviour diagrams for igneous rocks. *Mineralogical Magazine* **53**: 25–42.

Deer, W.A., Howie, R.A. & Zussman, J. (1963) *Rock Forming Minerals* Vol. 4: *Framework Silicates.* Longman, London.

Smith, J.V. & Brown, W.L. (1988) *Feldspar Minerals* Vol. 1: *Crystal Structure, Physical, Chemical and Microtextural Properties.* Springer-Verlag, Berlin.

Tuttle, O.F. & Bowen, N.L. (1958) Origin of granite in the light of experimental studies in the system $NaAlSi_3O_8$–$KAlSi_3O_8$–SiO_2–H_2O. *Geological Society of America Memoir* **74**.

feldspathoid minerals A group of aluminosilicate minerals with some similarities to the **feldspars** but containing less **silica** in their formulae, and characteristic of **silica**-undersaturated alkaline **igneous rocks**. Because rare, relative to **feldspars**, they are therefore diagnostic of a distinctive and important igneous family. There is no clear definition of feldspathoid minerals, and they have a variety of framework structures, but on the criteria above the following minerals may be included:

Leucite	$KAlSi_2O_6$ (limited ss. of Na)
Nepheline–Kalsilite	$NaAlSiO_4$–$KAlSiO_4$ (ss.)
Analcime	$NaAlSi_2O_6 \cdot H_2O$ (H_2O content is variable)

Sodalite group

Sodalite	$Na_8(Al_6Si_6O_{24})Cl_2$
Nosean	$Na_8(Al_6Si_6O_{24})SO_4$
Haüyne	$(Na,Ca)_{4-8}(Al_6Si_6O_{24})(SO_4,S_{1-2})$
Cancrinite	$(Na,Ca)_{7-8}(Al_6Si_6O_{24})$ $(CO_3,SO_4,Cl)_{1.5-2.0} \cdot 1$–$5H_2O$ (ss. with *Vishnevite*, which contains K and has $SO_4 > CO_3$)

In addition, although they are not framework silicates, the *melilite* group bear a similar relationship to the **feldspars**:

Melilite group

Gehlenite–Akermanite	$Ca_2Al_2SiO_7$–$Ca_2MgSi_2O_7$ (ss.)
Soda *melilite*	(limited ss. with *gehlenite*)
$NaCaAlSi_2O_7$	

Note that **leucite** is equivalent to **orthoclase** minus $1SiO_2$ ($KAlSi_2O_6 + SiO_2 = KAlSi_3O_8$) and that **nepheline** is equivalent to **albite** minus $2SiO_2$($NaAlSiO_4 + 2SiO_2 = NaAlSi_3O_8$). The sodium equivalent of **leucite**, $NaAlSi_2O_6$, is not a feldspathoid, but is the **pyroxene**, **jadeite**. **Sodalite** is essentially 6 **nepheline** +2 sodium chloride, $6NaAlSiO_4 + 2NaCl = Na_8(Al_6Si_6O_{24})Cl_2$. **Sodalite** and **cancrinite** are important repositories for volatile components (Cl, CO_3, SO_4, S and H_2O) which reach high concentrations in strongly alkaline **magmas**.

The feldspathoids have a variety of framework structures, with Al–O and Si–O tetrahedra linked in different three-dimensional configurations. The alkali and alkaline earth ions, and the larger CO_3^{2-} and SO_4^{2-} ions and water molecules, when present, occupy large cavities in the (Si,Al)–O framework. Feldspathoids have lower density than **feldspars**, and have properties intermediate between **feldspars** and the very open-structured **zeolite** group of minerals. It is relatively easy to exchange cations between feldspathoids and molten salts and salt solutions, but they do not show the reversible low temperature hydration–dehydration characteristic of **zeolites**.

Leucite is a characteristic mineral of potassic and ultra-potassic mafic and felsic volcanic and hypabyssal igneous rocks. **Potassium feldspar** melts incongruently, at water vapour pressures of <0.25 GPa, to **leucite** plus liquid, but congruently above, and **leucite** is not stable at moderate or high P_{H_2O}. In the absence of water it is stable up to 3 GPa, and can thus occur in the **upper mantle**. Like ice, and unique among silicates, it has the unusual property of being less dense ($D = 2.47$–2.50 Mg m^{-3}) than a melt of its own composition. At room temperature **leucite** has tetragonal symmetry, but it becomes cubic at about 625°C. Although white in colour, **leucite** *phenocrysts* usually have a well-developed icositetrahedral shape (Fig. F12a), which serves to distinguish them from **feldspar**. Refractive indices are 1.508–1.511. Because the symmetry is close to cubic, birefringence is very low. Euhedral crystals have octahedral outlines in thin section and distinctive repeated **twinning** on {110}. This is slightly reminiscent of **microcline**, except that as many as four, not two, sets of twins can be seen (Fig. F12b).

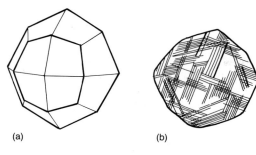

Fig. F12 Feldspathoid minerals. (a) Icositetrahedral morphology of leucite and analcime. (b) Appearance in thin section (crossed polarizers) showing multiple twinning in several directions.

Leucite crystals in plutonic and many high-level rocks are replaced by a vermiform intergrowth of **nepheline** and potassium **feldspar**, known as *pseudoleucite*. Various possible origins have been proposed for *pseudoleucite*: subsolidus breakdown of **leucite** to **sanidine** and **nepheline**; reaction of **leucite** with sodium-rich liquids; and breakdown of **potassium**-bearing **analcime**.

Nepheline occurs as an interstitial or euhedral phase in **phonolites** and **nepheline syenites**, and is the main mineral in the volcanic rock **nephelinite** and its plutonic equivalent **ijolite**, in which it plays the role normally undertaken by **feldspar**. It forms a *solid solution* series with **kalsilite** but the **potassium**-rich members are extremely rare, occurring only in certain extremely sodium-poor lavas. **Nepheline– kalsilite** intergrowths resembling **perthites** occur but are also very rare. Many natural **nephelines** in plutonic rocks are slightly richer in **silica** than the ideal formula, and commonly have compositions near $Ne_{74}Ks_{22}Qz_{3.9}$ (wt.%). Charge balance may be achieved by the presence of empty alkali sites (voids) in the structure. In H_2O- and SiO_2-poor environments **nepheline** is stable to about 4 GPa; **kalsilite** may be a stable phase to very high pressures (12 GPa).

Nepheline looks like **feldspar** and is usually turbid, white or pink in colour. In thin section its low relief, poor **cleavage** and uniaxial negative character are diagnostic. It is often even more turbid than coexisting **feldspar** in plutonic rocks. When occurring as *phenocrysts* in volcanic rocks, less commonly in plutonic rocks, it may have an oily **lustre** and the hexagonal shape of appropriate sections may be distinctive. **Quartz** lacks the turbidity and is uniaxial positive. The density of **nepheline** is 2.56–2.66 $Mg\ m^{-3}$.

Analcime (*analcite* in older texts) is sometimes placed with the **zeolite** minerals, rather than the feldspathoids. It is a cubic framework silicate which crystallizes at low temperatures as an interstitial phase in some basic rocks, as a major phase in rare, strongly **alkaline** (agpaitic) **igneous rocks**, where it often replaces other feldspathoids, and as apparently primary *phenocrysts* in *blairmorite*, a rare type of **phonolite**. It can also occur as a cement in some **sandstones**. Its low temperature stability relative to **albite**

is obscure, and probably depends on **pH**, among other factors. The water content is variable. Like **leucite**, **analcime** may crystallize in well formed icositetrahedra (Fig. F12) and, although cubic, it may be slightly anisotropic and can exhibit similar **twinning**. The two minerals are hard to distinguish, both in hand specimen and thin section, although the *paragenesis* normally provides a clue. **Analcime** has a slightly lower **refractive index** (1.479–1.493) and is less dense (2.24–2.29 $Mg\ m^{-3}$).

The **sodalite** group are uncommon minerals that occur in some **nepheline syenites** and **phonolites**, and are most abundant in the rare strongly **alkaline**, agpaitic **igneous rocks**. They are cubic, with low **refractive index** (1.48– 1.50) and density (2.27–2.50 $Mg\ m^{-3}$). They have poor **cleavages** (in contrast to **fluorite**) and lack the twins of **leucite** and **analcime**. *Nosean* sometimes has a curious dark brownish reaction rim in thin section; there are several chemical tests for distinguishing the **sodalite** minerals from each other, but no simple means. **Sodalites** can be brightly coloured. Usually they are pale grey, pink, green or blue, and they can be pale pink or blue in thin section. Sometimes the colouring is intense. *Hackmanite* is a variety with the entertaining property of changing from a distinct beetroot colour to pale green after a few seconds' exposure to light, and the **beryllium** sodalite *tugtupite* is intense pink. *Lazurite* is the blue constituent of the semiprecious stone **lapis lazuli**, which occurs in metamorphosed **limestone**, and is the original constituent of the pigment *ultramarine*.

Cancrinite occurs as a late-crystallizing or replacive mineral in some **nepheline syenites** and **ijolites**. It has distinctive optical properties with hexagonal symmetry and a well-developed **cleavage**, and can be white, pale blue, yellow or red. It has moderate birefringence (maximum 0.25) and all refractive indices are less than balsam.

The *melilite* group of minerals are often not considered to be feldspathoids because they lack a framework structure. However, they occur in high CaO, low SiO_2 igneous rocks, mostly, although not exclusively, volcanic or subvolcanic, and Ca–Al *melilite* (*gehlenite*) bears much the same relationship to basic **plagioclase** as **leucite** and **nepheline** do to **alkali feldspar**. It proxies for **plagioclase** in the rare rock type *melilitite*. *Melilite* is a tetragonal orthosilicate in which pairs of SiO_4 tetrahedra are linked by one shared oxygen to give Si_2O_7 groups, and these groups are linked by Al–O and/or Mg–O tetrahedra. It is colourless or brown in hand specimen, colourless or faintly brown in thin section. The refractive indices are higher than balsam (1.62–1.67). A distinctive feature is the anomalous blue interference colour, caused by strong **dispersion** (variation with wavelength) of the birefringence. Density is 2.95–3.05 $Mg\ m^{-3}$. [IP]

Brown, W.L. (ed.) (1984) *Feldspars and Feldspathoids. Structures, Properties and Occurrences.* NATO ASI Series, Series C, Vol. 137. Reidel, Dordrecht.

Deer, W.A., Howie, R.A. & Zussman, J. (1963) *Rock Forming Minerals* Vol. 4: *Framework Silicates.* Longman, London.

[245]

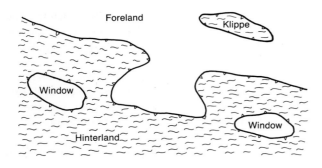

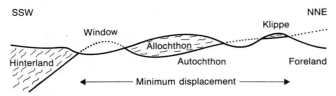

Fig. F13 Map and cross-section showing outcrop pattern of a low-angle thrust forming klippe and windows (**fenesters**). (After Roberts, 1982.)

felsic rock An **igneous rock** with one or more of the following minerals as a major component of the mode: *fe*ldspar + *le*nad + *si*lica (e.g. respectively **feldspars**, **feldspathoids** and **quartz**) (*lenad* is a term based on the **feldspathoid minerals**, e.g. **Leucite** and *ne*pheline). [RST]

fenester (*fenster*) Tectonic 'window'; an area where the rocks below a *thrust sheet* have been exposed by **erosion**. The rocks exposed in the fenster may belong to a lower *thrust sheet* or to the **autochthonous** basement. The converse structure is termed a **klippe** (pl. 'klippen'), which is an erosional remnant of a higher *thrust sheet* (see Fig. F13). [RGP]

Roberts, J.L. (1982) *Introduction to Geological Maps and Structures*. Pergamon, Oxford.

fenestrae (*birdseyes*) Voids, larger than intergranular pores, which may be partially or completely filled by sediment, or by some cementing material. They are usually small, only a few millimetres in long axis, and vary from rounded (bubble) to planar to irregular. The round and irregular forms are predominantly the result of either desiccation or air entrapment during flooding, typically on tidal flats. The planar (or laminoid) fenestrae are usually associated with *microbial mats* and result from gas entrapment beneath cohesive mats or by the shrinkage and/or **oxidation** of organic-rich laminae covered by sediment. Generally fenestrae are commonly seen in ancient carbonate intertidal to supratidal deposits. On modern tidal flats they form where the sediment is frequently exposed. Similar fenestrae also occur in lake deposits.

Bubble-like fenestrae also occur in **sands** and are termed *key-hole vugs*. In all cases the preservation of fenestrae requires early lithification. (See **stromatolites**.) [VPW]

Shinn, E.A. (1983) Tidal flat environment. In: Scholle, P.A., Bebout, D.G. & Moore, C.H. (eds) *Carbonate Depositional Environments*. American Association of Petroleum Geologists Memoir **33**: 171–210.

fenite A term applied to rocks which have been **metasomatized** as a result of emplacement of alkaline or **carbonatite** intrusive rocks. *Fenitization* therefore characteristically occurs within *country rocks* around high-level alkaline and **carbonatite** intrusions in continental settings such as the East African Rift Valley (and at the type locality of Fen in south Norway). There are two principal styles of *fenitization*;
1 *sodic fenitization* (or simply '*fenitization*') involving the formation of sodium-rich minerals such as **nepheline**, together with **alkali feldspar** and alkali **pyroxene** (such as *aegirine*), and
2 *potassic fenitization* (or '*feldspathization*') which involves the growth of **alkali feldspar** with minor **nepheline**, and alkali **pyroxene**. [RST]

feral relief A landscape in which the main valley sides are dissected by many *insequent streams*. Such relief is related to rapid **runoff** and intense dissection. [NJM]

ferberite ($FeWO_4$) The iron end-member of the **wolframite** (*wolfram*) series of minerals. [DJV]

fergusonite (($Y,Er,Ce,Fe)NbO_4$) An **ore** mineral of the **rare earth elements**. [DJV]

fermentation The *anerobic* process by which *bacteria* metabolize oxygen containing organic matter as an energy source, liberating hydrogen and carbon dioxide (CO_2) as by-products:

$$CH_2O + H_2O \rightarrow CO_2 + H_2O \qquad (1)$$

In the absence of sulphate the carbon dioxide is reduced by *methane*-generating *bacteria*:

$$4H_2 + CO_2 \rightarrow CH_4 + 2H_2O \qquad (2)$$

Fermentation takes place during the shallow **burial diagenesis** of organic matter and combining equations (1) and (2) provides a source of bicarbonate and *methane* in the **pore-waters**

$$2CH_2O + H_2O \rightarrow CH_4 + HCO_3^- + H^+$$

Fermentation reactions effectively reduce detrital mineral oxidations and lead to the **precipitation** of iron-rich carbonates.

$$7CH_2 + 2Fe_2O_3 + 6H_2O \rightarrow 3CH_4 + 4FeCO_3 + 4H_2O$$

(See **organic matter diagenesis, early diagenesis**.) [SDB]

Claypool, G.E. & Kaplan, I.R. (1974) The origin and distribution of methane in marine sediments. In: Kaplan, J.R. (ed.) *Natural Gases in Marine Sediments*, pp. 99–139. Plenum, New York.

Curtis, C.D. (1983) Geochemistry of porosity enhancement and reduction in clastic sediments. In: Brooks, J. (ed.) *Petroleum Geochemistry and Exploration of Europe*, pp. 113–25. Geological Society Special Publication No. 12. Blackwell Scientific Publications, Oxford.

ferricrete An **iron**-rich **duricrust** often formed at the top of **deep weathering** profiles in humid tropical and sub-tropical environments. The term **laterite** is sometimes used as a synonym for ferricrete. However, **laterite** is a broader term applied to surficial accumulations that are rich in **iron** and **aluminium** owing to the loss of alkalis and **silica** through chemical mobilization. In ferricretes, the **precipitation** of **iron oxides** and hydroxides from **groundwater** of during **pedogenesis** initially forms pea-sized nodules called *pisoliths*. As ferricretization proceeds, the *pisoliths* aggregate and may eventually form a thick, hard sheet of ferricrete — called a **plinthite**. [AW]

McFarlane, M.J. (1983) Laterites. In: Goudie, A.S. & Pye, K. (eds) *Chemical Sediments and Geomorphology*, pp. 7–58. Academic Press, London.

ferrimagnetism The form of **ferromagnetism** (*s.l.*) in which the magnetic lattices within a grain are oppositely magnetized but of different strength. This results in a **spontaneous magnetization** such as that of **magnetite**. [DHT]

ferrimolybdite ($Fe_2(MoO_4)_3 . 8H_2O$) A canary yellow, soft molybdate mineral produced by **alteration** of **molybdenite**. [DJV]

ferroactinolite ($Ca_2Fe_5Si_8O_{22}(OH)_2$) A dark green **amphibole mineral**. [DJV]

ferrochrome An alloy of **iron** and **chromium** (60–72%) used in the manufacture of very hard steel. [AME]

ferromagnetism There are two senses in which ferro-magnetism is used. In the widest sense (*s.l.*), it comprises all materials that exhibit a **spontaneous magnetization**, i.e. internal quantum-mechanical exchange or super-exchange forces cause the electron spins to become coupled but quench the coupling between the magnetization associated with the electron orbits. This results in strong magnetic linkages within a magnetic lattice, forming **magnetic domains**. Within an individual grain, exchange forces may couple all electron spins in the same direction giving rise to a strong **spontaneous magnetization** even in the absence of an external **magnetic field** (Fig. F14). This behaviour is ferromagnetic (*s.s.*) and is typical of **iron**, steel, **nickel** and their alloys. If super-exchange forces are involved, the electron spin couplings are antiparallel giving rise to two oppositely magnetized lattices. If these are exactly anti-parallel, the substance has no external field and is termed

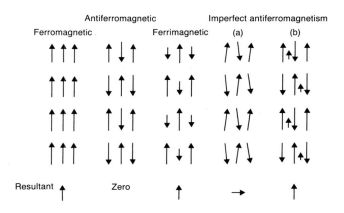

Fig. F14 Ferromagnetism. Schematic representation of spontaneous magnetization in crystals. The arrows represent the elementary moments, the resultants being given at the bottom. (After Jacobs, 1984.)

antiferromagnetic, while if the magnetizations are op-posite, but unequal, then the substance will have an external field, but weaker than that of a ferromagnet (*s.s.*), and its magnetic behaviour is termed **ferrimagnetic**. [DHT]

Jacobs, J.A. (1984) *Reversals of the Earth's Magnetic Field*. Adam Hilger, Bristol.
O'Reilly, W. (1984) *Rock Magnetism and Mineral Magnetism*. Blackie, Glasgow.

ferropseudobrookite ($FeTi_2O_5$) A rare **iron–titanium** oxide mineral. (See **oxide minerals**.) [DJV]

fersmannite ($Na_4Ca_4Ti_4(SiO_4)_3(O,OH,F)_3$) A rare **titanium**-bearing silicate (*island silicate* or *nesosilicate*) mineral. [DJV]

fetch The extent of open water across which a **wave**-generating wind blows. This helps determine the height and energy of **waves** and thus their **erosional** and depositional effect on the coast since increases in fetch, as well as in wind-speed and duration, results in larger wind-**waves**. The growth of wind-**waves** may be limited by a short fetch as in enclosed seas or lakes but waves in the open ocean may travel great distances outside their fetch areas if the wind stops blowing or changes direction. Many features of coastal deposition orientate to face the direction of maximum fetch. (See **beach**.) [JDH]

Komar, P.D. (1976) *Beach Processes and Sedimentation*. Prentice-Hall, Englewood Cliffs, NJ.

field capacity An area of *soil* is said to be at field capacity after draining under the influence of **gravity** has ceased. The water that has drained out is called the *specific yield* of the *soil* and that remaining the *specific retention*. [AME]

figure of the Earth The Earth is not a sphere, but its shape can be closely approximated by considering the

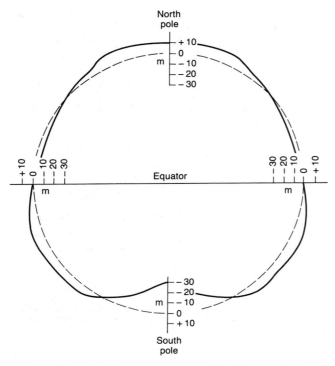

Fig. F15 Figure of the Earth. Height of the geoid (solid line) relative to a spheroid of flattening 1/298.25 (broken line) assuming the Earth to be axially symmetrical about the polar axis. Note exaggeration. (After Bott, 1982.)

Earth to be a rotating body in hydrostatic equilibrium acted upon by its own gravitation and such a shape can be mathematically discribed by the **spheroid** equation, assuming the density distribution in the Earth to be laterally homogeneous (Fig. F15). Satellite precessions can be analysed to provide precise determination of the Earth's gravitational field from which the **equipotential surface** of the Earth, the **geoid**, can be determined. In detail the Earth is irregular in shape, being approximately pear-shaped. Satellite observations have shown that the shape of the Earth is gradually changing due to mass redistributions consequent upon **isostatic rebound** following the decrease in the polar ice sheets and consequent changes in sea-level during the last 12 000 years, combined with the gradual decrease in the rate of rotation of the Earth. [DHT]

Bott, M.HP. (1982) *The Interior of the Earth*. Arnold, London.
Jeffreys, H. (1976) *The Earth, Its Origin, History and Physical Constitution*. Cambridge University Press, Cambridge.
Lambeck, K. (1980) *The Earth's Variable Rotation*. Cambridge University Press, Cambridge.

filtering The process of suppressing certain information (usually **noise**) in a waveform. The filtering process can discriminate on the basis of frequency, wavelength, **move-out** (velocity) or particle motion. Filtering can be implemented via electric circuits on analogue records or via computer programs on digital records — digital filtering.

Frequency filtering is designed to select or pass certain bands of frequencies and attenuate others. Filters may be classified according to their characteristics: low-pass pass low frequencies, high-pass pass high frequencies; band-pass pass a range of frequencies; band reject attenuate a continuous range of frequencies. In velocity filtering seismic signals are discriminated on the basis of their **moveout** or apparent velocity across a **seismic array** of detectors and is particularly used to eliminate source-generated **noise**, such as *surface wave* arrivals, from **seismic reflection** records. Particle motion filtering is used to discriminate between **noise** and signals on the basis of the three-dimensional pattern of their particle motion.

Inverse filtering is otherwise known as **deconvolution**. Coherence filtering **attenuates** where certain coherence or **correlation** tests are not satisfied. (See Fig. E4, p. 196.) [GS]

Kulhanek, O. (1976) *Introduction to Digital Filtering in Geophysics*. Elsevier, Amsterdam.

fineness The quality or fineness of native **gold** is usually expressed in parts per million (1000 is pure **gold**). The common alloy metals are **silver** and **copper**. The fineness of **placer gold** is about 500 to 999, that of **veins** approximately 500–850. The fineness of **placer gold** increases with the distance transported and with decreasing grain size, the alloy metals apparently being leached out. Because of its softness, **gold** for jewellery is generally alloyed with **copper, silver, nickel** or **palladium**. Its fineness or purity is then expressed in **carats**; one **carat** means one part **gold** in twenty-four. [AME]

finite rotation The actual rotation, measured in degrees, about a *Euler pole*, required to rotate one point, or series of points such as a coastline or *oceanic magnetic anomaly*, on the Earth's surface to be superimposed on another. The rotational movement, obtained this way, does not necessarily represent the actual path of the mobile entity. (See **instantaneous rotation**.) [DHT]

Le Pichon, X., Francheteau, J. & Bonnin, J. (1973) *Plate Tectonics*. Elsevier, Amsterdam.

finite strain The total change in shape (**strain**) of a deformed body relative to its shape before **deformation**. Finite strain can be measured in one, two or three dimensions. It gives no indication of the **strain** history, i.e. the sequence of **deformation** resulting in the final shape. In a three-dimensional coordinate reference frame, finite strain is represented by the three by three matrix that transforms the initial coordinates of any point to the co-ordinates of their final position, after **deformation**. Any combination of three by three matrices that multiply together to equal this matrix represents a geometrically possible sequence of **strain** increments, i.e. a potential *strain history*.

Some authors (e.g. Means, 1978) use *total strain* to refer to finite strain as defined here. Finite strain then refers to any measurable, as distinct from an infinitesimal, **strain**. The usage defined above is more common. [SB]

Means, W.D. (1978) *Stress and Strain*. John Wiley & Sons, New York.

fire fountaining A style of **volcanic eruption** in which **magma** is continuously erupted and rises to a height of up to a few hundred metres as an incandescent jet, before falling to the ground as **tephra**, which may immediately become reconstituted and flow away as **lava flow**. Fire fountains occur in **magmas** of low **viscosity**, typically **basalts**, and are a characteristic feature of *Hawaiian eruptions* (see **volcanic eruption**). The **magma** rises into the fire fountain as a result of its velocity of ascent through the subsurface conduit and/or as the result of the subsurface *vesiculation* of the **magma**. A line of coalescing fire fountains simultaneously active along a fissure (see **fissure eruption**) is known as a *curtain-of-fire*. [PTL]

fireclay/refractory clay An **underclay** that is rich in **kaolinite** and can be used commercially as a refractory clay. For this purpose clays must have low **mica** and **iron** contents. Fireclays now face many competing materials for the manufacture of refractory bricks, so they are now increasingly used for non-refractory products such as facing bricks, pipes and ceramic stoneware. [AME]

first arrival (*first break*) The earliest recorded seismic signal from a particular **shot**. Interpretation of **seismic refraction** data in terms of Earth structure principally relies on the picking of the refracted **head waves**, which are most easily discriminated as first arrivals. [GS]

fish/Pisces Water-living vertebrate animals, cold-blooded and with persistent gills.

The fish are divided into a number of classes: the *Agnatha* which have no biting jaws, and the others, which have hinged jaws equipped with teeth of one sort or another. There does not yet seem to be any known intermediate between these two main divisions of fish-like creatures.

A pre-fishlike condition may be subsumed under the term *Chordata*, which implies that these forms have the very basic structures seen in all other backboned animals. Thus such a form as *Branchiostoma* (the Lancelet) has a cartilagenous stiffening rod or notochord running the length of the body, a tubular nerve cord lying dorsal to this stiffening rod, gill slits in the front part of the body opening into the oesophagus, a post-anal tail and V-shaped muscle blocks or myotomes. All these characters are unique to the *Chordata* and are seen in one form or another right through to the **Mammalia**.

The origin of the fish-like primitive condition is generally

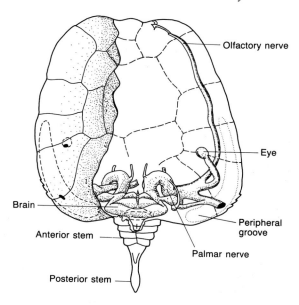

Fig. F16 Fish. *Mitrocystella* (Ordovician) from above with right-hand plates removed showing nervous system as interpreted by Jeffries (1968). (After Clarkson, 1979.)

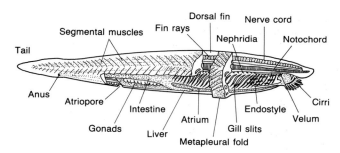

Fig. F17 Fish. The cephalochordate *Branchiostoma* (Amphioxus). This living animal provides another model for the ancestors of vertebrates that lived approximately 600 Ma.

held to lie within the *Tunicata* which are mostly sessile filter-feeding animals as adults, but which produce a larva with the characters seen in an amplified form in *Branchiostoma*. These larvae may in turn be compared with the early development stages of some **echinoderms**, so that of all the invertebrates it is generally recognized these come closest in relationship to the *chordates*.

The current debate as to which group within the **echinoderms** is closest to the *chordates* has lead to one group being singled out by some in preference to the others. This is the *Calcichordata* (**Cambrian–Devonian**), but the homologies drawn between them and the primitive *chordates* are not totally clear and the debate is still active (Fig. F16).

For instance, the classical view is that the earliest *chordate* was superficially fish-like, but with only a rudimentary enlargement at the anterior end of the nerve cord. Likewise the tail region merged smoothly into the head region, and the gills were each quite distinct and lay between adjacent muscle blocks. Each gill arch muscle was supplied in a similar manner to a normal muscle block with a 'dorsal' and a 'ventral' branch of a segmental nerve. The skeleton is entirely cartilagenous. *Branchiostoma* is taken to represent such a level of organization (Fig. F17).

The *calcichordates* (Fig. F16) have many characters which have been interpreted as being more akin to the true vertebrate condition than those seen in *Branchiostoma*. These include a very distinct locomotory tail, as well as the more normal chordate characters of gill openings, notochord and dorsal nerve cord. In the head region the organs are not arranged symmetrically, and the 'brain' and cranial nerves are developed at a very much higher level than is seen in *Branchiostoma*. The skeleton is, as the name implies, made of calcite. In all other chordates with a mineralized skeleton, the mineral is **apatite**.

Lamprey larvae develop in a manner very similar to *Branchiostoma*, with an inherent asymmetry of the anterior of the body. Thus the left gill openings appear first. It is also recognized that the vertebrate body has two main regions; the cephalic and the somatic, each part with its own pattern of innervation, and this combination of characters is seen by some to be most closely matched among the invertebrates by the *Calcichordata*. If this is so,

then they may represent some distant link between the early **echinoderms** and the *chordates* and the *tunicates* are then an offshoot (sister group) of the main stem.

In any case, the true fishes are all organized at a very much higher level than is seen in *Branchiostoma*. There is always a true brain and a greater or lesser condensing of the anterior segments of the body to form a distinct head region. Associated with these developments are the organs of special sense: eyes, nose and balance and pressure sensing organs. Even the earliest fishes have some sort of lateral outgrowths from the body wall in the form of fins of one sort or another to assist in locomotion and stability. If this is the case, then *Branchiostomata* may be little more than a 'degenerate' representative of early forms.

The first *chordate* known in the fossil record is *Pikaia* from the **Cambrian** of western Canada, the *Burgess Shale*. It seems to be superficially *Branchiostoma*-like, but is a rare component of the fauna and has problems of interpretation. However, by the **Ordovician**, and in several parts of the world (United States, Australia) a flourishing fauna of jawless fishes had come into existence. They are representatives of the *Diplorhina*, of which the modern hagfish, *Myxine* may be a surviving relative.

Organized at much the same level as the diplorhinids, but contrasting with them in several major characters

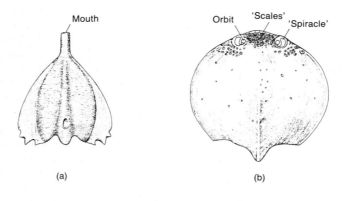

(a)

(b)

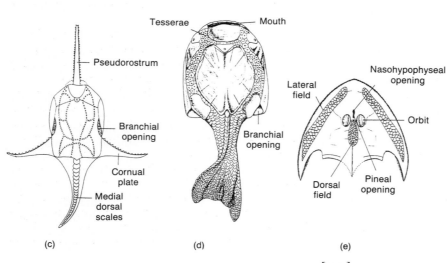

(c)

(d)

(e)

Fig. F18 Fish. Dorsal view of the head shield or carapace of jawless fishes to show a range of outline shapes. (a) and (b) Representatives of the family Amphiaspididae. (a) *Eglonaspis*. The eyes are lost and the mouth opens at the end of a long tube. (b) *Gabreyaspis*. Openings lateral to the orbits may be similar to spiracles of skates and rays. (c) *Doryaspis*, a peculiar pteraspidomorph. (d) The psammosteid *Drepanaspis*. (e) *Cephalaspis*. (After Moy-Thomas & Miles, 1971.) Not to scale.

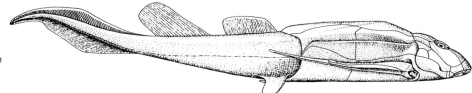

Fig. F19 Fish. *Bothriolepis*, a placoderm from the Upper Devonian, about 40 cm long. A freshwater, bottom-dwelling fish.

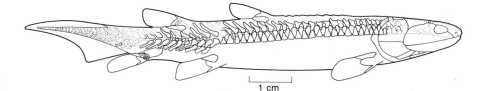

Fig. F20 Fish. *Osteolepis* reconstruction of endoskeleton. (After Andrews & Westoll, 1970.)

1 cm

of the head region, are the *Monorhina* (Fig. F18). Their modern representatives are held to be the lampreys (e.g. *Lampetra*) whose fossil record extends back to the **Carboniferous** (*Mazon Creek* Fauna) or even as far as the Upper **Silurian** in the form of *Jamoytius*, an eel-like animal with a suctorial mouth apparently very similar to that of a modern lamprey.

Both the above subclasses had an extensive external bony armour, but retained a cartilagenous internal skeleton. They lasted into the **Devonian** and are used as zone fossils in the **Devonian** of the Baltic Shield and equivalent strata in Britain and Nova Scotia. A separate radiation of these *agnathans* is recorded from China, many with bizarre-shaped head shields (Fig. F18).

Representatives of all the major groups of jawed fishes appear almost simultaneously in the **Silurian**. These include the *Placodermi* (Fig. F19), the *Osteichthyes* and the *Chondrichthyes*. The first group is now extinct, unless represented by the 'rabbit fishes' or *Holocephali*. The *Chondrichthyes* (sharks, skates and rays) have had a continuous successful history as open water predators over their entire existence. They differ from the *Osteichthyes* in not having a swim bladder, or other buoyancy organ other than oil in their tissues. Their anatomy is directed towards adaptations which require them to be moving so that they can rise clear of the seafloor, their specific gravity being slightly greater than seawater. They are hydrodynamically sophisticated in shape, using their stiff fins and their body shape to generate lift and control forces. Pitching moments can also be generated by flexing or extending the tail on the vertebral column.

The *Osteichthyes* (bony fish) comprise the *Acanthodii* (a primitive group with bony spines along the leading edges of their fins and no bony internal skeleton), the *Actinopterygii* (ray-fin fishes) and the *Sarcopterygii* (lobe-fin fishes). The latter may be further divided into the extant *Dipnoi* (lungfish) and the *Crossopterygii* (Fig. F20). It is from this group that the **Amphibia**, and hence the **Reptilia**, **Aves** and **Mammalia** arose.

The bony fish seem to have had the ability to live in freshwater from the beginning of their existence, as opposed to the majority of the other forms which seem to have been limited to brackish water at best. Many of their earliest representatives are found in non-marine deposits in areas of the world which at that time were in tropical or subtropical latitudes and which gave rise to *red bed* deposits. The belief is that living in water under these conditions may have been physiologically stressful, the heat and seasonality of rainfall leading to high rates of evaporation and periods of drying out of the rivers and lakes. It was under these conditions that two adaptations make their appearance and which have left an indelible imprint particularly on the *Sarcopterygii* and their descendants. The first and most widespread was the appearance of accessory respiratory organs and the second was the development of limbs (from fins) to enable their owners to travel from pool to better pool.

In the agnathans and other fish without neutral buoyancy, such as is given by a swim bladder, depth control is by a subtle combination of body shape, fins and a heterocercal tail (a tail with uneven distribution of the 'fringe' above and below the body axis). Such a tail is the mark of a fish with no swim bladder. The *Osteichthyes* with their swim bladder modify their tails so that they are no longer required to assist depth control, and so become symmetrical. Thus one of the resulting advantages is that all the power developed by a bony fish can be used for propulsion. Depth control is now by secreting or excreting gas into or out of the swim bladder. As the fins are no longer required to provide direct lift forces, they can be folded away against the body and so reduce drag. Their use is now to increase manoeuvrability, to assist turning and to provide a braking action. These fish can remain static in the water unless their fins are beating; in which case they can move forwards and backwards, unlike sharks.

Evolving in parallel with these changes in fin shape and function to a state where precise control is achieved, is a series of changes in jaw morphology. The first jawed fishes had a simple hinged jaw with rows of teeth of varying size with which to cut lumps out of their prey, or to kill and subdue smaller prey. Thus the majority of early fishes were macrophage carnivores; subtleties of feeding such as filter-feeding do not seem to have arisen until the **Jurassic**.

The increased manoeuvrability of the bony fish enabled them to exploit more confined spaces than can the sharks. New feeding niches became available where, for instance, fish could move slowly through the restricted passages in a coral reef grazing on individual branches in a colony, or taking individual **brachiopods** from their massed beds. **Molluscs** could 'escape' by burrowing, but the *Teleostei*, the most recently evolved of the bony fish, have so loosened the bony articulations round their face and jaw bones that their mouths can be used as powerful suctorial pumps to extract their prey from even quite deep in the sediment.

There is thus a whole range of body and fin shapes and proportions seen in bony fish from the late **Paleozoic** to the Recent. These shapes can be related to their overall way of life, and reflect a change from a feeder on epi-benthonic invertebrates or open water carnivores, to one capable of great precision of selection of food items.

Whereas the *Actinopterygii* emphasized the buoyancy-keeping role of their gas bladder, the *Sarcopterygii* maintained the original emphasis on respiration in theirs. However, in the Recent marine genus *Latimeria* and perhaps by implication in other later (**Mesozoic**?) *crossopterygians* the swim bladder/lung has been invaded by a mass of fatty tissue, which itself being less dense than water acts as a buoyancy organ. Their modern freshwater relatives, the lungfish, retain a respiratory function for their gas bladder.

The most significant group of *sarcopterygians* from a long-term evolutionary point of view were the *rhipidistians* (Fig. F20). These fish have been accepted for a long time as the model for the group from which arose the first land-living tetrapods, the **Amphibia**. Modification of the lateral fins allowed a supporting function for them, which in turn allowed them to be used as fulcra while the fish struggled overland, 'swimming'. Ultimately the combination of support and lateral undulation lead to a true walking limb. (See also Table F2.) [ARIC]

Alexander, R.McN. (1981) *The Chordates* (2nd edn). Cambridge University Press, Cambridge.
Andrews, S.M. & Westoll, T.S. (1970) The postcranial skeleton of rhipidistian fishes excluding *Eusthenopteron*. *Transactions of the Royal Society, Edinburgh* **68**: 391–489.
Carroll, R.L. (1987) *Vertebrate Paleontology and Evolution*. W.H. Freeman, Chicago.
Clarkson, E.N.K. (1979) *Invertebrate Palaeontology and Evolution* (1st edn). Allen & Unwin, London.

Table F2 Fish/Pisces

Phylum Chordata
Subphylum Vertebrata
SUPERCLASS PISCES

Class Agnatha
  Subclass Monorhina (= Cephalaspidomorpha)
    Order Osteostraci
      Anaspida
      Cyclostomata
  Subclass Diplorhina (= Pteraspidomorpha)
    Order Heterostraci
      Coelolepida

Class Placodermi
  Order Petalichthyida
    Rhenanida
    Arthrodira
    Phyllolepida
    Pyctodontida
    Antiarchii

Class Chondrichthyes
  Subclass Elasmobranchii
    Order Cladoselachii
      Pleurocanthodii
      Selachii
      Bradyodonti
  Subclass Holocephali

Class Osteichthyes
  Subclass Acanthodii
    Order Climatiformes
      Ischnacanthiformes
      Acanthodiformes
  Subclass Actinopterygii
    Order Chondrostei
      Holostei
      Teleostei
  Subclass Sarcopterygii
    Order Crossopterygii
      Dipnoi

Moy-Thomas, J. & Miles, R.S. (1971) *Palaeozoic Fishes* (2nd edn). Chapman & Hall, London.
Romer, A.S. (1966) *Vertebrate Paleontology* (3rd edn). Chicago University Press, Chicago.
Storer, T.I. & Usinger, R.L. (1965) *General Zoology* (6th edn). McGraw-Hill, New York.

fissility Surfaces of weakness along which a fine-grained rock splits easily. Fissility is a more general term than **cleavage**, and may be produced by both tectonic and sedimentary processes. [TGB]

fission track dating A dating method applied to **glassy** volcanic **rocks**, plutonic rocks, sedimentary rocks, man-made glasses and baked clay artefacts. It is dependent on all these materials containing small quantities of **uranium** either in the glass or in, e.g. **zircon** crystals in baked clay. The method is based on the following principle: a small proportion of the isotope **uranium**-238 decays by spontaneous fission of the atom into two heavy nuclei with atomic weights of between 70 and 160. When fission occurs a damage or 'fission' track is left within the material. The concentration of fission tracks in a sample will depend on both the amount of **uranium** present in the material and on the length of time during which the tracks have accumulated. Fission tracks will increase with time from a start date corresponding to the last time of heating of the sample (e.g. eruption and cooling of a lava, firing of pottery). In order to determine this date, the concentration of **uranium** is measured and the number of fission tracks is counted under a microscope. Since the rate of decay of uranium-238 by fission is known, the age can be calculated.

The method has been used to date **obsidian** tools and sources and fired clay objects (pottery and tiles). Manmade glasses are less easy to date because of the small number of fission tracks (due to the fairly recent date of manufactured glasses), except for some recent glasses in which **uranium** has been added as a colouring agent. Fission track counting can also help to determine whether an **obsidian** artefact has been heat-treated in antiquity to improve its flaking qualities (compare **electron spin resonance**).

The fission tracks in minerals resulting from nuclear fission anneal (i.e. close) at different rates dependent on the subsequent falling temperature. Consequently the fission track method can be used to study the thermal history and rates of **uplift** of rocks (Wagner & Reimer, 1972). The method has recently been used extensively in the study of the **uplift** history of mountain belts and the inversion history of sedimentary **basins**, principally by making use of the fission tracks in **apatites** and **zircons** in **igneous** and metamorphic **rocks**. [PK/OWT]

Storzer, D. & Wagner, A.W. (1982) The application of fission track dating in stratigraphy: a critical review. In: Odin, G.S. (ed.) *Numerical Dating in Stratigraphy*, Part 1, pp. 200–21. John Wiley & Sons, Chichester.
Wagner, G.A. & Reimer, G.M. (1972) Fission-track tectonics: A tectonic interpretation of fission track apatite ages. *Earth and Planetary Science Letters* **14**: 263–8.

fissure eruption Eruption of **magma** more or less simultaneously at several points along the length of an elongate volcanic conduit. Fissure eruptions differ in this respect from *central eruptions*, in which **magma** is erupted from a single pipe-like **volcanic vent**. (See **volcanic eruption**.) [PTL]

fissure vein An old term for **veins**. In the past, some workers made a genetic distinction between fissure veins and *lodes*, or simply between **veins** and *lodes*. Fissure veins were considered to have resulted mainly from the infilling of pre-existing open spaces, whilst the formation of *lodes* was held to involve the extensive **replacement** of pre-existing *host rock*. Such a genetic distinction has often

proved to be unworkable. It is nowadays recommended that all such **orebodies** be called **veins** no matter how they were formed and the term *lode* be dropped. [AME]

fixed carbon The **carbon** found in the material left after all the volatile matter has been expelled from **coal**. Fixed carbon is not determined direct but is the difference in an air-dried **coal** between the sum of the other components and 100%. It is used as an index of the amount of **coke** that will be produced on carbonization, as a measure of the solid combustible material in **coal** and, with a correction for **ash** content, as an index of **coal rank**. [AME]

flake graphite A commercial term that denotes flat, platy grains disseminated through metamorphic rocks. The common *host rocks* are quartz–mica **schists**, feldspathic or micaceous **quartzites**, **marble** and **gneiss**. [AME]

flake/ground/scrap mica Fine-grained **mica** used for various industrial purposes, e.g. paper, paint, rubber. It is a by-product of various operations such as the processing of *sheet mica*, *pegmatite*, **granite**, micaceous **schists** and **kaolin**. [AME]

flame structure Structure at the base of a bed associated with *load casts*. Down-current pointing fingers or wedges of sandy or silty sediment, formed by partial loading of wet sediment into a cohesive/semi-cohesive finer-grained substrate. **Shear stresses** during sediment loading cause the generally consistent sense of overturning in flame structure. (See **wet sediment deformation**.) [KTP]

Flandrian transgression The rise in sea-level which occurred in *Holocene* times and in the Late *Pleistocene* as the ice sheets of the last *glacial* melted. World sea-levels rose by as much as 170 m, flooding the continental shelves and transforming *glacial* and fluvial valleys into *fjords* and **rias** respectively. This transgression combed up large amounts of sediment to produce major depositional coastal features including shingle ridges (e.g. Chesil Beach). The transgression was more or less complete by 6000 years ago, and at its peak may have reached a few metres above its present level. [ASG]

Kidson, C. (1982) Sea-level changes in the Holocene. *Quaternary Science Reviews* **1**: 121–51.

flap Type of recumbent **syncline** created by **gravity collapse**. [RGP]

flaser and lenticular bedding/lamination The two main categories in a spectrum of types of interbedding/interlamination of mud deposited by fall-out from suspension in slack water, with aqueous current and/or **wave ripple**

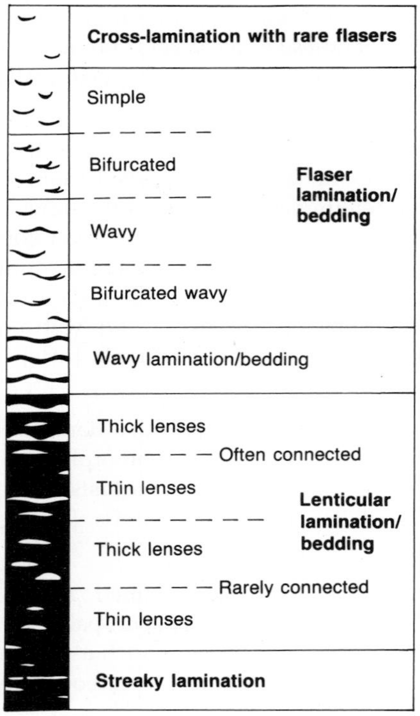
Cross-lamination with rare flasers
Simple
Bifurcated
Wavy
Bifurcated wavy
Flaser lamination/ bedding
Wavy lamination/bedding
Thick lenses
Often connected
Thin lenses
Thick lenses
Rarely connected
Thin lenses
Lenticular lamination/ bedding
Streaky lamination

Fig. F21 The spectrum of **flaser and lenticular bedding/lamination**. (After Reineck & Wunderlich, 1968.)

cross-laminated **sand** (Fig. F21). *'Flasers'* are mud lenses preserved in ripple troughs. The lenses of lenticular (or *'linsen'*) bedding are of **sand** within mud. Three other categories within the spectrum are 'wavy' and 'streaky bedding' or 'lamination' and 'cross-lamination with rare flasers'.

The structures form in environments where there is an alternation of slack water and wave or current activity, such as areas of **tidal** influence, or where sediment supply is episodic, such as river **floodplains**. [DAR]

Reineck, H.-E. & Wunderlich, F. (1968) Classification and origin of flaser and lenticular bedding. *Sedimentology* **11**: 99–104.

flash A lake or other water-filled depression formed through surface subsidence. The term is commonly applied to lakes in parts of Cheshire, north-west England, which have formed through **suffosion** resulting from natural **dissolution** and mining of **halite** in the *Keuper* Saliferous Beds which underlie unconsolidated surface deposits of *glacial till* and outwash. [AW]

Bell, F.G. (1975) Salt and subsidence in Cheshire, England. *Quarterly Journal of Engineering Geology* **9**: 237–47.

flat A horizontal or subhorizontal replacement **orebody**. Flats commonly branch out from **veins** and lie in carbonate *host rocks* beneath an impervious cover such as **shale**. (See also **thrust belts**.) [AME]

Evans, A.M. (1993) *Ore Geology and Industrial Minerals: An Introduction* (3rd edn). Blackwell Scientific Publications, Oxford.

flatiron A small, steep-sided, triangular-shaped **mesa**. Flatirons often occur in groups where residual blocks of an **erosion**-resistant stratum protect weaker, underlying rocks from **erosion**. [AW]

flattening/flattening strain A shape change (**strain**), or the process producing a shape change in which there is **contraction** along only one **principal strain** direction. The **strain ellipsoid** for a flattening strain is disc-shaped; the maximum and intermediate **principal strains** (measured as **stretches** or **quadratic elongations**) are greater than 1. On a **Flinn diagram**, the *flattening field* ($k < 1$) represents these ellipsoids. Any anisotropic volume change (**dilation**) will affect the **finite strain** seen in the **strain ellipsoid**. True flattening refers to **deformation** accompanied by volume change in which the **tectonic** component of the net **strain** is of flattening type, in contrast to **plane strain** or *constriction*. [SB]

flexure of the lithosphere The bending of an **elastic lithosphere** *plate* to support a load. In mountain belts the load is made up of uplifted **crust** and **mantle** material. Where the **lithosphere** behaves as an elastic plate, its flexure results in depression beneath the load, and hence a **basin** adjacent to the load, together with a 'peripheral bulge' further towards the **foreland**. Some **foreland basins**, such as the Swiss Molasse Basin, are examples of flexural depression ahead of a mountain belt load, in this case the Alps. The essential part of lithospheric flexure is that the lateral strength of the **lithosphere** partly supports the load. This is responsible for the departure of some active mountain belts from **isostatic** equilibrium.

The width and positions of the areas of **uplift** and depression depend on the *flexural rigidity* of the **lithosphere**, as well as the density distribution of the load. A strong **lithospheric** *plate* has high *flexural rigidity*, and leads to long wavelength, shallow **uplifts** and depressions. A weak **lithospheric** *plate* (low *flexural rigidity*) forms relatively narrow, deep **uplifts** and depressions. The *flexural rigidity* is often expressed as an *'effective elastic thickness'*, the thickness of an ideal elastic plate that behaves in the required manner. This has high values for high *flexural rigidity*. For ideal **Airy isostasy**, the *effective elastic thickness* is zero, and all the load is supported by the rock immediately beneath it.

Lithospheric flexure has a strong influence on sediment distribution in the **foreland** parts of mountain belts. Changes in the configuration of the load (basement *thrust sheets*) alter the distribution of areas of **uplift** and sediment deposition. The changing nature and extent of sediment distribution ahead of a mountain belt provides valuable information on the evolution of the *thrust* geometry with time. Flexure of the **lithosphere** also affects the geometry of **extensional basins** and **rifts**. These subtle effects are seen most clearly in the sedimentary **basin** fill as local **unconformities** and facies changes. [SB]

Watts, A.B., Cochran, J.R. & Selzer, G. (1975) Gravity anomalies and flexure of the lithosphere. *Journal of Geophysical Research* **80**, 1391–8.

Flinn diagram A graph developed by Flinn (1962) on which each point represents an ellipsoid of a particular shape; the plot is used to illustrate and analyse the shapes of **strain ellipsoids**. The graph uses ratios of the **principal strains** (expressed as **quadratic elongations**) of the ellipsoid:

The ordinate is $\quad a = \dfrac{\text{maximum principal strain}}{\text{intermediate principal strain}}$

and the abcissa is $\quad b = \dfrac{\text{intermediate principal strain}}{\text{minimum principal strain}}$

Both a and b are always greater than unity, so each axis starts at a value of one. Each point on the plot represents a particular ellipsoid shape, e.g. a sphere lies at the point $(1,1)$. The parameter k is defined as the gradient of a line joining any point to the point $(1,1)$ (the origin), thus:

$$k = \frac{a - 1}{b - 1}$$

k is used to represent ellipsoid shapes and to subdivide all possible ellipsoids into three geometrically significant and convenient groups.

Where $k = 1$, ellipsoids represent **plane strain**, i.e. no **elongation** or **shortening** takes place normal to a particular plane. Below the line $k = 1$, $k < 1$ defines the '*flattening field*', ellipsoids are oblate, and no two of the **principal strains** equal. On the axis where $k = 0$, ellipsoids are oblate (disc-shaped) with equal maximum and intermediate **principal strains**.

In the upper part of the Flinn diagram, the ellipsoids represented have maximum and intermediate **principal strains** equal when they lie on the axis where $k = \infty$; these are prolate, with the minimum and intermediate **principal strains** equal. Ellipsoids with $k > 1$ have three unequal **principal strains**, and represent *constrictional* **strains**; this defines the *constrictional field*.

A modification of the Flinn diagram was suggested by Ramsay (1967) in which the natural logarithms of the ratios of **principal strains** are used, to facilitate analysis of **strain** increments. These two diagrams can be used to show the range of **strain ellipsoid** shapes over an area, or to show the *strain history* in areas where successive stages in a progressive **deformation** can be identified. It can also be useful where there has been a change in volume. Contours of percentage volume loss or gain can be algebraically superimposed on particular ellipsoid families, for example $k = 1$, to model possible combinations of **dilation** and **tectonics**. Use of these techniques is illustrated

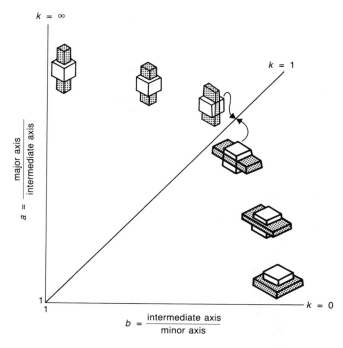

Fig. F22 The **Flinn diagram**, showing flattening and constriction fields, and the shapes of cubes representing the strain ellipsoids. (After Hobbs *et al.*, 1976.)

by the work of Ramsay & Wood (1973) and of Mawer (1983). Finally it should be noted that this diagram illustrates only the shapes of **strain ellipsoids** without reference to spatial orientation; the attitudes of the **principal strain** directions may change with **strain**, and should be considered in any **tectonic** analysis. (See Fig. F22 and **Hsu diagram**.) [SB]

Flinn, D. (1962) On folding during three-dimensional progressive deformation. *Quarterly Journal of the Geological Society, London* **118**: 385–433.

Hobbs, B.E., Means, W.D. & Williams, P.F. (1976) *An Outline of Structural Geology* (2nd edn). John Wiley & Sons, New York.

Mawer, C.K. (1983) State of strain in a quartz mylonite, Central Australia. *Journal of Structural Geology* **5**: 401–9.

Ramsay, J.G. (1967) *Folding and Fracturing of Rocks*. McGraw-Hill, New York.

Ramsay, J.G. & Wood, D.S. (1973) The geometrical effect of volume change during deformation processes. *Tectonophysics* **16**: 263–77.

flint (SiO_2) An impure, poorly crystalline form of **silica**. Flint was one of the most important raw materials of prehistoric stone age cultures, and was utilized world-wide wherever natural sources are present. Its fine-grained, homogeneous texture made it an ideal material for the manufacture of stone tools, which were generally produced by flaking techniques. It is particularly important in stone assemblages of the Paleolithic in Europe and in Africa, and it played a crucial part in tool manufacture throughout the Mesolithic, Neolithic, and into the Bronze Age. The earliest classifications of flint types were made in northern France, and sites in this area such as Abbeville and St Acheule have given their names to subdivisions of Paleolithic cultures world-wide. While the earliest stone tools (*c.* 2–2.5 Ma) were simple pebble tools, followed by choppers and then handaxes, later in prehistory implements were produced by fine flaking and by reworking of thin blades of flint. Thus beautifully flaked flint knives and points were made as early as the Lower and Middle Paleolithic periods, while in the Neolithic period sickles and arrowheads of several different shapes were produced. During the Mesolithic, Neolithic and Bronze Age, flint was used for **axes**, mainly flaked but including some polished **axes** and distinctive perforated maces from northern Europe. Flint was obtained in antiquity by collecting pieces from river gravels and sea-shore, and also by **flint mining**. [OWT/DJV]

Watson, W. & Sieveking, G. de G. (1968) *Flint Implements*. The Trustees of the British Museum, London.
Sieveking, G. de G. & Hart, M.B. (eds) (1986) *The Scientific Study of Flint and Chert*. Cambridge University Press, Cambridge.

flint mine/mining Prehistoric **flint** mines are known in western Europe, notably in Poland, England, Belgium and France. A good example is the Grimes Graves complex in Norfolk, UK, where *c.* 350 mineshafts up to 9 m deep have been discovered, some with galleries extending laterally into the **flint** deposits. The **flint** was extracted using deer antler picks, and lamps made in chalk vessels provided light. The site produced a large number of **axe** heads which were traded within Britain. Some flint mine products are distinctive in appearance, for example the 'Chocolate' and 'Banded' flints of Poland; others, such as those in the south of England, look similar to each other. Attempts chemically to characterize flint mines and their products in order to trace the extent of trade from a particular mine have been largely successful. [OWT]

flinty crush-rock An old term (e.g. see Peach *et al.*, 1907) for a fine-grained or cryptocrystalline **cataclastic** rock with a flinty appearance and often exhibiting intrusive relationships to the *host rock*, simulating **tachylite**. Such rocks are now known as **pseudotachylite** (if there is evidence for melting) or *ultramylonite* (if there is not). [RGP]

Peach, B.N., Horne, J., Gunn,, W., Clough, C.T. & Hinxman, L.W. (1907) *The Geological Structure of the North-west Highlands of Scotland*. Memoirs of the Geological Survey of Great Britain, HMSO.

flood routing Estimation of the shape of a **hydrograph** at any point along the river during the course of a flood event. Flood routing follows the passage of a **hydrograph** downstream from an upstream point where the **hydrograph** is known. Thus the problem is to estimate the **hydrograph** shape and timing at points for which no gauging records exist. Two properties of the **hydrograph** change during its passage downstream. The peak discharge obviously occurs later at points downstream; this is

known as *translation*. More importantly, peak discharge declines downstream and the **hydrograph** flattens out, a process known as **attenuation**. Routing methods seek to incorporate these two effects. Flood routing is required for large drainage **basins** where movement of the flood discharge through the channel system takes a significant amount of time; routing is not normally undertaken for small **basins**. Routing is needed for *flood forecasting* downstream, and is used by design engineers for the design of bridges and other river structures, and for estimating the effects of floods produced by dam bursts. Two main categories of flood routing are commonly used. Empirical methods, such as the *Muskingum method*, rely on the continuity equation and on a relationship between flow rate and the temporary storage of water in the channel during the flood. Since such methods are essentially trial-and-error, more recently engineers have considered the hydraulics of open channel flow. Such methods are based upon the *Saint Venant equations* which govern gradually varying unsteady flow in open channels. Flood routing is most efficiently applied to single lengths of river channel and is less successful where a number of tributaries join the main channel along the routed section. [TPB]

floodplain The area of land across which a river flows that may be periodically inundated by floodwater from that river. The floodplains of adjacent rivers may overlap to form an alluvial plain. The width of a floodplain is controlled by channel size, maximum flood discharge and topography. [JA]

Florida Bay A triangular-shaped marine bay, famed for its carbonate sediments, between the Everglades at the southern tip of Florida, the Gulf of Mexico, and a line of *Pleistocene reef*-**limestone** islands, the Florida Keys. Salinities range from 10–55 p.p.t. and temperatures average 25°C. The primary areas of sediment production are a network of carbonate mud mounds. These originated as localized areas of seagrass-stabilization on transgressed **karstic** surfaces of *Pleistocene* **limestones**, and developed into topographic highs by the accumulation principally of disintegrated codiacean *algae* and seagrass epibionts. Unlike the **Bahama Banks**, the sediments contain significant calcitic fossil debris in addition to **aragonite**. [IJF]

flos ferri A variety of **aragonite** that resembles **coral**. (See **carbonate minerals**.) [DJV]

flotation (*froth flotation*) A method of concentrating minerals in mineral processing plants by selective flotation. Ground **ore** is mixed with water and a frothing reagent in a vat or cell. The mixture (pulp) is stirred and air is blown in at the bottom. The valuable mineral becomes attached to bubbles to rise and form a surface froth that is skimmed off. The chemistry of the pulp can be so carefully controlled that **chalcopyrite** can be picked up but **pyrite** left to sink. Flotation is the most important method of concentrating metal sulphides and some **industrial minerals**. [AME]

flow Used in structural geology for permanent **deformation** that has a continuous **strain** distribution. (See **ductile deformation**.) Flow is thus essentially a description of the state of **strain** in a rock, but it also specifies that the **stress–strain** relationship is non-recoverable, in distinction to **elasticity**. Because **strain** continuity depends on the scale of observation, flow is also a scale-sensitive concept, for which a scale should ideally be given, e.g. flow at a macroscopic scale, which is also known as *ductile flow*. *Cataclastic flow* is permanent, continuously distributed **strain** achieved by **fracture** and sliding of rigid particles.

Flow banding is a layering produced by flow. The term is applied to both igneous and metamorphic rocks. *Flow foliation* (or *flow lineation*) is a more general term describing a surface or line produced by flow. [TGB]

flow cleavage A type of slaty **cleavage** in which the rock **fabric** is presumed to reflect recrystallization accompanied by solid-state **flow**, so causing partial or total obliteration of traces of the original sedimentary **bedding**. [RST]

flow folding Formation of **folds** in layering by **shear** or bulk **flow** of rock in a direction oblique or normal to the layering. Minor **folds** in intrusive **igneous bodies** formed by variations in flow rate during emplacement are examples of this type of folding. The term has been used to describe similar **folds** (see **fold**) in solid rock, but **folds** with a particular geometry can form by a variety of mechanisms. Such genetic terminology must be used with caution. (See **fold mechanism**.) [SB]

flow regime Classification of flows based on the frictional resistance which they experience, originally developed by American hydraulic engineers. The *lower flow regime* is characterized by increasing **shear stress** on the bed as flow velocity increases: this is accompanied by the development of **ripples**, then **dunes** as the stable equilibrium bedforms. The presence of **ripples** and **dunes** causes **flow separation** and therefore an increased resistance to flow. However, at a critical flow strength, **shear stress** on the bed abruptly decreases as the bedforms are washed out and replaced by a **plane bed**. **Flow separation** no longer takes place, thereby reducing the flow resistance. This is the start of the *upper flow regime*, and **shear stresses** on the bed start to increase again as first **standing waves**, then **antidunes** are formed under the progressively increasing velocity of flow. [PAA]

flow separation The detachment of the **boundary layer** from a surface through generation of adverse (positive) pressure gradients close to that surface, commonly as a result of abrupt changes in the bed geometry. Flow separation creates a region of recirculating flow in which fluid pressures and velocities are less than in the surrounding freestream flow. The separation zone is bounded by a **shear layer** which terminates at a *reattachment point* where the freestream fluid once again impinges upon the bed.

Flow separation zones are present at a wide range of scales: behind individual grains, over small pits in erosive beds, in the **lee side** of many transverse bedforms, behind boulders and bank protrusions, at tight meander bends and acute angle river confluences. The influence of flow separation on *turbulence* generation is profound and as such it is of fundamental importance in the formation of many sedimentary structures. (See **ripple**.) [JLB]

flow visualization The techniques and processes utilized in making the structure of a flow visible for either direct observation or photographic record. In general, flow visualization techniques may be split into those which produce either qualitative or quantitative results and those which can be used in air or water flows. Visualization techniques should seek to minimize any possible disturbance to the flow and are easiest to use if non-contaminative so that long observation periods are possible.

The introduction of smoke into airflows or dyes into water flows are common techniques. Evaporation of paraffin in air and its subsequent condensation produces a dense smoke whilst in water flows either potassium permanganate, methyl blue or water-based poster paints provide excellent dyes. These methods usually provide a qualitative assessment: however, photographic records may allow subsequent quantitative analysis, for example measurements of eddy size and growth. These methods, especially dyes in water, are often contaminative; however, the use of particles suspended in the flow is not, and effective visualization may be achieved through use of neutrally buoyant particles. In water, various highly reflective plastics are of great use and can provide excellent records of the movement of individual particles for **streamline** mapping. Additionally, in water or higher **viscosity** fluids, aluminium flakes, which align their long axes parallel to the flow, are suitable. Quantitative analysis of suspended particles is possible although care should normally be taken to ensure the two-dimensionality of the recorded image.

A third common method of visualization generates a flow marker through electrical methods. A popular method is the non-contaminative hydrogen bubble technique which relies upon electrolytic **precipitation** of hydrogen at a fine **platinum** wire cathode which normally has its anode as the channel wall. Passage of a small current through the wire produces hydrogen: if the wire is speck insulated or the voltage pulsed, then patches of dye can be generated and their downstream path quantitatively analysed. Similar markers can be produced using **tellurium** as a cathode or **tin** as an anode in solutions containing sodium carbonate. Another electrical technique relies upon the change in **pH** of an electrolyte produced at an electrode: if the fluid is titrated to near its end point of an appropriate indicator, then the application of a voltage at the electrode will produce a change in the colour of the fluid. A widely used indicator is thymol blue which changes colour from amber (neutral) to dark blue (basic).

A final group of visualization techniques relies upon optical methods utilizing **refractive index** variations that occur within a fluid due to density variations associated with a temperature or concentration field. The **schlieren**, shadowgraph and interferometer methods all fall within this group. Simple methods using candle floats or the reflection of light paths from a water surface may also provide adequate surface flow markers. [JLB]

Clayton, B.R. & Massey, B.S. (1967) Flow visualisation in water: a review of techniques. *Journal of Scientific Instruments* **44**: 2–11.
Grass, A.J. (1983) The influence of boundary layer turbulence on the mechanics of sediment transport. In: *Mechanics of Sediment Transport, Proceedings of Euromech 156*, pp. 3–17. A.A. Balkema, Rotterdam.
Honji, H., Kaneko, A. & Matsunaga, N. (1980) Flows above oscillatory ripples. *Sedimentology* **27**: 225–9.
Merzkirch, W. (1974) *Flow Visualisation*. Academic Press, New York.
Tritton, D.J. (1977) *Physical Fluid Dynamics*. Van Nostrand Reinhold, Wokingham.

fluidization (*aqueous fluidization of sediment*) Process where the vertical escape of fluid from a granular aggregate such as **sand** exerts sufficient **drag forces** on the grains to support them momentarily against the force of **gravity**. At high rates of fluid escape, the bulk volume and therefore **porosity** of a sediment rapidly increases, with the net result that there is a change from grain-support to fluid-support — at this stage the sediment is described as fluidized. Fluidization is a grain-support process and should not be used to describe a type of current as is commonly used in 'fluidized flow'. **Cohesive** forces between grains become more important in resisting fluidization as grain sizes decrease and the clay content increases. The optimum grain sizes for fluidization range from 0.1 to 0.5 mm. Fluidization typically involves the turbulent escape of **pore fluids** thereby destroying primary sedimentary structures, reduces the **viscosity** of the sediment, elutriates any finer grains and clays and imparts a negligible *yield strength* to the sediment. The local elutriation of finer grain sizes and clays from fluidization pathways, or pipes, together with their concentration in zones of low-velocity fluid-escape produces '**dish-structures**'. Elutriated sediment may form subaqueous '**sand volcanoes**' on the top surface of beds. Fluidization may result from **earthquakes**, cyclic loading of sediment by storm **waves**, and high rates of sediment accumulation causing **overpressuring** at depth with **failure**. (See **sediment gravity flow**.) [KTP]

fluorapatite ($Ca_5(PO_4)_3F$) A variety of **apatite**. [DJV]

fluorine test dating Method of dating buried bone by determining the amount of fluorine present in the sample. Fluorine content in the bone increases with time as fluorine in the surrounding environment gradually replaces calcium. Used mainly for relative dating of bones from the same site. (Compare **radiometric assay**.) [OWT]

fluorite (CaF_2) A mineral with important industrial uses as a flux in smelting. [DJV]

fluorspar Material containing sufficient **fluorite** to make it of commercial interest. Acid grade fluorspar contains a minimum of 97% CaF_2, less than 1.5% SiO_2 and must be low in **sulphur**. Ceramic grade fluorspar of No. 1 **grade** must carry 95–96% CaF_2 and No. 2 80–90% CaF_2, other specifications limit the **silica**, Fe_2O_3, **calcite** and **base metal** contents. Metallurgical grade fluorspar must run at least 60% CaF_2 and be low in **sulphur** and **lead**.

Fluorspar is often associated with **lead** and **zinc veins** in **carbonate-hosted base metal deposits**. In some of these deposits fluorite may be the most important economic mineral. Fluorspar also occurs in stratiform deposits and in **carbonatite–alkaline igneous rock** complexes. [AME]

Harben, P.W. & Bates, R.L. (1984) *Geology of the Nonmetallics*. Metal Bulletin Inc., New York.

fluviokarst A **limestone** landscape produced by the combined action of fluvial **erosion** and 'true' **karst** processes (i.e. **limestone dissolution**). Roglic (1960) coined the term and applied it to the inland part of the Yugoslavian **karst** area where river action and *slopewash* are important and only a limited range of **karst** landforms is present. Surface drainage becomes important within **limestone** areas where there is a cover of impermeable material, or a shallow impervious basement rock, or where **allogenic rivers** flow into the area. In all these cases fluviokarst forms, including gorges and **dry valleys**, may be produced. [HAV]

Roglic, J. (1960) Das Verhältnis der Flusserosion zum Karstprozess. *Zeitschrift für Geomorphologie* **4**: 116–28.

fluxgate magnetometer A continuous-reading field instrument for measuring the strength of the **geomagnetic field**. It is based on the time taken for two **ferrimagnetic** cores to reach **magnetic saturation** when affected by an alternating **magnetic field** which reinforces the field of one of the cores and opposes the other. By pointing the elements in the appropriate direction, the horizontal and vertical elements of the **geomagnetic field** can be measured. Total field measurement requires less stringent orientation. Largely superseded by the **proton magnetometer**.

This type of **magnetometer** may be used to detect buried features on archeological sites, such as kilns, which may be expected to show **magnetic anomalies** in comparison with the surrounding *soil*. Two fluxgate magnetometers may be used, one vertically above the other to form a **magnetic gradiometer**, and the difference in their response measured to provide the vertical magnetic gradient. [PK/OWT]

Dobrin, M.B. (1976) *Introduction to Geophysical Prospecting* (3rd edn). McGraw-Hill, New York.

fluxion structure A finely lensoid or banded **structure** characteristic of **mylonites** and *ultramylonites*, and often expressed as a streaky colour-lamination. It is produced by the **elongation** of grains and grain aggregates of contrasting composition at high **shear strains**. [RGP]

focal mechanism solution (*fault plane solution*) The identification of the direction and sometimes **magnitude** of the mechanical forces that have given rise to an **earthquake**.

In practice, virtually all **earthquakes** are found to be consistent with Reid's **elastic rebound theory**, and have a 'double-couple' mechanism; two orthogonal pairs of couples causing **failure** on one of two perpendicular planes (the **fault** and *auxiliary planes*). Focal mechanism solution is synonymous with 'fault plane solution', the determination of the orientation of the **fault** plane and by inference that of the pressure and tension axes causing the **earthquake** (see Fig. F23).

The double-couple mechanism predicts that the first motion of the *P wave* will alternate between compressional ('positive') and rarefactional ('negative') in the four quadrants enclosed by the **fault** and *auxiliary planes*, corresponding to rock movement toward or away from the observer. This alternation is exploited in the classic method of fault plane solution: *P wave* first motions are picked from **seismograms** at a widely distributed set of **teleseismic** stations. The polarities are plotted on a **focal sphere**. The positive and negative first motions should fall into separate areas, bounded by two orthogonal planes. Geological evidence is used where possible to choose which is the **fault** plane. This widely used method is actually only of use for **earthquakes** of body-wave **magnitude** of about 5.5 or larger, for which the *P waves* will be substantially larger than background noise and so whose polarities will be unambiguous.

Some **earthquakes** have been recognized as having mechanisms other than the double couple — for example, an explosive component caused by **magma** injection or an implosive component caused by a chemical phase change. In these cases, the pattern of **seismic waves** will differ and more general descriptions of a source must be used.

Numerous other focal mechanism techniques have been developed both to improve double-couple determination and to allow identification of non-double couple sources. Some exploit the amplitude variations of the double-couple mechanism by comparing the amplitudes of various *P* and/or *S wave* arrivals, such as the direct *P wave* and the

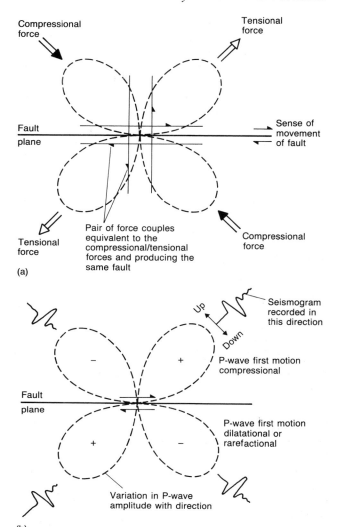

Fig. F23 Schematic view of the double-couple earthquake **focal mechanism**. (a) The compression and tension caused by tectonic forces cause a failure along a fault plane at 45° to them, where shear stress is at a maximum. Two force-couples are equivalent to the compressions and tensions. (b) The seismic waves radiated by the double-couple mechanism: in two quadrants their first motion is compressional ('positive'), in the other two dilational ('negative') and their amplitude varies considerably with direction. Four schematic seismograms are shown: note how the 'polarity' of their first motion changes from ground 'up' to ground 'down' in the four quadrants.

P wave reflected from the Earth's surface above the **earthquake** focus. Others use different waves types entirely, such as the *Rayleigh* and *Love surface waves*. By modelling quantitatively the *P* waveforms and amplitudes recorded, others combine focal mechanism solution for double-couple or more general sources with a determination of the **seismic moment**, and even the 'source time function' (how fast and for how long slip took place).

Focal mechanism solutions have played a central part in **seismology**, especially since 1962–3 when the first world-wide network of standardized *seismographs* was installed. The solutions soon revealed a coherent picture of directions of movement consistent with the formation and collision of tectonic *plates*. Focal mechanism solutions are now routinely computed by several agencies for most large **earthquakes**. (See **elastic rebound theory**, **seismic source**.) [RAC]

Cox, A. (1973) *Plate Tectonics and Geomagnetic Reversals*. W.H. Freeman, San Francisco.
Stauder, W. (1962) The focal mechanism of earthquakes. *Advances in Geophysics* **9**: 1–76.

focal sphere A hypothetical sphere centred on an **earthquake** *focus*, used to represent data on the characteristics of **seismic wave** radiation from the *focus*. Measurements made on **seismograms** are projected back onto the focal sphere's surface, using standard Earth models, to determine where the appropriate ray intersects the focal sphere. Representation of the focal spere on a *stereogram* is the most commonly used method for the determination of **focal-mechanism solutions**. (See **focal mechanism solution**.) [RAC]

fold Describes the curved or angular shape of an originally planar geological surface; folding refers to the formation of such curved surfaces. Folds result from bulk **strain** most commonly associated with regional metamorphism at relatively deep **crustal** levels. Folding also occurs at high **crustal** levels, for example associated with displacements on **faults** or **décollement** surfaces. **Soft-sediment deformation**, particularly **slumping**, can form complex folds, prior to full lithification of the rocks. Description and discussion of folded rocks depends on the use of geometrical terms discussed by Fleuty (1964) and Ramsay (1967), among others, which are summarized below. Most planar surfaces in geology occur in sub-parallel groups, such as *bedding* planes. Fold geometries are described in terms of the deformed shapes of individual planes, and of the relationships between these shapes.

Consider first an isolated surface, curved to form alternately concave upward and downward regions. This forms a *fold train* (Fig. F24). The areas of greatest curvature on this folded surface are the *hinge zones*, and the more planar areas separating them are the *fold limbs*. More precisely, the *hinge lines* lie in the folded surface, linking points of maximum curvature. *Hinge zones* that are concave upwards are called *fold troughs*, those concave downwards *fold crests*. The separation of two adjacent crests or troughs measured along the length of the fold train is the *fold wavelength*; the separation of a crest and a trough measured across the *fold train* is twice the *fold amplitude* (see **fold system**). Concave upwards folds are called **synforms**, and convex upwards folds, **antiforms**. (See also **anticlines** and **synclines**.)

Fold symmetry relates to the similarity of size and shape of the limbs of a fold. A fold is symmetrical when the two opposing limbs have mirror symmetry across the axial surface. In other words, the axial surface is planar, and bisects the angle between the *fold limbs*, the *interlimb angle* (see below). Asymmetrical folds occur in a variety of forms, reflecting differences in limb length or limb thickness. These variations are often systematic. For example, the folds in a *fold train* may consistently show the western limb shorter than the eastern limb. In such cases, asymmetric folds are described as 's' or 'z' folds, depending on the sense of rotation of the shorter limb. This notation is useful in mapping large-scale folds on the basis of *parasitic fold* asymmetries, as well as in defining **vergence**. Changes in limb thickness also define asymmetric folding, and often occur together with limb length changes, for example in **flow folds** resulting from bulk **simple shear**. Although the axial surface of asymmetrical folds can bisect the *interlimb angle*, this is not the general case.

Certain geometrical features are strictly defined in terms of the **fold profile**, which is the trace of a folded surface in a plane perpendicular to the *fold hinge* at any point along its length. This plane is the *fold profile plane*. The **fold profile** can be described in terms of its shape, so that a *sinusoidal fold* has a profile fitting a sine curve. If the *fold profile plane* has the same orientation everywhere along the hinge, and the **fold profile** is identical wherever it is seen, then the fold is *cylindroidal* or *cylindrical*. Hobbs *et al.* (1976) reserve the term *cylindrical* for folds with a constant profile consisting of arcs of a circle, making *cylindroidal* the more general term. However, Turner & Weiss (1963),

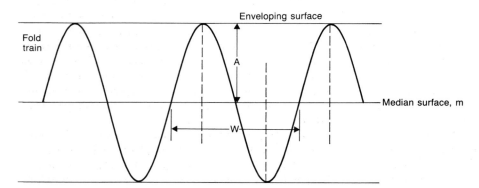

Fig. F24 A **fold** train showing median and enveloping surfaces, fold amplitude A and wavelength W. (After Ramsay 1967.)

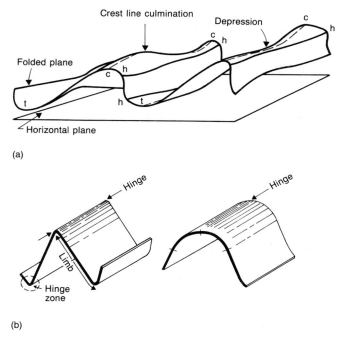

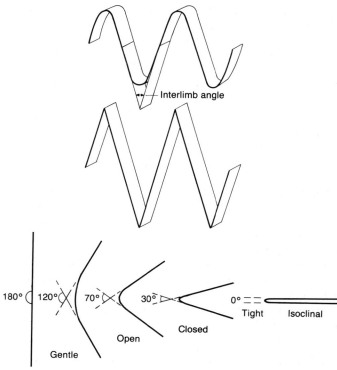

(a)

(b)

Fig. F25 (a) **Folds** in an isolated layer showing crest (c), trough (t) and hinge lines (h), with culminations and depressions along the length of the folds. These folds are non-cylindrical. (After Ramsay, 1967.) (b) Cylindroidal and cylindrical folds showing different proportions of hinge zone and limbs. (After Fleuty, 1964.)

Fig. F26 The interlimb angle can be defined for both angular and rounded **folds**; folds are described in terms of their interlimb angles, using the divisions shown. (After Ramsay, 1967 and Fleuty, 1964.)

Fleuty (1964) and Ramsay (1967) use *cylindrical* as the general term. *Non-cylindrical folds* show different profiles at different positions along their hinge lines. The **fold system** shown in Fig. F25 is *cylindrical*, with all hinge lines straight and parallel. Similar parallel straight lines can be drawn anywhere on the fold surface. The orientation of these lines is the axial direction of the folds, or the *fold axis*. For *cylindrical folds* the orientation of the *fold axis* coincides with the hinge lines. Changes in hinge line **plunge** along the length of a *non-cylindrical fold* define *culminations*, where the hinge zone is elevated, and *depressions*, where it is lowered. Examples of these fold types are **domes** and **basins**. If the **fold profile** changes along the hinge line such that the amplitude decreases to zero, and the folded layer is shaped like part of a cone, then the fold is *conical*.

These geometrical features apply equally to folds with angular or curved *hinge zones*. The distinction between angular and curved **fold profiles**, discussed by Hobbs *et al.* (1976), can be quantified in terms of the extent of the *hinge zones* relative to the *fold limbs*. This is negligible for angular folds such as **chevron folds** (see below), and large for curved folds. The *fold tightness* is defined in terms of the angle between *fold limbs*, specifically the *interlimb angle*, shown in Fig. F26. *Gentle* folds have *interlimb angles* greater than 120 degrees, *open* folds have angles between 70 and 120 degrees, etc. Ideal *isoclinal folds* have an

interlimb angle of zero degrees, i.e. the limbs are parallel. The term is often used for folds with *interlimb angles* up to 10 degrees.

Most examples of folds in geology affect many originally subparallel layers, rather than one folded surface. When several rock layers are folded together, the *fold core* refers to that part of the multilayer closest to the *hinge zone* (or axial surface, discussed below) in a section such as the plane of a map cutting through the fold. For example, an **anticline** is distinguished from an **antiform** by the occurrence of demonstrably stratigraphically older rocks in the *fold core*. The geometrical features described above can be applied to parts of these **fold systems**, but the relationships between adjacent folded layers is significant for both the resultant geometry and for the **fold mechanisms** involved in their formation. *Fold style* describes both the geometry of an individual folded surface and the shapes of folds resulting from the relationship between adjacent folded layers. In many cases, the shapes of successive layers are closely related. This is not so for **disharmonic folds**, in which there is little or no geometrical relationship between adjacent layers, e.g. the *wavelength* and *amplitude* may change.

Where successive folded surfaces are related, the geometrical terms above can be supplemented; successive positions of *fold hinges* through the layered sequence define *axial surfaces*, which may be planar, i.e. forming the

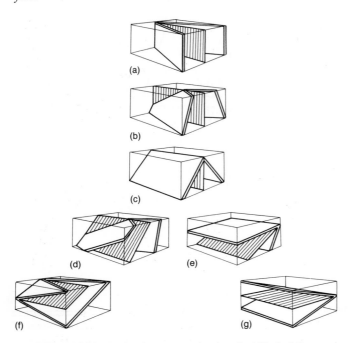

Fig. F27 Ideal **folds** showing the terms used to describe fold orientations. (a) Vertical. (b) Upright plunging. (c) Upright horizontal. (d) Inclined plunging. (e) Inclined horizontal. (f) Reclined. (g) Recumbent. (After Fleuty, 1964.)

The variety of fold shapes possible is described within a system of fold classification based on features unaffected by fold orientation, to facilitate comparisons between folds in different **tectonic** settings. A classification based on the changes in layer thicknesses relates to the distribution of **strain** in that layer and to the **fold mechanism**. Ramsay (1967) suggests a classification of this type based on the curvature of the two folded surfaces defining a layer. **Dip isogons**, lines joining points of equal **dip** on the these surfaces, have distinctive patterns in different *fold classes*, described in profile section, as shown in Fig. F28.

In the fold classification based on **dip isogons**, Class 1 folds have greater curvature on the inner arc than on the outer, so that the **dip isogons** converge downwards. The *fold limbs* have the same or greater orthogonal thickness than *fold hinges*. Class 2 folds have identical curvatures on inner and outer arcs, so that the **dip isogons** are parallel, and parallel to the *axial plane*. The orthogonal thickness of the *fold limbs* is less than that in the *hinge zone*. Class 3 folds have greater curvature on the outer arc, so that the **dip isogons** diverge downwards. Again, the orthogonal limb thicknesses of the limbs are less than that in the *hinge zones*.

Class 1B and 2 folds are common in natural **fold systems**. Class 1B folds maintain the layer thickness everywhere perpendicular to the layer surface; they are called *parallel folds*, as the inner and outer arcs remain parallel. *Concentric folds* are Class 1B folds in which outer and inner layer boundaries form concentric arcs of circles. Class 2 folds are called *similar folds*, as the inner and outer arcs of a layer have mathematically similar, i.e. identical, shapes, so that the thickness of the layer is unchanged parallel to

axial plane, or more generally, curved. The orientation of these surfaces, together with the **plunge** of the *fold axis* defines *fold orientation* in three dimensions, as in Fig. F27. Natural folds often have complex sets of hinge lines, with multiple and branching *axial surfaces*.

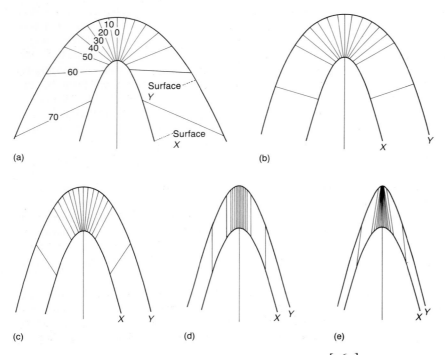

Fig. F28 Classification of **folds** based on their dip isogons. (a) Fold 1, class 1A. (b) Fold 2, class 1B (parallel). (c) Fold 3, class 1c. (d) Fold 4, class 2 (similar). (e) Fold 5, class 3. Dip isogons have been drawn at 10° intervals from the lower to the upper surfaces X and Y. (After Ramsay & Huber, 1987.)

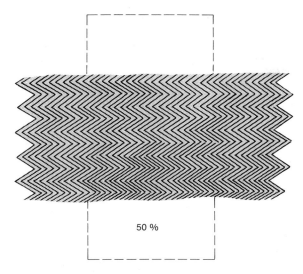

Fig. F29 Kink bands and chevron **folds** developed in experimental deformation (50% strain) of a multilayer. The dashed line shows the original shape of the rock. (After Hobbs *et al.*, 1976.)

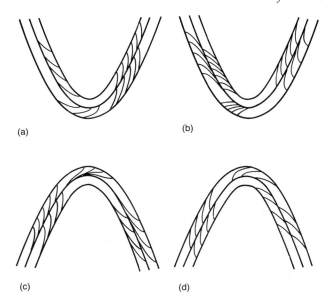

Fig. F30 **Fold** orientation using younging information to identify anticlines and antiforms, etc. (a) Synformal syncline. (b) Synformal anticline. (c) Antiformal anticline. (d) Antiformal syncline. (After Ramsay, 1967.)

the *axial plane*. The thickness perpendicular to the layer surface is less in the *fold limbs* than in the *hinge zones*. Ideal *similar* (Class 2) *folds* are the only type of folds that can occur throughout a multilayer sequence, without discontinuities along layer surfaces or change of *fold class*. Any other folds in a multilayer must change class along the *axial plane*. The **rheology** of the layers of rock is clearly of importance in the formation of folds. More and less **competent** layers in a multilayer often form folds of distinct classes which vary through the system.

A special case of a folded multilayer is that of **chevron folds**, **box folds** and **kink bands**, as shown in Fig. F29. These are angular folds (with negligible *hinge zones*) found in closely spaced multilayers. Experimental work (Paterson & Weiss, 1966) has demonstrated the evolution of **chevron folds** from **kink bands** and **box folds** during *layer-parallel shortening*.

The *orientation* of a *fold* is defined in terms of geometrical features such as the *axial surface* and *hinge line*. The terms **antiform** and **synform** describe whether a fold is convex upwards (*closing* upwards) or concave upwards (closing downwards) respectively. *Neutral folds* are sideways closing. When the stratigraphic **younging** direction is clear, **anticlines** and **synclines** can be defined. As shown in Fig. F30, an **anticline** can have **antiformal** or **synformal** orientation. Figure F27 also shows some of the common descriptions of *fold orientation* based on *axial surface* and *hinge line* orientation. Fleuty (1964) reviews the nomenclature used for describing folds in greater detail. When the *hinge line* is close to horizontal, the *axial surface* may be vertical, dipping or horizontal. Such folds are *upright*, *inclined* (*tilted*), or *recumbent* in turn. An *overfold* is a fold structure in which the *axial surface* and both *fold limbs* **dip** in the same direction, i.e. one limb is overturned. If the *hinge line* is not horizontal, but has moder-

ate **plunge**, then folds are described as *plunging inclined*, etc. If the *hinge line* plunges steeply, or is vertical, folds are described as *reclined* or *vertical*.

In general, in natural examples, the shapes of folded layers can be approximated to the ideal geometries discussed here. Departures from these geometries can then be recognized, and their significance assessed. Understanding of the geometrical consequences of folding is essential for analysis of complex **fold systems**, **fold interference patterns**, and **superposition of strata**. (See **buckling**, **sheath folds**, **strain**.) [SB]

Fleuty, M.J. (1964) The description of folds. *Proceedings of the Geological Association* 75: 461–92.
Hobbs, B.E., Means, W.D. & Williams, P.F. (1976) *An Outline of Structural Geology*. John Wiley & Sons, New York.
Paterson, M.S. & Weiss, L.E. (1966) Experimental deformation and folding in phyllite. *Geological Society of America Bulletin* 77: 343–74.
Ramsay, J.G. (1967) *Folding and Fracturing of Rocks*. McGraw-Hill, New York.
Ramsay, J.G. & Huber, M.I. (1987) *The Techniques of Modern Structural Geology* Vol. 2: *Folds and Faults*. Academic Press, London.
Turner, F.J. & Weiss, L.E. (1963) *Structural Analysis of Metamorphic Tectonites*. McGraw-Hill, New York.

fold axial trace The line of intersection of the **fold** *axial plane* or *axial surface* with the topographic surface, or with some other defined surface. The term is most often used to mark the trend of a **fold** on a map. [RGP]

fold belt A large-scale linear piece of **crust** characterized by a group of related **folds**. The term is sometimes used synonymously with *orogenic belt*, but is more usually applied to a part of an *orogenic belt* characterized by a

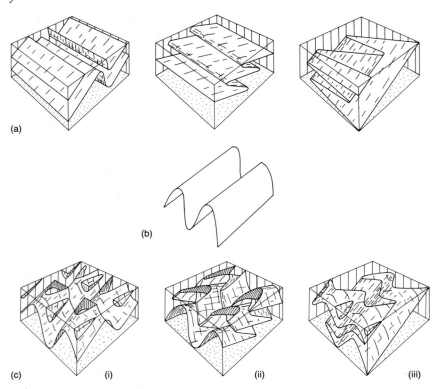

Fig. F31 Fold interference patterns generated by the superimposition of upright folds on previous folds of varying attitude. (a) F1 folds. (b) F2 folds. (c) Interference structures. (After Ramsay, 1967.)

continuous set of **folds** and separated from adjoining folded zones by less deformed areas or by stable blocks. The term does not imply a lack of other structures within the belt, and many fold belts exhibit *thrusts*, for example, that are genetically related to the folding. The Alpine fold belt of southern England, and the Jura fold belt of the Western Alps, are examples of Alpine fold belts that are separated both from each other and from the main Alpine belt by tracts of unfolded strata. [RGP]

fold closure Used to indicate the direction in which the *hinge* of a **fold** lies in relation to its *limbs* or *axial plane*, i.e. the direction in which the *limbs* converge. Thus a **fold** is said to close westwards, or upwards, for example. (See **facing**.) [RGP]

fold culmination An elevated zone on a *fold hinge* or *fold crest* of variable height; or a high point on a folded surface showing a *dome-and-basin* type of structure. Conversely *fold depressions* are the lowest parts of a *fold hinge* or folded surface. [RGP]

fold interference pattern/structure When a folded surface is deformed by a later set of superimposed **folds**, the resulting structure is termed an interference structure, which produces a characteristic outcrop pattern, the interference pattern. Ramsay (1967) developed a geometrical analysis of fold interference structures based on

the superimposition of later oblique *shear (flow) folds* on an already folded surface. Figure F31 shows the three basic types of interference structure that may be produced in this way. These are: *dome-and-basin*, *crescent-and-mushroom* and *double zigzag*. The geometry depends on the spatial relationships between the two **fold** sets, in particular between the attitudes of the two *fold axes* and *axial planes*, and on the **shear** or **flow** direction of the second **deformation**. Other possible geometries exist, but the types shown in the figure are those commonly recognized in the field. The figure demonstrates the effect of superimposing a vertical **flow** direction, producing upright F2 **folds**, on three types of F1 geometry: (a) upright F1 **folds** with *axes* and *axial planes* making a large angle with F2, producing the *dome-and-basin* structure; (b) F1 *overfolds* with *axes* and *axial planes* making a large angle with those of F2 but where the *axial planes* are oblique to the F2 **flow** direction, producing the *crescent-and-mushroom* structure; and (c) F1 *overfolds* whose *axial planes* make a large angle with those of F2, but whose axes make a small angle with those of F2, and where the F2 **flow** direction is oblique to the F1 *axial planes*, producing the *double zigzag structure*.

The term *interference pattern* implies **fold** superimposition, but identical patterns may be produced by *heterogeneous strain* during progressive **deformation**, e.g. in the production of **sheath folds**, where the *fold axis* is in a direction of **shortening** in the **finite strain ellipsoid**. [RGP]

Ramsay, J.G. (1967) *Folding and Fracturing of Rocks*. McGraw-Hill, New York.

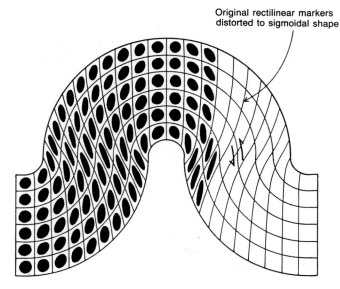

Original rectilinear markers distorted to sigmoidal shape

Fig. F32 Fold mechanism. Internal strain shown by the shapes of strain markers in a layer deformed by flexural flow. (After Ramsay, 1967.)

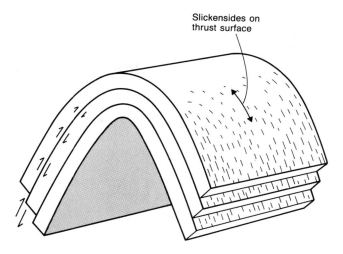

Slickensides on thrust surface

Fig. F33 Fold mechanism. Localized slip between layers undergoing deformation by flexural slip forms minor faults, here thrusts. (After Ramsay, 1967.)

fold mechanism The way in which a **fold** forms, i.e. the **deformation** history and distribution during folding. The **finite strain**, i.e. the final shape of a folded layer, does not uniquely define the *strain history*, nor can it necessarily give any information about **stress**. Overall, **folds** reflect contraction of a layer in at least one direction. However, this can occur in a variety of ways. Patterns of **strain** in naturally and experimentally folded layers, seen as **cleavage** development, or layer thickness changes, suggest three important mechanisms of accommodating folding: *flexural slip*, *flexural flow*, and *tangential longitudinal strain* (Figs F32, F33 and F34).

Flexural flow is a mechanism of folding consisting of *layer-parallel shear* in *fold limbs*, without distortion in *hinge zones*. *Flexural flow* folding implies distributed **simple shear** across the *fold limbs*. The intensity of **shear strain** depends on **fold** shape, and on position within the **fold**. This **strain** distribution ideally results in parallel, Class 1B **folds** (see **fold**). Natural examples of such **folds** show intense **cleavage** development and **vein** formation in the *fold limbs*, which dies away towards the *fold hinge zones*.

Flexural slip is a similar process to *flexural flow* folding, but occurs by discontinuous layer-parallel **shear**, distributed between rock layers, and *interlayer slip* on their bounding surfaces. *Flexural slip* is indicated by **slickensides** and fibrous mineral growth (*slickencrysts*) on

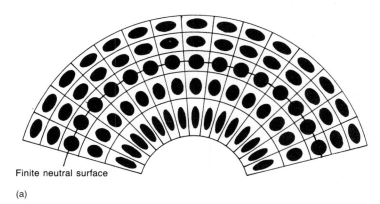

Finite neutral surface

(a)

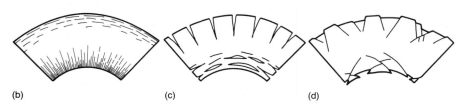

(b)　　　　　(c)　　　　　(d)

Fig. F34 Fold mechanism. (a) Internal strain shown by the shapes of strain markers in a layer deformed by tangential longitudinal strain. Strain can be accommodated by: (b) cleavage development; (c) opening of fissures; (d) minor fault populations. (After Ramsay & Huber, 1987.)

[265]

bedding planes within the *fold limbs*. Again, there is less **strain** in the *hinge zones*.

Tangential longitudinal strain is a fold mechanism recognized by a pattern of layer-parallel extensional **strain** on the outer arc of a folded layer, and layer parallel contractional **strain** on the inner arc. The maximum **principal strain** direction thus lies tangential to the **fold** surface. If the **strain** is distributed so that the perpendicular layer thickness is unaffected, the **folds** formed are parallel, Class 1B. In this folding mechanism, **strain** is most intense in the *hinge zone* of the **fold**, and weakly developed in the *fold limbs*. This **strain** distribution can result from **buckling deformation** of a layer isolated within rock of differing **rheology**, or from bending of a rock layer.

Another common fold mechanism occurs during **simple shear**, when there are irregularities in the **foliation** making it non-parallel to the **shear plane**. The **foliation** is then in contractional or extensional orientation (see **strain ellipse**) and *asymmetrical folds* form and grow. Continued **shear strain** can rotate and stretch the **foliation** until only the *hinge zones* of these **folds** remain, as *intrafolial folds*, to indicate the *strain history* of otherwise parallel-layered rocks. *Heterogenous simple shear*, in which the intensity of **shear** varies through the rock, can also form and amplify **folds**. *Flow folding* refers to formation of **folds** in layered rock by **shear** or bulk **flow** of rock in a direction oblique or normal to layering. Minor **folds** in intrusive **igneous bodies** formed by variations in **flow** rate during emplacement are examples of this type of folding. It should be noted that combinations of these idealized fold mechanisms with other **strain** patterns (**pure shear, simple shear**, *volume loss* or **dilation**), especially when these are heterogeneous, can produce a variety of **fold** geometries, and it can be impossible to deduce the sequence of **deformation** from **fold** shape alone.

Folds can also result from **faulting** processes. Where a **fault** with finite displacement ends, the **strain** continuity between faulted and unfaulted rock is maintained by **folds** called *tip folds* (see Coward & Potts, 1983). Movement of **faults** over non-planar **fault** surfaces also results in **folds** in the upper rock, termed *fault-bend folds*. In general, it is not possible to deduce the mechanism of formation of a **fold** solely from its geometry and/or internal **strain** distribution; a combination of this and its tectonic setting is necessary, but still may not be sufficient information. [SB]

Coward, M.P. & Potts, G.J. (1983) Complex strain patterns developed at the frontal and lateral tips to shear zones. *Journal of Structural Geology* 5: 383–401.

Ramsay, J.G. (1967) *Folding and Fracturing of Rocks*, pp. 391–460. McGraw-Hill, New York.

Ramsay, J.G. & Huber, M.I. (1987) *The Techniques of Modern Structural Geology* Vol. 2: *Folds and Faults*. Academic Press, New York.

fold nose The *hinge zone* of a **fold**. [RGP]

fold plunge Plunge of a **fold axis**. [RGP]

fold profile The trace of a **folded** surface in a plane perpendicular to the *fold hinge* at any point along its length. For a *cylindrical fold*, the profile plane will have the same orientation everywhere along the hinge, and the fold profile will be identical wherever it is seen. *Non-cylindrical folds* show different profiles along their hinge lines. [SB]

Ramsay, J.G. (1967) *Folding and Fracturing of Rocks*. McGraw-Hill, New York.

fold style Collective term for various geometrical features of the **fold profile** generally considered useful in classification and in indicating the mechanism of formation of **folds**. (See **fold mechanism**.) [RGP]

fold system A group of related **folds**, often of different sizes or geometries, that formed together. A common type of fold system is where folding affecting several rock layers is **disharmonic**. In such a system, **décollement** surfaces must be present between adjacent layers, and the **fold** shape and *fold class* will vary between layers. *Polyclinal folds* are characterized by variably inclined *axial surfaces* along and between layers. A common difference between **fold** shapes in a multilayered folded sequence is in the scale of folding. Narrow layers tend to form **folds** of shorter wavelength than thicker layers (see **buckling**).

In many cases, fold systems consist of **folds** on several scales affecting the same layers. Geometrical analysis of these **folds** is simplified by the construction of **enveloping surfaces**, which link *crests* or *troughs* of **folds** of similar scales. The order of **folds** can be defined thus: *first order folds* are larger than *second order folds*, and fold the **enveloping surfaces** of *second order folds*, and so on. *Second order folds* often show a particular pattern of asymmetry depending on their position with respect to first order *fold hinges*. These **folds**, called *parasitic folds*, or *satellite folds* are referred to as having 's' or 'z' asymmetry, or being symmetrical 'm' **folds**. As shown in Fig. F35, *s* and *z folds* form on the opposite sides of *major fold hinge zones*, which are marked by 'm' **minor folds**. (Notice that all observations in a particular area must be made in the same frame of reference with respect to the **major folds**, e.g. looking down **plunge**.) This method is of importance in mapping regions of large-scale **deformation**. (See also **vergence, facing**.)

Care must be taken in this method of mapping to distinguish *parasitic folds* from **folds** resulting from *flexural slip*, and from the superposition of one or more **fold systems**. *Flexural slip* **folds** would be preferentially developed on the *fold limbs*, and symmetrical 'm' **folds** would not be expected in the *hinge zones*. The latter process, discussed in detail by Ramsay (1967), results in **minor folds** of the same shape and geometry on opposing sides of a main *fold hinge*, i.e. **folds** with opposing **facing**, but the same **vergence** on opposing sides of a major *fold hinge*. [SB]

Ramsay, J.G. (1967) *Folding and Fracturing of Rocks*. McGraw-Hill, New York.

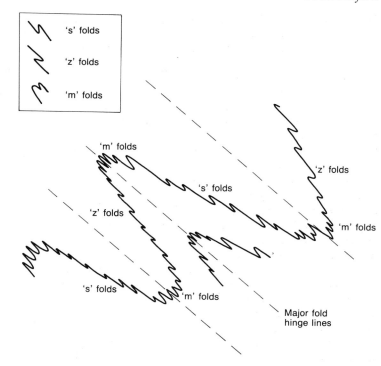

Fig. F35 Fold system. The relationship between major fold structures and asymmetrical parasitic folds: minor 's' folds and 'z' folds, defined as shown, occur on opposing major fold limbs, while symmetrical 'm' folds occur in their hinge zones.

Ramsay, J.G. & Huber, M.I. (1987) *The Techniques of Modern Structural Geology* Vol. 2: *Folds and Faults*. Academic Press, New York.

fold trend A rather imprecise term for the orientation of a *fold axis* or *fold hinge*, considered in the horizontal plane. The term may be applied either to the **strike** of the *axial plane* (or *axial surface*), or to the **plunge** direction of the *hinge*. The term is more used in a general sense to describe the mean geographic orientation of a set of **folds**, or of a **fold belt**. [RGP]

foliation A repeated or penetrative planar feature in a rock mass. This may be defined by, for example, a planar grain-shape **fabric**, a compositional layering or pervasive **fracture** set. The term is most commonly used for **fabrics** in metamorphic rocks without a specific genetic implication and thus includes **cleavage**, **schistosity** and **gneissosity**. Such 'metamorphic' foliations may have a sedimentary origin (e.g. **bedding**) or an igneous origin (e.g. compositional layering, magmatic **veins**); they may also be originally **tectonic** (e.g. **cleavage**). *Transposed foliations* are developed by **deformation** of a primary surface, commonly by the process of folding accompanied by *diffusive mass transfer*. *Igneous foliations* may also be defined by compositional variations or changes in crystal size or shape; they can be produced by the process of *differentiation* or by **flow** in a **magma**. Several types of foliation in several orientations may be present in a rock. These may be irregular or curved in form and may be **folded** as a result of subsequent **deformation**. The surfaces

to which they are parallel are called *S-surfaces* and rocks with **tectonic** foliations are *S-tectonites*, or *L–S tectonites* when both a **lineation** and **tectonic** foliation are present. [TGB]

Williams, P.F. (1977) Foliation: a review and discussion. *Tectonophysics* **39**: 305–28.

fool's gold A term sometimes applied to **pyrite** or **chalcopyrite** or mixtures of these minerals which can be mistaken by the eager for **gold**. [AME]

footwall The wall and, by extension the body of rock, lying beneath a horizontal or inclined **fault** or **orebody**. [AME]

Foraminifera An order of single-celled protozoans whose largely marine fossil record is of major importance to **biostratigraphy** and environmental analysis of **Devonian** to *Holocene* sediments. The group is related to the familiar, naked *Amoeba* and to the skeletal marine **radiolaria**. The order Foraminiferida is characterized by an outer ectoplasm of fine, granular pseudopodia and an inner endoplasm enclosed within a skeleton (test) of varying composition, consisting of a single chamber or many chambers, each connecting through an opening (foramen). Test composition and architecture are of primary importance in classification of the group (e.g. organic-walled Allogromiina, agglutinated Textulariina, high-**magnesium calcite** Miliolina, low-**magnesium calcite** or **aragonite**

Rotaliina) and this feature has proved to be a valuable tool in the interpretation of *paleosalinity* and seawater chemistry. The group is largely studied from bulk sampling of marine sediments, by washing disaggregated clays, marls or **chalks** through 1 mm to 0.06 mm sieves. The microscopic tests are picked out under reflected light and are usually studied in detail using scanning **electron microscopy**. Remains in indurated rocks, especially **limestones**, are studied in 0.03 to 0.06-mm-thick thin sections or stained acetate peels.

Foraminiferal remains have become increasingly common in sea-floor sediments since they first appeared in the **Cambrian** Period and they have, from time to time, been important as **limestone** producers. Deposits notably rich in their microscopic remains include **Carboniferous–Permian limestones**, *Cretaceous* to **Tertiary chalks**, deep sea '*Globigerina oozes*' and tropical **limestones** (incuding **nummulitic limestones**) as well as temperate carbonates. Marine argillites of **Mesozoic** and younger age also abound in foraminiferid remains and such is their abundance and ease of study that they have been used to provide **biostratigraphy** and environmental indices for many **wells** and cores into marine rocks over the last half century. Thus foraminiferid expertise has been developed by *hydrocarbon exploration* companies, geological surveys, universities and the **deep sea drilling projects**.

Three informal groupings of Foraminifera are often distinguished by experts using these microfossils: planktonics, smaller benthics (or benthonics) and larger benthics. Planktonic Foraminifera are the remains of small, free-floating forms that lived in the top 1000 m layer of the oceans. These first appeared during late **Jurassic** times and flourished especially from mid **Cretaceous** times onward, to provide a standard **biostratigraphic** scale for marine rocks that can be applied over much of the globe. At first, planktonic Foraminifera were intensively studied by oil company biostratigraphers but they have latterly taken a key role in the analysis of the drill sites through the deep sea floor (see **deep sea drilling project**). As well as contributing to zonal schemes (based on a wide variety of taxa, such as **Cretaceous** *Globotruncana* spp., Paleogene *Morozovella* spp. and *Neogene Globorotalia* spp.) these forms have also provided important information on *paleoceanography*. Studies of their lateral distribution can be used to plot temporal shifts in warm or cool water masses, for example, while the stable isotope chemistry of their skeletons provide valuable indications of changes in seawater temperature, salinity, productivity and changing polar ice volumes through the **Cretaceous** to *Holocene*. Studies of taxonomic evolution show interesting patterns too: the group suffered abrupt and almost total **extinction** at the *Cretaceous–Tertiary boundary* (perhaps in response to a *meteorite* impact) and their descendants suffered decline during the late **Eocene** and **Oligocene** (at a time of falling temperature and sea-level).

The term 'larger benthic Foraminifera' applies not to single ecological or taxonomic groups but to a variety of forms whose tests are mostly larger than 1 mm with endoskeletal structures. At present, such forms are often found in shallow, illuminated waters of subtropical to tropical latitudes, associated with carbonates and *reefs*, and they often contain endosymbiotic *algae*. Although living taxa are seldom larger than 10 mm across, warm episodes in the *Maastrichtian*, middle **Eocene** and late **Oligocene–Miocene** encouraged the evolution of larger forms approaching 150 mm across. The complex internal structure of larger benthic Foraminifera is best seen in orientated thin sections, and the species often show evidence of gradualistic, iterative evolutionary trends in test architecture. This, coupled with relatively high rates of evolutionary turnover, has enabled the group to occupy an unrivalled position in the **biostratigraphy** of warm shallow water **Tethyan limestones**. Thus, fusulinids (e.g. *Schwagerina*, *Neoschwagerina*) provide indices for the *Pennsylvanian* to **Permian** rocks; lituolids and orbitolinids (e.g. *Orbitopsella*, *Orbitolina*) are used for **Jurassic** to *Tertiary* **limestones**; miliolids (e.g. *Alveolina*) are useful for late **Cretaceous** and *Tertiary* **limestones**, especially of restricted facies; orbitoids and rotaliids (e.g. *Lepidocyclina* and *Nummulites*) are important for ramp and bank edge **limestones** of similar age.

Smaller benthic Foraminifera comprise the remaining genera and species of the order and some 30 000 of these have been described. This broad category provided the rootstock from which the groups discussed above were derived. Their **biostratigraphic** utility is relatively limited; forms such as simple *Hyperammina* spp. may range from **Ordovician** to *Holocene* though others, such as ornately sculptured *Uvigerina* spp., provide zonal stratigraphy on a regional scale. These smaller benthic forms are more greatly valued as *paleoenvironmental* indicators, with distinctive assemblages now recognized from the whole spectrum of marine environments from estuaries and lagoons to *abyssal plains*. An approximate indication of *paleosalinity* can be achieved by analysing the relative proportions of suborders in an assemblage, if taken in combination with species diversity: low diversity, agglutinated Textulariina dominant (brackish); low diversity, Miliolina dominant (hypersaline); higher diversity, Rotaliina dominant (various open marine conditions). Assemblage diversity also tends to be higher on continental shelves and declines down towards the *abyssal plain*, where primitive agglutinated forms dominate below the calcium **carbonate compensation depth** (*c.* 3000 to 5000 m). Thus a ratio of calcareous : agglutinated smaller benthic Foraminifera provides some indications of $CaCO_3$ availability. Another ready reckoner to *paleodepth* has been the planktonic : benthonic (P : B) ratio, with the latter dominating on the inner shelf and the former increasing in abundance with depth (or access to the ocean) until planktonic tests dominate in the *Globigerina oozes* of bathyal depths.

These *paleoenvironmental* measures decrease in value when applied to progressively more ancient assemblages. This is because many genera and higher groups have gradually changed their habitat preference through time. The simple agglutinated test of *Ammodiscus*, for example, is a typical abyssal form today but in **Mesozoic** times was commonly living in brackish or **anoxic** shelf facies, and its earliest preference was for **Silurian** reefs. There is also some evidence that the first planktonic Foraminifera largely lived over continental shelves and progressively shifted into the open ocean through time. [MDB]

Bolli, H.M., Saunders, J.B. & Perch-Nielsen, K. (eds) (1985) *Plankton Stratigraphy*. Cambridge University Press, Cambridge.
Brasier, M.D. (1980) *Microfossils*. Allen & Unwin, London.
Haynes, J.R. (1981) *Foraminifera*. Macmillan, London.
Jenkins, D.G. & Murray, J.W. (1981) *Stratigraphical Atlas of Fossil Foraminifera*. Ellis Horwood, Chichester.

force That which produces motion in a body. Force is defined as mass times acceleration. The SI unit of force is the **newton** (N) defined as the force required to produce an acceleration of 1 metre per second per second in a mass of 1 kg, i.e. $1 \, \text{kg} \, \text{m} \, \text{s}^{-2}$. Force is the fundamental mechanical parameter governing **deformation**, but is more usually considered in terms of **stress**. [RGP]

forced oscillation A vibration that does not arise from natural resonances within a solid or quasi-solid body. [DHT]

forces acting on plates Lithospheric *plates* are driven by **forces** applied to their peripheries. These **forces** are opposed by various resistances and when there is a balance between driving and resistive **forces** a constant *plate* velocity is maintained. These **forces** are shown in Fig. F36.

The driving forces are the **ridge-push** force at an *ocean ridge* and the **slab-pull** force at a **subduction zone**, the latter deriving from the negative buoyancy of the downgoing slab with respect to the **asthenosphere**. Both these forces act on the edges of the driven *plate*. **Mantle drag** applied to the base of the *plate* may drive or resist motion depending on the relative velocities of the *plate* and the **asthenosphere**. The overriding *plate* at the **subduction zone** may be thrown into tension by the **subduction suction** or *trench suction* force, which is one of the mechanisms of **back-arc spreading**.

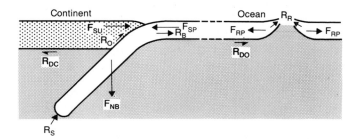

Fig. F36 Some of the **forces acting on plates**. (After Kearey & Vine, 1990, developed from Forsyth & Uyeda, 1975 by Bott, 1982.)

Resistance to motion is provided by the ridge resistance force at an *ocean ridge*, the resistance to bending of the downgoing slab, the frictional resistance of the overriding *plate* and the viscous resistance of the **asthenosphere** to the downgoing slab. Not shown in the figure are the resistances to motion caused by friction at **transform faults** and at *continent–continent collisions*.

It is now generally accepted that the **ridge-push** and **slab-pull** forces are dominant in driving *plates*, with the **mantle drag** force generally resistant to motion. The **lithosphere** is viewed as the top of the **convection** cell driving *plate* motion, and this type of **convection** termed **Orowan–Elsasser convection**. [PK]

Bott, M.H.P. (1982) *The Interior of the Earth, its Structure, Constitution and Evolution* (2nd edn). Arnold, London.
Forsyth, D.W. & Uyeda, S. (1975) On the relative importance of the driving forces of plate tectonics. *Geophysical Journal of the Royal Astronomical Society* **43**: 163–200.
Kearey, P. & Vine, F.J. (1990) *Global Tectonics*. Blackwell Scientific Publications, Oxford.

forearc basin An elongated **basin**, between an *oceanic trench* and the outer, non-volcanic part of a **subduction zone**, in which sediments can accumulate. Commonly lying to the landward side of the **accretionary prism**. [DHT]

foredune A linear coastal **dune** which is found behind, and parallel to, the **backshore** zone of **beaches**. Foredunes are essentially *stationary dunes* stabilized by vegetation, for example *Ammophila arenaria* (marram grass) and *Agropyron pungens* (couch grass). Conditions for their development are an available supply of **sand** and on-shore winds. In many countries foredune growth is promoted by **sand** fences and planting in order to provide a natural defence from marine flooding and **erosion**, a fine example being the Les Landes **dunes** in south-west France. The term has also been applied to **dunes** which develop upwind of stationary objects. [RDS]

foreland The undeformed marginal region bordering an *orogenic belt*. For example the Alpine foreland during the **Mesozoic** and *Tertiary* is the stable part of Europe west and north of the Alpine front. [RGP]

foreland basin (*foredeep basin*) Tectonic **basin** formed on the **foreland**, immediately adjacent to the margin of an *orogenic belt* and genetically related to the belt. The origin of sedimentary **basins** is discussed by Bally (1982). Foreland basins have been ascribed to **flexure of the lithosphere** caused by the increased **load** of the thickened orogenic pile. Examples of well-studied foreland basins include the Molasse trough of the Alps and the Alberta basin of the Canadian Cordilleran belt. [RGP]

Bally, A.W. (1982) Musings over sedimentary basin evolution. *Philosophical Transactions of the Royal Society, London* **A305**: 325–38.

foreshock One of a series of small-to-moderate **magnitude earthquakes** which precede some 40–50% of large, shallow **earthquakes**. They occur in the same general area as the mainshock, and often increase in number and strength over days or weeks before it. They indicate that the subsurface **stress** level is locally becoming high enough to cause rupture. These precursory events can be very valuable in **earthquake prediction**. (See **aftershock, earthquake prediction, earthquake swarm**.) [RAC]

form line Line drawn on a map to indicate the general direction of the **strike** of a **folded** surface. A tick on the line indicates the **dip**. A set of form lines will illustrate the geometry of the folding in a similar way to the outcrop pattern of the strata, for example, but form lines are more precise, and are not affected by topography. They can therefore be used in areas where individual formations have not been mapped. A set of *form line contours* can be drawn in a precise manner such that the spacing is proportional to the **dip**. A contoured map constructed by this means will illustrate the shape of a folded surface in the same way that a topographic contour map displays topographic relief (see Ragan, 1973). A *form surface* is any planar surface that intersects the ground surface as form lines and which may be used for structural mapping. [RGP]

Ragan, D.M. (1973) *Structural Geology: An Introduction to Geometrical Techniques* (2nd edn). John Wiley & Sons, New York.

formation evaluation The gathering of physical data on rocks during oil exploration. This is most commonly achieved by **well logging**. The basic log is an electric one which measures **self potential** and various types of **resistivity**. **Sonic**, *dipmeter* and *radioactivity logs* are also recorded. (See **geophysical borehole logging, logging boreholes and wells**.) [AME]

forsterite (Mg_2SiO_4) The magnesian end-member of the **olivine** *solid solution* series of minerals. (See **olivine minerals**.) [DJV]

foskorite Magnetite–olivine–apatite rock found in some carbonate-alkaline igneous complexes as at Palabora, RSA and Kovdor, Russia. These rocks may be mined for their **iron** and **phosphorus** contents. [AME]

fossil placer A term for a lithified **placer** such as the Witwatersrand **bankets**. [AME]

fossil preservation The incorporation of evidence of once-living organisms into the geologial record. It is influenced by several factors which may or may not be dependent upon each other. The most important of these are:
1 Preservation potential of the environment.
2 Preservation potential of the organism.

3 Post-mortem history of the organism up to the time of burial.
4 Post-burial history of the remains and their enclosing sediments.

Regardless of any other factors, the preservation of an organism is ultimately dependent on the preservation of the environment in which it is buried. Environments within an erosive regime are far less likely to be preserved than those in a depositional setting. Consequently aquatic environments have a greater preservation potential than terrestrial ones, with marine continental shelf environments being most frequently preserved. However, fossil examples of deep oceanic sediments become increasingly rare with age due to their loss through **subduction** of **oceanic crust**.

Assuming that the environment itself is preserved then several other factors control the preservation of the biota. The life habits of the organism will influence its preservation potential, an infaunal mode of life being more favourable than an epifaunal or **pelagic** one. Organisms which periodically shed certain structures during growth, such as exuviae of **arthropods** or deciduous leaves, further enhance their chances of representation in the fossil record.

The preservation of **trace fossils**, sedimentary structures formed by the activities of organisms, depends primarily on the sediment having a sufficient colour or texture contrast in order that such fossils may subsequently be detected by paleontologists. X-rays may be useful for detecting *bioturbation*.

The study of the post-mortem history of the organism is known as *taphonomy* or, if only considered up to the time of burial, *biostratinomy*. In normal circumstances it is strongly influenced by the structure and preservation potential of the organism. In an oxygenated environment microbial decay and the activities of scavengers destroy the soft tissues soon after death so that most fossils comprise only the hard parts of organisms, such as bones, teeth or shells, on which the agencies of wind, water or biological activity may then operate. They may be broken up or eroded, or may undergo selective sorting according to size, shape, weight or, in cases of biological agencies, species. Inorganic materials, such as **calcite** or phosphate, have a greater preservation potential than do proteinaceous materials, such as *chitin* or *cellulose*, although under certain conditions, such as in deep water, **calcite** and **silica** are unstable and pass into solution following death.

The structure and robustness of these hard parts have a strong influence on their preservation, though dependent on the nature of the environment. Multi-element skeletons in which the skeletal elements are united only by soft tissues will rapidly disarticulate under normal conditions whereas compact crystalline single-element skeletons are more resistant to post-mortem destruction even in quite high energy environments.

Under exceptional circumstances organisms may be incorporated into the fossil record with certain features

[270]

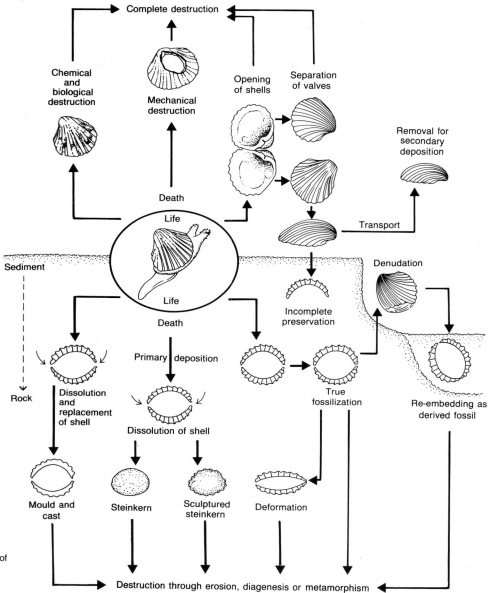

Fig. F37 Fossil preservation. The fate of skeletal material after the death of the animal. (After Müller, 1979.)

preserved which are ordinarily lost through *biostratinomic* processes. Such exceptional fossil faunas have been termed *Lagerstätten* or *conservation deposits*. (See **exceptional fossil deposits**.)

Following burial, the fossil remains may subsequently undergo considerable modification or even complete destruction (Fig. F37). They may retain their original composition and form, particularly if they have a compact structure composed of stable material, but more frequently they undergo some degree of modification in composition or structure. **Compaction** may cause the fossil remains to be compressed to a degree dependent on the nature of the sediment and the structure of the remains. The original material may be replaced by different minerals, such as **calcite**, **silica** or iron **pyrite**, or different crystal forms,

such as in the conversion of **aragonite** to **calcite**. If an unfilled shell is subsequently removed by **dissolution** then an external mould is left which, if subsequently filled by material, forms an external cast. If the shell was originally sediment-filled then **dissolution** of the shell produces internal and external moulds, subsequent infilling of the cavity between them forming an internal and external cast. The sediment infilling is known as a *steinkern* which, if the ornament of the external mould is impressed upon its surface following shell **dissolution**, forms a sculptured *steinkern*. Destruction may occur through the **erosion** of a rock unit and its contained fossils, through thermal metamorphism or by **diagenetic** processes, such as the selective **dissolution** of **aragonitic** shells in certain sediments or the destruction of all **calcitic** remains by **dolomitization**.

Fossils surviving the **erosion** of their enclosing rock may be incorporated into later sediments as derived or *remanié fossils*. [MJS]

Babin, C. (1980) *Elements of Palaeontology*. John Wiley & Sons, Chichester.

Brett, C.E. & Baird, G.C. (1986) Comparative taphonomy: a key to paleoenvironmental interpretation based on fossil preservation. *Palaios* **1**: 207–27.

Müller, A.H. (1979) Fossilization (Taphonomy). In: Robison, R.A. & Teichert, C. (eds) *Treatise on Invertebrate Paleontology, Part A* (*Introduction*), A2–A78. Geological Society of America & University of Kansas, Boulder, CO & Lawrence, KS.

Seilacher, A., Reif, W.E. & Westphal, F. (1985) Sedimentological, ecological and temporal patterns of Lagerstätten. *Philosophcal Transactions of the Royal Society, London* **B311**, 5–24.

fossil record and evolution Darwin sought confirmation of his theory of evolution in the fossil record, expecting to find in this storehouse of ancestors proof of the transmutation of species by the existence of graded series of

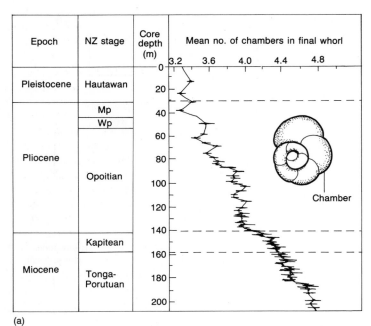

(a)

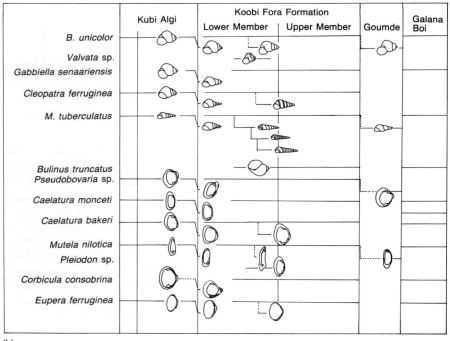

(b)

Fig. F38 Fossil record and evolution. (a) Gradual transformation ('phyletic gradualism') in a planktonic foramineral lineage from a deep-sea core. The horizontal bars are 95% confidence limits. (After Malmgren & Kennett, 1981.) (b) 'Punctuated equilibria' in lacustrine molluscs from E. Africa. (After Williamson, 1981.)

intermediates linking old and new forms. At the time of *The Origin* (1859) no such series were known and the fossil record seemed to present additional difficulties for the theory through the sudden appearance of diverse, complex organisms at the base of the **Cambrian** system. Darwin could only suggest in explanation that the fossil record provided a woefully incomplete sample of past life.

The fossil record *is* incomplete but there are enormous disparities in the quality of the record between groups and attempts to disprove evolution through an absence of intermediate forms ('missing links') have often been based on the most unsatisfactory material. Large terrestrial *vertebrates* leave very few fossils because of small population sizes and low potential for permanent entombment in sediment in an environment where **erosion** generally outweighs deposition. Thus the fossil record of *man*, for example, cannot be expected to provide convincing evidence of transformation. By contrast, the mineralized hardparts of certain marine invertebrates (including *bivalves, gastropods,* **brachiopods**, *echinoids*) stand a quite good chance of escaping physical and chemical destruction and of being preserved in sediment as fossils. It is not surprising that the most convincing evidence of gradual transformation comes from mineralized marine micro-organisms (Fig. F38) which have vast populations

and thus yield numerous fossils. The adaptive significance of such evolution is unclear — we know too little of the biology of the organisms involved — but for rather less complete sequences of forms in other groups (e.g. oysters, sea urchins and, surprisingly, horses) it is possible to advance adaptive explanations which bear out the action of natural selection.

The fossil record of the appearance of more advanced grades of organization (as recognized by biologists) is also consistent with their origin by evolution. Thus **mammals (Triassic)** do not appear until the **reptilian** grade (**Carboniferous**) has been emplaced, and this in turn follows the appearance of the **amphibian (Devonian)** and **fish (Ordovician)** grades (Fig. F39). The same is true of the now quite well-known *Precambrian* history of life. Thus complex multicellular organisms with nucleate, eukaryotic cells (*metazoans, metaphytes*) appear after the origin of single-celled *eukaryotes* (c. 1.4 Ga); these in turn succeed anucleate *prokaryotes* (over 3 Ga), some of whose varied forms have been widely regarded by biologists as the precursors of eukaryotic cell organelles. There are inconsistencies in this evidence: **sponges**, representing a tissue level of organization intermediate between the cellular level of early *eukaryotes* and the organ level of *metazoans*, are present in the **Cambrian** but are now

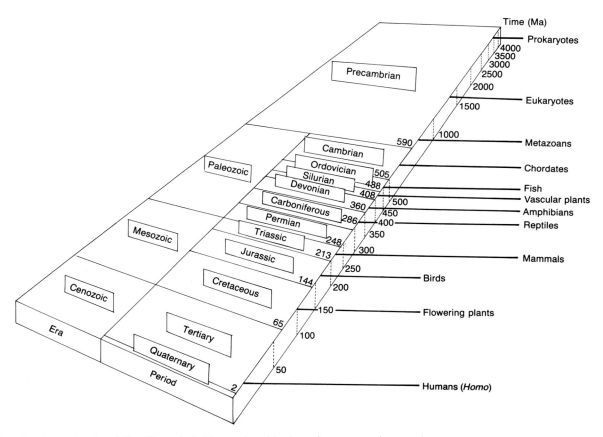

Fig. F39 Fossil record and evolution. The geological time-scale and the times of appearance of some major groups.

known to be preceded by more complex animals in the 'Ediacaran' fauna (c. 670 Ma). Such anomalies may perhaps be due to an incomplete record or they may reflect incorrect inferences about the likely evolutionary pathway to higher organisms.

The *Ediacaran fauna*, named after the place of its first discovery (in southern Australia) in 1947, is known from a number of cases of unusual soft-body preservation. In its low diversity assembly of fairly simple organisms (apparently mainly jellyfish with some worms) it helps to illuminate Darwin's 'abominable mystery' of the appearance of diverse *metazoans* at the base of the **Cambrian** (*c.* 590 Ma). However, the relatively short interval before the latter implies that evolution must nevertheless have proceeded at a rapid rate. *Precambrian*/**Cambrian** metazoan diversification has recently been resolved as an approximately logistic expansion. Part of this apparent increase in diversity may be an artefact of the acquisition of mineralized hard parts — increasing preservation potential — by organisms of much greater antiquity. However, the progressive increase in the complexity of hard parts themselves, the diversification of **trace fossils** (marks in the sediment left by the activities of both hard and soft-bodied organisms) and the very high diversity of soft-bodied organisms in the fauna of the *Burgess Shale* (British Columbia), a Middle **Cambrian** example of exceptional preservation, all point to a genuine increase in diversity over the *Precambrian–Cambrian boundary*. (See **fossil preservation, exceptional fossil deposits**.)

The *Burgess Shale* fauna is of especial significance for its inclusion of the most advanced phylum (*Chordata*) and also of a number of bizarre organisms which are impossible to classify in existing phyla. These indicate the extent to which evolution can proceed in an ecological vacuum resulting from a lack of competitors. The same pattern of 'over'-diversification (followed by elimination of seemingly unsuccessful morphologies) is also evident in evolution *within* phyla (e.g. **echinoderms**, cephalopod **molluscs**) during the early **Paleozoic**. Later episodes of diversification have generally occurred at lower taxonomic levels (no new phyla with fossil representatives appear after the early **Ordovician**) but rapid expansion, generally associated with the invasion of a new habitat or the **extinction** of a competing group, is also characteristic. Rapid diversification of higher plants and vertebrates in the almost virgin terrestrial realm was facilitated by the key adaptations of, respectively, vascular transport of solutes (late **Silurian**) and mechanisms of air-breathing (**Devonian**). By contrast, the early *Tertiary* radiation of **mammals** attended the decline of the ruling **reptiles** of the **Mesozoic**, including the *dinosaurs*. The failure of **mammals** to expand during their long **Mesozoic** history may evince a principle whereby prior occupants of an ecological niche may not be displaced by competitive superiors until they suffer some externally-mediated reduction in diversity. In the case of the **reptiles** this was the *mass extinction* event at the close of the **Mesozoic** which severely affected both terrestrial

and marine biotas. In the case of bivalve **molluscs**, which did not replace the supposedly inferior **brachiopods** as the dominant element in marine shelly assemblages until the **Mesozoic**, the *mass extinction* event at the end of the **Paleozoic** seems to have been critical in resetting the balance of diversity.

All the above are uniquely paleontological insights into the nature of large-scale evolution. In recent years fossil evidence has also been adduced in support of novel theories of evolution at the species level. It has been noted that even in well-preserved groups such as *bivalves* and **brachiopods** there is rather little evidence of gradual transformation. This led Eldredge & Gould to suggest in 1972 that the norm for most species is morphological stasis and that evolutionary changes (when they occur) take place at a rate too fast for stratigraphic resolution. This latter may imply nothing exceptional, given the enormous intervals of time which may be represented by the shortest of sedimentary sequences. However stasis, if it is real, may imply the existence of constraints on morphological evolution whose importance, if not probable occurrence, had been largely unsuspected by neontologists. An example illustrating the Eldredge–Gould model of evolution ('*punctuated equilibria*', as opposed to '*phyletic gradualism*') is provided by Williamson's (1981) study of **Cenozoic** lacustrine **molluscs** in East Africa (Fig. F39). In this case morphological stasis occupied several millions of years while evolution occurred within 5–50 000 years. [ALAJ]

Darwin, C. (1859) *On the Origin of Species by Means of Natural Selection, or the Preservation of Favoured Races in the Struggle for Life*. John Murray, London.
Eldredge, N. & Gould, S.J. (1972) Punctuated equilibria: an alternative to phyletic gradualism. In: Schopf, T.J.M. (ed.) *Models in Paleobiology*. W.H. Freeman, San Francisco.
Glaessner, M.F. (1984) *The Dawn of Animal Life. A Biohistorical Study*. Cambridge University Press, Cambridge.
House, M.R. (ed.) (1979) *The Origin of Major Invertebrate Groups*. Academic Press, London.
Malmgren, B.A. & Kennett, J.P. (1981) Phyletic gradualism in a Late Cenozoic planktonic foraminiferal lineage; DSDP Site 284, southwest Pacific. *Paleobiology* **7**: 230–40.
Raup, D.M. & Stanley, S.M. (1978) *Principles of Paleontology* (2nd edn). W.H. Freeman, San Francisco.
Stanley, S.M. (1979) *Macroevolution, Pattern and Process*. W.H. Freeman, San Francisco.
Whittington, H.B. (1980) The significance of the fauna of the Burgess Shale, Middle Cambrian, British Columbia. *Proceedings of the Geologists' Association* **91**: 127–48.
Williamson, P.G. (1981) Paleontological documentation of speciation in Cenozoic molluscs from Turkana Basin. *Nature* **293**: 437–43.

foundry sand Sand suitable for forming moulds into which molten metal may be poured to produce metal castings. Such **sands** should be refractory to withstand melting; cohesive to take and hold their shape, so some clay content is necessary; porous to allow gases to escape and low in **iron**. [AME]

fowlerite A **zinc**-rich variety of the mineral **rhodonite**. [DJV]

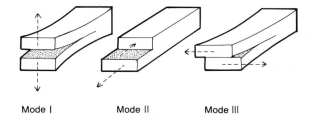

Fig. F40 Modes of **fracture**.

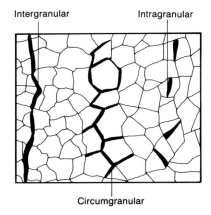

Fig. F41 Fracture. Inter-, circum-, and intragranular cracks, all showing mode I displacements.

fracture (1) In structural geology, a discontinuity across which there has been separation. Fractures thus include all cracks, **joints** and **faults**, but exclude **stylolites**. Fractures are conveniently classified into three modes by the direction of the displacement vector between the two sides of the fracture relative to the surface and perimeter of the fracture (Fig. F40). Mode I is also described as **tensile** or opening mode; Mode II as in-plane **shear** or sliding mode, and Mode III as anti-plane **shear** or tearing mode.

Fracture modes may be combined on a single fracture to define a *mixed mode* or *hybrid fracture*. Fractures in granular materials may be classified into *intergranular* (between grains), *circumgranular* (or *grain boundary*) (around grains) and *intragranular* (within grains) (Fig. F41). Fracture propagation rates are described as dynamic if they are fast enough to involve significant kinetic energy terms; dynamic fractures accelerate towards a terminal velocity at some value below the speed of sound in the material, resulting in a seismic phenomenon. Slower rates of fracture propagation are described as subcritical or 'slow'. **Creep** (aseismic) **deformation** of many materials and of the **crust** can be understood by subcritical crack propagation.

The study of fracture surfaces (*fractography*) can reveal much about fracture propagation. Figure F42 illustrates some of the more common features observed on rock fracture surfaces. The curving surface traces are known as *plumes*, and have been subdivided into S, C and rythmic C types. Fractures are initiated in all cases at the con-

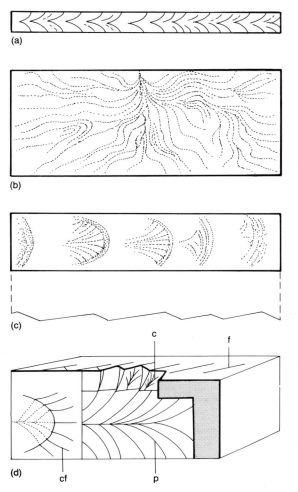

Fig. F42 Fracture surface features. (a) S plumes. (b) C plumes. (c) Rhythmic C plumes. (d) Rib structure: c, cross joints; f, fringe joints; p, plumes; cf, conchoidal fringe joints. (After Price, 1966; Bahat & Engelder, 1984 and Ramsay, 1986.)

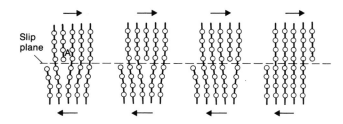

Fig. F43 Fracture. Glide. Movement of a dislocation (A) through a crystal lattice along a slip plane produces a displacement of one part of the lattice with respect to another.

vergence of the *plume* structure, and propagate away with the fracture surface approximately normal to the *plume*. In cross-section, *plumes* may have a saw-tooth profile called *rib-structure* (Fig. F43). This may be divided into subsidiary fractures at the edge of the main fracture surface, known as *en echelon fractures* (also known as *fringe-* or *f-joints*, *feather fractures*, and *pinnate fractures* (see below)) (Fig. F43). Shorter fractures cross-linking the *en echelon fractures* are known as *cross joints* or *c-joints*, and a final class of minor fracture consists of *conchoidal fringe joints*, formed radially in advance of convex rib structure (Fig. F43c).

A *hydraulic fracture* is a fracture caused by **pore-fluid pressure**. **Pore-fluid pressure** causes **tensile stresses** which may be sufficiently large to exceed the *compressive stress* plus **tensile strength** of the rock. Artificial *hydraulic fracture* (the *hydrofrac technique*) is induced in boreholes by sealing off part of the borehole, which is pumped with fluid until fracture. The fracture will occur in the plane normal to the least *principal stress*; this may be vertical, leading to a horizontal *hydraulic fracture*, or horizontal, giving a vertical fracture, which can be used to determine both the orientation and magnitude of the minimum horizontal **stress** if the **tensile strength** of the rock is known. The main use of the technique, however, is to create access to *hydrocarbon* **reservoirs**, or more recently, to **geothermal** heat supplies by increasing **permeability**. *Hydraulic fracture* has been inferred as a mechanism for the formation of natural *extension fractures*. High **pore-fluid pressures** may be created naturally by **overpressuring** due to compaction of sediments, an effect known from drill holes in sedimentary **basins** in several parts of the world.

A *pinnate fracture* is a minor fracture intersecting a larger fracture at an acute angle. *Pinnate fractures* often occur in a dense group of many short fractures, which intersect the larger fracture at 45° and are often found only on one side of it. The direction of movement of the *fault block* containing the fractures may be given by the direction of closure of the fractures on the main fracture plane. They may form due to **tensile stresses** created by frictional sliding on the main fracture. *Pinnate fractures* are also known as *feather fractures*.

Glide, translation gliding, is the movement of a **dislocation** through a *crystal lattice* along a plane (the *slip plane*). One part of the *crystal lattice* can slip with respect to another by the motion of a **dislocation** through the lattice, so that the lattice on either side of the *slip plane* remains undistorted, resulting in *intragranular glide*. The **slip** produced by the propagation of a single **dislocation** on the lattice is shown in Fig. F43. The *slip plane* and direction are defined by their *slip system*. This is the simplest way in which a crystal can deform by intracrystalline **plasticity**. *Glide* is promoted by **stress** and temperature.

(2) **Fracture** in minerals refers to the breaking of minerals that does not follow particular crystallographic directions (unlike **cleavage** or **parting**). It may, nevertheless, be distinctive and an aid to identification. Examples of fracture include:

(a) *Conchoidal*, the type of fracture observed in **quartz** or in the glass used to make bottles, whereby smooth curving surfaces resembling the interior surface of a shell result;

(b) *Hackly*, whereby the fracture surfaces are jagged with sharp edges;

(c) *Fibrous*

(d) *Splintery*

(e) *Uneven* or *irregular*, in which the surfaces produced are rough and irregular. [TGB/DJV]

Atkinson, B.K. (ed.) (1987) *Fracture Mechanics of Rock*. Academic Press, New York.

Bahat, D. & Engelder, T. (1984) Surface morphology on joints of the Appalachian Plateau, New York and Pennsylvania. *Tectonophysics* **104**: 299–313.

Hancock, P.L. & Barka, A.A. (1987) Kinematic indicators on active normal faults. *Journal of Structural Geology* **9**: 573–84.

Hull, D. (1975) *An Introduction to Dislocations*. Pergamon, Oxford.

Jaeger, J.C. & Cook, N.G.W. (1979) *Fundamentals of Rock Mechanics* (3rd edn), pp. 226–7. Chapman & Hall, London.

Lawn, B.R. & Wilshaw, T.R. (1975) *Fracture of Brittle Solids*. Cambridge University Press, Cambridge.

Nicolas, A. & Poiries, J.P. (1976) *Crystalline Plasticity and Solid State Flow in Metamorphic Rocks*. Wiley Interscience, London.

Price, N.J. (1966) *Fault and Joint Development in Brittle and Semi-Brittle Rock*. Pergammon, London.

Ramsay, J.G. (1986) *The Surface Fractures of Joints in the Jura Mountains, Switzerland*. Abstract, Tectonic Studies Group A.G.M., University of Hull.

fracture porosity **Porosity** developed as a result of the **fracturing** of rocks. **Fracturing** renders rocks more porous and usually more permeable to gas, oil and water. A number of **reservoir** rocks are only in this category by virtue of **fracturing**. Given a favourable association with source sediments and a **cap rock**, almost any rock type except **evaporites** can become a fractured-rock **reservoir**. The best examples are, however, fractured **limestones** and **dolomites**. The most famous example of these is the Asmari Limestone of Iran. [AME]

fracture zone A zone of past or present *transcurrent* or **transform** motion, usually within the **oceanic crust** and forming a marked topographic break extending for several hundreds or thousands of kilometres along which matching linear *oceanic magnetic anomalies* are offset, and also the present-day spreading *ocean ridge*. Only very weakly seismic along most of its length, mainly due to differential rates of cooling on either side of it, but very strongly active between the offset *oceanic ridge* crests. The fracture zones are also zones in which the **oceanic crust** is usually more strongly *serpentinized* due to *hydrothermal* circulation through it. [DHT]

fragipan An acidic, cemented horizon between the base of the *soil* zone and the underlying parent material or bedrock. Fragipans are normally compact and **brittle**, and the binding matter includes clays, **silica**, **iron**, **aluminium** and organic matter. Many are the result of *periglacial* processes, and they are widespread in Europe and North America. [ASG]

Grossmann, R.B. & Carlisle, F.J. (1969) Fragipan soils of the eastern United States. *Advances in Agronomy* **21**: 237–79.

franklinite ($(Zn,Fe,Mn)(Fe,Mn)_2O_4$) A rare **zinc**-rich **spinel** mineral found at Franklin, New Jersey, USA. (See **oxide minerals**.) [DJV]

Frasch process A method of extracting underground **sulphur**. A well of about 250 mm diameter is driven through overlying rocks to the **sulphur** deposit. Inside the metal casing is a 200 mm hot water pipe and within it a return **sulphur** pipe, itself containing an air pipe. Water heated to about 150°C is pumped into the **sulphur** to melt it and the molten **sulphur**–water mixture is blown up the return **sulphur** pipe with hot compressed air. [AME]

free air anomaly A **gravity** measurement to which has been applied a **latitude correction** and **free air correction**. Free air anomalies are frequently used when interpreting marine gravity data from deep water, as the **Bouguer anomalies** from such areas are unrealistically large. Free air anomalies give some idea of the degree of **isostatic** compensation of a region. [PK]

free air correction (FAC) Correction to a **gravity** measurement for the decrease of **gravity** with height in free air because of the increased distance from the centre of mass of the Earth.

$$FAC = 3.086\,g.u.\,m^{-1}$$

(See **reduction**.) [PK]

free face Wood's (1942) discussion of slope development began with consideration of the retreat of a rock cliff, or free face. **Weathering** of the cliff may occur by a variety of processes: slab **failure**, rock *avalanches*, *rock falls* and granular disintegration. **Talus** accumulates at the foot of the free face, and the upper part is gradually attacked by creep and *rainsplash* (see **raindrop impact erosion**). As the slope evolves through time, the free face is gradually consumed by enlargement of these two slope elements. [TPB]

Wood, A. (1942) The development of hillside slopes. *Proceedings of the Geologists' Association* **53**: 128–40.

free oscillations The Earth, as any other body, if struck will resonate most freely at particular harmonics. Generally an energy impulse equivalent to an **earthquake** of **magnitude** of ≥7.5 is required before such resonances can be detected unambiguously. Both torsional and spheroidal oscillations occur, i.e. motions perpendicular to the Earth's radius and both radial and tangential occur, the analysis of which provides critical information on the physical properties of its interior. [DHT]

Smith, S.W. (1966) Free oscillations excited by the Alaskan earthquake. *Journal of Geophysical Research* **71**: 1187–93.

free-milling gold Gold in an **orebody** that occurs in the native form and is easily amalgamated or cyanided for recovery purposes. Where it occurs in minerals that are difficult to treat it is called *refractory gold*. [AME]

freestone Any fine-grained stone, especially **limestone** or **sandstone**, that can be cut and worked in any direction without **fracturing**. [AME]

freibergite A **silver**-bearing **ore** mineral of the **tetrahedrite** group. [DJV]

French chalk A compact variety of **talc** used to mark cloth or remove grease stains from materials. In a ground form it may be used for absorbing tacky substances. [AME]

frost weathering The disintegration of rock by frost. The process produces angular fragments, which may form such features as **screes** and *blockfields*, promotes **tor** formation by acting along **joints** on exposed faces of susceptible rocks, and comminutes rock into material that can be silt-sized. To operate effectively the rock must contain moisture and attain sub-zero temperatures, but there is some dispute as to whether frequent cycles of small magnitude and duration are more effective than longer cycles of deeper intensity. There is also some controversy as to the precise mechanism involved in frost weathering. Although the traditional explanation is that the 9% volumetric expansion caused when water changes into ice sets up **stresses** that break rock, it is likely that an alternative or additional mechanism involves volume expansion caused by water migration and ice segregation processes within rock pores. [ASG]

Lautridou, J.P. & Ozouf, J.C. (1982) Experimental frost shattering. *Progress in Physical Geography* **6**: 215–32.
McGreevy, J.P. (1981) Some perspectives on frost shattering. *Progress in Physical Geography* **5**: 56–75.

Froude number The influence of **gravity** upon a flow is given by the ratio of inertial to **gravity** forces as expressed by the dimensionless Froude number, *Fr*, where

$$Fr = U/\sqrt{gL}$$

in which U is the mean flow velocity, g is *acceleration due to gravity* and L is a length term, usually taken as mean flow depth in open channel flows. The denominator is equivalent to the *celerity* (phase velocity) of a shallow water **gravity** wave $C = (\sqrt{gL})$. When the Froude number is less than unity and the influence of **gravity** is more pronounced, the flow is termed subcritical. When above unity the flow is termed supercritical and inertial forces predominate. At $Fr = 1$, water surface **waves** should appear to be standing still (**standing waves**), since $C = U$.

An important modification of the Froude number,

the *densimetric Froude number* Fr_d, expresses the velocity of gravity waves in a system with fluids of differing densities, such as in **turbidity currents**, and is given by addition of terms for the density difference between two layers, $\Delta\rho$, and the mean density, ρ, in the equation

$$Fr_d = U/\sqrt{Lg\Delta\rho/\rho}$$

Large values of this number indicate the greater mixing of momentum and density differences between the two flows. [JLB]

Chow, V.T. (1959) *Open-Channel Hydraulics*. McGraw-Hill, London.
Francis, J.R.D. (1975) *Fluid Mechanics for Engineering Students*. Arnold, London.

fuchsite A **chrome**-rich variety of the mineral **muscovite**. (See **mica minerals**.) [DJV]

fulje Deep parabolic depressions between closely interlocking crescentic **dunes**. One of the earliest detailed descriptions is given by the English explorer Wilfred Blunt following an expedition across the Nafud, northern Saudi Arabia, in 1879. The term is used in Australia to describe small **parabolic dunes**. [RDS]

Blunt, A. (1985) *A Pilgrimage to the Nejd*. Century Publishing, London.

Fuller's earth A natural absorbent clay used for decolorizing oils and fats, fulling cloth, etc. Mineralogically it is calcium **montmorillonite**, the **clay mineral** of non-swelling **bentonite**. In US parlance Fuller's earth refers to **attapulgite** and **sepiolite**, both of which can be similarly utilized. [AME]

fumarole A **volcanic vent**, apart from active **volcanic eruption** vents, from which **volcanic gases** escape. Fumaroles commonly occur as clusters in volcanically active areas, but also occur where there was no eruption in the immediate past. Fumaroles may persist for thousands of years after the last eruption at many volcanoes. Fumaroles emit a complex mixture of gases, the most common constituent usually being steam, with CO_2, CO, sulphur gases, N_2, HF, HCl, H_2 and NH_3 also present. The composition of fumarolic gases can vary widely, even within individual volcanic areas. The temperature of fumaroles is greater than about 100°C. At lower temperatures, the water-dominated nature of the mixture will usually ensure emission as a *hot spring*. Precise analysis of fumarolic gases is very difficult, because of practical problems encountered in the collection of samples not contaminated by air. [PTL]

Macdonald, G.A. (1972) *Volcanoes*. Prentice-Hall, Englewood Cliffs, NJ.

G

gabbro A coarse-grained **igneous rock** of basic chemical composition composed essentially of **plagioclase** which contains more than 50% of the **anorthite** component, **pyroxene** (*clinopyroxene > orthopyroxene*) and accessory **olivine**, **quartz** and **nepheline**. Gabbros form the lower part ('Layer 3') of the **oceanic crust**, form major intrusions within **continental crust** (e.g. the Skaergaard, Stillwater and **Bushveldt** intrusions), and form intrusions within **island arcs**. [RST]

gadolinite ($YFeBe_2SiO_4)_2O_2$) A **rare earth element ore** mineral. [DJV]

gahnite ($ZnAl_2O_4$) A **zinc aluminium spinel** mineral. (See **oxide minerals**.) [DJV]

gaining stream A river that receives water from an **aquifer**. This occurs where the bottom of a river valley lies below the **water table**. Discharge of water may occur through the bank or bed of the river and therefore be unseen. Nevertheless these discharges are responsible for the largest proportion of flow out of **aquifers**. [AME]

Gal The cgs unit of acceleration in $cm\,s^{-2}$. Named after Galileo. The average value of **gravity** on the Earth's surface is 980 Gal. [PK]

galaxite ($MnAl_2O_4$) A **manganese–aluminium spinel** mineral. (See **oxide minerals**.) [DJV]

galena (PbS) The major **ore** mineral of **lead**. [DJV]

gallium (Ga) At. no. 31, at. wt. 69.72, m.p. 29.78°C, d. 5.907. Metallic in appearance, it is used in doping semiconductors, light-emitting diodes, electrical devices and phosphors. It is produced as a by-product from some lead–zinc **ores**. [AME]

gamma (γ) The cgs subunit of **magnetic field** strength equal to 10^{-5} **gauss**. [PK]

gangue The unwanted material, minerals or rock with which **ore** minerals are usually intergrown. Mines commonly possess mineral processing plants in which the raw ore is milled before the separation of the **ore** minerals from the gangue. The plants provide **ore** concentrates, and tailings which are made up of the gangue. [AME]

ganister A **sandstone** or siltstone underlying a **coal seam**. These rocks have usually been highly leached by plant roots and can be used for making refractory **silica** bricks for lining furnaces. [AME]

garnet minerals Minerals particularly characteristic of metamorphic rocks but also found in some **igneous rocks** and as detrital grains in sediments. They crystallize in the cubic system and have a strong power of crystallization, tending to occur as 12-sided rhombic dodecahedra or 24-sided icositetrahedra, or a combination of both. The group has a general formula $A_3^{2+}B_2^{3+}Si_3O_{12}$ where A may be **magnesium**, **iron**, **manganese** or calcium and B is **aluminium**, **iron** or more rarely **chromium**.

Almandine, $Fe_3^{2+}Al_2Si_3O_{12}$, is the most common species in the group, though it generally contains appreciable amounts of both **magnesium** and **manganese** partially replacing Fe^{2+}. The distribution of **magnesium** and **iron** between coexisting garnet and **biotite** or *clinopyroxene* may be of use in **geothermometry**. Experimental work has indicated that at, for example, 200 GPa fluid pressure the stability field for **almandine** ranges from 540 to 900°C. In hand specimen, **almandines** are dark red to black, but they are colourless to pale pinkish red in thin section and are isotropic. **Almandine** is the typical garnet of the garnet–mica **schists** resulting from the regional metamorphism of argillaceous sediments and as such it is used as a zonal mineral in regions of progressive metamorphism of these rocks. It also occurs in higher grade rocks of the **granulite** facies and as a product of thermal metamorphism, e.g. in garnet–**cordierite hornfels**. It is only occasionally found in **igneous rocks**.

Pyrope, $Mg_3Al_2Si_3O_{12}$, generally occurs with appreciable amounts of other garnet molecules (chiefly **almandine** and subsidiary **grossular**) in *solid solution*. The typical **pyrope** of high-grade metamorphic rocks contains around 40–70% of this molecule. **Pyrope** garnets with Cr_2O_3 up to 8% are common, many such garnets having a characteristic greenish violet or purple hue and occurring in association with **diamond**-bearing **kimberlite**. Direct synthesis of euhedral **pyrope** in the presence of water is achieved at near 2500 GPa and 1000°C. **Pyrope** occurs in certain ultrabasic rocks such as mica **peridotites** or **kimberlites** and in associated **serpentinites**. The well-known Bohemian garnets of **gem** quality are **pyrope** and occur in the debris of a **basaltic** breccia derived from a **peridotite**. In *eclogites*, in which garnet is an essential constituent, the garnet is dominantly **pyrope**.

Spessartine, $Mn_3Al_2Si_3O_{12}$, generally occurs in *solid solution* with **almandine** $\pm$ **grossular**. It is typically orange-red in colour, but ranges to darker red when appreciable **almandine** substitutes. It occurs in some **skarn deposits**, in **manganese**-rich assemblages of **metasomatic** origin, and in some **granite** *pegmatites* and **aplites**.

Grossular, $Ca_3Al_2Si_3O_{12}$, may occur in virtually pure end-member composition but it forms a continuous series with **andradite**. It is especially characteristic of thermally or regionally metamorphosed impure calcareous rocks.

Andradite, $Ca_3Fe_2^{3+}Si_3O_{12}$, frequently occurs with more than 90% of this molecule, but more commonly falls in the **andradite–grossular** series; this ranges in colour from yellowish to dark brown. Many **andradites** in **skarn deposits** show compositional (and colour) zoning. **Andradites** of primary origin in **alkaline igneous rocks** may contain appreciable amounts of **titanium** and are black in hand specimen (**melanite**). Like **grossular**, **andradite** is typical of the metamorphism of impure calcareous rocks but is particularly to be found in **metasomatic skarn deposits** involving the addition of Fe_2O_3 ($\pm SiO_2$).

Uvarovite, $Ca_3Cr_2Si_3O_{12}$ is the rarest of the six common garnet species and is typically dark green to emerald green. It is found in **serpentinite** and in association with **skarn orebodies** in which the **chromium** is introduced **metasomatically** from basic or ultrabasic **igneous rocks**. [RAH]

Deer, W.A., Howie, R.A. & Zussman, J. (1966) *An Introduction to the Rock-forming Minerals.* Longman, London.

garnierite $((Ni,Mg)_3Si_2O_5(OH)_4)$ A green **nickel**-rich **serpentine group** mineral exploited as an **ore** in nickeliferous **laterites**. (See **clay minerals**.) [DJV]

gas cap A volume of gas lying above an oil accumulation within an oil **reservoir**. When the gas content of **crude oil** in a **reservoir** is sufficient to saturate the oil under the existing **reservoir** pressure, then any excess gas will form a free gas cap above the oil. In a **reservoir** where there was originally no gas cap and **depletion drive** is the mechanism of removal of the oil then, when the **reservoir** pressure falls to the saturation pressure, free gas will come out of solution to form a secondary gas cap. [AME]

gas cap drive When a free **gas cap** exists in an oil **reservoir**, the cap will expand as oil is withdrawn from the **reservoir** exerting a downward pressure on the oil driving it in the direction of the pressure gradient, i.e. towards the **well** or **wells**. A gas cap drive provides an effective recovery mechanism and production does not decline so rapidly as when **depletion drive** alone is employed. If **permeability**

is good, high rates of oil production can be maintained for many years giving oil recoveries of 30–75%. Gas–oil ratios increase with time and eventually most of the production is gas.

If during production oil is stripped of its gas at the wellhead and this is injected under pressure into the top of the **reservoir**, then an artificial **gas cap** is created which may produce greatly increased recoveries. This is a very common practice in **oilfields**. [AME]

North, F.K. (1985) *Petroleum Geology.* Allen & Unwin, Boston.

gas pool A discrete accumulation of natural gas in a single **reservoir** with a single trap. Several pools may occur side by side or overlap laterally to constitute a field. [AME]

gas trap A geometrical arrangement of strata that can accumulate gas (or oil) beneath it. The most common traps are **domes** and **anticlines**. (See **oil trap**.) [AME]

gas–oil ratio (GOR) The volume of gas in a **reservoir** in relation to the volume of oil. GORs in North America are expressed as cubic feet of gas per barrel *as they exist in the* **reservoir**. A *barrel* contains less than $6\,ft^3$ of oil so figures of 1000 or greater for the GORs of pools may seem at first sight to be in error. The explanation lies in the fact that gas is highly compressible and large volumes can be held in solution in oil pools in deep **reservoirs** that are under high **confining pressures**. [AME]

gauss (G) The c.g.s. unit of **magnetic field** strength. [PK]

Gauss' theorem The outward flux of the force of attraction over any closed surface in a gravitational field is equal to 4π times the mass enclosed by the surface. It can be shown that Gauss' theorem gives the result that for a local **gravity anomaly** digitized on a two-dimensional grid in which Δg_i represents the average **gravity anomaly** in a grid square of area Δa_i, the **excess mass** (M) is given by:

$$M = \Sigma \, \Delta g_i \Delta a_i / 2\pi G$$

where G is the **gravitational constant**. [PK]

Ramsey, A.S. (1964) *An Introduction to the Theory of Newtonian Attraction.* Cambridge University Press, Cambridge.

gaylussite $(Na_2Ca(CO_3)_2 . 5H_2O)$ An **evaporite** mineral deposited from saline lakes. [DJV]

gedrite $(Na_{0.5}(Mg,Fe)_2(Mg,Fe)_{3.5}(Al,Fe^{3+})_{1.5}Si_6Al_2O_{22}(OH)_2)$ An **amphibole mineral**. [DJV]

geikielite ($MgTiO_3$) A rare **titaniferous** oxide mineral, isostructural with **ilmenite**. (See **oxide minerals**.) [DJV]

gemstone/gem A mineral or organic substance which has an intrinsic value based on its beauty, durability or rarity. Many gems possess all these attributes. Most gemstones are minerals formed in the **crust** or, more rarely, in the **upper mantle**; **jadeite**, **nephrite** and **lapis lazuli** are rocks composed of one or more minerals. Gems of organic origin include **amber** and **jet** (fossilized resin and wood respectively), *pearl*, **coral** and *ivory*. About one hundred minerals have been classed as gemstones.

Rarity may take several forms. Some gem minerals are rare in occurrence, either containing the less abundant chemical elements, such as **beryl** (Be) and **topaz** (B), or requiring unusual formation conditions, for example **diamond**, which is **carbon** crystallized at extremely high temperatures and pressures. Many gems are unusual forms of common minerals, possessing an exceptional colour or optical effect, such as **amethystine quartz** or *iridescent* **labradorite** feldspar. Exceptional size or relative flawlessness also influence desirability and value.

Durability is a function of a gemstone's ability to resist chemical attack, **abrasion** and breakage, and its stability at surface temperatures and pressures. Most gem minerals are **corrosion**-resistant, stable silicates (**beryl**, **topaz**, **tourmaline**, **quartz**, **feldspars**, etc.) or oxides (**corundum**, **spinel**, **chrysoberyl**). Some porous gems are less resistant, notably **opal**, which may crack due to dehydration, and **turquoise**, which can become discoloured by absorption of oils, etc. *Pearl*, shell and much **coral** are composed of carbonates (**calcite** or **aragonite**) and react with dilute acids; **amber** melts at 250–300°C and **jet** will burn like **coal**.

Ideally gems should have a hardness of 7 or more on Mohs' scale, sufficient to resist **abrasion** by common dust and grit, most of which is **quartz** (7). Organic gems are relatively fragile, having hardnesses of 4 or less. Gems must also be tough to resist breakage during fashioning and wear. Despite their hardness some gems, notably **zircon** (7.5) and **emerald** (7.5) are **brittle** and chip easily. Other gems cleave readily in one or more crystal directions, for example, **diamond** (10) with 4 octahedral **cleavages** and **topaz** (8) with a basal **cleavage**. The toughest gems are **nephrite**, **jadeite** and **agate** which, although they all have hardnesses of 7 or less, derive great strength from their polycrystalline structure of microscopic interlocking fibres or grains.

Beauty in gemstones derives from their interaction with light. Colour is the most important gem characteristic and arises in several ways. The 'body colour' seen in most gemstones is due to absorption of certain wavelengths of white light within the gem, the unabsorbed wavelengths being transmitted as the body colour. The causes of absorption are complex, generally involving the presence of certain transition elements and damage or irregularities in the *crystal lattice*. The most important colouring elements are **chromium**, **iron**, **manganese**, **titanium**, **copper** and **vanadium**. In a few gems they are present as essential elements in the mineral's chemical composition and typically give rise to a limited range of stable colours as in **peridot** (**iron**) and **turquoise** (**copper**). In most gems the colouring elements are present as impurities and give rise to a wide range of colours as in **corundum** (ruby — **chromium**, sapphire — **iron**, **titanium**) and **beryl** (**emerald** — **chromium**, **aquamarine** and *heliodor* — **iron**, **morganite** — **manganese**). These colours tend to be less stable and may be enhanced or altered by heating or various forms of irradiation; such treatments are commonly practised today. Some gemstones are markedly pleochroic, displaying either shades of the same colour or two or three quite different colours, as in **iolite**, **tanzanite** and **andalusite**.

'Fire' (flashes of spectrum colour) is due to **dispersion** of light by refraction, and is best seen in some colourless gems, notably **diamond** and **zircon**. The play of spectrum colours seen in precious **opal**, fire **agate** and some **labradorite** is due to interference of light, which arises by **diffraction** or reflection at very small-scale regular mineral structures.

Lustre is defined as the amount and quality of white light reflected from the surface of a gem and is a function chiefly of its **refractive index**. **Lustre** is described qualitatively: starting with the highest **lustres** the range extends from metallic (e.g. **hematite**) and *adamantine* (e.g. **diamond**) to vitreous (most gems) and waxy (**turquoise**), with various intermediate categories. The spangling seen in **aventurine quartz** and **feldspar** is due to reflection of light from numerous metallic or **mica** inclusions in these gems. Reflection from sets of fibrous inclusions or hollow tubes, aligned parallel to specific crystal directions, gives rise to the star (**asterism**) or cat's-eye (**chatoyancy**) effects that can occur in many gem species, most notably **corundum** and **chrysoberyl**.

Skilled cutting and polishing are required to display a gemstone's beauty to best effect. Directional crystalline properties such as pleochroism, **cleavage** and variations in hardness impose constraints in orienting the material to obtain the best colour and quality of cut. The cabochon is the oldest style of cutting, producing a round or oval gem with a simple curved top surface. This cut is still used to display colour and texture in opaque gems and optical effects such as **asterism** and **chatoyancy**. Faceted styles, most commonly the brilliant and step cuts and their variants, are used for transparent gemstones where the **facets** are angled to reflect light from both the surface and interior of the stones. The modern brilliant cut was devised to display the **lustre** and fire of **diamond**, while step cuts best display the body colour of a gem. The ideal **facet** angles vary according to the optical properties of each gem species.

The origins of gem minerals are exceedingly diverse. **Diamonds** form at depths of 100–200 km and are brought to the surface in **kimberlite magmas**, typically in ancient stable **shield** areas such as southern Africa. High grade regional metamorphism of aluminous sediments produces **corundum**, **spinel**, **chrysoberyl** and **garnets**, as in Sri Lanka. **Corundum** formed at depth may occur as **xenocrysts** in **basalts**, as in Thailand and Kampuchea. **Jadeite** forms at the high pressures and relatively low temperatures associated with **subduction zones**. **Peridot** is the rare gem-quality **olivine** that occurs in some **basalts** and **peridotites**. **Skarns** are an important source of **corundum** and **spinel** (Burma and Afghanistan), **zircon**, **lapis lazuli** (Afghanistan, Siberia), **spessartine** and **grossular garnet**. Large and fine-quality crystals of gem **quartz** varieties, **amazonite** and **moonstone feldspar**, **topaz**, **tourmaline**, most **beryl** varieties, **spodumene**, **chrysoberyl** and **spessartine garnet** originate in *pegmatites*; notable localities include Brazil, Madagascar and California. Most **emerald** occurs as exomorphic crystals in mica **schists**, formed by reaction of **granitic magmas** with the **chromium**-bearing *host rock*. In Colombia, source of the finest **emeralds**, crystals occur in **hydrothermally** altered calcareous **shales** and **limestones**. Some gem minerals are of secondary origin; **malachite** and **azurite** form in the upper oxidized zones of **copper** deposits by reaction of **copper** minerals with circulating carbonated water, and **turquoise** forms by the action of near-surface waters on aluminous igneous or sedimentary rocks. The biggest commercial sources of **amethyst** and **agate** occur as secondary deposits in gas **vesicles** in **basalts**, as in the Rio Grande do Sul area of Brazil; precious **opal** forms as secondary deposits in volcanic rocks (Central America) or **sandstones** (Australia).

Gem mining falls into two major categories of contrasting scale and sophistication: **diamond mining** and *'coloured stone' mining*. All **diamonds** are valuable, in industry or in jewellery, so exploration and mining of both alluvial and **kimberlite** pipe sources are systematic and employ the most modern techniques. This huge industry grew up after the discovery of the comparatively large **kimberlite** pipe sources in South Africa in the 1870s. Until this time all **diamonds** had been mined from **alluvial deposits** in India, Kalimantan and Brazil. **Diamonds** are meticulously graded and their supply and marketing is closely controlled.

The production of other gem species is more haphazard. The source deposits are commonly small and in remote and difficult terrain, with patchy distribution and unpredictable yields of gem-quality material; consequently mining operations tend to be small-scale and employ simple techniques and equipment. Gem **gravels**, alluvial and eluvial, are among the most abundant and easily worked sources of fine gems: in Sri Lanka and Burma they have been worked by hand for many centuries. Modern mechanized methods are used in some recently discovered deposits as in the **sapphire** workings at Kings Plains, New South Wales.

Gem identification is mainly by optical properties. **Refractive index** and birefringence values are characteristic for many gem species and mostly lie within the range measured by a standard refractometer; however, accurate results are obtained from polished, flat surfaces only. In some species the presence of certain colouring elements gives rise to characteristic absorption spectra which can be observed in both rough and worked material with a simple hand-held spectroscope. Under a hand lens or microscope many gems display inclusions or growth structures which may be sufficiently distinctive to identify the gem species, or to distinguish natural material from the ever increasing variety of sophisticated modern synthetics and imitations. Specific gravity, measured by heavy liquid or hydrostatic weighing techniques, will distinguish between **jadeite**, **nephrite** and their simulants. [CW]

Anderson, B.W. (1976) *Gemstones for Everyman*. Faber, London.

Bauer, J. & Bouska, V. (1983) *A Guide in Colour to Precious and Semi-precious Stones*. Octopus Books, London.

Webster, R.A. (1979) *Practical Gemmology* (6th edn). NAG Press, London.

geo A linear narrow cleft or ravine running inland from a **sea cliff**. This Scottish word, used especially in the well-jointed *Old Red Sandstone* cliffs of Caithness and Orkney, has been adopted to describe coastal clefts eroded at the base of the cliff usually along a line of weakness, such as a **fault**, **dyke** or **joint** plane. Initially, a narrow **cave** develops which is driven inland. Continued **erosion** weakens the roof which may then collapse in stages leaving natural bridges temporarily spanning the cleft until they too collapse. [JDH]

geobarometry of ores The use of *fluid inclusions* or the chemical compositions of mineral systems to establish the pressure at the time of deposition of minerals or that of later changes (e.g. metamorphism) to which they have been subjected. In a few simple cases *fluid inclusions* can be used to infer the depth of formation and hence the **confining pressure**. One such case is when boiling of the **hydrothermal solution** occurred at the time of trapping. If boiling did not occur, then the **confining pressure** must have exceeded the vapour pressure of the fluid, which can be calculated from the salinity and temperature data, giving a minimum pressure value.

Certain mineral systems, particularly sulphide, can be used as geobarometers. For example the iron content of **sphalerite** *in equilibrium* with **pyrite** and **pyrrhotite** is strongly pressure dependent above 300°C and this geobarometer has been applied to the study of a number of deposits. [AME]

Barton, P.B. & Skinner, B.J. (1979) Sulfide mineral stabilities. In: Barnes, H.L. (ed.) *Geochemistry of Hydrothermal Ore Deposits*, pp. 278–403. John Wiley & Sons, New York.

Roedder, E. & Bodnar, R.J. (1980) Geologic pressure determinations from fluid inclusion studies. *Annual Review of Earth and Planetary Sciences* 8: 263–301.

geocronite ($Pb_5(Sb,As)_2S_8$) A sulphosalt mineral of **lead**. [DJV]

geodesy The science of the measurement and mapping of the surface of the Earth. Geodetic techniques can be used to study active processes of **deformation**, and can also be applied to the periodic measurement of the separation of continents to monitor **continental drift**. (See **satellite radiopositioning**, **satellite laser ranging** and **very long baseline interferometry**.) [PK]

geodynamics The study of dynamic processes which have affected, or are affecting, the solid Earth. The term usually refers to **tectonic** processes rather than those only affecting the surface of the Earth, such as sedimentation. Many geodynamic processes can be related directly or indirectly to the **lithospheric stresses** generated by **plate tectonics**. [PK]

geoid The **equipotential surface** of the Earth's gravitational field represented by the sea-level surface, and defined on land as the level water would reach in a canal open to the sea. The geoid is usually taken as the datum plane in **gravity reduction**. [PK]

geologic time-scale The time-scale composed of standard stratigraphic divisions based on rock sequences and calibrated in years. The standard stratigraphic divisions comprise a *chronostratigraphic* scale now conceived as a scale of rock sequences with standard reference points selected in sections, each particularly complete at the boundary and known as *boundary stratotypes*. Such a scale is a convention which in detail has yet to be agreed and pending such agreement the construction of a geologic time-scale is concerned with the calibration in years of the traditional stratigraphic scale. Despite the lack of any grand design in the establishment of the traditional stratigraphic scale, its components — the geologic *systems* — have been discriminated and utilized world-wide on the basis of distinctive fossils. Given the fuzziness of the *system* boundaries, however, it follows that the definition of the time equivalent of a *system* — the *period* — cannot be exact, furthermore the assignment of ages, expressed in years, to the limits of a *period* or lesser *chronostratigraphic* unit almost always requires interpolation and extrapolation, so adding to the uncertainty. Any time-scale must thus be seen as ephemeral and the ages assigned to the *period* limits, etc., are best viewed as conventions.

The calibration of the traditional stratigraphic scale is primarily based on the establishment of tie points that relate to particular rock samples in which a fortunate combination of characters allows *radiometric age determination(s)* on rocks closely related to those with fossils, which can be used to correlate with *stratotypes*. Such tie points may be established by the direct dating of **authigenic** minerals in fossiliferous sediments, by the dating of volcanic rocks interbedded with fossiliferous sediments, or by the dating of *intrusive* **igneous rocks** the time range of which in relation to fossiliferous sediments can be tightly restricted.

HISTORY OF DEVELOPMENT. Prior to the discovery of *radioactivity*, estimates of the age of the Earth were made on the basis of **denudation** phenomena — the time taken for the oceans to reach their present level of salinity, the time taken for the known maximum thickness of sediments to accumulate — or the energy source of the Sun and the cooling history of the Earth. Such approaches failed either because they simplified highly complex *systems* (as in the **denudation** approaches), or because there were additional factors not taken into account of which the scientific world of the time was unaware (the phenomena of nuclear fusion and *radioactivity*). Following the discovery of *radioactivity*, the first *radiometric age determinations* were published in 1907 and indicated that the Early **Paleozoic** rocks were some 430 Ma old. Several progressively improved time-scales were constructed by the British geologist Arthur Holmes and were based on the *uranium–lead method* during the first half of the present century, while the first 'modern' scale, which made extensive use of **potassium–argon** as well as *uranium–lead age determinations*, was published in 1961 by J.L. Kulp.

SOURCES OF INFORMATION. Since the publication of Kulp's time-scale two major reviews of the problem have been made by the Geological Society of London (Harland *et al.*, 1964; Harland & Francis, 1971); one by the International Subcommission on Stratigraphic Classification (Cohee *et al.*, 1978); one by the Geological Society of London and the Subcommission on Geochronology (Snelling, 1985); and two by independent groups of workers, namely, Harland *et al.*, (1982) and Odin (1982). These works include important data banks and reviews which together with more recent data form the basis of the time-scale presented here.

CONSTRUCTION OF A TIME-SCALE. The construction of a geologic time-scale is an iterative process which may be visualized as follows. A graph is constructed with a numerical horizontal (x) axis graduated in units of millions of years. The vertical or (y) axis is a stratigraphic time-scale showing *periods*, epochs and ages, the precise duration of which are initially unknown but which can be 'guessed' on the basis of existing *radiometric age* data, or on the number of faunal zones, subjective rates of evolution of faunal zones, rates of deposition of pelagic sediments, etc. Tie points, for which an age with its attendant analytical errors and an estimate of its stratigraphic position with its attendant uncertainty can be indicated, will define a straight line (within the limits of the aforementioned errors and uncertainties) provided the relative lengths of the *chronostratigraphic* divisions of the vertical axis have been estimated correctly. A best-fit solution is thus arrived at by progressive adjustments of the *chronostratigraphic* units on the vertical axis. Where sufficient data are available either side of a *chronostratigraphic* boundary, statistical methods may be used to give a more objective estimate of the boundary position (see Harland *et al.*, 1982).

A GEOLOGIC TIME-SCALE. The time-scale presented here is largely based on the most recent review sponsored by the Geological Society of London and the Subcommission on Geochronology (Snelling, 1985). The interim time-scale presented therein has been only slightly modified on the basis of more recent information.

The important datum level of the *Precambrian–Cambrian boundary* is the subject of exhaustive discussion in Snelling (1985) by various contributors. The boundary is now accepted as being marked by the incoming of small shelly fossils, often of uncertain affiliation, which are indicative of the *Tommotian* Age. The *Tommotian* Age is succeeded by the *Atdabanian* during which the first *trilobites* appeared. *Tommotian* fossils are particularly well developed in parts of the former Soviet Union and in China. Numerous age determinations have been reported from China, mainly by the *rubidium–strontium method* on the fine-grained **clay minerals** from **shales**. Unfortunately, this approach is prone to ambiguity; the ages could be too old because of the presence of detrital grains, too young because they record the time of **diagenesis** and not actual sedimentation, and by accident correct because of a fortuitous combination of the aforementioned phenomena! The best that can be said of the ambiguous data is that they may indicate maximum ages for the *Tommotian* of about 590 Ma and for the *Atdabanian* 570 Ma. Elsewhere in the world — France (Vire-Carolles), Morocco (Anti-Atlas), the Arabian-Nubian massif, Newfoundland (Holyrood and Burin Peninsula), and Massachusetts, less ambiguous age determinations are confounded by stratigraphic ambiguities, particularly the absence of strata containing *Tommotian* small shelly fossils. However, Odin *et al.* (in Snelling, 1985) attempted an assessment of these data and concluded that the early *Tommotian* could be set at about 540 Ma. There is thus a significant and unresolvable difference of opinion as to the date of the *Precambrian–Cambrian boundary* which obviously affects the dates assigned to the **Cambrian** Epoch in general since the *Cambrian–Ordovician boundary* is restricted by a fairly secure upper limit of about 490 Ma for the beginning of the **Arenig** Epoch. Pending resolution of the arguments summarized above, a conventional age of 570 Ma is proposed for the date of the *Precambrian–Cambrian boundary*.

The ages assigned to the other *chronostratigraphic* boundaries through the **Paleozoic** Era are summarized in Fig. G1. This scale is based on those presented in the 1985 Geological Society of London: Subcommission on Geochronology publication (Snelling, 1985) and the subsequent discussion, both published and unpublished. The graphic–iterative method outlined above was the prime method of deriving the scale but, in the absence of limiting data, many boundaries have been interpolated on the

Period	Epoch	Age	Ma	
Triassic	Late	Rhaetian	210	
		Norian	220	
		Carnian	230	
	Middle	Ladinian	235	
		Anisian	240	
	Early	Scythian	250	(235)
Permian	Late	Tatarian	255	
		Kazanian	260	
	Early	Kungurian	270	
		Artinskian Sakmarian	280	
		Asselian	300	(286)
Carboniferous	Silesian	Stephanian	305	
		Westphalian	315	
		Namurian	325	
	Dinantian	Visean		
		Tournaisian	355	(352)
Devonian	Late	Famennian		
		Frasnian	375	
	Middle	Givetian		
		Eifelian	390	
	Early	Emsian Siegenian Gedinnian	410	(412)
Silurian	Pridoli Ludlow		424	
	Wenlock		428	
	Llandovery		438	(433)
Ordovician	Ashgill		446	
	Caradoc		455	
	Llandeilo		460	
	Llanvirn		470	
	Arenig		490	
	Tremadoc		510	(509)
Cambrian			530	
		Atdabanian	550	
		Tommotian	570	

Fig. G1 The **geologic time-scale**, Cambrian to Triassic. The numbers in brackets are the ages proposed by Kulp (1961). They have been adjusted to take into account the adoption, in 1976, of conventional decay constants. Ages to beginning of chronostratigraphic units.

Period	Epoch	Age	Ma	
Neogene	Pliocene		5.3	
Neogene	Miocene		23	
Paleogene	Oligocene	Chattian	27	
Paleogene	Oligocene	Rupelian	34	
Paleogene	Eocene	Bartonian	39	
Paleogene	Eocene	Lutetian	45	
Paleogene	Eocene	Ypresian	53	
Paleogene	Paleocene	Thanetian	59	
Paleogene	Paleocene	Danian	65	(65)
Cretaceous	Late	Maastrichtian	72	
Cretaceous	Late	Campanian	83	
Cretaceous	Late	Santonian	86	
Cretaceous	Late	Coniacian	88	
Cretaceous	Late	Turonian	91	
Cretaceous	Late	Cenomanian	95	
Cretaceous	Early	Albian	107	
Cretaceous	Early	Aptian	114	
Cretaceous	Early	Sarremian	116	
Cretaceous	Early	Hauterivian	120	
Cretaceous	Early	Valanginian	128	
Cretaceous	Early	Berriasian	135	(138)
Jurassic	Late	Tithonian	139	
Jurassic	Late	Kimmeridgian	144	
Jurassic	Late	Oxfordian	152	
Jurassic	Middle	Callovian	159	
Jurassic	Middle	Bathonian	170	
Jurassic	Middle	Bajocian	176	
Jurassic	Middle	Aalenian	180	
Jurassic	Early	Toarcian	188	
Jurassic	Early	Pliensbachian	195	
Jurassic	Early	Sinemurian	201	
Jurassic	Early	Hettangian	205	(185)

Fig. G2 The **geologic time-scale**, Jurassic to Pliocene. See comments for Fig. G1.

basis of the epoch intervals derived by the statistical analysis of the data by Harland *et al.* (1982).

The **Jurassic–Cretaceous** time-scale (Fig. G2) follows that of Hallam and Hancock (in Snelling, 1985) which is based on the carefully researched and very fully documented scale derived by Kennedy and Odin (in Odin, 1982). It is preferred to the scale proposed by Harland *et al.* (1982) — **Jurassic** 213–144 Ma, **Cretaceous** 144–65 Ma — because of the excellent database used by Kennedy and Odin and because of the equally good database used by Forster and Warrington (in Snelling, 1985) to define the end of the **Triassic** Period.

The **Paleogene** time-scale (Fig. G2) is taken from Curry and Odin (in Odin, 1982). It, too, is a carefully researched and fully documented scale and is preferred to the scales presented by Harland *et al.* (1982) and Berggren *et al.* (in Snelling, 1985) which interpolate Epoch and Age boundaries from a few, rather poorly defined *radiometric age determinations* using *oceanic magnetic anomalies* as secondary clocks. Apart from the *Oligocene–Miocene boundary* for which an age of 23 Ma is here adopted, the remaining *Neogene* time-scale is that proposed by Berggren *et al.* (in Snelling, 1985). Thus, 5.3 Ma is adopted for the *Miocene–Pliocene boundary* and 1.6 Ma for the *Pliocene–Pleistocene boundary*. [NJS]

Cohee, G.V., Glaessner, M.F. & Hedberg, H.D. (eds) (1978) *Contributions to the Geologic Time Scale*. Papers given at the Geological Time Scale Symposium 106.6. 25th International Geological Congress Sydney, Australia, August 1976. Published by The American Association of Petroleum Geologists, Tulsa, OK.

Harland, W.B. & Francis, E.H. (eds) (1971) *The Phanerozoic Time-scale: A Supplement. Part 1, Supplementary Papers and Items*. Geological Society Special Publication No. 5. Blackwell Scientific Publications, Oxford.

Harland, W.B., Gilbert-Smith, A. & Wilcock, B. (1964) *The Phanerozoic Time-scale. A Symposium*. A supplement to the *Quarterly Journal of the Geological Society, London*, **120S**.

Harland, W.B., Cox, A.V., Llewellyn, P.G. *et al.* (1982) *A Geologic Time Scale*. Cambridge University Press, Cambridge.

Kulp, J.L. (1961) Geologic time scale. *Science* **133**: 3459, 1105–14.

Odin, G.S. (ed.) (1982) *Numerical Dating in Stratigraphy* (2 vols). John Wiley & Sons, Chichester.

Snelling, N.J. (ed.) (1985) *The Chronology of the Geological Record*. Memoirs of the Geological Society No. 10. Blackwell Scientific Publications, Oxford.

geological maps and mapping A geological map is a precisely-oriented scaled-down representation of the rocks of the Earth's **crust**. Maps may be of the surface geology or of sections of the subsurface at a precise and stated datum. A wide variety of styles and types of maps at a whole range of scales from 1 in several million (e.g. 1:5 million for the Metallogenic Map of the USA) to 1 in 100 or even greater is produced. Many published geological maps also contain interpretative **cross-sections**, a sequence of strata and in some cases a summary diagram of the rock relationships and a brief account of the geology.

Geological mapping is the process of making a geological map. Through mapping, the geologist is able to appreciate the spatial distribution of many aspects of the rocks and also features of geology too large or too diffuse to be appreciated on the ground. The whole exercise develops into a three-dimensional appraisal of the geology with surface geology being projected to depth. Through an understanding of the age relationships of the rocks it is possible to establish a geological history of the area.

Surface geological mapping is carried out at various scales. Reconnaissance mapping in many countries is done at a scale of 1:250 000 and detailed mapping at 1:10 000, the latter being the standard mapping scale for the British Geological Survey. Commercial and industrial mapping is often done at a much larger scale in critical areas. As well as showing surface features, geological maps should have geographic coordinates or a metric grid and if the latter it should, where possible, be a national rather than a local grid.

The geological map is the basic tool in geology from which subsequent studies can be made and accuracy in the production of such maps is of paramount importance. There are many unpublished procedural manuals on geological mapping for use by company or survey personnel and many university geology departments have produced handbooks on the techniques of geological mapping for their students. Barnes (1981) and large sections of the books by Compton (1985) and Moseley (1981) provide good accounts of the techniques of geological mapping.

SURFACE MAPPING. For field mapping a geologist needs base maps and *aerial photographs*. In addition to standard geological equipment he may also require an altimeter, a pedometer and an alidade level, particularly when the base maps lack detail. A good camera is now regarded as an essential tool for field work and binoculars are often useful for viewing inaccessible cliffs. The use of equipment is outlined in Barnes (1981) and Compton (1985).

Before going into the field, *aerial photographs* of the area should be examined and a basic *photogeological map* made. Moseley (1981) outlines the principles but for more detailed appraisal refer to Allum (1986). The use of *satellite imagery* is becoming more important in field mapping, particularly where no base map or *aerial photograph* cover exists. These photographs provide an excellent base for field mapping, especially when enlarged to a workable scale (1:10000+) since they show much more terrain detail than topographic maps and should be viewed in the field as a stereopair with a pocket stereoscope. (See **remote sensing**.)

In the field, geological observations are recorded on the base map or transparent photographic overlay at the exact locality where they are made. The field geologist should observe everything of significance about the rocks such as lithology, **structure**, fossil content, weathered profile, etc., and even where there is no solid rock the nature of the float, vegetation, *soil* colour, surface features, etc., should be noted. Colours should be used for lithologies and a set of conventional symbols for other features such as structural elements. As much data as possible should be recorded on the map without actually cluttering it, the additional detail, such as lengthy descriptions, sketches and enlargement maps or section logs, being recorded in a notebook. All notes should be given a locality number, marked on the map, and accompanied in the notebook by a grid reference in order to facilitate cross-referencing. The field map should be self-explanatory with a clear keÿto symbols and colours used and it should be seen to 'grow' as mapping progresses, enabling the geologist to formulate ideas as to the geology of the area as working hypotheses.

Finding an exact location on the map is not always an easy task, particularly if the base map lacks detail. Some base maps can be little better than a sheet of blank paper with a few streams, stylistically marked crags and a few spot heights. Under these conditions some of the techniques such as re-sectioning, pace and compass traverses and altimetry will have to be employed and all of these are described in the main texts cited. It is much easier to locate one's position on an *aerial photograph* since large-scale photographs show virtually every exposure. However, these are not without their attendant problems such as projection differences between map and photograph, distortion away from the centre of the photograph, differences in scale with changes in the height of the ground and different oblique views of particular features (parallax) during the flight traverse. These all have to be overcome (Moseley, 1981, Chapter 5 and Compton, 1985, Chapter 7). Maps give a more exact scale than photographs, but photographs show more detail than maps and consequently the best approach is to use both in conjunction.

All detailed geological mapping should start with a reconnaissance of the area in order to determine which mapping method will be most suitable, to note areas of good exposure and unexposed ground, in order to determine where the most suitable place to start would be, to get an overall impression of the geology and to establish a plan for the mapping programme. Several methods can be employed (Barnes, 1981).

1 *Traverse mapping*. This is the method used for regional (scales 1:250000 to 1:50000) and often for reconnaissance mapping. It is particularly suitable for featureless country or for thickly wooded areas where location of position is difficult, even with *aerial photographs*. Traverses may be systematically spaced along compass-bearing lines or irregularly spaced following roads or tracks, rivers or ridges and all should, where possible, be across the geological grain of the area. The spacing between traverse lines is dictated partly by the scale of the mapping and partly by the nature of the geology, more complex terrain demanding closer spacing than vast areas of uniform geology. Traverses need not be exclusively in straight lines, some may zigzag from feature to feature and others may be a loop returning to starting point. The geology is recorded along the line and traverses often provide a good framework on which to build the intervening geology.

2 *Contact or boundary mapping*. Normally used for more detailed mapping (scales of 1:50000 to 1:10000). This method involves following a geological contact and is a modified traverse in that the route taken zigzags along the contact. This method can be employed where exposure is good and the contact can be seen easily in the field or where the contact follows a definite topographic feature or vegetation change. *Aerial photographs* are particularly useful for this type of mapping since subtle changes not obvious on the ground can often be detected on them. In poorly exposed ground a *soil* auger should be used to determine the rock below the superficial cover, par-

ticularly when the lithological boundary is placed along a topographic feature. The use of features for boundary mapping has its limitations where glacial deposits are the superficial cover. *Solifluction* and **cambering** can also cause problems.

3 *Exposure mapping* (sometimes called *outcrop mapping*). This is a technique of detailed mapping at scales of 1:10 000 and larger. Every exposure is visited and its limits plotted. It is a good technique where exposure is limited since it records only fact and is a sound base for a subsequent interpretation of the geology. Topographic features which may have a geological control are marked as they may represent a geological boundary. Rock types on the exposures are normally coloured in the field according to a chosen code, and structural data recorded, since this aids interpretation. Areas between exposures of the same lithology can be shaded in the same colour as the exposures but not as heavily, since this distinguishes between interpretation and observation. It is advisable to extrapolate boundaries whilst in the field but it should be obvious on the map as to what is a definite boundary (marked by a solid line) and an inferred boundary (marked by a broken line).

4 *Line mapping*. This is detailed mapping at scales of 1:2500 or more. It involves using a measured base line on a bearing (as for a traverse) and taking offsets at right angles from the line to the exposure. This can be a very satisfactory way of recording the details of complex geology which are beyond the scale of the base map, but which do not merit a complete survey of the area. It is employed in **surface pit** and **underground mapping**.

5 *Detailed grid mapping*. This is ultra-detailed mapping at scales in the order of 1:100 to 1:10. In some respects it could be regarded as an accurate means of producing a field sketch of a critical exposure. In it a squared grid of cord is constructed over the exposure and the detail of each square is plotted on graph paper. A similar technique using a combination of cord and survey poles can be used to map the details of road cuttings or other similar sections. Those skilled in the use of a plane table may find that a more satisfactory technique for mapping horizontal surfaces in detail.

6 *Mapping superficial deposits*. These are often ignored during bedrock surveys but in *glaciated* or *periglacial* areas it is important that these deposits are mapped. They include large **screes**, boulder-clay *moraines* and other deposits of *till* such as *drumlins*, drainlines, outwash **sands** and **gravels** and post-glacial **peat bogs**. Mapping these deposits requires careful attention to surface features, the use of soil augers or even the excavation of temporary exposures by way of pitting and trenching (Moseley, 1981, Chapter 13).

In desert areas, **sand dunes**, **talus** slopes, **alluvial fans** and cones resulting from flash-floods are all important aspects of the recent history of the area and should not be omitted from the map. Since some of these features change with time, old *aerial photographs* may not show them in their present stage but will allow studies of their development to be made.

Final maps. These are produced from the field maps and are an interpretation of the geology of the area mapped, based on the field data. Government surveys and large industrial companies employ draughtsmen to carry out the production of these maps. The form of the final map, even if published on the same scale as the field map, will be a simplification of it, though it should show as much detail as possible without crowding the map. Reductions in scale will obviously lead to a reduction in detail, but these smaller-scale maps will have the advantage of covering a larger area and give a better impression of the regional situation. Corporate policy normally dictates the style of published maps, the colour and symbol scheme and the amount of data to be included. In many cases (e.g. UK) Drift and Solid editions of each 1:50 000 sheet are prepared, the former showing the extent of the superficial deposits and the latter the interpreted sub-outcrop pattern below the superficial deposits. If there is a great deal of cover the two maps will look very different and the solid edition may not have great detail on it because of the lack of firm data over much of the area. [DER]

Allum, J.A.E. (1986) *Photogeology and Regional Mapping*. Pergamon, Oxford.

Barnes, J.W. (1981) *Basic Geological Mapping*. Geological Society of London Handbook Series. Open University Press, Milton Keynes.

Compton, R.R. (1985) *Geology in the Field*. John Wiley & Sons, New York.

Moseley, F. (1981) *Methods in Field Geology*. W.H. Freeman, San Francisco.

geomagnetic dipole field Some 80% of the **geomagnetic field** can be accounted for by a single geocentric dipole inclined at 11.3° to the Earth's axis of rotation. The remaining field is the **geomagnetic non-dipole field**. [DHT]

Parkinson, W.D. (1983) *Introduction to Geomagnetism*. Scottish Academic Press, Edinburgh.

geomagnetic field The **magnetic field** of the Earth. [PK]

geomagnetic non-dipole field That part of the **geomagnetic field** (*c.* 20%) that is not accounted for by the main **geomagnetic dipole field**. [DHT]

Parkinson, W.D. (1983) *Introduction to Geomagnetism*. Scottish Academic Press, Edinburgh.

geomagnetic polarity time-scale The **geomagnetic field** changes polarity at various time-scales, sometimes remaining of constant polarity for 50–60 Ma and at other times changing several times within 1 Ma. The last well-substantiated reversal was 730 000 years ago, although it is suspected that there may have been intervals of 100–300 years (*polarity excursions*) during which the field has attempted to reverse but still retained its present polarity.

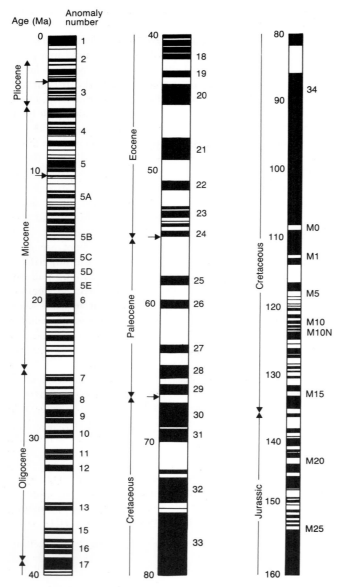

Bott, M.H.P. (1982) *The Interior of the Earth.* Arnold, London.
Jacobs, J.A. (1984) *Reversals of the Earth's Magnetic Field.* Adam Hilger, Bristol.
Khramov, A.N. (1987) *Paleomagnetology.* Springer-Verlag, Berlin.
Tarling, D.H. (1983) *Palaeomagnetism.* Chapman & Hall, London.

the last 5 Ma can be well dated using **igneous rocks** that preserve such polarity changes and can also be dated by radiometric methods. For earlier times the errors in *radiometric dating* become too large and conventional relative **biostratigraphic** dating methods must be used. Conveniently, the polarity changes of the **geomagnetic field** during the last 180 Ma are preserved in the **igneous rocks** of the **ocean crust** as these were injected and cooled sequentially at the *oceanic ridges* and, as they cooled, retained a record of the polarity of the **geomagnetic field** that can be readily determined from a study of their associated **magnetic anomalies** as determined at the ocean surface (see **Vine–Matthews hypothesis**). Using such records, a polarity scale can be produced that is dated by the age of the igneous ocean floor. To some extent, this itself is dated by the *oceanic magnetic anomalies* themselves, but such a record can be cross-checked using *radiometric dating* of polarity changes recorded in rocks on the continents, and also by determining the age of the sediments immediately overlying the igneous **oceanic crust**, mainly using samples drilled as part of the *JOIDES* and subsequent deep ocean drilling programmes. For even older times, the polarity sequences can only be determined using rock sequences preserved on the continents and dated mainly by their relative positions in the stratigraphic column, supplemented by *radiometric* and **biostratigraphic** *dates.* (See Fig. G3). [DHG]

Fig. G3 The **geomagnetic polarity time-scale** for the last 160 Ma. Periods of normal polarity are shown in black and the numbers assigned to prominent magnetic anomalies are shown to the right of each column. Note that the scale from 80 to 160 Ma is condensed by a factor of two. Calibration points back to 90 Ma are marked by arrows. (After Bott, 1982.)

Studies of the last reversal suggests that the actual polarity change takes place over some 3000 years, is preceded by a period when the intensity of the field decreases to about 20% of its usual value, and is followed by a similar period as the intensity regains its original magnitude. However, any intensity behaviour of the **geomagnetic field** is particularly difficult to determine, both in sediments and volcanic rocks. Polarity changes are particularly valuable as they provide global time markers that are extremely brief compared with all other global time markers, such as those based on **fossil evolution**. The changes during

geomagnetic/magnetic elements Descriptors used to describe the strength and direction of the **geomagnetic field**. The total field (*B*) has a vertical component (*Z*) and a horizontal component (*H*) in the direction of **magnetic north** (Fig. G4). The **dip** of *B* is known as the *inclination* (*I*) and the angle between *H* and true north the *declination*. *I* is downwards to the north in the northern magnetic hemi-

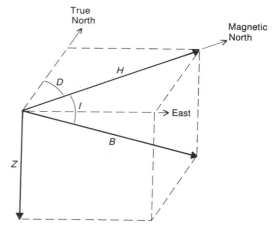

Fig. G4 The **geomagnetic elements**.

sphere and vertical downwards at the north magnetic pole. *I* is upwards to the north in the southern magnetic hemisphere and vertical upwards at the south magnetic pole. *I* is zero at the *magnetic equator*, which approximates the geographic equator. For a geocentric dipole approximation, *I* is related to latitude ϕ by

tan *I* = 2 tan ϕ

and this relationship is utilized in **paleomagnetic** studies to determine *paleolatitude* from the **dip** of the **natural remanent magnetization** as the *declination* (*D*) vanishes to zero when averaged over periods of a few thousand years. All geomagnetic elements are subject to **secular variation**. [PK]

geomagnetism The study of the Earth's **magnetic field**. Measurements of **geomagnetic field** strength, after **reduction** to remove all variations caused by non-geological effects such as latitude/longitude and **diurnal variation**, provide **magnetic anomalies** which can be interpreted in terms of bodies of anomalous magnetization. **Magnetic surveys** can be undertaken on land, at sea and in the air (see **airborne geophysical surveys** and **shipborne geophysical sureys**). The term is also applied to those methods which measure the time-varying **magnetic field** associated with **magnetotelluric** currents. [PK]

geomorphological magnitude and frequency The frequency distributions of variables such as peak river discharge, daily rainfall intensity and **wave** heights are typically positively skewed, such that small events occur frequently whilst large events are much rarer. For river systems, Wolman & Miller (1960) argued that medium-sized flood events achieved the bulk of fluvial work over the long term so that such events represent the channel-forming discharge. They showed that **convolution** of the frequency curve of flood discharge together with the curve of the work achieved by floods of a given magnitude yields a curve which peaks for floods of moderate frequency and magnitude, roughly accordant with **bankfull discharge** or the mean annual flood. The notion of a **dynamic equilibrium** for channel form related to floods of such recurrence interval stems directly from this argument. More recently it has been argued that the concept outlined by Wolman & Miller requires modification to cope with response to catastrophic floods which are so large that major **threshold** changes occur within the fluvial system. Not all major floods produce lasting effects, with channel form reverting to its former state over years or decades — this period of recovery being termed the '*relaxation time*' of the *system*. However, in some *systems*, particularly in semi-arid areas, extreme floods appear to produce more drastic responses, perhaps because the flood frequency curve is very right-skewed, so that the channel modifications persist. (See **river flood**). [TPB]

Wolman, M.G. & Miller, J.P. (1960) Magnitude and frequency of forces in geomorphic processes. *Journal of Geology* **68**: 54–74.

geomorphological threshold A change in a landform initiated by changes in the morphology of the landform itself through time. Such changes are thus inherent or intrinsic to the landform rather than initiated by changes in external controls. Such intrinsic thresholds have been extensively described by Schumm (1977, 1979). Examples may include: headcuts on **alluvial fans**; **arroyo** incision and river pattern changes. The existence of geomorphological thresholds stresses the operation of a **dynamic** metastable **equilibrium** especially at the early stage of landform evolution. [TPB]

Schumm, S.A. (1977) *The Fluvial System*. Wiley-Interscience, New York.
Schumm, S.A. (1979) Geomorphic thresholds: the concept and its applications. *Transactions of the Institute of British Geographers* **NS4**: 485–515.

geomorphology A term of obscure origins that probably has its roots in the United States Geological Survey of the 1880s. In 1891 W.J. McGee wrote: 'The phenomena of degradation form the subject of geomorphology, the novel branch of geology. . . . It has come to be recognized that the later history of world growth may be read from the configuration of the hills as well as from the sediments and fossils of ancient oceans. . . . The field of science is thereby broadened by the addition of a co-ordinate province — by the birth of a new geology which is destined to rank with the old. This is *geomorphic geology*, or geomorphology'.

This reconstruction of the **erosional** history of the Earth through the study of **erosional** landforms formed the core of **denudation chronology**, but since its definition the scope of the term geomorphology has become wider. Most geomorphologists now regard the core of the subject as being the comprehension of the form of the ground surface and the processes which mould it. In recent years a concern with understanding the processes of **weathering**, **erosion**, transport and deposition, together with determination of the rates at which such processes operate, has become central. Geomorphology now has many component branches (e.g. **anthropogeomorphology**, **applied geomorphology**, **climatic geomorphology**). [ASG]

Chorley, R.J., Schumm, S.A. & Sugden, D. (1984) *Geomorphology*. Methuen, London.
McGee, W.J. (1891) The Pleistocene history of northeastern Iowa. *11th Annual Report of the United States Geological Survey*, pp. 189–577.
Selby, M.J. (1985) *Earth's Changing Surface*. Clarendon Press, Oxford.

geopetal fabric Fabric formed by the partial infill of a cavity with sediment. These structures are particularly common in **limestones**, with sparry **calcite** occluding the cavity above the internal sediment. The flat upper surface of the internal sediment represents an originally horizontal level (Fig. G5). Hence geopetal structures can be used to estimate the amount of post-depositional tilting of strata or the presence of original depositional **dips**, as well as being useful *way-up indicators*. This interpretation assumes that

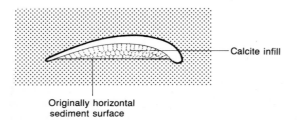

Fig. G5 **Geopetal fabric** developed beneath a shell.

sediment has not been banked in a cavity by currents, and that the sediment was fully lithified prior to tilting. [DMP]

geophone An instrument used on land, marsh or shallow water to detect seismic energy. Most geophones convert the seismic energy to electric voltage using a moving coil technique, like a microphone. A coil is suspended by springs in the **magnetic field** of a magnet attached to the geophone case, which is coupled with the Earth via a spike in the ground (Fig. G6). The **seismic wave** moves the case producing relative motion between the magnet and the coil due to the latter's inertia. This movement induces a voltage in the coil proportional to its velocity relative to the magnet. Geophones usually have a natural frequency between 4 and 100 Hz. [GS]

Anstey, N.A. (1981) *Seismic Prospecting Instruments* Vol. 1: *Signal Characteristics and Instrument Specifications*. Gebrüder Borntraeger, Berlin.

geophysical anomaly Any perturbation from the norm in a measured field, usually resulting from a change in the physical properties of the underlying rocks, and applying particularly to **gravity** and **magnetic fields**. The Earth's **gravity** and **magnetic fields** vary over the surface as a result of various phenomena, including location, elevation, topography and the presence of subsurface bodies of rock with anomalous density or magnetization. The aim of **reduction** is to remove all variations in the measured field which do not result from the latter cause. Consequently

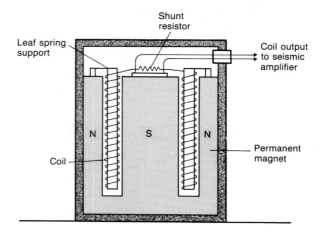

Fig. G6 Schematic cross-section through a moving coil **geophone**.

gravimeter drift, **latitude**, **free air**, **Bouguer**, **terrain**, *Earth tide* and **Eötvös corrections** are applied to gravity observations, and **diurnal** and **geomagnetic corrections** are applied to magnetic data. The resulting **gravity** and **magnetic anomalies** then reflect changes in the density or magnetization of the subsurface, and these can be processed or modelled to provide information about the causative body. [PK]

geophysical borehole logging (*geophysical well logging*, *wire-line logging*, *downhole geophysical survey*) The recording against depth of any of the characteristics or properties of the rock formations traversed by a measuring apparatus in a borehole. The collection of a complete **core** from a borehole is time-consuming and expensive, and the cuttings flushed to the surface by rotary or percussion drilling are difficult to interpret. Consequently it is common practice to obtain information on the material traversed by a borehole *in situ* by making use of apparatus mounted on a **sonde** and lowered down the borehole. The technique is widely employed in *hydrocarbon exploration* as it provides important information on the properties of **reservoir** rocks.

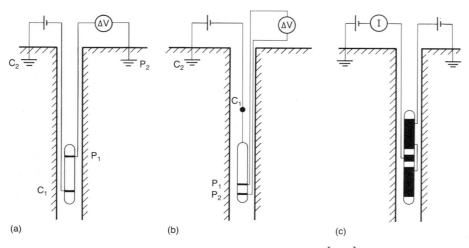

Fig. G7 **Geophysical borehole logging**. Resistivity logging devices. (a) Normal log. (b) Lateral log. (c) Laterolog. C — current electrode, P — potential electrode, ΔV — potential difference, I — current.

The main parameters provided by borehole logging are lithology, bed thickness, **porosity**, water and *hydrocarbon* saturation, **permeability**, temperature and **dip**. This is accomplished by making use of the various type of log described below, either alone or in combination.

1 Electric (*resistivity*) **logs** are used to determine water and *hydrocarbon saturation*, **porosity** and the presence of permeable zones. **Resistivity** is measured between various configurations of electrodes (Fig. G7). They include the following types:

(a) *Normal log* (Fig. G7a). Provides information on a thick shell around the borehole.

(b) *Lateral log* (Fig. G7b). The zone tested reaches much farther from the well.

(c) *Laterolog* (Fig. G7c). A focused log in which the zone tested is a circular disc.

(d) *Microlog*. Small electrodes are mounted on a pad which is pressed firmly against the side of the well. Provides information on the material immediately adjacent to the electrodes.

2 *Induction logs* are used in wells filled with insulating mud which prevents the use of **electrical logs**. Measurements are made by **electromagnetic induction**, and may be focused or unfocused.

3 *Self potential* (SP) *logs* measure the SP effects generated when **pore-fluids** of different ionic concentration are in contact. They are particularly useful in identifying shale sequences.

4 *Radioactivity logs* make use of natural or induced *radioactivity*.

(a) *Gamma logs* measure natural gamma radiation, which is most pronounced in the presence of **clay minerals**.

(b) *Gamma–gamma* (*density*) *logs* use artificial gamma rays, whose scatter by the formation is a measure of its density.

(c) *Neutron* (*neutron–gamma ray*) *logs* use neutron bombardment to stimulate non-radioactive elements to emit gamma rays. This effect is most pronounced in the presence of hydrogen ions, whose concentration depends on **porosity**.

5 *Sonic* (*continuous velocity*) *logs* use ultrasonic sound pulses to measure the seismic velocity of the formation, which can be used to estimate **porosity**.

6 *Temperature logs* measure downhole temperature gradients. Combined with **thermal conductivity** measurements, these can be used to calculate **heat flow**.

7 *Dipmeter logs* take four horizontal microresistivity readings at 90° intervals to determine formation **dip** and **strike**.

Other less frequently used methods include *magnetic* and *gravity logs* using specialized **magnetometers** and **borehole gravimeters**.

Some examples of downhole logs illustrating the use of some of these techniques are shown in Fig. G8. [PK]

Ellis, D.V. (1987) *Well Logging for Earth Scientists*. Elsevier, Amsterdam.

Labo, J. (1986) *A Practical Introduction to Borehole Geophysics*. Society of Exploration Geophysicists, Tulsa, OK.
Robinson, E.S. & Cahit, C. (1988) *Basic Exploration Geophysics*. John Wiley & Sons, New York.

geophysics The application of the methods and techniques of physics to the study of the structure of the Earth and the processes affecting it. [PK]

geostrophic current A current in which the pressure gradient force (PGF) and the **Coriolis force** (CORF) are in balance. Pressure gradients in the ocean result mainly from slope of the surface relative to horizontal or horizontal variations in water density. Currents established to redress these imbalances (*gradient currents*) do not flow along the gradient but, due to the action of the **Coriolis force**, tend to flow at right angles to the pressure gradient along lines of equal pressure (*isopycnals*) with the PGF and CORF acting in opposite directions. When the PGF and CORF are in balance the resulting current is geostrophic. Geostrophic theory (which assumes that frictional forces can be neglected) can be used to calculate the magnitude and variation of currents with depth, given latitude and vertical density profiles. The concept of geostrophic flow derives from the equations of motion of fluid particles on a rotating sphere (see Pond & Pickard, 1983). The situation in the oceans is complicated by the action of frictional forces and wind **stress** but many major oceanic currents such as the Gulf Steam are essentially geostrophic. [AESK]

Harvey, J.G. (1976) *Atmosphere and Ocean: Our Fluid Environments*. Artemis Press, London.
Pond, S. & Pickard, G.L. (1983) *Introductory Dynamical Oceanography* (2nd edn). Pergamon, Oxford.

geosyncline The concept of the geosyncline was central to **orogenic** theory from the early days of geology until the **plate tectonic** revolution. The term was first used by Hall (1859) and Dana (1873) to describe an elongate trough in which great thicknesses of sediment had accumulated in a zone that subsequently became part of an *orogenic belt*. The geosyncline was contrasted with adjoining much thinner continental shelf sequences. Whereas Hall considered that the depression of the trough was due to the sedimentary **load**, other early workers realized that some other mechanism of subsidence was necessary. Subsequently, geosynclines became classified into a number of categories such as mono-, poly-, meso-, ortho- and para-geosynclines. However, the most useful subdivision was into *eugeosyncline* and *miogeosyncline* (Stille, 1940) referring respectively to **basins** with oceanic affinities and abundant magmatism, and to continental slope sequences without related magmatism. Argand (1924) explained geosynclines as the result of **crustal attenuation** preceding the separation of continents, but this idea was not generally accepted until the **continental drift** theory was universally accepted

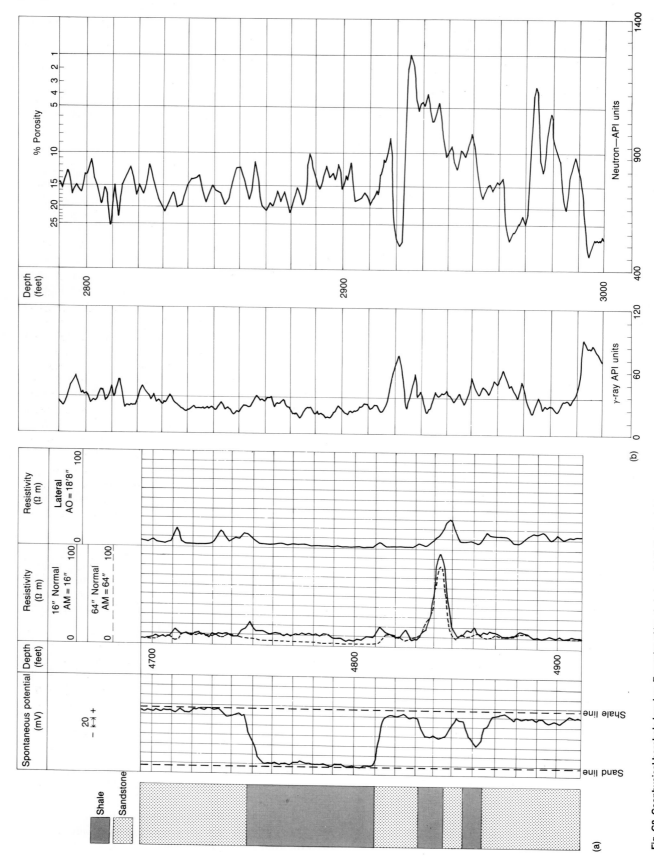

Fig. G8 Geophysical borehole logging. Examples of borehole logs. (a) Self potential, long and short normal and lateral logs over a sandstone–shale sequence. (b) Gamma and neutron logs over a sequence of dolomite and shale.

in the 1960s. According to **plate tectonic** theory, there is no place for the geosyncline concept, and no genetic link between the formation of a deep sedimentary **basin** and subsequent **orogenesis**. *Orogenic belts* are attributed to *plate collision*, and the fact that pre-existing sedimentary accumulations become involved is accidental, except in the sense that all **active-** or **passive-margin basins** will eventually become accreted to the continental margin. (See **plate tectonics**.) [RGP]

Argand, E. (1924) La tectonique de l'Asie. *Comptes rendus du congrès géologique international, Belgique, 1922,* pp. 171–372.

Dana, J.D. (1873) On some results of the Earth's contraction from cooling, including a discussion of the origin of mountains and the nature of the Earth's interior. *American Journal of Science, Series 3,* **5**: 423–43; **6**: 6–14, 104–15, 161–71.

Hall, J. (1859) *Natural History of New York, Part VI, Paleontology,* Vol. 3. Albany, New York.

Miyashiro, A., Aki, K. & Sengör, A.M.C. (1982) *Orogeny.* John Wiley & Sons, Chichester.

Stille, H. (1940) *Einführung in den Bau Amerikas.* Gebrüder Borntrager, Berlin.

geotectonic Relating to major Earth structure. 'Geotectonics' is the branch of geology dealing with this topic, and is the study of large-scale **crustal** structures (e.g. *orogenic belts*, **cratons**, blocks and **basins**) rather than individual **structures** such as **folds** or **faults**. It also includes the study of processes (e.g. **plate tectonics**) controlling major Earth structure. [RGP]

geotherm Curve representing the variation of temperature with depth. The geotherm is a straight line if the *geothermal gradient* is constant (i.e. for a constant **heat flow** and uniform **thermal conductivity**), but deviations from linearity are common. Even in the same lithological unit,

Table G1 Concentrations in p.p.m. of some of the elements in two **geothermal brines**. (1) Salton Sea, California, (2) Cheleken, Turkmen Republic

Element	1	2
Cl	155 000	157 000
Na	50 400	76 140
Ca	28 000	19 708
K	17 500	409
Mg	54	3 080
B	390	—
Br	120	526.5
F	15	—
NH_4	409	—
HCO_3^-	>150	31.9
H_2S	16*	0
SO_4^{2-}	5	309
Fe	2 290	14.0
Mn	1 400	46.5
Zn	540	3.0
Pb	102	9.2
Cu	8	1.4

* Sulphide present; all S reported as H_2S.

significant deviations are produced by *radioactive* elements producing heat, changes in conductivity due to temperature, effect of past climatic variations diffusing into the Earth, surface topography, etc. [ATB]

geothermal Concerned with the heat flowing from the interior of the Earth to the surface. (See **heat flow**). [ATB]

geothermal brine The saline solution present in **geothermal systems** and *hot springs*. The data for two geothermal brines are given in Table G1. It can be seen that the major constituents are sodium, **potassium**, calcium, **magnesium** and chlorine. They may also contain significant quantities of **base** and other **metals** of economic importance and, in the right circumstances, could give rise to mineral deposits. (See **brine**.) [AME]

geothermal energy Heat energy obtained by tapping **geothermal systems**, by drilling down to deep **aquifers** containing hot water and pumping some to the surface or by drilling holes into hot rocks and pumping water through these rocks to heat it and thence to the surface.

Geothermal systems utilizing natural **groundwater** can be broadly classified into two: 'high enthalpy' systems utilizing steam at temperatures >200°C for electricity generation, and 'low enthalpy' systems (temperatures <100°C) more commonly used for space heating (e.g. apartment blocks in the Paris Basin) or other industrial applications. Geothermal power generation has been in existence in Ladarello in Italy from the beginning of this century, and at present New Zealand, the Philippines, Japan, East Africa, and areas of central America and the United States can all boast successful power production in high enthalpy fields.

Deep **aquifers** in regions of high heat flow may contain water at a high enough temperature for district heating systems for houses, factories and/or glasshouses for producing fruit and vegetables. Generally this water is recycled by pumping it back down into the **aquifer**.

The third system involves the drilling of holes into hot rocks at suitable distances from each other and the creation of permeable zones between the bottoms of the holes through which water can be pumped. The permeable zones are created by producing *hydraulic fractures* in the hot rocks with water under high pressure. (See **hot dry rock concept**.) [AME]

Armstead, H.C.H. (1978) *Geothermal Energy.* Spon, London.

Armstead, H.C.H. & Tester, J.W. (1987) *Heat Mining.* Spon, London.

Butler, E.W. & Pick, J.B. (1982) *Geothermal Energy Development.* Plenum, New York.

Downing, R.A. & Gray, D.A. (1986) Introduction. In: Downing, R.A. & Gray, D.A. (eds) *Geothermal Energy — The Potential in the United Kingdom.* British Geological Survey, HMSO, London.

geothermal system Circulating **groundwater** system set in motion by a high *geothermal gradient*. Geothermal

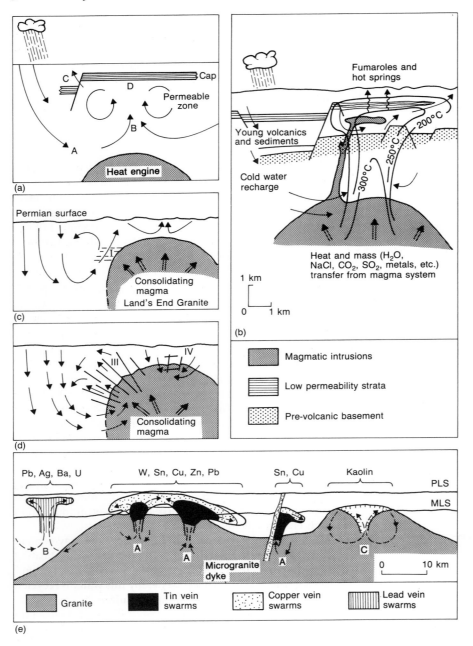

Fig. G9 (a) Schema showing some of the features of a **geothermal system**: (A–D) are explained in the text. (b) Schema showing the structure of a geothermal system like that of the Taupo Volcanic Zone, New Zealand. (c and d) Schemata illustrating the evolution of some of the mineralization in a flank of the Land's End Granite. In detail, these show: (c) initial emplacement of the pluton with the development of an H_2O-saturated carapace enclosing still-consolidating magma. It also shows formation of tin- and magnetite-bearing skarns (I) in aureole rocks, by aqueous solutions of a dominantly magmatic origin. Time: about 290–270 Ma. (d) Further crystallization of the pluton has taken place, and joints and fractures have formed in the crystallized carapace. With the formation of a water-rich phase that has separated from the H_2O-saturated melt, an extensive geothermal system has come into being. This has produced the main stage mineralization (III and IV) of tin- and copper-bearing quartz veins. Time: about 270 Ma. (Type II mineralization is that of pegmatites.) (e) Schema of possible fossil geothermal systems associated with the granite batholith of south-west England, illustrating the different types and settings of mineralization in that region, and the district zoning developed there: (A) Dines (1956) type emanative centres; (B) cross course mineralization; (C) kaolin deposits (weathering may have played a part in their formation).

systems from where a heat engine (usually **magmatic**) at depths of a few kilometres sets deep **groundwaters** in motion (Fig. G9). These waters are usually **meteoric** in origin but in some systems **connate** or other saline waters (Salton Sea) may be present. Systems near the coast may be fed wholly or partially by seawater (Svartsengi, Iceland). **Magmatic** water may be added by the heat engine, and some ancient systems appear to have been dominated by **magmatic** water, at least in their early stages, e.g. **porphyry copper** and **disseminated molybdenum deposits**. Dissolved constituents (**geothermal brine**) may be derived by the circulating waters from a **magmatic** body at depth, or from the *country rocks* which contain the

system. These may be altered by the solutions to mineral assemblages identical to those found in some **wall rock alteration** zones associated with **orebodies**. Common sulphides, such as **galena** and **sphalerite**, occur in a number of modern systems, and at Broadlands, NZ, an amorphous **antimony–arsenic–mercury**–thallium sulphide precipitate enriched to **ore grade** in **gold** and **silver**, has been formed.

The principal features of a geothermal system are shown in Fig. G9(a). **Meteoric water** sinking to several kilometres depth (A) enters a zone of high **heat flow**, absorbs heat and rises into one or a succession of permeable zones (BD). There may be outflow at an appreciable rate along a path such as (C) or much slower outflow by permeation of the **cap rock** (**mudstone**, **tuff**, etc.). Outflow through (C) depends on the **permeability** of the rocks and the pressure at the top of the zone (BD). If the outflow rate does not exceed that of the inflow, an all liquid system will prevail. With a higher outflow rate, a steam phase will form in BD and the steam pressure will decrease until the mass outflow through (C) is reduced to equal the mass inflow. A dynamic balance then obtains with a lowered water level in the permeable horizon, boiling water and the development of **convection** currents in the water.

Figure G9(b) illustrates the structure of a geothermal system in volcanic terranes like that of the Taupo Volcanic Zone, New Zealand. Note that the hot waters are circulating through, reacting with and probably obtaining dissolved constituents from both the **magmatic** intrusion and the *country rocks*. In Fig. G9(c and d), geothermal systems are postulated to explain **vein tin** and **copper** mineralization in and adjacent to the Land's End Granite in south-west England. In Fig. G9(e) we have a broader picture, with geothermal systems being invoked to explain some of the different types of mineralization in south-west England and the zoning of metals that is one of the well-known features of this orefield.

Modern geothermal systems such as those of North Island, NZ are an important source of **geothermal energy** which can be used for generating electricity, district heating and other purposes. [AME]

Dines, H.G. (1956) The metalliferous mining region of south-west England. *Memoirs of the Geological Society of Great Britain* **1**.

Ellis, A.J. (1979) Explored geothermal systems. In: Barnes, H.L. (ed.) *Geochemistry of Hydrothermal Ore Deposits*, pp. 632–83. John Wiley & Sons, New York.

Henley, R.W. & Ellis A.J. (1983) Geothermal systems ancient and modern: a geochemical review. *Earth Science Reviews* **19**: 1–50.

Weissberg, B.G., Browne, P.R.L. & Seward, T.M. (1979) Ore metals in active geothermal systems. In: Barnes, H.L. (ed.), *Geochemistry of Hydrothermal Ore Deposits*, pp. 738–80. John Wiley & Sons, New York.

geothermometry of mineral deposits Mineral deposits are formed at temperatures and pressures ranging from very high, at deep **crustal** levels, to atmospheric, at the surface. Some *pegmatites* and **magmatic segregation deposits** have formed at temperatures around 1000°C and under many kilometres of overlying rock, whilst **placer** deposits and sedimentary **ores** have formed under surface conditions. Most **orebodies** were deposited between these two extremes. Clearly, knowledge of the temperatures and pressures obtaining during the **precipitation** of the various minerals will be invaluable in assessing their probable mode of genesis, and such knowledge will also be of great value in formulating exploration programmes. In this entry it is only possible to touch on a few of the methods that can be used.

The most useful method is the study of *fluid inclusions* which can be used on many transparent minerals such as **fluorite**, **baryte**, **quartz**, **cassiterite** and **sphalerite**. Other methods are the use of inversion points in **polymorphism** and the resolution of **exsolution** textures. More useful than these last two are: (i) the study of sulphide and **oxide mineral** systems, where certain mineral assemblages in equilibrium with each other possess one or more minerals whose composition reflects the temperature of equilibration; and (ii) the use of **stable isotopes** particularly of oxygen and **sulphur**. [AME]

gersdorffite (NiAsS) A relatively rare **ore** mineral of **nickel**. [DJV]

geyser A vent from which hot water and steam is periodically and violently ejected at the Earth's surface. Geysers occur in volcanically active areas. **Groundwater** in a system of deep, interconnecting conduits is heated by underlying **magma** to near boiling point for the appropriate pressure. Initiation of boiling creates an upwelling of the water in the conduits, and therefore further boiling owing to the reduced pressure (see **retrograde boiling**). A runaway situation occurs in which the water rises more and more rapidly as it continues to boil, and the mixture of water and steam is erupted from the geyser as a jet which can rise up to about 70 m high. Cooler water draining back into the deep conduits requires a period of time to become heated to near boiling point again. The periodicity of the famous Old Faithful geyser in Yellowstone, Wyoming, averages 65 min, ranging from 32 to 120 min. [PTL]

geyserite A variety of **opal** deposited by *hot springs* and **geysers**. [DJV]

giant gas field Fields with very large **natural gas** reserves. Nearly one hundred fields each contain more than $140 \times 10^9 \, m^3$ but so far only fifteen have each produced more than $100 \times 10^9 \, m^3$. Some of these giant fields are also giant **oilfields**, e.g. Prudhoe Bay in Alaska and A.J. Bermudez in Mexico. The Urengoi Field in Siberia is capable of supplying 15% of the present world demand for **natural gas**. [AME]

gibber A desert plain, especially in Australia, which is mantled with a layer of pebbles or boulders. It is in effect a type of **stone pavement**. The pebbles are often made of highly resistant material, including fragments of **silcrete**. [ASG]

gibbsite (Al(OH)$_3$) A mineral exploited as an **ore** of **aluminium** in **bauxite** deposits. [DJV]

gilgai A form of micro-relief characterized by groups of small knolls or hummocks separated by shallow troughs or **basins**. On sloping surfaces, ridges and furrows develop parallel to the direction of slope. In arid regions, the formation of gilgai is probably related to volumetric expansion of **clay minerals** — especially **smectites** — during periodic wetting and drying cycles. This process may also move coarse material to the surface (see **desert pavement**), filling the troughs of the gilgai with stones. In areas of *permafrost*, similar micro-relief and polygonal patterned ground is created by volumetric changes during freezing and thawing of the *soil* zone. [AW]

Ollier, C.D. (1966) Desert gilgai. *Nature* **212**: 581–3.
Verger, F. (1964) Mottureaux et gilgais. *Annales de Géographie* **73**: 413–30.

gipfelflur A summit plane identified by mountain peaks of similar elevation; an imaginary plane which intersects the summits of the main peaks of a mountain region. A gipfelflur is related to neither a structural uniformity nor an ancient **peneplain**. [AW]

girdle distribution Pattern of arrangement of points (usually **poles of a plane**) along a **great circle** on a *stereogram*. (See **stereographic projection**.) [RGP]

glacial deposition Glaciers produce large amounts of debris. According to its location with respect to the **glacier**, such debris transported by an ice mass may be divided into three main categories : englacial debris (which occurs within the **glacier**), supraglacial debris (which occurs on the **glacier** surface), and subglacial debris (which occurs at the base of the **glacier**). Deposition of the transported material is a complex process, but that fraction deposited directly from the ice is called *till*. It consists of a wide range of grain sizes, and is thus often called *boulder clay*. It also possesses very little stratification, frequently contains far-travelled *erratic* material, tends to have *clasts* with edges and corners blunted by **abrasion**, and often has its larger particles showing a **preferred orientation** or alignment.

Traditionally, two types of *till* have been recognized: lodgement *till*, which is laid down subglacially when debris is released directly from the sole of the ice, and ablation *till*, which accumulates initially in a supraglacial position and is later lowered to the ground surface by undermelting. Ablation *till* can be further subdivided into meltout *till* and flow *till*; the former is the direct product of ablation continuing beneath a cover of detritus, while the latter consists of debris that has built up on the ice and after saturation with meltwater becomes so unstable that it flows or slumps into nearby hollows.

Till may be deposited as a series of distinctive landforms. Ridges of glacially deposited material occurring along the terminal margins of **glaciers** and *ice sheets* are called end *moraines*. On the sides of **glaciers** there may be prominent ridges called lateral *moraines*.

In some areas, particularly where a **glacier** opens out on

to a plain, the *till* is deposited as swarms of rounded hummocks, called *drumlins*.

Closely related to *drumlins* are a whole series of other streamlined forms described under the broad heading of fluted ground *moraine*. Some of the largest flutes are up to 20 km long, 100 m wide and 25 m high, but they are generally much smaller that this and are found just beyond the glacial front.

There is also a series of landforms and deposits that, while not strictly speaking made of *till*, owe their form and origin to glacial agency. Included in this category are *kames* and *eskers*. These are the product of glacial rivers. *Kames* are features produced at the margins of the ice and *eskers* are primarily forms developed beneath the ice. *Kames* consist of irregular undulating mounds of bedded **sands** and **gravels** that are essentially a group of alluvial cones or **deltas** deposited unevenly along the front of a stagnant or gradually decaying *ice sheet*. Some *kames* occur as *kame-terraces* and were formed along the trough between the **glacier** and the valley side, forming narrow flat-topped terrace-like ridges. Perforation or moulin *kames* are formed in hollows and perforations in decaying ice, and if they develop along major *crevasses* they may have an angular dog-leg form.

Eskers are elongated ridges of stratified **gravel**, usually thought to be the casts of streams formed either on the ice or beneath it. The largest examples may be 100 m high and extend for tens of kilometres.

The meltwater that flows from an ice mass creates large outwash plains which are sometimes called after their Icelandic name of *sandur*. These are characterized by multi-thread channels called *braids*, by abundant coarse debris and by marked variations in discharge, including catastrophic floods (e.g. Jökulhlaup).

The term *glaciofluvial* is used for meltwater activity. Sediments resulting from the interaction of **glaciers** and the sea are called *glacimarine* sediments. [ASG]

Brodozikowski, K. & van Loon, A.J. (1991) *Glacigenic Sediments.* Elsevier, Amsterdam.
Dowdeswell, J.A. & Scourse, J.D. (1990) *Glacimarine Environments: Processes and Sediments.* Geological Society, London.
Menzies, J. & Rose, J. (ed) (1987) *Drumlin Symposium.* Balkema, Rotterdam.
Van der Meer, J.J.M. (ed) (1989) *Tills and Glaciotectonics.* Balkema, Rotterdam.

glacial erosion Recent studies of the quantities of material being carried away by glacial meltwater streams indicate that **erosion** in a glaciated **catchment** may frequently be equivalent to a ground surface lowering of 2000–3000 mm per thousand years, perhaps ten times the norm for ordinary fluvial **catchments** without **glaciers**. Such **erosion** is achieved in a variety of ways. First, **glaciers** can in some respects be likened to a conveyor belt. If a rockfall puts a vast amount of coarse debris on to a **glacier** surface, for example, or if *frost shattering* sends down a mass of angular rock fragments on to the **glacier** surface, it can then be transported, almost whatever its size, down valley.

Second, beneath **glaciers** there is often a very considerable flow of meltwater. This may flow under pressure through tunnels in the ice at great speed, and may be charged with coarse debris from the bed of the **glacier**. Such subglacial streams are highly effective at wearing down the bedrock beneath a **glacier**. Third, although **glacier** ice itself might not cause marked **erosion** of a rock surface by **abrasion**, when it carries coarse debris at its base some **abrasion** can occur. This grinding process has been observed directly by digging tunnels into **glaciers**, but there is other evidence for it; rock beneath **glaciers** may be striated or scratched and much of the debris in **glaciers** is ground down to a fine mixture of silt and clay called *rock flour*.

Glaciers also cause **erosion** by means of *plucking*. If the bedrock beneath the **glacier** has been **weathered** in pre-glacial times, or if the rock is full of **joints**, the **glacier** can detach large particles of rock. As this process goes on, moreover, some of the underlying **joints** in the rock may open up still more as the overburden of dense rock above them is removed by the **glacier**. This is a process called **pressure release**.

As debris-laden ice grinds and plucks away the surface over which it moves, landforms are produced which give distinctive character to glacial landscapes. Of the features resulting from glacial quarrying, one of the most impressive is the *cirque*. This is a horseshoe-shaped, steep-walled, glaciated valley head. It is called a *cwm* in Wales, and a *corrie* in Scotland.

As *cirques* develop they eat back into the hill mass in which they have developed. When several *cirques* lie close to one another, the divide separating them may become progressively narrowed until it is reduced to a narrow, precipitous ridge called an *arête*. Should the **glaciers** continue to whittle away at the mountain from all sides the result is the formation of a pyramidal *horn*.

Some of the most spectacular erosional effects of glaciation observable today are those of valley glaciation. The lower ends of spurs and ridges are blunted or truncated; the valleys assume a U-shaped configuration; they become more linear; and hollows or troughs are excavated in their floors. Many high-latitude coasts are flanked by narrow troughs, called *fjords*, which differ from land-based glacial valleys in that they are submerged by the sea.

A further erosional effect of valley **glaciers** is the breaching of watersheds, for when ice cannot get away down a valley first enough — perhaps because its valley is blocked lower down by other ice, or because there is a constriction — it will overflow at the lowest available point, a process known as *diffluence*. The result of this **erosion** is the creation of a *col*, or a gap in the watershed.

The long profile of a glacial trough is characteristically irregular. The factors determining where pronounced deepening of the valley floor takes place have been the subject of many arguments. For example, it has been proposed that where two **glaciers** join there is a great increase in erosive capacity. Elsewhere it can be shown that the

glaciers have plucked preferentially at areas of non-resistant and closely jointed rock. Sometimes the basins may be areas where the rock was deeply **weathered** by chemical action or by freeze–thaw in preglacial times. Alternatively, if the valley is for some reason laterally constricted, say by a zone of resistant rock, the **glacier** enlarges its cross-sectional area by eroding its floor more deeply.

Tributary valleys to the main glacial trough have their spurs ground back and truncated. Furthermore, the floor of a trunk **glacier** is deepened more effectively than those of feeders from the side or at the head, so that after a period of prolonged glaciation such valleys are left as *hanging valleys* high above the main trough.

The development of an *ice sheet* tends to scour the landscape. In Canada there are vast expanses of empty territory where the *Pleistocene* **glaciers** scoured the land surface, removing almost all the superficial deposits and exposing the **joint** and **fracture** patterns of the crystalline rocks beneath. Streamlined and moulded rock ridges develop, including *roches moutonnées*. They are interspersed with scoured hollows which may be occupied by small lakes when the *ice sheet* retreats. In Scotland, relief that is dominated by this mixture of rock ridges and small basins is called *knock and lochan* topography.

Glaciers produce a great deal of meltwater, some of which flows on the **glacier**, some within, and some over the bedrock. It often flows very swiftly; moreover, the meltwater that moves in tunnels at the base of a **glacier** may be subject to very great pressure, and under certain conditions it can flow uphill. Some streams may also contain large quantities of debris that may give them further erosive power. In Denmark and Germany great subglacial meltwater rivers flowing southwards towards the wasting margin of the last Scandinavian *ice sheet* are thought to have cut the so-called *tunnel valleys*, some of which are over 70 km long and in places over 100 m deep. Some such channels have humped sections in their long profiles, indicating that the subglacial streams accomplished much **erosion** while flowing uphill under great pressure.

Lakes are a widespread feature of many areas that have been glaciated, and they result from a great variety of erosional and depositional causes. [ASG]

Eyles, N. (ed) (1983) *Glacial Geology*. Pergamon Press, Oxford.
Syvitski, J.P.M., Burrell, D.C. & Skei, J.M. (1987) *Fjords: Processes and Products*. Springer Verlag, New York.
Sugden, D.E. & John, B.S. (1976) *Glaciers and Landscape*. Arnold, London.

glacial proximal trough Trough resulting when a steep rock body, such as a rock *drumlin*, exposed to **glacier** ice, running water or wind, increases the velocity of the moving element as it flows round the obstruction. This increased velocity is often sufficient to initiate **erosion** around the body and a proximal depression is therefore created (Lassilla, 1986). [ASG]

Lassila, M. (1986) Proximal troughs and ice movements in Gotland, southern Sweden. *Zeitschrift für Geomorphologie* NF **30**: 129–40.

glacier A mass of snow and ice which, if it accumulates to sufficient thickness, deforms under its own weight and **flows**.

The biggest glaciers are called *ice sheets*. These have a flattened dome-like cross-section and are hundreds of kilometres in width. The most famous *ice sheets*, because of their size, are those of the two polar regions — Antarctica and Greenland. Those dome-shaped masses with a smaller area, less than about $50\,000\,km^2$, are called *ice caps*. Non-valley glaciers can be summarized as follows:

Ice sheet: more than $50\,000\,km^2$, with a flattened dome, which buries underlying relief.

Ice cap: a small *ice sheet*, with an area less than $50\,000\,km^2$, but which still buries the landscape.

Ice dome: the central part of an *ice cap* or *ice sheet*.

Outlet glacier or ice: a stream of ice that drains part of an *ice sheet* or *ice cap*, and which often passes through confining mountains.

Ice shelf: a thick, floating *ice sheet* which is attached to a coast.

Ice field: a relatively flat and extensive mass of ice.

The other main class of glacier is the valley glacier. These occupy basins or valleys in upland areas, and can be further subdivided into miscellaneous types as follows:

Valley glacier: a body of ice that moves down a valley under the influence of **gravity** and is bounded by cliffs.

Cirque glacier: a small ice body that occupies an armchair-shaped hollow in mountains which has been cut into bedrock; the hollows are sometimes called *cwms* or *corries*.

Niche glacier: a small upland ice body resting upon a sloping rock face or in a shallow hollow that the glacier itself has modified only slightly.

Diffluent glacier: a valley glacier that diverges from a trunk glacier and crosses a drainage divide through a *diffluence col*.

Piedmont glacier: a glacier that leaves its confining rock walls and spreads out to form an expanded foot glacier; formed on a lowland or at a mountain foot.

The movement of the glacier takes place in three main ways: by sliding over bedrock, by internal **deformation** (**creep**) of the ice, and by alternate compression and extension of the ice mass in response to changes in the bedrock surface below the ice.

Alpine glaciers have generally been recorded as moving at velocities between 20 and 200 m per year, but may accelerate to rates in excess of 1000 m per year down the steeper slopes. Some of the fastest rates are found on outlet glaciers of the polar *ice sheets*, where velocities as high as 7000 m per year have been recorded. Some glaciers are prone to surging, or flowing for short periods at rates of over 10 km per year. Glaciers in the Himalaya and the United States are known to have moved at rates over 300 m per day at the height of a surge. [ASG]

Drewry, O. (1986) *Glacial Geologic Processes*. Arnold, London.
Sharp, M. (1988) Surging glaciers: behaviour and mechanisms. *Progress in Physical Geography* **12**: 349–70.

glaciotectonism Structures and landforms (e.g. displaced mega-blocks) produced by **deformation** and dislocation of pre-existing soft bedrock (e.g. **chalk**) and **drift** masses as a direct consequence of glacier ice movement. [ASG]

Aber, J.S. (1985) The character of glaciotectonism. *Geologie en Mijnbouw* **64**: 389–95.

glacis The French term for a **pediment** which is generally applied to **erosional pediments** in arid regions, especially in North Africa. Several generations of glacis can give the footslopes of mountains and **escarpments** a stepped appearance. In North Africa, these multiple pediments — the highest being the oldest — have been attributed to increased **erosion** during wet periods in the **Quaternary** (Coque, 1962). [AW]

Coque, R. (1962) *La Tunisie Présaharienne: Étude Géomorphologique*. Armand Colin, Paris.

glass sand **Sand** for glassmaking. The **silica** content ranges from 95 to 99.8%. Alumina up to 4% is acceptable for common glass production but it must be <0.1% for optical glass although a small amount does help to prevent devitrification. **Iron oxides** can only be tolerated for green and brown bottle glass. The **sand** grains should be even-sized and less than 20 but more than 100 mesh (ASTM). [AME]

glassy rock A rock, usually volcanic, which consists partly or wholly of glass. Glassy rocks occur when molten rock cools too fast to allow crystallization. Volcanic glass of **basaltic** composition occurs as rinds on **pillow lavas**, as **hyaloclastite** and as fine-grained **tephra**. Silicic volcanic glass is more common, and most silicic **tephra**, and many entire silicic **lava flows**, are glassy. Volcanic glass is unstable and is susceptible to hydration — the low temperature absorption of water — which is usually associated with chemical changes. **Basaltic** glass is normally hydrated and chemically altered to **palagonite**. **Silicic** glass is normally hydrated to **perlite**, and subsequently experiences devitrification.

Glassy metamorphic rocks occur as dynamically metamorphosed rocks in **faults** (**pseudotachylites**) and as certain thermally metamorphosed rocks (**buchites**). [PTL]

glaucodot ((Co,Fe)AsS) A relatively rare **ore** mineral of **cobalt**. [DJV]

glauconite $((K, Na, Ca)_{0.5-1}(Fe^{3+}, Al, Fe^{2+}, Mg)_2(Si, Al)_4O_{10}(OH)_2 \cdot nH_2O)$ A **clay mineral** similar in composition to **biotite** and found as an **authigenic** mineral in sedimentary rocks. Glauconite was used in antiquity for green pigment, and as such used by the Romans in wall paintings. [OWT/DJV]

glaucophane ($Na_2Mg_3Al_2Si_8O_{22}(OH)_2$) A blue to black **amphibole mineral** found in metamorphic rocks. [DJV]

gleitbretter Layering produced by **deformation**. **Microlithon** is a more acceptable term for the same structure. (See also **cleavage**.) [RGP]

glory hole A large open pit, often without benches, from which **ore** has been, or is being extracted. [AME]

glory-hole mining **Open pit mining** in which **ore** is worked from the surface downwards in a conical excavation and is removed through underground workings beneath the **orebody**. [AME]

gloss coal The highest **rank lignite**. It is black, compact with a *conchoidal fracture* and glossy **lustre**. [AME]

gmelinite ($Na_2,Ca)(Al_2Si_4O_{12}).6H_2O$) A **zeolite mineral**. [DJV]

gnamma Basin produced in a rock surface, especially in **igneous rocks** and **sandstones** by **weathering** processes. They often occur on the surfaces of **inselbergs** and **tors**. [ASG]

gneiss/gneissosity A gneiss is a metamorphic rock characterized by gneissosity, which is a type of **foliation** characterized by compositional layering (*gneissic layering*) or lensoid structure and found in high-grade metamorphic or deformed **igneous rocks**. The layers or lenses may be produced either by **deformation** of a pre-existing texture or structure, or by *metamorphic segregation*, or, most commonly, by a combination of both. **Augen gneisses** are typical of deformed coarse-grained **igneous rocks**, particularly those containing *phenocrysts* or *megablasts*. These deform to produce the augen. In many gneisses, the origin of the layered structure is obscure, and in the past there has been much debate centred around the origin of banded granitoid gneisses, with some geologists favouring a metasedimentary and others a meta-igneous origin. Field observations in well-exposed **Precambrian shield** terrains, such as south Greenland, have demonstrated how heterogeneous **igneous bodies** may produce regularly banded gneisses under high **strains** in **shear zones**. [RGP]

goethite ($\alpha FeO.OH$) The orange 'rusty' hydrated oxide of **iron**, the typical **weathering** product of **iron** minerals. [DJV]

gold (Au) A native metal mineral, the main **ore** of the metal (see **native elements**, **precious metal**).

Gold was one of the earliest metals to be used by man, because of its occurrence naturally in the native, state, its ease of working and its attractive appearance. Gold artefacts are known from the 5th millenium BC onwards and many artefacts from this period may have been subsequently melted down and reworked. Gold was collected as **nuggets** from river **gravels** and obtained from **vein** deposits by crushing and washing the **ore**. The Egyptians and the Romans used the latter method at mines in Egypt and Nubia, and in south Wales. Working was by hammering, and later melting. Gold was used for jewellery, coins and decorative purposes, being too soft for working tools.

Gold was used extensively in the early Egyptian, Near Eastern and Aegean civilizations, and later by the Romans and Greeks in Europe and the pre-Hispanic Indians of central and south America. Some of the most famous examples of gold working in antiquity include the jewellery of Tutankhamun's Tomb, the Mycenean 'Mask of Agamemnon', and the British Early Bronze Age Rillaton Cup. Efforts to trace gold artefacts to their source by **chemical analyses** have been partially successful, for example in relating some Greek gold and **silver** coins to the Laurion mines near Athens, Greece. [OWT/DJV]

gold dust The fine specks of **gold** found in **gold placer** deposits. The very finest **placer gold** is called *flour gold*. Its extremely fine-grained nature explains its wide distribution throughout the entire thickness and width of auriferous stream gravels. [AME]

Gondwanaland The southern *supercontinent*, comprising South America, Africa, Arabia, Malagasy, the Indian continent, Sri Lanka, Australia, New Zealand and Antarctica — together with minor continental fragments within the Indian and South Atlantic Oceans, such as the Seychelles, Falklands, Kerguelen, etc. Originally named following observations of fossil similarities in late **Carboniferous** and **Permian** age deposits recognized in India in the Kingdom of the Gonds. It is probable that South China and parts of south-eastern Asia also formed part of this *supercontinent* which began to break up some 180 Ma ago. It is not clear when the *supercontinent* originally formed, but probably more than 2 Ga ago. [DHT]

Audley-Charles, M.G. & Hallam, A. (1988) *Gondwana and Tethys*. Geological Society Special Publication No. 37. Oxford University Press, Oxford.

goshenite A colourless **gem** variety of the mineral **beryl**. [DJV]

gossan A Cornish word used to describe the cellular mass of **limonite** and **gangue** developed by the **oxidation**

of outcrops of sulphide deposits. Gossans are generally characterized by **boxworks**.

Surface waters percolating down the outcrops of sulphide **orebodies** oxidize many **ore** minerals and yield solvents that dissolve other minerals. **Pyrite** is almost ubiquitous in sulphide deposits and this breaks down to produce insoluble iron hydroxides (**limonite**) and sulphuric acid:

$$2FeS_2 + 15O + 8H_2O + CO_2 \rightleftharpoons$$
$$2Fe(OH)_3 + 4H_2SO_4 + H_2CO_3$$

and

$$2CuFeS_2 + 17O + 6H_2O + CO_2 \rightleftharpoons$$
$$2Fe(OH)_3 + 2CuSO_4 + H_2CO_3$$

Copper, **zinc** and **silver** sulphides are soluble and thus the upper part of a sulphide **orebody** may be oxidized and generally leached of many of its valuable elements right down to the **water table**. This is called the *zone of oxidation*. The ferric hydroxide is left behind to form a **residual deposit** at the surface and this is the gossan or *iron hat* — such features are eagerly sought by prospectors. As the water percolates downwards through the *zone of oxidation*, it may, because it is still carbonated and still has oxidizing properties, **precipitate** secondary minerals such as **malachite** and **azurite**. However the bulk of dissolved material is carried on down to the zone of **supergene enrichment**. [AME]

grab sample A random sample, often hurriedly collected, of mineralized ground. It has no statistical validity and is often collected merely to check the nature of the mineralization. [AME]

graben Depressed *fault block*. Graben are often bounded by a series of parallel or sub-parallel **step faults** with the same sense of **displacement**. The **faults** on each side of the graben form **conjugate** sets and **dip** inwards. The opposite type of structure is a **horst**, which is an elevated **fault** block. (See **rift**.) [RGP]

grade in geomorphology A term used in **geomorphology** to describe the condition of **dynamic equilibrium**. The term was first used by G.K. Gilbert to explain the graded condition of a river where an exact balance exists between the load provided to the channel and the amount of material that can be moved by the river. Without change in the external supply of discharge or debris, the river will not erode or deposit sediment. Gilbert recognized that negative feedback operates to maintain grade: if a graded river encounters a reach of lower gradient, deposition will occur to steepen the bed gradient; on a steeper channel reach, **erosion** takes place to lower the gradient. Gilbert saw the smooth **long profile** of rivers as the visible manifestation of grade. W.M. Davis incorporated the concept of grade into his **cycle of erosion**: the smooth profiles of both slopes and rivers indicate that grade is achieved during the mature stage of landform development. [TPB]

Gilbert, G.K. (1877) *The Geology of the Henry Mountains*. United States Geographical and Geological Survey, Washington, DC.

grade of coal A little-used **coal** classification based on the degree of purity, i.e. the amount of **ash** present. [AME]

grade of ore The concentration of a metal in an **orebody**. Grade (or *tenor*) is usually expressed as a percentage or in parts per million (p.p.m.). The process of determining these concentrations is called **assaying**.

Grades vary from **orebody** to **orebody** and, clearly, the lower the grade, the greater the tonnage of **ore** required to provide an economic deposit. The tendency in metalliferous mining during this century has been to mine lower and lower grade **ores**. This has led to the development of more large scale operations with outputs of 40 kt of **ore** per day being not unusual. [AME]

graded bed Layer in which the grain size distribution is organized into a gradual, abrupt or step-wise vertical and/or lateral change throughout or in part of a bed. Where there is an upward change from coarser to finer grain sizes, the grading is described as 'normal' or 'positive', the converse being defined as *'inverse'*, *'reverse'* or *'negative'*. There may be a gradual change in the entire grain size distribution, as *'distribution'* or *'content' grading*. Alternatively, only the larger grain sizes may grade as in *'coarse-tail' grading*. If only one gradation of grain sizes occurs in a bed, it is called 'simple' or 'single' grading whereas 'multiple', 'repeated' or 'recurrent' grading involves stacked graded layers. Graded bedding develops when changes in flow velocities or concentrations allow the deposition of different grain size populations from either aeolian or fluid systems. [KTP]

gradiometer An instrument measuring the gradient of a **potential** field. A magnetic gradiometer consists of two sensing elements of **fluxgate**, **cesium-vapour** or **proton magnetometer** type, kept at a fixed separation in a vertical or horizontal plane at a distance that is small with respect to the distance of the causative magnetic body. Gradient readings tend to enhance the delineation of anomalies caused by shallow structures and resolve complex anomalies into their individual components. Regional and temporal trends in the **geomagnetic field** are automatically removed. Magnetic gradiometers are widely used in archaeological surveys. [PK]

grain size coarsening Recrystallization of aggregates of grains to produce an overall larger grain size. This is a common process when temperature is elevated, for example during metamorphism. Increasing the grain size results in a smaller proportion of rock volume being taken up by grain boundaries; there is a decrease in surface energy for the rock as a whole, which can drive the process. Grain size coarsening may be accompanied by metamorphic reactions, such as in **contact metamorphism**, or occur solely by grain boundary migration in rock such as **quartzite**. [SB]

granite A coarse-grained **igneous rock** composed of quartz (over 20%) and feldspar in which **plagioclase** and **alkali feldspar** are present in approximately equal proportions. Rock widely used as a **building stone** in antiquity, in Europe, the Near East, Russia, Egypt, South Africa, in India and Nepal for Hindu temples and by the Incas and Mayas. Egypt possesses probably the most famous granite used in antiquity, the 'Rose Syenite' from Syene, used for tombs, monuments and pavements. Other early uses of granite include stele from the Sudan, lamps from Alaska, statues and vases from Egypt and occasional **millstones**. [OWT]

granite–greenstone terrain One of the two main types of **Archean crust**, the other being the *high-grade gneiss terrain*. Granite–greenstone terrains are characterized by the association of **granite batholiths** and **greenstone belts**, which exhibit low metamorphic grades, typically *greenschist* facies. Such terrains are found in most of the major *Precambrian* **shield** regions of the world. Well-studied examples include the Barberton and Rhodesian **Cratons** of southern Africa, the Superior Province of Canada and the Yilgarn and Pilbara **Cratons** of Western Australia. Although it has been suggested that granite–greenstone and *high-grade gneiss terrains* represent fundamentally different types of **crust**, there is considerable evidence in many of the **Archean** regions of a transition between the two types of terrain. Thus in the Superior Province, for example, typical granite–greenstone terrain of the southern and central parts of the Province appear to merge into typical high-grade terrains in the north-west and north-east, which represent deeper levels of **erosion** and higher degrees of **deformation**. Moreover, many of the **granite batholiths** appear on closer observation to be areas of reworked older high-grade **gneiss** basement injected by younger **granite plutons**.

Granite–greenstone terrains that have not been subjected to subsequent lateral compression exhibit a characteristic outcrop pattern produced by rounded **batholiths** separated by irregular **greenstone** areas with cuspate contacts. The **deformation** of the **greenstones** appears to be largely contact-parallel and has been attributed to gravity-driven **diapiric** emplacement of solid **granitic** basement into the supracrustal cover, due to the negative buoyancy of the denser mafic **greenstones**.

Some of the older granite–greenstone terrains are

thought to represent the earliest period of permanent formation of the **continental crust**, the associated **greenstone** sequences being apparently floored by **oceanic crust**. These early terrains may have formed during a long period of **Archean** time from *c.* 3.8 Ga to *c.* 3.3 Ga. Younger terrains appear to be floored by **gneissose** basement indicating their formation on pre-existing **continental crust**. Isotopic evidence from the **Archean** of Western Greenland suggests that the **continental crust** grew significantly by lateral **accretion** between 3.8 and 3.3 Ga. [RGP]

Nisbet, E.G. (1987) *The Young Earth*. Allen & Unwin, Boston.

granodiorite A coarse-grained **igneous rock** composed of **quartz** (over 20%), and **feldspar** in which **plagioclase** forms over 67% of the total **feldspar** content. Granodiorite is a major component of **batholiths**. [RST]

granophyre A fine to medium-grained **felsic rock** (commonly **porphyritic**) of acid-chemical composition which forms minor intrusions and is characterized by the presence of a groundmass containing intergrown **quartz** and **alkali feldspar**. The intergrowths may contain isolated hieroglyphic-like crystals of **quartz** within **alkali feldspar**, or radiating discontinuous fibre-like growths of **quartz** which have grown normal to the crystal faces of **feldspar** *phenocrysts*. The 'hieroglyphs' or fibres are in parallel optical orientation, and this relationship between **quartz** and **alkali feldspar** is generally interpreted in terms of simultaneous crystallization of these two minerals from a cotectic/**granite** minimum melt. The intergrowth of **quartz** and **alkali feldspar** may be referred to as a *micropegmatite* or a *micrographic texture* (cf. **graphic texture**). [RST]

granulite A metamorphic rock crystallized within the high-T–high-P granulite facies, so characterized by a mineral assemblage of **plagioclase** and **pyroxene** ± **garnet**, **quartz**, anhydrous aluminosilicates (e.g. Al_2SiO_5), **alkali feldspar**, **calcite** and **forsterite**-rich **olivine**. Granulites commonly have a characteristic crystalloblastic mineral **fabric** in which the stable grain-boundary configuration shows an equilibrium dihedral angle of *c.* 120°. Hydrous mineral phases such as **amphibole** or **mica** are rare or absent. The **feldspar** may be dark in colour (and may contain *antiperthite*) and **alkali feldspar** may be **perthitic**. **Quartz** may also be dark in colour due to the presence of **rutile** (TiO_2) inclusions.

Granulite-facies rocks (including **charnockite**) are common in **Archean shield** areas and commonly exhibit compositional layering (e.g. **gneissose** texture) which may be subhorizontal on a regional scale and may be associated with *migmatites* and migmatitic **granites**. They are hence interpreted as products of regional metamorphism under high-T–high-P conditions at the base of the **continental crust**. [RST]

Park, R.G. & Tarney, J. (eds) (1987) *Evolution of Lewisian and Comparable Precambrian High Grade Terrains.* Geological Society Special Publication No. 27. Blackwell Scientific Publications, Oxford.

graphic texture An intergrown texture between **quartz** and **alkali feldspar** (generally **microcline–microperthite**). The intergrowth is analogous to that developed within **granophyre** but is visible in hand specimen. Such intergrowth is believed to result from simultaneous crystallization of **quartz** and **alkali feldspar**, as in the case of the development of **granophyric** (*micropegmatite*) texture. [RST]

graphite (C) The naturally occurring (low pressure) form of **carbon**. (See **native element**.) [DJV]

graptolites A distinctive class of extinct colonial animals, known almost exclusively from their skeletal hardparts. The class is placed with increasing confidence in the phylum *Hemichordata*, an extant wholly benthonic phylum with its origins in the Middle **Cambrian**. Graptolites occurred as benthos at least until the upper **Carboniferous**, possibly the **Permian**, but had a major planktonic component, the graptoloids, from the lowest **Ordovician** until the latest Lower **Devonian** or earliest Middle **Devonian**.

Benthonic graptolites are variable in form from minutely encrusting to upright conical colonies in excess of 30 cm. Most are, however, less than 5 cm long. They may be attached to shells or seaweed with a holdfast, or 'rooted' in the bottom sediment. A majority occurred in shelf environments, including inshore deposits, and they commonly occur in association with a varied benthos, to which they may be attached.

Planktonic graptolites, whilst typically preserved in deeper slope or offshore deposits (the classic, **black**, graptolitic **shale**), also occur in varied sedimentary environments, including inshore deposits into which they drifted after death. For this reason they have, unlike benthonic graptolites, high value in stratigraphic correlation, most especially in the **Ordovician** and **Silurian** *systems*. In size they range from about 5 mm to in excess of 1000 mm. Colonies with a span of more than 400 mm are not uncommon and in addition to being the first macro-zooplankton on Earth, planktonic graptolites may well have been the largest zooplankton in the history of the Earth. The largest colony known was a multibranched form with a diameter of 1200 mm.

Each colony has a number of branches (stipes), usually 1–2 mm wide, of various lengths in different species, and numbering as few as one or several hundred. Each stipe itself is a sequence of small cups (thecae) which housed the colonies' zooids during life (Rickards & Stait, 1984). The cups, indeed the whole of the graptolite periderm, are composed of the protein *collagen* in various fibrillar arrangements (Rickards & Dumican, 1984). Each theca is constructed with tiny growth increments (full rings and half rings dovetailing along zigzag commissures) and then overlain, and sometimes lined, with *collagen* fibrils in the form of criss-crossing cortical bandages.

Graptolites are classified according to the nature and disposition of the stipes, and according to the nature of the thecae. Most benthonic graptolites, especially the order *Dendroidea*, have two basic types of thecae along the stipes, so-called autothecae and diminutive bithecae associated with them. Planktonic graptolites have only one thecal type (theca; equivalent to the autotheca of dendroids) the bithecae having been lost in evolution early in the **Ordovician**, at least in graptoloids.

Within the benthonic dendroids neither bithecae nor autothecae vary much from species to species and classification of them is effected largely by examination of stipe disposition. This is to some extent also true of **Ordovician** graptoloids, but in the **Silurian** and **Devonian** the thecae show enormous variation in form, on generally simpler colonies, and this affords the most precise means of erecting a classification at species level.

Little is known of evolutionary changes in benthonic graptolites. Those genera which occur in the **Cambrian** also occur, apparently unmodified, in the **Carboniferous**. However, the planktonic graptoloids show great species diversity (Rickards, 1978) with several conspicuous diversity peaks, notably in the *Arenig, Caradoc, Llandovery,* and *Ludlow* series, each being lower than the previous one, and leading to **extinction** of the planktonic forms, probably in the earliest Middle **Devonian**. The patterns of evolution conform in general to that of the *punctuated equilibrium model.* The evolutionary explosions seem to originate from very few lineages; that at the base of the **Silurian** System from one only. Over a period of a few million years there is spectacular divergence typified by the appearance of new morphological types, and characterized in detail by patterns indicating that *mosaic evolution* is dominant. Thereafter is a period of quiescence, with parallel evolution *sensu lato* the dominant mode, coupled with gradual decline (reflected in **extinctions** outpacing originations). The main pattern is then repeated by one or more lineages becoming involved in explosive evolution.

Understanding of the mode of life of graptolites, benthonic forms apart, is in its infancy. Whilst there is a most spectacular fossil record and appreciation of their evolution, whether the graptoloids were passive drifters or automobile is unresolved (Kirk, 1969; Rickards, 1975). *Paleobiogeography* is better understood, because the morphological diversity allows specific identification and accurate plotting. It is known that in the *Arenig* there were two major provinces, the one tropical, the other at higher latitudes (so that the diversity peak referred to above is a composite of two provinces). There is some evidence of provincialism at later times, but to nothing like the same extent.

The reasons for the **extinctions** of the graptolites —

benthos in the **Carboniferous**, plankton in the **Devonian** — is discussed at length by Koren' & Rickards (1979): it is possible that the final act in the case of the planktonic forms was determined by the increased amount of lime in the world's oceans (organic-walled plankton having difficulties at such times); and that the benthonic graptolites were adversely affected by competition in the changing sea margins, especially if the developing **land plants** on the *Old Red Sandstone* continent effected changes in the food supply of the benthonic graptolites. (See Fig. G10.) [RBR]

Kirk, N.H. (1969) Some thoughts on the ecology, mode of life, and evolution of the Graptolithina. *Proceedings of the Geological Society, London* **1659**: 273–92.

Koren', T.N. & Rickards, R.B. (1979) Extinction of the graptolites. *In* Harris, A.L., Holland, C.H. & Leake, B.E. (eds). *The Caledonides of the British Isles–Reviewed*, pp. 457–66. Geological Society Special Publication No. 8. Scottish Academic Press, Edinburgh.

Rickards, R.B. (1975) Palaeoecology of the Graptolothina, an extinct class of the phylum Hemichordata. *Biological Review* **50**: 397–436.

Rickards, R.B. (1978) Major aspects of evolution of the graptolites. *Acta Palaeontologica Polonica* **23**: 585–94.

Rickards, R.B. & Dumican, L.W. (1984) The fibrillar component of the graptolite periderm. *Israel Journal of Earth Science* **6**: 175–203.

Rickards, R.B. & Stait, B. (1984) *Psigraptus*, its classification, evolution and zooid. *Alcherringa* **8**: 101–12.

gravel Loose sedimentary material having a grain size greater than **sand**. For geologists this boundary lies at 2 mm but in engineering circles it is put at 4.75 mm. Gravel fragments are usually of variable composition depending on their provenance. The ideal economic gravel deposit consists of a wide range of grain sizes of both **sand** and gravel so that a large number of screened sizes can be produced for different uses. Gravel deposits usually contain some clay which means that washing will be necessary. Undesirable weak materials such as **shale** can be removed by passing the screened gravel through very turbulent water in which only the sound fragments sink, or heavy media separation may be used employing a water suspension of **magnetite** or ferrosilicon having a density of about $2.5 \, \text{Mg m}^{-3}$, again the sound fragments sink. All this is necessary as buyers such as highway authorities and the construction industry have stringent specifications, especially for **aggregate** that goes into concrete. [AME]

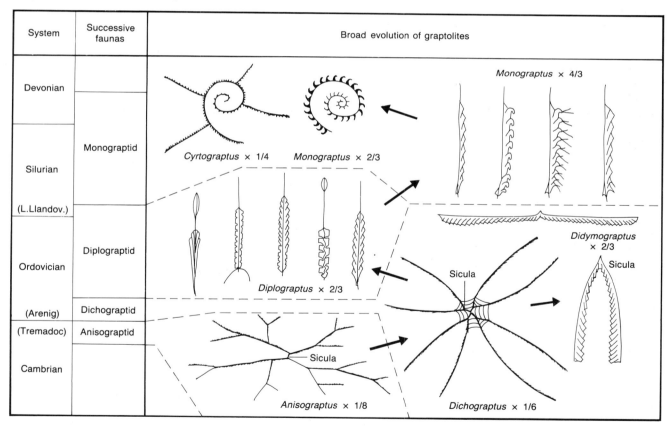

Fig. G10 Broad evolution of **graptolites**. Four successive, evolving and overlapping faunas are known leading from the Dendroidea (through the family Anisograptidae) to the Graptoloidea (Dichograptina; Diplograptina: Monograptina). Many evolutionary lineages are known and some typical faunal elements are depicted here. In the change from Anisograptid to Dichograptid we see a reduction in the number of stipes (and hence of thecae), a reduction in the thickness of the cortical layer of the periderm, a loss of bithecae and the beginnings of a tendency towards a scandent position of the stipes, that is they tend to come upwards in later species, rather than hanging down as in earlier species. In the Diplograptid Fauna this last tendency reaches a peak with two stipes back to back as in *Diplograptus* and *Climacograptus*, enclosing the nema for much of the rhabdosome length. Diplograptids, hydrodynamically 'heavy', tend to have distal floats. In the Monograptid fauna only one stipe remained in most forms and that uniserial scandent although some did develop branches or become coiled.

Constructional and Other Bulk Materials. Block 4 of The Earth's Physical Resources. Open University Press, Milton Keynes.

Harben, P.W. & Bates, R.L. (1984) *Geology of the Nonmetallics.* Metal Bulletin Inc., New York.

gravimeter (*gravity meter*) A small, protable instrument for measuring changes in **gravity** (but see **absolute gravity instrument**). Stable, or static, gravity meters are based on the principle that the weight of a small mass suspended from a spring changes with a change in gravity, and causes a change in the length of the spring according to **Hooke's law**. The change in length is very small, and requires amplification by optical, mechanical or electronic means. The spring serves the dual purpose of supporting the mass and acting as the measuring device, but the relatively strong spring required for the former purpose acts to the detriment of the latter.

Modern gravimeters make use of additional spring *systems* which exert an additional force that acts in the same sense as the spring movement resulting from a change in **gravity**. Such instruments are known as unstable or astatic gravimeters. The unstable principle is exemplified by the LaCoste & Romberg gravimeter (Fig. G11). The meter consists of a hinged beam carrying a small mass which is supported by a spring attached immediately above the hinge. With an increase in **gravity**, the beam moves down, and the angle between the spring and beam decreases. This causes the torque exerted by the spring on the beam to decrease, and hence to amplify the extension. The spring used in modern meters is the so-called '*zero-length spring*' which is pretensioned during manufacture so that the force exerted is proportional to the actual length of the spring rather than its extension. This results in an increased sensitivity over a wide range. All gravimeters have mechanisms to compensate for temperature changes, which would cause expansion or contraction of the spring to the detriment of the **gravity** measurement.

All gravimeters have to be accurately levelled before reading. This is accomplished manually during a land survey. The development of automatic levelling devices and methods of compensating for extraneous accelerations have made it possible for gravimeters to be used at sea and in the air (see **shipborne geophysical surveys** and **airborne geophysical surveys**). [PK]

Kearey, P. & Brooks, M. (1991) *An Introduction to Geophysical Exploration* (2nd edn). Blackwell Scientific Publications, Oxford.

gravimeter drift The change in reading with time of a **gravimeter** left at a fixed location. It arises from anelasticity of the mainspring. Drift is corrected by repeated readings at a **base station** during the course of the survey. [PK]

gravitational constant (*universal gravitational constant*) The constant of proportionality in **Newton's law of gravitation**, equal to $6.67 \times 10^{-11}\,\mathrm{m^3\,kg^{-1}\,s^{-2}}$. It has been suggested that the gravitational constant has decreased over geological time, causing a gradual increase in the Earth's radius. (See **expanding Earth**.) [PK]

gravitational load/pressure The pressure exerted by **gravity** acting on a piece of the Earth's subsurface of a given thickness and unit cross-sectional area. In a *triaxial stress* field, the gravitational load is the main contributor to the vertical *principal stress*, and is given by:

$$\sigma = \rho g z$$

where ρ is the density of the overlying rock, g is **gravity** and z the depth. [RGP]

gravitational potential (U) The scalar defined:

$$U = GM/r$$

where G is the **gravitational constant** and r the distance to the gravitating body of mass M. The first derivative of U in any direction gives the **gravity anomaly** in that direction. [PK]

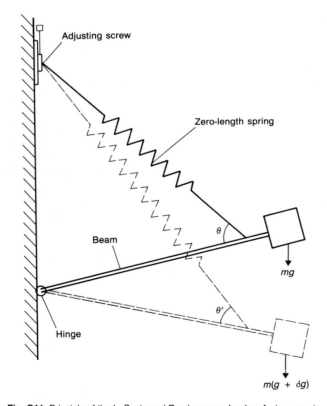

Fig. G11 Principle of the LaCoste and Romberg **gravimeter**. An increase in gravity δg causes an extension of the spring which is amplified by the reduced torque exerted on the beam as θ reduces to θ'. (After Kearey & Brooks, 1991.)

gravity (*acceleration due to gravity*) The acceleration experienced by an object in a gravitational field according to

Newton's law, on Earth resulting from the attraction of the mass of the Earth on the object. Using a spherical approximation of the shape of the Earth, gravity (g) is related to the mass (M) and radius (R) of the Earth by:

$$g = GM/R^2$$

where G is the **gravitational constant**.

Small variations in gravity, known as **gravity anomalies**, which result from changes in the density of the underlying rocks can be isolated by **reduction** and used to derive the possible depth and form of the causative body. The technique is extensively used in academic studies for delineating subsurface bodies of rocks on medium to large scales, and in investigations of **isostasy**. It is not widely used in exploration as it is a relatively expensive technique, requiring accurate heighting and an involved **reduction** procedure. It is used, however, in estimating **ore** tonnages via an **excess mass** calculation, and can be adapted to the location of subsurface voids in an **engineering geology** or archeological context (see **microgravity**). Gravity has applications in **geodesy** and *inertial navigation*. It also has military implications as the flight-path of a missile is affected by **gravity anomalies** along its track. [PK]

Kearey, P. & Brooks, M. (1991) *An Introduction to Geophysical Exploration* (2nd edn). Blackwell Scientific Publications, Oxford.

gravity anomaly The variation in the Earth's gravitational field, expressed in **gravity units** or **milligals**, caused by variation in the density of the subsurface rocks. Gravity anomalies should be isolated by the **reduction** procedure. [PK]

gravity collapse structure Structure produced by purely gravitational movement down a slope. A group of **structures** of this type are described by Harrison & Falcon (1936) and attributed to **slip** or gliding of **competent** sheets of rock (**limestone**) over *incompetent* **anhydrite**-bearing **shales** on the slopes of **anticlinal** ridges breached by **erosion**. These different types, illustrated in Fig. G12, are: **cascade folds**, *slip sheets* and **flaps**. [RGP]

Harrison, J.V. & Falcon, N.L. (1936) Gravity collapse structures and mountain ranges. *Quarterly Journal of the Geological Society, London* **92**: 91–102.

Gravity Formula *Clairault's formula* with constants agreed in 1967, which describes how **gravity** varies over the reference **spheroid**. The most accurate form of the formula (Mittermayer, 1969) is

$$g_\phi = 9780318.5(1 + 0.005278895 \sin^2 \phi + 0.000023462 \sin^4\phi) \text{ g.u.}$$

where g_ϕ is the predicted value of **gravity** at latitude ϕ. The Gravity Formula is used to compute the **latitude correction** in gravity **reduction**. [PK]

Mittermayer, E. (1969) Numerical formulas for the Geodetic Reference System 1967. *Bolletino di Geofisica Teorica ed Applicata* **11**: 96–107.

gravity gliding (*gravity sliding*) Movement of a sheet of rock down an inclined surface under **gravity**. The base of the displaced sheet usually rests on a **detachment horizon** or **décollement** plane parallel to **bedding**. This is the *sole fault*. At the proximal end of the sheet, a *listric normal fault* cuts up to the contemporary land surface, thus enabling the sheet to become detached from the unmoved **autochthon**. Some very large sheets are attributed to this process, e.g. the Bearpaw Mountains of Montana and Wyoming near the eastern margin of the North American Cordillera. This block is approximately 50 km across, and lies on undeformed horizontal strata about 200 km east of the margin of the mountain belt. The basal detachments of these large sheets have very small inclinations (*c.* 2°) and the ability of the sheet to overcome the high frictional resistance on the **detachment fault** is attributed to high fluid pressure (Hubbert & Rubey, 1959).

Many of the great **nappes** in *orogenic belts* have been attributed to gravity gliding, e.g. the Helvetic **nappes** of the Swiss Alps (see Fig. N1, p. 415). These **nappes** are composed of folded massive **limestones** resting on a **detachment horizon** of **ductile shales**, and have been attributed to sliding off the uplifted Aar Massif. However, caution is required in interpreting such structures, since a compressional **fold-thrust belt** may be deformed by movement on a lower **thrust** to produce a locally **foreland**-dipping **nappe**. Such a **nappe** may or may not be subjected to gravitational gliding. Thus the attitude of the detachment plane is not a reliable indicator. The presence of **extensional** structures in the proximal part of the sheet and compressional in the distal part are considered to be diagnostic of gliding. [RGP]

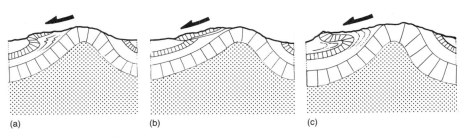

Fig. G12 Gravity collapse structure. (a) Cascade. (b) Slip sheet. (c) Flap.

(a)   (b)   (c)

Hubbert, M.K. & Rubey, W.W. (1959) Role of fluid pressure in mechanics of overthrust faulting. *Geological Society of America Bulletin* **70**: 115–206.

Suppe, J. (1985) *Principles of Structural Geology.* Prentice-Hall, Englewood Cliffs, NJ.

gravity unit (g.u.) The SI subunit applicable to gravitational fields, equal to an acceleration of $1\,\mu m\,s^{-2}$. [PK]

Great Barrier Reef A huge *reef* belt running for over 2000 km along the north-east coast of Australia, constituting the world's largest *reef* province. Extensive *reef* growth occurs along the continental shelf edge with the development of linear or ribbon *reefs*. The width of the shelf increases southward down the Queensland coast and reaches 300 km at its widest, defining a wide shelf **basin** up to 40 m deep, in which a belt of large 'inner' *reefs* are developed, still some 150 km from the mainland. In this belt some of the reefs cover areas of 100 km². The main growth phase began in the **Pliocene**-*Pleistocene* but has been episodic as a result of *Pleistocene* sea-level changes. The present growth phase began some 8000 years ago. [VPW]

Stoddart, D.R. (1978) The Great Barrier Reef and the Great Barrier Reef Expedition 1973. *Philosophical Transactions of the Royal Society, London* **A291**: 5–22.

great circle A circle on the Earth's surface that corresponds to a circumference, e.g. all lines of longitude and the Equator are great circles. (See also **small circle**.) [DHT]

Great Salt Lake The modern Great Salt Lake, Utah, USA, is a perennial, shallow (<12.5 m deep), wide, stratified saline lake (Fig. G13), a remnant of the formerly more extensive Lake Bonneville, itself a continental lake formed in an arid intermontane **basin**. Although evaporation exceeds precipitation in the area, lake water-levels are maintained by three rivers draining mountain ranges to the east; depth therefore varies seasonally and significant fluctuations in lake levels have been recorded during modern times with over 6 m change in the last one hundred years. For example, in 1850 the lake covered 4500 km² at an average depth of 4 m, but a later rise of water-level flooded an additional 1100 km², increasing the volume of the lake nearly four times. In contrast, during the drought of 1930–1935, a thick layer of **halite** was **precipitated** over the lake floor and, had the drought continued, the Great Salt Lake would have become ephemeral. Much of the present solute load may be a product of **dissolution** of the salt produced during this drought.

Today both **halite** and **gypsum** are formed both in parts of the lake and within the underlying and surrounding sediments. These surrounding arid mudflats are former lake bottom sediments, deposited during a higher water-level stand in the lake. In these mudflats **halite** occurs as

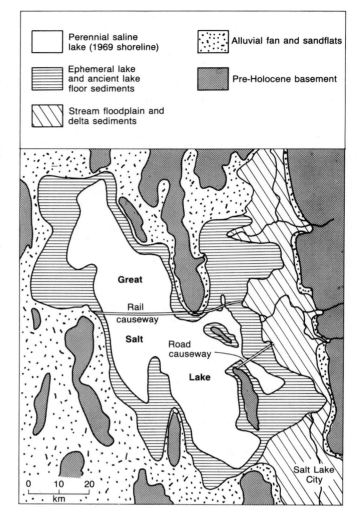

☐ Perennial saline lake (1969 shoreline)	⦂⦂ Alluvial fan and sandflats
▤ Ephemeral lake and ancient lake floor sediments	▨ Pre-Holocene basement
◿ Stream floodplain and delta sediments	

Fig. G13 Summary of modern depositional environments around **Great Salt Lake**, Utah, USA. (After Hardie *et al.*, 1978.)

an ephemeral crust, or in the form of hopper crystals; **gypsum** forms lensoid crystals and, more rarely, crusts. In the 1950s a railroad causeway built across the lake divided the water masses into two parts (Fig. G13). In the northern area where it is isolated by the causeway from major freshwater input, the water commonly exceeds **halite** salinity (up to 32.5% salinity from about 25%). Here **halite** is formed at the **brine**–air interface, where it floats as individual suspended crystals and as thin rafts before coalescing into larger rafts which sink to the lake bottom. South of the causeway the lake water is dominated by the fluvial input and the lake water is fresher, with little **evaporite precipitation**. A causeway built later to the south has reduced salinities further (to 8.2%) in the south-eastern arm of the lake (Fig. G13).

Although dominated by siliciclastic sedimentation from the surrounding **alluvial fans** and distal fan sediments, carbonate sediments are **precipitated** as **aragonite** needles

[306]

and as fine **calcite** crystals within the lake. Moreover, wind-induced **wave** action within the Great Salt Lake has led to the formation of **ooids**, both **aragonitic** and of high magnesian **calcite** in the recent past. High magnesian **calcite ooids** present contain a radial structure, whereas **aragonitic ooids** have a concentric structure; more rarely **ooids** demonstrate a composite mineralogy. In addition, in the recent past **aragonitic** *algal* mounds have built up within areas of the lake and are now found to depths of 4 m. These mounds occur over 100 km^2 and are circular or oval in plan, up to 1.5 m in diameter and 0.5 m high and are commonly surrounded by *ripple-laminated ooid* **sands**. [GMH]

Hardie, L.A., Smoot, J.P. & Eugster, H.P. (1978) Saline lakes and their deposits — a sedimentological approach. In: Matter, A. & Tucker, M. (eds) *Modern and Ancient Lake Sediments*, pp. 7–41. Special Publication of the International Association of Sedimentologists No. 2. Blackwell Scientific Publications, Oxford.

greenalite ((Fe,Mg)$_3$Si$_2$O$_5$(OH)$_4$) A hydrated **iron** silicate mineral of the **serpentine** group exploited as an **ore** in certain *'iron formations'*. (See **clay minerals**.) [DJV]

greenockite (CdS) An **ore** mineral of **cadmium**. [DJV]

greenstone A general term applied to a dark-green altered, low- or medium-grade metamorphosed basic **igneous rock** such as a **basaltic lava** (e.g. **spilite**) or minor intrusion (e.g. **dolerite**). The green colour reflects the *greenschist* facies mineral assemblage which includes fine-grained aggregates of **epidote**, *prehnite*, *pumpellyite*, **chlorite**, **actinolite**, **albite**, **sphene** and possible **carbonate minerals**. These minerals may replace primary **magmatic** minerals such as **pyroxenes** and **plagioclase** which record the former presence of igneous textures. With increasing metamorphic grade, greenstones pass into **plagioclase–hornblende** bearing rocks characteristic of the *amphibolite* metamorphic facies.

Greenstone is a loose term used by archeologists to refer to a group of metamorphosed basic **igneous rocks** such as *amphibolites*, **schists**, **gabbros** and *epidiorites*; also sometimes used by archeologists to include **serpentinites** and **jades**. Use of the term may be partly justified since prehistoric man may have used many of these rocks interchangeably. Greenstones were mainly used in antiquity for polished stone **axes**. [RST/OWT]

greenstone belt Linear to irregular-shaped volcano-sedimentary outcrop occurring within a predominantly **granitic** terrain of early *Precambrain* age (see **granite–greenstone terrain**). Greenstone belts have been identified in all the major low-grade **Archean** provinces, e.g. the Superior and Slave Provinces of Canada, the Rhodesian and Kaapvaal Provinces of southern Africa, and the Pilbara and Yilgarn **Cratons** of Western Australia. The belts extend for distances of several hundred kilometres and are very variable in width. They contain predominantly mafic to ultramafic volcanic sequences with subordinate sediments, typically *greywackes*, but including characteristic assemblages of **banded iron formation, chert** and carbonate. The supracrustal sequences, around 10–15 km in thickness in the average greenstone belt, are considered to be laid down in **extensional basins** either on thinned **continental** or on **oceanic crust**. The origin and tectonic evolution of greenstone belts is discussed under **granite–greenstone terrain**. [RGP]

greisen An aggregate of **quartz** and white **mica** (commonly **muscovite** or **lepidolite**) with accessory **cassiterite, fluorite, rutile, topaz** and **tourmaline** formed as a result of **alteration** of **granite** by fluorine-rich **hydrothermal solutions**. Greisens may therefore form in association with **granites** and may be associated with **tin–tungsten** mineralization, and may also occur in association with **kaolin** deposits. [RST]

greisen deposits These are mainly important for their production of **tin** and **tungsten**. Usually one element is predominant but there may be by-product output of the other. **Greisens** are usually developed at the upper contacts of **granite** intrusions and are sometimes accompanied by **stockwork** development.

Deposits of this type have been worked in the Erzgebirge on either side of the Czech–German border for many years, and cross-sections of some typical deposits are given in Fig. G14 together with a plan and section of the East Kemptville **orebody**. The Erzgebirge deposits occur as massive **greisens** and **greisen**-bordered **veins** in the uppermost endocontacts and exocontacts of small **lithium** mica–albite granite cupolas. These coalesce at depth to form a large **batholith**. The mineralization at East Kemptville is, however, developed beneath an inflection in the **granite**–metasedimentary contact of a discrete **pluton** that is part of the South Mountain Batholith. [AME]

Evans, A.M. (1993) *Ore Geology and Industrial Minerals: An Introduction.* (3rd edn). Blackwell Scientific Publications, Oxford.

grèze litée Bedded **scree** of angular rock debris, with the inclination of the layers paralleling that of the slope. In contrast to ordinary *gravitational* debris *slides*, the deposits, which in France may be up to 40 m thick, show a striking predominance of fines in their distal parts. The layers may dip as much as 40°. Snow patches (*nivation*) may play a role in their formation and *downwash* is an important process. The deposits have a rhythmic nature which suggests that under cold conditions the following happens: first, freezing of rocks on a cliff face causes disintegration, thereby releasing coarse, angular debris which slides downward over frozen subsoil; second, the following phase of thaw causes a mantle of half-fluid

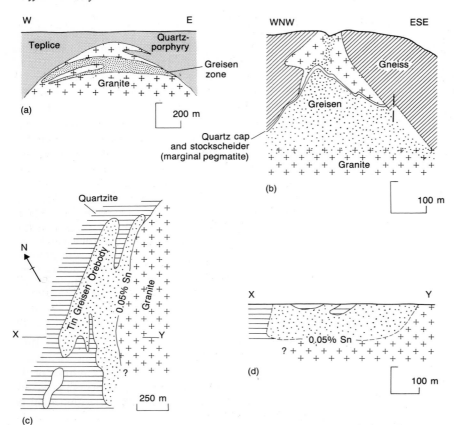

Fig. G14 Some tin **greisen** orebodies. (a) Section through Cínovec, Czechoslovakia. (b) Sadisdorf, Germany. (c) and (d) Map and section of East Kemptville, Nova Scotia, Canada. (After Evans, 1993.)

material rich in fines to spread over the stony layer (Guillen, 1964). [ASG]

Guillen, Y. (1964) Les grèzes litées comme dépôts cyclothémiques. *Zeitschrift für Geomorphologie* Supp. **5**: 53–8.

Griffith crack/flaw A crack around which **tensile stresses** concentrate leading to crack propagation and **failure**. Griffith (1924) observed that the **strength** of glass fibres was much less than theoretically predicted, and postulated that this was due to submicroscopic flaws. This led to an analysis of **stress** around an isolated flaw, which showed that **tensile stresses**, concentrated around the flaw, could cause **failure** at the lower values. Any cracks, flaws or inclusions in a material which act as **stress** concentrators to nucleate cracks and to reduce **strength** are hence referred to as Griffith cracks or flaws. [TGB]

Griffith, A.A. (1924) The theory of rupture. In: Biezeno, C.B. & Burgers, J.M. (eds) *Proceedings of the First International Congress for Applied Mechanics*, pp. 54–63. Technische Boekhandel en Drukkerij J. Waltman Jr., Delft.

Jaeger, J.C. & Cook, N.G.W. (1979) *Fundamentals of Rock Mechanics* (3rd edn), Chapter 4. Chapman & Hall, London.

grike Term originating from northern England and used to describe troughs separating blocks (or **clints**) in **limestone pavements**. The troughs, or runnels, develop along **joints** and other planes of weakness within the rock by solutional widening. The equivalent German term is *Kluftkarren*. **Solutional** widening may occur under a *soil* cover or due to the direct influence of rainfall. [HAV]

groove cast Elongate, relatively straight gutter-like scours or **sole marks** ranging in width from about 1 mm to decimetres, with lengths up to at least tens of metres. Groove casts are best preserved as **sand**-grade sediment in more cohesive finer grained siltstones and **mudrocks**. Grooves are produced when tools or clasts suspended in a current, such as pebbles, shells or plant material, erode a linear trackway into a soft **cohesive** substrate. The tool may become grounded and preserved at the end of the gouge or scour, or it may be resuspended in the current. Grounded icebergs on the continental shelf of Newfoundland are documented to have produced grooves 6 m deep, 30 m wide and several kilometres in length. Generally grooves are most common in *flysch* successions. [KTP]

grossular ($Ca_3Al_2Si_3O_{12}$) A garnet group silicate mineral. (See **garnet minerals**.) [DJV]

ground subsidence The relative sinking of the ground surface, especially that produced by human actions. Major causes of humanly induced ground subsidence include drainage of organic *soils* (as in the English Fenland), disturbance of the thermal equilibrium of *permafrost* (to produce *thermokarst*), removal of **groundwater** for municipal, industrial or agricultural supply (e.g. in Mexico City, where there has already been about 9 m of subsidence), mining or solution of **salt** beds (as in Cheshire, England), the abstraction of oil and **natural gas** (e.g. Long Beach, California, where around 9 m of subsidence has occurred), and the mining of **coal** and other minerals. There are examples of ground subsidence being sudden and catastrophic, as for example when **dewatering** of **aquifers** in **limestone** regions (e.g. Georgia, USA and the Transvaal, South Africa) leads to rapid **sinkhole** development. [ASG]

Coates, D.R. (1983) Large scale land subsidence. In: Gardner, R. & Scoging, H. (eds) *Mega-geomorphology*, pp. 212–34. Oxford University Press, Oxford.

groundwater The water present in water-saturated loose or indurated materials (rocks) beneath the **water table**. [AME]

grout/grouting A fine-grained slurry with high water content that is poured or injected into rocks to fill **joints** and other fissures before sinking a shaft or driving a tunnel through them, or building a dam on them. The aim is to strengthen the rocks and/or reduce their **permeability**. [AME]

growan Partially **weathered** or decomposed **granite** or other crystalline rock — especially as found on Dartmoor, south-west England. Growan may be formed through the **alteration** of **granite** by metamorphic processes, or it may be a product of **deep weathering**. The **erosion** of growan from around unweathered **corestones** may result in the formation of **tors** through exhumation. [AW]

growth anisotropy The tendency of certain minerals to grow at varying rates in different directions in a **stress** field. This can result in a strong bulk **anisotropy** following metamorphism. [SB]

growth fibre Fibrous crystal growing in response to rock **strain**. Growth fibres are found in **pressure shadows** forming *pressure fringes*, along **fault** surfaces (forming *slickencrysts*) and in **veins** and *tension gashes*. The fibres form in response to small **strain** increments by repeated **fracturing** followed by sealing of the crack with more of the wall mineral, or with a different mineral. This mechanism is known as 'crack-seal' (Ramsay, 1980). When this occurs, the long axis of the fibres follows the maximum **extension** direction at any one time. Thus curvature of fibres indicates changes in orientation of **principal strain** axes. (See **crack-seal mechanism**.) [SB]

Ramsay, J.G. (1980) The crack-seal mechanism of rock deformation. *Nature* **284**: 135–9.

grunerite ($Fe_7Si_8O_{22}(OH)_2$) A light brown **amphibole mineral**. [DJV]

grus An accumulation of poorly sorted, angular rock fragments formed during the **weathering** of crystalline rocks especially **granites**. Grus is a mixture of **gravel**, **sand**, silt and clay. It often occurs around **granite** outcrops such as **tors** or **inselbergs**, particularly in arid and semi-arid regions where the mechanical breakdown of rock predominates over chemical **weathering** processes. Grus is generally coarser in texture than **growan** which is the product of chemical **weathering** processes. [AW]

guano A calcium phosphate deposit formed by a reaction of the excreta of birds with **limestone** rock. Deposits of this type on small *oceanic islands*, e.g. Nauru, Christmas Island in the Pacific Ocean were once the world's main suppliers of phosphate. Similar deposits of bat excrement in caves have been worked in Malaysia. [AME]

gull A fissure or crack, sometimes sediment filled, which opens up on **escarpments** developed in sedimentary rocks, e.g. the Cotswolds in central England, as a result of the tensions produced by **cambering**. [ASG]

gully erosion The **erosion** of steep-sided channels and small ravines in bedrock and weakly consolidated surficial materials by ephemeral streams. Gullying is initiated when surface **runoff** erodes the *soil* zone and incises the underlying materials. Often, **erosion** commences when the vegetation cover is reduced or removed as a result of climatic change or human interference. Gullying may occur in bedrock such as **shale** or clay, or in weakly **cohesive** sediments such as volcanic **ash**, **loess** and **colluvium**. While gullying results from **accelerated erosion** when such materials loose their **cohesion** (see **badlands**), the residual materials must retain some **cohesion** in order to preserve the nearly vertical walls of the channels.

In addition to incision and **erosion** by periodic streamflow, **piping** and **tunnel erosion** at *headwalls* may also contribute to gully growth. Water moving beneath the surface erodes soluble material and fine-grained particles, excavating cavities which eventually collapse. In some cases, the surface *soil* may protect the underlying, erodible materials from incision. However, if the *soil* zone is breached, rapid **erosion** may result producing steep-sided gullies.

Once gullying is initiated, catastrophic rates of *headwall erosion* can occur. Rates approaching 650 m a^{-1} have been reported from areas of irrigated farmland (Akhmadov, 1979). However, some evidence suggests that rapid gullying takes place only during the first 25% of the gully

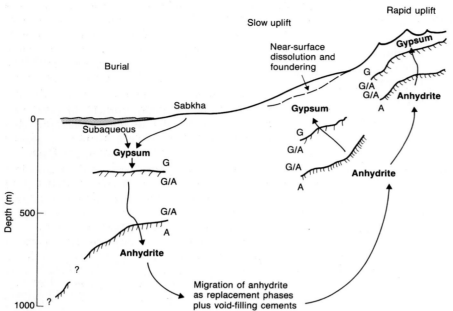

Fig. G15 Gypsum–anhydrite transitions on burial and subsequent uplift. G, all gypsum; A, all anhydrite; G/A, mixture of gypsum and anhydrite. (After Murray, 1964.)

system's life. After the phase of rapid growth, the rate of **erosion** decreases until the gully reaches its maximum dimensions; thereafter, vegetation begins to colonize the eroded surface and the gully becomes stabilized. The morphology of individual gully systems is related to their stage of development — juvenile gullies are generally linear, while more mature systems are often dendritic. Nevertheless, the physical and chemical properties of the eroding materials — especially their grain-size characteristics and dispersivity — along with climatic, geologic, and topographic factors, vegetation and land use, can also influence the morphology of gully *systems* (Stocking, 1979; Korotina, 1981). In parts of southern Africa, for example, the morphology of some **donga** *systems* follows exactly the pattern of footpaths and cattle tracks identifiable on *aerial photographs* which predate the present phase of gullying. The most likely reason for this is that the destruction of grass cover along tracks allowed **erosion** of the *soil* which protected the highly erodible material beneath. Once exposed, these materials are prone to rapid incision to depths of 30 m or more, and gullies up to 3 km in length have formed. [AW]

Akhmadov, H.M. (1979) On recent growth of gullies in Tadjikistan. *Geomorfologia* **1979–4**: 51–5.

Korotina, N.M. (1981) Rate of gully growth in the Ulyanovsk Region (right Bank of the Volga River). *Geomorfologia* **1981–4**: 78–83.

Stocking, M.A. (1979) Catena of sodium-rich soil in Rhodesia. *Journal of Soil Science* **139**: 517–22.

Watson, A., Price Williams, D. & Goudie, A. (1987) Reply to 'Is gullying associated with highly sodic colluvium?' Further comment to the environmental interpretation of southern African dongas. *Palaeogeography, Palaeoclimatology, Palaeoecology* **58**: 123–8.

Gutenberg discontinuity The **seismic discontinuity** at 2900 km depth which corresponds to the **core/mantle** boundary. [DHT]

guyot A flat-topped, submarine peak. Guyots originated as volcanoes which were elevated above sea-level. The flat top to guyots is presumed to have been caused by subaerial and marine planation during exposure. The truncated guyots then sank to their present submarine positions. Many guyots are now 1500 m below sea-level.

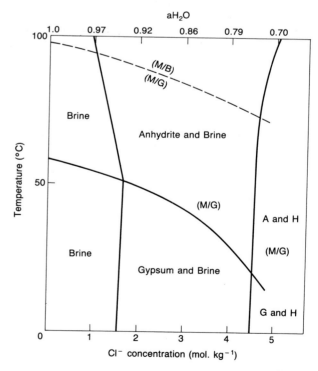

Fig. G16 Gypsum–anhydrite stability equilibrium in diluted seawater, seawater and seawater brines. G, gypsum; B, bassanite; A, anhydrite; M/G, metastable gypsum. (After Kinsman, 1966.)

Guyots are particularly numerous in the Pacific and their distribution pattern is related to **plate tectonics**. [HAV]

Menard, H.W. (1986) *Islands*, pp. 126–7. Scientific American Library, New York.

gypsum ($CaSO_4 . 2H_2O$) An important **evaporite** mineral, also known as *alabaster*, exploited for use in '*plaster of Paris*' and other industrial applications. **Alabaster** was widely used in antiquity. [OWT/DJV]

gypsum–anhydrite equilibrium Gypsum and **anhydrite** are both calcium sulphate minerals, **gypsum** being the hydrated form ($CaSO_4 . 2H_2O$), and anhydrite the dehydrated form ($CaSO_4$). **Gypsum** is commonly found near the surface, whereas **anhydrite** characterizes the higher temperatures and pressures associated with burial of a kilometre or more; rarely, however, **gypsum** is found down to depths of 1200 m, suggesting neither temperature nor pressure is the prime agent of anhydritization. There is a continuum of change between the two minerals, as **gypsum** is converted to **anhydrite** on burial and, conversely, **anhydrite** is hydrated to **gypsum** once **uplifted** to the shallow subsurface (Fig. G15). Until the 1950s **anhydrite** had not been found to form at surface temperatures and pressures, although it was common at depth. Its discovery in the **sabkha** environments of the Arabian Gulf and, later, Baja California albeit after a **gypsum** precursor, greatly aided comprehension of the transformation from **gypsum** to **anhydrite** and has complemented earlier laboratory studies.

Modern surface **anhydrite** is produced by dehydration of a **gypsum** precursor, although *bassanite*, or *hemihydrate* ($2CaSO_4 . H_2O$), may form an intermediate phase and, indeed, is a necessary intermediate phase in the absence of water. Laboratory experiments on gypsum–anhydrite equilibrium have been hampered by difficulties in nucleating, and therefore **precipitating**, anhydrite from supersaturated solutions at temperatures below 60°C and have shown metastable **gypsum** to be **precipitated** well within the calculated stability fields of **anhydrite** (Fig. G16). Thus **gypsum** is likely to be the calcium sulphate mineral most widely formed as a primary **brine precipitate** in nature, whereas **anhydrite** forms only from a precursor **gypsum** where surrounding **brines** are hygroscopic or where subaerial desiccation of **gypsum**, via *bassanite*, takes place.

Rehydration of **anhydrite** to **gypsum** requires the addition of **meteoric waters**, and therefore some **porosity** in the precursor **anhydrite**. This reaction, if no calcium sulphate is removed in solution, produces a 61% volume increase, although the **lithostatic pressure** of 60–75 m overburden can prevent expansion. Volume increases are indicated by contortions of the **gypsum** beds and by numerous fibrous, cross-cutting **gypsum** veins, commonly parallel to the surface. Many secondary **gypsum** occurrences show no evidence of an increase in volume, indicating removal of some calcium sulphate in solution in addition to overburden pressure. [GMH]

Kinsman, D.J.J. (1966) Gypsum and anhydrite of recent age, Trucial coast, Persian Gulf. In: Rau, J.L. (ed.) *Second Symposium on Salt*, pp. 302–26. North Ohio Geological Society, Cleveland, OH.
Murray, R.C. (1964) Origin and diagenesis of gypsum and anhydrite. *Journal of Sedimentary Petrology* **34**: 512–23.

H

half-graben Type of asymmetrical *fault block* **structure** common in **extensional fault systems**. The structure is formed by a dipping **fault** on one side of a block of tilted strata, creating a wedge-shaped zone that may become filled with younger sediments. **Extensional fault** zones typically exhibit a series of half-graben (Fig. H1). [RGP]

half-life The time taken for the mass of a *radioactive* element to decrease by a half, used in *radioactive dating*. [PK]

half-spreading rate *Oceanic magnetic anomalies* allow the age of the ocean floor to be determined at different places and hence to calculate the rate of movement from each *oceanic ridge*. This is the half-spreading rate as the **oceanic crust** on both sides of the ridge are or were spreading at identical rates, therefore the ocean is or was opening at twice the rate of the individual oceanic basin on each side. [DHT]

halite (NaCl) The mineral also known as *rocksalt*, occurring most commonly in **evaporite** sequences and exploited for use in the chemical industry and for culinary purposes. [DJV]

halloysite ($Al_4Si_4O_{10}(OH)_8 . 8H_2O$) A **clay mineral** made up of an irregular sequence of **kaolinite** layers plus inter-layer water. [DJV]

haloclasty The disintegration of rock or building material as a result of the action of salts, which can cause disruption through salt crystallization, salt hydration, or the thermal expansion of salts. The process appears to be especially important in arid areas, but it can also occur in polluted cities, in coastal situations, or in areas where the rocks themselves may weather to produce soluble salts. See also **salt weathering**. [ASG]

Goudie, A.S. (1985) Salt weathering. *School of Geography, University of Oxford, Research Papers Series* No. 32.

hammada Flat or gently dipping bare rock desert surface, sometimes with a thin veneer of gravel or a boulder *lag*. Surficial debris is often angular, its origin being **weathered** surface rock. From the arabic word 'Hammadat', meaning 'unfruitful'. Covers large areas of desert in many arid regions. [RDS]

Hammer chart A graticule devised by Hammer (1939) used in calculating **topographic corrections** for **gravity reduction**. The graticule (Fig. H2) is divided into some 130 compartments. The centre is placed on a topographic map of the area under consideration and the mean elevation determined for each compartment. Standard tables provide the gravitational contribution of each compartment, and these are summed to give the *terrain correction*. [PK]

Hammer, S. (1939) Terrain corrections for gravimeter stations. *Geophysics* **4**: 184–94.

hamra A red, sandy *soil* with a high clay content as found in some desert regions. The term is from the Arabic 'hamra', 'red'. [AW]

hangingwall The wall and, by extension, the body of rock lying above a horizontal or inclined **fault** or **orebody**. [AME]

hard coal A high **rank coal**, e.g. coals with a *calorific value* above $23.86\,MJ\,kg^{-1}$; this includes most **bituminous coals** and **anthracite**. [AME]

hardground An indurated surface caused by synsedimentary lithification of the seafloor sediment. The term has also been used in the past in a more general way for any hard substrate, including **unconformity** surfaces, especially when colonized by a hard substrate biota.

There has been considerable interest in present-day hardground surfaces as sites of active marine **cementation**. Fossil examples have also been a focus of attention because of the often well-preserved biotas associated with them and because of their sedimentary significance. A hardground reflects a period of induration and, since most **cementation** is a relatively slow process, the hardground must represent a pause or decrease in rate of sedimentation. This is often confirmed by a study of the associated biota.

Fossil hardgrounds are recognized by a variety of criteria. They are often identified by having an irregular morphology reflecting local **erosion** and, because hardgrounds are generally rather thin, they are readily reworked to form **intraclasts** (*intraformational conglomerates*). However, they are more commonly recognized because of evidence of having been colonized by a hard substrate biota of borers and encrusters. Identifying a true hardground from its fossil biota requires care because other

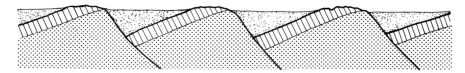

Fig. H1 Half-graben.

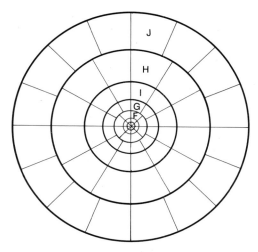

Fig. H2 A typical **Hammer chart**. A series of these with zones varying in radius from 2 m to 21.9 km are used with topographic maps of varying scale.

types of substrate may be colonized by similar biotas mimicking hardgrounds, and a succession of phases of colonization by different communities may have occurred as the substrate became lithified. Many hardgrounds have mineral encrustations, typically of **iron** and **manganese** oxides, especially if they represent extended periods of exposure on the seafloor.

They occur in a wide variety of environments from tidal flats to the deep ocean. They cover large areas of present-day subtropical shallow water carbonate environments where the seawater is supersaturated with respect to calcium carbonate. In such settings stabilized surfaces may become rapidly cemented by **aragonite** or high magnesian **calcite** cements and stabilization may be facilitated by sediment binding by *cyanobacteria*.

The term strictly refers to an indurated seafloor and does not cover such phenomena as **beachrocks**. [VPW]

Fursich, F.T. (1979) Genesis, environments, and ecology of Jurassic hardgrounds. *Neues Jahrbuch für Geologic und Palaeontologie Abhandlung* **158**: 1–63.
Goldring, R. & Kaźmierczak, J. (1974) Ecological succession in intra-formational hardground formation. *Palaeontology* **17**: 949–62.

hardness of water The property of water resulting from the presence of dissolved alkaline earth salts which prevents the formation of a lather with soaps. Temporary hardness is due to the presence of calcium bicarbonate and can be removed by boiling. Permanent hardness due to other calcium and **magnesium** salts such as sulphates can

be removed by *ion-exchange* processes or by detergents. The total hardness of a water sample is usually expressed in p.p.m., i.e. total hardness = 2.497 Ca + 4.115 Mg. [AME]

harmotome ($Ba(Al_2Si_6O_{16}) \cdot 6H_2O$) A **zeolite mineral**. [DJV]

harzburgite An **ultramafic**/ultrabasic **rock** composed of **olivine** (over 40%) and *orthopyroxene* (5–60%) with minor *clinopyroxene* (below 5%). Formation of **basaltic magma** from *lherzolite* **mantle** (*olivine–orthopyroxene–clinopyroxene* ± **plagioclase**, **spinel** or **garnet**) leaves a harzburgite residue. Harzburgites are therefore common among ultramafic **xenoliths** contained in **alkali basalts**, and also form the **mantle** sequence of **ophiolite** complexes. [RST]

hastingsite ($NaCa_2Fe_4Al,Fe)Al_2Si_6O_{22}(OH)_2$) A calcic **amphibole mineral**. [DJV]

hauyne (($NaCa)_{4-8}(AlSiO_4)_6(SO_4)_{1-2}$) A **feldspathoid mineral**. [DJV]

hawaiite A basic alkaline volcanic rock containing **plagioclase** of **andesine** composition (An_{30}–An_{50}) together with **pyroxene**, Fe–Ti oxide and/or minor **quartz** or **nepheline**. [RST]

head of water (*hydraulic head*) The height of the surface of water above a specific point especially when considered as a measure of the pressure of water at that point. Where there are two bodies of water at different levels then the difference in height of their surfaces is known as the *head difference*. When the water level in an **aquifer** (or in other situations) is lowered this is known as the *head loss*. [AME]

heat flow In the continents, the temperature measured at different depths within the upper 1–3 km, yet below some 200 m (the depth below which climatic temperature effects, including those of the last **ice age**, can be neglected) increases at some 20 to $40°C\,km^{-1}$. Similar increases are also observed in oceanic sediments, although the depth of study is usually only a few metres. The average global *geothermal gradient* is some $24°C\,km^{-1}$, which would correspond to partial melting of **mantle** rocks at some 50 km depth. As seismic studies do not indicate any **plasticity**, never mind actual melting, until much greater depths, both continental and oceanic *geothermal gradients* are assumed to decrease in the lower **crust** and upper **mantle** so that they intersect the melting point of **mantle** rocks at the top of the seismic **low velocity zone** which is commonly considered to be at a temperature of *c*. 1200°C as the fluids present at these depths are usually considered to be the result of a 0.1% partial melt. **Xenoliths** brought up from the **lower continental crust** indicate ambient temperatures in such regions of about 900°C, although the effect of impurities, particularly carbon dioxide and **sulphur**, on temperature estimates for such depths are not clear. Below the **low velocity zone**, the *geothermal gradient* must be below that of the **adiabatic** increase in melting point with depth, *c*. 0.33°C km^{-1}, as seismic evidence does not indicate the presence of partial melting. The actual distribution of conductive heat flow in the oceans indicates a decrease with distance from the *oceanic ridges* that is proportional to the square root of the age of the **oceanic crust**, varying from *c*. 100 mW m^{-2} near the crest to less than 50 mW m^{-2} in oceanic areas more than 100 Ma old. Virtually all of this heat must originate in the underlying **mantle** as the **oceanic crust** is very poor in *radioactive* elements. A similar square root of age dependence is seen in the continents, with a heat flow of 60–75 mW m^{-2} associated with *Tertiary* and **Mesozoic orogenic** areas and *c*. 38 mW m^{-2} in **cratonic** areas. The average continental heat flow is about 57 mW m^{-2} and is mostly accounted for by radiogenic heat produced in the **upper continental crust**, which contains some 2.8 p.p.m. **uranium**, 10.7 p.p.m. **thorium**, and 2.8 wt% **potassium**, and the concentration of radiogenic heat sources in the **lower continental crust** is about an order of magnitude less (0.3 p.p.m. **uranium**, 1.1 p.p.m. **thorium**, 0.3 wt% **potassium**). The annual global heat loss is some $4.1–4.3 \times 10^{13}$ W (1.3×10^{21} J a^{-1}), the uncertainty arising mainly from difficulty in estimating the amount of heat loss due to **convective** transfer of heat, particularly at *oceanic ridge* systems. Internal heat sources are radiogenic heat (mainly from **uranium**, **thorium** and **potassium**), gravitational (particularly separation of the **core**), exothermic chemical reactions, latent heat of crystallization and exothermic phase changes, kinetic heat (particularly frictional) and any remaining primordial heat. Locally, surface heat flow may be raised by tectonic events, such as **diapirs**, carrying heat upwards. Most of the heat flow in **cratonic** areas can be accounted for by radiogenic heating from surficial radiogenic elements, implying little or no contribution from the lower **crust** or **lithospheric mantle**. Approximately 75% of the global heat loss is through the **oceanic crust** (Fig. H3). [DHT]

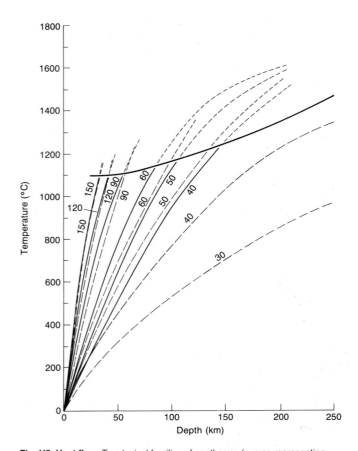

Fig. H3 Heat flow. Two typical families of geotherms (curves representing the variation of the Earth's temperature with respect to depth) are shown here for a continental province (– – –) and an oceanic province (——). The members of each family of geotherms are labelled according to the heat flow (in milliwatts per square metre) produced at the Earth's surface. As might be expected, the temperature found at a given depth in a region of high heat flow will be higher than the temperature at the same depth in a region of low heat flow. The heavy black curve that most geothermal curves intersect represents the temperature at which rock will begin to melt in the mantle. The depth at which such melting is observed is variable, depending on the heat flow and the geotherm for that region. Under some continental regions with low heat flow (in particular the stable Precambrian shields) there is probably no melting. Many geologists and geophysicists believe that the base of the rigid lithosphere is defined by the onset of partial melting. (After Pollack & Chapman, 1977 (a) and (b).)

Pollack, H.N. & Chapman, D.S. (1977b) The flow of heat from the Earth's Interior. *Scientific American* **237**: 60–76.

heavy-metal pollution The introduction of heavy metals into the environment, usually by human activities, particularly industrial processes, in quantities that will adversely affect the environment or the living things within it. Heavy metals are any of the metals that react readily with dithizone (C_6H_5N) and include **arsenic**, **cadmium**, **copper**, **lead**, **nickel** and **silver**. Since they are elements, heavy metals cannot be broken down into non-toxic forms and thus represent a long-term threat. [NJM]

Bott, M.H.P. (1982) *The Interior of the Earth*. Arnold, London.
Brown, G.C. & Mussett, A.E. (1981) *The Inaccessible Earth*. Allen & Unwin, London.
Pollack, H.N. & Chapman, D.S. (1977a) On the regional variation of heat flow, geotherms, and the thickness of the lithosphere. *Tectonophysics* **38**: 279–96.

hectorite (Na$_{0.3}$(Al,Mg,Li)$_3$Si$_4$O$_{10}$(OH)$_2$.4H$_2$O) A **clay mineral** belonging to the **montmorillonite** (or **smectite**) family and characterized by being **lithium**-bearing. [DJV]

hedenbergite (CaFeSi$_2$O$_6$) A *clinopyroxene* mineral. (See **pyroxene minerals**.) [DJV]

heliotrope A green and red variety of **chalcedony**, the mineral that is itself a cryptocrystalline variety of **silica**. (See **gem**.) [DJV]

hematite (Fe$_2$O$_3$) A major **ore** mineral of **iron** and **accessory mineral** in many rocks. (See **oxide minerals**.) Hematite was used to give red pigment for European Paleolithic cave art. [OWT|DJV]

hemimorphite (Zn$_4$Si$_2$O$_7$(OH)$_2$.H$_2$O) A mineral that occurs in the oxidized zone of **zinc ore** deposits and that has constituted a minor **ore** of **zinc** in several deposits. [DJV]

hemipelagite Abbreviation for *hemipelagic deposit*. Sediment, generally in the finer-grained spectrum, formed from the settling of grains out of suspension and from low density, low **viscosity turbidity currents**, **nepheloid layers** and other semi-permanent/permanent ocean currents (see **thermohaline current**). The biogenic component ranges from 5 to 75%, with >40% of the terrigenous component as silt-grade sediment. Hemipelagites typically appear structureless, occur in layers up to decimetres thick and may show thorough to partial *bioturbation*. [KTP]

hercynite (FeAl$_2$O$_4$) A member of the **spinel** group of minerals. (See **oxide minerals**.) [DJV]

hessite (Ag$_2$Te) A relatively rare **ore** mineral of **silver**. [DJV]

heulandite (CaAl$_2$Si$_7$O$_{18}$.6H$_2$O) A **zeolite mineral**. [DJV]

hiddenite A green **gem** variety of the mineral **spodumene**. [DJV]

high energy window The suggestion, first introduced by Neumann (1972) that in the mid-*Holocene* on tropical coasts there was a period when **wave** energy was higher than at present. This occurred during the phase when the present sea-level was being first approached by the **Flandrian** (*Holocene*) **transgression** and prior to the protective development of **coral** *reefs*. The 'window' may have operated on a more local scale on individual *reefs* with **waves** breaking not on margins of an extensive *reef* flat as at the present time, but more extensively over a shallowly submerged *reef*

top prior to the development of the *reef* flat (Hopley, 1984). [ASG]

Hopley, D. (1984) The Holocene 'high energy window' on the central Great Barrier Reef. In: Thom, B.G. (ed.) *Coastal Geomorphology in Australia*, pp. 135–50. Academic Press, Sydney.
Neumann, A.C. (1972) Quaternary sea level history of Bermuda and the Bahamas. *American Quaternary Association Second National Conference Abstracts*, pp. 41–4.

high sulphur crude oil **Crude oil** containing more than 1.7 wt% **sulphur**. The **sulphur** must not be in the form of H$_2$S, oils with high H$_2$S contents are called *sour crudes* (see **crude oil**). Intermediate sulphur crudes run 0.6–1.7% **sulphur** and low sulphur crudes less than 0.6% **sulphur**. Very '*sweet*' crudes have a **sulphur** content as low as 0.1–0.2%. There has long been a demand by refineries for *sweet crudes* and excessive consumption of them so that they now only form a small proportion of the world's oil reserves. [AME]

Himalayan mountain belt The Himalaya are a complex of mountains, including large uplifted **plateaux** as well as deformed areas, most of which are still currently active. It is generally recognized as resulting from a combination of a series of *plate* collisions, commencing at least 100 Ma ago, between the Indian subcontinent and various **island arcs** and relatively small continental blocks, such as Lut and Dzungaria, and eventually mainland Asia. More recently, the main features have arisen as a consequence of the underthrusting of the Indian subcontinent crustal rocks beneath those of Asia, causing local doubling of the crustal thickness and associated **isostatic** adjustments. (See also **indentation tectonics**.) [DHT]

hinge fault **Fault** where the displacement decreases to zero at its termination, so that the **fault** acts as a hinge. [RGP]

history of the Earth sciences The history of the Earth sciences comprises the dual and interrelated themes of the utilitarian and philosophical understanding of the Earth. These themes have for long periods been independent of one another, although at certain times (as for example between 1920 and 1950 when **geophysics** found application as the principal technique for finding new *hydrocarbon reservoirs*) there has been a fruitful interchange of ideas and technologies. In the account that follows, the history is considered under four headings: the structure of the Earth; the dynamics of the Earth; the record of the past; and the exploitation of resources.

THE STRUCTURE OF THE EARTH. The philosophical understanding of Earth the planet can be said to have begun in the 6th century BC with Pythagoras, native of the Aegean island of Samos but banished to the Greek colony in southern Italy, whose followers taught that the Earth was round because at eclipse it cast a curved shadow on the Moon. Eratosthenes, the second director of the Library at Alexandria, calculated the diameter of the Earth one mid-summer day in the late 3rd century BC from the difference in angle of the midday sun between Alexandria and Aswan (on the Tropic of Cancer) 750 km to the south. Subsequently little advance was made in understanding the form or properties of the Earth until after the Renaissance.

The rise of the British Navy through the 16th century and the exploration of the Americas inspired research into improving location-finding in the open ocean. In 1576 Robert Norman, purveyor of nautical instruments in London, carefully measured the inclination of a magnetic needle from the horizontal, an observation that was refined in 1580 by William Borough (1536–1599), a retired naval commander. This same interest in magnetism encouraged William Gilbert (?1540–1603), physician to Queen Elizabeth, to model the *Earth's magnetic field* by placing magnetic needles around a sphere of *lodestone* (**magnetite**), the inclination of the needles indicating latitude. Yet when Norman and Borough's measurements were repeated in 1622 and 1635 the magnetic inclination at London had changed by more than seven degrees in fifty years, a remarkable demonstration that the **geomagnetic field** was dynamic.

The determination of longitude was only possible through the exact measurement of time. In 1670 the Académie Royale des Sciences funded an astronomer, Jean Richer (1630–1696), to travel to Cayenne in French Guiana where he found that a pendulum clock, adjusted for Paris, ran 148 seconds a day slow. Isaac Newton's (1642–1727) *Principia* appeared in 1687, in which it was claimed that the Earth was held together by the force of **gravity** and predicted that, as a result of centrifugal force, the diameter of the Earth at the equator must be one part in 229 longer than at the poles. Newton saw that Richer's observations were in accord with the anticipated reduction of **gravity** at the Equator.

The **geodetic** survey of France from the Mediterranean to Channel coasts was completed in 1718 when the survey director Jacques Cassini (1677–1756) announced that, contrary to Newton's prediction, the Earth was shaped like an egg (prolate ellipsoid). In 1735 three French mathematicians (Pierre Louis Moreau de Maupertuis (1698–1759), Pierre Bouguer (1698–1758) and Alexis-Claude Clairaut (1713–1765)) obtained funding from l'Académie Royale des Sciences, to survey comparative degrees of latitude in Peru and Lapland, to find the true shape of the Earth. The Lapland survey was completed in only 14 months, while the Peruvian expedition took nine years. The results confirmed Newton's predicted flattening at the poles. Through their work the three Frenchmen had mastered the basics of terrestrial **gravity** observation: Clairaut established the relationship between **gravity** variation and latitude, and Bouguer (on the Peru survey)

discovered the influence of the density of near-surface rocks on the relationship between **gravity** and altitude.

In 1774 the British Astronomer Royal Nevil Maskelyne (1732–1811) measured the offset from the vertical (in the reference frame of the fixed stars) of two plumb lines located to either side of an isolated mountain (Schiehallion) in Scotland. From their relative gravitational attraction, Maskelyne estimated 'the mean density of the Earth to be double that of the hill'. A more accurate determination was made by Henry Cavendish (1731–1810) at the end of the 18th century through the determination of Newton's **gravitational constant** by the sensitive measurement of the attraction between two masses of known weight. He obtained a figure of the average density of the Earth as 5.48 times that of water: almost twice the density of the common rocks found at the surface.

As a result of the initiative of Baron Friedrich von Humboldt (1769–1859), explorer, geographer and climatologist, *magnetic observatories* were installed at a number of locations around the world in the 1830s. The great mathematician, Karl Ferdinand Gauss (1777–1855) was director of the Observatory at the University of Göttingen between 1807 and 1855, and inventor of the highly sensitive bi-filar **magnetometer**, in use for more than a century. Gauss also initiated the Magnetischer Verein (Society of Magnetic Research), which coordinated simultaneous magnetic observations throughout Europe from 1836 to 1841. In 1835 Gauss discovered a small non-dipolar component of the **geomagnetic field** that comprised a series of apparently random highs and lows.

The introduction of steam-powered pumps in the late 18th century allowed mines to be excavated deeper (down to 500 m by the early 1800s) and consequently allowed measurements of the increase of temperature with depth into the Earth. In 1797 in Scotland, James Hall (1761–1832) began experiments on the fusing and crystallizing of rock melts. By 1830 it was realized that the extrapolated temperature increase measured in mines would intercept the melting curve of the common rocks, at least by a depth of 80 km. The concept of an Earth comprising a thin solid **crust** over a large reservoir of molten rock became widely discussed, but, as demonstrated in 1838 by the Cambridge mathematician William Hopkins (1793–1866), the absence of significant *Earth tides* and the shifts of the orientation of the Earth's axis from the influence of the Moon and the Sun proved that the Earth must be a solid, at least to a depth of 1600 km.

The greatest advance in investigating the Earth's interior came at the end of the 19th century, after the invention of *seismographs* capable of recording distant **earthquake** vibration, first by a British physicist James Ewing (1856–1935), working in Tokyo in 1880. In 1889, a Japanese **earthquake** was accidentally recorded on a sensitive **gravimeter** in Germany: proof that **earthquake** waves travelled through the Earth. In 1895 the Englishman John Milne (1850–1913), having spent fifteen years studying **earthquakes** in Japan, returned to England to organize a global seismic network. Milne's recordings (allied with information from Italian seismic stations) were employed by Richard Dixon Oldham (1858–1936), former head of the Indian Geological Survey, who in 1906 reported that secondary 'transverse waves' (*shear-waves*) radiating from **earthquakes** located on the opposite side of the Earth, were slowed in passing through an internal **core** of different material. Oldham also saw from the relative speeds and hence densities of the outer shell of the Earth that the '**crust** cannot be more than a score of miles thick'.

The presence of a metallic **core** to the Earth had been anticipated for more than a century to explain Cavendish's observations of the high average density of the planet. The German seismologist Emil Wiechert (1861–1928) had already proposed the existence of an **iron–nickel core** from analogy with the composition and abundances of *meteorites*. In 1909 a pupil of Wiechert's, Beno Gutenberg (1889–1960), from a detailed study of distant **earthquakes**, determined the exact diameter of the **core** (at 7000 km) which had been underestimated by Oldham. In the same year a Yugoslav physicist Andrij Mohorovičić (1857–1936), using records of a local **earthquake**, defined Oldham's predicted discontinuity between the lighter '**crust**' and the denser underlying **mantle** (subsequently known as the **Mohorovičić Discontinuity** or *Moho*). In 1926 the English mathematical geophysicist Harold Jeffreys (1891–1989) proved that the **core** was completely opaque to *shear-waves*, and hence that the dense metal was in the liquid state. Ten years later the Danish seismologist Inge Lehmann was studying the times taken for vibrations from Western Pacific **earthquakes** to be recorded in Europe, and found that she could only interpret some wave-impulses if at the heart of the **core** there lay a denser inner **core**, subsequently proved to be solid. In 1946 a German physicist, Walter Elsasser, working in the US, proposed that the liquid metal **core** was a self-exciting dynamo. The shifting **geomagnetic field**, observed since the end of the 16th century, reflected flow within the **core**.

The more detailed study of the times taken for **seismic waves** to traverse different paths through the Earth's interior has helped map additional, but less significant, depth-dependent density boundaries. By the early 1980s, through the use of rapid computer processing facilities, full three-dimensional **seismic tomographic** velocity models for the **mantle** had been constructed from the records of thousands of **earthquakes** recorded through the Earth. The lateral distribution of low and high density regions in the **mantle** probably reflected the rising and falling columns of **convection**. Mapping the mineralogy of the **mantle** has been assisted by observations of **mantle xenoliths** contained within deep-sourced **magma** pipes and experimental and theoretical investigation of high pressure phase changes.

THE HISTORY OF EARTH DYNAMICS. From the birth of modern science at the end of the 17th century, for more than a century, scientific observation was principally em-

ployed to accommodate and embroider the story in Genesis of Creation and the Great Noachian Flood. Perhaps most significant among these early accounts was that of Thomas Burnet (1636–1715), who wrote in 1681 his much reprinted *Theory of the Earth* in which the original smooth surface of the planet (on which was cultivated the Garden of Eden) comprised a layer of dust that had foundered into the subterranean oceans at the time of the Flood.

Yet as Burnet was writing, Robert Hooke (1635–1703), in charge of experiments at the newly formed Royal Society of London, gave a series of lectures detailing how the skeleta of marine animals have been lifted out of the sea, to be stranded as fossils in the hills and mountains, by the action of '**earthquakes**' (a word used to describe many forms of activity including **volcanic eruption**). Hooke sought to substantiate this theory within accounts of recent **earthquakes**. These ideas were lost on Hooke's death, but rediscovered and publicized by Rudolf Erich Raspe (1737–1794), a German naturalist, educated at the University of Göttingen, who came to live in England after being caught stealing coins in the collection he was superintending in Hesse. Through the mediation of Raspe, Hooke's ideas became known to the English lawyer and geologist Charles Lyell (1767–1849), who made them fundamental to the first volume of his great work *The Principles of Geology*, published in 1830. Lyell employed the eye-witness accounts of land-**uplift** in two recent **earthquakes** in Chile and India to substantiate Hooke's theory that land was suddenly raised out of the sea. The best evidence for such a process came five years later from the observations of coastal **uplift** made by Charles Darwin (1809–1882) and Robert Fitzroy (1805–1865) (Captain of the British survey vessel *Beagle*) in the aftermath of the great February 20th 1835 **earthquake** at Concepción, Chile.

In 1817 the Parisian anatomist Georges Cuvier (1769–1832) accommodated the Noachian deluge as just the latest in a series of catastrophes that had occurred through the long history of the Earth. This concept, later termed '**catastrophism**', was well received, and was elaborated by Jean-Baptiste-Armand-Louis-Leonce Elie de Beaumont (1798–1874), who in 1829 announced that each catastrophe was associated with escaping vapours and the sudden upheaval of colossal mountain ranges. All mountains formed in the same cataclysm shared the same orientation, thus allowing ages of mountain building to be identified by the points of the compass, from which episodes of mountain building could be globally correlated: both ideas that survived deep into the 20th century.

Those who opposed **catastrophism** (known as *uniformitarians*), such as Charles Lyell, generally had no fully-formed explanation for mountains, except that they somehow resulted from the cumulative effects of **earthquake**-induced **uplift** repeated over very long time periods. Around 1860 observations from the Appalachians revealed that mountains formed where the sediments were thickest and hence that deep sedimentary piles, later termed '**geosynclines**', predestined mountain ranges. At the same period, mapping in the Alps revealed far more complexity than could be derived from simple repeated **uplift**. In the 1880s massive horizontal overthrust sheets of sediments (**nappes**) were identified and soon recognized in other older mountain ranges. Such crustal shortening was seen to be consistent with the model of an Earth in thermal contraction. The greatest Alpine geologist, Edmund Suess (1831–1914), denied all observations of land **uplift** in **earthquakes**, and in *Das Antlitz der Erde*, a compendious review of the geology of all the continents, begun in 1885, he fitted all his geological observations into the contraction theory.

There remained at the end of the 19th century a few geologists who opposed the model of the **contracting Earth**, most significantly the Reverend Osmond Fisher, a mathematical geologist at Cambridge, who believed that the interior of the Earth was engaged in **convection**. In his book *The Physics of the Earth's Crust* published in 1880, this convective flux was considered capable of raising mountain ranges and creating **rifts**. Fisher believed that the upwelling substratum (**mantle**) rose along the mid-ocean plateaux and that at the margins of the Pacific, as beneath **earthquake**-ridden Japan, 'the sub-**oceanic crust** is in a very unstable condition, as it would be if it were thus sinking'.

Soon after 1900, as Suess was completing the final volume of his masterwork, the **contracting Earth** model was beginning to seem less secure. The discovery of *radioactivity*, and of radioactive heat-releasing elements in crustal rocks, had suggested by 1904 that the Earth might have its own internal heat source, although the implications of this for geology were not appreciated for twenty years. The remarkable 6 m horizontal **fault** displacement that accompanied the great 1906 San Francisco **earthquake**, the identification of major **extension** in **rifts** such as those in East Africa, and new measurements of the amount of crustal shortening in the Alps all strained the the contraction theory towards its elastic limit.

However, the greatest body of inconsistent data had emerged from the study of the global geography of floral and faunal populations (first recognized around 1860) and the fossil record of such populations within the geological past: most strikingly in the identical **Carboniferous** 'Gondwana' flora distributed through a number of dispersed continents around the southern hemisphere. In 1909 a German meteorological geophysicist, Alfred Wegener (1880–1930), struck by the similarity in the Atlantic coastlines, noted the extraordinary geological correspondence between the rejoined continents once the intervening ocean had been removed. The idea was written up as a book that was translated into four languages. Between 1921 and 1926 Wegener's theory of **continental drift** became widely discussed. Wegener claimed that Greenland had been attached to northern Europe as late as the last **ice age**, and believed he could find confirmation of the continuing opening of the northern Atlantic, at rates as fast as 30 m per year, first from lunar

and subsequently radiodistance observations. The theory was generally, but not universally, dismissed, attacked by physicists such as Harold Jeffreys, because it made impossible demands on motion in the **mantle**, and by geologists primarily because it was alien in both scope and origin.

In 1928 the English geologist and pioneer of radioisotope dating, Arthur Holmes (1890–1965), proposed that the concentration of *radioactive* elements in the Earth's substratum (**mantle**) was sufficient to cause **convection**, with the consumption of old ocean floors in advance of the moving continents and the creation of new ocean floors in their wake. This in turn could drive **continental drift** at more credible speeds of a few centimetres a year. In 1932 the Dutch geophysicist Felix Vening-Meinesz (1887–1966) employed his own submarine **gravity** observations of a marked negative anomaly over deep-ocean trenches around the East Indies, combined with the evidence for associated deep **earthquakes**, to suggest that the ocean trench lay above a region where the ocean floor was being dragged down by convective overturn.

The theory of **continental drift** remained outcast until the early 1950s when some British geophysicists explained the changes in the location of the **magnetic poles**, preserved in ancient rocks, as resulting from moving continents. The post-war programme of oceanographic surveys, made possible by the wartime technological advances in marine geophysics (**reflection seismology**, **magnetics** and **gravity**) soon revealed the extraordinary simplicity of the geology of the oceans. The mid-Atlantic ridge was located exactly at the centre of the ocean, and was part of a 54 000 km mid-ocean mountain-chain that garlanded the planet, everywhere associated with a narrow corridor of enhanced **seismicity** and a high **heat flow**. In 1959 Harry Hess (1906–1969), a marine geologist at Princeton, who had formerly worked with Vening-Meinesz, proposed that **convection** currents in the **mantle** were rising under the **ocean ridges**, and generating new **ocean crust**: a model (termed **seafloor spreading**) closely comparable with those of Fisher in 1880 and Holmes in 1928.

That certain rocks were magnetized contrary to the present-day **geomagnetic field** had been known since the late 19th century and, from work undertaken on Japanese **lavas** in the 1920s, it was shown (at least in part) to be determined by the age of the rock. In the early 1960s, an approximate (**potassium–argon dated**) chronology was established by researchers at Menlo Park, California for the last few reversals in the Earth's magnetic polarity.

In 1963 Lawrence Morley in Toronto and Fred Vine with his supervisor Drummond Matthews in Cambridge, independently proposed that the strip-like *oceanic magnetic anomalies* mapped parallel to sections of the mid-Atlantic and Carlsberg (Indian Ocean) ridges could have formed, symmetrically to either side of the ridge, as a result of continuous **seafloor speading** during episodes of normal and reversed geomagnetic polarity. In 1965 John Tuzo Wilson, a geophysicist from Toronto, met Vine and Hess in Cambridge and conceived of the idea of **transform faults**, **strike-slip faults** that offset spreading ridges. He termed the broad rigid blocks of the Earth's outer shell that were separated by the narrow zones of seismic activity: *'plates'*. Together Wilson and Vine recognized that the pattern of seafloor topography, *oceanic magnetic anomalies* and **seismicity** off the West Coast of North America could be explained by a combination of **transform faults** (like the San Andreas Fault) and spreading ridges. The **magnetic anomalies** mapped in the ocean floor revealed the past pattern of movements; **seismicity** maps those movements taking place today (see **Vine–Matthews hypothesis**). By 1966 other magnetic profiles across spreading ridges had been found with a perfect bilateral symmetry that corroborated the seafloor spreading model, and in 1967 Dan McKenzie at Cambridge provided the spherical geometrical understanding of the possible **plate boundary** motions as constrained by the movement of rigid sections of a spherical shell.

At the end of the 1950s a number of marine geophysicists had supported a model proposed by Samuel Warren Carey, Professor of Geology in Tasmania, in which the separation of the continents was accomplished by rapid **Earth expansion**, an idea first suggested around 1890. However, the existence of '**subduction zones**' where **oceanic crust** was consumed, and ocean area approximately conserved, was demonstrated as a modification of Vening-Meinesz's model of 1932, from observations of the **Benioff–Wadati zone** of deep **earthquakes** dipping away from the ocean. Where the Benioff zone intersected the surface lay the deep ocean trenches with their large negative **gravity anomalies**. The measured velocities of **earthquake** waves showed that the subducting **oceanic crust** remained cold to great depths. The mechanics of the **underthrusting** of the oceanic *plate* beneath the continent had been observed in the two 'super-great' **earthquakes** in southern Chile and Alaska in 1960 and 1964.

The theory of **plate tectonics** had a revolutionary impact on many aspects of the Earth sciences. By 1970 it had been summoned to explain the origin of mountain-ranges, through the collision of continents. 'Geosynclines' were simply the great piles of sediment that had built up along the margins of continents prior to collision. Other important features of global tectonics also emerged, including: the creation of *marginal seas* through **back-arc spreading** above **subduction zones**, the formation of sedimentary **basins** above stretched **continental crust**, and the comparative simplicity of oceanic **plate boundaries**, relative to the width and complexity of zones of *continent–continent collision*. By the 1980s satellite **gravity** and **geoidal** observations had been combined with **seismic tomography** to map the internal **convection** currents that drive the *plates*, and the motion of the continents had begun to be measured directly (at speeds one thousandth of those proposed by Wegener) from repeated satellite surveys.

THE RECORD OF THE PAST. The idea that there was any rival history to that recorded in ancient books emerged from a debate about the meaning of fossils (a term employed to describe any regularly shaped stone) that continued from the Renaissance to the end of the 18th century. When Leonardo da Vinci (1452–1519), in the late 15th century, wrote that former marine life could be found preserved in rocks high above sea-level, he left the observation unpublished in his private notebooks. The favoured view at the time was that fossils were 'sports of nature' that had grown in the rock. The confrontation with the Church over new observations and interpretations of the world began after the publication in 1543 of *De Revolutionibus* by Nicholas Copernicus (1473–1543), a book that championed the heliocentric solar system, subsequently publicized by Tycho Brahe (1546–1601), and confirmed in the observations of Johannes Kepler (1571–1630) on the elliptical orbits of the planets, and Galileo Galilei (1564–1642) of the moons of Jupiter. Copernicus' work was placed on the *Index of Prohibited Books* in 1616 where it remained until 1835. Galileo's trial by the Vatican in 1632 drove scientific discussion into a fertile samizdat, where other heretical theories concerned with the origin of fossils were gaining currency.

In the diluvial concept that emerged in Italy to reconcile Leonardo's observations with the account in the Bible, fossils were all laid down in the great Noachian flood, and contained species that had not survived that inundation. This was revolutionary in acknowledging that history was contained in the rocks, although requiring some passing neglect of the statement that 'two of all flesh, wherein there is the breath of life' went into the ark, and hence that fossils without living equivalents could not be organic in origin. In the collections that were privately established through the 16th century, and in particular in their catalogues, the range of varieties of minerals and fossils became widely known.

When a young Danish doctor, Niels Stensen (1638–1686) was invited to Medici Florence in the middle of the 17th century, he saw that 'tongue-stones' were formerly shark's teeth and visited many quarries in Tuscany collecting fossils. In the summary of his field observations, published (after much scrutiny and debate from the censors) in 1669, Steno (his surname latinized) reasoned that any solid body must be older than the rock within which it is enclosed and that the stratum below had to exist before the overlying layer was laid down: the fundamental laws of **superposition**. At the end of the 17th century, in the newly founded Royal Society of London, Robert Hooke recognized that the term 'fossil' was being applied to both organic and inorganic (mineral) remains and required better definition.

Over the course of the next 150 years the battle between scientific rationalism (inspired by the works of Newton) and biblical literalism continued to be fought, with most published opinions attempting some diluvial reconciliation. In Switzerland, Johann Jacob Scheuchzer

(1672–1733) devoted a lifetime to uncovering fossilized remains (lizard or **fish** skeletons) of those early human sinners who had perished in the Flood.

In 1740 an Italian priest, Anton-Lazzaro Moro (1687–1764), divided those rocks found in mountains into Primary, unstratified rocks of igneous origin, and the layered Secondary rocks. The *Tertiary* rocks, 'lying in the valleys or slopes' of the mountains were named by a Venetian mining engineer, Giovanni Arduino (1714–1795) around 1770. At the newly created Mining Academy at Freiberg, in the last two decades of the 18th century, Abraham Gottlob Werner (1750–1817), expanded his practical background in mineralogy to encompass Moro's ideas on Primary and Secondary formations to create his grand theory called 'geognosy'. The Primitive crystalline rocks (**granite** and **gneiss**) were the first precipitate of a primordial ocean, and were subsequently overlain by 'Floetz' formations filled with organic petrifactions. The youngest material in valleys and coastal plains was 'Alluvial' in age.

Werner's theories were well-received in the academies, but became progressively undermined in the field. The volcanic **craters** and recent eruptive products of central France were first identified in 1751, but it was not until 1763 when Nicholas Desmarest (1725–1815), Inspector of Manufactures at Limoges, found a volcanic **lava flow** formed into prismatic columns, that the igneous origin of **basalt** could be demonstrated. Cracks, formerly claimed as evidence of dessication, were unquestionably the product of contraction on cooling. In 1764 Desmarest began a detailed mapping project of the area from Volvic to Mont Dore in the Massif Central, the first detailed field investigation ever undertaken to resolve a general principle. Over the next two decades he showed that the lavas had flowed down river valleys that had subsequently had to cut their way through the lavas anew, a process that was continuing. The oldest **lava flows** now formed the tops of the hills although these also had simply followed the original topography. All in turn had been shaped by running water, as was happening today. This reconstruction was the best demonstration both of the complexity of the history contained in rocks and the significance of great lengths of time. The account of Earth history preserved in the rocks was becoming irreconcilable with the story of creation contained in the Bible, and by the end of the 18th century it was generally recognized among natural philosophers that the Earth was tens or even hundreds of thousands of years old.

Desmarest's careful observations were in harmony with the philosophy espoused by an Edinburgh philosopher and industrialist, James Hutton (1726–1797). He saw, in the **erosion** of rocks by water, the need for a corresponding creation of rocks through the agency of heat beneath the ocean floor, to sustain a balance of land and sea that had existed for an indefinite period. The new self-confidence in the significance of their science encouraged the establishment of geology as its own discipline, more utilitarian

than philosophical, as marked by the foundation of the Geological Society of London in 1807.

To make sense of history demanded a geological chronology. The 18th century attempts to establish the order in which sediments were deposited, as those of Werner, had characterized ages of rocks simply according to their character. However, between 1791 and 1799 an English canal engineer William Smith (1769–1839), noticed in the lengthy rock-cuts excavated through the sediments of Somerset that seemingly identical **limestones** could be distinguished by their fossils. Around 1799 he mapped the geological strata close to Bath, and by 1815 had extended this map over much of England. William Smith's method of stratigraphic mapping soon became recognized as the historical key to the elucidation of the geological record.

The subdivision of geological time that was established through the 19th century itself reflects every stage of the historical development of *stratigraphy*. The 18th century Italian organization of Primary, Secondary, *Tertiary* (and **Quaternary**) provided the broader frame, the first three being given alternative Greek names according to their fossils: **Paleozoic** ('old-life'), **Mesozoic** ('middle-life') and **Cenozoic** ('new-life'). The *Tertiary* was subdivided by Charles Lyell, in his *Principles of Geology* of 1830, into **Eocene** ('dawn of the new'), **Miocene** ('smaller new') and **Pliocene** ('greater new') from the proportion of modern species found in each sub-period, based on observations made by Gerard Paul Deshayes (1797–1875) working at the Jardin des Plantes in Paris. The **Cretaceous** and **Carboniferous** are French names that survived from the 18th century belief in the universal character of sedimentation. The remaining **Mesozoic** periods, **Jurassic** (from the Jura) and **Triassic** (from its threefold division), reflect the French and German influence while the **Paleozoic** was subdivided by English geologists. The **Permian** (unfossiliferous in Britain) was named by the English gentleman-geologist Roderick Murchison (1792–1871) following a visit to Perm in the Russian Urals in 1841. The sediments older than the **Carboniferous** were termed 'Devonian' by Murchison and the Cambridge Professor of Geology Sedgwick (1785–1873) in 1840. Murchison had already named the older sediments of the Welsh borderland after a local tribe, active in Roman times, '**Silurian**', while, further to the north, Sedgwick was mapping still older rocks he termed **Cambrian** (after the Roman name for Wales). A battle developed between Sedgwick and Murchison, both trying to extend their own formation over the other's territory; a battle that Murchison appeared to be winning after his appointment as Director of the Geological Survey in 1855 allowed him to impose his expanded **Silurian** on official maps. The battle outlived the two original antagonists, and was only resolved in 1879 when Charles Lapworth (1842–1920) proposed an intervening 'Ordovician' period (after another obscure Welsh tribe) separating Sedgwick's **Cambrian** from Murchison's **Silurian**.

The use of the fossil record as a means of stratigraphic

identification naturally emphasized the significance of the changing pattern of life-forms through geological time. While Charles Lyell had demanded the uniformity of nature's laws as the means of geological explanation, he considered life itself to lie beyond such strictures: new life-forms had emerged or been created throughout prehistory. Lyell had no sympathy for the views of the French naturalist Lamarck, who believed that species could transform themselves by an act of will. In 1844, in his *Vestiges of the Natural History of Creation*, published anonymously, Robert Chambers, a self-taught Scottish naturalist and publisher, attempted to unify Lyell's *uniformitarianism* with the development of living forms (by some process of transmutation and recapitulation — the idea that higher biological forms develop through all the lower forms that preceded them). Chambers' book was reprinted many times, and prepared the way 'as a valuable lightning rod for channelling off the initial thunders of orthodoxy' for Charles Darwin's *On the Origin of Species* of 1857, the final reconciliation, through the medium of natural selection, of Lyell's *uniformitarian* philosophy with the diversity of life and the fossil record.

On his global voyage on the survey vessel *Beagle*, from 1831 to 1835, Darwin had been a keen student of Lyell's *Principles of Geology* and between 1838 and 1841 had been secretary to the Geological Society of London. Darwin's *On the Origin of Species* marked the final separation of geology from Genesis, and also the beginning of the split of biology from geology.

While Lyell had cautiously called for 'indefinite' lengths of time to account for Earth history, Darwin was more specific in proposing 300 Ma for the **erosion** of the Weald, a large whaleback ridge of sediments in south-east England. This figure became the focus for Darwin's critics, and although by the third edition in 1860 Darwin had removed this estimate, the attack against Evolution concentrated on starving the theory of time. For the next forty years the debate was dominated by the pronouncements of William Thomson (Lord Kelvin) (1824–1907) over the age of the Earth (from calculations of the rate of cooling from a molten state) and the age of the Sun (assumed to be heated by the collapse of meteors). By the end of the century there was a consensus amongst most physicists and geologists (employing estimates based on such figures as the salt-content of the oceans relative to that provided by rivers) that the Earth was 20–100 Ma old; a consensus that began to fragment in 1904 after the discovery of *radioactive* heating in rocks and *soil*. Immediately before World War I, Arthur Holmes (1890–1965), a British geologist trained under the physicist R.J. Strutt (Lord Rayleigh) (1842–1919), measured the **lead** content (formed by radioactive decay) of **uranium ores** and found ages of hundreds of millions of years. The methods and results of *radioisotope dating* became generally accepted around 1920, freeing geological history of the constraint imposed by Kelvin. In 1931 Holmes estimated 500 Ma since the lower **Cambrian** and between 1600 and 3000 Ma for the age of the Earth. The birth of the Earth was accurately dated (at around 4550 Ma) in 1956 by Claire Patterson, from a study of **uranium/lead** isotope ages of *meteorites* corroborated from the relative isotopic abundances of terrestrial **lead** deposits.

Radioisotope age-dating finally allowed the history of those rocks older than the **Cambrian** to be appropriately scaled. From the accident that the oldest British sediments coincide approximately with the emergence of hard skeleta in marine organisms, early 19th century geological chronology had relegated all earlier periods (more than 85% of Earth history) to a formless 'pre-Cambrian' void. The finding of pre-Cambrian fossils of life-forms going back almost as far as the oldest known sediments has placed the origin of life at the very beginning of recorded crustal history.

Since the 18th century the sedimentary record has been interpreted as reflecting the history of past climate, and in 1830 Lyell proposed that global climatic change was primarily influenced by the disposition of polar landmasses. Around 1820 Ignatz Venetz (1788–1859), a Swiss civil engineer, discovered traces of ice **erosion** far from the high mountains, and in 1836 convinced the young Swiss geologist Jean Agassiz (1807–1873), who found widespread evidence of a recent lowland *glaciation*. Over the next five years Agassiz publicized the idea of an **ice age**, travelled to northern Britain where the products of glacial **erosion** and deposition were found to be commonplace, and wrote a highly influential work on the subject. At the height of the 'heroic age of geology' the **ice age** (along with the discoveries of the bones of giant extinct *dinosaurs*) captured the popular imagination. For Agassiz, the **ice age** was simply the latest in Cuvier's long series of cataclysms, that had been the cause of **extinction** throughout the geological record.

In 1909 Alfred Wegener reinterpreted both the pattern of past *climates* reflected by the sediments and the recent **ice age** as resulting from the disposition of the continents. Wegener's *paleoclimatic* reconstructions, based on the work of his father-in-law Wladimir Köppen (1846–1940) are now generally accepted; while his belief that the last **ice age** originated in Greenland, joined to Europe until a mere 100 000 years ago, is shown to be fantastic.

The possibility of global correlation of variations in **eustatic** sea-level was emphasized by Edmund Suess in 1885, and also by the great American geologist Thomas Chrowder Chamberlain (1843–1928), who in 1904 considered that the Earth, created from a pile of asteroids, underwent sudden episodes of consolidation, contraction and **orogeny**, each of which marked a sharp change in sea-level. The concept of a falling sea-level throughout much of the *Tertiary* was used by a number of **geomorphologists** from 1930 to 1960 to explain the perceived (but undated) 'marine erosion surfaces' claimed to exist at elevations up to 1000 m. This attempt at global correlation of elevation was never without controversy and was abandoned when local vertical movements were found to be widespread in the theory of **plate tectonics**. Global correlation of changes in **eustatic** sea-level was re-attempted from the **seismic** stratigraphy of onlap and offlap sequences in a number of sedimentary basins and continental margins, by Peter R. Vail and co-workers in 1977. Periods of high or low eustatic sea-levels have been recognized as of great importance, not only in affecting the style of ocean-margin sedimentation, but also the biological diversity and biomass of continental shelves.

The reconstruction of the past motion of the continents from *paleoclimatological* data became reinforced in the 1950s from the use of paleomagnetic evidence, by Edward Irving in London and Keith Runcorn in Cambridge and Newcastle. The last few million years of the geological record of variations in the polarity of the magnetic dipole generated by the core were patiently reconstructed from recent (potassium–argon dated) volcanic materials by Richard Doell and Allan Cox (1926–1987) at Berkeley in 1963. Their chronology provided the all-important key for unravelling the mystery of the stripe-like oceanic magnetic anomalies within the model of seafloor spreading.

The theory of plate tectonics, constructed between 1965 and 1967, has provided the framework within which all past major tectonic processes can be comprehended. In the 1970s the geological record was once again under scrutiny for the detailed understanding of the drift of genes within populations and the explanation of the sudden global episodes of *mass extinction*. Lyell's *uniformitarian* (or 'actualistic') philosophy has everywhere been vindicated (as in the local geological reconstructions made possible through understanding the processes of vulcanism, faulting, metamorphism, erosion and sedimentation), except within the apparently 'catastrophic' episodes by which the early 19th century divisions of the geological record into discrete and universal periods had been made possible.

THE EXPLOITATION OF RESOURCES. Man's understanding of the properties and spatial associations of surface rocks is extremely old. The advances in technology represented by the Stone, Bronze and Iron Ages were won by practical geological knowledge: the ability to find and quarry stones and mineral ores. The locations of these materials provided a major reason for trade: tin for Greek bronze had to be brought from as far away as Cornwall, England. Other geological materials such as silica for making glass, clay for pottery and marble for statuary, were sought and excavated from at least 2000 years BC. Rare deposits of native silver and gold (used since at least 3000 BC), or of gemstones, remained the most closely guarded geological secrets. Diamonds from India were known from about 1000 BC and, as the hardest of all materials, soon became the most concentrated form of wealth. By the third century BC Indian diamonds were being traded from China to Italy.

Prospecting and exploiting the Earth required skills of field-observation, skills that were not written down. The only records of this early enterprise are to be found inscribed into the Earth in the pits and shafts of early excavations. In Yugoslavia, at around 2500 BC, copper was being mined to depths greater than 20 m. About 314 BC Theophrastus of Athens wrote a short treatise *Peri Lithon* (On the Stones), a catalogue of the numerous minerals traded in Athens, a number of which, such as amethyst, cinnabar, gypsum and crystal (quartz) have held the same names through to the present day.

The introduction of printing at the end of the 15th century and the consequent expansion of literacy for the first time encouraged the creation of a mining literature. The mining industry in central Europe was almost entirely Germanic. Through providing the silver for coinage, wealthy mining communities attracted talented engineers and physicians, such as Georg Bauer (1494–1555), who in the middle 16th century wrote in Latin (under the name Agricola) a number of works of geology, most importantly *De re Metallica*, an encyclopedic account of mining lore accumulated over the centuries. By the middle 18th century in the face of dwindling supplies at home, and greater competition overseas, the German mining industry reinvigorated itself through founding higher education Mining Academies, the first in Prague in 1762.

The industrial revolution, which began in England at the end of the 18th century, served to expand the market for minerals (principally iron and coal) while the mines themselves provided the first customers for steam-power. The development of steam-driven pumps in the 18th century allowed mine-shafts to be dug deeper (down to 500 m by 1810) and together with some of the earliest railways and haulage systems allowed mine output to be rapidly expanded.

In the first few decades of the 19th century the major impetus to geological mapping was economic in origin and closely bound up with industrialization. Deposits of iron ore, salt, brick clay and building stone were shown to be confined to a particular age of strata. Around 1840 Murchison made the important economic discovery that the earliest plants were found in the upper sediments of the Silurian and hence that there was no possibility of finding coal in lower (and hence older) sediments.

In the middle of the 19th century, with the invention of the kerosene lamp, rock oil (petroleum) rapidly switched from being a patent medicine to an important fuel. The earliest oil provinces were in Russia and the eastern states of the USA. Initially employing techniques used for the extraction of *rocksalt* brine, technological improvements extended the maximum depth of drilling from 150 m to 500 m between 1860 and 1880. However, a geological understanding of the formations and structures wherein oil was likely to be found did not begin to be developed until the end of the century. Oil geology was rarely taught at university and the American Association of Petroleum Geologists was not founded until 1916.

In the 1920s, an enormous demand for petroleum followed the mass-production of the motor-car. At this same period, geophysics began to be harnessed in the search for hidden **petroleum** traps. The first geophysical discovery of an **oilfield** was in 1924 at a Texas **salt dome** using an Eötvös **torsion balance gravimeter**. However, by far the most important geophysical technique came from **seismology**. Physicists and seismologists were attached to the German, British, French and American armies in World War I, to locate artillery positions by employing arrays of microphones. In the 1920s the American team found themselves in direct competition with the leading German sound-ranger Dr Ludger Mintrop, in attempting to employ seismic methods for the mapping of underground structure in the southern states of USA. The seismic anomalies provided by **salt domes** and hidden **anticlines** were mapped with a fan-shaped array of seismic ray-paths between the explosion source and a distant line of receivers (see **fan shooting**). The first productive field was found by the technique in 1924 and by the late 1920s the enormous benefits and savings offered by **seismic prospecting** caused a rapid expansion in the number of field crews and seismic contracting companies. Yet by the end of the 1930s there were few courses and no textbooks on seismological prospecting; the techniques themselves were proprietary. The first journals on **applied geophysics** emerged in Germany in 1924 (*Zeitschrift für angewandte Geophysik*); in the US the Society of Petroleum Geophysicists was founded in 1932 and renamed the Society of Exploration Geophysicists in 1937.

Almost all early **seismic prospecting** involved **seismic refraction**. True reflection **seismology** required the new magnetic recording capability, developed during World War II, and the **common depth point stacking** method, introduced in 1957 in which **noise** and reverberation is suppressed by **stacking** numerous shotpoint to detector configurations, all sharing the same mid-point subsurface reflector. **Stacking** in turn required all the storage and processing capability of the early generation of digital electronic computers, and the multi-billion dollar geophysical industry has subsequently been a major customer for every new innovation in computer technology. Early reflection **seismology** was mostly undertaken at sea where the **seismic source** and string of **hydrophones** were readily moved. Land-based **seismic reflection** using vibrating trucks in place of dynamite charges was developed around 1960.

The other major innovation in *hydrocarbon prospecting* has come from the development of **borehole logging** techniques. Two French professors, Conrad and Marcel Schlumberger, were pioneers in the use of electrical and **magnetic survey** methods between 1913 and 1922. In 1927 the two brothers employed **electrical resistivity** to log a borehole in the Pechelbronn **oilfield** in France and over the next decade consistently pioneered new **borehole logging** techniques and expanded Schlumberger, making their company the dominant force in an enormous oil industry market.

Geophysics also enhanced *mineral prospecting*, allowing mineral deposits buried beneath a thick sedimentary cover

to be identified from their magnetic or gravimetric signature. The airborne **magnetometer**, developed at the end of World War II for the detection of submerged enemy submarines, allowed vast areas of remote territory to be rapidly surveyed (see **airborne geophysical survey**). Reflection **seismology** has also evolved from the search for oil to be employed in defining the structure and extent of **coal seams**.

All these new techniques, pioneered and developed for industry, have in turn brought their rewards for the understanding of Earth structure. Magnetic and gravimetric surveys have assisted in mapping the **crust**, while deep **seismic reflection** (with extended reflector recording times of up to 30 s), first used in the USA, has been able in the 1980s to image major structures in the lower **crust** and **upper mantle**. [RMW]

Hjulström curve An empirical curve describing the critical condition for **erosion** of river channel bed sediments in terms of mean flow velocity. Hjulström (1935) predicted that medium **sand** (0.25–0.5 mm diameter) is the most easily eroded grain size, with higher velocities being required to entrain coarser or finer materials. Clays are normally eroded as aggregates; it is their **cohesive strength** that resists entrainment. Once eroded, clay particles are transported at a wide range of flow velocities. [TPB]

Hjulström, F. (1935) Studies of the morphological activity of rivers as illustrated by the river Fyris. *Bulletin of the Geological Institute, University of Uppsala* **25**: 221–527.

hogback A long ridge, consisting of a sharp crest and steep slopes on both flanks. The ridge is produced by differential **erosion** of steeply inclined rock *strata*. [NJM]

hollandite ($Ba_2Mn_8O_{16}$) An **ore** mineral of **manganese**. [DJV]

holmquistite ($Li_2(Mg,Fe)_3(Al,Fe^{2+})_2Si_8O_{22}(OH)_2$) A mineral of the **iron–magnesium amphibole** group. (See **amphibole minerals**.) [DJV]

holokarst A term coined by Cvijic to describe the coastal Dinaric **karst** belt of Yugoslavia where the full suite of **karst** landforms are found and **limestone** solution processes dominate the landscape. Cvijic contrasted this landscape with the less well-developed **karst** landscape further inland which he termed **merokarst**. The term holokarst has been generally used to describe any **limestone** landscape with a fully developed range of **karst** features. [HAV]

Cvijic, J. (1893) Das Karstphänomen. *Geographische Abhandlung* **5**: 217–329.
Cvijic, J. (1925) Types morphologiques des terrains calcaires. Le holokarst. *Comptes Rendus de l'Académie des Sciences* **180**: 592–4.

holomictic The status of a lake which undergoes a complete circulation of the water column, thereby breaking down the seasonal stratification, commonly at times of winter cooling. (See **meromictic**.) [PAA]

Hutchinson, G.E. & Löffler, H. (1956) The thermal classification of lakes. *Proceedings of the National Academy of Sciences, Washington* **42**: 84–6.

hoodoo Erosional pillar of rock or weakly consolidated sediment. Hoodoos are often irregularly shaped because of variations in the **erodibility** of the horizontal *strata* in which they are developed. They are most common in arid and semi-arid regions. [AW]

Hooke's law The extension of an elastic spring is proportional to the force applied to it, the constant of proportionality being the *elastic constant*. Hooke's law underlies the functioning of modern **gravimeters**. [PK]

horn silver A colloquial name for the mineral **chlorargyrite** (AgCl). [DJV]

hornblende ((Ca,Na)$_{2-3}$(Mg,Fe,Al)$_5$Si$_6$(Si,Al)$_2$O$_{22}$(OH)$_2$) An important and widespread rock-forming mineral, a monoclinic calcic **amphibole**. (See **amphibole minerals**.) [DJV]

hornfels A rock resulting from thermal metamorphism and solid-state recrystallization within a *contact aureole* around an igneous intrusion. Hornfelses are commonly fine to medium-grained pelitic rocks which may contain **porphyroblasts** of minerals such as **biotite, andalusite** and **cordierite** and commonly show traces of sedimentary/tectonic structures which have resisted metamorphic recrystallization. [RST]

horse-tailing The host of mineralized **fractures** that sometimes occurs at or near the ends of **veins**. These clusters of **veins**, when drawn in plan or section, resemble the hairs of a horse's tail. Classic examples of horse-tailing were encountered during the mining of the **copper veins** at Butte, Montana. [AME]

horst Elevated *fault block*. Horsts may be bounded by a set of parallel or near-parallel **step faults** with the same sense of displacement; these **faults** form **conjugate** sets dipping outwards on each side of the block. The converse type of structure is a **graben**, which is a depressed **fault** block. Horsts and **graben** often occur in association in **extensional fault systems**. [RGP]

hot dry rock (*HDR*) **concept** A method of extracting **geothermal** energy from rocks of little natural **permeability**. The technique involves the creation of a twin borehole

system penetrating several kilometres of rock, the depth depending upon the subsurface temperatures in the region (Fig. H4). One borehole is used for injecting cold water into the ground and the other for recovering water that has been heated by the rocks at depth. Efficient heat exchange is achieved by using any naturally-occurring **joint** system or artificially-induced fracture system, through which the water has to percolate between the injection and recovery wells. The hot water recovered can be used for electricity generation if the temperature is high enough, or else for space heating or other commercial applications such as in agriculture (Fig. H4).

The pioneering work in HDR was carried out in the early 1970s in New Mexico by the Los Alamos National Laboratory in the United States. The boreholes were drilled into granitic basement and the subsurface temperatures encountered were *c.* 200°C at 3 km depth and *c.* 320°C at 4.5 km. Electricity generation schemes have operated there for several years without any apparent depletion of the **geothermal** resource.

Research into HDR technology in the UK has been concentrated at the Camborne School of Mines. At their test-site on the Carnmenellis **granite** a temperature of 79°C has been encountered at 2 km depth and there has been

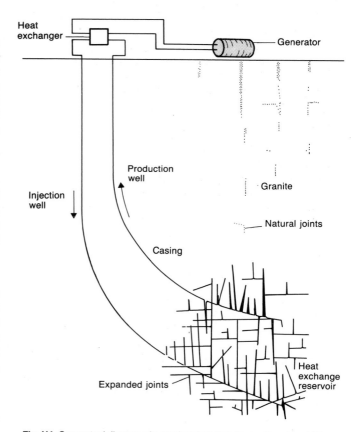

Fig. H4 Conceptual diagram of a two borehole **hot dry rock** system. (After Lee, 1986.)

sustained heat extraction from their twin borehole system for a few years. It is envisaged that their new borehole system will operate at *c.* 6 km depth, where the temperature is expected to be around 240°C.

Generally speaking, the best prospects for HDR projects are in high heat production **granites**, with the associated high subsurface temperatures at shallow depths. Electricity generation requires a minimum rock temperature of *c.* 200°C, economical viability of the projects depending critically on drilling costs. (See **geothermal energy**.) [ATB]

Lee, M.K. (1986) Hot dry rock. In: Downing, R.A. & Gray, D.A. (eds) *Geothermal Energy — The Potential in the United Kingdom.* British Geological Survey, HMSO, London.

Smith, M.C., Nunz, G.J. & Ponder, G.M. (1983) *Hot Dry Rock Geothermal Energy Development Program. Annual Report for Fiscal Year 1982.* Los Alamos National Laboratory, Los Alamos, USA.

hotspot While this term can refer to any positive heat anomaly, it normally refers to areas of volcanicity that appear to have been persistent over at least several tens of millions of years, such as the Hawaiian hotspot that has been continuously active for more than 100 Ma, giving rise to the Emperor Seamount–Hawaiian Ridge volcanic chain as the Pacific *plate* has moved over this active centre. Other possible hotspots are Iceland, Tristan da Cunha, Amsterdam Island, possibly the southern part of the 90° Ridge, etc. It has been suggested that these hotspots are a result of localized streaming of volcanic materials, possibly originating from the **core–mantle** interface. This has led to hotspot locations being considered as a **mantle** framework against which to measure the absolute motions of tectonic *plates*, although small relative movements do occur between hotspots. (See **absolute plate motion**.) [DHT]

Morgan, W.J. (1971) Convective plumes in the lower mantle. *Nature* **216**: 42–3.

Hsu diagram A method of illustration of the shapes of **strain ellipsoids** suggested by Hsu (1966) (Fig. H5). It is used in similar situations to the **Flinn diagram**, and employs a parameter v (nu) based on the **principal natural strains** thus:

$$v = \frac{2\bar{e}_2 - \bar{e}_1 - \bar{e}_3}{\bar{e}_1 - \bar{e}_3}$$

In terms of Flinn's parameter k,

v = −1, $k = \infty$ prolate
v = 0, $k = 1$ **plane strain**
v = 1, $k = 0$ oblate   [SB]

Hsu, K.J. (1966) The characteristics of coaxial and non-coaxial strain paths. *Journal of Strain Analysis* **1**: 216–22.

Hobbs, B.E., Means, W.D. & Williams, P.F. (1976) *An Outline of Structural Geology.* John Wiley & Sons, New York.

huebnerite (MnWO$_4$) End-member of the **wolframite** group of minerals. [DJV]

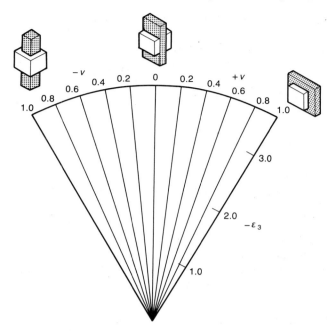

Fig. H5 Hsu diagram showing the shapes of cubes representing strain ellipsoids of varying shapes. Compare with the Flinn diagram (Fig. F22, p. 255). (After Hobbs *et al.* 1976.)

hum A residual **limestone** hill. Often found rising from the floors of **poljes** in Yugoslavia. The term originates from the village of Hum in the Yugoslavian **karst** region. [HAV]

humite (Mg$_7$(SiO)$_3$(F,OH)$_2$) A member of the **humite** group of island silicate (*nesosilicate*) minerals. (See **chondrodite**.) [DJV]

hummocky cross-stratification A type of **bedding** structure formally recognized by Harms *et al.* (1975, p. 87) consisting of gently curved, low-angle **cross-stratification**. The undulating laminae may be both convex-up, forming a hummock, and more commonly concave-up, forming a **swale** (Fig. H6). The hummocks and **swales** are commonly highly three-dimensional in plan view and radially symmetrical. The hummocks have a spacing in the range of 1 to 6 m.

There has been considerable debate as to the precise origin of the structure. This debate has centred on whether hummocky cross-stratification is produced solely by the **oscillatory flow** under *storm* **waves**, for example in hurricanes, or by a flow regime involving a unidirectional component. Those who believe the structure to be singularly diagnostic of the former action of *storm* **waves** place its environment of formation between fairweather and *storm* **wave** base on the continental shelf, implying

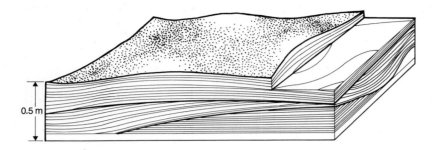

Fig. H6 Block diagram showing **hummocky cross-stratification**.

0.5 m

a precise environmental bracketing. Others believe the structure to be produced wherever the appropriate conditions of high sediment fluxes and mixed unidirectional-reversing/pulsating flow exist. This latter view would allow, for example, hummocky cross-stratification to form under density currents (see **turbidity currents**).

A form of hummocky cross-stratification with a smaller crest to crest spacing has recently been interpreted as due to the passage of **antidunes** in the **upper flow regime**.

The term *swaley cross-stratification* has been introduced to distinguish a variant of hummocky cross-stratification in which the hummocks are rare or absent and the **swales** are preferentially preserved. [PAA]

Harms, J.C., Southard, J.B., Spearing, D.R. & Walker, R.G. (1975) *Depositional Environments as Interpreted from Primary Sedimentary Structures and Stratification Sequences.* Short Course No. 2. Society of Economic Paleontologists and Mineralogists, Tulsa, OK.

hyalite A clear colourless variety of the mineral **opal**, with a globular or botryoidal surface. [DJV]

hyaloclastite Volcanic rock composed of quench-fragmented glass. Hyaloclastite is poorly sorted, consisting of very angular and blocky non-vesicular fragments. The shards formed by spalling of glassy crusts of lavas when they come into contact with water. Hyaloclastite is usually non-bedded and commonly fills gaps between adjacent **pillow lavas**. Most hyaloclastites are **basaltic**, but some are silicic in composition (Furnes *et al.*, 1980). Most hyaloclastites are formed when **basaltic** lava is erupted in the oceans, but some are formed by effusion of lava into lakes or beneath **glaciers**. **Basaltic** hyaloclastites are composed of two varieties of glass, a brown glass called *sideromelane* and a black glass, which contains tiny quench microcrysts, called **tachylite** (Macdonald, 1972). (See **glassy rock, palagonite**.) [PTL]

Cas, R.A.F. & Wright, J.G. (1987) *Volcanic Successions Modern and Ancient.* Allen & Unwin, London.
Furnes, H., Fridleifsson, I.B. & Atkins, F.B. (1980) Subglacial volcanics — on the formation of acid hyaloclastites. *Journal of Volcanology and Geothermal Research* **8**: 95–110.
Macdonald, G.A. (1972) *Volcanoes.* Prentice-Hall, Englewood Cliffs, NJ.

hyalophane ((K,Ba)(Al,Si)$_2$Si$_2$O$_8$) A relatively rare **barium feldspar mineral**. [DJV]

hydration layer dating Method of dating of **obsidian** artefacts manufactured in antiquity. **Obsidian** absorbs water at a rate dependent on the chemical composition of the **obsidian** and on the ambient temperature. The hydration layer thus formed can be measured under a microscope. The newly flaked surface of an artefact will begin formation of a hydration layer immediately, and measurement of the thickness of this layer (if undamaged) and a knowledge of the rate of hydration will give a date of manufacture of the artefact. Because of difficulties in determining this rate of hydration for all cases, this method of dating has been more successful in determining relative dates of artefacts of the same chemical type and burial conditions than absolute dates. [OWT]

hydraulic cement Cement prepared by calcining argillaceous **limestone**. A process made known by John Smeaton in 1756 who discovered that it would set under water. Argillaceous **limestones** vary too much in composition for the manufacture of modern cement and hydraulic cement production ceased early in the 19th century. [AME]

hydraulic equivalence A term which expresses the size of a grain of a given density to the equivalent size of a **quartz** grain (density = 2.65 Mg m^{-3}) which has an identical **settling velocity**. The **settling** or *fall* **velocity** of a grain will depend upon its size, shape and density and how these interact with its hydrodynamic behaviour. The use of hydraulic equivalence enables the behaviour of different size, shape and density grains to be compared, this being of use not only in settling experiments but also in considerations of **placer** mineral accumulations. [JLB]

hydraulic geometry The study of channel form in relation to external controls of discharge and sediment. Four elements of hydraulic geometry can be identified: the cross-sectional form of the channel, the bed configuration, the channel planform and the **long profile** of the river. River systems are characterized by self-regulation, with the hydraulic geometry adjusting to variations in the magnitude of the external controls. Whilst true stability never exists for river channels, **dynamic equilibrium** has been identified for many river systems in humid areas, particularly since empirical work on hydraulic geometry particularly since empirical work on hydraulic geometry by Leopold *et al.* (1964). A mean channel state is assumed to exist in response to a dominant or channel-forming discharge. Thus, downstream variations in channel dimensions are made with respect to a reference discharge, usually **bankfull discharge** which reflects the dominant discharge. At-a-station changes in hydraulic geometry relate to temporal changes in discharge. In both cases, empirical relationships suggest a consistent relationship between discharge and channel dimensions. Bed-forms and channel pattern have also been successfully related to discharge variations. Since **dynamic equilibrium** implies some variation about an average state, recent interest has focused on transient departures from the average, particularly on the effects of extreme flood events. In some rivers the effects of large floods is ephemeral, and the channel geometry gradually recovers over time. In other systems, particularly in semi-arid areas, large floods produce major changes in channel form which persist; such adjustments fit less well into the established view of an equilibrium hydraulic geometry. [TPB]

Leopold, L.B., Wolman, M.G. & Miller, J.P. (1964) *Fluvial Processes in Geomorphology.* W.H. Freeman, San Francisco.

hydraulic gradient The loss of pressure (**head**) along the direction of flow when water flows at constant velocity through a horizontal pipe of constant diameter. This results from a conversion of potential energy into heat, which is lost from the system. If the *head loss* is divided by the distance between two points where it has been measured, this provides a quantity known as the hydraulic gradient. Water (or oil) always flows down the hydraulic gradient whether this is in a pipe or an **aquifer**. [AME]

hydraulic jump The abrupt increase in flow depth caused where a fast flowing supercritical flow (**Froude number** > 1) rapidly changes to the slower velocity subcritical state (**Froude number** < 1). Flow depth before the jump is always less than that after the jump. Hydraulic jumps commonly occur at abrupt changes in bed relief within alluvial channels and are associated with significant energy losses at these sites. [JLB]

hydraulic radius The ratio of a stream's wetted perimeter to the cross-sectional area of **flow** in a channel. It measures the efficiency of a section in conveying **flow**, as it is the area of **flow** per unit length of water–solid contact. [ASG]

hydraulically smooth and rough Terms that express the relative size of sediment grains to the thickness of the **viscous sublayer** of a flow. Where sediment particles lie within the **viscous sublayer** the bed is termed hydraulically smooth and the particle **Reynolds number** is less than approximately 5. Transitional boundaries are created where the roughness elements begin to exceed the sublayer thickness. Fully hydraulically rough boundaries are generated where the diameter of the sediment particles is approximately five times the thickness of the **viscous sublayer** and above particle **Reynolds numbers** of around 70. Once a bed is hydraulically rough, enhanced *turbulence* generation and vertical mixing are encouraged within the flow. For instance, in water flows a **sand** bed ceases to behave as a hydraulically smooth surface at the threshold for sediment movement at grain diameters above approximately 0.6 mm. This threshold grain size is also linked with the onset of **lower-stage plane bed** conditions. [JLB]

Leeder, M.R. (1982) *Sedimentology: Process and Product.* Allen & Unwin, London.
Richards, K.S. (1982) *Rivers: Form and Process in Alluvial Channels.* Methuen, London.

hydroboracite ($CaMgB_6O_8(OH)_6.3H_2O$) A borate mineral. [DJV]

hydrocarbon generation The formation of the *hydrocarbons* present in **natural gas** and **crude oil** accumulations. Most workers nowadays believe that these *hydrocarbons* were generated from disseminated organic matter in sedimentary rocks; the particular rocks in which the generation occurred are known as *source rocks*. A lower limit of about 0.4% organic **carbon** (in *kerogen*) is generally considered necessary for a rock to act as a source bed, but most recognized source beds contain 0.8–2% and the best as much as 10%. The most important *kerogen* results from the accumulation and decomposition of large amounts of planktonic, aquatic organisms which thrive where oxygen-rich bottom waters rise to the surface. When dead organic matter (OM) reaches the seafloor the constituents necessary for hydrocarbon generation are only preserved if the water is essentially **anoxic**. The change from OM to *kerogen* takes place from shallow depths down to about 1000 m with temperatures of up to about 50°C. *Kerogen* is the intermediate product, formed by **diagenetic** transformation of the OM, which gives rise to **petroleum** and it is the organic matter of **oil shales**. With deeper burial and heating (1000–6000 m and 50–175°C) the large molecules in *kerogen* break down (crack) to form smaller, lower molecular weight *hydrocarbons*. Oxygen is lost rapidly by dehydration and loss of CO_2 so that H_2O and CO_2 are the initial products. At higher temperatures volatiles (*hydrocarbon* and *methane*) and liquid **petroleum** products (C_{13}–C_{30}) develop. Mild transformation produces dominantly liquid products, intense transformation produces mainly gas. Thus there will be a critical point beneath which no commercial oil accumulations will be found although gasfields may occur. (See also **immaturity of oil**.) [AME]

North, F.K. (1985) *Petroleum Geology.* Allen & Unwin, Boston.

hydrogen (H) At. no. 1, at. wt. 1.0079. The lightest element. It occurs free in some natural gases and is a widespread constituent of water, many minerals and natural organic materials. [AME]

hydrogeology The study of underground water, i.e. **groundwater**. [AME]

hydrogeomorphology Also termed *fluvial geomorphology*. The study of landforms produced by the action of fluvial processes. [TPB]

hydrograph A graph of the discharge of a river or stream; this is more properly called a *discharge hydrograph*. A *stage hydrograph* is a graph of river *stage*. The *stage* is the height of the water surface above a reference level and it can be measured using a vertical graduated post (a staff gauge) or with an automatic recorder. Using a graph or formula to relate discharge to *stage* for the entire range of flows that may occur at a particular point on a river, a hydrologist can obtain the discharge value simply by reading the river stage value. [AME]

hydrogrossular $(Ca_3Al_2O_8(SiO_4)_{1-m}(OH)_{4m})$ A hydrous silicate mineral of the garnet group. (See **garnet minerals**.) [DJV]

hydroisostasy The **isostatic** reaction of the Earth's **crust** to the removal or application of a mass of water. For example, climatically controlled fluctuations in the level of water in a closed **basin** (e.g. Lake Bonneville in Utah) may create warped shorelines. More generally, however, **eustatic** changes of sea-level have affected the depth of water over the continental shelves causing the **crust** to be depressed at times of high sea-level and to be elevated at times of low sea-level. [ASG]

Bloom, A.L. (1967) Pleistocene shorelines: a new test of isostasy. *Geological Society of America Bulletin* **78**: 1477–94.
Crittenden, M.D. (1963) *New Data on the Isostatic Deformation of Lake Bonneville.* Professional Paper of the United States Geological Survey **454E**.

hydrological cycle The sum total of processes whereby water evaporates from the seas to fall as precipitation on land surfaces and thence to flow both through **aquifers** and rivers back to the sea. [AME]

hydrology The study of the distribution, conservation, use, etc., of the water of the Earth and its atmosphere. [AME]

hydrophone An instrument used to detect seismic energy in water. A hydrophone is sensitive to pressure variations set up by a **seismic wave**, in contrast to a **geophone** which measures velocity of movement. A piezoelectrical material, such as barium titanate or lead zirconate converts changes in pressure produced by the passing of a **seismic wave** to electrical voltage variations. At sea hydrophones are deployed in arrays in a '*streamer*' towed behind a ship. Modern devices have two piezoelectrical crystals wired in parallel such that horizontal accelerations produced by the ship's movement through the water are minimized on the records. [GS]

hydrothermal alteration The **alteration** produced by **hydrothermal solutions**. Frequently alongside **veins** or around irregularly shaped **orebodies** are found **alteration** of the *country rocks*. This may take the form of colour, textural, mineralogical or chemical changes, or any combination of these. **Alteration** is not always present, but when it is, it may vary from minor colour changes to extensive mineralogical transformations and complete recrystallization. Generally speaking, the higher the temperature of deposition of the **ore** minerals the more intense is the **alteration**, but it is not necessarily more widespread. This **alteration**, which shows a spatial and usually a close temporal relationship to **ore** deposits, is called **wall rock** or hydrothermal alteration.

The areal extent of the **alteration** can vary considerably, sometimes being limited to a few centimetres on either side of a **vein**, at other times forming a thick halo around an **orebody** and then, since it widens the drilling target, it may be of considerable exploration value. The drilling targets in the **uranium** field of the Athabaska Basin in Saskatchewan are enlarged by a factor of ten to twenty times by the **alteration**. The spatial and temporal relationships suggest that **wall rock alteration** is due to reactions caused by the mineralizing *hydrothermal fluid* permeating parts of the **wall rocks**. Many *alteration haloes* show a zonation of mineral assemblages resulting from the changing nature of the **hydrothermal solution** as it passes through the **wall rocks**.

The different **wall rock alteration** mineral assemblages can be compared with metamorphic facies as, like these, they are formed in response to various pressure, temperature and compositional changes. They are not, however, generally referred to as **alteration** facies but as types of **wall rock** or hydrothermal alteration and they are described under separate entries, e.g. **advanced argillic alteration, seritization**.

The study of **hydrothermal solutions** has shown that they are commonly weakly acidic, but may become neutral or slightly alkaline by reaction with **wall rocks** (or by mixing with other waters, e.g. **groundwater**). The solutions contain dissolved ions that are important in ion-exchange reactions, and the composition of a particular **hydrothermal solution** will have an important bearing on the nature of the **wall rock alteration** it may give rise to. Since the chemistry of **wall rocks** can also vary greatly according to their petrography, it is clear that predictions as to the course of **wall rock alteration** reactions are fraught with difficulties. Nevertheless, there is, despite a variety of controls, a considerable uniformity in the types of **wall rock alteration** which facilitates their study and classification. The controls of **wall rock alteration** fall into two groups governed respectively by the nature of the *host rocks*; other factors of importance are their grain size, physical state (e.g. sheared or unsheared) and **permeability**, and for the **hydrothermal solution** important properties are the pressure, temperature, chemistry, **pH** and Eh.

Although most rock-forming minerals are susceptible to attack by acid solution, **carbonates**, **zeolites**, **feldspathoids** and calcic **plagioclase** are least resistant; **pyroxenes**, **amphiboles** and **biotite** are moderately resistant, and sodic **plagioclase**, **potash feldspar** and **muscovite** are strongly resistant. **Quartz** is often entirely unaffected. [AME]

hydrothermal deposit Any mineral deposit formed by **precipitation** from a **hydrothermal solution**. This term has been used more commonly in the past in discussing deposits of economic interest, but with the increasing knowledge of the importance of **hydrothermal solutions** in the late stage development and **alteration** of **igneous rocks** it is now frequently used to describe non-economic mineral deposits, e.g. **veins** of **calcite**, **zeolites**, etc. [AME]

hydrothermal solutions and processes Hot aqueous solutions have played a part in the formation of many different types of mineral and **ore** deposit, for example **veins**, **stockworks** of various types, **massive sulphide deposits** and others. Such fluids are usually called hydrothermal solutions and many lines of evidence attest to their important role as mineralizers. Homogenization of *fluid inclusions* in minerals from **hydrothermal deposits** and other **geothermometers** have shown that the depositional range for all types of deposit is approximately 50–650°C. Analysis of the fluid has shown water to be the common phase and usually it has salinities far higher than that of seawater. Hydrothermal solutions are believed to be capable of carrying a wide variety of materials as diverse as **gold** and **muscovite**, showing that the physical chemistry of such solutions is complex and very difficult to imitate in the laboratory.

SOURCES OF THE SOLUTIONS AND THEIR CONTENTS. The widespread occurrence of **brines** shows that saline hydrothermal solutions are, and have been, very active and widespread in the **crust**. In some present-day **geothermal systems** the circulation of hydrothermal solutions is under intensive study. Whence the water of these solutions? Data from water in mines, tunnels, drill holes, hot **springs**, *fluid inclusions*, **minerals** and rocks suggest that there are five sources of subsurface hydrothermal waters. The five are:

1 Surface water, including **groundwater**, commonly referred to by geologists as **meteoric water**.
2 Ocean (sea) water.
3 Formation and deeply penetrating **meteoric water**.
4 Metamorphic water.
5 Magmatic water.

Most formation **water** may have been originally **meteoric water** but long burial in sediments and reactions with the rock minerals give it a different character. Present evidence appears to show that similar deposits can be formed from detectably different types of water and, on the other hand, that waters of at least two parentages have played an important role in the formation of some **orebodies**.

Many data are now available on the chemistry of hydrothermal solutions from studies of *fluid inclusions* (which provide most of our data), modern **geothermal brines**, hot **springs**, and **brines** encountered during deep drilling operations in **oilfields** (see **oilfield water**). Some data from these sources are given under **geothermal brine**. The major constituents are **sodium**, **potassium**, calcium and *chlorine*; other elements and radicals are usually present in amounts less than 1000 p.p.m. Little can be said about the source of any of the **base metals** in solution except **lead**. **Lead** isotopic studies have shown that it may have been derived from a number of very different sources. **Lead** in deposits in south-east Missouri appears to have been leached from a **sandstone** underlying the **orefield** whereas that in some leadfields of Utah was derived from associated igneous intrusions. On the other hand **lead** in the deposits of Pentivy, France and Broken Hill, Australia appears to have come directly from the **mantle**.

The evidence from **lead** isotopic studies thus suggests that **ore** fluids may collect their metals from a **magma**, if that was their source. Alternatively, they may collect more metals from the rocks they pass through, or obtain all their metallic content from the rocks they traverse, which can contain (in trace amounts) all the metals required to form an **orebody**. Our present knowledge indicates that most rocks can act as a source of geochemically scarce elements which can be leached out under suitable conditions by hydrothermal solutions.

Because of the spatial relationship that exists between many **hydrothermal deposits** and **igneous rocks**, a strong school of thought holds that consolidating **magmas** are the source of many, if not all, hydrothermal solutions. The solutions are considered to be low temperature residual fluids left over after *pegmatite* crystallization, and containing the **base metals** and other elements which could not be accommodated in the crystal lattices of the silicate minerals precipitated by the freezing **magma**. This model derives not only the water, the metals and other elements from a hot body of **igneous rock**, but also the heat to drive the mineralization system. The solutions are assumed to move upwards along **fractures** and other channelways to cooler parts of the **crust** where deposition of minerals occurs.

In many orefields, however, e.g. most **carbonate-hosted base metal ore deposits**, there are no **plutonic** intrusions that might be the source of the **ores**. For some of these deposition of the **ore** minerals from hydrothermal solutions generated by expulsion of **connate water** from **shale**-rich **basins** has been postulated. These are believed to have carried **base metals** in hot **brine** into **anhydrite**-bearing **limestones** where **base metal** sulphides were **precipitated**.

Some mineralizing hydrothermal solutions were part of a **geothermal system** which gave rise to **epigenetic deposits** such as **veins**, whilst others reached the seafloor to form **syngenetic massive sulphide deposits**.

MEANS OF TRANSPORT OF MINERAL MATERIAL. Sulphides and other minerals have such low solubilities in pure water that it is now generally believed that the metals were transported as complex ions, i.e. the metals are joined to complexing groups (ligands). The most important are HS^- or H_2S, Cl^- and OH^-; it is possible that organic ligands may also play a part. Thus **silver** and **lead** might be transported as $AgCl_2^-$ and $PbCl_3^{3-}$. On the other hand, **gold** in the Broadlands (NZ) **geothermal system** is probably in solution as a bisulphide complex.

MINERAL DEPOSITION. Theoretical considerations and geological observations show that many different mechanisms, acting in combination or singly, can cause **precipitation** of sulphides and oxides from hydrothermal solutions. For example, in a saturated solution with $PbCl_2$ as the dominant **lead** complex in solution we may write:

$$PbS + 2H^+ + 2Cl^- = PbCl_2 + H_2S$$

This equation shows that dilution *or* addition of H_2S would precipitate **galena**. Dilution might result from the rising solution encountering **groundwater**, addition by meeting another solution carrying H_2S. Another very effective mechanism is boiling which for the above solution would increase the **pH** and thereby induce **precipitation**. Solubility can also be decreased by the cooling which a solution may suffer when it moves towards the surface. [AME]

Barnes, H.L. (ed.) (1979) *Geochemistry of Hydrothermal Ore Deposits*. John Wiley & Sons, New York.

Evans, A.M. (1993) *Ore Geology and Industrial Minerals: An Introduction* (3rd edn). Blackwell Scientific Publications, Oxford.

Guilbert, J.M. & Park, C.F., Jr. (1986) *The Geology of Ore Deposits*. W.H. Freeman, New York.

Henley, R.W., Truesdall, A.H., Barton, P.B. & Whitney, J.A. (1984) *Fluid–Mineral Equilibria in Hydrothermal Systems*. Reviews in Economic Geology 1. Society of Economic Geologists, Tulsa, OK.

Skinner, B.J. (1979) The many origins of hydrothermal mineral deposits. In: Barnes, H.L. (ed.) *Geochemistry of Hydrothermal Mineral Deposits* (2nd edn), pp. 1–21. John Wiley & Sons, New York.

hydroxylapatite ($Ca_5(PO_4)_3(OH)$) One of the **apatite** group of phosphate minerals. [DJV]

hydrozincite ($Zn_3(CO_3)_2(OH)_6$) A secondary mineral produced by **alteration** of **zinc ores**. [DJV]

hypersthene ((Mg,Fe)SiO_3) An *orthopyroxene* mineral common in certain igneous and metamorphic rocks. (See **pyroxene minerals**.) [DJV]

hypocentre (*focus*) The location of an **earthquake** or explosion in space and time: i.e., latitude, longitude, depth, and time of occurrence ('origin time'). It usually refers, in the case of **earthquakes**, to the point of initiation of rupturing. The term '**epicentre**' refers just to the latitude and longitude of the point at the surface vertically above the hypocentre.

Hypocentres are determined, for **earthquakes** recorded at **teleseismic** distances, by comparing the predicted **arrival times** of *P waves* from an estimated hypocentre with the actual **arrival times**; the hypocentre is revised iteratively until the best match is obtained. Most routine hypocentre estimates have an uncertainty of some 10–20 km in latitude and longitude but rather more in depth, and a few seconds in origin time. (See **deep-focus earthquakes**, **seismicity**, **teleseismic**.) [RAC]

hypogene mineralization/alteration Mineralization or **alteration** caused by ascending **hydrothermal solutions**. [AME]

hypolimnion The lower, cold layer of a stratified lake undisturbed by daily or seasonal mixing. As a result of the stagnation or very slow circulation of the water mass in the hypolimnion, it has a great tendency to *anoxia*, allowing the preservation of organic **carbon**-rich material on the lake floor derived by fall-out from the **epilimnion**. The top of the hypolimnion is a zone of rapid temperature change, or *metalimnion*, the plane of maximum temperature gradient being the **thermocline**. [PAA]

hypothermal deposit **Epigenetic** mineral **deposit** formed at high temperature (300–600°C) and considerable depth (3–15 km). Generally found in or near deep-seated acid **plutonic** rocks in deeply eroded areas of *Precambrian* or **Paleozoic** terrane, rarely in younger rocks. The **ore** zones consist of both **fracture**-filled and **replacement bodies** and they are particularly important as **ores** of **gold**, **tin** and **tungsten**. Mineralogical changes with increasing depth are very gradual over thousands of metres and **bottoming** of the mineralization is never abrupt. The **gold** deposits of

Kirkland Lake, Ontario and the **tin** deposits of Cornwall are examples. [AME]

hypsometry The area–altitude distribution of land, i.e. the measurement of the elevation of the land surface or seafloor above or below a given **datum**, usually mean **sea-level**. [ASG]

Cogley, J.G. (1985) Hypsometry of the continents. *Zeitschrift für Geomorphologie* Supp. 53.

I

Iapetus The hypothetical ocean that is assumed to have lain between North America and Europe–Africa some 500 Ma ago. The closure of this ocean, by *subduction* of its margins, is considered the cause for the collision between these continents, resulting in the formation of the Caledonian mountain systems in **Siluro-Devonian** times (*c.* 450 Ma). Remnants of Iapetus are thought to be represented by **obducted** fragments, particularly in Newfoundland and Scotland. [DHT]

ice ages Major changes have occurred in the Earth's climate system through geological time. Of these, the periodic incidence of ice ages is explained by boundary conditions internal to the Earth's system, and external forcing functions. An important external factor has been an increase in global insolation through time as the result of a 20–30% increase in *solar luminosity*. Earlier, the interaction between low luminosity and a reduced *greenhouse effect* may be part of the explanation for *Precambrian* ice ages. On time-scales greater than a million years, boundary conditions provided by ocean–continent configurations, seaways and mountain ranges are important. *Orbital forcing* is important on time-scales of 1000 to 100 000 years. But at all time-scales, feedback mechanisms within the atmosphere–ocean–biosphere–solid-Earth climate system modulate these internal and external controls.

'Ice age' is used in different ways. On the one hand it refers to a long period of *glaciation* in time, such as the **Permo-Triassic** ice age; or on the other hand, to the **Cenozoic** ice ages, where it may have a specific connotation as the geochronological equivalent of the *chronostratigraphic stage*. Late **Cenozoic** ice ages after about 900 000 years ago typically lasted 100 000 years, and were separated by **interglacials** lasting on average about 10 000 years.

Ice ages occur when large *ice sheets* grow on the Earth's surface. These need not necessarily be polar in location, because, provided temperatures are low enough and precipitation adequate for ice nourishment, they may occur in mid-latitudes, as in the *Pleistocene*; or apparently even in lower latitudes if an appropriate combination of low *solar luminosity* and greenhouse gases obtain, as during some of the *Precambrian* ice ages. The late **Cenozoic** ice ages are reconstructed on a wealth of inderdisciplinary data unsurpassed in earlier geological history. But information for earlier ice ages decreases in quality with increasing age. Thus the evidence for the **Permo-Triassic** and late **Ordovician** ice ages is clear in terms of present-day analogues, but evidence for *Precambrian* ice ages is less so.

It relies mainly on the recognition of *tillites*, **dropstones** and **striations**.

At least four major *Precambrian* ice ages occurred. The first evidence for *glaciation* occurs some 2.3 Ga ago, which coincides with a rapid expansion of **stromatolites**. These may have been the means by which **carbon** was withdrawn from the atmosphere. Thus a reduced *greenhouse effect* and the low luminosity of the Sun at that time combined to give the conditions for an ice age. Between 900 and 600 Ma ago three ice ages, with *glaciation* that was continental in scale, appear to have affected all continents. Every region with *Precambrian* rocks shows evidence for this (Hambrey & Harland, 1981); but because of the range of *radiometric ages*, and uncertain position of *plates*, they cannot be regarded as showing synchronous *glaciation* everywhere. The causes of these ice ages are not fully understood. **Paleomagnetic** evidence suggests that they occurred in low latitudes which may have been possible because the luminosity of the Sun was at least 10% below that of the present. Late *Precambrian* mountain building episodes may also have been a contributory cause.

In contrast, the two major **Paleozoic** ice ages were both located across the South Pole. Both show evidence for ice movement, more or less synchronous in time, and consistent with **paleogeographical reconstructions** of the southern continents. The earlier occurred during the Late **Ordovician** (about 450 Ma), and evidence is found in the Sahara Desert of North Africa and also in Saudi Arabia. A variety of indicators shows that ice movement flowed outward from the present equator. At that time, however, North Africa was close to the South Pole (Fig. I1).

The later **Permo-Triassic** ice age (250–320 Ma) was also related to a grouping of the southern continents around the South Pole. Evidence for this **Gondwanaland** *glaciation* is found as *tillites* and polished and striated bedrock surfaces in South America, Africa, India and Australia.

It is difficult to determine the causal factors of these pre-**Cenozoic** ice ages other than in general terms. But the evidence for the **Cenozoic** allows for an understanding of the basic requirements for ice age conditions. At the simplest level, these are: a combination of low temperatures (provided by *orbital forcing* and reduced *greenhouse effect*), and an abundant supply of precipitation to grow the *ice sheets*. This depends on **paleogeography**. Thus, the movement of Antarctica to its location over the southern pole seems to have been a major prerequisite for the **Cenozoic** ice ages. A further reinforcement of this was the closure of the Drake Passage, between Antarctica and South America 30–40 Ma ago, which effectively isolated

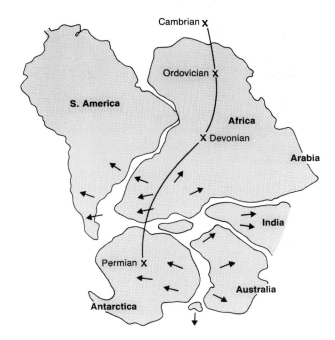

Fig. I1 Ice ages. The path of Gondwanaland across the South Pole during the Paleozoic. Arrows show the pattern of ice-movement for the Permo–Triassic glaciation. (After Eicher & McAlester, 1980.)

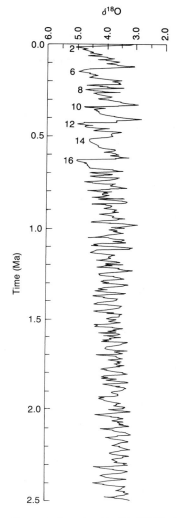

Fig. I2 Ice ages. Oxygen isotope signal for Site 607 (North Atlantic). Before 700 000 years ago the signal is dominated by the 41 000 year tilt frequency, and afterwards by the 100 000 year frequency, when large ice sheets grew in North America and Europe. The Matuyama–Brunhes reversal occurs in Stage 19, and the Gauss–Matuyama reversal just before 2.4 Ma. (After Ruddiman *et al.*, 1986.)

Antarctica when the circum-Antarctic cold current developed. **Strontium** ratios (see **strontium isotope analysis**) from deep-sea sediments offshore from Antarctica indicate that *glaciation* commenced during the **Oligocene**, but the large *ice sheet* of the present-day did not develop until the **Pliocene**. Prior to the closure of the Panama seaway, 3.1 Ma ago, and even after the earlier closure of the **Tethys** seaway, one of the main characteristics of global ocean circulation was that of a zonal flow from Atlantic to Pacific. Closure of the Panama seaway, however, deflected warm tropical surface waters northwards via the Gulf Stream and North Atlantic Drift. The establishment of this meridional flow of surface water was essential to provide a moisture source for high latitudes in the northern hemisphere. Soon after, ice grew in Iceland, and at 2.4 Ma major *glaciation* was taking place in the northern hemisphere when *ice sheets* had grown large enough to launch ice floes in tidewater environments.

The study of the *Pleistocene* ice ages was placed on a sound basis by Emiliani's (1955) work on *oxygen isotope stratigraphy* (see **stable isotopes of carbon and oxygen in sediments, oxygen isotope analysis**). The oxygen isotope signal (variability in the $^{18}O/^{16}O$ ratio in planktonic and/or benthonic **Foraminifera**) is a first-order indicator of continental ice volumes. Subsequent work by Shackleton & Opdyke (1973) and others extended Emiliani's work. *Oxygen isotope stratigraphy* shows that the former fourfold classification of *Pleistocene* ice ages was oversimplified (Kukla, 1977). Oxygen Isotope Stages are numbered by

counting backwards in time: thus Stage 1 represents the *Holocene* (present-day), and odd numbers represent **interglacials**. Even numbered *stages* correspond with ice ages. For example, Stage 2 corresponds with the last *glaciation*, while Stage 100 corresponds with the first major excursion in the oxygen isotope signal, which indicates extensive northern hemisphere *glaciation* 2.4 Ma ago.

Age estimates for the timing of the ice ages are based on the recognition of *geomagnetic reversals* in deep-sea **cores**. This allows the timing of the ice ages and **interglacials** to be estimated (Fig. I2). The frequencies show the regular rhythms or *cycles* attributed by *Milankovich* and others to *orbital forcing* (Hays *et al.*, 1976, Imbrie & Imbrie, 1979). These are: eccentricity of the orbit (100 000 year cycle), variation in axial tilt (41 000 year cycle), and precession,

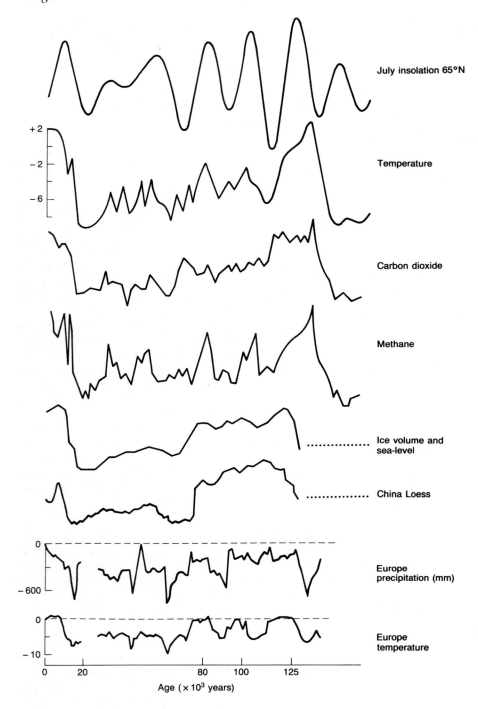

July insolation 65°N

Temperature

+2

−2

−6

Carbon dioxide

Methane

Ice volume and sea-level

China Loess

0

−600

Europe precipitation (mm)

0

−10

Europe temperature

0   20   80   100   125

Age (× 10³ years)

Fig. I3 Ice ages. Signals of climatic change for the last major glacial cycle. (After Rutter, 1990.)

which varies the distance between the Earth and the Sun, notably on June 21st (23 000 and 19 000 year cycles). These vary the amount of solar radiation received at the top of the Earth's atmosphere at different latitudes. Tilt is dominant north of 65° N, and precession south of that latitude. The 100 000 year eccentricity frequency remains a mystery because it lacks enough power to drive major climate changes. Some disagreement obtains on the timing of the 'last **interglacial**': *oxygen isotope stratigraphy* and

orbital considerations place this at 125 ka, which is the average age of a global sea-level high-stand at that time. But the estimate from the Vostok ice core in Antarctica places it somewhat earlier at about 135 ka, as does the analysis of a **calcite vein** in Nevada. These, however, may be phase effects within the climate system which is highly diachronous.

The ice age record for the last 2.4 Ma shows that it is dominated by the tilt (41 000 year) frequency between

[334]

2.4 Ma and about 0.9 Ma. Then, after a transitional period, the 100 000 year eccentricity frequency becomes dominant (Fig. I2). There is no general agreement as to what caused this. What may be significant is that this coincides with the growth of large mid-latitude *ice sheets* in North America and Europe. These may have been caused by changes in the position of the jet stream which brought cold air and moisture-bearing storm tracks to mid-latitudes. One theory suggests this was caused by **uplift** of the Tibetan Plateau, and the Sierra Nevada and Colorado Plateau, which deflected high-level wind-systems to cause major waves in the circum-polar vortex of cold air which sits astride the North Pole. The increased **weathering** such **uplift** caused would uptake carbon dioxide from the atmosphere, further decreasing global temperatures by reducing the '*greenhouse effect*'.

Land–sea correlations depend almost entirely on fixed points in the continental evidence provided by age estimates using a variety of methods, for example, *carbon-14*, **uranium**-series, **thermoluminescence**, *amino acid racemization*, **fission track dating**, etc. Time-series data on different parts of the climate system display the orbital frequencies: for example, sea-level change, vegetational history, **magnetic susceptibility** in **loess**, and, for the last ice age, the greenhouse gases carbon dioxide and *methane* (Fig. I3). Thus it is evident that all the major components of the Earth's climate system were pulsing to the same orbital beat, thereby demonstrating the coupled nature of the Earth's climate system.

Some events cannot be explained by *orbital forcing*. The Younger Dryas cooling (11–10 ka) is such a non-linear event. This, and other short-lived events, may be related to adjustments within the climate system, in which the role of deep-ocean circulation seems critical, especially in the production of North Atlantic deep water. Similar 'internal' rhythms, or modulations, of *orbital forcing*, may account for the strong 100 000 year cycle, which could be due to interactions between climate and *glacial* **isostasy** of the Earth's **crust**. [DQB]

Broecker, W.S. & Denton G.H. (1990) What drives glacial cycles? *Scientific American* **262**(1): 43–50.
Crowley, T.J. (1983) The geologic record of climatic change. *Reviews of Geophysics and Space Physics* **21**: 828–77.
Eicher, D.L. & McAlester, A.L. (1980) *History of the Earth*. Prentice-Hall, Englewood Cliffs, NJ.
Emiliani, C. (1955) Pleistocene temperatures. *Journal of Geology* **63**: 538–78.
Hambrey, M.J. & Harland, W.B. (1981) *Earth's Pre-Pleistocene Glacial Record*. Cambridge University Press, Cambridge.
Hays, J.D., Imbrie, J. & Shackleton, N.J. (1976) Variations in the Earth's orbit: pacemaker of the ice-ages. *Science* **194**: 1121–32.
Imbrie, J. & Imbrie, K.P. (1979) *Ice-ages: Solving the Mystery*. Macmillan, London.
Kukla, G.J. (1977) Pleistocene land–sea correlations, I, Europe. *Earth Science Review* **13**: 307–74.
Ruddiman, W.F., McIntyre, A. & Raymo, M. (1986) Palaeoenvironmental results from North Atlantic sites 607 and 609. *Initial Report, Deep Sea Drilling Project* **94**: 855–78.
Rutter, N. (1990) What significant climatic and environmental changes have occurred in the past, and what were their causes? In: International Geosphere Biosphere Programme (eds) *Global Change* Report 12, 7.1.2–7.1.30. Swedish Academy of Sciences, Stockholm.
Shackleton, N.J. & Opdyke, N.D. (1973) Oxygen isotope and paleomagnetic stratigraphy of Pacific core V28-238: oxygen isotope temperatures and ice volume on a 100 000 and 1 000 000 year scale. *Quaternary Research* **3**: 39–55.

Iceland spar A varietal name for the clear colourless form of the mineral **calcite**. [DJV]

idocrase A name sometimes used for the mineral **vesuvianite**. [DJV]

igneous body A volume of **igneous rock** separated by discrete boundaries from the adjoining '*country rock*' or '*host rock*' into or onto which it has been emplaced. Igneous bodies are divisible into *intrusive* (emplaced at depth) and *extrusive* (emplaced at the surface). *Intrusive* bodies are further divisible into sheet-like bodies (**dykes** and **sills**), cylindrical (**pipes**, **volcanic vents** and *stocks*) and volumetrically large bodies such as **batholiths**.

The emplacement of intrusions may be controlled by the external **stress** field, or the intrusion may itself produce a **stress** field, which induces **deformation** in the *country rock* and/or controls the emplacement of subsequent intrusions from the same source. *Emplacement* may thus be either *forceful* or *permitted* depending on whether the *country rock* is actively deformed or passively displaced. Emplacement of **dykes** and **sills** is commonly **dilational**, i.e. *permitted*, and the passive displacement of the **wall rocks** is controlled by the regional **stress** field, the sheets tending to lie normal to the minimum *principal stress axis*. Secondary **stress** fields created around *forceful* intrusions may give rise to **cone sheets** or **radial dykes**. [RGP]

igneous rock A rock which has solidified from molten or partially molten rock material (i.e. **magma**). Igneous rocks include rocks which have crystallized at depth within the Earth's **crust** (or **mantle**) and/or those which have crystallized from molten rock during ascent to the Earth's surface as extrusive **lava flows** or **pyroclastic rocks**. [RST]

ignimbrite A **pyroclastic rock** consisting predominantly of **pumice** and **ash**, transported by a pyroclastic flow mechanism, and of rather large volume (1 km^3–2000 km^3). Ignimbrites are pumiceous and large-volume varieties of **ash-flow tuffs**. Ignimbrites may or may not be welded.

Ignimbrites consist of **pumice**, **ash**, and separated (commonly broken) *phenocrysts*, all derived from **magma** of silicic or intermediate composition, and lithic blocks and *clasts*. The latter are derived from the eruption vent walls, or are plucked into the pyroclastic flow during transport. Most ignimbrites are poorly sorted and consist of particles sizes in the range <0.1 mm to >1 m. The term ignimbrite refers to both the rock type and to the whole pyroclastic sheet. Ignimbrites range from about 1 m to over 100 m thick, and commonly consist of more than one flow unit, each flow unit having a fine-grained basal layer, normal grading of lithic *clasts* and *reversed grading* of **pumice** *clasts*. Because of their large volume and pyroclastic flow mechanism of transport and deposition, ignimbrites fill pre-existing valleys, and tend to flatten out topography, sometimes producing nearly-flat **ash-flow** plains.

Ignimbrites having volumes greater than about 10 km³ are usually associated with the formation of **calderas**. Such ignimbrites may be divided into a **caldera**-fill portion (the inflow sheet), and a more widespread portion outside the **caldera** (the outflow sheet). Inflow sheets are commonly thick, coarse-grained, densely welded, and interbedded with landslide deposits. They are rarely well exposed, but may constitute half the total volume of the ignimbrite. Outflow sheets show many vertical and lateral facies variations. Degree of welding increases towards the **volcanic vent**, and welding is most intense at the base of thick valley-ponded sections, where the weight of overlying **tuff** is greatest. Some outflow sheets are entirely non-welded. Close to the vent (the proximal facies), ignimbrites contain a high proportion of large lithic blocks (up to 3 m across) which locally form lithic-rich breccias (called lag breccias) which are relatively depleted in **ash** and **pumice** and up to 25 m thick. Such breccias show intimate gradational and interbedded contacts with more **ash**-rich, lithic-poor ignimbrite. The breccias are commonly bedded, giving the impression of having been deposited by an air-fall mechanism. Beyond the proximal facies, the greatest area covered by most ignimbrite outflow sheets consists of rather monotonous, **ash**-rich, poorly-sorted, non-bedded pyroclastic deposits, with most **pumice** and lithic *clasts* in the size range up to 20 cm. This facies is commonly strongly valley-ponded, and may be called normal ignimbrite. It may be associated with veneer deposits deposited as thin (10 cm–2 m thick) mantles on the crests of hills poking above the general level of the top of ignimbrites. The distal facies of ignimbrite outflow sheets, deposited perhaps 20–150 km from the vent, consists of thin, **ash**-rich deposits with few **pumice** or lithic *clasts*. They are relatively well sorted and resemble fine-grained air-fall **tuffs**.

Probably all ignimbrites possess mineralogical and chemical zonations. Crystal content of **pumice** *clasts* usually increases in later-erupted flow units. This is thought to indicate that the ignimbrite eruptions tapped progressively deeper levels of **magma** chambers in which *phenocryst* content of the **magma** increased with depth. In some ignimbrites, certain **phenocryst** phases appear or disappear, or show systematic chemical variation, through the erupted stratigraphy. Chemical zonation is usually characterized by decrease in abundances of silicon and incompatible trace elements (e.g. rubidium, **niobium** and **thorium**) in progressively later-erupted flow units. There are corresponding progressive increases in compatible elements such as calcium, **magnesium** and **titanium**. These chemical zonations are thought to mean that ignimbrite

eruptions usually tap chemically zoned **magma** chambers in which abundances of silicon and incompatible trace elements increase roofward. [PTL]

Cas, R.A.F. & Wright, J.V. (1987) *Volcanic Sucessions Ancient and Modern.* Allen & Unwin, London.
Fisher, R.V. & Schmincke, H.-U. (1984) *Pyroclastic Rocks.* Springer-Verlag, Berlin.
Wilson, C.J.N. (1985) The Taupo eruption, New Zealand II. The Taupo ignimbrite. *Philosophical Transactions of the Royal Society, London* **A314**: 229–310.
Wilson, C.J.N. & Walker, G.P.L. (1985) The Taupo eruption, New Zealand I. General aspects. *Philosophical Transactions of the Royal Society, London* **A314**: 199–228.

ijolite A **plutonic** rock containing over 90% **nepheline** and mafic minerals, generally **pyroxene** and also **amphibole**, *sphene*, **apatite** and **melanite**. Ijolite rocks form a series depending upon the content of mafic minerals, with limits of 0–30–70–90–100% corresponding respectively to *urtite*, ijolite, *melteigite*, and alkali *pyroxenite*. Most ijolites have normal igneous textures, mainly *subophitic*, and some are intruded by **veins** which are mineralogically similar to the *host rock* but with **pyroxene**, **wollastonite** and any **feldspar** exhibiting a prismatic to acicular habit with the long axes at right angles to the walls of the **vein**. This is termed '*comb-structure*' and is common in ijolites and **fenites**. Ijolites form concentric intrusions and **dykes** which form part of **carbonatite**–ijolite–**nephelinite** complexes in a variety of continental areas. [RST]

illite $(K_{1.5}Al_2(Al_{1.5}Si_{2.5}O_{10})(OH)_2)$ A **clay mineral** similar in composition to **muscovite** mica. [DJV]

illitization The **alteration** of a precursor mineral to **authigenic illite**. Most commonly referred to in the case of illitization of **smectites** and **illite**–**smectite** clays during burial, but other known precursor minerals include **feldspar**, **muscovite** and **kaolinite**. (See **authigenesis**, **burial diagenesis**, **diagenesis**, **neoformation of clays**, **replacement**.) [SDB]

Hower, J. (1981) Shale diagenesis. In: Longstaff, F.J. (ed.) *Clays and the Resource Geologist.* Mineralogical Association of Canada 7: 60–77.
Srodon, J. & Eberl, D.D. (1984) Illite. In: Bailey, S.W. (ed.) *Micas*, pp. 495–544. Reviews in Mineralogy No. 13. Mineralogical Society of America, Washington, DC.

illuviation The accumulation of material within the lower *soil* zone. Leaching and **eluviation** of the upper *soil* zone carries fine-grained materials (silt and clay) and water-soluble minerals to the lower horizons. If the materials in **suspension** and solution are not flushed from the *soil* zone, they accumulate as illuvial *clay pans* or *soil crusts*. In arid regions, some **pedogenic calcretes** and **gypsum** crusts are of illuvial origin having accreted through evaporation of *soil* moisture containing salts leached from the surface. [AW]

ilmenite (FeTiO₃) An **iron–titanium ore** mineral also found as an **accessory mineral** in many rocks. (See **oxide minerals**.) [DJV]

ilvaite (CaFe₃²⁺Fe³⁺O(Si₂O₇)(OH)) A double-island silicate mineral (*sorosilicate*). [DJV]

imaginary component (EM) That part of the secondary field that is 90° out of phase with the primary field in **electromagnetic induction methods**. It is also referred to as the *out-of-phase* or *quadrature component*. At low frequencies (and in highly resistive bodies), induction effects being less significant than resistive effects for **eddy currents**, the secondary field will be predominantly 'imaginary' and at high frequencies (and large conductivities) the secondary will be almost in phase with the primary. [ATB]

imbricate structure A set of subparallel overlapping slices of rock bounded by closely spaced **faults** which join together to form the same **fault** at depth. The closely spaced **faults** are called *imbricate faults*. Rock between two *imbricate faults* is known as an *imbricate slice*. The term *schuppen structure* has been used synonymously with imbricate structure in *orogenic belts*. Identification of imbricate structure does not depend on scale; examples are found on scales varying from centimetres to tens of kilometres.

Imbrication is widely recognized in **thrust belts**, in which it can occur in both *piggyback* and *overstep propagation* sequences, to form *imbricate stacks*. A *piggyback thrust* sequence refers to imbrication in which new *thrust segments* develop closer to the **foreland** than existing *thrusts*. Thus early-formed **faults** and associated **folds** are carried '*piggy-back*' in the **hangingwall** to later-formed *thrusts*. In an ideal *piggyback sequence*, all the **shortening** occurring at any one time is taken up on the lowest, latest-formed *thrust*. Figure I4 shows a structure developed by *piggyback propagation*. This sequence can be identified where movement on a **fault** folds or tilts an existing **fault** surface. This **fault** sequence is most commonly observed in **thrust belts** overall, and is implied when '*in-sequence*' thrust development is described. It is necessary to identify *out-of-sequence faults* and **strains** when constructing **balanced sections**, and in the analysis of **thrust belts** generally. This *thrust* sequence is the converse of an **overstep fault** sequence, shown in Fig. I5, in which successively later-formed *thrusts* form closer to the hinterland than existing structures. Early-formed *thrusts*, and *thrust*-related **folds** can be truncated by later-formed **faults**. This order of propagation is less common than a *piggyback sequence* in **thrust belts**.

Imbrication in **thrust belts** forms two groups of structures, *imbricate fans* and *duplexes*, shown in Fig. I6. In an *imbricate fan*, Fig. I6(a), the imbricate **faults** branch from a

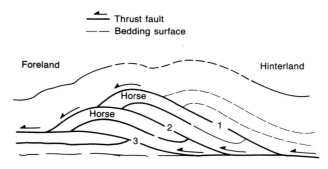

Fig. I4 A piggyback **imbricate thrust structure** (a duplex) in which the faults developed in the order 1,2,3, from hinterland to foreland. The fault-bounded packages of rock are known as 'horses'.

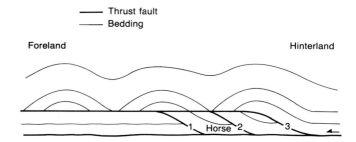

Fig. I5 An overstep **imbricate thrust structure**, in which the faults developed in the order 1,2,3, from foreland to hinterland.

floor thrust or *sole thrust*, and terminate in **folds** or **strain** zones in the overlying rock. The terminations of the *thrust* surfaces are known as *tip lines*, or *tips*, and **faults** which do not reach the surface are called *blind thrusts*. In a *duplex* (Fig. I6b–d) the **faults** branch from an underlying floor *thrust*, but also join a common higher level *thrust*, called the *roof thrust*. A *duplex*, like a simple **ramp**, transfers displacement from lower to higher stratigraphic levels. Each package of rock surrounded and isolated by **faults** is known as a *horse*. Where *thrust faults* cut through bedded or banded rocks, *horses* often contain gentle or open **anticline–syncline** pairs, formed by the curved shapes of the *thrust fault* at **ramps**. Different amounts of imbricate *thrust* displacements relative to **fault** separations result in the imbricate slices in a *duplex* dipping to the hinterland, to the **foreland**, or in both directions, which forms an *antiformal stack*.

In real examples, where the upper parts of imbricate stacks are eroded away, some *duplexes* cannot be distinguished from *imbricate fans*. The distinction is important, however, because the net displacement represented by the structure differs in the two cases. A significant feature of imbricate systems is their usefulness in identification of the order of **fault** movement and in the calculation of minimum displacement for the *thrust belt* as a whole. For a branching *thrust* system, displacement identified in *imbricate stacks* sums to give a minimum for that system as a whole, taking into account any distributed **strain** around

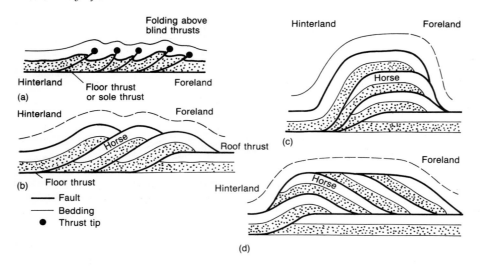

Fig. I6 **Imbricate structures** in a thrust regime. (a) An imbricate fan in which each fault dies out to a tip upwards (blind thrusts). (b) Duplex in which the horses dip towards the hinterland. (c) An antiformal stack duplex in which the horses dip towards both the hinterland and the foreland. (d) A duplex in which the horses dip towards the foreland.

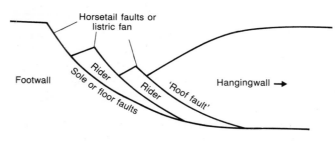

Fig. I7 **Imbricate structure**. Imbrication on listric faults in an extensional fault regime, showing 'riders'. (After Gibbs, 1984.)

the **fault** surfaces. Where **faults** branch, at *branch lines*, displacement is transferred between different **fault** segments. The displacements measurable on *imbricate thrusts* must also have occurred on the floor *thrust* in an *imbricate fan*. In a *duplex*, the sum of the imbricate **fault** displacements gives a minimum for the roof *thrust* as well as the floor *thrust*. Where much of a *thrust sheet* has been removed by **erosion**, reconstruction of the *thrust* displacement pattern in this way provides evidence of the **shortening** represented by the system.

Imbrication is also recognized in **extensional fault systems**, where repeated branching of closely spaced **faults** from an underlying *detachment* forms arrays of tilted *fault blocks*. These **faults** are usually emergent, so that *imbricate fans* are formed. The **faults** are often *listric*. This structure, shown in Fig. I7, is known as a *listric fan*, or as *horsetail faults*; the blocks separated by these **faults** are called *riders*. Extensional systems can be analysed in terms of order and magnitude of **fault** movements in similar ways to that described for *thrust fault* systems.

In **strike-slip fault systems**, **en echelon** segments of *wrench faults* are often linked by shorter **fault** segments of differing trend. These form *releasing* or *restraining bends*, whether they have extensional or compressional displacement components respectively. Where several subparallel bend segments exist, *strike-slip duplexes* are formed. Figure

I8 shows imbrication on **strike-slip faults**, where the imbricate **fault** segments branch onto the main **fault** at depth, as well as joining along the length of the **fault**. In cross-section, this **fault** pattern can be seen to be one way of producing positive (compressional) and negative (extensional) *flower structures*. [SB]

Butler, R.W.H. (1987) Thrust sequences. *Journal of the Geological Society, London* **144**: 619–34.
Gibbs, A.D. (1984) Structural evolution of extensional basin margins. *Journal of the Geological Society, London* **141**: 609–20.
Woodcock, N.H. & Fischer, M. (1986) Strike-slip duplexes. *Journal of Structural Geology* **8**: 725–37.

immaturity of oil A heavy oil (see **API gravity**) that has not undergone sufficient natural cracking (see **hydrocarbon generation**) to convert it into a light oil. It will be strongly naphthenic or asphaltic. **Crude oils** from very young (immature) to mature develop by the natural breakdown of large molecules. This process, if not interrupted, increases the API value of the oil and its contents of lower molecular weight *hydrocarbons* and **aromatics**. [AME]

impactogen A continental *rift valley* system in the distal region of a *continent–continent collisional* orogen resulting from tensional **stresses** associated with **indentation tectonics**. Typical examples are the Shansi Graben, developed some 3000 km behind the Himalayan mountains, and the Rhine Graben behind the Alps. [PK]

Şengör, A.M.C. (1976) Collision of irregular continental margins: implications for foreland deformation of Alpine-type orogens. *Geology* **4**: 779–82.

***in situ* combustion** A method of recovering heavy, viscous oils from a **reservoir** when primary methods (e.g. **depletion drive**) have failed. The oil in the **reservoir** is heated by burning it. Combustion is kept going by pumping air into the formation to create a front of burning oil which, as it advances, breaks down the oil into **coke** and

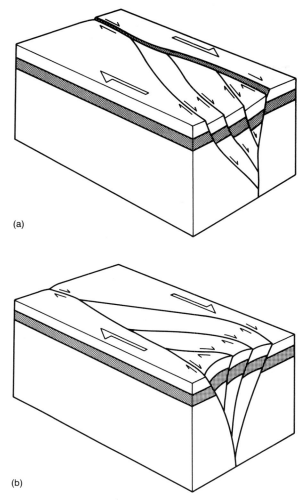

(a)

(b)

Fig. I8 Imbricate structure. Imbrication at (a) a releasing bend, and (b) a restraining bend, on a strike-slip fault. The oblique slip faults formed have extensional and compressional components, and form negative and positive flower structures respectively. (After Woodcock & Fischer, 1987.)

light oil. The latter is pushed ahead to producing **wells**. [AME]

***in situ* mining** Extraction of wanted materials of a mineral deposit without using physical mining methods, e.g. **Frasch process**, acid leaching of **uranium**. [AME]

inclusion trail A linear array of microscopic inclusions within a crystal. The inclusions are commonly fluids, **iron oxides**, hydroxides or other minerals; they may originate in a variety of ways. Many fluid and mineral inclusions are incorporated in the growth of a crystal particularly when growth is episodic, as for example in the **crack-seal mechanism**. In sedimentary rocks, clastic grains coated by

iron oxides or hydroxides and subsequently cemented may be delineated by inclusion trails. *Fluid inclusions* commonly mark cemented **microfractures**, and may also form in association with *crystal plasticity* during **recovery**, when they are found on **dislocation** networks. [TGB]

indentation tectonics (*escape*, *extrusion tectonics*) The **deformation** affecting the **lithosphere** of the overriding continental *plate* during a *continent–continent collision*. The theory of indentation of plastic materials has been developed in a mechanical engineering context, and allows the prediction of the configuration of lines of **failure** for simple shapes of both plastic medium and die (Fig. I9). From an examination of satellite imagery (see **remote sensing**) and **focal mechanism solutions** of **earthquakes** in the Himalayan region (Fig. I10), it has been proposed (Tapponier & Molnar, 1976) that indentation theory could be applied to this region, with India acting as the die and mainland Asia the plastic medium. Simple mechanical modelling of this process (Tapponier *et al.*, 1982), in which a block of laminated plasticene, confined at one edge, is deformed laterally by a rigid indenter (Fig. I10), has produced a configuration of slip lines which appears to mimic many of the **strike-slip faults** of the Himalayan region, and suggests that Indochina owes its position to extrusion from the unconfined, eastern part of the indented **lithosphere**. Also, the region some 3000 km north of the Himalaya is thrown into tension by the indentation, thus explaining the presence of the Shansi Graben and Baikal Rift, which are structures termed **impactogens** (Fig. I9). A criticism of the indentation model is that it takes no account of thickening (by *thrusting*) or thinning (by **extension**) of the plastic layer. An alternative model suggested by England (1982) views the **deformation** of Asia as the response of a continuous (i.e. unfaulted) viscous medium to both the edge forces of *continental collision* and the internal forces generated by different **crustal** thicknesses. Mathematical analysis of this situation provides a solution which can explain, in broad terms, the northerly progression from *thrust* faulting in the Himalaya through **strike-slip faulting** to *normal faulting* in the distal part of the mountain range.

Neither of these models can be completely successful in simulating the pattern of **deformation** associated with a *continental collision*, as both make simplifying assumptions. They do, however, provide insight into the development of this process, and it is possible that, in the future, theories encompassing both models will lead to a fuller understanding of *continental collision*. [PK]

England, P.C. (1982) Some numerical investigations of large scale continental deformation. In: Hsu, K.J. (ed.) *Mountain Building Processes*, pp. 129–41. Academic Press, London.
Tapponier, P. & Molnar, P. (1976) Slip-line field theory and large-scale continental tectonics. *Nature* **264**: 319–24.
Tapponier, P., Peltzer, G., LeDain, A.Y., Armijo, R. & Cobbing, P. (1982) Propagating extrusion tectonics in Asia: new insights from simple experiments with plasticene. *Geology* **10**: 611–16.

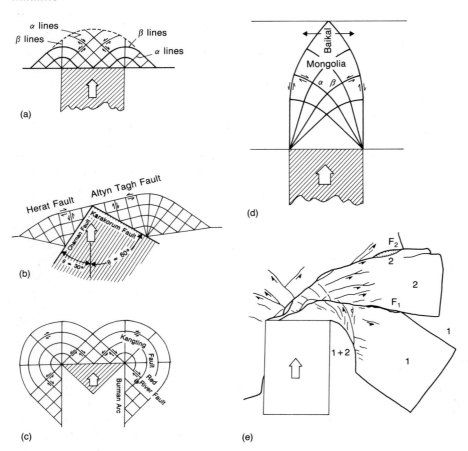

Fig. I9 Indentation tectonics. Plane indentation of semi-infinite rigid–plastic media by differently shaped rigid dies, with slip-line names corresponding to Himalayan faults. (a) Flat die. (b) Wedge. (c) Flat triangular die. (d) Plane indenter, bounded medium. (After Tapponier & Molnar 1976.) (e) An indentation experiment using plasticene. F = major fault, number = sequence of events. (After Tapponier *et al.*, 1982.)

indialite ($(MgFe)_2Al_4Si_5O_{18} \cdot nH_2O$) A ring silicate (*cyclo-silicate*) mineral that is the high-temperature **polymorph** of **cordierite**. [DJV]

indicolite Name for the blue variety of the mineral **tourmaline**. [DJV]

induced magnetization The magnetization possessed by a magnetic material in the presence of an ambient **magnetic field**, usually in the field direction. The other main type of magnetization is **remanent magnetization** which a rock acquires primarily as it cools through the **Curie temperatures** of its magnetic minerals and later in its history due to viscous relaxation, chemical processes, etc. (See also **rock magnetism**, **paleomagnetism**.) The net magnetization of a body is the vector sum of all the magnetizations. Although the effect of the induced magnetism usually dominates, there are notable exceptions (see **Koenigsberger ratio**) and it is unwise to disregard remanence in magnetic interpretation. [ATB]

induced polarization Exploration method widely used in the search for sulphides. The method is based on the empirical observation that certain mineralized regions, on the termination of direct current flowing through them, show a gradually decaying pattern of potential difference, rather similar to the decay pattern of an RC circuit. This transient time-domain pattern has a frequency-domain equivalent: the **apparent resistivity** measured at different frequencies is different. These properties are powerful indicators of sulphide mineralization, especially the disseminated type. The quantitative parameters describing them are **chargeability** and **percentage frequency effect**. (See also **overvoltage**, **membrane polarization**.) [ATB]

Keller, G.V. & Frischknecht, F.C. (1966) *Electrical Methods in Geophysical Prospecting*. Pergamon, Oxford.
Sumner, J.S. (1976) *Principles of Induced Polarisation for Geophysical Exploration*. Elsevier, Amsterdam.

inductively-coupled plasma emission spectrometry (ICPES) An analytical technique used in geology and archeology for the determination of many major, minor and trace elements in silicate rocks. The sample is digested in, typically, a mixture of hydrofluoric acid, nitric acid and perchloric acid. The sample solution is transformed into an aerosol of droplets and injected into a plasma of high-temperature ionized argon gas. The atoms of the sample

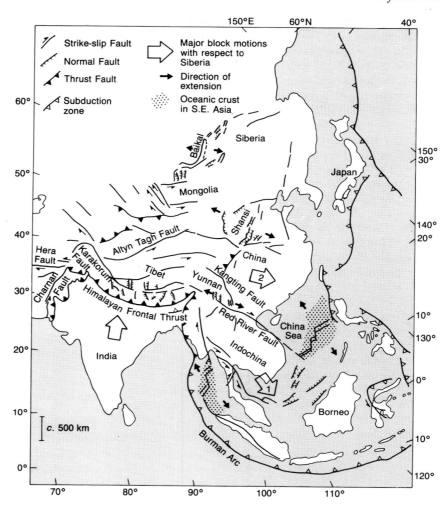

Fig. I10 Schematic map illustrating **indentation tectonics** in south-eastern Asia. Numbers = sequence of extrusions. (After Tapponier *et al.*, 1982.)

are excited and emit radiation including light of wavelengths characteristic of the elements present in the sample. This radiation is measured by a spectrometer and compared with that emitted from standard (known composition) solutions to give a quantitative analysis. Detection limits are typically 0.2–1 p.p.m. [OWT]

industrial diamonds Diamonds of less than **gem** quality because of inferior colour, size, shape or structural defects. These are used for industrial purposes such as drilling, cutting and lapping with a great variety of uses and scale. For example, industrial diamonds are used in a dentist's drill and an oil drilling rig. About 75% of industrial diamonds are used as grains or powder in grinding wheels and saw blades. The varieties of industrial diamond include industrial stones — large, below **gem** quality stones, **bort** — small irregular and imperfect stones used mainly in drilling, **ballas** and **carbonado**. [AME]

industrial minerals A term covering minerals of economic importance that are normally used by industry in the form

of the mineral itself. Thus they are not sought after as a source of one of the metallic elements they contain, as are the **ore** minerals such as **galena** from which **lead** is separated in a smelter. Two examples of industrial minerals are **kaolinite**, used as a filler in paper and many other industrial purposes and **baryte**, used as a filler in paint and asbestos. [AME]

infiltration Entry of water into the *soil* generally by downward flow through all or part of the *soil* surface. The rate of entry, relative to the rate of supply, determines how much of the water, if any, will pond on the surface possibly to form infiltration-excess **overland flow**. The *infiltration rate* is defined as the volume flux of water flowing into the *soil* per unit area of *soil* surface. If water is freely available, the *infiltration rate* reaches a maximum value, termed '**infiltration capacity**' by R.E. Horton (1940), and more recently '*infiltrability*' by some authors. Thus if water supply exceeds the **infiltration capacity**, infiltration is profile-controlled, whilst if the supply rate is less than the **infiltration capacity**, the infiltration is supply controlled. In unstructured *soils*, infiltration follows a 'classic' pattern with the development of surface saturation and the gradual penetration of a wetting front down into the profile. In structured *soils*, where water supply exceeds the **infiltration capacity** of the individual *soil beds*, water may infiltrate down the structural voids, sometimes called *macropores*, and rapidly flow deep into the *soil profile*. In *soils* close to saturation, or where deep percolation is limited by an impermeable horizon within the *soil profile*, infiltration may be limited so that the profile becomes saturated and surface **runoff** may be generated. **Infiltration capacity** is greatly influenced by *soil* texture, the amount of *soil* organic matter and by vegetation cover. The presence of structural voids, old root holes and other biopores may raise the *infiltration capacity* greatly. Typical values of **infiltration capacity** are $100 \, \text{mm h}^{-1}$ for a woodland loam, 10 to $20 \, \text{mm h}^{-1}$ for sandy and silty *soils* under pasture, and below $5 \, \text{mm h}^{-1}$ for heavy clay *soils*. Rainbeat on bare *soils* and surface compaction by overgrazing or by use of heavy farm machinery can lower infiltration by several orders of magnitude. *Soil* moisture levels also influence **infiltration capacity** which falls as the wetted zone deepens eventually to approach a constant minimum rate which is characteristic for the soil profile. Numerous attempts have been made to devise mathematical curves to describe the infiltration process: some like *The Green–Ampt* (1911) or *Horton* (1940) *equations* are empirical; others, like the *Philip* (1957) *equation* have a theoretical basis. [TPB]

Green, W.H. & Ampt, G.A. (1911) Studies on soil physics, 1. The flow of air and water through soils. *Journal of Agriculture and Soils* 4: 1–24.
Horton, R.E. (1940) An approach towards a physical interpretation of infiltration capacity. *Proceedings of the Soil Science Society of America* 5: 399–417.
Philip, J.R. (1957) The theory of infiltration. *Soil Science* 83: 345–57.

infiltration capacity The maximum rate at which water can enter the *soil* of an area by **infiltration**. [AME]

infrastructure That part of an *orogenic belt* deformed and metamorphosed at deeper levels of the **crust**, in contrast with the **suprastructure**, which consists of those parts deformed at a relatively shallow level. The distinction between the two is often difficult to make, and is of very limited usefulness. [RGP]

injection well A **well** in a gas or **oilfield** through which water, gas or steam is pumped into the **reservoir** formation to maintain the **reservoir** pressure so as to enhance recovery. (See **enhanced oil recovery**.) [AME]

inlier Area of older rocks surrounded by younger. The term is mainly used to describe outcrops of older *strata* isolated by **erosion**, e.g. in a valley cut through younger *strata*. Inliers may be created by the interaction of **structures** with topography, as in the case of an **anticline** or a **horst** crossing a valley, causing respectively *fold* and *fault inliers*. The converse is **outlier**. [RGP]

INPUT® An airborne **electromagnetic induction method** which employs transient currents and measures the decaying secondary field from the Earth after the primary field has been switched off. The energizing pulse in the INPUT system is a half sine wave through a horizontal loop around the aircraft and a towed vertical coil is the receiver. At least four time-slices of the secondary are recorded, the earliest representing poorer conductors and the later ones better and larger conductors. (See **airborne geophysical survey**.) [ATB]

Barringer, A.R. (1963) The use of audio and radio frequency pulses for terrain sensing. *Proceedings of the Second Symposium on Remote Sensing of Environment*, pp. 201–14. The University of Michigan.
Keller, G.V. & Frischknecht, F.C. (1966) *Electrical Methods in Geophysical Prospecting*. Pergamon, Oxford.

inselberg A large, steep-sided outcrop of solid rock or of huge boulders rising abruptly from an otherwise flat landscape. In German, the term refers to an 'island mountain'. Smaller rock and boulder outcrops are called **tors** or **kopjes**, and large, domed inselbergs are called *bornhardts*. Considerable debate has focused on the origin of inselbergs. It has been held that they are residuals produced by **parallel retreat** of bedrock slopes and, hence, are similar to **mesas**; others have argued that they are remnants following scarp retreat on deeply **weathered** land surfaces.

The reason for the preservation of masses of relatively unweathered rock within **deep weathering** profiles may be structural, since rocks with fewer **joints** are less susceptible to **weathering**. Alternatively, variations in the geochemistry of the rock may cause differential **weathering**. With **granites**, for example, rocks with higher **potassic feldspar** contents appear to be less prone to chemical **weathering** (Pye *et al.*, 1986).

Though similar to **monadnocks** in temperate regions, inselbergs are most common in arid and semi-arid regions of the tropics and subtropics. They may be developed in **sandstones** (as at Ayers Rock in Australia) or in crystalline rocks such as **granites** and **gneisses** (as in the Matopos of Zimbabwe). Inselberg landscapes, comprising extensive, flat plains dotted with massive outcrops of bare rock, are common on the ancient continental **shield** rocks of Africa and the Arabian Peninsula. [AW]

Bremer, H. & Jennings, J. (eds) (1978) Inselbergs. *Zeitschrift für Geomorphologie* Supp. **31**.
King, L.C. (1975) Bornhardt landforms and what they teach. *Zeitschrift für Geomorphologie* NF **19**: 299–318.
Pye, K., Goudie, A.S. & Thomas, D.S.G. (1984) A test of petrological control in the development of bornhardts and koppies on the Matopos Batholith, Zimbabwe. *Earth Surface Processes and Landforms* **9**: 455–67.
Pye, K., Goudie, A.S. & Watson, A. (1986) Petrological influence on

differential weathering and inselberg development in the Kora area of central Kenya. *Earth Surface Processes and Landforms* **11**: 41–52.

insolation weathering The shattering or granular disintegration of rock on the land surface caused by rapid expansion and contraction resulting from marked temperature fluctuations. Though insolation weathering used to be regarded as an important process in desert areas where there are large diurnal variations in temperature, there is little empirical evidence to support its occurrence. Indeed, attempts to simulate insolation weathering in laboratories have met with little success. The shattering of rocks in deserts is more commonly achieved by moisture or salt rather than heating and cooling alone. [AW]

Schattner, I. (1961) Weathering phenomena in the crystalline of the Sinai in the light of current notions. *Bulletin of the Research Council of Israel* **10G**: 247–66.

instantaneous rotation The motion of a tectonic *plate* at any given instant relative to another. It is defined in terms of its *Euler pole* and the rate of angular rotation about that axis. The integration of instantaneous rotations provides the actual path of the plates relative to each other. (Compare with **finite rotation**.) [DHT]

Le Pichon, X., Francheteau J. & Bonnin, J. (1973) *Plate Tectonics*. Elsevier, Amsterdam.

intensity of magnetization (J) *Induced intensity of magnetization* (J_i) refers to the magnetization induced by a weak external **magnetic field** (usually the **geomagnetic field**) and expressed in A m^{-1}. It arises from the alignment of initially randomly orientated elementary dipoles parallel to the external **magnetic field**, and its strength is proportional to that of the field, the constant of proportionality being the **magnetic susceptibility**. **Remanent magnetization** (J_r) refers to the **spontaneous magnetization** retained in the absence of an external field caused by the **preferred orientation** of the elemental dipoles. (See **natural remanent magnetization**.) [PK]

interconnectedness An index of the proportion of individual sediment bodies of a particular type that are in touch with each other in a given succession. Interconnectedness is in most common usage with respect to the coarse members of **alluvial deposits** where it refers to the proportion of individual channel-belt **sandstone** bodies that are in contact with each other. [JA]

Leeder, M.R. (1978) A quantitative stratigraphic model for alluvium with specific reference to channel deposit density and interconnectedness. In: Miall, A.D. (ed.) *Fluvial Sedimentology*. Canadian Society of Petroleum Geologists Memoir **5**: 587–96.

interdune An area of topographically low and generally rather flat ground separating aeolian **dunes** whose surface generally comprises a gravel or coarse lag. Interdunes may be erosional, depositional, or stable. Because of their low position and depending on the level of the **water table** and amount of **deflation**, they may be wet (i.e. flooded, and known as *slacks*), damp, dry, or cemented by **duricrusts** or **evaporites**. Some may have *algal* mats, or higher forms of vegetation, which may or may not have associated **nebkhas**, shadow **dunes** and aeolian scours. They can act as temporary conduits for ephemeral rivers sourced outside the **dune** system. Alternatively, small ephemeral streams may drain from one interdune to another. Dry interdunes can be flat, with small **wind ripples** or aeolian **plane beds** (see **aeolian stratification**), or they may have larger bedforms such as *granule ripples* (see **wind ripple**), slipfaceless *zibars* (see **aeolian sand sheet**), or even small **slip-faced dunes**. Therefore, interdune deposits include carbonates and **evaporites**, as well as clastic sediments with sedimentary structures such as low angle, dry surface, **aeolian stratification**, **deflation** structures, *desiccation cracks* (see **mud crack**), subaqueous **wave** and current generated structures, **adhesion structures**, **paleosols**, fluvial erosional structures (channels and scours) and the deposits of small **deltas**. Interdune deposits are consequently significant in ancient sequences as sensitive recorders of *paleoenvironments*, as barriers to *hydrocarbon migration* within aeolian **reservoirs**, as possible *hydrocarbon source rocks* and as fossiliferous horizons for **biostratigraphy**.

It will be obvious from the above that the nature of an interdune depends ultimately on factors such as climate, tectonism and regional geology, the main mechanism of control being the form taken by the adjacent **dunes**, itself a function of wind regime and **sand** supply. Where, due to a limited rate of net **sand** accumulation, the only **dunes** are isolated *siefs* or **barchans**, interdunes are commonly either *deflation* lags ('*serir*', '**reg**', or '**gibber**' plains') with wind-facetted *clasts*, or bare bedrock. Where there is more net deposition, different **dune** geometries develop and interdunes are predominantly sandy, either because they are eroded in earlier **dune** deposits, or because they are depositional. Their extent and plan form are controlled by **dune** geometry. The position of the **water table** will affect the nature of their deposits both in the ways explained earlier and by controlling **dune** geometry.

Because of their topographically low sites of deposition, interdune deposits have a higher preservation potential than **dune** sands, so the ratio of the preserved thicknesses of the two does not represent the aerial importance of the two environments during deposition. Where angles of **dune** climb are low, amalgamated interdune deposits may accumulate. One must therefore be cautious when attempting to distinguish interdune from **aeolian sand sheet** deposits and when assessing the significance of vertical sections through ancient interdune deposits in terms of the aerial extent of individual interdunes and of associated **aeolian bounding surfaces**. [DAR/RDS]

Ahlbrandt, T.S. & Fryberger, S.G. (1981) Sedimentary features and significance of interdune deposits. In: Ethridge, F.G. & Flores, R.M.

(eds) *Recent and Ancient Nonmarine Depositional Environments: Models for Exploration*. Special Publication of the Society of Economic Paleontologists and Mineralogists **31**: 293–314.

Breed, C.S. & Grow, T. (1979) In: McKee, E.D. (ed.) *A Study of Global Sand Seas*. Professional Paper of the United States Geological Survey **1052**: 253–304.

Kocurek, G. (1981) Significance of interdune deposits and bounding surfaces in aeolian dune sands. *Sedimentology* **28**: 753–80.

interference ripple (*ripple mark*) A **bedding** surface sedimentary structure consisting of **ripples** arranged in two or more orientations at a high angle so as to define a polygonal pattern. Some authors restrict the term to the interfering patterns of trochoidal crested **ripples** produced by the interference of **waves** moving simultaneously in two different orientations. Others use it for any pattern of overlapping *current* and/or **wave ripples** produced either simultaneously or one after the other. Care must be taken not to confuse interference ripples with biogenic structures or with those formed beneath curled clay flakes (see **mud crack**).

'*Ladder (or ladder-back) ripples*' are a category of interference ripples consisting of long wavelength **wave ripples** with smaller **wave ripples** in their troughs, the two sets being orientated at right angles. The structure is observed to form during a fall of water level, e.g. in an intertidal or evaporating ephemeral lake setting, where late stage **waves** form **ripples** in the earlier troughs, but not on the crests which are too shallow or exposed. However, a recent report of possible subtidal examples has made their diagnostic significance dubious (Reddering, 1987). [DAR]

Reddering, J.S.V. (1987) Subtidal occurrences of ladder-back ripples: their significance in palaeoenvironmental reconstruction. *Sedimentology* **34**: 253–7.

interflow Subsurface flow within the *soil* layer. Use of the word seems to imply that interflow is rapid and contributes to the storm **hydrograph**, and so equates with **overland flow**. *Subsurface stormflow* or *throughflow* are analogous terms. [TPB]

interfluve The area of high ground which separates two adjacent river valleys. [ASG]

interglacial A phase of relative warmth between cold phases (*glacials*) when the great *ice sheets* retreated and decayed, and *tundra* conditions were replaced by forest over the now temperate lands of the northern hemisphere. The *Holocene* is an interglacial and has seen a massive rise in sea-level (the **Flandrian transgression**) as the ice caps have released large amounts of meltwater into the oceans. Some of the **Quaternary** interglacials may have been warmer than today. Analysis of the record preserved in deep-sea cores suggests that there may have been around 17 *glacial*/interglacial cycles in the last 1.6 Ma. [ASG]

intergranular displacement **Displacement** along grain boundaries, between adjacent grains. Contrast *intragranular displacement* (within-grain) and *transgranular displacement* (across grain boundaries). (See **deformation mechanism**.) [RGP]

interlayer slip **Deformation** achieved by **displacement** on **bedding** or **foliation** surfaces. This is a necessary part of **fold mechanisms** such as *flexural slip*. It can also occur when sequences of layered rocks undergo **simple shear** parallel to **bedding** or **foliation**. **Strain** due to interlayer slip can be difficult to quantify where no markers cross **bedding** or **foliation** surfaces. The occurrence of this **deformation** must be taken into account when measuring **strain**. [SB]

intermediate argillic alteration A form of **wall rock alteration** characterized by the development of **kaolin**- and **montmorillonite**-group minerals mainly as **alteration** products of **plagioclase**. It is an intermediate to low grade form of **alteration** which merges into either **propylitic alteration** or fresh rock. [AME]

internal wave A **wave** motion occurring along a density interface within the body of a stratified fluid. Internal waves most commonly occur in the oceans and in lakes along the permanent or seasonal **thermocline**. They may also occur along the boundaries between water masses, in **estuaries** along the interface between dense seawater and lighter freshwater (*halocline*) as well as along less pronounced density steps. Because of the smaller density contrast involved, internal waves generally have larger amplitudes than surface **waves** (from a few metres in coastal waters to over 100 m in the open ocean) but are slower, with periods of a few minutes to several hours. Although internal waves do not directly set the surface in motion, their effects may be observed by the presence of bands or 'slicks' of calm or smooth water alternating with rougher water. This is caused by surface convergence over internal waves or interactions of **waves** with the sea bed in shelf areas. When internal waves impinge on the sea bed, for example at continental margins and on the shelf, they break and can be responsible for **suspension** of sediment. [AESK]

LaFond, E.C. (1962) Internal waves. In: Hill, M.N. (ed.) *The Sea* Vol. 1, pp. 731–55. Wiley-Interscience, New York.

Munk, W.H. (1982) Internal waves and small-scale mixing processes. In: Warren, B.A. & Wunsch, C. (eds) *Evolution of Physical Oceanography*, pp. 264–91. MIT Press, Cambridge, MA.

International Geomagnetic Reference Field (*IGRF*) Formula used to correct for latitude, longitude (*geomagnetic correction*) and **secular variation** in the **reduction** of magnetic measurements. The formula expresses the theoretical undisturbed **geomagnetic field** as a **spherical harmonic** equation to a large number of harmonics. Since records of the **geomagnetic field** are limited to a few, fixed, *magnetic observatories*, the formula is less well known than the corresponding **Gravity Formula**, which is based on world-wide measurements with portable **gravimeters**. [PK]

Peddie, N.W. (1983) International Geomagnetic Reference Field — its evolution and the difference in total field intensity between new and old models for 1965–1980. *Geophysics* 27: 691–713.

International Gravity Formula Formula used for the **latitude correction** in **gravity reduction** in the period 1930 to 1967. It was superseded by the **Gravity Formula**, whose constants are known with greater accuracy. [PK]

International Gravity Standardization Net (*IGSN*) World-wide network of locations set up in 1971 where the absolute value of **gravity** is known. The locations, or their derivatives, are used as **base stations** to convert the relative field readings of a **gravimeter** into absolute values. [PK]

Morelli, C., Gantar, C., Honkasalo, T. *et al.* (1971) *The International Gravity Standardisation Net 1971*. Special Publication No. 4. Bureau Central de l'Association Internationale de Geodesie, Paris.

interpluvial A relatively dry phase interspersed with the wetter phases (*pluvials*) of the *Pleistocene* and *Holocene*. In many parts of the tropics the period at the end of the Late-Glacial maximum (between 18 000 and 13 000 years ago) was especially dry, permitting **dune** fields to expand from desert cores and to spread out over exposed continental shelves. Lake levels in areas like the *Rift Valley* of East Africa fell at this time, and rivers like the Niger and the Nile were greatly diminished. [ASG]

Goudie, A.S. (1992) *Environmental Change* (3rd edn). Clarendon Press, Oxford.

intersection point The point on the longitudinal profile of an **alluvial fan** above which channel incision into older fan sediments occurs and below which there is deposition. The position of the intersection point will change with discharge conditions, differential subsidence associated

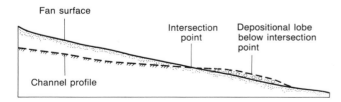

Fig. I11 Radial profile of an alluvial fan showing the position of the **intersection point**. This point will move up and down the fan surface in response to phases of incision and aggradation, probably related to tectonic activity. (After Hooke, 1967.)

with **faulting** and tilting of the fan surface. The existence of an intersection point often relates to active tectonic **deformation** (Fig. I11). [JA]

Hooke, R. Le B. (1967) Processes on arid-region fans. *Journal of Geology* 75: 438–60.

interstadial A phase of lesser *glaciation* and relatively greater warmth during the course of a major *glacial* phase, which was not of sufficient magnitude and/or duration to be classed as an **interglacial**. [ASG]

inter-arc basin An elongated area between the outer arc and volcanic arc of a **subduction zone**, mainly filled or being filled by sediments derived from the volcanic **island arc**. [DHT]

intraformational clast (*intraclast*) Sedimentary particle produced by the **erosion** of sediment soon after its deposition, and incorporated in deposits only slightly younger, certainly of the same episode of **basin** fill.

Intraclasts of mud and carbonate are the most common types because of the **cohesion** achieved by the former as soon as it is deposited and because of the common early lithification of the latter. In **sandstone**-dominated sedimentary successions, such as some fluvial sequences, the presence of *mud clasts*, sometimes termed '*rip-up clasts*', may be the only indication that muddy sediment was available for deposition, albeit temporarily, in low energy sub-environments, such as **bar** tops.

Sandstone intraclasts are significant in that they indicate that either early, near-surface **cementation** of **sand** by **evaporites** or by **pedogenesis** was taking place, or that tectonically controlled channel incision or **faulting** had re-exposed **sands** which had just undergone **burial diagenesis**. [DAR]

intrenched meander A meander in a river channel that has become incised into the surrounding landscape as a result of local **tectonic uplift**. [ASG]

inverse grading An upward increase in mean grain size of sediment within a bed, typically over centimetres. Also referred to, less commonly, as *reverse* or *negative grading*. Inverse grading commonly occurs in **beach** lamination and in the coarser-grained **sediment gravity-flow** deposits. Inverse grading suggests that the sedimenting grains were free to move relative to each other and that they were deposited under the dominant influence of grain-grain interactions. In **sediment gravity-flows**, inverse grading best develops where there are large grain concentrations, poor sorting and coarse grain sizes in a dense grain dispersion. (See **graded bed**, **kinematic sieving**.) [KTP]

inverse problem Problem relating to the interpretation of **potential** fields which states that it is possible to compute the anomaly of any given body, but the determination of the causative body from any given anomaly cannot be performed without ambiguity. This ambiguity can often be decreased by using external controls on the causative body, such as are provided by the surface geology and other, complimentary, geophysical techniques. [PK]

inversion The reversal of the sense of vertical movement of a block or region. Inversion occurs both in *orogenic belts* and in **cratons**. Widespread inversions in *orogenic belts* are often caused by reversal of motion on major **dip-slip fault** systems. **Extensional fault systems** may be reactivated in compression causing block **uplift**, or the reverse may occur; *thrust systems* may become reactivated in **extension** causing the formation of **basins**. Certain long-lived **fault** systems exhibit several inversions. For example, the Moine Thrust and Outer Isles Fault are members of a **fault** system in north-west Scotland that was probably active in **extension** during the Late **Proterozoic** when the Torridonian and Moine **basins** were formed; they became reactivated during the Caledonian thrust movements, and reactivated yet again during Mesozoic **extension**. Each reactivation gave rise to inversion.

Inversions are marked by breaks in the stratigraphic record. Major regional inversions are probably caused by abrupt changes in relative *plate* movement **vectors**. [RGP]

inverted relief A topographic phenomenon resulting from the transformation of areas of high relief — such as **anticlines** — into valleys or depressions as a result of enhanced **erosion** of the uplands. Landscapes of inverted relief are common in deserts where gravelly stream beds can become ridges, or salt-encrusted lake beds can be transformed to flat-topped hills, when the surrounding terrain is **eroded** by **deflation** of unconsolidated, fine-grained materials. [AW]

invisible gold Gold that cannot be seen even with the aid of an optical microscope. This gold is either sub-microscopic native **gold** or is present in interstitial *solid solution* in minerals such as **pyrite** or **arsenopyrite**. [AME]

inyoite ($CaB_3O_3(OH)_5 . 4H_2O$) A borate mineral. [DJV]

iodargyrite (AgI) A rare supergene **silver ore** mineral. [DJV]

iodine (I) At. no. 53, at. wt. 126.9045, m.p. 113°C. Occurs in **brines**, **calcrete**, seawater and seaweeds. The main sources are **brines** in Japan, USA and the former USSR. Iodine is also recovered from Chilean nitrate deposits. [AME]

iodobromite ($Ag(Cl,Br,I)$) A rare supergene **silver ore** mineral. [DJV]

iolite A **gem** variety of the mineral **cordierite**. [DJV]

iridium (Ir) A very rare, naturally occurring, **platinum** group metal. (See **native element**.) [DJV]

iridosmine (Ir,Os) A very rare but valued **platinum** group metal mineral. (See **native element**.) [DJV]

iron (Fe) A rare mineral. (See **native element**.) Native and particularly *meteoritic* iron was prized by early man. Native iron was occasionally used for tools by stone age peoples, for example in Greenland where flakes broken from pieces of native iron were used as cutting tools. True iron technology involving smelting was developed first by the Hittites of Anatolia and Syria in the 2nd millenium BC, heralding the Iron Age in Europe and the Near East. By *c.* 500 BC iron working had spread over most of Europe, and had been developed by the Chinese, who were the first to begin casting it (as opposed to hammering at red heat). The greater hardness of iron resulted in the abandonment of the earlier **bronze** weapons in these areas. Iron technology was not developed in the pre-Columbian Americas. [OWT/DJV]

iron oxide A general term for the range of iron oxides used in industry. An ancient and continuing use is as pigments. Apart from **titanium** oxide pigments, natural and artificial iron oxide pigments are the most widely used throughout the world. There are three categories of natural iron oxide pigments: **ochres** (red and yellow), siennas (orange) and **umbers** (dark brown). In general **hematite** contributes red shades, **limonite** yellow and **magnetite** brown to black. *Micaceous iron oxide* (MIO), which is prepared from specular **hematite**, cannot be produced synthetically. It is used as a corrosion-protective pigment in paints for engineering structures. [AME]

island arc Most of the calc-alkaline volcanoes that border the Pacific have curved traces that are convex towards the ocean. Similarly, the Scotia and Lesser Antilles arcs in the Atlantic–Caribbean are convex towards the **oceanic lithosphere** being **subducted**. The island arc volcanics overlie the **Benioff–Wadati zones** and the composition of the volcanoes appear to be directly related to the depth of these planes as they show an increase in **potassium** relative to sodium (normalized for **silica** content) with increasing depth of seismic activity, and similarly systematic variations in the ratios of **rare earth elements**. The volcanic island arcs are commonly bordered by sedimentary arcs on their oceanic side which are underlain by an older volcanic substrate. [DHT]

Gill, J. (1981) *Orogenic Andesites and Plate Tectonics*. Springer-Verlag, Berlin.

Toksöz, M.N., Uyeda, S. & Francheteau, J. (eds) (1980) *Oceanic Ridges and Arcs*. Elsevier, Amsterdam.

isochron A line joining points of equal age, usually based on *radiometric dating*. Also a line of a constant ratio of radioactive isotopes used to determine the radiometric age of a rock or mineral. [DHT]

isopachyte Line joining points of equal stratigraphic thickness of a formation or group of *strata*. An isopachyte map is contoured to indicate the three-dimensional shape of a unit of variable thickness. The technique is used, for example, in the study of sedimentary **basins** and in portraying the geometry of stratigraphic units cut off by **unconformities**. Isopachyte maps may be prepared from borehole data from which thicknesses are directly obtainable, or by geometric construction using *stratum contours* for the top and base of the unit, and subtracting the lower from the higher values where they intersect. [RGP]

isoseismal A line on a map separating zones of equal **earthquake intensity**. The first known use of this concept was by Robert Mallett, a British scientist who travelled to Italy and compiled observations of damage caused by the 16 December 1857 Naples **earthquake**. The form of the isoseismal contours can give some insight into the focal mechanism (for example, they may be elongated along the strike of a **fault** plane) but are very susceptible to distortion because of local *soil* and bedrock conditions. The **earthquake**'s **epicentre** can (but need not) fall within the maximum isoseismal contour, and hence isoseismals estimated from historical archives are useful for obtaining approximate locations for historic (pre-instrumental) **earthquakes**. For roughly circular isoseismals, various empirical relationships have been given which estimate the depth of *focus*. (See **earthquake intensity**, **hypocentre**.) [RAC]

isostasy While undertaking measurement of the **figure of the Earth** in the Andes in the 18th century, Bouguer noted that the Andes were characterized by a lower gravitational attraction than the Earth should have if it were homogeneous, even after allowing for the density variation in the immediate vicinity. A similar deflection from the vertical was subsequently found in India in the mid-19th century, attributable to the Himalaya. This regional discrepancy was modelled in two different ways by **Airy** and **Pratt** and subsequently termed isostasy by Dutton in 1889. The basic idea of isostasy is that below a certain depth (the **depth of compensation**) all pressures exerted by the rocks above this level are equal. Pratt proposed that the average density of **crustal** rocks is lower

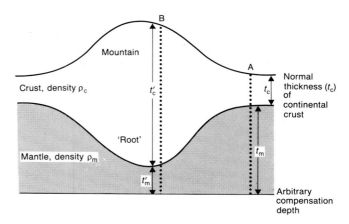

Fig. I12 Airy **isostasy**: the weight of rock above the compensation depth is constant.

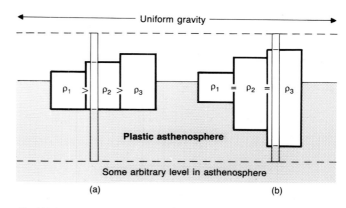

Fig. I13 Isostasy. Isostatic compensation. (a) According to Pratt's hypothesis. (b) According to Airy's hypothesis. If compensation is perfect, the masses in similar columns through the lithosphere down to the same level in the asthenosphere are identical. Gravity is uniform along the top line, since mass in a column below the line is the same everywhere.

where the elevation is greatest, while Airy proposed that the higher topography was compensated by the presence of **continental crustal** roots (Figs I12 and I13). Both hypotheses adequately model the observed phenomenon, although the physical reality probably lies between the two models, i.e. most recent *orogenic belts* tend to have deep crustal roots, but some areas of marked elevation, such as the **Basin-and-Range** Province of western North America, are characterized by hotter, and therefore lower density, underlying **mantle** rocks than normal. However, the fact that isostatic anomalies are persistent features of the Moon's **gravity** field, being associated with mascons (mass concentrations) that formed more than 3 Ga ago, but transient in terms of 1 Ga or so on the Earth, has been fundamental in developing the concept of a **lithosphere** and **asthenosphere** (Barrell, 1914). As the Earth is loaded, such as by *ice sheets*, volcanoes or even oceanic **tides**, the

lithosphere gradually **flexes** with associated flow of the lower plastic **asthenospheric mantle** rocks. The very gradual disappearance of isostatic anomalies on the Earth can therefore be used to determine the **viscosity** of the **mantle** and the *flexural rigidity* of the **lithosphere**. (See **isostatic rebound**.) [DHT]

Barrell, J. (1914) The strength of the Earth's crust. *Journal of Geology* **22**: 635–83.
Bott, M.H.P. (1982) *The Interior of the Earth*. Arnold, London.
Dutton, C.E. (1889) On some of the greater problems of physical geology. *Bulletin of the Philosophical Society of Washington* **23**: 51–64
Mörner, N. (ed.) (1980) *Earth Rheology, Isostasy and Eustasy*. John Wiley & Sons, Chichester.

isostatic rebound Excess loads on the Earth's **lithosphere**, such as *ice sheets*, volcanic piles, etc., cause a depression of the **lithosphere**, with consequent flow of the **asthenosphere** in order to compensate for the **flexure of the lithosphere**. After the load has been removed, e.g. the ice melts or the rocks are eroded, the flexural forces attempt to restore the **lithosphere** to its original position but this is delayed by the elastic response of both the **lithosphere** and **asthenosphere**. This allows the *flexural rigidity* of the **lithosphere** and also the **viscosity** of the **asthenosphere** to be determined. The decrease in the size of the **Great Salt Lake** in Utah or the **uplift** of Fennoscandia following deglaciation are examples of unloading, while the Kariba Dam and associated reservoir in southern Africa is an example of new **stresses** being applied. [DHT]

Mörner, N. (ed.) (1980) *Earth Rheology, Isostasy and Eustasy*. John Wiley & Sons, Chichester.

isothermal Having the same temperature (spatially). Isothermal surfaces (or *isotherms*) have zero temperature gradients tangential to them, so the **flow** of **heat** will be normal to the surface at every point on the surface in an isotropic medium. The *Curie isotherm*, below which the temperatures are too high for minerals to sustain magnetism (**Curie point**), is a significant concept, defining the lowest depth at which **magnetic anomalies** can originate. [ATB]

isovol A line of equal volatile content in **coal**. Isovol maps are one of the many types of **coal quality maps**. [AME]

J

J-type lead An isotopically **anomalous lead** whose **lead** isotopic ratios yield a negative model **lead** age for the deposit in which they occur, i.e. an age some time in the future. These leads must have had a complex crustal history, during which they acquired extra amounts of radiogenic **lead**. They are named J-type after Joplin, Missouri from where the first examples were collected. **Leads** from the orefields of the Mississippi Valley are all J-type leads whereas those in most similar **carbonate-hosted base metal deposits** are ordinary or **common leads**. [AME]

jacobsite ($MnFe_2O_4$) A rare mineral of the **spinel** group. (See **oxide minerals**.) [DJV]

jade The term used for the semi-precious stone that includes both the mineral **jadeite** and the mineral **nephrite**. Jades were highly prized in antiquity, as today, for objects of decoration and personal adornment. Despite the difficulties of working these hard materials, jade objects were produced in the early civilizations of China and Japan, in central and south America, and in many parts of Europe. The Chinese used jade for sculptures, figurines, chisels, knives and coins, whilst in Europe polished **axes** were produced. [OWT/DJV]

jadeite ($NaAlSi_2O_6$) A green, compact, single-chain silicate mineral (*inosilicate*) of the **pyroxene** group found only in metamorphic rocks and prized as the semi-precious stone 'jade'. (See **pyroxene minerals, gems**.) [DJV]

jamesonite ($Pb_4FeSb_6S_{14}$) A minor **ore** mineral of **lead**. [DJV]

jargon Name given to yellowish or smoky varieties of the mineral **zircon**. [DJV]

jarosite ($KFe_3(SO_4)_2(OH)_6$) A secondary mineral found as crusts and coatings on **iron ores**. [DJV]

jasper A granular microcrystalline variety of the mineral **quartz**, usually coloured red by included **hematite**. Jasper was used for tool manufacture on some south African Stone Age sites and in Sri Lanka. [OWT/DJV]

jasperoid A form of **wall rock alteration** characterized by the development of fine-grained **hematite**-stained **silica**.

This is a common **alteration** type in some **epithermal deposits**, e.g. **Carlin-type**. [AME]

jet A lustrous variety of **lignite** that occurs as isolated masses in some bituminous **shales**. It is used for jewelry and other ornamentation.

Jet, probably mainly from the Yorkshire coast of England, was employed by the Bronze Age inhabitants of Britain for the manufacture of beads and buttons. Jewelry of **lignite** and **shale**, superfically similar in appearance to jet, may have been an alternative for those who could not afford or obtain jet. [AME/OWT]

jimthompsonite (($Mg,Fe)_{10}Si_{12}O_{32}(OH)_4$) A silicate mineral belonging to a group closely related to the **amphiboles** and known as the *biopyriboles*. (See **pyribole**.) [DJV]

johannsenite ($CaMnSi_2O_6$) A **pyroxene mineral**. [DJV]

JOIDES Joint Oceanographic Institutions for Deep Earth Sampling, originally established to undertake deep-sea drilling. The programme later evolved into the **Deep Sea Drilling Program** (**DSDP**) and then the *International Program for Ocean Drilling* (*IPOD*). [DHT]

joint A **fracture** on which any **shear displacement** is too small to be visible to the unaided eye. Joints are a ubiquitous feature of rocks exposed at the surface of the Earth. Joints usually occur in a group known as a *joint set*, defined as a collection of joints with a common orientation; joints belonging to a set are *systematic*. **Fracture** sets in folded rocks often have quite consistent geometrical relationships to the **fold** orientation which is given orthogonal reference axes *a*, *b* and *c* (Fig. J1). *ac* joints (*cross joints*) are perpendicular to the *fold axis* and *bc* joints (*longitudinal joints*) are parallel to the axis and perpendicular to the folded layer at the *hinge* of the **fold**. **Fractures** parallel to the *fold axis* that remain perpendicular to the folded layer on *fold limbs* are known as *radial joints*.

Conjugate joints are joints that intersect each other, often with an acute angle of 60°.

En echelon joints are parallel or subparallel joints slightly offset from each other in the direction perpendicular to the joint surface. *Feather joints* or *pinnate fractures* are minor joints adjacent to a larger **fracture** and intersecting it at an acute angle. The V shape defined by the intersection

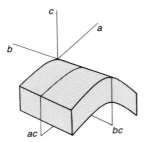

Fig. J1 Common **joint** orientations with respect to folds.

Table J1 The **Jurassic** system

Major subdivisions	Traditional German classification	Stages
Upper	Malm	Tithonian* Kimmeridgian Oxfordian
Middle	Dogger	Callovian Bathonian Bajocian Aalenian
Lower	Lias	Toarcian Pliensbachian Sinemurian Hettangian

* Roughly equivalent to Volgian in boreal realm.

is often taken to point in the direction of **displacement** on the larger **fracture**. Joints are common features of igneous intrusions, in which they characteristically form perpendicular to the cooling surface. Such *cooling joints* may form a network around elongate prisms of the intrusive rock (especially in tabular-shaped intrusions such as **dykes** and **sills**); this is described as *columnar jointing*. The columns may have approximately hexagonal cross-sections and may be cut perpendicular to their axes by smaller convex joints. A structure common in **granitic** intrusions is a set of **fractures** parallel to the ground surface; this is *sheeting*, defined by *sheet joints*.

A *master joint* is a joint whose extent is appreciably greater than the average of the set in which it occurs, and against which less prominent joints terminate.

Joints may be open, or filled by a variety of cementing minerals, for example **quartz**, **calcite** and iron oxides or hydroxides. Joint surfaces may show a variety of surface markings, including *plume* or *feather structure*, *augen* or *rib structure*, *cross fractures*, *fringe joints*, *en echelon joints*, and *conchoidal fringe joints* (see **fracture**).

Because there is no visible **displacement** on a joint, many speculative inferences have been made about the relationship of joints to the **stress** system in which they originated. *Shear joints* are considered to form by **shear failure** in orientations similar to *shear faults*, i.e. at acute angles to the maximum *principal stress*. *Extension* (or *tension*) *joints* are inferred to form by **tensile failure** perpendicular to the least *principal stress*, and *hybrid joints* are intermediate between the two.

There have been many hypotheses to account for the origin of the **stresses** responsible for joint formation. A commonly invoked mechanism is the development of high **pore-fluid pressures** due to **compaction** of sediments leading to jointing by *hydrofracture*. Large compressional horizontal **stresses** may develop during regional tectonism, which may explain some associations between joints and **folds**. The development of both compressional and **tensile stresses** during **uplift** of rocks can be predicted, depending on initial **stress** conditions. In a laterally confined situation, relief of the vertical load can lead to a minimum *principal stress* approximately perpendicular to the ground surface. Extension **fractures** would then form parallel to

the ground surface and appear as *sheeting*. Alternatively, horizontal tensile **stresses** may arise through the change in radius of curvature of a sedimentary **basin** as it moves further from the Earth's centre, forming vertical *extension joints*. It has been suggested that joints are formed in poorly consolidated sediments due to dynamic loading from **earthquakes**. Finally, thermal **stresses** induced by cooling are clearly responsible for all *cooling joints*.

Joint frequencies are noticeably variable. **Fracture** frequencies are considered to be controlled by **strain** (*hinge areas* of **folds** may have higher joint frequencies), lithology (**sandstones** commonly have higher joint frequencies than adjacent **shales**, for example) and **bed** thickness (thinner **beds** have higher frequencies). It also appears that **fractures** may form localized **deformation** zones of higher fracture frequency as an inherent aspect of their development. [TGB]

Hancock, P.L. (1985) Brittle microtectonics: principles and practice. *Journal of Structural Geology* **7**: 437–57.

Price, N.J. (1966) *Fault and Joint Development in Brittle and Semi-brittle Rock*. Pergamon, Oxford.

Jurassic The stratigraphic system taking its name from the Jura Mountains of Switzerland; its subdivision into eleven stages and many more zones is based on *ammonites* (Table J1). Because of pronounced faunal provinciality, correlation of the youngest Jurassic stage presents greater difficulties than the earlier ones, and three terms have been widely used: *Tithonian*, *Volgian* and *Portlandian*. The bases of the *Tithonian* and *Volgian* are essentially coincidental, but the *Volgian–Ryazanian boundary* in the boreal realm is slightly younger than the *Tithonian–Berriasian boundary* in the **Tethyan** realm, so the generally accepted *Jurassic–Cretaceous boundaries* in the different realms do not exactly coincide. In the absence of *ammonites*, the most useful fossils for correlation are **dinoflagellates** in marine and *pollen* and *spores* in non-marine facies.

There were only two major continents, **Laurasia** in the north and **Gondwanaland** in the south. During the early part of the period these were united at their western end and separated by a major equatorial ocean, **Tethys**, which widened towards the east. In the mid-Jurassic that section which we now know as north-western Africa moved away from North America, so creating a narrow ocean, the precursor of the modern Atlantic, which continued to widen throughout the remainder of the period. Antarctica, South America and Africa began to separate at the end of the Jurassic. **Seafloor spreading** also took place in the proto-Pacific (**Panthalassa**), and various islands were accreted onto the Asian and American margins.

Sea-level was comparatively high for most of the Jurassic, with up to a quarter of the present continental area being inundated at its highest stand late in the period (*Oxfordian* to early *Tithonian*). Most of what is now western Europe was covered by sea for almost the entire period, as were the Pacific margins, but much of **Gondwanaland** was never flooded, the sea being restricted to shallow bays or straits. Most of North America also remained uncovered, but a shallow sea spread from the Pacific into the western interior of the United States and Canada during the middle and early part of the late Jurassic. Another area of persistent land was what we now know as eastern Asia.

The Jurassic climate was appreciably more equable than that of the present day. There were probably no polar ice caps, temperate conditions extending as far as the Arctic and Antarctic. Similarly, the climatic conditions which today characterize the tropics extended well north and south spreading as far as, for example, western Europe. Continental interiors in low and middle latitudes were arid, and there was a spread of arid conditions in southern Eurasia late in the period, bound up with the creation of a rain shadow by the newly-formed Cimmeride mountains.

The vertebrate terrestrial life of the Jurassic period was dominated by **reptiles**. The *dinosaurs* had first appeared late in the **Triassic** from a *thecodont* stock which also gave rise to *pterosaurs* and, later, **birds**. From small bipedal animals there evolved the huge, spectacular creatures familiar to us all, such as the herbivorous *Brontosaurus*, *Diplodocus* and *Stegosaurus* and the carnivorous *Allosaurus*. Flying animals include the truly **reptilian** *pterosaurs* and the first organisms that could be called **birds** as distinct from **reptiles**, as represented by the pigeon-sized

Archeopteryx from the Upper Jurassic *Solnhoffen Limestone of Bavaria*. There were two important groups of **reptiles** which lived in the sea, the dolphin-like *ichthyosaurs* and the long-necked *plesiosaurs*; both groups were beautifully adapted to marine life. *Turtles* and *crocodiles* are also found in Jurassic deposits. Jurassic **mammals** were small and subordinate and obviously did not compete directly with the *dinosaurs*. The **fish** faunas were dominated by the *holosteans*, characterized by heavy rhombic scales.

Because they are far more abundant, the marine invertebrate fossils are of more importance to stratigraphers and **paleoecologists**. As already noted, the *ammonites*, which are abundant in many marine deposits, are by far the most valuable stratigraphically because of their high rate of evolutionary turnover. In shallow marine deposits another **molluscan** group, the *bivalves*, dominates, and gives important information on *paleoenvironments*. Also significant in many deposits are such groups as **corals**, *gastropods*, *belemnites*, *crinoids*, **echinoids**, **Bryozoa**, **Foraminifera** and *ostracods*. A large-scale provinciality, expressed as the **Tethyan** and boreal realms, is recognizable in some groups, most notably the *ammonites* and *belemnites*. As this seems broadly related to *paleolatitude* a number of paleontologists have sought a climatic explanation, but the matter is undoubtedly more complicated and various paleogeographic factors should not be overlooked.

With regard to **land plants**, the Jurassic might well be called the age of the *gymnosperms*, including *conifers*, *gingkos* and their relatives, the *cycads*. *Ferns* and *horsetails* make up the remainder of the land flora. [AH]

Arkell, W.J. (1956) *Jurassic Geology of the World*. Oliver & Boyd, Edinburgh.

Cope, J.C.W., Getty, T.A., Howarth, M.K., Morton, N. & Torrens, H.S. (1980) *A Correlation of Jurassic Rocks in the British Isles Part 1: Introduction and Lower Jurassic*. Geological Society of London Special Reports No. 14. Blackwell Scientific Publications, Oxford.

Cope, J.C.W., Duff, K.L., Parsons, C.F., Torrens, H.S., Wimbledon, W.A. & Wright, J.K. (1980) *A Correlation of Jurassic Rocks in the British Isles Part 2: Middle and Upper Jurassic*. Geological Society of London Special Report No. 15. Blackwell Scientific Publications, Oxford.

Hallam, A. (1975) *Jurassic Environments*. Cambridge University Press, Cambridge.

Haq, B.U., Hardenbol, J. & Vail, P.R. (1987) Chronology of fluctuating sea-levels since the Triassic. *Science* **235**: 1156–67. (This article contains biostratigraphic zonal schemes as well as a sea-level curve based on seismic stratigraphy.)

K

K-cycle The name given to a concept of landscape development involving the cyclic **erosion** of *soils* on upper hillslopes during unstable climatic phases and *soil* development during stable phases. The term, which is much used in Australia, is broadly comparable to the idea of **rhexistasy** and **biostasy**. [ASG]

kainite ($KMg(Cl,SO_4) \cdot 2\frac{3}{4}H_2O$) An **evaporite** mineral. [DJV]

kalsilite ($KAlSiO_4$) A **feldspathoid mineral**. [DJV]

kamacite (Fe,Ni) An **iron–nickel** alloy mineral found in *meteorites*. (See **native element**.) [DJV]

kamenitza Solution **pan**, or solution **basin**, formed on exposed **limestone** surfaces by **dissolution** of calcium carbonate. According to Bögli (1960) they are a type of **karre** which form on **karst** surfaces which are partly covered by sediment. Fine sediment and *algae* often collect in the bottom of kamenitzas, which may be flat-bottomed and have undercut rims. [HAV]

Bögli, A. (1960) Kalklösung und Karrenbildung. *Zeitschrift für Geomorphologie* Supp. **2**: 4–21.

kaolin Also known as 'china clay', an important mineral product largely made up of the mineral **kaolinite**. (See **clay minerals**.) [DJV]

kaolinite ($Al_2Si_2O_5(OH)_4$) A **clay mineral**, the main component of **china clay**. (See **kaolin**.) [DJV]

kaolinitization The **alteration** of a precursor mineral to **authigenic kaolinite**. Precursor minerals commonly comprise mica (**muscovite**, **biotite**) or **feldspar**, and **alteration** proceeds via a **dissolution–precipitation** mechanism. (See also **authigenesis**, **diagenesis**, **neoformation of clays**, **replacement**.) [SDB]

Keller, W.D. (1978) Kaolinisation of feldspars as displayed in scanning electron micrographs. *Geology* **6**: 184–8.
Loughnan, F.C. (1982) Genesis and synthesis: kaolins in sediments. In: van Olphen, H. & Veniale, F. (eds) *Proceedings of International Clay Conference 1981*, pp. 187–94. Developments in Sedimentology No. 35. Elsevier, Amsterdam.

Karman–Prandtl velocity law A term which expresses the velocity profile within a turbulent **boundary layer** as a logarithmic function of the distance from the bed. This velocity law is only valid in the lower 15–20% of the flow above the **viscous sublayer**. The law is based upon considerations of turbulent mixing within the **boundary layer** and upon the assumptions (i) that the mixing length of eddies within a **turbulent flow** is directly proportional to the flow depth as expressed by the proportionality of **von Karmans constant** and (ii) that the **shear stress** is constant. The law may be given by:

$$U = \frac{1}{\kappa} \cdot \left(\frac{\tau}{\rho}\right)^{0.5} \cdot \ln \frac{y}{k}$$

where U is the velocity, κ is the **von Karman constant**, τ is the **shear stress**, ρ the fluid density, y the height of the point at which U is to be determined and k is the roughness constant. (See **roughness height**, **turbulent flow**.) [JLB]

Francis, J.R.D. (1975) *Fluid Mechanics for Engineering Students*. Arnold, London.
Middleton, G.V. & Southard, J.B. (1984) *Mechanics of Sediment Transport* (2nd edn). Short Course No. 3 Lecture Notes. Society of Economic Paleontologists and Mineralogists, Tulsa, OK.

karren (sing. karre) A term of German origin used to describe minor solutional features developed on carbonate rocks. They are formed largely by the **dissolution** of **carbonate minerals** by rainwater, soilwater or underground flows. The French term *lapiés* is also used to describe such features, but there is no comparable English word. Bögli (1960) classified karren features into those developed on bare, partially covered and covered **karst** surfaces. Bare surface forms include *rain pits* and *solution flutes (Rillenkarren)*. *Rillenkarren* are commonly found on steep to nearly vertical surfaces and have sharp ribs between the flutes. Another common form on bare surfaces are *Trittkarren*, which are heel-shaped pits. The distribution of bare surface features is influenced by water flow and slope angle, although organisms and other factors may influence some forms. **Solution pans** (**kamenitza**, which is a Slav term) and **solution** notches are forms found on partly covered surfaces. Subsoil karren include rounded solution runnels (*Rundkarren*) and **solution pipes**. Karren-type forms may also develop on non-carbonate rocks, in which case they are usually known as *pseudokarren*. [HAV]

Bögli, A. (1960) Kalklösung und Karrenbildung. *Zeitschrift für Geomorphologie* Supp. **2**: 4–21.

Jennings, J.N. (1985) *Karst Geomorphology*, pp. 73–83. Basil Blackwell, Oxford.

karst A term used to describe terrain with distinctive landforms and drainage arising mainly from *solutional erosion*. The word karst is a germanicized form of the Slav word 'krs' which was originally used as a name for the large area of **limestone** with distinctive scenery in Western Slovenia (the Dinaric karst). Since then, the term has been expanded to describe similar areas of terrain. Karst terrain has been identified in many areas of the world, large areas being found in Europe, China, Malaysia, the United States and Australia. In Britain, karst areas are found on the Mendip Hills, in the Craven district of North Yorkshire and elsewhere. There has been much debate over the conditions necessary for the development of karst terrain. True karst (as opposed to **pseudokarst**) development requires an easily soluble rock, such as one of the carbonate rocks (e.g. **limestone**, **dolomite**), or an **evaporite** rock like **gypsum**. The **weathering** process of **solution** occurs when natural waters make contact with these rocks, leaving very little insoluble residue. Karst terrain development is also favoured by the presence of a considerable thickness of well-**jointed** rock. In some parts of Yugoslavia, for example, the **limestone** is over 4000 m thick. High relief and high annual precipitation also encourage the development of karst landscapes.

An essential component of karst terrain is the development of underground drainage systems, through *solutional erosion*, which leads to the production of often spectacular **cave** systems, as well as a range of closed depressions on the surface through which water is chanelled underground (e.g. **dolines**). The concentration of **erosion** underground leads to the preservation of upstanding residual features. A long-standing debate in karst **geomorphology** has concerned the role of climate in the formation of different karst types. Distinctive karst landscapes have been identified in tropical and temperate areas. It has been hypothesized that karst development is maximized in humid tropical climates (producing **tower** and **cockpit karst** forms) due to high precipitation levels, faster chemical reaction rates, and greater *soil* CO_2 concentrations. As with many climatic control hypotheses in **geomorphology**, however, this is an inadequate explanation of the variability of karst terrain. Structural factors are also very important to the nature of karst landscapes. Recent geomorphological work in karst areas has elucidated the rates and nature of *solutional erosion*, and provided more accurate information upon the long term development of karst landscapes. [HAV]

Gunn, J. (1986) Solute processes and karst landforms. In: Trudgill, S.T. (ed.) *Solute Processes*. John Wiley & Sons, Chichester.
Jennings, J.N. (1985) *Karst Geomorphology*. Basil Blackwell, Oxford.

karst bauxite Bauxite developed on a highly irregular karstified **limestone** or **dolomite** surface. Karst bauxites include the oldest known **bauxites** — those in the lands just north of the Mediterranean which range from **Devonian** to mid-**Miocene**. Other major deposits are the *Tertiary* ones of Jamaica and Hispaniola. Texturally, karst bauxites are quite variable. The West Indian examples are **gibbsitic ores** with a structureless, earthy, sparsely concretionary texture. European karst bauxites, on the other hand, are generally lithified and texturally *pisolitic*, **oolitic**, fragmental or even bedded. Mineralogically they are predominantly **boehmitic ores**. These and other facts indicate that the West Indian **bauxites** have strong affinities with upland (high-level) deposits, whilst the European karst bauxites are more reminiscent of **peneplain** (low-level) deposits. [AME]

kavir A **playa**, continental **sabkha**, or other saline desert **basin** which may be periodically flooded by rainfall, **runoff**, or rising **groundwater**. The term is used mainly in Iran to describe saline marshes and salt deserts such as the Kavīr Desert (Dasht-e-Kavīr) which is nearly 400 km wide. [AW]

keatite (SiO_2) A name given to a synthetic tetragonal form of **silica**. [DJV]

kegelkarst A German term for a **limestone** landscape characterized by conical hills, or *kegels*, interspersed with closed depressions. This landscape type is common in tropical **karst** areas, for example Jamaica and Java, and is also known as *cone karst* or **cockpit karst**. [HAV]

Lehmann, H. (1936) *Morphologische Studien auf Java*. Engelhorn, Stuttgart.

kernite ($Na_2B_4O_6(OH)_2 . 3H_2O$) An important chemical mineral and source of **boron** compounds. [DJV]

kersantite A **calc-alkaline lamprophyre** containing **biotite**, **plagioclase** (generally **oligoclase–andesine**) and **augite**, with or without **diopside** and **olivine**. [RST]

kieserite ($MgSO_4 . H_2O$) An **evaporite** mineral. [DJV]

kilobar Unit of pressure equal to 1000 bar or 100 megapascal. See **bar**, **pascal**. [RGP]

kimberlite A *serpentinized* and carbonated *porphyritic* **mica-peridotite** comprising **phenocrysts** of **olivine** and **phlogopite** in a fine-grained groundmass containing **olivine**, **phlogopite**, **pyrope**, Fe–Ti oxide, **perovskite**, with **serpentinite**, **chlorite** and carbonates. Kimberlites are commonly extremely brecciated and with large crystal and/or rock fragments within a fine-grained matrix. They commonly occur in *volcanic pipes* (and occasional **dykes**) and are characterized by a wide range of **xenolithic** inclusions which include (with increasing depth of origin), angular blocks of *country rock*, more rounded blocks of high-grade metamorphic rocks (middle and lower **crust**) and rounded blocks of **ultramafic rocks** of **mantle** origin. Kimberlites are the main economic source of **diamond**. [RST]

Ross, J., Jaques, A.L., Ferguson, J. *et al* (eds) (1989) *Kimberlites and Related Rocks* Vol. 1: *Their Composition, Occurrence, Origin and Emplacement.* Proceedings of the 4th International Kimberlite Conference, Perth 1986. Geological Society of Australia Special Publication No. 14. Blackwell Scientific Publications, Melbourne.

Ross, J., Jaques, A.L., Ferguson, J. *et al* (eds) (1989) *Kimberlites and Related Rocks* Vol. 2: *Their Mantle/Crust Setting, Diamonds and Diamond Exploration.* Proceedings of the 4th International Kimberlite Conference, Perth 1986. Geological Society of Australia Special Publication No. 14. Blackwell Scientific Publications, Melbourne.

kinematic Refers to motion. 'Kinematics' is the study of motion. The term is used to describe **structures**, properties or characteristics that convey information regarding the relative motion of an object, e.g. a *thrust sheet* or a *plate*. Thus *'kinematic indicators'* are **structures** that can be used to determine the direction and sense of **displacement** of a **fault** or **shear zone**. [RGP]

kinematic sieving A process invoked partly to explain **inverse** or *reverse* **grading**, in which 'vibration-strain' during flow promotes the downward filtration or percolation of smaller grains between relatively larger grains, thereby preferentially displacing larger particles towards the free sedimentation surface of the flow. (See **inverse grading**.) [KTP]

kinematic viscosity The resistance of a substance to a change in shape as given by the ratio of the molecular **viscosity** divided by the density of the medium. Use of kinematic viscosity enables comparison between substances of different densities in relation to their resistance to change in shape. [JLB]

kink Type of **fold** with a characteristic angular profile; *kinking* is the **fold** process producing such **structures**. (See **kink band**, **chevron fold**.) [RGP]

kink band A localized band defined by sharp boundaries where the orientation of a **structure** changes abruptly. The edges of the band are known as *kink band boundaries* or *kink planes*. By analogy with **faults**, displacement of the upper side of the kink band with respect to the lower defines *reversed* or *contractional kink bands* (Fig. K1a) or *normal* or *extensional kink bands* (Fig. K1b). Kink bands may occur on a variety of scales from the **microscopic**, where kink bands in crystal lattices are responsible for **twinning** and for producing bands of *undulose extinction* in **quartz**, to mesoscopic. Kink bands that occur in two directions with

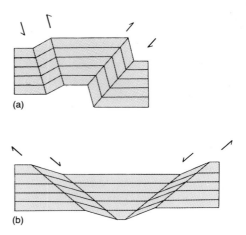

Fig. K1 A pair of conjugate **kink bands**. (a) Reversed or contractional kink bands. (b) Normal or extensional kink bands. (After Ramsay & Huber, 1984).

the same sense of **displacement** are known as *conjugate kink bands* (Fig. K1a and b), which can be viewed as a **box fold** when considered together. If equally sized kink bands are adjacent and parallel, they describe **chevron folds**. The growth of two isolated **conjugate** kink bands can lead to the formation of **box folds** and **chevron folds** in a continuous sequence. Kink bands are particularly common structures in finely foliated rocks. The formation of kink bands can be thought of as a response to **stress** in a similar way to **buckling**, and is predicted by theoretical analysis of the viscous **deformation** of materials with high **anisotropies**, for example **slates** or thinly bedded sediments. [TGB]

Ramsay, J.G. & Huber, M.I. (1987) *The Techniques of Modern Structural Geology* Vol. 2: *Folds and Fractures.* Academic Press, London.

klippe (pl. klippen) *Thrust* **outlier**. (See **fenster**.) [RGP]

Koenigsberger ratio The ratio between the intensity of **natural remanent magnetization** and the magnetization induced by the local **geomagnetic field**. In cgs and SI units, a ratio greater than 1.0 indicates that **remanent magnetization** would be significant in the interpretation of **magnetic anomalies**. It is now commonly measured as Q_n, i.e. the ratio between the **remanent magnetization** and the magnetization induced at room temperature by an applied field of about $50\,\mu$**Tesla**. [DHT]

komatiite An volcanic **ultramafic rock** characterized by the morphological features of subaerial and submarine **basaltic lava** flows (including **pillow** and *hydroclastite* structure), and evidence of rapid cooling such as volcanic glass and quench crystallization textures. Komatiites have a distinctive feature termed *spinifex texture*. (Spinifex is a variety of spiky grass.) Spinifex is a texture characterized by large, randomly oriented or subparallel skeletal, plate or lattice **olivine** grains, or acicular **pyroxene** grains, and is formed by rapid crystallization of ultramafic liquids. Komatiites are defined as having >18% MgO and are composed essentially of **olivine** and **pyroxene** (± **chromite**) in a glassy or devitrified groundmass. They are associated with komatiitic **basalts** which have *spinifex textures* but lower MgO contents (10–18%). The association of komatiites and komatiitic **basalts** is termed the komatiite suite or series.

The komatiite suite is characteristic of **Archean** terranes but also occurs within certain **Phanerozoic** volcanic areas, such as the fore-arc regions of the western Pacific, where **boninite lavas** have affinities with komatiitic **basalts**. The komatiite suite is important because it demonstrates the occurrence of ultramafic high-MgO melts, indicating higher degrees of *partial melting* and hence a higher **heat flow** within the **Archean**, and locally within the **Phanerozoic**. The rocks are also important because of their association with **Archean nickel** deposits in Australia, Zimbabwe and Canada, and **gold** and **antimony** mineralization in other **Archean** areas. [RST]

Arndt, N.T. & Nisbet, E.G. (eds) (1982) *Komatiites*. Allen & Unwin, London.

kopje (*koppie*) A rocky hill or outcrop of large boulders. In southern Africa, the term is applied to **tors** and small **inselbergs**. Crystalline rocks — especially **granites** and **dolerites** — commonly form kopjes which are probably the product of exhumation of relatively unweathered rock and **corestones** from within **deep weathering** profiles. [AW]

Gibbons, C.L.M.H. (1981) Tors in Swaziland. *Geographical Journal* **147**: 72–8.

krennerite (AuTe$_2$) A rare **gold ore** mineral. [DJV]

kunzite A clear lilac-coloured **gem** variety of the mineral **spodumene**. [DJV]

Kupferschiefer Probably the world's best known **copper-rich shale**. It is of late **Permian** age and has been mined at Mansfeld (eastern Germany) for almost 1000 years. The Kupferschiefer underlies about 600 000 km^2 in Germany, Poland, Holland and England (Fig. K2). **Copper** concentrations greater than 0.3% occur in about 1% and **zinc** concentrations greater than 0.3% in about 5% of this area. Thus, although all the Kupferschiefer is anomalously high in **base metals**, **ore grades** are only encountered in a few areas. The most notable recent discoveries have been in southern Poland where deposits lying at a depth of 600–1500 m have been found during the last two decades. Here, the Kupferschiefer is 0.4–5.5 m thick. Average **copper** content is around 1.5% and the reserves at 1% **copper** amount to some 1500 Mt, making Poland the leading **copper** producer in Europe.

The Kupferschiefer consists of thin alternating layers of carbonate, clay and organic matter with **fish** remains

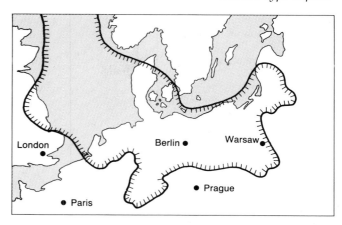

Fig. K2 Extent of the Zechstein sea in Central Europe. The **Kupferschiefer** occurs at the base of the Upper Permian (Zechstein). (After Evans, 1993.)

which give it a characteristic dark grey to black colour. The Kupferschiefer is the first marine transgressive unit overlying the non-marine Lower **Permian** Rotliegendes. The Kupferschiefer may represent a tidal marsh (**sabkha**) environment which developed as the sea transgressed desert **sands**. The **copper** and other metals are disseminated throughout the matrix as fine-grained sulphides and they appear to have been developed during **diagenesis**. [AME]

Evans, A.M. (1993) *Ore Geology and Industrial Minerals: An Introduction* (3rd edn). Blackwell Scientific Publications, Oxford.
Haranczyk, C. (1986) Zechstein copper-bearing shales in Poland. In: Friedrich, G.H., Genkin, A.J., Naldrett, A.J. *et al.* (eds) *Geology and Metallogeny of Copper Deposits*, pp. 461–76. Springer-Verlag, Berlin.
Schmidt, F.-P., Schumacher, C., Spieth, V. & Friedrich, G. (1986) Results of recent exploration for copper–silver deposits in the Kupferschiefer of West Germany. In: Friedrich, G.H., Genkin, A.J., Naldrett, A.J., *et al.* (eds) *Geology and Metallogeny of Copper Deposits*, pp. 572–82. Springer-Verlag, Berlin.

Kuroko-type deposit A type of volcanic-associated **massive sulphide deposit** which is usually an **ore** of **copper**, **zinc** and **lead** with or without **gold** and **silver** values. They are named from the **Miocene** Kuroko deposits of Japan. The volcanic association is with bimodal suites: **tholeiitic basalts**, **calc-alkaline lavas** and **pyroclastics**; but the deposits themselves are developed within or at the top of acid volcanic series and commonly show a spatial relationship to intrusive **rhyolite domes** and explosive volcanic activity (Fig. K3). The **plate tectonic** setting is that of **back-arc spreading** and their principal development was in the early **Proterozoic** and **Phanerozoic**. Although the similar, dominantly **copper–zinc** deposits of the **Archean** were long considered to be a variant of the Kuroko-type, they are now considered to be a separate Primitive-type.

Zoning is often apparent in the **orebodies**. **Galena** and **sphalerite** are more abundant in the upper half of the **orebodies** whereas **chalcopyrite** increases towards the **footwall** and grades downward into **chalcopyrite**

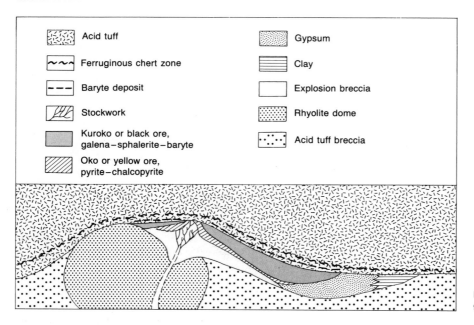

Acid tuff		Gypsum	
Ferruginous chert zone		Clay	
Baryte deposit		Explosion breccia	
Stockwork		Rhyolite dome	
Kuroko or black ore, galena–sphalerite–baryte		Acid tuff breccia	
Oko or yellow ore, pyrite–chalcopyrite			

Fig. K3 Schematic section through a **Kuroko-type deposit**. (After Evans, 1993.)

stockwork ore (see Figs M4 and M5, **massive sulphide deposit**, p. 387.)

Textures vary with the degree of recrystallization. The dominant original textures appear to be colloform banding of the sulphides with much development of framboidal **pyrite**, perhaps reflecting colloidal deposition. Commonly, however, recrystallization, often due to some degree of metamorphism, has destroyed the colloform banding and produced a granular **ore**. This may show banding in the **zinc**-rich section, whereas the **chalcopyrite ores** are rarely banded.

The Kuroko deposits of the Kosaka district, Japan, are a good example. All the Japanese Kuroko deposits are associated with **Miocene** volcanics and fossiliferous sediments developed along the eastern margin of a major **geosyncline**. Mineralization occurred during a limited period of the Middle **Miocene** over a strike length of 800 km in the Green Tuff volcanic region. Within this region more than a hundred Kuroko-type occurrences are known, but most are clustered into eight or nine districts. Their *host rocks* are acid **pyroclastic** flows which are also centred on **domes**, particularly those showing evidence of explosive activity. Kuroko ores have a consistent stratigraphical succession of **ore** and rock types, and an idealized deposit (Fig. K3) contains the following units:

1 hangingwall: upper volcanic and/or sedimentary formation;
2 ferruginous quartz zone: chiefly **hematite** and **quartz**;
3 baryte ore zone;
4 Kuroko or black ore zone: **sphalerite–galena–baryte**;
5 Oko or yellow ore zone: cupriferous **pyrite ores**; about this level, but often towards the periphery of the deposit, there may be the Sekkoko zone of **anhydrite–gypsum–pyrite**;
6 Keiko or siliceous ore zone: **copper**-bearing, siliceous, **disseminated** and/or **stockwork ore**;
7 footwall: silicified **rhyolite** and **pyroclastic rocks**. [AME]

Evans, A.M. (1993) *Ore Geology and Industrial Minerals: An Introduction* (3rd edn). Blackwell Scientific Publications, Oxford.
Ohmoto, H. & Skinner, B.J. (1983) The Kuroko and related volcanogenic massive sulphide deposits. *Economic Geology* 5.

kutnahorite (CaMn(CO$_3$)$_2$) A **manganese**-rich **dolomite** mineral. (See **carbonate minerals**.) [DJV]

kyanite (Al$_2$SiO$_5$) An important mineral in rocks formed by regional metamorphism of pelites and valued commercially for its refractory qualities. An island silicate (*nesosilicate*) mineral. [DJV]

L

labradorite A **plagioclase feldspar** mineral. (See **feldspar minerals**.) [DJV]

laccolith A concordant minor igneous intrusion with a known or presumed flat floor and a convex upper surface, so yielding an overall lens-like form. Laccoliths are roughly circular in plan, with diameters of up to c. 8 km with thicknesses from a few to hundreds of metres and are commonly of basic (**basaltic**) composition. [RST]

lag gravel A feature comprising a residual accumulation of coarser particles from which the finer material has been removed by wind. [NJM]

lahar A flow of volcanic debris, intimately mixed with water, that moves and is deposited as a mass, also the rock formed from such a flow. Lahars follow pre-existing valleys and are very mobile, travelling at velocities of up to about $40 \, m \, s^{-1}$ for distances of up to 120 km (Janda *et al.*, 1981). They form poorly sorted deposits, commonly containing blocks over 1 m in diameter, and have a fine-grained (silty) matrix. Lithic *clasts* in lahars are usually angular, in contrast to the rounding of *clasts* in **alluvial deposits** with which they may be interstratified. Lahars range from <0.5 m to >50 m in thickness, and usually possess no significant internal stratification, although locally there may be normal (and, in some cases, *reverse*) *grading* of large lithic blocks. Commonly, **preferred orientation** of ellipsoidal *clasts* is not observed, or only locally present at the base of the flow (Fisher & Schmincke, 1984). Lahars can occur in almost any volcanic setting, but are most common on steep-sided **andesitic** *stratovolcanoes*, especially those covered by snow for at least part of the year, e.g. the Cascade volcanoes, USA.

Lahars are thought to move by **laminar flow**, or possibly as a plug riding on a basal zone of **laminar flow**. They are able to carry large blocks for considerable distances. The flow mechanism is regarded as different from that of streams hyperconcentrated in clastic debris, and moving lahars are thought to have over about 60% by volume of solids (Fisher & Schmincke, 1984). Lahars are more mobile than dry **debris flows** and dry rock *avalanches*, implying that the water lubricates the flow of lahars. Many lahars originate as a direct result of **volcanic eruptions**; e.g. by deposition of hot **tephra** onto snow or ice, or by heavy falls of rain during or immediately after an eruption, or by flowage of **ash flows** into rivers or lakes. Lahars can also originate in ways not directly related to **volcanic eruption**; e.g. mobilization of unconsolidated **tephra** by heavy rain, or sudden release of water stored in **crater** lakes. (See **mudflow, pyroclastic rock**.) [PTL]

Fisher, R.V. & Schmincke, H.-U. (1984) *Pyroclastic Rocks*. Springer-Verlag, Berlin.
Janda, R.J., Scott, K.M., Nolan, K.H. & Martinson, H.A. (1981) Lahar movement, effects, and deposits. In: Lipman, P.W. & Mullineaux, D.R. (eds) *The 1980 Eruptions of Mount St Helen's, Washington*. Professional Paper of the United States Geological Survey **1250**: 461–78.

Lahn-Dill iron deposit Syngenetic, conformable, iron-bearing layer or lens which is dominantly siliceous at the bottom and calcareous at the top. These deposits occur in the **Devonian** and the Lower **Carboniferous** of Central Europe and the **Devonian** of the Taurus Mountains, Turkey. **Grades** are around 35% Fe. The **orebodies** which are found with rocks of the **spilite**–*keratophyre* group, are generally small — about 5 Mt; but the combined **orebodies** around one particular volcanic centre can amount to as much as 100 Mt. The principal **iron** minerals are **hematite, magnetite, siderite** and **limonite**. There are many similarities with **Algoma-type iron formation** and these deposits are also believed to be of volcanic-exhalative origin. [AME]

Quade, H. (1976) Genetic problems and environmental features of volcanosedimentary iron-ore deposits of the Lahn-Dill type. In: Wolf, K.H. (ed.) *Handbook of Strata-bound and Stratiform Ore Deposits* Vol. 7, pp. 255–94. Elsevier, Amsterdam.

Lamé constants (λ and μ) The most convenient **elastic moduli** to use in theoretical work when it is required to express **stress** in terms of **strain**. [GS]

laminar flow Flow characterized by the absence of turbulence and in which the mean flow velocity and instantaneous velocities at a point are always exactly the same. Consequently, laminar flows can be represented solely by the downstream vector of velocity. Laminar flows are characterized by the dominance of viscous forces in relation to inertial forces and are present at flow **Reynolds numbers** of less than 500. The **streamlines** in laminar flows are smooth with the appearance of infinitesimally thin layers of fluid sliding parallel to adjacent layers. Laminar flow boundary layers possess velocity gradients that are less steep than **turbulent flow** boundary layers (parabolic rather than logarithmic) due to the lack of mixing between the fluid close to the boundary and that above it. Examples of laminar flows in geological environments are the flow of thick muds and **debris flows**, the movement of **glaciers** and the flow of **lavas** and **magmas**, although it should be noted that these all also display **non-Newtonian flow** behaviour. [JLB]

Allen, J.R.L. (1985) *Principles of Physical Sedimentology*. Allen & Unwin, London.
Tritton, D.J. (1977) *Physical Fluid Dynamics*. Von Nostrand Reinhold, Wokingham.

laminite A sediment laminated on a millimetre scale (see also **rhythmite**). Laminites are common in glacial and non-glacial lakes, and can also occur as extensive blankets in certain marine environments lacking *bioturbation* and turbulence, such as off an ice margin. Lacustrine laminites may involve a variety of mineralogies. For example, the 'fish-beds' of the **Devonian** Orcadian Basin of northern Britain consist of triplets consisting of (i) *micritic* or microspariitic carbonate (**calcite** or **dolomite**), (ii) silt grade clastic material consisting mostly of **quartz**, and (iii) organic **carbon**-rich layers. The stratification of the dilute **Devonian** lake is thought to have promoted the preservation of the delicate lamination beneath an oxygen-deficient **hypolimnion**. In other lacustrine sequences which were deposited in hydrologically closed **basins**, the laminites contain **evaporite** mineral layers (such as **gypsum**) as components, and laminites accumulating on the floor of the **Dead Sea** consist of couplets of **aragonite** and **gypsum**, although the **gypsum** is mostly reduced by sulphate-reducing *bacteria*. Laminites are therefore extremely valuable indicators of *paleoclimates*. [PAA]

lamproite A **potassium**- and **magnesium**-rich mafic to ultramafic alkaline **lamprophyric** rock of volcanic or hypabyssal origin composed dominantly (5–30%) of **titanium**-rich **phlogopite**, *clinopyroxene* (commonly **diopside**), alkali **amphibole** (commonly K–Ti *richterite*), **olivine**, **leucite** and **sanidine**, together with **accessory minerals** (5–10%) which include chrome **spinel**, *priderite*, *wadeite*, **nepheline**, **ilmenite**, *shcherbakovite*, *jeppeite*, **apatite**, **perovskite**, *sphene* and *armalcolite*. **Xenoliths** and **xenocrysts**, including **olivine**, **pyroxene**, **garnet** and **spinel** of **upper mantle** origin, may be present and **diamond** as a rare **accessory mineral**. Lamproites have greater mineralogical and textural variations than **kimberlites** and, although they do contain some minerals characteristic of **kimberlites**, the chemical compositions of these two minerals are often distinctly different. Lamproites commonly show **alteration** to secondary phases that include **analcime**, **chlorite**, **quartz**, TiO_2 **polymorphs**, **zeolites**, **chlorite**, **serpentine**, **barite**, **albite** and **clay minerals**. In chemical composition, lamproites have over 5% K_2O and MgO, high K_2O/Na_2O and K_2O/Al_2O_3 ratios. They form **lava** flows and **pyroclastic rocks**, minor intrusions and **diatremes**, and occur within a variety of continental regions. Lamproitic **craters** are generally wider and shallower than those of **kimberlites**.

Lamproite **magmas** are produced by the *partial melting* of **mantle peridotite** that was enriched in **potassium**-bearing and other incompatible-element-enriched phases, such as **phlogopite** and **apatite**, most probably as a result of a **metasomatic** event prior to melting. By contrast with **alkali basaltic** and **kimberlitic** melts which are apparently produced from the *partial melting* of a CO_2-enriched **mantle peridotite** (i.e. a source with a relatively high CO_2/H_2O ratio), water is probably the major volatile species involved in the petrogenesis of lamproites.

Fertile lamproites, i.e. those carrying **diamonds**, appear to be the **silica**-saturated varieties *orendite* and *madupite* which carry **sanidine** rather than **leucite**. The richest known diamondiferous pipe is the lamproitic Argyle AK1 in Western Australia. Proven reserves are 61 Mt grading $6.8\,ct^{-1}$ with a **gem** content of 5%, cheap **gem** 40% and **industrial diamonds** 55%. No known **kimberlite** is as rich as this in **diamonds**. (See **diamond deposit**.) [AME/RST]

Bergman, S.C. (1987) Lamproites and other potassium-rich igneous rocks: a review of their occurrence, mineralogy and geochemistry. In: Fitton, J.G. & Upton, B.G.J. (eds) *Alkaline Igneous Rocks*. Geological Society Special Publication No. 30. Blackwell Scientific Publications, Oxford.
Kornprobst, J. (ed.) (1984) *Kimberlite I. Kimberlites and Related Rocks*. Elsevier, Amsterdam.

lamprophyllite ($Na_3Sr_2Ti_3(Si_2O_7)_2(O,OH,F)_2$) A rare **titanium**-bearing silicate, an island silicate (*nesosilicate*). [DJV]

lamprophyre Minor intrusion (**dyke** or **sill**) of mesocratic and melanocratic mineral composition which contains essential (*phenocryst* or groundmass) **biotite** (or **phlogopite**) and/or **amphibole**, with *clinopyroxene* and **olivine** (and *melilite* in certain types), with groundmass **feldspars** and/or **feldspathoids**. The mineralogy and classification of lamprophyres is summarized in Table L1. Lamprophyres commonly show **hydrothermal alteration** of the primary phases, and may contain **calcite**, **zeolites** and **clay minerals** as primary constituents. **Calc-alkaline** lamprophyres occur in association with *subduction*-related and post-orogenic granitoid intrusions and as post-orogenic intrusions within areas of *continent–continent collision*, and alkaline and **melilitic** lamprophyres tend to occur in association with anorogenic alkaline complexes and/or **carbonatites**. [RST]

Rock, N.M.S. (1984) Nature and origin of calc-alkaline lamprophyres: minettes, vogesites, kersantites and spessartites. *Transactions of the Royal Society of Edinburgh, Earth Sciences* **74**: 193–227.

land plants Plants that spend all or most of their lives in subaerial environments. Although the term includes mosses, liverworts, lichen, fungi, *algae* and *bacteria*, it is often used to refer exclusively to plants with a fluid-conducting vascular system.

Colonization of the land surface may have begun as

Table L1 Mineral compositions of **lamprophyres**

Felsic constituents				Predominant mafic minerals	
Feldspar	Foid	Biotite, diopside, augite, ± olivine	Hornblende, diopside, augite, ± olivine	Amphibole (barkevikite, kaersutite), olivine, biotite	Melilite, biotite ± titanaugite ± olivine ± calcite
or > pl	—	Minette	Vogesite*		
pl > or	—	Kersantite	Spessartite		
or > pl	fsp > foid			Sannaite[†]	
pl > or	fsp > foid			Camptonite	
—	Glass or foid			Monchiquite	Polzenite
					Alnoite

* Termed calc-alkaline lamprophyres; [†] termed alkaline lamprophyres.
foid, feldspathoid; fsp, feldspar; or, orthoclase; pl, plagioclase.

early as the *Precambrian*, but good evidence for terrestrial plants first appears in the early **Silurian** in the form of dispersed tube-like cells, fragments of cuticle and spores with triangular trilete marks and walls made of sporopollenin (Pratt *et al.*, 1978). Initially restricted to damp near-ground environments, early subaerial plants were probably thalloid (sheet-like). Selection in favour of wide *spore* dispersal by wind, and competitive mutual overgrowth, gave rise to an erect habit. Vertical growth above the relatively static humid boundary layer of air near the substrate was accompanied by specialized features to combat desiccation: a vascular system composed of lignified tube cells (tracheids) for fluid transport and mechanical support, a waxy external coating (cuticle) to reduce evaporation and external injury, pores (stomata) to allow gas exchange through the cuticle while minimizing water loss, underground organs (roots or rhizoids) for anchorage and water and mineral absorption. The evolution of land vegetation had profound effects on **erosion** and sedimentation, the **hydrologic** and **carbon** cycles, and global climate (Beerbower, 1985), effects that continue to the present.

Vascular plants probably evolved from the green *algae*. The earliest record of a demonstrable vascular plant (*Cooksonia*) is from the **Silurian**. Present evidence suggests that early vascular plants were simple, dichotomously branched, erect, naked, herbaceous axes supporting terminal globose sporangia. **Silurian** and early **Devonian** *vascular plants* are conventionally divided into three groups: Rhyniopsida (or Rhyniophytina) with dichotomous branching and terminal sporangia, Zosterophyllopsida (Zosterophyllophytina) with dichotomous or 'H' type branching and lateral sporangia, and the Trimerophytopsida (Trimerophytina) with monopodial branching (main erect axes with lateral side branches) and sporangia in clusters on specialized fertile branches. The Rhyniopsida may have contained the ancestral group of all vascular plants. The Zosterophyllopsida appear to have given rise to the

lycophytes while the Trimerphytopsida ultimately gave rise to all other vascular plant groups.

A number of early land plants produced small outgrowths (enations) from their axes which presumably functioned as rudimentary leaves. These enations are thought to have evolved into the microphylls found on members of the Lycopsida. True leaves (megaphylls) are thought to represent reduced, planated and webbed branching systems.

Early land plants produced only one kind of *spore* which, after release and landing in a suitable (wet) environment, germinated to form a thin sheet of cells (a gametophyte) bearing the sex organs (antheridia-producing motile sperm cells and archegonia containing sessile egg cells). Water was required for fertilization, which gave rise to the vascular spore-producing (sporophyte) plant. This type of reproductive cycle (homospory) is seen in most modern ferns. During the **Devonian** heterospory evolved in which two different kinds of *spore* were produced: small microspores (containing the male gametes) and larger ($\geq$200 μm) megaspores containing the egg cells and food reserves. Heterospory allowed for greater outcrossing and the evolution of dormancy mechanisms, limited the dependence on water for fertilization, and provided food reserves for the young sporophytes. The competitive advantages of these characteristics, enhanced by enclosure of the sporangium by a protective integument, were further exploited by the development of heterospory into the seed habit.

By the end of the **Devonian** full forest ecosystems, true leaves and seeds, and all classes except the Angiospermopsida, had evolved.

The **Carboniferous** saw the diversification of the *lycophytes* and *sphenophytes* (some of which attained tree stature), the *ferns*, seed ferns and *cordaites*. The first *conifers* appeared during the **Carboniferous**. Euramerican **coal swamp** forests were dominated by arborescent (tree form) members of the Lepidodendrales (e.g. *Lepidodendron*) while the arborescent sphenopsid *Calamites* occupied more disturbed riparian sites. Understorey elements included a variety of pteridosperms and *ferns*. Drier areas may have supported *cordaites* and a variety of *ferns* and pteridosperms. Well-defined phytogeographic provinces became established in the late **Devonian** and early **Carboniferous**.

Changes in global climate and sea-level rise at the end of the **Paleozoic** contributed to the demise of the arborescent *lycophytes*. In the southern hemisphere Glossopteridales and seed ferns belonging to the morphologically diverse genus *Dicroidium* became ecologically dominant in the **Permian** and **Triassic**, while in **Laurasia** *conifers*, *ginkgophytes*, *ferns*, *cycadophytes* and seed ferns were ubiquitous throughout much of the **Mesozoic**. At low latitudes *conifers* belonging to the Cheirolepidiaceae were dominant.

Angiosperms (flowering plants — the most widespread and diverse class today), first appear in the fossil record in the late early **Cretaceous** as pollen and leaves. Early morphologic trends show rapid diversification suggesting an early **Cretaceous** origin of the group (Doyle & Hickey, 1976) but their ancestry is unclear. From early appearances in disturbed and somewhat arid environments at low latitudes, the *angiosperms* spread polewards. By the **Paleocene** the first *angiosperm*-dominated forest ecosystems had evolved (Wolfe & Upchurch, 1986).

It is generally agreed that the archetypal primitive *angiosperm* bore a bisexual entymophilous (insect pollinated) blossom. Early flower and *angiosperm* pollen morphology suggests primitive *angiosperms* were of this type with radially symmetrical open blossoms. Subsequent evolution of flower form was strongly linked with that of the biotic pollinators (mostly *insects*). Co-evolution of flowers and biotic pollinators gave rise to specific pollen targeting which, in conjunction with specialized seed dispersal by animals (especially **birds**), resulted in communities with high species diversity and led to the break-up of low diversity communities typical of predominantly anemophilous (wind pollinated) *gymnosperms*. This in turn provided the framework for further habitat and morphological diversification in both animal and plant groups. In spite of the advantages of entymophily many *angiosperms*, including some early forms, adopted anemophily.

Apart from the formation of **coal**, land plants have great geological significance. Dispersed organs such as leaves (and isolated cuticles), *spores* and *pollen*, and *seeds* provide powerful **biostratigraphic** tools particularly if used in conjunction with analyses of depositional regimes. In vegetation that is in equilibrium with its environment plant morphology, anatomy, leaf and cuticle architecture, and physiognomy reflect strongly the prevailing climate (Wolfe, 1979). This is especially so for *angiosperms*. Land plants contain important *paleoclimatic* signals, particularly as regards atmospheric temperatures, precipitation/evaporation, and seasonality which in turn may be used to predict the likely sites of formation of climate-related mineral and *hydrocarbon* resources. **Diagenesis** of land-plant-derived resins, waxes and fats may also play a role in oil and gas generation. (See Table L2.) [RAS]

Beerbower, R. (1985) Early development of continental ecosystems. In: Tiffney, B.H. (ed.) *Geological Factors and the Evolution of Plants*, pp. 47–91. Yale University Press, New Haven, CT.

Doyle, J.A. & Hickey, L.J. (1976) Pollen and leaves from the mid-Cretaceous Potomac Group and their bearing on early angiosperm evolution. In: Beck, C.B. (ed.) *Origin and Early Evolution of Angiosperms*, pp. 139–206. Columbia University Press, New York.

Pratt, L.M., Phillips, T.L. & Dennison, J.M. (1978) Evidence of non-vascular land plants from the Early Silurian (Llandoverian) of Virginia, USA. *Review of Palaeobotany and Palynology* 25: 121–49.

Stewart, W.N. (1983) *Paleobotany and the Evolution of Plants*. Cambridge University Press, Cambridge.

Thomas, B.A. & Spicer, R.A. (1987) *The Evolution and Palaeobiology of Land Plants*. Croom Helm, London.

Tiffney, B.H. (ed.) (1986) *Geological Factors in the Evolution of Plants*. Yale University Press, New Haven, CT.

Wolfe, J.A. & Upchurch, G.R. (1986) Vegetation, climatic and floral changes at the Cretaceous–Tertiary boundary. *Nature* 324:148–52.

Wolfe, J.A. (1979) *Temperature Parameters of Humid to Mesic Forests of Eastern Asia and their Relation to Forests of Other Areas of the Northern Hemisphere and Australasia*. Professional Paper of the United States Geological Survey 1106.

landslide Rapid movement of *soil* downslope *en masse* so that there is little or no **deformation** of the *soil structure*. Landslides may have curved or planar **failure** planes. Rotational **failures** along curved **shear planes**, sometimes called **slumps**, occur in soft cohesive sediments such as clay or **shale**. Single or multiple rotational **slides** may occur. Translational **slides** with a straight **failure** plane occur in non-cohesive materials where only frictional **strength** controls the occurrence of **failure**. The distinction between landslides and **flows** is often difficult to make and depends on the amount of **deformation** of the *soil* during the **failure**. [TPB]

Anderson, M.G. & Richards, K.S. (eds) (1987) *Slope Instability*. John Wiley & Sons, Chichester.

langbeinite ($K_2Mg_2(SO_4)_3$) An **evaporite** mineral. [DJV]

lanthanum (La) At. no. 57, at. wt. 138.9055, d. 6.14. A metal used in steels and in **magnesium** and **aluminium** alloys. La_2O_3 is used in glass, as a glass polish, as a high temperature refractory and as a host matrix for fluorescent phosphors. Lanthanum is recovered from **bastnäsite** from **carbonatite** deposits and from **placer deposits** of **monazite**. The chief producers are the United States, Australia and China. [AME]

lapis lazuli A highly prized ornamental stone made up of a mixture of **lazunite**, **calcite**, **pyroxenes** and other silicates.

Lapis lazuli was valued in antiquity in the Old World for its attractive blue colour which made it suitable for making jewellery and small decorative objects such as seals and small vessels. It was also used for inlay on statues

Table L2 A classification of **land plants** (To ordinal level excluding fungi, algae, and bacteria)

Kingdom Plantae

SPORE PRODUCING PLANTS
Division Bryophyta (mosses, liverworts, hornworts)

Division Tracheophyta (vascular plants)
Class Rhyniopsida (primitive vascular plants)
    Order Rhyniales†
Class Zosterophyllopsida (ancestors of microphyllous plants)
    Order Zosterophyllales†
      Asteroxylales†
Class Lycopsida (club mosses and related plants)
    Order Drepanophycales†
      Protolepidodendrales†
      Lycopodiales
      Selaginellales
      Lepidodendrales†
      Isoetales (quillworts)
Class Trimerophytopsida† (ancestors of megaphyllous plants)
Class Sphenopsida (horsetails and their relatives)
    Order Pseudoborniales†
      Sphenophyllales†
      Equisetales
Class Filicopsida (ferns and their relatives)
    Order Cladoxylales†
      Stauropteridales†
      Zygopteridales†
      Coenopteridales†
      Marattiales
      Filicales
      Salviniales
Class Progymnospermopsida (possible ancestors of some gymnosperms —
some may have produced seeds)
    Order Aneurophytales†
      Archaeopteridales†
      Protopityales†

SEED PRODUCING PLANTS
Class Gymnospermopsida ('naked seeded' plants — an artificial group)
    Order Cycadeoidales† ('Cycadophytes')
      Cycadales (cycads)
      Pteridospermales† (seed ferns)
      Caytoniales†
      Glossopteridales†
      Pentoxylales†
      Czekanowskiales†
      Ginkgoales (*Ginkgo* and extinct related plants)
      Cordaitales†
      Voltziales†
      Coniferales (conifers) ('Coniferophytes')
      Taxales (Yews)
Class Gnetopsida (an artificial group consisting of *Gnetum*, *Welwitschia*, and
Ephedra)
Class Angiospermopsida (flowering plants)
    Subclass Monocotyledonae
      Order Alismatales (Water plantains)
        Hydrocharitales (Canadian pond weed)
        Najadales (*Potomogeton*)
        Triuridales
        Commelinales
        Eriocaulales
        Restionales
        Poales (Grasses)
        Juncales (Rushes)
        Cyperales (Reeds, Sedges)
        Typhales (Bulrushes)
        Bromeliales (Bromeliads)

        Zingiberales (Bananas, Ginger)
        Arecales (Palms)
        Cyclanthales
        Pandanales (Screw pines)
        Arales (Duckweeds, Aroids)
        Liliales (Daffodils, Yams)
        Orchidales (Orchids)
    Subclass Dicotyledonae
      Order Magnoliales (Magnolias)
        Illicales
        Laurales (Laurel, Avacado)
        Piperales (Pepper)
        Aristolochiales (Birthworts)
        Nymphaeales (Water lilies)
        Ranunculales (Buttercups, Curare)
        Papaverales (Poppies)
        Sarraceniales (Pitcher plants)
        Trochodendrales
        Hamamelidales (Planes, Witch hazels)
        Eucommiales
        Leitneriales
        Myricales
        Fagales (Birches, Beeches, Oaks)
        Casuarinales (She oaks)
        Caryophyllales (Cactus, Beets, Spinach)
        Batales
        Polygonales (Rhubarb)
        Plumbaginales (Thrift)
        Dilleniales (Peonies)
        Theales (Tea)
        Malvales (Lime, Cotton, Balsa)
        Urticales (Elms, Nettles)
        Lecythidales (Brazil nuts)
        Violales (Violets, Pumpkins)
        Salicales (Willows)
        Capparales (Capers)
        Ericales (Heathers, Rhododendrons)
        Diapensiales
        Ebenales (Chicle, Persimmons)
        Primulales (Primroses)
        Rosales (Roses, Sundews)
        Fabales (Peas)
        Podostemales
        Haloragales
        Myrtales (Mangroves, Cloves)
        Cornales (Dogwoods)
        Proteales (Proteas, Banksias)
        Santalales (Mistletoes)
        Rafflesiales
        Celastrales (Holly)
        Euphorbiales (Box, Pointsettia)
        Rhamnales (Buckthorn)
        Sapindales (Maples, Citrus)
        Juglandales (Walnut, Hickory)
        Geraniales (Flax, Coca)
        Polygalales
        Umbellales (Ivy, Ginseng, Carrots)
        Gentianales (Gentians, Olives)
        Polemoniales (Potato, Morning Glory)
        Lamiales (Teak, Mint)
        Plantaginales (Plantains)
        Scrophulariales (Foxgloves, Sesame)
        Campanulales (Bellflowers)
        Rubiales (Coffee, Quinine)
        Dipsacales (Elders, Teasel)
        Asterales (Composites)

† Known only from the fossil record.

and tablets. Much of the lapis lazuli used in the ancient Near East may have come from the important mine of Badakshan in Afghanistan, from where it was traded as far as Egypt. The most famous ancient object of this material is probably the **gold** and lapis 'Ram in the Thicket' statue from Ur in southern Iraq (mid-third millenium BC), while the earliest use of lapis is thought to be in the 4th millenium BC. Later, lapis lazuli was used by the Romans, and was powdered to give a dark blue pigment. [OWT/DJV]

Laplace's equation Relationship obeyed by all **potential** fields which, in Cartesian coordinates, states that the sum of the rates of change of the field's gradients in three orthogonal directions is zero. Thus, for a coordinate system with horizontal axes x, y and a vertical axis z, and a potential field A

$$\frac{\mathrm{d}^2 A}{\mathrm{d}x^2} + \frac{\mathrm{d}^2 A}{\mathrm{d}y^2} + \frac{\mathrm{d}^2 A}{\mathrm{d}z^2} = 0$$

Solution of Laplace's equation allows the computation of what the potential field would be at any level above (*upward continuation*) or below (*downward continuation*) the plane of observation. [PK]

Gunn, P.J. (1975) Linear transformations of gravity and magnetic fields. *Geophysical Prospecting* **23**: 300–12.
Ramsey, A.S. (1964) *An Introduction to the Theory of Newtonian Attraction*. Cambridge University Press, Cambridge.

larsenite ($PbZnSiO_4$) A rare mineral of the olivine group. (See **olivine minerals**.) [DJV]

lateral secretion The derivation of mineral-forming materials from the **wall rocks** in the immediate vicinity of a **vein** or other mineral deposit. An example of deposits formed in this way has been described by Boyle from the Yellowknife Goldfield of the Northwest Territories of Canada where the deposits occur in **quartz**-carbonate lenses in extensive **chloritic shear zones** cutting **amphibolites** (*metabasites*). The deposits represent concentrations of **silica**, carbon dioxide, **sulphur**, water, **gold**, **silver** and other metallic elements. The principal minerals are **quartz**, **carbonates**, **sericite**, **pyrite**, **arsenopyrite**, **gold** and *aurostibnite*. *Alteration haloes* of carbonate–**sericite**–**schist** and **chlorite**–**carbonate**–**schist** occur in the *host rocks* adjoining the deposits.

The dominant mineral of the **veins** is **quartz** and the profile of **silica** alongside the lenses is shown in Fig. L1. Traced from a value of about 52% in the **wall rocks** the **silica** content decreases dramatically in the **wall rock** *alteration haloes* before becoming nearly 100% in the **quartz** lens; and this of course obtains on both sides of the lens. By making use of mine plans Boyle was able to show that more **silica** has been removed from the **wall rocks** than is present in the **quartz** lenses. It is highly probable that all the other materials in the **quartz** lenses were also derived from the **wall rocks**. [AME]

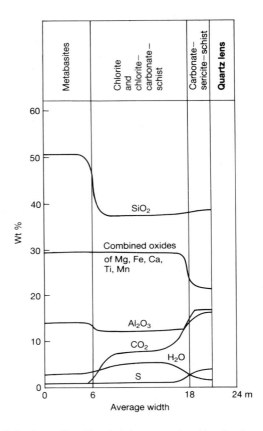

Fig. L1 Lateral secretion. Chemical changes produced by alteration of metabasites. Yellowknife gold deposits, Canada. (Modified from Boyle, 1959.)

Boyle, R.W. (1959) The geochemistry, origin and role of carbon dioxide, sulfur and boron in the Yellowknife gold deposits, Northwest Territories, Canada. *Economic Geology* **54**: 1506–24.
Brimhall, G.H. (1979) Lithologic determination of mass transfer mechanisms of multiple-stage porphyry copper mineralization at Butte, Montana: vein formation by hypogene leaching and enrichment of potassium-silicate protore. *Economic Geology* **74**: 556–89.

laterite A **residual deposit** formed by the **weathering** of rocks under humid tropical conditions and consisting mainly of **iron** and **aluminium** hydroxides. With both metals present in considerable amounts these rocks are of no value as a source of either.

Occasionally, however, laterites derived from basic or ultrabasic rock may be sufficiently rich in **iron** to be workable. These deposits, which may be as much as 20 m thick but are usually less than 6 m, consist of nodular red, yellow or brown **hematite** and **goethite** which may carry up to 20% alumina. Deposits of this type form mantles on **plateaux** and are worked in Guyana, Indonesia, Cuba, the Philippines, etc.

A good example is the Conakry deposit in Guinea which is developed on **dunite**, the change from laterite to **dunite** being sharp. Most of the laterite consists of a hard crust, usually about 6 m thick. The **ore** as shipped contains 52% **iron**, 12% alumina, 1.8% silica, 0.25% **phosphorus**,

0.14% **sulphur**, 1.8% **chromium**, 1.15% **nickel**, 0.5% TiO_2 and 11% combined water. This last figure illustrates one of the drawbacks of these **ores** — their high water content, which may range up to 30%. This has to be transported and then removed during smelting. [AME]

latite A volcanic rock with a similar **silica** content to **andesite** but with a higher K_2O and K_2O/Na_2O ratio in comparison with **calc-alkaline andesite** such that *phenocrysts* of **plagioclase** and **biotite** ± **sanidine** are more abundant than **pyroxene** and **hornblende**. [RST]

latitude correction The correction applied to **gravity** measurements during **reduction** to take account of the variation of **gravity** with latitude. The **Gravity Formula**, based on **Clairault's formula**, is used for this purpose. The variation in the **geomagnetic field** with latitude is corrected using the *geomagnetic correction*. [PK]

laumontite $((CaAl_2Si_4O_{12}) . 4H_2O)$ A **zeolite mineral**. [DJV]

Laurasia The northern *supercontinent*, prior to the formation of the Atlantic Ocean, which comprised North America, Greenland, Europe and Asia. It is now known that most of Asia should not be included in this *supercontinent* as Siberia only joined in the **Triassic** (*c.* 200 Ma ago), and many of the smaller blocks in southern and eastern Asia have only become accreted to Siberia since then. A more useful term is therefore **Laurentia**. [DHT]

Laurentia A *supercontinent* comprising the Canadian Shield, Greenland and parts of north-western Europe and some other parts of North America. This configuration existed in slightly modified form before the formation of the *Caledonian* Mountains, but the term usually applies to the western 'Atlantic' continent during and subsequent to the formation of the Variscan Mountains. [DHT]

lava flow A dense mass of molten rock moving as a stream on the Earth's surface. Lava flows usually issue directly from a **volcanic vent**, although they can form by very rapid reconstitution of **magma** explosively fragmented by **fire fountaining**. The forms of the lava flows, especially their surface texture, are influenced by many factors, including their **viscosity**, eruption rate and the slope of the ground. Very fluid, commonly rather fast-moving flows form *pahoehoe* lavas, which have thick, smooth, wavy surfaces, and tend to be rather thin (*c.* 10 cm–2 m). More viscous, slower-moving flows tend to be thicker (*c.* 10–30 m thick), and have either very rough, clinkery surfaces, in the case of *aa* lavas, or surfaces consisting of a layer of blocks of lava with rather smooth surfaces, in the case of blocky lavas. Single lava flows may have parts with different surface texture types. *Pahoehoe*

flows are largely restricted to basaltic compositions, but blocky and *aa* surfaces occur in lavas having a range of chemical compositions.

Individual **volcanic eruptions** may produce several flows, and these may be superimposed, forming a compound lava flow. Compound lava flows usually consist of *pahoehoe* flows. In contrast, simple lava flows consist of a single massive flow. (See **dome, pillow lava, vesicle**.) [PTL]

law of accordant junctions The law which states that tributary rivers enter a main river at the same level as that river, without any sudden drop as in a glaciated *hanging valley*. The idea was first proposed by Playfair (1802) and is known as '*Playfair's law*'. The law has been widely accepted by fluvial geomorphologists, but field evidence does not always support the law. The channel size of the tributary relative to the main river may be an important factor affecting whether the confluence is accordant or not (Kennedy, 1984). [NJM]

Kennedy, B.A. (1984) On Playfair's law of accordant junctions. *Earth Surface Processes and Landforms* **9**: 153–73.
Playfair, J. (1802) *Illustrations of the Huttonian Theory of the Earth*. Reprinted 1964, Dover, New York.

lawsonite $(CaAl_2(Si_2O_7)(OH)_2 . H_2O)$ A double island silicate (*sorosilicate*) mineral commonly found in **gneisses** and **schists**. [DJV]

lazurite $((Na,Ca)_8(AlSiO_4)_6(SO_4,S,Cl)_2)$ A rare framework silicate (*tectosilicate*) mineral. [DJV]

lead (Pb) At. no. 82, at. wt. variable depending upon source, approx. 207.2. Occurs naturally as **galena** and a number of secondary minerals. **Galena** is won from many different types of metalliferous deposits but especially from **carbonate-hosted base metal deposits** and **massive sulphide deposits**. Lead is used extensively in alloys, storage batteries, petrol, radiation and sound shields, paints and high quality glass.

Lead was commonly added to **tin bronzes** during the Late Bronze Age in Europe, even though this would reduce the hardness of the resultant alloy; perhaps added to economize on **tin** use. Lead was also occasionally added to early glasses. Its use for water pipes, tanks and in tableware by the Romans is well known and this, along with the preferred use of lead vessels for the preparation of wine, must have resulted in some lead poisoning in the population. [AME/OWT]

lechatelierite (SiO_2) A name given to naturally occurring fused **silica** or **silica** glass. [DJV]

lee side/slope The downstream side/slope of a body or obstruction that is sheltered from the dominant flow direction. Lee sides are present at many different physical scales from sheltered slopes downwind of **mountain ranges**, to large lee side **slip-faces** of aeolian and aqueous bedforms to lee sides of individual sediment grains. Many lee side localities are associated with the phenomenon of **flow separation** and its consequent sedimentological effects. Lee sides in most bedforms are associated with flow velocities that are much reduced compared to those of the surrounding freestream. Consequently these lee slopes of bedforms are preferred sites of sediment deposition. [JLB]

lepidocrocite (γFeO . OH) A secondary **iron** mineral, a **polymorph** of **goethite**. [DJV]

lepidolite $(K(Li,Al)_{2-3}(AlSi_3O_{10})(O,OH,F)_2)$ A **lithium-**bearing **mica mineral**, commonly pink or lilac in colour. [DJV]

leucite $(KAlSi_2O_6)$ A **feldspathoid mineral**. [DJV]

leucitite Fine-grained, frequently porphyritic **lavas** and minor intrusions, essentially composed of **leucite** (>30–50%, dependent upon definition) and *clinopyroxene* (*titan-augite*, **diopside** and/or *aegirine*–**augite**) together with accessory **nepheline**, Fe–Ti oxides and **apatite**. They occur within continental **rift** areas (e.g. the African Western Rift), within volcanic arcs (e.g. Indonesia) and within other complex post-tectonic continental settings, including the famous Leucite Hills of Wyoming (USA).

Leucitite was one of the three main types of **igneous rock** used for **millstones** in the Roman Mediterranean (the others are **rhyolite** and **basalt**). Leucitite is particularly suitable for milling because of the abrasive surface produced by the leucite crystals. Leucite-bearing lavas were also an important **building stone** in parts of Roman Italy, e.g. at Pompeii. [RST/OWT]

Gupta, A.K. & Yaki, K. (1979) *Petrology and Genesis of Leucite-bearing Rocks*. Springer-Verlag, Berlin.

leucoxene A fine-grained yellow to brown **alteration** product of **titanium**-rich minerals (**ilmenite**, **titanite**, **perovskite**) comprised mainly of **rutile**. [DJV]

levée A raised bank bordering a terrestrial or subaqueous channel that results from rapid deposition of sediment from water escaping from the confines of the channel. Levée deposits are generally finer than the associated channel deposit and thin and fine away from the channel. They are composed of interbedded coarse and fine sediments deposited in varying **hydrological** conditions. In an alluvial setting, levées allow the level of the river to rise above its **floodplain** and when breached will then result in sudden diversion of the current causing major flooding or channel **avulsion**. Such levée breaches are termed *crevasses*. [JA]

lherzolite A phaneritic rock (**peridotite**) containing 40–90% **olivine**, with >5% of each of *orthopyroxene* and *clinopyroxene* together with accessory **plagioclase**, **spinel** and/or **garnet**. Lherzolite is capable of yielding **basaltic magma** during *partial melting* and so is commonly taken as a model for the composition of the **upper mantle**. Lherzolite forms the ultramafic component of some **ophiolite** complexes (cf. **harzburgite**) and also occurs as **xenoliths** within **alkali basalts**, **lamproites**. and **kimberlites**. **Garnet** lherzolites may contain accessory **phlogopites**, sulphides, carbonates and chrome **spinel**. Partial melting of **mantle** lherzolite first leaves residual **harzburgite** (as occurs in many **ophiolite** complexes) and then **dunite**. [RST]

Nixon, P.H. (ed.) (1987) *Mantle Xenoliths*. John Wiley & Sons, Chichester.

lichenometry A technique first proposed by Beschel in 1950 to find either a relative or an absolute date for exposure of a surface using lichen measurements. The most long-lived species suitable for the technique form concentric growths on rocks. The 'relative dating' of substrata exposed or deposited at different times may be achieved by mapping contact lines between lichen populations with distinctly different maximum diameters: it enables the ranking of deposits in order of deposition. 'Absolute dating' depends on the accurate construction of growth curves from which growth rates can be estimated and size–age correlations established.

Two main approaches have been suggested for growth curve construction; each requires that maximum diameters of correctly identified species on substrata of known date are plotted on graphs with size–time axes. The dated surfaces of graveyard memorials are frequently used to provide a range of dates within a reasonably uniform environment but other surfaces which fill these criteria may also be used.

One approach is to fit an envelope curve round the measurements of largest single growths providing maxima through time. Confidence in this curve is limited since it is likely to be based on the growth of very few specimens. In the other approach a mean growth rate is provided, in addition to an error term, by tracing the curve through the mean of the five largest growths on each surface carrying maximum diameters through time. The problem with this approach is that as a population ages, individuals remaining from the original colonization period are likely to become scarce and it may be increasingly difficult to find the required five specimens on the limited surface areas of older tombstones. Whichever approach is used it must be rigorously maintained when dating is attempted and sampling areas at both the growth curve construction site and the dating site should be as nearly equivalent as possible.

After a period of colonization and subsequent rapid growth, the duration of which depends on local conditions, most crustose lichens adopt a slower constant growth rate which varies over the long term about a mean according to seasonal variations and other factors such as competition. For example, average long-term growth of *Rhizocarpon geographicum*, the species most frequently used by geomorphologists, has been found to be approximately 3 mm in 10 years in a wide range of environments.

Extrapolation of growth curves can only be fully justified if there is independent dating evidence to support the extrapolation. Thus, the technique is best used where other dating evidence needs additional confirmation. However, if no other evidence is available, dates may be more firmly established by using the different growth curves of several species as independent evidence for a likely dating period. The best estimate for a date is then based on the mean of the different species' dates and a standard error is provided.

The dating range depends on both species specific and environmental factors. The range is probably unlikely to exceed 500 years in most temperate environments. In harsh climates rock **weathering** rates could determine life span so that the possible maximum range in these conditions may depend on this. Some earlier estimations for dating in the Arctic have credited the technique with a range between 4500 and 9000 years. Absolute dating is based on oldest surviving lichens so only minimum estimates for dates can be established and many other factors affecting lichen establishment and survival mean that surfaces may have been exposed for longer. New lichen growths can start at any time on surfaces freshly exposed as a result of the activities of man, animal interference, vegetation changes and rock **weathering**.

The major use of lichenometry so far has been to date *glacial* deposits; other uses include the dating of *mass-wasting* events, **seismological** movements, and changes in lake, river or **glacier** levels marked by lichen trimlines. The technique has also been used for assessing the detailed treatment of historic or prehistoric structures.

Problems in the past have mainly concerned the effects on growth of environmental variation (including pollution and delay in colonization), the measurement of composite growths, differences in species' life spans and an inconsistent methodology. Despite these problems if lichenometry is applied with sufficient care it can provide a useful additional dating method covering the most recent period of active change in areas where other dating evidence is lacking or needs support. [VW]

Innes, J. (1985) Lichenometry. *Progress in Physical Geography* **9**: 187–254.

Locke, W.W. III, Andrews, J.T. & Webber, P.J. (1979) A manual for lichenometry. *British Geomorphological Research Group Technical Bulletin* **26**: 1–47.

Webber, P.J. & Andrews, J.T. (eds) (1973) Lichenometry. *Arctic and Alpine Research* **5**: 293–432.

Winchester, V. (1985) A proposal for a new approach to lichenometry. *British Geomorphological Research Group Technical Bulletin* **33**: 3–20.

lift force The force acting upon a sediment grain that acts in a direction normal to the flow direction and is generated by pressure differences over the surface of the grain. Flow over a sediment grain is forced to increase in velocity over the top of the grain where the **streamlines** converge and the fluid pressure consequently drops. However, as the base of the grain near the bed flow is decelerated, **streamlines** diverge and the fluid pressure is consequently relatively high compared to the freestream. The difference between higher pressures acting upon the base of the grain and lower pressures acting upon its top surface generates a lift force acting in a direction normal to the bed. The lift force, F_L, may be given by:

$$F_L = C_L(\rho . U^2/2)d$$

where C_L is a dimensionless lift proportionality coefficient, ρ is the fluid density, U is the mean flow velocity and d is a measure of grain size, such as cross-sectional area exposed to the flow.

The lift force together with the **drag force** resolve a resultant fluid force acting upon a sediment grain that is instrumental in the initiation of sediment movement (See **Shields diagram**, Fig. S29, p. 566.) [JLB]

Middleton, G.V. & Southard, J.B. (1984) *Mechanics of Sediment Transport* (2nd edn). Short Course No. 3 Lecture Notes. Society of Economic Paleontologists and Mineralogists, Tulsa, OK.

lignite A soft, low **rank coal** normally referred to in Europe as **brown coal**. It is a dull, earthy material, brown to black in colour. It may take a massive **sapropelic** form, but is more commonly made of humic material with recognizable wood and plant remains in a finer grained, organic groundmass. By contrast *sub-bituminous coal* is generally black and harder, and some may have a vitreous **lustre**. Lignites and *sub-bituminous coal* tend to crack and fall apart on drying out after mining, the process usually known as *slacking*.

Lignite was used in Bronze Age Britain for the manufacture of small beads. [AME/OWT]

limburgite A volcanic rock (or minor intrusion) composed of **olivine** and **pyroxene** crystals within **basaltic** glass (see **tachylite**). Such rocks are commonly taken to be alkali-rich and/or silica-undersaturated although (alkali-poor) **komatiites** from southern Africa were originally termed limburgites. [RST]

lime (CaO) Substance produced by the calcining of high purity **limestone**. A part of the lime is converted to the more stable hydrated lime $Ca(OH)_2$. Lime has a vast number of industrial applications, e.g. in the production of soda ash using the Solvay process, for the production of exceptionally pure calcium carbonate, etc. [AME]

limestone Rock consisting of more than 50% calcium carbonate $(CaCO_3)$. Limestones constitute significant vol-

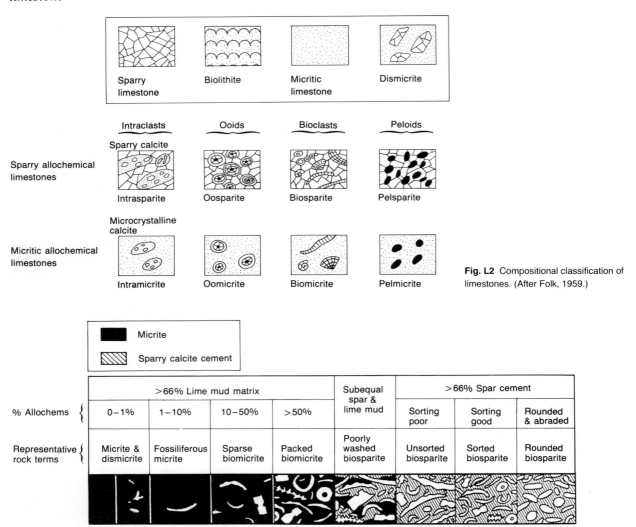

Fig. L2 Compositional classification of limestones. (After Folk, 1959.)

Fig. L3 Textural classification of **limestones**. (After Folk, 1962.)

umes of the sedimentary record and are the repositories for much of the fossil record. In addition, limestones and **dolomites** contain about 60% of the world's known *hydrocarbon reserves*, host major metalliferous deposits and are used widely in the construction and chemical industries.

Most limestones, at least since early **Cambrian** times, are partly or wholly biogenic in origin. Carbonate production in present-day seas is mainly by organisms such as **molluscs**, **corals**, **echinoderms**, **Foraminifera** and **calcareous algae**, while in the past other organisms were the main contributors. Biological production is greatest in very shallow waters, less than 20 m in depth, and most limestones in the geological record represent such shallow water deposits. However, extensive carbonate production also occurs in the open ocean by planktic organisms such as **Foraminifera**, *coccolithoporoids* and *pteropods*, but at lower production rates than in shallow water.

Carbonate sediments forming today consist of three forms of calcium carbonate — **aragonite**, high **magnesium** (or magnesian) **calcite** (HMC) with >5 mol.% Mg, and low **magnesium** or (magnesian) **calcite** (LMC) with <5 mol.% Mg. The former two types are metastable under most **diagenetic** conditions and are stabilized to the low **magnesium** form. This instability is one of the main driving forces of **diagenesis** and results in major textural and geochemical changes. Looking at ancient limestones can be said to be like looking 'through a glass darkly' because of the **diagenetic** veil. Such instability-controlled processes are especially important during the early stages of **diagenesis** while during deeper burial pressure- and temperature-dependent processes become more significant.

Limestones, in comparison to other rock types, are mineralogically simple but are nevertheless highly varied in their composition. Three main components can be recognized: grains (or **allochems**), matrix and cement.

Depositional texture recognizable					Depositional texture not recognizable
Original components not bound together during deposition				Original components were bound together	
Contains mud (i.e. clay and fine silt-size carbonate)		Grain-supported	Lacks mud and is grain-supported		Crystalline carbonate
Mud-supported					
<10% grains	>10% grains				
Mudstone	Wackestone	Packstone	Grainstone	Boundstone	Crystalline

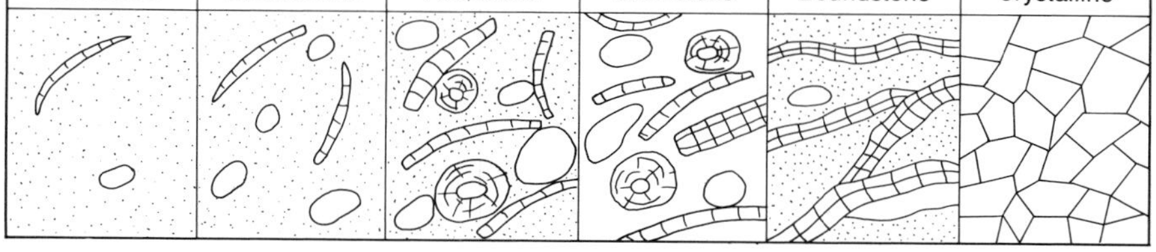

Fig. L4 Dunham's (1962) textural classification of **limestones**.

Two types of grains can be recognized: skeletal and non-skeletal. The former consist of skeletal material (*bioclasts*), either whole or fragmented, of animals or calcareous plants. The latter represent grain-types such as **ooids**, *peloids* (sand-sized grains of carbonate mud of polygenetic origin but commonly either the cemented fecal pellets of mud-ingesting organisms, or micritized grains), *calcareous aggregates* (carbonate grains which have been agglutinated, often by microbial processes, forming irregular masses referred to as *grapestones* if the incorporated grains are **ooids**), and *clasts* (reworked carbonate lithologies either locally-derived (*intraclasts*) or from a lithology not associated with the main depositional system (*lithoclast*). If the resulting rock type has predominantly **sand**-sized grains it is known as a *calcarenite*. Where the grains exceed 2 mm (or 1 mm in some classifications) the rock is known as a *calcirudite*.

The matrix in limestones consists of very fine-grained carbonate referred to as *micrite*. The term is an abbreviation for microcrystalline **calcite** and should strictly be used where the crystal or grain size is less than 4 μm in size. *Micrite* appears dark in thin section. This material is regarded as being the equivalent of clay-grade matrix in **sandstones** and its presence in rocks provides a clue to low energy levels during deposition. Rocks with such a matrix are said to be texturally immature. The origins of such muds are still uncertain but some represent the fine-grained remains of *algally*-produced carbonate while others possibly represent mud **precipitated** directly from the water column. Some *micrite* in *reefs* is actually a cement. *Micrite* is susceptible to recrystallization to form *microspar* (4–20 μm in size) or *pseudospar* (>20 μm in size). The term *calcilutite* is used for mud-grade limestones.

Cements in limestones are represented by sparry **calcite** or *sparite* which is generally clear in thin section. The cement is **precipitated** and nucleates from a substrate, such as a grain, and grows into a pore space. A large variety of types of cement fabrics have been recognized and many are characteristic of certain **diagenetic** environments.

There are several systems for classifying limestones. The simplest is based on grain size with the terms *calcirudite*, *calcarenite* and *calcilutite* being used. Two widely used classifications, by Folk and Dunham, are based on the concept of *textural maturity* and are comparable to **sandstone** classifications. The amount of matrix present, and the degree of sorting of the grains, provides a clue to the energy level during deposition. Folk (1959) introduced a simple division of limestones depending on whether they have a matrix (*micritic limestones*) or a sparry cement (*sparry limestones*) (Fig. L2). He also recognized several other forms where the rock was wholly crystalline, or where there was evidence of organic binding (*biolithite*). The limestone could be further classified on the types of grains (**allochems**) present and Folk introduced a series of terms such as *oosparite* (an *oolitic limestone* with a spar cement) and *biomicrite* (a *bioclastic limestone* with a *micritic* matrix). Folk (1962) later introduced a more detailed classification based on the matrix-cement content and also on the degree of *sorting* and rounding (Fig. L3).

The most widely used and simplest classification is that of Dunham (1962) (Fig. L4). This recognizes divisions based on depositional texture or where the texture is purely crystalline (**diagenetic**). Within the depositional divisions five types are recognized in which one category has grains which were bound by some organism or organisms (*boundstone*), and four divisions in which no binding occurred. These four are differentiated by the means of fabric support (matrix or grain supported) and by the amount of matrix. Those rock types lacking a matrix are the most mature, having been deposited in higher energy conditions. The Dunham classification also has been extended to include a variety of *reef* rocks.

There are problems in applying both classifications and there are many limestones where the fabric is **diagenetic** in origin and the simple concept of *textural maturity* is not applicable. It is usually necessary to 'filter-out' the **diagenetic** overprint before the rock can be interpreted regarding the energy-level of its environment of deposition.

The morphometric properties of limestones, i.e. grain size, shape, roundness and *sorting*, are less reliable as indicators of depositional processes than they are in siliciclastic rocks. Grain size in limestones is often more a function of biological factors than energy level. Grain-roundness and sphericity properties also have to be interpreted with care; for example, **ooids** grow into spheroids rather than being abraded into that shape and *peloids* are commonly rounded because of their passage through the colon of an organism. It is the types of grains, rather than their morphometric properties, which are more reliable environmental indicators in limestones.

Limestone was important in antiquity mainly as a **building stone**; it was a major building material of the Maya and figures largely in the Mediterranean and in Egyptian remains. The Sphinx and early Egyptian pyramids are of **nummulitic limestone**. Limestone was also used more rarely for various smaller objects including quernstones and carvings (the famous Head of Nefertiti is of painted limestone), and even for tools of the Lower Paleolithic people of Africa. [OWT/VPW]

Dunham, R.J. (1962) Classification of carbonate rocks according to depositional texture. In: Ham W.E. (ed.) *Classification of Carbonate Rocks*. American Association of Petroleum Geologists Memoir **1**: 108–21.
Folk, R.L. (1959) Practical petrographic classification of limestones. *American Association of Petroleum Geologists Bulletin* **43**: 1–38.
Folk, R.L. (1962) Spectral subdivision of limestone types. In: Ham, W.E. (ed.) *Classification of Carbonate Rocks*. American Association of Petroleum Geologists Memoir **1**: 62–84.

limestone pavement A glacially stripped **limestone** platform dissected by solutional **weathering** into blocks and runnels. The blocks are known in the north of England as **clints** and the runnels as **grikes** and the whole assemblage of features forms a suite of **karren**. Good examples of limestone pavements are found in the Craven district in Yorkshire, and in the Burren, County Clare, Eire. Processes of *glacial* scour acting upon a bedded **limestone** have produced the characteristically flat surfaces, and the presence of **joints** leads to the development of runnels and other topographic features. Solutional widening of **joints** may occur both before and after *glacial* scour, and involves both subaerial and subsoil processes. Controversy exists over the role of *soil cover*, and *soil erosion*, in the formation of pavement topography. The spacing of **grikes** is related to slope and **joint** frequency. Limestone pavements often support rare floras and have also suffered much damage from human removal of **limestone** blocks and plants. [HAV]

Drew, D.P. (1983) Accelerated soil erosion in a karst area: The Burren, western Ireland. *Journal of Hydrology* **61**: 113–126.

limiting depth The maximum depth at which an anomalous body could lie and still produce the observed **geophysical anomaly**. [PK]

limonite ($FeO \cdot OH \cdot nH_2O$) A field term used for a hydrated **oxide mineral** of (ferric) **iron**, the precise structure of which is uncertain. [DJV]

lineament A major linear topographic feature of geological significance on the Earth's surface, such as the trace of a long-lived **fault** system. Abrupt edges of mountain chains, long, narrow, **fault**-controlled valleys or ridges, or lines of volcanic activity can all be described as lineaments. *Fault zones* seen as lineaments often have a complex history of reactivation, i.e. a series of **fault** displacements, not necessarily in the same direction, over long periods of geological time. Localization of **deformation** in younger rocks above basement lineaments is a characteristic feature. [SB]

lineation A repeated or penetrative linear structure in a rock mass. This may be defined by, for example, a linear grain **shape fabric**, or elongate crystal habit (*mineral lineation*), axes of small **folds** or **boudins**. The term is most commonly used for fabrics in metamorphic rocks without a specific genetic implication. Some lineations arise from high **penetrative strains** leading to **elongation** of strained features (e.g. grain shape, pebbles); these are termed *elongation lineations* (see also **stretching lineation**). In the latter case, the lineation may define the direction of maximum **extension**, which in **simple shear** becomes nearly parallel to the **shear** direction at high **strains**; such lineations are often assumed to be in the **shear** direction in **mylonites**.

Lineations are also common features of *fault planes*, where they are often taken to indicate the direction of the **fault** *slip vector*, e.g. **slickenside** striations (see **fault**). *Intersection lineations* are linear features produced by the intersection of two planes. A common example is the line produced by the intersection of **cleavage** with a folded surface such as **bedding**; in the case of **axial planar cleavage**, this lineation is parallel to the *fold axis*. Several types of lineation with different orientations may be present in a rock and earlier lineations may be folded by later **folds**. Rocks with well-developed lineations are termed *L-tectonites*.

Lineations due to primary igneous processes are found in **igneous rocks** (for example due to aligned elongate crystal shapes) and also due to sedimentary processes in sedimentary rocks (for example *flutes*, *tool marks* and primary *current lineations* caused by movement of *clasts* or fluid). [TGB]

linnaeite (Co_3S_4) An **ore** mineral of **cobalt** with the **spinel** crystal structure. [DJV]

liquation deposit Oxide–sulphide deposit formed by *liquid immiscibility*. In exactly the same way that oil and water will not mix but form immiscible globules of one within the other, so in a mixed sulphide–silicate **magma** the two liquids will tend to segregate. Sulphide droplets separate out and coalesce to form globules which, being denser than the **magma**, sink through it to accumulate at the base of the intrusion or **lava flow**. Iron sulphide is the principal constituent of these droplets which are associated with basic and ultrabasic rocks, because **sulphur** and **iron** are both more abundant in these rocks than in acid or intermediate rocks. Chalcophile elements such as **copper** and **nickel** also enter these droplets and sometimes the **platinum** group metals. All the world's major **nickel** sulphide deposits were formed by this process. [AME]

liquefaction A sudden loss of **shear resistance** associated with a collapse of the grain-supported framework in *under-consolidated sediment*. Framework collapse gives rise to a temporary increase in the **pore-fluid pressure**, as the transitory fluid-supported suspension resediments from the base upwards until the entire sediment is again grain-supported but more tightly packed. Fluid escape is rapid but not sufficiently turbulent to destroy the continuity of sedimentary **laminations**. **Convolute lamination** and over-steepened *cross-bedding* are formed typically during sediment liquefaction. Liquefaction may be caused by **earthquakes**, cyclic loading of sediments during severe *storms*, and other processes that shock underconsolidated sediments. Sediment liquefaction may lead to **sediment gravity flows**, **slides** and other *mass flows*. [KTP]

liquefied natural gas (*LNG*) *Methane* liquefied at −160°C and atmospheric pressure. This reduces its volume by more than 600 times for oceanic transport in specially designed tankers. [AME]

liquefied petroleum gas (*LPG*) Liquefied propanes and **butanes** extracted from *wet gas*. The liquid fraction of the *wet gas* is known as **condensate**. [AME]

litharge (γ-PbO) The red **oxide mineral** of lead. [DJV]

lithiophilite (Li(Mn,Fe)PO_4) A mineral occurring in *pegmatites*. [DJV]

lithium (Li) At. no. 3, at. wt. 6.941, m.p. 180.54°C, d 0.531. The lightest alkali metal, it is used in low melting alloys and as a heat transfer medium. Lithium is recovered from some **pegmatite deposits** and from lithium-rich **brines** in the USA and Chile. [AME]

lithosphere The upper layer of the Earth, comprising **continental** and **oceanic crust** together with those upper-most parts of the **mantle** that have **viscosities** significantly greater than 10^{21-22} Pa s (generally $>10^{24}$ Pa s). It is conventionally recognized as showing **brittle deformation** if subjected to stresses of some 100 MPa although **earthquakes** in the **continental crust** are generally confined to the uppermost 10–12 km. The tectonic *plates* are comprised entirely of lithosphere, the thickness of which is usually defined in the oceans by the depth to the seismic **low velocity zone** which is within 2–3 km of the surface immediately below the central rift zone in the *oceanic ridges*, but rapidly increases to 60–80 km within some tens of kilometres of the crest, and then increases to 120–180 km beneath the older parts of the **oceanic crust**. Beneath the continents, its thickness is mainly defined by the evidence for both the persistence of the sources for **carbonatitic** intrusions, particularly **kimberlites**, and the depletion in radiogenic **heat flow** producing elements in both the **lower crust** and underlying **mantle** required to account for the low **heat flows** in **cratonic** areas. Both observations suggest that the continental lithosphere beneath **cratonic** areas is at least 250 km thick, and possibly as much as 450–500 km thick in these areas. [DHT]

Bott, M.H.P. (1982) *The Interior of the Earth*. Arnold, London.
Pollack, H.N. & Chapman, D.S. (1977) On the regional variation of heat flow, geotherms, and the thickness of the lithosphere. *Tectonophysics* **38**: 279–96.
Tozer, D.C. (1972) The concept of a lithosphere. *Geofysica Internacionale* **13**: 363–88.

lithostatic pressure/stress The vertical **stress** due to the force of **gravity** acting on a rock mass. The value of the lithostatic stress σ is simply calculated by the density (ρ) multiplied by depth (z) and the acceleration due to **gravity** (*g*):

$$\sigma = \rho g z$$

The density is usually taken to be the bulk density of both rock and any fluids contained. This is also known as the **load pressure** or **load stress**. [TGB]

lithostratigraphy The subdivision and correlation of stratal sequences by means of rock type. Because there are only a few types of commonly recurring sedimentary rocks, which rarely persist laterally for any great distance, this method is much less reliable for correlation than **biostratigraphy**, but in unfossiliferous strata it is the only one practical to use. For purposes of **geological mapping** and stratal logging, it is necessary to establish a hierarchical scheme of units named after geographic localities, in ascending order **bed**, *member*, *formation*, *group* and *supergroup*. The *formation* is generally regarded as the smallest unit mappable on a reasonable scale. In the extensive subsurface exploration undertaken by oil companies, much use is made of methods such as *gamma ray* and *resistivity logging*, which depend on particular strata registering a distinctive signature recorded by geophysical instruments. (See **geophysical borehole logging**.) [AH]

Boggs, S., Jr. (1987) *Principles of Sedimentology and Stratigraphy*. Merrill, Columbus, Ohio.

lithotypes in banded coal The four basic constituents of **banded coal** that can be distinguished in hand specimen and used in **coal classification**.
1 *Vitrain* — black, vitreous material that forms the most obvious constituent of **bituminous coal**. It is more **brittle** than other constituents, often breaks with a conchoidal **fracture** and occurs as thin (6–8 mm), closely jointed bands.
2 *Clarain* — bright to semi-bright bands of finely laminated **coal** having a silky **lustre** and which often contain thin *vitrain* bands alternating with a duller **attrital** groundmass.
3 *Durain* — grey to black bands with a dull to greasy **lustre**. Harder than other lithotypes with a tendency to fracture into large blocky fragments.
4 *Fusain* — when fairly pure is a soft friable material resembling charcoal which easily disintegrates into a black fibrous powder. It may be impregnated with minerals and is then hard. *Fusain* is only a minor constituent of most **bituminous coals** and in these it occurs as thin lenses only a few millimetres thick.

Various terms such as *clarodurain, clarofusain, duroclarain, fusoclarain, vitrodurain* are used to describe materials with intermediate characteristics.

There are difficulties in the field use of these terms and other methods of logging **coals** are employed, e.g. see **bright coal**. [AME]

Ward, C.R. (ed.) (1984) *Coal Geology and Coal Technology*. Blackwell Scientific Publications, Melbourne.

lizardite ($Mg_3Si_2O_5(OH)_4$) A serpentine mineral. (See **serpentine group**.) [DJV]

load pressure/stress Applied **stress** in experimental **deformation**; also used as a synonym for **lithostatic pressure/stress**. [RGP]

loess Silt of aeolian derivation. Loess is composed of the fine-grained silt-grade material transported in suspension as a result of winnowing from a coarser grained, sand grade bed-stock. The thick and extensive loess deposits of the world are thought to be derived by winnowing of *glacial* terrains, but deserts far removed from any *glacial* activity can also provide much silt-grade material. Primary wind-deposited loess is commonly weakly laminated or structureless. [PAA]

Pye, K. (1984) Loess. *Progress in Physical Geography* **8**: 176–217.

logan stone A large, exposed boulder which is balanced such that it is easily rocked while remaining in place. [AW]

logging boreholes and wells The recording of information obtained during the examination of boreholes and **wells**. Obtaining the maximum amount of pertinent data from boreholes is important in all aspects of geology but particularly in applied geology whether it be concerned with mineral, **coal** or oil exploration; **site investigation** for the erection of engineering structures or for waste disposal; water supply and so on. The logging may be by visual and optical examination of drill **core** or chippings, when these are available, or by lowering various instruments down the boreholes to measure some of the physical properties of the strata penetrated by the borehole (see **geophysical borehole logging**). These include a number of electrical properties, *radioactivity*, **dip** of the strata, etc. Cameras may be lowered to photograph the rocks and in shallow holes a periscope can be used for visual inspection. [AME]

Erickson, A.J. Jr. (ed.) (1984) *Applied Mining Geology*. American Institute of Mining, Metallurgical and Petroleum Engineers, New York.
North, F.K. (1985) *Petroleum Geology*. Allen & Unwin, Boston.
Ward, C.R. (1984) *Coal Geology and Coal Technology*. Blackwell Scientific Publications, Melbourne.

long profile, river The altitude of the river channel plotted against distance downstream. Long profiles are usually concave upwards though they are rarely smooth. Much work has attempted to treat the profile as an ideal shape and to fit functional relationships to the observed long profile. Convex sections are taken to indicate that **rejuvenation** has taken place; attempts have been made to extrapolate the concave sections of such profiles in order to estimate former sea-levels, such as the classic paper of Jones (1924) on the river profiles of central Wales. Increasing discharge and decreasing size of bed material downstream provide a general explanation for long profile concavity. [TPB]

Jones, O.T. (1924) The longitudinal profiles of the Upper Towy drainage system. *Quarterly Journal of the Geological Society, London* **80**: 568–609.

longshore drift The net longshore (shore parallel) sediment transport in one direction resulting from the summation of the longshore sediment transport by all the **waves** incident on the coastline. The majority of the sediment in the nearshore zone moves alongshore and results in the rapid lateral migration of many coastal morphological features.

The driving force for longshore sediment transport is oblique **wave** approach to the coastline providing a longshore energy vector; hence, the lower the angle between the coastline and the **wave** approach, the greater the longshore transport. Although longshore transport occurs throughout the nearshore zone, it is on the **beach** face that the largest percentage occurs. Two processes are responsible. The first is a complex longshore current produced within the *surf zone*. The sediment transport rate is proportional to the product of the longshore current velocity (which is greatest in the *surf zone*) and the bed **shear stress** (which is maximum at the breaker line). The maximum longshore transport of sediment thus occurs directly landward of the *breaker zone*. The second process is the zigzag motion of sediment in the *swash zone*, where the **swash** of waves incident at an angle to the **beach** moves sediment up the **beach** at an oblique angle. The return gravity-induced **backwash** drives the sediment back down the **beach**, resulting in a net longshore transport of sediment. [RK]

Komar, P.D. (1976) *Beach Processes and Sedimentation*. Prentice-Hall, Englewood Cliffs, NJ.

lonsdaleite (C) A hexagonal **polymorph** of **diamond** found in iron *meteorites*. [DJV]

lopolith A saucer-shaped igneous intrusion, commonly of mafic composition in which both upper and lower contacts are concave upwards. Lopoliths include some of the largest basin-shaped masses of **igneous rock** such as the **Bushveldt** intrusion (South Africa), and the Sudbury (Ontario, USA) and Duluth (Minnesota, USA) intrusions. The rocks in such intrusions commonly display layered and banded textures. [RST]

losing stream A stream that loses water to permeable rocks when flowing across them because the stream bed is higher than the **water table**. [AME]

low velocity zone The study of seismic **arrival times** from the Bikini and Eniwetok **atolls** in the Pacific Ocean provided the first unequivocal evidence that *P waves* travelling at depths between some 60 and 200 km within the Pacific were slowed by some 10% compared with *P waves* travelling at higher levels and that *S waves* were both slowed and their energy dissipated. Such behaviour indicates the presence of fluids within this part of the **mantle** and is commonly interpreted as being due to some 0.1% *partial melting* at the level at which the *geothermal gradient* lies close to the melting curve of **mantle** minerals. Alternatively, the fluid could be an **exsolution** phenomenon, such as the dehydration of **amphiboles** or **serpentines** at temperatures of *c.* 800°C (or even **serpentines** at *c.* 600°C), or to other fluids. The low velocity zone is well established in all oceanic areas and its top surface is widely adopted as the boundary between the **lithosphere** and **asthenosphere**, which varies from a few kilometres to 40–60 km thick beneath the crests of the *oceanic ridges*, to some 120 to 180 km beneath the older parts of the ocean basins. The low velocity zone is only poorly evidenced beneath the younger parts of the continents, i.e. areas that have undergone **orogenesis** during the last 600 Ma, and appears to be absent beneath the oldest parts of the continents. The base of the low velocity zone is poorly defined but is generally considered to be at about 250 km in the oceanic areas. [DHT]

Bird, J.M. (1980) (ed.) *Plate Tectonics*. American Geophysical Union, Washington, DC.
Dziewonski, A.M. (1971) Upper mantle models from 'pure path' dispersion data. *Journal of Geophysical Research* **82**: 1985–2000.
Le Pichon, X., Francheteau, J. & Bonnin, J. (1973) *Plate Tectonics*. Elsevier, Amsterdam.

lower continental crust That part of the **continental crust** that lies between 10–12 km depth and the **Mohorovičić discontinuity**. It is formed of rocks of **granodioritic** composition in *granulite* grade metamorphism and has an average density of about $3.0 \, \text{Mg m}^{-3}$. [DHT]

lower mantle That part of the Earth's **mantle** that lies below 700 km and above the Earth's **core** at 2900 km depth. It is characterized by very uniform physical properties, e.g. its **viscosity** is 10^{21-22} Pa s. Very slight vertical and lateral variations in properties have been reported using **seismic tomography**, but most evidence for such variations is only slightly greater than the noise level of their detection, i.e. they are very small indeed. It is considered to be composed of close packed oxides with a **perovskite** structure. (See also **mantle**.) [DHT]

lower stage plane bed A flat sediment bed produced at low **shear stresses** in **sands** coarser than approximately 0.6 mm diameter. These beds, over which there is sediment transport, are essentially flat except for shallow grooves and scours, 2–3 grain diameters high, that are aligned parallel to the flow. No transverse bed undulations are present. At these relatively low **shear stresses** in coarse **sands** this bedstate replaces the current **ripples** present in finer grained sediment. (See **bedform theory, hydraulically smooth and rough**.) [JLB]

Leeder, M.R. (1980) On the stability of lower stage plane beds and the absence of ripples in coarse sands. *Journal of the Geological Society, London* **137**: 423–30.

lunette A **dune** formed as an arcuate mound on the **lee side** of a **deflated** *lagoon*, lake **basin**, or river bed. Lunettes may occur singly or as two or three more-or-less parallel ridges. They are composed either of normal **quartz dune sand** or of clay aggregates and **evaporite** minerals. They are a feature of semi-arid areas, and while most examples are generally only 20 m or less in height, some examples are known from Tunisia which are over 140 m high. The classic locality is Australia (Bowler, 1973) but excellent examples occur around salt lakes in the High Plains of Texas, in the Pampas of Argentina, in Algeria and in the Kalahari. [ASG]

Bowler, J.M. (1973) Clay dunes: their occurrence, formation and environmental significance. *Earth Science Reviews* **9**: 315–38.

Goudie, A.S. & Thomas, D.S.G. (1986) Lunette dunes in southern Africa. *Journal of Arid Environments* **10**: 1–12.

lustre A property of **minerals** that arises from the inter-action of light with the mineral surface. The two main types of lustre are metallic and non-metallic. In the former, which is characteristic of metals, alloys and certain com-pounds of metals such as many sulphides, much of the (visible) light incident on the surface is reflected back. There is not a sharp division between these two types of behaviour and some minerals with an intermediate lustre are described as submetallic. Examples of minerals exhibit-ing a metallic lustre are native **gold**, **galena** and **pyrite**; whereas **hematite** and **wolframite** exhibit a submetallic lustre.

Minerals with a metallic lustre are opaque to light and their optical properties can be explained in terms of an electronic structure in which large numbers of vacant energy levels occur just above the highest energy levels containing electrons. Thus, photons of energies within the visible light-range are absorbed by exciting these electrons, but most of this absorbed energy is immediately re-emitted as visible light that returns to the observer. Minerals with a metallic lustre therefore reflect much of the light incident upon them.

Minerals with a non-metallic lustre generally transmit visible light, at least through thin edges; those that transmit most effectively are said to be *transparent* and those where the quality of transmission is poorer are said to be *translucent*. In this case the electronic structure is such that large numbers of vacant energy levels do not occur at energies such that excitation can be caused by photons of visible light. However, photons of a particular energy (wavelength) may be absorbed and give rise to coloured minerals. The lustre of non-metallic minerals is further described using the following terms, some of which are self-explanatory.

Adamantine The brilliant sparkling lustre of **diamond**; due to the mineral being transparent and having a high refractive index. Other examples include **cerussite** and **anglesite**.

Vitreous The lustre of broken glass; shown by **quartz** and many of the silicate minerals.

Resinous Having the lustre of resin; shown, for example, by **sphalerite** and by native **sulphur**.

Greasy Having an appearance as if covered by a thin layer of oil, although actually caused by the light being scattered by a surface that is rough on a microscopic scale, e.g. **nepheline**.

Silky Having a silk-like appearance and due to reflection of light from a fine aggregate of parallel fibres. Examples include fibrous **gypsum** and **chrysolite**.

Pearly An iridescent pearl-like lustre shown mostly by **cleavage** surfaces of layer lattice silicates such as **talc**. [DJV]

luzonite (Cu_3AsS_4) A comparatively rare **ore** mineral of **copper**. [DJV]

lysocline The depth in the ocean which separates well-preserved (above) from poorly-preserved (below) assem-blages of a given calcareous microfossil group. The most common usage is the foraminiferal lysocline — referring to preservation of **Foraminifera** — although the term has also been applied to **coccoliths**. The term sediment lysocline refers to the preservation of assemblages in the sediment whereas the hydrographic lysocline refers to preservation of assemblages in the water column. The lysocline marks the level within the water column at which the rate of **dissolution** of calcium carbonate starts to increase rapidly with depth and lies above the **calcite compensation depth**. [AESK]

Berger, W.H. (1976) Biogenous deep sea sediments. In: Riley, J.P. & Chester, R. (eds) *Chemical Oceanography* Vol. 5 (2nd edn), pp. 260–388. Academic Press, London.

Kennett, J. (1982) *Marine Geology*. Prentice-Hall, Englewood Cliffs, NJ.

M

maar A type of **tuff ring** in which the centre of the **crater** lies below the surrounding ground surface, having experienced down-**faulting** or sagging. [PTL]

macerals The basic organic constituents of **coal** that can be recognized under the microscope. They can be regarded as the 'minerals' that make up the group of organic rocks called **coal**. Macerals are the coalified remains of the various plant materials that existed at the time of **peat** formation. They differ from each other in their morphology, hardness and optical properties. Thin sections of **coal** can only be prepared with great difficulty as they must be about one third the thickness of rock thin sections. All modern observations are therefore made using polished sections and reflected light illumination with oil immersion as in metallurgy and **ore microscopy**.

Macerals are classified according to the *Stopes–Heerlen system* into three groups, see Table M1, on the basis of their physical appearance, chemical characteristics and biological affinities. All maceral names in this classification have the suffix -inite. Two macerals make up the bulk of the **lithotypes in banded coal** called *vitrain* and *fusain* and they have been named *vitrinite* and *fusinite* respectively. The two lithotypes *clairain* and *durain* are composed of mixtures of many macerals.

Vitrinite is the principal maceral in most **coals**. It is mainly formed of the remains of trunks, branches, stems, roots and leaves of **land plants**. When the cell structure is visible microscopically the term *telinite* is used. *Collinite* is the structureless component of *vitrinite*. In reflected light *vitrinite* is a medium grey in contrast with the lighter *inertites* and the darker *exinites*.

The *exinite* group is composed of a diverse assemblage of small organic particles such as *algae*, *spores*, cuticles and resin bodies. This group is characterized, especially in low **rank coals**, by high **hydrogen** and volatile contents. The most common member of this group in **banded coal** is *sporinite* — coalified *spores* and *pollen*. *Cutinite* represents the waxy cuticular coatings of leaves and other tissues. *Alginite* consists of coalified *algal* remains and is characteristic of *boghead coals* which are themselves composed dominantly of *algal* material. It is also abundant in some **oil shales**, but rare in **banded coals**. *Boghead coals* (*torbanites*) are thought to reflect depositional environments in which clear, aerated surface waters, free of humic matter, permitted *algal* colonies to flourish. On dying the remains of these colonies accumulated in **euxinic** bottom oozes, to form deposits of *alginite* or *bituminite*.

Intertite group macerals have higher **carbon** and lower **hydrogen** contents than macerals in **coals** of equivalent **rank**. They are harder than the other macerals and therefore tend to have a high relief. The name of this group implies their relative inertness during **coke** manufacture and other industrial processes. In some *coalification* processes plant materials are not transformed into *vitrinite*, but into highly reflectant *fusinite*. Similar material of lower reflectance is called *semifusinite*. *Macrinite* resembles *fusinite* but lacks cell structure. *Micrinite* forms very small rounded grains about a micrometre across. It is a ubiquitous constituent of **bituminous coal**, but is usually only present in small quantities. *Inertodetrinite* consists of broken fragments of *inertite* macerals and *sclerotine* the coalified remains of fungal material.

A different maceral classification is generally used for low **rank** coals such as **lignite** and *sub-bituminous coal* as these show more variability in their macerals and appear more complex under the microscope. (See **microlithotypes of coal**.) [AME]

International Committee for Coal Petrology (1963) (1971) (1976) *International Handbook of Coal Petrography* (2nd edn and supplements). Centre National de Recherche Scientifique, Paris.

Stach, E., Mackowsky, M.-Th., Teichmuller, M., *et al.* (1982) *Stach's Textbook of Coal Petrology*. Gebrüder Borntraeger, Berlin.

Ward, C.R. (1984) *Coal Geology and Coal Technology*. Blackwell Scientific Publications, Melbourne.

Table M1 Stopes–Heerlen classification of **macerals**

Maceral Group	Maceral
Vitrinite	Telinite
	Collinite
Exinite	Sporinite
	Cutinite
	Suberinite
	Resinite
	Alginite
	Liptodetrinite
	Fluorinite
	Bituminite
	Exudatinite
Inertinite	Fusinite
	Semifusinite
	Macrinite
	Micrinite
	Sclerotinite
	Inertodetrinite

machair A term used in the highlands of Scotland and Hebrides to describe coastal areas of calcareous sandy *soils* carpeted by rich grassland. A summary of the major characteristics of machair landscape is given in Ritchie (1976). [RDS]

Ritchie, W. (1976) The meaning and definition of Machair. *Transactions of the Botanical Society, Edinburgh* **42**: 431–40.

macroscopic Describes **structures** or other features of large size, measured in kilometres to hundreds of kilometres. This term refers broadly to features ranging from those seen on **geological maps** at 1:10 000 scale up to regional and world-scale **structures** such as *orogenic belts*, **lineaments**, and **lithospheric** *plates*. [SB]

magma A melt that generally contains suspended crystals and dissolved gaseous constituents or volatiles. Magmas are formed by *partial* or total *melting* of pre-existing solid **mantle** and/or **crustal** rocks. They are initially under-saturated with volatiles but ascent into regions of lower pressure (and temperature) within the **crust** may cause crystallization and volatile-saturation. Such magmas may be erupted on the surface as **lava flows**. However, **exsolution** and subsequent expansion of gas in magmatic systems may cause explosive **volcanic eruption**. Magmas comprise polymers of interconnected but disordered Si–O tetrahedra, with other cations such as **magnesium**, **iron**, calcium, sodium and **potassium** which occur in looser co-ordination with the oxygens, and magmas are hence less ordered in comparison with silicate crystals. The history of a magma body from formation to crystallization depends upon the composition, size, geometry and volatile content.

Magmas range from rare ultramafic types (e.g. **komatiite**) through **basalt**, **andesite**, **dacite** and **rhyolite** in composition. The temperatures of these magmas near to the Earth's surface range from over 1500°C (**komatiite**) to more common temperatures between *c.* 1200° and 700°C. Overall densities of magma bodies range from *c.* 2.20 to $3.00\,\mathrm{Mg\,m^{-3}}$. The behaviour of magmas during uprise and crystallization is very sensitive to volatile (e.g. H_2O, CO_2) content. The concentration of water in a melt is largely dependent upon the pressure, increasing from low values at atmospheric pressure (1 bar) to a maximum of *c.* 15–20% at 1 GPa (the maximum concentration depending upon the availability of water). Dissolved water has the effect of depolymerizing melt structure so causing rocks to melt (to form magma) at lower temperatures with increase in pressure (causing greater solution of H_2O within the melt).

The physical properties of magmas include **viscosity** and **shear strength**. The **viscosity**, or resistance of the liquid to **flow**, reflects the rearrangement of the polymerized silicate melt structure during **deformation**. Since basaltic melts are less polymerized than **rhyolite** melts (see above), they are less viscous; **viscosities** of **basalt** melts are *c.* $10–10^2\,\mathrm{Pa\,s}$ in comparison with *c.* 10^{3-6} for **rhyolite** melts. Also for a given melt composition, a higher H_2O content causes depolymerization and hence gives a lower **viscosity**. Most magmas exhibit **non-Newtonian** plastic behaviour in their crystallization range, causing apparent **viscosities** to be greater than idealized crystal- and bubble-free melts.

The behaviour of a magma body during uprise, *vesiculation* and possible eruption depends upon temperature, volatile content, specific heat, latent heat of crystallization, apparent **viscosity**, **thermal conductivity** density, dimensions and shape as well as equivalent properties of the *country rocks*. Large static magma bodies (measured in terms of tens of kilometres) may take of the order of a million years to crystallize. Convecting magma bodies may cool faster than static ones and produce anisotropic **fabrics** and may permit significant chemical fractionation. (See **magma diversification**.) [RST]

magma diversification The diversification processes acting upon a **magma** body leaving the source region that generally change the overall bulk chemical composition before final crystallization into solid rock. These processes include:

1 *magmatic differentiation*; the separation of initially homogeneous (parent) **magmas** into two or more (daughter) **magmas** of contrasted chemical composition. This may occur by:

(a) *crystal–liquid fractionation* or *fractional crystallization*. Such separation of crystals from the melt may occur (for example) by settling (or floating) of crystals that are denser (or less dense) than the coexisting melt, by separation of crystals from liquid by **flow** in a **magma** conduit (e.g. a **dyke**), or by crystallization of early-formed minerals on the walls of a **magma** body, leaving **magma** of contrasted composition within the centre and upper part of the **magma** body. Such *crystal–liquid fractionation* processes are widely regarded as representing the most important mechanism of magmatic diversification.

(b) *liquid immiscibility*; the separation of a homogeneous **magma** into two contrasted liquids. Such separation appears to be restricted in significance to certain high-Fe **basaltic** melts and highly *alkaline* melts.

(c) *vapour transport*, in which certain constituents (e.g. sodium, and possibly also **potassium** and silicon) become concentrated in a separate vapour phase which migrates within, or escapes from a **magma** body. Such a process is also of local significance and may apply to differentiation of volatile-rich **silica**- and alkali-rich **magmas**.

2 Assimilation or *contamination*; the chemical incorporation of solid material from **wall rock** (or **xenoliths**) into the **magma** to form a contaminated or *hybrid rock*. Such a process is of undoubted local significance but the large-scale significance of **assimilation**/*contamination* is a matter of debate.

3 *Magma mixing;* the mixing of two **magmas** of contrasted compositions within a **magma** body, to form a hybrid daughter product compositionally intermediate between them. *Magma mixing* certainly occurs on a local scale, but the large-scale significance of **assimilation**/*contamination*, is controversial. [RST]

Best, M.G. (1982) *Igneous and Metamorphic Petrology.* W.H. Freeman, New York.

magnesiochromite ($MgCr_2O_4$) A mineral with the **spinel** crystal structure. (See **oxide minerals**.) [DJV]

magnesioferrite ($MgFe_2O_4$) A mineral with the **spinel crystal structure** (See **oxide minerals**.) [DJV]

magnesite ($MgCO_3$) A **carbonate mineral**, the source of *magnesia* (MgO). [DJV]

magnesium (Mg) At. no. 12, at. wt. 24.305, d 1.74. An alkaline earth metal. Used in alloys and castings, for deoxidizing and desulphurizing metals, glass, ceramics, fillers ($MgCO_3,MgCl_2$), in flocculating agents, catalysts, refractories and many other industrial processes. Recovered from **magnesite** deposits, **dolomite**, **evaporites**, **brines** and seawater. [AME]

magnetic anisotropy A **ferromagnetic** particle will normally be more easily magnetized by an applied field in one direction than another, giving rise to magnetic anisotropy which deflects the original direction in proportion to the magnitude of the anisotropy. In **magnetite**, such anisotropy is almost entirely due to the actual shape of the grain, being most easily magnetized along its long axis, while **hematite** has a very strong crystalline anisotropy, being virtually only magnetized within its basal plane. Measurements of the magnetic anisotropy can thus provide rapid estimations of the shape and crystalline alignment of the magnetic minerals within a rock. [DHT]

magnetic anomaly The variation in the **geomagnetic field**, expressed in **nanoteslas** or **gammas**, caused by variations in the magnetic properties of the subsurface rocks. Magnetic anomalies should be isolated by the **reduction** procedure. [PK]

magnetic domain A small volume of a **ferromagnetic** material within which the electron spins are coupled to result in a unidirectional **magnetic field** within that volume which is commonly around 1 μm diameter in natural **iron oxides**. [DHT]

magnetic epoch A time interval of constant geomagnetic polarity, usually much greater than 10 000 years in duration. [DHT]

magnetic field The magnetic field (*B*), or *magnetic induction*, due to a pole of strength *m* at a distance *r* from the pole is given by

$$B = \mu_o m/4\pi\mu_R r^2$$

where μ_o and μ_R are constants corresponding to the **magnetic permeability** of vacuum and the relative **magnetic permeability**. The units of *B* are **nanotesla** (nT). [PK]

magnetic moment For a dipole, the product of the pole strength of one of the poles and the distance between the poles. Units are $A\,m^2$. [PK]

magnetic permeability (μ) Magnetic permeability terms contribute to the constant of proportionality in the relationship between **magnetic field**, pole strength and distance. Magnetic permeability describes the magnetic properties of the medium between the field point and pole. μ_o refers to the **permeability** of vacuum (in $Vs\,m^{-2}$, $Wb\,m^{-2}$ or $H\,m^{-1}$, Wb = Weber, H = Henry), μ_R to the relative **permeability** (dimensionless).

$$\mu = \mu_o\mu_R$$

where μ is expressed in $H\,m^{-1}$. [PK]

magnetic pole Magnetic dipoles are surrounded by lines of magnetic force which can be followed with a small compass needle. The points at which the lines of force converge are known as the poles of the magnet. Dipoles are characterized by two poles of equal magnitude and different sign. The geomagnetic poles are similarly located where the lines of force of the *Earth's magnetic field* converge. They are displaced from the geographic poles and not exactly antipodal. A magnetic pole can only very rarely exist in isolation without the association of an equal pole of opposite polarity. [PK]

magnetic potential The magnetic potential *V* at a distance *r* from a pole of strength *m* is given by

$$V = \mu_o m/4\pi\mu_R r$$

where μ_o is the **magnetic permeability** of vacuum and μ_R the relative **permeability** of the medium between the point of interest and the pole. Differentiation of the **potential** in any direction provides the **magnetic field** in that direction. [PK]

magnetic survey In addition to the search for magnetic rocks and minerals, magnetic survey methods can be used to locate buried archeological features such as pottery kilns, ditches, pits. These features show **magnetic anomalies** in comparison with the surrounding *soil* and subsoil. There are two main ways in which such anomalies are acquired. First, fired objects such as baked clay and pottery kilns acquire **thermoremanent magnetization** as a result of the realignment of the direction of magnetization of some grains of **iron oxides**, parallel to the **geomagnetic field**. The specific **remanent magnetization** resulting varies from 10^{-10} to $10^{-7}\,\text{Wb}\,\text{m}\,\text{kg}^{-1}$ ($1\,\text{Wb}\,\text{m}\,\text{kg}^{-1} = 10^7/4\pi$ $\text{emu}\,\text{g}^{-1}$). Second, archeologically accumulated *soil* and organic remains show an enhanced **magnetic susceptibility** in comparison with the underlying subsoil. This enhancement is due to past heating of the *soil* by fires, and is also associated with the decay of organic material during alternating humid and dry conditions when **reduction** and re-**oxidation** of the **iron oxides** present take place. The enhancement is particularly marked in archeological rubbish pits and ditches. Two frequently used instruments are the **proton** and **fluxgate magnetometers**. [OWT]

magnetic susceptibility (k) The dimensionless constant of proportionality in the relationship between **intensity of induced magnetization** and the **magnetizing force** of the inducing field. [PK]

magnetite (Fe_3O_4) A mineral with the **spinel crystal structure**, being strongly magnetic (**ferrimagnetic**), it is the major contributor to the magnetic properties of rocks. (See **oxide minerals**.) [DJV]

magnetizing force (H) When an electric current flows through a coil of wire a **magnetic field** (B) flows through and around the coil as the result of a magnetizing force. B is proportional to H, the constant of proportionality being the **magnetic susceptibility**. H is expressed in $\text{A}\,\text{m}^{-1}$. [PK]

magnetohydrodynamics The explanation of the **geomagnetic field** is dependent on direct observations combined with theoretical considerations of the behaviour of electrically conducting materials that are in motion within a **magnetic field**. The principle basis is that a good electrical conductor will carry a line of magnetic force with it as it moves, thus a **poloidal field** can be transformed into a **toroidal field**, or vice versa, by motions within the electrically conducting outer **core** of the Earth. In practice, such lines of force gradually decay, giving rise to slow changes in the **poloidal field** observed at the Earth's surface. [DHT]

Hide, R. (1982) On the role of rotation in the generation of magnetic fields by fluid motions. *Philosophical Transactions of the Royal Society, London* **A303**: 223–34.

Jacobs, J.A. (1987) *The Earth's Core* (2nd edn). Academic Press, London.

magnetometer An instrument for measuring **magnetic fields**. **Fluxgate**, **proton** and **cesium vapour magnetometers**

are mainly used for measurement of the **geomagnetic field** (and, in the case of the **cesium vapour** instrument, the fields of other planetary bodies). **Spinner** and **cryogenic magnetometers** are used on samples of rock for **paleomagnetic** purposes. [PK]

magnetostratigraphy The use of geomagnetic polarity **reversal** events recorded in stratal sequences for the purpose of correlation. Sensitive **magnetometers** can recognize these events in many sediments and a reliable magnetic time-scale has been established back to about 170 Ma. The method is especially valuable where fossils are sparse or absent and has been widely used for *Tertiary* and **Quaternary** correlation in deep-sea cores. It is the only means for establishing reasonably precise correlation between marine and terrestrial sequences. (See **geomagnetic polarity time-scale**.) [AH]

Boggs, S., Jr. (1987) *Principles of Sedimentology and Stratigraphy*. Merrill, Columbus, OH.

magnetotelluric (MT) **Electromagnetic induction method** using naturally occurring electromagnetic fields to investigate **electrical condutivities** at depth. The method is based on the theory of diffusion of plane electromagnetic waves into a conductor. Since the **depth of penetration** increases with period, the range of investigation can be some tens of kilometres provided the long periods of the geomagnetic spectrum are used, allowing the **lower crust/upper mantle** region to be explored, for example.

The '*wave impedance*', Z_{xy}, defined as the ratio of the amplitudes of orthogonal horizontal electric and **magnetic field vectors** (E_x/H_y) at a given frequency, is proportional to the **apparent resistivity**, ρ_a, (Cagniard, 1953) and is the parameter determined from field data. Usually the electric and **magnetic fields** are recorded continuously and the amplitude information obtained by **power spectrum** analysis.

For horizontally layered media, standard curves of ρ_a (or wave impedance) against period allow curve-matching in MT depth-sounding, complementing the d.c. method (**vertical electrical sounding**). The advantage of MT in this application is that electromagnetic waves, unlike d.c., penetrate highly resistive media virtually unimpeded. Also, the source is freely available.

In both the former USSR and North America, MT has been used in **petroleum** exploration: to determine the depth to basement in sedimentary **basins**, and to map thicknesses of sediments and delineate **structures** overlain by high (seismic) velocity **basalts**. In *mineral exploration* the target is shallower and here MT uses the Schumann frequency range (**sferics**). [ATB]

Cagniard, L. (1953) Basic theory of the magnetotelluric method of geophysical prospecting. *Geophysics* **18**: 605–35

Keller, G.V. & Frischknecht, F.C. (1966) *Electrical Methods in Geophysical Prospecting*. Pergamon, Oxford.

Vozoff, K. & Asten, M. (eds) *Magnetotellurics in Oil Exploration in the USSR*. Translated by Keller, G.V. Society of Exploration Geophysicists, Tulsa, OK.

major fold Relative term to describe the larger (**macroscopic**) **folds** in a complexly folded terrain. Major **folds** are those on a map scale with wavelengths of the order of kilometres. The term is contrasted with '**minor fold**', which is a **fold** distinguishable at outcrop scale downwards. [RGP]

malachite ($Cu_2CO_3(OH)_2$) A bright green mineral found in the oxidized portions of **copper ore** deposits.

Apart from being one of the **ores** used for **copper** production in antiquity, malachite was employed to make green pigment (used for painting and as a cosmetic), and for small ornaments and jewellery. Deposits were exploited in Sinai and Zimbabwe. [OWT/DJV]

mammals/Mammalia Vertebrate animals with only one set of replacement teeth and a dentary-squamosal contact in the jaw hinge.

The traditional definition of a mammal is of an animal which has 'warm blood' and which suckles its young. This is totally inadequate as a paleontological definition, as neither of these two characters is preserved in fossils, except that in some cases soft tissue can be mummified. These mummies are always of remains far too late in time to help resolve the critical *reptile–mammal transition*, or even help elsewhere, so soft-tissue characteristics cannot be used taxonomically.

As teeth are by far the most resistant part of the vertebrate body, they have come to be used almost universally in the classification of mammals (Table M2). Many mammals are known almost exclusively from teeth alone, particularly from the late **Mesozoic** when so much of importance was happening in this critical phase of vertebrate evolution. As far as teeth are concerned, mammals differ from all other tooth-bearing vertebrates in that they have only one replacement generation of teeth. **Fish**, **amphibians** and **reptiles** have continual replacement of teeth, where they are present, and even in the toothed **birds** of the **Cretaceous** there is evidence for a **reptilian** mode of tooth replacement. In mammals the first generation of teeth is known as the 'milk dentition', and only those teeth which are *not* replaced, i.e. the posterior cheek teeth, are known as 'molars'. The series of teeth in front of the molars are the 'premolars' and have a set of precursors, the 'milk molars'. These cheek teeth tend to have more than one root, but the presence or absence of more than one root does not make a tooth a molar, or not. Only care in description and detailed knowledge of mammal dentitions can be used to distinguish between the cheek tooth series.

In the late 19th century a system of homologizing tooth cusps across the *reptile–mammal boundary*, and within the mammals themselves, was devised, and is known as the

Table M2 Classification of **mammals**

Class Mammalia
  Subclass Prototheria
    Order Monotremata
      (?) Docodonta
      (?) Triconodonta

  Subclass Allotheria
    Order Multituberculata

  Subclass Theria
    Infraclass Trituberculata
      Order Symmetrodonta
        Pantotheria
    Infraclass Metatheria
      Order Marsupialia
    Infraclass Eutheria
      Order Insectivora
        Tillodonta
        Taeniodonta
        Chiroptera
        Primates
        Creodonta
        Carnivora
        Condylarthra
        Amblypoda
        Proboscidea
        Sirenia
        Desmostyla
        Embrithopoda
        Notoungulata
        Astrapotheria
        Lipopterna
        Perissodactyla
        Artiodactyla
        Edentata
        Pholidota
        Tubulidentata
        Cetacea
        Rodentia
        Lagomorpha

'*Cope–Osborne theory*'. The assumption is that the single pointed tooth of the reptile grew elaborate extra cusps in order to become the mammalian tribosphenic molar. Thus the principal cusp in a mammal tooth is assumed to be representative of this original uni-cusped tooth, and was termed the 'protocone'. Other terms such as 'paracone', 'hypocone' and so on were coined to describe the accessory cusps, and the system came to be recognized as the best method of describing and comparing teeth in almost every order of mammals (Fig. M1).

However, if it is assumed that the protocone represents the original '**reptilian**' tooth, then it should follow that this cusp should be the first to appear embryologically, in the developing young. Unfortunately this is not the case. Also, when the *reptile–mammal transition* is examined, advanced cynodont *Therapsida* with a jaw-joint approaching that of the true mammal are seen to have teeth in which there is no true 'protocone' analogue,

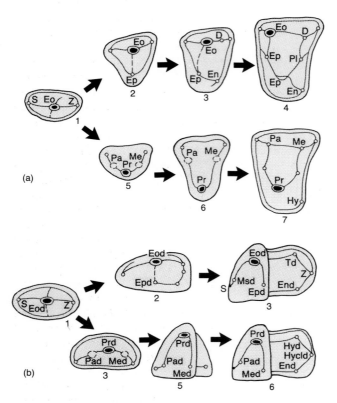

Fig. M1 Mammalia. Evolution and nomenclature of dental cusps. (a) Origin of the upper teeth; 1,2,3,4 after Vandebroek. Eo = eocone; Ep = epicone; S = mesiostyle; Z = distostyle, D = distocone; En = endocone; Pl = plagiocone. 5,6,7 after Cope and Osborne. Pr = protocone; Pa = paracone; Me = metacone; Hy = hypocone. (b) Origin of the lower teeth; 1,2,3 after Vandebroek. Eod = eoconide; S = mesiostylide; Z = distostylide; Epd = epiconide; Msd = Mesioconide; Td = teloconide; End = endoconide. 4,5,6 after Cope and Osborne. Prd = protoconide; Pad = paraconide; Med = metaconide; Hyd = hypoconide; End = entoconide; Hycld = hypoconulide. (After Chaline, 1987.)

but which have post-canine teeth with three or four substantial cusps in a straight line accompanied by a row of small cusps in a cingulum on the inner side of the tooth. One of these small cusps is in fact the true analogue of the 'protocone', and may be seen to be slightly larger than its neighbours. A contrasting system of comparisons of tooth cusps is useful for comparing the teeth at the crucial *reptile–mammal transition*, and forms part of Fig. M1.

Dental differentiation had of course been achieved by the cynodont stage of *Therapsida* evolution, and incisors, canines and 'molars' are clearly noted in these forms. The first mammals had triconodont post-canine teeth, but at the end of the **Jurassic** or the beginning of the **Cretaceous** the tribosphenic molar tooth-pattern had appeared, which characterizes both *marsupial* and *placental* mammals. The term 'tribosphenic' describes the action of the 'protocone' of the upper molars which crushes against the 'talonid' of the lower molar in a similar manner to the action of a mortar-and-pestle. However the real difference between

a reptilian tooth and that of a mammal concerns their relationship when the jaw is closing and does not rely on the presence of a talonid. In a **reptile** no tooth-to-tooth occlusion takes place, whereas even in the earliest mammal (and probably in some *Therapsida*) a shearing action occurred between the opposing 'protocones' of the cheek teeth. A comparison with the action of the closing jaws of a pair of garden rose-trimmers has been made for the **Triassic** triconodont type of tooth, with the peak of the major cusp closing opposite the valley between the opposing cusps, and so forcing the food to be trapped in the wedge-shaped space between the cusps while the opposing peaks shear against one another and then penetrate what is held there. This kind of action is most useful for instance in puncturing the fluid-filled body spaces (coeloms) of animals like worms, and so incapacitating them, before they are further cut up. As the vast majority of **Mesozoic** mammals were very small, the size of shrews (*Insectivora*), this method of feeding is quite logical. As body size increases, and less energy-rich foods become used, more efficient breaking up of foodstuffs becomes more important, and the combined slicing and crushing action of the tribosphenic molar is seen to be crucial.

Aegialodon of the Wealden is the first really well-known mammal to show the presence of a true 'protocone'. Succeeding *Aegialodon* and similar forms, *Holoclemensia* and *Pappotherium* of the Albian of Texas are potential common ancestors to both the *marsupials* and *placentals*, i.e. the *Metatheria* and *Eutheria*, and have been referred to as being of *metatherian–eutherian* grade. The Deltatheriidae and *Endotherium* of Asia are probably similar, but their true relationships are still obscure.

Differentiation of the *marsupials* and *eutherians* is seen to have taken place by the Upper **Cretaceous** in North America. The former have a complex biogeographical history, being typical of the South American *Tertiary*, and they penetrated the Australasian Realm by way of the Antarctic during the **Oligocene**. Recent fossil discoveries have confirmed this route, as there was an alternative suggested by way of Northern Europe and Asia which held some attraction for its protagonists for a long time. The original *marsupial* immigrant to Australia was probably an animal very similar to the modern oppossum, *Didelphys*, but from this founder-stock has evolved the entire *marsupial* fauna known today. *Marsupial* equivalents to 'lions', 'tigers', 'wolves', 'sheep', 'cows', 'rats' and 'mice' have all evolved from the original stock. Unique to Australasia however are the specialized bipedal, hopping, jumping kangaroos and wallabies which fulfil the same function in the Australian savannah as do the antelopes and deer of the 'Old' and 'New' Worlds.

The *Eutheria* (*placentals*) are numerically and taxonomically the most important mammalian infraclass. They originated at about the same time as the *marsupials* in the Upper **Cretaceous** of North America, and much of their early history is contained in rocks of that age in both America and Mongolia. Early development of all the mammals took place alongside the later history of the *dinosaurs*, and they seem to have been night-adapted forms from the beginning of their history, thus keeping out of the way of the large, active **reptilian** predators of the time.

By the end of the **Cretaceous** the following orders are known: *Condylarthra, Primates, Insectivora, Carnivora,* and both orders of *ungulates*, the *Perissodactyla* and *Artiodactyla*. Bats, dermopterans (flying lemurs), rodents and cetaceans (whales, porpoises) appear by the earliest *Tertiary*, but a whole range of taxa appear much later in time and have no known point of origin among the recognized taxons. These last enigmatic types include the xenarthrans (armadillos, sloths, giant anteaters), pholidotids (pangolins), tubulidentids (aardvarks), lagomorphs (rabbits and hares), hyracoids, proboscideans and sirenians. The last group of three are probably quite closely related and are represented today by *Hyrax,* elephants and sea-cows.

The *Insectivora* represent the basic stock from which all other *eutherian* mammal orders can be derived. The smaller representatives of the order do rely to a very great degree on insects, slugs, worms and similar small animal life for their diet, but many are in fact quite cosmopolitan in their diet. Hedgehogs are omnivorous, taking eggs, mammals smaller than themselves, bird nestlings and carrion, as well as more exotic dishes such as bread-and-milk when living in suburban gardens. Water-shrews are semi-aquatic, as their name implies, and feed on a range of aquatic food. They are also much larger than the normal 'shrew'. Their dentition seems to have been evolved to subdue small, mobile, prey. Early insectivores have a tooth count in each jaw probably close to the primitive number of about 11, i.e. a total of 44; comprising incisors (3), canines (1), premolars (4) and molars (3).

The *Carnivora* are probably closest to the *insectivores*, and on the other hand reduce the tooth count dramatically so that often there are only three premolars and a maximum of two molars. Either the last premolar or the first molar is expanded into a 'Carnassial' or shearing tooth, adapted to remove large quantities of flesh from the prey carcase. The canines often also undergo great enlargement, as in the sabre-tooth cats. A great deal of the evolution seen in the mammals is iterative, similar adaptations being seen in unrelated lineages within the same broad group, and so in the **Oligocene, Miocene** and **Pliocene,** different groups of the cats evolved enlarged canines to deal with the corresponding thick-skinned ungulates living alongside them. The most famous sabre-tooth is the North American *Pleistocene* genus *Smilodon*, but there were even marsupial sabre-tooths living in South America before the two continents became rejoined in the **Miocene/Pliocene.**

The two major, successful, ungulate (hoofed) orders are the *Perissodactyla* (horses and other odd-toed herbivores) and the *Artiodactyla* (cows, sheep and other even-toed herbivores). Their success is attributed to their ability to detoxify alkaloids in *angiosperm* plant tissues, where other groups were very much less successful. Two main strategies are adopted; foregut fermentation and hindgut fermentation. In the former, the stomach is divided into several chambers in which the ingested foodstuff is exposed to the action of *bacteria* and protozoa, which break down the plant tissues and extract a proportion of the energy in the form of fatty acids. The partly digested material is returned to the mouth as 'cud', where additional digestive enzymes are added to the cellulose bolus, which is then returned to the stomach and further digested. What is capable of being extracted further is then taken out while the foodstuff is passed down the gut. In all herbivores the gut is very much lengthened so as to provide the maximum surface area of absorption. This expansion of the gut is secondarily seen in the very bulky bodies typical of all herbivores, and reflected in their expanded rib cages. A strategy which allows part of the digestive system to act without the owner actually having to be in the open and feeding, i.e. when the cud is being chewed, must have added to its selective advantage.

The dental adaptations seen in *artiodactyls* involve the loss of the upper incisor teeth and the elaboration of the cheek teeth to provide a battery of flat-crowned, but roughened surfaces to comminute the food. The tongue is roughened to catch and pull large hanks of vegetation into the mouth; this food is passed directly to the stomach for fermentation. Thus a large meal can be acquired quickly, but digested at leisure in the security of close cover.

Associated with the elaboration of the *artiodactyl* digestive strategy was the evolution of large and sometimes complex horns. These can be related to their owner's way of life: as a browser on leaves or grazer on grasses, but is a much more complex study than can be summarized here.

The *Perissodactyla* have similar batteries of roughened cheek teeth to comminute their food but they retain a set of incisors in both jaws. The cheek teeth are typically much increased in height to combat wear brought on through grazing on silica-rich grasses. Their gut is elongated in a similar manner to the *artiodactyls*, but they use an expanded cecum, a diverticulum of the hindgut, in which to store their food and in which the major part of the digestive process takes place. The *perissodactyls* are therefore nominally less efficient at digesting their food, but they can process it much faster than can the *artiodactyls* and so make up for this apparent deficiency in bulk intake and speed of processing.

All *ungulates* modify their limbs to become very fast runners. The femur is shortened, and the lower limb bones are elongated. In all but the very earliest *ungulates* the bones of the wrist and ankle are also modified and lengthened, so that the limbs are very much elongated and the stride increased. This set of adaptations seems to have coincided with the spread of grassland during the *Tertiary*, particularly in the western interior of North America. For instance it is believed that the earliest representative of the horses was adapted to a leaf-eating diet in forested areas growing on soft ground. Therefore their teeth are adapted to eating relatively soft tissue, and their feet have toes splayed out to counteract any tendency to sink into soft ground. This stage is typified by *Eohippus* of the European **Eocene**. With the presumed growth of the mid-continental grasslands, a series of horses evolved which developed high-crowned teeth and gradually longer legs, concomitantly with the reduction of the number of functional toes in both feet. Not all changes are seen in all representatives of the horse line all at the same time, a process known as *mosaic evolution*. However, in general, each stage of horse evolution can be related to specific horizons within the *Tertiary*, particularly in North America.

In addition, the *Tertiary* mammal faunas of the Great Plains of North America are so rich that, in the absence of marine sediments, they can be used to correlate terrestrial deposits over long distances, and so define '*Land Mammal Ages*'. These sequences take their place along with the Baltic fish-zones and the South African **Permo-Triassic reptile** horizons as being among the relatively few vertebrate-bearing zones capable of being correlated by these usually scarce fossils.

One other 'success' story within the mammals is the evolutionary history of the *primates*. The earliest *primate* is known from the **Cretaceous** of North America, but similar primitive forms (*Pleiadapus*) are also known from Europe and North Africa in the lower *Tertiary*.

A sequence of genera with increasing size of brain, binocular vision and locomotory improvements leads through to the **Oligocene**, when the apes and monkeys diverged. The New World monkeys are descended from Old World ancestors which migrated across the Atlantic prior to the final breakup of **Gondwanaland**, and retain several primitive characteristics of their lower *Tertiary* ancestors.

The apes underwent their principal evolutionary radiation in Africa, with several 'waves' emmigrating north and east across through the then densely forested Arabian peninsula, to Asia. The gibbons may have been the first, in the early **Miocene**, followed by the ancestors of the orang-outan. Other extinct, giant genera are also known, such as *Gigantopithecus* which is found in the Siwaliks of northern India. At the close of the **Miocene** the forests of the Afro-Arabian region disappeared and a series of more *man-like apes* evolved in what is now East and South Africa.

The conventional view is that the first real *man-like ape* appeared some 5 Ma ago in the form of *Kenyapithecus* (=*Ramapithecus*?) and then by the way of *Australopithecus* to *Homo habilis*, the first of the tool-makers. Dates for early australopithecines range through the upper **Pliocene** and the first true *men* are also probably of this age (*c.* 3.2 Ma).

A further elaboration of the brain, associated with the development of an upright stance and concomitant social evolution is presumed in *Homo erectus* and *H. sapiens neanderthalensis*. The final stage is represented by *H. s. sapiens*, whose earliest fossil remains are found in southern Africa dated to more than 250 000 years, at Border Cave, Zululand. [ARIC]

Bowler, P.J. (1987) *Theories of Human Evolution: A Century of Debate 1844–1944*. Basil Blackwell, Oxford.
Carroll, R.L. (1987) *Vertebrate Paleontology and Evolution*. W.H. Freeman, Chicago.
Chaline, J. (1987) *Paléontologie des Vertébrés*. Dunod, Paris.
Romer, A.S. (1966) *Vertebrate Paleontolgy* (3rd edn). Chicago University Press, Chicago.
Savage, R.J.G. & Long, M.R. (1986) *Mammal Evolution: An Illustrated Guide*. British Museum (NH), London.

man-made earthquakes There are four classes of **earthquakes** caused by human activity: mining-induced, reservoir-induced, fluid-injection, and explosion **aftershocks**. They are caused by local modification of the natural **stress** patterns in the Earth's **crust** which then reactivate pre-existing **faults**.

1 *Mining-induced*. Excavation of cavities, mine shafts and tunnels changes the regional **stress** pattern. This may cause small **earthquakes** on existing **faults** around the mine. In deep mines, these '*rockbursts*' can be quite large, up to **magnitude** 4 or more, and cause fresh faulting. Other effects related directly to mining, such as collapses and breakouts, may also be registered as **seismic events**.

2 *Reservoir-induced*. There is a link between the filling of some large (>100 m high) dams/reservoirs and seismic activity beneath and around the reservoirs. Well-known examples are Lake Kariba, Zambia/Zimbabwe border (2000 **earthquakes** culminating in a **magnitude** 5.8 shock less than a year after filling), and the Koyna Dam, India (a main shock of **magnitude** 6.5, causing damage and fatalities, and continuing seismic activity of tens-to-hundreds of **earthquakes** monthly, correlated with local rainfall patterns). Other well-documented cases include the Hoover Dam in the USA, the Hsingfengkiang Dam in China and the Nurek Dam in the former USSR. The seismic activity usually declines once a main shock has happened.

The **earthquakes** occur on pre-existing **faults**. The existing tectonic **stress** patterns are modified by the weight of water and, more importantly, because of the injection of water under pressure into the subsurface (see **4** below). Careful surveys of seismic activity and **fault** locations are vital in preparing the design of large reservoirs. See **hydroisostasy**.

3 *Explosion aftershocks*. Underground nuclear explosions are often followed by **aftershocks**. Most result from the collapse of the cavity left by the explosion; some are small **earthquakes**. The outward pressure of the explosion allows regional **stress** to cause a small **earthquake** directly underneath the explosion and simultaneous with it ('*tectonic release*'). Similarly, nearby **faults** can be reactivated and small shallow (<5 km depth) **earthquakes** can happen later. Empirically, only **faults** within a radius of $300Y^{1/3}$m (where Y is *explosion yield* in kilotons TNT equivalent) are activated. Detailed studies at the Nevada Test Site and of

tests in the Aleutian Islands have shown no proof that natural *earthquakes* are stimulated on any larger scale.

4 *Fluid injection.* When waste water from a U.S. military arsenal in Denver, Colorado was injected down a 3.7 km deep borehole, frequent small **earthquakes** (magnitude < 4.3) began at a rate of over 40 per month. This seismic activity could be controlled by the starting or stopping of injection. It demonstrated the role of **pore-fluids** in **earthquake** initiation, as perfectly dry rocks would have greater **strength**. The Denver data have been confirmed by experiments in oilfields. Fluid injection has been proposed as a way of mitigating **earthquake** hazards; water pumped into a **fault** could release **stress** as numerous small **earthquakes** rather than letting it 'lock up' and cause a large event. (See **reservoir-induced seismicity, nuclear-explosion seismology**.) [RAC].

Bolt, B.A. (1978) *Earthquakes – A Primer*. W.H. Freeman, San Francisco.
Glasstone, S. & Dolan, P.J. (1977) *The Effects of Nuclear Weapons*. US Government (DOD and ERDA), Washington, DC.
Judd, W.R. (1974) Seismic effects of reservoir impounding. *Engineering Geology* **8**: 1–212.

manganese (Mn) At. no. 25, at. wt. 54.938, d 7.2. Transition element of Group VII. A soft grey metal, it is the most important ferro-alloy element in steels. Also used in other alloys. Manganese compounds are used in batteries, dyes, paints, chemical processes, etc. The principal sources are sedimentary manganese deposits and their metamorphosed equivalents together with **residual deposits**.

Manganese was added to ancient glasses in the Roman, Byzantine and Dark Age periods in Europe as a decolorant (cf. **antimony**) or to produce a purple colour (depending on the composition of the glass and the oxidizing/reducing conditions of production). [AME/OWT]

manganese nodules Nodules of *ferromanganese oxides* found mainly on modern ocean floors. *Ferromanganese oxides* appear in various forms in marine environments and there is no accepted classification that differentiates clearly between objects such as fragments of **coral**, shark's teeth, that are stained or thinly encrusted with **manganese** minerals and objects that are termed manganese nodules. The problem is that some 'nodules' have rock fragment cores comprising 50% or more of the nodule whereas others may have no such clearly defined core at all. The general size range of nodules is 5–200 mm and the outer surfaces are rounded, common descriptions being spheroidal, ellipsoidal, botryoidal. These surfaces may be smooth or granular and are usually reddish brown to black. The interiors generally show a zoning produced by a variation in mineral content of the growth layers. The minerals can be divided into three groups: **manganese, iron oxide** and **accessory minerals**. The principal **manganese** minerals are todorokite (Mn, Ca, Mg)$Mn_3O_7 \cdot H_2O$, *birnessite* $Na_4Mn_{14}O_{27} \cdot 9H_2O$ and *vernadite* (δ-MnO_2). The **iron oxides** are various forms of

FeOOH and are often amorphous to **X-ray diffraction**. The **accessory minerals** include **feldspars**, clays, silicas, the **zeolite phillipsite** and **barite**. Biogenic **apatite, calcite** and **aragonite** also occur.

Apart from oxygen the major elements are **manganese** and **iron** with **manganese** generally predominant. It can however vary from 1 to 40%. **Iron** generally runs 1–25%. Other elements of potential economic interest are **nickel** 0.1–2%, **copper** <0.05–2%, **cobalt** <0.05–2.50%, **zinc** <0.05–0.09%, **vanadium** <0.005–0.3% and **molybdenum** <0.005–0.15%.

Manganese nodules are found over most of the deep ocean floor and in shallower seas such as parts of the Baltic. They form by the extraction of metals from sea-water and the **pore-waters** of seafloor muds but, as can be seen from the above figures, their metal contents can be very variable. Economic interest has centred mainly on those with a combination of about 3% **nickel, cobalt** and **copper** and there are only a few areas where such high **grades** are known. The main area is in the north-eastern equatorial Pacific Ocean; less important areas lie in other parts of the Pacific Ocean and in the Indian Ocean. These are areas of high biological productivity in the surface waters. Organisms here extract metals from the water and sink after death to decay on the seafloor, liberating metals into the bottom waters and the *pore-waters* of the sediments where they are scavenged by *ferromanganese oxide* phases.

During the late 1960s and the 1970s there was a general expectation that nodule mining was only a few years away. For the time being, however, most of the exploration by industrial concerns has ceased for two main reasons. The first is the depressed state of most metal prices, particularly for those metals that occur in metal-enriched nodules, and the second is the deep-sea mining regime proposed under the *Law of the Sea Convention*. This would place all the seafloor beyond 200 miles of coastal states under the jurisdiction of an *International Seabed Authority* which would be responsible for all mineral deposits in the oceans and perhaps mine them itself. [AME]

Aplin, A.C. & Cronan, D.S. (1985) Ferromanganese oxide deposits from the central Pacific Ocean, II. Nodules and associated sediments. *Geochimica et Cosmochimica Acta* **49**: 437–51.
Cronan, D. (1985) A wealth of sea-floor minerals. *New Scientist* **June**: 34–8.
Haynes, B.W., Law, S.L. & Barron, D.C. (1985) An elemental description of Pacific manganese nodules. *Marine Mining* **5**: 239–76.

manganite (γ-MnO(OH)) An **ore** mineral of **manganese**. [DJV]

manganotantalite ((Mn,Fe)Ta_2O_6) An **ore** mineral of **tantalum**. [DJV]

mangrove swamp An intertidal association of mudflats and mangrove vegetation found in the tropics. Mangrove swamps are similar features to **salt marshes**, although the latter are largely restricted to temperate coastlines. Mangrove swamps usually develop along sheltered, low-energy coastlines, including **estuaries**, where there is a ready supply of sediment. Large areas of mangrove swamp are found in Florida, along the Malay peninsula and on the Pacific coast of Colombia. There has been much debate over the influence of the mangrove trees upon sedimentation within **swamps**, which seems to be great in many areas depending upon tidal regimes and other factors. Mangrove swamps exhibit zonations of habitats, related to height above sea-level. Some mangrove trees colonize early and are highly tolerant of inundation, such as the *Avicennia* genus. Other mangroves, such as *Rhizophora* and *Bruguiera* prefer higher substrates. All mangroves have developed special adaptations to allow them to survive in such conditions, including prop roots and pneumatophores. Creek systems similar to those developed in **salt marshes** occur in mangrove swamps, but have not been widely investigated. Cores taken through mangrove swamps can provide much *paleoenvironmental* information. [HAV]

Smith, T.J. & Duke, N.C. (1987) Physical determinants of inter-estuarine variation in mangrove species richness around the tropical coastline of Australia. *Journal of Biogeography* **14**: 9–19.
Woodroffe, C.D., Thom, B.G. & Chappell, J. (1985) Development of widespread mangrove swamps in mid-Holocene times in northern Australia. *Nature* **317**: 711–13.

manjiroite $((Na,K)Mn_8O_{16} \cdot nH_2O)$ A secondary **ore** mineral of **manganese**. [DJV]

Manning equation Perhaps the best-known formula of its type for estimating stream velocity on the basis of *channel roughness*, slope and **hydraulic radius**.

$$v = k \cdot R \cdot s \cdot n$$

where R = **hydraulic radius** (cross-section area divided by wetted perimeter), s = channel gradient, k = a constant, n = the Manning roughness coefficient.

The constant k is taken to be 1 in SI units, 1.49 in fps units and 4.64 in cgs units. This means that Manning's n is a dimensionless constant for a given channel regardless of the units used to measure R. Manning's n may take a wide variety of values depending on the calibre of bed and bank material, and on the amount of vegetation present in the channel or, if relevant, on the **floodplain**. n may be around 0.02 for fine gravels, and between 0.03 and 0.05 for coarser gravel with a few boulders; cobbles with large boulders have n values between 0.04 and 0.07. Sluggish reaches with deep weedy pools have n values between 0.05 and 0.08. **Floodplains** with dense stands of trees may yield n values above 0.1 and rarely as high as 0.2. Manning's n is often used as the basis for reconstructing flood discharge. [TPB]

mantle The zone of the Earth between the **Mohorovičić discontinuity** (at *c.* 11 km below the ocean floor and averaging about 35 km depth beneath the continents) and the **core** at a depth of *c.* 2900 km (Fig. M2). It is solid to the transmission of **seismic waves**, i.e. its *shear modulus* is everywhere greater than zero, so that *S waves* are always transmitted through it. The average composition of the mantle is generally modelled on the basis of the composition of oceanic **upper mantle** minerals, found in association with **ophiolites**, and of continental **upper mantle**, based on **xenoliths** within **kimberlite** intrusions, and comprises a mixture of some 75% **dunite** and 25% **basalt**. The crystal forms, rather than total composition, change as a function of increased pressure, with the silicates eventually forming close-packed oxides at depths greater than 700–800 km. The uppermost few kilometres of mantle beneath the **oceanic crust** are anisotropic to **seismic wave** propagation, indicating that **olivine** crystals within them are oriented relative to the orientation of the *oceanic ridge*. Beneath all oceans, **seismic waves**, particularly *S* waves, are slowed and their energy reduced as they pass through the uppermost part of the mantle, mostly between 60 and 160 km down to about 250 km, thereby form-

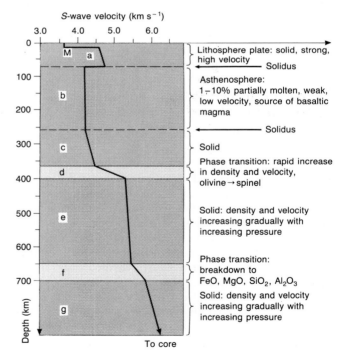

Fig. M2 Mantle. A modern view of the structure of the outermost 700 km of the Earth illustrated by a plot of S-wave velocity against depth. Note how changes in velocity mark the important zones: lithosphere, partially molten asthenosphere, transitions to more dense molecular structures. (After Press & Siever, 1982.)

ing the **low velocity zone**, the thickness of which depends approximately on the square root of the age of the overlying oceanic **lithosphere**. However, the **low velocity zone** is absent beneath most, probably all, of the continental **cratonic** shields and is suppressed beneath old continental *orogenic belts*. Similar variations in **electrical conductivity** at this level also confirm the marked lateral inhomogeneity of the mantle at these depths. Indeed, the entire mantle above 700 km depth has a very complex structure, with numerous vertical and lateral density discontinuities as indicated by changes in **seismic velocities**, and is commonly divided into an **upper mantle**, between the **Mohorovičić discontinuity** and a depth of some 350–370 km, and a deeper **transition zone** between 350 and 700 km (although the term, **upper mantle**, is sometimes used to include the **transition zone**). The strongest vertical discontinuities in seismic velocities usually occur within a zone only some 2 to 20 km thick and are thought to represent increases in average density due to the collapse of lower density crystalline silicate structures to higher density forms, with little or no change in the average chemical composition of the mantle rocks. The most important of these transitions are attributed to **olivine** and **pyroxene** structures compressing to **garnet** and **spinel** forms, with an increased density of *c.* 9% at about 250 km depth, and the **garnet** and **spinel** structures compressing to **perovskite** structures at about 550 km, with a 7.5% increase in density. Below about 700 km the mantle has a particularly uniform velocity structure both laterally and vertically and is defined as the **lower mantle**. This zone is believed to comprise silicate oxides that gradually increase velocity with depth, reflecting the increased compression of the oxides by the overlying rocks (self-compression) without significant change in their composition or structure. Detailed analyses of seismic **arrival times** suggest that there may be a slight increase in the concentration of **iron oxide** with depth and even more sophisticated studies of **delay times**, *seismic tomography*, indicate minor lateral variations in temperature that appear to be continuous between the **upper** and **lower mantle** and may reflect **convection** current redistribution of heat within the entire mantle. At the base of the **lower mantle**, there is a zone some 200 km thick immediately overlying the **core** in which there appear to be significant variations in seismic velocity but the causes of such variations are entirely speculative at the moment. It is similarly disputed whether **convective** motions within the mantle are continuous throughout the mantle, as indicated by its uniform **viscosity**, below the **lithosphere**, of 10^{21-22} Pa s, or two or more levels of partially linked **convective** activity, mainly in the **upper** and lower **mantle**, which have been suggested to account for geochemical variations in mantle-derived volcanic rocks. [DHT]

Anderson, D.L., Sammis, C. & Jordan, T. (1971) Composition and evolution of the mantle and core. *Science* **171**: 1103–12.
Bott, M.H.P. (1982) *The Interior of the Earth*. Arnold, London.
McElhinny, M.W. (ed.) (1979) *The Earth: Its Origin, Structure, and Evolution*. Academic Press, London.
Pollack, H.N. & Chapman, D.S. (1977) On the regional variation of heat flow, geotherms, and the thickness of the lithosphere. *Tectonophysics* **38**: 279–96.
Press, F. & Siever, R. (1982) *Earth* (3rd edn). W.H. Freeman, San Francisco.
Ringwood, A.E. (1975) *Composition and Petrology of the Earth's Mantle*. McGraw-Hill, New York.

mantle drag If either the **asthenospheric** part of the **mantle** is moving faster than the **lithosphere**, or vice versa, then there is either a positive or negative frictional drag between them, depending on the **shearing stress** and **viscosity** of the **mantle** at the boundary. Some tectonic models assume that the **lithosphere** must be mechanically decoupled from the **asthenosphere** because the boundary marks the level at which **mantle** *partial melting* occurs. In this model, **mantle** motions are caused by the **flow** required to compensate for the destruction of **lithosphere** at **subduction zones** and the creation of new **lithosphere** at *oceanic ridges*. In the **Orowan–Elsasser** model, the **ridge** 'push' and **subduction suction** forces on the **lithospheric** *plates* simply decrease with depth while other models, usually involving whole **mantle convection**, consider that **mantle** motions can be transmitted across the **asthenosphere–lithosphere** interface and that these drive the tectonic *plates*. This model requires that the top of the seismic **low velocity zone** is not a zone of decoupling, and the presence of fluids is not due to *partial melting* at this level. (See **forces acting on plates**.) [DHT]

Bott, M.H.P. (1982) *The Interior of the Earth*. Arnold, London.
Davies, P.A. & Runcorn, S.K. (eds) (1980) *Mechanisms of Continental Drift and Plate Tectonics*. Academic Press, London.

mantle transition zone The location in the **mantle**, approximately between 370 and 700 km, within which several sharp increases in seismic velocity occur. Each increase takes place over some 10–20 km, and results from a rapid increase in density with depth, attributed to phase transitions in the constituent silicates and oxides without major change in the overall chemical composition. (See **mantle**.) [DHT]

mantled gneiss dome Type of **structure** first described by Eskola (1949) from the **Precambrian** of Finland. The **structure** comprises a dome-shaped core of **granitic gneisses** overlain by a supracrustal **mantle**, or cover succession, of metasedimentary and metavolcanic rocks. The rocks of the core typically show igneous textures in the central parts of the dome, but are deformed near the core–cover interface, showing a strong **foliation** parallel to the interface. The surface of the domes varies in attitude from shallowly to steeply outward-dipping, to overturned in some cases. Many such **structures** have been interpreted as deformed **unconformities**, and there are *manto*

examples of unconformable relationships at the base of the cover sequences. There is also evidence in many cases of intrusion of core material into the cover. No single explanation will satisfy all the evidence, but it is now widely believed that the explanation for many of the mantled gneiss domes, particularly in **granite–greenstone terrains**, is the gravitational instability of less dense **granitic** basement when overlain by a thick supracrustal sequence dominated by mafic to ultramafic volcanic rock. The **granitic** substratum rises through the cover as a solid **diapir**, and may form a range of structures from simple **domes** and stocks to overturned mushroom shapes.

Evidence of intrusion of core **igneous rocks** in many examples may relate either to pre-doming emplacement of igneous sheets, or to syn- or post-doming **magmatic** activity. [RGP]

Eskola, P. (1949) The problem of mantled gneiss domes: William Smith lecture. *Quarterly Journal of the Geological Society, London* **104**: 461–76.

Gorman, B.E., Pearce, T.H. & Birkett, T.C. (1978) On the structure of Archaean greenstone belts. *Precambrian Research* **6**: 23–41.

Schwerdtner, W.M., Stone, D., Osadetz, K., Morgan, J. & Stott, G.M. (1979) Granitoid complexes and the Archaean tectonic record in the southern part of northwestern Ontario. *Canadian Journal of Earth Science* **16**: 1956–77.

manto A horizontal or subhorizontal tubular **orebody**. A misuse of the Spanish which means a sheet, blanket, etc., and the term is sometimes used to describe tabular **orebodies**. [AME]

mapping surface/openpit mines and quarries Mapping in this environment enables a progressive three-dimensional picture of the geology to be constructed since there is continuous exposure with limited debris. Fresh new exposures are constantly being produced with only those of the higher levels being **weathered**.

The basic purpose of this mapping is to record and describe the geology, to provide data for **ore grade** calculations and quality control and to enable **fracture** patterns to be determined in order to facilitate geotechnical studies, all of which are used in planning future development of the mine. There are some drawbacks, mainly in regard to **fracture** pattern analysis, since there are many blast-induced **fractures** present which must be distinguished from natural **fractures**. Rock textures and structures are not emphasized as is often the case with **weathered** rocks.

Location is little problem in this environment since there are permanent survey stations and well-surveyed borehole

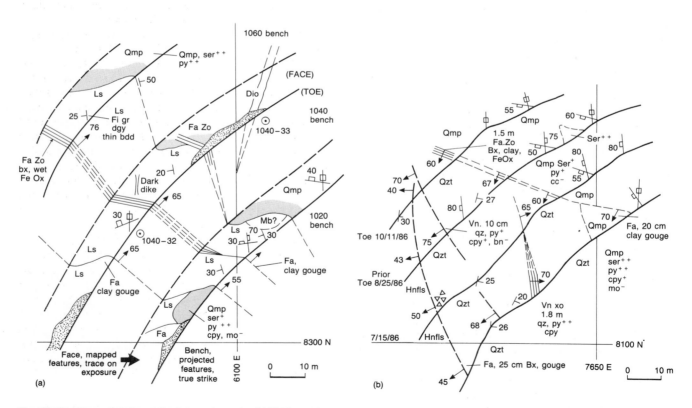

Fig. M3 Mapping surface/openpit mines and quarries (a) Field note sheet with openpit mine geology recorded by the face method. (b) Field note posting sheet, with openpit mine geology recorded on one level by the toe method. (After Peters, 1987.)

positions which provide a very good basis for tape and compass or plane table surveys. Whilst there is always a team of surveyors available at the mine who have the skills to plot locations accurately for the geologist, there may not be time to call them in before critical exposures are destroyed by normal production. Consequently **geological mapping** often has to be executed with speed and without specialist survey assistance. A problem can arise for compass work in pits since there is often a lot of metallic plant and even the deposit itself may deflect the compass needle. Under these circumstances angles between lines may be a more accurate way of recording positions rather than a compass bearing.

Aerial photographs taken at regular intervals are often used to monitor the development of the pit, as are ground photographs of working faces taken with a telephoto lens from standard stations. Both of these are of more use if stereo-pairs are taken.

The scale of **geological maps** of open pits is normally quite large. For example, Bingham Copper Mine in Utah is mapped at a scale of 1:1200 or 1:600 (1 inch to 100 feet and 1 inch to 500 feet respectively). It is quite common for imperial measurements to be used in mines in the United States at the present day and certainly any reference to older maps of the US or UK will be of this nature. A very common scale used by surveyors in the past was 2 chains to the inch (1:528). Where the metric system is used, scales of 1:2500, 1:2000, 1:1000 or 1:500 are the most common for both plans and sections. It is completely impractical to change from imperial to metric scale during the working of a mine since old plans and new plans would cease to be compatible.

Geological data are plotted by either the *face method* or the *toe method*. In the *face method* all geological data on the face are plotted in plan view or at a reference datum measured accurately near the *toe* (base of working face) and those of higher levels on the face are estimated. In a multi-bench open pit this results in a step-like appearance of the geology and gives a good three-dimensional impression (Fig. M3a). The only true **strike** directions of geological contacts are present on the horizontal benches, the rest being **apparent dip** and **strike**.

In the *toe method* (it is difficult and dangerous to map the *crest*, which is the top of the working face) the base of the face is mapped as a line and mine level plans show a succession of *toe* positions. It is in fact a version of detailed line-mapping on a horizontal plane with excellent control for interpolation between lines. All features in a face are recorded (and sometimes sketched) and projected down to the reference level, the final map representing a single horizontal plane on which all contacts are shown with their true strike direction (Fig. M3b). This method is ideal for single-bench operations where the deposit is fairly flat lying and most of the mineralization occurs on one bench. However, if the operation is multi-bench it is necessary to produce a separate plan for each level.

In many mines both methods are used. Pit mapping may be carried out by the *face method* with all working plans being of this type. The data collected are then transferred to a series of level-by-level *toe* plans and to a series of cross-sections. Some mine survey departments produce simplified plans of the salient features of the deposit on transparent material such as Perspex or Plexiglas and assemble these one above the other at the correct interval to give a 'see-through' model of the deposit. This is a particularly good way of enabling the complete three-dimensional nature of the mine to be appreciated and of conveying geological information to the non-geologist. [DER]

Peters, W.C. (1987) *Exploration and Mining Geology* (2nd edn). John Wiley & Sons, New York.

marble A metamorphosed **limestone** formed by recrystallization under conditions of thermal or regional metamorphism. Marbles derived from pure **limestones** are composed of recrystallized **calcite** but the presence of other minerals (e.g. **quartz** and *phyllosilicates*) may cause the formation of **dolomite** marble and calc-silicate **schists** and gneisses, which may contain **quartz**, **diopside**, **tremolite**, **talc**, **wollastonite-plagioclase**, **forsterite**, **idocrase** and **grossular–andradite garnet** in addition to **calcite** and **dolomite**. Marble may form an attractive **building**/sculptural **stone** although it is important to note that some stonemasons apply the term marble to rocks that may be easily polished, including particularly attractive, but unmetamorphosed, **limestones**.

Marble has been very widely used by man for buildings, statues and decoration. It provided temples and statues in India (the Taj Mahal is built mainly of marble), and was used for building in Russia, Europe, China, Morocco, the Sudan, South Africa and in South America. In Britain **limestones** sometimes erroneously referred to as marbles, including the Sussex and Derbyshire 'marbles' were used extensively by the Romans, while both Romans and later the Saxons made use of the **Jurassic** Purbeck 'marble'. The most famous use of marbles is in the Classical Greek and Roman temples and statues of the Mediterranean. The quarries of high-quality marble in Greece, Italy, the Aegean islands and in Asia Minor were widely exploited from the 6th and 5th centuries BC onwards. One of the best known is the Pentelic marble quarry near Athens which provided the building material for the Parthenon.

Much work has been done on the relating of marble buildings and statues from the Mediterranean area to their sources, to judge the extent of marble trade. Petrographic and **chemical analyses** proved insufficiently discriminating between different marble quarries, but more recent work on the isotopic composition (particularly **carbon**, **oxygen** and **strontium** isotopes) of marbles has been more successful in this characterization. [RST/OWT]

marcasite (FeS_2) A relatively common sulphide mineral formed at low temperatures. [DJV]

March analysis Analysis of the behaviour of passive markers in an homogeneous body deforming by *viscous flow*. The analysis predicts that elongate markers initially parallel to the axis of **shortening** will rotate symmetrically away from this direction to form a bimodal distribution of **preferred orientations** in any plane with the angle between the **shortening** direction and the **preferred orientation** increasing with **strain**. It has been suggested that *phyllosilicates* may be rotated passively during **deformation** according to March analysis to produce the **preferred orientation** of *slaty cleavage*. [TGB]

margarite ($CaAl_2(Al_2Si_2O_{10})(OH)_2$) A brittle **mica mineral**. [DJV]

marialite ($Na_4(AlSi_3O_8)_3(Cl_2,CO_3,SO_4)$)   A metamorphic mineral. (See **scapolite series**.) [DJV]

marine snow A general term for the discarded mucus feeding sheets or strands of gelatinous zooplankton such as *pteropods*, *salps* and *appendicularians*. These delicate mucus structures act as traps for both organic and inorganic suspended particulate matter. They are periodically discarded in response to clogging, disturbance or predation, and settle. Marine snow is responsible for transporting fine particulate matter to the ocean floor and may be locally a more important agent in this process than the fecal pellets of planktonic organisms (Asper, 1987). Because of its delicate structure and composition, marine snow has no preservation potential and its widespread occurrence has only recently been recognized with the introduction of more refined sampling techniques. [AESK]

Aldridge, A.L. (1976) Field behavior and adaptative strategies of appendicularians (cordata: tunicata). *Marine Biology* **38**: 29–39.
Asper, V.L. (1987) Measuring the flux and sinking speed of marine snow aggregates. *Deep-Sea Research* **34**: 1–17.

martite A name given to the mineral **hematite** or an intergrowth of **hematite** and **magnetite** replacing **magnetite** along octahedral **cleavage** planes. [DJV]

mass deficiency The difference in mass between a body of relatively low density and the higher density surrounding rock that would otherwise occupy its volume. (See **Gauss' Theorem**.) [PK]

mass movement types Most classifications are based on the work of Sharpe (1938). His basic division was between 'slide' and 'flow'. Slow flows include *solifluction* and *soil creep*, whilst more rapid flows include *debris avalanches*, *earth flows* and *mud flows*. Slides include *rock falls*, *rock slides*, and *planar* and *rotational slumps*. 'Heave' processes, such as *frost heave*, *freeze–thaw* movements and **cambering**, are also included in a number of classificatory schemes describing mass movement types. [TPB]

Sharpe, C.F.S. (1938) *Landslides and Related Phenomena*. Columbia, New York.

mass solute transfer The transport of dissolved ionic or molecular species from reaction sites within donor sediments where **dissolution** takes place (solute source) to reaction sites in acceptor sediments where **precipitation** takes place (solute sinks). Transport is usually inferred to take place on a scale larger than individual sediment packages and is driven through permeable conduits by either a *thermobaric flow* out of a compacting **basin** or by a hydrologic **head** during **meteoric water** ingress. On a much smaller, local scale, diffusive mass solute transfer has also been suggested for several specific **diagenetic** reactions (e.g. see **pressure dissolution**). However, in most **diagenetic** systems it is extremely difficult to quantify the extent to which mass solute transfer has occurred in any given setting and to identify the transfer mechanism. Mass solute transfer is an extremely controversial subject and remains one of the unsolved enigmatic problems of **diagenesis**.

There are several trends within many deep burial sequences and associated lines of circumstantial evidence that suggest mass solute transfer occurs between **mudrocks** and **sandstones** during **burial diagenesis**. Some of the most convincing evidence is provided by mineralogical and chemical documentation that the **smectite** to **illite** reaction in **mudrocks** during burial, which involves large-scale liberation of (sodium, calcium) (**magnesium, iron**) and silicon together with dehydration water, is spatially and temporally associated with the addition of ions to interbedded **sandstones** in the form of **quartz** and carbonate cements. This interpretation is supported by the comparison of bulk chemical analyses of shallow and deeply buried **sandstone–mudrock** pairs which document a loss of **iron** and **potassium** (normalized against Al_2O_3, assumed immobile) from **mudrocks** and a corresponding gain with depth in associated **sandstones**. Furthermore, the light isotopic character of **carbon** in late carbonate cements typical of many deep **sandstones** is widely considered to indicate the mass transfer of the products of thermal **decarboxylation** of organic matter from the adjacent **mudrocks** (see **organic matter diagenesis**). Where exotic cements (such as **barite**, sulphates, **ankerite**, etc.) are abundant or diagenetic sulphide mineralization is developed in **sandstones** with no obvious detrital source, some mass solute transfer is clearly required. On a more tenuous argument, the common observation of either leaching or **cementation** in **sandstones** juxtaposed with thick **mudrock** sequences may also provide evidence of mass solute transfer from **mudrocks**. The development of **overpressuring** in **sandstones** clearly indicates water migration into porous lithologies from enclosing **mudrocks** whilst the very existence of *hydrocarbon reservoirs* is proof

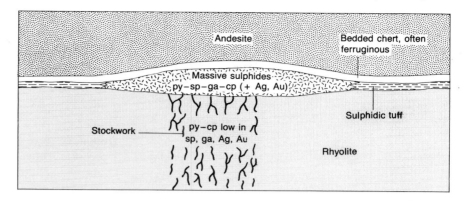

Fig. M4 Schematic cross-section through an idealized volcanic-associated **massive sulphide deposit** showing the underlying feeder stockwork and typical mineralogy. py = pyrite; sp = sphalerite; ga = galena; cp = chalcopyrite. (After Evans, 1993.)

that fluids migrate from **mudrocks** to **sandstones** in the subsurface.

To précis the opposition to mass solute transfer between **mudrocks** and **sandstones** during burial, several authors have correctly pointed out that the amount of water available from compaction and dehydration in a subsiding **basin** is finite, and probably insufficient to allow significant mass transfer of low solubility species, particularly with respect to **quartz**. Moreover, it has recently become apparent that many of the ions released from burial **clay mineral** reactions and **organic matter diagenesis** in **mudrocks** are reprecipitated locally within the source **mudrocks** as replacive and pore-filling cements. The enigma remains. [SDB]

Bjorlykke, K. (1980) Cementation of sandstones. *Journal of Sedimentary Petrology* **49**: 1358–9.
Curtis, C.D. (1978) Possible links between sandstone diagenesis and depth-related geochemical reactions occurring in enclosing mudstones. *Journal of the Geological Society, London* **135**: 107–17.

mass spectrometry Method for determining the ratio and/or concentration of isotopes within materials including rocks and stone artefacts. Procedures depend on the element to be analysed and the sample may be taken into solution and deposited on a filament, or may be in the form of a gas. In a conventional mass spectrometer atoms of the sample are ionized and ions accelerated by a potential difference (voltage); a **magnetic field** is applied perpendicular to the direction of motion of the ions, which are thus constrained to move in an arc, whose curvature is dependent partly on the ion mass. By varying the voltage applied, ions of different mass may be detected in turn by a detector and the relative proportion of isotopes present measured. The method is important particularly in isotope determinations for **potassium–argon dating** and in **accelerator radiocarbon dating**, and in archeological applications of **oxygen** and **strontium isotope analyses**. [OWT]

massive sulphide deposit A large, generally **stratiform**, conformable **orebody** composed dominantly of iron

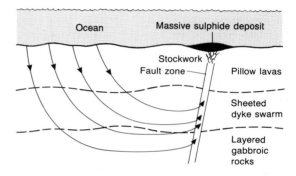

Fig. M5 Diagram to show how seawater circulation through oceanic crust might give rise to the formation of an exhalative **massive sulphide deposit**. (After Evans, 1993.)

sulphide. These deposits are now becoming generally known as *volcanic-associated massive sulphide deposits*. They often consist of over 90% iron sulphide usually as **pyrite** ± **pyrrhotite** in lenticular to sheetlike **orebodies** at the interfaces between volcanic units or at volcanic–sedimentary interfaces. Beneath them there is usually a **stockwork** that may be **ore grade** and which appears to have been the feeder channel up which mineralizing fluids penetrated to form the overlying massive sulphide deposit (Fig. M4). These **orebodies** commonly occur in groups and in any one area they are found at one or just a few stratigraphical horizons. On economic grounds they may be divided into three classes: (a) Zn–Pb–Cu, (b) Zn–Cu, (c) Cu. There may also be **gold** and **silver** values and the **pyrite** is frequently used for sulphuric acid production. A few deposits are worked for the **pyrite** alone. The major minerals are **pyrite**, (**pyrrhotite**), **sphalerite**, **galena**, **chalcopyrite**. The **gangue** is principally **quartz**. Some deposits have considerable developments of **barite** or **gypsum**. Using data such as geological occurrence, geochemistry, etc., they are divided into **Besshi**, **Cyprus**, **Kuroko** and *Primitive* types.

Once thought to be **epigenetic** in origin, they are now considered to be **syngenetic** volcanic-exhalative deposits. Deposits of this sort are now forming around *hydrothermal* *vents* (**black smokers**) along various **oceanic ridge** systems such as the East Pacific Rise and the Mid-Atlantic Ridge. They show an intimate relationship with the volcanism and are generally thought to have formed from circulating seawater that becomes a concentrated **brine** at depth, dissolving **copper** and other metals from the rocks it traverses and carrying these up to the surface where they are **precipitated** as sulphides with **sulphur** derived from the seawater, see Fig. M5. [AME]

Edwards, R. & Atkinson, K. (1986) *Ore Deposit Geology*. Chapman & Hall, London.
Evans, A.M. (1993) *Ore Geology and Industrial Minerals: An Introduction* (3rd edn). Blackwell Scientific Publications, Oxford.

material (adj.) Relating to matter or substance. In a geological context, 'material' is used as a noun for a substance, normally rock: thus, 'siliceous material'. The term is also used as an adjective to describe properties or characteristics relating to rock as a substance: thus 'material properties'. In the study of **strain**, a *'material line'* is a line joining fixed points in a body of rock material, to distinguish it from a geometrical construction such as a *fold axis*, or *strain axis*, which may vary in relation to material points or lines in a rock. [RGP]

Mayen lava Lavas of the Mayen region in the Eifel area of western Germany are one of the most famous sources of milling rocks in Europe. The highly vesicular grey **nepheline tephrite** was quarried for **millstones** apparently

continously from the Hallstatt Period (Bronze Age to Iron Age) until the last century. Mayen **millstones** were traded widely in north-west Europe from the Iron Age onwards and were termed *'Cullen'* or *'Koln' stones* (derived from the nearby town of Cologne) in recent centuries. The lava is sometimes referred to as *'Niedermendig'* or *'Andernach' lava*, after other nearby towns. [OWT]

meander belt The area of land a meandering channel occupies between successive **avulsions**. As individual meanders migrate, the channel position within the channel belt changes such that **channel deposits** are found over a greater area than occupied by the channel at any one time. Progressive lateral migration of a channel may lead to the formation of an asymmetrical meander belt of which only part is active. [JA]

Schumm, S.A. (1977) *The Fluvial System*. John Wiley & Sons, New York.

meandering stream A single channel stream with planform **sinuosity** (channel length divided by channel reach length) greater than 1.3 (Leopold & Wolman, 1957). The choice of 1.3 as the class boundary is arbitrary and not related to a natural threshold. Meandering channels migrate by selective channel bank **erosion** and **point bar** deposition, meander cutoff and **avulsion**. Meander amplitudes range from tens to hundreds of metres controlled by a combination of factors of which slope and

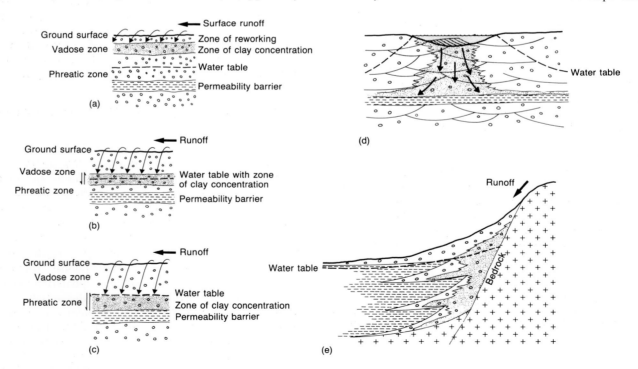

Fig. M6 Mechanical infiltration of fines. Processes resulting in the concentration of mechanically infiltrated fines. (a) Concentration in vadose zone. (b) Concentration about fluctuating water table. (c) Concentration above impermeable barriers. (d) Concentration beneath active fluvial channels. (e) Concentration in marginal conglomerate adjacent to source of influent seepage. (After Walker, 1976.)

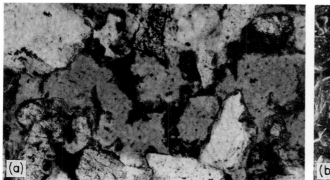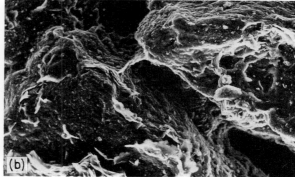

Fig. M7 Mechanical infiltration of fines. (a) Thin section photomicrograph showing development of clay coats around detrital quartz grains. Note the pore-bridging and geopetal fabrics. (b) Scanning electron microscopy of mechanically infiltrated clay showing the tangentially orientated clay coatings to detrital grains and pore-bridging fabric.

discharge are of major importance. Meandering streams occur in alluvial, **tidal** and submarine environments. [JA]

Leopold, L.B. & Wolman, M.G. (1957) *River Channel Patterns — Braided, Meandering and Straight*. Professional Paper of the United States Geological Survey **282B**.

mechanical infiltration of fines The introduction of clay and silt grade material into coarse-grained, porous and permeable sediments where surficial influent **seepage** accompanies *alluviation* in areas where **water tables** are low. Clay and silt are not normally deposited with high-energy alluvium but remain as **suspension** load. However, during waning flow, water infiltrates into the **vadose zone** of the surficial sediment above the **water table**. The infiltrated fines are deposited within the interstitial pores of the sediment as the influent flow velocity is reduced by either permeability barriers within the sediment or upon attainment of the **water table** level (Fig. M6). As coarse-grained alluvium is typically deposited clay-free, the process is extremely important for producing secondary matrix in older proximal continental clastic sediments where clay may amount to 25% of the sediment. Mechanically infiltrated clay is usually concentrated in the surficial layers of sediment where it forms **geopetal textures**, occurring on the upper surface of pebbles, filling larger interstices and bridging across narrow *pore throats* (Fig. M7). It characteristically displays a clastic **fabric** and develops tangentially orientated clay coatings to detrital grains that are readily distinguished from true **authigenic** clay precipitates. (See **depositional environment related diagenesis, neoformation of clays, vadose zone**.) [SDB]

Walker, T.R. (1976) Diagenetic origin of continental Red Beds. In: Falke, H. (ed.) *The Continental Permian in Central West and South Europe*, pp. 240–82. NATO Advanced Study Institute, Series C No. 22. D. Reidel Publishing Co., Dordrecht, Holland.

megaturbidite *(seismoturbidite)* Very thick-bedded carbonate/siliciclastic *turbidite* (>1 m thick), for example the **Miocene** Contessa Bed in the Italian Apennines that

reaches approximately 15 m in thickness. Typically, megaturbidites are laterally continuous over entire **basins**, and may show internal **cross-stratification**, or *ripple-lamination*, that indicates ponding of the **turbidity current** in the **basin**, resulting in flow reflection from the **basin** slopes. Megaturbidites may show a distinct lower **sand**-rich division capped by a relatively thick mud-rich division formed by rapid deposition from a dense mud suspension. These beds provide potentially useful correlation or *marker beds* for erecting a high-resolution *lithostratigraphy* in a **basin**. The origin of megaturbidites involves enormous volumes, of the order of 10–100 km^3, of sediment **failure** from **basin** margins. If seismically triggered mass failure produces such beds, they are referred to as 'seismoturbidites', although since beds of any thickness may have the same origin, this latter term should be used only where the genesis of a *turbidite* is known. [KTP]

meionite (Ca$_4$(Al$_2$Si$_2$O$_8$)$_3$(Cl$_2$CO$_3$SO$_4$) A metamorphic mineral. (See **scapolite series**.) [DJV]

mélange Rock body composed of disrupted fragments, of metre to kilometre in scale, of pre-existing continuous rock bodies; e.g. composed of blocks of **competent** strata in a matrix of finer-grained material. There are two distinct types of mélange, tectonic and sedimentary. A sedimentary mélange is termed an **olistostrome**, and is produced by sedimentary processes such as *gravity* slumping or *sliding*. A tectonic mélange is produced by tectonic **deformation**; i.e. the disruption and mechanical mixing of pieces of older rock as a result of post-sedimentary **deformation**. There is clearly some overlap in terminology with *'fault breccia'*, but the blocks in a mélange may be very much larger than in a typical *fault breccia*. Many examples of tectonic mélange sequences described in the literature have subsequently been shown to be deformed **olistostromes**.

The term was first used by Greenly (1919) for the

Gwna group in Anglesey, North Wales, and was subsequently applied to the Ankara mélange in Turkey (Bailey & McCallien, 1950) and the Franciscan mélanges in California (Bailey *et al.*, 1964). Sedimentary mélanges are considered to be typical of *ocean trench* sedimentation, and when such units subsequently become involved in **underthrusting** at an **active margin**, the distinction between sedimentary and tectonic processes becomes difficult to maintain. It has been suggested that the Franciscan mélanges are the product of the **underthrusting** process itself. *Offscraping* at the toe of an **accretionary prism** is likely to generate sequences that are partly tectonic and partly sedimentary in origin. [RGP]

Bailey, E.B. & McCallien, W.J. (1961) Structure of the northern Apennines. *Nature* **191**: 1136–7.
Bailey, E.H., Irwin, W.P. & Jones, D.L. (1964) Franciscan and related rocks, and their significance in the geology of western California. *Californian Division of Mines and Geology Bulletin* **183**.
Greenly, E. (1919) *The Geology of Anglesey*. Geological Survey of England and Wales Memoir **1**.
Scholl, D.W., von Huene, R., Vallier, T.L. & Howell, D.G. (1980) Sedimentary masses and concepts about tectonic processes at underthrust ocean margins. *Geology* **8**: 564–8.

melanite ($Ca_3Fe_2(SiO_4)_3$) A black variety of the mineral **andradite**. (See **garnet minerals**.) [DJV]

melanterite ($FeSO_4 . 7H_2O$) A green-blue secondary **iron** mineral. [DJV]

membrane polarization (also known as **electrolytic polarization**) An **induced polarization** effect which occurs in many rocks, irrespective of the presence of metallic minerals, owing to constrictions in the pore spaces through which ionic conduction takes place. Charges build up on the walls of the pores, and where the walls narrow, the flow of ions is inhibited, resulting in an accumulation of charge. Once the imposed current is switched off, the charge gradually disperses, producing a measurable transient potential difference. **Clay minerals** are often associated with strong membrane polarization effects. (See **induced polarization**.) [ATB]

Keller, G.V. & Frischknecht, F.C. (1966) *Electrical Methods in Geophysical Prospecting*. Pergamon, Oxford.

membrane tectonics As **lithospheric** *plates* are thin, rigid sheets, relative to the size of the Earth, they will be distorted by the actual figure of the Earth, particularly the equatorial bulge, as they move over its surface. This results in tensional forces on the edges of the *plate*, and compressional forces at its centre, as it passes from a short to long Earth radius area (the Benue Trough in Nigeria has been attributed to this mechanism), and in compressional forces when a *plate* moves from long to short radii regions. [DHT]

Turcotte, D.L. (1974) Membrane tectonics. *Geophysical Journal of the Royal Astronomical Society* **36**: 33–42.

meneghinite ($CuPb_{13}Sb_7S_{24}$) A rare **ore** mineral of **copper** and **lead**. [DJV]

mercury (Hg) A naturally occurring liquid metal known colloquially as '*quicksilver*'. (See **native elements**.) [DJV]

mere A small lake, the origin of which is often unclear, that occurs on *glacial* outwash deposits and other superficial materials in areas like the Cheshire Plain and the Breckland of East Anglia in England. Some may be relict *thermokarst* features, some may be the result of **solution** of underlying rocks (e.g. **chalk**), but others may have other origins. [ASG]

merokarst A **limestone** landscape characterized by only partial development of typical **karst** landform assemblages. Cvijic used this term to describe the inland part of the Yugoslavian **karst** region where the **limestone** beds are thinner and fewer **karst** landforms are developed than in the coastal zone, which he called **holokarst**. [HAV]

Cvijic, J. (1893) Das Karstphänomen. *Geographische Abhandlung* **5**: 217–329.
Cvijic, J. (1925) Types morphologiques des terrains calcaires. Le merokarst. *Comptes rendus de l'Académie des Sciences* **180**: 757–8.

meromictic The status of a lake which is permanently stratified with a well-developed **epilimnion** and **hypolimnion**. Atmospheric disturbances such as wind **stress** and seasonal heating are insufficient to break down the temperature (and occasionally chemical) stratification of the water mass, so that only the surface layers undergo seasonal mixing. (See **holomictic**.) [PAA]

Hutchinson, G.E. & Löffler, H. (1956) The thermal classification of lakes. *Proceedings of the National Academy of Sciences, Washington, DC* **42**: 84–6.

mesa A steep-sided, flat-topped **plateau** or promontory of rock surrounded by a flat **erosional** plain. Large mesas may be called *table mountains* (Spanish *mesa*, 'table'); small mesas are called **buttes**. They may form as a result of slope **erosion** by **backwearing** (**parallel retreat**) or may be islands of relatively soft rock protected from **erosion** by a capping of harder material such as a **cuirasse**. [AW]

mesoscopic Describes structures or other features of intermediate size measured in metres. Mesoscopic features are seen on a cliff section (500 m), in outcrop (10 m), or in a hand specimen (0.1 m). Sediment channels, **ripples**, *tension gashes*, and some **foliations** are mesoscopic; observations at this scale provide useful information about macroscopic features, and form a framework for **microscopic** analysis. [SB]

mesosphere The name, now largely discontinued, for a layer immediately below the **lithosphere** and broadly equivalent to the seismic **low velocity zone** in the Earth's **mantle**, i.e. equivalent to the **upper mantle**, but excluding the **transition zone**. It has also been used for the **mantle** zone immediately below the **asthenosphere**. [DHT]

mesothermal deposit Epigenetic mineral **deposit** intermediate between **epithermal** and **hypothermal deposits**. Such deposits were formed at 200–300°C at depths of 1200–4500 m. Generally found in or near intrusive **igneous rocks**, they consist of **fracture**-filled and **replacement bodies** and are important as **ores** of **gold, silver, copper, lead, zinc, tungsten**, etc. **Orebody zoning** is well developed and the deposits have a good vertical range. [AME]

Mesozoic The collective term for the **Triassic** to **Cretaceous**. [AH]

metacinnabar ($Hg_{1-x}S$) The high-temperature form (>344°C) of the mineral **cinnabar**. [DJV]

metal factor parameter An empirical parameter measured in **induced polarization** methods indicating the presence of certain types of mineralization. Both **chargeability** and **percentage frequency effect** vary with current density and hence **apparent resistivity** of the *host rock*. The metal factor parameter is introduced in an attempt to normalize these parameters by dividing them by the d.c. **resistivity** so that the effect of variations in **resistivity** are eliminated to some extent. For frequency effect the metal factor is defined as

$$\frac{2\pi 10^5 (\rho_0 - \rho_\infty)}{\rho_\infty \rho_0}$$

where ρ_0 and ρ_∞ are the **apparent resistivities** at d.c. and infinite frequency. [ATB]

Telford, W.M., Geldart, L.P., Sheriff, R.E. & Keys, D.A. (1976) *Applied Geophysics*. Cambridge University Press, Cambridge.

metallogenic provinces and epochs Regions of the **crust** characterized by relatively abundant mineralization of one type are called metallogenic provinces. When certain types of mineralization are formed over large areas during definite periods in the Earth's history, then these are known as metallogenic epochs.

Metallogenic provinces can be delineated by reference to a single metal or to several metals or metal associations. In the latter case the province may show a zonal distribution of the various metallic deposits as in the Variscan **orogenic** belt of north-western Europe and its northern **foreland**, Fig. M8. In this province the principal epochs of **epigenetic** mineralization were *Hercynian* (end **Carboniferous** to early **Permian**) and *Saxonian* (middle **Triassic** to **Jurassic**).

Tin deposits are an excellent example of an element restricted almost entirely from the economic point of view to a few metallogenic provinces, of which the most important is the so-called *tin girdle* of south-east Asia. Even more striking is the fact that most **tin** mineralization is post-**Precambrian** and confined to certain well marked

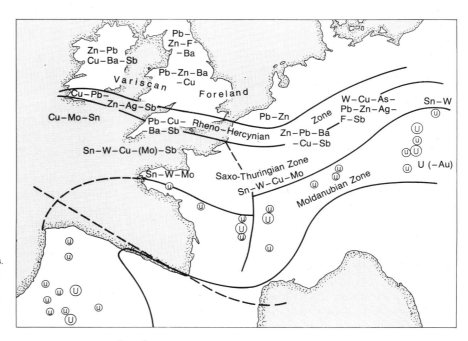

Fig. M8 Metallogenic provinces and epochs. Regional metal zonation of epigenetic deposits in the Variscan Metallogenic Province of north-west Europe. Sizes of symbols in Moldanubian Zone and Spain indicate relative sizes of uranium deposits. (After Evans, 1993.)

epochs. Equally striking is the strong association of these deposits with post-tectonic **granites**. Among **tin** deposits of the whole world, 63.1% are associated with **Mesozoic granites**, 18.1% with *Hercynian* (late **Paleozoic**) **granites**, 6.6% with *Caledonian* (mid-**Paleozoic**) **granites** and 3.3% with *Precambrian* **granites**. The reasons for the restricted development of **tin** deposits in space and time are still not clearly understood. [AME]

Evans, A.M. (1993) *Ore Geology and Industrial Minerals: An Introduction* (3rd edn). Blackwell Scientific Publications, Oxford.
Schuiling, R.D. (1967) Tin belts on continents around the Atlantic Ocean. *Economic Geology* **62**: 540–50.

metamorphic aureole (*contact aureole*) Metamorphism caused by emplacement of host **magma** into cooler *host* (or '*country*') *rocks*. Such metamorphism may cause **hydrothermal** and/or chemical changes (**metasomatism**) as a result of heating of **groundwater** within *country rocks* (± magmatic water), as well as recrystallization and development of metamorphic minerals. The metamorphic effects are co-linear around the contact between the igneous body and the *country rocks*, but may show variations as a result of variations in rock type and the subsurface attitude of the contact. **Contact metamorphic** effects may be minor in anhydrous rocks such as **sandstones** but may be more important in sedimentary rocks such as **shales** (**mudrocks**) and impure carbonates.

The aureole width depends upon a range of factors such as intrusion temperature and the *country rock* temperature, composition and volatile content, and may range from centimetres to kilometres in width. Metamorphic aureoles are generally zoned in terms of **fabrics** and mineralogy, reflecting the thermal gradient and *country rock* characteristics noted above. Such zonation reflects the decrease in metamorphic grade away from the igneous contact. [RST]

metamorphic banding Banding resulting from recrystallization of original **fabrics** such as stratification and other sedimentary (and igneous) structures to yield new planar **fabrics** such as **foliation** and **gneissose** banding. In terms of increasing metamorphic grade, metamorphic banding may result from:
1 **Brittle deformation** of rocks in planar **fault** and **shear** zones near to the Earths' surface (see **mylonite**).
2 Development of a **penetrative fabric** by **pressure-solution** processes, mineral reorientation/recrystallization and *ductile flow* (see **foliation**).
3 Solid-state recrystallization (see **gneissose banding**). [RST]

metamorphic differentiation Separation of components or phases of a rock by metamorphic processes; also sometimes termed *metamorphic segregation*. The process of diffusion, which may involve a fluid phase (*solution transfer*) can concentrate components or phases in particular areas, producing a heterogeneous distribution, often in the form of layering. This may be assisted by tectonic **strain**. A common example is the differentiation of **mica** and **quartz** into discrete layers in a **slate** or **phyllite**. (See **pressure solution**.) [TGB]

metapedogenesis Human alteration of a *soil*'s properties. This may result from deliberate actions or may be unintentional; the changes may be beneficial or detrimental. Agricultural practices such as fertilizing the *soil*, preventing **erosion**, and irrigation can be beneficial. However, mismanagement of cultivated land may lead to depletion of nutrients through overgrazing or overcropping, *soil erosion* by wind or water, or salinization as a result of inadequate drainage or the use of saline irrigation water. [AW]

Bidwell, O.W. & Hole, F.D. (1965) Man as a factor of soil formation. *Soil Science* **99**: 65–72.
Yaalon, D.H. & Yaron, B. (1966) Framework for man-made soil changes — an outline of metapedogenesis. *Soil Science* **102**: 272–7.

metasomatism A metamorphic process involving a significant change in the chemical composition of the rock. Although **contact** and regional **metamorphism** involves dehydration as well as recrystallization, metasomatism is generally taken to involve major chemical gains and/or losses, usually as a result of fluid flow through fissures (and/or **porosity**) in rocks around an igneous intrusion. Metasomatic processes include *fenitization* and seawater–rock interaction. [RST]

meteoric diagenesis in carbonates The **diagenesis** caused by rainfall-derived **groundwater** in carbonate rocks. It is important in **limestone** formation because of two factors. Firstly, most carbonate sediments are deposited in very shallow water and are prone to subaerial exposure by minor sea-level falls. Secondly, carbonate sediments contain two forms of calcium carbonate, **aragonite** and **magnesium calcite**, which are unstable in **meteoric waters**, and are replaced by the more stable, and less soluble, low-**magnesium calcite**. **Meteoric waters** act as weak acids and contain dissolved carbon dioxide to form carbonic acid as well as organic acids derived from the *soil* as rainwater percolates into the underlying sediment.

Besides **dissolution** and *stabilization* (mineral transformation) another important process is **cementation** where calcium carbonate is reprecipitated as the stable low-**magnesium** form.

Two broad **alteration** zones can be defined: the **vadose zone** above the **water table** and the **phreatic zone** below it. These two zones have different styles of **dissolution**, *stabilization* and **cementation**. Water drains through the **vadose zone** and is locally held both at *pore throats* by capillary action in meniscate geometries and on the undersides of grains as pendant droplets. In both types each wetting and drying cycle may result in the **precipitation** of minute amounts of **calcite** cement but multiple cycles will create cement zones with meniscus-like or microstalactitic (or gravitational) forms. **Cementation** is generally highly irregular in amount and distribution in the **vadose zone**.

In the **phreatic zone** all pores are filled with water. The upper part of the zone is typically mobile and **cementation** is more uniform in style and extent. The lower parts may be stagnant with slow **alteration** rates. The petrographic and geochemical distinctions between these two zones are sufficiently marked for them to be recognized in **limestones** exposed to meteoric diagenesis as far back as the **Paleozoic**.

The style of **alteration** ultimately depends on the climate, and 'humid' and 'arid' styles have been documented. A number of exposed *Pleistocene* carbonates have been studied and a developmental sequence recognized from unconsolidated, fresh carbonate sediments to lithified, stabilized **limestones**. Complete *stabilization* and partial **cementation** can occur in 100 000 years. [VPW]

James, N.P. & Choquette, P.W. (1984) Diagenesis 9 — limestones — the meteoric diagenetic environment. *Geoscience Canada* **11**: 161–94.

meteoric water Groundwater resulting from rainfall and **infiltration** within the **hydrological cycle**. **Meteoric groundwater** is modified chemically during its passage through rocks and some may become hard, some decidedly saline. (See **hardness of water**.) [AME]

meteoric water diagenesis (*telogenesis*) Diagenesis taking place where the existing *eogenetic* (see **depositional environment related diagenesis**) or *mesogenetic* (see **burial diagenesis**) pore-waters are flushed from the host sediment by the ingress of **meteoric waters** (that is, **groundwaters** that originate from atmospheric precipitation and reach the **phreatic zone** by **infiltration**). **Meteoric waters** are driven by a pressure **head** from the water column above sea-level, usually provided by an effective topographic relief in the **catchment** area. A downward flow is developed perpendicular to equipotential lines and meets **pore-waters** expelled from a compacting **basin** at a **hydrological boundary**. Ingress usually results from one or a combination of three main processes; tectonic **uplift** and **inversion**, **eustatic** sea-level fall or through the seaward progradation of a coastline. Excluded from the *telogenetic* realm are continental systems where the depositional (*eogenetic*) **pore-waters** are **meteoric** in origin. Theoretically, the depth of **meteoric water** penetration below sea-level is in the order of 40 times the *hydraulic head* developed given contrasting seawater and freshwater densities of $1.025\,\mathrm{Mg\,m^{-3}}$ and $1.00\,\mathrm{Mg\,m^{-3}}$ respectively, although in practice the ingress will be less, due to the opposing compactive flow. However, given suitable **hydrogeological** conditions, freshwater may extend for hundreds of kilometres and even flush intracratonic **basins**; freshwater **aquifers**, for example, have been reported 120 km off the west Florida coast in *Tertiary* carbonates.

Sediments exposed to meteoric water diagenesis com-

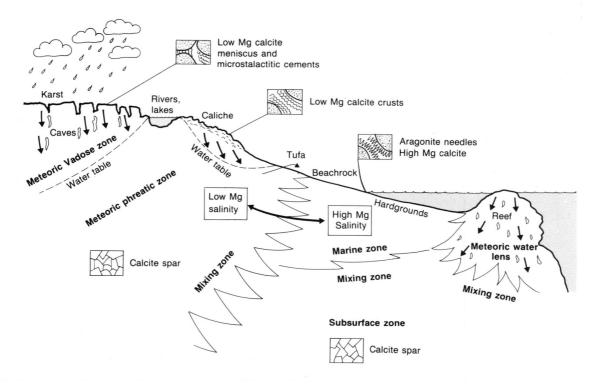

Fig. M9 Meteoric water diagenesis. Meteoric water ingress in Pleistocene limestones with the resultant mixing between marine and subsurface waters. (After Folk, 1974.)

monly display extensive **alteration**. **Meteoric waters** are characteristically very dilute, **oxidizing**, saturated in CO_2 and may be acidic in nature. They are thus highly reactive towards sediments deposited in the marine environment or to those that have experienced deep **burial diagenesis** prior to **uplift** and **inversion**. *Telogenetic* modifications are documented both from a range of rocks being altered at or near the present Earth's surface and from ancient sediments beneath paleo-**unconformities**.

Quaternary glacio-**eustatic** sea-level fluctuations have periodically lowered sea-level and exposed a variety of *Pleistocene* marine carbonates flanking the main oceans to meteoric water diagenesis (Fig. M9). Considerable interest has been devoted to the study of these carbonates. Many of their original components have undergone extensive **dissolution** whilst others have been neomorphically replaced and recrystallized (see **neomorphism**), and the original **fabric** modified by varying types and styles of **cementation**. Much of the **dissolution** is fabric-selective and occurs on a very fine scale. The most soluble crystals are preferentially dissolved creating micropores which eventually coalesce to form larger pores, moulds and *vugs*. The most intense **dissolution** occurs near to the surface in the **vadose zone** where percolating rainwaters, highly undersaturated with respect to **calcite**, dissolve all **carbonate minerals**. In such instances the sediment **fabric** may eventually collapse, leaving behind a *residual breccia*.

Neomorphic transformation of the primary constituents can occur in either the **vadose** or **phreatic zones** although the resulting **fabrics** are often quite distinctive. The replacement of **aragonite** by low-**magnesium calcite** produces **neomorphic fabrics** with detailed preservation of primary ultrastructures only in the **vadose zone**; such relict textures are rare in phreatic **neomorphic** replacements.

Cementation fabrics are also often indicative of either **vadose** or **phreatic zones** (Table M3). In general, meteoric cements comprise blocky, non-ferroan, low-**magnesium** calcitic spar. However, in the **vadose zone**, meniscus

Table M3 Meteoric water diagenesis. Comparison of vadose and phreatic diagenesis in carbonate sediments

Vadose diagenetic environment	Phreatic diagenetic environment
Input of highly carbonate under-saturated waters with rapid flow waters	Rapid attainment of saturation with respect to carbonate in relatively slow water flow
Presence of CO_2 in soil zone promotes intense carbonate mineral dissolution	Mouldic or vuggy porosity development close to water table from dissolution of aragonite
CO_2-loss to atmosphere enables minor cement precipitation typically forming meniscus or microstalactitic cements	Rapid, often extensive cementation with equant or isopachous calcite and syntaxial calcite overgrowths on echinoderms

cements, occurring as small low-**magnesium calcite** crystals, coat all free grain surfaces (Fig. M10). On the lower surface of **allochemical** grains vadose cements can develop gravitational *micro-stalactitic structures*. Vadose cements typically fill pores of less than $100\,\mu m$ diameter but rarely occlude larger pores. As a result, fine-grained carbonate sediments become well cemented whilst coarser-grained carbonates often are only lightly cemented. Phreatic cements, by contrast, consist of coarse low-**magnesium calcite** crystals typically $100-250\,\mu m$ in diameter (Fig. M10). Unlike vadose cements they occlude pores of all sizes, with the size of crystals increasing away from the initial substrate due to competitive crystal growth. Near-surface meteoric phreatic cements may appear morphologically similar to deep burial carbonate cements, but they will tend to be non-ferroan **calcites** that contrast with the ferroan nature of most deep carbonate cements.

An understanding of the early meteoric water diagenesis of **limestones** is economically important because subaerially exposed and *telogenetically* altered shallow water **limestones** frequently retain their high

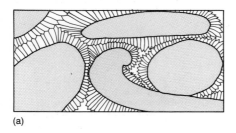
(a)

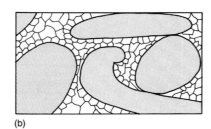
(b)

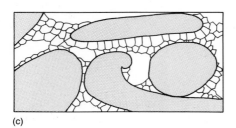
(c)

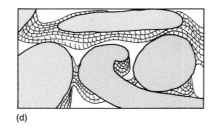
(d)

Fig. M10 Meteoric water diagenesis. Fabric of meteoric water carbonate cements. (a) Marine phreatic aragonite cement. (b) Fresh water phreatic calcite cement. (c) Vadose meniscus calcite cement. (d) Vadose microstalactitic calcite cement.

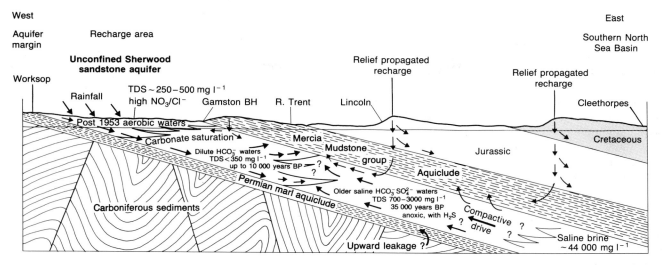

West

Aquifer
margin   Recharge area

**Unconfined Sherwood
sandstone aquifer**

East

Southern North
Sea Basin

Fig. M11 Meteoric water diagenesis. Schematic W–E cross-section through the southern North Sea Basin margin aquifer showing possible groundwater flow paths and mixing zones in the Sherwood Sandstone aquifer. (After Bath *et al.*, 1987.)

porosity and **permeability** characteristics, even during deep burial, and thus make excellent **aquifers** or **petroleum reservoir** rocks. The *Pleistocene* **limestones** provide examples of sediments which have undergone meteoric water diagenesis but their *telogenetic* fabrics have not been overprinted by **burial diagenesis**. The features displayed in their **fabrics** assist with the diagenetic interpretation of ancient **limestones** which experienced meteoric **alteration** prior to burial.

The effects of *telogenetic* modifications can also be important in permeable clastic lithologies. There are many large sedimentary **basins** in which **meteoric water** flushing of **sandstone aquifers** has been well documented to penetrate down-dip for tens of kilometres to depths exceeding 3 km of burial. In the Western Canadian Basin of northern Alberta, for example, **meteoric waters** are in hydraulic communication in several **sandstone aquifers** throughout a 1 km thick wedge of **Mesozoic–Paleozoic** sediments despite the presence of interbedded low **permeability mudrock** sequences. The invading **meteoric water** is generally held to be responsible for a variety of diagenetic reactions, including mineral **dissolution** and **cementation** in addition to *hydrocarbon accumulation*.

In general, however, since the **solubility of silica** is so low, dilute **meteoric waters** very rapidly attain saturation with respect to **quartz**. Consequently there is often little potential for reaction with the great majority of detrital **sandstones**. **Alteration** of siliciclastic sediments by **meteoric water** is thus typically restricted to the more mineralogically immature **sandstones** or earlier formed **authigenic** minerals. In the arkosic **Triassic** Sherwood Sandstone **aquifers** of south Devon, the English Midlands and along the southern North Sea Basin margin, for example, **meteoric waters** have penetrated some 20 km down the **aquifer** gradient (Fig. M11). **Feldspar** grains have been extensively altered to **authigenic kaolinite**

whilst carbonate and sulphate cements have undergone widespread **dissolution** in the meteoric invasion zone.

The **alteration** of sediments in the **weathering** zone can also be included in the *telogenetic* regime (see **diagenesis**). In this sense, diagenetic processes overlap with those of **pedogenesis**. Studies of *soil* types and *soil*-forming processes are particularly relevant because the **alteration** of sediments at or near the surface under the influence of **meteoric water** in *soil profiles* takes place by similar processes which may subsequently affect detrital grains and **authigenic** minerals during later burial. An understanding of the ways in which detrital minerals are altered or of the mechanism by which **pedogenic** minerals form in *soils* may enable, therefore, constraints to be placed on the conditions of similar mineral reactions during **diagenesis**. (See **meteoric diagenesis in carbonates**.) [SDB]

Bath, A.H., Milodowski, A.E. & Strong, G. (1987) Fluid Flow and Diagenesis in the East Midlands Triassic Sandstone Aquifer. In: Goff, J.C. & Williams, B.P.J. (eds) *Fluid Flow in Sedimentary Basins and Aquifers*. Geological Society Special Publication No. 34. Blackwell Scientific Publications, Oxford.

Folk, R.L. (1974) The natural history of crystalline calcium carbonate: effect of magnesium content and salinity. *Journal of Sedimentary Petrology* **44**: 40–53.

Land, L.S. (1970) Phreatic versus vadose meteoric diagenesis of limestones: evidence from fossil water table. *Sedimentology* **14**: 175–85.

Toth, J. (1980) Cross-formational gravity-flow of groundwater: a mechanism of the transport and accumulation of petroleum. In: Roberts, W.H. & Cordell, R.J. (eds) *Problems of Petroleum Migration*, pp. 121–67. Studies in Geology No. 10. American Association of Petroleum Geologists, Tulsa OK.

miarolitic cavity A small, crystal-lined cavity in an intrusive rock. Such cavities result from segregation of small pockets of gas into an irregular cavity (or *vug*) bounded by the outlines of surrounding crystals. In some

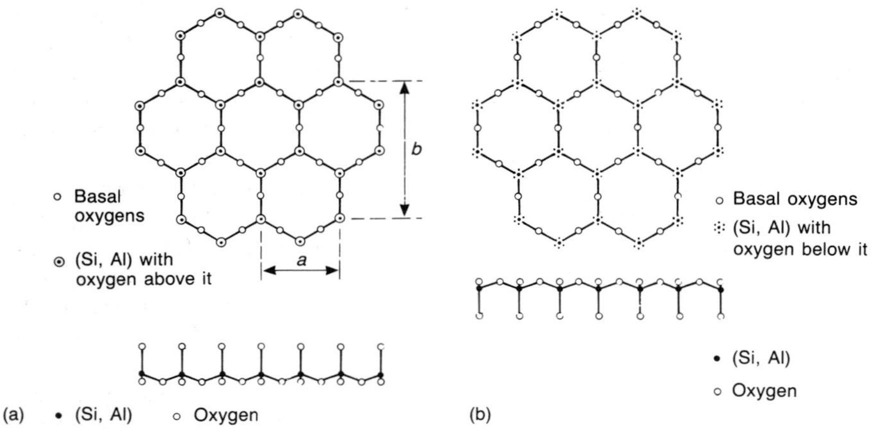

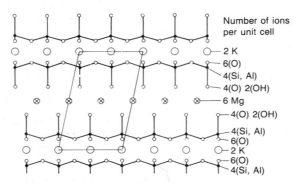

○ Basal
oxygens

⊙ (Si, Al) with
oxygen above it

• (Si, Al) ○ Oxygen

(a)

○ Basal oxygens

⠿ (Si, Al) with
oxygen below it

(b)

• (Si, Al)

○ Oxygen

Fig. M12 Mica minerals. (a) Plan of tetrahedral layer $(Si,Al)_4O_{10}$ with tetrahedra pointing upwards, and end view of layer looking along the *y* axis. (b) Plan and elevation of tetrahedral layer with tetrahedra pointing downwards.

Number of ions
per unit cell

— 2 K
— 6(O)
— 4(Si, Al)
— 4(O) 2(OH)
— 6 Mg
— 4(O) 2(OH)
— 4(Si, Al)
— 6(O)
— 2 K
— 6(O)
— 4(Si, Al)

Fig. M13 Mica minerals: phlogopite, $K_2Mg_6[Si_6Al_2O_{20}](OH)_4$. Elevation of Fig. M12 (a) and (b) superimposed and linked by a plane of octahedrally coordinated cations. Composite layers are shown linked by potassium ions and the simplest unit cell is outlined. View is along the *y* axis.

cases late-stage, euhedral **hydrothermal** crystals grow freely from the walls into the gas pocket. Miarolitic cavities, or *miarolitic fabrics*, are therefore analogous to spherical/ellipsoidal **vesicular** cavities/**fabric** in volcanic rocks. [RST]

mica minerals Although these sheet silicates show a considerable variation in chemical and physical properties, they are all characterized by a platy morphology and perfect basal **cleavage** which are a consequence of their layered atomic structure. Of the micas, **muscovite**, **phlogopite** and **lepidolite** are of considerable economic importance. The most common micas which are dealt with here include **muscovite**, **glauconite**, **lepidolite**, **phlogopite** and **biotite**; the first three are distinct mineral species but **phlogopite** and **biotite** are separated merely for convenience in dealing with otherwise so large a group.

The basic structural feature of the micas is a composite sheet in which a layer of octahedrally coordinated cations is sandwiched between two identical layers of linked $(Si,Al)O_4$ tetrahedra. Two of these tetrahedral sheets [of composition $(Si,Al)_2O_5$] are illustrated in Fig. M12, with a sheet on the left in which all the tetrahedra point upwards

(as shown in the elevation drawing below it) and a sheet on the right with the tetrahedra pointing downwards. The two sheets superimpose and are linked by a plane of cations as shown in Fig. M13. Additional hydroxyl ions, and the apical oxygens of the inwards-pointing tetrahedra, complete the octahedral coordination of the sandwiched cations. The central *Y* ions (Mg^{2+} in Fig. M13) determine the positions of the two tetrahedral sheets so that they are displaced relative to one another by *a*/3 in the [100] plane, the expected overall hexagonal symmetry being thus reduced to monoclinic. The hexagons may be superimposed, however, in six different ways (Smith & Yoder, 1956) giving stacking sequences leading to detailed variations in symmetry. The layers have a net negative charge which is balanced by planes of *X* ions (**potassium**, sodium, etc.) lying between them. Chemically the general formula of the micas can be given as $X_2Y_{4-6}Z_8O_{20}(OH,F)$, where *X* is mainly **potassium** or sodium (calcium in the '*brittle micas*' not dealt with here), *Y* is mainly **aluminium**, **mangesium** or **iron** but also **manganese**, **chromium**, **titanium**, **lithium**, etc., and *Z* is mainly silicon or **aluminium** but perhaps also Fe^{3+}. The micas can be subdivided into so-called di-octahedral and tri-octahedral classes in which the number of *Y* ions is 4 and 6 respectively. Further subdivisions of the common micas are made according to the principal constituents in the categories *X*, *Y* and *Z*, and the approximate formulae are given in Table M4. For the di-octahedral common micas, substitution of Si^{4+} for Al^{3+} in tetrahedral coordination is generally balanced electrostatically by equivalent substitutions of divalent ions for Al^{3+} in octahedral sites. In addition to the ions listed in the table, all micas contain hydroxyl and fluorine ions, with OH generally dominant except in **lepidolites**.

The optical properties of mica cover a wide range but all have negative optical sign and have α approximately perpendicular to their perfect (001) **cleavage**. Most are biaxial, with the axial angle moderate for di-octahedral and generally small for tri-octahedral micas. Birefringence is generally very weak in the plane of **cleavage** plates

[396]

but strong in transverse sections. In the coloured micas pleochroism is strong, the absorption being greatest for vibration directions parallel to the **cleavage** (and may partially mask the birefringence colour under crossed polars). In thin sections the relief is low for **lepidolite** and **phlogopite** but moderate to high for other micas. The di-octahedral micas generally have their optic axial planes perpendicular to (010) while the others have the optic axial plane parallel to (010). Well-formed crystals often show {110} faces forming 'books' of mica with a pseudohexagonal outline. The mica minerals occur in igneous, metamorphic and sedimentary rocks and are frequently the origin of **schistosity** in such rocks as *mica schists*.

Muscovite is commonest in regionally metamorphosed pelitic rocks; in low-grade environments it is found in **albite–chlorite–sericite schists** or in **sericite phyllites** (**sericite** being a name given to fine-grained *white mica*). **Paragonite** is the sodium-bearing analogue of **muscovite**. At higher grades of metamorphism **muscovite** is often associated with **biotite** and **garnet** but at the highest grades it dissociates to **potassium-feldspar** and **sillimanite**. In acid **igneous rocks muscovite** is less common than **biotite** but occurs in **muscovite granites** and **muscovite–biotite granites**; it is also the commonest mica in **aplites**. **Muscovite** with its low **iron** content is sought after in blocks capable of yielding **cleavage** sheets of relatively large area for use in electrical insulation; the absence of structural defects and inclusions is important. It is the characteristic product of fluorine **metasomatism** (*greisenization*) at **granite–slate** contacts. **Muscovite** is resistant to **weathering** and may occur in some **sandstones**, enhancing their **fissility** as in '*flagstones*'.

Glauconite is a mica mineral which occurs almost exclusively in marine sediments, particularly in *greensands*. It is generally found in rounded fine-grained aggregates of poorly-formed platelets. Its green colour is typical, normally with a greater birefringence than the **chlorites**. It is generally accepted that **glauconites** are formed from a variety of starting materials by marine **diagenesis** in shallow water. They have thus been used in **potassium– argon dating** to give direct evidence of the date of sedimentation. Possible relationships with *celadonite* have been much discussed (Odom in Bailey, 1984).

Phlogopite and **biotite** form a continuous series of tri-octahedral micas, the nomenclature here adapted being: **phlogopite**, Mg:Fe > 2:1, whereas for **biotite**, Mg:Fe < 2:1. The two principal occurrences of **phlogopite** are in metamorphosed impure **dolomitic limestones** and in ultrabasic rocks such as **kimberlite** and **leucite lamproites**. **Phlogopite** in large blocks can be used to yield undistorted sheets for use in thermal and electrical insulation. **Biotite** occurs in a greater variety of geological environments than any of the other micas. In metamorphic rocks it is formed under a range of temperature and pressure conditions and is abundant in many contact and regionally metamorphosed sediments. In thermal metamorphism it forms

at an early stage from **chloritic** sediments and persists in medium-grade **hornfelses** until at the highest grades it is ultimately replaced by potassium-**feldspar**, *orthopyroxene* and **sillimanite**. In regional metamorphism the appearance of **biotite** marks the onset of the T–P conditions of the **biotite** zone, typical rocks of which include **biotite schist**, **biotite–sericite schist**, **biotite–chlorite schist** and **albite–biotite schist**. It continues to be stable in the **garnet** zone and is an important constituent of many **garnet– mica schists**. In **igneous rocks** it is commonest in the intermediate and acid plutonic rocks; it occurs much less frequently in extrusive rocks and when present it is commonly partially altered.

Lepidolite is the most common **lithium**-bearing mineral and occurs almost exclusively in **granite pegmatites** associated with other **lithium** minerals (**amblygonite**, **spodumene**), **tourmaline**, **topaz**, **cassiterite**, **beryl** and **quartz**. It is mainly derived by **hydrothermal** replacement of **muscovite** or **biotite**. *Zinnwaldite* is somewhat similar, contains appreciable **iron** and is generally greyish brown in hand specimens as opposed to the colourless, pink or purple **lepidolite**. [RAH]

Bailey, S.W. (ed.) (1984) *Micas*. Reviews in Mineralogy Vol. 13. Mineralogical Society of America, Washington, DC.
Deer, W.A., Howie, R.A. & Zussman, J. (1966) *An Introduction to the Rock-forming Minerals*, pp. 193–221. Longman, London.
Foster, M.D. (1960) *Interpretation of the Composition of Tri-octahedral Micas*. Professional Paper of the United States Geological Survey **354B**.
Smith, J.V. & Yoder, H.S. (1956) Experimental and theoretical studies of the mica polymorphs. *Mineralogical Magazine* 31: 209–35.

micritization A form of degradation in which coarse calcareous material is replaced or reduced in size, eventually to *micrite* (average grain size < 4 µm). The most significant form of micritization is that resulting from biological activity. Centripetal micritization (from the outside inwards) of skeletal and non-skeletal carbonate grains appears to be a consequence of the boring activity of organisms such as **algae** and fungi. Borings, formerly occupied by living tissue, become filled with precipitates of *micrite*. Repetition of this boring and filling process results initially in the development of a thin *micrite* rind which is characteristically dark in thin sections. Continued

Table M4 Mica minerals. Approximate chemical formulae of common micas

	X	Y	Z
Di-octahedral			
Muscovite	K_2	Al_4	Si_6Al_2
Glauconite	$(K,Na)_{1.2-2.0}$	$(Fe,Mg,Al)_4$	$Si_{7-7.6}Al_{1.0-0.4}$
Tri-octahedral			
Phlogopite	K_2	$(Mg,Fe^{2+})_6$	Si_6Al_2
Biotite	K_2	$(Mg,Fe^{2+},Al)_6$	$Si_{6-5}Al_{2-3}$
Lepidolite	K_2	$(Li,Al)_{5-6}$	$Si_{6-5}Al_{2-3}$

micritization can result in the complete destruction of the original **fabric** of the grain.

The *paleoenvironmental* significance of micritization has been overstated. Many examples of micritization have been documented from the shallow, *lagoonal* areas of carbonate shelves and platforms, where photosynthetic **algae** are active borers. However, light-independent organisms such as *bacteria* and fungi can cause micritization, thus invalidating the argument that micritization is restricted to the **photic zone**. (See **microboring**.) [DMP]

Bathurst, R.G.C. (1966) Boring algae, micrite envelopes and lithification of molluscan biosparites. *Geological Journal* **5**: 15–32.

microboring A small (submillimetre) diameter boring made by a micro-organism (algae or fungi) into a substrate. The organisms bore by chemical **dissolution** and are important in the *bioerosion* of carbonate substrates. The borings are typically tubular and are up to 1 mm long. Carbonate grains may be intensively bored by algae to burial depths of 370 m, and by fungi to depths exceeding 5000 m. These infestations both dissolve the grains and mechanically weaken them. The borings may be filled with fine-grained precipitated calcium carbonate to form a *micrite* envelope. (See **micritization**.) [VPW]

microcline ($KAlSi_3O_8$) The triclinic (low temperature) form of **potash feldspar**. (See **feldspar minerals**.) [DJV]

microcontinent A small fragment of **continental crust**, varying from the size of Malagasy to that of Fiji, the Seychelles, etc. Usually a broken-off continental remnant, separated during **continental splitting**, but then becoming completely isolated following a jump in the location of **seafloor spreading** which left it surrounded by oceanic rocks. Such microcontinents do not have independent motion but simply move with the oceanic *plate* within which they now lie. [DHT]

microfracture A microscopic discontinuity across which there has been separation. Microfractures may be classified into Modes I, II and III or *mixed mode*, and *inter*, *intra* and *circumgranular* using the same conventions given for **fractures**, and may be divided into *microfaults* and *microcracks* on the same basis as **fractures** (i.e. depending on whether or not there is visible **displacement**).

Microfractures are universally observed in the precursor stage of **faulting**, during *triaxial compression* experiments; they have a good **preferred orientation** perpendicular to the *least principal stress* and dominantly Mode I (extensional) **displacements**. They may link points of impingement between grains. Natural *microcracks* are common in deformed rocks, and they too may have a good **preferred orientation**, and Mode I **displacements**; this can be used to infer the **stress** system under which

they formed. The experimental formation of *microcracks* in the pre-**faulting** stage has been used as an analogy for **earthquake** precursory phenomena, from which it has been shown that changes in **stress, strain, pore-fluid pressure** and *seismic velocity* can be accounted for by *microcracking*; this is known as *dilatancy–diffusion theory* (see **earthquake prediction**).

The formation of *microcracks* is responsible for a number of other changes (such as **dilatancy** and **dilatancy** *hardening*) which occur before **faulting**. The **faulting** itself is due to the linkage of microfractures formed in the precursory stages.

Seven mechanisms of *microcrack* formation may be identified. Impingement-induced *microcracks* form due to **tensile stresses** created at points of impingement between grains. *Shear fault*-induced *microcracks* are associated with **shear** displacement along **faults**, and flaw-induced *microcracks* grow from **stress** concentrations around pre-existing flaws. Elastic mismatch-induced *microcracks* are due to differential **elastic strains** along a boundary, while plastic-induced *microcracks* are similar, but due to *plastic* **strains**. **Cleavage**-induced *microcracks* along **cleavage** planes are common in minerals with good **cleavages**, and finally pre-existing flaw-induced *microcracks* are opened along former flaws such as grain boundaries. [TGB]

Kranz, R.L. (1983) Review of microcracks. *Tectonophysics* **100**: 449–80.

microgravity The technique of measuring **gravity** to *microgal* (10^{-2} g.u.) accuracy. Specialized **gravimeters** are necessary, and the LaCoste–Romberg Model D meter has been widely used. Microgravity surveys are used in the search for subsurface voids, the monitoring of the movement of underground water and in measuring rates of **neotectonic** movements. [PK]

microlite ($Ca_2Ta_2O_6(O,OH,F)$) A mineral found in *pegmatites*. [DJV]

microlithon A tabular-shaped body of rock bounded by **cleavage** surfaces. The **cleavage** is often a *fracture cleavage*, but may also be a **crenulation-cleavage** or **pressure-solution cleavage**. During **buckling** of a multilayer, microlithons commonly form along the hinges of the buckles between the highly attenuated or sheared limbs. The spacing of the microlithons is controlled by the wavelength of the initial buckles, and is therefore larger for thicker layers. The microlithons may be either perpendicular or inclined to the original layering, and preserve the primary layering at a state of lower **strain** than the sheared **cleavage** surfaces which bound them. The structure of microlithons bounded by sheared **cleavage** surfaces is also known as *gleitbretter*. [TGB]

de Sitter, L.U. (1964) *Structural Geology* (2nd edn). McGraw-Hill, New York.

microlithotypes of coal Microscopically determined **maceral** associations. **Coal macerals** rarely occur alone; they are usually associated with **macerals** of the same or of the two other **maceral** groups. A microlithotype is termed monomaceral, bimaceral or trimaceral according to the number of groups present. By convention a microlithotype is only recognized as such if its thickness is at least 50 μm. Microlithotypes are given the suffix -ite to distinguish them from **macerals**.

There are seven microlithotype groups: *vitrite, inertite* and *liptite*, which are monomaceral; *clarite, vitrinertite* and *durite*, which are bimaceral and the trimaceral *trimacerite*. *Vitrite* is composed of the *vitrinite* **macerals** and makes up at least 40–50% of most of the **Carboniferous** deposits of the Northern hemisphere. In **Gondwanaland coals** its proportion is rarely greater than 20–30%. *Vitrite* layers are usually 1–10 mm thick. The *inertite* group is composed of *inertinite* group **macerals**; the members *fusite*, composed of *fusinite*, and *semifusite* are the most common.

Clarite and *vitrinertite* consist of *vitrinite* and one other **maceral**, *durite* of *inertinite* plus *exinite* **macerals**. *Clarite* contains *exinite* group **macerals** and forms up to 20% of **bituminous coals**. *Vitrinertite* consists of *vitrinite* and *inertinite*; it is characteristic of high **rank coals** for in these, *exinite* group **macerals** have been converted into *vitrinite*-like material and are no longer recognizable. *Durite*, when present, may be most common in **coals** near the margins of **coal basins**, where it may make up as much as 30% of a **seam**. Other **coals** may contain little or no *durite*.

After *vitrite*, *trimacerite* is the most frequently occurring group of microlithotypes in many **coals**. It contains, by definition, more than 5% of **macerals** from each of the three **maceral** groups in various combinations. [AME]

Stach, E., Mackowsky, M.-Th., Teichmuller, M. *et al.* (1982) *Stach's Textbook of Coal Petrology.* Gebrüder Borntraeger, Berlin.

microperthite An intergrowth of the **alkali feldspar** minerals **albite** and **potash feldspar** visible only under the optical microscope. (See **feldspar minerals**.) [DJV]

microscopic Describes **structures** or other features of small size, measured in fractions of millimetres. Broadly, these are visible using an optical microscope, having a size range of 0.005–2 mm. Examples include **mortar structure** and **twinning**. Features smaller than this, imaged using techniques such as **electron microscopy**, are generally called '*submicroscopic*'.

The term *microstructure* is used to describe the microscopic features of a rock. Microstructural analysis examines the processes that formed the observed rock textures, and relates these to conditions of **deformation** and the pattern of **strain** over a wider area. [SB]

microseism The most prominent, naturally occurring **seismic noise**. Microseisms are long-duration *Rayleigh* (*surface*) *waves* of 5–20 s period, and amplitude 0.1–10 000 nanometres (largest at coasts and shores of large lakes and smallest in continental interiors). Their main component (5–9 s period) is thought to arise from pressure on the sea-floor from standing **waves** on the sea surface in *storm* centres, near coasts or in lakes. A subsidiary 12–18 s period component comes from direct **wave** action against shelving seafloor. Separate short- and long-period *seismometers* are used to avoid recording microseisms. An example of microseisms is shown with **earthquake magnitude** (see Fig. E4, p. 196). (See **seismic noise**.) [RAC]

Longuet-Higgins, M.S. (1950) A theory of the origin of microseisms. *Philosophical Transactions of the Royal Society, London* **A243**: 1–35.

migration of oil and gas The migration of oil or gas from its **source rock** into the **reservoir rock** in which it now occurs beneath an **oil** or gas **trap**. When source sediments are compressed by burial the *hydrocarbons* migrate out, mainly across the stratification as shown by geological reconstructions of the migration path, into a porous *host rock*; this is known as primary migration. If the *host rock* is permeable as well as porous then the *hydrocarbons* will migrate through it to collect beneath a trap. They may also pass from this **reservoir** rock to another one. These are secondary migrations and they are the process by which oil and gas collect into commercial pools. The migration takes place largely along the stratification and it is probably controlled by **buoyancy** differences between gas, oil and water which are all competing for space in the **reservoir**. [AME]

North, F.K. (1985) *Petroleum Geology*. Allen & Unwin, Boston.

millerite (NiS) A relatively rare **ore** mineral of **nickel**. [DJV]

milligal (mgal) The cgs unit of **gravity** corresponding to an acceleration of 10^{-3} cm s^{-2}. [PK]

millstone Millstones and simple grinders and querns have been used for grinding grain, plants, olives, pigments, salt and metal **ore** throughout the world since early prehistoric times. The simplest consist of two stones, the lower one slightly concave and the upper (rubber) cylindrical or cigar-shaped. These are known as saddle querns, and from these gradually developed more sophisticated millstones with holes in the upper stone for insertion of grain into the mill (known as hopper-rubbers), followed by mills operated by rotary motion, driven by hand, animals and later water and wind. The rocks used for millstones were carefully chosen for their roughness, efficiency in cutting, hardness, and degree of contamination of whatever was being ground. **Limestones** and **travertines**, **sandstones**, **granites** and lavas were all used in the Old World and the New, the **sandstones** being one of the most common types. In Europe young fresh **vesicular** lavas were often used in antiquity, including **basalts**, **rhyolites** and **leucitites**. Millstones from the most important quarries of these rocks were traded hundreds of kilometres from the Iron Age onwards. Two of the best-known millstones types, both exported world-wide during recent centuries, are *Cullen stones* (cf. **Mayen lava**), and the Millstone Grit from Derbyshire in the UK. [OWT]

mima mound Earth mound that takes its name from Mima Prairie, Thurston County, Washington, USA. Such mounds are characteristically up to around 2 m in height, 25–50 m in diameter, and occur at a density of 50–100 or more to the hectare. There are many hypotheses for their origin (Cox & Gakahu, 1986), including that they are **erosional** residuals, that they result from depositional processes around vegetation clumps, that they are the product of frost sorting, and they have been formed by communal rodents. [ASG]

Cox, G.W. & Gakahu, C.G. (1986) A latitudinal test of the fossorial rodent hypothesis of Mima mound origin. *Zeitschrift für Geomorphologie* NF **30**: 485–501.

mimetic crystal growth Growth of a mineral during metamorphism which mimics in shape or orientation pre-existing textural features. Thus platy minerals such as **mica** can grow in **slates** in orientations governed by the arrangement of minerals in **mudstone**. In higher grade metamorphism, it is possible to find, e.g., **hornblende** fibres forming aggregates with the typical square shapes of **pyroxene**. Hydration of **pyroxenes** in **granulites** in this case results in mimetic growth of **hornblende**. Orientation on recrystalization of many minerals (e.g. **quartz**, **calcite**) is governed by that of the pre-existing grain in the absence of a strong **deformational** control. [SB]

mimetite ($Pb_5(AsO_4)_3Cl$) A secondary **ore** mineral of **lead** found in the **oxidized** portions of **lead** deposits, and forming a complete *solid solution* to **pyromorphite**. [DJV]

mineral A term that is difficult to define both accurately and succinctly. A generally accepted working definition is that 'A mineral is a naturally occurring homogeneous solid with a defined (but not necessarily fixed) chemical composition and a highly ordered atomic arrangement.' Examining this definition we find 'naturally occurring' includes all those substances not produced by human activity, although materials that crystallize on artefacts discarded by man, such as utensils of various kinds, have been regarded as minerals. The term 'homogeneous solid' implies that we are dealing with a single substance that cannot be physically subdivided, even if the particular mineral may only be homogeneous on the scale observed under the optical or **electron microscope**. 'Solid' excludes all gases and liquids so that H_2O as ice is a mineral, but

as water it is not. Also, **mercury** as the native metal would not be a mineral, although this liquid metal occurs naturally and is commonly regarded as a mineral. The 'defined (but not necessarily fixed) chemical composition' indicates that we can give a chemical formula for a mineral such as SiO_2 for **quartz** or ZnS for **sphalerite**. By 'not fixed' it is possible to include the many examples of minerals in which two or more elements substitute for one another, e.g. $(Mg,Fe)_2SiO_4$ **olivine**; $(Fe,Mn)WO_4$ **wolframite**. It does not entirely take account of the fact that at the level of parts per million (or sometimes even fractions of a per cent) all natural solids contain numerous impurities making individual mineral specimens from different regions very different in composition when studied in great detail. The 'highly ordered atomic arrangement' refers to the ordered **crystal structures** of most minerals, whereby the atoms (or ions) are linked together to produce a regular three-dimensional pattern that is repeated through the crystal, i.e. minerals are regarded as crystalline. Again, there are examples of poorly crystalline or even amorphous substances that are commonly regarded as minerals, e.g. various crypto-crystalline forms of **silica** such as **chalcedony**. The above definition also rightly excludes, if only by implication, such organic solids as **coal**, **lignite** and *bitumen* although minerals may certainly be formed by processes that involve living organisms such as *algae* or *bacteria*. [DJV]

mineraloid A term sometimes used to refer to naturally occurring substances that fall outside the 'strict' definition of the term **mineral** being applied by the writer involved. Thus, certain authors would use the term 'mineraloid' in reference to a non-crystalline solid phase such as a natural glass, or a non-solid such as **mercury**. These mineraloids are generally regarded as falling within the domain of study of the science of mineralogy. [DJV]

minette A calc-alkaline lamprophyre comprising *phenocrysts* of **biotite**, *clinopyroxene* (e.g. **diopside**, **augite**), ± **hornblende** and **olivine**, with **alkali feldspar** > **plagioclase**. [RST]

minette ironstone A **siderite**–**chamosite**–**calcite** rock. The **chamosite** is often **oolitic**, the **iron** content is around 30% while lime runs 5–20% and **silica** is usually above 20%. The high **lime** content forms one contrast with **banded iron formation** and often results in these ironstones being self-fluxing **ores**.

Minette ironstones are particularly widespread and important in the **Mesozoic** of Europe, examples being the ironstones of the English Midlands, the minette **ores** of Lorraine and Luxembourg, the Salzgitter **ores** of Saxony and the **iron ores** of the Peace River area, Alberta. The only mining on a significant scale is limited to Lorraine and Luxembourg. [AME]

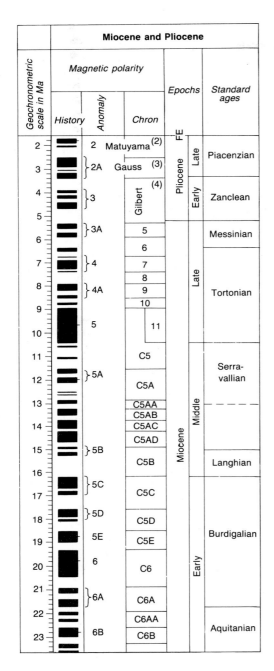

Fig. M14 Miocene.

Miocene and Pliocene					
Geochronometric scale in Ma	Magnetic polarity			Epochs	Standard ages
	History	Anomaly	Chron	FE	
2		2	Matuyama (2)	Late	Piacenzian
3		2A	Gauss (3)	Pliocene Late	
4		3	Gilbert (4)	Pliocene Early	Zanclean
5					
6		3A	5	Late	Messinian
			6		
7		4	7		Tortonian
8		4A	8 / 9		
			10		
9		5	11		
10					
11			C5	Middle	Serra-vallian
12		5A	C5A		
13			C5AA / C5AB		
14			C5AC		
15		5B	C5AD	Miocene	
16			C5B		Langhian
17		5C	C5C		
18		5D	C5D		
19		5E	C5E	Early	Burdigalian
20		6	C6		
21		6A	C6A		
22			C6AA		Aquitanian
23		6B	C6B		

minium (Pb_3O_4) A brownish-red **oxide mineral** of **lead**. [DJV]

minnesotaite ($Fe_3Si_4O_{10}(OH)_2$) The **iron**-rich analogue of the mineral **talc**, found in the low-grade metamorphic **iron** deposits known as *Precambrian Iron Formations*. [DJV]

minor fold, structure Fold or other **structure** distinguishable at outcrop scale. Contrast **major fold**. [RGP]

Miocene The Miocene has a duration of *c.* 17 Ma from the end of the **Oligocene** *c.* 23 Ma to the base of the **Pliocene** Epoch at *c.* 5 Ma and has been divided into six stages (Berggren *et al.*, 1985) (see Fig. M14).

The amelioration of climate in the Late **Oligocene** continued into the Early Miocene, but by the Middle Miocene the world climate took on a modern aspect. There is some evidence that at 15 Ma, East Antarctica was in a deep freeze with a permanent *ice sheet*. A major sea-level fall occurred at 10 Ma and the earliest known **glaciers** of recent times appeared in Alaska and West Antarctica and a permanent *ice sheet* formed at 6.6 Ma, when sea-level fell again, and **glaciers** appeared in South America.

With the progressive disappearance of the Ural Sea at the end of the **Oligocene** and in the Miocene, and the collision of India and Eurasia which resulted in the **uplift** of the Himalayas, the Eurasian continent took on a modern appearance. The **Tethyan** region was, however, still an area of complex change at *c.* 18 Ma. In the Early Miocene at *c.* 18 Ma, the Middle East region began to close, separating the **Tethyan** from the Indo-Pacific Provinces and linking Africa with Asia and Europe thus allowing African *proboscidians* and bovids (antelope and cattle, etc.) to enter Eurasia. Several million years later, the western end of the **Tethys** was closed and became separated from the Caribbean Province. This closure joining Africa to Europe in the region of Gibraltar occurred in the Middle/Late Miocene. The circum-Antarctic current assumed its modern deep-water circulation at *c.* 22 Ma. In the Late Miocene at *c.* 6 Ma the enclosed Mediterranean Sea became a large evaporating **basin**, similar to the **Dead Sea**; continuous evaporation led to a drop in its surface area by as much as 2000 km². **Eustatic** sea-level changes probably influenced by **interglacials** at high latitudes in Antarctica breached the barrier of Gibraltar periodically and flooded the **basin** with oceanic waters. Clays with planktonic **Foraminifera** and **coccoliths** were thus intercalated with Late Miocene **evaporites**. During periods of desiccation water courses cut deep canyons which today are preserved filled with sediments. There was a dramatic increase in volcanic activity at *c.* 14 Ma including the Hawaiian **hotspot** and Central America.

Apes and monkeys shared common ancestors until they diverged at *c.* 21 Ma and the early *hominid Ramapithicus* appeared *c.* 14 Ma. During a global volcanic upheaval *c.* 14 Ma land **mammals** lost much of their diversity, with the number of genera falling by 30%. During the next few million years the cool climate was favoured by grass which in turn stimulated a burst in the evolution of grazing **mammals**. Modern cats (*Felis*) and the first elephants evolved about *c.* 8 Ma and 7 Ma and respectively, as the chilling of the planet continued.

In the oceans the planktonic **Foraminifera** underwent another radiation reaching a peak of species diversity at *c.* 15 Ma followed by a steady decline. Calcareous nannofossil diversity steadily declined until the end of the Lower Miocene when just over 30 known species have been recorded (Tappan, 1980). A radiation in taxa occurred during the Middle Miocene resulting in a total of over 50 species.

On land, minor fluctuations in temperature have been recorded and there is evidence that some cooling even affected the tropical areas; elder and spruce were growing in Borneo, while extensive coniferous forests developed in high northern latitudes and open grassland appeared at middle and high latitudes (Thomas & Spicer, 1987). The following extant orders appeared: Lamiales, Plumbaginales, Dilleniales, Ranunculales, Capparales, Primuleales, Alismales, Nijales and Areles (Thomas & Spicer, 1987). (See **land plants**.) [DGJ]

Berggren, W.A., Kent, D.V. & van Couvering, J.A. (1985) The Neogene: Part 2. Neogene geochronology and chronostratigraphy. In: Snelling, N.J. (ed.) *The Chronology of the Geological Record*, pp. 211–60. Memoirs of the Geological Society No. 10. Blackwell Scientific Publications, Oxford.

Tappan, H. (1980) *The Paleobiology of Plant Protists*. W.H. Freeman, San Francisco.

Thomas, B. & Spicer, R. (1987) *Evolution and Palaeobiology of Land Plants*, pp. 1–304. Croom Helm, London.

mise-à-la-masse **Resistivity method** where one of the current electrodes is placed virtually at infinity, and the other within the conductive body. The method is feasible only when the conductor is accessible: usually as an outcrop, or within a borehole. The potential electrode pair is moved about either on the surface (commonly in a radial pattern around the current electrode), or in the borehole, with the object of mapping **equipotential** lines. The direct injection of the current into the **orebody** allows a better definition of the spatial extent of its three-dimensional form. [ATB]

Kearey, P. & Brooks, M. (1991) *An Introduction to Geophysical Exploration* (2nd edn). Blackwell Scientific Publications, Oxford.

Telford, W.M., Geldart, L.P. & Sheriff, R.E. (1990) *Applied Geophysics* (2nd edn). Cambridge University Press, Cambridge.

mixing corrosion The phenomenon whereby two **karst** waters which are saturated with calcium carbonate mix to form an undersaturated water, which then becomes capable of further **dissolution** of calcium carbonate. The geomorphological importance of this phenomenon was first recognized by Bögli. The renewed **aggressivity** of the water occurs because the relationship between calcium carbonate equilibrium saturation and carbon dioxide partial pressure is an exponential one. Mixing of two saturated waters which contain different levels of carbon dioxide may cause the resultant water to fall below the exponential curve, and become aggressive. Mixing may also lead to a change in kinetic regime, which also may encourage further **solution** of calcium carbonate. The phenomenon is often important at some depth within a **limestone** mass. [HAV]

Bögli, A. (1964) Mischungskorrosion — ein Beitrag zur Karstmorphologie. *Erdkunde* **18**: 83–92.

mobile belt Large-scale linear belt or zone of the **continental crust** characterized by **tectonic** activity during a given period of Earth history. Such a zone is contrasted with a **craton**, which is the adjoining relatively **stable tectonic zone**. 'Mobile belt' is sometimes used synonymously with '*orogenic belt*' but carries no implication of mountain belt creation. It is therefore used by many authors in describing **Precambrian** belts of thermo-tectonic activity that may not be precisely analogous to **Phanerozoic** *orogenic belts*. [RGP]

modal composition The composition of a rock sample in terms of the volumetric proportions of the different **minerals** in the whole rock. Modal compositions may be determined from inspection of a petrographic thin-section. [RST]

Modified Mercalli Scale The most widely used **earthquake intensity** scale is the 12-point Modified Mercalli (MM) Scale. The original scale was devised by the Italian seismologist and volcanologist Mercalli in 1902 as a refinement of the 1878 10-point **Rossi-Forel Scale**. Because it is based on how people perceive the **earthquake** and its effect on structures, it has been revised twice to allow for changing design and construction standards: first by Wood and Neumann in 1931 to be applicable to Californian buildings, then generalized by Richter in 1958 to include several standards of masonry. The formal definitions of the scales are given in Table M5. An approximate relation between MM **intensity**, i (from IV to X only), and peak horizontal ground acceleration A (in $\mathrm{cm\,s^{-2}}$) is:

$$\log_{10}(A) = 0.01 + 0.30\,i$$

In the UK, a similar relationship has been given for *surface-wave* **magnitude** M_s at R kilometres from the source:

$$i = 1.36 + 1.80\,M_s - 1.18\ln R - 0.0019\,R$$

The definitions of points on the MM scale are given in Table M5. (See **earthquake intensity**.) [RAC]

Bolt, B.A. (1988) *Earthquake*. W.H. Freeman, San Francisco.

mogote A residual **limestone** hill found within tropical **karst** terrain. The name comes from the local word for such hills in Cuba where they are found for example in the Sierra de los Organos. They are generally steep-sided features, scattered over plains. [HAV]

Day, M. (1978) Morphology and distribution of residual limestone hills (mogotes) in the karst of northern Puerto Rico. *Geological Society of America Bulletin* **89**: 426–32.

Mohorovičić discontinuity (*Moho*) A **seismic discontinuity** between the **crust** (both oceanic and continental) and the underlying **mantle**. It occurs at some 11–12 km depth

Table M5 An abbreviated definition of the 1931 **Modified Mercalli** Intensity **Scale**. (After Bolt, 1988.)

I Not felt except by a very few under especially favourable circumstances. (I Rossi-Forel scale)

II Felt only by a few persons at rest, especially on upper floors of buildings. Delicately suspended objects may swing. (I to II Rossi-Forel scale)

III Felt quite noticeably indoors, especially on upper floors of buildings, but many people do not recognize it as an earthquake. Standing automobiles may rock slightly. Vibration like passing of truck. Duration estimated. (III Rossi-Forel scale)

IV During the day felt indoors by many, outdoors by few. At night some awakened. Dishes, windows, doors disturbed; walls make creaking sound. Sensation like heavy truck striking building. Standing automobiles rocked noticeably. (IV to V Ross-Forel scale)

V Felt by nearly everyone, many awakened. Some dishes, windows, and so on broken; cracked plaster in a few places; unstable objects overturned. Disturbances of trees, poles, and other tall objects sometimes noticed. Pendulum clocks may stop. (V to VI Rossi-Forel scale)

VI Felt by all, many frightened and run outdoors. Some heavy furniture moved; a few instances of fallen plaster and damaged chimneys. Damage slight. (VI to VII Rossi-Forel scale)

VII Everybody runs outdoors. Damage negligible in buildings of good design and construction; slight to moderate in well-built ordinary structures; considerable in poorly built or badly designed structures; some chimneys broken. Noticed by persons driving cars. (VIII Rossi-Forel scale)

VIII Damage slight in specially designed structures; considerable in ordinary substantial buildings with partial collapse; great in poorly built structures. Panel walls thrown out of frame structures. Fall of chimneys, factory stacks, columns, monuments, walls. Heavy furniture overturned. Sand and mud ejected in small amounts. Changes in well water. Persons driving cars disturbed. (VIII + to IX Rossi-Forel scale)

IX Damage considerable in specially designed structures; well-designed frame structures thrown out of plumb; great in substantial buildings, with partial collapse. Buildings shifted off foundations. Ground cracked conspicuously. Underground pipes broken. (IX + Rossi-Forel scale)

X Some well-built wooden structures destroyed; most masonry and frame structures destroyed with foundations; ground badly cracked. Rails bent. Landslides considerable from river banks and steep slopes. Shifted sand and mud. Water splashed, slopped over banks. (X Rossi-Forel scale)

XI Few, if any, (masonry) structures remain standing. Bridges destroyed. Broad fissures in ground. Underground pipelines completely out of service. Earth slumps and landslips in soft ground. Rails bent greatly

XII Damage total. Waves seen on ground surface. Lines of sight and level distorted. Objects thrown into the air

below the ocean floors and at much variable depths below the **continental crust** where it averages some 35–40 km thickness. Originally discovered by Mohorovičić in the European **continental crust** who observed that *P waves* were refracted at the base of the **continental crust**, and then travelled through the **mantle** immediately below it, at some 8.1 km s^{-1}. Energy refracted back upwards from this

head wave eventually arrived at the surface before *P waves* travelling directly through the **continental crust**. The **arrival times** and velocities can be used to determine the depth of the discontinuity using the standard techniques of **seismic refraction**. The depth can also be calculated using **reflected seismic waves** and the boundary was originally considered to be very sharp, but recently **Vibroseis**® studies indicate that it is complex and probably layered. [DHT]

Mohr diagram Graphical representation of a state of **stress** or **strain** employing circles. Mohr diagrams can be constructed to show **stress** and **strain** (*infinitesimal* or **finite**) in both two and three dimensions, and the diagram for **stress** is widely used in studies of **failure** and yield while the **strain** versions are useful as methods of **strain** determination. The construction can be made for any second-order tensor quantity.

Mohr's original construction was for **stress**. Figure M15 has a vertical axis on which **shear stress** (τ) is plotted, and a horizontal axis for *normal stress* (σ_N). A general state of **stress** is represented by a circle, centre $C = (\sigma_1 + \sigma_2)/2$ on the σ_N axis, and radius $(\sigma_1 - \sigma_2)/2$. In geological literature, compressive **stresses** are conventionally taken as positive. The *normal* and **shear stress** acting on any plane, *P*, can be read off from its coordinates of σ and τ, where the orientation of the normal to *P* with respect to τ is given by half the angle between *P* and the σ_N axis (2θ). This can also be expressed algebraically as

$$\sigma_N = \tfrac{1}{2}(\sigma_1 + \sigma_2) + \tfrac{1}{2}(\sigma_1 - \sigma_2)\cos2\theta$$
$$\tau = \tfrac{1}{2}(\sigma_2 - \sigma_1)\sin2\theta$$

The corresponding values for a plane normal to *P* can also be deduced, and the inverse problem of deducing the *principal stresses* from measurements of σ_N and τ in two or more planes can be solved. Many other properties of the state of **stress** are also evident from this simple diagram, such as the fact that maximum **shear stress** is on a plane with $2\theta = +/-90°$, or $\theta = +/-45°$ to σ_1, and that the **stress** tensor is symmetrical. Due to this latter property, it is common to represent only the upper half of the diagram.

The sequence of Mohr circles in the two-dimensional plot representing states of **stress** at **failure** defines the *Mohr envelope*, which is widely used in deriving *failure criteria*. The *Mohr envelope* is a line showing the relationship between **shear** and *normal* **stress** at **failure** on a Mohr diagram. The *Mohr envelope* is obtained experimentally by plotting Mohr circles representing the state of **stress** at **failure** for various **confining pressures**, and constructing the bounding surface (Fig. M16). This may also be described by a formula derived from fitting a curve to the data or from theory. Mohr's hypothesis was that **failure** occurred when the **shear** and *normal* **stress** acting on the **shear** plane reached a critical value given by the *Mohr*

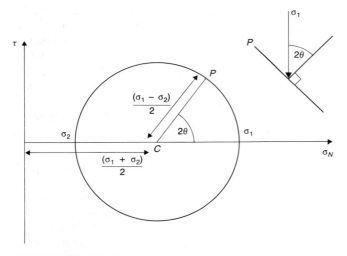

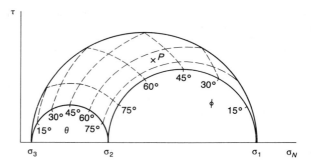

Fig. M17 The **Mohr diagram** for stress in three dimensions. The three principal stresses (σ_1, σ_2, and σ_3) lie along the normal stress (σ_N) axis. The shear stress (τ) and normal stress acting on any plane with a normal at ϕ to σ_1 and θ to σ_2 (e.g. point P) can be read from the two axes. (After Jaeger & Cook, 1979.)

Fig. M15 The **Mohr diagram**. Representation of the normal stress (σ_N) and shear stress (τ) acting on a plane P with normal orientated at 2θ to σ_1. The principal stresses are σ_1 and σ_2.

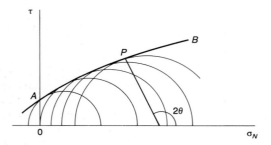

Fig. M16 The **Mohr** envelope (AB) joins points representing the state of shear (τ) and normal stress (σ_N) at failure under various confining pressures. P is one such point, and the angle 2θ is twice the angle between the normal to the shear plane and the maximum principal stress, σ_1, as in Fig. M15.

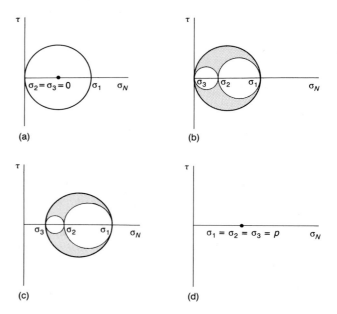

Fig. M18 The **Mohr diagram** for various states of three dimensional stress. Shear stress (τ) is plotted against normal stress (σ_N). (a) Uniaxial stress; σ_2 and σ_3 are zero. (b) Biaxial stress; σ_3 is zero. (c) Triaxial stress. The general state of stress in the Earth's crust. (d) Hydrostatic pressure. All three principal stresses are equal, in this case positive. (After Hobbs *et al.*, 1976.)

envelope; this predicts that **failure** will occur when the *Mohr circle* just touches the *Mohr envelope*, and the angle of the **shear plane** to the maximum *principal stress* will be given by the tangent to the *Mohr envelope* (Fig. M16), which is generally observed in experiments.

One of the simplest theoretical *Mohr envelopes* is the *Coulomb criterion* that **shear stress** is a linear function of normal **stress**; this predicts a straight line *Mohr envelope*. However, almost all observed *Mohr envelopes* are concave downwards, and such a shape is predicted by other **failure** criteria such as the *Griffith criterion* and all its derivatives.

In three dimensions, the three *principal stresses* σ_1, σ_2 and σ_3 are also represented as points on the σ_N axis (Fig. M17). The state of **stress** on a principal plane of **stress** is shown by a *Mohr circle* connecting the two *principal stresses* acting in the plane, such as the circle between σ_1 and σ_2. The *normal* and **shear stresses** on any plane P are its coordinates as before, and the orientation of P can be

defined by angles ϕ and θ within two of the *principal stress* planes (Fig. M17). P is constrained to lie on or within the three *Mohr circles*. This area can be contoured for ϕ and θ, as shown, to locate any plane P.

The three-dimensional Mohr construction is useful for the above calculation of *normal* and **shear stress** on any plane, and can also be used to illustrate various **stress** states as shown in Fig. M18. *Uniaxial stress*, Fig. M18(a), is the state in which there is only one non-zero *principal stress*, σ_1, compared with *biaxial stress*, where there are two non-zero *principal stresses*, σ_1 and σ_2 (Fig. M18b). The most general case of triaxial **stress**, with all *principal stresses*

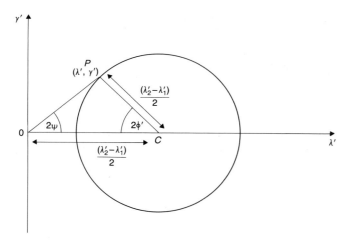

Fig. M19 Mohr. Representation of the two dimensional strain parameters λ' and γ' in any plane P orientated at $2\phi'$ to λ_1'. The angular shear strain at P is 2ψ.

compressive, as in all the deeper parts of the Earth, is as shown in Fig. M18(c). *Hydrostatic pressure*, positive or negative with three equal *principal stresses*, is shown by a single point in Fig. M18(d).

Finite strain in two dimensions can be represented by *Mohr circles* in coordinates of λ' and γ', where $\lambda' = 1/\lambda$ and $\gamma' = \gamma/\lambda$ are the reciprocal **quadratic elongations** and reciprocal **shear strains** respectively. These two quantities are defined in such a way that **strain** can be represented by a *Mohr circle* with principle **quadratic elongations** λ_1 and λ_2. The *Mohr circle* is centred at $C = (\lambda_2' + \lambda_1')/2$ and has a radius $(\lambda_2' - \lambda_1')/2$ (Fig. M19). The values of λ' and γ' for any plane P can be read off from its coordinates, and its orientation is given by $2\phi'$, the angle between P and λ_1' in the strained state.

The definitions of λ' and γ' in terms of the reciprocal principal **quadratic elongations** (λ_1' amd λ_2') are thus:

$$\lambda' = \frac{(\lambda_1' + \lambda_2')}{2} - \frac{(\lambda_2' - \lambda_1')}{2} \cos 2\phi'$$

$$\gamma' = \frac{(\lambda_2' - \lambda_1')}{2} \sin 2\phi'$$

A useful feature of this diagram is that the *angular shear strain* of $P(\psi)$ is given by half the angle between OP and the λ' axis (2ψ). The diagram is therefore particularly convenient for calculating the principal **quadratic elongations** from measurements of deformed objects.

The Mohr diagram for three-dimensional **strains** has the three reciprocal principal **quadratic elongations** λ_1', λ_2' and λ_3' as points on the axis, connected by *Mohr circles* which give the values of λ' and γ' on the principal planes of the **strain ellipsoid**, the values for a general direction, P, can be read from plotting its position on or within the *Mohr circles* using the angles from the *principal strain axes*. Angular **shear strain** at P is again represented by the angle between OP and the λ' axis. The three-dimensional Mohr

diagram is useful for depicting various types of **strain ellipsoid** and also for deriving the principal **quadratic elongations** and orientations from field **strain** data. For the latter, it is convenient to plot a *Mohr locus*: this is a closed line representing the **strain** on a general section of the **strain ellipsoid**, such as might be measured on a **bedding** plane, for example.

The Mohr diagram can be used in both two- and three-dimensional forms to represent *infinitesimal strains* (ε) or **strain rates** ($\dot{\varepsilon}$). In these cases, the construction and use of the diagram follows exactly the same procedure as given for **stress**, with the quantities of ε or $\dot{\varepsilon}$ substituted for **stress**. [TGB]

Hobbs, B.E., Means, W.D. & Williams, P.F. (1976) *An Introduction to Modern Structural Geology*. John Wiley & Sons, New York.

Jaeger, J.C. & Cook, N.G.W. (1979) *Fundamentals of Rock Mechanics* (3rd edn). Chapman & Hall, London.

Means, W.D. (1976) *Stress and Strain*. Springer-Verlag, Berlin.

Ramsay, J.G. & Huber, M.I. (1983) *The Techniques of Modern Structural Geology*. Vol. 1: *Strain Analysis*, Sessions 6, 8. Academic Press, New York.

Treagus, S.H. (1986) Some applications of the Mohr Diagram for three-dimensional strain. *Journal of Structural Geology* **8**: 819–30.

molecular orbital (MO) A wave function which describes the behaviour of an electron in the presence of many nuclei. The simplest representation for an MO is a linear combination of **atomic orbitals** (AO), $\theta_i = \Sigma a_{ri}\phi_r$, where a_{ri} describes how much the rth AO, ϕ_r contributes to the ith MO θ_i. The energies of these MOs can be found by the *variation method* in which the values of the coefficients a_{ri} for stable MOs are determined by minimizing the MO energy (E) with respect to each coefficient in turn ($\partial E/\partial a_{ri} = 0$). This generates a series of secular equations, $\Sigma a_r(H_{rs} - E.S_{rs}) = 0$. If there are n AOs they will generate n MOs and there will be n equations corresponding to the n possible values of s and each equation will have n terms (n possible values of r). H_{rr} is the *Coulomb integral* for AO r (usual symbol α_r), which is closely related to the ionization energy for that AO. H_{rs} ($r \neq s$) is the *resonance integral* and measures the energy of interaction between the AOs r and s (symbol β_{rs}). Hückel proposed that $\beta_{rs} = 0$ unless r and s are adjacent. S_{rs} is the *overlap integral* and measures the spatial overlap between orbitals r and s. If $r = s$ $S_{rr} = 1$ for a normalized AO. In the widely used Neglect of Overlap (NoO) Approximation S_{rs} ($r \neq s$) = 0. The Hückel and NoO approximations allow the secular equations to be considerably simplified. The possible values of E can then be calculated and the set of coefficients appropriate to each MO found. In symmetric molecules, group theory can be used to simplify the calculations.

This simple Hückel MO theory can be used to give a very useful first approximation to the MO structure of a molecule. The squares of the coefficients indicate the charge distribution in the orbital and the product of coefficients on adjacent atoms measures *bond order* which is related to *bond strength*. In most molecules such MO calculations will yield the energies of both occupied and unoccupied orbitals so that electronic transitions can be discussed and the charge distributions in both ground and excited states calculated.

Implicit in the simple MO theory outlined above is the idea that the mutual repulsions between electrons can be averaged out, so that one electron moves in an attractive field due to the nuclei and a repulsive field due to the other electrons — hence the term 'one-electron' MO theory. The deficiencies inherent in this approach can be demonstrated by calculating the effect of placing two electrons in a single MO. The overall two-electron wave function (ψ) is based on the product of the individual electron wave functions. Thus for two electrons (1 and 2) in the bonding orbital of dihydrogen (ϕ_A, ϕ_B are H $1s$ orbitals on atoms A and B) $\psi = [(2)^{-0.5}(\phi_A(1) + \phi_B(1))]$. $[(2)^{-0.5}(\phi_A(2) + \phi_B(2))]^{so}$ $\psi = (0.5).(\phi_A(1).\phi_A(2) + \phi_B(1).\phi_B(2) + \phi_A(1).\phi_B(2) + \phi_B(1).\phi_A(2))$. The first two terms both represent the *two* electrons at one end of the molecule, i.e. $H_A^-H_B^+$ and $H_A^+H_B^-$. Thus half of the terms in ψ are 'ionic' and only the remaining two terms are 'covalent' indicating an equitable electron distribution. For this reason the simple MO method is said to overemphasize the 'ionic character' of a bond. This anomalous picture can be rectified either by the inclusion of other terms based on excited wavefunctions (*configuration interaction*) or by reducing the importance of the ionic terms (see **valence bond theory**). Other calculation methods consider electron–electron repulsions explicitly, e.g. CNDO, MINDO, etc. (Dewar, 1973) or dispense with hydrogen-like AOs altogether and use functions based on collections of Gaussian functions — *ab initio* methods. Yet another approach (Johnson) considers the way in which electrons might be scattered by a molecule to devise appropriate wave functions for the electrons of the molecule — scattered wave X_α. [DSU]

Ballhausen, C.J. & Gray, H.B. (1965) *Molecular Orbital Theory.* Benjamin, New York.

Barrett, J. (1970) *Introduction to Atomic and Molecular Structure.* John Wiley & Sons, London.

Dewar, M.J.S. (1973) The role of semi-empirical SCF MO methods. In: Price, W.C., Chissick, S.S. & Ravensdale, T. (eds) *Wave Mechanics — The First Fifty Years*, pp. 239–54. Butterworths, London.

Johnson, K.H. (1972) *Scattered Wave Theory of the Chemical Bond. Advances in Quantum Chemistry* Vol. 7. Academic Press, New York.

Murrell, J.N., Kettle, S.F.A. & Tedder, J.M. (1965) *Valence Theory.* John Wiley & Sons, London.

Urch, D.S. (1979) *Orbitals and Symmetry.* Macmillan, London.

molluscs/Mollusca A diverse phylum typified by possession of a calcareous shell and in which the soft parts can be regarded as various modifications of a hypothetical 'archimollusc' body plan (Fig. M20). The shell may consist of one, two or, more rarely, seven or eight parts; the components are typically some form of coiled, hollow cone produced by varying rates of accretion around the aperture. Archimollusc, a concept of comparative anatomists, possesses a ventral muscular foot beneath a visceral mass containing the organs and a dorsal cap-shaped shell secreted by a layer of cells known as the mantle. The gut runs from an anterior mouth, provided with a rasping radula, to a posterior invagination (the mantle cavity) containing paired gills used for respiration. There are six living mollusc classes but a further extinct class, *Rostroconchia*, is now generally recognized and some authorities consider that others may have existed in the early history of the phylum. *Gastropoda, Bivalvia* and *Cephalopoda* have by far the largest fossil records. The seven main classes exhibit the following modifications of the archimollusc ground plan.

Monoplacophora (**Cambrian**–Recent) Multiple paired gills, kidneys, gonads and shell attachment muscles. In most forms the shell is only slightly coiled.

Amphineura (**Cambrian**–Recent) In subclass Polyplacophora dorsal surface covered by seven or eight calcareous plates. Subclass Aplacophora lacks calcification and has no fossil record.

Gastropoda (**Cambrian**–Recent) Anterior part of foot developed into a head with sense organs. In the post-larval stage visceral mass twisted (torted) through 180° about a dorso-ventral axis such that the anus comes to lie above the head. The shell is typically helically coiled. The gills are replaced by a lung in the air-breathing Pulmonata.

Cephalopoda (**Cambrian**–Recent) Differentiated head region provided with well-developed eyes. Visceral mass bent into a 'U' but not torted. The shell is internally partitioned by calcareous septa, the chambers being connected by a fleshy siphuncle. It is external (and commonly planispirally-coiled) in a large proportion of forms but in the subclass *Coleoidea* a calcareous shell is only developed internally and may be lacking entirely (many squids, octopodids). Where present it is blade-like in form.

Bivalvia (**Cambrian**–Recent) Shell consisting of two components (valves) of generally low inflation, joined dorsally by a partly or wholly organic ligament. Shell closed by adductor muscles running between the valves. Head region and radula lacking. Gills adapted for filter feeding as well as respiration.

Rostroconchia (**Cambrian**–**Permian**) Pseudobivalved shell; 'valves' connected dorsally by a continuous calcareous strip. The soft parts of this recently recognized extinct class were probably arranged much as in the *Bivalvia*, although ligament and adductors were lacking.

Scaphopoda (**Ordovician**–Recent) Univalved tusk-shaped shell, open at both ends, formed by ventral fusion of an initially pseudobivalved shell followed by continued accretion at the posterior aperture and resorption anteriorly. Differentiated head region and radula but gills rudimentary.

The above scheme is based on the work of Runnegar & Pojeta (1974). They consider that the uni- and bivalved classes can be linked to a monoplacophoran ancestry by a series of principally **Cambrian** intermediates. However, the earliest **Cambrian** (*Tommotian*) Mollusca, whilst in many cases possessing a simple conical shell, have not yet

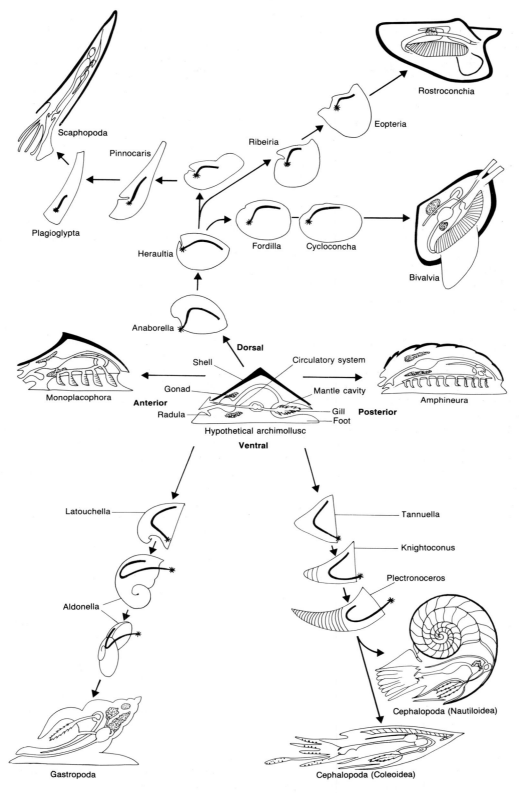

Fig. M20 Mollusca. Derivation of molluscan classes from a hypothetical archimollusc ancestor. Intermediate forms are early Paleozoic genera with extent of shell apertures shown by thickened outline; inferred position of mouth (starred) and gut also indicated. (After Runnegar & Pojeta, 1974 and Clarkson, 1986.)

supplied the indications of metameric segmentation (e.g. multiple muscle scars) present in *Monoplacophora*. It may be safer to assume that the soft parts were closer to those of the hypothetical archimollusc ancestor proposed by zoologists. In this view monoplacophorans and their presumed close relatives, polyplacophoran *amphineurans* (likewise exhibiting traces of metameric segmentation in their multiplated shells), would be regarded as offshoots of a primitively unsegmented molluscan line rather than intermediates between distant segmented ancestors ('worms') and the other molluscan classes. Recent discoveries of diverse ?late *Precambrian Polyplacophora* in China lend credence to the idea of segmentation as a primitive molluscan character, although the co-occurrence of univalves with slight helical coiling (?*Gastropoda*) should be noted. In Fig. M20 the fossil intermediate forms identified by Runnegar & Pojeta are interposed into an essentially zoological derivation scheme, thus giving a composite model. A very different, but controversial, view of molluscan phylogeny has been put forward by Yochelson (1979).

The divergence and subsequent evolutionary histories of the classes identified above can be explained in terms of their feeding habits and the incidence of predation. *Monoplacophora* are epifaunal grazers which use the radula as a rasp. Once regarded as an extinct **Paleozoic** class, a few genera (e.g. *Neopilina*) have been found in the modern deep sea. They have been supplanted in shallow water niches largely by *Gastropoda*, whose characteristic visceral torsion permits complete withdrawal of the head region into the shell and thus affords protection from predators. *Bellerophontids*, an important **Paleozoic** group of non-septate planispirally-coiled forms, include some examples with paired muscle scars. They cannot have undergone torsion and are therefore to be regarded as monoplacophorans rather than *gastropods*. The fossil record of *Amphineura*, also grazers (e.g. the modern *Chiton*), is poor owing to disarticulation of the hard parts. Their relatively low diversity at the present day (600 species) may relate to competition with grazing *gastropods* or greater susceptibility to predation.

Gastropoda are the most diverse molluscan class at present and include, as well as grazers, numerous predators and infaunal deposit and suspension feeding forms. The predatory habit is known from drill holes in other shells as far back as the **Trias**, but first became important in the mid-**Cretaceous**. Infaunal habitats were probably colonized in the **Paleozoic** when there are indications from apertural modifications of the existence of siphons to facilitate oxygenation of the gills when buried. The deposit- and suspension-feeding strategies represented did not, however, lead to the same diversification as occurred in the similarly-adapted *Bivalvia*. Gastropods, like *bivalves*, invaded the freshwater realm in the **Carboniferous**. They are the only molluscan group also having terrestrial representatives.

The shell form of the *Bivalvia* is adapted to an infaunal mode of life in that with the foot extended, closure of the valves displaces body fluids into the foot, distending it and permitting its use as an anchor against which the shell can be drawn into the sediment by the action of retractor muscles. The elastic ligament opens the valves on relaxation of the adductors and allows the cycle to be repeated. *Fordilla* (L. **Cambrian**), the earliest generally recognized bivalve, has a shell shape similar to many modern burrowing forms, thus infaunality is probably the primitive bivalve condition. It provides protection from predators and physical disturbance but inhibits movement, thus bivalves have abandoned the radular feeding method and become essentially sedentary, using the gills as a filtering device to extract suspended food. One order, the Nuculoida, reverted at an early stage to a more active, deposit feeding mode of life within the sediment, making use of a large foot and employing extensions of the mouth as feeding organs. Other suspension-feeding groups rapidly evolved epifaunal forms, anchored by a fibrous byssus, but these and later-evolving reclining and cemented forms are now less widespread, generally occurring in habitats where the incidence of predation is relatively low (e.g. oysters in brackish waters). A major post-**Paleozoic** radiation of infaunal bivalves occurred in conjunction with the development of siphons and of gills (eulamellibranch) with a high pumping capacity. This allowed occupation of deeper levels in the sediment than had been available to non-siphonate forms and afforded increased protection from predators. The life habits of *Rostroconchia* are uncertain. Their gills may not have been adapted for suspension feeding and in the presumed absence of a radula they were probably deposit feeders. In the early part of their history they were restricted to calcareous facies in which bivalves were rare and their ultimate **extinction** may have been due to competition with *bivalve* deposit feeders.

Scaphopoda appear to feed mainly on benthic **Foraminifera** which they suck into the mouth by means of water currents set up by contractions of the body. Although probably represented in the **Ordovician** they are not recorded again until the **Devonian**. Their eventual radiation may be related to diversification of their foraminiferal prey.

Through evolution of a septate shell whose chambers are pumped out after being formed, the *Cephalopoda* have been able to attain neutral buoyancy and thus free themselves from the substrate. The living, externally shelled *Nautilus* is able to vary its buoyancy by altering the water content of the chambers and has been observed to make diurnal journeys from deep to shallow water. The highly developed tentaculate head region appears adapted for a predatory mode of life but movement (through rhythmic expulsion of water from the mantle cavity) is relatively sluggish and it appears that only slow-moving organisms are taken, together probably with dead animals. The internally shelled (if at all) squids and cuttlefish are capable of much faster swimming, taking **fish**, but the octopodids have reverted to a benthic hunting existence.

These coleoids are much the most abundant cephalopods at the present day but in the past externally shelled forms have been at least as abundant. They were the first to evolve and included many straight and slightly coiled forms. The subclass *Nautiloidea* gave rise to the *Ammonoidea*, characterized by septal folding, and to the *Coleoidea*, in the **Devonian**. The ammonoids rapidly became the dominant externally shelled forms, evolving more complicated septal folding through a history punctuated by episodes of *mass extinction*; that at the end of the **Cretaceous** seeing the demise of the subclass. The coleoids suffered the **extinction** of the important Order *Belemnitida* early in the Tertiary but have susequently fared better than the nautiloids, represented now solely by *Nautilus*. [ALAJ]

Clarkson, E.N.K. (1986) *Invertebrate Palaeontology and Evolution* (2nd edn). Allen & Unwin, London.

Moore, R.C. (ed.) (1957–1971) *Treatise on Invertebrate Paleontology: Mollusca* (Parts I,K,L,N). Geological Society of America and University of Kansas Press, Lawrence, KS.

Peel, J.S., Skelton, P.W. & House, M.R. (1985) Mollusca. In: Murray, J.W. (ed.) *Atlas of Invertebrate Macrofossils*. Palaeontological Association and Longman, London.

Purchon, R.D. (1968) *The Biology of the Mollusca*. Pergamon, Oxford.

Raup, D.M. (1966) Geometrical analysis of shell coiling: general problems. *Journal of Paleontology* 40: 1178–90.

Runnegar, B. & Pojeta, J., Jr. (1974) Molluscan phylogeny: the paleontological viewpoint. *Science* 186: 311–17.

Yochelson, E.L. (1979) Early radiation of Mollusca and mollusc-like groups. In: House, M.R. (ed.) *The Origin of Major Invertebrate Groups*. Academic Press, London.

molybdenite (MoS_2) The major **ore** mineral of **molybdenum**. [DJV]

molybdenum (Mo) At. no. 42, at. wt. 95.94. A Group VI element. Used extensively in steels, particularly cutting steels and as a filament material. MoS_2 is used as a solid lubricant and molybdenum compounds are used in dyes. The production of molybdenum is from **disseminated molybdenum deposits** and as a by-product from some **porphyry copper deposits**. [AME]

monadnock An upstanding rock, hill or mountain which is isolated in an otherwise flat plain. In the *Davisian cycle of erosion* it is a product of the late stages and rises above a **peneplain**. The type example is Mount Monadnock in New Hampshire, USA. [NJM]

monalbite ($NaAlSi_3O_8$) A monoclinic, high-temperature form of the mineral **albite**. (See **feldspar minerals**.) [DJV]

monazite (($Ce,La,Y,Th)PO_4$) A rare earth phosphate occurring as an **accessory mineral** in **granites** and related rocks and concentrated as an **ore** mineral (chiefly of **thorium**) in **beach sands**. [DJV]

monocline Asymmetric **fold** produced by a local steepening of **dip**. The term is used normally for **macroscopic** structures only. [RGP]

montebrasite (($Li,Na)Al(PO_4)(OH,F)$) A variety of the mineral **amblygonite**. [DJV]

monticellite ($CaMgSiO_4$) An **olivine mineral**. [DJV]

montmorillonite (($Al,Mg)_8(Si_4O_{10})_3(OH)_{10} \cdot 12H_2O$) A **clay mineral**. [DJV]

moonstone A variety of **albite** or **oligoclase** exhibiting an opalescent play of colours and valued as a semiprecious **gemstone**. (See **feldspar minerals**). [DJV]

morganite A pink coloured **gem** variety of the mineral **beryl**. [DJV]

morphological mapping The mapping of surface relief and landforms. Such maps are used for regional planning, engineering, particularly for route decisions and potential construction hazards and **resource** surveys. Morphological maps combine basic mapping of a region, often from *aerial photographs*, with field surveys of landforms and surficial deposits. In central Europe they have been used by geomorphologists to reconstruct the evolution and origin of landforms. [RDS]

Brunsden, D., Doornkamp, J.C. & Jones, D.K.C. (1979) The Bahrain surface materials resources survey. *Geographical Journal* 145: 1–35.

mortar structure/texture An optical **microstructure** consisting of relatively large **strained** grains surrounded by smaller recrystallized **strain**-free grains. It is typical of **mylonites** developed from monomineralic rocks, e.g. **quartzite** or **limestone**. **Strain** is localized in the grain margins because of their irregular shapes. The texture consists of relict grain 'cores', and 'mantles' of finer grains produced by dynamic recrystallization (White, 1976). This texture does not result from **cataclasis**. [SB]

White, S.H. (1976) The effects of strain on the microstructures, fabrics and deformation mechanisms in quartzites. *Philosophical Transactions of the Royal Society, London* **A283**: 69–86.

moss agate A variety of the **silica mineral agate** with moss-like patterns, valued as a semi-precious stone. (See **gem**.) [DJV]

Mössbauer spectroscopy The recoilless emission and resonant absorption of gamma rays by nuclei in solids. Although observable in about 40 elements, by far the majority of work has involved **iron** with some studies on **gold**, **tin**, **tellurium** and **antimony**. The Mössbauer spectrum provides information on the nature (**oxidation** state, spin state) and environment (coordination in a crystal, whether magnetically coupled, etc.) of the atom concerned and has been much used to study crystal chemistry of **iron** in minerals (e.g. Fe site populations), magnetic ordering, or (in both minerals and bulk rock or soil samples) ferrous: ferric ratios. The method has been used in archeology to determine the temperature and atmospheric conditions of firing of medieval British pottery, and to characterize sources and prehistoric artefacts of **obsidian**. [DJV, OWT]

Hawthorne, F.C. (1988) *Mössbauer Spectroscopy*. In: Hawthorne, F.C. (ed.) *Spectroscopic Methods in Mineralogy and Geology*. Reviews in Mineralogy Vol. 18. Mineralogical Society of America, Washington DC.

mound spring Artesian spring, which forms preferentially along **fault** lines, and which forms a small mound. Solutes and colloids are precipitated to form **travertines** of calcium carbonate, together with various siliceous and ferruginous deposits. Wind-blown **sand** and accumulated plant debris, together with mud and **sand** carried up with the **spring** water, assist in their formation. Where **springs** display high rates of water flow there tends to be little or no mound formation — they are too erosive. However, **springs** with low discharge rates and **laminar flow** experience high rates of evaporation (especially in arid environments) and have a greater possibility of accumulating chemical precipitates. [ASG]

Ponder, W.F. (1986) Mound springs of the Great Artesian Basin. In: de Deckker, P. & Williams, W.D. (eds) *Limnology in Australia*, pp. 403–20. Junk, Dordrecht.

moveout (*stepout*) The difference in **arrival time** of a signal at two separate **geophones** along a survey line. Normal moveout is the moveout of a **seismic reflection** from a horizontal reflector. Dip moveout is similarly defined but isolates the component of moveout due to the dip of a reflector. In the **common depth point** (CDP) technique the normal moveout correction is applied to a set of CDP traces so that they can be summed and this **stack**ed trace plotted at the location above the CDPs. [GS]

mud crack Vertical to subvertical *shrinkage crack* produced by the contraction of cohesive muddy sediment beds. Mud cracks are preserved in rocks if they have been infilled by a different sediment, usually **sand**, from the bed above (Fig. M21). Subaerial drying of mud produces '*desiccation cracks*'. Subaqueous shrinkage cracks, often referred to as '*syneresis cracks*', form in response to **dewatering** of a mud layer as it compacts under its own weight, as flocs break down, or as certain **clay minerals** shrink due to increased inter-particle attraction consequent on a change in the salinity of the waters above. Cracks can develop with a **preferred orientation** when there is an additional source of

stress such as a slope. They are usually parallel-sided in plan and taper downwards in vertical section. Ancient crack fills are often folded due to differential compaction of the surrounding mud.

One of the most important controls on the geometry of *desiccation cracks* is the thickness of the desiccating mud layer; thicker layers produce wider and more widely spaced cracks. The spacings of *desiccation cracks* vary from centimetres to metres. Widths and depths can reach several centimetres and decimetres respectively. Subaqueous processes cannot cause as much shrinkage of mud as desiccation, so the resultant cracks tend to be small and unjoined (Donovan & Foster, 1972).

Desiccation of thin mud layers can produce concave-up curled clay flakes due to greater shrinkage of the top surface. Such flakes may be preserved if sand is gently blown or washed beneath them. Reticulate patterns of ridges on **sandstone bedding** surfaces may have formed in this way and might be confused with **interference ripples**. [DAR]

Allen, J.R.L. (1987) Desiccation of mud in the temperate intertidal zone: studies from the Severn estuary and eastern England. *Philosophical Transactions of the Royal Society, London* **B315**: 127–56.
Donovan, R.N. & Foster, R.J. (1972) Subaqueous shrinkage cracks from the Caithness Flagstone Series (Middle Devonian) of northeast Scotland. *Journal of Sedimentary Petrology* **42**: 309–17.

mud ripple A structure on mudstone **bedding** surfaces, and **sandstone** casts of them, which is similar in form and scale to the straighter crested type of *current ripple*. Such structures grade in form towards *flute marks* (see **sole mark**) and, like them, are probably an **erosional** feature. They can be produced experimentally by subaqueous currents over weakly cohesive mud. [DAR]

Allen, J.R.L. (1984) *Sedimentary Structures: Their Character and Physical Basis* (2nd edn). Developments in Sedimentology No. 30. Elsevier, Amsterdam.

mudflow A mass flow of debris, intimately mixed with water, and containing a high proportion of mud (fine **sand**, silt and clay). Mudflows are essentially the same phenomenon as **lahars**, and show very similar depositional and lithological features to those rocks. The nomenclature of water-rich **debris flows**, and the rocks they form after deposition, is in a state of confusion. The terms mudflow and **lahar** are both deeply entrenched in the literature, and their usual (and sensible) use is that both are varieties of **debris flows**, rich in water, but not sufficiently rich so as to flow as a hyperconcentrated stream; both are usually (exclusively?) of volcanic origin, and that mudflows are rather richer in fine-grained material than **lahars**. Mullineaux & Crandell (1962) and Janda *et al*. (1981) thought that **lahar** should encompass all water-rich volcanic *mass flows*, and that **debris flows** and mudflows should be varieties of **lahar** respectively relatively poor and rich in muddy matrix material (fine **sand**, silt and clay). Macdonald (1972) thought that **lahar** and mudflow

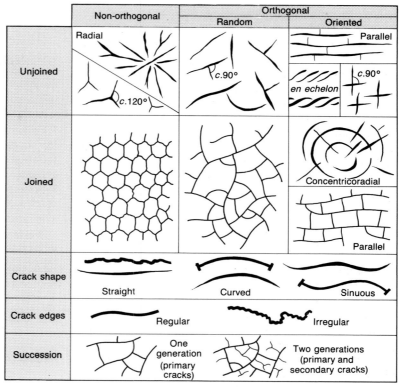

(a)

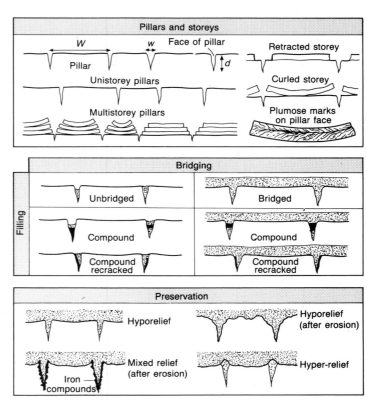

(b)

Fig. M21 Mud crack. (a) Outline classification of desiccation cracks viewed in plan. (b) Characterization and outline classification of desiccation cracks, their fillings and modes of preservation as seen in section. (After Allen, 1987.)

mudlump

should be synonyms, but Fisher & Schmincke (1984) thought that **lahars** should be a variety of **debris flow**, and that the rock-name mudflow should be discarded. (See **pyroclastic rock, tuff ring**.) [PTL]

Fisher, R.V. & Schminke, H.-U. (1984) *Pyroclastic Rocks*. Springer-Verlag, Berlin.
Janda, R.J., Scott, K.M., Nolan, K.H. & Martinson, H.A. (1981) Lahar movement, effects and deposits. In: Lipman, P.W. & Mullineaux, D.R. (eds) *The 1980 Eruptions of Mount St Helen's, Washington*. Professional Paper of the United States Geological Survey **1250**: 461–78.
Macdonald, G.A. (1972) *Volcanoes*. Prentice-Hall, Englewood Cliffs, NJ.
Mullineaux, D.R. & Crandell, D.R. (1962) Lahars from Mount St Helens, Washington. *Geological Society of America Bulletin* **73**: 855–69.

mudlump Small-scale landform associated with the Mississippi **delta** of the USA. The mechanism of formation is thought to be as follows (Morgan *et al.*, 1968): rapid forward growth of distributary channels causes deposition of **deltaic sand**, mud and organic material on top of unstable prodelta clay; this causes **loading** which in turn causes **diapiric** intrusions of plastic clays through the overlying **sands**; updoming or extrusion occurs, producing the mudlumps. [ASG]

Morgan, J.P., Coleman, J.M. & Gagliano, S.M. (1968) Mudlumps: diapiric structures in Mississippi delta sediments. *American Association of Petroleum Geologists Memoir* **8**: 145–61.

mudrock Sedimentary rock composed primarily of **clay minerals** (average of about 60%) with lesser amounts of **quartz** (30%) and still smaller amounts of **feldspar**, rock fragments, carbonate, **hematite** or other iron minerals and organic matter (all less than 5%). Mudrocks comprise at least 65% of sedimentary rocks.

Mud particles are defined as grains less than 62 µm in size, but most are in the clay size range of less than 4 µm. Clay particles tend to adhere together in the presence of organic matter and in saline water rich in electrolytes. This clumping together of clay particles into randomly oriented aggregates is termed *flocculation*. These floccules are consistently in the silt size range of 4 to 62 µm. (See **clastic sedimentary rock**.) [PAA]

mullion structure Type of linear, cylindrical structure consisting of elongate rods or columns, which may be either complete or incomplete in **cross-section**. Mullions vary in size from 20 mm across to surfaces with curvatures of 2 m and may be up to 100 m long. They may have smooth or corrugated surfaces and define a **lineation** that may be parallel to *fold axes*. They are most common in strongly deformed metamorphic rocks of all grades.

Mullion structures may be formed in a variety of ways. One of the commonest types of mullion is formed by **buckling** of the interface between **competent** and *incompetent* **beds**, producing a cuspate **cross-section**. Mullions may also be formed by the intersection of

bedding and **cleavage**, particularly when the **deformation** has accentuated linear **sole-marks** on the *bedding plane*. *Fold mullions* are formed by *fold hinges*. Mullions are also commonly formed by **boudinage**, so that the mullions correspond to linear **boudins**. Careful examination of mullions is necessary to determine their origin before deductions are made about **strain** implications.

Mullions have been classified:
1 *Fold mullions*: these have regular curved surfaces formed from a previous **foliation**, and represent *fold hinges* of that **foliation**.
2 *Bedding mullions*: polished or striated *bedding planes*.
3 *Cleavage mullions*: polished or striated cylinders bounded by **cleavage** planes.
4 *Irregular mullions*: cylinders with irregular **cross-sections**.
Many mullions are considered to have formed by **deformation** in *constrictional strain* with **extension** parallel to their length. [TGP/RGP]

Ramsay, J.G. & Huber, M.I. (1987) *The Techniques of Modern Structural Geology* Vol 2: *Folds and Fractures*. Academic Press, London.
Wilson, G. (1953): Mullion and rodding structures in the Moine series of Scotland. *Proceedings of the Geologists' Association* **64**: 118–51.

mullite ($Al_6Si_3O_{15}$) A rare aluminosilicate mineral. [DJV]

multiple intrusion An igneous intrusion characterized by successive emplacement of **magmas** of distinctive but similar (e.g. basic) composition, but distinguished by internal contacts and/or degrees of **phenocryst** content and crystallinity. Examples of such intrusions include **dykes** and **sills**. (See **composite intrusions**.) [RST]

multiple reflection (*multiple*) Seismic energy that has been reflected from more than one reflector in the subsurface before being recorded. Short period multiples arrive immediately after the primary **seismic reflection** producing a complex elongated waveform, whereas long period multiples form distinct events with a greater **moveout** than primary **seismic reflections** with a similar **arrival time**. As such these multiples can be attenuated by the application of the correct normal **moveout** during the **common depth point stack**ing technique. Predictive **deconvolution** is often applied to simplify 'ringing' events produced by short period multiples, by 'predicting' the ringing events and **attenuating** them. [GS]

multiplexing The process whereby multiple input channels of information are interleaved and transferred to a single output channel. The input channels are usually rapidly sampled in sequence at constant intervals in time and this digital information is then fed to a single output channel for recording on a computer-readable medium. Transmission of information can be undertaken by frequency domain multiplexing by allocating different channels of input data to various modulating frequencies within the bandwidth of the carrier. *De-multiplexing* is the process by which the multiplexing is reversed. [GS]

muscovite (KAl$_2$(AlSi$_3$O$_{10}$)(OH)$_2$) A common and important member of the **mica** group of **minerals**, also known as 'white mica'. [DJV]

muskeg The water-logged *marsh* and **bog** land of the north-west of Canada, or a **peat**-filled **basin** characterized by *Sphagnum* moss. [ASG]

mylonite Fine-grained **foliated** rock with recrystallized texture and a strong **lineation** resulting from **shear** in a major *ductile* **fault** or **shear zone** (see also **fault rock**). *Mylonitization*, the process of forming a mylonite, generally consists of *crystal plastic strain* by *dislocation climb* and **recovery**, forming new **mineral** grains by *dynamic recrystallization*. Other **deformation mechanisms** such as *grain boundary sliding* may also operate in the formation of a mylonite. Development of a mylonite is a gradual process of grain size reduction in which three stages, *protomylonite*, mylonite and *ultramylonite*, are distinguished on the basis of the proportions of relict and new grains (see **fault rocks** and Sibson, 1977). The terms *mylonite gneiss* and *mylonite schist* refer to mylonites with well-developed **gneissosity** and **schistosity** respectively. A *blastomylonite* is a type of mylonite that contains, within a fine-grained groundmass, some crystals of considerably larger size that grow during or after **deformation**. These **porphyroblasts** indicate metamorphic grade, and help to define **deformation** conditions when syntectonic. Where deformed, such crystals can be useful indicators of the sense of **shear** of the **deformation**, e.g. where they show *pressure fringes* (see **pressure shadow**), or where they contain *inclusion trails*. Differences in original composition lead to a strong compositional layering by streaking out **minerals** such as **quartz** to form ribbons, and the **fractur**ing of **minerals** such as **feldspar** (depending on conditions of **deformation**). *Intrafolial folds*, often curvilinear (see **sheath folds**), are common in mylonite zones. The **strain** processes of the different **minerals** will depend on the temperature and pressure states at the time of **deformation** (see **deformation mechanism**). The lattice **deformation** characteristic of *dynamic recrystallization* means that most mylonites develop strong crystallographic **preferred orientations** (see Law *et al.*, 1986). **Quartz** and **calcite** **fabrics** have been extensively studied in terms of *c*-axis distributions, which can indicate the sense of **shear** in *non-coaxial deformation*.

Mylonite zones are belts or zones of mylonite along major *fault zones*. They can outcrop over lengths of hundreds of kilometres, and can be kilometres thick. They can form in **strike-slip**, **extensional** or compressional tectonic regimes, and are characteristic of moderate–high temperature **deformation** (depending on mineralogy). *Mylonite zones* are rarely homogeneous, and can contain all types of mylonites, often arranged as high and low **strain** domains. They generally represent **deformation** at depths of greater than 10–15 km. The **uplift** resulting from *thrust faulting* means that they commonly are found in **thrust belts**. [SB]

Law, R.D., Casey, M. & Knipe, R.J. (1986) Kinematic and tectonic significance of microstructures and crystallographic fabrics within quartz mylonites from the Assynt and Eriboll regions of the Moine thrust zone. *Transactions of the Royal Society of Edinburgh, Earth Sciences* **77**: 99–125.
Sibson, R.H. (1977) Fault rocks and fault mechanisms. *Journal of the Geological Society, London* **133**: 191–213.

myrmekite A vermicular intergrowth of **quartz** and sodic **feldspar** adjacent to an **alkali feldspar** crystal. The intergrowth may be within part of a euhedral **plagioclase** crystal, or may form within a lobate or cauliflower-like part of such a crystal. The origin is assumed to be secondary, caused by release of **quartz** as a result of replacement of **alkali feldspar** by **plagioclase** (i.e., K-feldspar + CaO → **anorthite** + K$_2$O + 4SiO$_2$). [RST]

N

nacrite ($Al_2Si_2O_5(OH)_4$) A **clay mineral** of the same composition as the common mineral **kaolinite** but different crystal structure. [DJV]

nagyagite ($AuTe.6Pb(S,Te)$) A rare telluride mineral found in **hydrothermal veins**. [DJV]

nanotesla (nT) Subunit equal to 10^{-9} tesla ($V\,s\,m^{-2}$) used in measuring **magnetic field** strength. [PK]

nappe A body of rock that has undergone considerable horizontal **tectonic** transport in an *orogenic belt*. There is generally considerable **folding** and **ductile deformation** within nappes, with a downwards increase in **strain** intensity. A *fold nappe* comprises a large *asymmetric fold* structure with a subhorizontal *axial surface* and *hinge line*, (a *recumbent fold*). The term nappe has been used synonymously with *thrust sheet*, but nappes are now generally understood to show higher intensities of internal **deformation** than *thrust sheets* (see discussion by Dennis *et al.*, 1981). In *fold nappes*, the lower overturned limb is most highly strained and sheared out. A **shear zone** is often developed at the base of a nappe, which can take up much of the displacement of the nappe as a whole. Nappes form an important part of the internal zones of *orogenic belts*; the Helvetic Nappes of Switzerland are good examples (Fig. N1). [SB]

Dennis, J.G. *et al.* (1981) What is a thrust? What is a nappe? In: McClay, K.R. & Price, N.J. (eds) *Thrust and Nappe Tectonics*, pp. 7–12. Geological Society Special Publication No. 9. Blackwell Scientific Publications, Oxford.
Hobbs, B.E., Means W.D. & Williams, P.F. (1976) *An Outline of Structural Geology*. John Wiley & Sons, New York.

native element A **mineral** that is made up of chemical elements in an uncombined or 'native' state. About 24 elements are known to occur in their native state in rocks; all of these are relatively rare, with none constituting large masses of rocks. However, included in this group are some of the most economically important **ore** minerals (native **gold**, **silver**, **platinum**) and **gemstones** and **industrial minerals** (**diamond**, native **sulphur**).

The important native elements are commonly classified as follows:
Metals:
  Gold (Au) (and **gold–silver** alloys)
  Silver (Ag)
  Copper (Cu)
  Platinum (Pt) (and other platinum group metals)
  Iron (Fe) (and (Fe,Ni) alloys)
Semi-metals:
  Arsenic (As)
  Antimony (Sb)
  Bismuth (Bi)
Non-metals:
  Sulphur (S)
  Graphite (C)
  Diamond (C)

This classification arises essentially from the differences in bonding in the members of the three groups, and hence in the characteristic properties of those with metallic bonding (opaqueness, high **electrical** and **thermal conductivity** and metallic **lustre**), as against those in which bonding is essentially covalent and hence have properties distinctly different from those of metals, or those of intermediate character that are termed *semi-metals*.

Further subdivision of the members of the above groups may be made on the basis of **crystal structure**. **Gold**, **silver**, **copper** (and the very rare native **lead**) all have cubic close-packed structures, as do **platinum** and **palladium**. Specifically, the structures are based on the face-centred cubic lattice, whereas **iron** has a body-centred cubic lattice. Hexagonal close packing occurs in *osmium* and in native **zinc**.

Many of the metals that occur as native elements also occur in nature as alloys, and such phases are also classed with the native elements. A complete *solid solution* exists between **gold** and **silver**, and native **gold** commonly contains up to 10–15% of **silver**. The name **electrum** is applied to **gold** with >20% **silver**. Natural alloys are particularly important amongst the **platinum** group metals, **platinum** itself forming alloys with **iron** and with **iridium**. The *osmium*-**iridium** alloy phases have a hexagonal close-packed structure. **Gold**, **silver**, the **platinum** group metals and alloys, although rare in occurrence, are all of great importance as **ores** of the **precious metals**. **Iron** (α-Fe) always contains some **nickel** but is very rare in the Earth's crust, although the alloys *awaruite* (Ni_3Fe) and *wairauite* ($CoFe$) occur as rare phases in **peridotites**. A particular area of interest, however, is the occurrence of **iron** and **nickel–iron** alloys in *meteorites*, commonly as lamellar intergrowths of **kamacite** (Fe with *c.* 5.5 wt% Ni and a body-centred cubic structure) with **taenite** (Fe with *c.* 27–65 wt% Ni and a face-centred cubic structure).

The *semi-metals* **arsenic**, **antimony** and **bismuth** belong to an isostructural group (space group R$\bar{3}$m), with trigonal

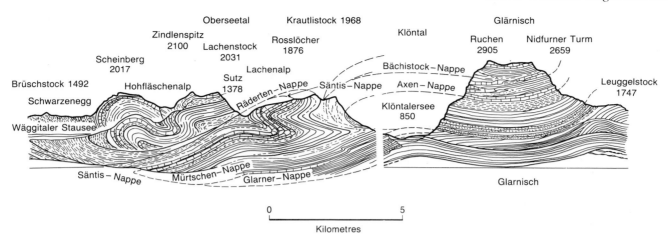

Fig. N1 Heim's (1919) cross-section through the Swiss Alps showing major **nappes** and fold nappes, displaying discontinuities at their bases, and significant internal deformation. (After Hobbs *et al.*, 1976.)

structures in which groups of four atoms form pyramidal groups that in turn link together to form layers parallel to {0001} (giving a good basal **cleavage**). All three have similar physical properties, being **brittle** and showing poorer **electrical** and **thermal conductivity** than the metals. They are rare minerals, being found as minor phases in metalliferous **ores**.

Sulphur occurs in several forms as a native element but the commonest is α-sulphur or rhombic **sulphur**, which has the atoms bonded into puckered rings of eight atoms. As well as occurring in areas of recent volcanic activity, it is found in sedimentary sequences from which it is extracted.

Carbon occurs in the dramatically contrasting forms of **graphite** and **diamond**. In **diamond**, the **carbon** atoms are all linked tetrahedrally to four neighbouring **carbon** atoms by strong covalent bonds to give a cubic phase of great hardness. In **graphite** (hexagonal) **carbon** atoms are linked into hexagonal arrays forming sheets parallel to {0001} with weak bonding (and large spacing) between the sheets, and therefore a very well-developed basal **cleavage**. **Diamond** is exploited largely for use in abrasives and also, of course, as a **gemstone**. [DJV]

natroalunite $((Na,K)Al_3(SO_4)_2(OH_6)$ A sulphate mineral, the sodium-rich (Na > K) equivalent of **alunite**. [DJV]

natrolite $(Na_2Al_2Si_3O_{10} . 2H_2O)$ A **zeolite mineral**. [DJV]

natural arch A bridge or arch of rock joining two rock outcrops, which has been produced by natural processes of **weathering** and **erosion**. They may occur in coastal situations as a result of **joint** enlargement, in **karstic** areas as a result of **solutional** attack, and in other sedimentary rock areas as a result of differential **erosion** by fluvial processes. [ASG]

natural gas The gaseous members of the materials that constitute **petroleum**. These are the first four members of the paraffin series (see **crude oil**), *methane*, ethane, propane and *n*-**butane**. *Dry gas* is composed almost entirely of *methane*. *Wet gas* contains significant amounts of the other gases. *Dry gas* can be of three different origins:
1 as a by-product of the generation of **crude oil** (see **hydrocarbon generation**) or through its thermocatalytic modification at high temperatures;
2 by the modification of **coal**, perhaps often due to **contact metamorphism**, and
3 by *bacterial* **alteration** of organic matter at or near the Earth's surface — *marsh gas* — which, of course, has no commercial importance. [AME]

natural remanent magnetization The magnetic remanence carried by rocks that has been acquired by natural geological processes — the study of which is termed **paleomagnetism**. Five main processes are involved. **Thermal remanence**, is associated with all rocks that have, at some time in their history, been heated and is the magnetization acquired as they cooled from ambient temperatures at or below their **Curie temperature**. **Chemical remanence** is acquired as new magnetic grains grow in the **geomagnetic field** and is thus acquired at any time that such chemical changes occur, including both early changes, such as **diagenesis** in sediments, and changes during metamorphism or **weathering**. Already magnetized grains, as they are deposited in water or air, are also partially aligned by the **geomagnetic field** and thus acquire a **depositional (detrital) remanence**. As all magnetic materials lying in a **magnetic field** very gradually lose their original magnetization, they also acquire a time-dependent **viscous magnetization**. Occasionally, rocks may be struck by lightning, in which case they acquire a very complex magnetization, most of which is an **isothermal remanent magnetization**. [DHT]

[415]

natural strain A linear measurement of shape change during **deformation** based on the integral of instantaneous **strain** values. For an instantaneous **strain** measurement, the small increment of increase in length, dl, of a line of initial length l, form the ratio, which is summed thus:

$$n = \int_{l_u}^{l_d} \frac{dl}{l}$$

where l_d = deformed length
l_u = underformed length

This integral reduces to:

$$n = \tfrac{1}{2} \ln(\lambda) = \ln(1 + e)$$

where λ = **quadratic elongation**
and e = **elongation**

The natural strain can thus be measured from the **finite strain**; it is always a smaller value than the **elongation** in similar circumstances. [SB].

Means, W.D. (1979) *Stress and Strain*. Springer-Verlag, Berlin.

Navier–Stokes equations A series of equations of linear momentum for a moving viscous compressible fluid, particularly relating **stress** and the state of **strain**. [DHT]

Tritton, D.J. (1988) *Physical Fluid Dynamics*. Clarendon Press, Oxford.

nebkha An aeolian bedform consisting of wind-blown **sand** collected and stabilized in and behind vegetation. Alternative names are *coppice* or *shrub-coppice dunes* or *mounds*. **Sand** accumulates because of the local reduction in wind speed caused by the vegetation, and may reach heights of over 3 m. Seasonally or episodically active **slip faces** may or may not be present, as may **lee-side** 'shadows' of unvegetated **sand**. **Erosion** may occur between clumps of vegetation by wind channelling. Internally, depositional stratification is obscured or destroyed by *bioturbation* by the deep roots of desert plants' roots, producing a sedimentary structure termed '*dikaka*' (Glennie, 1970). Preferential early **cementation** may occur around such roots. Nebkha deposits and *dikaka* indicate that non-saline **groundwater** was present in an ancient sedimentary environment. [DAR/RDS]

Glennie, K.W. (1970) *Desert Sedimentary Environments*. Developments in Sedimentology No. 14. Elsevier, Amsterdam.
Gunatilaka, A. & Mwango, S. (1987) Continental sabkha pans and associated nebkhas in southern Kuwait, Arabian Gulf. In: Frostick, L.E. & Reid, I. (eds) *Desert Sediments: Ancient and Modern*, pp. 187–203. Geological Society Special Publication No. 35. Blackwell Scientific Publications, Oxford.

necking Localized thinning of a structure during **extension**. Necking is common in the **competent** units of multilayers stretched parallel to the layering, and may occur periodically to give **pinch-and-swell structure**, with less **competent** adjacent units flowing into the thinned

regions. With increasing **strain**, the neck **fractures** to separate discrete **boudins**. The amount of necking before **fracture** depends on the conditions of **deformation** and the competence contrast between layers. A lower competence contrast promotes necking over **fracture**. (See **boudin**.) [TGB]

Néel point/temperature The temperature at which the coupling between anti-parallel magnetic lattices within the same **ferrimagnetic** or **antiferromagnetic** material breaks down. [DHT]

neocatastrophism A term introduced in the mid-20th century by paleontologists concerned with sudden and massive **extinctions** of life forms, such as that which afflicted the great **mammals** at the end of the *Pleistocene*. It has also been extended into **geomorphology** by Dury (1980), where there is increasing concern with the role of events of great magnitude and low frequency in moulding the Earth's surface. [ASG]

Dury, G.H. (1980) Neocatastrophism? A further look. *Progress in Physical Geography* **4**: 391–413.

neoformation of clays In its broadest sense, the **authigenesis** of **clay minerals**. This may be the result of direct **precipitation** from a **pore-fluid**, the **neomorphism** of one **clay mineral polymorph** to another or the **replacement** of a precursor mineral by a newly formed **clay mineral**.

Clay minerals may form in a variety of geological environments, ranging from **diagenetic** (including all **diagenetic regimes** — *eogenesis*, burial and *telogenesis/weathering* through **hydrothermal**) to low-grade metamorphic. With evolving physico-chemical conditions (P, T, **pore-fluid** chemistry) one **clay mineral** may transform via **replacement** processes into another.

Although there is a voluminous quantity of data available concerning the origin of **clay minerals** in diverse geochemical environments, there is little coherency or agreement as to the exact geochemical conditions of **clay mineral** formation or to the controlling factors which may influence their **precipitation** or stability.

Formerly, it was widely held that clay mineral neoformation could take place when detrital clays, derived from continental **weathering**, entered the marine environment (see Millot, 1970), although it is now generally accepted that detrital clays do not undergo fundamental structural modifications in depositional **pore-waters**. It is, however, vitally important to consider the processes and conditions that were ultimately responsible for the formation of detrital clays, often in the source **weathering** areas, when investigating the origin of **clay mineral** assemblages.

Indeed, much of our current understanding of **clay mineral** stability and their response to the geochemical

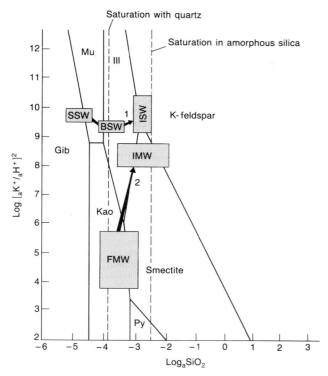

Fig. N2 Neoformation of clays. Combination diagram showing stability relations in the system $K_2O-Al_2O_3-SiO_2-H_2O$. Pathway 1 depicts evolution of normal marine seawater during early diagenesis by interstitial reaction with pore waters. SSW = surface seawater, BSW = bottom seawater, ISW = interstitial seawater. Pathway 2 depicts equilibration of fresh meteoric water (FMW) with the detrital assemblage to produce average interstitial meteoric water (IMW). Gib = gibbsite; Mu = muscovite; Ill = illite; Kao = kaolinite; Py = pyrophyllite.

environment has developed from studies of rock **weathering** and *soil* science and, in many cases, concepts derived from **weathering** can be applied to *eogenetic* and shallow **burial diagenetic** reactions. Differences in **clay mineral** occurrences can often be related directly to the chemistry of the enclosing **pore-waters**; these relationships can be

shown graphically on ion activity diagrams and used to depict possible **pore-water evolution** (Fig. N2).

Clay minerals are sheet silicates related to the **micas**. **Authigenic** clays are typically fine-grained which makes their structural and chemical characterization difficult. The classification scheme generally accepted is based on the **micas** which form at higher temperatures and pressures. Four major groups of **authigenic** clays (and their inter-stratifications) are most commonly reported in sediments: *kandites*, **illites**, **smectites** and **chlorites**. No individual analytical technique can be employed comprehensively to characterize any one particular **clay mineral** species; usually a combination of petrography (thin section or scanning **electron microscopy**), **X-ray diffraction** and analytical transmission **electron microscopy** or *microprobe* data are required.

Clay minerals may be of detrital or **authigenic** origin in sediments. Their distinction is often difficult, particularly in finer-grained sediments, as both detrital and **authigenic** clays may display a variety of textural modes and **fabrics** (Figs N3, N4 and N5; see also **mechanical infiltration of fines**). In the most general terms, detrital clays tend to be very fine-grained, poorly crystalline and of indistinct morphology, whilst **authigenic** clays tend to be well crystallized, of distinct, delicate morphology (often with euhedral crystal habit) and occur as grain coatings or pore-fillings.

The formation of **authigenic** clays involves the presence of an aqueous pore medium and, consequently, the type of **clay mineral** precipitated and its exact chemical composition reflects that of the enclosing **pore-waters**.

Precipitation of the *kandite* group of minerals (**halloysite**, **kaolinite** and **dickite**) is generally considered to be favoured by dilute, molecularly dispersed low-**pH** aqueous solutions (see Fig. N2). They are often associated with **granitic** or *arkosic* host lithologies rich in **feldspar**, although this need not be the case as many **groundwaters** in the tropics and humid temperate climatic zones are often sufficiently saturated with alumina and silica to precipitate *kandites*. **Halloysite**, the fully hydrated **polymorph**

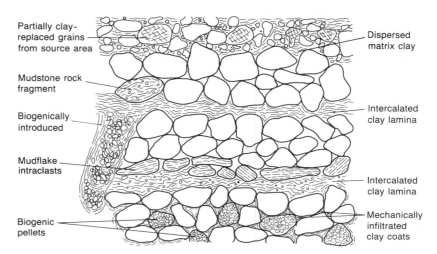

Fig. N3 Neoformation of clays. The modes of occurrence of detrital clay in sandstones. (After Wilson & Pittman, 1977.)

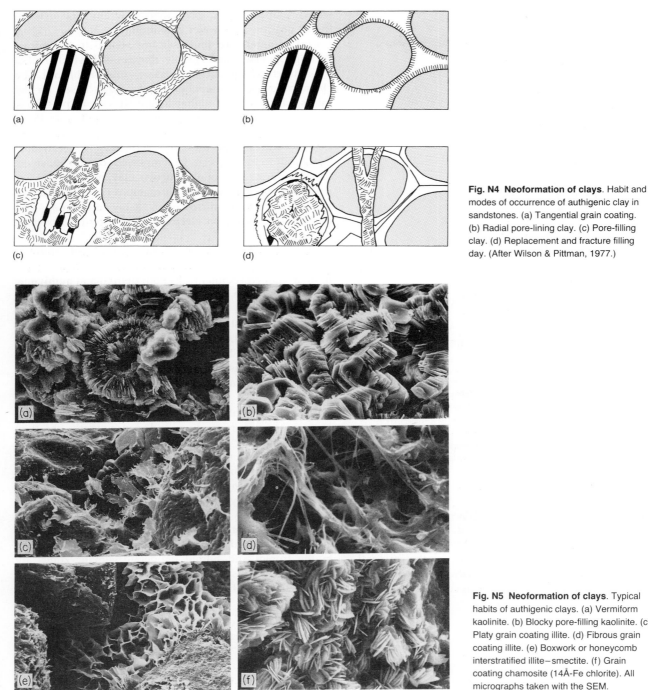

Fig. N4 Neoformation of clays. Habit and modes of occurrence of authigenic clay in sandstones. (a) Tangential grain coating. (b) Radial pore-lining clay. (c) Pore-filling clay. (d) Replacement and fracture filling day. (After Wilson & Pittman, 1977.)

Fig. N5 Neoformation of clays. Typical habits of authigenic clays. (a) Vermiform kaolinite. (b) Blocky pore-filling kaolinite. (c) Platy grain coating illite. (d) Fibrous grain coating illite. (e) Boxwork or honeycomb interstratified illite–smectite. (f) Grain coating chamosite (14Å-Fe chlorite). All micrographs taken with the SEM.

of **kaolinite**, is widely reported as a **weathering** product of K-**feldspar** in *soil* environments but **kaolinite** is the dominant *kandite* described in **diagenetic** regimes, both as an **alteration** product of **feldspar** and as a direct precipitate within intergranular **porosity**. It is unlikely that all **kaolinite** in **diagenetic** regimes is derived from the transformation of **halloysite** (although the reaction is certainly possible on thermodynamic grounds). Rather, there is

probably some fundamental kinetic or geochemical control on the differential **precipitation** of **halloysite** and **kaolinite** in *soil* and **diagenetic** regimes. Similarly, little is known of the conditions governing the stability of **dickite** relative to **kaolinite**. Dickite is often reported from **hydrothermal deposits** and is also known as a pore-filling mineral in some **sandstones**; this has led to the idea that **dickite** may be the high temperature **polymorph** of **kaolinite**, a suggestion supported by the high degree of order typical of **dickite**.

The occurrence and morphology of **authigenic** *kandites* in **diagenetic** settings also varies considerably and can be related to the geochemistry of the **pore-waters**. *Kandites* are commonly recorded as **alteration** products of both K-**mica** and **feldspar**, e.g.

$$2K\,AlSi_3O_8 + 2H^+ + H_2O = Al_2Si_2O_5\,(OH)_4 + SiO_2 + 2K^+$$
K-feldspar   kaolinite

For such reactions to proceed, the dissolved reaction products (Si, K) have to be continually removed or locally precipitated within other minerals (such as **quartz** overgrowths or illitic clays, etc.). This has led to the suggestion that a continuous flux of dilute acidic **pore-waters** is required to extensively 'kaolinize' feldspars. **Meteoric water** is usually ascribed as being responsible for the required freshwater flush. Pore-filling *kandites* may occur either as very coarse, vermiform aggregates comprising many stacked **kaolinite** plates or as dense, blocky aggregates of small, randomly oriented euhedral crystals (Fig. N5a and b). This variation in the morphology of **authigenic** *kandites* may be controlled by growth rate, with the vermiform *kandites* having grown more rapidly than the small euhedral *kandites*. The rapid input of nutrient ions implied with the vermiform *kandites* is consistent with a through-flow of meteorically derived **pore-waters** whilst slower *kandite* growth in the absence of a **pore-water** flow is typical of more closed reaction systems.

Illite, I-S clay and **smectite** are, in contrast to the *kandite* group, generally considered to form in **pore-waters** of alkaline **pH** and high ionic strength, specifically high K^+/H^+ or $(FeMg)Mn^{2+}/H^+$ ratios. In general, **illite** end-member **precipitation** is favoured by high K^+ activity and the **smectitic** end-member is favoured by higher (FeMg) activity, a view borne out by the common **alteration** of K-feldspar by **illite** and basic (FeMg) minerals by **smectite**. Both **illite** and **smectite** can form in moderately reducing to highly **oxidizing** environments, as reflected in the variation of Fe^{3+} contents from <5% to *c*. 15%. **Smectites** display a wide variation in interlayer cation chemistry (Ca, Na, K, Mg) which largely reflects either parent material or **pore-water** chemistry. **Illites**, by contrast, are much more restricted in their interlayer cation composition; despite the dominance of NaCl in sedimentary aqueous solutions, most diagenetic **illites** are K-dominant. Although dioctahedral **montmorillonitic smectites** are typical of alkaline conditions, numerous studies have reported the widespread development of trioctahedral **beidellite**-like **smectites** in highly acidic (pH < 4.5) *soils*. **Illites** and **smectites** (and interstratified mixtures of these two end-members) can, therefore, form in a wide range of geochemical conditions as reflected in their diverse chemical composition. As a consequence they may also occur in a wide variety of geological settings. In the **weathering** environment **smectites** are found in both poorly-drained *soils* with high cation retentivity, typically associated

with the **alteration** of basic rock lithologies, and also in well-drained *soils* in semi-arid climates with little **runoff**. **Illite** is a common **weathering** product of the lower part of many *soil* profiles, particularly in the presence of K-**feldspar**, although with increasing degree of **weathering illites** evolve towards **smectitic** products, often via a **vermiculite** precursor. In the marine environment, **illite** is commonly reported as a major component of the **clay mineral** assemblage but most **illite** in Recent marine sediments is considered to be detrital in origin, derived from the **weathering** of continental material. **Smectites**, on the other hand, do form **authigenically** on the seafloor in Recent marine sediments as a result of seawater **alteration** of volcanic rocks, low temperature combination of biogenic **silica** with Fe-oxyhydroxides and direct **precipitation** from *hydrothermal fluids*.

The reaction of **smectite** to **illite** with increasing burial during *mesogenesis* is well known and has been documented by many workers in several sedimentary **basins**. The process was first recognized in *Tertiary* sediments of the Texas Gulf Coast but has since been studied in great detail. The reaction is complex and does not simply take place by cation exchange but also involves **aluminium** substitution for silicon, reduction of octahedral **iron** and eventually loss of both octahedral **magnesium** and **iron**. These changes may take place in more than one way. The original **smectite** 2 : 1 layers may remain intact and the reaction may largely involve ionic substitution and expulsion:

$$4.5K^+ + 8Al^{3+} + KNaCa_2Mg_4Fe_4Al_{14}Si_{38}O_{100}(OH)_{20}.10H_2 \rightleftharpoons$$
$$K_{5.5}Mg_2Fe_{1.5}Al_{22}Si_{35}O_{100}(OH)_{20} + Na^+ + 2Ca^{2+} +$$
$$2.5Fe^{3+} + 2Mg^{2+} + 3Si^{4+} + 10H_2O$$

In this reaction, **aluminium** is a mobile component and 1 mole of **smectite** is used to produce 1 mole of **illite**. However, the reaction may actually involve collapse of the 2 : 1 **smectite** structure to liberate **aluminium** for the **illitization**:

$$3.93K^+ + 1.57KNaCa_2Mg_4Fe_4Al_{14}Si_{38}O_{100}(OH)_{20}.10H_2O \rightleftharpoons$$
$$K_{5.5}Mg_2Fe_{1.5}Al_{22}Si_{35}O_{100}(OH)_{20} + 1.57Na^+ + 3.14Ca^{2+} +$$
$$4.28Mg^{2+} + 4.78Fe^{3+} + 24.66Si^{4+} + 57\,O^{2-} + 11.40(OH)^- +$$
$$15.7H_2O$$

This reaction suggests that approximately 3 moles of **smectite** transform to 2 moles of **illite** resulting in an overall reduction in the amount of 2 : 1 **clay mineral** material present in the **mudrock** as a result of **smectite** collapse.

Current ideas suggest that both these two proposed mechanisms may be too simple. Detailed high resolution *transmission electron microscopy* studies indicate that the reaction involves the breakdown of major parts of both octahedral and tetrahedral units with a fundamental change in composition and structure of these sheets as they become reconstituted to form defect-free 2 : 1 **illite** layers. Both **iron** and **magnesium** can diffuse to new sites or move out of the system to be incorporated into other mineral components. Transport of reactants (K, Al) and products (Na, Si, Fe, Mg and H_2O) are inferred to take place through the surrounding **smectite** matrix along lattice dislocations by diffusion. These studies suggest that the **illitization** reaction proceeds by at least partial internal utilization of components common to both **smectite** and **illite** to form the new **illite** layers.

The relationship between the **illitization** of **smectite** in **mudrocks** and **precipitation** of **authigenic illites** in porous **sandstones** is also not yet fully clear. Pore-filling, hairy or filamentous **illites** are commonly reported from deeply buried *hydrocarbon reservoir* **sandstones** (see Fig. N5c). Many of these occurrences are specific to the water zone below the *hydrocarbon reservoir*. The **illite** laths range up to $50\,\mu m$ in length, typically between $0.1\,\mu m$ and $0.5\,\mu m$ in width and up to $200\,\text{Å}$ thick. Most **illite** laths possess perfect crystal morphology, displaying (110) prism face terminations. In general, the chemistry of these **illites** approaches that of *phengite*, with low (FeMg) contents. The available morphological, textural and chemical evidence suggests that these **illites** were direct precipitates from the interstitial **pore-fluids**. There is, however, very little data on the associated formation waters in these deep **reservoirs**.

The **illitization** of *kandite* is commonly reported in similar deep burial *hydrocarbon* settings. In all these cases the **illitization** takes place at considerable depth and at temperatures around 100°C. However, temperature alone is clearly not sufficient to promote the conversion of **kaolinite** to **illite** as an influx of **potassium** is clearly required:

$$8Al_2Si_2O_5(OH)_4 + 3K^+ = 2K_{1.5}Al_4Si_{6.5}O_{20}(OH)_2 \\ + 3Si(OH)_4 + 8Al^{3+} + 16OH^-$$

Thus the **pore-water** chemistry is likely to be an important control on the stability of **kaolinite**. Given a suitable supply of K^+, **illitization** of **kaolinite** should, therefore, proceed spontaneously. However, although **illitization** is commonly recorded from deep burial **sandstones**, **kaolinite** decomposition is not always associated with **illitization** and neither do **kaolinites** always decompose when **illitic** clays form. Indeed, **illitic** minerals may coexist with **kaolinite** over a wide range of T–P conditions and probably in a wide range of **porewater** compositions. Thus, whilst **pore-water** chemistry is an important control on the **illitization** of **kaolinite**, kinetic or specific saturation conditions may influence the reaction and **kaolinite** may remain metastable within the stability field of **illite**.

Chlorite authigenesis during **weathering** or *eogenesis* is generally held to take place under extremely reducing conditions in low-**pH** solutions with relatively high (FeMg) activity but correspondingly low sulphide and bicarbonate activity. In oxidizing environments most **chlorites** are unstable. **Magnesium-chlorites** generally alter towards a **smectitic** clay whilst **iron**-rich **chlorites** commonly produce interstratified *vermiculite*-**chlorite** or *vermiculite* daughter clays. However, in some settings **chlorites** do remain stable in **oxidizing** environments, particularly **iron**-rich varieties that contain significant proportions of ferric **iron**. *Berthierine* is recorded from modern *soil* profiles and **iron**-rich $14\,\text{Å}$ **chlorites** occur in *Pleistocene glacial* sediments. Thus under **oxidizing** conditions **iron**-rich **chlorites**, mixed-layer **chlorite-smectite** or **chlorite**-*vermiculites* may persist.

Within *eogenetic* environments **authigenic chlorite** is reported from both marine and fluvial systems associated with organic matter and reducing conditions. *Berthierine* ($7\,\text{Å}$-chlorite) characteristically forms **ooids**, often in the vicinity of **deltas**, suggesting clay **authigenesis** takes place via reaction of detrital **iron**-coated **ooids** and seawater (see **depositional environment related diagenesis**). The **oxidation** state of the seafloor and shallow sediment is critical for the **precipitation** of *berthierine*; if the environment becomes oxidizing *glaucony* forms in preference to *berthierine*. **Chamosite** ($14\,\text{Å}$ 2:1 layer Fe-rich **chlorite**) is described as a grain-coating cement in many porous **sandstones** from marine, **deltaic** or vegetated fluvial systems and is interpreted to reflect **anoxic**, low-**pH** conditions (Fig. N5f). Additionally, (FeMg)-**chlorite** is a widespread early cement in many volcaniclastic **sandstones** where a local source of (FeMg) from the **alteration** of volcanic detritus is usually inferred.

During gradual burial, **chlorite** undergoes a variety of chemical and structural changes. Initially, $7\,\text{Å}$ **chlorite** evolves towards more stable $14\,\text{Å}$ structures. With increasing burial there is a trend towards increasing **magnesium** content at the expense of **iron**, an increase in polytype stability from disordered stacking types ($1b_d$) through intermediate species (16, $\beta = 97°$) to well-ordered types (1b, $\beta = 90°$).

In the Salton Sea **geothermal** field of California, **authigenic chlorite** shows a decrease in Al^{VI} and an increase in total (FeMg) with depth. Other studies have documented the reaction of **authigenic kaolinite** to **magnesium**-rich aluminous **chlorite** with depth. As with the reaction involving **illitization** of **kaolinite**, neither temperature nor burial depth are the overriding factors governing the stability of **kaolinite**; more important is the ready source of suitable (FeMg) cations either from the breakdown of detrital or **authigenic** minerals present in the sediment or introduced during **pore-water** migration. Indeed, in **kaolinite**-rich environments (**coal** *tonsteins*, quartzose **sandstones** with **kaolinite** cements, etc.) where a source of **potassium**, **magnesium** or **iron** is lacking, *kandites* may persist to great depths and burial temperatures, evolving from disordered to ordered and ultimately to **dickite**. Clearly, the chemistry of the **pore-waters** during burial is of prime importance in controlling **clay mineral** stability. [SDB]

Hurst, A. & Irwin, H. (1982) Geological modelling of clay diagenesis in sandstones. *Clay Minerals* **17**: 5–22.

Millot, G. (1970) *Geology of Clays: Weathering, Sedimentology, Geochemistry.* Springer-Verlag, New York.

Velde, B. (1977) *Clays and Clay Minerals in Natural and Synthetic Systems*. Elsevier, Amsterdam.

Wilson, M.D. & Pittman, E.D. (1977) Authigenic clays in Sandstones: recognition and influence on reservoir properties and paleoenvironment analysis. *Journal of Sedimentary Petrology* **47**: 3–31.

neoglacial A small-scale *glacial* advance that has occurred during the *Holocene* after the time of maximum *glacial* shrinkage characteristic of the present **interglacial**. Fluctuations appear to have been frequent, but there appear to be sparse temporal correlations between different areas (Grove, 1979). The latest neoglacial advance was the so-called Little **Ice Age** that caused valley **glaciers** to advance between around AD 1550 and 1850. [ASG]

Grove, J.M. (1979) The glacial history of the Holocene. *Progress in Physical Geography* **3**: 1–54.

neomorphism According to the original definition (Folk, 1965, pp. 20–1), a comprehensive term for all **diagenetic** transformations between one mineral and itself or a **polymorph**, whether the new crystals are larger or smaller or simply differ in shape from the previous ones. Excluded, however, are simple pore space filling processes; during neomorphism, pre-existing crystals or mineral grains must have been gradually consumed and their place simultaneously occupied by new crystals of the same mineral or a **polymorph**.

As later summarized by Bathurst (1971, pp. 475–6), neomorphism includes two distinct, but complex, *in situ* processes: (i) *polymorphic transformations* ('inversion' of Folk), and (ii) *recrystallization*, in which the gross chemical composition remains unchanged. These two processes in turn embody several different reaction mechanisms, but in the **diagenetic** environment presumably all involve an aqueous phase and do not proceed as dry, solid-state reactions that are typical of the metamorphic environment. Primary recrystallization and grain growth are thus unlikely to occur during **diagenesis**. Neomorphism is a particularly appropriate term to describe **diagenetic fabrics** produced in carbonate sediments as a result of the **aragonite–calcite** transformation or the processes whereby *micrite* and microcrystalline skeletal **calcite** are replaced by a coarse, sparry **calcite**. The latter process, referred to as aggrading neomorphism, produces a diverse array of **diagenetic** fabrics in **limestones** that are often difficult to distinguish from those produced by passive **cementation** direct from a **pore-fluid**.

Although the concept of neomorphism was developed in carbonate sediments, it is also applicable to the diagenetic **silica** system common in **silcretes**, precious **opal** deposits, **cherts** and marine **siliceous oozes**. During progressive **diagenesis** this **silica** system evolves from **opal**-A (biogenic **silica**) through **opal**-CT (a disordered cristobalite–tridymite mixed phase) to **chalcedony** and, ultimately, microcrystalline **quartz**. As in carbonate sediments, the transformation must take place in the presence of an aqueous phase during **diagenesis** and presumably involves a **dissolution**–*reprecipitation* mechanism, even though **dissolution** rates are probably very slow and proceed on a grain-by-grain scale with very local **precipitation**. Such a system is governed by the aqueous solubility of the component phases, primarily a function of *crystal structure* and particle size. [SDB]

Bathurst, R.G.C. (1971) *Carbonate Sediments and their Diagenesis*. Developments in Sedimentology No. 12, Chapter 12. Elsevier, Amsterdam.

Folk, R.L. (1965) Some aspects of recrystallization in ancient limestones. In: Pray, L.C. & Murray, R.C. (eds) *Dolomitization and Limestone Diagenesis: A Symposium*. Special Publication of the Society of Economic Paleontologists and Mineralogists **13**: 14–48.

neotectonics 'The study of late **Cenozoic deformation** . . . it undoubtedly reflects widespread and lively interest in crustal movements which are recent enough to permit detailed analysis and the assessment of rates of change' (Vita-Finzi, 1986, pp. 14–15). Methods employed in the establishment of recent crustal movements include the presence of **faults** and other **structures** in recent geological materials, **deformation** of geomorphological features (e.g. terraces, shorelines) of recent age, archeological and historical evidence, **geodetic** surveys, and **seismological** monitoring. [ASG]

Vita-Finzi, C. (1986) *Recent Earth Movements: An Introduction to Neotectonics*. Academic Press, London.

nepheline ((Na,K)AlSiO$_4$) A rock-forming mineral of the **feldspathoid** group. (See **feldspathoid minerals**.) [DJV]

nepheline syenite A plutonic **igneous rock** composed essentially of **nepheline** and feldspar (both **albite** and **microcline**). It is high in alumina (*c.* 25%) and soda (*c.* 9%) and is therefore suitable for the manufacture of container and sheet glass as well as vitreous whiteware, glazes and other ceramic uses. Minor uses are as a filler in paint, plastics and rubber and as an **abrasive**. For commercial uses the **iron** content must be very low and few nepheline syenites combine this requirement with proximity to markets. Three big commercial deposits are exploited outside the former USSR and these are at Blue Mountain, Ontario; Sternoy, Norway and Canaan, Brazil. [AME]

Harben, P.W. & Bates, R.L. (1984) *Geology of the Nonmetallics*. Metal Bulletin Inc., New York.

nephelinite A fine-grained **igneous rock** containing more than 10% modal **nepheline** and little or no **alkali feldspar** (cf. **basanite**). Such rocks may be further classified as indicated in Table N1.

Nephelinites may be aphyric or **porphyritic** lavas containing *phenocrysts* of euhedral **nepheline** and *clinopyroxene* (**diopside**–*aegirine*–**augite**) ± Fe–Ti oxide in a groundmass comprising small Fe–Ti oxide, *aegirine*–**augite**,

Table N1 Classification of **nephelinite**

Mineralogy	Rock term
Nepheline > mafic minerals	Nephelinite
Nepheline < mafic minerals	Melanephelinite
Melilite + nepheline > mafic minerals	Melilite-nephelinite
Melilite + nepheline < mafic minerals	Melilite-melanepheline

nepheline, **perovskite** and **biotite**. The **nepheline** *phenocrysts* are frequently altered to *analcite*, **calcite** or *natrolite*. Nephelinites occur in association with **carbonatites** and other **alkaline igneous rocks** (e.g. **ijolites**) in a variety of continental areas. [RST]

Le Bas, M.J. (1977) *Carbonatite–Nephelinite Volcanism*. John Wiley & Sons, New York.

nepheloid layer A layer of water, generally occurring on the sea bed, containing increased amounts of suspended sediment. Nepheloid layers have been identified by the scattering of light by fine particles measured with *nephelometers* or *transmissometers*. Most of the scattering which these instruments record is caused by the clay sized fraction. Nepheloid layers generally occur in the deep ocean and can be from a few hundred to about 1500 m thick. The bottom nepheloid layers with the greatest concentrations of suspended sediment occur in regions of strong bottom current activity, particularly in the zones of western boundary currents (see **thermohaline current**). Nepheloid layers at intermediate depths within the water column commonly occur adjacent to the shelf edge or continental slope where material has been brought into suspension or along density interfaces such as the **thermocline**. Nepheloid layers are also associated with **submarine canyons** and are a common feature on muddy shelves. [AESK]

Biscay, P.E. & Eittreim, S.L. (1977) Suspended particulate loads and transports in the nepheloid layer of the abyssal Atlantic Ocean. *Marine Geology* **23**: 155–72.
McCave, I.N. (1986) Local and global aspects of the bottom nepheloid layers in the world ocean. *Netherlands Journal of Sea Research* **20**: 167–81.

nephrite A tough, compact variety of the mineral **tremolite** that provides much of the material known as **jade**. (See **amphibole minerals**.) [DJV]

neptunian dyke A vertical sheet of sediment infilling a fissure in an older rock formation. Cracks and fissures in a substrate may be formed by tectonic flexure, **weathering** or **solution**. Sedimentary infills of these cavities are referred to as neptunian deposits (**sills** are more or less horizontal, **dykes** vertical), and may record periods of sedimentation not preserved in the standard vertical sequence of strata. Organisms associated with these deposits are often of reduced size (dwarf faunas), reflecting the restricted habitat.

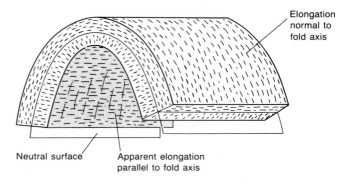

Fig. N6 Neutral surface of a fold.

Neptunian dykes should not be confused with injection structures, in which older sediment is forced upwards into younger deposits. (See **wet-sediment deformation**.) [DMP]

neptunite ($KNa_2Li(Fe,Mn)_2TiO_2(Si_4O_{11})_2$) A complex *inosilicate* mineral. [DJV]

neutral surface of a fold The surface within a folded layer, usually parallel to the layer boundaries, that separates regions of compression from regions of **extension**. A neutral surface separates an extensional region in the outer arc of a **fold** from a compressional region in the inner (see Fig. N6). Such regions swap over between **antiforms** and **synforms**, and the neutral surface is thus not continuous from one **fold** in a set to the next. Neutral surface folding is typically associated with **buckling**. [RGP]

Hobbs, B.E., Means, W.D. & Williams, P.F. (1976) *An Outline of Structural Geology*. John Wiley & Sons, New York.

neutron activation analysis (INAA) Method of analysis for a small number of major elements and many trace elements normally present in silicate rocks, including the **rare earth elements**. A sample of typically *c*. 300 mg is bombarded with thermal neutrons in a nuclear reactor where some isotopes form new *radioactive* isotopes as a result of the addition of a neutron to the nucleus. These decay in processes involving the emission of a gamma ray, the energy of which is ultimately characteristic of the original isotope. Thus the presence in the sample of certain isotopes (and hence elements) is determined, while the number of each type of gamma ray emitted indicates the amount of that element present. In this way elements present at levels of less than 1 p.p.m. can be analysed. Detection limits can be improved further by pre-separation of elements by chemical means. INAA is widely used in geology, mainly for the determination of **rare earth elements**, and is also extensively used by archeologists for the characterization of, in particular, pottery and **obsidian** (Fig. N7). [OWT]

Tite, M.S. (1972) *Methods of Physical Examination in Archaeology*. Seminar Press, London.

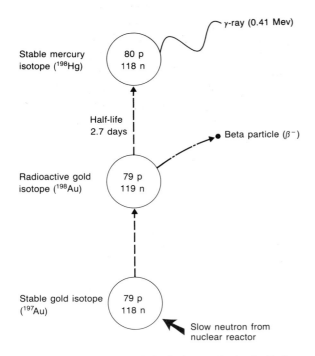

Fig. N7 Neutron activation analysis. Nuclear reaction involved in the emission of a gamma ray whose energy (0.41 MeV) is effectively characteristic of the element, gold. The numbers in the circles refer to the number of protons (p) and neutrons (n) in the associated nuclei.

neutron diffraction The **diffraction** of neutrons by a crystal or crystalline materials. Thermal neutrons from a nuclear reactor, collimated and passed through crystal monochromators, give beams of wavelength 0.11 nm. Unlike X-rays, neutrons are scattered by the nuclei of atoms, except that atoms with **magnetic moments** give additional scattering. Neutron scattering does not increase steadily with atomic number (as does X-ray scattering), absorption is generally low, different isotopes of the same substance scatter differently, and nuclear spin can also contribute. Therefore, although neutron diffraction has lower resolution than **X-ray diffraction**, it is advantageous when studying magnetic materials, solids consisting of light atoms, etc. [DJB]

Newton Standard (derived SI) unit of **force**, defined as the force required to move a mass of 1 kg with an acceleration of 1 m per second per second, i.e. $1 \, kg \, m \, s^{-2}$. [RGP]

Newton's law of gravitation The force of attraction (F) between two masses m_1 and m_2, whose dimensions are small with respect to the distance between them (r), is proportional to their product and the inverse square of r. The constant of proportionality is the **gravitational constant** G.

$$F = Gm_1 m_2 / r^2$$

[PK]

Newtonian behaviour Flow in which the **shear strain rate** is a linear function of **shear stress**. (See **Newtonian fluid**, **non-Newtonian fluid**.) [TGB]

Means, W.D. (1979) *Stress and Strain*. Springer-Verlag, Berlin.

Newtonian fluid A fluid whose **viscosity** is constant regardless of the external **shear** force that is applied to it and which obeys the **Navier–Stokes equations**. In tensor notation, the matrix of coefficients that relate **shear stress** to **shear strain rate** are known as the *viscosity coefficients*. Water is a Newtonian fluid, whereas water-saturated muds are non-Newtonian in their behaviour. Fluids which require an initial *yield* stress before motion occurs are known as *Bingham plastics* and include some debris and **lava flows**. (See **non-Newtonian fluid**, **Newtonian behaviour**.) [JLB/TGB/DHT]

Johnson, A.M. (1970) *Physical Processes in Geology*. Freeman Cooper, San Francisco.
Tritton, D.J. (1988) *Physical Fluid Dynamics*. Clarendon Press, Oxford.

nickel (Ni) At. no. 28, at. wt. 58.69. A transition element of Group VIII. Used extensively in alloys, particularly steels and cast iron, as a coinage metal, in glass (green) and catalysts. It is produced from nickel sulphide deposits. (See **orthomagmatic deposit**, **nickel laterite deposit**.) [AME]

nickel laterite deposit Laterite enriched in **nickel**. Residual **nickel** deposits are formed by the intense tropical **weathering** of rocks rich in trace amounts of **nickel** such as **peridotites** and **serpentinites**, running about 0.25% Ni. During the laterization of such rocks, **nickel** passes (temporarily) into solution but is generally quickly reprecipitated either on to **iron oxide minerals** in the **laterite** or as **garnierite** and other nickeliferous phyllosilicates in the weathered rock below the **laterite**. **Cobalt** too may be concentrated, but it is usually fixed in a **wad**. **Grades** of economic deposits range from 1.5 to 3% Ni.

The first major **nickel** production in the world came from nickeliferous **laterites** in New Caledonia where mining commenced in about 1876. It has been calculated that there are about 64 Mt of economically recoverable **nickel** in land-based deposits. Of this, about 70% occurs in lateritic deposits, although less than a half of current **nickel** production comes from these **ores**. [AME]

Golightly, J.P. (1981) Nickeliferous laterite deposits. *Economic Geology 75th Anniversary Volume*, pp. 710–35.

nickeline (NiAs) An **ore** mineral of **nickel**, also known as *niccolite*. [DJV]

niobium (Nb) (*columbium*) At. no. 41, at. wt. 92.9064. A Group V metal used in special steels and with **titanium** in superconductors. It is obtained commercially from the minerals **pyrochlore** and **columbite**, the former from some **carbonatites** and the latter from some **placer** and **pegmatite deposits**. [AME]

nitre (KNO$_3$) A nitrate mineral used as a source of nitrogen compounds and in fertilizer manufacture, also known as *saltpetre*. [DJV]

nitratine (NaNO$_3$) A nitrate mineral found as a salt bed (**caliche**) in desert areas. [DJV]

nivation This is a term used in two senses (Thorn, 1988): as a collective noun it is used to identify an assemblage of weathering and transport processes that are intensified by a late-lying snow patch; as an adjective it is used as a term to describe landforms substantially modified by the presence of snow-patch related processes. The term was introduced by Matthes (1900). Nivation processes include mechanical and chemical **weathering**, creep and *solifluction*, and resultant landforms may include nivation hollows and *cryoplanation* (*altiplanation terraces*). [ASG]

Matthes, F.E. (1900) Glacial sculpture of the Bighorn Mountains, Wyoming. *US Geological Survey 21st Annual Report 1899–1900*, pp. 167–90.
Thorn, C. (1988) Nivation: geomorphic chimera. In Clark, M.J. (ed) *Advances in Periglacial Geomorphology*, pp. 3–31. Wiley, Chichester.

nodule Irregular, spherical, ellipsoidal, flattened to cylindrical body, composed of a variety of materials; common forms being **calcite**, **siderite**, **pyrite**, **gypsum** and **chert**. The term *concretion* is also used but is restricted by some workers to refer to nodules with a concentric structure. Nodules with internal irregular cracks or **veins** are termed *septarian nodules*. Nodules are common in a variety of *soils* and **evaporite deposits**, and also form in sediments during early burial but such nodules may continue to grow during deeper burial. Early **diagenetic calcite** nodules are widespread in ancient **mudrock** sequences and the geochemistry of these nodules has been the subject of much study. [VPW]

Hesse, R. (1986) Diagenesis II, Early diagenetic pore water/sediment interaction: modern offshore basins. *Geoscience Canada* **13**: 165–96.

noise Any unwanted signal or disturbance on a record which does not represent the primary information from a source. Noise is usually divided into random noise and coherent or source-generated noise. Random or ambient noise is produced by background movements or instrument effects and in theory an increase in **signal to noise ratio** of $\sqrt{N}$ can be achieved in the presence of random noise if N independent signals are summed. Coherent noise on **seismic records** includes **multiple reflections** and *surface waves*. [GS]

non-conformity Type of **unconformity** where younger *strata* rest on an **erosion** surface cut across non-bedded **igneous rocks**. [RGP]

non-contacting conductivity measurement The measurement of ground **electrical conductivity** using **electromagnetic induction methods**. No ground contact is required so that measurements can be made at walking pace and the subsurface volume sampled is averaged in such a way that resolution is considerably improved over standard **resistivity methods**. [PK]

Zalasiewicz, J.A., Mathers, S.J. & Cornwell, J.D. (1985) The application of ground conductivity measurements to geological mapping. *Quarterly Journal of Engineering Geology* **18**: 139–48.

non-Newtonian fluid/flow Mixtures of fluids of different **viscosities**, fluids that behave as both viscous fluids and elastic solids (i.e. they show elastic **recovery** after being stressed), and fluids that show a time-dependent response to an imposed **stress** have motions, when stressed, that are difficult to quantify, except under very simple situations. Most **convective** motions within the Earth involve non-Newtonian fluids, as in the **mantle**. (See **Newtonian fluid**.) [DHT]

Tritton, D.J. (1988) *Physical Fluid Dynamics*. Clarendon Press, Oxford.

non-polarizing electrode Electrode designed to avoid the accumulation of charges on metal electrodes by **electrolytic polarization** effects. This is of no consequence for current electrodes, but should be avoided on potential electrodes as it distorts current flow in their vicinity and produces errors in potentials measured. The use of non-polarizing electrodes, which eliminate this effect, is essential in polarization studies (**self-potential** and **induced polarization**).

A typical non-polarizable electrode would be a metal (e.g. **copper**) rod immersed in a saturated solution of one of its own salts (e.g. **copper** sulphate), and housed in a (ceramic) pot with a porous base; the solution seeping out through the base makes the ground contact. [ATB]

Keller, G.V. & Frischknecht, F.C. (1966) *Electrical Methods in Geophysical Prospecting*. Pergamon, Oxford.

nontronite (Fe$_2$(Al,Si)$_4$O$_{10}$(OH$_2$. Na$_{0.3}$(H$_2$O)$_4$)) A **clay mineral** of the **montmorillonite group**. [DJV]

norbergite (Mg$_3$(SiO$_4$)F,OH)$_2$) A mineral of the **humite** group. [DJV]

normal grain growth An increase in grain size of a polycrystalline aggregate during recrystallization. This can occur after grain size reduction during **deformation**, or can affect an originally fine-grained rock such as a **chert**.

Energy is required to maintain the bounding surfaces of a grain; normal grain growth occurs to minimize the amount of surface area, for a given volume of aggregate. In isolation, a sphere has minimum surface area for a given volume. In an aggregate, equant polyhedral grains are formed, with volumes greater than the initial grain volumes, by normal grain growth. [SB]

normal polarity A **remanent magnetization** which has the same direction as the present **geomagnetic field**, i.e. the north **paleomagnetic pole** lies in the northern hemisphere. (See also **reversed polarity**.) [DHT]

normative composition The composition of an **igneous rock** sample expressed in terms of the weight proportions of idealized (or 'normalized') anhydrous minerals that crystallize from **magma**. The minerals include the end-member composition of *solid solution* series (e.g. **olivines**, **pyroxenes** and **plagioclases**) and are calculated in a specified sequence from a **chemical analysis** of a rock. The normative compositions (or 'norms') may be used for comparison of fine-grained or glassy rocks that may not be amenable to **modal analysis**, and may be used to compare anhydrous and hydrous (e.g. **mica**- or **amphibole**-bearing) rocks. (However, they are not entirely appropriate for rocks containing high proportions of hydrated minerals). The norm emphasizes certain subtle distinctions in composition such as the degree of *silica-saturation* (see **basalt**) and *alumina-alkali balance* (see **alkaline igneous rock**). [RST]

nosean $(Na_8(AlSiO_4)_6)SO_4$ A rare **feldspathoid mineral**. [DJV]

notch Landform that develops at the base of a cliff, platform or *reef* flat, especially in **limestones** and on tropical coasts. Deep narrow notches are characteristic of areas with a low **tidal** range. Their positions are related to lithological and structural controls, **tidal** characteristics and sea-level history. In general, the higher the amplitude of the **waves** and the higher the **tidal** range, the greater is the difference in elevation between the notch roof and the floor. Notches in the humid tropics may be 1–5 m in depth. Although mechanical action of **waves** may contribute to their development, most investigators now believe that chemical or biochemical **corrosion**, or biological boring and grazing activities, are important. [ASG]

Woodroffe, C.D., Stoddart, D.R., Harmon, R.S. & Spencer, T. (1983) Coastal morphology and Late Quaternary History, Cayman Islands, West Indies. *Quaternary Research* **19**: 64–84.

nubbin A small rounded or elongate earth lump, a few centimetres in diameter, produced by heaving associated with the growth of needle ice. [NJM]

nuclear explosion seismology **Seismic waves** generated by underground nuclear explosions may be recorded thousands of kilometres from the explosion. Since 1963, when the Partial Test-Ban Treaty came into force requiring all nuclear tests to be carried out underground, **seismology** has been the main method of verifying compliance with test-ban treaties. Many advances in **seismology** and the earth sciences have been brought about by the needs of nuclear explosion seismology. Nuclear explosion seismology, or *'forensic seismology'*, has three stages:
1 *Detection.* All possible **seismic events** are identified and located. The routine detection capability, at **teleseismic** range, of the current world-wide networks of *seismographs* is about m_b 3–4 in the northern hemisphere, with the best performance coming from **seismic arrays**.
2 *Identification.* The seismic signals are used to determine whether the event was an **earthquake** or an underground nuclear explosion. The simplest 'discriminant' is location; only shallow events beneath land are candidate explosions because of the limitations in drilling technology. Distinguishing between explosions and most shallow **earthquakes** is feasible using the '$m_b : M_s$ ratio': explosions are inefficient sources of *surface waves* compared with P waves, whereas shallow **earthquakes** are the opposite. Other criteria, mostly based on the frequency spectrum of P and other wave types, can also be used.
3 *Yield estimation.* The size or *'yield'* of an *explosion* (in kilotonnes (kt); 1 kt is equivalent to the energy released by 1000 tonnes of TNT) is estimated from the **magnitude** of the **seismic waves** it produces, using previously established 'magnitude-yield' curves of explosions of known yield. The **magnitude** produced by a given *explosion yield* varies greatly depending on what rock type it is detonated in and the regional **attenuation**. 'Soft' rocks such as dry sediments give **magnitudes** at least 0.5 units smaller than 'hard' or water-saturated rocks (where 1 kt gives about m_b 4 and M_s 2). Firing explosions in large cavities can 'decouple' the explosion and reduce **magnitudes** by as much as 1.8 magnitude units, so providing a possible means of evading detection.

Until the 1980s most nuclear-explosion seismology concentrated on **teleseismic**-range monitoring because the former USSR refused to allow siting of *seismographs* within its territory. Routine capability was sufficient to monitor the 1974 Threshold Test-Ban Treaty which banned explosions larger than 150 kt: most well-coupled explosions of kilotonne-range yield could be detected and those larger than some 10 kt identified and a yield accurate to a factor of *c.* 2 assigned. With the changed political situation in the former Soviet Bloc, so-called 'in-country' *seismographs* are now feasible and research efforts are directed at regional-range monitoring, lowering the detection and identification thresholds, and improving the precision of yield estimates. [RAC]

Barnaby, F. (1990) *A Handbook of Verification Procedures*. Macmillan, London.

Bolt, B.A. (1976) *Nuclear Explosions and Earthquakes: The Parted Veil.* W.H. Freeman, San Francisco.

Dahlman, O & Israelson, H. (1977) *Monitoring Underground Nuclear Explosions.* Elsevier, Amsterdam.

nuée ardente A laterally moving, turbulent, hot cloud composed of air, **volcanic gases** and suspended fine-grained **tephra** generated by volcanic activity. Nuées ardentes are a feature of *peléean eruptions* (see **volcanic eruption**), and are thought to be generated by upward **convection** of hot gas and strained **tephra** from an active **pyroclastic** flow. In the type example, the 1902 eruption of Mount Pelée, Martinique, the nuée ardente became detached from the underlying block and **ash flow** and travelled some 3 km as a separate unit (Fisher & Heiken, 1982). Nuées ardentes form fine-grained deposits, less than a few metres thick, characterized by low-angle *cross-bedded* sedimentary structures similar to those observed in **base surges**. This deposit is overlain by a thin (few centimetres) air-fall layer from **tephra** suspended in the upper, buoyant part of the cloud. These nuée ardente deposits overly the block and **ash** deposit of the associated **pyroclastic** flow, but are distributed over a wider area. The potentially destructive character of nuées ardentes is demonstrated by the devastation of the town of St Pierre during the 1902 eruption, when virtually all the 30 000 inhabitants were killed when it was overrun by a nuée ardente.

Nuées ardentes were observed rising above small-volume scoriaceous **pyroclastic** flows during the post 18th May 1980 eruptions of Mount St Helens, Washington. [PTL]

Fisher, R.V. & Heiken, G. (1982). Mt Pelée, Martinique: May 8 and 20, 1902, pyroclastic flows and surges. *Journal of Volcanology and Geothermal Research* **13**: 339–71.

Lipman, P.W. & Mullineaux, D.R. (eds) (1981) *The 1980 Eruptions of Mount St Helens, Washington.* Professional Paper of the United States Geological Survey **1250**.

nugget A lump of **gold** in its natural state. The majority of nuggets have been found in **alluvial deposits** where they generally range in size from 0.5–3 cm; but larger nuggets have been found, the largest being the Welcome Stranger from Ballarat, Australia which weighed 2280 troy oz. Native **platinum** occasionally occurs in nugget form. [AME]

nummulitic limestone Limestone containing nummulites (fossil **Foraminifera**); widespread in Egypt and quarried for some early Egyptian monuments. [OWT]

Nusselt number This hydrodynamic parameter corresponds to 1 for the possible initiation of **convective** flow and higher values increase the probability of **convective** motion. Related to **Rayleigh number**. [DHT]

Tritton, D.J. (1988) *Physical Fluid Dynamics.* Clarendon Press, Oxford.

Nyquist frequency (*folding frequency*) Half the sampling frequency of a digitized signal. Frequencies higher than the Nyquist that are present in the signal that is sampled will **alias** as lower frequencies in the sampled signal. Thus the Nyquist frequency is the highest frequency reliably restored from a digital signal. Care must be taken either to choose the correct Nyquist frequency or sampling interval for the frequency content of a given signal to be sampled so that the sampled signal is not distorted. An *anti-alias filter* is often used to **attenuate** frequencies higher than the Nyquist frequency before a signal is sampled. [GS]

O

obduction In certain areas, e.g. Cyprus, Iran, New Zealand, remnants of oceanic **lithosphere** (**ophiolites**) overlie **continental crust**. Such occurrences show that oceanic **lithosphere** is not always subducted, but can occasionally be either carried onto or *underplated* by **continental crust**. This process by which dense oceanic **lithosphere** is transferred onto **continental crust** is termed obduction. There is some evidence suggesting that obduction of true **oceanic crust** may be very rare and that most obducted fragments are derived from the lithosphere of *marginal basins.* [DHT]

Dewey, J.F. & Bird, J.M. (1971) Origin and emplacement of the ophiolite suite: Appalachian ophiolites in Newfoundland. *Journal of Geophysical Research* **76**: 3179–206.

obsidian A glassy volcanic rock of intermediate to acid chemical composition, formed by the chilling of rhyolitic lava (cf. **rhyolite**). Obsidian is associated with young (generally less than 10 Ma) volcanic activity, including **island arcs** which produce mainly **calc-alkaline** and, more rarely, peralkaline obsidians, with volcanic activity at constructive **plate boundaries**, and with intra-plate activity, commonly producing peralkaline obsidian. Obsidian is generally grey or black, sometimes green (mainly peralkaline obsidian with its higher **iron** content) and occasionally red; it is often transparent in normal light and may contain darker bands of microcrystallites, which form slowly as the glass devitrifies.

Obsidian is of great importance archeologically as it was widely used in prehistoric times for the making of small tools; its predictable **fracture** and sharp cutting edges made it ideal for sharp blades, and it was also used for arrowheads. Its shiny attractive appearance made it suitable also for decorative articles such as bowls and mirrors. Among the most important sources used in antiquity are the Mediterranean islands (Melos, Lipari, Sardinia, Pantelleria), the Carpathians, central and eastern Turkey (Çiftlik, Acigöl, Lake Van), east Africa, New Zealand (Mayor Island), the Pacific north-west, western USA, Central America, and the south-west Pacific. Within all these areas studies have been carried out to characterize the obsidian sources and trace the extent of trade in their products. These studies indicate overland and seaborne trade in obsidian over distances of more than 500 km in Neolithic Europe. The most successful methods of characterization have proved to be chemical analysis for **trace**, particularly incompatible, **elements**, often determined by **neutron activation analysis**, **X-ray fluorescence** analysis and **optical emission spectroscopy**. Obsidians can be dated by **fission track dating** giving another means of characterizing sources and artefacts, while the date of working on an artefact can sometimes be determined by **hydration layer dating**. [OWT]

Cann, J.R. (1983) Petrology of obsidian artefacts. In: Kempe, D.R.C. & Harvey, Anthony P. (eds) *The Petrology of Archaeological Artefacts*, pp. 227–55. Clarendon Press, Oxford.

oceanic crust This is formed of four layers, of which three are of igneous origin and are the fundamental components (Fig. O1). The uppermost layer comprises sediments that are of variable thickness, being virtually non-existent on the crests of the *oceanic ridges*, and increase fairly uniformly in thickness to some 2 or 3 km with increasing distance (age) from the ridge, although thicknesses of 10–12 km may occur where the oceanic sedimentation processes are supplemented by detritus derived from the continents flowing into the ocean basins. The age of the basal sediments also increases away from the ridge crest. The uppermost igneous layer (also known as layer 2) is largely composed of **basaltic lava flows**, with underlying feeder **dykes**, forming a layer about 2 km thick with a seismic *P wave* velocity of about 5 km s^{-1} which is underlain by the major oceanic crustal layer (layer 3), formed of **gabbroic** rock some 5–6 km thick with *P wave* velocities of 6.7 km s^{-1}. This is underlain by a thin layer (layer 4), probably about 0.5 km thick, with *P wave* velocities of about 7.4 km s^{-1}, referred to as the layered complex and formed of cummulate **gabbros**, which immediately overlie **mantle** rocks with *P wave* velocities of 8.1 km s^{-1}. A remarkable feature of the igneous part of the oceanic

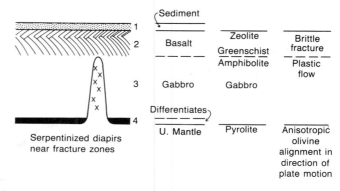

Fig. O1 A model of the **oceanic crust**. The seismic layering (1–4) can be interpreted in terms of composition, metamorphic grade and rheology, but all give similar zonation. (After Tarling, 1978.)

[427]

crust is its highly consistent thickness over vast areas. The linear **oceanic magnetic anomalies**, measured at the sea surface, appear to be associated with the **dyke** complex in layer 2 because of the sharpness of the edges of the anomalies and the magnetic instability of the **gabbros** of layer 3. The igneous part of the oceanic crust originates at *oceanic ridges* as a result of **magmatic** differentiation of **mantle** rocks. These accumulate as a **gabbroic magma** that feeds the overlying **dykes** and **lavas**. As the *magma chamber* becomes sealed by the now cooled overlying **dykes** and **lavas**, the denser crystals convect downward to form a crystal mush on the bottom of the *magma chamber* which eventually forms layer 4. As the **igneous rocks** continue to cool, both at the surface and sides, the composition and structure becomes 'frozen in' and forms the basement onto which later sediments accumulate. The lower part of the oceanic crust, the **gabbros**, are thought to **deform plastically** as most **earthquake** activity appears to be in the upper layers. The difference in mechanical behaviour may partly reflect increased metamorphism of the lower rocks, and particularly *serpentinization* along **fracture zones** related to the circulation of hydrothermal fluids. Occasional mineralization appears to be concentrated where these fluids expel into the oceanic waters, presently seen as **black smokers**, but the mineralization associated with such activity appears to be mainly concentrated at the sediment/ **igneous rock** interface and is probably related to the escape of volatile-rich minerals that may have originated within the **gabbro** layer. (See **ophiolite**.) [DHT]

Bott, M.H.P. (1982) *The Interior of the Earth*. Arnold, London.
Cann, J.R. (1974) A model for oceanic structure developed. *Geophysical Journal of the Royal Astronomical Society* **39**: 169–87.
Tarling, D.H. (ed.) (1978) *Evolution of the Earth's Crust*. Academic Press, London.

ochre Ochre had a number of uses in antiquity; the Romans record it as a source of **iron**, as colouring for pottery decoration and painting, as a cosmetic and as medicine. It was used in Paleolithic cave art and there are several examples of red ochre painted skeletons found in Upper Paleolithic burials in Europe, presumably as part of funeral rites. [OWT]

Oersted (Oe) The cgs unit of *magnetic induction*, numerically equal to the **gauss**.[PK]

offlap Structure formed where successive wedge-shaped **beds** do not extend to the margin of the underlying **bed**, but terminate within it (see Fig. O2). Such a structure is typical of sedimentary sequences in contracting **basins** with regressing shorelines. The converse situation produces **onlap**. [RGP]

Ohm's law An empirical law of physics which states that the electric current that flows between two points is proportional to the potential difference between them, the constant of proportionality being the resistance between the points. Although widely used, it is not universally true; there are many materials which do not obey the law, and most materials obey it over only limited ranges of potential differences, and resistances change with temperature. Ohm's law for continuous media,

$$E = \rho J$$

with E and J as electric field and current density, and ρ as the **electrical resistivity** of the medium, is much used in geophysics. [ATB]

oil basin Sedimentary **basin** containing commercial accumulations of oil. Whereas it has been estimated that there are about 700 distinguishable sedimentary **basins** in the world, about 95% of the world's commercial oil accumulations occur in only fifty of them and 70–75% in only ten, with almost half the total present in the **Persian Gulf** Basin alone. [AME]

oil shale Sedimentary rock containing substantial amounts of organic matter (OM) that has reached only a low level of *maturation* such that liquid oil has never been expelled from the *kerogen* (see **hydrocarbon generation**). Most so-called oil shales are bituminous, non-marine **limestones** or marls containing *kerogen* and only a few marine examples can properly be termed **shales**. Their OM content is much higher than in most **oil source rocks** and in commercially significant deposits is at least 30%. Apart from **coals**, oil shales are the most abundant fossil fuel **resource**, but the obstacles to their exploitation are immense and the contribution of shale oil to world oil supplies during this century has been negligible. [AME]

North, F.K. (1985) *Petroleum Geology*. Allen & Unwin, Boston.

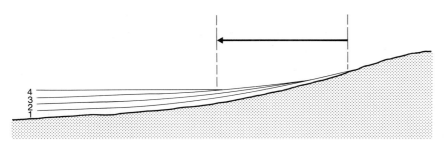

Fig. O2 Cross-section showing **offlap** of successive beds 1–4 associated with a regressive shoreline.

oil source rock Rock within which **hydrocarbon genera-tion** produces **natural gas** and **crude oil**. Sufficient organic matter (OM) to form source rocks has been concentrated in certain **shales** and **limestones**. Not all **shales** are poten-tial source rocks. Rich source sediments of different ages are much alike in their gross appearance. They are most commonly brown, often dark brown but may also be black or green. Lamination is generally developed with pale silty bands alternating with bands packed with organic remains and the rocks may be strongly *bituminous*. In addition most source rocks are phosphatic and uraniferous; sulphides, especially **pyrite**, are common. Rich, **Albian** age, source rock **shales** occur in the Kazdhumi Formation of the Iraq–Iran Oilfields. On the other hand two of the most prolific source sediments known are **limestones**: the **Jurassic** Hanifa Formation of Saudi Arabia and the Cretraceous La Luna Formation of Venezuela. The latter is a dark brownish-grey to black *bituminous* **limestone**, thinly laminated to very well bedded with lenticular bands of dark **chert**, **fish** remains and phosphate pellets.

Source sediments are early deposits in most **oil basins** and are normally older than the main **reservoir** rocks, from which they may be separated by whole formations devoid of oil accumulation. The combination of high OM, undisturbed laminar layering, **pelagic** flora and fauna,

presence of **phosphorus** and the necessity for *anerobic* conditions to preserve the OM, indicates a depositional environment below a **thermocline**, probably in the deepest part of the **basin**, although the absolute depth need not have been great.

An oil source rock may also produce gas and **associated gas** is very common. Deep environments may produce non-associated gas and so may other source rocks such as **coal**. In north-western Europe enormous gas reserves occur in Lower **Permian sandstones** extending from the River Elbe westwards into the British sector of the southern North Sea. Here the source rock, **Carboniferous coal**, lies beneath the porous, desert-deposited, **reservoir sandstones** of the Rotliegendes Formation which are perfectly sealed by thick **Permian** *Zechstein* **evaporites**. [AME]

North, F.K. (1985) *Petroleum Geology.* Allen & Unwin, Boston.

oil trap Any geometrical arrangement of strata that leads to the accumulation of oil or gas or both. *Hydrocarbon traps* have been divided into three types: structural, stratigraphical and combination traps, the last named being a combination of stratigraphical and structural trap-ping mechanisms. The most common traps are **anticlinal**,

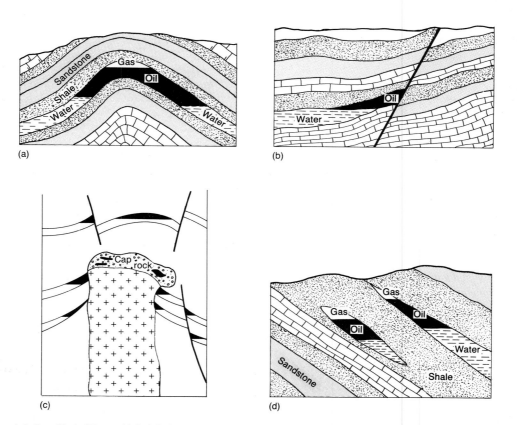

Fig. O3 Structural and stratigraphical **oil traps**. (a) Anticlinal trap developed in a sandstone reservoir in an open, asymmetrical fold. (b) Oil trapped by a fault seal. (c) Schematic diagram of salt dome traps, in supercap, cap rock, and flank sandstones (abutting, fault-sealed and pinch-out). (d) Two types of stratigraphical traps. Right, sandstone wedge-out; left, sandstone lens.

if we take the broadest interpretation of the term. Other important structural traps are developed by **faulting** and piercement structures, e.g. **salt domes**. Stratigraphical traps are created by variations in the stratigraphy that are not due to structural **deformation**.

Anticlinal traps are the easiest to locate because they are large and extensive structures which can often be located by surface mapping, or at depth by geophysical investigations. Because tectonic **anticlines** have considerable vertical continuity of shape there may be many separate gas or oil pools stacked one above the other. An elongate, open **anticline** is an ideal **petroleum** trap since its extended limbs facilitate **migration**, its hinge zone permits accumulation, provided a **cap rock** is present, and its crest generally contains **periclinal** areas in which oil pools can form. In asymmetrical **folds** the gentler-dipping limb often contains the most oil, Fig. O3.

Fault traps may form wherever an inclined **reservoir** is cut by a **fault** that brings impermeable rocks against that portion of the dislocated **reservoir** which dips downwards away from the **fault**, Fig. O3.

Salt dome traps may be present in those **oil basins** that are underlain by **evaporite** beds rich in **halite** as in the Gulf Coast of the USA. The **salt diapirs** are quite impermeable and where the upward-moving salt plug has domed and pierced the overlying and surrounding strata, then many different oil traps may be formed (Fig. O3).

Stratigraphical traps result from lateral and vertical variations in thickness, texture, lithology and **porosity** of **reservoir** rocks. An inclined **sandstone reservoir** intercalated in **shale** may wedge out up **dip**, as **beach** and **bar** sands do, thus forming a wedge-out trap. A variation of this is the **sandstone** lens trap which may or may not be inclined, Fig. O3. *Shoestring sands* are long narrow bodies of **sandstone** enclosed by **shales**; they may have been formed as offshore **bars** or by meandering rivers. Small oil pools generally characterize these **sandstones** but some of the Texan occurrences have very large oil accumulations. Porous **limestone** *reefs* when buried by impermeable sediments form potential *reef* traps, but these are generally small by comparison with other traps.

Unconformities can also produce important oil traps. Underlying tilted **reservoir** rocks may be sealed at the **unconformity** to form an **unconformity** trap.

The Prudhoe Bay Field in Alaska (Fig. O4) is a striking example of a *combination trap*. Here the overall structure is a huge **anticline** but the actual trap for the principal oil pool occurs where the homoclinally dipping **reservoir** rock is truncated up **dip** by a *normal fault*, which brings down impermeable **shales** against the **reservoir** rock, and where it is cut across by a major **unconformity**. [AME]

oilfield A region of the Earth's **crust** containing a number of oil accumulations (pools). During the whole history of oil and gas exploration it is estimated that over 30 000 oil and gas fields have been discovered, but some of these

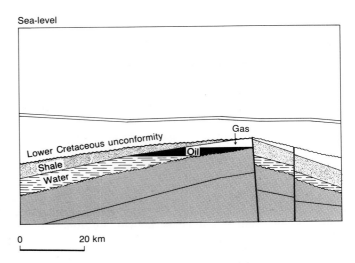

Sea-level

Lower Cretaceous unconformity — Gas — Oil — Shale — Water

0   20 km

Fig. O4 Oil trap. North–south section through the Prudhoe Bay Oilfield, Alaska. See text for description. (Vertical exaggeration of eight times.)

were extremely small and have long since been abandoned. [AME]

North, F.K. (1985) *Petroleum Geology*. Allen & Unwin, Boston.

oilfield water The term given to the water found in *hydrocarbon*-bearing **reservoir** rocks. These waters are usually grouped into four classes: (a) sulphate–sodium waters, (b) bicarbonate–sodium, (c) chloride–**magnesium**, (d) chloride–calcium. Classes (c) and (d) also contain sodium. Most oilfield waters are of class (d); sodium is the dominant cation but Ca^{2+} and Mg^{2+} are present. Chloride is practically the sole anion, sulphate being absent. Class (c) water is dominant in many **evaporite**-bearing sequences and classes (a) and (b) occur in near-surface **artesian** conditions. Classes (a) and (b) are usually oxygen-bearing and may give rise to **oxidation** of the oils in contact with them. [AME]

North, F.K. (1985) *Petroleum Geology*. Allen & Unwin, Boston.

Oligocene The Oligocene has a duration of *c*. 13 Ma from the end of the **Eocene** at *c*. 37 Ma to the base of the **Miocene** at 23 Ma and has been divided into two stages (Berggren *et al.*, 1985; see Fig. O5).

East Africa domed and **rifted** to produce a series of *rift valleys* flanked by volcanoes and these were flooded by lakes. Arabia began to break with Africa along the line of the Red Sea. A sea gap opened between Antarctica and South America at *c*. 30 Ma and between Antarctica and the southernmost Australian continental shelf at about *c*. 30 Ma, and Australia finally separated from Antarctica and moved northwards. The waters around Antarctica were free to circulate with the development of the circumAntarctic Current or West Wind Drift. The current flowed around Antarctica cutting the continent off from

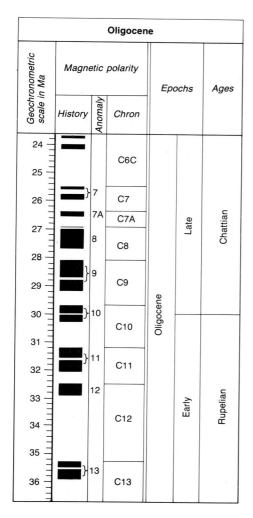

Fig. O5 Oligocene.

occurred at *c.* 35 Ma. Hang-nose primates evolved at *c.* 29 Ma. With the development of the circumAntarctic current, the first of the modern whalebone whales, which feed by filtering plankton from the sea, evolved in the Southern Ocean. The newly created oceanographic regime became favoured by whales because of the high biotic productivity of the waters. As the distance between Australia and Antarctica increased, and the current grew progressively, the whales became larger, to become the largest marine mammals of all time (Calder, 1983).

In the oceans the species diversity of planktonic **Foraminifera** had fallen drastically by the beginning of the Oligocene and only recovered slowly as ocean temperatures increased toward the Late Oligocene.

During the mid-Oligocene period the climate remained cool and dry with the result that vegetation was less dense in some areas. During this time large areas of grassland occurred on the plains of North America.

In the seas calcareous nannofossil diversity was low throughout the Oligocene. Some **Eocene** species disappeared in the lower part of the Oligocene, while new *sphenoliths* appeared in the Late Oligiocene. No disc-shaped *discoasters* are found and *D. flandrei*, *D. tanii nodifer* and *D. tanii* are usually the only species found.

At the end of the **Eocene** there had been a sudden floral change: broad-leaved deciduous forests expanded southwards in the northern hemisphere and middle and high latitude diversity declined while later in the Oligocene diversity increased (Thomas & Spicer, 1987).

The following extant orders appeared: Nelumboniales, Violales, Cucurbitales, Salicales, Rosales, Rhamnales, Oleales, Elaeagnales, Campanulales and Asterales (Thomas & Spicer, 1987). (See **land phants.**) [DGJ]

Berggren, W.A., Kent, D.V. & Flynn, J. (1985) Jurassic to Paleogene: Part 2. Paleogene geochronology and chronostratigraphy. In: Snelling, N.J. (ed.) *The Chronology of the Geological Record*, pp. 141–95. Memoirs of the Geological Society No. 10. Blackwell Scientific Publications, Oxford.

Calder, N.C. (1983) *Timescale, An Atlas of the Fourth Dimension*. Chatto & Windus, London.

Thomas, B. & Spicer, R. (1987) *Evolution and Palaeobiology of Land Plants*, pp. 1–304. Croom Helm, London.

oligoclase A **plagioclase feldspar** mineral. (See **feldspar minerals.**) [DJV]

olistolith/olistostrome A subaqueously-emplaced mineral block, rock or biogenic *clast* (e.g., *reef*) of pebble, cobble or boulder size that is exotic in terms of petrography, composition and texture with respect to the surrounding stratigraphy. An olistostrome is a bed or layer of olistoliths. The size of blocks may reach dimensions of hundreds of metres, with gravity-driven transport over at least tens of kilometres. Common olistoliths are volcanics, igneous intrusives, carbonates, **sandstones**, **cherts** and metamorphic *clasts* within relatively deep-water mixed **sandstone**–mudstone or carbonate successions.

warmer waters from the north. Consequently Antarctic ice accumulated more extensively and there was a fall in sea-level at *c.* 29 Ma. Japan separated from the Asia mainland with the development of a small oceanic basin, the Japan Sea.

The first major decline in **Cenozoic** temperatures occurred at the beginning of the Oligocene with wintry conditions beginning on land at high latitudes and a corresponding drop in seawater temperatures. A second major cooling occurred at *c.* 29 Ma with an associated very large fall in sea-level and a possible temporary accumulation of ice on Antarctica. Temperatures gradually recovered during the Late Oligocene.

Many species of the early **Cenozoic mammals** died out, including primitive insectivores and some hoofed animals. The appearance of the first cats, dogs and rhinoceroses occurred at *c.* 35 Ma, and by *c.* 30 Ma early pigs and bears had evolved (Calder, 1983). The horse was now 0.61 m tall and had three toes. The appearance of the forerunners of the broad-nose (platyrrhine) New World monkeys

Magnetic polarity table:

Geochronometric scale in Ma	Magnetic polarity			Epochs	Ages
	History	Anomaly	Chron		
24–36			C6C	Late	Chattian
		7	C7		
		7A	C7A		
		8	C8		
		9	C9		
		10	C10	Early	Rupelian
		11	C11		
		12	C12		
		13	C13		

Oligocene

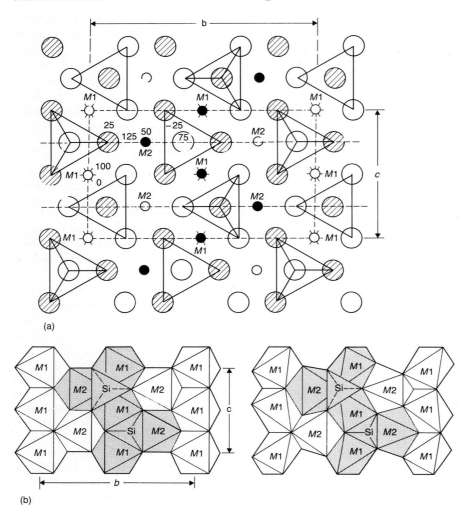

(a)

(b)

Fig. O6 Olivine minerals. The crystal structure of olivine. (a) Idealized olivine structure parallel to (100) plane (Si atoms are at centres of tetrahedra and not shown; small open circles are Mg atoms at x = 0, small solid circles are Mg atoms at x = $\frac{1}{2}$). (b) Olivine structure parallel to (100) plane showing ideal hexagonal close packing (left) and the actual structure (right).

Olistostromes, with constituent olistoliths, tend to develop where submarine **fault** scarps are subject to *mass wastage* resulting in rockfall, individual block gliding, **slides** and/or *mass flows* such as **debris flows**. They are particularly well documented from forearc **accretionary prisms** and submarine *normal fault* scarps along **passive continental** and oblique-slip **margins**, where steep slopes develop. Sheared or tectonized olistostromes form one type of **mélange**. [KTP]

olivine minerals A group of orthosilicate ('island silicate') rock-forming **minerals** of general formula M_2SiO_4 where M = **magnesium**, **iron**, **manganese** and calcium (although other metals, notably **nickel**, may substitute in the M positions but in relatively minor or trace quantities in natural olivines). All olivine minerals have the same crystal structure (orthorhombic; space group *Pbnm*) in which isolated SiO_4 tetrahedra are linked by (divalent) metal cations occurring in equal numbers of six-fold co-ordinate sites known as the M_I and M_{II} sites (see Fig. O6). Natural olivines are dominated by the phases in the

solid solution series from **forsterite** (Mg_2SiO_4) to **fayalite** (Fe_2SiO_4), an unbroken series in which intermediate members may be precisely designated by the mol% **forsterite** (Fo) component or **fayalite** (Fa) component (e.g. as Fo_{88}, equivalent to Fa_{12}). Intermediate members of this very important series are also given names, the complete series being **forsterite** (Fo_{100}–Fo_{90}), **chrysolite** (Fo_{90}–Fo_{70}), *hyalosiderite* (Fo_{70}–Fo_{50}), *hortonolite* (Fo_{50}–Fo_{30}), *ferro-hortonolite* (Fo_{30}–Fo_{10}) and **fayalite** (Fo_{10}–Fo_0). Other olivine minerals, all of which are much rarer, include **tephroite** (Mn_2SiO_4), *knebelite* $(Mn,Fe)_2 SiO_4$, **monticellite** ($CaMgSiO_4$), *glaucochroite* ($CaMnSiO_4$) and *kirschsteinite* ($CaFeSiO_4$). There is also a complete *solid solution* between **fayalite** and **tephroite**.

The chemistry of the olivine minerals has been studied in great detail. The M_I and M_{II} sites are both distorted octahedra (Fig. O6). Spectroscopic and **X-ray diffraction** studies have shown that Fe^{2+} favours the M_I sites and will tend partially to order into them, the degree of order being a function of thermal history of the sample. Phase relations between members of the olivine group and between

[432]

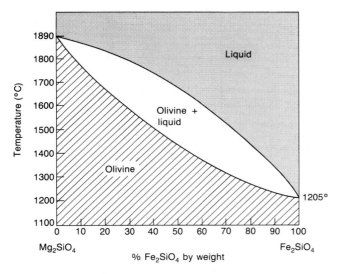

Fig. O7 Olivine minerals. Phase diagram for the system forsterite–fayalite.

olivines and other rock-forming **minerals** have also been much studied. The binary system **forsterite–fayalite** (Fig. O7) is a classic example of continuous reaction in which the earlier crystallizing phases are **magnesium**-rich relative to the melt and immediately react, becoming more **iron**-rich as crystallization proceeds. Disequilibrium can lead to *zoned crystals*, or *fractional crystallization* with early **magnesium**-rich olivines separating to leave relatively **iron**-rich residual melts. The compositions of olivines are often, therefore, a useful indication of the differentiation stage of the parent **magmas** in which they crystallized.

In the **forsterite–fayalite** series, refractive indices vary linearly with composition and the optic axial angle also varies systematically and increases from $2V_\gamma = 82°$ for Mg_2SiO_4, to $2V_\gamma = 134°$ for Fe_2SiO_4, changing sign at approximately Fa_{12}. There are also systematic variations in cell parameters and density.

The more magnesian olivines of the **forsterite–fayalite** *solid solution* series are the most important minerals in **ultramafic igneous rocks** (**dunites** and **peridotites**) and very common constituents in mafic rocks (**gabbros, dolerites** and **basalts**). More **iron**-rich olivines are less common, but occur in **iron**-rich **dolerites**, *ferrogabbros* and **granophyres**. **Fayalite** is fairly common in **quartz**-bearing **syenite** and important in **fayalite** *ferrogabbro*. **Fayalite** also occurs in regionally (or thermally) metamorphosed **iron**-rich sediments, whereas thermally metamorphosed magnesian **limestones** and **dolomites** contain the most **magnesium**-rich olivines. **Monticellite** is a rare mineral found in some **ultramafic rocks** and metamorphosed and **metasomatized limestones**. The **manganese**-rich olivines occur in **iron–manganese ore** deposits and **skarns**. The magnesian olivines are widely regarded as the most important minerals in the upper regions of the **mantle** of the Earth (as evidenced by **peridotite nodules** from deep volcanic sources and the **peridotites** associated with

ophiolite complexes). For this reason, the high pressure mineralogy of olivine and other aspects of its **mineral physics** have been much studied. The long predicted transformation of Mg_2SiO_4 from the olivine to the **spinel** structure was eventually confirmed experimentally, although it does involve an intermediate step (the formation of the so-called β-phase). Olivines are very susceptible to **hydrothermal alteration**, low-grade metamorphism and **weathering**, common **alteration** products being **serpentine, chlorite, amphibole, talc,** carbonates and **iron oxides**. [DJV]

Deer, W.A., Howie, R.A. & Zussman, J. (1982) *Rock-Forming Minerals.* Vol. 1A; *Orthosilicates* (2nd edn). Longman, London.
Ribbe, P.H. (ed.) (1982) *Orthosilicates* (2nd edn). Reviews in Mineralogy Vol. 5. Mineralogical Society of America, Washington, DC.

omphacite $((Ca,Na)(Mg,Fe,Al)Si_2O_6)$ A **pyroxene mineral** found in *eclogites*. [DJV]

oncoid (*oncolith*) A type of carbonate *coated grain*. The term is generally used if the grain is believed to have been coated by microbial mats which were either calcifying or later became calcified. As in the case of **ooid** and **pisoid**, the use of the term oncoid is fraught with problems. It is rare that any fossil oncoid can be confidently shown to have been formed by the activities of ancient microbial communities. The term is usually given to *coated grains* where the laminae are irregular in thickness, relief and continuity, suggesting a biological influence on growth.

The geological significance of oncoids depends on ascertaining their exact origin. True microbial oncoids can form in marine and non-marine settings, from very shallow, even periodically exposed settings, to the deep ocean.

A more reasonable use of the term would be for *coated grains* with irregular coatings, reserving the terms **ooid** and **pisoid** for regularly *coated grains*. A rock made up largely of oncoids should be termed an *oncolite*. [VPW]

Peryt, T. (ed.) (1983) *Coated Grains*. Springer-Verlag, Berlin.

onlap (*overlap*) **Structure** formed where successive wedge-shaped **beds** extend further than the margin of the underlying **bed**, such that they lie partly on older basement (see Fig. O8). Such a structure is typical of sedimentary

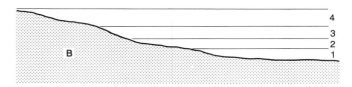

Fig. O8 Cross-section showing **onlap** of successive beds 1–4, each resting partly on older basement B.

sequences in expanding **basins** with transgressive shore-lines. [RGP]

Roberts, J.L. (1982) *Introduction to Geological Maps and Structures.* Pergamon, Oxford.

onyx A layered microcrystalline variety of **silica** valued for decorative purposes. (See **silica minerals**.) [DJV]

onyx marble A banded **calcite** or **aragonite** used for decorative purposes. (See **carbonate minerals**.) [DJV]

ooid (*oolith*) One type of carbonate *coated grain*. The terminology for this group of features is rife with inconsistencies and problems. The term refers to their resemblance to fish eggs. They are small (<2 mm in diameter) *coated grains* with a calcareous cortex and a nucleus, typically a shell fragment or **sand** grain. The cortex exhibits fine laminae which are commonly very regular in thickness and so are uniformly concentric. The laminae commonly exhibit a *micritic* or radial micro-structure but lack any obvious biogenic features. The term 'superficial' ooid refers to ooids whose cortex width is less than the radius of the nucleus.

A rock largely composed of ooids is called an *oolite*. Forms with diameters of more than 2 mm are called **pisoids**.

Ooids are polygenetic in origin and many are abiogenic, simply formed by the physico-chemical **precipitation** of calcium carbonate on a nucleus, but others are microbially **precipitated**.

They are not diagnostic of any particular environment and can form in **caves**, calcareous *soils*, streams, lakes and marine settings. They require an environment where $CaCO_3$ can be **precipitated** and where there is sufficient movement for the grains to be coated concentrically. For multiple coatings to form, the growing grain must be retained in the environment.

Most ancient ooids formed in very shallow warm marine waters and extensive ooid formation occurs today on **beaches**, tidal inlets and tidal banks in areas such as the **Bahama Banks**, **Shark Bay** and the **Persian Gulf**. [VPW]

Flügel, E. (1982) *Microfacies Analysis of Limestones.* Springer-Verlag, Berlin.
Peryt, T. (ed.) (1983) *Coated Grains.* Springer-Verlag, Berlin.

opal ($SiO_2.nH_2O$) An essentially amorphous form of **silica** that includes common varieties as well as the **gemstone**, precious opal. (See **silica minerals**.) [DJV]

open pit/opencast mining Mining by excavating from the surface. The cheapest form of hard rock mining. A method used whenever possible in **coal**, **industrial minerals**, **bulk materials** and metal mining. [AME]

ophiolite An association of ultrabasic/ultramafic–basic/mafic rock types, commonly *serpentinized* forming a distinctive stratigraphic sequence, which is believed to represent a section through the oceanic **lithosphere**, tectonically emplaced at an **active** plate **margin** onto older continental, **island arc** or **oceanic crust** by obduction. A complete ophiolite sequence is a distinctive set of rock types as shown in Fig. O9. These comprise (from bottom to top), an ultramafic unit of **harzburgite** (with minor **dunite** and **lherzolite**), characterized by metamorphic textures and a complex **deformation** history; cumulate **ultramafic rocks** including **dunite**, **wehrlite**, *pyroxenite* and *leucogabbro*; **gabbro** with 'isotropic' texture with irregular pods and **veins** of **trondhjemite** near the top, and with

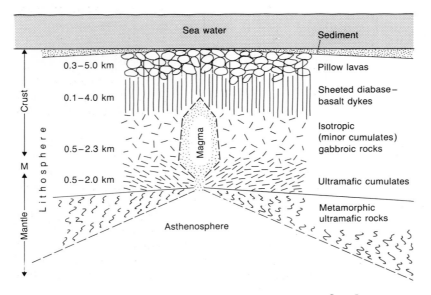

Fig. O9 Highly idealized section through a typical **ophiolite** sequence in its oceanic setting — either mid-ocean rift or dilating marginal basin. Range of thickness of units within the ophiolite is given on the left. The M-discontinuity marking the seismic boundary between the crust and the higher-velocity mantle occurs near the top of the ultramafic layer. The lower part of the ophiolite sequence is the alpine ultramafic complex. Overlying marine sediments are usually found with ophiolite.

dolerite (*diabase*) **dykes** increasing in abundance upwards; a *sheeted* **dolerite** *dyke complex*; and a layer of **pillow lavas**. These igneous units may be overlain by deep-sea sedimentary rocks.

The sequence of rock types within an ophiolite complex is correlated with the layers of rock forming the **oceanic crust** (Fig. O9), and has been used to interpret the formation of oceanic **lithosphere**. The ophiolite sequence comprises products of **mantle** *partial melting*, **magma** segregation, intrusion and eruption, formed during extension of oceanic **lithosphere** at an *ocean ridge*. These products include **harzburgite**, which is interpreted as a refractory residue of *partial melting* of more fertile (**lherzolite**) **mantle**, overlain by cumulate **gabbros** formed in *magma chambers*, which fed the **pillow lava** through the *sheeted dyke complex* (Fig. O9) [DHT/RST]

Coleman, R.G. (1977) *Ophiolites*. Springer-Verlag, Heidelberg.
Gass, I.G. & Smewing, D. (1980). Ophiolites: obducted oceanic lithosphere. In: Emiliani, C. (ed.) *The Sea*, Vol. 7. *The Oceanic Lithosphere*. John Wiley & Sons, New York.

ophitic texture A type of **poikilitic texture** in slowly cooled basic **igneous rocks** (e.g. **gabbro** and **dolerite**) in which euhedral, randomly-oriented **plagioclase** crystals ('laths') are partially or completely enclosed within plates of **pyroxene** (partial enclosure = *subophitic texture*). Ophitic texture is particularly characteristic of minor intrusions such as **dolerite**. [RST]

optical emission spectroscopy Method of analysis of rocks for major and **trace elements**. The sample is powdered, mixed with **graphite**, inserted into a carbon arc and vaporized. The light emitted is characteristic of the elements present in the sample, and the density of lines from light of different wavelengths recorded on a photographic plate is proportional to the concentration of elements present. The method has been used in geology and archeology, particularly for early analytical characterization of **obsidian**, but is now largely superseded by **X-ray fluorescence analysis**, **neutron activation analysis**, and **inductively-coupled plasma emission spectrometry**. [OWT]

Ordovician The second of three systems in the Lower *Paleozoic*. It is thought to span a period of about 67 Ma from 505 Ma to 438 Ma. It was founded in 1879 by Charles Lapworth based on rocks in the Arenig to Bala region of North Wales (home of the early British tribe, the Ordovices) which Sedgwick and Murchison had formerly referred to the **Cambrian** and **Silurian** Systems respectively. British workers have usually regarded the base of the *Arenig* Series as being the base of the Ordovician, but international opinion would now include the underlying *Tremadoc* Series within the Ordovician. There are six series in the standard British succession; the *Tremadoc Arenig*, *Llanvirn* and *Llandeilo* Series all have their type

sections in Wales, the *Caradoc* Series is based on rocks in Shropshire and the *Ashgill* on the sequence in the Cautley Inlier of northern England (Fig. O10). The base of the *Tremadoc* Series has never been formally defined but a suitable level for the boundary is at the base of a mudstone band yielding the lowest examples of *Rhabdinopora flabelliforme s.l.* (Whittington *et al.*, 1984). The *Tremadoc* Series has a distinctive fauna which in the type area is characterized by olenid *trilobites*, anisograptids and **molluscs**. The succeeding *Arenig* Series, in its type area, lies with slight **unconformity** on the *Tremadoc*, but continuous sequences from the *Tremadoc* through into the *Arenig* are found in South Wales.

The boundary between the Ordovician and overlying **Silurian** has been chosen at the base of the *Parakidograptus acuminatus* Biozone in the Dob's Linn section in the Southern Uplands of Scotland.

The Ordovician System has commonly been divided into Lower, Middle and Upper divisions, but these units have never been formally defined and are used in different ways in different places, so should not be regarded as established stratigraphic units. Outside Britain it is commonly difficult to recognize the standard British Series, and local series are commonly used. In the USA the system is commonly divided into a lowermost *Canadian* or *Ibex* Series, followed by the *Champlainian* which is often replaced by two series, the *Whiterockian* and *Mohawkian*, and then by the *Cincinnati* Series. Three Series, the *Ichangian*, the *Neichiashanian* and the *Chientangkiangtgian* are recognized in China (Fig. O10). In Australia the Ordovician is divided into 10 stages based on **graptolite** faunas; these resemble those in the USA and provide a basis for intercontinental correlation.

Biostratigraphic correlation in the Ordovician is best effected where there are **graptolites**, which usually occur in more off-shore grey to **black shales**. The system can be divided into about 17 **graptolite** zones, though further subdivisions can sometimes be made locally. Correlation using **conodonts** is also effective, particularly in carbonate facies; a sequence of 12 faunas have been recognized in the middle and upper Ordovician of the midcontinent USA. *Chitinozoa* and **acritarchs** have proved helpful in fine-grained clastic facies. Certain bathypelagic *trilobites* such as the cyclopygids have a wide distribution which makes them useful for correlation, while the careful use of concurrent ranges of *trilobites* and **brachiopods** can be effective locally.

The diversity of marine life increased rapidly from the start of the *Tremadoc* and many new groups appeared in the early Ordovician. In particular, groups of epifaunal suspension feeders with calcareous skeletons became common. Amongst the **brachiopods**, the strophomenids, orthids and pentamerids diversified early in the period and the rhynchonellids and spiriferids rather later. *Crinoids* and *cystoids* were present throughout the Ordovician but increased in abundance and diversity from the *Caradoc* onwards, as did the rugose and tabulate **corals** which first appeared in the middle Ordovician. **Bryozoa** appeared near the base of the Ordovician and progressively diversified. *Trilobites* which had been common in the **Cambrian** diversified in the early Ordovician into a variety of new forms represented by four new orders and five new sub-orders. Several new morphological types appeared including the Trinucleina, Harpina, Cheirurina, and Odontopleurida. Amongst the plankton, *nautiloids*, **graptolites** and **conodonts** all increased in abundance and diversity in the early Ordovician.

The result of this Ordovician faunal diversification is that assemblages of fossils become more common and varied throughout the system. The success of the shelly biota is reflected in the development of **coral–stromatoporoid** *reefs* with an associated **brachiopod**, **bryozoa**, *crinoid* fauna, from middle Ordovician onwards. During the Ordovician benthic communities dominated by articulate **brachiopods** occur in progressively deeper water shelf facies until from *Caradoc* times onwards Ordovician clastic shelves are roughly partitioned into about five depth/facies belts, each

Ordovician System

Series: *Period*		*Epoch*	British Stages: *Age*	*British Biozones*	*Ma*	*Estonia*		*Kazakhstan*	*China*		*Australia*	*North America*		
S		Llandovery	Rhuddanian	*Parakidograptus acuminatus*	438									
Ordovician (O)	Bala	Ashgill	Hirnantian			Porkuni	Harju	Tolen	Wufeng	Jiangtangjiang (Chientangkiang)	Bolindian	Richmond	Cincinnatian	
			Rawtheyan	*Dicellograptus anceps*										
			Cautleyan	*Dicellograptus complanatus*		Pirgu								
			Pusgillian											
		Caradoc	Onnian	*Pleurograptus linearis*	448	Vormsi		Zharyk	Linxiang			Maysville		
			Actonian	*Dicranograptus clingani*		Nabala		Dulankara	(Linh-siang)		Eastonian	Eden		
			Marshbrookian			Rakvere								
			Longvillian			Oandu Keila		Anderken	Baota (Pagoda)			Sherman	Trenton	Mohawkian
			Soudleyan	*Climacograptus wilsoni* (*Diplograptus multidens*)				Yerkebidaik						
			Harnagian	*Climacograptus peltifer*		Johvi Idavere						Kirkfield		
												Rockland		
			Costonian				Viru		Hule	Aijiashan (Neichiashan)	Gisbornian	Black River	Champlainian	
		Llandeilo		*Nemagraptus gracilis*	458	Kukruse		Tselin-ograd				?		
												Chazy		Whiterockian
				Glyptograptus teretiusculus		Uhaku								
		Llanvirn		*Didymograptus murchisoni*	468	Lasnamagi Aseri		Karakan	Niushang		Darriwilian	Whiterock		
				Didymograptus bifidus				Kopaly						
		Arenig		*Didymograptus hirundo*	478	Kunda	Oeland	Kogashyk	Chongyi		Yapeenian			
				Isograptus gibberulus (*Didymograptus extensus*)		Volkhov					Castle-mainian	Beekmantown	Canadian = Ibex	
				Didymograptus nitidus							Chewtonian			
				Didymograptus deflexus		Latorp		Rakhmet	Ningguo (Ningkuo)	Yichang (Ichang)	Bendigonian			
				Tetragraptus approximatus							Lancefieldian			
		Tremadoc		*Apatokephalus serratus*	488	Olentian								
				Clonograptus heres & Shumardia pusilla										
				Symphysurus incipiens					Xinchang		Warendian	Gasconada		
				Dictyonema flabelliforme	505						Datsonian			
€			Dolgellian								Payntonian			

Fig. O10 Stratigraphic table showing the division of the **Ordovician** System into the Series, Stages and Biozones in Britain, and series (printed vertically) and stages (printed horizontally) elsewhere.

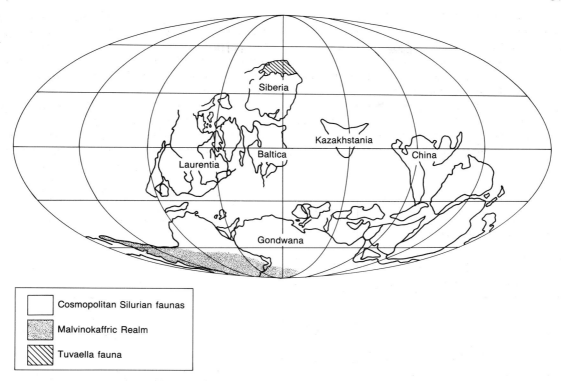

Fig. O11 A possible distribution of continents in the Lower **Ordovician** and the main biogeographic provinces based principally on trilobite faunas. (After Scotese *et al.*, 1979.)

occupied by a separate **brachiopod** association (e.g. the *Dinorthis, Howellites, Dalmanella, Nicolella* and *Sericoidea* association in the mid *Caradoc* of Wales).

The successful diversification of Ordovician faunas was interrupted by a major episode of **extinction** near the close of the Ordovician. A first phase of **extinction** at the beginning of the Hirnantian Stage (the topmost Ordovician Stage) reduced the **graptolites** to a few species, caused the disappearance of several *trilobite* families and many tens of genera, and greatly reduced the diversity of **brachiopods**, *cystoids* and *ostracods*; **acritarchs** and **conodonts** also became rare in some regions. The residual fauna is represented in many places throughout the world by the *Hirnantia* fauna (*Hirnantia* is an orthid **brachiopod**) which usually has a low diversity. The second phase of **extinction** at the end of the Hirnantian seriously affected the **conodonts**, **acritarchs**, **corals** and **brachiopods** and the development of *reefs* was temporarily inhibited. It is estimated that in total 22 families became extinct in the late Ordovician which makes it one of the largest **extinction** events in the fossil record.

The distribution of the main continental blocks in the Ordovician is not well constrained. It appears likely that in early Ordovician times there was a large **Gondwanan** continent in the southern hemisphere stretching from the pole to the equator. **Laurentia**, Siberia and *Kazakhstania* were in tropical latitudes. There is disagreement about the latitude of *Baltica* in the southern hemisphere. By the end of the Ordovician, Saharan Africa lay at the South

Pole and supported a large ice cap while **Laurentia**, Siberia and *Kazakhstania* were still in the tropics. *Baltica* probably lay close to the tropics, but Iberia, which on some reconstructions is associated with *Baltica*, has *glacimarine* sediments and must have been in relatively high latitudes.

The distribution of Ordovician faunas helps to identify continental blocks. Isograptid faunas in the lower Ordovician occur at the margins of continents (Fig. O11) while non-isograptid faunas occur internally. In the lower Ordovician the **Gondwanan** continent had an *Asaphopsis* *trilobite* fauna, **Laurentia** a bathyurid fauna, *Baltica* an asaphid fauna and central Europe a *Selenopeltis* fauna. The faunas of Siberia and *Kazakhstania* are of bathyurid aspect.

In middle Ordovician times both the *trilobite* and **brachiopod** faunas of **Laurentia**, *Baltica* and Central Europe, including England, were still distinct but progressively through the upper Ordovician the faunas of the three blocks became more alike. It has been suggested that *Baltica* and the rest of Europe had joined by the late *Caradoc* and Europe and **Laurentia** were close together by the end of the Ordovician.

Sea-level stood relatively high throughout most of the Ordovician covering large areas of the continents. Regressions occurred in the lower and upper *Tremadoc*, at the end of the *Arenig*, in the *Llandeilo*, and generally during the *Ashgill* but with a particularly marked regression at the end of the *Ashgill*. This last regression and the ensuing transgression were probably related to the growth and

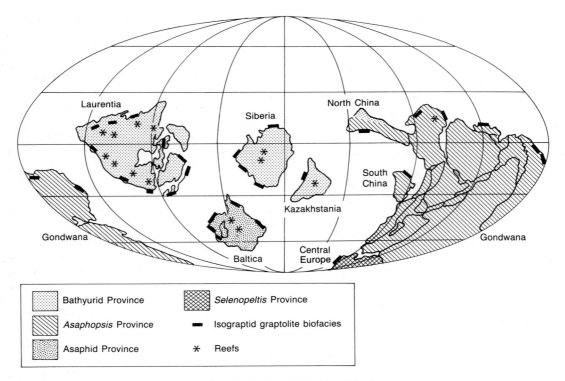

Fig. O12 The distribution of isograptid graptolite biofacies and reefs in the Lower **Ordovician**. (After Scotese *et al.*, 1979.)

decay of an *ice cap* centred on Saharan Africa on the **Gondwanan** plate.

The shallow epeiric seas on the continental platforms and continental interiors produced widespread carbonate sequences in tropical regions, e.g. midcontinent USA, Arctic Canada, Greenland, north-west Scotland, the Russian platform and Siberian platform. In several of these regions there are associated *reef* **limestones** (Fig. O12). In more temperate regions such as Iberia, France, Czechoslovakia and England, platforms had a clastic cover. Mixed carbonate/clastic sequences, occur in Scandinavia, Australia, and north-eastern parts of the former USSR.

Many shelves and slopes peripheral to the continents had a **shale** facies, often with **graptolites** (the isograptid biofacies on Fig. O12). **Mobile belts** with associated volcanics occurred in western South America, the Appalachians, south-east Ireland, Wales, the Lake District of England and the Southern Uplands of Scotland, the Sakmaro-Leinvinskaya zone of the Urals and *Khazakhstania*, north-west of the Omuler Mountains in the north-east of the former USSR. [PJB]

Bassett, M.G. (ed.) (1976) *The Ordovician System, Proceedings of the Palaeontological Association Symposium, Birmingham, September 1974.* University of Wales Press and National Museum of Wales, Cardiff.

Bruton, D.L. (ed.) (1984) *Aspects of the Ordovician System.* Palaeontological Contributions from the University of Oslo, No. 295. Universitetsforlaget, Oslo.

Lockley, M.G. (1983) A review of the brachiopod dominated palaeo-communities from the type Ordovician. *Palaeontology* **26**: 111–45.

Ross, R.J. Jr. (1984) The Ordovician System, progress and problems. *Annual Review of Earth and Planetary Sciences* **12**: 307–35.

Ross, R.J. Jr., Adler, F.J., Amsden, T.W. *et al.* (1982) *The Ordovician of the United States of America, Correlation Chart and Explanatory Notes.* International Union of Geological Sciences Publication No. 12, 73 pp. + 3 sheets (there are IUGS correlation charts for most parts of world).

Scotese, C.R., Bambach, R.K., Barton, C., Van der Voo, R. & Ziegler, A.M. (1979) Paleozoic base maps. *Journal of Geology* **87**: 217–77.

Sepkoski, J.J. & Sheehan, P.M. (1983) Diversification, faunal change and community replacement during the Ordovician radiations. In: Tevesz, C. & McCall, P.L. (eds) *Biotic Interactions in Recent and Fossil Benthic Communities*, pp. 675–717. Plenum, New York.

Whittington, H.B., Dean, W.T., Fortey, R.A., *et al.* (1984) Definition of the Tremadoc Series and the Series of the Ordovician System in Great Britain. *Geological Magazine* **121**: 17–33.

ore A metalliferous mineral or an aggregate of metalliferous minerals, more or less mixed with **gangue**, which can be exploited at a profit. The profit element is crucial and material that cannot be worked at a profit should not be referred to as ore. Strictly speaking ore is material from which a metal or metals are extracted. There is, however, no equivalent word to describe crustal material from which **industrial minerals** are separated and therefore the term ore is sometimes applied to commercial deposits of **fluorspar**, **gypsum**, etc. [AME]

ore microscopy The technique of examining opaque minerals in reflected light with the use of a polarizing microscope. [AME]

Craig, J.R. & Vaughan, D.J. (1981) *Ore Microscopy and Ore Petrography.* John Wiley & Sons, New York.

ore reserve (*measured*, *indicated* and *inferred reserve)* The calculable tonnage of **ore** in an **orebody**, including some that is believed, even though not conclusively proved, to be present, is known as the ore reserve. Mining geologists classify ore reserves into three classes: proved, probable and possible. Proved **ore** has been so thoroughly sampled that we can be certain of its outline, tonnage and average **grade**, within certain limits. Elsewhere in the **orebody**, sampling from drilling and development workings may not have been so thorough, but there may be enough information to be reasonably sure of its tonnage and **grade**; this is probable **ore**. On the fringes of exploratory workings there may be enough information to infer that **ore** extends for some way into only partially explored ground and that it may amount to a certain volume and **grade** of possible **ore**. In most countries, these, or equivalent, words have nationally recognized definitions and legal connotation. [AME]

orebody A volume of rock that can be exploited commercially for its metal content. Various types of orebody are shown in figures illustrating **massive sulphide deposit**, p. 387; **Kuroko-type deposit**, p. 355; **carbonate-hosted base metal deposit**, p. 90; **Carlin-type deposit**, p. 94; **diamond deposit**, p. 170; **greisen deposit**, p. 307; **Bushveld Complex**, p. 76. [AME]

organic matter diagenesis Most sediments, when deposited, contain a mixture of inorganic detrital **mineral** grains, fine-grained amorphous **mineral** detritus, dead organic material (carbohydrates, proteins, lipids, lignin, fats, waxes, resins, *chitin*, etc.), numerous living micro-organisms (*algae*, *bacteria*, etc.) and **porosity**, the latter being variously wetted and oxygenated depending upon the depositional environment. Such assemblages are intrinsically unstable; dead organic matter, an extremely effective reducing agent, is brought into intimate contact with fine-grained **mineral** detritus and amorphous detrital material — the products of subaerial **weathering** — effective oxidizing agents. On deposition, therefore, there is an inherent tendency for organic matter to react with, and be oxidized by, fine-grained detrital materials (principally **iron** and **aluminium** sesquioxides) and dissolved oxidizing agents (e.g. dissolved oxygen, nitrate, sulphate) through the processes of **diagenesis**. The presence of micro-organisms that utilize the organic matter as a food source further catalyses this potential reactivity.

The type and extent of **alteration** of organic matter depends upon several variables, principally the amount and type of organic matter present, the depositional environment and the rate of deposition (Fig. O13). In highly oxidizing continental environments with slow rates of deposition (and therefore burial) *all* the organic matter

may be oxidized to CO_2 and H_2O by aerobic **bacteria** that use molecular oxygen for respiration at or close to the depositional surface. By contrast, very rapid burial in a marine environment results in a significant proportion of the organic matter surviving initial oxidative **diagenesis**. In such environments reducing conditions develop after burial of a few centimetres if the sedimentation rate is sufficient to carry the organic matter away from the influence of molecular oxygen. A regular depth zonation of **diagenetic** processes that degrade organic matter is then established within the uppermost metres of the sedimentary succession (Fig. O14). *Anaerobic* sulphate-reducing *bacteria* (e.g. *Desulfovibrio*) utilize oxygen from dissolved SO_4^{2-} in the interstitial **pore-waters**, reducing the **sulphur** to S^{2-}. Energy for this *sulphate reduction* is provided by the degradation of organic matter which is converted to CO_2, NH_4, H_2O and residual *humin* (or *kerogen*). Sulphate is replenished by downward diffusion from the depositional **pore-waters** so that the lower limit of *sulphate reduction* is determined by the supply of SO_4^{2-}. Reduced **sulphur** is not incorporated into the bacterial cell but normally combined with **iron** to form *hydrotroilite* which slowly converts to **pyrite**. Beneath the zone of *sulphate reduction* further degradation of organic matter takes place through *fermentation*. This is an *anaerobic* process by which *bacteria* use oxidized forms of organic matter as an energy source to liberate acetate, bicarbonate and *methane*. **Fermentation** is dependent upon a supply of suitable organic substrates which, as a consequence of activity, are gradually reduced with increasing depth.

Residues not used by aerobic and *anaerobic* micro-organisms recombine through condensation and polymerization to produce macromolecules of soluble organic acids and insoluble *humin* compounds. Organic acids are, in turn, consumed by *bacteria* but may persist until temperatures of around 80°C when they undergo thermal destruction. With increasing burial, however, the proportion of insoluble *humin* gradually predominates over organic acids until at depths of around 1 km virtually all the remaining organic matter is acid insoluble *humin*, the precursor of *kerogen*.

At greater burial depths the **alteration** of *kerogen* takes place largely as a result of increasing temperature, and is generally referred to as thermal **decarboxylation**. Structural rearrangements take place in the *kerogen* in an attempt to produce a higher degree of ordering and thus greater stability. The adjustment to increasing temperature is achieved through a progressive elimination of functional groups and linkages between **carbon** chains, liberating a wide range of compounds, including CO_2, H_2O, liquid *hydrocarbons* and, ultimately, dry *methane* gas. **Petroleum** generation is gradual through high to low molecular weight *hydrocarbons* and can also be considered as an attempt of *kerogen* to adjust to its changing geochemical environment by attaining higher degrees of ordering with increasing temperature and burial depth. The ultimate end-product of this response to increasing temperature

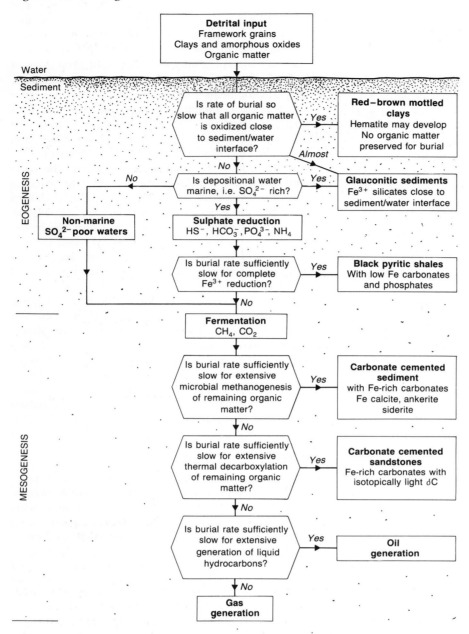

Fig. O13 Influence of depositional environment and burial rate on the type of **organic matter diagenesis** and resulting reaction products. (After Curtis, 1987.)

is **graphite** as low-grade metamorphic conditions are reached.

As a result of these depth-related **alteration** processes, the composition of the organic matter becomes fundamentally and irreversibly altered. In its initial state, immature *kerogen* falls into three main compositional groups; sapropelic algal *kerogen* (type I), rich in lipids; marine planktonic *kerogen* (type II), rich in ester groups; and humic *kerogen* (type III), derived from **land plants** and largely comprising woody, vegetal material. A cross plot of **kerogen** composition in terms of oxygen:**carbon** ratio against **hydrogen**:**carbon** ratio (termed a *van Krevelen*

diagram after the first scientist to use this plot) highlights these primary depositional differences well (Fig. O15). Each *kerogen* type initially has a distinct oxygen:**hydrogen** ratio. During **burial diagenesis** the composition of each *kerogen* type changes in terms of its three major elemental components (C, H, O). On the *van Krevelen diagram* this change is expressed as a distinct trend of decreasing oxygen and **hydrogen** contents for each *kerogen* type. These trends, termed *evolution paths*, eventually converge together during deep burial as the composition of the *kerogen* approaches 100% **carbon**.

The relative proportion of the various products of

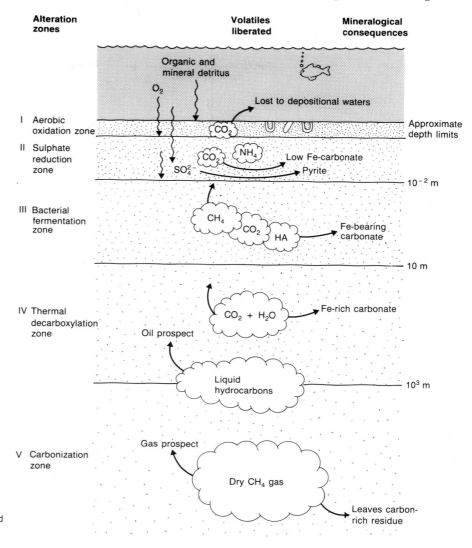

Fig. O14 Organic matter diagenesis. Depth-related zones of organic matter degradation and the main reaction products. HA = humic acids.

kerogen maturation are strongly dependent upon the initial *kerogen* composition (Fig. O16). *Sapropelic kerogen* with high initial **hydrogen:carbon** but low initial **oxygen:carbon** atomic ratios, typical of *algal* material, characteristically produces a high yield of paraffinic oil. By contrast, humic *kerogen* (low H/C ratio, high O/C ratio) typical of plant and woody material characteristically produces little liquid *hydrocarbon* but large volumes of CO_2 and *methane*.

The progressive **alteration** of organic matter cannot be considered in isolation of the host or enclosing sediments. Reaction products of organic matter diagenesis, including CO_2, H_2O, organic acids, *hydrocarbons* and *methane*, may all interact with the inorganic constituents of the associated sediments. Expulsion of H_2O may contribute to the interstitial **pore-waters**, providing water for migration and the development of **overpressures**. Organic acids may complex metal cations and enhance **mass solute transfer** (see also **chelation**). CO_2 dissolves·in H_2O to produce

bicarbonate which may subsequently be incorporated into carbonate cements if accompanied by **iron** or **magnesium** reduction. In the absence of equalizing alkali-generating reactions, excess carbonic acid may be available for dissolving unstable minerals in the subsurface (see **secondary porosity**). **Oxidation** of organic matter results in a concomitant reduction of ferric **iron**, causing a rise in **pore-water pH** and Fe^{2+} available for incorporation into **carbonates** and **chlorite minerals**. (See **burial diagenesis**.) [SDB]

Curtis, C.D. (1987) Mineralogical consequences of organic matter degradation in sediments: inorganic/organic diagenesis. In: Leggett, J.K. & Zuffa, G.G. (eds) *Marine Clastic Sedimentology*, pp. 108–23 Graham & Trotman, London.

Hunt, J.M. (1979) *Petroleum Geochemistry and Geology*. W.H. Freeman, San Francisco.

Surdam, R.C. & Crossey, L.J. (1985) Organic-inorganic reactions during progressive burial: key to porosity and permeability enhancement and preservation. *Philosophical Transactions of the Royal Society, London* **A315**: 135–56.

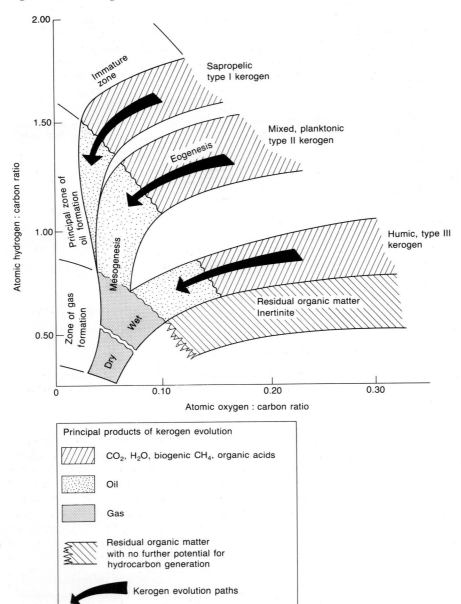

Fig. O15 Organic matter diagenesis.
Evolution of kerogen composition during diagenesis expressed as a function of oxygen: carbon and hydrogen: carbon ratios (the van Krevelen diagram). Successive stages of alteration are indicated by the principal products generated. (After Tissot & Welte, 1978.)

Tissot, B. & Welte, D.H. (1978) *Petroleum Formation and Occurrence.* Springer-Verlag, Berlin.

organic weathering The disintegration and decomposition of rocks through the action of micro-organisms, plants, animals and decaying organic matter. Organic influences upon **weathering** processes have often been neglected by geomorphologists, although they are of general importance. Organisms may participate in the physical **weathering** of rocks through root wedging, for example, especially in **jointed** rocks which encourage root penetration. Living organisms and decaying organic material also play an important biochemical role in **weathering**, involving a complex set of processes such as cation exchange,

chelation, solution by root exudates and the effects of organic acids. Humic acids, for example, are active in **chelation**, decompose silicates and, in particular, **amphiboles**. *Bacteria* and microflora such as *algae*, fungi, and lichens often play a particularly important role in rock **weathering**. On bare rock surfaces such organisms grow as pioneer communities, and within *soils* they may play a key role in the acidification of *soil* waters. Organisms may also help protect rock surfaces from other **weathering** processes by forming a living 'skin' and so reducing the effects of rain and temperature changes. [HAV]

Berthelin, J. (1983) Microbial weathering processes. In: Krumbein, W.E. (ed.) *Microbial Geochemistry*, pp. 223–62. Basil Blackwell, Oxford.

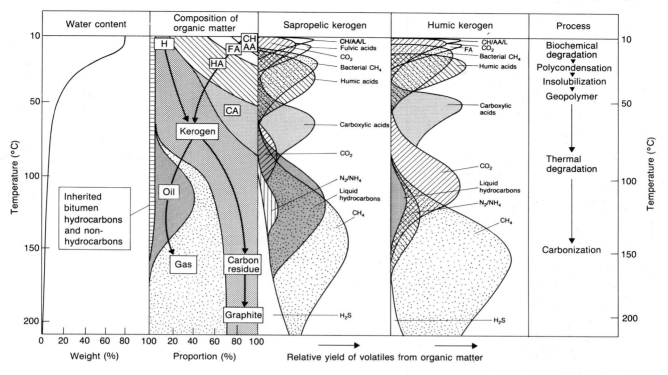

Fig. O16 Organic matter diagenesis. The relative proportions of volatiles released from sapropelic and humic organic matter during burial diagenesis. H = humic; CH = carbohydrates; AA = amino acids; L = lipids; FA = fulvic acids; HA = humic acids; CA = carboxylic acids; HC = hydrocarbons. (Data sources: Hunt, 1979; Surdam & Crossey, 1985; Tissot & Welte, 1978.)

Eckhardt, F.E.W. (1985) Solubilisation, transport and deposition of mineral cations by microorganisms — efficient rock weathering agents. In: Drever, J.I. (ed.) *The Chemistry of Weathering*, pp. 161–73. Reidel, New York.

oriented core Usually referring to rock samples that have been drilled in the field and have been oriented relative to present-day horizontal and **true north**. [DHT]

oriented lake Lake which possesses a preferred long axis orientation. Such lakes occur in *permafrost* areas, where they may be referred to as *'thaw lakes'*, or in arid areas, where they are generally called **pans**. The former are mainly caused by *thermokarst* subsidence, while the latter are primarily caused by *deflation*. Both are shaped into characteristic shapes (clams, kidneys, ovals, D-shapes) by wind action, and in most cases the long axis is at right angles to prevailing winds. [ASG]

Carson, C.E. & Hussey, K.M. (1963) The oriented lakes of Arctic Canada: a reply. *Journal of Geology* **71**: 532–3.

orogen Total mass of rock deformed during an **orogeny** (mountain-building period). [NJM]

orogeny/orogenesis Process of creation of mountain belts by **tectonic** activity. An orogenic belt is a major

linear segment of the Earth's **crust** affected by orogenesis. Orogeny is contrasted with **epeirogeny**, which affects more stable **cratonic** parts of the Earth's **crust**. Orogenic belts are distinguished by **folding**, **faulting**, regional metamorphism and igneous activity, as well as by vertical movements that are generally greater than those of adjoining **cratons**. Thus although vertical **uplift** is responsible for the creation of a mountain belt topography, it is the origin of the **uplift** that is important in orogenesis.

In terms of **plate tectonic** theory, orogenic belts mark sites of *continent–continent* or *continent–island arc collision* zones. These are situated at *destructive plate boundaries* characterized by convergent *plate* motion. Where one of the converging *plates* is oceanic, **subduction** takes place to accommodate the convergence. However where two continental parts of the converging *plates* come into contact, **subduction** is inhibited by the buoyancy of the **continental crust**. Continued convergence may take place only by thickening the **crust**. The process of crustal thickening, which is central to the formation of an orogenic belt, may take place by crustal-scale overlap, as has been suggested for the Himalaya by Bird (1978), or by a process of **tectonic** thickening of one or both of the opposed continental slabs. Thus the **deformational** aspects of orogeny may be explained by *plate* convergence. The igneous and metamorphic effects associated with orogeny are partly related to the **subduction** process that precedes collision. At least one, and typically several **subduction zones** become welded together in the final collisional event of an orogeny. The effects of **subduction** on the upper slab of a **subduction zone** are often known as '*subduction orogeny*' and include the formation of **paired** high-temperature/low pressure and low-temperature/high pressure **metamorphic belts**, and also the **calc-alkaline** *volcanic* activity characteristic of such zones. The effects of collision may bring several **subduction** complexes together in a single belt, and also superimpose new thermal effects resulting from the collision process. **Crustal** thickening may lead to melting at the base of the thickened **crust**, and movements on deep thrust-sense **shear zones** may generate additional heat. The process of collision and **uplift** therefore generates a major tectono-thermal event, usually with a characteristic thermal age that is geochronologically detectable because of the regional resetting of isotopic systems. This thermotectonic age may be used to date 'old' orogenies such as those in the **Precambrian** where detailed stratigraphic age control is lacking.

The major orogenic belts of the **Phanerozoic** are usually grouped into three world-wide orogenic systems: the *Alpine (Alpides)*, *Hercynian (Hercynides)* and *Caledonian (Caledonides)*. These systems result from the amalgamation of many individual **subduction** and *collision* orogenic belts. Major systems of this type have been termed '*orogenic cycles*' by Sutton (1963) in the belief that orogenic activity is cyclic over geological time. There is no doubt that major changes in *plate* configuration take place at intervals of several hundred Ma, and that these changes follow major culminations of tectono-thermal activity. However the periods between these climactic events do not appear to be of uniform length. In the **Precambrian**, major changes are related to widespread *cratonization* events that occurred at *c.* 2.5, *c.* 1.6, and *c.* 1.0 Ga, and have been used to define the end of the **Archean**, and of the Early and Mid-**Proterozoic** respectively, Thus while orogenies are generally **diachronous** over large sectors, reflecting the complexity of *plate* configurations, the climax or ending of the major periods of orogenic activity may be broadly coeval. [RGP]

Bird, J.M. (1978) Initiation of intracontinental subduction in the Himalaya. *Journal of Geophysical Research* **83**: B10, 4975–8.
Miyashiro, A., Aki, K. & Sengor, A.M.C. (1982) *Orogeny*. John Wiley & Sons, Chichester.
Sutton, J. (1963) Long term cycles in the evolution of the continents. *Nature* **198**: 731–5.

Orowan–Elsasser convection A convective model for the interior of the Earth in which low density **mantle** material rises at the *oceanic ridges* to form new **lithosphere**, and dense, cool **lithosphere** descends at the *oceanic trenches*. Such **convection** could be restricted to the **upper mantle** and implies the **lithosphere** travels faster than the underlying **mantle**. [DHT]

Bott, M.H.P. (1982) *The Interior of the Earth*. Arnold, London.

orpiment (As_2S_3). A rare **ore** mineral of **arsenic**. [DJV]

orthite A synonym for the mineral **allanite**. [DJV]

orthoclase ($KAlSi_3O_8$) A **feldspar mineral**. [DJV]

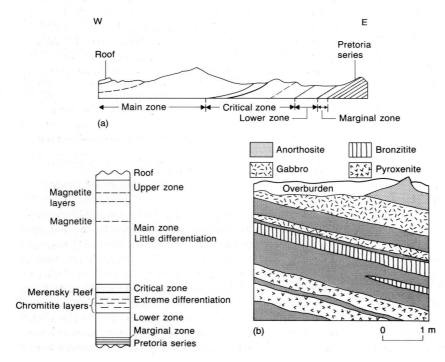

Fig. O17 **Orthomagmatic deposit**. Sections showing the occurrence of economic minerals in the Bushveld Complex. (After Evans, 1993.)

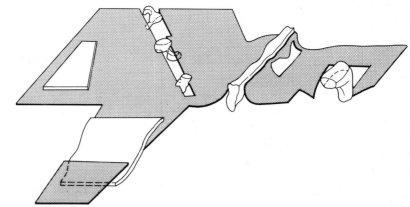

Fig. O18 Orthomagmatic segregation deposit. Diagram showing the shapes of podiform chromite deposits in New Caledonia and their relationships to the plane of the foliation in the host peridotites. (The foliation is nearly always parallel to the compositional banding). Note the disturbance of the foliation around the two deposit types on the right. The tabular deposits are of the order of 50–100 kt. (After Evans, 1993.)

orthoferrosilite (FeSiO₃) An end-member *orthopyroxene* mineral. (See **pyroxene minerals**.) [DJV]

orthomagmatic/magmatic segregation deposit Ore deposit, apart from *pegmatite*, that has crystallized direct from a **magma**. The type formed by *fractional crystallization* is usually found in ultrabasic and basic **plutonic** rocks; those produced by **liquation** are associated with either **plutonic** or volcanic ultrabasic and basic rocks.

The first group belongs to the class of **igneous rocks** known as **cumulates**, but as well as the normal bands of **olivine**, **pyroxene** and **plagioclase**, layers rich in **ore** minerals such as **chromite** and **magnetite** are now present as in the **Bushveld Complex** which exemplifies the stratiform or Bushveld-type of **chromite** deposit. In this deposit type the layers of massive **chromite**, a few millimetres to over 1 m in thickness, occur in the lower portions of stratified igneous complexes of either funnel-shaped (**Bushveld**; Great Dyke of Zimbabwe) or **sill**-like intrusions (Stillwater, USA; Kemi, Finland) associated with ultrabasic differentiates of the parent **magma**. The **chromite** and other layers can be traced laterally for tens of kilometres. In addition to the **chromite** layers there may be layers enriched in **platinum** metals, such as the Merensky Reef, and deposits of this type are the world's major producers of these metals. Near the top of such intrusions economically mineable layers of vanadiferous **magnetite** may occur (Fig. O17).

Insignificant in terms of **reserves** but important producers at the present time are the podiform **chromite** deposits of the former USSR, Albania and Turkey. The morphology of these **orebodies** is irregular and unpredictable (Fig. O18) although the basic form is sheetlike to podlike. They usually contain less than 1 Mt of **ore** except for those in the southern Urals which can range up to 90 Mt. Podiform **chromite** deposits occur in **harzburgite** sections of **peridotite** masses or **peridotite–gabbro** complexes of Alpine-type (**ophiolites**) or in parts of dismembered **ophiolites**.

A different association is that of the large **ilmenite** segregations found in **anorthosites**. To compete with the low cost beach **placer** producers of **titanium** these deposits have to be large tonnage **orebodies** that can be mined by **opencast** methods and thus there are only a few such operations in the world. The biggest is that of Tellnes, Norway with reserves of 300 Mt of 18% TiO₂.

The most important **liquation** formed deposits are the Cu–Ni–Fe(–Pt) **orebodies**. These are essentially **nickel** producers, but some have co-product or by-product **copper** and sometimes **iron** and/or **platinum**. Only Canada, the former USSR and Australia have important deposits of this type. Their principal mineralogy is basically simple: **pyrrhotite**, **pentlandite**, **chalcopyrite** and **magnetite**. These deposits are associated with both volcanic and plutonic ultrabasic and basic **igneous rocks**. The volcanic association is mainly with **komatiites** in **Archean greenstone belts** but to a minor extent with **tholeiites**. The plutonic association is with large, stratiformly layered complexes such as **Bushveld**, South Africa and Sudbury, Canada or with smaller intrusions related to flood **basalts**, e.g. Noril'sk, Russian Republic emplaced in **cratons**. The latter association occurs either in areas that have suffered a catastrophic release of energy, e.g. an *astrobleme* (Sudbury), or in intracontinental **rift** zones (Noril'sk). [AME]

Edwards, R. & Atkinson, K. (1986) *Ore Deposit Geology*. Chapman & Hall, London.

Evans, A.M. (1993) *Ore Geology and Industrial Minerals: An Introduction* (3rd edn). Blackwell Scientific Publications, Oxford.

oscillatory flow An aqueous flow which reverses periodically, usually restricted to the flow set up by the passage of *gravity waves* with periods in the range 1–20 s. Wind **stress** on the sea surface causes the formation of water **waves** with a characteristic wavelength, height and phase velocity (or *celerity*). The passage of these **waves** causes an orbital motion of water particles in the underlying water column. The nature of the orbits depends on the **wave** conditions but also critically on the depth of water. In deep water, where water wavelengths (L) are small compared with water depth (*h*), the orbits of water particles are circles which decrease in diameter exponentially with depth. Deep water **waves** ($h/L > 0.5$) are therefore not capable of forming a **wave**-rippled bed. In intermediate water depths ($0.5 > h/L > 0.05$) the orbits of the water particles decrease in diameter downwards through the water column but they become flattened ellipses rather than circles. In shallow water ($h/L < 0.05$) the orbits retain their surface diameter with depth but are strongly flattened ellipses and the water movement immediately above the bed resolves itself into a purely to-and-fro horizontal motion. Intermediate and shallow water **waves** are responsible for the formation of **wave ripple** marks in a cohesionless substrate.

Under oscillatory flows the **wave boundary layer** is thin compared with its unidirectional current counterpart. This is because the short period of the reversals prevents the **boundary layer** from developing fully. Because the **wave boundary layer** is thin, steep velocity gradients and large **shear stresses** result. This means that **waves** are very efficient at stirring up sediment from the bed.

Progressive *gravity waves* produce not only a purely oscillatory flow, but also various superimposed mass transports. The mass transport may be essentially constant and in the direction of **wave** propagation (sometimes called a steady streaming), or may be itself periodic and related to the effects on the flow of the curvature of the bed or to the existence of a **standing wave** caused, for example, by a nearby **beach** or harbour wall. Ignoring other components to the near-bed flow, such as those due to *storm set-up*, these mass transports cause an asymmetry to the velocity field and may be responsible for the widespread occurrence of a unidirectional *cross-lamination* within **wave ripple marks**. (See **ripple marks**.) [PAA]

outlier An isolated hill or group of rocks lying at some distance from the main body of similar rocks, and immediately surrounded by rocks of an older age. [NJM]

overbank deposit Deposition of suspended sediment and the finer fractions of **bedload** by flood water at times when **bankfull discharge** has been exceeded and the river is flowing across the **floodplain**. Significant amounts of material may be deposited by this mechanism so that overbank deposits can form an important constituent of alluvium if flooding is common, although usually within-channel deposition is much more important. Since the transporting capacity of the flow tends to decrease away from the channel margin, so the size and amount of deposition tends to vary in similar fashion; in this way natural **levées** are formed. Vertical **accretion** by overbank deposition provides at most only a centimetre or two of material during each flood episode. The deposit may fine upwards. However, laminated sediments are not commonly found on **floodplains**, perhaps because of the lateral reworking of deposits by meander migration, so the importance of overbank deposition is therefore hard to assess. (See **river flood**.) [TPB]

overland flow Downslope flow of water across the *soil* surface. Such **runoff** is usually produced during rainfall or as the result of snowmelt, although *exfiltration* of *soil* water, often at the foot of steep slope sections, can maintain surface **runoff** even in the absence of rainfall. Two main mechanisms are responsible for overland flow. **Infiltration**-excess overland flow occurs when the rainfall intensity exceeds the **infiltration capacity** of the *soil* surface (sometimes called 'Hortonian' *overland flow*). Saturation-excess overland flow can occur either when rain falls on to a saturated surface and being unable to **infiltrate** forms 'direct' **runoff**; or, when water moving downslope through the *soil* (*throughflow*) exceeds the *transmission capacity* of the *soil* so that the excess must exfiltrate as 'return' flow. **Infiltration**-excess overland flow may occur widely over a **basin** of uniform soils, whereas saturation-excess overland flow is usually produced from limited areas of the **basin**. Overland flow of either type is often instrumental in causing the surface **erosion** of *soils*. [TPB]

overpressuring (*geopressuring*) Occurs in subsurface sediments as a result of restricted fluid mobility either within the overpressured formation or in the enclosing sediments. Normally the pressure of **pore-waters** increases systematically with burial depth as a function of the weight of the overlying water column up to the **water table** or sea-level and is referred to as *hydrostatic pressure* (Fig. O19). If, however, the flow of fluids out of a formation is impeded then the fluid pressure will increase above the *hydrostatic gradient* and may reach and even exceed **lithostatic pressures** (that is, the pressure resulting from the total weight of the overlying rock sequence).

Overpressuring is common in actively subsiding sedimentary **basins** where compaction is taking place, particularly in those **basins** associated with thick **mudrock** sequences and flow of **geothermal** waters. Excess **pore-fluid pressures** may be related to expulsion of fluids from compacting **mudrocks** (causing fluid influx into adjacent **sandstones**), to thermal **decarboxylation** of organic matter, to depth-related dehydration reactions in **mudrock** sequences or to the thermal expansion of fluids and rock matrix associated with upward moving, hot, migrating fluids. High **sinuosity** fluvial and deltaic sandbodies entirely enclosed in **mudrock** sequences or more continuous sandbodies juxtaposed against **mudrocks** by **faulting** are particularly susceptible to overpressuring. Fluid loss from such overpressured formations is often achieved along **faults** and is episodic, being triggered by fluid pressure increase above lithostatic load. Thick

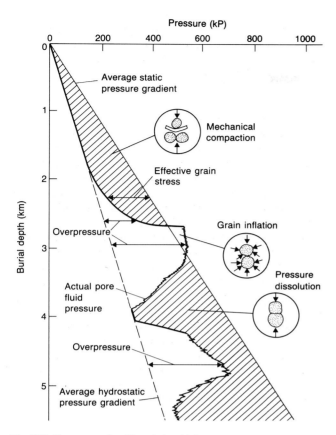

Fig. O19 Overpressuring. The relationship between burial depth and pressure for the average hydrostatic and lithostatic load in an imaginary sediment package illustrating the development of overpressure and its effect on grain stress. (After Gretener, 1977.)

packages of **sand**-dominated continental **sandstones** are less commonly overpressured because of the stacked multistorey nature of the sandbodies that allows unrestricted fluid flow.

Distinct overpressure zones may develop within an actively subsiding sedimentary **basin** coincident with any one of the fluid-releasing **diagenetic** reactions. Associated with the interaction between *hydrostatic* and geostatic *pressure* regimes are major phase changes of fluids and changes in mineral and *hydrocarbon* solubility. There is, therefore, great potential for **mass solute transfer** of soluble constituents from the deeper to shallow parts of the **basin** and their upward redistribution as cements, liquids (**brines** and oil) and gases (*methane*).

Under normal **lithostatic pressures** mechanical compaction of the sediment results in **porosity** loss and **fabric** reorientation (Fig. O19). Overpressuring reduces the **effective stress** on detrital framework grains within a sediment and thus reduces the **porosity** loss by mechanical compaction and **pressure dissolution** (see also **burial diagenesis, compaction**). If overpressures exceed lithostatic load then sediment inflation may occur (Fig. O19). Regional compressional **tectonics** may also induce overpressuring. [SDB]

Gretener, P.E. (1977) *Pore pressure: Fundamentals, General Ramifications and Implication for Structural Geology*. Continuing Education Course Notes, Series 4. American Association of Petroleum Geologists, Tulsa, OK.
Jones, P.H. (1980) Role of geopressure in the hydrocarbon and water system. In: Roberts, W.H. & Cordell, R.J. (eds) *Problems of Petroleum Migration*, pp. 207–16. Studies in Geology No. 10. American Association of Petroleum Geologists, Tulsa, OK.

overprinting Superimposition of a younger **structure** on an older, usually considered on a **mesoscopic** to **microscopic** scale. Evidence of overprinting is important in establishing relative age relationships and **deformation** sequences in complexly deformed terrains. The best evidence of overprinting is where a later **deformation** causes crenulation, **folding** or displacement of an earlier **foliation**. [RGP]

overstep The stratigraphic relationship between **beds** at an **unconformity** when the oldest unit of the younger sequence is in contact with more than one of the older rock sequence. This **unconformity** geometry indicates that tilting or **folding** occurred before deposition of the younger beds. [SB]

overthrust Used as a noun, synonymous with *thrust*. In a **plate tectonic** context, the term may be used as a verb to distinguish the displacement of a *thrust sheet* over the edge of an opposing *plate* rather than under (**underthrusting**). An alternative term for overthrusting in this context is **obduction** (Dewey & Bird, 1971), or *flake tectonics* (Oxburgh, 1972). [RGP]

Dewey, J.F. & Bird, J.M. (1971) Origin and emplacement of the ophiolite suite: Appalachian ophiolites in Newfoundland. *Journal of Geophysical Research* **76**: 3179–206.
Oxburgh, E.R. (1972) Flake tectonics and continental collision. *Nature* **239**: 202–15.

overvoltage (*electrode polarization*) An **induced polarization** effect due to unequal rates of electrochemical (**oxidation**/reduction) reactions involving metallic **minerals** in rocks. Charges build up at metal–electrolyte contacts as a result, necessitating an extra **potential** gradient (or, overvoltage) to drive a current through the **mineral**. After the current is switched off, the charges diffuse away, producing a decaying voltage (see **chargeability**). Metallic sulphides and oxides and **graphite** are **minerals** which exhibit this behaviour, and dissemination of **minerals** enhances the effect. Although the aim of **induced polarization** methods is measurement of electrode polarization, **membrane polarization** will also contribute, indistinguishably, to the results. [ATB]

Keller, G.V. & Frischknecht, F.C. (1966) *Electrical Methods in Geophysical Prospecting*. Pergamon, Oxford.

oxbow A curved lake that has been isolated from a stream after an acute meander has been cut off from the channel.

Table O1 Oxide minerals. Spinels grouped by composition

	Spinel series (Al)	Magnetite series (Fe)	Chromite series (Cr)	Titano spinels (Ti)
Mg	Spinel ($MgAl_2O_4$)	Magnesioferrite ($MgFe_2O_4$)	Magnesiochromite ($MgCr_2O_4$)	Qandilite (Mg_2TiO_4)
Fe^{2+}	Hercynite ($FeAl_2O_4$)	Magnetite (Fe_3O_4)	Chromite ($FeCr_2O_4$)	Ulvöspinel (Fe_2TiO_4)
Zn	Gahnite ($ZnAl_2O_4$)	Franklinite ($ZnFe_2O_4$)		
Mn	Galaxite ($MnAl_2O_4$)	Jacobsite ($MnFe_2O_4$)		
Ni		Trevorite ($NiFe_2O_4$)		
		Maghemite (γ-Fe_2O_3)		

The term is also used for the acute meander itself, before the cut-off stage, in which there is only a narrow neck of land between adjacent reaches, or the land within the bend. The term derives from the characteristic plan shape that resembles the U-shaped frame used to harness an ox. [NJM]

oxidation (1) The loss of electrons from an ion or atom. (2) More specifically, the loss of **hydrogen** or the addition of oxygen to an element through chemical reaction. Hence, when **carbon** is burnt in the presence of oxygen it oxidizes to carbon dioxide. When **iron** is oxidized it turns to rust:

$$4Fe + 3O_2 \rightarrow 2Fe_2O_3$$

When this occurs during the **weathering** of rocks or during *soil* formation, the materials typically become reddened (*rubefied*). The process of oxidation is often coupled with *chemical reduction* in a balanced exchange of electrons; this is termed a *redox reaction*. [AW]

oxide minerals The oxide minerals are commonly found as **accessory minerals** in many rocks and occasionally reach rock-forming concentrations.

Spinel gives its name to a large group of **minerals** having the same structure. These **spinels** may be grouped by composition as shown in Table O1.

The **spinel** group has cubic structure with a unit cell which contains 32 oxygen atoms arranged in layers perpendicular to an axis passing through opposite corners of the cube (the triad axis). In between these oxygen layers there are 24 cations of which eight are divalent (typically Mg, Fe^{2+}, Zn, Mn) and 16 are trivalent (Al, Fe^{3+}, Cr). Variations to this generalization are provided by *maghemite* and **ulvöspinel**. In *maghemite* some or all of the spaces which could be occupied by divalent cations are unoccupied, giving a cation-deficient structure. The trivalent cations in **ulvöspinel** are replaced by Fe^{2+} + Ti^{4+} which maintain the electrical charge balance. The **spinel** structure is able to accommodate a wide range of common elements and many intermediate compositions are possible. **Pleonaste** is the name given to compositions intermediate between **spinel** and **hercynite**. Pure **spinel** is transparent and red in colour and is also known as **ruby spinel**.

The **spinel** group occurs frequently as **accessory minerals** in igneous and metamorphic rocks and in sediments as detrital grains. In metamorphic rocks the **spinel** composition is generally related to the bulk composition of the rock. **Magnetite** is a common **accessory mineral** in **igneous rocks** whilst **chromite** occurs more frequently in basic and ultrabasic rocks where it may form layers which are an economic source of **chromium**. In similar environments **magnetite** may be **titanium**- and **vanadium**-rich and form an important source of **vanadium**. Extensive layers of **chromite** paralleled by layers of **vanadium**-bearing *titanomagnetite* are notable features of the layered ultrabasic rocks of the **Bushveld Complex** of South Africa. Rocks in which **chromite** is the dominant phase are known as *chromitites*. Where **zinc**, **manganese** or **nickel** are locally abundant with **iron**, **franklinite**, **jacobsite** or *trevorite* may be found and provide an economic supply of those metals as at Sterling Hill and Franklin, New Jersey, USA.

In more oxidized rocks **hematite** (α-Fe_2O_3) occurs as the dominant **iron oxide**, there being an **oxidation** process from **magnetite** to **hematite** via *maghemite*. **Hematite** occurs in several forms such as *kidney ore* which has a fibrous, radial structure internally, leading to rounded external surfaces which are red in colour resembling both the shape and colour of kidneys. Crystalline, platy **hematite** has a high metallic **lustre** and is known as specular **hematite** or **specularite**. When **hematite** forms octahedral *pseudomorphs* after **magnetite** it is known as **martite**.

Where **titanium** is abundant, usually in more basic rocks, **magnetite** may become **titanium**-bearing when it is known as *titanomagnetite*. There is complete *solid solution* between **magnetite** and **ulvöspinel** via *titanomagnetite* compositions since they share the **spinel** structure. **Ilmenite** ($FeTiO_3$) has a trigonal structure similar to **hematite** and at high temperature there is complete *solid solution* between **hematite** and **ilmenite**. These relationships can conveniently be illustrated on a triangular diagram (Fig. O20). *Maghemite*, and the **titanium**-bearing equivalent *titanomaghemite*, occupy the shaded field in Fig. O20 depending on the degree of **oxidation**. Under hot, oxidizing conditions such as those found at volcanic **fumaroles**, **pseudobrookite** (Fe_2TiO_5) may be found. **Pseudobrookite** has an orthorhombic structure and extends in composition across to $FeTi_2O_5$. *Armalcolite* ([Fe, Mg]Ti_2O_5) was initially discovered in samples collected on the first Lunar landing and has now been found in

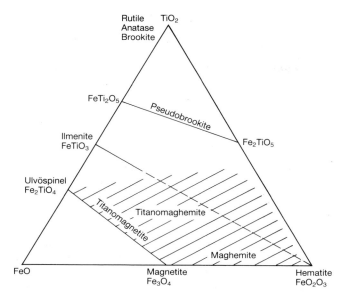

Fig. O20 Iron–titanium **oxide minerals**. Pseudobrookite and titanomagnetite have a range of composition along solid–solution series. Complete solid–solution between ilmenite and hematite only occurs at high temperatures. The alteration minerals maghemite and titanomaghemite can occur within an ill-defined range of compositions, shown shaded.

meteorites and on Earth. *Armalcolite* has the same structure as **pseudobrookite** and is found in highly reducing environments.

Rutile, anatase and **brookite** all have the formula TiO_2 but different structure; that is they are **polymorphs. Rutile** is the most common, occurring widely as an **accessory mineral** in metamorphic rocks and *pegmatites* and as a detrital mineral in sediments. It also occurs as fine oriented needles in **quartz** where it forms attractive specimens (*sagenite*). **Anatase** and **brookite** usually form as secondary minerals derived from **alteration** of other **titanium**-rich phases; **anatase** being produced by low-temperature processes (**weathering** and **hydrothermal alteration**) whilst **brookite** appears to result from high-temperature **alteration. Leucoxene** is an **alteration** product of **ilmenite** which usually consists of finely divided **rutile** or **brookite** but may in natural examples be intimately associated with **hematite** or **pseudobrookite**.

The iron oxyhydroxides **goethite, limonite** and **lepidocrocite** form as **weathering** products of the **iron oxides** and provide the yellow and brown colouration associated with **iron**-rich weathered material.

Corundum (Al_2O_3) occurs chiefly in **aluminium**-rich igneous and metamorphic rocks. It is hard (9 on Mohs' scale) and survives well as a detrital mineral. **Corundum** may be coloured when trace metals are present and the valuable **gemstones, ruby** and **sapphire** are transparent **corundum**. The red of **ruby** is caused mainly by **chromium** and **vanadium**, whilst **iron** and **titanium** provide the blue of **sapphire. Gibbsite** ($Al[OH]_3$), **diaspore** (α-$AlO[OH]$) and **boehmite** (γ-$AlO[OH]$) are **weathering** products of

aluminium-rich silicate rocks which have formed clays from which the **silica** has been removed by **groundwater** leaching. This occurs mainly under tropical **weathering** conditions to form **bauxite** or **laterites** in which **gibbsite** is usually the dominant phase.

Periclase (MgO) has a cubic structure and is generally greyish white with a perfect cubic **cleavage**. It forms by high-temperature metamorphism of **magnesium**-rich **limestones** and **dolomites** and it finds industrial use as a **refractory** material. **Brucite** ($Mg[OH]_2$) is an **alteration** product of **periclase** forming during **contact metamorphism** or low-temperature **hydrothermal alteration**.

The **tin** oxide **cassiterite** (SnO_2) is the most important **ore** from which **tin** is obtained. It is found in **veins** and *pegmatites* associated with **granites** and it concentrates readily in **alluvial deposits** such as those which occur in Malaysia and Yakutia, in the former USSR. A significant proportion of the world's **tin** supply is obtained from dredging these deposits. Underground mining for **cassiterite** usually follows **veins** extending out from the **granite** into the *country rock* (e.g. Cornwall) where it occurs with other **ores** and **stannite** (Cu_2SnFeS_4). **Wood tin** is a colloform variety of **cassiterite**, with **hematite** and **silica**, formed by secondary **alteration** in an oxidizing environment and leading to concentrically banded material.

Perovskite ($CaTiO_3$) is an **accessory mineral** which occurs in basic **igneous rocks** and **kimberlites. Cerium** and the other **rare earth elements** may be present to a significant degree, replacing the calcium whilst **niobium** and **tantalum** may replace **titanium** to an appreciable amount. **Perovskite** has an orthorhombic structure which approximates to a distorted cubic close-packed arrangement. Other minerals with the **perovskite**-type structure are known and because of the close-packed nature the structure is stable at high pressures. Thus **perovskite** in **kimberlites** and **carbonatites** is considered to have crystallized at considerable depth and ($Mg,Fe)SiO_3$ with the **perovskite**-type structure may be an important constituent of the deep **mantle**.

Uraninite is cubic and occurs as a primary **accessory mineral** of **igneous rocks**, particularly **granites**. It is *radioactive* and forms an important contribution to the radiation hazard and *radon* production associated with such rocks. Because some of the **uranium** is *radioactive* it breaks down eventually to form **lead** and most **uraninites** contain some radiogenic **lead** produced in this way; approximately 1% Pb for every 100 Ma. **Uraninite** probably approximates to the formula UO_2 when formed and analyses of young **uraninites** often fit this formula, but with *radioactive* decay and attendant **alteration** the formula is more usually close to U_3O_8 indicating some **oxidation**. The fine-grained colloform variety of **uraninite** is called **pitchblende** and this may be more oxidized and contain some water. **Uraninite** may also contain some **thorium** but **thorianite** (ThO_2) is the naturally occurring oxide.

Columbite ($[Fe,Mn]\ Nb_2O_6$) and **tantalite** ($[Fe,Mn]$

Ta_2O_6) form a series with almost complete variability of the Fe/Mn and Nb/Ta content. When the proportion of **iron** is high the mineral is black in colour whilst **manganotantalite** is red. These minerals are found chiefly in **granite** *pegmatites* especially those containing **albite** and there is often an association with **lithium** minerals.

Pyrolusite (MnO_2) generally forms as a black or dark blue-grey *concretion* under highly oxidizing conditions although prismatic crystals have been found. It frequently forms a dark blue-grey surface coating on **joint** or **fracture** surfaces and occasionally these show a dendritic pattern. **Manganite** (MnO[OH]) forms in low-temperature **hydrothermal veins** or during secondary **alteration** which if taken further produces a series of hydrated **manganese** oxides which may contain **barium** (**psilomelane**, ($BaH_2O)Mn_5O_{10}$) or **potassium** (**cryptomelane**) or **lead**. These are usually very fine-grained and used to be thought of as amorphous but are now known to be monoclinic. Much-altered material consisting of an assemblage of altered **manganese** oxides is known in the field as **wad**. [JFWB]

Deer, W.A., Howie, R.A. & Zussman, J. (1962) *Rock-forming Minerals* Vol. 5: *Non-silicates*. Longmans Green, London.
Rumble III, D. (ed.) (1976) *Oxide Minerals*. Reviews in Mineralogy, Vol. 3. Mineralogical Society of America, Washington, DC.

oxygen isotope analysis Determination of the ratio of oxygen isotopes (^{18}O, ^{16}O) by **mass spectrometry**. It has been applied to spondylus shells (used for jewellery and traded in Neolithic Europe) to determine the source of the shells, on East African **chert** artefacts and to determine the source of raw materials in 4th century AD glasses from the Near East. [OWT]

oxygen minimum layer/zone The level in the ocean where concentrations of dissolved oxygen are at a minimum, which generally occurs between depths of 150 m and 1000 m. The bounding limit of the oxygen minimum zone is generally taken to be $0.2\,ml\,l^{-1}$. Where the flux of organic matter is high and the oxygen minimum is intense and intersects the sea bed, **anoxic** sediments may accumulate. [AESK]

P

Pacific-type coast A coast which lies parallel to the trend of topography and geological **structure**, also called a *longitudinal* or *concordant coast*. This gives rise to a smooth linear coastline except where marine inundation of the intervening valleys has produced a series of coast-parallel promontaries and island chains characteristic of a **dalmatian coast**. Examples of such coasts are found in British Columbia in Canada. At the other extreme is the *Atlantic-type coast* (also *transverse* or *discordant coast*) where topography and geological **structure** is at right angles or oblique to the coastline as occurs in north-west Spain. [JDH]

paired metamorphic belt Two adjacent linear belts of different metamorphic grade. Commonly a high pressure/low-temperature belt (*blueschist* or *prehenite–pumpellyite* zone) adjacent to a low pressure/high-temperature zone as in the present Japanese **island arc** as described by Miyashiro (1972). The metamorphic grades reflect the pressure, temperature and hydrous conditions that occur within a **subduction zone** and so the distribution of such metamorphic belts can be used to assess the previous direction (polarity) of subduction. This concept has been challenged by Barber (1982), who proposes that the juxtaposition of paired belts was accomplished by subsequent strike-slip tectonics. [PK/DHT]

Barber, A.J. (1982) Interpretations of the tectonic environment of southwest Japan. *Proceedings of the Geological Association* **93**: 131–45.
Miyashiro, A. (1972) Metamorphism and related magmatism in plate tectonics. *American Journal of Science* **272**: 629–56.

palagonite A volcanic rock consisting of hydrated and chemically altered **basalt hyaloclastite**. **Hyaloclastite** consists of fine-grained, quench-fragmented, glassy material which is susceptible to hydration. During the hydration and associated chemical **alteration** — a process known as *palagonitization* — the glass is converted to a yellow or reddish-brown isotropic material. Gas & Wright (1987) reviewed the chemical affects of the **alteration: potassium** and H_2O are gained, and silicon, sodium and **manganese** are lost from the glass. **Iron, titanium, magnesium,** calcium and **aluminium** are also mobile, but show less systematic behaviour. *Palagonitization* is normally associated with intergranular deposition of **quartz, zeolites** and **calcite**. The process of *palagonitization* and deposition of such secondary minerals contribute to the lithification of **hyaloclastite**. *Palagonitization* can occur at any tempera-ture, in the presence of either seawater or **meteoric water**. The process is speeded up by elevated temperatures.

Palagonite is perhaps most spectacularly exposed in Iceland, where large volumes of **basaltic magma**, erupted beneath **glaciers** during the *Pleistocene*, formed **hyaloclastite**, subsequently altered to palagonite, and locally known as '*Moberg*'. [PTL]

Cas, R.A.F. & Wright, J.V. (1987) *Volcanic Successions Modern and Ancient*. Allen & Unwin, London.
Jakobsson, S.P. (1979) Outline of the petrology of Iceland. *Jokul* **29**: 57–73.

Paleocene The first epoch of the **Cenozoic** Era (Schimper, 1874). It has a duration of *c.* 14 Ma from the end of the Late **Cretaceous** at 66.4 Ma to the beginning of the **Eocene** epoch at *c.* 52 Ma (Berggren *et al.*, 1985). The Paleocene consists of two stages, the *Danian* and the *Thanetian* separated by an unnamed interval (see Fig. P1)

At the beginning of the Paleocene, the North Atlantic was open between Gibraltar and Newfoundland but North America and Europe were still joined by Greenland. Madagascar had separated from Africa, but Arabia was still attached to Africa. India was moving northwards but had not yet collided with Asia.

The migration of the **mammals** between America and Europe took place via Greenland in Paleocene times and between North and South America via a narrow land connection which had been established at the beginning of the Paleocene. The climate warmed through the epoch and there were no *glaciations*.

As a result of the major biotic changes in the Late **Cretaceous** times, which included the **extinction** of the *ammonites* and a dramatic reduction in the numbers and diversity of the **brachiopods**, the most common fossil marine invertebrates in the Paleocene were *bivalves* and *gastropods*. Many groups of **Mesozoic** *bivalves* continued into the Paleocene; oysters, mussels (*Mytilus*) and cockles (*Protocardia*) were common and fossil wood was frequently bored by the bivalve *Teredo* (ship-worm). A radiation of a genetically new group of planktonic **Foraminifera** began in the Paleocene, having replaced the *Maastrichtian* species in the lowermost *Danian*; this radiation peaked in the Late Paleocene (Cifelli, 1969).

After the **extinction** of the larger **reptiles** in the Late **Cretaceous**, the **mammals** evolved rapidly and radiated into most environments. In South America a great diversity of *marsupials* evolved while elsewhere, **mammals** in the Early Paleocene included small *insectivores* and *marsupials*

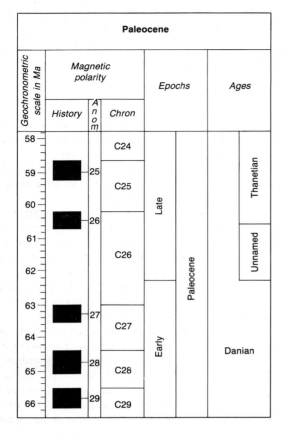

Fig. P1 Paleocene

Poales and Typhales (Thomas & Spicer, 1987). (See **land plants**.) [DGJ]

Berggren, W.A., Kent, D.V. & Flynn, J. (1985) Jurassic to Paleogene: Part 2 Paleogene geochronology and chronostratigraphy. In: Snelling, N.J. (ed.) *The Chronology of the Geological Record*, pp. 141–95. Memoirs of the Geological Society No. 10. Blackwell Scientific Publications, Oxford.

Cifelli, R.J. (1969) Radiation of Cenozoic planktonic foraminifera. *Systemic Zoology* **18**: 154–68.

Hickey, L.J. (1981) Land plant evolution compatible with gradual not catastrophic change at the end of the Cretaceous. *Nature* **292**: 529–31.

Perch-Nielsen, K. (1982) Maastrichtian coccoliths in the Danian: survivors or reworked 'dead bodies'? *Abstract, IAS Meeting, Copenhagen*, 122.

Schimper, W.P. (1874) *Traité de Paléontologie Végétale* Vol. 3. J.B. Baillière, Paris.

Thomas, B. & Spicer, R. (1987) *Evolution and Palaeobiology of Land Plants*, pp. 1–304. Croom Helm, London.

paleoecology The study of the interactions of organisms with one another and with their environment in the geologic past. It differs from the ecology of extant organisms in that the incomplete preservation of the fossil record prevents the direct observation of many aspects of the biota. *Uniformitarianism* is the most important principle in paleoecology, although evolutionary differences between the habits and requirements of extinct organisms and their extant representatives, if any, may need to be considered. Frequently data from paleoecologic analyses are used in *paleoenvironmental* studies. Conversely, information from non-paleontological sources, such as sedimentology and geochemistry, may aid the interpretation of fossil communities or individuals. Two major subdivisions of paleoecology can be recognized, *paleoautecology* and *paleosynecology*.

PALEOAUTECOLOGY. This is the study of the ecology of individual organisms or taxonomic groups. Where extant representatives of a fossil group are known, aspects of the anatomy, physiology and life history of the fossil group can be deduced from a direct comparison with their living relatives. This becomes progressively less reliable earlier in the geological record, where extinct groups predominate, but several other approaches can be adopted in such cases.

The adaptive functional morphology of an organism may be determined by *homology* (comparison with the same structure in living representatives), analogy (comparison with similar morphology in extant but unrelated organisms) or the paradigm approach, which involves an engineering-type analysis of the function of particular structures. The possible influence of phylogenetic or architectural constraints on morphology should also be considered in such analyses.

Direct evidence of activity, in the form of **trace fossils** associated with resting, locomotion or feeding, is a valuable source of information on life habits although it may prove difficult to relate particular traces to specific fossil organisms. The orientation of a fossil, or of epizoans on a host organism, may indicate its life position though it is

which had survived from the **Cretaceous**. The rise and increase in the diversity of the **mammals** was dramatic and by *c.* 55 Ma the evolution of bats, early whales and the ancestors of the hoofed animals, including horses which initially were only the size of dogs, took place. Other new forms included the ancestors of *rodents* and *primates*.

In the oceans some 15 to 18 calcareous nannofossil genera survived the *Cretaceous–Tertiary boundary* event (Perch-Nielsen, 1982). From this relatively small genetic pool, new taxa evolved slowly at first and then rapidly during the Late Paleocene. The **coccoliths** of the family Coccolithacea developed in the lower Paleocene and the important *Tertiary* lineage of star and disc-shaped nannofossils, the *discoasters*, evolved in the Late Paleocene at *c.* 60 Ma. Paleocene palynoflora are strongly differentiated from those of the Upper **Cretaceous**.

Precipitation on land increased after the Late **Cretaceous** and there was an increase in the abundance of toothed-margin deciduous leaved plants in the Paleocene (Hickey, 1981); there was a low diversity deciduous forest at this time in a wet environment (Thomas & Spicer, 1987). The first firm records of grass *pollen* occur in the Paleocene.

The following extant floral orders appeared: Didymeliales, Hamamelidales, Eucommiales, Casuarinales, Polygonales, Rutales, Polygalales, Cornales, Gentianales, Restionales, important to discriminate between biological and purely mechanical orientation. Skeletal structure may provide indications of life habits and environmental factors. Growth line spacing in accretionary skeletons may show evidence for seasonality, trauma or stunting. The biogeochemistry of the fossil remains may provide evidence for temperature and salinity tolerance of the organism.

Every organism represents part of a population of interacting individuals of the same species. A number of quantitative techniques are available to the paleontologist for analysing different aspects of fossil populations. Population age structure can be determined from the size-frequency distribution of the individuals. This can then be used to construct a *survivorship curve*, a graph of the number of survivors of a single age group versus the age of the group, or to identify seasonal spawning or mortality patterns, growth rates, maximum ages or current sorting. Species which show marked fluctuations in abundance at successive horizons may be opportunistic species whereas those with a more stable population structure are probably equilibrium species. Several other criteria to test this are listed in Dodd & Stanton (1981).

The nature of the associated sediments is an important source of information concerning the organism's mode of life. Benthic organisms characteristically occur in a more restricted range of facies than do **pelagic** or pseudopelagic ones. Morphologic variation within a population or lineage may be ecophenotypic and so demonstrate the effect of environmental changes on the organism.

PALEOSYNECOLOGY. This is the study of the ecology of fossil communities as a whole. A direct comparison between living and fossil communities is rarely possible since much of the fossil biota is unlikely to be preserved at all. For this reason the terms 'fossil community', for associations of fossils ecologically related amongst themselves, and 'fossil assemblage' for associations without ecological relations to each other or their enclosing sediment, are preferable to the terms used by ecologists for living biotas.

Paleoecologists can use several analytical techniques to interpret these fossil associations. Natural groupings of organisms into communities can be identified by one of two methods. In *Q-mode analysis* compared samples are grouped together on their relative similarity. In *R-mode analysis* the distribution of individual taxa are compared and those that co-occur are grouped together whilst those that are mutually exclusive are placed in different communities. Community boundaries are more clearly defined in the former method but the latter emphasizes patterns of co-occurrence and mutual exclusion and is thus useful in delimiting fossil communities.

The community structure can be analysed in terms of its diversity and its trophic structure. Biotic diversity is strongly influenced by time, environmental stability and resources. *Richness, S,* the number of species or taxa present, is the simplest measure of diversity though it is strongly dependent on sample size. The *dominance*

diversity, the relative abundance of taxa within a sample, or its inverse, equitability, can be calculated using one of several different equations, such as the *Shannon–Weaver dominance diversity equation,*

$$H = -\sum_{i=1}^{s} P_i \log P_i$$

where P_i is the proportion of the *i*th species in the sample. The maximum value of H occurs when all the species are equally abundant.

Trophic structure is most conveniently analysed by plotting the various taxa on ternary diagrams of feeding habit, substrate-niche or other parameters.

Direct biotic interactions between different elements of the biota are an important aspect of synecology and may take the form of parasitism, predation or symbiosis of one sort or another. A less direct form of interaction may occur through *taphonomic* feedback, the buildup of skeletal hard-parts, which inhibits the mobile infauna. The presence of deposit feeders is also an important regulatory process as they may ingest the larvae of other benthic forms.

An extension of *paleosynecology* is *paleoenvironmental interpretation*, this being very much dependent on paleoecological data as well as information from sedimentology and geochemistry. Direct comparison with modern analogues is important although frequently not possible in earlier parts of the geological column. In such cases *Taxonomic Uniformitarian Analysis* is the best approach. With this technique the tolerance for various environmental parameters of each fossil taxon in the community is estimated from those of a comparable living taxon. The value of each environmental parameter may then be deduced from the overlap of the values from all taxa or by comparing the *fossil community* with the most comparable modern analogue. In both instances the results have much greater reliability than if only one or a small number of taxa were used. [MJS]

Ager, D.V. (1963) *Principles of Palaeoecology*. McGraw-Hill, London.
Craig, G.Y. (1966) Concepts in Palaeoecology. *Earth Science Reviews* **2**: 127–55.
Dodd, J.R. & Stanton, R.J. (1981) *Paleoecology, Concepts and Applications*. John Wiley & Sons, Chichester.
McKerrow, W.S. (1978) *The Ecology of Fossils*. Duckworth, London.
Raup, D.M. & Stanley, S.M. (1978) *Principles of Paleontology*. W.H. Freeman, San Francisco.
Tevesz, M.J.S. & McCall, P.L. (eds) (1983) *Biotic Interactions in Recent and Fossil Communities*. Plenum, London.

paleogeography The broad-scale study of ancient geographies, in terms of such features as distribution of land and sea, upland and lowland, ocean shallows and deeps, and climate. [AH]

paleogeological map Map constructed to show the outcrop pattern of an area at a given time. Such a map may be constructed for a time period corresponding to the base of an unconformable sequence, for example, by restoring the surface of **unconformity** to the horizontal and mapping the **subcrop** traces (**feather edges**) of underlying formations. [RGP]

Roberts, J.L. (1982) *Introduction to Geological Maps and Structures.* Pergamon, Oxford.

paleohydrology The reconstruction of **flow regimes** for abandoned, often ancient, river channels. Parameters such as **bankfull discharge** and flow **shear stresses** are estimated through study of exposed sections of old channels, where channel geometry and channel sediments can be measured, or sometimes by measurement of planform dimensions of old channels preserved in old **floodplain** or terrace deposits and visible on *aerial photographs*. [TPB]

Gregory, K.J. (ed.) (1983) *Background to Palaeohydrology.* John Wiley & Sons, Chichester.

paleomagnetic pole The location of the **geomagnetic** pole calculated from the mean of magnetic directions determined at a specific location and assuming that this mean direction represents an average **geomagnetic field** that was both geocentric and dipolar. [DHT]

paleomagnetism The study of the magnetic properties of rocks. It is primarily concerned with the directional properties of **natural remanent magnetizations**, but as the interpretation of such observations also requires an understanding of their physical properties, it also includes **rock magnetism**. Over archeological time-scales, the subject is termed **archeomagnetism** and includes the study of remanences acquired by human processes that have affected both archeological and geological materials. The natural remanence of rocks may be acquired in the direction of the **geomagnetic field** by processes associated with the formation of the rock itself, i.e. the natural remanence often includes primary components, such as a **thermal remanent magnetization** acquired by **igneous rocks** as they cool and **depositional remanent magnetization** acquired by sediments as they are deposited, although most lithified sedimentary rocks are characterized by a **chemical remanent magnetization** acquired during **diagenesis**. The **natural remanent magnetizations** almost invariably also include a secondary time-dependent **viscous remanent magnetization**, mostly associated with the **geomagnetic field** during the last few 100 000 years including, in some rocks, magnetizations acquired since the rock samples were collected. The **natural remanent magnetizations** of rocks may also include, or be dominated by, other secondary magnetizations associated with thermal or chemical changes such as those accompanying burial, natural mineralogical **exsolution, weathering**, etc. Most paleomagnetic studies attempt to isolate particular components acquired at known geological times, usually the primary magnetizations as these can be assigned to a specific time, the age of the rock. Secondary magnet-

izations can be similarly isolated and interpreted, but determining their age is usually much more difficult. Such component analysis is usually undertaken by means of partial *demagnetization*, applied in incremental steps of either temperature or alternating **magnetic field** strength, both techniques being applied while the present **geomagnetic field** is cancelled. Such methods preferentially remove the lower *coercivity* or least thermally stable components, i.e. that part of the *coercivity* or thermal spectra that is most likely to be associated with any viscous remanences. The behaviour of the magnetic **vector** during such incremental *demagnetization* is usually analysed by computational methods to identify identical **vectors**, isolated over at least three consecutive incremental steps. The age of each component isolated in this way is then assessed, mainly by determining whether the magnetization was acquired before or after a particular geological event, usually the **folding** or differential tilting, affecting the same sequence of rocks. If the component is pre-**folding**, then it is generally assumed to be primary, although the ultimate test is whether such directions make geological sense for such a time. The fundamental problem in paleomagnetic analyses is that secondary magnetizations, particularly the formation of **hematite** at some later time, can result in secondary magnetizations that are possibly more resistant to *demagnetization* procedures than any remaining primary remanence. After such linear, stable magnetic components have been isolated they can be used for both relative and absolute *magnetic dating*, using the fact that the **geomagnetic field** changes on three different time-scales: (i) **secular variations**, 10^2–10^4 years, (ii) polarity change, 10^4–10^7 years, and (iii) **polar wandering**, 10^6–10^9 years. Comparison of directions of remanence acquired at similar times in different areas can be used to determine whether such areas have moved relative to each other since the magnetization was originally acquired, i.e. determine whether the areas were originally part of the same **tectonic** unit and the relative rotation of each block relative to the other. If both areas have remained part of the same **tectonic** unit since their magnetizations were originally acquired, then the mean direction of remanence of rocks of similar age in both areas will point towards the same ancient **geomagnetic** pole position, i.e. their mean direction will correspond to the same **paleomagnetic pole**. Areas that were originally part of the same **tectonic** unit but have since moved relative to each other, or were never part of the same **tectonic** unit, will have different **paleomagnetic poles**. These can be matched with each other to determine the original **paleogeographic** relationships between the two areas. For any one locality in which the average **geomagnetic field** direction can be determined for time intervals corresponding to a few 100 000 years, i.e. averaging out geomagnetic **secular variations**, the mean *inclination* can be used to calculate the ancient latitude (*paleolatitude*) assuming the *axial geocentric dipole hypothesis*, and the mean *declination* provides the orientation of the tectonic unit relative to the **paleo-**

magnetic pole. As the **paleomagnetic pole** corresponds to the Earth's rotational pole for that time, a major contribution of paleomagnetism is to provide **paleogeographic reconstructions** of the **continents**. However, paleomagnetism can be used to study all chemical and thermal changes that may have occurred to a rock, as each of these processes will, to varying degrees, affect the pre-existing remanence and usually provide a component of magnetic remanence that can be used to study the geological process itself. [DHT]

Collinson, D.W. (1983) *Methods in Rock Magnetism and Palaeomagnetism*. Chapman & Hall, London.
Khramov, A.N. (1987) *Paleomagnetology*. Springer-Verlag, Berlin.
Tarling, D.H. (1983) *Palaeomagnetism*. Chapman & Hall, London.

paleosol A *soil* that formed on a landscape of the past. Such *soils* may occur within sequences of sediments (buried paleosols) or may have been buried and then exposed (exhumed paleosols) or have not been buried but formed under a set of earlier conditions no longer operating (relict paleosol). While most studies of paleosols have concentrated on **Quaternary** examples found in **loess** and **alluvial deposits**, they are now widely recognized in older sedimentary sequences and most continental deposits in the geological record contain paleosols. *Precambrian* paleosols are well documented. (See **pedogenesis**.) [VPW]

Wright, V.P. (ed.) (1986) *Paleosols: Their Recognition and Interpretation*. Blackwell Scientific Publications, Oxford.

Paleozoic The collective term for the **Cambrian** to **Permian** systems. [AH]

palinspastic reconstruction Geometrical reconstruction, in the form of a map or **cross-section**, of the original geometry of a set of rocks before **deformation**. (See **balanced section**.) [RGP]

palladium (Pd) A very rare native metal mineral. (See **native element**.) [DJV]

palynology The scientific study of carbonaceous microorganisms, whose remains can be extracted from siliceous rocks by means of treatment with hydrofluoric acid. The subject matter includes *pollen* and *spores* produced in the terrestrial environment and **dinoflagellates** and **acritarchs** in the marine environment. [AH]

pan A closed depression, often with a characteristic shape (see **oriented lake**), that can occur in great numbers in arid and semi-arid areas such as Patagonia, the High Plains of Texas, the Kalahari and Transvaal, and Western Australia. Pans may result from such processes as **solution** of **limestone** or **calcrete** and animal activity (buffalo or hog wallows), but the prime cause of their development is deflational activity acting on surfaces of susceptible materials (e.g. **shales**, fine **sandstones**, unconsolidated aeolian **sands**, lake sediments, etc.). On their **lee sides** they may possess **clay dunes** or **lunettes**. [ASG]

Goudie, A.S. & Thomas, D.S.G. (1985) Pans in southern Africa with particular reference to South Africa and Zimbabwe. *Zeitschrift für Geomorphologie* NF **29**: 1–19.

panfan A landform or landscape in an arid region in which hills and ridges have been worn down, **pediments** are extensive and coalescing and **basins** have been infilled. [NJM]

Pangea A hypothetical single continent which is thought to have comprised all of the **continental crust** for much of the Earth's history. This continent is thought to have split into two *supercontinents*, **Laurasia** and **Gondwanaland**, some 300 Ma ago, each of which subsequently broke up to form our present continents. Pangea was surrounded by the ocean **Panthalassa**, and may have existed through most of the **Proterozoic** and **Paleozoic**. It is known that several continental blocks, e.g. the North and South China blocks, did not form part of this single continent during the **Phanerozoic**, but such exceptions are small compared to the size of Pangea. The hypothesis implies that **continental splitting** did not occur during the **Proterozoic** but became increasingly common during the **Phanerozoic**. Some authorities consider that Pangea only existed for a brief time during the **Permo-Triassic**, i.e. *c*. 300 to 250 Ma ago. [DHT]

Piper, J.D.A. (1987) *Palaeomagnetism and the Continental Crust*. Halsted Press, New York.

pantellerite A type of peralkaline **rhyolite**. By definition, *peralkaline rocks* have molecular excess of $(Na_2O + K_2O)$ over Al_2O_3, and **rhyolites** have more than 10% **normative quartz**. Peralkaline **rhyolites** can be divided into pantellerites and *comendites*. Fresh, unaltered peralkaline silicic **glassy rocks** can be divided into pantelleritic and *comenditic* types on the basis of contents of **normative quartz** and total **normative** femic minerals, the pantellerites having more than about 12.5% **normative** femic minerals (Macdonald & Bailey, 1973). Macdonald (in Bailey *et al.*, 1974) proposed a simpler classification of peralkaline, silicon-saturated volcanic rocks based on abundances of total iron and Al_2O_3 abundances, which can also be used for hydrated and slightly chemically altered rocks. Pantellerites are normally weakly **porphyritic**, with the *phenocryst* assemblage **alkali feldspar**, sodic *ferrohedenbergite*, **fayalite**, **aenigmatite**, **amphibole**, **quartz** and Fe–Ti oxides (Sutherland, in Bailey *et al.*, 1974).

Most pantellerites occur in continental **rift** tectonic settings (e.g. Kenya, Ethiopia), and in broad, extensional regimes, such as the **Miocene** of the western United States. They are rare on *oceanic islands*, but occur in Iceland and on the Island of Pantelleria, Italy (the type locality). They also occur rarely in active volcanic arcs (e.g. Mayor Island, New Zealand). (See **trachyte**.) [PTL]

Bailey, D.K., Barberi, F. & Macdonad, R. (eds) (1974) Oversaturated peralkaline volcanic rocks. *Bulletin Volcanologique* **38**: 497–860.
Macdonald, R. & Bailey, D.K. (1973) *The Chemistry of the Peralkaline Oversaturated Obsidians*. Professional Paper of the United States Geological Survey **440-N-1**: 1–37.

Panthalassa The hypothetical ocean that surrounded **Pangea**. This ocean is commonly considered to be the ancestral Pacific Ocean. [DHT]

parabolic dune A crescentic **dune** with arms tethered by vegetation pointing upwind. Movement is through **deflation** of **sand** from the central bowl of the upwind face, normally with a slope angle between 2° and 15°, which is then deposited a short distance downwind in a mound. In arid climates parabolic dunes are found on desert margins where a sparse vegetation cover restricts the **sand** flow. Upward growth of the **sand** mound in these areas is normally limited by the lack of vegetation. In wetter areas, where there is more vigorous vegetation growth, the **dunes** are commonly higher, sometimes up to 50 m, with more steeply inclined slopes. Parabolic dunes commonly develop from *blow-outs*. [RDS]

paragenetic sequence The order of deposition (crystallization) of minerals. Not to be confused with *paragenesis* which is a synonym for mineral association. The term paragenetic sequence is most commonly used in connection with **epigenetic, hydrothermal deposits**. World-wide studies of these indicate that there is a general order of deposition of minerals in these deposits. Exceptions and reversals are known but not in sufficient number to suggest that anything other than a common order to deposition is generally the case. A simplified, general paragenetic sequence is as follows:
1 silicates;
2 **magnetite, ilmenite, hematite**;
3 **cassiterite, wolframite, molybdenite**;
4 **pyrrhotite**, *löllingite*, **arsenopyrite, pyrite, cobalt** and **nickel** arsenides;
5 **chalcopyrite, bornite**, *sphalerite*;
6 **galena, tetrahedrite, lead** sulphosalts, tellurides, **cinnabar**.
Of course, not all these minerals are necessarily present in any one deposit and the above list has been drawn up from evidence from a great number of **orebodies**. [AME]

paragonite ($NaAl_2(AlSi_3O_{10})(OH)_2$) A **mica mineral** isostructural with **muscovite**. [DJV]

parallel retreat A mode of slope evolution in which the form and angle of the entire slope remains constant as **erosion** ensues. Only the concavity at the slope base (the **waning slope**) increases in length and reduces in angle. Thus ground loss measured in a horizontal direction occurs at the same rate over the whole slope, or over those slope elements subject to parallel retreat. Parallel retreat may occur where **erosion** is **weathering**-limited, as on a cliff. Gilbert (1909) argued that uniform **regolith** thickness over a convex or rectilinear slope implied that **weathering** and ground loss were also uniform. Parallel retreat was a central assumption of the model of slope evolution proposed by Walther Penck in 1924. [TPB]

Gilbert, G.K. (1909) The convexity of hilltops. *Journal of Geology* **17**: 344–51.
Penck, W. (1924) *Die morphologische Analyse*. English translation by Czech, H. & Boswell, K.C. (1953) *Morphological Analysis of Landforms*. Macmillan, London.

paramagnetism The magnetic property of certain materials which, when placed in a **magnetic field**, acquire a magnetization that is in the same direction as the applied field, but disappears as soon as the field is removed. This property is superimposed on **diamagnetism** and may itself be superimposed by **ferromagnetism**. [DHT]

parasitic antiferromagnetism The magnetization that arises if the two equally and oppositely magnetized lattices in an **antiferromagnetic** material are slightly distorted so that they no longer cancel each other; such materials therefore have a **spontaneous magnetization** even in zero applied field. **Hematite** has a parasitic antiferromagnetism. [DHT]

pargasite ($NaCa_2Fe_4(Al,Fe)Al_2Si_6O_{22}(OH)_2$) An **amphibole mineral**. [DJV]

parna Aeolian clay deposit (Butler, 1956) which occurs in Australia either as a discrete **dune** or as a thin, discontinuous, widespread sheet. Clay pellets make up the greater proportion of the material, though there may be some '*companion sand*'. They may be derived from the **deflation** of material from unvegetated, saline lake floors, or from other *soil* or alluvial surfaces. Parna is in effect a **loessic** clay. [ASG]

Butler, B.E. (1956) Parna — an aeolian clay. *Australian Journal of Science* **18**: 145–51.

partial area model A concept proposed by Betson (1964) to account for the low ratio of **runoff** to rainfall in **catchments** where **infiltration**-excess (or '*Hortonian*') **overland flow** is presumed to be the dominant *storm* **runoff** generating mechanism. Betson argued that only small parts of the **basin** contributed surface **runoff** whilst in the rest of the **basin** the **infiltration capacity** is high enough to allow complete **infiltration** to occur. Such partial areas might be fields with no vegetation or those disturbed by overgrazing or excessive use of heavy machinery, for example. The term is often used erroneously to include surface **runoff** produced by saturation-excess **overland flow**; such contributing areas are more properly called *'variable source areas'* in order to distinguish the distinctive **runoff** process occurring. [TPB]

Betson, R.P. (1964) What is watershed runoff? *Journal of Geophysical Research* **69**: 1541–52.

particle cluster A grouping of particles on a bed generated through the mutual hydrodynamic interference between the constituent *clasts*. Such clusters are characteristic of **gravel** bed rivers and may be split into three components: (i) a principal large *clast* with its long axis transverse to flow causes the accumulation of (ii) imbricated *clasts* on its upstream **stoss side** and (iii) a much finer-grained tail or wake deposit in the **flow separation** zone generated on its downstream side. As such these structures may prove excellent *paleocurrent indicators* in **gravels**. [JLB]

Brayshaw, A.C. (1984) *Characteristics and origin of cluster bedforms in coarse-grained alluvial channels*. Canadian Society of Petroleum Geologists Memoir **10**: 77–85.

parting The breaking along planes of weakness in **minerals** such as *twin planes* or planes resulting from **deformation**. It resembles **cleavage** because the planes lie parallel to crystallographic planes. However, whereas **cleavage** planes are common to all crystals of a particular **mineral**, parting planes are not shown by all crystals, only those that are twinned or that have been subject to the appropriate **deformation**. Examples of parting include the octahedral parting of **magnetite**, rhombohedral parting of **corundum** and basal parting of **pyroxene**. [DJV]

pascal Standard (derived SI) unit of **pressure** or **stress**, defined as the reaction of a surface of 1m^2 in area to a **force** of 1 **newton** (N). Thus 1 pascal (Pa) $= 1 \text{kg m}^{-1} \text{s}^{-2}$. For geological purposes, the *megapascal* $(= 10^6 \text{Pa})$ is commonly used. [RGP]

passive continental margin (*trailing continental margin*) A continental edge, originally formed by **continental splitting**, which was left behind as **seafloor spreading** carried it, and the older parts of the **oceanic crust** bordering it, away from the active **seafloor spreading** ridge. These margins, sometimes also known as *quiescent continental margins*, still show some **seismicity**, mostly associated with cooling and **isostatic** adjustment to both differential cooling between ocean and continent, and to the effect of the sedimentary loading on the continental edge and fore-slope. [DHT]

patronite (VS$_4$) An **ore** mineral of **vanadium**. [DJV]

patterned ground A general term for features developed in *soil*, sediments and **fractured** bedrock, including more or less symmetrical forms (e.g. circles, nets, stripes, polygons). They are characteristic of areas subject to frost processes, and *cryoturbation*, heaving and thermal desiccation and contraction processes are probably important in their formation. Patterned ground also occurs in some lower latitude areas where seasonal extremes of *soil* moisture levels causes expansion and contraction phenomena to develop. In Australia patterned ground forms are called *gilgai*. [ASG]

Washburn, A.L. (1979) *Geocryology: A Survey of Periglacial Processes and Environments*. Arnold, London.

pause plane A surface within a **cross-stratified** bed representing a period of minimum **flow** velocity. Pause planes may be **erosional** or non-**erosional** and occur between structures due to slackening (below) and reactivation (above). Pause planes may be draped with fine sediment. They are particularly common in tidal **sandwave** deposits. (See **reactivation surface**.) [PAA]

Boersma, J.R. & Terwindt, J.H.J. (1981) Neap-spring tide sequences of intertidal shoal deposits in a mesotidal estuary. *Sedimentology* **28**: 151–70.

peat A mass of fibrous plant debris that is partly decomposed and often dark brown in colour. The peats that accumulate in certain **deltaic** and coastal areas have long been recognized as the precursors of **coal**. For the formation of peat there must be (i) substantial growth of vegetation, (ii) sufficient standing water around the accumulating dead vegetation to prevent **oxidation** or bacterial destruction of the organic components and (iii) no introduction of detrital sediment into the site of peat accumulation. With favourable conditions, as in north-west Borneo, peat growth may reach $3–4 \text{ mm a}^{-1}$ and thicknesses of over 17m have been found. [AME]

Péclet number A dimensionless number which describes the relative importance of diffusion compared to **flow**. If D is the *diffusion coefficient* $(\text{m}^2 \text{s}^{-1})$, L is the average distance of migration (m) and U is the interstitial water **flow** velocity in a sediment (m s^{-1}),

$$Pe = \frac{D}{LU}$$

If $D \gg LU$, **diffusion** is the main transporting agent to a site, otherwise **flow** is the dominant process. (See **diffusion in sediments**.) [PAA]

pectolite (Ca$_2$NaH(SiO$_3$)$_3$) An *inosilicate* mineral of the pyroxenoid group. (See **pyroxenoid minerals**.) [DJV]

pediment A gently sloping **erosion** surface, generally inclined at about 0.5° to 7.0°, becoming steeper upslope (concave upwards). Pediments occur on the flanks of steep-sided hills and mountains, and are common at the foot of **mesas** and **inselbergs**. They are cut in bedrock but are often mantled with a veneer of alluvium or **gravel**. Notwithstanding the morphological similarity to depositional fans, pediments are **erosional** landforms.

Pediments are characteristic of many arid and semi-arid landscapes (in North Africa they are known as **glacis**). It has been suggested that they are the product of **erosion** by **sheetfloods** or of **backwearing** of hillslopes by **parallel retreat**. Since pediments are defined by the break-in-slope between the steeper hillslope at the top of the pediment, it is likely that specific **weathering** processes and **erosional** mechanisms produce the landform. There must be a thorough breakdown of solid rock to weathered material in order for the **erosional** mechanisms to strip the debris from the surface. Though pediments have been recognized in temperate and humid tropical landscapes, their morphology is often masked by vegetation and *soil* accumulation. In such instances, their classification as true pediments may be inappropriate. [AW]

Adams, G. (ed.) (1975) *Planation Surfaces*. Dowden, Hutchinson & Ross, Stroudsburg, PA.

pedogenesis The process of *soil* formation. While the term *soil* is defined differently by different specialists a widely used definition is that *soil* is a natural body consisting of horizons of mineral matter and/or organic constituents which differ from the parent material in their morphological, physical, chemical, mineralogical and biological characteristics (modified from Birkeland, 1984).

Soil is a product of the interplay of several factors: parent material, climate, organisms, topography and time. Parent material includes both lithified and unconsolidated material. Climate exerts a fundamental control on *soil formation* and affects not only **weathering** and other *soil forming* processes, but also influences the types of organisms present. The biological components include plants and animals in a complex web of interactions involving nutrient recycling and decomposition, as well as in varied symbiotic interrelationships. Topography influences the *soil* in many ways, such as drainage characteristics and stability. There are sufficient variations in *soil* processes down slopes to result in distinctive changes in *soil* type down-gradient. Such slope-related sequences of *soils* are called *catenas* or *toposequences*. Time is also important and *soils* develop through various stages (*chronosequences*) although some *soil* features achieve a steady state during their evolution. *Soils* take long periods to form but climates and landscapes are dynamic and many *soils* have had complex histories.

The processes responsible for *soil formation* are varied and do not all operate uniformly. It is the variations in the types and degrees of these processes which result in the development of many different *soil* types.

The main *soil forming* processes include:
Weathering (the physical and chemical **alteration** of rocks and minerals at or near the Earth's surface).
Leaching (the removal from the *soil* of the soluble constituents).
Eluviation (the removal of material from a horizon of a *soil*, either in solution or suspension).
Illuviation (the movement and accumulation of material produced by **eluviation**).
Podsolization (the downward movement of **iron**, **aluminium** and organic matter resulting in a prominent light coloured eluvial horizon with residual **silica**).
Rubifaction (reddening due to the formation of **hematite**).
Ferrallitization (accumulation of sesquioxides of **iron** and **aluminium** under humid tropical climates).
Gleying (waterlogging and the development of anerobic conditions resulting in slow rates of organic decomposition and the formation of ferrous **iron**).
Calcification (accumulation of **calcite** or **dolomite**, usually in *soils* with a moisture deficit).
Salinization (accumulation of salts in the *soil*).
Biological processes such as *humification* (conversion of organic matter into *humus*).
Pedoturbation (the churning of the *soil* caused by physical processes, such as clay shrinkage and swelling, and biological activity).

As a result of these processes, changes occur in the *soil* with losses (leaching), additions (from **weathering** and rainfall), transformations (by the formation of clays, iron compounds) and translocations (by eluvial and illuvial processes and *podsolization*, *humification*). All these processes result in the most striking features of soils, *horizonation*.

Soil profiles are divided into several prominent horizons recognized by their different colours, structures and compositions:
O horizon — mainly organic layer.
A horizon — mixed humified organic matter and mineral material.
E horizon — horizon beneath the A horizon which contains less organic matter, sesquioxides of **iron** and clay than the underlying horizon. Effectively material has been removed by **eluviation**.
B horizon — a horizon which underlies an O, A or E horizon but is different from the parent material. It may exhibit a variety of features including concentrations (of organic matter, clays, **iron** or **aluminium**) representing an illuvial horizon or concentrations of soluble compounds such as calcium carbonate.
C horizon — a horizon which lacks properties of the other horizons but includes material in various stages of **weathering**. It may also contain accumulations of various soluble materials. If the *soil* develops over consolidated material, a zone of **weathered** rock occurs which may be relatively thin, or under conditions of intense or prolonged **weathering**, may be very thick. Such zones are referred to as **regolith** or **saprolite**.

R horizon — bedrock.

These are the main (master) *soil* horizons and a variety of other subhorizons can be designated, resulting in the detailed subdivision of a profile.

A number of classifications are available for *soils* and the resulting terminology is rather bewildering. Two widely used systems are the international FAO World Soil map system and Soil Taxonomy, based on the United States Department of Agriculture handbook. Both these systems recognize major *soil* 'orders' by the presence of diagnostic horizons which are strictly defined. The latter system uses a complex series of original names while the former incorporates traditional terms. [VPW]

Birkeland, P.W. (1984) *Soils and Geomorphology*. Oxford University Press, New York.
Duchaufour, P. (1982) *Pedology, Pedogenesis and Classification*. Allen & Unwin, London.
Fitzpatrick, E.A. (1983) *Soils and Their Formation, Classification and Distribution*. Longman, London.

pegmatite deposit The bulk composition of most pegmatites of economic interest is close to that of **granite** and some of these contain high contents of volatile components (OH,F,B) and a wide range of **accessory minerals** containing rare lithophile elements. These include **beryllium, lithium, tin, tungsten**, rubidium, **cesium, niobium, tantalum**, the **rare earth elements** and **uranium** for which pegmatites are mined as well as **industrial minerals** such as **feldspar** and **mica**.

Pegmatitic deposits of **spodumene, petalite, lepidolite** and other **lithium** minerals are exploited throughout the world for use in glass ceramics, fluxes in **aluminium** reduction cells and the manufacture of numerous **lithium** compounds. These deposits often yield by-product **beryllium**, rubidium, **cesium, niobium, tantalum** and **tin**. They may be internally zoned, or unzoned as are the highly productive **lithium** pegmatites of King's Mountain, North Carolina. The largest known pegmatitic **lithium** resources are in Zaïre in two **laccoliths** each about 5 km long and 0.4 km wide. **Reserves** have been put at 300 Mt and **tantalum, niobium, zirconium** and **titanium** values have been reported. A recent challenge to pegmatitic **lithium** production comes from the exploitation of **lithium**-bearing **brines**.

The traditional source of **beryllium** (beryl) has now been overtaken by *bertrandite*, which occurs on a commercial scale in **hydrothermal deposits**. The main source of **beryl** is pegmatites, with the former USSR ($2\,kt\,a^{-1}$) being the world's largest producer, followed by Brazil with less than $400\,t\,a^{-1}$. Pegmatites are also important as a source of **tantalum**, but it must be noted that the largest **reserves** — about 7.25 Gt of **tantalum** — are in slags (running about 12% Ta_2O_5) produced during the smelting of **tin ores** in Thailand. Significant **reserves** are present in the Green-bushes Pegmatite, Western Australia together with **tin** and most of the **tin** mined in Thailand is won from pegmatites. Uraniferous pegmatites and pegmatitic **granite**

have been exploited in a number of localities. Among the more important deposits are those of Bancroft, Ontario and the enormous Rössing Deposit in Namibia. [AME]

Černý, P. (ed.) (1982) *Short Course in Granitic Pegmatites in Science and Industry*. Mineralogical Association of Canada, Winnipeg.
Evans, A.M. (1993) *Ore Geology and Industrial Minerals: An Introduction* (3rd edn). Blackwell Scientific Publications, Oxford.

pelagic Of or pertaining to the deep sea, e.g. pelagic sediment. (See **pelagite, calcareous ooze, siliceous ooze, red clay**.) [AESK]

pelagite Abbreviation for **pelagic** sediment, formed by deposition mainly from the slow settling from suspension of calcareous and siliceous biogenic material that generally comprises finer grain-size populations, and in which the biogenic component is >75%. **Pelagic** sedimentation is favoured in **basins** protected from high rates of terrigenous sediment accumulation, in silled **basins**, slope-basin floors and *abyssal plains*. Typical pelagites include **radiolarian cherts, red clays, chalk**, nannofossil, **Foraminifera**, *diatomaceous* and *pteropod* oozes. Pelagites typically occur as structureless layers up to decimetres thick and may show *bioturbation* to varying degrees. Typical rates of **pelagic** sedimentation range up to about $4\,mm\,a^{-1}$, but commonly are between 0.02 and $0.15\,mm\,a^{-1}$. [KTP]

Penck and Brückner model Developed during the first decade of this century by A. Penck and E. Brückner, working primarily in Bavaria, to provide a framework for understanding the *Pleistocene* history of the Alps, this model was very widely adopted, and applied all over the world. Four main *glacial* phases were initially identified — *Günz, Mindel, Riss* and *Würm* — together with various **interglacials**, including the so-called 'Great Interglacial' between the *Mindel* and the *Riss*. A more complex *Pleistocene* history, with many more *glacial*–**interglacial** cycles, has now been determined on the basis of deep-sea core studies, and this means that the Penck and Brückner model is now superseded. [ASG]

Penck, A. & Brückner, E. (1909) *Die Alpen in Eiszeitalter*. C.H. Tauchnitz, Leipzig.

peneplain A low-angle surface, almost a plain or plane, which occurs during the old age stage of a cycle of **erosion** in which slope angles decline through time. [ASG]

Davis, W.M. (1899) The geographical cycle. *Geographical Journal* **14**: 481–504.

penetrative Applies to a feature such as a **fabric** or a **microstructure** that is developed throughout the rock on the scale of observation. For example, a *penetrative cleavage* implies alignment of platy minerals throughout the specimen, rather than only in parts of the rock. However, a *penetrative cleavage* on outcrop or hand specimen scale may be *non-penetrative*, i.e. only present in certain areas of the rock, when it is examined in greater detail, such as with a microscope. This is a useful term for comparative rock descriptions, bearing in mind the importance of the scale of the observations. [SB]

pennantite A variety of the mineral **chlorite**. [DJV]

pentlandite ($(Fe,Ni)_9S_8$) The most important **ore** mineral of **nickel**. [DJV]

percentage frequency effect (PFE) The parameter measured in frequency-domain **induced polarization** methods. Certain mineralization allows charge build-up in the ground during current flow, rather similar to capacitors, which gives rise to a frequency-dependent **apparent resistivity** in the medium. PFE, which quantifies this behaviour, is defined as

$$\frac{100\,(\rho_0 - \rho_\infty)}{(\rho_\infty)}$$

with the subscripts of the **apparent resistivity**, ρ, indicating frequencies. In field measurements, ρ_0 is ρ at some very low frequency, typically *c.* 1 Hz or less, and ρ_∞ is ρ at a few tens of Hz. PFE is equivalent to **chargeability** in time-domain measurements. [ATB]

Keller, G.V. & Frischknecht, F.C. (1966) *Electrical Methods in Geophysical Prospecting.* Pergamon, Oxford.

perched aquifer/water table A perched aquifer consists of a locally developed, water-saturated body of material that lies above the regional **water table**. It owes its existence to an impermeable layer that lies immediately beneath it. The upper limit of a perched aquifer will form a perched water table. [AME]

perennial head The highest point along the course of a river from which a year-round flow of water occurs. Above this point the flow may be intermittent as a result of seasonal rise and fall of the **water table**. [AME]

perennial stream A stream or river that flows all year and every year. [AME]

periclase (MgO) An **oxide mineral** found in **contact metamorphosed limestones**. [DJV]

pericline Fold, usually with elliptical or circular outcrop, in which the **dips** vary in direction around the structure. Such **folds** include **domes** and **basins**, with radial *limb* **dips**, and *brachyanticlines* and *brachysynclines*, which are elongate with varying axial **plunge** (Fig. P2). [RGP]

peridot A clear green **gem** variety of **olivine**. (See **olivine minerals**.) [DJV]

peridotite An ultramafic rock containing 40–90% of **olivine** and **pyroxene**. Such peridotites include **harzburgite** (<5% *clinopyroxene*) and **wehrlite** (<5% *orthopyroxene*) but the dominant type of peridotite is **lherzolite**. [RST]

periglacial geomorphology Although the term was used first by Losinski (1909) to describe **frost weathering** conditions in the Carpathian Mountains and developed as a concept refering to conditions peripheral to *ice sheets* and **glaciers** (present or *Pleistocene*), modern usage refers to a wide range of conditions regardless of their proximity to **glaciers**, either in time or space. *Periglacial* areas may be underlain by **permafrost**, but it is not a necessary feature of *periglacial* environments. Frost processes are, however, a fundamental component. Nonetheless, periglacial environments are often characterized by **permafrost**-related landforms: *ice mounds* (*pingos* and *palsas*), subsidence phenomena (*thermokarst*), and various types of patterned ground (ice, wedge, polygons, etc). Active zone processes are also significant (e.g. *frost shatter*ing, congelifraction, **scree** and *blockfield* formation, protalus ramparts, etc). Frost and mass movement processes may produce *cryoplanation* terraces, while spring **runoff** may cause deep fluvial incision. Some *periglacial* areas have rock **glaciers** (lobate or tongue-shaped bodies of frozen debris which move downslope or down-valley). In the cold phases of the *Pleistocene*, *periglacial* environments were greatly expanded in extent and large portions of the mid-latitudes show relict *periglacial* features. [ASG]

Clark, M.J. (ed) (1988) *Advances in Periglacial Geomorphology.* Wiley, Chichester.
Lozinski, W. von (1909) Über die mechansche Verwitterung der Sandsteine im gemässigten Klima. *Bulletin International de l'Académie des Sciences et des Lettres de Cracovie, Classe des Sciences Mathématiques et Naturellels* 1: 1–25.

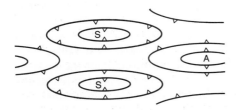

Basin

Dome

Fig. P2 Types of **pericline**: brachyanticline (A), brachysynclines (S), basin and dome.

peristerite The name given to very fine scale intergrowths of t o **plagioclase feldspar minerals** (overall composition in the range An$_2$ to An$_{15}$) not visible to the naked eye although possibly detectable due to *irridescence* in specimens termed **'moonstone'**. (See **feldspar minerals**). [DJV]

perlite Hydrated silicic volcanic glass, characterized by curved, concentric, 'onion-skin' cracks, thought to have been formed during volume changes associated with hydration. Water contents are typically 2–5% by weight. Used commercially this term includes any volcanic glass that will expand or decrepitate on heating to form a light-weight frothy material. Such material has a low thermal conductivity, high sound absorption and high resistance to heat which makes it suitable for insulation board manufacture and other constructional uses. Because of its absorbency, it is also used as a rooting medium, *soil* conditioner, filter material and filler in paints, plastics, etc. Hydration is usually associated with chemical changes, notably gain of **potassium** and loss of sodium. Perlite deposits occur as flows, **dykes**, **sills** and **domes** of hydrated **obsidian** and **pitchstone** in young volcanic regions. Due to devitrification, few deposits older than **Oligocene** are of commercial interest. The USA, Greece, the former USSR and Hungary are the biggest producers. (See **glassy rock**.) [AME/PTL]

Stewart, D.B. (1979) The formation of siliceous potasic glassy rocks. In: Yoder, H.S. (ed.) *The Evolution of the Igneous Rocks*, pp. 339–50. Princeton University Press, Princeton, NJ.

permafrost A thermal condition of rock, *soil* and sediment where temperatures remain below 0°C for at least two consecutive winters and the intervening summer. It is essentially permanently frozen ground (Miller, 1947). The surface layer, called the active layer is subject to seasonal thaw. Permafrost underlies about one quarter of the Earth's land surface, and is widespread in Siberia, northern Canada, Alaska and Tibet. Thicknesses may exceed 500 m. Permafrost also underlies some coastal areas. According to climatic and material conditions permafrost may be continuous, discontinuous or sporadic. [ASG]

Muller, S.W. (1947) *Permafrost or Permanently Frozen Ground and Related Engineering Problems*. J.W. Edwards, Ann Arbor, MI.

permanence of continents Long before **plate tectonics**, the evidence of a prolonged stratigraphic record on the continents led to the supposition that the continents were fairly permanent features of the Earth's surface, although suffering occasional marine inundations. It is now realized that this is mainly because they are much less dense than the oceanic rocks and are therefore not subducted, and consequently persist to eventually collide with other continental blocks. [DHT]

Wise, D.U. (1973) Freeboard of continents through time. *Geological Society of America Memoir* **132**: 87–100.

permeability The coefficient linking the flow rate through a medium to the pressure gradient in the *Darcy equation*,

$$Q = \frac{KAdP}{dx}$$

where Q is the volume of transmitted flow per unit time (**flow** rate), A is the cross-sectional area of the porous medium, and dP/dx is the pressure gradient over distance x (*hydraulic gradient*) and permeability K is measured in Darcys. In **sands**, permeability of unconsolidated **sand** decreases as grain size becomes finer and as *sorting* becomes poorer. For wet-packed samples, permeability varies over more than an order of magnitude between extremely well sorted **sands** and very poorly sorted **sands**. [PAA]

Permian System named after the ancient kingdom of Permia in Russia, and suggested by Sir Roderick Murchison in 1841 after his tour of Imperial Russia. He found a thick succession of sediments, now known to include beds from the *Asselian* to the *Tartarian* (Fig. P3), overlying the **Carboniferous** in the region between the Volga and the Urals; he correctly interpreted them as lying between the **Carboniferous** and the **Triassic** *New Red Sandstone*. Prior to this (1822) workers in Britain had realized that the post-**Carboniferous** Penrith Sandstone and Magnesian Limestone were equivalent to the Rotliegendes and Zechstein of Germany and by 1853 a wide belt of Permian rocks had been recognized between the Mississippi and Rio Colorado in the USA. Further, the two divisions of the Permian, first described in Germany, were found to be widely present elsewhere and have been retained as epochs in the current divisions of the Period. In this classification the smaller standard divisions are based on Russian localities, but only the ages are given in Fig. P3 which is modified from Harland *et al.* (1982). **Biostratigraphical** subdivision is primarily by marine fusulinacean **Foraminifera**, the base of the Permian being marked by the entrance of *Triticites californicus* in the USA, but there are boundary problems with the underlying **Carboniferous** at many other places in the world. **Brachiopod** biozones have also been established. On land the base of the Permian is widely marked by the entrance of the seed fern *Callipteris conferta*.

Early Permian world geography is a continuation of **Carboniferous** trends. Dynamic approaching and coalescing continents quickly completed the assembly of the **Pangea** *supercontinent* which was accomplished early in the Period (Fig. P4). Initially the major continents of **Laurasia**, *Angaraland* and **Gondwanaland** were still separated by seaways. **Laurasia** and *Angaraland* became linked after the *continental collision* that formed the Ural chains in the early Permian and **Laurasia** and **Gondwanaland** drifted and rotated progressively into closer configuration during the late **Carboniferous** and early **Permian** to form the Appalachian chains on the

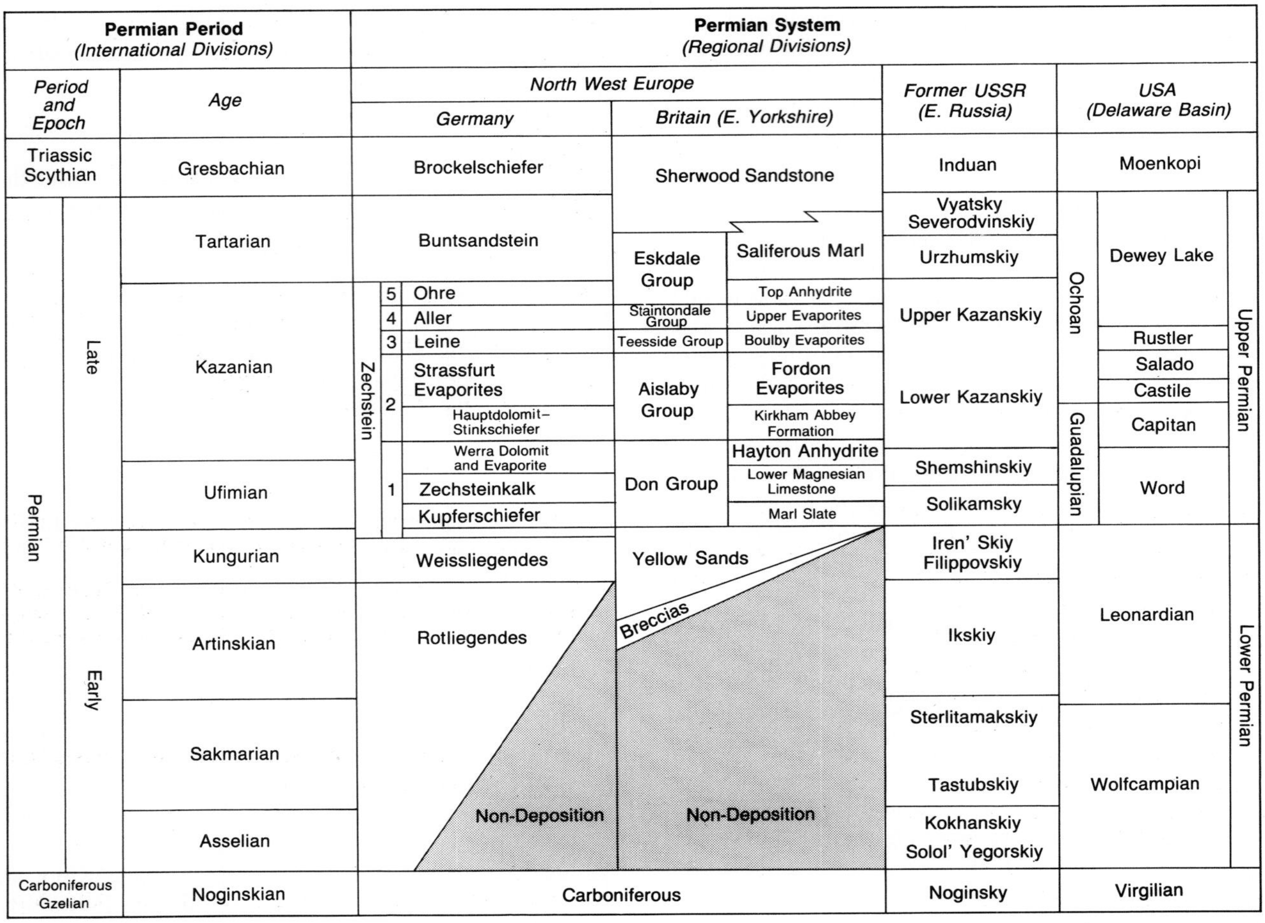

Permian Period (International Divisions)		Permian System (Regional Divisions)				
Period and Epoch	Age	North West Europe — Germany	North West Europe — Britain (E. Yorkshire)	Former USSR (E. Russia)	USA (Delaware Basin)	
Triassic Scythian	Gresbachian	Brockelschiefer	Sherwood Sandstone	Induan	Moenkopi	
Permian — Late	Tartarian	Buntsandstein	Eskdale Group / Saliferous Marl	Vyatsky Severodvinskiy / Urzhumskiy	Dewey Lake	Upper Permian
	Kazanian	Zechstein 5 Ohre; 4 Aller; 3 Leine; Strassfurt Evaporites 2 Hauptdolomit–Stinkschiefer	Staintondale Group / Teesside Group / Aislaby Group — Top Anhydrite, Upper Evaporites, Boulby Evaporites, Fordon Evaporites, Kirkham Abbey Formation	Upper Kazanskiy / Lower Kazanskiy	Rustler; Salado; Castile; Capitan (Ochoan / Guadalupian)	
	Ufimian	Zechstein 1 Werra Dolomit and Evaporite; Zechsteinkalk; Kupferschiefer	Don Group — Hayton Anhydrite, Lower Magnesian Limestone, Marl Slate	Shemshinskiy / Solikamsky	Word	
Permian — Early	Kungurian	Weissliegendes	Yellow Sands	Iren' Skiy Filippovskiy	Leonardian	Lower Permian
	Artinskian	Rotliegendes	Breccias	Ikskiy		
	Sakmarian	Non-Deposition	Non-Deposition	Sterlitamakskiy / Tastubskiy	Wolfcampian	
	Asselian			Kokhanskiy / Solol' Yegorskiy		
Carboniferous Gzelian	Noginskian	Carboniferous		Noginsky	Virgilian	

Fig. P3 The **Permian** system.

eastern margin of USA. First freshwater links were established between **Laurasia** and **Gondwanaland** allowing the aquatic **reptile** *Mesosaurus* to reach South Africa and Brazil. Almost at the same time *labyrinthodont* **amphibians** migrated south and east to that part of **Gondwanaland** that is now northern India. By the middle of the Permian, terrestrial tetrapod faunas from **Laurasia** were able to reach *Angaraland* and spread eastwards to Siberia. The main southwards and eastwards migration of tetrapods, particularly the *therapsid* **reptiles**, took place in the Late Permian when they spread southwards to South Africa and India and eastwards to China. By this time the tetrapods were able to migrate by land routes to all the linked continents of **Pangea**. The origin of the world distribution of their descendants dates from this event.

In the British Isles and much of continental Europe the Early Permian was a period of interior hinterland subaerial **erosion**, during which the newly raised *Hercynian* mountains were gradually reduced to a **peneplain**. The sedimentary record is fragmentary and consists of **sands** and breccias laid down partly in **sand** deserts and partly in elevated rocky deserts under dry tropical conditions. At the beginning of the Late Permian a rise in sea-level,

perhaps caused by the beginning of retreat and melting of the vast Antarctic continental *glaciation*, caused the flooding from the north of a large sub-sea-level drainage **basin** which covered north-east England, the North Sea and much of north-west Europe. This *Zechstein Sea* filled rapidly to considerable depths leaving the aeolian **dune sands** on the floor of the **basin** little disturbed. The first deposit, the *Kupferschiefer* of Germany and the *Marl Slate* of north-east England, is very uniform and extensive. This deposit is the beginning of five major cycles of sedimentation, each of which has a carbonate, sulphate and chloride phase (Fig. P3). The control of these **evaporite/carbonate** cycles is believed to be **eustatic** with major changes in sea-level causing changes in deposition from carbonates in a restricted marine **basin** to thick **evaporites** in an isolated salt **basin**. Important reserves of **dolomite**, **anhydrite**, **halite** and potash salts are present in the five Zechstein sedimentary cycles. In north-east England a shelf-edge **barrier reef** complex can be traced for more than 30 km along the margin of the *Zechstein Sea*.

Similar shelf-edge *reefs* are developed on the margins of carbonate/**evaporite basins** in Texas and New Mexico, USA. In the Delaware Basin the Capitan Limestone *reef* of

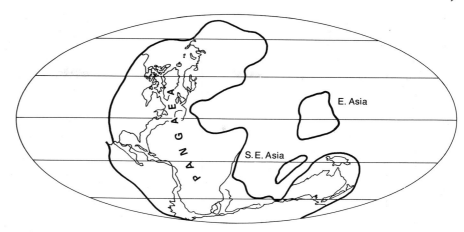

Fig. P4 The **Permian** continents.

Guadeloupian age separates shallow-water shelf carbonates and nearshore **sabkha evaporites** from stagnant water deep **basins** where organic-rich **limestones, shales** and **sandstones** were laid down. These deep **basins** are important *source rocks* for **petroleum** now found in **reservoirs** on the surrounding shelf areas. In much the same way as in the *Zechstein Sea*, circulation of seawater became restricted soon after the beginning of the Late Permian and with rapid evaporation in a hot dry climate about 600 m of **anhydrite, halite** and **potassium** salts were laid down in the Delaware and adjacent **basins**. A major feature of middle Permian times throughout the warmer climatic zones of the world is the widespread development of carbonate/**evaporite basins**.

In the upper part of the Late Permian continuously lowering sea-level and increasing aridity of the hot dry interior of the vast **Pangea** *supercontinent* led to progressive draining and desiccation of virtually all the major seas and the development of desert conditions of the *New Red Sandstone*. Continental red **sandstones** and siltstones occur over the Zechstein **basin** and date from the top of the Permian; very similar deposits occur in West Texas over the Delaware and associated **basins**.

Land plant communities were restricted in arid tropical regions during the Permian, but under temperate conditions in **Gondwanaland** widespread **peat** deposits formed from a broad-leaved deciduous *Glossopteris flora* in what are now South Africa, India and Australia. These deposits gave rise to important reserves of **coal**.

The marine invertebrate faunas of the world changed dramatically at the end of the Permian. A high rate of **extinction** in the **Foraminifera** (*fusulinaceans*), **corals** (*rugosans*), **brachiopods** (productids), **Bryozoa** (fenestellids), *ammonoids* (*goniatites*), **echinoderms** (*blastoids* and many *crinoids*) and **arthropods** (*trilobites*) greatly altered the composition of shallow water marine life, though the terrestrial plants and vertebrates show no marked changes at this time. This faunal change was recognized as of much significance by early geologists who made it the basis for the separation of the **Paleozoic** and **Mesozoic** eras. [GALJ]

Harland, W.B., Cox, A.V., Llewellyn, P.G. *et al.* (1989) *A Geologic Time Scale.* Cambridge Earth Science Series. Cambridge University Press, Cambridge.
Smith, D.B., Brunstrom, R.G.W., Manning, P.I. & Shotton, F.W. (1974) *A Correlation of Permian rocks in the British Isles.* Geological Society of London Special Report No. 5. Scottish Academic Press, Edinburgh.

perovskite (CaTiO$_3$) An **oxide mineral** found in **nepheline syenites** and **carbonatites**. [DJV]

Persian Gulf (*Arabian Gulf*) One of the world's largest *hydrocarbon provinces* and an area of extensive carbonate and **evaporite** deposition. The Gulf is 1000 km long and 200–300 km wide. It has an asymmetric form with a gentle slope on the Arabian side, down to 100 m, and a steeper slope on the Iranian side. The Arabian coastal zone, especially along the southerly extent, is an area of extensive shallow water carbonate deposition and it constitutes a modern *carbonate ramp* (see **carbonate buildup**). The seafloor topography in this zone is complicated by local highs around **salt domes** which are associated with shoals and **coral** *reefs*.

The climate is arid subtropical with high evaporation and with little **runoff** from the surrounding arid areas. As a result much of the water of the Gulf is above normal salinity. Another consequence is that the input of land-derived siliciclastics is very low, favouring the accumulation of carbonates. Along the south-east of the Gulf (the Trucial Coast area) extensive **barrier islands** occur, sheltering hypersaline *lagoons* and tidal flats in their lee. As a result of rapid progradation during the last few thousand years an extensive supratidal zone has formed. The high evaporation rate results in **evaporite precipitation** on and within these sediments to form a **salt flat** or **sabkha**. [VPW]

Purser, B.H. (ed.) (1973) *The Persian Gulf: Holocene Carbonate Sedimentation and Diagenesis in a Shallow Epicontinental Sea.* Springer-Verlag, Berlin.

perthite A relatively coarse intergrowth (observable with the naked eye) of the minerals **potash feldspar** and **albite**. (See **feldspar minerals**.) [DJV]

petalite ($Li(AlSi_4O_{10})$) A framework silicate (*tectosilicate*) mineral of importance as an **ore** of **lithium**. [DJV]

petroleum In geological usage this term includes the entire spectrum of naturally occurring *hydrocarbons* whether they be gaseous (**natural gas**), liquid (**crude oil**) or solid (**asphalt**, *bitumen* or tar). [AME]

pH A parameter related to the abundance of available hydrogen ions and defined as $pH = -\log_{10}(a_{H^+})$ where a_{H^+} is the activity of the hydrogen ion H^+ (e.g. pH is 8 if activity of hydrogen ions is 10^{-8}). Hydrogen ions vary in abundance inversely with hydroxyl ions (OH^-). The equilibrium constant for the dissociation of water is 10^{-14} at 25°C and equals the product of the activity of hydrogen ions and hydroxyl ions. A *neutral solution* is one where the activities of the two species are equal: $pH = 7$ at 25°C. *Acid solutions* have excess H^+ hence lower pHs; *alkaline solutions* have excess OH^-, hence higher pH values.

The range of values of pH for natural waters were summarized by Baas Becking *et al.* (1960). Rainwater is acid because of dissolved **carbon** dioxide. In unpolluted air the pH would be 5.7, but a major environmental hazard is posed by '*acid rain*' with pH often less than 4 owing to dissolved gases from industrial processes. Soilwaters range from acid in humic *soils* to mildly alkaline where dominated by neutralizing reactions with **minerals**. Lake waters are often slightly alkaline because of dissolved bicarbonate. *Alkaline lakes* have pH up to 12 related to dissolved borates or sodium carbonate. Seawater has a pH around 8–8.2 in equilibrium with atmospheric **carbon** dioxide. Atmospheric carbon dioxide pressures are thought to have been rather higher at many times in the geological past, so pHs of surface waters would have been lower. **Pore-waters** may be acid or alkaline. [IJF]

Baas Becking, L.G.M., Kaplan, I.R. & Moore, D. (1960) Limits of the natural environment in terms of pH and oxidation–reduction potentials. *Journal of Geology* **68**: 243–84.

Phanerozoic The term given to the interval of time that has elapsed since the *Precambrian*, and the strata formed during that time. For a best estimate of the age of the component periods see **geologic time-scale**. [AH]

phenakite (Be_2SiO_4) A rare *nesosilicate* mineral found in *pegmatites*. [DJV]

phenocryst A relatively large, generally euhedral, mineral grain within a fine-grained (or glassy) matrix/ groundmass of an intrusive or extrusive **igneous rock**. Common phenocrysts in **igneous rocks** include **olivine**, **pyroxenes**, **feldspars** and **quartz**. An **igneous rock** containing phenocrysts is termed *porphyritic* in texture, and such a texture may reflect slow cooling/growth of large crystals at depth, followed by cooling (as a result of uprise or extrusion) accompanied by the crystallization of the finer-grained matrix. [RST]

phillipsite ($KCa(Al_3Si_5O_{16}) \cdot 6H_2O$) A **zeolite mineral**. [DJV]

phlogopite ($KMg_3(AlSi_3O_{10})(OH)_2$) A mineral of the mica group. (See **mica minerals**.) [DJV]

phonolite A fine-grained commonly *porphyritic* **felsic igneous rock** containing over 10% modal **feldspathoid** (e.g. **nepheline**) together with **alkali feldspar** (e.g. **sanidine** and/or **anorthoclase**) and minor sodium-rich **amphibole** (e.g. **riebeckite**) or **pyroxene** (e.g. *aegirine*). With decrease in **alkali feldspar**, phonolites grade into **nephelinite** and with increase in **alkali feldspar** they grade into **trachyte**. Phonolite may contain **leucite** (cf. **leucitite**). Phonolites occur in association with **alkali basalt–trachyte** associations within varied continental *alkaline* volcanic provinces and on *oceanic islands*. [RST]

phosgenite ($Pb_2CO_3Cl_2$) A rare **carbonate mineral** of **lead**. [DJV]

phosphate analysis Determination of the concentration of *phosphate* in the occupation layers of archeological sites. Decay of animal or human remains leaves a residue of phosphates in the *soil*. The method is used to confirm human or animal occupation of **cave** sites, and to pinpoint human burial areas. [OWT]

phosphate rock A general term used to describe rocks with a high concentration of phosphate **minerals**, commonly in the *francolite*–**apatite** series, whether the rocks are of igneous or sedimentary origin. Most commercial producers exploit sedimentary deposits, but **apatite**-rich igneous intrusions are exploited in the former USSR, South Africa, Brazil and Finland; they provide about 16% of world production. A commercial **grade** rock usually contains at least 60% BPL (bone phosphate of lime, also known as triphosphate of lime or TPL), although lower **grades** are used by some producers in their own chemical processing plants.

Apart from **guano**, which provides about 2% of world production, the bulk of production (82%) is from **phosphorites** of which the biggest producers are the USA and Morocco. Most **phosphorites** occur as **beds**, centimetres to tens of metres thick, composed of grains (frequently termed pellets) of cryptocrystalline carbonate *fluorapatite* or collophane and variable amounts of detrital sedimentary material.

Most **phosphorites** formed under marine conditions at low *paleolatitudes*, i.e. within 40° of the Equator, and probably at water depths of less than 500 m, i.e. shelf environments. One of the critical factors was probably the influx of nutrient-rich water, generally by upwelling from oceanic depths leading to the development of a prolific biota, which gave rise to the initial concentration of **phosphorus**. [AME]

Nriagu, J.O. & Moore, P.B. (eds) (1984) *Phosphate Minerals*. Springer-Verlag, Berlin.
Slansky, M. (1986) *Geology of Sedimentary Phosphates*. North Oxford Academic, London.

phosphorite A **phosphate** rock. [DJV]

phosphorus (P) At. no. 15, at. wt. 30.97376. A non-metal of Group V. Phosphorus compounds are important as fertilizers but are also important in matches, pesticides, special glasses and chinaware, alloys (steels, phosphor bronze), metal treating, detergents, electrical components, foods and drinks. Commercial production is from **phosphate rocks**. [AME]

photic/euphotic zone The zone at the surface of the ocean in which the intensity of light penetration is sufficient to support *photosynthesis*. The intensity of light penetration decreases exponentially with depth in the ocean due to **attenuation** by absorption and scattering. The depth of the photic zone therefore varies significantly from 100 to 150 m in the open ocean to a few metres in turbid coastal or estuarine waters. The lower limit of the photic zone is marked by the compensation level at which the rates of photosynthetic production and respiration consumption of oxygen are equal. This level is generally taken to be that reached by 1% of the surface incident light. [AESK]

Parsons, T.R., Takahashi, M. & Hargrave, B. (1984) *Biological Oceanographic Processes* (3rd edn). Pergamon, Oxford.

phreatic zone The subsurface zone beneath the **water table** in which the intergranular pores and fissures are completely filled with water at hydrostatic pressures greater than atmospheric. Also known as the *zone of permanent saturation*. (cf. **vadose zone**.) [SDB]

phreatic/phreatomagmatic eruption Phreatic eruption is the explosive ejection of **tephra** consisting of mud and lithic material from a **volcanic vent**, from which juvenile magmatic material is absent. The eruptions result from magmatic heat causing explosive boiling of **groundwater**; there is no direct interaction between **magma** and erupted material. Phreatic deposits commonly form small cones consisting of weakly-bedded breccia and mud deposits.

Phreatomagmatic eruption is equivalent to *hydrovolcanic eruption*. It includes any explosive interaction between **magma** and water (from sea, lake, or **groundwater**). The proportion of erupted juvenile, magmatic material to brecciated and disaggregated *country rock* in phreatomagmatic deposits varies widely.

Phreatomagmatic eruptions occur in all **magma** compositions. The most common phreatomagmatic volcanic form for **basaltic magmas** is a **tuff ring**. Phreatomagmatic eruptions typically generate very fine-grained **tephra** which forms air-fall and **base surge** deposits. Bomb sags and bedding slumps are also characteristic. *Pyroclastic flow* deposits are rare. **Mudflow** deposits interbedded with the air-fall and **pyroclastic** surge beds may represent reworked **tephra**. The water content of phreatomagmatic **tephra**, when deposited, varies widely. Some is erupted and deposited hotter than about 100°C, so that condensed water is not present. Such deposits are called 'dry'. At lower temperatures, condensed water is present, and the deposit is called 'wet'.

Very widespread, very fine-grained air-fall **tephra** resulting from the interaction of silicic **magma** and water are called *phreatoplinian deposits* (Self & Sparks, 1978). (See **volcanic euption**). [PTL]

Fisher, R.V. & Schminke, H.-U. (1984) *Pyroclastic Rocks*. Springer-Verlag, Berlin.
Self, S. & Sparks, R.S.J. (1978) Characteristics of widespread pyroclastic deposits formed by the interaction of silicic magma and water. *Bulletin Volcanologique* 41: 1–17.

phyllic alteration A type of **alteration** in the *host rocks* of **porphyry copper** and **molybdenum** deposits. When present in other **hydrothermal** mineral **deposits** it is referred to as **sericitization**. [AME]

Evans, A.M. (1993) *Ore Geology and Industrial Minerals: An Introduction* (3rd edn). Blackwell Scientific Publications, Oxford.

phyllite A regionally metamorphosed **foliated** pelitic rock, i.e. one chemically similar to a **shale**. Grain sizes in phyllites are broadly intermediate between **slates** and **schists**. Thus *phyllosilicate* crystals, often abundant, are coarse enough to be seen with a lens, and give a distinct sheen to the **cleavage** surfaces. The **cleavage** surfaces are less planar than in slate, and often show **crenulation cleavage**. [SB]

phyllonite A dynamically metamorphosed *phyllosilicate*-rich rock, formed under similar conditions to **mylonites**, i.e. moderate to high temperatures and pressures, or slow **deformation** rates, at depths typically greater than 10 km. Phyllonites are typically formed by retrogressive metamorphism of higher grade **schists** or **gneisses**. Recrystallization during **deformation** results in **preferred orientation** of *phyllosilicate* flakes. The grain alignment can be less well developed than in a regionally metamorphosed rock, and so phyllonites may be correspondingly less **fissile**. [SB]

physiography A word that has obscure origins (Stoddart, 1975) but which came into regular usage in the English-speaking world in the 19th century, notably in the work of T.H. Huxley (1877). It was defined by Dana in 1863:

Physiography, which begins where geology ends — that is, with the adult or finished Earth — and treats (1) of the Earth's final surface arrangements (as to its features, climates, magnetism, life, etc.), and (2) its systems of physical movements and changes (as atmospheric and oceanic currents, and other **secular variations** in heat, moisture, magnetism, etc.).

[ASG]

Dana, J.D. (1863) *Manual of Geology: Treating on the Principles of the Science*. Bliss, Philadelphia.
Huxley, T.H. (1877) *Physiography: An Introduction to the Study of Nature*. Macmillan, London.
Stoddart, D.R. (1975) 'That Victorian science': Huxley's *Physiography* and its impact on geography. *Transactions of the Institute of British Geographers* **66**: 17–40.

phytogeomorphology The study of the important relationships between **land plants** and landforms, two of the most visible elements of landscape. They are directly and readily recognized and recorded by modern methods of **resources** survey, which include the increasing use of **remote sensing** (Howard & Mitchell, 1985). [ASG]

Howard, J.A. & Mitchell, C.W. (1985) *Phytogeomorphology*. John Wiley & Sons, New York.

phytokarst A **karst** landform, usually small-scale, which has been predominantly shaped by organic action (strictly speaking by plants). **Biokarst** is a wider term. The term was coined by Folk *et al*. (1973) to describe some complexly dissected pinnacles from an inland swampy area on Grand Cayman Island. These pinnacles were covered with *blue-green algae*, which Folk and co-workers concluded had attacked the **limestone** to produce the complex pitting. Small pinnacles in **cave** entrances, which are oriented to the light and covered with *algal* growths, have also been given the name **phytokarst**. The processes involved with the formation of such features are still the subject of debate. [HAV]

Bull, P.A. & Laverty, M. (1982) Observations on phytokarst. *Zeitschrift für Geomorphologie* **26**: 393–416.
Folk, R.L., Roberts, H.H. & Moore, C.H. (1973) Black phytokarst from Hell, Cayman Islands, British West Indies. *Geological Society of America Bulletin* **84**: 2351–60.

pi (π) diagram/pole On a *stereogram*, the π pole or axis is the normal to the best-fit **great circle** through a **girdle distribution** of **poles** to a folded surface. The π pole thus represents the mean *fold axis*. The π diagram is preferable to the **beta diagram** in cases where there are a large number of measurements, since it avoids the problem of the large number of meaningless intersections inherent in the β method, as shown by Ramsay (1967). [RGP]

Ramsay, J.G. (1967) *Folding and Fracturing of Rocks*. McGraw-Hill, New York.

picrite A dark-coloured volcanic or minor intrusive rock characterized by abundant **olivine** (with other ferromagnesian minerals, totalling over 90%) and minor **plagioclase** (up to 10%).

Picrite was used for the manufacture of some British Neolithic stone **axes**, including battle-axes and **axe** hammers. The main archeological source of the rock is believed to be near Hyssington, Shropshire, UK. [OWT/RST]

piemontite ($Ca_2MnAl_2O(SiO_4)Si_2O_7)(OH)$) A *sorosilicate* mineral isostructural with **epidote**. [DJV]

piercement fold Fold formed as a result of the **diapiric** intrusion of **evaporite** material (most commonly salt). The **fold** is developed on the upper surface of the **diapir**, which cuts through the surrounding strata, hence 'piercement'. [RGP]

piezoremanent magnetization The magnetization acquired by rocks when subjected to sudden impact, such as by a *meteorite*, or by the application of prolonged **stresses**. This form of **natural remanent magnetization** is poorly understood but is of potential value for studying changing seismic **stress** by determining accompanying changes in the local **geomagnetic field**. [DHT]

pigeonite (*c*. $Ca_{0.25}(Mg,Fe)_{1.75}Si_2O_6$) A **pyroxene mineral**. [DJV]

pillow lava A volcanic rock consisting of many rounded, sack-like parts, about 0.2–2 m in diameter, separated from each other by fine-grained (originally glassy) rinds. Pillow lavas form when **lava flows** are erupted into water (or mud or ice), or when **lava flows** enter water from land. Not all submarine **lava flows** are pillow lavas. The individual sacks, or pillows, may be tube-like or pillow-like in three-dimensional shape. The pillows are rather poorly fitted together, and the spaces between the pillows, which may form a considerable proportion of the whole rock, are filled by **hyaloclastite** and sediment. The pillows are usually jointed, with a set of **joints** radiating from a central point.

During an eruption in Hawaii, **lava flows** were observed to flow into the sea, and during underwater flow the **lavas** were seen to advance partly by extrusion of pillow-like lobes (Moore *et al*., 1973). The lobes were extruded from a main lava tube, like paste being squeezed from a tube. The surface of the growing lobes consisted of chilled lava, which was hot to touch, but presumably below boiling point, as steam was not generated at the contact. The incandescent, molten nature of the lava in the centre

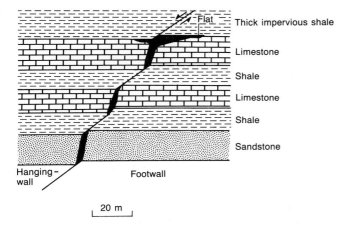

Fig. P5 Vein occupying a normal fault and exhibiting **pinch-and-swell** structure, giving rise to ribbon ore shoots. The development of a flat beneath impervious cover is also shown. (After Evans, 1993.)

of the lobes was only observed briefly when the chilled surface cracked. These observations in Hawaii proved the non-explosive nature of pillow lava formation, which contrasts to the violently explosive **phreatomagmatic eruptions**.

Pillow lavas seem to only occur in lavas of **basaltic** composition. They can form at all water depths including the lowest parts of the *ocean ridges*. The vescularity of pillow lavas is approximately related to the depth of water at the time of extrusion. (See **vesicle**.) [PTL]

Cas, R.A.F. & Wright, J.V. (1987) *Volcanic Successions Modern and Ancient*. Allen & Unwin, London.
Moore, J.G., Phillips, R.L., Grigg, R.W., Peterson, D.W. & Swanson, D.A. (1973) Flow of lava into the sea, 1969–1971, Kilauea volcano, Hawaii. *Geological Society of America Bulletin* 84: 537–46.

pinch-and-swell structure Any repetitive thinning and thickening of a body of rock material. This structure normally has an underlying structural origin. One mechanism giving rise to large-scale developments of this structure is the formation of mineral **veins** in **faults**. A typical **vein** displaying pinch-and-swell structure is shown in Fig. P5 from which it will be seen that the pinches and swells correspond to variations in the stratigraphical succession. This pinch-and-swell structure can create

difficulties during both exploration and mining, often because only the swells are workable and if these are imagined in a section at right angles to that of the figure it can be seen that they form ribbon *ore shoots*.

The origin of pinch-and swell structure is shown in Fig. P6. An initial **fracture** in rocks changes its attitude as it crosses them according to the changes in physical properties of the rocks and these properties are in turn governed by changes in lithology (Fig. P6a). When movement occurs producing a *normal fault* then the less steeply dipping sections are held against each other to become bearing surfaces, and open spaces (*dilatant zones*) form in the more steeply dipping sections. Then, should **minerals** be deposited in these cavities, a **vein** will be formed. Of course with reverse movement on the initial **fracture** the steeper parts of the **fault** now act as bearing surfaces and the *dilatant zones* are formed in the less steeply dipping sections. (See also **boudinage**.) [AME]

Evans, A.M. (1993) *Ore Geology and Industrial Minerals: An Introduction* (3rd edn). Blackwell Scientific Publications, Oxford.

pinnate fracture Minor **fracture** intersecting a larger **fracture** at an acute angle. Pinnate fractures often occur in a dense group of many short **fractures**, which intersect the larger **fracture** at 45° and are often found only on one side of it. The direction of movement of the **fault** block containing the **fractures** may be given by the direction of closure of the **fractures** on the main **fracture** plane. They may form due to **tensile stresses** created by frictional sliding on the main **fracture**. Pinnate fractures are also known as *feather fractures*. (See **feather joint**.) [TGB]

Hancock, P.L. & Barka, A.A. (1987) Kinematic indicators on active normal faults. *Journal of Structural Geology* 9: 573–84.

pipe (1) Subsurface channel ranging in size from a large crack several millimetres in width to a diameter up to three metres. Pipes may be distinguished from tunnels which form by **eluviation** of particles from inadequately designed earth-filled dams under high hydraulic gradients. Natural pipes usually form in *soil* or **peats** subject to cracking. Often such soils also have a very low *hydraulic conductivity* so that flow is concentrated in the cracks and not absorbed into the *soil* matrix. Pipe development is

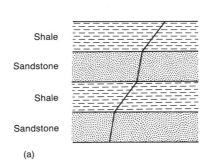

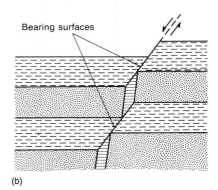

Fig. P6 Formation of **pinch-and-swell** structure in veins. (a) Initial fracture showing refraction as it crosses beds of different competency. (b) Movement along the fracture gives rise to open spaces. (After Evans, 1993.)

therefore encouraged by factors which cause cracking, such as *soil* shrinkage or desiccation. Pipes may also form from biopores or from holes produced by mass movement. Although pipes were first described in semi-arid areas, much recent research has been concerned with their occurrence in the organic *soils* of the uplands of north-west Europe. The **hydrological** and **geomorphological** significance of pipes is not yet totally clear. In **badlands**, flow in the pipes is ephemeral but are clearly of **erosional** significance both in relation to *gullying* and to mass movement. In humid areas, pipes may be both perennial and ephemeral. There they can contribute significantly to *storm* **runoff** and may be important sites for the extension of *gullies* on blanket **peat** (Jones, 1981). (See **gully erosion**.)

(2) A vertical or subvertical tubular **orebody**. Pipes and **mantos** are often found in association, the pipes frequently acting as feeders to the **mantos**. Sometimes **mantos** pass up from bed to bed in a flat lying area by way of pipe connections, often branching as they go. [TPB/AMT]

Jones, J.A.A. (1981) *The Nature of Soil Piping: A Review of Research*. Geo Books, Norwich.

pisoid (*pisolith*) A type of carbonate *coated grain* similar to an **ooid** but larger than 2 mm in diameter. Much confusion exists over its usage and some workers restrict its use to refer to non-marine **ooids**. The origins and occurrences are broadly the same as for **ooids**. A rock composed of pisoids is called a *pisolite*. [VPW]

Peryt, T. (ed.) *Coated Grains*. Springer-Verlag, Berlin.

pitch The orientation of a line, measured as an angle from the horizontal, in a specified non-vertical plane. Compare **plunge**, which is the orientation measured in a vertical plane. A measurement of pitch must give the **strike** and **dip** of the plane of measurement, plus the angle of pitch and the **strike** direction from which the pitch angle is measured (since there are two possible directions in a given plane for the same pitch angle), see Fig. P7.

The method is useful in the field where precise measurements of angles within inclined **joint**, **foliation** or **bedding** planes are more convenient than direct measurement of the **plunge**. The plunge may be easily derived using a *stereogram*. (See **stereographic projection**.) [RGP]

pitch coal (*bituminous brown coal*) A brittle, lustrous **bituminous coal** or **lignite** having a conchoidal fracture. [AME]

pitchblende Name given to the massive variety of the mineral **uraninite** (UO_2), a major **ore** of **uranium**. [DJV]

pitchstone Hydrated, and incipiently recrystallized, silicic volcanic glass. Pitchstones typically have an irregular

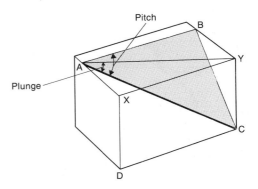

Fig. P7 Plunge and **pitch** of an inclined line: the plunge of the line AC is the angle YAC: the pitch is the angle BAC.

fracture and a dull, resinous appearance. The hydration and incipient crystallization are secondary, low-temperature effects, similar to those of **perlite**. Some pitchstones have a **spherulitic** texture. (See **glassy rock**.)

Pitchstone was used by prehistoric man in the UK for making small tools including blades and arrowheads. The main source exploited was the island of Arran in west Scotland. [PTL/OWT]

placer deposit Sedimentary accumulation of economic minerals formed by mechanical processes. Usually this natural gravity separation is accomplished by moving water, though concentration in solid and gaseous mediums may also occur. The dense or heavy minerals so concentrated must first be freed from their source rock and must possess a high density, chemical resistance to **weathering** and mechanical durability. **Placer** minerals having these properties in varying degrees include: **cassiterite**, **chromite**, **columbite**, **copper**, **garnet**, **gold**, **ilmenite**, **magnetite**, **monazite**, **platinum**, **ruby**, **rutile**, **sapphire**, *xenotime* and **zircon**. Since sulphides readily break up and decompose they are rarely concentrated into placers. There are, however, some notable *Precambrian* exceptions and a few small recent examples.

Placer deposits have formed throughout geological time, but most are of **Tertiary** and Recent age. The majority of placer deposits are small and often ephemeral as they form on the Earth's surface usually at or above the local base level, so that many are removed by **erosion** before they can be buried. Most placer deposits are low **grade**, but can be exploited because they are loose, easily worked materials which require no crushing and for which relatively cheap semi-mobile separating or hydraulic mining plants can be used. Mining usually takes the form of dredging, about the cheapest of all mining methods. Older **placers** are likely to be lithified, tilted and partially or wholly buried beneath other lithified rocks. This means that exploitation costs are much higher and then the deposits, to be economic, must contain unusually valuable

Table P1 A classification of **placer deposits**

Mode of origin	Class
Accumulation *in situ* during weathering	Residual placers
Concentration in a moving solid medium	Eluvial placers
Concentration in a moving liquid medium (water)	Stream or alluvial placers Beach placers Offshore placers
Concentration in a moving gaseous medium (air)	Aeolian placers

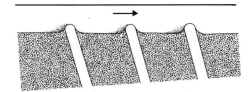

Fig. P9 Placer deposit. Quartzite ribs interbedded with slate serving as natural riffles for the collection of placer gold. (After Evans, 1993.)

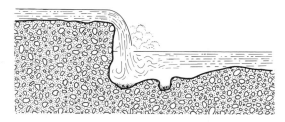

Fig. P10 Placer deposit. Plunge pools at the foot of waterfalls and potholes can be sites of heavy mineral accumulations. (After Evans, 1993.)

minerals (e.g. **gold**) or be of high **grade**. The traditional genetic classification of **placers** is given in Table P1.

Residual placers accumulate immediately above a bedrock source (e.g. **gold** or **cassiterite** vein) by the chemical decay and removal of the lighter rock materials; they may grade downwards into weathered **veins**. In *residual placers* chemically resistant light minerals (e.g. **beryl**) may also occur. *Residual placers* only form where the ground surface is fairly flat; when a slope is present, **creep** will occur and *eluvial placers* will be generated (Fig. P8). *Residual placers* formed over **carbonatites** are important as producers of **apatite** and as sources of **niobium**, **zircon**, *baddeleyite*, **magnetite** and other **minerals**.

Stream or *alluvial placers* were once the most important type of placer deposit and primitive mining made great use of such deposits. The ease of extraction made them eagerly sought after in early as well as in recent times and they have been the cause of some of the world's greatest **gold** and **diamond** rushes.

Our understanding of the exact mechanisms by which concentrations of heavy minerals are formed in stream channels is still incomplete. The concepts of fall velocity, **hydraulic equivalence**, bed configuration and grain density have all been invoked; but these concepts apply in situations where the mineral grains are being transported either in suspension or by **saltation**. In most fluvial and littoral marine situations the transport of **sand** and larger particles is largely as part of a traction carpet, in which case the important processes are (i) **entrainment equivalence** and (ii) *interstice entrapment*. What we do know is that heavy mineral concentrations are developed when

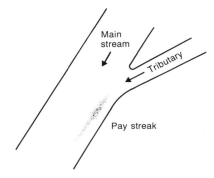

Fig. P11 Placer deposit. A pay streak may be formed where a fast-flowing tributary enters a master stream. (After Evans, 1993.)

we have irregular flow and this may occur in a number of situations — always provided a source rock is present in the **catchment** area.

The first example is that of emergence from a canyon. In the canyon itself net deposition is zero. Again, where there is fast-moving water passing over projections in the stream bed, the progress of heavy minerals may be arrested (Fig. P9). **Waterfalls** and **potholes** form other sites of accumulation (Fig. P10) and the confluence of a swift tributary with a slower master stream is often another site of concentration (Fig. P11). Most important of all, however, is deposition in rapidly flowing meandering streams. The faster water is on the outside curve of meanders and slack water is opposite. The junction of the two, where **point bars** form, is a favourable site for deposition of heavies. With lateral migration of the

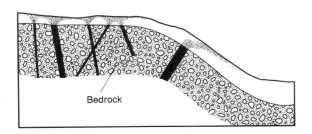

Fig. P8 The formation of residual (left) and eluvial (right) **placer deposits** by the weathering of cassiterite veins. (After Evans, 1993.)

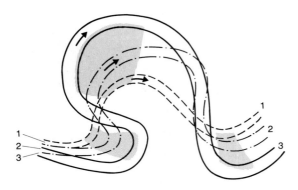

Fig. P12 Placer deposit. Formation of pay streaks (grey areas) in a rapidly flowing meandering stream with migrating meanders. 1 = original position of stream; 2 = intermediate position; 3 = present position. Note that pay streaks are extended laterally and downstream. Arrows indicate the direction of water flow.

meander (Fig. P12), a *pay streak* is built up which becomes covered with alluvium and eventually lies at some distance from the present stream channel. The middle reaches of a river are most likely to contain such placer deposits where we have well-graded streams in which a balance has been achieved between **erosion**, transportation and deposition. Gradients measured on a number of placer **gold** and **tin** deposits average out at a little under 1 in 175. Much of the world's **tin** is won from *alluvial placers* in Brazil and Malaysia.

Beach placer deposits are described elsewhere. Offshore placers occur on the continental shelf usually within a few kilometres of the coast. They have been formed principally by the submergence of alluvial and/or **beach placers** (*drowned placers*). Good examples are the **tin** placers of Indonesia. The most important *aeolian placers* have been formed by the reworking of **beach placers** by winds, e.g. the large *titanomagnetite* **iron sand** deposits of North Island, New Zealand that are estimated to contain more than 1000 Mt of *titanomagnetite*, 300 Mt of which are present in the Taharoa deposit.

Among ancient placer deposits the most outstanding examples are the **gold–uranium**-bearing *conglomerates* of the late Archean and early **Proterozoic**. The principal deposits occur in the Witwatersrand Goldfield, South Africa. The *host rocks* are *oligomic conglomerates* (**vein quartz** pebbles) having a matrix rich in **pyrite**, **sericite** and **quartz**. The **gold** and **uranium minerals** (principally **uraninite**) occur in the matrix together with a host of other heavy **minerals**.

The **orebodies** appear to have been formed around the periphery of an intermontane, intracratonic lake or shallow water inland sea at and near entry points where sediment was introduced into the **basin**. The individual mineralized areas formed as fluvial *fans* which were built up at the entry points. Each *fan* was the result of sediment deposition at a river mouth that discharged through a canyon and flowed across a relatively narrow piedmont plain before entering the **basin**. These processes led to the formation of the world's greatest goldfield, which between its discovery in 1886 and 1983, produced over 35 000 t of **gold** with an average **grade** of $10 \, \mathrm{g \, t^{-1}}$. Average mined **grades** are now below this figure. [AME]

Edwards, R. & Atkinson, K. (1986) *Ore Deposit Geology*. Chapman & Hall, London.
Evans, A.M. (1993) *Ore Geology and Industrial Minerals: An Introduction* (3rd edn). Blackwell Scientific Publications, Oxford.
Guilbert, J.M. & Park, C.F. (1986) *The Geology of Ore Deposits*. W.H. Freeman, New York.
Macdonald, E.H. (1983) *Alluvial Mining*. Chapman & Hall, London.
Slingerland, R.S. & Smith, N.D. (1986) Occurrence and origin of water-laid placers. *Annual Review of Earth and Planetary Sciences* **14**: 113–47.

plagioclase The name given to the series of **feldspar minerals** with compositions between $NaAlSi_3O_8$ and $CaAl_2Si_2O_8$. [DJV]

plagionite ($Pb_5Sb_8S_{17}$) A rare **ore** mineral of **lead**. [DJV]

plancheite ($Cu_8(Si_4O_{11})_2(OH)_4 \cdot H_2O$) A blue secondary hydrated silicate (*inosilicate*) mineral of **copper**. [DJV]

plane bed Flat sediment bed over which sediment transport is occurring but which is essentially featureless, any bed irregularities being less than 2–3 grain diameters in height although they may be many diameters in length. The term may therefore be applied to **sand** grade sediments in both lower and upper **flow regimes** as well as to **gravel** beds in which **particle clusters** are developed. (See **lower stage plane beds**, **upper stage plane beds**, **primary current lineation**.) [JLB]

plane strain A shape change (**strain**) in three dimensions in which the intermediate **principal strain** is equal to 1, i.e. there is no net **extension** or **contraction** in the intermediate **principal strain** direction. This type of **strain** distribution has been extensively studied as it can be realistically dealt with in two dimensions, in the plane containing the maximum and minimum **principal strains**. Approximate plane strain is one of the conditions necessary for **strain** analysis using **balanced sections**. [SB]

planèze A wedge-shaped **lava flow** on the slopes of a dissected volcano that protects the underlying volcanic core from **erosion**. [NJM]

plasticity **Deformation** causing permanent, continuous **strain** that does not involve **brittle failure** or significant change in total volume. In a plastic material, any **stress** above a critical value known as the *yield stress* causes continuous, permanent **strain**.

The existence of a positive *yield stress* distinguishes plastic behaviour from fluid **flow**. At **stresses** below this

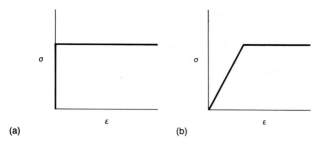

Fig. P13 Plasticity. Stress (σ) and strain (ε) relationships for (a) rigid-plastic behaviour and (b) elastic-plastic behaviour.

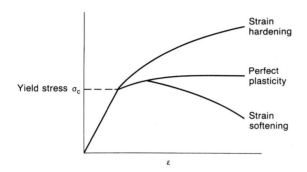

(a)

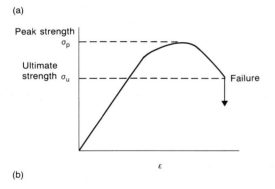

(b)

Fig. P14 Definition of some of the terms used in **plasticity** theory.

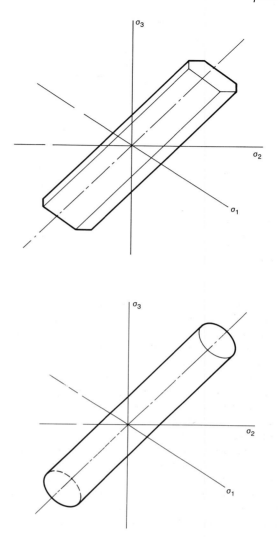

Fig. P15 Plasticity. Yield surfaces of the Tresca and Von Mises criteria in coordinates of principal stress (σ_1, σ_2, σ_3).

value, the material is *rigid-plastic* if no **deformation** occurs; commonly, however, there will be **elastic deformation** and the material can be described as *elastic-plastic* (Fig. P13). Many materials depart from perfect **plasticity** in that progressive **strain** is accompanied by either an increase or decrease in **strength**. An increase in **strength** is known as *strain hardening* or *work hardening*; a decrease as **strain softening** or *work softening* (Fig. P14), which may be either a linear or non-linear function. The slope of the stress–strain curve gives the *coefficient of work hardening* or *work softening*. In many materials, only a limited amount of plastic **strain** may occur before **failure**; the **stress** at which this occurs is the *ultimate strength* (Fig. P14).

The onset of **plasticity** is given by a *yield criterion*, which is the relation between the *principal stresses* at the yield point. On a diagram showing the three *principal stresses* along each of three perpendicular axes, the *yield criterion* can be represented by a *yield surface*, defining the **stress** conditions at which yield occurs (Fig. P15). The criterion is often expressed in terms of the *stress invariants* (see **stress**). For an isotropic substance, the *yield criterion* is symmetrical about the line $\sigma_1 = \sigma_2 = \sigma_3$. Two common *yield criteria* have been proposed.

1 The *Tresca criterion*. Yield occurs at a limiting value ($\frac{1}{2}\sigma_0$) of the maximum **shear stress**. This can be expressed as:

$$\sigma_1 - \sigma_3 = \sigma_0$$

The *yield surface* for the *Tresca criterion* is a hexagonal cylinder with the line $\sigma_1 = \sigma_2 = \sigma_3$ as its axis (Fig. P15); it is independent of the *hydrostatic stress* ($\sigma_1 + \sigma_2 + \sigma_3$)/3. The criterion predicts yield **stress** of σ_0 in uniaxial compression and tension, which is suitable for metals but not rocks.

2 The *Von Mises criterion*. Yield occurs at a limiting value of the second *stress invariant*, $J_2 = \sigma_0^2/3$. This can be expressed as:

$$(\sigma_2 - \sigma_3)^2 + (\sigma_3 - \sigma_1)^2 + (\sigma_1 - \sigma_2)^2 = 2\sigma_0^2$$

Its yield surface is a circular cylinder with the line $\sigma_1 = \sigma_2 = \sigma_3$ as its axis; like the *Tresca criterion*, it is independent of *hydrostatic stress*, and also it predicts uniaxial *yield stresses* of σ_0 in compression and tension (Fig. P15). Versions of the *Tresca* and *Von Mises criteria* have been proposed in which the yield **stress** is also a function of *hydrostatic stress*; these *yield surfaces* are conical surfaces with the $\sigma_1 = \sigma_2 = \sigma_3$ as their axis.

The theory of plasticity is developed in a similar manner to the theory of **elasticity**, but with **strain rates** replacing **strains**.

The fundamental law of plasticity is that the **strain rates** ($\dot{\varepsilon}$) are proportional to the *deviatoric stress* /S/

$$/S/_1 = 2\phi\dot{\varepsilon}_1 \quad /S/_2 = 2\phi\dot{\varepsilon}_2, \quad /S/_3 = 2\phi\dot{\varepsilon}_3$$

This is one form of the *Levy–Mises equations*, assuming no volume change. The value of ϕ is determined by the *yield criterion*; it is not a physical constant analogous to **the elastic constants**. For viscous **Newtonian fluids**, the same equations also hold with the **viscosity** substituted for ϕ. The *Levy–Mises equations* can be modified to incorporate a component of elastic behaviour, in which case they become the slightly more complex *Prandtl–Reuss equations*.

Plastic **deformation** may localize on *slip lines*, which are curves showing the direction of maximum **shear stress** at any point. For a perfectly plastic material, these will also be the directions of maximum **strain rate**, and they are easily calculated from the maximum or minimum principal *stress trajectories*, to which they form an angle of 45°; there are two orthogonal sets of *slip lines*. Velocity components may change across *slip lines*; such velocity discontinuities have been suggested to mark the locus of **faults** in plastically deforming materials, and the concept has been applied with some success to explain crustal **fault** systems, for example, assuming that Asia deformed as a plastic material following collision with India. (See **indentation tectonics**.)

'Plasticity' is also used in a wider sense to mean a **deformation mechanism**. *Crystal* or *intracrystalline plasticity* is **deformation** by movement of *dislocations* by *glide* and/or *climb*. It is a thermally activated process that produces a continuous **strain** distribution at all scales above the atomic. Plastic **deformation mechanisms** are sometimes taken even more generally to refer to all mechanisms which can produce a **mesoscopically** continuous **strain** including *crystal plasticity, diffusive mass transfer*, and *grain boundary sliding*. These have also been called 'quasi- plastic' mechanisms. [TGB]

Hill, R. (1950) *The Mathematical Theory of Plasticity*. Clarendon Press, Oxford.

Jaeger, J.C. & Cook, N.G.W. (1979) *Fundamentals of Rock Mechanics* (3rd edn). Chapman & Hall, London.
Nadai, A. (1950) *Theory of Flow and Fracture of Solids*. McGraw-Hill, New York.

plate boundary/margin The edge of a tectonic *plate* is defined by the occurrence of localized lines of persistent seismic activity. Three types of boundary are recognized: (i) constructive, or accretionary, where new plates form by igneous intrusions along the crest of the *oceanic ridges*, (ii) destructive, where the oceanic *plate* is subducted into the **mantle**, i.e. at oceanic trenches (**subduction zones**), and (iii) transform boundaries, where adjacent plates move tangentially along **transform faults**. [DHT]

Le Pichon, X., Francheteau, J. & Bonnin, J. (1973) *Plate Tectonics*. Elsevier, Amsterdam.

plate tectonics The recognition in the late 1960s that most (*c.* 80%) of the Earth's seismic activity was concentrated in narrow, continuous zones, commonly less than 10 km wide along the crest of the *ocean ridges* and **transform faults**, although much more diffuse within **subduction zones**, indicated that vast areas of the Earth's surface were in relative motion to each other. The relative motions of these areas were also indicated by **focal mechanism solutions** which indicated tensional **faulting** on the *oceanic ridges*, lateral motion along the **transform faults** between the *oceanic ridges*, and complex motions in **subduction zones**, within which discrete patterns of seismic activity, the **Benioff–Wadati zones**, could be recognized. This distribution pattern and first motions, combined with the dating of the ocean floor by means of *oceanic magnetic anomalies*, demonstrated that some 10 to 12 major **lithospheric** *plates* could be defined (Fig. P16), with numerous microplates within present and past **subduction** and *continent–continent collision* zones in particular. The *plates* within the Pacific comprise entirely oceanic **lithosphere**, with all other *plates* including both continental and oceanic **lithosphere**, but all partaking in rigid body motion relative to each other, i.e. exhibiting high torsional rigidity. The relative motion of the *plates* was found to be consistent with the arguments for **continental drift** and **continental reconstruction** if the motions were combined with the generation of new oceanic **lithosphere** at the *oceanic ridges*, and its destruction within **subduction zones** (Fig. P17). It was also recognized that most of the world's **calc-alkaline** volcanic activity could similarly be linked with **plate margins**, while the accretionary boundaries, at the *ocean ridge* crests, comprised mainly **basaltic** eruptions. However, not all volcanic activity is confined to *plate margins*. The lateral extent of *plates* is thus well defined by seismic activity (although sometimes difficult to determine precisely in **subduction zone** regions) and consistent with the distribution of volcanoes, but the definition of the vertical extent of a *plate* is still disputed and various definitions exist based on different Earth models. Virtually all models define the base of the **lithospheric** *plate* as the **brittle–ductile transition** zone, and most models assume that this is located where the *geothermal gradient* approaches the melting point of the **mantle** rocks so that *partial melting* occurs, giving rise to the seismic **low velocity zone**, which is thus equivalent to the **asthenosphere**. However, the **brittle–ductile transition** zone must occur some 200° or more below the melting temperature of the **mantle** rocks as the onset of **plastic deformation**, particularly on the scale of the **mantle**, will occur well below the actual melting point. Similarly, the decrease in seismic velocity can be accounted for by other explanations than *partial melting*. The thickness of the continental *plate* is determinable from the depth at which consistent volcanic activities are generated, e.g. the **kimberlite** pipes that characterize the edge of the **Archean cratons** are generated from depths of at least 250 km and the consistent location of this form of volcanic activity demonstrates that at least this thickness of **lithosphere** must travel with the **continental crust**. Similarly, depleted **heat flow** from the **mantle** beneath the **continental crust** requires depletion in radiogenic heat-producing elements within the uppermost 300–350 km of the continental **lithosphere** beneath **Archean** terrains, which is consistent with the absence of the seismic **low velocity zone** in such areas. The physical definition of the thickness of the **lithosphere** is thus not clearly established and this is reflected in the uncertainty

concerning the relative importance of the different **forces acting on plates**. It is widely considered that **slab pull**, supplemented by **ridge push**, are the most significant, but these assume that the **lithosphere** is mechanically decoupled from the **asthenosphere**, while **mantle drag** forces could be predominant if the **lithosphere** and **asthenosphere** are mechanically coupled.

The motions of individual *plates* can be determined directly using **geodetic** observations, particularly the *Global Positioning System* and **very long baseline interferometry**, and, over geological time, by analysis of linear *oceanic magnetic anomalies*, **paleomagnetism**, *paleoclimatology* and geological matching between continental units. Such methods allow the definition of the **instantaneous plate motion**, in some instances, but more commonly define the **finite plate motion**. These methods also indicate that the continental part of tectonic *plates* has been in motion, moving relative to the Earth's axis of rotation (based on *paleoclimatic* analyses) and to the **paleomagnetic pole** (based on **paleomagnetic** analyses). Such continental *plates* are not sufficiently dense to be subducted, giving rise to the **permanence of the continents**, but it is unclear to what extent **continental splitting** occurred in **Proterozoic** times, and it is conceivable that most of the continents formed a single unit, **Pangea**, and that inter-*continental collision* was thus inhibited, particularly in the early **Proterozoic**. However, the fact that **continental drift**

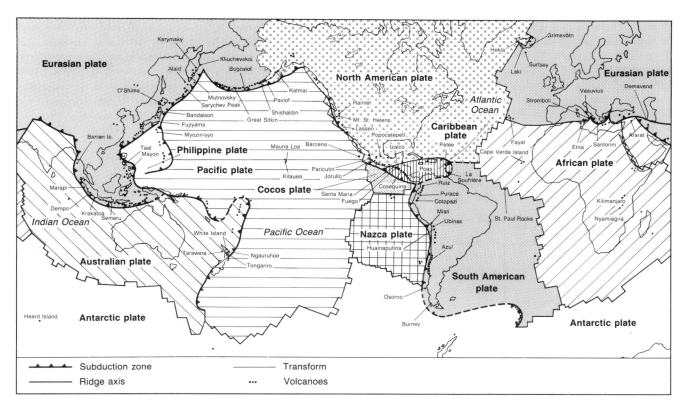

Fig. P16 Plate tectonics. The active volcanoes of the world are not distributed randomly on the Earth's surface; they tend to be associated with plate boundaries. (After Press & Siever, 1982.)

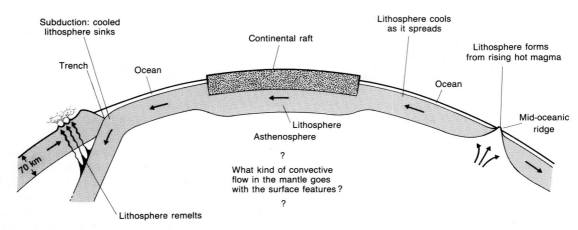

Fig. P17 Plate tectonics. The motion of plates, spreading from mid-ocean ridges and sinking in subduction zones, is the surface manifestation of convection currents in the interior. The nature of the flow in the interior is uncertain. (After Press & Siever, 1982.)

occurred suggested that plate tectonic forces were operating, but not necessarily causing *continent–continent collisions* until later in the **Proterozoic**. It appears that plate tectonic activity has therefore been continuous throughout geological time on the Earth, and there is some evidence for possible plate tectonic activity on Mars, but not on the Moon.

Although the term 'plate tectonics' has structural connotations, the fact that most geological processes can be better understood in terms of this model means that it is commonly applied to non-tectonic features, such as biogeographic reconstructions. In this sense, the term applies to a model of the evolution of the Earth's surface and hence has implications for all surface features, including the biosphere, atmosphere and oceans. It is also probable that the surface expression of plate tectonic events may have been different at different times, reflecting changes in the continental *geothermal gradient* and hence changes in the **lithospheric** thickness. [DHT]

Bird, J.M. (1980) (ed.) *Plate Tectonics*. American Geophysical Union, Washington, DC.
Burke, K. & Sengör, A.M.C. (1979) Review of plate tectonics. *Reviews of Geophysics and Space Physics* **17**: 1081–90.
Condie, K.C. (1989) *Plate Tectonics and Crustal Evolution* (2nd edn). Pergamon, Oxford.
Kearey, P. & Vine, F.J. (1990) *Global Tectonics*. Blackwell Scientific Publications, Oxford.
Le Grand, H.E. (1988) *Drifting Continents and Shifting Theories*. Cambridge University Press, Cambridge.
Le Pichon, X., Francheteau, J. & Bonnin, J. (1973) *Plate Tectonics*. Elsevier, Amsterdam.
Press, F. & Siever, R. (1982) *Earth* (3rd edn). W.H. Freeman, San Francisco.

plateau Any relatively flat area of large extent at high elevation. It may be bounded by steep slopes falling away or rising to mountain ridges, is often deeply dissected, and may be of **tectonic**, volcanic or residual origin. [NJM]

plateau lava One of a number of **lava flows**, collectively of great volume, which cover earlier topography and create a new, nearly flat, and areally extensive plateau. The **lavas** (also called *flood lavas*) form **plateaux** larger than 1000 km^2, and up to about 300 000 km^2 in area. Most lava plateaux consist overwhelmingly of **basalt**, usually **tholeiitic** in composition. The well-known early **Mesozoic** Karoo province, Southern Africa (Erlank, 1984) is typical of lava plateaux. It covers 140 000 km^2 and consists dominantly of **tholeiitic basalt**, with some **rhyolite** and **dacite** lavas (which give the province a compositional bimodality with respect to **silica** content). There are rare associated **alkaline igneous rocks**. The **lavas** were erupted during an episode of **lithospheric** tension during the break-up of **Gondwanaland**. Other **basalt**-dominated plateau lava provinces include the Deccan, India, and Columbia River Plateaux, Washington and Oregon, and the Parana, Brazil.

The **Miocene** plateau **phonolite** lavas associated with the early stages of development of the Kenya **rift** valley originally covered about 100 000 km^2, and appear to be the only example of a large lava plateau dominated by salic and/or undersaturated lavas (Williams, 1982).

The only other volcanic rocks that can be extruded in sufficient volume, and that are capable of forming plateaux, are **ignimbrites** (Macdonald, 1972). Volumes of individual **ignimbrites** associated with formation of **calderas** may exceed 1000 km^3. Successive eruption of **ignimbrites** from closely-spaced or overlapping **calderas** may cover areas of up to about 20 000 km^2 with flat-lying, gently-dipping **pyroclastic rocks**. Such deposits may swamp previous topography, and produce new, gently-dipping or nearly flat plateaux, variously called *ignimbrite plains, ignimbrite plateaux, ash-flow plains*, or *ash-flow fields* (e.g. Smith, 1960; Macdonald, 1972). Because of the unconsolidated nature of much of this **pyroclastic** material, *ignimbrite plains* are more susceptible to **erosion** than lava plateaux. There are many large examples of *ignimbrite plains* in the western United States (e.g. Yellowstone, Wyoming, and the Bishop Tuff, California) and in the Andes (Baker, 1981). The central part of the North Island of New Zealand can be considered to be an *ignimbrite plain*, and there are many smaller examples wherever **caldera**-related **ignimbrites** have been erupted. (See **shield volcano**.) [PTL]

Baker, M.C.W. (1981) The nature and distribution of upper Cenozoic ignimbrite centres in the central Andes. *Journal of Volcanology and Geothermal Research* **11**: 293–315.

Erlank, A.J. (ed.) (1984) *Petrogenesis of the Volcanic Rocks of the Karoo Province*. Special Publication of the Geological Society of South Africa **13**.

Macdonald, G.A. (1972) *Volcanoes*. Prentice-Hall, Englewood Cliffs, NJ.

Smith, R.L. (1960) Ash flows. *Geological Society of America Bulletin* **71**: 795–842.

Williams, L.A.J. (1982) Physical aspects of magmatism in continental rifts. In: *Continental and Oceanic Rifts*, pp. 193–222. Geodynamics Series 8. American Geophysical Union.

plateau uplift Mechanism that produces the large uplifted **plateau** regions of the **continental crust**. Major **plateaux** are many hundreds of kilometres across and about 2 km in height. They are normally in **isostatic** equilibrium since their excess height is compensated either by thickened **crust** or by thinner **lithosphere**. Possible mechanisms for the creation of such structures include magmatic emplacement or *underplating*, **tectonic** thickening, thermal heating causing volume increase, and metamorphic changes in the lower **crust** involving hydration or phase changes. Some **plateaux** are clearly associated with thermal and related igneous activity, for example the Ethiopian and East African **plateaux** situated on the African **rift** system. Others are more plausibly attributed to **tectonic** thickening (e.g. the Tibetan Plateau). The former are related to **extensional** tectonic regimes, the latter to compressional. [RGP]

Park, R.G. (1988) *Geological Structures and Moving Plates*. Blackie, Glasgow.

platinum (Pt) A very rare native metal **ore** mineral. (See **native element**.) [DJV]

playa lake (*salina*) A part of a saline lake depositional complex comprising a shallow, continental, **brine** lake, which may be ephemeral. Such lakes are also described as **alkali lakes** when wet, or, when dry, as playa, *dry lake*, *alkali flat*, **salt flat** or inland **sabkha**. These lakes form in enclosed inland drainage basins and are part of a larger depositional complex (Fig. P18) which includes **alluvial fans**, sandflats, mudflats, aeolian **dune** fields and ephemeral stream **floodplains** in addition to the lake itself. Modern ephemeral saline lakes may extend over an area as large as $8000 \, km^2$ at maximum stand, as in the case of Lake Eyre, Australia, or may be less than $1 \, km^2$. Although many modern playa lakes have a perennial fluviatile input, for example **Great Salt Lake**, major lake **recharge** is by surface **runoff** during infrequent heavy storms and, in many lakes, by **groundwater springs** (Fig. P18).

Within the drainage basin, coarse detritus is trapped within the **alluvial fans** and surface water movement is restricted to few ephemeral or perennial streams and rivers. **Sheetfloods**, during *storms*, sweep clastic **sands** and, away from the **alluvial fans** themselves, clays onto the exposed mud flats. These sediments become cemented by carbonate (here **precipitating** as an **evaporite** mineral) or **gypsum**, produced in part from the evaporating flood

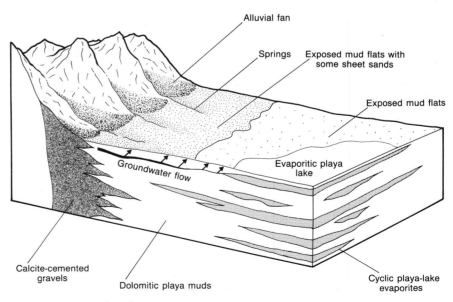

Fig. P18 Depositional complex of **playa lakes**. (After Eugster & Hardie, 1978 and Kendall, 1984.)

Alluvial fan

Springs

Exposed mud flats with some sheet sands

Exposed mud flats

Groundwater flow

Evaporitic playa lake

Calcite-cemented gravels

Dolomitic playa muds

Cyclic playa-lake evaporites

waters but mostly from already-concentrated **ground-waters** (Fig. P18). These **groundwaters** may be drawn to the surface by capillary rise, **evaporative pumping** or by plant root systems, tapping **groundwater** some depth below the suface; **springs**, or a seasonally higher **water table**, may directly tap the **water table**. Where no **groundwater** discharge occurs at the surface, **evaporites** are rare.

The **evaporite** minerals which form within, and around, a playa are a reflection of the **groundwater** composition and the **groundwater** source of that region. Thus, although many continental **groundwaters** are poor in sulphate, where modern **groundwaters** are derived from **dissolution** of ancient **gypsum/anhydrite** formations, playa lakes **precipitate gypsum** which may be reworked during lake lowstands to produce aeolian **gypsum dunes** (Fig. P19). The geographical and geological setting of the drainage basin, together with its size, will also control the amount and extent of playas, and hence the magnitude of **evaporite** formation. Steep-sided basins with perennial lakes (e.g. **Dead Sea**) have restricted extent and rarely become ephemeral; in contrast, wide basins contain shallow perennial lakes (e.g. **Great Salt Lake**) which may become ephemeral during drought and produce large expanses of **evaporite** crusts, albeit surrounded by wider areas of mudflats. Climate is also important, both because drought may produce **evaporites** and because a subsequent **pluvial** period may result in **dissolution** of former **precipitated evaporites**, thus enhancing the salinity of the playa lake. This continued **dissolution** of **precipitated** salts means that playa **evaporites** have a low preservation potential; indeed, continental **evaporites**, particularly before the *Tertiary*, are rare in the geological record. [GMH]

Eugster, H.P. & Hardie, L.A. (1978) Saline lakes. In: Lerman, A. (ed.) *Chemistry, Geology and Physics of Lakes*, pp. 237–93. Springer-Verlag, New York.
Kendall, A.C. (1984) Evaporites. In: Walker, R.G. (ed.) *Facies Models* (2nd edn), pp. 259–96. Geoscience Canada Reprint Series 1.

Playfair's law The assumption, following Playfair (1802), that fluvial junctions will be 'accordant', i.e. both river channels and their valleys will form confluences which join without any abrupt change in elevation between them. Playfair (1802, p.102) states: 'Every river appears to consist of a main trunk, fed from a variety of branches, each running in a valley proportional to its size, and all of them together forming a system of vallies communicating with one another, and having such nice adjustment of their declivities, that none of them join the principal valley, either on too high or too low a level; a circumstance which would be infinitely improbable if each of these vallies were not the work of the stream that flows in it'. Since then, most geomorphologists have assumed that discordant junctions are therefore exceptional in some way, and represent disruption of the normal state of affairs, perhaps by **tectonic** activity or as the result of **glacial erosion**, which often causes *hanging valleys* to form. Little research has been conducted on stream junctions, however; recent work by Kennedy (1984) shows that discordant **thalwegs** are more common than Playfair's law would suppose, although discordant valleys remain uncommon where fluvial **erosion** has been dominant. [TPB]

Kennedy, B.A. (1984) On Playfair's Law of accordant junctions. *Earth Surface Processes and Landforms* 9: 153–73.
Playfair, J. (1802) *Illustrations of the Huttonian Theory of the Earth*. Edinburgh.

pleonaste A green variety of **spinel mineral** intermediate in composition between **spinel** and **hercynite**. (See **oxide minerals**.) [DJV]

plinthite A thick, well-cemented horizon formed by **illuvial accretion** of **ferricrete** or other hard *soil crusts*. Plinthites are often **lateritic** crusts which protect softer underlying materials from **erosion**. In arid and semi-arid environments (notable in Australia and Africa), **ferricrete** and **silcrete** plinthites form **cuirasses** over broad tracts of the landscape. [AW]

Goudie, A.S. (1973) *Duricrusts in Tropical and Subtropical Landscapes*. Oxford University Press, Oxford.
McFarlane, M.J. (1983) Laterites. In: Goudie, A.S. & Pye, K. (eds) *Chemical Sediments and Geomorphology*, pp. 7–58. Academic Press, London.

Pliocene The Pliocene epoch had a duration of *c.* 4 Ma from the end of the **Miocene** epoch at *c.* 5.4 Ma to the base

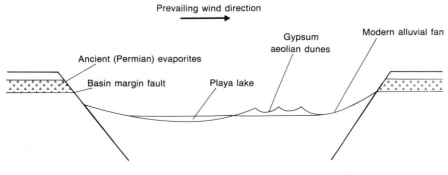

Fig. P19 Formation of gypsum-precipitating **playa lake** with gypsum aeolianite dunes from ancient evaporite formations. Based on White Sands National Park and surrounding area, New Mexico, USA.

of the **Pleistocene** at *c.* 1.6 Ma and has been divided into two stages (Berggren *et al.*, 1985). (See Fig. M14, p. 401.)

Oxygen isotope analyses have shown that the first major pulse of ice-rafted debris in the North Atlantic occurred at 2.4 Ma, and was preceded by a minor pulse at about 2.5 Ma. North Atlantic deep water, similar to that of the present day, has been established at least since *c.* 3.5 Ma (Shackleton *et al.*, 1984).

The collision between India and Asia climaxed in the building of the Himalayas and at *c.* 4.5 Ma the major Andean **uplift** raised the land to 4000 m. At *c.* 3.9 Ma the water route between the Pacific and the Atlantic became dammed in the Central American region. This promoted the development of the warm Gulf Stream current which traversed the North Atlantic. The Panama bridge became fully open to land animal passage at *c.* 3.0 Ma and northern **mammals** crossed the isthmus while many of their South American counterparts became **extinct**. The first **oceanic crust** evolved in the Red Sea at *c.* 3.5 Ma. In the North Atlantic the cold Labrador Current began transporting cold water southwards along the east coast of North America at *c.* 5 Ma. The Great Apes of Africa divided into parties that were the ancestors of the gorillas and chimpanzees. The gorillas evolved in the dense forest, the chimpanzees in the more open forest; baboons evolved having adapted to the open country and using trees only for refuge. At *c.* 4.5 Ma modern camels, bears and pigs evolved. At *c.* 3.7 Ma the modern horse appeared in North America. The first small stone tools made by an unknown human hand date from 2.4 Ma, from eastern Africa. As forest changed to open savannah, fleet-footed grazing animals flourished. Modern big cats, bison, sheep and wild dogs evolved at *c.* 1.8 Ma. Most fossil marine invertebrates are barely distinguishable from modern species.

In the oceans calcareous nannofossil diversity gradually decreased throughout the Pliocene culminating in the **extinction** of the *discoasters* at *c.* 1.9 Ma. Nannofossil size decreased in the Pliocene; Late Pliocene **coccolith** flora are dominated by small (<6 μm) placoliths. The first species with slits in its shield evolved in the Pliocene, while the planktonic foraminiferal faunas took on their modern appearance and were distributed in latitudinal belt faunal provinces.

Hominid evolution became diversified with *Australopithicus* living between *c.* 4 Ma and 1 Ma in the Early **Pliocene** and both *Homo habilis* and *Homo erectus* appeared in the Late Pliocene at *c.* 2.5 Ma and *c.* 1.7 Ma respectively. [DGJ]

Berggren, W.A., Kent, D.V. & van Couvering, J.A. (1985) The Neogene: Part 2 Neogene geochronology and chronostratigraphy. In: Snelling, N.J. (ed.) *The Chronology of the Geological Record*, pp. 211–60. Memoirs of the Geological Society No. 10. Blackwell Scientific Publications, Oxford.

Shackleton, N.J., Backman, J. & Zimmerman, H. *et al.* (1984) Oxygen isotope calibration of the onset of ice-rafting and history of glaciations in the North Atlantic region. *Nature* **307**: 620–3.

plis de couverture Folds of a cover sequence above a basal **detachment horizon** or **décollement**. **Folds** in the basement below the **décollement** horizon are termed '*plis de fond*'. [RGP]

plume structure A radial structure formed by surface relief on a **fracture** surface. Plume structure is common on the **fracture** surfaces of many fine- and intermediate-grained rocks, and has been recently classified into three types (Fig. P20):
1 S (straight)-type plumes, consisting of crescentic shoulders or barbs disposed symmetrically about an axis commonly along the centre of a bed.
2 C (curved)-type plumes consisting of wavy barbs radiating from a single point.
3 Rythmic C-type plumes, consisting of cycles of fans C-type plumes. Plume structure originates during **fracture propagation**; the direction of propagation is from the convex to the concave side of the barbs. Synonyms occasionally used are *herringbone structure*, *feather structure* and *hackle mark*. [TGB]

Bahat, D. & Engelder, T. (1984). Surface morphology on joints of the Appalachian plateau, New York and Pennsylvannia. *Tectonophysics* **104**: 219–313.

plunge The orientation of a linear structure (e.g. *fold axis*) measured as the angle between the line and the horizontal in the vertical plane. The plunge is given as an angle and a bearing (azimuth), which is the direction of plunge, thus 30° to 45° or 30° north-east. Compare **pitch**. [RGP]

plus-minus method (*Hagedoorn method*) A **seismic refraction** interpretation method using **reversed profiles** to delineate refractors which vary laterally both in their depth from the surface and velocity. If the **reciprocal time** for a **head wave** to travel from one **shot** to the other on a **reversed profile** is t_r, and t_{g_1} and t_{g_2} are the **arrival times** of the **head wave** from the same refractor at an intermediate **geophone** location s_g from the two **shots**, then the minus times are defined as $t_{g_1} - t_{g_2} - t_r$, whilst the plus times are $t_{g_1} + t_{g_2} - t_r$. The minus times are plotted against distance and define the true refractor velocity corrected for lateral changes in depth. The plus times relate to depth variations of the refractor from the surface and are converted to depth by multiplying them by a factor equal to

$$\frac{1}{2}\left[\frac{v_0}{\sqrt{\left(1 - \frac{v_0^2}{v_r^2}\right)}}\right]$$

where v_0 is the overburden velocity and v_r the velocity of the refractor. [GS]

Hagedoorn, J.G. (1959) The Plus-Minus method of interpreting seismic refraction sections. *Geophysical Prospecting* **7**: 158–82.

pluton A large, relatively thick body of magmatic rock with steep contacts which has been emplaced and crystal-

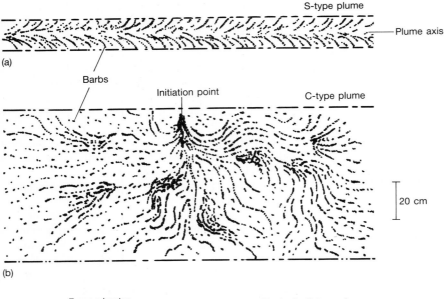

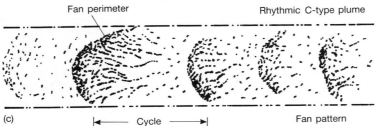

Fig. P20 Various **plume structures**.
(a) Straight plume. (b) Curving plume.
(c) Rhythmic plume. Barbs on each pattern
are individual traces or fine lines within the
plume that mark the direction of local fracture
propagation. Fan perimeters in (c) designate
loci of arrest lines. They are convex toward
the direction of fracture propagation. (After
Bahat & Engelder, 1984.)

lized beneath the surface of the Earth, and now exposed
with irregular polygonal outcrop (cf. minor intrusions such
as **dyke**, **laccolith**, **lopolith** and **sill**). A **batholith** is a
multiple intrusion composed of many plutons. Plutons
may show sharp or diffuse concordant or discordant contacts with the surrounding, *country rock*, and are rarely
homogeneous in composition. Internal variation may result
from **wall rock assimilation**, from *differentiation* processes
(see **magmatic diversification**) or from compositional inhomogeneities in the source region for the different **magmas**. A common feature of plutons of **diorite**, **granodiorite**
and **granite** composition is the occurrence of inclusions of
rock that differ in composition and/or **fabric** from the *host
rock*. These also may have sharp or diffuse margins and
include **xenoliths** of *country rock* and concentrations of
early-formed crystals from the **magma** itself; these are
termed *autoliths*.

The depth of emplacement of plutons ranges from subvolcanic locations just below the surface to near the base
of the **crust**. The features of plutons emplaced at varied
depths are varied. High-level plutons show sharp, locally
discordant contacts against unmetamorphosed or weakly-
metamorphosed sedimentary and volcanic rocks, and are
associated with minor intrusions (see above) and extrusive
rocks. By contrast, plutons emplaced at a deep level within
the **crust** may show diffuse, often concordant, contacts
against metamorphic rocks (cf. *migmatites*), and may be

characterized by metamorphic **fabrics**. High-level plutons
are termed post-**tectonic** in nature, whereas low-level
plutons are often emplaced in association with **orogenic**
activity so may be termed syn-**orogenic**. [RST]

pluvial (*lacustral*) A time of greater moisture availability
caused by increased precipitation and/or reduced evapotranspiration levels. Pluvials caused lake levels to rise,
river systems to be integrated, and **groundwater** to be
recharged in the arid and seasonally humid tropics in the
Pleistocene and early *Holocene*. Pluvials used to be equated
in a simple temporal manner with *glacials*, but this point of
view is no longer acceptable, for in low latitudes one of
the most important pluvials occurred in the *Holocene*
interglacial at around 9000 years ago. [ASG]

Street, F.A. (1981) Tropical palaeoenvironments. *Progress in Physical
Geography* **5**: 157–85.

poikilitic texture An inequigranular texture observed in
gabbroic and **ultramafic rocks** in which a single large
mineral grain, termed an *oikocryst*, encloses several
randomly oriented smaller grains of another phase or
phases. **Ophitic texture** is a special case of poikilitic
texture. Such texture may result from sparse nucleation of
a phase in the interstices of earlier-formed crystals which
are then enclosed by the growing *oikocryst*, or by dis-

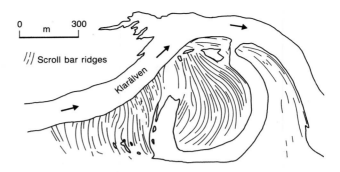

(a)

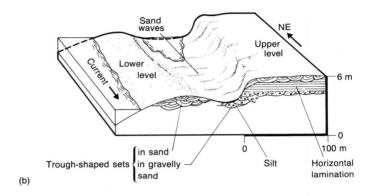

(b)

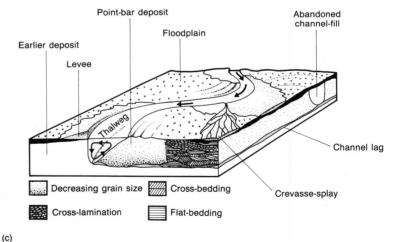

Fig. P21 (a) **Point bar** in plan view showing accretion topography of scroll bars showing former positions of point bar in River Klarälven, Sweden. (After Sunborg, 1956.) (b) Point bar with benches, showing the well developed stepped profile, and associated internal structures observed in trenches, Beene point bar, Red River, Louisiana. (After Harris *et al.*, 1963.) (c) The classical point bar model of a meandering stream. (After Allen, 1964.)

(c)

ruption and invasion of earlier-crystallized minerals by a melt which then crystallizes to form *oikocrysts*. [RST]

point bar A channel bar that may range in composition from mud to coarse *conglomerate* occurring on the convex side of a channel bend. Deposition on the point bar results from conditions of reduced flow in a helical system or **flow separation**. The form of the point bar depends upon the sediment calibre, channel dimensions and discharge characteristics. Sedimentary surfaces dip towards the channel and may be gradual slopes or stepped surfaces. The grain size varies such that there is a general reduction in grain size in the downstream direction and with height above the base of the channel. This grain size variation is frequently associated with changes in bedform type. Migration of channel meanders by **erosion** of the concave bank and deposition on the point bar leads to lateral migration structures known as **epsilon cross-stratification** within the point bar (Fig. P21). [JA]

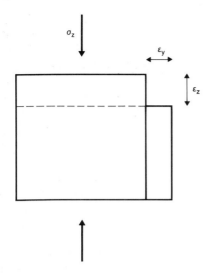

Fig. P22 Poisson's ratio $\nu = -\varepsilon_y/\varepsilon_z$.

Allen, J.R.L. (1964) Studies in fluviatile sedimentation: six cyclothems from the Lower Old Red Sandstone, Anglo-Welsh basin. *Sedimentology* **3**: 163–98.

Bridge, J.S. (1977) Flow, bed topography and sedimentary structure in open channel bends: a three-dimensional model. *Earth Surface Processes* **2**: 401–416.

Harris, J.C., MacKenzie, D.B. & McCubbin, D.G. (1963) Stratification in modern sands of the Red River, Louisiana. *Journal of Geology* **71**: 566–80.

Smith, D.G. (1987) Meandering river point bar lithofacies models: Modern and ancient examples compared. In: Ethridge, F.C. & Flores, R.M. (eds) *Recent Developments in Fluvial Sedimentology*. Special Publication of the Society of Economic Paleontologists and Mineralogists **39**: 83–91.

Sundborg, A. (1956) The River Klarälven: A study of fluvial processes. *Geografiska Annaler* **38**: 127–316.

Poisson's equation The relationship between **gravitational** and **magnetic potentials** arising because both are controlled by the inverse of the distance to the attracting object. The equation allows the transformation of **magnetic fields** into **gravity** fields and vice versa, which can be of use in the combined interpretation of **magnetic** and **gravity anomalies**. [PK]

Poisson's ratio The ratio between a lateral **strain** (ε_y) and a longitudinal **strain** (ε_z) in an elastic body due to a uniaxial longitudinal **stress** (σ_z). The **stress** and **strains** on a rectangular body are shown in Fig. P22.

Poisson's ratio $\nu = -\varepsilon_y/\varepsilon_z$

This ratio is an important material property, which is between 0.2 and 0.3 for many rocks and has a value of 0.5 for incompressible materials and liquids. The value is 0.25 for a *Poisson solid* — a homogeneous elastic solid for which the *shear modulus* and *Lamé constants* are equal. The reciprocal of Poisson's ratio is known as *Poisson's number*. Poisson's ratio is one of the fundamental elastic

properties of a material, and therefore is relevant to all aspects of **elastic deformation** and **seismic wave** propagation.

The ratio of compressional *P wave* velocity to shear *S wave* velocity is given by

$$\sqrt{\frac{2(1 - \sigma)}{(1 - 2\sigma)}}$$

[TGB/GS]

Jaeger, J.C. & Cook, N.G.W. (1979) *Fundamentals of Rock Mechanics* (3rd edn), p. 110. Chapman & Hall, London.

polar wander The Earth's axis of rotation precesses in space with a periodicity of some 25 700 years. It also shows fluctuations in its precise location on the Earth's surface by up to about 3 m, most of which are termed the **Chandler wobble** and have a characteristic time constant of some 40 years. There has also been a net motion of the pole, during the last 80 years, of some 6 m (3.5 marcs a^{-1}) towards Newfoundland and is attributed to changes in sea-level. For earlier periods of time, there is clear evidence from *paleoclimatic* indicators that the Earth's climatic pole changes, relative to any given tectonic *plate*, and an identical change is shown by the **paleomagnetic pole** positions for similar times from the same *plate*. These motions are termed polar wander, but it is not yet established whether such motions are entirely attributed to motions of the *plates* relative to the rotational pole, or whether a small component of **true polar wander** is also involved. [DHT]

Goldreich, P. & Toomre, A. (1969) Some remarks on polar wandering. *Journal of Geophysical Research* **74**: 2555–67.

Lambeck, K. (1980) *The Earth's Variable Rotation*. Cambridge University Press, Cambridge.

polar wobble The Earth's axis of rotation is approximately fixed in space, but shows both regular and irregular changes. The main systematic change is the annual wobble with an amplitude of *c.* 100 marcs associated with both planetary gravitational fields and changes in oceanic and atmospheric mass distributions. Its amplitude is very similar to that of the **Chandler wobble**, which has a 430 day periodicity and appears to be excited by unknown forces at this time. Another wobble, the *Markowitz wobble*, has a 30 years periodicity and an amplitude of 25 marcs, with a retrograde motion, but this does not appear to decay and is still of unknown origin. [DHT]

Lambeck, K. (1980) *The Earth's Variable Rotation*. Cambridge University Press, Cambridge.

polarity epoch A period of constant **geomagnetic field** polarity, usually much greater than 10 000 years. (A *geomagnetic event* or *excursion* corresponds to a short period of constant polarity, usually lasting less than 10 000 years.) [DHT]

polder A flat, low-lying area of land which has been reclaimed from the sea or a lake by artificial drainage. In The Netherlands, there are large tracts of polderland which are below sea-level. Much of the former Zuider Zee has been reclaimed by enclosing large areas behind dykes and draining them by pumping out the water — thereby creating a polder. Once the salinity of the polder sediments has been reduced, fertile *soils* develop and the land can be brought into agricultural production. [AW]

pole figure *Stereogram* showing the distribution of crystallographic orientations, usually in the form of poles to crystallographic planes. The distributions are usually shown on an *equal-area plot* in order that statistical analysis of the distribution may be made. **Quartz** and **calcite** *c*-axes are commonly analysed in this way. (See **stereographic projection**.) [RGP]

pole of a plane The normal to a plane; i.e. the line to which all lines in a plane are perpendicular. The pole is used in **stereographic projection** as a more convenient way of plotting planes, especially where there are a large number of measurements. (See **pi (π) diagram**.) [RGP]

polje A large, usually flat-floored, closed depression found in **karst** areas. The term comes from the Slav word for field, and in many cases poljes provide very good agricultural land. They are usually covered with alluvium and streams may flow across them and sink into **ponors**. In many poljes periodical flooding occurs forming shallow lakes over the polje floor. There are many different types of polje found within the Yugoslavian **karst** (famous examples are Planina polje and Popovo polje) and there is, accordingly, much debate over how they form. [HAV]

Gams, I. (1978) The polje: the problem of definition. *Zeitschrift für Geomorphologie* 22: 170–81.

pollucite (CsAlSi$_2$O$_6$. H$_2$O) A rare **feldspathoid mineral** found in *pegmatites*. [DJV]

poloidal field The **geomagnetic field** as determined at the Earth's surface is a poloidal field, i.e. it has both radial and tangential components. Compare with **toroidal field**. [DHT]

Parkinson, W.D. (1983) *Introduction to Geomagnetism*. Scottish Academic Press, Edinburgh.

polybasite (Ag$_{16}$Sb$_2$S$_{11}$) A rare **silver ore** mineral. [DJV]

polygonal karst **Karst** landscape produced by a closely packed assemblage of closed depressions, the divides between which form polygons. This landscape was first described by Williams (1972) from New Guinea where the residual hills along the polygonal divides are the dominant landscape features, and has since been reported from the Waitomo district of New Zealand. [HAV]

Williams, P.W. (1972) Morphometric analysis of polygonal karst in New Guinea. *Geological Society of America Bulletin* 83: 761–96.

polygonization The formation of polygonal grains or **subgrains**. Polygonization is one of the processes that contribute towards **recovery**, the restoration of a **deformed** material to an undeformed state. By forming polygonal grain or **subgrain** boundaries, the surface energy of the boundaries is reduced and a more stable, lower-energy configuration is attained. The process occurs by the *climb* and *cross slip* of **dislocations** towards the boundary; it is therefore thermally activated and can occur only at relatively high temperatures. In the resulting microstructure there is often only a small misorientation of the *crystal lattice* across the boundaries, which are straight and define regular shapes such as hexagons or rectangles in cross-section. [TGB]

polyhalite (K$_2$Ca$_2$Mg(SO$_4$)$_4$. 2H$_2$O) An **evaporite mineral** valued as a source of **potassium**. [DJV]

polymorphism The ability of a chemical compound to exist in the solid state as two or more forms with different **crystal structures**. Just as materials can generally exist as solid, liquid or gas, depending on pressure and temperature, so they can sometimes occur in more than one form as a crystalline solid. The polymorphs will have the same chemical composition but their physical properties, such as density, **refractive indices**, specific heat, **elastic moduli**, will be different.

There are numerous examples of polymorphism in minerals: **diamond** and **graphite** (C); **calcite** and **aragonite** (CaCO$_3$); **sphalerite** and **wurtzite** (ZnS); **pyrite** and **marcasite** (FeS$_2$); **rutile**, **anatase** and **brookite** (TiO$_2$), **quartz**, **tridymite**, **cristobalite**, **coesite** and **stishovite** (SiO$_2$); and **kyanite**, **andalusite** and **sillimanite** (Al$_2$SiO$_5$).

The first stage in studying polymorphism is to determine the pressure and temperature conditions under which each polymorph is stable, i.e. to obtain the phase diagram. As would be expected, the polymorphs with the higher density and higher coordination number tend to occur at higher pressures and lower temperatures. The occurrence of a particular polymorph in a rock may provide an indication of the pressure and temperature conditions under which it crystallized or was metamorphosed. For instance, **kyanite**, **sillimanite** and **andalusite** may be used as indicators of the conditions under which metamorphism occurred. **Andalusite** is stable only at pressures below about 350 MPa and its presence implies metamorphism at correspondingly shallow depth (less than about 10 km). The formation of **sillimanite** requires temperatures above about 500°C, since below that temperature it is unstable relative to **andalusite** or **kyanite** depending on the pressure. (See **geobarometry of ores, geothermometry of mineral deposits**.)

Transformation between polymorphs may occur rapidly and reversibly, as between α-**quartz** and β-**quartz** at 573°C, or it may be slow, as between **quartz** and **tridymite** at 870°C or **tridymite** and **cristobalite** at 1470°C. The slowness of the transitions allows **quartz, tridymite and cristobalite** to exist metastably for long periods and the polymorph present shows how high a temperature was reached. Polymorphic transformations will proceed in steps from the unstable form through a series of metastable states of decreasing free energy until the stable form is reached (*Ostwald step rule*). Thus colloidally **precipitated** $CaCO_3$ first forms the unstable **aragonite**. It may thus be possible to get an idea of the cooling history of a rock, i.e. the time it spent in a particular temperature range, from the progress of polymorphic transitions.

Polymorphic transformations were classified by Buerger according to the differences in **crystal structure** between the polymorphs:

1 Transformations of first coordination, i.e. changes in the number or arrangement of the nearest neighbours of an atom. In the transformation from **aragonite** to **calcite**, each calcium atom is linked to 6 oxygen atoms and each oxygen atom to 2 calcium in **calcite**, while in **aragonite** each calcium is linked to 9 oxygens and each oxygen to 3 calcium.

2 Transformations of second coordination, i.e. between polymorphs which have the same bonding to first neighbours but different relationships between second nearest neighbours. The transitions between **quartz–tridymite–cristobalite** are examples: silicon is linked to 4 oxygen atoms in all three polymorphs but they differ in the way the SiO_4 tetrahedra are linked together.

3 Transformations of disorder. The transformations between the high and low temperature forms of the **feldspars** are examples. In the high temperature forms, the **aluminium** and silicon atoms are distributed randomly over the tetrahedral sites but at low temperatures the

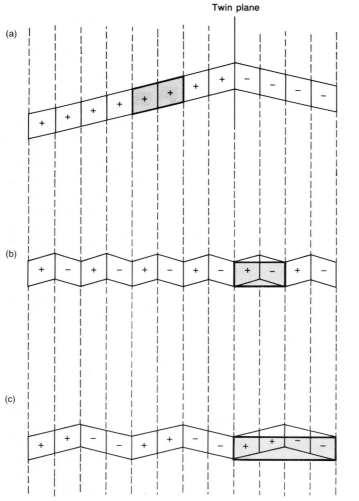

Fig. P23 Polytypism. Stacking of structural layers in the enstatite polytypes. The layers parallel to (100) are marked with dashed lines. The parallelograms represent structural units and show the direction of displacement at each layer which is also marked as + or −. The stacking sequences are (a) clinoenstatite + + + + + + . . . (or the twin-related − − − − − . . .), (b) protoenstatite + − + − + − . . . , (c) orthoenstatite + + − − + + − − . . . . In each case the unit cell is shaded.

aluminium atoms all occupy a single set of sites, the others containing only silicon.

4 Transformations of bond type, in which the nature of the bonding changes. These are usually accompanied by drastic changes in physical properties, e.g. **graphite–diamond**, metallic white **tin**–covalently bonded grey **tin**.

The kinetics of a transformation between polymorphs reflect the structural changes. *Displacive transformations* are those which occur by the expansion, distortion or rotation of coordination polyhedra without bonds being broken. The energy barrier is low and such transformations are rapid. *Reconstructive transformations* are those in which bonds must be broken and new bonds formed; the energy required to break the bonds initially is high and such transformations take place very slowly. (See also **polytypism**.) [JEC]

Megaw, H.D. (1973) *Crystal Structures: A Working Approach*, Chapter 15. W.B. Saunders, Philadelphia.

Verma, A.R. & Krishna, P. (1966) *Polymorphism and Polytypism in Crystals*. John Wiley & Sons, New York.

polytypism A particular kind of **polymorphism** in which a compound can exist with two or more layer-like **crystal structures** which differ in their layer stacking sequences. The **crystal structures** are made up of layers which are similar but need not be crystallographically identical. Polytypism differs from **polymorphism** (as strictly defined) in allowing small differences in chemical composition, strictly up to 0.25 atoms per formula unit. (For large chemical differences, the term *polytypoid* should strictly speaking be used.)

The differences between polytypes and their layer stacking sequences arise from the relationships between successive layers. A layer may be displaced in two or more different directions relative to the previous one or by different amounts in the same direction. Or a layer may be rotated by different amounts relative to the previous one, with or without changes in relative displacement.

The polytypes of the **pyroxene** $MgSiO_3$, **clinoenstatite**, *orthoenstatite* and *protoenstatite*, provide a convenient example. The **crystal structure** can be described in terms of layers parallel to (100) having a monoclinic cell. Successive layers are displaced by an irrational fraction of *c*-axis repeat and this displacement may be in either direction, denoted + or −. In **clinoenstatite**, successive layers are displaced in the same sense, the sequences +++++... and −−−−−... being twins related by reflection in a (100) plane (Fig. P23a). In *protoenstatite*, alternate (100) layers are displaced in opposite directions, i.e. the stacking sequence is +−+−+−... and the unit cell is orthorhombic (Fig. P23b). *Orthoenstatite* has the sequence ++−−++−−++... and is also orthorhombic but with the *a* cell edge doubled (Fig. P23c).

The three **enstatite** polytypes, **clino-**, *proto-* and *ortho-*, with sequences +++++..., +−+−+... and ++−−++−−++... can be used to illustrate the nota-

tions used to describe polytyes. The simplest *notation* is that of *Ramsdell* which gives the number of layers in the repeat unit of the stacking sequence and a letter to describe the symmetry of the polytype: in the **enstatite** examples, 1M, 1OR and 2OR. Letters used for the symmetry are triclinic = A (anorthic) or TC, monoclinic = M, orthorhombic = OR or O, trigonal = T, tetragonal = Q (quadratic) or TT, hexagonal = H, cubic = C.

The fuller and more flexible modified *Gard notation* describes the unit cell relationship more completely; it begins with a letter for the symmetry and then gives the polytype cell edge repeats as multiples of the single-layer cell repeats. **Clino-**, *proto-* and *ortho-***enstatite** are respectively **enstatite**-M*abc*, OR*abc* and OR2*abc*.

A complete description of polytypism requires a description of the stacking sequence and the relationships between successive layers, and these are not always the same as in the **enstatite** example. In **wollastonite** ($CaSiO_3$), successive (100) layers may be either superimposed directly (undisplaced) which we shall denote *T*, or displaced by +*b* which we shall denote *G*. The two simplest polytypes have the sequences *TTTT*... and *TGTGTG*... but other more complex sequences such as *TTGTTG*..., *TTTGTTTG*..., *TTTTGTTTTG*... and *TTGTGTGTTGTGTG*... have also been identified.

The **micas** are the classic example of polytypism in the **clay minerals**. Successive (001) layers may be displaced in one of six directions at 0, 60, 120, 180, 240 and 300° to each other, giving a variety of sequences with 1, 2, 3 or 6 layers in the repeat unit of the sequence and monoclinic, orthorhombic, trigonal or hexagonal symmetry.

Generally, polytypes of a mineral differ very little in their free energy but the activation energy required to bring about transformation may be large. Which polytype of a mineral occurs in nature can provide an indication of pressure and temperature conditions if the phase diagram is known and equilibrium has been reached. The extent to which a phase transition between polytypes has taken place and the degree of structural order may give a guide to changing P–T conditions and cooling or heating rates. (See **geobarometry of ores**, **geothermometry of mineral deposits**.) [JEC]

Bailey, S.W. (1980) Structures of layer silicates. In: Brindley, G.W. & Brown, G. (eds) *Crystal Structures of Clay Minerals and their X-ray Identification*. Mineralogical Society Monograph **5**: 1–123.

Buseck, P.R., Nord, G.L. & Veblen, D.R. (1980) Subsolidus phenomena in pyroxenes. *Reviews in Mineralogy* **7**: 117–211.

Henmi, C., Kawahara, A., Henmi, K., Kusachi, I. & Takeuchi, Y. (1983) The 3*T*, 4*T* and 5*T* Polytypes of wollastonite from Kushiro, Hiroshima Prefecture, Japan. *American Mineralogist* **68**: 156–63.

Henmi, C., Kusachi, I., Kawahara, A. & Henmi, K. (1978) 7*T* wollastonite from Fuka, Okayama Prefecture. *Mineralogical Journal* **9**: 169–81.

International Mineralogical Association (1977) Report of the International Mineralogical Association (IMA) — International Union of Crystallography (IUCr) Joint Committee on Nomenclature. *Acta Crystallographica* **A33**: 681–4.

Smith, J.V. & Yoder, H.S. (1956) Experimental and theoretical studies of the mica polymorphs. *Mineralogical Magazine* **31**: 209–35.

ponor A Slav term for a stream **sinkhole** or **swallow hole** found in **limestone** areas. Ponors are found in Yugoslavian **poljes** and **blind valleys**. [HAV]

pool and riffle sequence The development of alternating deep (pool) and shallow (riffle) sections along the course of a river channel. This is a feature of both straight and meandering channels. Pools are found on the meander bend with riffles on the straight sections between the bends. The pool is usually associated with an asymmetrical channel cross-section formed by **erosion** of the outer bank and deposition on the inner bank (the **slip-off slope**). Riffles tend to be wider and shallower at all stages of flow. Successive riffles and pools are regularly spaced at five to seven times the channel width. This rhythmical change in the bed topography of streams is thought to be related to the initiation of the meandering pattern itself. Langbein & Leopold (1966) argued that the energy loss in the shallower riffle is balanced by the higher velocity of the pool section. The overall effect is to produce a more uniform dissipation of energy through a meandering channel section compared to a straight one. These links are, however, incompletely understood since meanders can develop without pools and riffles, and some straight channels with pool and riffle sequences are evidently stable and do not form meanders. Nevertheless, it is clear that the combination of meanders and pool and riffles represents two interacting sources of flow resistance capable of modifying the rate and distribution of energy loss at the reach scale. [TPB]

Langbein, W.B. & Leopold, L.B. (1966) *River Meanders and the Theory of Minimum Variance*. Professional Paper of the United States Geological Survey, **442H**.

porcellanite Term used by archeologists for deposits of rock formed by thermal metamorphism of a bole or *soil* horizon within **basalt**. Two **axe factories** providing Neolithic polished **axes** were found within a porcellanite deposit in Northern Ireland. [OWT]

pore-fluid pressure The pressure exerted by fluids within the pores of a rock. This pressure consists of *normal stress* only with no *deviatoric* components, and equal values of all three *principal stresses*, $\sigma_1 = \sigma_2 = \sigma_3 = P$. If the rock is saturated by **pore fluid**, the pressure (P) at any depth (z) is calculated simply as the pressure due to a column of fluid, $P = \rho g z$ where ρ is the density of the fluid, often water, and g is the acceleration due to **gravity**. This is *hydrostatic* pore-fluid pressure.

However, natural pore-fluid pressures are known to exceed hydrostatic pressures, for example in **oilfields**, due to overconsolidation of impermeable sediments, applied tectonic **stresses** or dehydration reactions; conversely they may be less than hydrostatic in underconsolidated sediments. The pore-fluid pressure is often measured by the parameter λ, which is the ratio of pore-fluid pressure to **lithostatic pressure** due to the weight of rock; λ may have values between 0 and 1, and may approach the higher value in highly **overpressured** rocks.

Pore-fluid pressure has a very important influence on rock **failure** in experiments, where it has been observed that the applied *normal stresses* are reduced by an amount corresponding to the pore-fluid pressure, causing **failure** at lower values of **differential stress** (see **effective stress**). This has been inferred to apply to natural **deformation**, suggesting for example that large *thrust sheets* can be emplaced under conditions of low **shear stress** due to high pore-fluid pressures, and variations in pore-fluid pressure play an essential role in some theories about **earthquake** precursor events. (See **earthquake prediction**.) [TGB]

Hubbert, M.K. & Rubey, W.W. (1959) Role of fluid pressure in mechanics of overthrust faulting. *Geological Society of America Bulletin* **70**.
Jaeger, J.C. & Cook, N.G.W. (1979) *Fundamentals of Rock Mechanics* (3rd edn) Chapter 8. Chapman & Hall, London.

pore-waters/fluids and their evolution Pore-waters are the aqueous solutions that occupy the pore spaces in *soils*, sediments and subsurface rocks generally. The composition of pore-waters often reflects their ultimate origin although their present composition is usually modified as a result of mixing with other pore-fluids and by **diagenetic** interaction with their host sediments at source, during migration or at their final site of residence. Pore-waters are responsible for providing nutrients to terrestrial **plants** in *soil* systems, are the ubiquitous medium through which **diagenetic** reactions take place, are associated with *hydrocarbon* **migration** and can potentially transport large quantities of solute in the subsurface (see **mass solute transfer, migration of oil and gas**). The important role that pore-waters play during **diagenesis** cannot, therefore, be overstressed. They are, however, difficult to study; water samples from the subsurface cannot be obtained easily without contamination (Collins, 1975). Indirect methods of study, via inferences from **electric logs**, **diagenetic** mineralogies, variation in the composition of *concretions* within sediments, *fluid inclusions* entrapped in **authigenic** minerals or by squeezing sediments and extracting pore-fluids, provide important data on pore-water salinity, composition and evolution with time.

Many interstitial subsurface pore-waters originate from the depositional environment, having been trapped with their host sediment during burial. In the shallow subsurface, therefore, the composition of pore-fluids often reflects that of the depositional environment, and in general terms, three extremes of composition can be recognized (see also **depositional environment related diagenesis**): marine derived pore-waters, originally with moderate sulphate; continental, dilute, low-**pH** waters associated with vegetated temperate climates; and continental, high salinity, alkaline evaporitic waters associated with hot dry desert environments (Table P2). In all

Table P2 Pore-fluids and their evolution. The variation in the generalized composition of pore-waters from extremes of depositional environments, compared with deep representative sedimentary basinal brines. TDS, total dissolved solids in mg l^{-1}. Other concentrations in p.p.m.

	Normal marine	Desertic interstitial	Surface river (Mississippi)	Basinal brines (Tertiary, Texas)	
pH	*c.* 8	*c.* 8.5	*c.* 6.5	*c.* 5.8	*c.* 6.2
Eh	+ve	extremely +ve	slightly +ve	−ve	−ve
TDS	35 000	2 370	120	202 016	107 375
Ca^{2+}	411	402	15	6 800	23 200
Mg^{2+}	1 290	904	4	596	117
Na$^+$	10 800	85	6	53 800	17 870
K$^+$	392	43	3	520	160
[Al]	tr	0.05	0.88	7.8	0.3
[Fe]	0.0034	—	0.67	41	22
[Mn]	0.0004	0.0002	0.2	0.0004	0.6
Si(OH)$_4$	1.5	8	13	30	144
HCO$_3^-$	142	560	58	225	61
SO$_4^{2-}$	1 356	38	11	820	4
Cl$^-$	19 400	396	8	98 200	68 500

cases these pore-fluids react with their enclosing sediments and gradually evolve with time. The extent of reaction will depend upon the original pore-fluid composition (its reactivity), the composition of the host sediment (percentage of unstable components), amount of organic matter and pore-water residence time. In general terms, pore-water evolution in marine sediments is dominated by the reaction with, and products of, **organic matter diagenesis** whilst pore-water evolution in continental desert systems usually reflects the high concentration of dissolved species and detrital composition; pore-water evolution in vegetated temperate climates typically lies intermediate between these two extremes. **Meteoric**-derived **waters** will almost invariably increase in salinity during **diagenesis** by reaction with unstable minerals in the host sediment.

It is more difficult to generalize on the composition and evolution of pore-waters during deeper burial. Historically, increased burial was considered to be associated with increasing salinity. However, salinity reversals in depth profiles are now commonly recorded in deep sedimentary **basins**. Pore-water salinities may range from around 8000 mg l^{-1} to in excess of 300 000 mg l^{-1}, whilst compositions cover a diverse spectrum of NaCl, SO$_4^{2-}$, and HCO$_3^-$ dominated waters. The pore-waters from the Texas Gulf Coast of the USA have received by far the most detailed attention (Table P2) and are known in some detail. Chemically distinct pore-waters throughout the Gulf Coast area can be used to define hydrochemical regions and are inferred to reflect separate pore-water origins. Highly saline (>250 000 mg l^{-1}) Ca–NaCl **brines** result from deep-seated **evaporite dissolution**; low salinity, high alkalinity waters, commonly present in **overpressured** zones and associated with **secondary porosity**, result from **mudrock** compaction and dehydration reactions. Depth-related changes in pore-water composition indicate mixing of waters of different origin and/or **diagenetic** reaction with host sediments. How far

these observations can be applied to other large sedimentary **basins** has yet to be determined. Bearing in mind the potential for widespread **subsurface fluid migration** and pore-water mixing, together with a diverse variety of pore-fluid sources (depositional, **meteoric**, compactional, dehydration, **decarboxylation**) and the subsequent pore-water/sediment interaction during **diagenesis**, this remains a challenging and exciting area of research. The isotopic composition (O, D and radiogenic/stable isotopes of dissolved constituents) can be expected to play an increasing role in determining the origin of pore-fluids and their evolution. [SDB]

Collins, A.G. (1975) *Geochemistry of Oilfield Waters*. Developments in Petroleum Science No. 1. Elsevier, Amsterdam.

Morton, R.A. & Land, L.S. (1987) Regional variations in formation water chemistry, Frio Formation (Oligocene), Texas Gulf Coast. *American Association of Petroleum Geologists Bulletin* **71**: 191–206.

porosity Generally applied as a term to include all the openings or interstices within the framework of a *soil*, sediment or other rock lithology. The interstices within the rock framework are individually referred to as pores. Porosity, therefore, can be considered as the property a material has of containing pores, and is normally expressed as a percentage of the bulk volume of the material occupied by pore space.

The term pore is derived from the Greek word *poros* meaning 'passageway'. It is, however, now normal practice in sedimentology to restrict the use of the term pore to describe discrete enlargements within the **fabric** of the porosity and distinguish these from the more restricted openings that connect adjacent pores together, which are referred to as *pore throats*. Collectively, pores and *pore throats* constitute the pore system. Defined in this manner, pores will usually make up the bulk of the porosity in most sediments whilst the interconnecting *pore throats* will contribute only a very small percentage of the total pore volume. *Pore throats*, however, are important in their own right in that they exert a controlling influence on the fluid transmissibility of a pore system (see **permeability**).

The porosity system in any sediment may be extremely complex and heterogeneous in its form and distribution. Individual pores may be of significantly different origins, may vary in their connectivity, display a wide range of shapes and sizes and are modified with time during **diagenesis** by the opposing processes of **dissolution** (which enhances porosity) and **cementation** (which reduces porosity). A useful genetic classification, valid for both carbonate and siliciclastic sediments, is to divide **porosity** into *primary* and **secondary**. *Primary porosity* is used to define all the initial pore space that is present in a sediment at deposition. This varies from as much as 90% of the sediment bulk volume in argillaceous and carbonate muds on the seafloor to a theoretical 47% for an open packed, perfectly sorted, clay-free, coarse-grained sediment composed of ideal spherical particles. In reality, the maximum *primary porosity* for well-sorted **sandstones** is around 40% because detrital grains are rarely true spheres and initial packing densities lie intermediate between open and closed end members. *Primary porosity* is reduced during **diagenesis** by the combined processes of **cementation** and **compaction**. (See also **burial diagenesis**.)

Secondary porosity is distinguished from *primary porosity* in being generated within sediments after their deposition and, therefore, falls within the scope of **diagenesis**. **Secondary porosity** can be created, modified or destroyed at any stage in the diagenetic evolution of any sediementary lithology. It is extremely useful to be able to place the main phases of porosity generation (or destruction) in the context of **basin** development and, therefore, the three **diagenetic regimes**, *eogenesis*, *mesogenesis* and *telogenesis*, provide a useful genetic framework for subdividing the timing of **secondary porosity** formation.

These concepts of *primary* and **secondary porosity** simply depend upon the time when the porosity was created and are independent of both their mode of origin and **fabric**. It is equally important to be able to classify porosity systems in terms of descriptive types, either for the purpose of evaluating the physical properties that influence the fluid content of sediments or for elucidating the geological origin of the sediment and its **fabric**. There are many such classification systems in the literature, but the essential element in each system is the recognition of a series of physically or genetically distinctive pore types that can be defined in terms of size, shape, connectivity, relationship to particular framework constituents and/or mode of origin. In coarse-grained **carbonates** (*grainstones*) and *siliciclastics* (**sandstones**, *conglomerates*) most *primary porosity* occurs between the framework grains and is termed intergranular. Additionally, where bioclastic material is a major detrital component of the sediment, pore space within skeletal components (such as **foraminiferal** tests, **coral** cavities or *gastropod* chambers, etc.), termed intragranular porosity, may be significant. Sediments containing either an argillaceous or a carbonate mud matrix generally lack intergranular porosity. However, significant porosity may be present within the matrix material, although pore aperture radii are often less than $0.5\,\mu m$, and the porosity is generally referred to as *microporosity*. **Secondary porosity** results from either the leaching of detrital framework grains, termed grain-dissolution porosity, or the removal of earlier precipitated **authigenic** cements, termed cement-dissolution porosity. Fracture porosity, although quantitatively a minor component of the bulk porosity volume, is often commonly developed and is important in increasing the connectivity and hence **permeability** of many pore systems. [SDB]

Choquette, P.W. & Pray, L.C. (1970) Geologic nomenclature and classification of porosity in sedimentary carbonates. *American Association of Petroleum Geologists Bulletin* **54**: 207–50.
Pittman, E.D. (1979) Porosity, diagenesis and productive capability of sandstone reservoirs. In: Scholle, P.A. & Schluger, P.R. (eds.) *Aspects of Diagenesis*. Special Publication of the Society of Economic Palaeontologists and Mineralogists **26**: 159–73.

porphyritic texture An inequigranular texture characterized by the occurrence of larger crystals, termed *phenocrysts*, set in a finer-grained (or glassy) matrix or groundmass. Porphyritic texture may result from two-stage cooling of a **magma**, causing slow cooling and slow growth of larger crystals, followed by rapid cooling (caused, for example, by extrusion) forming the finer-grained matrix. Such textures may also result from more complex histories, some of which may involve a single-stage cooling history. [RST]

porphyroblast (*metacryst*) A relatively large mineral grain formed during metamorphic recrystallization within a fine-grained matrix. Porphyroblasts are euhedral or subhedral and may include minerals such as **feldspars**, **garnet**, **staurolite** and **kyanite**. [RST]

porphyry copper deposit Large low-grade **stockwork** to **disseminated deposit** of **copper** which may also carry minor recoverable amounts of **molybdenum**, **gold** and **silver**. Usually they are **copper–molybdenum** or **copper–gold** deposits. They must be amenable to bulk mining methods, that is **open pit** or, if underground, **block caving**. Most deposits have grades of 0.4–1% **copper** and total tonnages range up to 1000 million with a few giants being even larger. Selective mining is of course impossible and *host rock*, **stockwork** and disseminated mineralization have to be extracted *in toto*. In this way, some of the largest man-made holes in the **crust** have come into being.

The typical porphyry copper deposit occurs in a cylindrical, stock-like, **composite intrusion** having an elongate or irregular outcrop about $1.5 \times 2\,km$, often with an outer shell of equigranular medium-grained rock. The central part is *porphyrite* — implying a period of rapid cooling to produce the finer-grained groundmass — the porphyry part of the intrusion. The most common hosts are passively intruded, acid, **plutonic** rocks of the **granite** clan

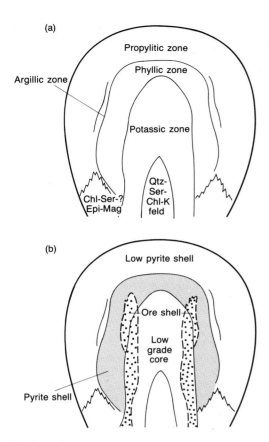

(a)

Propylitic zone

Phyllic zone

Argillic zone

Potassic zone

Qtz-
Ser-
Chl-K
feld

Chl-Ser-?
Epi-Mag

(b)

Low pyrite shell

Ore shell

Low
grade
core

Pyrite shell

Fig. P24 (a) Hydrothermal alteration zoning pattern and (b) schematic diagram of the principal areas of sulphide mineralization in the Lowell–Guilbert model of **porphyry copper deposits**. Solid lines represent the boundaries of the alteration zones shown in (a). (After Evans, 1993).

ranging from **granite** through **granodiorite** to **tonalite**, *quartz monzodiorite* and **diorite**. However, **diorite** through *monzonite* (especially *quartz monzonite*) to **syenite** (sometimes alkalic) are also important *host rock*-types.

The mineralization occurs within a series of zones of **wall rock alteration** and is best described by reference to its position relative to these zones. There are two main types of porphyry copper deposits — the *Lowell–Guilbert Model* and the *Diorite Model*. In the former the **alteration** pattern is as shown in Fig. P24. The *potassic zone* is characterized by **potassic alteration** of the host stock, the *phyllic zone* by **sericitization**, the *argillic zone* (not always present) by **intermediate argillic alteration** and the *propylitic zone* by **propylitic alteration**. The **ore** zone may be (i) totally within the host stock, (ii) partially in the stock and partially in the *country rocks* or (iii) (rarely) in the *country rocks* only. The most common shape for the **orebody** in the *Lowell–Guilbert Model* is that of a steeply walled cylinder coinciding more or less with the boundary between the *potassic* and *phyllic zones*, Fig. P24.

The *Diorite Model* differs from the *Lowell–Guilbert* in a number of ways, the most important being the absence of the *phyllic* and *argillic* zones. Important amounts of **gold**

may now be present and the **molybdenum copper** ratio low.

The majority of porphyry copper deposits are associated with **Mesozoic** and **Cenozoic** *orogenic belts* in both **island arc** and *continental margin* settings. The main exceptions are many of the deposits in the former USSR and those of the Appalachians which are of **Paleozoic** age. All these deposits appear to have formed from **hydrothermal solutions** derived from the parent **magmas**. Late in the mineralization process **meteoric water** was frequently drawn into the zone of mineralization and played a part in redistributing and concentrating the **ore**. In some porphyry copper deposits **supergene enrichment** was very important in developing the **orebodies**. [AME]

Evans, A.M. (1993) *Ore Geology and Industrial Minerals: An Introduction* (3rd edn). Blackwell Scientific Publications, Oxford.
McMillan, W.J. & Panteleyev, A. (1980) Ore deposit models: 1. Porphyry copper deposits. *Geoscience Canada* 7: 52–63.
Titley, S.R. (ed.) (1982) *Advances in Geology of the Porphyry Copper Deposits*. University of Arizona Press, Tucson, AZ.
Titley, S.R. & Beane, R.E. (1981) Porphyry copper deposits. *Economic Geology 75th Anniversary Volume*: 214–69.

porphyry tin deposit A **stockwork tin** deposit. These deposits have some similarities with **porphyry coppers** but are of much lower tonnage. They occur in both **plutonic** and subvolcanic acid intrusions but most are of too low **grade** to be exploited at the present day. [AME]

portal Exit location of a meltwater stream at the snout or front of an ice mass. [ASG]

Portland cement A fine-grained, grey powder made by calcining a mixture of **limestone** and clay or **shale**. It is used with water and **sand** to prepare mortar or with water, **sand** and **aggregate** to make concrete. When the **limestone** and clay are heated together a cement *clinker* is formed which contains a number of anhydrous compounds. The most important is tricalcium silicate (Ca_3SiO_5), which reacts rapidly with water to form hydrated silicate grains. These interlock and give the mortar or concrete its strength. [AME]

Constructional and Other Bulk Materials. Block 4 of The Earth's Physical Resources. Open University Press, Milton Keynes.

Portland stone A yellowish white **oolitic limestone** from the Isle of Portland (a peninsula in southern England) which is widely used as **building stone**. [AME]

potash/potassium feldspar ($KAlSi_3O_8$) A general name given to the potassium end-member of the **alkali feldspar** series. (See **feldspar minerals, orthoclase, microcline, sanidine**.) [DJV]

potassic/potassium silicate alteration A form of **wall rock alteration**, of which **biotitization** is one form, characterized by the formation of new **potash feldspar** and/or **biotite**. **Anhydrite** is also often developed and can form up to 15% of the altered rock. [AME]

potassium (K) At. no. 19, at. wt. 39.0983. An alkali metal widely distributed in silicate rocks, e.g. as **orthoclase**. Extensively used in fertilizers. KOH is used as an electrolyte in batteries, in glass and in ceramics. Obtained commercially from **evaporite deposits**. [AME]

potassium–argon dating Geological dating method based on the following principles: rocks containing **potassium** (K) contain a small amount of *radioactive* **potassium**-40, present at a constant fraction (0.012%) of the total **potassium**. 11% of the potassium-40 decays to argon (argon-40) by electron capture (the remaining 89% decaying to calcium-40) at a known rate. Most argon is expelled from the rock on eruption and argon-40 now present is partly radiogenic. By determination of the concentrations in the rock of total **potassium** (by chemical methods), **potassium**-40 (by calculation) and argon-40 (by **mass spectrometry**) and a knowledge of the rate of decay of **potassium**-40 by electron capture, an age since formation of the rock can be calculated. Corrections must be made for the presence in the rock of absorbed atmospheric argon, and errors can occur through loss of **potassium** and argon-40 subsequent to formation, and the presence of small amounts of argon-40 in the rock at formation. The method has been used to date levels of early *hominid* occupation at Olduvai Gorge, East Africa. [OWT]

potential The potential at any point is defined as the energy required to bring a unit quantity from infinity to that point against the ambient potential field. A potential field is any field that obeys **Laplace's equation**, such as **gravity**, **magnetic** or electrical. (See **gravitational potential**, **magnetic potential**.) [PK]

potentiometric/piezometric surface The imaginary surface that indicates the static **head of water** in an **aquifer**. (See **artesian aquifer**.) [AME]

pothole (1) A deep, circular hole in the bed of a river, or **cave** stream, caused by **abrasion**. These are also called *swirlholes*.
(2) A range of vertical shafts in **limestone** areas which connect underground **cave** systems with the surface. The equivalent French term is **aven**. [HAV]

potter's clay (*pot earth*) Any clay or earth that can be used for making pottery. When **iron oxides** are present the typical red-brown colour of much cheap pottery is developed on firing; other colours are produced by adding coloured oxides. [AME]

powellite (CaMoO$_4$) A molybdate mineral showing partial *solid solution* to **scheelite**. [DJV]

power spectrum A representation of a time or distance function in the frequency domain as the relationship between the power of the function at various frequencies or wavenumbers. The power spectrum is the square of the amplitude spectrum. [GS]

pozzolan A pumiceous **ash** which the Romans mixed with lime to make cement. Today pozzolanic material is mixed with cement, generally in the proportion of 10–30% by weight, to form a material often used in the construction of concrete dams. [AME]

Prandtl number A dimensionless coefficient equal to $C\mu/R$, where C is specific heat, μ **kinematic viscosity**, and R is **thermal conductivity**. [DHT]

Tritton, D.J. (1988) *Physical Fluid Dynamics*. Clarendon Press, Oxford.

prase A dull green microcrystalline variety of **silica**. [DJV]

praseodymium (Pr) At. no. 59, at. wt. 140.91. A lanthanide metal used in glasses and ceramics to produce a characteristic yellow colour and in thermoelectric materials. Recovered commercially from **monazite** and **bastnäsite**. [AME]

Pratt hypothesis A model to account for **isostatic** compensation in which it is assumed that at a constant depth, usually taken as 113.7 km, the **depth of compensation**, the total pressure due to the overlying rocks is the same, irrespective of their elevation, i.e. the greater the regional elevation, the lower the density of rocks in the column beneath. [DHT]

pre-diagenetic controls on diagenesis Includes all those variables which influence the course and extent of **diagenesis** both prior to its beginning and also during any of the subsequent **regimes of diagenesis**.

The fundamental importance of an understanding of the pre-diagenetic variables on a **diagenetic** system is illustrated in Fig. P25, which shows some of the complex interrelationships that climate and **tectonic** setting may impart on the mineralogical composition and **fabric** of a particular sediment package and its interstitial **pore-waters**. On the largest scale, provenance is largely a function of the overall **tectonic** setting. This strongly influences detrital mineralogy as a consequence of the actual material available and subsequent mineralogical partitioning during transport. Climate in the source area may play a significant role in affecting the degree of surficial **weathering** in a particular environment which will

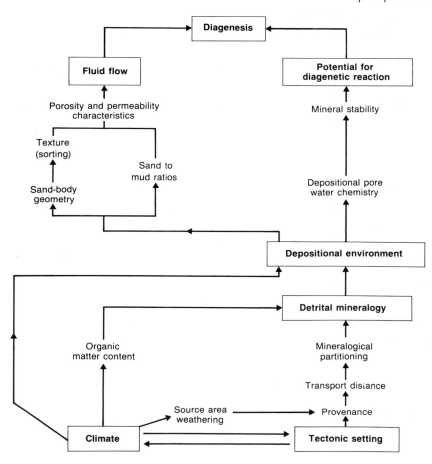

Fig. P25 Pre-diagenetic controls on diagenesis.
Simplified graphic flow chart illustrating some of the
potential interrelationships between diagenesis and
prediagenetic conditions.

determine the stability of **minerals** and hence both detritus available for transport and chemistry of **pore-waters**.

At a second level, once in the **basin** of deposition, not only will the composition of the initial sedimentary assemblage exert a controlling influence on the **diagenetic** style, but also the gross biological, chemical and physical environment. For example, whether the sediment was deposited in an aqueous medium or by aeolian processes, the chemistry, Eh and **pH** of the ambient medium, the height of **water tables**, prevailing temperatures and concentration of organic matter in the environment are just some of the factors that can be expected to impart some influence on **diagenetic** processes. The depositional environment is, therefore, likely to be an important controlling factor on the style and extent of **diagenesis** (see **depositional environment related diagenesis**).

Unfortunately, the influence of such pre-diagenetic variables does not end with the depositional environment, nor even the initial **pore-water** system. A specific depositional environment will essentially define sediment texture, **sand** to mud ratios and overall sandbody geometry. Hence, **porosity–permeability** characteristics, resultant fluid **flow** properties, interconnectivity of sandbodies in the subsurface and general sediment–interstitial fluid interaction may all be related to the original de-

positional system. Furthermore, the style and rate of burial will also be related to the sedimentation rate, itself a function of **tectonic** setting. Clearly, therefore, both depositional environment and structural framework not only exert a profound influence on initial **diagenetic** effects, but may also affect the **diagenetic** history of a sediment deep into the subsurface. [SDB]

precious/noble metal Gold, silver and **platinum** group metals. [AME]

precipitation The deposition of an **authigenic mineral** from a **pore-fluid** in either solid form by crystallization or as a gel by flocculation (with subsequent crystallization) resulting in the **cementation** (or partial **cementation**) of the **porosity** of the *host rock*.

Precipitation must take place from a supersaturated solution. The type of **authigenic** mineral precipitated depends upon the type of chemical species in solution and rate of introduction of dissolved species into the **pore-waters**. Precipitation of a solid substance can be divided into two fundamental processes, *nucleation* and *crystal growth* (Berner, 1981). *Nucleation* invariably precedes *crystal growth*, and the two processes are separated by an energy barrier as a result of the developing interface between the crystal nucleus and the solution. Once this free energy barrier has been overcome, *crystal growth* takes place spontaneously with a net decrease in free energy until sufficient material is removed from solution so that supersaturation ceases and a state of equilibrium is attained.

Both the rate of *nucleation* and *crystal growth* are a function of the degree of supersaturation of the solution. At low levels of supersaturation, crystal *nucleation* is very slow and excess dissolved material in solution is consumed by crystal growth on a limited number of nuclei. This results in a high degree of crystallinity in the **authigenic mineral**, and will tend to favour growth on existing mineral phases as **syntaxial** overgrowths. By contrast, at high solute concentrations and a high level of supersaturation, the rate of *nucleation* may be so rapid that virtually all the excess solute is precipitated instantaneously as very small (<100 Å) nuclei, effectively decreasing the saturation so that little solute remains for *crystal growth*. The resulting crystal forms will tend to be very poorly crystallized, often fine-grained and with a high degree of disorder.

Energetically, the precipitation of overgrowths can be considered as an attempt for detrital grains to become more stable and crystal-like. Initial rounded or angular detrital grains are effectively anhedral crystals, possessing high surface area to mass ratios. The energetic response during **diagenesis** is to lower the surface free energy and surface area by precipitation of stable **authigenic** overgrowths so that the detrital grain approaches the low surface free energy state of a euhedral crystal. [SDB]

Berner, R.A. (1981) Kinetics of weathering and diagenesis. In: Lasaga, A.C. & Kirkpatrick, R.J. (eds) *Kinetics of Geochemical Processes*, pp. 111–33. Reviews in Mineralogy Vol. 8. Mineralogical Society of America, Washington, DC.

preferred orientation Statistical concentration of linear or planar structural or **fabric** elements in a particular attitude or group of attitudes. The term is most frequently applied to three-dimensional distributions represented on *stereograms*, e.g. **pole figures**. A *crystallographic preferred orientation* refers to the alignment or preferred orientation of the axes of symmetry of crystal lattices in a deformed rock. This results from crystal **plastic deformation** and affects all crystallographic axes of a **mineral** grain. Usually only one axis is measured with a universal stage on an optical microscope. Most measured examples are from *uniaxial* **minerals** such as **quartz**, **calcite**, and **olivine**.

The example shown is a **quartzite** from close to the Moine Thrust in north-west Scotland, showing an asymmetrical distribution about the **foliation** plane (Fig. P26). The *fabric skeleton*, lines linking the highly populated parts of the distribution as shown on the right, is used to indicate the **shear** sense of the **deformation**. [SB]

Law, R.D., Casey, M. & Knipe, R.J. (1987) Kinematic and tectonic significance of microstructures and crystallographic fabrics within quartz mylonites from the Assynt and Eriboll regions of the Moine thrust zone. *Transactions of the Royal Society of Edinburgh, Earth Sciences* **77**: 99–125.
Schmid, S.M. & Casey, M. (1986) Complete fabric analysis of some commonly observed quartz c-axis patterns. In: *Mineral and Rock Deformation: Laboratory studies — The Paterson Volume*. American Geophysical Union, Geophysical Monograph **36**: 263–86.

prehnite ($Ca_2Al(AlSi_3O_{10})(OH)_2$) A green secondary silicate mineral found lining cavities in **basalt** and related rocks. [DJV]

preselite/prescellite Archeologists' term for spotted **dolerite** outcropping in the Preseli Hills, south Wales, UK. Its most famous use was for many of the 'Bluestones' of Stonehenge, UK but it was also occasionally used for Neolithic stone **axes**. [OWT]

pressure Three-dimensional **stress** state in which the magnitude of the **stress** is equal in all directions; more accurately known as *hydrostatic pressure* or *stress*. The *hydrostatic pressure* created by **gravitational load** at a given depth in the **crust** is termed **lithostatic pressure**. The term pressure is also frequently used to refer to the equidimensional **stress** created by a fluid (e.g. **pore-fluid pressure**, or *magma pressure*). [RGP]

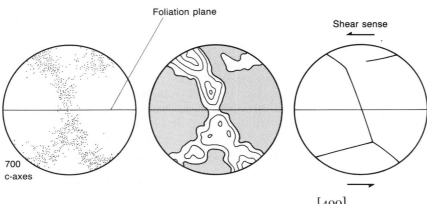

Foliation plane

Shear sense

700 c-axes

Fig. P26 Quartz c-axis **preferred orientation** from close to the Moine Thrust, north-west Scotland, sinistral sense of shear. (a) Lower hemisphere projection. (b) Contoured plot. (c) Fabric skeleton. (After Law *et al.*, 1987.)

pressure release The **weathering** mechanism which results in spalling and **exfoliation** of **plutonic** rocks and other rocks which have been deeply buried when they are exposed at the land surface. The layers of rock which are detached from the masses of bedrock may be from several millimetres to over 5 m thick. Pressure release occurs when the rocks, especially **granites**, are exhumed through the **erosion** of the overlying rocks. As the **lithostatic pressure** under which **plutonic igneous rocks** solidified is released — or as compressional forces resulting from **diastrophic** processes are relaxed — the rock masses expand. In response to this **unloading**, the rocks fracture parallel to the surface or along a plane perpendicular to the direction of compression. When domed **plutons** of **granite** are exposed, pressure release can bring about the **exfoliation** of huge sheets of rock. Many **granite inselbergs** exhibit these characteristic **weathering** features. [AW]

Bradley, W.C. (1963) Large-scale exfoliation in massive sandstones of the Colorado Plateau. *Geological Society of America Bulletin* **74**: 519–27.
Twidale, C.R. (1973) On the origin of sheet jointing. *Rock Mechanics* **5**: 163–87.
Watson, A. & Pye, K. (1985) Pseudokarstic micro-relief and other weathering features on the Mswati Granite (Swaziland). *Zeitschrift für Geomorphologie* NF **29**: 285–300.

pressure shadow The region around a relatively rigid body in a deformed rock which has undergone **extensional strain**. This volume is commonly defined by the growth of new minerals such as **calcite** or **quartz**. These new **minerals** form the pressure shadow; their growth is related to **pressure solution** elsewhere, the supply of material for the pressure shadow depending on diffusion through fluid-filled grain boundaries or fissures. *Pressure fringes* are marked by growth of fibrous **minerals** on certain parts of the margins of the more rigid body, which is often a **pyrite** crystal. Fibres in pressure fringes grow with long axes parallel to the incremental maximum **principal strain**. Hence the variation in the orientation of fibres within *pressure fringes* indicates changes in orientation of **strain** increments during the **strain** history. An example of pressure fibres showing changing **strain** increments is shown in Fig. P27. Ramsay & Huber (1985) discuss the method in detail, using this and other examples. The orientations of *pressure fringes* (or pressure shadows) can be mapped out to deduce **strain** distributions over large areas.

Orientations of pressure shadows around objects such as **porphyroblasts** can be used to indicate the sense of **shear**. Figure P28 shows the geometries of *pressure fringes* resulting from **deformation** with and without rotation of the more rigid crystal with respect to the matrix. [SB]

Etchecopar, A. & Malavielle, J. (1987) Computer models of pressure shadows: a method for strain measurement and shear sense determination. *Journal of Structural Geology* **9**: 667–77.
Ramsay, J.G. & Huber, M.I. (1985) *The Techniques of Modern Structural Geology* Vol. 1: *Strain analysis*. Academic Press, New York.

pressure-solution cleavage A **cleavage** developed as a result of preferential solution occurring at the contact surfaces of grains/crystals where the external pressure exceeds the fluid pressure of interstitial fluid. Such solution may occur during burial and **diagenesis** of sedimentary rocks (see **stylolites**) and is termed a **cleavage** where it leads to the development of a planar **foliation** as a result of orientated sedimentary and/or **tectonic stress** at higher P–T conditions. **Pressure solution** causes an increase in grain/crystal contact area and a reduction in pore/cavity space so may lead to the formation of a strongly recrystallized **foliated** metamorphic rock. [RST]

pressure solution/dissolution The enhanced rate of transfer of material from a **mineral** grain into the intergranular fluid with increasing **stress** when the external pressure exceeds the hydraulic pressure exerted by interstitial **pore-fluids**; thus **overpressure** inhibits pressure solution. Dissolved material moves through fluid lying on grain boundaries by diffusional mass transfer. This occurs at a rate higher in general than the rate of transfer in the absence of fluid (*Coble creep*). The term pressure solution also refers to the process of **strain** by **dissolution**, diffusion through fluid on grain boundaries and redeposition of **minerals**. In a **tectonic**, non-hydrostatic **stress** system, **dissolution** is enhanced on surfaces normal to the maximum *principal stress* direction, and deposition

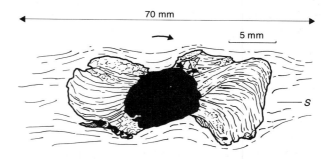

Fig. P27 Curved quartz fibres forming a **pressure shadow** around a framboidal pyrite from Lagrave (French Alps). The asymmetry indicates a dextral shear. (After Etchecopar & Malavielle, 1987.)

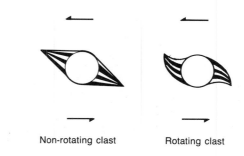

Non-rotating clast   Rotating clast

Fig. P28 Pressure shadow. Sketches of different end-member fibre patterns developed in pressure fringes around clasts that rotate with the overall deformation, or remain fixed with respect to the matrix.

favoured on surfaces normal to the minimum *principal stress*, in the ideal case. In most rocks, pore space and **rheological** contrasts between minerals affect the distribution of pressure solution and redeposition.

One of the most common occurrences of pressure solution is during **diagenesis** as a response to burial of sedimentary rocks. The increasing vertical load leads to **dissolution** on subhorizontal surfaces, and deposition in pore spaces, and results in **compaction**. This pressure solution early in the history of the sediment can be a source for the intergranular cement. **Tectonic** occurrences of pressure solution are common. When the rock affected by pressure solution is a bioclastic **limestone** containing fragments of known initial shape and size (Fig. P29), it is possible to estimate the volume of material movement. This can occur on surfaces called **stylolites**. These are generally marked by deposits of insoluble minerals, usually clays, left behind after diffusional mass transfer of **calcite** along the **stylolite**. These surfaces vary in shape from near planar or wavy, to strongly stepped and dog-toothed. The shape depends on the relative rates of **dissolution** of individual grains; thus interpenetration of grains, and sutured grain–grain contacts are characteristic of pressure solution **deformation**. When wavy or planar **dissolution** surfaces occur subparallel to each other, closely-spaced, and spread over a wide area, they can form a **macroscopic cleavage**. Pressure solution also acts to redistribute minerals during **crenulation** of existing **cleavage**.

The occurrence of diffusional mass transfer along mesoscale surfaces, such as **cleavage**, **stylolites** or **bedding**, indicates the possibility of bulk volume change, if there is not equivalent local redeposition of material. Volume losses of 10–30% have been documented for the **tectonic** process of **cleavage** formation. Volume loss during **diagenetic** compaction of carbonates can be even larger; **stylolites** subparallel to bedding can form at very early stages in burial. Thus pressure solution can form an integral part of **diagenesis**. as well as being an important

Fig. P29 Conglomerate pebbles that indent each other through **pressure solution** removing material. (Photograph courtesy of R.W.H. Butler, University of Leeds.)

mechanism for **strain**. (See **pressure shadow, deformation mechanism**.) [SB/SDB]

De Boer, R.D. (1977) Pressure solution: theory and experiment. *Tectonophysics* **39**: 287–301.
McClay, K.R. (1975) Pressure solution and Coble creep of minerals: a review. *Journal of the Geological Society, London* **134**: 57–70.
Rutter, E.H. (1983) Pressure solution in nature, theory and experiment. *Journal of the Geological Society, London* **140**: 725–40.

primärrumpf A tectonically upwarped **dome** of rock which, though still undergoing **uplift**, is **eroding** at a rate equal to the rate of **uplift**. [AW]

primary current lineation Low, linear ridges of grains aligned parallel to the flow direction on the surface of **upper stage plane beds**. These ridges are 2–3 grain diameters high and can be hundreds or thousands of diameters in length. Their origin is linked to the existence of high velocity **sweep** events within the **viscous sublayer** which entrain sediment and then subsequently deposit it as ridges parallel to the flow. Consequently, the transverse spacing of **sweeps** and primary current lineation ridges are directly related. [JLB]

Allen, J.R.L. (1985) *Principles of Physical Sedimentology*. Allen & Unwin, London.
Weedman, S.D. & Slingerland, R. (1985) Experimental study of sand streaks formed in turbulent boundary layers. *Sedimentology* **32**: 39–48.

primary structure Structure produced during the initial formation of a rock, rather than by **deformation**. Such **structures** are divisible into sedimentary and igneous, depending on the *host rock*. Certain primary structures, e.g. *cross-bedding*, *slump folds*, or igneous *flow foliations*, may be confused with deformational structures in highly deformed and/or metamorphosed rocks. [RGP]

principal strains The relative sizes (measured as **stretches** or **quadratic elongations**) of the three axes of a **strain ellipsoid**. The *principal strain planes* are perpendicular to each of the axes. The three **stretches**, S1, S2, S3, or *X*, *Y*, *Z*, are respectively the maximum, intermediate and minimum principal **strains**. Their orientations define the *strain axes*. For two-dimensional **strains**, a **strain ellipse** and two principal **strains** are used. The **strain ellipsoid** represents the deformed shape of a unit sphere, so that the principal **strains** are ratios, which are often quoted normalized to a minimum principal **strain** of 1. This facilitates comparison between different examples. (See **strain, pure shear, simple shear, plane strain, finite strain**.) [SB]

prod and bounce mark Type of *sole structure* or *tool mark*. Prod marks are asymmetric, elongate, semicircular to triangular depressions with the downcurrent part being deeper and broader relative to the upcurrent end. Prod marks are formed when a tool or *clast* in a sediment-flow impacts the sediment surface at an acute to oblique angle and is momentarily stopped before being resuspended upward into the current. Where the impacting tool hits the sediment surface at an acute angle and is effectively instantaneously resuspended into the current, a more symmetrical depression is formed that tapers similarly in the up- and down-current direction, and is called a bounce mark. Prod and bounce marks commonly occur together, and are found typically in *turbidite* successions. [KTP]

propagating rift A fracture, or system of fractures, that is continuing to extend its length. In the ocean **basins**, this term usually refers to a ridge, with associated **rift** features, that is growing at the expense of another. On the continents, the term usually applies to subparallel fractures, associated with a **rift** valley, that are still extending their lengths, usually comprising ancient lines of weakness being opened by **tectonic** forces, such as those associated with **membrane tectonics** and **mantle diapirs**. Compare with quiescent **rift** valleys, called **aulacogens**. [DHT]

Hey, R., Duennebier, F.K. & Morgan, W.J. (1980) Propogating rifts on mid-ocean ridges. *Journal of Geophysical Research* **85**: 3647–58.

propylitic alteration A complex **alteration** characterized by **chlorite**, **epidote**, **albite** and carbonate (**calcite**, **dolomite** or **ankerite**). The propylitic alteration zone is often wide and can be a useful guide in *mineral exploration*, e.g. for **porphyry copper deposits**. [AME]

Proterozoic The younger eon of the *Precambrian*, that part of the Earth's history before the appearance of life forms with hard parts that gave rise to body fossils at the beginning of the **Cambrian** period, and extending from 2500 Ma until the start of the **Cambrian** period (*c.* 550 Ma). The older eon of the *Precambrian* is the **Archean**. There are proposals to divide Proterozoic time into early, mid and late eras at 1600 Ma and 900 Ma, although these divisions cannot be recognized by any changes within the rocks. The only major changes in biota that occur in the Proterozoic are the incoming of *eukaryotic* life forms about 1300 Ma (all previous forms were *prokaryotes* which have no cell walls around their nuclei) and the first *Metazoa* at *c.* 700 Ma. There was a gradual evolution of **stromatolites** throughout the Proterozoic and it is likely that changes in their shape and diversity will enable **stromatolite**-based stratigraphic subdivisions of global application to be erected in the near future. By far the most important paleontological feature of the Proterozoic, however, was the sudden emergence of a very diverse fauna of *metazoans*, exemplified by the *Ediacara fauna* of South Australia. These are all soft bodied but include many classes of the *coelenterates*, *annelids* and **arthropods**. These fossils typify the latest division of the *Precambrian*, the **Vendian** System. Another feature of the Late *Precambrian*, which may also allow regional correlations in these essentially unfossiliferous rocks, is the presence of several periods of *glaciation*, at about 600 and 800 Ma and a less widespread *glaciation* near the boundary of middle and late *Precambrian* at 950 Ma. Whether these world-wide occurrences are due to world-wide *glaciations* or to very rapid **continental drift** is still a matter of dispute. Earlier *glaciations* are sporadically recorded throughout the Proterozoic.

As much as 85% of the **continental crust** is thought to have been formed by the end of the **Archean**, but much of this was either reworked or covered by sedimentary rocks during the Proterozoic and a large proportion of the world's *Precambrian* **shield** areas is now represented by Proterozoic rocks.

Although the *greenstone belts* and *granulite–gneiss belts* of **Archean** times continued to form in a limited way in the early Proterozoic, there was a dramatic change about 2500 Ma with the development of linear *orogenic belts* with all the features of modern **plate tectonic orogenies**. There are no particular features of the Proterozoic that are characteristic of the period although some igneous suites are almost unique. One of the features of Proterozoic time, as of **Archean** time, is the widespread development of **dyke swarms** that penetrate the whole stable **crust** but also extend syntectonically into the contemporaneous **mobile belts**. Such widespread **dyke swarms** only seem to develop in later times during stretching of the **crust**. This is probably due to a thinner **lithosphere** in **Archean** and Proterozoic time. **Archean** crustal evolution culminated in the stabilization of large areas of **crust** and on these stable crustal areas major sedimentary **basins** developed, starting with the Witwatersrand in **Archean** time but continuing for hundreds of millions of years in many **shield** areas with extensive and thick deposits of **quartzite**–carbonate–**shale** assemblages, often with very large deposits of **banded iron formation**. The latter indicate very stable depositional environments as individual bands only millimetres thick can be correlated laterally for 300 km. Another feature of importance in the sedimentary record of these **basins** is the presence of *red beds*. From *c.* 2000 Ma **hematite** coatings on sedimentary grains began to be found and this time also marks the last great episode of banded *ironstone* deposition, suggesting that a major change in seawater and atmospheric chemistry had taken place, namely the change from an oxygen-rich hydrosphere to an oxygen-bearing atmosphere. Other effects of a reducing atmosphere in the early Proterozoic may be the formation of detrital **uranium** deposits and extensive **manganese** deposits, while in the later Proterozoic major **phosphorites** and **evaporites** become more common. All these sedimentary records are still extant because of the stable crustal regime which resulted in these thick sedimentary **cratonic** deposits being laid down but subsequently neither deformed nor metamorphosed.

Elsewhere in the **crust** Proterozoic **mobile belts** were developing as linear **orogenies**. Many display all the characteristics of modern **plate tectonic** *continental-margin* **orogenies**. The Wopmay Orogen of Northwest Territories, Canada, developed between 2100 and 1800 Ma and was initiated by a shelf sequence interrrupted by continent-directed **rifts** (with associated alkaline intrusives). This was succeeded by the foundering of the continental margin to give deep water **shales** and *flysch* and continentward thrusting of an uprising cordillera from the core of the **orogen**. Late **plutons** and further collision tectonics with *molasse*-type sediments complete the cycle. Other collision **orogenies** are recognized in most **shield** areas, but many early and mid-Proterozoic linear belts reveal only reworked **Archean gneisses** of high metamorphic grade, with the only Proterozoic crustal additions being synorogenic **dyke swarms** and late granitic complexes. Such belts have been ascribed to an 'ensialic' type of **orogeny** although many now regard these belts only as very deep-seated examples of *continental collision* with all surface effects expunged by **erosion**. Many of these belts seem, however, to be developed by **shear** regimes with very large lateral translation across the belt, the **shear** couple also affecting the **craton** and being marked by the widespread **dyke swarms**. Such structures may also be tied, by the same **stress** system, to large tensional trough systems on adjacent margins and such whole-continent **strain** seems to be a characteristic feature of this period.

The late Proterozoic of Africa has such a continent-wide feature. The *Pan-African Orogeny* of 900–600 Ma is a series of zones of activity in a reticulate pattern over the whole of the African Shield and beyond. Certain features in Arabia, Mauritania and even southern Britain seem to be continental margin **orogenies** of conventional type and others in the centre of the Shield (e.g. the *Hoggar*) seem to be collision **orogenies** due to the closure of small oceans. Many features, however, cross the continent and cause **deformation** and metamorphism in such a complex network that normal opening and closing of ocean **basins** seems an unlikely explanation.

There are two major phases of igneous activity that are in many ways unique to the Proterozoic. During the Early and mid-Proterozoic there were many enormous basic complexes intruded into the stable **cratonic** areas. The *Stillwater Complex* in Montana and the *Great Dyke of Zimbabwe* (480 km long but only 6 km wide) are probably very late **Archean** in age but represent intrusions of this type. The largest of these intrusions, the **Bushveld Complex** in South Africa, is of 6600 km² and is up to 9 km thick and both it and the *Sudbury Irruptive*, Canada were intruded about 2100–2000 Ma. Such large eruptions of basic **magma** into single intrusions were not to occur again, although the Muskox and Duluth Intrusions of *c.* 1100 Ma are still very large layered basic bodies.

In the mid-Proterozoic most of the continental landmass was, for a time at least, probably in the form of a large *supercontinent*, **Pangea**. This large mass was traversed at this time by **rift** zones which gave rise to suites of characteristic intrusives and volcanics. *Rapakivi granites* and *anorthosite* **plutons** are the most typical of these and occur in a swathe from California to Ontario through Sweden to the Ukraine and in a southern belt from Australia to East Africa and India. Late stages of the activity included **dykes** and bimodal volcanicity followed by alkaline and **carbonatitic** intrusions. Continental, **rift**-controlled sedimentation gives the key to the explanation of this diverse suite of largely igneous features. Although usually occurring between 1400 and 1100 Ma, some of the igneous activity of this type seems to be as early as 1750 Ma indicating **rifting** activity over enormous lengths of time in one area of the **crust**.

Large proportions of Proterozoic rocks were thus largely controlled by the **Archean** stabilization of the continents. There was a great deal of continental anorogenic and **rift**-related igneous activity and large and long-lasting sedimentary **basins** formed on these stable **cratons**. However, there is also extensive evidence for *continental-margin* **subduction** and **plate tectonic** development of cordilleran and collision **orogenies** and even where there is no evidence of Proterozoic sedimentation it now seems likely that the reworked Archean of many Proterozoic **orogenies** marks the lower **crustal** representation of major *plate* collisions. [AEW]

Punnkollu, S.N. & Andrews-Speed, C.P. (eds) (1984) Proterozoic — evolution, mineralisation and orogenies. *Precambrian Research* **25**: 1–348.
Windley, B.F. (1977) *The Evolving Continents*. John Wiley & Sons, Chichester.

proton magnetometer (*nuclear precession magnetometer*) A **magnetometer** which measures a **magnetic field** by the precession of protons (hydrogen ions) in a liquid about the field direction. The frequency of precession is proportional to the strength of the field. Little orientation of the sensor is necessary, so that the **magnetometer** can be rapidly read and used on land, at sea and in the air. It is also the standard instrument used in *magnetic observatories*.

In addition to the search for magnetic **minerals**, proton magnetometers and **gradiometers** are widely used in archeological prospecting to detect buried features such as walls and pottery kilns. [PK/OWT]

Kearey, P. & Brooks, M. (1991) *An Introduction to Geophysical Exploration* (2nd edn). Blackwell Scientific Publications, Oxford.

protore Mineral material in which an initial but uneconomic concentration of metals has occurred that may, by further natural processes, be upgraded to the level of **ore**. [AME]

proustite (Ag_3AsS_3) A sulphosalt **ore** mineral of **silver** found in **vein** deposits, also known as *light ruby silver*. [DJV]

pseudobrookite ($FeTiO_5$) An **oxide mineral** found in **igneous rocks**. [DJV]

pseudokarst Landforms and landscapes which resemble those found in areas of **limestone** (or other calcium carbonate rocks) but which are developed on other rocks. The term has been used in two ways: first, to describe features such as **caves**, **dolines**, and **karren** developed through **dissolution** of water-soluble rocks and **minerals** other than calcium carbonate. Examples include, solution and collapse features (**dolines**) in **halite** rock, caves in some volcanic rocks and **karren** and solution pits in **granites**. However, according to this definition, features produced by **dissolution** of **gypsum** (calcium sulphate) should also be classified as pseudokarstic. Otvos (1976) proposed that all landforms produced by the **dissolution** of rock should be termed **karstic**. The second way in which the term pseudokarst has been used is that put forward by Otvos. In his definition, only those landforms and terrains resembling **karstic** features but formed by processes other than rock **dissolution** are termed pseudokarstic. Examples of such features include **pans** resembling **dolines** but which are formed by **deflation**, and some *lava caves* which form as molten volcanic rocks solidify. [AW]

Naum, T. & Butnaru, E. (1967) Le volcano-karst des Calimani (Carpathes roumaines). *Annales de Spéléologie* **22**: 727–55.
Otvos, E.G. (1976) 'Pseudokarst' and 'pseudokarst terrains': problems of terminology. *Geological Society of America Bulletin* **87**: 1021–7.
Watson, A. & Pye, K. (1985) Pseudokarstic micro-relief and other weathering features on the Mswati Granite (Swaziland). *Zeitschrift für Geomorphologie* NF **29**: 285–300.

pseudoleucite The name given to a mixture of the minerals **nepheline**, **orthoclase** and **analcime pseudomorphous** after **leucite**. [DJV]

pseudophite A massive variety of the mineral **chlorite** sometimes used as a substitute for **jade**. [DJV]

pseudotachylite Extremely fine-grained or glassy-looking rock found along **fault** zones, representing rapid displacement and melting of material by **shear** heating. Fused rock has been found at the base of major recent **landslide** sheets, and can be made experimentally by friction between rock surfaces, but natural examples rarely show glassy textures. Pseudotachylite often shows injection features and **veins**, indicating that it was in a fluid state; thus devitrification may have occurred to remove the glassy texture. Many rocks classified as pseudotachylites may have resulted from very intense grain size reduction by **cataclasis**. (See **fault rock**, **glassy rock**.) [SB]

pseudowollastonite ($CaSiO_3$) A high temperature (>1120°C) form of the mineral **wollastonite**. (See **pyroxenoid minerals**.) [DJV]

psilomelane The name given to botryoidal masses of **manganese oxide minerals** now known to be a mixture of several minerals of which **romanechite** is a major constituent. [DJV]

ptygmatic fold A rounded **fold** in which the *fold amplitude* is large with respect to the thickness of the folded layer, and the *fold wavelength* is small with respect to the trough-to-trough separation along the layer. Ptygmatic folds generally come close to an **elasticas fold profile**, i.e. one in which the fold interlimb angle has a negative value. In such a **fold**, the limbs of an **anticline**, for example, converge downwards due to the excessive curvature of the hinge. The ptygmatic folds in Fig. P30 show some **elasticas** profiles. The **folds** are generally concentric, i.e. without change in individual layer thickness. They are most common in layers of higher **competence** isolated in a less competent matrix, and can result from **buckling**. [SB]

Ramsay, J.G. & Huber, M.I. (1987) *The Techniques of Modern Structural Geology* Vol. 2: *Folds and Fractures*. Academic Press, London.

puddingstone Popular name for coarse *conglomerate*. Applied particularly to the silicified *conglomerate* of the **Eocene** Reading Beds of Hertfordshire, England ('Hertfordshire puddingstone') which was used for **millstones** in pre-Roman Britain. [OWT]

pull-apart basin **Extensional basin**, usually of relatively small size, formed in a **strike-slip zone**. [RGP]

pumice and scoria Pumice is a light coloured, highly **vesicular**, acidic volcanic glass and scoria is its rusty red to black mafic counterpart. By virtue of their porous nature these materials have low density and insulating properties. They are used in the construction industry, in light-weight structural concrete, plaster aggregate, loose-fill insulation and concrete blocks. [AME]

pumping test A method of measuring **aquifer** properties. Water is pumped out of one **well** and the manner in

23 mm

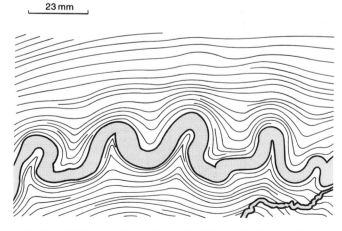

Fig. P30 Ptygmatic folds in a vein within granitic gneiss showing strong hinge curvature and high amplitude with respect to the layer thickness. (After Ramsay & Huber, 1987.)

which the **cone of depression** expands is measured using observation **wells** around the pumping **well**. The steepness of the cone depends upon the **hydraulic gradient** which in turn depends on the pumping rate and the **transmissivity** and *storage coefficient* of the **aquifer**. [AME]

Price, M. (1985) *Introducing Groundwater*. Allen & Unwin, London.

pure shear A shape change (distortional **strain**) without volume change in which the *strain axes* do not rotate, i.e. an irrotational **deformation**. A cube would become a cuboid with sides parallel to the *strain axes* under this **deformation**, as shown in the Fig. P31. If pure shear is used to refer to continuing **deformation**, then the incremental *strain axes* must not rotate with respect to material lines. This is thus a **coaxial deformation** geometry. [SB]

pyrargyrite (Ag_3SbS_2) A sulphosalt **ore** mineral of **silver** found in **vein** deposits, also known as *dark ruby silver* [DJV]

pyribole (*biopyribole*) The term 'pyribole' was originally coined to encompass the common features of **pyroxene** and **amphibole** mineralogy. It now includes with these the recently characterized non-classical chain silicates. The term 'biopyribole' is used to include sheet silicates: 'bio-' is from '**biotite**' and does not signify biological origin.

The sheet silicates (**micas**, **talc**), **amphiboles** and **pyroxenes** have many structural features in common. The characteristic building block of **amphiboles** and **pyroxenes** is the 'I-beam' formed by two silicate chains, held together by cations at sites formed mainly by the non-bridging oxygens. There is a logical progression from the single-width **pyroxene** I-beam, through the **amphibole** double chain, to the two-dimensional structure of the sheet silicates. The octahedral cation sites of the **micas** correspond to those within the pyribole I-beam structure, and the interlayer sites to those between the backs of adjacent I-beams in **amphiboles**.

Alternatively, the relationship can be described by envisaging the **amphibole** stacking structure as alternate 'slices' parallel to (010) of **pyroxene** (the edge sections of adjacent **amphibole** units) and mica (the central portion of the **amphibole** chains). This approach, outlined in articles

by Thompson and by Veblen, in *Amphiboles and Other Hydrous Pyriboles — Mineralogy* edited by Veblen (1983), can be generalized to describe the non-classical pyriboles.

Non-classical pyribole structures were first observed in *orthoamphiboles* during the study of silicates by high resolution **electron microscopy**. As well as frequent **dislocations** of the chain structure, ordered regions were observed with regular chains equivalent to triple the **pyroxene** chain (where the **amphibole** chain is double) and with alternate double and triple chains. In addition, there were many domains ('finite sheets') showing multiple chains up to and over 300 chain units wide. The interested reader is referred to the report by Veblen *et al.* (1977).

Naturally occurring minerals with non-classical chain structures were first identified from a metamorphic formation near Chester, Vermont, USA; the minerals were probably derived by **alteration** from **anthophyllite**. Two of these, **jimthompsonite** (orthorhombic in symmetry) and *clinojimthompsonite* (monoclinic), contain triple chains; a third, **chesterite** (orthorhombic), contains alternate strips of double and triple chains. A fourth phase, probably 'clinochesterite', was present in too small a quantity to be fully characterized.

Wide chain domains are likely to be important as intermediates in the reactions by which sheet silicates are formed from **amphiboles**. (See **mica minerals**, **electron microscope**.) [ADL]

Veblen, D.R. (ed.) (1983) *Amphiboles and Other Hydrous Pyriboles — Mineralogy*. Reviews in Mineralogy Vol. 9A. Mineralogical Society of America, Washington, DC.
Veblen, D.R., Buseck, P. & Burnham, C.W. (1977) Asbestiform chain silicates: new minerals and structural groups. *Science* 198: 359–65.

pyrite (FeS_2) The most common **sulphide mineral**. [DJV]

pyrochlore (($Ca,Na)_2(Nb,Ta)_2O_6(O,OH,F)$) An **oxide mineral** containing **niobium** and **tantalum** and found associated with **alkaline igneous rocks**. [DJV]

pyroclastic rock A rock formed by accumulation of material generated by explosive fragmentation of **magma**, and/or previously solid rock, during the course of a **volcanic eruption**. The material forming the rock is called **tephra**. A **tuff** is a pyroclastic rock that is coherent because

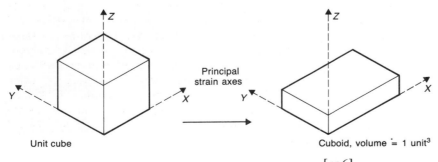

Fig. P31 Deformation of a unit cube by a **pure shear**; there is no rotation of principal strain axes.

[496]

of welding (see **welded tuff**) or any other lithification process. During **volcanic eruptions**, extruded **magma** forms **lava flows** and pyroclastic deposits in various proportions. Most eruptions generate at least some pyroclastic deposits, and, in some eruptions, all the extruded **magma** is explosively fragmented to form pyroclastic deposits, as in the case of most **ignimbrite** eruptions.

Pyroclastic deposits consist of different proportions of **pumice**, **scoria**, **ash**, lithic fragments and crystals. **Pumice** is strongly vesiculated (see **vesicle**) and fragmented **magma** of silicic composition. **Scoria** is vesiculated and fragmented **magma** of mafic to intermediate composition. **Ash** is fine-grained magmatic material mostly formed from the glass walls between adjacent bubbles in vesiculating **magma**. Lithic *clasts* are non-vesicular fragments, most of which were obviously derived from previously solid rock. Most lithic *clasts* are of a kind called *accessory lithics* — those which were torn from the vent walls during eruption, and which may or may not be volcanic rocks from the superstructure of the volcano. *Accidental lithics* are those which were plucked from the ground during transport of **tephra** as **ash flows** or *base caps*. *Cognate lithics* are those which consist of juvenile material, derived from the same **magma**, or one very closely related to that represented by the **pumice** or **scoria** *clasts*. The origin of *cognate lithics*, which are sometimes glassy (see **glassy rock**), is problematic. In some cases, they represent material derived by explosive disruption of a plug filling the **volcanic vent**, or of **dykes** and conduits situated between a *magma chamber* and the vent. Crystals are released from **porphyritic magmas** as they explosively fragment into **ash** and **pumice** or **scoria**.

Pyroclastic fragments can be divided according to size into *juvenile bombs* and *lithic blocks* (both >64 mm), *lapilli* (2–64 mm) and **ash** (<2 mm). Descriptive classifications of pyroclastic rocks have been devised based on the relative proportions of (blocks + bombs), *lapilli*, and **ash**; or of lithic fragments, crystals and (**pumice** + **ash**) (Schmid, 1981; Fisher & Schmincke, 1984).

Pyroclastic deposits erupted on land can be classified according to the way the **tephra** was transported immediately prior to deposition. Several transport mechanisms are recognized: air-fall, pyroclastic flow (see **ash flow**), pyroclastic surge (see **base surge**) and **mudflow** or **lahar**. Each transport mechanism generates deposits with different degrees of lateral continuity, average grain size, and degree of *sorting* (Wright *et al.*, 1980; Cas & Wright, 1987). Pyroclastic deposits erupted and/or deposited below water are either accumulations from showers of **tephra** through the water, or from various types of *gravity flows*, ranging from **ash-flows** which entered the water from land, to water-saturated **turbidity currents**. (See **phreatic eruption**, **volcanic eruption**). [PTL]

Cas, R.A.F. & Wright, J.V. (1987) *Volcanic Successions Modern and Ancient*. Allen & Unwin, London.
Fisher, R.V. & Schmincke, H.-U. (1984) *Pyroclastic Rocks*. Springer-Verlag, Berlin.

Schmid, R. (1981) Descriptive nomenclature and classification of pyroclastic deposits and fragments: recommendations of the IUGS subcommission on the systematics of igneous rocks. *Geology* 9: 41–3.
Wright, J.V., Smith, A.L. & Self, S. (1980) A working terminology of pyroclastic deposits. *Journal of Volcanology and Goethermal Research* 8: 315–36.

pyrolusite (MnO_2) The most important **manganese ore** mineral. [DJV]

pyromorphite ($Pb_5(PO_4)_3Cl$) A supergene mineral found in the oxidized portions of **lead veins**. [DJV]

pyrope ($Mg_3Al_2Si_3O_{12}$) The **magnesium** garnet, deep red to black in colour and valued as a **gem** when clear and transparent. (See **garnet minerals**.) [DJV]

pyrophanite ($MnTiO_3$) An **oxide mineral** related to **ilmenite**. [DJV]

pyrophyllite ($Al_2Si_4O_{10}(OH)_2$) A comparatively rare *phyllosilicate* mineral found in metamorphic rocks. [DJV]

pyrophyllite deposit Pyrophyllite has many physical properties in common with **talc** and many similar uses. Deposits of economic importance are the result of the **hydrothermal alteration** of acidic volcanic rocks as in Japan and Korea, the leading world producers. It has also been formed during the metamorphism of **tuffs** as in the USA. [AME]

pyroxene minerals Silicate minerals whose internal structure consists of a single chain of linked silicate tetrahedra with cations occupying sites formed between oxygen ions at the edges of the chains.

STRUCTURE. The essential feature of the pyroxene structure (Fig. P32a) is the linkage of SiO_4 tetrahedra which each share two of their four corners to form a continuous chain. This chain is not straight, but staggered or kinked. The 'bases' of the tetrahedra form an approximate plane parallel to the chain length; the apices of the tetrahedra lie in a second, approximately parallel, plane. The repeat along the chain comprises two tetrahedra and is approximately 0.52 nm in length. The pyroxene structure is equivalent to half the **amphibole** double chain and there are strong structural similarities between **amphiboles** and pyroxenes. In both, the chains are held together by metal ions at sites formed mainly by the non-bridging oxygens (those linked only to one silicon). Strong bonds between apical oxygens and cations hold chains together in pairs, forming the characteristic I-beam building blocks (Fig. P32b). In the overall structure, the remaining cations are located at sites formed by oxygens at the edges of adjacent I-beams.

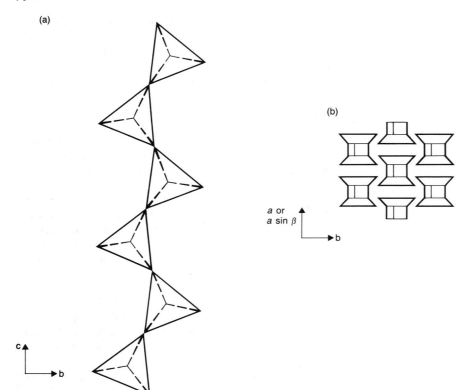

Fig. P32 Structure of **pyroxene minerals**: (a) Chain structure viewed along the length of the chain. (b) Chain structure viewed end-on, showing the stacking of I-beams and the location of cations within the structure.

The direction of polymerization of the chain defines the Z- or c-axis of the **crystal structure** according to the normal convention. There are two crystallographically distinct sites for metal ions and, depending on the precise details of the chain structure, either one or two distinct tetrahedral sites. The available ionic species are ordered to varying degrees among these sites.

Most naturally occurring pyroxenes are monoclinic (*clinopyroxene*). The Mg–Fe pyroxenes **enstatite** and *ferrosilite*, and intermediate compositions, are commonly orthorhombic (*orthopyroxene*) but monoclinic **polymorphs** are also known. There is a second orthorhombic form ('*protopyroxene*') which has not been found in nature but which has been synthesized in laboratory experiments and whose natural occurrence at high temperatures has been inferred from detailed study of pyroxenes thought to have changed from this form as they cooled.

CHEMISTRY AND NOMENCLATURE. The overall chemistry of the pyroxenes can be represented as

$$A_{1-p}(B,C)_{1+p}T_2O_6$$

where in common pyroxenes A is Na or Ca; B is Mg or Fe^{2+}; C is Al or Fe^{3+}; and T is Si or Al.

Nomenclature of pyroxenes was rationalized in 1988 on the basis of crystal chemistry. The principal groups now recognized are: Mg–Fe pyroxenes; Ca pyroxenes; Ca–Na (or Na–Ca) pyroxenes; Na pyroxenes; Mn–Mg pyroxenes; and Li pyroxene. Classification details are given in Table

P3. The interested reader is referred to Morimoto (1988) for fuller details including the names now recognized and those formally abandoned.

In the Fe^{2+}- and Mg-containing pyroxenes, there is wide variation in the Mg/Fe^{2+} ratio. There is complete *solid solution* between Mg and Fe^{2+} end members in the major groups but pyroxenes with high Fe/Mg ratios ($Fe^{2+}/(Mg + Fe^{2+})$ over 85%) in the Ca–Mg–Fe system characteristically contain significant amounts of Mn. Ca and Na can replace Mg and Fe^{2+} at the larger cation site between the edges of the I-beams in the structure (A in the formula). Because they are significantly larger than either Mg or Fe^{2+}, there is a limit to the extent to which either can be introduced into the Mg–Fe pyroxenes without a change in the structure. There is a miscibility gap between these and the calcic and sodic varieties, except at elevated temperatures where the components may mix. As in **amphiboles**, **exsolution** may occur during cooling of a mineral which forms at high temperature as a single phase. For example, **augite** occurs widely as **exsolution** lamellae in *orthopyroxene*, as a product of the 'inversion' of **pigeonite** on cooling.

The stability relations between the various pyroxene compositions are complex and have been widely studied, both by laboratory experiments and by investigation of naturally occurring assemblages. The interested reader is referred to articles in Prewitt (1980) for details.

Replacement of a divalent ion by Na (at A in the formula) can only take place in a linked substitution, commonly involving the introduction of Al or Fe^{3+} at C to maintain

Table P3 Pyroxene minerals. The major pyroxene groups. (From Morimoto, 1988.)

General formula: $A_{1-p}(B,C)_{1+p}T_2O_6$

Ca–Mg–Fe²⁺ pyroxenes
The end member compositions for this group are: $Mg_2Si_2O_6$ (enstatite); $Fe_2Si_2O_6$ (ferrosilite); and $Ca_2Si_2O_6$ [the pyroxenoid wollastonite]

Where $Ca/(Ca + Mg + Fe^{2+}) < 5\%$, the pyroxene is:
[clino]enstatite, if $Mg/(Mg + Fe^{2+}) < 50\%$
[clino]ferrosilite, if $Mg/(Mg + Fe^{2+}) > 50\%$
(prefix 'clino' if monoclinic, no prefix if orthorhombic)

Where $5\% < Ca/(Ca + Mg + Fe^{2+}) < 20\%$, the pyroxene is pigeonite

Where $20\% < Ca/(Ca + Mg + Fe^{2+}) < 45\%$, the pyroxene is augite

Where $45\% < Ca/(Ca + Mg + Fe^{2+}) < 50\%$, the pyroxene is:
diopside, if $Mg/(Mg + Fe^{2+}) < 50\%$
hedenbergite, if $Mg/(Mg + Fe^{2+}) > 50\%$

$Ca/(Ca + Mg + Fe^{2+})$ should not exceed 50%

Ca–Na and Na-pyroxenes
The end members for this group are $NaAlSi_2O_6$ (jadeite); $NaFe^{3+}Si_2O_6$ (aegirine); and the combined members of the Ca–Mg–Fe²⁺ group

Where $Na_A > 80\%$, the pyroxene is:
jadeite, if $Al/(Al + Fe^{3+}) < 50\%$
aegirine, if $Al/(Al + Fe^{3+}) > 50\%$

Where $80\% > Na_A > 20\%$, the pyroxene is:
omphacite, if $Al/(Al + Fe^{3+}) < 50\%$
aegirine-augite, if $Al/(Al + Fe^{3+}) > 50\%$

Where $Na_A < 20\%$, the pyroxenes are classified in the Ca–Mg–Fe system.

Others
Other recognized pyroxene end members include:

Mn-pyroxenes:
johannsenite, $CaMnSi_2O_6$
kanoite, $MnMgSi_2O_6$

Fe³⁺-pyroxenes:
acmite, $NaFe^{3+}Si_2O_6$
fassaite, $CaFe^{3+}AlSiO_6$

charge balance. There is generally some substitution of Al for Si in the tetrahedral chain (T in the formula), charge balance again being maintained by introduction of higher valent ions such as Fe^{3+} (at C in the formula). The charge coupled substitution of Al both for divalent ions at B and for Si at T is often referred to as *Tschermak's component*. In particular, the composition CaAl.AlSiO₆ is called 'calcium Tschermak's molecule' or 'calcium Tschermak's component'; this is not known as a natural pyroxene and is not formally named, but its composition represents an important theoretical end member in pyroxene composition space. The composition $Ca_2Si_2O_6$ (corresponding to the **pyroxenoid**, **wollastonite**, but not to any known natural pyroxene) is also used as a compositional component.

As in **amphiboles**, the Li-containing species (**spodumene**, $LiAlSi_2O_6$) is distinct because of the small size of the Li^+ cation which enters at A in the formula. **Spodumenes** in most cases contain almost the formula proportions of Li and Al.

Of less common substituents, trivalent and tetravalent **titanium** have been found in pyroxenes in a *meteorite* which fell in Allende, Chile, in 1973.

OCCURRENCE. Pyroxenes are stable over a wide variety of pressure and temperature conditions. They play a major role in igneous geology and in a wide range of metamorphic events.

The melting point for most pyroxenes is between about 1200 and 1600 °C at the Earth's surface, although at many compositions melting is complicated by the phase relationships between different pyroxene **polymorphs** and by incongruous melting. For example, **enstatite** melts to **olivine + quartz** at 1577 °C:

$$Mg_2Si_2O_6 \rightarrow Mg_2SiO_4 \text{ (solid)} + SiO_2 \text{ (liquid)}$$

Melting point generally decreases with increasing iron content. The sodic pyroxene **acmite**, $NaFe^{3+}Si_2O_6$, melts incongruently at the low temperature of 990 °C with separation of **hematite** (Fe_2O_3).

Magnesium-rich *orthopyroxenes* are an early crystallization product of basic **magmas**. They form at early and intermediate stages in igneous processes, both by direct crystallization and by reaction of earlier-formed phases (such as **olivine**) with the melt: they are common constituents of some ultrabasic rocks. *Orthopyroxenes* may also result from inversion, on cooling, of Fe–Mg *clinopyroxenes* crystallized from more **iron**-rich differentiates of basic **magmas**. They occur extensively in *layered intrusions* such as Skaergaard or Stillwater. They are common in metamorphic rocks of medium and higher-grade metamorphism. It is thought that high-**magnesium** *orthopyroxenes* are also important constituents of the **upper mantle**.

Monoclinic **polymorphs** of the *orthopyroxenes*, *clinoenstatite* and *clinoferrosilite*, and a second orthorhombic form (sometimes called *protoenstatite*), all occur naturally. It is generally believed that the orthorhombic forms are the thermodynamically stable ones at the Earth's surface, but the **polymorphism** between the species is complex and not fully understood.

Clinopyroxenes of the **diopside–hedenbergite** series occur in a variety of metamorphic rocks, the more **magnesium**-rich varieties being characteristic of rocks derived from calcium-rich sediments (e.g. siliceous **dolomites**). **Augites**, by contrast, occur mainly in **igneous rocks**, e.g. **gabbros** and **basalts**; *aegirines* and *aegirine-augites* occur most commonly as the later products of crystallization of alkaline **magmas**.

Pyroxenes are commonly the most abundant **iron–magnesium** phase of **basaltic** rocks, and can accommodate all the major elements of **basalts** except **potassium**. As well as being common terrestrial **igneous rocks**, **basalts** occur both on the Moon and in *meteorites* and their study provides the means to compare the different characteristics of planetary formation and history. Thus knowledge of pyroxenes provides a powerful planetary probe.

PYROXENES IN GEOTHERMOMETRY AND GEOBAROMETRY. Equations used to represent phase equilibria can often be rewritten to allow the temperature (or pressure) of equilibration to be calculated from mineral composition and known thermodynamic constants. The relative simplicity of the pyroxene structure (it contains only two crystallographically distinct cation sites) means that it is also possible to model the distribution of cations over the two sites, and between coexisting pyroxenes or between pyroxenes and other phases, in terms of *solid solution* mixing. This, and the widespread occurrence of pyroxenes, gives them considerable value in estimating the temperatures or pressures of equilibration of natural pyroxene-bearing assemblages, i.e. as **geothermometers** or **geobarometers**.

RECOGNITION. Pyroxenes, like **amphiboles**, exhibit a range of colours but are normally pale to dark green or brown. They normally have vitreous **lustre** and good **cleavages** which can be seen in hand specimen or thin section. Hardness is normally 5 or 6 on Mohs' scale, and density is between 3.0 and 3.6 Mg m^{-3}. In hand specimen, *orthopyroxenes* may appear as irregular grains or coarse **cleavage** fragments; otherwise, pyroxenes form short or, in some cases, longer prismatic crystals.

In thin section, pyroxenes range in colour from colourless to green or brown. Except for Fe^{3+}-containing species (*acmites*) and *orthopyroxenes* close to *ferrosilite*, they are typically only weakly pleochroic. Crystals show a characteristic nearly square cross-section, which distinguishes them from **amphiboles** (diamond-shaped). Most optical properties such as pleochroism and birefringence are composition dependent. [ADL]

Deer, W.A., Howie, R.A. & Zussman, J. (1978) *Rock Forming Minerals* Vol 2A: *Single Chain Silicates* (2nd edn). Longman, Harlow.
Morimoto, N. (1988). Nomenclature of pyroxenes. *American Mineralogist* **73**: 1123–33.
Papike, J.J. (ed.) (1969) *Pyroxenes and Amphiboles: Crystal Chemistry and Phase Petrology.* Special Paper of the Mineralogical Society of America **2**.
Prewitt, C.T. (ed.) (1980) *Pyroxenes.* Reviews in Mineralogy Vol. 7, Mineralogical Society of America, Washington, DC. [Course notes from a short course arranged by the MSA and containing authoritative material covering all aspects of pyroxene mineralogy.]
Saxena, S.K. (1973) *Thermodynamics of Rock-forming Crystalline Solutions.* Springer-Verlag, Berlin.

pyroxenoid minerals A group of silicate **minerals** which are analogous in formula to **pyroxenes**, having a Si:O ratio of 1:3, but do not crystallize in the **pyroxene** structure.

The pyroxenoid group includes **wollastonite**, CaSiO$_3$; **rhodonite**, MnSiO$_3$; *bustamite*, CaMnSi$_2$O$_6$; *pyroxmangite*, (Mn,Fe)SiO$_3$; and **pectolite**, Ca$_2$NaHSi$_3$O$_9$. The triclinic chain structures are conditioned by the large size of the calcium or **manganese** cation. In **wollastonite**, **pectolite** and *bustamite* the chain repeat is three tetrahedra (approximately 0.73 nm); in **rhodonite** it is five (1.22 nm) and in *pyroxmangite* seven (1.74 nm).

The most important pyroxenoid is **wollastonite**. Although not a **pyroxene**, it is used as a notional end-member to describe **pyroxene** compositions. It is formed by metamorphism of impure **limestones** in conditions of high grade regional metamorphism, and in igneous contact zones where silicon may be present in the sediment or introduced by **metasomatism**. It is white or greyish, may be fibrous, and has vitreous or (in fibrous varieties) silky **lustre**. It dissolves in dilute hydrochloric acid with separation of **silica**.

Particularly at high temperatures, the **wollastonite** structure can accommodate considerable quantities of **iron** and **manganese** replacing calcium. Because ions of intermediate size, such as Fe^{2+} and Mn, may behave as 'small' cations at low temperature and high pressure but as 'large' cations at low pressure, pyroxenoids containing substantial **iron** or **manganese** may be **polymorphic** with **pyroxenes**. For example, small interlocking grains of **iron**-rich **pyroxene** observed in rocks of the Skaergaard intrusion may have originated by inversion of **iron**-rich **wollastonite**.

The **manganese**-containing pyroxenoids **rhodonite** and *bustamite* are usually associated with **manganese orebodies** and often result from **metasomatic** activity. They are typically pinkish; **rhodonite** in particular alters readily to black **manganese** oxide which appears as **veins** in the **mineral**. **Rhodonite** can contain a significant amount of calcium but *solid solution* between **rhodonite** and *bustamite* is limited.

Pectolite forms aggregates of sharp needle-like crystals (it should be handled with care). In composition, most **pectolites** approach the formula Ca$_2$NaHSi$_3$O$_9$, but Mn^{2+} can replace Ca which increases the density and changes the optical properties. Like **wollastonite**, it is decomposed by hydrochloric acid with separation of **silica**. It occurs typically as a **hydrothermal** mineral, for example in association with **zeolites** in cavities in basic **igneous rocks**. [ADL]

Deer, W.A., Howie, R.A. & Zussman, J. (1978) *Rock Forming Minerals* Vol. 2A: *Single Chain Silicates* (2nd edn). Longman, Harlow.

pyroxferroite (Ca$_{0.15}$Fe$_{0.85}$SiO$_3$) A **pyroxenoid mineral** found in lunar lavas. [DJV]

pyroxmangite ((Mn,Fe)SiO$_3$) A **pyroxenoid mineral**. [DJV]

pyrrhotite (Fe$_{1-x}$S) A widely occurring **sulphide mineral**. [DJV]

Q

quadratic elongation A linear measure of shape change (distortional **strain**) based on changes in line lengths thus:

$$\text{quadratic elongation} = \left[\frac{\text{deformed length}}{\text{undeformed length}}\right]^2$$

Quadratic elongation is the square of the **stretch**, and simply related to the **elongation**. It is used to define the values of the **principal strains**, and so to define shapes of **strain ellipsoids**. It is also useful in theoretical **strain** calculations. (See **Mohr diagram**.) [SB]

quartz (SiO_2) The most commonly occurring **silica mineral**, important in a wide variety of rocks and mineral deposits.

Quartz pebbles and quartz quarried from **veins** were both used in antiquity for tool manufacture. Quartz pebbles were particularly common in early Paleolithic sites of China and Africa and are frequent finds at the early *hominid* site of Olduvai Gorge. In northern parts of Sweden where there is no locally available **flint**, quartz was the dominant tool raw material, with indications of prehistoric quarrying of nearby quartz **veins**. [OWT/DJV]

quartzite Quartzite was widely used by early man for tool manufacture, despite its poor flaking qualities compared with, e.g. **flint**. However, because of their frequent availability, particularly in river **gravels**, quartzite pebbles were very commonly used in early Paleolithic settlements in Europe, Asia, India and in Africa at such important sites as Olduvai Gorge. Since quartzite can occur as large masses, it was also used as a **building stone**, especially by the ancient Egyptians who used it in their tombs and statues. The Colossi of Memnon at Thebes are carved from a block of quartzite of *c.* 720 tonnes. [OWT]

Quaternary The most recent of the geological periods, during which large-scale climatic oscillations occurred on a world-wide basis, leading to *glacials* and **interglacials** in middle and high latitudes of both hemispheres and to **pluvials** and *interpluvials* at more equatorial latitudes. These two responses to climatic change are not necessarily synchronous.

Originally the Quaternary was equated with the last million years of geologic time but, with the recognition that *glacial*-**interglacial** cycles continue back further than this, the Quaternary period now begins at *c.* 2 Ma. It is subdivided into two unequal parts; the *Pleistocene* from the beginning of the Quaternary to 10 000 years ago and the *Holocene* dating from 10 000 years ago to the present day.

The concept of an **Ice Age** was put on a scientific basis by Agassiz in 1837 when he showed that evidence in Switzerland could only be explained if **glaciers** had once been very much more extensive than they are today. Almost immediately geologists such as Buckland and Lyell extended the application to northern Europe. In 1909, Penck and Brückner recognized that there were at least four separate *glacial*–**interglacial** cycles from their studies of the river terraces of the alpine rivers and their relationship to the *moraines* upstream. These four glacial periods were called, in order of decreasing age, *Gunz, Mindel, Riss* and *Würm* after the alpine rivers involved in their investigations. Similar fourfold *glaciations* were then recognized in northern Europe and named *Elster, Saale, Warthe* and *Weichsel*. In North America a similar sequence of *glaciations* was named: *Nebraskan, Kansan, Illinoian, Iowan* and *Wisconsin*. Several episodes of *glaciation* are recognized in the southern hemisphere in New Zealand and the southern Andes.

In the second half of this century, this sequence of *glacials* and their intervening **interglacials** has been largely superseded by an elaborate local nomenclature that takes into account that there were many more major climatic oscillations than had hitherto been suspected. More up-to-date classifications and correlations of Quaternary deposits may be found in Sibrava *et al.* (1986).

The reasons for this revolution in Quaternary sciences stemmed, in the beginning, from the work of Emiliani (1966) who published the results of an investigation of the **oxygen isotope** ratio (between $\delta^{18}O$ and $\delta^{16}O$) in the shells of **Foraminifera** at various depths down a core in marine Quaternary deposits in the Caribbean. Interpreting the changes in this ratio as reflections of changing temperatures of the ocean water, he recognized that there had been many *glacial*/**interglacial** cycles and that *glacials* were on the whole much longer than **interglacial** periods. This contrasts with the view that prevailed since Penck and Brückner, that the *glacial* events were relatively short. Shackleton (1967) pointed out that much of the **oxygen isotopic** ratio was a reflection not of ocean temperature changes but of global ice volume. The changes in ratio should then be universal throughout the Earth's oceans and thus long Quaternary cores could be correlated using isotope stages. There has been a widespread tendency to utilize these isotopic stages as a basis of classification of terrestrial Quaternary sequences. Caution, however should be exercized in this application because the deposits reflect local environmental conditions which may differ markedly from the globally integrated signal of the isotopic ratios.

Application of **magnetic** and **radiocarbon dating** techniques have set the isotopic stages in a secure time dimension, so that it has been possible to calculate the periodicities of the major climatic fluctuations. It has become apparent that the *glacial*/**interglacial** cycles match to a remarkable degree the changes in insolation in the middle latitudes of the northern hemisphere caused by the perturbations in the Earth's orbit round the Sun in turn caused by the presence of the other planets in the solar system (Hays *et al.*, 1976). It would seem that changes in eccentricity, precession of the equinoxes and changes in the angle of tilt of the Earth's axis relative to the plane of the ecliptic, all contribute to the periodicity of *glaciations*. Since these astronomical variations can be predicted, as so-called *Milankovich cycles*, it is possible to say with certainty that the present day represents the final phase of an **interglacial** and that fully *glacial* conditions will be reimposed about 20 000 years in the future.

Deep ice cores taken through the world's *ice cap* reach beyond the last **interglacial** and record changes in a variety of components of the atmosphere including volcanic and nuclear fall-out. Most important is the preservation in the ice of samples of the atmospheric carbon dioxide during the past *glacial*/**interglacial** cycle. It is apparent that the carbon dioxide has an effect on the timing and intensity of *glaciation*, high values being associated with warmer periods. If the demonstrable increase in atmospheric carbon dioxide caused by the burning of fossil fuels and the destruction of forests has a similar warming effect, we should expect an episode of global increase in temperature in the coming century.

The effect of Quaternary climatic changes has been dra-matic on the flora and fauna in changing their geographic distributions. For the most part, species remained the same throughout this period with the conspicuous exception of the **mammals**. Many of the latter evolved rapidly and there is abundant evidence of **extinctions**, especially amongst the larger species. It has been suggested that the extermination of many of these species may be due to human exploitation, though climatic factors probably contributed to their vulnerability.

The rise of *Homo* to a position of dominance on the Earth takes place within the Quaternary; *Homo erectus* emerged from Africa and rapidly colonized most of Eurasia during the Lower *Pleistocene*. The use of stone tools manufactured to a set pattern and the control of fire even in the early *Pleistocene* shows considerable sophistication in the exploitation of the environment. The appearance of *Homo sapiens* and the invasion of Australia took place during the Last *Glaciation* and of North America probably rather later towards the end of the same *glacial* period. Only during the *Holocene* has human culture so dominated the Earth so as to alter the environment in ways that have no analogue in Quaternary time. [GRC]

Emiliani, C. (1966) Palaeotemperature analysis of Caribbean cores P6304-8 and P6304-9 and a generalized temperature curve for the past 425 000 years. *Journal of Geology* **74**: 109–24.

Hays, J.D., Imbrie, J. & Shackleton, N.J. (1976) Variations in the earth's orbit: pacemaker of the ice ages. *Science* **194**: 1121–32.

Imbrie, J. & Imbrie, K.P. (1978) *Ice Ages, Solving the Mystery.* Macmillan, London.

Nilsson, T. (1983) *The Pleistocene, Geology and Life of the Quaternary Ice Age.* Reidel, Dordrecht.

Sibrava, V., Bowen, D.Q. & Richmond, G. (1986) Glaciations of the Northern hemisphere. *Quaternary Science Reviews* **5**: 1–513.

Shackleton, N.J. (1967) Oxygen isotope analysis and Pleistocene temperatures reassessed. *Nature* **215**: 15–17.

West, R.G. (1968) *Pleistocene Geology and Biology.* Longman, London.

R

radar A method of electromagnetic distance measurement based on the reflection time of a pulse of radiation. The technique can be adapted to geophysical exploration, and a ground-based radar transmitter used to provide a depth section very similar to a **seismic reflection seismogram**. The method is extensively used in small-scale **site investigation** and archeological surveys. [PK]

Davis, J.L. & Annan, A.P. (1989) Ground-penetrating radar for high-resolution mapping of soil and rock stratigraphy. *Geophysical Prospecting* **37**: 531–51.

radial dykes A set of steep **dykes** arranged in a radial pattern around a central **pluton**. The radial pattern may be explained by the local **stress** field generated by the plutonic body. Examples of radial dyke sets are found around several of the *Tertiary* igneous complexes of north-west Scotland, e.g. Skye. These sets become regional **dyke swarms** when traced outwards from their centres. (See **igneous body**.) [RGP]

radiocarbon dating A method of obtaining absolute dates for certain types of organic material. Also known as *carbon-14 dating*, it is one of the most important and successful dating methods so far developed. It has been applied to recent geological deposits which can be dated by associated layers of charcoal (for example ages of eruptions of Thera (Santorini) volcano, Greece). However, its main application has been in archeology for the dating of charcoal, and to a lesser extent, of bone and antler.

Principles of the methods are as follows: all living organisms take up **carbon** from the environment, and this **carbon** includes a proportion of the radioactive isotope, carbon-14 (formed from nitrogen-14 as a result of cosmic ray entry into the Earth's atmosphere). The amount of **carbon** in living things reaches an equilibrium value until the animal or plant dies, after which it no longer takes up **carbon**, and the already present carbon-14 begins to decay at a known rate. By measuring (from its β emission) the amount of carbon-14 present now in a once-living organism, and with a knowledge of the decay rate of carbon-14 and the equilibrium value, the length of time elapsed since the organism's death can be calculated. Thus a date with an associated error is obtained for the archeological horizon which contained the sample dated.

Problems of the methods include changing estimates of the rate of decay of carbon-14 and, more important, variation over time of the carbon-14 reservoir in the Earth's environment, due largely to changes in the cosmic ray flux. This has been shown by the dating of long-living bristle-cone pine trees in California by both radiocarbon and tree-ring counting (**dendrochronology**) methods; the **dendrochronology** indicated dates significantly earlier than the radiocarbon dates for periods of time before *c.* 1000 years BC, indicating a higher amount of carbon-14 in the Earth's atmosphere in the past. By comparing **dendrochronological** dates and radiocarbon dates for wood samples of up to 6500 years old, a calibration has been produced to correct radiocarbon dates for this factor.

Radiocarbon dating is normally suitable only for samples of less than *c.* 50 000 years old because of the small amounts of carbon-14 remaining to be measured in older samples. The recent development of **accelerator radiocarbon dating** has reduced the sample size required and may in the future extend the range of datable material back beyond 50 000 years BC.

Radiocarbon dating radically altered many theories on prehistoric development of ideas and technology. An important example of this is the dating of the development of domestication of plants and animals; radiocarbon dating showed that this took place in the Near East (in the Levant and Anatolia) soon after 8000 BC and spread to south-east Europe around 6000 BC. The spread of agriculture to western Europe, previously thought to be after 3000 BC, was put back to *c.* 5000 BC. Similarly, it was demonstrated that megalithic (large stone) architecture was practised in Neolithic western Europe before its supposed prototypes in the eastern Mediterranean, thus undermining the established idea of major developments diffusing from the East, and paving the way for a more enlightened view of independent development of similar ideas in different parts of the Old World. (See Fig. R1.) [OWT]

Tite, M.S. (1972) *Methods of Physical Examination in Archaeology.* Seminar Press, London.

Radiolaria A subclass of marine *protozoans* whose **silica** skeletons make an important contribution to deep-sea sediments and ancient radiolarian **cherts**. This group of free-floating, single-celled *protozoans* has roughly spherical cells with stiff, thread-like pseudopodia that extend radially over a delicate silica endoskeleton. These pseudopods are used to catch prey whilst buoyancy is regulated through modulation of the frothy ectoplasm. Ectoplasm and endoplasm are characteristically separated by a tough central capsular membrane.

The radiolarian skeleton is generally of solid **opaline** silica and consists, in the simplest forms, of either radial

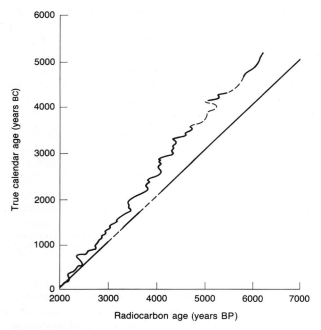

Fig. R1 Radiocarbon dating. Relationship between the radiocarbon age (years BP calculated using the 5568 year half-life) of tree-ring dated wood and the true calendar tree-ring age (years BC), as determined from tree-ring studies on the bristlecone pines and other long-lived trees. The diagonal straight line represents the situation for coincidence of the radiocarbon and true calendar ages. The calibration curve is shown as a dashed line for those periods for which there is a scarcity of data. (After Tite, 1972.)

elements (spicules, spines, bars) that serve to support the radiating pseudopods, or tangential elements (lattice shell) that serve to protect the bulk of the cytoplasm, or of both elements in combination. The form of the skeleton is used to distinguish the different orders of radiolarians: Spumellaria (spherical or discoidal lattice, plus radial elements); Nasellaria (a primary spicule, a spinose ring or an elongate lattice shell showing bipolar or heteropolar symmetry). The aforementioned polycystine radiolarians with solid skeletal elements can be differentiated from Phaeodaria with hollow elements; these are rarely preserved.

Radiolarian remains preserved in **cherts** and in **flints** may be studied in petrographic thin-sections or isolated by the use of weak hydrofluoric acid. Weaker acids may be used on calcareous rocks and boiling with various chemicals may free the remains from shales, **chalks** or deep-sea oozes. Specimens are then studied by transmitted light or scanning **electron microscopy**.

At the present day, radiolarians are mainly to be found seaward of the continental slope, in the upper few hundred metres of the water column and especially in regions of upwelling enriched in food and **silica**. They are generally most abundant and diverse in equatorial latitudes where their silica skeletons sink through the water column and contribute to an equatorial belt of siliceous *radiolarian oozes*. These tend to accumulate below

the calcium **carbonate compensation depth** (*c.* 3000 to 5000 m) and may contain as many as 100 000 skeletons per gram of sediment. Since the oceans are undersaturated with respect to silica, delicate radiolarian skeletons often suffer **dissolution** where sedimentation rates are slow and their remains are less common than might be expected. The red muds of the *abyssal plains* contain only those parts of radiolarian skeletons most resistant to solution. *Radiolarian oozes* and **cherts** appear to have been characteristic of oceanic sediments throughout much of the **Phanerozoic**, though the group is thought to have flourished mainly over the continental shelves in **Paleozoic** times.

The **dissolution** suffered by these delicate siliceous skeletons means that their fossil record is rather patchy. Even so, they are valued as **biostratigraphic** indices of oceanic or deeper shelf siliceous sediments and can provide useful zonal schemes for **Mesozoic** and **Cenozoic** sediments, such as those encountered in the **Deep Sea Drilling Project**.

The earliest radiolarians were spumellarians of middle **Cambrian** age and this group developed gradually through the course of the **Paleozoic**. Nasellarians appeared by late **Triassic** times and the whole group reached its peak of diversity and abundance during the **Cretaceous** Period, at a time of relatively high sea-levels and widespread **pelagic** sedimentation. The subsequent decline through the course of the **Cenozoic** may be related to competitive pressure for silica brought about by the *diatom* radiation. [MDB]

Anderson, O.R. (1983) *Radiolaria*. Springer-Verlag, New York.
Bolli, H.M., Saunders, J.B. & Perch-Nielsen, K. (eds) (1985) *Plankton Stratigraphy*. Cambridge University Press, Cambridge.

radiometric assay Determination of the concentration of **uranium** in buried bone, achieved by measuring the *radioactivity* of the sample. The **uranium** is gradually absorbed into the bone from percolating **groundwater**, so that this measurement gives an estimate of the date of burial. Generally used for relative dating of bones from the same site (cf. **fluorine test**). [OWT]

radiometric surveying The geophysical detection of *radioactive* elements. There are about twenty naturally occurring *radioactive* elements, but the majority are very rare or only weakly *radioactive*. The only elements common and *radioactive* enough to be of relevance in exploration are **uranium** (U), **thorium** (Th) and **potassium** (^{40}K). **Uranium** and **thorium** are important as sources of nuclear fuel and occur in several minerals such as **pitchblende** and **monazite**. They decay to **lead** via three decay paths.

There are three types of radiation emitted during decay. α-particles correspond to helium nuclei ejected from the nucleus during disintegration. β-rays correspond to the electrons emitted when neutrons decay to protons. γ-rays represent the electromagnetic radiation emitted during decay, and can also originate by *K*-capture when an electron from the innermost orbit enters the nucleus, forming a different element. α-particles are very weak, and β-rays can only propagate a few metres. γ-rays, however, can travel several hundreds of metres, and are the main target in surveying.

Two forms of instrument are used to detect radiation. *Geiger counters* monitor β-rays and are thus restricted to ground surveys. β-rays cause the ionization of a gas, and the positive ions and electrons so formed are accelerated, cause a discharge pulse across an anode resistor and register their presence as a series of audible clicks. The instrument has to be very close to the outcrop of *radioactive* rock for its β-rays to be detected. The *scintillation counter* (*scintillometer*) measures γ-rays by counting the scintillations they produce on a screen made of some substance such as **zinc** sulphide or sodium iodide treated with thallium. The *gamma-ray spectrometer* is an extension of the *scintillometer* which is capable of separating the characteristic γ-rays of **uranium**, **thorium** and **potassium-40** so that the source can be identified. Interpretation of the data is usually only qualitative, with readings significantly above the background radiation level indicating the presence of *radioactive* material, which is then investigated further using other techniques. Both the *scintillation counter* and *gamma-ray spectrometer* are commonly mounted in aircraft. (See **airborne geophysical surveys**.)

Radiometric techniques involving both natural and induced *radioactivity* are routinely used in **geophysical borehole logging**. [PK]

Durrance, E.M. (1986) *Radioactivity in Geology*. Ellis Horwood, Chichester.

radium (Ra) At. no. 88, at. wt. 226.0254. A *radioactive* metal used in self-luminous paints, as a neutron source and in radiotherapy (being supplanted by artificial radioisotopes). Isolated from **uranium ores**. [AME]

raindrop and hailstone-generated structure Sedimentary structure formed by rain or hail on unconsolidated sediment.

Rain, hail or spray impact marks consist of craters up to a centimetre in diameter, several millimetres deep with raised rims. They are circular in plan, or elliptical if the precipitation is driven by wind or the sediment surface has a slope. When they completely cover a surface, they interfere to produce a polygonal pattern of narrow ridges separating irregular, partly connected depressions. Some pits contain delicate inner rings of sediment which has been attracted by the surface tension of a melting hailstone or an impacting raindrop. Such surface-tension structures can form without impact craters when hailstones melt after impacting very compacted sediment or after transport by water or rolling by wind (Rubin & Hunter, 1984). Surface tension structures in **sand** may help distinguish impact marks from subaqueous gas-bubble escape struc-tures with which they may otherwise be confused, but gas bubble surfaces can also attract at least mud-grade sediment. Rain and hail impact marks may also be confused with foam impressions.

Rain impact ripples, best described by Clifton (1977), are produced when large raindrops are driven obliquely onto fine **sand** by a strong wind. They are similar in appearance and size to **adhesion ripples** except that their steeper, often overhanging sides and migration directions are downwind.

All these structures have a much higher preservation potential in cohesive mud than in **sand**, which is likely to be reworked after the surface dries. [DAR]

Clifton, H.E. (1977) Rain impact ripples. *Journal of Sedimentary Petrology* **47**: 678–9.
Rubin, D.M. & Hunter, R.E. (1984) Sedimentary structures formed in sand by surface tension on melting hailstones. *Journal of Sedimentary Petrology* **54**: 581–2.

raindrop impact erosion Movement of *soil* particles on the *soil* surface as a result of raindrop impact by a combination of rebound, dislodging downslope and undermining. On level ground *rainsplash erosion* may cause crusting of the *soil* surface, thereby lowering the **infiltration capacity**, though otherwise it achieves little more than rearrangement of particles. On a slope, however, there is a net transport of material downslope both because the splash trajectory is longer in the downslope direction, and because of the downslope component of the impact force. An exception to this may occur, however, if the raindrop trajectory is not vertical but is towards the slope; in this unusual case the net splash transport could then be upslope. The size of particles dislodged is related to the momentum of the raindrops. It is therefore likely that the *soil* particles moved by rainsplash will be the finer fraction of the *soil* material. The effect of raindrop impact erosion is much reduced by the presence of **overland flow** of depths greater than about 5 mm. It is therefore likely that splash is most effective when no surface **runoff** occurs or when it is close to the divide where flow depths are necessarily limited. Rainsplash erosion contributes to the convex shape of the drainage divide observed on many **badland** slopes in semi-arid areas. [TPB]

raised beach **Beach deposits** stranded at altitude by a fall in relative sea-level. It may be part of a raised shoreline which may also include **estuarine** muds and silts, shell beds and **shore platforms** backed by former **sea cliffs**. Many factors cause sea-level change but the principal ways in which beaches become elevated are global sea-level change or **eustasy**, local depression of the land surface or **isostasy**, and **tectonic** elevation of the land surface. Raised beaches in lower latitudes outside the former extent of *Pleistocene* ice are generally related to changes in global sea-level. The raised beaches of the Mediterranean Sea are good examples of this, although

hydroisostasy, the loading of continental shelves by water, may be a complicating factor. The limited shelf and **tectonic** stability of some **coral** islands lead to their identification as '*Pleistocene* dip-sticks' from which a clearer **eustatic** signal could be derived. Raised beaches in **isostatically** loaded areas show different elevations dependent on their distance from the loading centre. Spectacular and youthful raised beaches and shorelines occur in tectonically active areas like New Zealand and Japan. [JDH]

Pethick, J.S. (1984) *An Introduction to Coastal Geomorphology*. Arnold, London.

ramp The part of a **fault** that cuts across datum surfaces, most commonly **bedding**, in *thrust* and **extensional fault systems**. In *thrust systems*, *frontal ramps* trend normal to transport direction, *lateral ramps* parallel (i.e. with **strike-slip** geometry) and *oblique ramps* at an angle between. **Hangingwall** *ramps* and *footwall ramps* are distinguished by the truncation of **bedding** or datum surfaces such as a **foliation**, which is seen in the **hangingwall** to the **fault** in **hangingwall** *ramps* and in the **footwall** in *footwall ramps* (Fig. R2). Segments of **faults** that are parallel to **bedding** surfaces, or close to horizontal, are known as *flats*. A **fault** surface composed of *flats* and *ramps* has a *stairstep* or *staircase* trajectory. In **strike-slip fault systems**, analogous structures are called *restraining* or *releasing bends*. (See **thrust belts, imbricate structure**.) [SB]

Boyer, S.E. & Elliott, D. (1982) Thrust systems. *American Association of Petroleum Geologists Bulletin* **66**: 1196–230.
Butler, R.W.H. (1982) Structural evolution in the Moine of northwest Scotland: A Caledonian linked thrust system? *Geological Magazine* **123**: 1–11.

random fabric/structure Fabric or set of structures with no order or **preferred orientation**; sometimes termed *isotropic*. [TGB]

rank The rank of a **coal** indicates the degree of *coalification* it has reached. *Hilt's law* of 1873 states that rank increases with depth; this is in response to the concomitant increase in temperature which might also arise at shallow depths in the neighbourhood of igneous intrusions. The transition from low rank **brown coal** to high rank **anthracite** is marked by an increase from 70% to 95% in **carbon**, a decrease from 5% to 3% in **hydrogen** and from 20% to about 3% in oxygen. With increasing rank the woody cells of **brown coals** are homogenized and compacted giving rise to the formation of the *vitrinite* and other **macerals** of **bituminous coals** and **anthracite**. The rank of a **coal** can be measured by determining the *reflectance* of its *vitrinite* which increases from about 0.3 to 2.5%. [AME]

rapakivi texture A porphyritic texture in which rounded *phenocrysts* of **potassium feldspar** (often a few centimetres in diameter) are surrounded by a rim of sodium-rich **plagioclase** (usually **oligoclase**). Rapakivi-textured **granites** are characteristic of **Proterozoic** *anorthosite–granite* associations and occur in Scandinavia, South Greenland and the western USA. [RST]

rapids A section of a river channel in which the **flow** is locally faster and more **turbulent** than on other reaches. Such a section is usually characterized by a steep incline, though without sufficient break of slope to form a **waterfall**, and resistant rock outcrops which obstruct the water's flow. [NJM]

rare earth elements Elements with atomic numbers from 57 to 71 plus scandium and **yttrium**. The rare earth elements are important for archeological work in the charac-

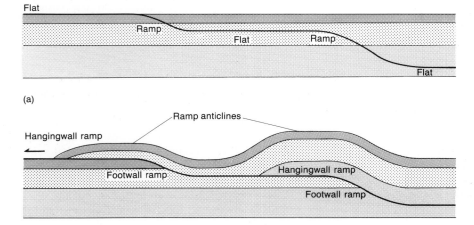

Fig. R2 Sketches showing a thrust fault with a stairstep trajectory (a) before and (b) after displacement. **Ramps** in the footwall and hangingwall show where the fault cuts across bedding surfaces, i.e. they are marked by bedding cutoffs. In (b), footwall and hangingwall ramps are separated after displacement; folds are seen above hangingwall ramps, and where the hangingwall has moved above footwall ramps.

Fig. R3 The **reactivation surface** of a dune bedform. The dune migrated from left to right over upper stage plane bed horizontal laminae with the reactivation surface being picked out by the concentration of dark, heavy minerals along this surface. The knife is approximately 15 cm long.

terization of stone artefacts and their sources. They have been used successfully in the provenancing of **obsidian** artefacts and to distinguish between sources of American **steatite** objects. [OWT]

Rayleigh number A dimensionless number used in fluid dynamics, *Ra*, depending on **gravity** (*g*), the depth of the liquid (*d*) involved, the difference in temperature (ΔT) between the top and bottom (mainly the temperature gradient above that of the **adiabatic** gradient), the **thermal diffusivity** (*K*), density (ρ), volume coefficient of thermal expansion (α) and the **kinematic viscosity** (η) of the liquid.

$$Ra = \frac{d^3\, g\, \Delta T\, \rho\, \alpha}{K\, \eta}$$

It is used mainly to estimate when **convection** commences in a fluid. [DHT]

Tritton, D.J. (1988) *Physical Fluid Dynamics*. Clarendon Press, Oxford.

reactivation surface A surface which records **erosion** of a bedform foreset slope caused by either a rise or fall in the level of the flow. Such surfaces are common in subaqueous **dune cross-stratification** and may represent **erosion** on the falling stage of a flood event, reworking at lower stages or **erosion** on the rising stage of the subsequent flood wave. The surface is commonly at a lower angle than the *angle of repose* foreset slopes on either side of it (Fig. R3). [JLB]

Collinson, J.D. & Thompson, D.B. (1982) *Sedimentary Structures*. Allen & Unwin, London.

real component (EM) That part of the secondary field that is in phase (or 180° out phase) with the primary field in **electromagnetic induction methods**. It is also referred to as the *in-phase component*. At low frequencies (and in highly resistive bodies), induction effects being less significant than resistive effects for **eddy currents**, the secondary field will have a prominent out-of-phase component, but at high frequencies (and large conductivities) the secondary will be almost entirely 'real'. [ATB]

realgar (AsS) A red **arsenic** sulphide formerly used as a pigment and found in **veins** and deposits from volcanic exhalations and *hot springs*. [DJV]

recharge Precipitation that reaches the **water table** and helps to replenish the supply of **groundwater**. The area receiving this precipitation is termed the recharge area, see Fig. A8, p. 31. [AME]

reciprocal time The travel time between two common points on a **reversed seismic profile**. Often relates to the travel time of a **head wave** from the surface at one end **shot** to the surface at the other on a **reversed profile**. The reciprocal time is an important component in the analysis of **seismic refraction** data using techniques such as the **plus-minus** and **time-term** methods. [GS]

recoverability The percentage of the total metal contained in an **ore** that is present in the concentrate prepared by processing the **ore**. Recoveries vary considerably. **Copper ores** usually lie in the range 80–90%, but **vein tin** deposits may be as low as 40% and average about 65%. [AME]

recovery The change from a **strained** to an unstrained state during or after **deformation**. It is an important factor in the **deformation mechanism** known as *dislocation climb*. During such **deformation**, recovery reduces the **dislocation** density and reorders **dislocations** into more stable arrangements. Walls of **dislocations**, surrounding volumes of the *crystal lattice* containing few **dislocations**, are characteristic of *dislocation climb*, and are seen optically as **subgrains** in minerals such as **quartz**. Continual recovery during **deformation** means that areas of crystal lattice free of **strain** are continually generated, so that **deformation** continues. Recovery is not a feature of **deformation mechanisms** in which **strain hardening** occurs, such as *dislocation glide*. [SB]

red clay (*brown clay*) A reddish brown **pelagic** sediment composed predominantly of **clay minerals** with minor components of very fine **quartz** and other minerals, volcanic **ash**, cosmic dust and occasional **fish** teeth. Red clays are the product of very slow ($1000\,\text{mm}\,\text{a}^{-1}$) deposition in the deeper (>4 km) central parts of the ocean basins below the **calcite compensation depth**. [AESK]

red ochre A name given to the red, earthy variety of **hematite**. [DJV]

reduction The procedure used in processing **gravity** and **magnetic** data in which all causes of the field's variation other than those originating from the underlying rocks are removed. In **gravity** processing the method is often referred to as reduction to the **geoid**, as sea-level is usually the most convenient datum to which to relate measurements. (See **geophysical anomaly**.) [PK]

Reeh calving An explanation for the calving of *icebergs* from the front of **glaciers** and *ice sheets*, whereby it is maintained that the **stresses** are greatest at a cross-section of floating **glacier** at a distance of about the ice thickness from the ice front and **stresses** are of a magnitude sufficient to cause **fracture** (Reeh, 1968). [ASG]

Reeh, N. (1968) On the calving of ice from floating glaciers and ice shelves. *Journal of Glaciology* 7: 215–32.

reflected seismic wave The energy that has returned to the surface by reflection at an **acoustic impedance** contrast (reflector) or series of contrasts in the Earth. Reflection is the process whereby a signal incident on an **acoustic impedance** contrast at an angle θ to the normal is returned on the same side of the contrast at an angle θ. Reflected waves are used extensively in **seismic exploration** for *hydrocarbon* **resources** to image geological structures forming potential **traps** for **oil** and gas. [GS]

reflection coefficient The ratio of the amplitude of the **reflected wave** to that of the wave incident on an **acoustic impedance** contrast. For a wave impinging the contact at a right angle and the **reflected wave** returning by the reversed path, the reflection coefficient can be calculated under the assumption of continuity of **stress** and displacement across the interface as

$$\frac{A_2 - A_1}{A_2 + A_1}$$

where A_1 is the **acoustic impedance** of the layer in which the wave travels and A_2 is **acoustic impedance** of the layer on the other side of the contrast. Reflection coeffients vary from -1 for a wave incident on the sea surface from below (the negative sign indicates phase reversal, i.e. a peak becomes a trough) to 0.2 for a normal strong reflector. In the more general case of a wave incident at an angle to a contact, both shear and compressional waves are reflected and transmitted. The amplitudes of these various waves are described by **Zoeppritz' equations**. [GS]

refracted seismic wave The wave that is transmitted through an **acoustic impedance** contrast, often with a change of direction dependent on the velocity contrast. *Snell's law* controls the change in direction of the refracted wave and states that

$$\frac{\text{sine of the incident angle}}{\text{velocity of first medium}} = \frac{\text{sine of the refracted angle}}{\text{velocity of the second medium}}$$

Head waves are developed when the refracted angle becomes 90°, i.e. the refracted wave propagates along the interface between the two media and the angle of incidence is known as the **critical angle**. [GS]

refraction wave The process by which **wave** crests are bent until they become parallel to the submarine contours. In water depths of approximately half the wavelength, the velocity of a **wave** decreases with decreasing depth so that the section of the **wave** in shallow water travels more slowly than its deep water section leading to bending of the **wave** crest. *Wave rays* or *orthogonals*, drawn normal to the **wave** crests, are lines between which **wave** energy is constant and due to the bending of the crests, these *orthogonals* converge and diverge at the coast creating

Fig. R4 Refraction wave. Refracted wave crests in Kilmurrin Cove, Co. Waterford, Ireland. Bending of the crests continues until the crest is parallel to the depth contours and coastline.

concentrations of **wave** energy on headlands and spreading of energy in bays (Fig. R4). [JDH]

King, C.A.M. (1972) *Beaches and Coasts*. Arnold, London.

refractive index The most important and fundamental optical property of a material, and, in the Earth Sciences, of a **mineral**. Refractive index (RI) is the ratio of the velocity of light in a vacuum (V) to that in the material concerned (*v*):

$$RI = V/v$$

As a standard, the velocity of light in a vacuum is taken as unity and other light velocities related to this; since in air $v = 0.9997$, this is also generally taken as unity. As described above, refractive index refers to a material through which light can be transmitted. If light is reflected from a flat polished surface of an opaque mineral (at normal incidence) there is a complex refractive index (N) where

$$N = n + ik$$

and *n* is the ratio of velocities in air and the medium, *k* is the absorption coefficient, and *i* is the complex conjugate. For translucent minerals, various methods have been devised for measuring refractive index, including immersion in liquids of varying refractive index or the use of instruments such as *refractometers*. In crystals of **anisotropic** materials, the refractive index varies as a function of orientation as well as the wavelength of the light.

Refractive index is occasionally used in archeological work as a method of characterizing **glassy rocks** and relating artefacts to a particular source. Egyptian **obsidians** and Cypriot **pumice** artefacts have been studied in this way. [DJV/OWT]

Gribble, C.D. & Hall, A.J. (1985) *A Practical Introduction to Optical Mineralogy*. George Allen & Unwin, London.

Wahlstrom, E.E. (1969) *Optical Crystallography* (4th edn). Wiley, New York.

reg A stony desert floor, generally bare of vegetation, surficially comprising **gravels** but underlain by a mixture of sediment sizes. A combination of wind and water **erosion** removes the finer sediment sizes from the surface. [RDS]

regimes of diagenesis A useful genetic classification of the various stages of **diagenesis** developed by Choquette & Pray (1970) for carbonate rocks subsequently applied to siliciclastic sediments by Schmidt & McDonald (1979). Three conceptual regimes are recognized, comprising *eogenesis*, *mesogenesis* and *telogenesis*, that relate **diagenesis** to the development and evolution of the sedimentary **basin** (Fig. R5). *Eogenesis* includes all the **diagenetic** responses that take place at or near the surface of sedimentation where the chemistry of the interstitial waters is mainly controlled by the depositional environment (see **depositional environment related diagenesis**). *Mesogenesis* begins when the sediment is sealed by effective burial from the influence of surface processes on the chemistry of the interstitial waters and includes all those **diagenetic** processes that modify the chemistry of interstitial **pore-waters** and **fabric** of the sediment as a result of burial (see **burial diagenesis**). *Telogenesis* represents the regime where **diagenesis** is influenced by the ingress of **meteoric water** from the surface once the **basin** has been subject to **uplift** or **inversion** subsequent to burial (see **meteoric water diagenesis**). **Inversion** may take place at any stage of the burial history of the **basin** and the burial–**uplift** cycle may be repeated several times so that the resulting **diagenetic** sequences may be extremely complex. (See **diagenesis**.) [SDB]

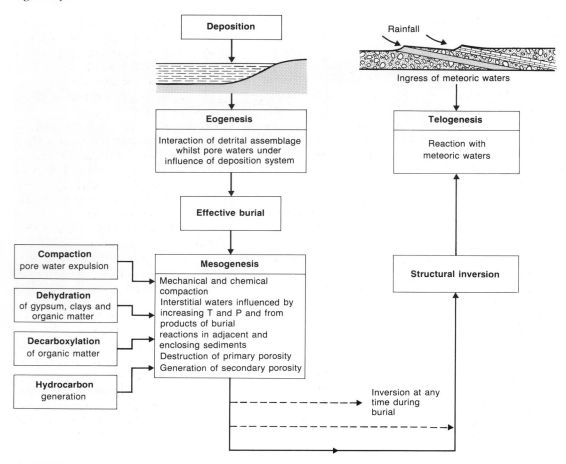

Fig. R5 The **regimes of diagenesis** and their relationships with each other.

Choquette, P.W. & Pray, L.C. (1970) Geologic nomenclature and classification of porosity in sedimentary carbonates. *American Association of Petroleum Geologists Bulletin* **54**: 207–50.

Schmidt, V. & McDonald, D.A. (1979) The role of secondary porosity in the course of sandstone diagenesis. In: Scholle, P.A. & Schluger, P.R. (eds) *Aspects of Diagenesis*. Special Publication of the Society of Economic Paleontologists and Mineralogists **26**: 175–207.

regional field The broad variation in a **gravity** or **magnetic field** over a region upon which local anomalies are superimposed. The regional field must be removed before interpretation of a *residual anomaly*; this can be accomplished by manual smoothing or by using an automatic method such as regression analysis or polynomial fitting. [PK]

Dobrin, M.B. & Savit, C.H. (1988) *Introduction to Geophysical Prospecting* (4th edn). McGraw-Hill, Singapore.

regolith The layer of loose, unconsolidated material that mantles bedrock over much of the Earth's land surface. The regolith may be made up of **weathered** bedrock (**saprolite**) which is *in situ*, or may comprise surface sediments such as volcanic **ash**, **colluvium**, alluvium, and *glacial* **drift** which have not been lithified. A *soil* may develop on the regolith but, strictly, the *soil* zone is distinct from it. [AW]

rejuvenation of river The stimulation of a river to increased **erosional** activity. This may result from a fall in *base level* or an increase in stream discharge. In general this will cause a stream to incise itself into its former valley, creating such features as knickpoints and river terraces. [NJM]

rejuvenation of structure Process of topographic renewal of *orogenic belts* by block **uplift**, long after the climax of the orogenic activity. For example, **Paleozoic fold belts** subsequently rejuvenated to give present-day mountain ranges include the Appalachians and the Urals. [RGP]

relative permeability The ratio of **effective permeability**, at a given fluid saturation, to the **absolute permeability**. [AME]

relative plate motion A tectonic *plate* is a part of the Earth's surface undergoing rigid body motion relative to other plates. Such motions can be determined, relative to each other, where the plate boundary between them is an *ocean ridge* by means of determining their **spreading rate** based on matching linear *oceanic magnetic anomalies* of identical age. Relative motions can also be determined using the same methods as for **continental reconstruction**, i.e. **paleomagnetic** methods, matching geological features, *paleoclimatic* and paleontological correlations, etc. It is now becoming possible to measure **absolute motions** based on **very long base line interferometry** and the *Global Positioning System*, in which the motions are determined relative to the Earth's orientation within space. [DHT]

Chase, C.G. (1978) Plate kinematics: the Americas, East Africa and the rest of the world. *Earth and Planetary Science Letters* **37**: 355–68.

remanent magnetization The magnetization, carried by **ferromagnetic** minerals, that is still present even when the samples are placed in zero external **magnetic field**. It arises spontaneously from the effects of quantum mechanical exchange and super-exchange forces that couple electron spin alignments so that such minerals have an internal magnetization even in the absence of an external magnetizing field. Such internal magnetizations can themselves be aligned to result in a strong external magnetization which, when it occurs in rocks, is termed **natural remanent magnetization**. Such net alignments are caused when the **geomagnetic field** is present during a series of natural processes, in particular heating, chemical change and deposition of already magnetized particles. Similar alignments can also be produced by laboratory processes, such as applying a direct **magnetic field** while simultaneously applying an alternating **magnetic field** (*anhysteretic magnetization*) or by rotating a sample within an alternating field (*gyromagnetic remanence*) as well as by heating, chemical change or deposition under laboratory conditions. [DHT]

Collinson, D.W. (1983) *Methods in Rock Magnetism and Palaeomagnetism*. Chapman & Hall, London.
O'Reilly, W. (1984) *Rock Magnetism and Mineral Magnetism*. Blackie, Glasgow.
Tarling, D.H. (1983) *Palaeomagnetism*. Chapman & Hall, London.

remote sensing In its broadest sense, the collection of information about something without being in physical contact with it. The term is usually restricted to the recording of an image of part of the Earth's surface using *electromagnetic radiation* (EMR). The recording system is normally carried by an aircraft or satellite at such a height that a complete image of a large area is shown. This provides a synoptic view revealing regional-scale features that are not otherwise readily apparent, and, during mapping, allowing extrapolation from known areas into areas that have not been studied in detail on the ground. Furthermore, the image can be recorded using wavelengths of EMR beyond the visible part of the spectrum and may reveal variations in surface composition that cannot be seen by the conventional field-based geologist.

Figure R6 shows the conventional divisions of the electromagnetic spectrum, the wavelength-dependence of the percentage transmission of EMR through the Earth's atmosphere, and the sources of the EMR detected in remote sensing. Techniques that produce images using reflected sunlight are considered first, and then thermal infrared and microwave remote sensing.

The earliest example of geological remote sensing was *aerial photography*, and this is still the most widely used technique. Photographs are taken in succession by a downward-pointing camera as an aircraft flies over the ground. Successive photographs overlap by about 60%, so they can be viewed through a *stereoscope* which allows people with normal binocular vision to see an apparently three-dimensional view of the terrain. This enables topographic contours to be drawn, and, if the **bedding** is sufficiently well expressed, **strike**, **dip** and other structural parameters can be measured. The scale of aerial photographs depends on the height at which the aircraft flies and the focal length of the camera lens; usually it is between 1:10 000 and 1:100 000. The sensitivity of photographic film is restricted to the visible and very near infrared (up to 900 nm wavelength) part of the spectrum. The most common (and cheapest) form of aerial photography uses normal black and white film (usually with a blue filter to reduce atmospheric haze). This is perfectly adequate for topographic, structural, and **geomorphological** mapping. Colour and false colour infrared films are useful to distinguish between different types of surface cover.

High-resolution photographic cameras have also been mounted on orbiting spacecraft or hand-held by astronauts. Pictures from the latter are especially valuable for recording transient phenomena such as *dust storms* or **volcanic eruption** plumes. However, most space-borne remote sensing systems take advantage of technological developments to record images in regions of the spectrum especially suitable for discriminating between different surface types. In these systems the image is recorded digitally by means of electronic sensors that scan the terrain and record the strength of the signal over adjacent small elements of the ground, known as picture elements, or *pixels*. Usually the image is scanned in successive lines that run at right angles to the ground-track of the satellite. Each line is scanned when the satellite is directly above the centre of the line, so any geometric distortion in the image is outwards from the central column of *pixels*. This is simpler to correct for than the radial distortion that affects photographs, but such geometry means that there is no stereoscopic effect when comparing successive images. Digital images are reconstructed by reassembling the *pixels* in the correct order and these are then enhanced on an image processing system, consisting of a computer, a colour monitor screen and a device for making 'hard copy' of the resulting image. Colour images are made by combining images recorded in several wavelength regions (or *spectral bands*). Because the *spectral bands* chosen

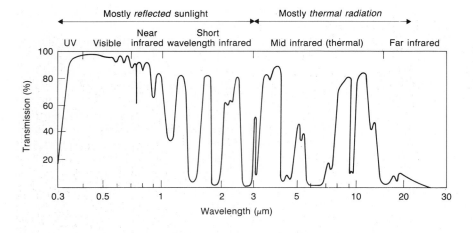

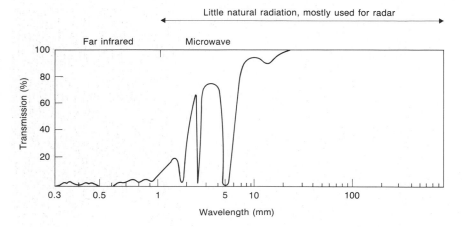

Fig. R6 Only those parts of the electromagnetic spectrum that are transmitted through the atmosphere are available for **remote sensing**. Moreover, the source of the radiation varies according to wavelength. (After Rothery, 1985.)

usually include at least one from beyond the visible spectrum, such an image is called a *false-colour composite*.

The achievements of satellite remote sensing in geology include: studies that have enabled detection of and recognition of the significance of major **faults** in poorly accessible regions like central Asia (e.g. Molnar & Tapponnier, 1975; Fig. R7); mapping of major rock types (e.g. Sultan *et al.*, 1986; Abrams *et al.*, 1988) and **alteration** zones (e.g. Abrams *et al.*, 1983); and detection and monitoring of **volcanic hotspots** and **eruption** plumes (e.g. Rothery, 1989). A widespread, but relatively unsung, use of satellite images is as base maps for logistic planning in *exploration* for, and development of, *hydocarbon* and **mineral** deposits.

For discrimination of rock types in well-exposed terrains, probably the most useful region of the spectrum is 2.0 to 2.5 μm, which is sensitive to **minerals** containing the hydroxyl group (OH^-) or the carbonate group (CO_3^{2+}) because these groups cause absorption features in the **mineral**'s reflectance spectrum. In *mineral exploration*, this can be useful for the detection of **hydrothermally altered** areas that are rich in hydroxyl-bearing **clay minerals**. Information about surface **iron** content comes from the visible part of the spectrum, especially at wavelengths

shorter than about 550 nm where oxidized **iron** (Fe^{3+}) causes strong absorption.

Vegetation reflects very strongly in the near infrared. Spectral anomalies in this band can be used to detect and map geobotanical effects, which may indicate a relationship between vegetation cover, *soil* type and the chemistry of the underlying rock. Comparison of images from several dates may show late production of leaves in the spring or premature yellowing or loss of leaves in the autumn, which are signs of vegetation stress due to poisoning, typically by heavy metals (such as **nickel** or **chromium**) or seepage of gases from *hydrocarbon reservoirs*.

Images from the *Landsat* series of *satellites* (originally called *ERTS*) have been used particularly widely for geological purposes. The first of the *Landsat* series was launched in 1972, carrying a *Multispectral Scanner* (MSS) imaging four spectral bands in the region 500–1100 nm, with a *pixel* size of about 80 m across. The second generation of *Landsat* satellites began operating in 1982, equipped with the MSS and a more powerful imaging system, the *Thematic Mapper* (TM), recording six spectral bands in the region 450–2350 nm with *pixels* about 30 m across and a thermal infrared channel with 120 m *pixels*. A complete

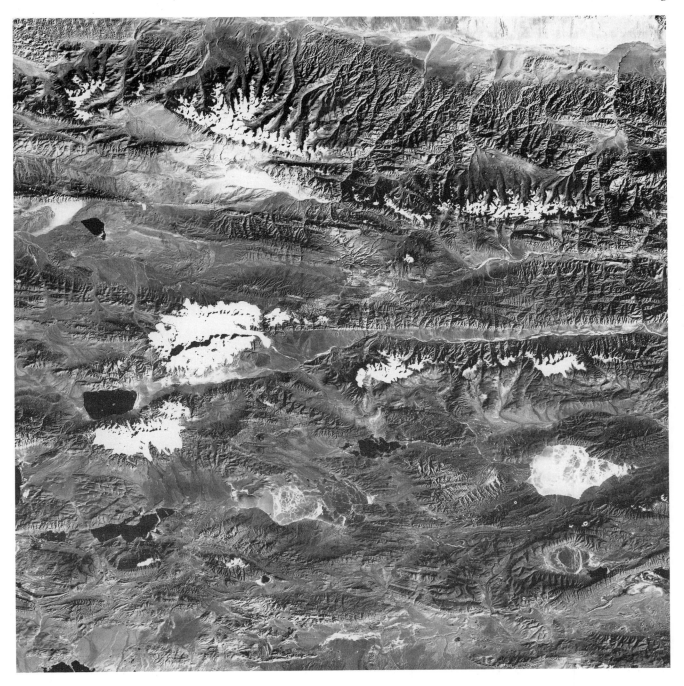

Fig. R7 Remote sensing. An extract from an enhanced Landsat MSS image showing the Kun Lun fault, a major east–west sinistral strike-slip fault that forms the northern boundary of the Tibetan plateau. Many other structural and geomorphological features are also visible.

Landsat image covers an area about 180 km across, and can usefully be printed at scales from 1:1 000 000 to about 1:100 000.

The other noteworthy satellite series began in 1986 with the launch of *SPOT-1*, the first of a series of French remote sensing satellites. *SPOT-1's* spectral range (500–890 nm) is less suited to lithologic mapping in well-exposed areas than that of the *Landsat* TM, but it has better spatial

resolution (20 m *pixels* in colour or 10 m *pixels* in a panchromatic mode) and it can be made to point slightly sideways, which enables it to give stereoscopic cover by using images of the same area recorded during different orbits.

Several exploration companies operate airborne scanner systems which can record images with higher spatial and spectral resolution than available from the current generation of satellites, and similar devices have been used to test proposed satellite instruments.

The thermal infrared part of the spectrum has received less attention. Although there is a thermal infrared sensor as part of the *Landsat* TM system and there have been experimental thermal systems in space with comparable spatial resolution, most geological applications of thermal infrared remote sensing have been based on airborne infrared scanners. The amount of thermal infrared radiation coming from the ground informs essentially about its surface temperature, so the most obvious uses are **geothermal** prospecting and detecting areas of **groundwater**, which are distinguishable (especially in night-time images) as a result of evaporative cooling. Another technique is to record thermal images at two times of the day (ideally pre-dawn and early afternoon) to estimate the diurnal temperature change; this can be related to the thermal inertia of the rock or *soil*, which is a bulk property of the upper 20 cm or so, and cannot be swamped by coatings of **weathering** minerals that can give misleading results in images recorded at shorter wavelengths. Thirdly, various studies have shown that suitably enhanced multispectral thermal images, i.e. images recorded in several channels in the thermal infrared (at a single time of day), can be used successfully for lithological mapping by capitalizing on subtle differences in emissivity between closely related rock types (e.g. Kahle & Goetz, 1983).

There is very little natural emission of microwaves by the Earth. Although passive microwave recording has its uses, for example in mapping the properties of polar ice, geologists use the microwave part of the spectrum mainly for **radar**. By processing the echoes from a continuous stream of **radar** pulses directed obliquely downwards from an aircraft or spacecraft, an image can be reconstructed that looks very similar to a strip of black and white photographs. **Radar** images show the topographic texture of the terrain very clearly and are an excellent substitute for aerial photographs or other images in parts of the world that are frequently cloud-covered, because clouds are transparent to microwaves of the wavelengths used in **radar**. Moreover, the properties of **radar** echoes relate to the roughness of the surface, so **radar** data are complementary to images of reflected sunlight (which relate to the chemical composition of the surface) and thermal images (which relate to temperature). In extremely arid conditions, **radar** can penetrate dry sand to a depth of a few metres, in which case the echoes give information on the underlying bedrock. Imaging **radar** experiments have been flown by the Space Shuttle (Ford *et al.*, 1983) and by Seasat (which operated for a few months in 1978), and airborne surveys are flown by a number of exploration companies.

The techniques described provide almost the only information available about the surfaces of the other planets, and remote sensing has other uses in the Earth sciences in general. For example, it reveals details of ocean circulation by showing the sea-surface temperature

(thermal infrared) and plankton distribution (reflected sunlight) on an ocean-wide scale and can map the dispersal of gases from large **volcanic eruptions** (backscatter of ultraviolet radiation).

Future developments in geological remote sensing will include more widespread use of imaging *spectrometers* capable of recording images in about 200 narrow channels in the reflected part of the spectrum. Whereas the wide spectral bands used by conventional scanners can suggest the presence of broad mineral types only, the higher spectral resolution of imaging *spectrometers* enables the identification of individual minerals due to slight composition-specific shifts in the wavelength of their absorption features. Imaging *spectrometers* and a variety of other remote sensing instruments will be carried by the *Earth Observation System* in the mid-late 1990s. This will be a collaborative, mutli-satellite effort between NASA, the European Space Agency and other countries, intended to give better information on the relationships between the **lithosphere**, hydrosphere and atmosphere on a planet-wide scale.

General introductions to geological remote sensing and the ways in which the images are processed are provided by Drury (1987) and Sabins (1986). Details of some of the suppliers from whom remote sensing data can be obtained are given by Carter (1986). [DAR]

Abrams, M.J., Brown, D., Lepley, L. & Sadowski, R. (1983) Remote sensing for copper deposits in southern Arizona. *Economic Geology* **78**: 591–604.

Abrams, M.J., Rothery, D.A. & Pontual, A. (1988) Mapping in the Oman ophiolite using enhanced Landsat Thematic Mapper Images. *Tectonophysics* **151**: 387–401.

Carter, D.J. (1986) *The Remote Sensing Sourcebook*. Kogan Page, London and McCarta Ltd, London.

Drury, S.A. (1987) *Image Interpretation in Geology*. Allen & Unwin, London.

Ford, J.P., Cimino, J.B. & Elachi, C. (1983) *Space Shuttle Columbia Views the World with Imaging Radar: The SIR-A Experiment*. JPL Publication 82–95, NASA Jet Propulsion Laboratory, Pasadena, CA.

Kahle, A.B. & Goetz, A.H.F. (1983) Mineralogic information from a new airborne Thermal Infrared Multispectral Scanner. *Science* **222**: 24–7.

Molnar, P. & Tapponnier, P. (1975) Cenozoic tectonics of Asia. *Science* **189**: 419–26.

Rothery, D.A. (1985) Remote sensing. *Geology Today* **5**: 105–8.

Rothery, D.A. (1989) Volcano monitoring by satellite. *Geology Today* **5**: 128–32.

Sabins, F.F. Jr (1986) *Remote Sensing* (2nd edn). W.H. Freeman, New York.

Sultan, M., Arvidson, R.E. & Sturchio, N.C. (1986) Mapping serpentinites in the Eastern Desert of Egypt by using Landsat thematic mapper data. *Geology* **14**: 995–9.

replacement As used in **diagenesis**, the growth of a chemically different **authigenic** mineral within the body of a pre-existing mineral, which may constitute the *host rock*, a detrital grain or an earlier formed **authigenic** mineral.

Conceptually, replacement can be considered in terms of **dissolution** and **precipitation** reactions (Pettijohn *et al.*, 1972). In this sense, replacement results from the **dissolution** of one mineral and the **precipitation** of another in its

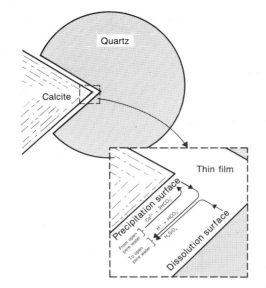

Dissolution: SiO_4 tetrahedra dissolve from quartz surface, hydrate→$H_4SiO_4°$ molecules in solution and diffuse into the surrounding pores

Precipitation: Ca^{2+} and HCO_3^- diffuse into the thin film separating quartz grains from the growing calcite crystal and more calcite precipitates according to the reaction:

$$Ca^{2+} + 2HCO_3^- \rightarrow CaCO_3 + H^+ + HCO_3^-$$

and to maintain precipitation H^+ must diffuse away and be lost to the thin film microenvironment

Fig. R8 The dissolution and precipitation mechanism of **replacement** via a thin film. This may be host grain or solution controlled, e.g. calcite replacement of detrital quartz grains. (After Pettijohn *et al.*, 1972, p. 403.)

place on a volume for volume or molecule for molecule basis. Replacement will, therefore, obviously occur when the **pore-fluids** are not in equilibrium with the minerals present in the host sediment and **dissolution** and **precipitation** reactions are thus initiated. It is not always possible, however, to determine from the resultant texture whether these events occurred simultaneously across an interface via a thin-film mechanism (Fig. R8) or if the **dissolution** and **precipitation** reactions were separated by a significant time gap.

The common preservation of relict internal **fabrics** and grain or mineral surface outlines within newly formed replacement minerals or replacive textures that mirror the morphology of the replacive mineral all suggest that replacement frequently proceeds via a thin-film mechanism. A good example of this type of process is often displayed in carbonate-**quartz** replacement **fabrics** (Fig. R8).

Two textural styles of replacement can often be discerned petrographically (Burley & Kantorowicz, 1986). Replacement that is selective or specific to sites of high surface free energy in the host mineral or grain is typical of solutions of low chemical reactivity and the replacement style is termed *host grain controlled replacement*. By contrast, non-specific replacement that is highly replacive of minerals regardless of their composition, structure or surface free energy characteristics is typical of highly concentrated, aggressive solutions and is termed *precipitate controlled replacement*.

Replacements are very common in **diagenetic** systems. Carbonate and sulphate cements are often highly replacive to detrital **quartz** and **feldspar** grains in **sandstones**. **Dolomite** commonly replaces **calcite** in **limestones** leading to the formation of *dolostones*. **Pyrite** is an almost ubiquitous replacement of organic matter in **mudrocks**. (See also **neomorphism**.) [SDB]

Burley, S.D. & Kantorowicz, J.D. (1986) Thin section and S.E.M. textural criteria for the recognition of cement dissolution porosity in sandstones. *Sedimentology* **33**: 587–604; and discussion–reply **33**: 605–14.

Pettijohn, F.J., Potter, P.E. & Siever, R. (1972) *Sand and Sandstone*, pp. 402–4. Springer-Verlag, Berlin.

replacement orebody A number of **epigenetic orebodies** have been formed by the **replacement** of pre-existing rocks, particularly carbonate-rich sediments; examples of these are **flats** and **skarn deposits**. Evidence of **replacement** can be found on every scale from **corrosion** textures on the **mineral** grain scale to three-dimensional evidence from detailed **underground mapping** which shows bold fronts of mineralization cutting across but not displacing the stratification of the *country rocks*; and **orebodies** containing isolated remnants of the *country rock* which are surrounded on all sides by **ore**, but which show no signs of rotation or forceful displacement. [AME]

reptation The process of grain transport by short trajectories following an aeolian grain impact event. The kinetic energy of an aeolian impact is distributed between the rebound of the impacting grain, dissipation into the granular bed and the ejection of reptating (Latin for 'crawling') grains. In contrast to the long trajectory **saltating** grains, the reptating population moves with low velocity. Spatial variations in the mass transport of the reptating population is thought to give rise to the growth and translation of aeolian *impact ripples*. The fastest growing *impact ripple* wavelength is roughly six times the mean reptation length. (See **saltation**, **aeolian stratification**.) [PAA]

Ungar, J. & Haff, P.K. (1987) Steady state saltation in air. *Sedimentology* **34**: 289–99.

Table R1 Classification of **reptiles**

Class Reptilia

Subclass Anapsida
   Order Cotylosauria
      Mesosauria
      Chelonia

Subclass Lepidosauria
   Order Eosuchia
      Squamata
      Sphenodontia*

Subclass Archosauria
   Order Rhynchocephalia (s.s.) = Rhynchosauria
      Thecodontia
      Crocodilia
      Pterosauria
      Saurischia
      Ornithischia

Subclass Euryapsida
   Order Araeoscelidia
      Sauropterygia
      Placodontia

Subclass Ichthyopterygia
   Order Ichthyosauria

Subclass Synapsida
   Order Pelycosauria
      Therapsida

* Retained in the lepidosaurs as a separate order, previously being part of the Rhynchocephalia s.l.

reptiles/Reptilia Class of vertebrate animals possessing a cleidoic egg and which are not **Mammalia** or **Aves**.

The original derivation of the name is from the Latin *reptilis*, which refers to their 'crawling' way of life. In classical times this would have been adequate; **amphibians** hopped, **mammals** suckled their young, **birds** flew and **fish** swam. However, in that the Reptilia are now known to comprise at least six subclasses (Table R1), some of which had life-styles not much different from the other vertebrate classes (e.g. *Pterosaurs* which flew and *Ichthyosaurs* which swam in the open sea), the preceding definition seems to be the best available.

The cleidoic egg is covered by a tough leathery or calcified shell and contains all the water and foodstuff necessary for the complete development of the embryo. Likewise other structures (membranes) are present which protect the embryo further and which receive waste products of metabolism. The hatchling is a miniature version of the adult and does not undergo metamorphosis. This pattern of egg is common to the **mammals** and **birds** as well as the reptiles and the three classes can be linked together as *Amniota*, as opposed to the **Fish** and **Amphibia** which are referred to as *Anamniota*. The possession of the amniote egg allowed the vertebrates to sever their ties with water for breeding purposes and enabled them to colonize new regions of the Earth, hitherto denied to the vertebrates.

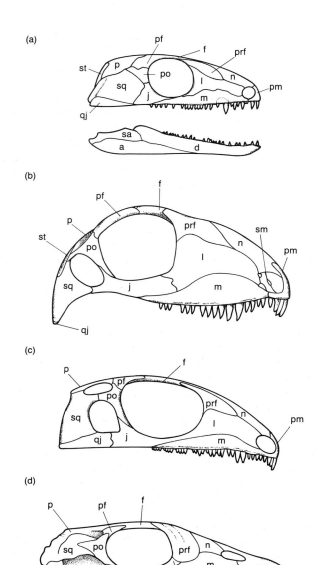

Fig. R9 Skulls of **reptiles** showing the pattern of temporal openings that distinguish the major groups. (a) The anapsid condition, illustrated by the protorothyrid *Paleothyris*. (b) The synapsid condition, exemplified by the early mammal-like reptile *Haptodus*. (c) The diapsid condition, shown by *Petrolacosaurus*. (d) The nothosaur *Neusticosaurus*, illustrating the parapsid or euryapsid condition. The diapsid and synapsid configurations are thought to have evolved separately from the anapsid condition. The euryapsid pattern has evolved from the diapsid pattern by loss of the lower temporal bar. a, angular; d, dentary; f, frontal; j, jugal; l, lacrimal; m, maxilla; n, nasal; p, parietal; pf, postfrontal; pm, premaxilla; po, postorbital; prf, prefrontal; q, quadrate; qj, quadratojugal; sa, surangular; sm, septomaxilla; sq, squamosal; st, supratemporal. (After Carroll, 1987.)

Within the reptiles the traditional way of subdividing the classes is by the pattern of the skull roof openings (Fig. R9). However, the reptiles which first appeared in the **Carboniferous** (*Mississippian = Dinantian*), had skulls which were pierced only for nose and eye openings. They were not very different from some **Amphibia** (e.g. the *Micro-sauria*, a subdivision of the *Lepospondyli*), and seem to have been very *lizard*-like in their structure and habits. These stem-reptiles or *Cotylosaurs* are brigaded together with the *Chelonia* (*turtles* and *tortoises*) (**Trias**–**Recent**) and several less well-known forms, e.g. *Mesosauria* (**Permo-Carboniferous**) as the subclass *Anapsida*. There is no certainty that they form a monophyletic taxon.

The *Mesosauria* were an aquatic group, seemingly adapted for catching small *crustaceans* with their long needle-like teeth. One genus, *Mesosaurus*, is found in sediments of **Permian** age in South Africa (Karoo Basin) and in Brazil (Parana Basin), which may have been laid down at the same time that the **Gondwanaland** terrestrial hinterland was still glaciated. Although classified as an anapsid, there are in fact small openings in the sidewall of the skull which may be homologous with similar openings in the skull of a later sister-group, the Millerosauria. These are sometimes considered to be modified anapsid reptiles, and at other times to be *Eosuchians*.

The *Chelonia* comprise the well-known *turtles* and *tortoises*. They occupy a wide range of ecological niches from the totally terrestrial to the totally aquatic. However, even the most aquatic of the turtles have to come ashore to lay eggs. Their trade mark is their bony carapace into which the limbs, neck and head can be drawn.

The *lizards* and related forms (*snakes*; amphisbaenians; the Tuatara *Sphenodon*) are sometimes classified with the *crocodiles* and their relatives as the *Diapsida*. The earliest diapsid or 'two-arched' reptile is *Petrolacosaurus* (*Pennsylvanian*) of Texas. It may represent a genuine common ancestor to both subclasses of modern reptile, but there is a long time-gap between *Petrolacosaurus* and the start of the great diapsid radiation in the uppermost **Permian** and the **Triassic**. It is probably safer therefore to treat the *Lepidosauria* and the *Archosauria* as separate groups only distantly related.

Of the *Lepidosauria*, *Youngina* (Upper **Permian**, South Africa) is the best known of their early representatives, the *Eosuchia*. However this term is now almost exclusively used for a brigade of poorly known fossil reptiles which may or may not be related and which may or may not represent the first radiation of diapsid types, including the *Euryapsida*. The first *lizards* are probably of Upper **Permian** age, but as they are part of the 'Eosuchian' problem, it leaves the first undisputed *lizards* to be of the **Jurassic** and later.

Sphenodon (New Zealand; Recent) is the most primitive living *lepidosaurian* and an example of a *living fossil*. Its skull differs from that in true *lizards* and *snakes* in having an 'unbroken' lower temporal bar, unlike some of its **Triassic** and **Jurassic** ancestors. The quadrate is therefore fixed in position and cannot be used to help in swallowing movements by rocking backwards and forwards in the same way that it can in most other squamates. However, as *Sphenodon* feeds on resistant invertebrate and plant material with a shearing action of its jaws, the need to swallow large mouthfuls does not arise. A skull capable of resisting the heavy pressures applied in processing food this way needs different adaptations from one which does not, and the apparently 'primitive' stage shown by *Sphenodon* may have been secondarily achieved from a form with a 'free' quadrate.

The *Archosauria* now include the *Rhynchosauria*, an aberrant group of Upper **Triassic** herbivores previously classified alongside the *sphenodontids*. They seem to represent a parallel radiation to the very primitive *archosaurs*, the proterosuchian *thecodonts*. The *rhynchosaurs* have batteries of ankylothecodont teeth, set in deep sockets which formed a powerful chopping mechanism, as opposed to the *sphenodontids* which have pleurodont teeth, or only bony cutting and shearing surfaces to the jaws.

The *Archosauria* proper contain the well-known forms such as the *Crocodilia* (**Trias**–**Recent**), the *Pterosauria* (**Trias**–**Cretaceous**) and both orders of 'dinosaur' (U. **Trias**–**Cretaceous**). The *crocodiles* arose as small, very active terrestrial *insectivores* in the Upper **Triassic**, derived from one of the many Middle **Triassic** thecodontian groups. They probably only became aquatic under competitive pressure from advanced *thecodonts*, such as the Rauisuchidae, but underwent several bursts of adaptation to water during the rest of the **Mesozoic** and later in the *Tertiary* and more recently, e.g. the *Callovian* thalattosuchians and modern freshwater forms.

Both the *dinosaur* orders, the *Saurischia* and the *Ornithischia*, probably arose from carnivorous ancestors such as the Rauisuchidae. The two types of *dinosaur* are distinguished primarily on the structure of their pelvis, but whereas the *Saurischia* comprise both carnivorous (*Theropoda*) and herbivorous (*Sauropoda*) forms, the *Ornithischia* were entirely herbivorous.

The *dinosaur* orders dominated the terrestrial ecology of the **Mesozoic** and many of them achieved great size and enormous mass. Some of the *dinosaur* faunas, e.g. Morrison (Middle **Jurassic**) were cosmopolitan, with identical genera being found as far apart as the central USA and Tanzania.

Among other factors affecting *dinosaur* **extinctions**, as they were presumably egg-layers, and as eggs are proportional in size to the body which laid them, developing eggs may have been liable to physiological stress if the atmospheric composition became adversely changed. During the **Cretaceous**, and associated with ongoing **seafloor spreading**, there was considerable volcanic activity which released very great amounts of carbon dioxide into the atmosphere. Developing *dinosaur* embryos may have found it more difficult to respire efficiently in an atmosphere enriched in carbon dioxide, in a similar way that chicks in modern hatcheries do. In the latter case, elevated mortality rates are noted, and the same may have held good for the **Cretaceous**. Thinner egg shells would permit an improvement in gas exchange for the embryo, but at the cost of becoming more fragile. This combination of phenomena in association with changes in the climate caused by a parallel *greenhouse effect* may well have been the cause of not only the **extinction** of the *dinosaurs*, but many of the other groups which faded from the scene at the same time. Those that survived (**mammals, birds,** *crocodiles, chelonians*) all lay 'small' eggs or are viviparous and may not have been as badly stressed as were the *dinosaurs* and others. A *bolide impact* may only have been the cause of removing the last few taxa from the record.

The *pterosaurs* had a successful history in parallel with the *dinosaurs*, but may themselves have been competitively eliminated by the success of the **birds**, rather than suffering the same stresses as their larger relatives.

The *Euryapsida* may or may not be a natural grouping of related forms. They are supposedly linked together by the possession of only one temporal opening, the upper, but whether the bones surrounding the temporal openings in all the components of the subclass are homologous is open to doubt. There is some good evidence that the skull of the *sauropyterygians* was originally *diapsid*, and an evolutionary sequence starting with something like *Youngina* in the Upper **Permian** led to the often-illustrated marine *plesiosaurs* and *pliosaurs* of the later **Mesozoic,** through an intermediate group, the *Nothosauria* of the **Triassic.** The *plesiosaurs* are exclusively aquatic animals with a variety of adaptations for both macrophagy and microphagy. Some must have been fearsome predators, particularly the large-headed *pliosaurs*; the long-necked *elasmosaurs* were probably **fish** eaters. Some were even filter feeders. All of them seem to have had to ballast their bodies with *gastroliths* in order to counteract the natural buoyancy of their air-filled lungs. These *gastroliths*

are among the pebbles now found some distance from the **Mesozoic** shore-lines and pose interesting problems for the sedimentologists working there.

The *Ichthyopterygia* are superficially dolphin-like reptiles which enjoyed great success as marine predators in the **Mesozoic.** They originated in the Lower **Triassic** and ran out in the Upper **Cretaceous.** Some were large enough to have been able to prey on the smaller *elasmosaurs*, but evidence from preserved stomach contents shows their preferred prey to have been *belemnites* and **fish.** Evidence also exists for 'live birth' in these forms and much new work is being directed towards reconstructing their fin outlines from exceptionally well-preserved specimens from, e.g., Holzmaden (Germany) and Barrow-on-Soar (UK). Their ancestry seems to have been entirely independent from an unknown group of *cotylosaurs*.

The *Synapsida* or **mammal**-like reptiles (**Carboniferous– Jurassic**), more than any one group of tetrapod, demonstrate the passage of one fossil group gradually evolving into another, in this case the *reptile–mammal transition*. There are only a very few characters which run through the entire *synapsid* lineage, but the possession of a true lower temporal opening in the first *pelycosaurs* distinguishes them from the contemporaneous *cotylosaurs*, which otherwise are very similar.

The first *pelycosaurs* are small *lizard*-like forms often preserved in the trunk infill of *Lepidodendron* trees in the *Pennsylvanian* of Nova Scotia. By the Lower **Permian** the *pelycosaurs* radiated into large carnivorous, piscivorous and even herbivorous forms. Some are characterized by a 'sail' on their backs supported by neural spines on the vertebrae

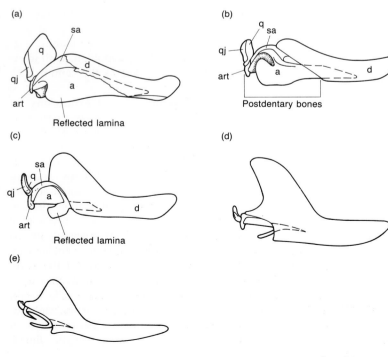

Fig. R10 Reptiles. Progressive changes in the structure of the jaw and elements of the middle ear from pelycosaurs to mammals. All in lateral view. (a) The sphenacodont pelycosaur *Dimetrodon*. All the elements of the lower jaw are suturally attached. The angular bears a reflected lamina. (b) Condition of an advanced therocephalian. Postdentary bones are no longer suturally attached to the dentary, but they remain very large. (c) *Thrinaxodon*, a primitive cynodont. (d) The advanced cynodont *Probainognathus*. (e) An early Jurassic mammal, *Morganucodon*. a, angular; art, articular; d, dentary; q, quadrate; qj, quadratojugal; sa, surangular. (After Allin, 1975.)

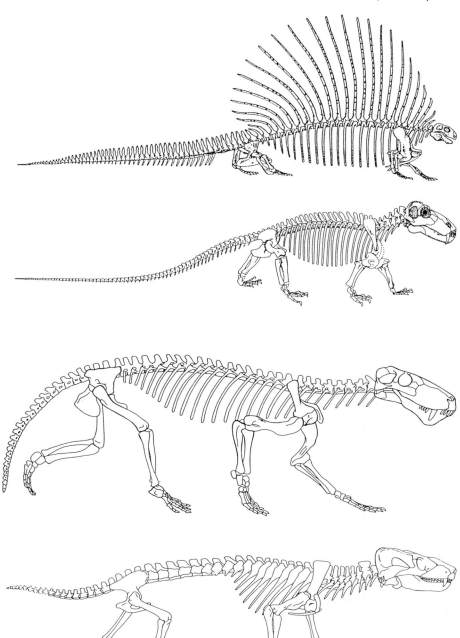

Fig. R11 Reptiles. *Edaphosaurus.* This herbivorous pelycosaur is from the late Pennsylvanian and early Permian. Skeleton, 3 m long. Unlike those of *Dimetrodon*, the spines bear transverse processes. (After Romer & Price, 1940.)

Fig. R12 Reptiles. *Titanophoneus.* Skeleton of a primitive dinocephalian therapsid from Russia, 2 m long. (After Orlov, 1958.)

Fig. R13 Reptiles. Skeleton of the gorgonopsid *Lycaenops*, original 1 m long. (After Colbert, 1948.)

Fig. R14 Reptiles. Skeleton of the lower Triassic cynodont *Thrinaxodon*, 0.5 m long. Greatly widened ribs probably served to support the trunk and to limit lateral undulation. (After Jenkins, 1984.)

and all show the start of a heterodont dentition. These forms are all typical of the south-western United States Lower **Permian**.

The story from this point onwards in the Upper **Permian** is followed firstly in Russia and then in the Karoo Basin of South Africa, where the **mammal**-like reptiles are so abundant that they are used as **biostratigraphic** markers. Improvements in limb posture and hence speed and

agility across the ground, with an enlargement of the temporal opening and further differentiation of the teeth mark the transition from the *pelycosaur* to the *therapsid*. The 'reptilian' phalangeal formula of 2-3-4-5-2(3) gives way to the **mammalian** formula of 2-3-3-3-3, showing that the feet were becoming pointed to the front of the animal and were no longer splayed outwards.

Within the *therapsids* the most marked tendencies are

Table R2 Reptilia and Mammalia homologies

Reptile		Mammal	
Stapes	Sound conducting/ quadrate support	Stapes	Sound conduction
Quadrate	Upper jaw hinge	Incus	Sound conduction
Articular	Lower jaw hinge	Malleus	Sound conduction
Angular	Pterygoideus insertion?	Tympanic	Support for ear-drum

towards an improved gait and towards a mammalian pattern of jaw-closing muscles, as is reflected in the bony anatomy of the lower jaw and teeth.

The lower jaw of a typical reptile has six or seven individual bones in each ramus, whereas a true **mammal** has only one (Fig. R10). A grey area exists in the evolutionary sequence from *reptile to mammal* where the principal hinge joint of the jaw is translating from the reptilian quadrate (upper jaw)–articular (lower jaw) hinge to the mammalian dentary–squamosal hinge, as in some cases all four bones may be involved. By the time the true **mammalian** state exists, the small supernumerary bones have largely been incorporated into the middle-ear structure. The homologies are as shown in Table R2.

The dentition of the carnivorous **mammal**-like reptiles becomes more and more **mammal** like, until by the Upper **Triassic** several forms of primitive **mammals** are known with tribosphenic-precursor teeth.

A sequence of **mammal**-like reptiles can be assembled to show this gradual acquisition of **mammalian** characters. It starts with a *pelycosaur* and progresses through a Russian biarmosuchid to a South African gorgonopsid and therocephalian and cynodont (Figs. R11–R15). Ultimately one cynodont must have given rise to the first **mammal** as seen in *Erythrotherium*, of Lesotho (Upper **Triassic**) or similar genera in South Wales and China, with all the characters recognizable in the modern orders, i.e. hair, warm blood, four-chambered heart, mammary glands and the care of the young. [ARIC]

Allin, E.F. (1975) Evolution of the mammalian middle ear. *Journal of Morphology* **147**: 403–38.

Carroll, R.L. (1987) *Vertebrate Paleontology and Evolution*. W.H. Freeman, Chicago.

Colbert, E.H. (1948) The mammal-like reptile Lycaenops. *Bulletin of the American Museum of Natural History* **89**: 353–404.

Jenkins, F.A. (1984) A survey of mammalian origins. In: Gingerich, P.D. & Badgley, C.E. (eds) *Mammals: Notes for a Short Course*. University of Tennessee Studies in Geology No. 8, pp. 32–47.

Jenkins, F.A. & Parrington, F.R. (1976) The postcranial skeletons of the Triassic mammals Eozostrodon, Megazostrodon and Erythrotherium. *Philosophical Transactions of the Royal Society, London* (B) **273**: 387–431.

Kemp, T.S. (1982) *Mammal-like Reptiles and the Origin of Mammals*. Academic Press, London.

Norman, D. (1985) *Illustrated Encyclopedia of Dinosaurs*. Salamander, London.

Orlov, J.A. (1958) Carnivorous dinocephalians from the fauna of Isheev (Titanosuchia). *Trudy. Pal. Inst. Akad. Nauk. USSR* **72**: 1–114.

Romer, A.S. (1966) *Vertebrate Paleontology* (3rd edn). University Press, Chicago.

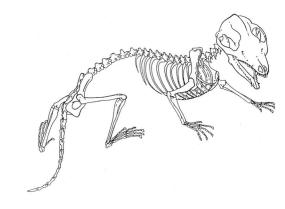

Fig. R15 Reptiles. Skeleton of a primitive mammal. The skeleton was 10 cm long to the base of the tail. (After Jenkins & Parrington, 1976.)

Romer, A.S. & Price, L.I. (1940) *Review of the Pelycosauria*. Special Paper of the Geological Society of America **28**: 1–538.

Smithson, T.R. (1989) The earliest known reptile. *Nature* **342**: 676–8.

reserves For **coal** deposits the term reserves refers to material that is actually known to be mineable under the technical and economic conditions prevailing at the time the assessment is made. Estimates of **coal** reserves may be of *in situ* reserves — the amount of **coal** in place, of recoverable reserves — the amount of **coal** that can probably be won from the deposit, or of marketable reserves — the amount of recovered **coal** that can ultimately be marketed after processing at the mine.

For oil geologists there is some confusion with regard to the categorization of reserves and there are various different systems in use. The USA has recently developed the following terminology for statistical purposes. Proved (measured) reserves are those which have been outlined by drilling and conservatively delineated. Probable (indicated) supply is oil which is likely to be present in extensions of existing fields but beyond the present known limits. Possible (or inferred) supply may come from future discoveries in already productive areas of formations. The former USSR used a sixfold classification and in 1983 a study group of the World Petroleum Congress prepared a variant of the American system. [AME]

North, F.K. (1985) *Petroleum Geology*. Allen & Unwin, Boston.

reservoir A rock containing subsurface oil and/or gas in exploitable quantities is called a reservoir rock or simply a reservoir. Any rock can be a reservoir provided it is both porous and permeable so that it can contain and yield oil or gas, or both, in commercial quantities. The common reservoir rocks are, however, the coarse clastic sediments, **sandstones** and *conglomerates*, that have retained a significant fraction of their original **porosity** and **permeability**. Certain carbonate sediments such as *calcarenites*, especially *grainstones*, may form equally good reservoirs.

Not all **sandstones** *s.l.* are likely to be good reservoir rocks. Suitable reservoir **sandstones** are those composed of hard, stable insoluble, equidimensional grains that will provide the rock with a strong framework. Clearly with these requirements **quartz** will be an important constituent. During their formation fine-grained, matrix-forming materials, e.g. clays, must have been excluded and **cementation** should be only partially developed. As a result the quality of the **sandstone** as originally formed depends on the nature of the source area, the depositional processes and the environment of deposition.

Carbonate rocks may also retain some original **porosity** but good **porosity** in carbonate reservoirs is often a result of *recrystallization* and most commonly of **dolomitization**. Other carbonate reservoirs have developed **fracture porosity**. [AME]

North, F.K. (1985) *Petroleum Geology*. Allen & Unwin, Boston.

residual deposit Deposit of economic interest formed by the action of **weathering** and **groundwater** processes on **protore**. For the formation of extensive deposits, intense chemical **weathering**, such as in tropical climates having a high rainfall, is necessary. In such situations, most rocks yield a *soil* from which all soluble material has been dissolved and these *soils* are called **laterites**. **Laterite** consisting of almost pure **aluminium** hydroxide is called **bauxite**. (See **nickel laterite deposit**.) [AME]

residual strength A lower value of **strength** measured in disturbed materials. Since **strength** depends partly on **porosity**, if a *soil* or rock initially has a low **porosity**, some part of its **strength** is not recoverable after **failure** since it undergoes a non-reversible increase in volume and **porosity** as **shear** takes place. The remoulded material therefore has a lower **strength** than the initial high peak **strength**. In granular material, this loss of **strength** from the undisturbed state may be of minimal importance, but in cohesive *soils*, where the **cohesion** forms a major part of the intact **strength**, residual strength may be only a small fraction of the *peak undisturbed strength*. This may be particularly important in 'quick' clays where the delicate 'house-of-cards' packing of clay particles may be completely destroyed upon **failure** causing almost total loss of **strength** and, possibly, **liquefaction**. Some *quick clays* may display *thixotropic* behaviour with partial **recovery** of **residual strength** occurring some time after the **failure**. [TPB]

resistivity method Exploration method to investigate the subsurface distribution of **electrical resistivity**, with wide-ranging applications (in *mineral exploration*, **hydrogeology**, **engineering geology**, **reservoir** evaluation, deep structure investigations, etc.).

Resistivity methods exploit the huge contrasts that exist in the **electrical resistivity** of Earth materials (e.g. some sulphides at 10^{-6} ohm m and dry crystalline rocks at 10^{6} ohm m), and the fact that the resistivity of rocks is controlled very strongly by **porosity** and **pore fluids** (see **Archie's formula**). In deep **crustal** and **mantle** studies, a significant feature is that, since most rocks are semiconductors, their resistivity decreases with temperature (i.e. with depth).

It is important to note that in all surface measurements of resistivity, it is the integrated effect of the subsurface **conductivity** structure to the **depth of penetration** that is measured. So the convention is to work with the concept of 'apparent resistivity', ρ_a, i.e. the resistivity of a homogeneous Earth which would produce the same response under the same conditions of measurement. Field data are usually expressed in terms of **apparent resistivity** before interpretation.

Resistivity methods fall into two broad categories: 'direct current' (or galvanic) methods where subsurface currents are initiated by electrodes in contact with the ground, and **electromagnetic induction methods** where the currents are induced by time variations of **magnetic fields**, natural or man-made. In the d.c. method, potential distributions produced at the surface by a very low frequency (*c.* 1 Hz) current of known strength, averaged over a number of full cycles to avoid polarization effects (see **electrolytic polarization**), are measured. The potential difference between two electrodes will be a function of the geometry of the potential and current electrodes as well as the subsurface resistivities. For each measurement a value of ρ_a can be calculated from the known electrode configuration, and from a suite of measurements of ρ_a, the **conductivity** structure is deduced.

d.c. exploration seeks to delineate either lateral resistivity contrasts (see **constant separation traversing**) or the vertical resistivity structure of a horizontally layered Earth (see **vertical electrical sounding**). The former is generally concerned with the location of contact zones, and quantitative interpretation is rarely attempted. The theory of **apparent resistivity** of layered structures has been studied extensively, however, and, despite the limitations of an inherently **inverse problem**, many interpretation schemes have been established. Until the 1970s these were mainly confined to curve-matching using a necessarily restricted number of standard curves, but with the advances in computer technology it is now possible to generate theoretical curves for specific multi-layered models and perform detailed curve-matching quickly and easily, even in the field. Furthermore, computer programs are readily available for the inversion of resistivity field data.

In archeology, resistivity methods are sometimes used in advance of excavation. Buried features such as walls and ditches show a change in resistance from the surrounding earth due to their different water content (Fig. R16). A smaller water content, such as may be expected in a buried stone wall, results in increased resistance. Several different arrangements of probes carrying the electrodes

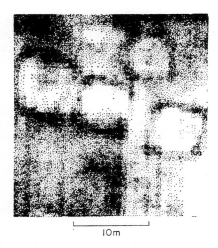

Fig. R16 Resistivity survey results, in the form of a dot density diagram, for a typical area of the Iron Age cemetery at Burton Fleming, Yorkshire. High dot density is associated with regions of low resistivity such as are produced by the rectangular enclosure ditches, four of which are clearly evident in the diagram. (After Tite, 1972.)

are used, but typically four probes may be put into the ground, two to carry the inserted current, two to detect the resistance. Probes separated horizontally by *c*. 1 m could be expected to detect features buried to approximately that depth. In practice, resistivity surveying is subject to complications of varying underlying geology which affect the measurements, as well as to problems of topography and vegetation cover. Locating small features may be very time-consuming but the method is very suitable for the tracing of linear features such as walls or earthworks once part is located. [ATB/OWT]

Ghosh, D.P. (1971) The application of linear filter theory to the direct interpretation of geoelectrical resistivity sounding measurements. *Geophysical Prospecting* **19**: 192–217.

Keller, G.V. & Frischknecht, F.C. (1966) *Electrical Methods in Geophysical Prospecting*. Pergamon, Oxford.

Koefoed, O. (1979) *Resistivity Sounding Measurements* (Geosounding Principles, 1). Elsevier, Amsterdam.

O'Neill, D.J. & Merrick, N.P. (1984) A digital linear filter for resistivity sounding with a generalised electrode array. *Geophysical Prospecting* **32**: 105–23.

Tite, M.S. (1972) *Methods of Physical Examination in Archaeology*. Seminar Press, London.

resource (1) Mineral. The total amount of a particular commodity (e.g. **tin**, **andalusite**) estimated for the world, or for a nation, but not for a company. Resources consist of **ore reserves**, known but uneconomic deposits, and hypothetical deposits not yet discovered. The estimation of the undiscovered potential of a region can be made by comparison with well-explored areas of similar geology.

Theoretically, world resources of most metals are enormous. Taking **copper** as an example, there are large amounts of rock running 0.1–0.3% and enormous volumes containing about 0.01%. The total quantity of **copper** in such deposits probably exceeds that in proven **reserves** by a factor of from 10^3 to 10^4. Nevertheless, the enormous amount of such material does not at present imply a virtually endless resource of metals. As **grades** approach low values, a concentration (the *mineralogical limit*) is reached, below which an element no longer forms a distinct physically recoverable mineral phase.

(2) Oil. Resources of oil are best expressed in a similar fashion to that for minerals.

Unfortunately many minerals economists, let alone informed laymen, confuse resources with **reserves** and this leads to extremely speculative figures being published as representing assured supplies of a particular commodity. On the other hand the pessimists will seize on world **reserve** figures for a particular commodity and, completely ignoring the possibility of new discoveries, will prophesy that the world will have no more of that particular material in a few decades or so. [AME]

retrograde boiling Boiling that occurs in a system in which the temperature is falling. In a stationary igneous intrusion the **confining pressure** will not fluctuate but cooling will promote crystallization. In an acid **magma** this will be largely of anhydrous minerals, so that the liquid **magma** is enriched in volatiles leading to an increase in the vapour pressure. If the vapour pressure rises above the **confining pressure**, then what is called retrograde boiling will occur and a rapidly boiling liquid will separate. When retrograde boiling occurs in a largely consolidated rock, the vapour pressure has to overcome the *tensile strength* of the rock as well as rising above the **confining pressure**. This will result in expansion and extensive and rapid development of **crackle brecciation**. The reason for this is that water released at a depth of about 2 km at 500°C would have a specific volume of 4 and, if 1% by weight formed a separate phase, it would produce an increase in volume of about 10%. At shallower depths the increase would be even greater and the degree of **fracture** intensity higher. Evidence for the development of retrograde boiling in **porphyry copper deposits** is common in the form of the widespread occurrence of liquid-rich and gas-rich *fluid inclusions* in the same thin section. [AME]

reversed magnetization A magnetization, usually the **natural remanent magnetization**, of rocks that is opposite to that of the present field, i.e. the north **magnetic pole** lies in the southern hemisphere. This definition becomes increasing impractical as the **paleomagnetic pole** approaches the present Equator. Normal and reversed fields are exactly 180° different in *declination* and of opposite sign of *inclination*. [DHT]

reversed profile A line of detectors that has **shots** at each of its ends such that **seismic waves** are recorded travelling in both directions along the profile. This term is sometimes used more specifically to describe the portion of profile recording waves from a particular target horizon from both directions. [GS]

reversing dune Dune with **slip-faces** on two opposing sides due to a bimodal wind regime more or less 180° apart. [RDS]

reworking Process whereby relatively deep **crustal** material, formed in a previous **orogeny**, becomes altered by **deformation** and metamorphism during a subsequent orogenic event. Effects of reworking are commonplace in **Precambrian shield** regions; for example, many areas of Early **Proterozoic crust** consist of reworked material of **Archean** age. Deeply dissected **Precambrian** terrains typically consist partly of reworked older basement and partly of deformed supracrustal cover. [RGP]

Reynolds number A dimensionless parameter that expresses the influence of viscous forces within a fluid in relation to the inertial forces acting upon the flow. The **flow** Reynolds number (*Re*) is given by

$$Re = \rho \cdot U \cdot l/\upsilon$$

in which ρ is the fluid density, U is the mean flow velocity, l is a length term and υ is the **kinematic viscosity**. In open channel flows the length term is usually expressed by the flow depth. Where viscous forces are predominant the flow is **laminar** (*Re* < 500) whereas at values above 2000 the flow is **turbulent** where the inertial forces become relatively far more significant. The **laminar–turbulent flow** transition occurs between values of 500 and 2000.

Another common form of the Reynolds number is the particle Reynolds number in which the **shear velocity** replaces the mean flow velocity and the particle diameter is used as the length term. This form of the Reynolds number expresses the relative inertia of the particle to the viscous forces of the fluid. (See **flow regime, laminar flow, turbulent flow**.) [JLB]

Chow, V.T. (1959) *Open-channel Hydraulics*. McGraw-Hill, London.
Middleton, G.V. & Southard, J.B. (1984) *Mechanics of Sediment Transport* (2nd edn). Short Course No. 3 Lecture Notes. Society of Economic Paleontologists and Mineralogists, Tulsa, OK.
Tritton, D.J. (1988) *Physical Fluid Dynamics*. Clarendon Press, Oxford.

rheology The study of the **flow** and **deformation** of matter. [DHT]

Tritton, D.J. (1988) *Physical Fluid Dynamics*. Clarendon Press, Oxford.

rhexistasy The term coined by Erhart (1956) to define the state of environmental and geomorphic instability which interrupts periods of **biostasy**. Rhexistasy is characterized by a deterioration in the vegetation cover, **erosion** of *soils*, and **denudation** of the landscape. Its onset may result from climatic change (aridification) or from **tectonic** disturbance. Once the environment has responded to the disruption which brought about the state of rhexistasy, and geomorphic forces are again in equilibrium with environmental conditions, **biostasy** is re-established. [AW]

Erhart, H. (1956) *La Genèse des Sols en tant que Phénomène Géologique*. Masson, Paris.

rhizocretion An accumulation of mineral matter, such as **calcite**, around roots. This accumulation may take place during the life of the plant root or after its death. A variety of types of concretionary root traces are found and Klappa (1980) has suggested the general term *rhizolith* be used for organo-sedimentary structures resulting from the accumulation and/or **cementation** around, **cementation** within, or **replacement** of, higher plant roots by mineral matter. Besides rhizocretions *sensu stricto* other forms of *rhizolith* include *root moulds* and *casts* left after the decay and later infilling of the sites of roots, *root tubules* which are cemented cylinders around root moulds, and *root petrifications* formed by the impregnation or replacement of organic matter. A rock showing evidence of having been formed largely by root activity is termed a *rhizolite* (Fig. R17). [VPW]

Klappa, C.F. (1980) Rhizoliths in terrestrial carbonates: classification, recognition, genesis and significance. *Sedimentology* **27**: 613–29.

rhodochrosite ($MnCO_3$) A comparatively rare **ore** mineral **of manganese** found in certain **hydrothermal veins**. (See **carbonate minerals**.) [DJV]

rhodolite A pale rose-red or purple **garnet mineral** corresponding compositionally to two parts **pyrope** and one **almandine**. [DJV]

Fig. R17 Rhizocretions in calcareous sands (Pleistocene) Coorong area, South Australia.

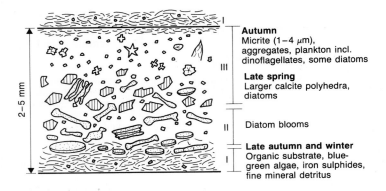

Autumn
Micrite (1–4 μm),
aggregates, plankton incl.
dinoflagellates, some diatoms

Late spring
Larger calcite polyhedra,
diatoms

Diatom blooms

Late autumn and winter
Organic substrate, blue-
green algae, iron sulphides,
fine mineral detritus

Fig. R18 Rhythmite. Schematic representation of a typical non-glacial varve from Lake Zürich, Switzerland. Units I and II form the late Autumn–Winter (dark) lamina; Unit III the late Spring–Summer (light) lamina. (After Kelts & Hsü, 1978.)

rhodonite (MnSiO₃) A **pyroxenoid mineral** found in **manganese** deposits and **manganese**-rich *iron formations*. [DJV]

rhourd A large pyramid-shaped **dune** formed by the intersection of smaller **dunes** in areas of multi-directional winds. [RDS]

rhyolite A fine-grained volcanic rock of acid chemical composition (hence compositionally equivalent to **granite**) which may be aphanitic or may contain *phenocrysts* of **feldspar** (**alkali feldspar** and/or sodium-rich **plagioclase**) and **quartz**, in a matrix composed of these minerals together with sparse mafic minerals. Rhyolites occur on *oceanic islands* and oceanic **island arcs** but are most abundant in continental arcs and continental *rift valleys* where they form extensive **ignimbrites** and rarer **domes** and **lava flows**.

Rocks of rhyolite composition were put to a number of uses in antiquity: a source of rhyolitic **ignimbrite** on Sardinia in the Mediterrranean provided many Roman **millstones**; in Britain, rhyolites were used by the Romans for some milestones, while earlier, in the Neolithic period, stone **axes** were occasionally made of this rock. Four of the 'bluestones' of Stonehenge, England, are of rhyolitic lava and **ignimbrite**. [RST/OWT]

rhythmite A finely laminated sediment (see **laminite**) in which different lithologies (usually two or three) are regularly repeated (Fig. R18). The regular repetitions constitute small cycles or rhythms. Rhythmites are particularly common in *glacial* lakes, and where an annual cycle of deposition can be demonstrated, they are termed *varves*. Rhythmites have therefore been widely described from *Pleistocene* lake sequences. In Glacial Lake Hitchcock, USA, three types of rhythmite were found (Ashley, 1975). Common to each type is a basal silty layer related to stream input into the lake in the 'summer' season, and a fine-grained upper layer deposited by fall-out from **suspension**, the so-called 'winter' layer. The details of these couplets, such as the thickness of the summer layer compared to the winter layer, and their constituent grain sizes appear to

vary as a function of distance from inflowing rivers and their **deltas**, and of the irregularities of the sub-lacustrine bathymetry. [PAA]

Ashley, G.M. (1975) Rhythmic sedimentation in Glacial Lake Hitchcock, Massachusetts-Connecticut. In: Jopling, A.V. & McDonald, B.C. (eds) *Glaciofluvial and Glaciolacustrine Sedimentation*. Special Publication of the Society of Economic Paleontologists and Mineralogists **23**: 304–20.

Kelts, K. & Hsü, K.J. (1978) Freshwater carbonate sedimentation. In: Lerman, A. (ed.) *Lakes: Chemistry, Geology, Physics*, pp. 295–323. Springer-Verlag, Berlin.

ria A submerged coastal valley originally formed by subaerial processes but inundated due to a rise in sea-level. Typically V-shaped in cross-section, the width and depth gradually and uniformly diminish inland. [NJM]

Richter earthquake scale This **earthquake** scale was formulated in 1935 by Charles Richter who, developing early ideas by the Japanese seismologist Wadati, used the amplitude of ground motion to provide a ranking of the size or strength of Californian **earthquakes**; their '**magnitude**', M, was defined as

$$M = \log(A) - Q(\Delta)$$

where A is the amplitude in millimetres of the **seismogram** written on a standard Wood–Anderson horizontal-component *seismograph* operating at a magnification of 2800, and Q(Δ) is a distance-correction term reducing the measurement to what would have been recorded at 100 km from the *epicentre*. Richter provided tables of Q(Δ) for **epicentral distances** of 10–600 km without allowances for varying depth of *focus* because all Californian **earthquakes** are less than 15–20 km deep. An equivalent definition in terms of ground displacement A (measured in nanometres) and **epicentral distance**, d (kilometres) is

$$M = \log_{10}(A) + 2.56 \log_{10}(d) - 4.67$$

This original Richter Scale is now known as the *Local Magnitude*, M_L, and is still widely applied in its original form even though it is strictly correct only for Californian **earthquakes**. Though it has limited theoretical basis, it is very useful and the original scale was extended by

Gutenberg & Richter in 1956 to recordings of **earthquakes** world-wide:

$$M = \log_{10}(A/T) + B(\Delta, h) - 3$$

where A and T are the amplitude (nanometres) and period (seconds) respectively of the *P wave* maximum measured on a short-period *seismometer*, and $B(\Delta, h)$ is a distance-correction term like $Q(\Delta)$ above but tabulated for distances Δ out to $110°$ and *focal depth h* up to 730 km. $B(\Delta, h)$ defines a globally averaged decay of amplitudes because of both geometric spreading and **anelastic attenuation**. Popularly called the Richter Scale, this is the *body wave* or *P wave* magnitude m_b. (See **earthquake magnitude**.) [RAC]

Boore, D.M. (1989) The Richter Scale: its development and use for determining earthquake source parameters. *Tectonophysics* **166**: 1–14.

Gutenberg, B. & Richter, C.F. (1956) Magnitude and energy of earthquakes. *Annali di Geofisica*. **9**: 1–15.

ridge push Mechanism whereby an *ocean ridge* exerts a force on the *plates* on either side, hence *'ridge push force'*. The origin and magnitude of this force are discussed by Bott & Kusznir (1984). The force arises from the buoyancy effect of the mass of warm, less dense **asthenosphere** that supports the ridge topography (the ridge is in **isostatic** equilibrium) and exerts a lateral **extensional stress** across the ridge. The magnitude of the resulting **stress** is estimated to be 20–30 MPa. The ridge push force is one of the main sources of renewable **stress** acting on the **lithosphere** *plates*, the others being **slab pull**, **subduction suction** and **plateau uplift**. (See **forces acting on plates**.) [RGP]

Bott, M.H.P. & Kusznir, N.J. (1984) Origins of tectonic stress in the lithosphere. *Tectonics* **105**: 1–14.

ridges and runnels Morphological highs and intervening lows trending parallel or subparallel to the coastline on certain **beaches**. They characterize gently sloping ($<0.6°$) sandy **beaches** with a high meso- or macro-**tidal** range, low to moderate **wave** energy and **fetch**-limited **waves**. The ridges are landwardly asymmetrical three-dimensional features with an amplitude from 0.1 to 1.5 m, and wavelength from 50 to 200 m. They are dissected by coast-perpendicular **runoff** channels.

Their genesis is related to the formation of **swash bars** whereby a steepened **swash** slope gradient is produced on an overall low gradient foreshore. A multiple ridged **beach** develops because the **tidal** height varies and the **swash** zone moves to new positions on the **beach** face. This process produces a series of ridges separated by intervening runnels. The position of the ridges corresponds roughly to the height at which the **tide** stands longest during the **tidal** cycle.

Ridge morphology has a well-defined annual cyclicity, with a reduced topography during the *storm* season, which builds under the fairweather **wave** climate. [RK]

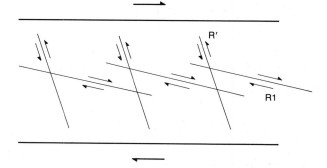

Fig. R19 Riedel shears (R1) are inclined at a low angle to the shear or fault zone and have the same sense. Conjugate Riedel shears (R') are at a high angle to the shear direction, with the opposite sense of displacement.

Orford, J.D. & Wright, P. (1978) What's in a name? — Descriptive or genetic implications of 'Ridge and Runnel' topography. *Marine Geology* **28**: M1–M8.

riebeckite ($Na_2Fe_3^{2+}Fe_2^{2+}Si_8O_{22}(OH)_2$) A bluish-coloured **amphibole** mineral, when asbestiform known as **crocidolite**.

An **axe factory** has been located in the riebeckite felsite of Shetland, UK, producing polished stone **axes** in the Neolithic period. [OWT/DJV]

Riedel shear A shear or **fault** within a **shear** or **fault zone** with the same sense as the zone, but inclined at a low angle (typically less than 15°) to the **shear** direction (Fig. R19). Riedel shears are also known as *synthetic* or *R1 shears*.

Riedel shears are ubiquitously observed in **simple shear** experiments and natural **shear** or **fault** zones. (e.g. Riedel, 1929; Tchalenko, 1968). They may be accompanied by a conjugate **shear** set which are referred to as *conjugate Riedel shears* (R') (Fig. R19). They are especially important elements of **strike-slip fault systems**. In **simple shear** experiments, Riedel shears form relatively early and may be rotated passively by the **shear** until they are no longer active; they may then be linked by new Riedel shears, *conjugate Riedel shears*, or *P shears* to form an anastomosing network of **shears** isolating **shear** lenses. This is also a very characteristic pattern of **faults** in **strike-slip** or *wrench fault zones*. On a smaller scale, Riedel and *conjugate Riedel shears* are also important features of both experimental and natural *fault gouges* [TGB]

Maddock, R.J., Hall, S.H. & White, S.H. (1986) Comparative microstructures of natural and experimentally produced clay-bearing fault gouges. *Pure and Applied Geophysics* **124**: 1–30.

Naylor, M.A., Mandle, G. & Sijpesteijn, C.H.K. (1986) Fault geometries in basement-induced wrench faulting under different initial stress rates. *Journal of Structural Geology* **8**: 737–52.

Riedel, W. (1929) Zur Mechanik geologisher Brucherscheinungen. *Zentralblatt für Geologie und Palaeontologie Abhandlung*. **13**: 354–68.

Tchalenko, J.S. (1970) Similarities between shear zones of different magnitudes. *Geological Society of America Bulletin* **82**: 1625–40.

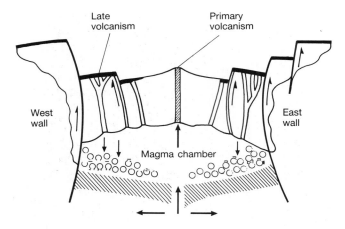

Fig. R20 Diagrammatic profile of the *c.*4 km-wide inner **rift** valley of the mid-Atlantic ridge, showing magma chamber in which lateral differentiation and accumulate deposition are taking place. Later volcanism on the flanks draws on differentiated magma, and crustal thickness increases towards the main inward-dipping boundary faults. (After Ballard & Van Andel, 1977.)

riffle A topographic high point in the undulating longitudinal profile of a **gravel**-bed river. Riffles have a characteristic downstream spacing of 5–7 channel widths and are separated by topographic low points or pools. Flow is divergent over the riffle and convergent in the pools. At high flow stages, lower velocities and **shear stresses** are present over the riffle than in the pool and commonly produce an accumulation of coarser material on the riffle than in the pool. (See **pool and riffle sequence**.) [JLB]

rift Major elongate down-**faulted** block (**graben**). The term was applied initially to relatively young structures expressed topographically as valley systems, such as the Rhine, African and Baikal rifts. However, it has now come to be regarded as a standard type of major **extensional graben** system of any age, whose original topographic expression may have been lost.

Rifts are found in both **continental** and **oceanic crust**. The major oceanic rifts are associated with the active crestal regions of *ocean ridges*, e.g. the mid-Atlantic ridge, where they are associated with seismic and volcanic activity. The major currently active rifts correspond to constructive **plate boundaries**, and are many thousands of kilometres in extent. The central mid-Atlantic rift (e.g. see Ballard & Van Andel, 1977) is about 100 km across and contains a volcanically active central ridge dominated by *extensional* **fissures** and their eruptive products (see Fig. R20). The flanks of the rift are marked by inward-dipping *normal* **step faults**. The depression of the rift floor is attributed to collapse following withdrawal of **magma** from the underlying chamber. Oceanic rifts are often segmented by **transform faults**, such that individual sections may be quite short.

Continental rifts are similar in structure and dimen-

sions to oceanic rifts, but vary widely in the nature and abundance of the associated vulcanicity. Moreover there are major changes in the associated magmatism along the length of individual rift zones. The origin of continental rifts has occasioned much debate. Many active rifts are directly connected to the present **plate boundary** network, e.g. the Gulf of Aden–Red Sea–African rift system; others may have an indirect connection via major crustal discontinuities by means of which displacements may be transferred to a **plate boundary**, e.g. the Rhine – Ruhr rift system, which is connected to the Alpine front to the south, and, formerly, through the North Sea rift system to the Atlantic rift in the north (see **impactogen**). The Atlantic and Indian Oceans clearly originated as rifts during the break-up of **Pangea** in the early **Mesozoic**, so that some rifts develop into oceans. Others have existed for long periods of geological time without developing into oceans, e.g. the African and Baikal rifts. Two opposed models have been proposed for rift generation: (i) **extension** of the **lithosphere**, where **crustal extension** precedes the volcanic effects, which are therefore secondary; and (ii) 'mantle activation' where a *mantle plume* causes crustal doming, vulcanicity and rifting — here the extensional effects are secondary (see Burke & Dewey, 1973). The East African rift is held to be an example of the mantle-activated type of rift, since the African Plate in which it is situated is in a state of general compression. Here doming and vulcanicity appear to pre-date the extensional rifting. However, in this case, as in others, the rift may be part of an older extensional network relating to the break-up of **Pangea** or **membrane tectonics**. (See **continental splitting**.) [RGP]

Ballard, R.D. & Van Andel, Tj.H. (1977) Morphology and tectonics of the inner rift valley at lat. 36′50°N on the mid-Atlantic ridge. *Geological Society of America Bulletin* **88**: 507–30.
Burke, K. & Dewey, J.F. (1973) Plume-generated triple junctions: key indicators in applying plate tectonics to old rocks. *Journal of Geology* **81**: 406–33.
Morgan, P. & Baker, B.H. (eds) (1983) Processes of continental rifting. *Tectonophysics* **94**.
Park, R.G. (1988) *Geological Structures and Moving Plates*. Blackie, Glasgow.

rigid body rotation Rotation in space without changing lengths or angular relationships of any material lines, and without displacement of the centre of rotation, i.e. an orientation change. Such a rotation can be applied to rock or to concepts such as the **strain ellipsoid**, *strain axes*, etc. Any **deformation** can be considered as a combination of translation (position change), internal or distortional **strain** (shape change), and rigid body rotation (orientation change). [SB]

rigid/rigidity Degree of resistance to **shear strain**. The rigidity or *modulus of rigidity* (G) is the ratio of **shear stress** (τ) to **shear strain** (γ) in **elastic deformation**.

$$G = \tau/\gamma$$

The SI unit of rigidity is $N\,m^{-2}$. [TGB]

rigidity of the Earth Seismic velocities of both *body waves* and *surface waves* are dependent on the **rigidity** of the rocks through which they pass, and to varying extents on the *shear* and *bulk moduli*. It is thus possible to use different **seismic waves** to determine the absolute **rigidity** of the Earth as a function of depth. Further constraints are provided by **free oscillations** of the Earth (solid *Earth tides*). [DHT]

Lambeck, K. (1980) *The Earth's Variable Rotation*. Cambridge University Press, Cambridge.
Mörner, N. (1980) (ed.) *Earth Rheology, Isostasy and Eustasy*. John Wiley & Sons, Chichester.

rill A small channel, generally less than a few centimetres across, that changes location with every **runoff** event and that can be obliterated by ploughing. [ASG]

rim syncline Annular **syncline** surrounding a **salt dome** or **diapir**. The structure is produced by the collapse of strata above the region evacuated by the salt, which has risen to form the **diapir**. [RGP]

ring complex/dyke/intrusion An annular tabular body with cylindrical form, so exposed at the surface with a ring-shaped outcrop. A ring dyke is an arcuate tabular body with a vertical axis and near-vertical or outward dipping contacts. **Cone sheets** are conical sheetlike bodies converging towards a buried apex. Both forms of ring intrusion are commonly mutually associated in a ring complex. Such complexes may be formed of almost any **igneous rock** and occur within *oceanic islands*, in continental margins (including **batholiths**) and in continental areas including the *Tertiary* volcanic province of north-west Scotland and the East African **rift** valley. [RST]

ringwoodite ($(Fe,Mg)_2SiO_4$) A mineral with the **spinel** crystal structure and the composition of an **olivine** found in stony *meteorites*. [DJV]

Fig. R21 Rip current. Longshore currents feed the rip currents which exit seawards through the breakers at zones of lower wave height. Atwick Beach, Holderness, east England.

rip current A strong narrow current which flows seaward from the **beach** through the *breaker zone*. Its strength makes it particularly dangerous to swimmers. Rip currents are fed by longshore currents which increase in velocity from a minimum midway between the rips to a maximum as the current turns seaward. A slow shoreward mass transport of water between the rips completes the nearshore cell circulation system. The spacing of the rips depends on variations in **wave** height alongshore. Opposite areas of high **waves**, the mean water level, or **wave** set-up, is enhanced. The resultant pressure gradient causes a current to flow from high to low **wave** set-up and this feeds the rip currents. Regular spacing of rip currents reflects systematic variation in **wave** height alongshore arising either from **wave refraction** or from interference caused by edge **waves** trapped within the nearshore zone. The nearshore cell circulation system of which rip currents are a part may be associated with cuspate and rhythmic shoreline features at a variety of scales and may also provide a means whereby the nearshore zone is flushed with freshwater and cleansed of pollutants (Figs R21 and R22). [JDH]

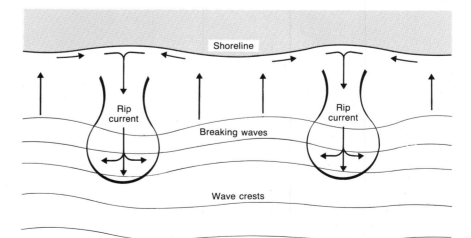

Fig. R22 Rip current.

[527]

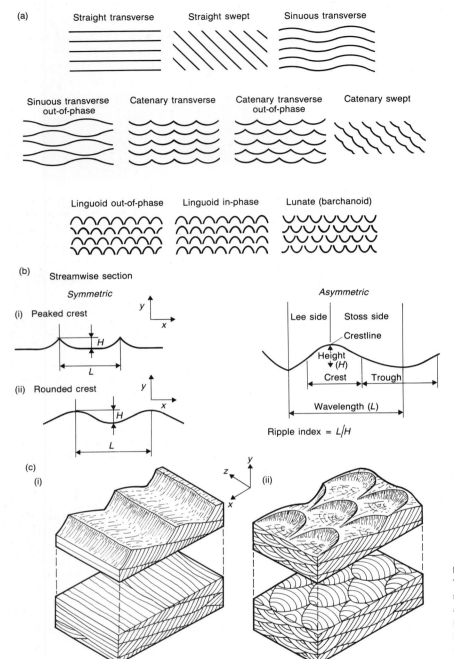

(a)

Straight transverse   Straight swept   Sinuous transverse

Sinuous transverse out-of-phase   Catenary transverse   Catenary transverse out-of-phase   Catenary swept

Linguoid out-of-phase   Linguoid in-phase   Lunate (barchanoid)

(b) Streamwise section

Symmetric

(i) Peaked crest

(ii) Rounded crest

Asymmetric

Lee side | Stoss side
Crestline
Height (H)
Crest | Trough
Wavelength (L)

Ripple index = L/H

(c)

(i)

(ii)

Fig. R23 Characteristics of **ripples**. (a) Variations in planform geometry. (b) Streamwise sections of symmetric and asymmetric ripples and their nomenclature. (c) Stratification produced from (i) straight crested current ripples producing planar cross-stratification and (ii) sinuous crested current ripples producing trough cross-stratification in the flow-normal plane.

Komar, P.D. (1976) *Beach Processes and Sedimentation.* Prentice-Hall, Englewood Cliffs, NJ.
Wright, L., Chappell, J., Thom, B., Bradshaw, M. & Cowell, P. (1979) Morphodynamics of reflective and dissipative beach and inshore systems, South Australia. *Marine Geology* **32**: 105–40.

ripple The stable, flow transverse bedform generated at relatively low **shear stresses** above the threshold for sediment movement in **cohesion**less **sands** of less than approximately 0.6 mm mean grain size. Ripple wavelength is generally less than 0.5 m and amplitude is commonly less than 0.03 m. The morphology of ripples displays a wide variation dependent upon the controlling fluid dynamic processes which are, in turn, related to the environment of deposition. Ripples can be described in terms of both their cross-sectional geometry and planform (see Fig. R23). The important distinction between asymmetrical and symmetical ripples can be assessed through comparison of *ripple indices* (wavelength : height ratio) and may, with care, be used to support the generality of the distinction between unidirectional flow and **oscillatory flow** structures. Unidirectional flows generate ripples with indices of between 10 and 40 whereas ripples produced under **oscillatory flows** have an index of commonly less than 4. A range of combined flow ripples also exists between these two end members.

The planform shape of ripples varies greatly. Ripple dimensions generally increase at higher bed **shear stresses** and ripple wavelengths vary with mean grain size, D, wavelength being approximated as $c.$ $1000\,D$ under unidirectional flows. Ripple dimensions are, however, independent of the flow depth and consequently have been linked to processes characteristic of the inner region of the turbulent **boundary layer**.

Current ripples are intimately associated with **flow separation** with separation occurring at the crest and reattachment in the trough. Ripple migration results from avalanching of grains down the ripple **lee side** and produces small-scale **cross-stratification**. In the flow-normal plane this may produce horizontal planar stratification with straight crested ripples or trough **cross-stratification** where sinuous crested ripples possess distinct scour troughs in their **lee sides**. Ripple-generated sedimentary structures are of the upmost importance in *paleoenvironmental analysis* and their dimensions may be used in *paleohydraulic reconstructions*. (See **climbing ripples, cross-stratification**.) [JLB]

Allen, J.R.L. (1968) *Current Ripples*. North Holland, Amsterdam.
Allen, J.R.L. (1984) *Sedimentary Structures: Their Character and Physical Basis*. Elsevier, Amsterdam.

river capture A process whereby one river, by more rapid vertical incision, effectively undercuts the drainage area of an adjacent river, thereby truncating its neighbour's **catchment** and enlarging its own. Evidence of capture is traditionally provided by an abrupt change of direction along the river channel, showing that a river which formerly flowed in one direction has had its headwaters captured by another river whose direction of drainage is quite different. For small **rills** and *gullies*, the term 'micropiracy' has been used to describe the way in which gradual migration of **rills** may occur back and forth across a hillslope, so effecting a relatively even **erosion** of the whole slope surface. [NJM]

river flood Water arriving in the river channel rapidly enough to produce a significant increase in discharge above *baseflow* levels so as to form a recognizable peak in **discharge**. Floods are produced by a variety of mechanisms, of which heavy rainfall and snowmelt are the most usual. Many methods of *flood prediction* now exist: these may be divided into statistical methods of predicting flood frequency; and methods of *flood forecasting*, often involving estimation of the whole flood **hydrograph** rather than just the flood peak discharge. Use of computers allows real-time *flood forecasting*, in combination with telemetry systems for precipitation and discharge monitoring. Such methods form the basis for the design of flood alleviation schemes and allow flood warnings to be issued. Floods are an important element of the natural behaviour of river systems. In humid areas, floods of moderate frequency and magnitude provide the dominant **discharge** with respect to the maintenance of channel form. Because the channel is completely full, **bankfull discharge** may best represent this range of morphologically significant floods. [TPB]

Ward, R.C. (1978) *Floods — A Geographical Perspective*. Macmillan, London.

Robin effect A mechanism which occurs in association with **glaciers**, where the removal of rock at the edge of steps above a subglacial cavity may be reinforced by a heat pump process proposed by Robin (1976). As ice moves over an irregular bed the **stress** distribution in the ice and at the **glacier** bed will change (due perhaps to variation in basal water supply). As Drewry (1986, p. 46) has explained, 'If the pressure increases the ice may melt with expulsion of any free water in veins and at crystal boundaries, with corresponding lowering of the pressure melting point. When the pressure is released the absence of the expelled water requires that the heat lost from the ice is restored by conduction through the rock leading to production of cold patches . . . The effect of a developing cold patch is both adhesion of ice to rock and an increase in rock hardness.' [ASG]

Drewry, D. (1986) *Glacial Geologic Processes*. Arnold, London.
Robin, G. de Q. (1976) Is the basal ice of a temperate glacier at the pressure melting point? *Journal of Glaciology* **16**: 183–96.

rocdrumlin (*tadpole rock*) Streamlined moulded bedrock form. Rocdrumlins are *drumlin*-shaped features developed in solid rock, whereas *drumlins* are usually made of *till* or **drift** with or without a rock core. They have a massive and high upstream side and an elongated and gently sloping downstream side. [ASG]

rock crystal A varietal name given to colourless crystals of the mineral **quartz**. [DJV]

rock magnetism The study of the magnetic properties of naturally occurring magnetic minerals, particularly the processes by which they become magnetized both naturally and in the laboratory. The minerals involved include the **iron oxides** (mainly *titanomagnetites* and *ilmenohematites*), the **iron** hydroxides (especially **goethite**), **iron** sulphides (mainly **pyrrhotite**) and extraterrestrial minerals, especially **iron**, **nickel** and their alloys, as found in *meteorites* and lunar samples. [DHT]

O'Reilly, W. (1984) *Rock Magnetism and Mineral Magnetism*. Blackie, Glasgow.
Tarling, D.H. (1983) *Palaeomagnetism*. Chapman & Hall, London.

roddon A sinuous silty ridge that snakes about above the general level of the **peat** Fens of East Anglia (England). Roddons are the remnants of ancient river systems that may initially have flowed between **levées** above the general level of the surrounding land. Alternatively they may subsequently have become relatively elevated as a consequence of **peat** wastage brought about by drainage activities. [ASG]

Fowler, G. (1932) Old river beds in the Fenlands. *Geographical Journal* **79**: 210–12.

rods/rodding Cylindrical structures composed of a single **mineral** in deformed rocks. Rods are similar in morphology to **boudins** and **mullions**, but are distinguished by their monomineralic composition which is separate from the enclosing rock. They are most commonly composed of **quartz**; rods of **calcite** and **pyrite** are also known. They define a **lineation** which may be parallel to *fold axes*, and may be formed by **deformation** of **mineral** segregations, including **veins**, sometimes by *detachment* and isolation of *fold hinges*, or by **elongation** of *conglomerate* pebbles. [TGB]

Wilson, G. (1953) Mullion and rodding structures in the Moine series of Scotland. *Proceedings of the Geological Association* **64**: 118–51.
Wilson, G. (1961) The tectonic significance of small scale structures and their importance to the geologist in the field. *Annales Société Géologique de Belgique* **84**: 424–548.

romanèchite ($BaMn^{2+}Mn_8^{4+}O_{16}(OH)_4$) An **ore** mineral of **manganese**. [DJV]

room and pillar mining A mining method normally used in flat lying or subhorizontal deposits in which an area is divided into a series of rooms, from which the **coal** or **minerals** are extracted, separated by a series of pillars that hold up the roof. If the pillars cannot be later removed then a considerable amount of economic material is lost in this method. [AME]

root zone of nappe Area from which a **nappe** has originated. Most **nappes** are now physically separated from their roots due to **erosion** and/or **deformation**. However, the **displacement** surface (*thrust* or **shear zone**) forming the base of the **nappe** usually cuts down through the **crust** to a major **detachment horizon**, within or at the base of the **crust**, to link eventually with a contemporary **plate boundary** (commonly a collisional **suture**). The root zone itself may correspond to a collisional **suture**. [RGP]

roscoelite ($KV_2(AlSi_3O_{10})(OH)_2$) An **ore** mineral of **vanadium** with a mica structure. (See **mica minerals**.) [DJV]

Rossi–Forel scale A 10-point **earthquake intensity** scale, probably the first formal scale of modern times. It was constructed by de Rossi of Italy and Forel of Switzerland in 1878. [RAC]

rotational slip A rapid mass movement in **cohesive** sediments in which the shape of the **failure** plane is roughly circular. A single **failure** may occur, or a sequence of retrogressive **failures** may develop. Such **failures** are common where river valleys incise into weak clay deposits, and on man-made embankments cut into clays. In some clays overlain by porous caprocks, especially in **sea cliffs** where *toe erosion* is continuous, a series of deep-seated retrogressive **failures** may be formed. [TPB]

Röthlisberger channels (*'R' channels*) A series of conduits or pipes at the bed of a **glacier** incised upwards into the ice (Drewry, 1986). Tunnel melting is produced by heat generated by **turbulent** water **flow**, and they are maintained by water descending to the bed via *moulins* and *crevasses*. They occur primarily along the central parts of the beds of valley **glaciers**.

Nye channels are water-discharging basal conduits associated with **glaciers** which, unlike **Röthlisberger channels** are incised into bedrock. [ASG]

Drewry, D. (1986) *Glacial Geologic Processes.* Arnold, London.

royalty A percentage of the revenue from the operation of a mining or quarrying enterprise, **coal**, gas, or **oilfield**, etc., paid to the owner of the mineral rights which may be a private individual, a company or a government. [AME]

rubellite A red to pink variety of the mineral **tourmaline**. [DJV]

ruby A red **gem** form of the mineral **corundum**. (See **oxide minerals**.) [DJV]

ruby copper A variety of the mineral **cuprite** (forming ruby-red transparent crystals. [DJV]

ruby spinel ($MgAl_2O_4$) A red **gem** spinel. (See **oxide minerals**.) [DJV]

rudite A clastic rock with more than 30% *clasts* of **gravel** grade, i.e. coarser than 2 mm, in a matrix of **sand** and/or mud grade (see **clastic rock**). Those with more than 85% clasts are termed *clast-* (or *particle-*, *grain-* or *framework-*) *supported*, since the *clasts* are mostly in contact, whereas *matrix-supported* rudites have 30–85% clasts. Rudites grade into pebbly (or *conglomeratic*) mudstones and **sandstones**. They are subdivided on the basis of *clast* morphology (and therefore *textural maturity* — see **clastic rock**) into *sedimentary breccias*, whose *clasts* are mostly angular and *conglomerates* (archeically '*puddingstones*') with mostly rounded to subangular *clasts*. The term *breccio-*, or *breccia-conglomerate* is sometimes used for intermediates. *Calcirudites* are *rudaceous* **limestones**. A *sedentary rudaceous deposit* is one produced *in situ* by **weathering**.

Conglomerates can be divided into those which are *oligomict*, or *monomict*, dominated by one *clast* composition, usually compositionally mature (see **sandstone** for the concept of *maturity*) assemblages of **quartz** and/or **chert**, and *polymict* or *petromict conglomerates*, which have *clasts* of a variety of compositions. *Intraformational conglomerates*, composed of **intraclasts**, contrast with *extraformational conglomerates* sourced mostly from outside the **basin** of deposition. The term *orthoconglomerate* has been used for *clast-supported conglomerates* inferred to have been deposited by aqueous currents, whereas *paraconglomerates* include matrix-supported *conglomerates*, *mixtites* and pebbly **mudstones** attributed to deposition by **sediment gravity flows**, or as glacial *till*, Genetic terms which have been coined for *conglomerates* include *tillite* and, for **alluvial fan** deposits, *fanglomerate*.

A typical site of deposition for a *sedimentary breccia* is a **scree**. 'Breccia', and 'brecciated' are also used for certain non-sedimentary fragmental rocks such as *fault breccias* and *volcanic* or *igneous breccias* (*agglomerates* and *explosion breccias*). *Solution*, or *collapse breccias* form as a result of the **dissolution** of underlying **limestone** or **evaporite** (see **evaporite-related collapse breccias**). [DAR]

Greensmith, J.T. (1978) *Petrology of the Sedimentary Rocks* (6th edn). Allen & Unwin, London.
Pettijohn, F.J. (1975) *Sedimentary Rocks* (3rd edn). Harper & Row, New York.

runoff A general term encompassing flow in streams and on hillslopes. Usually, however, it is applied to flow in small **basins** and so implies a hillslope source. Runoff may be divided into *baseflow* and *quickflow*. *Baseflow* is subsurface in origin and may be composed of either **throughflow**, **groundwater**, or both. *Quickflow* may be both surface and subsurface runoff — a mixture of **overland flow** and *subsurface stormflow*. [TPB]

runzelmarken (*wrinkle marks*) Parallel or reticulate ridges up to 1 mm high, with spacings of a few millimetres, on cohesive muddy sediment surfaces. They have been observed experimentally and in intertidal environments to form when strong wind **stress** affects such a surface when covered by only a very thin film of water. However, the presence of very similar wrinkled **mudstone** surfaces in ancient deep water deposits suggests that they may also form in response to subaqueous currents. Some at least may be the expression at the sediment–water interface of soft sediment loading in thinly interlaminated **sands** and muds. [DAR]

Allen, J.R.L. (1985) Wrinkle marks: an intertidal sedimentary structure due to aseismic soft-sediment loading. *Sedimentary Geology* **41**: 75–95.
Reineck, H.-E. (1969) Die Entstehung von Runzelmarken. *Senkenbergiana marit.* **1**: 165–8.
Teichert, C. (1970) Runzelmarken (Wrinkle marks). *Journal of Sedimentary Petrology* **40**: 1056.

rutile (TiO_2) An important **accessory mineral** in many rocks and an **ore** mineral of **titanium**. (See **oxide minerals**.) [DJV]

ruware A low, dome-shaped exposure of bedrock projecting from a cover of alluvium or **weathered** bedrock. It is either a relict or incipient **inselberg**. [ASG]

S

sabkha (*sebkha*) A broad plain or **salt flat** which contains **evaporites** and is formed in an arid or semi-arid climate; the sabkha level is dictated by the local level of the **water table** and forms an equilibrium geomorphological surface. The sabkha surface may be periodically inundated with water, either **meteoric runoff** or marine. Sabkhas may be either coastal or continental; the former are dominated by marine sediments and marine **groundwater**, whereas the latter are dominated by continental **groundwater** and sediments. A marine sabkha may grade laterally into a continental sabkha with no noticeable morphological discontinuity. Handford (1981) represented process associations of sabkhas on a triangular diagram, with end members comprising marine processes, fluvial–lacustrine processes and aeolian processes (Fig. S1). Evaporation at the sabkha surface gives rise to **evaporative pumping** and **evaporite** formation, both at the surface and within the upper metre of sediment.

Continental sabkhas may be dominated by fluvial–lacustrine processes or by aeolian processes (Fig. S1). The mud flats which surround **playa lakes** form continental sabkhas dominated by fluvial–lacustrine processes; these continental sabkhas are more prone to inundation by changing lake levels than other sabkha types, with clays settling from the ponded lake water. Such sabkhas are commonly developed in siliciclastic facies, either the aforementioned clays or intermixed clays and **sands**, derived from the adjacent **alluvial fans**. One typical example is Bristol Dry Lake, California where **evaporites** formed within these sabkha sediments are dominated by **halite**, both as displacive hopper crystals and chaotic displacive clay–**halite** mixes. Around the outer limits of the **salt flat gypsum** occurs in the form of regular and distorted laminae, **nodules**, coarse sand-sized crystals, some discoid, and as displacive **selenite** crystals. As with most playa-associated sediments the preservation potential of these sabkhas is low, and ancient examples are rare.

Aeolian processes may both create and modify sabkhas. The reworking of sediments by the wind produces aeolian **dunes** with intervening **deflation** surfaces, the level of the surface determined by the level of the **water table** within the area. One modern example is the Qattara Depression, Egypt, where constant **deflation** has lowered the surface to over 100 m below sea-level over a **basin** of some 18 000 km^2. Rapid migration of **dune** forms inhibits **groundwater** concentration by evaporation, but where **interdune** areas remain static **evaporites** may be **precipitated** within the surficial sediments. In these environments displacive

carbonates are a characteristic **evaporite** mineral, commonly **precipitated** within the upper few centimetres of the sediment.

Coastal plain, or marine, sabkhas (Fig. S1) develop as prograding supratidal plains on shallow stable shelves where they form an upper member of a **shallowing-upward cycle** (Fig. S2) in either carbonate or siliciclastic sediments. This environment is a product of both depositional and **diagenetic** processes, where depositional processes may be highly modified by the **diagenetic** growth of displacive calcium sulphate, either as **gypsum** or, more rarely, **anhydrite**, together with, in places, **halite**. Calcium sulphate development commonly occurs to a greater extent than in continental sabkhas, as evaporated seawaters provide an extensive source of both calcium and sulphate. Sulphates are enhanced in carbonate marine sabkhas where **dolomitization** releases additional calcium to form further **gypsum** and **anhydrite**. Where **dolomitization** does not occur, or occurs only to a small extent, as in siliciclastic marine sabkhas, sulphate abundances are relatively less and **magnesite** (MgCO$_3$) is common.

The continuum between marine and continental sabkhas is demonstrated, in part, by one modern sabkha, Laguna Salada, Baja California, which exhibits the evolution from a marine sabkha, formed as a prograding supratidal flat into the Gulf of California, to a **playa**-associated sabkha, now isolated more than 100 km from the Gulf as the Colorado River **delta** has prograded seawards.

This continuum is also developed within the sabkhas of the Trucial Coast, where the coastal margins are dominated by marine waters, whereas, a kilometre or more away from the coast itself, **groundwater** below the sabkha surface comprises continental **brines** (Fig. S3). It is these sabkhas which have dominated the literature on marine sabkhas, a result of research in the 1950s and 1960s on the Abu Dhabi sabkhas of the Trucial Coast, now unfortunately largely destroyed in the creation of new port facilities at Abu Dhabi. The discovery of penecontemporaneous **dolomitization** of the sabkha sediments together with the discovery of **anhydrite** as **nodules** and enterolithic layers (Fig. S2) within the upper metre of sediment (see **gypsum–anhydrite equilibrium**) resulted in a plethora of papers recognizing Trucial Coast-type analogues in the ancient sedimentary record, an abundance not representative of the variations in morphology and **hydrology** now recorded from modern sabkhas (Fig. S3). As an example of this popularization of sabkhas, one feature of sabkha sulphates is their *chicken wire texture*

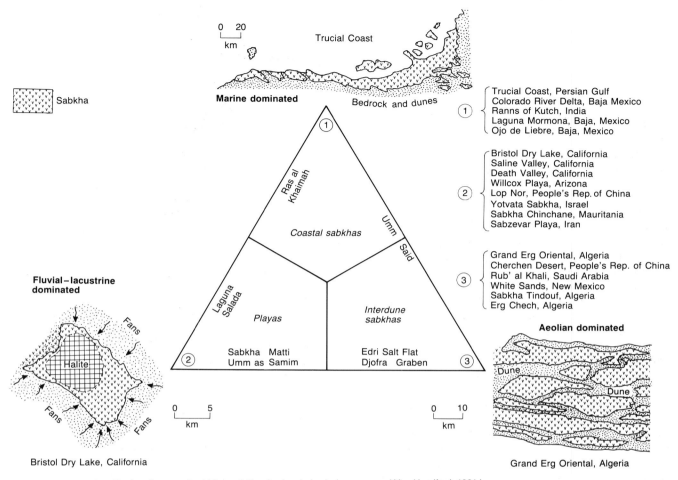

Fig. S1 Triangular classification diagram of **sabkhas** relating dominant physical processes. (After Handford, 1981.)

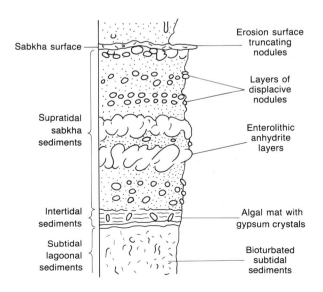

Fig. S2 Sabkha succession, based on the Abu Dhabi Sabkha, UAE Trucial Coast. (After Shearman, 1966.)

(Fig. S2). The existence of this texture in many ancient **evaporite** sequences was taken by many authors as evidence of a former sabkha environment. However, *chicken wire texture* is a common **diagenetic** product of sulphates and, as such, has been recorded from many ancient **evaporitic** depositional environments, demonstrating that one feature, taken in isolation, cannot be used to determine an ancient sedimentary environment.

There are several factors which control the development of marine sabkhas. Firstly, their size is controlled by the sedimentation rate and hence the development of a stable accretionary sediment wedge on a coastal plain. As the sabkha forms the upper surface of this accretionary wedge, a large sabkha can only be expected to develop where there are stable conditions and no steep offshore gradient. Moreover, large sabkha development necessitates a regional **groundwater** flow and hence a large hinterland with sufficient relief to form an adequate **recharge** area. The nature of the accretionary wedge and the **tidal** regime is also important. The modern large, planar Abu Dhabi Sabkha formed in a protected, microtidal

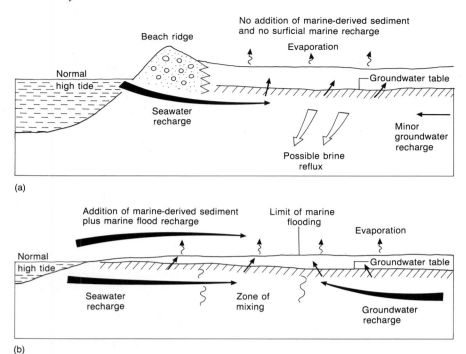

Fig. S3 Different modes of water and brine supply in **sabkhas**. (a) Based on Sinai coastal sabkha. (b) Based on Abu Dhabi sabkha. (After Kendall, 1984.)

environment where sea-level has been falling only slightly. Increased tidal action, progradation where there are no protective **barrier islands**, or more rapid changes in sea-level would result in a less regular planar sabkha surface and, possibly, less rapid progradation, features not yet recognized in ancient sequences described as sabkhas. A further important factor is the nature of the adjacent water body. Modern marine sabkhas develop adjacent to normal marine to slightly hypersaline waters. However, if a sabkha develops adjacent to a hypersaline water body where **evaporites** are being **precipitated**, the underlying sediments may be very different; *algal mats* may extend well into the subtidal environment where there is no browsing fauna, the facies therefore being dominated by *algal-laminated* **evaporitic** sediments. The additional presence of interbedded, *storm*-derived, clastic **evaporites** on the sabkha surface makes such sabkha sequences very difficult to recognize in ancient **evaporitic** sediments.

Although sabkhas incorporate a diverse range of sedimentary features and host sediments, all are prone to early **diagenetic alteration**, due to the rapid throughflow of **brines** and their evaporation at the sabkha surface. This constant, and rapid, evaporation gives rise to **evaporative pumping**, increasing both fluid throughput and penecontemporaneous **diagenesis**. The nature, the rate of flow and the origin of the fluids, together with the extent of their reaction with the host sediments, control the extent of this **diagenesis**. This can range from few displacive **evaporites**, as in many continental sabkhas, through displacive **evaporites** plus an **evaporite**-rich crust to large-scale displacive sulphates, as seen in the Abu Dhabi

sabkha, to large-scale displacement plus early **diagenetic alteration** and disruption of original **evaporitic** sediments. [GMH]

Handford, C.R. (1981) A process-sedimentary framework for characterizing recent and ancient sabkhas. *Sedimentary Geology* **30**: 255–65.

Kendall, A.C. (1984) Evaporites. In: Walker, R.G. (ed.) *Facies Models* (2nd edn), pp. 259–96. Geoscience Canada Reprint Series No. 1.

Shearman, D.J. (1966) Origin of marine evaporites by diagenesis. *Transactions of the Institution of Mining and Metallurgy* **B75**: 207–15.

saddle reef A saddle- to triangular-shaped development of **minerals** or **coal** in the *hinge* zone of a *fold* (Fig. S4). When **competent sandstones** are interbedded, and particularly when there are thin interbeds of **shale**, then there is a tendency for an opening to form between the base of one and the top of that beneath. The most famous of such mineralized dilatational zones are the auriferous saddle reefs of Bendigo, Australia where the main mineral is **quartz**. [AME]

salar A **basin** of inland drainage in an arid or semi-arid region. Salars may be inundated periodically by rainfall, **runoff** or **groundwater** rising above the floor. For most of the time, however, they are dry. Though **runoff** flowing into the **basin** carries with it fine-grained sediments, **deflation** of these deposits maintains the depression. The evaporation of floodwaters results in the accumulation of **evaporite** minerals in many salars. [AW]

Stoertz, G.E. & Ericksen, G.E. (1974) *Geology of Salars in Northern Chile*. Professional Paper of the United States Geological Survey **811**.

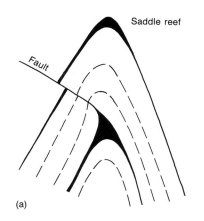
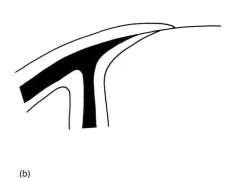

Fig. S4 (a) **Saddle reefs** of the type at Bendigo, Australia. (b) Queue anticlinal in coal, Belgium.

(a)   (b)

salcrete Surface crust of primarily sodium chloride which cements a **sand** surface through evaporation of moisture. The term was coined by Yasso (1966), and has been mainly used to describe crusts developed through evaporation of sea spray on **beaches**. [RDS]

Pye, K. (1980) Beach salcrete in North Queensland. *Journal of Sedimentary Petrology* **50**: 257–61.
Yasso, W.E. (1966) Heavy mineral concentrations and sastrugi-like deflation furrows in a beach salcrete at Rockaway Point. *Journal of Sedimentary Petrology* **36**: 836–8.

salic mineral A **normative mineral** comprised dominantly of silicon (**silica**) and **aluminium** (alumina), for example **quartz**, **feldspars** and **feldspathoids**. [RST]

salt diapir/dome A structure forming when less dense **halite** (with a density near $2.3\,\mathrm{Mg\,m^{-3}}$) becomes unstable under an increasing overburden of overlying sediments (density nearer $2.6\,\mathrm{Mg\,m^{-3}}$) and rises, buoyantly, through the sediment overburden. A salt dome contains a central salt plug, commonly 1–2 km in diameter, which may puncture 5–10 km of sediments; extreme salt movement (*halokinesis*), particularly where associated with active tectonism, may result in *piercement structures* and, in places, salt flowage on the surface.

The initiation of salt movement commonly starts with the formation of a *salt pillow*, with slight withdrawal of **halite** between adjacent pillows. Continued **halite** migration causes development of a diapir (Fig. S5), which may become detached from the original salt source as it rises. Salt domes deform their surrounding sediments as they rise, forming potential *hydrocarbon traps* in areas otherwise devoid of structural **deformation**. (See Fig. O3, p. 429.) [GMH]

salt flat A flat stretch of salt-encrusted ground, often the bed of a former salt lake. **Evaporite** salt flats are also found in arid supratidal zones adjoining intertidal flats, where they are known as **sabkhas**. A famous stretch of salt flats is the Bonneville Salt Flats, which are the remains of Lake Bonneville. [HAV]

salt heave A cause of damage to roads, runways, and foundations in desert areas as a result of the presence of soluble salts (including sodium chloride, **magnesium** sulphate and sodium sulphate). The process is akin to *frost heave*, in that the hydration and crystallization of salts plays a major role, but it is also akin to needle ice formation in that salt whiskers may grow vertically. The problem is especially severe when saline **groundwater** occurs near the ground surface. Possible techniques to deal with the problem have been developed (Horta, 1985), including brooming, embankments, barriers and the use of thick impervious surfacings. [ASG]

Horta, J.C. de O.S. (1985) Salt heaving in the Sahara. *Géotechnique* **35**: 329–37.

salt marsh A vegetated intertidal mud flat found along temperate coastlines. In the tropics they are generally replaced by **mangrove swamps**, although there are some tropical occurrences of salt marshes or salt marsh plants. Salt marshes are typically found on low-energy coasts, including **estuaries** and bays. Several different types of salt marsh can be recognized, depending upon the flora, tidal regime, sediment types, topography and **eustatic** and **tectonic** stability. In Western Europe and North America the most common colonizing genus is *Salicornia* (marsh samphire) or *Spartina* (marsh cord grass). Both of these are well adapted to the high salinities of the upper mud flat environment. Colonization of these species leads to increased deposition of sediment and a succession towards better adapted plants. In Britain and Western Europe, for example, *Salicornia* is succeeded by *Halimione portulacoides* (sea purslane) and *Aster maritima* (sea aster) at medium marsh heights. *Puccenelia maritima* (salt marsh grass), *Limonium vulgare* (sea lavender) and other species become dominant at the highest marsh levels. In North America a different species succession is found. Sediment **accretion** rates on salt marshes decrease over time. On young marshes (less than 100 years old), rates as high as $100\,\mathrm{mm\,a^{-1}}$ are found, whereas on older marshes rates of $0.01\,\mathrm{mm\,a^{-1}}$ are common. Salt marshes may be open or

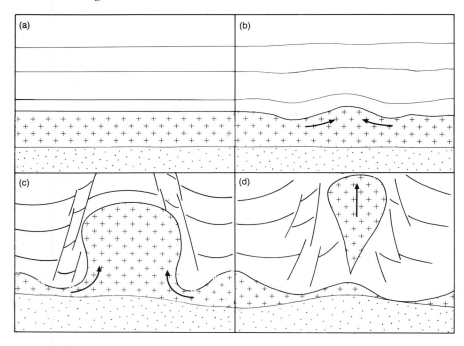

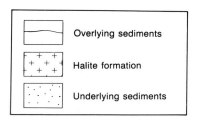

Overlying sediments

+ + +
+ + + Halite formation
+ + +

Underlying sediments

Fig. S5 Development of a **salt diapir**: (a) Progressive burial of halite formation. (b) Commencement of halokinesis with development of salt pillow. (c) Attached diapir, with salt withdrawal adjacent to diapir. (d) Detached diapir rising through overlying sediment column with complete salt withdrawal in some areas.

closed features, depending upon ease of access to open mudflats. Closed marshes develop behind **spits** and other barriers. Open marshes may end in a small cliff (or *microfalaise*) where **erosion** is occurring. Salt marshes are characterized by the presence of intricate creek systems. A high density of creeks per unit area of marsh surface is often found. *Salt pans* are also found on many marsh surfaces. [HAV]

Frey, R.W. & Basan, P.B. (1987) Coastal salt marshes. In: Dans, R.A. (ed.) *Coastal Sedimentary Environments* (2nd edn), pp. 225–301. Springer-Verlag, New York.

salt weathering **Weathering** caused primarily by physical changes produced by salt crystallization, salt hydration or the thermal expansion of salts (see also **haloclasty**). Substantial pressures can be produced as crystals grow from solutions in rock pores as salts take up water of crystallization during the hydration process, or as a result of the relatively high coefficients of linear expansion that many common salts possess. Laboratory studies show that among the most effective salts are sodium sulphate, sodium carbonate and **magnesium** sulphate. Further **weathering** may be promoted by chemical reactions involving the salts. Salt weathering has been recognized as an important **geomorphological** process in desert, coastal, polar and urban areas, and as a cause of decay in engineering structures, especially those made of concrete. [ASG]

Cooke, R.U. (1981) Salt weathering in deserts. *Proceedings of the Geologists' Association, London* **92**: 1–16.
Goudie, A.S. (1986) Laboratory simulation of 'the wick effect' in salt weathering of rock. *Earth Surface Processes and Landforms* **11**: 275–85.

saltation That component of **bedload transport** characterized by a series of ballistic jumps with grains adopting a steep ascent from the bed (>45°) to a height of several grain diameters followed by a low angle (<10°) descent. Individual saltation movements may be either uninterrupted, displaying a smooth parabolic-like trajectory, or interrupted by turbulent eddies or grain collisions at some point in the saltation path. Saltation height and its impact effects upon the bed are far greater in air than in water flows due to the much lower **kinematic viscosity** of the fluid. (See Fig. S6.) [JLB]

sand For geologists, mineral or rock grains of 0.625–2 mm diameter are of sand grade. Smaller particles are called fines and for most commercial purposes have to be removed by washing. Most sand is composed of **quartz** and the ideal sand deposit should contain a range of grain sizes so that a large number of different screened sizes can be marketed for various purposes. For many of these opaline **chert** or **shale**, chert, siliceous **limestone** and acid volcanic fragments are deleterious, as are mica, silt, organic matter, surface coatings, fissile **shale**, friable **sandstone** and other weak rock types. Sand is usually won from river, *glacial* or offshore deposits. It has many uses besides the obvious ones in the construction industries, e.g. glass making, foundry sand and so on. (See **gravel**.) [AME]

sand rose A circular figure (*rose diagram*) which depicts graphically the volume of **sand** moved by the wind from different compass directions, generally over a period of

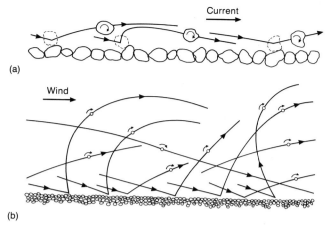

Current

Wind

(a)

(b)

Fig. S6 Saltation. Schematic representation of the paths of irregular particles saltating over a bed of similar grains. (a) In water. (b) In the wind.

one year. Sand roses may employ empirical data from **sand** traps at a specific location. Often, however, such information is not available and sand roses are drawn based upon potential rates of **sand** drift calculated from wind data from a specific meteorological station. In either case, the length of the arms of the sand rose indicates the volume of **sand** (actual or potential) moving from each direction. From this information it is possible to compute the resultant direction of net **sand** movement and the net volume of drift in that direction. A graphical representation of this vector usually accompanies the *rose diagram.* [AW]

Fryberger, S.G. & Ahlbrandt, T.S. (1979) Mechanisms for the formation of eolian sand seas. *Zeitschrift für Geomorphologie* NF **23**: 440–60.

sand sheet Area of predominantly aeolian **sand** where **dunes** with **slipfaces** are generally absent. Sand sheet surfaces can be rippled or unrippled, and range from flat to regularly undulatory to irregular (Kocurek & Nielson, 1986). They form in **ergs** where conditions are not suitable for **dunes**, or particular factors act to interfere with **dune** formation. These factors include a high **water table**, periodic flooding, surface binding or **cementation**, the presence of vegetation, and a significant coarse grain-size component. These same factors are also effective in promoting sand sheet accumulation where otherwise only sand transport without deposition would occur. [ASG]

Kocurek, G. & Nielson, J. (1986) Conditions favourable for the formation of warm-climate aeolian sand sheets. *Sedimentology* **33**: 795–816.

sand volcano A small mound of **sand** and mud generally less than 30 mm high and 25 to 50 mm in diameter. At the apex there is a conical depression from which water is emitted carrying fine-grained sediment in suspension. The deposition of this material causes the mound to accrete. Sand volcanoes develop on fluvial and lacustrine deposits

which extrude interstitial water at surface **springs** as the sediments settle and are compacted. [AW]

sandbed channel Channel composed of largely sandy material which is transported at a wide range of discharges. Such bed sediments are most readily moulded into different shapes, and a sequence of bed forms is associated with changing discharge (see **channel resistance**); such changes in bed configuration represent an important mechanism for the self-regulation of **channel resistance**. Such channels are likely to be wider and shallower than those formed in more **cohesive** sediments. [TPB]

sandstone A **clastic** sedimentary **rock** containing 25% or more by volume of *clasts* of **sand** grade, i.e. 0.0625 to 2 mm (see **clastic rock**). The most commonly used classification applicable to the majority of sandstones (or *arenaceous rocks*) is that of Pettijohn *et al.* (1973) shown in Fig. S7, in which terrigenous sandstones (or siliciclastic sandstones) are divided into *wackes* and *arenites* according to the proportion of matrix (see **clastic rock**) present. The classification therefore incorporates a major aspect of *textural maturity* (see **clastic rock**). Unfortunately, *arenite* has an alternative usage as a synonym of sandstone. Further subdivision is on the basis of *clast* composition using the proportions of **quartz**, **feldspar** and other rock fragments, which together make up the vast majority of **sand** grains. Polycrystalline **quartz** grains are usually considered as **quartz** rather than as rock fragments, except that some authors treat **chert** as a rock fragment. Outdated terms which are still used frequently are *orthoquartzite* and **quartzite** for **quartz** *arenite*. Some authors regard *greywacke* as outdated, preferring *wacke*. Further subdivisions and names have been used such as *phyllarenite*, *volcanic arenite* and *sedarenite* as terms for lithic *arenites* whose rock fragments are dominantly of metamorphic, volcanic and sedimentary origin respectively, and **ganister** for **quartz** *arenites* formed by *in situ leaching* in **Carboniferous** soils.

An aspect of the rationale for the terrigenous sandstone classification described above is the concept of *compositional maturity*, expressed numerically as the *compositional maturity index*, the ratio of **quartz** + **chert** grains to **feldspars** + rock fragments, i.e. of relatively stable grains resistant to physical and chemical processes of breakdown to relatively unstable, or 'labile' clasts. A compositionally-mature **sand** is one where prolonged and/or intense **weathering** and/or transport have removed most of the unstable grains which were present when the sediment was first eroded from an igneous or metamorphic source. Such **sands** and especially many supermature **quartz** *arenites* are likely to be polycyclic (or multicyclic), i.e. their grains have undergone many cycles of **erosion**, deposition and even lithification since they were last part of a non-sedimentary rock. Exceptions to this rule include **ganisters** (see above).

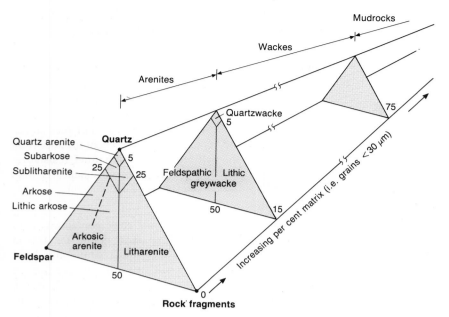

Fig. S7 Classification of **sandstones**. (After Pettijohn *et al.* 1973.)

Sandstone was a very important building material in the ancient world; it was used in many parts of Europe, in North Africa, in India (particularly for temple construction), in the Americas by the Mayas and Incas (for building and relief work), and in Egypt and the Levant. Some of the most spectacular uses are to be found in Egypt — the statues of Abu Simbel are of sandstone, as are many pyramids in the south of Egypt. The Egyptians quarried sandstone in large masses at several points along the Nile valley, particularly at the quarry of Silsila. Sandstone was also used for smaller objects, including the famous **millstone** grit querns and later water-**millstones** used in Britain and exported to many parts of the world. [DAR/OWT]

Pettijohn, F.J., Potter, P.E. & Siever, R. (1973) *Sand and Sandstone*. Springer-Verlag, Berlin.

sandstone–uranium–vanadium base metal deposit
Deposit found in terrestrial sediments, frequently fluviatile, which was generally laid down under arid conditions. As a result, the *host rocks* are often red in colour and for this reason **copper** deposits of this type are commonly referred to as *red bed coppers*. **Uranium**-rich examples are called *Colorado Plateau-type*, *roll-front-type*, *western states-type* or *sandstone uranium-type deposits*. The last term is now in common use.

In deposits of this type one or two metals are present in economic amounts, whilst the others may be present in minor or trace quantities. Thus **copper** mineralization (with **chalcocite**, **bornite** and **covellite**) is widespread in *red bed* successions though it is not often up to **ore grade** and the same applies to **silver** and **lead–zinc** mineralization. **Uranium** mineralization (±**vanadium**) may be accompanied by trace amounts of the above metals but usually occurs as separate deposits.

Conformable *red bed copper deposits* occur in some **sandstones** which were laid down under desert conditions. **Dune sands** are frequently porous and permeable and the **copper** minerals are generally developed in pore spaces. Examples of such deposits occur in the former USSR, in the **Trias** of central England, in Germany and in the southwestern USA. At the Nacimiento Mine in New Mexico a deposit of 11 Mt averaging 0.65% **copper** is being worked by **open pit** methods. Like other *red bed coppers*, this deposit has a high metal/**sulphur** ratio as the principal mineral is **chalcocite**. This yields a **copper** concentrate low in **sulphur** which is very acceptable to present-day custom smelters faced with stringent anti-pollution legislation.

Sandstone uranium-type deposits of this general type are widespread in the USA and they have provided over 95% of its domestic production of **uranium** and **vanadium**. From the global point of view they probably constitute a quarter of the world's **reserves** outside the Commonwealth of Independent States (CIS). In the USA these deposits are well developed in the Colorado Plateau region and in Wyoming. [AME]

Evans, A.M. (1993) *Ore Geology and Industrial Minerals: An Introduction* (3rd edn). Blackwell Scientific Publications, Oxford.

sandstorm The blowing of **sand**-sized particles into the air by a strong wind which reduces horizontal visibility to 1000 m or below. **Sand** grains are commonly raised a few metres above the ground surface and rarely above 15 m due to their mass. The predominant size of material differentiates a sandstorm from a *dust storm*. [NJM]

sandwave A large-scale periodic bedform. This term is used in conflicting ways. Some authors use the term to describe any large-scale periodic bedform, so that it is synonymous with **dune**. Others prefer to restrict its use to linear crested, two-dimensional, large wavelength/height structures. Finally, some workers use the term only for those large-scale bedforms which grow to equilibrium over time-scales much longer than individual steady or quasi-steady flows. Such sandwaves are typically fashioned by reversing flows of **tidal** frequency.

The sandwaves forming under reversing tidal **boundary layers** are thought to have a morphology and internal structure determined essentially by the relative sediment transports of the dominant and subordinate **tides**, or in other words the time–velocity asymmetry. Tidal sandwaves under highly asymmetrical flows are very much like the **dunes** found under unidirectional flows in having steep **lee slopes** dominated by avalanching. Under more symmetrical **tides**, however, the smaller inequality between the sediment transports on the dominant and subordinate **tides** means that the sandwave flanks are extensively reworked, causing low angle **lee** and **stoss slopes** and a complex internal geometry. [PAA]

Allen, J.R.L. (1980) Sand waves: a model of origin and internal structure. *Sedimentary Geology* **26**: 281–321.
Berné, S., Auffret, J.P. & Walker, P. (1988) Internal structure of subtidal sandwaves revealed by high resolution seismic reflection. *Sedimentology* **35**: 5–20.

sanidine ($KAlSi_3O_8$) A high-temperature form of **potash feldspar**. (See **feldspar minerals**.) [DJV]

sanitary landfill A land site where municipal solid waste is buried in a manner designed to cause minimal disturbance or pollution to the environment. The waste is typically compacted and periodically covered with earth. Controlled landfilling of municipal waste is cheaper than other disposal and treatment methods and the most important tool for municipal waste management in most developed countries. The major problem with this method, however, is the danger of *leaching* and **seepage** of waste substances which may enter **groundwater** or rivers, and careful site surveys (see **site investigation**) and strict controls are needed to minimize this risk. [NJM]

sannaite An alkaline **lamprophyre**, analogous to **camptonite** but with **alkali feldspar** more abundant than **plagioclase**. [RST]

sanukite A type of **hypersthene andesite**, black and glassy when fresh, used very commonly for stone tool production in prehistoric Japan. Artefacts of sanukite have been correlated with their source by means of *chemical analysis*. [OWT]

saponite ($(Mg,Fe)_3(Al,Si)_4O_{10}(OH)_2(Ca,Na)_{0.3}(H_2O)_4$) A **clay mineral** of the **montmorillonite** group. [DJV]

sapphire A blue **gem** variety of the mineral **corundum**. [DJV]

sapping The undermining of the base of a cliff, with subsequent **failure** of the cliff face, produced by processes which include **wave** action, lateral **erosion** by streams and **groundwater** outflow (*spring sapping*). Some **dry valleys**, canyons and **blind valleys** may be a consequence of sapping. [ASG]

Higgins, C.G. (1984) Piping and sapping: development of landforms by groundwater outflow. In: La Fleur, R.G. (ed.) *Groundwater as a Geomorphic Agent*, pp. 18–58. Allen & Unwin, Boston.

saprolite The material produced by the **deep weathering** of bedrock — especially crystalline rocks such as **granites** and **gneisses**. Saprolite is produced *in situ* by chemical **weathering**, particularly in humid environments in the tropics and subtropics. Thorough disintegration of the igneous and metamorphic rocks produces a fine-grained clayey material. Structures such as **quartz veins**, which are resistant to chemical decomposition, may remain relatively intact within the accumulation of saprolite. [AW]

Ollier, C.D. (1969) *Weathering*. Oliver & Boyd, Edinburgh.

sapropelic coal Fine-grained, faintly stratified to homogeneous massive **coal** deposit. These **coals** are generally dark in colour, tough and often show a marked conchoidal **fracture**. They may form complete **seams** of **coal** or occur as layers within **banded coal** seams, often being at the roof. The two principal types are *cannel coal*, composed of *spores* or fine organoclastic fragments, and *boghead coal*, composed largely of *algal* material. There exists, however, a large range of **coals** consisting of mixtures of these two end-members and intermediate forms are given names such as cannel-boghead and boghead-cannel. These coals are virtually impossible to distinguish from each other in hand specimen, microscopic analysis being necessary for correct identification. [AME]

Ward, C.R. (1984) *Coal Geology and Coal Technology*. Blackwell Scientific Publications, Melbourne.

sard A brown variety of the mineral **chalcedony**. [DJV]

sardonyx A variety of microcrystalline **silica** comprised of alternating layers of **onyx** and **sard**. (See **silica minerals**.) [DJV]

sarsen Blocks of **silica**-cemented **sandstone**, breccia or *conglomerate*, which are thought to have formed in the surface or near-surface environment of southern England during warm phases of the *Tertiary*, also called *grey wethers* or **puddingstones**. They are in effect masses of **silcrete** which have been broken down by *Pleistocene* frost action and mass movement. Comparable deposits, called *meulières*, are known from the *Tertiary* beds of the Paris Basin and elsewhere in France.

Much of the famous monument of Stonehenge, England, is formed of sarsens, assumed to be from the nearby Marlborough Downs. The outer circle and the famous trilithons (settings of two upright stones with one laid across the top of them) are of sarsens. These parts of Stonehenge are dated to within the Bronze Age (*c.* 2000 BC) but sarsens had been used earlier in southern Britain, for burial chambers of the preceding Neolithic period. The prehistoric monument of Avebury, Wiltshire, England also contains a stone circle and avenues of sarsens. [ASG/OWT]

Summerfield, M.A. & Goudie, A.S. (1980) The sarsens of southern England: their palaeoenvironmental interpretation with reference to other silcretes. In: Jones, D.K.C. (ed.) *The Shaping of Southern England*, pp. 71–100. Academic Press, London.

sastrugi Small wind-eroded irregular ridge up to 50 mm high commonly with a concave upwind face. Sastrugi form in moist or salt-cemented sandy surfaces and show parallel to the wind direction. [RDS]

satellite geodesy The use of artificial satellites or the Moon in making **geodetic** observations. See **satellite laser ranging**, **satellite radiopositioning** and **very long baseline interferometry**. [PK]

satellite laser ranging (*SLR*) **Geodetic** technique of determining the distance to an orbiting artificial satellite or a reflector on the Moon by measuring the two-way travel time of a reflected pulse of laser light. If two installations simultaneously track the same satellite, their relative location can be computed to an accuracy of about 80 mm. Repetition of the observations over a period of time can be used to measure the relative motion between the sites, and so to determine the rate of **continental drift**. The accuracy of the method is dependent upon knowledge of the satellite's orbit, whose major control is the Earth's gravitational field. [PK]

Christodoulis, D.C., Smith, D.E., Kolenkiewiez, R., *et al.* (1985) Observing tectonic plate motions and deformations from satelite laser ranging. *Journal of Geophysical Research* **90**: 9249–63.

satellite radiopositioning **Geodetic** technique of accurate three-dimensional positioning by radio interferometry from a constellation of artificial satellites known as the *Global Positioning System* (*GPS*). The simultaneous observation of multiple satellites allows extremely accurate positioning with a small, portable receiver. The present accuracy of distance measurements between two sites is about 1 p.p.m., and this may well increase to about

0.1 p.p.m. when the full constellation of GPS satellites is implemented.

Repeated measurements of the distance between two receivers can be used to measure the rate of **continental drift**. It is probable that satellite radiopositioning will be the standard method of determining location and elevation during geophysical surveys when its full implementation is realized. [PK]

satin spar A fibrous variety of the mineral **gypsum**. [DJV]

saturated zone The zone below the **water table** that is saturated with **groundwater**. [AME]

saturation magnetization The maximum intensity of *isothermal remanence* acquired by the application of strong ($\geqslant 100$ mT) **magnetic fields**, but measured after removal of the field. [DHT]

scabland A landscape of bare rock surfaces, thin *soil* cover and sparse vegetation that is underlain by flat **basalt** flows and dissected by dry channels formed by *glacial* floodwaters. A good example is the Columbia **lava plateau** of eastern Washington, USA. [NJM]

scapolite series ($3NaAlSi_3O_8 . NaCl$ to $3CaAl_2Si_2O_8 . CaCO_3$) A group of metamorphic minerals (*tectosilicates*) involving complete *solid solution* between **marialite** ($3NaAlSi_3O_8 . NaCl$) and **meionite** ($3CaAl_2Si_2O_8 . CaSO_4$ or $CaCO_3$). The name *wernerite* has been used for members intermediate in composition between **marialite** and **meionite**. [DJV]

scar A steep, rocky cliff in **limestone** areas, which is often characterized by outcrops of relatively bare, massively bedded **limestone**. The word is commonly used in northern England, and a famous example is Gordale Scar in Yorkshire. [HAV]

scar fold A **fold** formed by the **flow** of material surrounding a **boudin**. The less competent material around a growing **boudin** may flow into space created by the thinning and separation of **boudin** necks; the resulting **strain** can create a structure which may vary from a flexing of the layering to a *tight* or even *isoclinal fold*. [TGB]

scheelite ($CaWO_4$) A major **ore** mineral of **tungsten**. [DJV]

schistosity/schist **Foliation** produced by **deformation** consisting of the **preferred orientation** of tabular minerals with grain size coarse enough to be visible to the unaided eye. A schist is a rock possessing schistosity. Both *slaty*

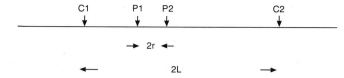

Fig. S8 Schlumberger configuration. P1 and P2 are the potential electrodes and C1 and C2 the current electrodes. In depth sounding, P1 and P2 will be symmetric about the mid-point.

cleavage and schistosity are tectonically produced planar **fabrics** which differ only in that the grain size of a schist is visible. In general, therefore, schists are metamorphic rocks of higher grade than **slates** (*greenschist* facies and higher) in which a coarser grain size has developed by metamorphic processes. [TGB]

Hobbs, B.E., Means, W.D. & Williams, P.F. (1976) *An Outline of Structural Geology*. John Wiley & Sons, New York.

schlieren Pencil-shaped, discoidal or blade-like inclusions composed of aggregates containing a higher concentration of mafic minerals than in the *host rock*. Schlieren may hence correspond to a type of *autolith*. [RST]

Schlumberger configuration An arrangement of electrodes in **resistivity methods** (Fig. S8). Here the potential electrode separation (2*r*) is commonly much smaller than the current electrode separation, (2*L*) so that the measured **potential** difference, *V*, divided by 2*r* approximates to the electric field. This array, with the electrodes symmetrically disposed about the point of investigation, is popular in **vertical electrical soundings** since only current electrodes need be moved for each new measurement; then, for a current *I*, the **apparent resistivity** ρ_a is

$$\rho_a = \pi \frac{(L^2 - r^2)}{2r} \frac{V}{I}$$

Comparison of the field data plotted as $\log(\rho_a)$ versus $\log(L)$ with theoretical curves, or the use of direct inversion techniques, yields resistivities and thicknesses of the subsurface layers. (See also **Wenner configuration**.) [ATB]

Schmidt net Type of stereographic net employing *equal-area projection*. Thus all 10° areas on a spherical surface plot as areas of equal size on such a net. The Schmidt net is used primarily for statistical analysis of orientation distributions. The alternative widely used type of **stereographic projection** produces the *equal angle* or **Wulff net**, in which all angles of the same size project to lines of the same length but areas are distorted. (See **stereographic projection**.) [RGP]

schollen Large blocks of *country rock* found in *till* sheets and resulting from large-scale *glacial* **erosion**. [ASG]

schorl The **iron**-rich (black) variety of the mineral **tourmaline**. [DJV]

scolecite ($CaAl_2Si_3O_{10} \cdot 3H_2O$) A **zeolite mineral**. [DJV]

scorzalite ($(Fe,Mg)Al_2(PO_4)_2(OH)_2$) A rare phosphate mineral found in high grade metamorphic rocks and *pegmatites*. [DJV]

scree/talus Any sloping accumulation of loose *clasts* of granule grade and above. Usually a wedge, metres to hundreds of metres high, of *clasts* located against a steeply sloping rock face from which they fall as a result of **weathering** and **erosion**. The *clasts* are generally angular, so their interlocking enables screes to exceed the normal *angle of repose* of unconsolidated sediment and reach 35° even where unstabilized by vegetation. Talus slopes are usually concave in form. The term is most often used for the isolated or laterally coalescing *fans* of **gravel** which occur in such subaerial settings as **fault** scarps in arid and semi-arid regions, or the steep sides of *glacial* valleys, but it also has been applied to their subaqueous analogues at *reef* fronts and **fault** scarps. Whilst a continuum exists between talus and **alluvial fans**, the latter usually slope at less than 11° and have a long profile which corresponds with that of the valley which provides the debris. Downslope transport processes include the falling and rolling, sliding, or creep of individual *clasts*. Also the impact of falling *clasts*, **earthquakes**, or *storm* **runoff** emerging from *gullies* above a scree may trigger *grain flows* (see **sediment–gravity flow**). Consequently, a downslope increase in *clast* size is characteristic and *imbrication* is also common with long axes parallel (rather than transverse) to the slope and dipping up to 30° upslope relative to its surface. Size sorting down screes is probably enhanced by a like-seeks-like mechanism whereby rolling *clasts* wedge into the spaces between stationary *clasts* of a similar size, but not between smaller ones. [TPB/DAR]

Statham, I. (1976) Debris flows on vegetated screes in the Black Mountain, Carmarthenshire. *Earth Surface Processes* **1**: 173–80.

scroll bar Ridge of sediment deposited parallel to the contours of a **point bar** within an alluvial channel. As the channel migrates by cut-bank **erosion** and **point bar** deposition, individual scroll bars are seen to move up the **point bar** relative to the channel position. On the **point bar** top surface the remnants of scroll bars may be observed as a ridge and **swale** topography that demonstrates the pattern of meander migration (Nanson, 1980) (see Fig. P21, p. 479). The term *meander scrolls* has also been applied to the topography of **exhumed point bars** but in this case the topography is the result of differential **erosion** of beds within a truncated *epsilon cross-stratified* complex. [JA]

Nanson, G.L. (1980) Point bar and floodplain formation of the meandering Beatton River, northeastern British Columbia, Canada. *Sedimentology* **27**: 3–29.

sea cliff Steep coastal slope whose break in slope is related principally to the interaction of marine and sub-aerial processes. Quarrying and **abrasion** by **waves** undercut and oversteepen the coastal slope whilst sub-aerial mass movement causes slope retreat and debris delivery to the slope foot. On the great majority of cliff coasts both sets of processes operate in different proportions depending on lithology, structure and climate. 'Active' sea cliffs depend upon constant **wave** removal of debris revealing a steep bedrock slope; the relative rates of delivery and removal together with geological structure determining the cliff form. At the other extreme, marine removal of debris may be so weak or infrequent as to term the cliffs 'inactive'. 'Relict' or former sea cliffs occur where all marine processes have been removed from the vicinity of the cliff by continual deposition at the cliff foot or by a fall in sea-level or, frequently, by a combination of both. As a result of the partial adjustment of cliff morphology to sea-level change, many cliffs show complex forms related to different combinations of processes. However, plunging cliffs (Fig. S9) are thought to be largely inherited forms little modified by present-day processes. [JDH]

Emery, K.O. & Kuhn, G.G. (1982) Sea cliffs: processes, profiles, classification. *Geological Society of America Bulletin* **93**: 644–53.
Trenhaile, A.S. (1987) *The Geomorphology of Rock Coasts*. Clarendon Press, Oxford.

sea-level datum The height of the mean sea-level surface to which all surface elevations are related, so that international consistency is maintained. It is frequently used as the reference level or datum for geophysical surveys. In the UK the sea-level datum (*Ordnance Datum* or *OD*) is taken as the mean sea-level at Newlyn or Liverpool. All *bench marks* (sites of known elevation) in the UK are related to *Ordnance Datum*. Sea-level corresponds to the **geoid** surface. [PK]

seafloor spreading The hypothesis that the oceanic **lithosphere** is generated in a narrow zone at the crest of each of the seismically active *oceanic ridges*. It is then carried perpendicularly away from its source, at identical rates on each side (see **half spreading rate**) but in opposite directions, and eventually sinks, or is carried back, into the **mantle** at **subduction zones** (Fig. S10a). This model is largely attributed to Hess in 1962, although Holmes in 1944 had much earlier proposed a similar model except that the new oceanic rocks were injected in a more dispersed manner, only moderately concentrated on the higher flanks of the ridge crest. The fundamental proof of the highly localized injection model was provided by the **Vine–Matthews hypothesis**, based on the dating of the ocean floor by means of the record of geomagnetic changes

Fig. S9 Sea cliffs. The plunging cliffs of Moher, Co. Clare, Ireland, are subject to intense wave quarrying and, in places, abrasion. The horizontally bedded sedimentary rocks help keep the cliffs vertical although it is likely that this form has been maintained over a considerable period of time spanning different sea-levels.

in polarity as preserved in the form of **thermoremanent magnetization** acquired by **magnetic minerals** in the **igneous rocks** as they originally formed and cooled at the *oceanic ridge* (Fig. S10b). The sharpness of the observed linear *oceanic magnetic anomalies* which this magnetization causes at the ocean surface means that the injection and locking in of the remanence must take place within extremely narrow zones, probably only a few metres in width. Deep-sea drilling to sample the sediments overlying the **magnetic anomalies** has not only confirmed the age of the magnetization but provided independent paleontological and *paleoclimatic* evidence for the systematic increase in age of the ocean floor away from the *oceanic ridges*. Subsequent to 1966, all of the ocean floors were dated by means of their linear *magnetic anomalies*, initially correlated against the *polarity time-scale* for the last 5 Ma (Heirtzler *et al.*, 1968), but since based on a combination of the polarity record in terrestrial rocks and the paleonto-logical dating of sediments immediately overlying the igneous ocean floor in which the anomalies reside. The nature of ancient ocean floor remnants that have been **obducted** onto the continents indicates that similar ocean **lithosphere** generation, i.e. by means of seafloor spread-

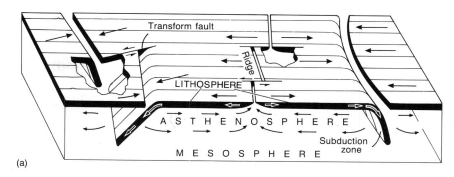

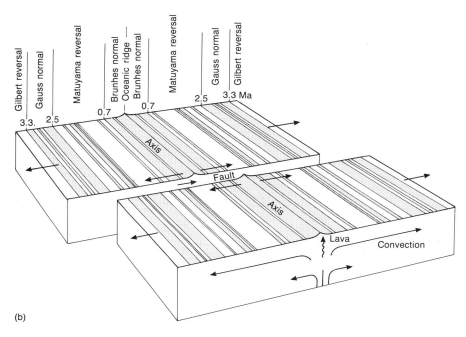

Fig. S10 (a) Schematic three-dimensional diagram showing the major features of **seafloor spreading**. (After Isacks *et al.*, 1968.) (b) The succession of normal and reversed magnetic anomalies on the seafloor provides strong support for the concept of seafloor spreading. Rocks of normal, or present-day, polarity are shown in shades of grey; rocks of reversed polarity are white. The symmetry of the magnetic stripes and the correlation with the time-scale of reversals worked out from lava flows on land suggest that molten rock upwelling along the ridge axis became magnetized as it cooled, was pushed out on both sides, and gradually moved outward with the separating plates. The separation of the two blocks represents a transform fault.

ing, has occurred throughout the history of the Earth, but this does not necessitate **continental splitting** occurring throughout geological time. It is possible that seafloor spreading also occurs within *marginal basins* overlying and adjacent to **subduction zones** (see **back-arc spreading**), but the irregularity of all **geophysical anomalies** in these areas suggests that either the injection of new **oceanic crustal** materials is more diffuse than at *ocean ridges* or that the anomalies are largely masked by great thicknesses of sediments in such areas. [DHT]

Elsasser, W.M. (1971) Sea-floor spreading as thermal convection. *Journal of Geophysical Research* **76**: 1101–12.

Isacks, B., Oliver, J. & Sykes, L.R. (1968) Seismology and the new global tectonics. *Journal of Geophysical Research* **73**: 5855–99.

Heirtzler, J.R., Dickson, G.O., Herron, E.M., Pitman, W.C. & Le Pichon, X. (1968) Marine magnetic anomalies, geomagnetic field reversals, and motions of the ocean floor and continents. *Journal of Geophysical Research* **73**: 2119–36.

Pitman, W.C. & Talwani, M. (1972) Sea-floor spreading in the North Atlantic. *Geological Society of America Bulletin* **83**: 619–46.

Press, F. & Siever, R. (1982) *Earth* (3rd edn). W. H. Freeman, San Francisco.

seal The barrier preventing the upward rise of *hydrocarbons* so that they remain in the **reservoir** rock to form a gas or oil pool. This barrier is also called the *roof rock*. The best seals are *ductile* sedimentary rocks such as clays or **shales**, **evaporites** being the ideal. (See **oil trap**, **cap rock**.) [AME]

seamount A submarine peak which does not rise above sea-level. Seamounts are volcanic features consisting of **basaltic** rocks. Those seamounts which are flat-topped, having been planed by marine and subaerial **erosion**, are known as **guyots**. There are well over 20 000 seamounts known to exist, many of which are found in the Pacific Ocean. In general, seamounts have elevations of over 1000 m and many rise *c.* 3000 m above the adjacent sea bed. According to Menard (1986) the total volume of undersea volcanoes is very approximately 10^7 km^3. There are far more subsea volcanoes than there are volcanoes on the continents, and they are in general much larger, due to the lack of **erosion** under the oceans. Many more new subsea volcanoes are being discovered by **side-scan sonar** systems. The distribution of seamounts is controlled by patterns of **seafloor spreading**. Seamounts often occur in chains, as for example the Emperor Seamount Chain to the north of the Hawaiian Islands. (See **hotspot**.) [HAV]

Crough, S.T. (1984) Seamounts as recorders of hot-spot epeirogeny. *Geological Society of America Bulletin* **95**: 3–8.
Menard, H.W. (1986) *Islands*. Scientific American Library, New York.

seat earth A thin horizon beneath a **coal** or **lignite** which contains fossil rootlets, and represents the original *soil* in which the vegetation grew. These horizons are often bleached in appearance and are rich in **kaolinite** clays. These features result from the *leaching* of various cations, including iron, by acidic *soil* waters, and the washing down (**eluviation**) of clays. The term **ganister** refers to **sandstones** beneath **coals** which have a very high **quartz** content and were used to manufacture refractory bricks. These formed from less **quartz**-rich parent materials because of extensive *leaching* and **eluviation**. *Tonsteins* are **kaolinite**-rich seat earths possibly formed by the **alteration** of either volcanic **ashes** or **illite**-rich clays. [VPW]

Percival, C.J. (1986) Paleosols containing an albic horizon: examples from the Upper Carboniferous of northern England. In: Wright, V.P. (ed.) *Paleosols: Their Recognition and Interpretation*, pp. 87–111. Blackwell Scientific Publications, Oxford.

secondary porosity The post-depositional formation during **diagenesis** of any pore space (see **porosity**) through the **dissolution** of detrital framework grains, matrix or earlier formed **authigenic** cements within sedimentary rocks (Fig. S11). In carbonate sediments dominated by a lime mud matrix, extensive **dolomite** replacement may also generate secondary porosity. Additionally, quantitatively subordinate amounts of secondary porosity may be generated or enhanced as a result either of tectonic fracturing or shrinkage accompanying dehydration of hydrous minerals (see Table S1).

In recent years the occurrence, recognition and origin of secondary porosity has been the subject of a lively debate (see Bjorlykke, 1984 and Schmidt & McDonald, 1979 for details of the opposing sides of the argument). Although most authors now accept that secondary porosity can be developed in sedimentary rocks, there is still considerable disagreement as to the extent to which it may be developed and which fluids are responsible for the generation of secondary porosity. Essentially two opposing mechanisms are currently favoured for the production of fluids capable of mineral **dissolution**:

1 Penetration of low **pH**, dilute **meteoric waters** driven by a topographic **head**.
2 Expulsion of low **pH**, highly **aggressive pore-waters** from organic-matter-rich **mudrocks** undergoing thermal *maturation* during deep burial. Sources of acidity include CO_2, organic acids or **clay mineral** reactions.

Additionally, mineral **dissolution** may take place when

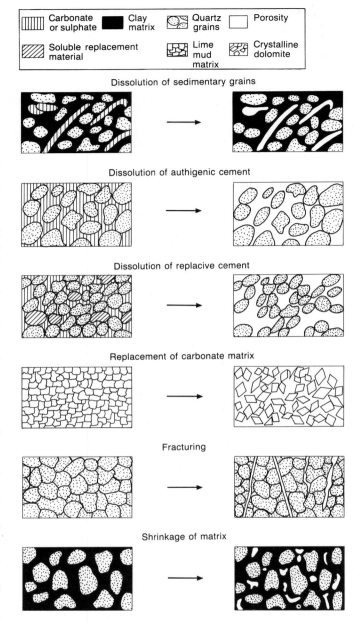

Fig. S11 The principal genetic types of **secondary porosity** in sediments.

two solutions of different composition mix in the subsurface. This process, known as *mixing corrosion*, is well known from **hydrogeological** regimes where a **meteoric water** flush encounters seawater-derived **connate groundwater** in coastal settings. However, the widespread applicability of this mechanism has not yet been considered in detail for deep burial regimes where, for example, dilute basinal fluids expelled from decompacting **mudrocks** may migrate and mix with concentrated formation **brines**.

The distribution of the pore system and **permeability** characteristics of secondary porosity are often funda-

Table S1 Summary of the main **secondary porosity** generating processes in sandstones and limestones and the principal minerals involved.

Porosity generating process	Mineral	Quantitative importance
Grain dissolution	Carbonates	Major in sandstones and limestones
	Silicates	Major in sandstones
	Other non-silicates	Minor in sandstones and limestones
Cement dissolution	Aragonite	Major in limestones
	Calcite	
	Dolomite	Major in sandstones and limestones
	Siderite	
	Sulphate	
	Other evaporites	Minor in sandstones and limestones
	Other non-silicates	
Matrix dissolution or replacement	Aragonite	Major in limestones
	Calcite	
Fracturing	Total rock	Locally important in sandstones and limestones
	Competent grains	
Shrinkage	Grains	Very minor in sandstones and limestones
	Matrix	

mentally different from those of primary **porosity**. An ability to recognize the textures of secondary porosity, quantify its presence and predict its distribution can, therefore, have important economic implications for both **aquifer** and *hydrocarbon* production. There is a series of petrographic-based criteria for both carbonate and *siliciclastic* sediments (Fig. S12). In general terms, grain **dissolution** secondary porosity is relatively easy to recognize, although identification of the detrital **mineral** or grain type

is not so obvious in cases of extreme or complete **dissolution**. Honeycombed grains, oversized pores, 'floating grain' textures, moulds, *vugs*, insoluble residues or other remnants within oversized pores are all **fabrics** indicative of framework grain **dissolution porosity**.

Mouldic fabrics are a specific type of grain **dissolution** porosity that result from the *leaching* of grains or replacive cements which have their characteristic morphology preserved in the newly formed void (e.g. **dissolution** of **aragonitic** *gastropod* shell fragments, **dissolution** of euhedral **anhydrite** crystals from a lime mud matrix, etc). In carbonate sediments peripheral solution of the enclosing cement or matrix around a mould result in the loss of the identity of the precursor grain or replacement and the irregular, oversized pore is referred to as a *vug*.

Recognition of cement-**dissolution porosity** can be more difficult, particularly if removal of the cement merely exhumes the primary intergranular **porosity**. However, many intergranular cements are replacive with respect to framework components and their *leaching* results in the formation of an enlarged intergranular **porosity fabric** as the replacive cement is dissolved. By contrast, reduced intergranular **porosity fabrics** may result if compaction took place either before initial **cementation** or after secondary porosity generation. Matrix secondary porosity occurs in **limestones** with a lime mud matrix through either the selective *leaching* of more soluble mineralogical components (e.g. **aragonite** needles) or partial **dissolution**. Replacement secondary porosity is associated with the **gypsum** to **anhydrite** (see **gypsum−anhydrite equilibrium**) and **calcite** to **dolomite** transformations. Both these reactions result in the **replacement** of a pre-existing **mineral** by

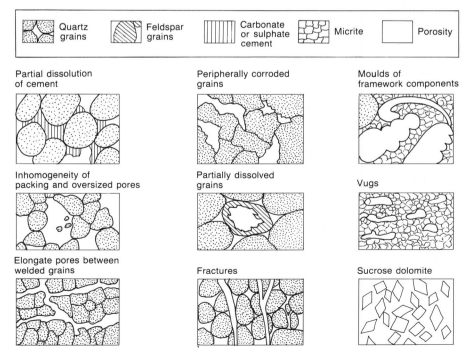

Fig. S12 Textural fabrics typical of **secondary porosity**.

a denser **mineral** and a corresponding volume decrease. Up to 13% of intercrystalline secondary porosity can be generated, for example, by the complete **dolomitization** of a **calcite** lime mud matrix.

Secondary porosity may be developed at any stage during the **diagenetic** evolution of a sedimentary sequence and can, therefore, be classified according to the **diagenetic** regime — *eogenesis, mesogenesis* or *telogenesis* — under which it developed. During *eogenesis*, for example, in strongly reactive environments (extremes of Eh, **pH** or **pore-water** chemistry), **pore-waters** may *leach* unstable detrital grains (such as **feldspars**) close to the depositional surface. Although the general trend during burial is for **porosity** to decrease with increasing depth, the invasion of hot, **aggressive** formation waters from deeper parts of a sedimentary **basin** into structurally higher sediments may result in **dissolution** of unstable framework grains or carbonate and sulphate cements. Thus, during *mesogenesis*, there is potential for localized increases in **porosity** with depth coincident with expulsion of **aggressive** formation waters from adjacent source rocks. Such secondary porosity generation may be multicyclical. Ultimately, however, as **cementation** and **compaction** continue, **porosity** will again be reduced with increasing burial. During *telogenesis*, secondary porosity may be generated in *reefal* carbonates as **meteoric waters** displace either *eogenetic* or *mesogenetic* waters and selectively dissolve the more soluble carbonates.

In all cases of secondary porosity generation there is only net effective gain in total **porosity** of a given sedimentary unit if the solute product of **dissolution** is removed from the **dissolution** site. If such **mass solute transfer** does not take place then reaction products will be **precipitated** and there is only **porosity** redistribution. (See **pore waters and their evolution, burial diagenesis**.) [SDB]

Bjorlykke, K. (1984) Formation of secondary porosity: how important is it? In: McDonald, D.A. & Surdam, R.C. (eds) *Clastic Diagenesis.* American Association of Petroleum Geologists Memoir **37**: 277–86.

Schmidt, V. & McDonald, D.A. (1979) The role of secondary porosity in the course of sandstone diagenesis. In: Scholle, P.A. & Schluger, P.R. (eds) *Aspects of Diagenesis.* Special Publication of the Society of Economic Paleontologists and Mineralogists **26**: 175–208.

secular variation Long-term, progressive and predictable variation in the **geomagnetic field**. It affects all the **geomagnetic elements**, and is probably caused by gradual changes in the circulation patterns of charged particles in the fluid outer **core**, where the main **geomagnetic field** originates. [PK]

sediment gravity flow An aggregate of grains moving with the aid of **gravity** but independently of the overlying medium. The friction between the particles may be overcome in four main ways, although there is considerable overlap between the different mechanisms:

1 In *grain flows* the particles are kept in suspension by grain collisions and near misses. *Grain flows* commonly occur where the sediment surface exceeds the *angle of repose*. Such a situation is common where **sand** accumulates on the upper parts of subaqueous and subaerial bedforms, eventually causing **failure** and avalanching of grains to the foot of the bedform. The weight of the particles in a *grain flow* is counterbalanced by *dispersive pressure*. This pressure or **stress** is greatest in the regions of largest **shear**, that is, just above the basal **shear plane** over which the *grain flow* is moving. *Inverse grading* is typical of *grain flows*. This may be due either to the action of **dispersive pressure** or to **kinematic sieving**. In the former case, the largest grains exert the greatest **dispersive pressure** and therefore move away from the basal zone of maximum **shear** in order to equalize the **stress** gradient. In the latter, small grains fall through the gaps between the larger grains.

2 Grains being transported in **debris flows** are supported by the strength of the matrix, usually a result of the inclusion of a clay–water slurry giving added **viscosity**. In order for a **debris flow** to be initiated, the internal **shear stress** must exceed the *yield strength*, which is itself controlled by the **cohesion** and the granular friction caused by particle interlocking. Even small amounts of clay–water slurry matrix (<1% by volume in some cases) can reduce the internal friction to very small values, allowing **debris flows** to move on surprisingly small slopes. The deposits of **debris flows** are typically badly *sorted* (Fig. S13) and commonly, but not exclusively, matrix supported.

3 *Liquefied flows* are kept in motion by the buoyancy imparted to the particles by the escape of **pore fluid**. The loss of the *pore fluid* from the base of the liquefied aggregate causes a progressive upward 'freezing' of the flow. The upward movement of **pore-fluid** may become concentrated into higher velocity pipes, causing **dish structures** and pillars to form (Fig. S13).

4 The particles in **turbidity currents** are kept in suspension primarily by the **turbulence** of the **flow**. As a result, **turbidity currents** generally involve relatively dilute suspensions of grains. **Autosuspension** is the condition whereby the grains are kept in suspension without any internal energy loss. *Turbidity flows* commonly evolve from sediment **failures** such as **slides** and **slumps** (Fig. S14), triggered in many cases by **earthquakes**. They may also occur, however, where relatively dense river waters enter lakes or seas. (See **Bouma sequence**.) [PAA]

sediment transport equation A great number and variety of equations have been proposed over many years to predict the sediment yield of a particular flow given a knowledge of the variables such as **shear velocity** or **shear stress**, grain size parameters, channel **hydraulic radius** and fluid characteristics. These equations have dealt with **bedload** and **suspended load** either separately or as one total sediment load. As sediment transport generally increases with channel size, the transport rate is usually

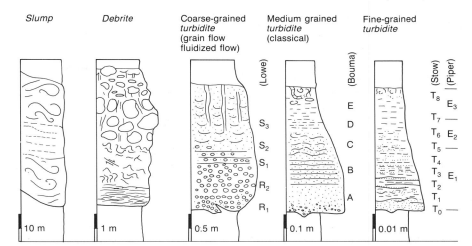

Fig. S13 Sediment gravity flow. Probable interrelationship of processes of initiation, long-distance transport and deposition of sediment in the deep sea. Post-depositional modification can involve current reworking, liquefaction and bioturbation.

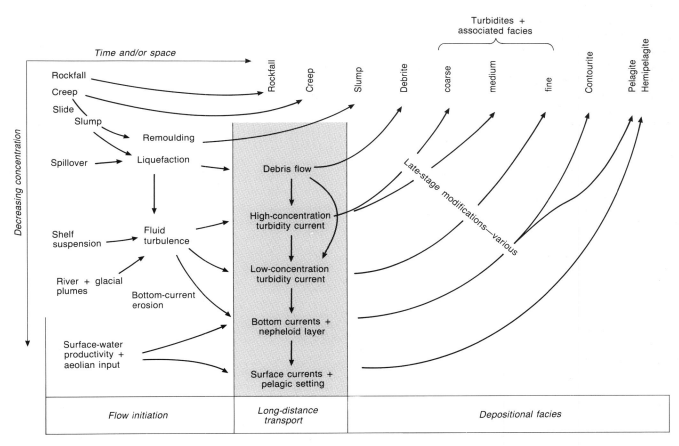

Fig. S14 Sediment gravity flow. Resedimented clastic facies models for slumps, debrites and turbidites, showing the idealized structural sequences. The scale bars give an indication only of typical unit thickness, which may vary widely in practice. Grain-size increases to the right for each column.

expressed as a volume, mass or weight per unit channel width per unit time.

Growing from the simple treatments of Duboys in 1879 who proposed the concept of **shear stresses** or tractive forces in relation to total sediment discharge, many subsequent theories have progressed to predict sediment discharge based upon tractive force assumptions. Many of these include the exceeding of a *threshold* flow velocity, **shear stress** or stream power although definition of this condition if often complex due to the effects of, among other things, flow depth:particle size ratios, **turbulent** fluctuations within the flow and sediment *sorting*. The *Meyer-Peter and Müller equation* and modifications of this have been widely used in many engineering problems although their application is generally limited to low transport stages in medium **sand–gravel bedload**.

Two other widely used transport equations departed from this tractive force approach. Bagnold (1966 and later developments) proposed an approach for both **bedload** and total load which viewed the sediment transport function as a transporting machine which has a rate of work, this rate of work being a function of the available power and an efficiency factor. Einstein in 1942 and later in 1950 developed a probabilistic approach to sediment transport. This appealed to considerations of individual particle movements within a bed and specifically to the **lift force** rather than the **shear stress**. Numerous other theories have been developed to deal with either different sediment types or in different sedimentary environments. [JLB]

Committee on Sedimentation of the Hydraulics Division (1971) Sediment transportation mechanics. H: Sediment discharge formulas. *Proceedings of the American Society of Civil Engineers, Journal of the Hydraulics Division* **97**: HY4.
Yalin, M.S. (1977) *Mechanics of Sediment Movement*. Pergamon, Oxford.

sediment yield The total amount of sediment exported from a drainage **basin** in a given period of time, usually expressed in units of tonne $km^{-2}a^{-1}$. Sediment yield is usually inversely correlated with **basin** area: the gross **erosion** rate measured on hillslopes is not reflected in yields from small **basins** because of sediment deposition on lower slopes; yields from large **basins** are usually less than from small ones because of sediment storage within the channel system. For this reason, the *sediment delivery ratio* has been designed as an empirical method of relating gross **erosion** to sediment yield for the entire **basin**. Sediment yields vary between **basins** for a variety of reasons including climate, lithology, relief and human impact. Langbein & Schumm (1958) identified a peak in sediment yield in semi-arid regions because of the combination of partial vegetation cover and seasonally intense rainfall. Sediment yields may also be high in *monsoonal* regions and in high-rainfall areas in the tropics. Human interference is notable as a source of **accelerated erosion** and greatly enhanced sediment yields, particularly as the result of agriculture and deforestation. [TPB]

Langbein, W.B. & Schumm, S.A. (1958) Yield of sediment in relation to mean annual precipitation. *Transactions of the American Geophysical Union* **39**: 1076–84.

sedimentary facies A body of sedimentary rock with specific and distinctive characteristics. The criteria used to define a facies may be biological (i.e. faunal or floral content), giving a *biofacies*, or physical and chemical, giving a *lithofacies*. The term facies is also used beyond a purely descriptive framework. For example, it can be used in a genetic sense for sediments deposited by a particular process, or in an evironmental sense for a range of sediment deposited in a particular **geomorphological** or oceanographical setting. [PAA]

seepage (1) An escape of gas or oil, generally to the surface but also into underground workings.
(2) A diffuse flow of water from an **aquifer** on to the Earth's surface, as compared with a more localized outflow which is termed a **spring**. [AME]

seiche A **standing wave** generated on the surface of enclosed or partially enclosed areas of water, such as gulfs, bays, **estuaries**, *fjords*, lakes or harbours. Seiches are caused by the action of an applied forcing mechanism such as winds, atmospheric pressure gradients, tides or, more rarely, **earthquakes** or sediment slides. Thus, many seiches are ephemeral features although coastal seiches generated by the impingement of oceanic surface waves or **internal waves** may be long-lasting and **tidal** seiches such as that of the Bay of Fundy are recurrent. The period of the seiche is largely dependent on the geometry and size of the **basin** concerned and can typically vary from a few minutes to several hours (e.g. Loch Earn, Scotland — 15 min; Bay of Naples, Italy — 48 min; Sea of Azov, former USSR — 24.5 h). An internal seiche is a standing **internal wave** which commonly affects the **thermocline**. (See also **tsunami**.) [AESK]

Wilson, B.W. (1972) Seiches. *Advances in Hydroscience* **8**: 1–94.

seif A variety of linear **dune** orientated in the direction of prevailing winds. Commonly found in a series of parallel ridges. Seifs are amongst the most impressive of **dune** types, sometimes reaching 100 m or more in height. Major areas of seif **dunes** are found in the deserts of south-west Africa, Sahara, Australia and Arabia. Active **sand** movement usually takes place on the **dune** crests, where the wind speeds are greatest. However there is usually little net **sand** movement in the less exposed areas between *dune ridges* across the **interdune** corridors. Winds approaching the **dune** are generally deflected parallel to the direction of **dune** orientation. [RDS]

Tsoar, H. (1983) Dynamic processes acting on a longitudinal (seif) sand dune. *Sedimentology* **30**: 567–78.

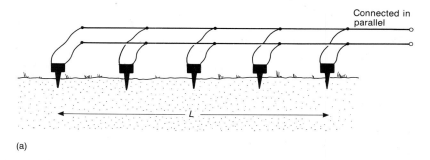

(a)

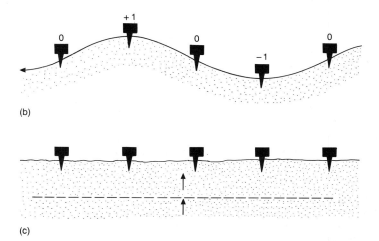

(b)

(c)

Fig. S15 Schematic illustration of how **seismic arrays** are used in seismic exploration to enhance and supress recording of seismic waves of different effective wavelengths. (a) Five geophones, equally-spaced over a distance *L*. (b) Response of array to wave of wavelength *L* is zero. (c) Response of array to vertically-travelling wave is +5. Waves of other wavelength, or arriving obliquely at the surface, give responses between these two extremes.

seismic array In **seismic exploration**, 'array' refers to groups of **seismic sources**, or (more usually) sensors, whose combined output can improve the **signal to noise ratio** of particular wavelengths and suppress others. Sources and/or phones are in linear patterns along the survey line. Their combined output will be zero for a wave whose wavelength is the same as the length of the array. Other wavelengths will be less suppressed. Arrays are most commonly used to minimize the amplitude of horizontally travelling *surface waves* and enhance the near-vertical **seismic reflections** used to image subsurface structures (see Fig. S15).

In **earthquake seismology**, arrays are groups of *seismometers* distributed over $100\,km^2$ or more. A seismic signal will arrive at each of *N* *seismometers* at a slightly different time, depending on the distance and direction of their source. Adding together the individual **seismograms** with suitable timeshifts to align the seismic signal enhances the signal and suppresses background **noise**, by a maximum factor of $\sqrt{N}$ when the **noise** is random. If the timeshifts are found empirically, they give an estimate of the **epicentre**; or for a given **epicentre** region the shifts are predicted and that region can be monitored. Arrays were introduced to improve detection of undergound nuclear tests and play a major role in that subject. In the 1960s and

1970s, they also contributed much to studies of Earth structure. [RAC]

Dahlman, O. & Israelson, H. (1977) *Monitoring Underground Nuclear Explosions*. Elsevier, Amsterdam.
Whiteway, F.E. (1965) The recording and analysis of seismic body-waves using linear cross arrays. *Radio Electronic Engineer* **29**: 33–46.

seismic belt On a global scale, the majority of **earthquakes** occur along the edges of **tectonic** *plates*, where the *plates* are created, collide, or slide past each other. On a world map these zones appear as linear 'seismic belts'. On a local scale too, **earthquakes** may be concentrated in belts along particular **faults**. (See **seismicity**.) [RAC]

seismic creep A form of non-instantaneous **deformation** in rocks. Following the removal (or application) of a **stress** *P*, most resultant **strain** *e(t)* disappears (or appears) at once but a further component decreases (or increases) over time *t*:

$$\mu \cdot e(t) = P \cdot (1 + \alpha(t))$$

where $\alpha(t)$ is known as the **creep** function and μ is the **rigidity** modulus. Laboratory experiments suggest the **creep** function is of the form $q \cdot \log(1 + at)$ (*a, q* constants).

Replacing P by the **stresses** due to a plane **seismic wave** allows seismic **attenuation** laws to be expressed in terms of the **creep** function, and show that q is related to the fractional loss of wave energy per wave cycle and a to the resonant frequency of grains in rocks. [RAC]

seismic discontinuity An abrupt boundary between two media of different **acoustic impedance** (=density × wavespeed). Some boundaries can be gradational, but if this distance is small compared to the wavelength of **seismic waves** it will 'appear' sharp. Note that, especially at solid–fluid interfaces such as the seafloor or the **core–mantle** boundary, the effects of a discontinuity may be different for P and S *waves*.

In the Earth's **crust**, there are many seismic boundaries between and within geological units. On a global scale, some major seismic discontinuities (some gradational over kilometres) are shown in Table S2 with their interpretations. [GS]

seismic event Any sudden disturbance on or within the Earth that generates **seismic waves**. This usually encompasses **earthquakes**, chemical explosions, underground nuclear explosions, and *rockbursts* and mine **failures**, but excludes sources of **microseisms** such as trees, heavy industry and traffic. Sonic booms from aircraft and *meteorites* can also register on *seismographs*, but are readily distinguished from seismic events because of the very slow velocity of sound in air ($0.33\,\mathrm{km\,s^{-1}}$) compared to P *waves* in rock (2–$6\,\mathrm{km\,s^{-1}}$ for the **crust**).

Regions of the Earth where much natural **earthquake** activity occurs are termed 'seismically active'. 'Aseismic' or 'seismically quiescent' regions are those where there are few or no **earthquakes**. [RAC]

seismic exploration The use of seismic techniques to find and delineate geological structures or bodies in the sub-surface that contain materials of economic importance. The commonest seismic exploration technique is the **seismic reflection** method which is used widely on land and sea to explore for geological structures containing *hydrocarbons*. The *mineral exploration* industry uses **seismic refraction** techniques to delineate sedimentary deposits containing minerals (**placer deposits**). [GS]

seismic gap An area within a **seismic belt** where no **earthquakes** have occurred over some period of time, usually tens of years. Over long periods (thousands of years) the slip rate along **faults** is believed to be near enough constant, thus seismic gaps are the anticipated sites of subsequent large **earthquakes**. Seismic gaps are one of the most effective means of long-term **earthquake prediction**; over 50% of successful forecasts have been made this way. Being aseismic zones within seismically active zones, the resultant 'hole' in the pattern of **earth-**

Table S2 Seismic discontinuties in the Earth's crust

Boundary	Approximate depth (km)
Base of crust (the Moho)	5–10 (oceans) 20–80 (continents)
Base of lithosphere	100–150
Phase change of olivine?	400
Phase change of spinel?	670
Outer core–mantle	2885
Inner core–outer core	5155

quake hypocentres has led to seismic gaps sometimes being called a '*Mogi Doughnut*' after the Japanese seismologist who first described them and their use in prediction. [RAC]

McCann, W.R., Nishenko, S.P., Sykes, L.R. & Krause, J. (1979) Seismic gaps and plate tectonics: seismic potential for major boundaries. *Pure and Applied Geophysics* **117**: 1082–147.

seismic head wave A **refracted wave** that impinges on the contact between low and higher velocity media at the **critical angle**, propagates along the contact, whilst sending head waves to the surface at the **critical angle**. Head waves give **first arrivals** on refraction surveys and are used for interpretation of the data, by methods such as the **plus-minus method**, in terms of layers of differing thicknesses increasing in velocity with depth. [GS]

seismic migration The process of putting dipping **seismic reflections** into their true spatial position on a seismic section rather than at the **common depth point**. Migration causes dipping events to move up-dip and steepen by an amount dependent on the initial dip and the migration velocity function assumed. With correct migration, **diffractions** are collapsed to a point and geological structures are imaged in their true geometric form. Seismic migration is commonly undertaken in two dimensions along the plane of a seismic section, though certain specialized techniques can consider the three-dimensional nature of Earth structure. [GS]

seismic moment An **earthquake** is a **shear failure** of a rock mass and the forces causing the **failure** can be represented as a set of couples. Seismic moment M_0 is the scalar size of those force couples. Hence it is a true measure of **earthquake** 'size' and is equally applicable to other types of sources (e.g. explosions) because they too can be represented by a system of forces. M_0 can be computed from broadband P wave **seismograms** at **teleseismic** range (either from their low-frequency spectral amplitude or integrating ground displacements). Seismic moment also has an equivalent definition as

$\mu \times$ (slip length along fault plane)

$\times$ (slip area on fault plane)

where μ is the **rigidity** modulus. M_0 is expressed in dyne cm or Newton metres ($1\,\mathrm{Nm} = 10^{-7}$ dyne cm). M_0 reaches

10^{28}–10^{30} dyne cm, and is routinely computed for most medium-to-large **earthquakes**. A moment magnitude M_w scale has been proposed:

$$M_w = \tfrac{2}{3} \log_{10}(M_o) - 10.7$$

(M_o in dyne cm) which does not saturate at large **magnitudes** as P or *surface wave* **magnitudes** m_b and M_s do. So, for example, the great Chile **earthquake** of 22 May 1960 gave M_s 8.3 but M_w 9.5, and the 28 March 1964 Alaska event was M_s 8.4 but M_w 9.2. (See **earthquake magnitude**.) [RAC]

Kanamori, H. (1977) The energy release in great earthquakes. *Journal of Geophysical Research*. **82**: 2981–7.

seismic noise The continuous motion of the Earth's surface which forms the background to specific seismic signals from **earthquakes** and explosions. In exploration **seismology**, noise may be used more loosely to include any unwanted signals.

Cultural noise (human activities) is generally at frequencies of 1 Hz and higher, and dies away rapidly from its source (factories, roads, etc). Noise at lower frequencies originates from stream/river flow, the oceans (5–20 s period **microseisms**, the main noise type) and atmospheric loading (periods over 30 s). The very low amplitudes of high-frequency (>10 Hz) Earth noise has only recently been appreciated now that *seismographs* with low internal noise have been developed. (See **microseism**.) [RAC]

Holcomb, L.G. (1989) Seismic noise. In: James, D.E. (ed.) *Encyclopedia of Solid-Earth Geophysics*. Van Nostrand, New York.

seismic prospecting The study of variations in the travel time and waveform of **seismic waves** from a known **seismic source** with the objective of locating geological structures that contain substances of economic importance such as oil, gas, **minerals** and water or the investigation of shallow geological conditions affecting civil engineering works and reserves of natural construction materials. (See also **seismic exploration**.) [GS]

seismic record A plot or recording of seismic energy released from a **seismic source** picked up by a detector or set of detectors over a particular time interval. The seismic record can be on paper or in computer-readable form. [GS]

seismic reflection The energy or signal waveform from a **reflected seismic wave** generated by a **seismic source**. This term is often used to describe the actual waveform of a **reflected wave** on a **seismic record**. The seismic reflection technique studies **reflected waves** and is used widely in **seismic exploration**. [GS]

seismic refraction The energy or signal waveform from a **refracted wave** or **head wave** generated by a **seismic**

source. This term is often used to describe the actual waveform of a **refracted wave** on a **seismic record**. The seismic refraction technique studies **refracted waves**, which are often **first arrivals** and are used to delineate the layered velocity structure of the Earth at various depths from shallow layers important for civil engineering **site investigations** up to the study of **crustal** variations on a continental scale. [GS]

seismic shooting The process of collecting seismic information for **seismic prospecting** and relates to the detonation of a **shot** in particular or **seismic source** in general to generate the seismic energy that is to be recorded. A seismic shooter is the person on a seismic surveying crew in charge of detonating the explosive to produce the **shot**. [GS]

seismic slip Slip on a *fault plane* accompanied by perceptible ground acceleration (**seismicity**). The slip rate during a **seismic event** is typically 0.1–$1\,\mathrm{m\,s^{-1}}$, and maximum ground displacements of several metres occur. Ground acceleration is from 0.1 to a few times the value of **gravity**. [TGB]

seismic source Any mechanism whereby **seismic waves** are generated. Specific sources used in solid-Earth geophysics include **earthquakes**, industrial or other chemical explosions and underground nuclear explosions. In exploration **seismology**, numerous forms of mechanical, electrical, and other artificial sources have been devised. Many natural and human activities are effectively seismic sources, e.g. industrial activity, traffic, wind action on topography and trees, but their effect is usually to create undesirable 'background noise'. [RAC]

seismic stratigraphy The study of stratigraphy and distinct, genetically related depositional units as interpreted from seismic data. The method involves the analysis of the non-structural information from **seismic reflections** and can be divided into four subheadings.

1 *Seismic sequence analysis* is used to divide the seismic section into units of common reflection/depositional characteristics by the identification of **unconformities** or changes in the seismic patterns. Each seismic sequence is thought of as a three-dimensional set of sediments deposited contemporaneously as part of the same geological system of depositional processes and environment. The package of **seismic reflections** from sedimentary deposits that form a **delta** is a good example of a seismic sequence. **Unconformities** are described by the way the reflections terminate as **onlap**, downlap, toplap, erosional truncation and concordance. Analysis of seismic sequences has lead certain workers to set up a time seismic stratigraphy by delineating apparent world-wide changes in relative sea-level which are though to be represented by the **unconformities**.

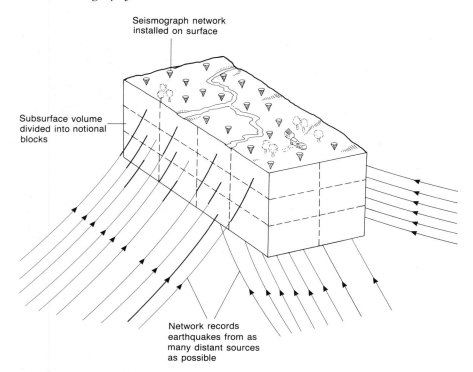

Seismograph network
installed on surface

Subsurface volume
divided into notional
blocks

Network records
earthquakes from as
many distant sources
as possible

Fig. S16 Schematic illustration of how **seismic tomography** is carried out. The relative arrival times (or amplitudes) of the seismic waves are used to determine the relative velocities (or attenuation) in each of the subsurface blocks that is 'hit' by a ray.

2 *Seismic facies analysis* concerns the distinctive characteristics which make one group of reflections look different from adjacent reflections. The reflections are described in terms of their continuity, amplitude, frequency and interval velocity and used to infer the depositional environment of the sediments. Thus the *paleoenvironment* can be reconstructed and depositional facies changes predicted so as to infer the presence of **reservoir** rocks holding *hydrocarbons* in *stratigraphical traps*.

3 *Reflection character analysis* involves the study of changes in a **seismic reflection** wave shape from one **seismic record** to another with the object of determining the nature of changes in stratigraphy or the fluid content and properties of a **reservoir** rock.

4 The study of direct *hydrocarbon* indicators in seismic data is the final seismic stratigraphy technique. [GS]

Sheriff, R.E. (1980) *Seismic Stratigraphy*. IHRDC Publications, Boston.

seismic tomography (literally, 'slice-picture') A means of mapping the seismic velocity variations in a subsurface volume of the Earth by comparing **arrival times** at a network of receivers of **seismic waves** traversing the volume in different directions (see Fig. S16). Its main elements are most easily described in an example.

A group of **seismographs** is located at surface over the volume of interest. Relative **arrival times** are measured of **seismic waves** from **earthquakes** at as many different distances and directions as possible. The subsurface 'target' volume is divided into a set of notional blocks.

Every source-to-station raypath is traced through the blocks to find which ones it traverses. Structure outside the volume is assumed constant. A system of simultaneous equations can then be set up with the relative arrival times and the path lengths in each block as knowns, and velocity anomalies as unknowns. Solving this can demand a great deal of computer power, depending on the number of blocks and individual raypaths. Tomographic methods can just as readily map out **attenuation** using wave amplitude data instead of **arrival times**.

Tomography can be applied two- or three-dimensionally and on any scale, from a borehole-to-borehole investigation of the uppermost 100 m, to the whole of the Earth's **mantle**, in all cases provided that a suitable geometry of sources and receivers is available and the 'ray-tracing' stage through the blocks is done accurately. Tomography has been used, for example, in the definition of subducting **lithosphere** as slabs of high velocity and the mapping of *magma chamber* shapes beneath volcanically active areas. [RAC]

Aki, K., Christoffersson, A. & Husebye, E.S. (1977) Determination of the three-dimensional seismic structure of the lithosphere. *Journal of Geophysical Research* **82**: 277–96.
Iyer, H.M. (1989) Seismic tomography. In: James, D.E. (ed.). *Encyclopedia of Solid-Earth Geophysics*, pp. 1133–51. Van Nostrand, New York.

seismic waves *Elastic waves* (i.e. which vibrate their host medium but cause no permanent **deformation**) propagating in the Earth (see Fig. S17). They may be

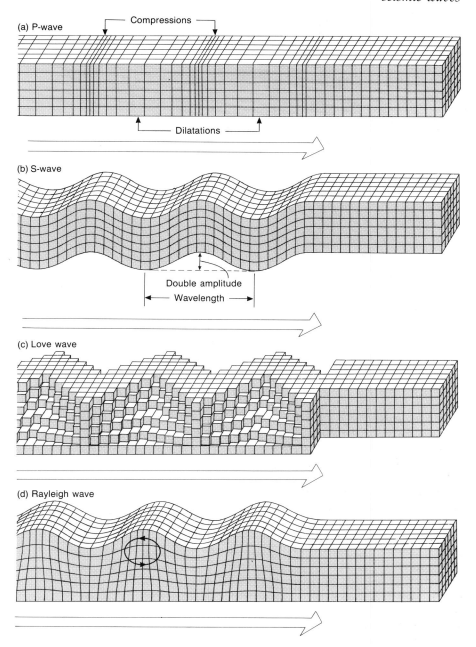

Fig. S17 Diagram to illustrate the forms of ground motion near the ground surface in four types of **seismic** (earthquake) **waves**. (After Bolt, 1976.)

divided into two general categories: *body* and *surface waves*. *Body waves*, whether their source is on the Earth's surface or within it, travel within the body of the Earth before emerging at surface. *Surface waves* travel exclusively around the surface of the Earth.

BODY WAVES.

P waves Longitudinal waves, i.e. compressions and rarefactions with particle motion in the direction that the wave is travelling. The fastest seismic wave, so the direct *P wave* path from source to receiver is *always* the first ('Primary') seismic wave to arrive at a given point.

P wave velocities are <2 km s^{-1} in unconsolidated and **weathered** rocks. **Crustal** velocities are 2–4 km s^{-1} in sedimentary rocks, 5–7 km s^{-1} in igneous and metamorphic rocks. **Mantle** velocities are 8–10 km s^{-1}: and up to 13 km s^{-1} at the centre of the inner **core**. Most **earthquakes** generate *P waves* at frequencies of 0.1–10 Hz.

S waves Tranverse ('Shear' or Secondary) waves with particle motion perpendicular to the direction the wave is propagating. For most solid materials, the velocity of *S waves* is about $1/\sqrt{3}$ of the *P wave* velocity. *S waves* are produced by **earthquakes** in greater amounts than *P*

[553]

Table S3 Seismic waves. In all cases, the type of wave (*P* or *S*) is self-evident except where stated and 'reflection' is taken to be from outside the boundary.

Code	Route taken
P, S	within crust or mantle
p, s	upward-travelling from source
c	reflection at outer core
K	P wave in outer core
i	reflection at inner core
I, J	P, S waves in inner core
LR	Rayleigh surface waves
LQ	Love surface waves

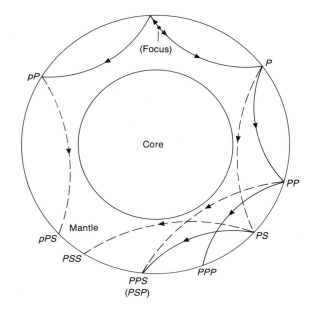

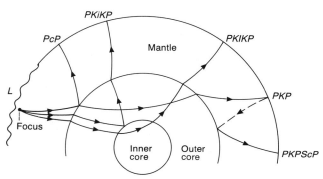

Fig. S18 Examples of **seismic wave** paths in the Earth's mantle (above), through the core (below), and the notation used to describe them. P wave paths are shown as solid lines, S waves as dashed. The inner core is omitted from the upper figure for clarity. (After Bullen & Bolt, 1985.)

because of the **shear** motion of the source, but *S waves* are **attenuated** more quickly than *P* so may not be so easily visible on **seismograms**. Just like light, *S waves* can be polarized.

Boundaries within the Earth cause *body waves* to be reflected as well as transmitted, and converted from *P* to *S* and vice versa. This means that seismic waves can take many paths as well as the direct route from source to a surface receiver, and a shorthand 'code' is used to describe a wave's path (see Table S3 and Fig. S18). For example, sPcP is an upward-travelling *S wave*, converted to *P* when reflected down from the surface, then reflected again at the **core–mantle** boundary before being recorded. Close to their source, the *body wave* arrivals will be a complex mixture of direct, reflected, and refracted arrivals.

SURFACE WAVES. Seismic *surface waves* travel, with velocities of 2.5–4.5 km s^{-1}, radially out from their source around the Earth's surface in a 'skin' of about one wavelength thickness. Because they exhibit **dispersion** they appear as an emergent wave packet rather than a discrete arrival. They are generated over a wide range of periods (up to 300 s or more), of which 15–20 s is the dominant period on **teleseismic** long-period **seismograms**.

The two types of *surface wave* are named after the mathematicians who first described their theory. *Rayleigh waves* have elliptical particle motion in a vertical plane along the direction the wave is travelling. *Love waves* are transverse waves in which particle motion is horizontal and perpendicular to the direction the wave is travelling (like a rope lying on the ground when one end is shaken from side-to-side). *Love waves* cannot travel in water, but are faster than *Rayleigh waves* of the same period.

OTHER TYPES OF WAVES. There are some special forms of *body* and *surface waves* which are important in **seismology**. Shallow (crustal) sources generate an *S wave* trapped in the **crust**, using it as a 'waveguide'. This is often the larges -amplitude arrival at local/regional range. It is termed the *Lg wave*. The *T wave* is similar but is a *P wave* trapped in the ocean. **Free oscillations** are *surface waves* of specific ultra-long wavelengths which match the natural

oscillation frequencies of the whole Earth and make it 'ring' like a bell. Only very large (**magnitude** >7.5) **earthquakes** can excite these waves appreciably. (See **dispersion, seismic discontinuity**.) [RAC]

Bolt, B.A. (1976) *Nuclear Explosions and Earthquakes: The Parted Veil*. W.H. Freeman, San Francisco.

Bolt, B.A. (1982) *Inside the Earth*. W.H. Freeman, San Francisco.

Bullen, K.E. & Bolt, B.A. (1985) *An Introduction to the Theory of Seismology* (4th edn). Cambridge University Press, Cambridge.

seismicity A term coined by Gutenberg & Richter in 1954 to describe the distribution of seismic activity in time, **magnitude**, location and depth. Though **seismology** has been an instrumentally-based science for only some 100 years, a longer-term view of seismicity can be assembled using descriptive records from China, Japan, Greece and elsewhere that date back as far as 2000 BC. Seismicity

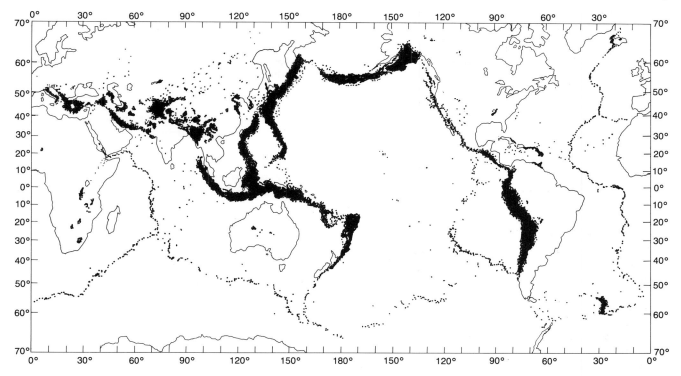

Fig. S19 Seismicity. Epicentres for earthquakes of magnitude greater than 4.5, in the period 1963–1977. (Map computer plotted by Peter W. Sloss of NGSDC.) (After Bolt, 1982.)

studies are vital in providing the statistical basis for **earthquake prediction** and hazard assessment as well as studies of the dynamic behaviour of the Earth.

The vast majority of **earthquakes** are located on the boundaries between **tectonic** *plates* — 'interplate' **earthquakes** (see Fig. S19). Some 70% occur around the margins of the Pacific Ocean, 25% in the Alpine–Himalayan Belt, and most of the remainder at *ocean ridges* and continental **rifts**. Most of these are linear zones apart from some areas of 'diffuse' seismicity (e.g. China, the Eastern Mediterranean), where the *plate* interactions are more complex. A few 'intraplate' events take place within *plates*. Some 75% of **earthquakes** are shallow (<70 km depth); deeper ones (up to 700 km) occur in only a few areas where *plates* are colliding (see **deep-focus earthquakes**). **Earthquake swarms** are also associated with volcanically active areas within or on the edge of *plates*.

The distribution of **earthquakes** in time is, averaged over hundreds of years, essentially constant with no periodicities or changes. Exceptions are (a) **foreshock** and **aftershock** swarms, (b) a few occasions where times of **earthquakes** have been correlated with the *tides* of the solid *Earth* and appear to be triggered by tidal forces (an accepted mechanism for seismic events deep within the Moon) and (c) intraplate **earthquakes**, relatively so infrequent that reliable statistics have yet to be built up. Overall, some 100 000 **earthquakes** strong enough to be felt happen every year. The average annual release of

seismic energy is 10^{17}–10^{18} J, over 90% of this coming from the few great **earthquakes** of **magnitude** 7 and larger.

The distribution of **earthquakes** by **magnitude** reveals a simple link between the **magnitude** M of **earthquakes** and the number $N(M)$ which happen at **magnitude** M, or the so-called *magnitude–frequency relationship*:

$$\log N(M) = a - b\,M$$

where a is a constant for the region under study and b is typically 0.8–1.2. This form of relationship has been proven in studies on all scales from local to global. Expressed in terms of cumulative number, where $N_c(M)$ is the number at **magnitude** M or larger, a representative result is

$$\log N_c(m_b) = 7.8 - 1.0\,m_b$$

for annual numbers of shallow **earthquakes** world-wide. This shows that the number of **earthquakes** increases by a factor of 10 for each **magnitude** unit less that is considered. For example, there are about 80 events per year of **magnitude** 6 or greater, but 800 per year of 5 or more and 8000 per year at 4 or larger. The almost universal value of 1 for b arises from the self-similar or fractal nature of **fault** lengths and the asperities (roughness) of **fault** planes.

Different configurations of *plate*-to-*plate* contacts have broadly characteristic **seismicity** patterns. **Transform faults**, where plates are slipping horizontally past each other have exclusively shallow **earthquakes** (maximum

[555]

depths 10–15 km) and mostly small-to-medium **magnitudes**, though large events can happen (e.g. 1906 San Francisco and 1989 Santa Cruz) if the boundary is locked. Where *plates* are diverging at *ocean ridges* and **rift** valleys, **earthquakes** are mostly shallow, small (**magnitude** <5) and caused by vertical movements as new **crust** is generated. In most places where two *plates* are converging, one bends and dives into the **mantle** beneath the other at a **subduction zone**. The world's largest **earthquakes** happen in the shallow part of the collision zone where one *plate* moves past the leading edge of the other. Intermediate and deep **earthquakes** happen within the subducted *plate* as it sinks into the **mantle**.

Intraplate **earthquakes** occur in continental and oceanic areas, mainly on old zones of weakness reactivated by *plate*-wide tectonic **stresses**. Though the largest have **seismic moments** only 1/100th as great as the largest **plate boundary earthquakes**, they can be extremely destructive, partly because **seismic waves** are not **attenuated** as quickly in intraplate regions and also because structures may not be designed to withstand **earthquakes**. (See **deep-focus earthquakes**.) [RAC]

Bolt, B.A. (1982) *Inside the Earth*. W.H. Freeman, San Francisco.
Isacks, B.L., Oliver, J. & Sykes, L.R. (1968) Seismology and the new global tectonics. *Journal of Geophysical Research* **73**: 5855–99.

seismogram A recording of the variation in ground displacement, velocity or acceleration with time. The amplified output of a **seismometer** is registered (usually along with a time signal) on smoked-paper drums, photographic paper, or ink traces on a chart recorder. More usefully, they may be analogue or digital recordings on magnetic tape, or even direct digital recordings onto computer disk or optical disk. In exploration **seismology**, a seismogram is more usually called a '**seismic record**' or '**trace**'. Data about the **seismometer**/amplifier system should be recorded too so that actual values of ground movement can be recovered. (See **seismometer**.) [RAC]

Simon, R.B. (1981) *Earthquake Interpretation: A Manual for Reading Seismograms*. Kauffmann, Los Altos, California.

seismology The study of **seismic waves** within the Earth, how those waves are generated by **earthquakes** and artificial explosions, how they propagate through the Earth and the information they hold about the structure of the Earth's interior. As well as being of value as a means of studying the Earth, seismology plays a part in human activities as the principal tool for *hydrocarbon* exploration, in trying to minimize **earthquake** hazards by **prediction** and building design and as the means of verification of nuclear testing treaties.

The techniques of seismology are now being extended within the solar system. Seismic experiments were among the first deployed on the Moon and Mars. 'Moonquakes' have revealed something of the Moon's structure. 'Helio-seismology' uses astronomical observations of seismic waves on the Sun to determine its internal structure. [RAC]

seismometer An electromechanical device for detecting ground movement and translating it into an electrical signal.

Coupled with a clock and recording system it forms a *seismograph*, allowing the **arrival times** and waveforms of **seismic waves** to be measured. A seismometer can be 'tuned' to respond to particular frequencies. Knowing its frequency response and the amount by which ground movement is magnified, actual ground displacements can be determined.

Seismic waves can have amplitudes from less than 0.1 nm to over 100 mm and periods of less than 0.02 s to about 1 h. **Arrival times** of waves need to be measured to better than 0.01 s. Only since the mid-19th century have suitably sensitive instruments (basically pendulums with varying means of suspension and magnification of their motion) capable of continuous recording been devised. The 1892 design of John Milne, an English seismologist working in Japan, was the first in widespread use. It had a magnification of about 100 for vibrations of about 10 s period. Later designs by Weichert and Galitzin used coils and permanent magnets mounted on a pendulum or suspended on springs (see **geophone**), producing an induced voltage proportional to ground movement and allowing the seismometer vibrations to be damped electromechanically so that it could operate at larger magnifications.

Galitzin-type designs typically have 1–3 s short-period (SP) or 15–20 s long-period (LP) responses, so as to record *body waves* and *surface waves* respectively without contamination by 5–10 s period **microseisms**. By the late 1950s some 700 *seismographs* were operating, though of disparate designs, often with timing errors of over 1 s and very few with known frequency reponses.

The *World-Wide Standardized Seismograph Network* (*WWSSN*), designed and installed in the early 1960s to provide the scientific and technical means of monitoring underground nuclear explosions, was the first globally-operating system to use a common seismometer design, timebase and frequency response, and to be calibrated daily. Six seismometers in a temperature-controlled vault recorded SP and LP vertical, north–south and east–west movements. Galitzin type seismometers drove galvanometers with small mirrors attached, reflecting a light beam onto a rotating drum of photographic paper. Over 120 were installed and provided data of immense value.

Coupled with digital recording, modern broadband (BB) seismometers can record almost the entire range of **seismic wave** amplitudes and frequencies with a single instrument. They sense the position of the suspended mass by incorporating the mass itself in a capacitance-effect circuit, using electrical feedback to provide damping and define the frequency response. They have a uniform magnification of 100 000 or more over a wide range of frequencies (typically 0.02–20 Hz; **microseisms** are subsequently filtered out by computer) sufficient to detect signals as small as ambient Earth noise. (See **microseisms, seismic noise, seismic waves, seismogram**.) [RAC]

Dewey, J.W. & Byerley, P. (1969) The early history of seismometry to 1900. *Bulletin of the Seismological Society of America* **59**: 183–277.

selenite A variety of the mineral **gypsum** that yields broad, colourless transparent **cleavage** folia. [DJV]

selenium (Se) At. no. 34, at. wt. 78.96. A stable grey form is metallic and is used as a photoconductor (xerography, photoelectric cells). It is also used as a steel additive and in glass. Selenium occurs as a significant impurity in some sulphide **ores** and is then recovered from smelter flue dusts. [AME]

self-potential method (*spontaneous polarization method*) Electrical exploration method based on the natural electric fields produced in the ground by electrochemical reactions involving certain minerals. The anomalous **potentials**, measured by simply planting electrodes at the surface, can be as high as hundreds of millivolts, and are always negative relative to unmineralized ground.

The important minerals which produce this effect are sulphides and **graphite**. The classical explanation for the phenomenon is that of Sato & Mooney (1960) and invokes currents set up within a mineralized zone due to a partial invasion by the **water table**. Below the **water table** the **pore-fluids** undergo **oxidation** reactions, releasing electrons which can pass through the conducting **ore** body, while at the top of the body the electrons take part in *reduction* reactions. The result is that below the **water table** an anodic half-cell is set up and above a cathodic one, the minerals themselves not participating in the reactions, merely facilitating the passage of electrons. This flow of current, electronic in the body and ionic in the fluid, establishes the negative **potential** at the top of the body.

Self-potential can be measured by any voltage measuring device that is sufficiently sensitive, but the **electrodes** used must be of the **non-polarizing** type. The usual practice is to run profiles across the area of interest, with one fixed electrode relative to which the **potential** of the other, which is moved along about 15 m each time, is measured. Long lengths of cable are obviously necessary to achieve this, and it may also be necessary to establish several 'sub-bases' depending upon the profile length. **Telluric currents** are a source of error in self-potential surveys.

The data presentation is in the form of profiles or contour maps, the information sought being usually of a qualitative nature, and the negative maxima lying more or less directly above zones of mineralization. Interpretation of a semi-quantitative nature, in particular relating anomaly-width to the depth of burial of the body, are also common. [ATB]

Kearey, P. & Brooks, M. (1991) *An Introduction to Geophysical Exploration* (2nd edn). Blackwell Scientific Publications, Oxford.
Sato, M. & Mooney, H.M. (1960) The electrochemical mechanism of sulphide self-potentials. *Geophysics* **25**: 226–49.
Telford, W.M., Geldart, L.P., Sheriff, R.E. & Keys, D.A. (1976) *Applied Geophysics*. Cambridge University Press, Cambridge.

self-reversal Some synthetic **ferromagnetic** minerals have the ability to acquire a **remanent magnetization** that has the opposite polarity to the external ambient **magnetic field** in which they were magnetized. This phenomenon can also occur in naturally occurring minerals, but self-reversing **natural remanent magnetization** is uncommon in rocks, being only fully substantiated in the Haruna **dacite** of Japan, although it has sometimes been claimed to occur in other rocks, but is usually not completely repeatable. The mechanisms to account for this behaviour by Néel (1955) are based on magnetic interaction between two **ferromagnetic** minerals with different **Curie temperatures**, with the material of lower **Curie temperature** having a greater **intensity of magnetization** than that of the higher. As such substances cool, the higher **Curie temperature** component becomes magnetized in the same direction as the ambient field, but as further cooling occurs, the lower **Curie temperature** component is primarily affected by the magnetization of the already cooled and magnetized high **Curie temperature** component and therefore acquires a magnetization parallel to this, rather than that of the ambient field. Such behaviour should be repeatable in the laboratory unless the high temperature component has since been destroyed by, for example, ionic migration of the **iron** atoms. [DHT]

Néel, L. (1955) L'inversion de l'aimantation permanente des roches. *Annales Géophysique* **7**: 90–102.
Tarling, D.H. (1983) *Palaeomagnetism*. Chapman & Hall, London.

semseyite ($Pb_9Sb_8S_{21}$) A rare **ore** mineral of **lead**. [DJV]

sepiolite ($Mg_4(OH)_2Si_6O_{15}H_2O$ + $4H_2O$) A clay-like, hydrous, secondary **magnesium** mineral found associated with **serpentine** (also known as '*meerschaum*'). [DJV]

sericite A fine-grained variety of the mineral **muscovite**. [DJV]

sericitization A form of **wall rock alteration** in which the dominant **alteration** products are **sericite** and **quartz**. **Pyrite** is often also present. When both the **potash** and **plagioclase feldspars** of **porphyry copper deposit** *host rocks* are altered in this way the **alteration** may be referred to as **phyllic**. [AME]

serpentine group ($Mg_3Si_2O_5(OH)_4$) A group of hydrated **silicate minerals** (*phyllosilicates*) with three common **polymorphs**, **antigorite**, **lizardite** and **chrysotile**. **Antigorite** and **lizardite** are commonly massive and fine-grained whereas **chrysotile** is fibrous (including **asbestiform** varieties). The serpentine minerals are common and widely distributed, occurring as an **alteration** of **magnesium** silicates, especially **olivine**, **pyroxene** and **amphibole**. [DJV]

serpentinite A rock composed dominantly of the **serpentine minerals** (**antigorite**, **lizardite** and **chrysotile**), with minor amounts of **brucite**, **talc**, Fe–Ti oxide and Ca–Mg carbonates. Such rocks are derived by **alteration** (*serpentinization*) of **ultramafic rocks** such as **dunite**, **peridotite** and *pyroxenite*, and there is complete gradation between unserpentinized **ultramafic rock** and serpentinite. There are two types of serpentinite **fabric**, the *massive* type which has relict textures of the ultramafic source rock and *sheared* serpentinite, in which the relict **fabrics** have been obliterated by intense **penetrative deformation**.

Although any deformed/metamorphosed **ultramafic rock** may be partially to completely serpentinized, many serpentinites are believed to have formed by **alteration** of oceanic **ultramafic rocks** derived from **ophiolite** complexes. Such rocks consist of **lizardite** + **chrysotile** + **brucite** + **magnetite** and probably formed below *c.* 300°C. Rarer **antigorite**-bearing serpentinites may have formed by *greenschist* or *amphibolite* facies metamorphism of **ultramafic rocks** at temperatures up to *c.* 500°C.

Serpentinite has long been a highly prized **stone** for **building** and sculptural use, popular because of its attractive colour and smooth texture. It has been used as a **building stone** in North Africa, India, Japan, Australia, and in the New World, and is most often seen in Europe in church architecture and decoration. It has also been used for smaller objects such as masks and bowls, particularly in pre-Hispanic central America where Olmecs, Aztecs and Maya all employed this stone. It may be referred to as 'greenstone' in archeological literature. [RST/OWT]

settling velocity The terminal velocity at which a grain will settle through a static fluid. Settling velocity is dependent upon the density, size, shape and bulk concentration of the sediment particles together with the density, **viscosity** and **rheological** properties of the fluid. Settling velocity measurements may be used to determine the grain size of sediments, especially with fine grain sizes. (See **drag force**.) [JLB]

sferic Natural electromagnetic field in the audiofrequency range, *c.*1–1000 Hz, with its origin in thunderstorm activity (unlike longer period variations which are of

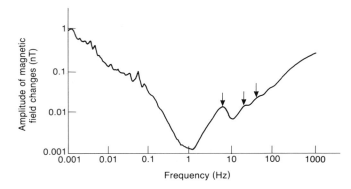

Fig. S20 Sferics. Typical amplitude spectrum of the geomagnetic field in the extremely low frequency range. Arrows indicate the Schumann resonance peaks. (Note the logarithmic scale of the axes.)

ionospheric or magnetospheric origin). Lightning strikes set up transients which are propagated in a waveguide mode in the Earth–ionosphere cavity, with the selective enhancement of the *Schumann resonance frequencies* at approximately 8, 14, 20, 25 . . . Hz, the exact values depending on the height of the ionosphere. The worldwide distribution of storms is such that sferics are always present, and they are the source field for **magnetotelluric** and **AFMAG** methods. (See Fig. S20.) [ATB]

Keller, G.V. & Frischknecht, F.C. (1966) *Electrical Methods in Geophysical Prospecting*. Pergamon, Oxford.

shadow zone A distance range from a **seismic source** within which some given type of **seismic wave** cannot emerge at surface because it is refracted away by some subsurface velocity contrast(s).

The most familiar example is the *epicentral* distance range 105–142° for *P* and *S waves* from shallow **earthquakes** (see Fig. S21). At closer ranges, the waves traverse only the **crust** and **mantle**. In travelling deeper to try to emerge further away, the wave enters the **core**, is refracted towards the vertical and emerges much further away. Similar but less marked shadow zones can exist at regional range because of **low velocity zones** in the **upper mantle**.

The term can also be loosely applied to the volume of the Earth not sampled by the **seismic waves**. (See **epicentral distance**.) [RAC]

Brown, G.C. & Mussett, A.E. (1981) *The Inaccessible Earth*. Allen & Unwin, London.

shakehole A northern English term for a **doline**, or closed depression occurring in **limestone** districts. The shakeholes of the Craven area, northern England, are conical subsidence **dolines** formed in *glacial moraine* overlying **limestone**. They are formed by a combination of collapse and solutional **weathering**. [HAV]

Waltham, A.C. & Sweeting, M.M. (eds) (1974) *The Limestone and Caves of North West England*. David & Charles, Newton Abbot, Devon.

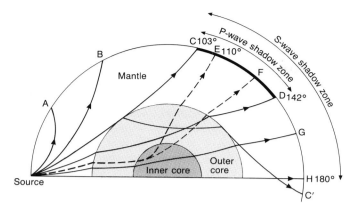

Fig. S21 Rays A–H leave their source at progressively steeper angles. The ray C grazes the outer core, but a ray only minimally steeper is refracted into the core and arrives at C′. The range C–D, the P wave **shadow zone**, would receive no rays at all except that the inner core deflects some energy into it (rays E, F). A little energy is also received in the shadow zone by diffraction around the core–mantle boundary (not shown). All epicentral distances beyond C are in the S wave shadow zone, because the fluid outer core does not transmit S. (After Brown & Mussett, 1981.)

shale Shale had a variety of uses in antiquity. The Romans in Britain made bowls and even tables out of **Jurassic** Kimmeridge shale, and shale armlets were produced in southern Britain. Shale is also reported to have been used as a fuel at this time. Earlier, the British Bronze Age people made shale beads and some Neolithic stone **axes** were made of baked shale from north Wales. It has also been used for tools in Stone Age African sites. [OWT]

shallowing upward carbonate cycle A common product of sedimentation on *carbonate platforms* and *shelves* resulting from carbonate production exceeding rates of relative subsidence.

Sedimentation in shallow-marine carbonate provinces is controlled by the complex interplay of **eustatic** sea-level changes, **tectonic** subsidence of the sedimentary **basin**, the internal generation of carbonate sediments, climate, and external sediment sources. Carbonate sediments are essentially produced *in situ*, and changes in the other factors will necessarily provoke a response in the rate and style of sedimentation on the platform. These changes could potentially produce a huge number of *lithofacies* associations, but in ancient and modern *carbonate platform* sequences it has been recognized that a limited number of cycles, repeated many times, accounts for the bulk of the sedimentary record. These *lithofacies* associations can be explained in terms of a shallowing (or shoaling) upward sequence which records the sedimentary response to changes in relative sea-level. Since the long-term vertical **accretion** rates for platform carbonates are many times greater than rates of relative sea-level rise, a sudden transgression is followed by a gradual build up to sea-level again, as tidal flats prograde on to the submerged parts of the platform top. Repeated many times, this process gives rise to cyclic *carbonate platform* sequences seen throughout the **Phanerozoic**.

The most common cycles, which record the sequences developed in high and low energy regimes following a transgressive event, are illustrated in Fig. S22. These cycles are typically a few metres thick and comprise a basal transgressive *lag* overlain by argillaceous **limestones** and then fossiliferous, *bioturbated* **mudstones** and *wackestones*. In a high energy setting, lime **sand** shoals (perhaps *oolitic*) may develop which spill over into lower energy areas, resulting in discrete *calcarenite* beds and lenses within finer-grained deposits. These subtidal units are overlain by cryptalgal **laminites** from the intertidal zone which themselves are succeeded by supratidal deposits containing evidence of exposure, desiccation or **meteoric carbonate diagenesis**. The cycles are rarely complete, and rapid lateral facies changes, reflecting the environmental complexity of the platform top, are common. In arid climates, **evaporites** (or evidence of their former presence) may characterize the supratidal unit. In some areas individual beds can be correlated over hundreds of kilometres, while cycles are typically repeated several hundred times in thick carbonate sequences.

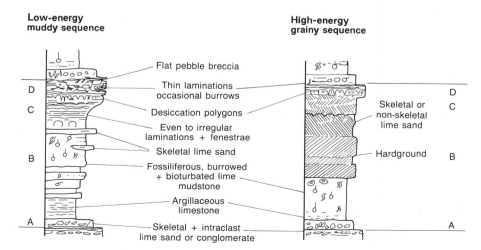

Fig. S22 Hypothetical **shallowing upward carbonate cycle** in low (left) and high (right) energy shelf environments. A = Basal transgressive lag. B = Subtidal, restricted to open marine. C = Inter-tidal. D = Supra-tidal. (After James, 1984.)

Low-energy muddy sequence

High-energy grainy sequence

Flat pebble breccia
Thin laminations occasional burrows
Desiccation polygons
Even to irregular laminations + fenestrae
Skeletal lime sand
Fossiliferous, burrowed + bioturbated lime mudstone
Argillaceous limestone
Skeletal + intraclast lime sand or conglomerate

Skeletal or non-skeletal lime sand
Hardground

Short-term cycles of this type commonly show a longer-term cyclicity when bed thicknesses and overall facies changes are examined, reflecting the different periodicities of the forcing mechanisms (Goldhammer *et al.*, 1987). In addition, platform-top cyclicity often results in rhythmic sedimentation in adjacent **basins** which are directly influenced by on-platform changes.

Numerous models have been suggested for generating cyclic platform deposits, and the whole issue remains contentious. The models are essentially of two types — *autocyclic* models, in which the cyclicity is forced by changes in the sedimentary dynamics of the system itself, and *allocyclic* models, which call on external forcing mechanisms. The *autocyclic* model recognizes that the growth of a platform is controlled by the size and productivity of the source area — the subtidal carbonate 'factory'. Progradation of the main sites of carbonate accumulation (the intertidal zones) into the source area will necessarily reduce the productivity. Ultimately, the source area may be eliminated and, in the absence of a source, the top of the platform may subside below sea-level, for the whole process to repeat itself. *Allocyclic* mechanisms (e.g. the *Punctuated Aggradational Cycles* of Goodwin & Anderson, 1985) call for geologically instantaneous **basin**-wide changes in relative sea-level. Glacio-**eustatic** fluctuations are the most favoured method for generating the short-period cyclicity seen in the geological record.

As the nature of the cyclic sequences is controlled by the interaction of numerous processes, the cycles provide a wealth of information about the **tectonic, eustatic** and **paleogeographic** setting of the carbonate province under investigation. [DMP]

Goldhammer, R.K., Dunn, P.A. & Hardie, L.A. (1987) High frequency glacio-eustatic sealevel oscillations with Milankovitch characteristics recorded in Middle Triassic platform carbonates in northern Italy. *American Journal of Science* **287**: 853–92.

Goodwin, P.W. & Anderson, E.J. (1985) Punctuated aggradational cycles: A general hypothesis for episodic stratigraphic accumulation. *Journal of Geology* **93**: 515–33.

James, N.P. (1984) Shallowing-upward sequences in carbonates. In: Walker, R.G. (ed.) *Facies Models* (2nd edn), pp. 213–28. Geoscience Canada Reprint Series No. 1.

shape fabric Planar or linear structure defined by a **preferred orientation** of the shapes of components of a material. Two factors must therefore contribute to forming a shape fabric: **anisotropic** shapes of individual components, and the overall alignment of this **anisotropy**. These may be, for example, ellipsoidal grains in a deformed **quartzite**, acicular crystals of a mineral defining a **lineation** in a metamorphic rock, or inequant pebbles in a *conglomerate*. The origin of the shapes of the components and their **preferred orientation** may be **tectonic** (due to an applied **stress** and resultant **strain**), metamorphic (due to mineral growth), igneous (due to flow in a lava or **magma**, or *crystal settling*), compactional (due to burial and compression of asymmetrical grains in a sediment), or

sedimentary (due to alignment of grains by sedimentary processes such as fluid **flow** or settling). [TGB]

Shark Bay A bay in Western Australia composed of two elongate embayments in which carbonate sediments are forming. The two inlets have partially restricted exchange with the Indian Ocean and as a result of the high evaporation rates and low **runoff**, have become hypersaline. The restriction is caused by the growth of sea-grass banks (sills). These isolate the Bay into three **basins**, Freycinet, Lharidon and Hamelin. As the barriers have developed these **basins** have become progressively more saline. The Freycinet Basin (salinity of 40–53%) has active sea-grass communities while the Hamelin Basin (Hamelin Pool) is more restricted (56–70% salinity) and has a restricted biota. The Lharidon Basin (Bight) is transitional between the two. Hamelin Basin is well known as the site of extensive **stromatolite** growth, both subtidally and in the intertidal zone. [VPW]

Hagan, G.M. & Logan, B.W. (1974) Development of carbonate banks and hypersaline basins, Shark Bay, Western Australia. In: Logan, B.W., Read, J.F., Hagan, G.M. *et al.* (eds) *Evolution and Diagenesis of Quaternary Carbonate Sequences, Shark Bay, Western Australia.* American Association of Petroleum Geologists Memoir **22**: 61–139.

shattuckite ($Cu_5(SiO_3)_4(OH)_2$) An *inosilicate* mineral found in the oxidized zones of **copper** deposits. [DJV]

shear Deformation involving changes in the angular relationship between material lines in a body, i.e. a *rotational* **stress** or **strain**. Shearing is often used broadly to describe sliding of planes parallel to their lengths, e.g. see **simple shear**. Any quantification of shear depends on a more precise description of the **stress** or **strain** involved. *Shear displacement* is the **displacement** produced by shear, i.e. a component of slip parallel to the plane of the discontinuity, on a **fault** or a **shear zone**. The *shear direction* is the direction in which the **displacement** takes place. A *shear fault* is a **fault** on which the **displacement** has been produced by shear rather than by **extension** normal to the **fault** surface, as in *extensional* **joints** or *fissures*. (See **pure shear, simple shear, shear strain**.) [SB]

shear band Narrow zone of localized **strain** developed in deforming anisotropic **foliated** rocks such as **mylonites** or **phyllonites**. **Deformation** in these zones folds the pre-existing metamorphic **foliation**, as shown in Fig. S23. Shear bands are useful in the field as *shear criteria*. They develop at angles of about 35° to the metamorphic **foliation** and are often *listric*, curving smoothly to become parallel to the early **fabric**. Their spacing is related to the scale of existing **fabric anisotropy**. In fine-grained rocks, shear band formation can be pervasive, destroying the earlier **foliation**.

Figure S24 shows possible shear band orientations; it is

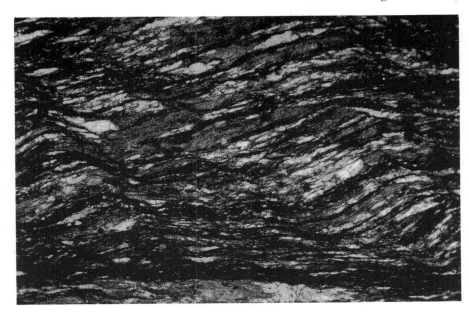

Fig. S23 **Shear bands**.

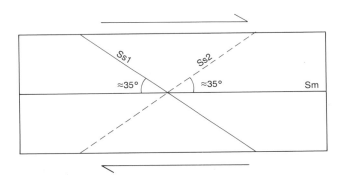

Fig. S24 Sketch showing the common orientations of **shear bands** (Ss1 and Ss2) developed within a zone of overall shear in which the mylonitic fabric (Sm) is as shown.

often the case that only one of the two possible orientations is developed, that which extends the earlier **foliation**. This geometry means that the overall **shear** sense can be deduced from the shear band orientation with respect to layering. When one set of shear bands is strongly developed, an *extensional crenulation cleavage* is formed. In cases where shear bands are developed in both of the possible orientations, the bulk **deformation** cannot be **simple shear** along the layering but must include a flattening or thickening component. [SB]

Platt, J.P. & Vissers, R.L.M. (1980) Extensional structures in anisotropic rocks. *Journal of Structural Geology* **2**: 287–410.
White, S.H. (1979) Large strain deformation: a report on a Tectonic Studies Group discussion meeting held at Imperial College, London on 14 November 1979. *Journal of Structural Geology* **1**: 333–9.

shear layer A zone of velocity gradient, principally directed normal to the **shear plane**, which is generated by the juxtaposition of two fluid bodies of differing velocity or density. Such shear layers are common in stratified flows, in regions of reversing currents and in association with **flow separation**. In low **viscosity** flows these regions are characterized by the generation of large-scale turbulence, the production of high turbulent **shear stresses** and the presence of large-scale eddies, often in pairs, which are responsible for the mixing of the two different velocity or density flows. [JLB]

shear plane The surface parallel to which points move in a rock undergoing **simple shear**, and therefore the plane of a **shear zone**. In the card-deck analogy for **simple shear**, in which playing cards slip past each other in a constant direction, the shear plane is parallel to the surfaces of the cards. [SB]

shear strain Deformation that involves changes in the intrinsic angular relationship between lines. The *angular shear strain* is the angular change produced in an original right angle during **strain**. The shear strain is defined as the tangent of this change in angle — see Fig. S25.

It is easy to calculate in cases of **simple shear** (shown in the Fig. S25) particularly where marker bands cross the shear plane and are offset, as in **shear zones**. In certain analyses of **strain**, discussed by Means (1979, p. 182), tensor shear strain is defined as half the value of γ. (See **strain, shear**.) [SB]

Means, W.D. (1979) *Stress and Strain*. John Wiley & Sons, New York.

shear strength/resistance The **strength** of a material under a **shear stress**. The shear strength is a function of

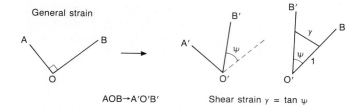

AOB→A'O'B'   Shear strain γ = tan ψ

For simple shear, on a unit cube

Fig. S25 Shear strain. Under a general strain, angle AOB → A'OB'. Shear strain γ = tan ψ. For simple shear on a unit square, γ is the offset of one side of the square with respect to the other.

normal stress and is therefore not a fundamental physical property. Shear strengths are measured experimentally by applying a *normal stress* to a surface and measuring the **shear stress** required for sliding or **failure**. This may be carried out directly in a **simple shear** experiment, or in a triaxial apparatus. The shear strength can be calculated by resolving the known **principal stresses** at **failure** onto the **failure** surface when the angle of the **failure** surface to the *principal stresses* has been measured.

The shear strength (τ) of a sliding surface is related to the *normal stress* (σ_n) in *Amonton's law of frictional sliding* by the *coefficient of friction*, μ and of an intact material by the equation of the *Mohr envelope*, which may be determined experimentally and is predicted by a *failure criterion*. [TGB]

shear stress The force per unit area that is exerted tangential to a given surface. A volume of rock (See Fig. S26) of cross-sectional area *A* has a force *F* applied to opposite faces. The **stress** θ applied is thus *F/A*. The shear stress on a given plane within the rock at an angle θ to

the forces is the **effective stress** across that plane, σ cos θ sin θ. (It should NOT be regarded as a simple couple, because that would cause the volume to rotate, but two equal and orthogonal couples). Shear stress is a maximum at 45° to the applied forces and so that is the geometry in which rocks actually fail — i.e. in **an earthquake**, the *fault plane* is at 45° to the compression and tension axes.

The shear stress in a fluid is caused by velocity gradients, **as in boundary layers**. [JLB/RAC]

Tabor, D. (1969) *Gases, Liquids and Solids*. Penguin, Harmondsworth, Aki, K. & Richards, P.G. (1980) *Quantitative Seismology; Theory and Methods* (2 Vols). W.H. Freeman, San Francisco.

shear velocity A velocity expressing the velocity gradient of the lowest regions of a **boundary layer**. The shear velocity measures the rate of increase in fluid velocity with distance away from the bed in the lowest 15% of the flow depth and as such has units of length/unit time. Shear velocity, U_*, is related to the fluid **shear stress** by $U_* = (\tau/\rho)^{0.5}$ where τ is the **shear stress** and ρ the fluid density. [JLB]

shear zone Zone of **ductile deformation** between two undeformed blocks that have undergone **shear** displacement relative to each other. In terms of relative **displacement**, a shear zone is the ductile analogue of a **fault** and the same terminology may be used; thus **strike-slip** shear zones may be classified into **sinistral** or **dextral**, and **dip-slip** into *thrust* or *normal*. Instead of the **shear displacement** being abrupt across the **shear plane** as in a **fault**, however, **displacement** is transferred across a zone either by continuous **strain**, or by a combination of continuous and discontinuous **strain** (in a *brittle–ductile shear zone*).

The geometry of shear zone displacements is summarized in Fig. S27. The **shear** or **displacement** direction is the linear movement **vector** for the **displacement**, and lies within the *displacement plane*, which is the plane normal to

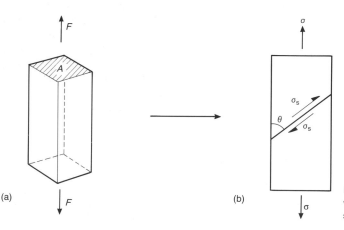

Fig. S26 Shear stress. (a) Rectangular bar of uniform cross-sectional area *A* with tensional force *F* applied: hence tensional stress σ = *F/A*. (b) Shear stress σ_s is given by $\sigma_s = \sigma \cos \theta \sin \theta$

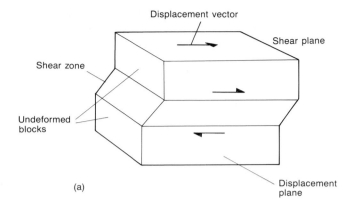

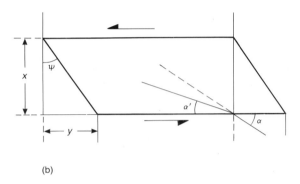

Fig. S27 (a) The **shear zone** is a zone of ductile deformation between two undeformed blocks which have been displaced relative to each other. Note the shear direction (displacement vector) lying within the shear plane which is parallel to the margins of the shear zone. (b) The displacement y is given by $x \tan \psi = x\gamma$ where x is the width, ψ the angular shear strain and γ the shear strain. If a line makes an initial angle α with the shear plane and an angle α' after shearing, $\gamma = \cot \alpha - \cot \alpha'$. (After Ramsay, 1980.)

the shear zone walls containing the *shear direction*. In ideal shear zones, **displacement** is effected by **simple shear**, i.e. two-dimensional **strain**, in which the maximum and minimum **principal strain** axes lie in the **displacement** plane and there is no change in the intermediate **strain** direction. The plane containing the **shear** direction, and perpendicular to the **displacement** plane, is termed the *shear plane*. The *shear plane* is parallel to the shear zone walls. In an ideal case of *homogeneous strain*, the displacement is simply related to the **shear strain** by the equation $y = \gamma x$, where y is the amount of **displacement**, γ is the **shear strain**, and x the width of the shear zone. The **shear strain**, $\gamma = \tan \psi$ where ψ is the angular **shear strain**. In most shear zones, the **strain** is heterogeneous across the shear zone, usually increasing to a maximum either in a central narrow high-**strain** zone, or in a series of such zones. In cases such as these, the **displacement** may be calculated as follows:

$$y = \int_0^x \gamma \, \mathrm{d}x$$

There are two main methods of determining **shear strain**, and hence **displacement**, in shear zones, (i) by measuring the progressive deflection of a line entering the shear zone at an angle α, where:

$$\gamma = \cot \alpha - \cot \alpha'$$

or (ii) by measuring the orientation or shape of the **finite strain ellipse**. The relationships between the **shear strain** and the geometry of the **strain ellipse** are summarized in Fig. S28.

Shear zones may be much more complex geometrically than the above model. It is common for shear zones to depart from the **simple shear** model by the addition of a **pure shear** component, such that the shear zone may be either **transpressional** or **transtensional**, by the addition of a *compressional* or *extensional* component of **strain** across the zone. Or the **strain** may depart from **plane strain** to become three-dimensional, e.g. by the addition of a *uniaxial flattening* across the shear zone. Like **faults**, shear zone **displacements** vary along their length producing **strain** incompatibilities. **Strain** may be distributed into the walls of the shear zone, or may be transferred into branching zones. Often complex shear zone networks are found enclosing blocks or lenses of undeformed material.

Shear zones are the characteristic method of achieving **displacements** at intermediate to deep levels of the **crust** and probably also in the **mantle** part of the **lithosphere**. Major **fault** displacements at high **crustal** levels are considered to transfer downwards into shear zones. Major continental **strike-slip fault systems** such as the San Andreas *fault zone* are thus visualized as the high-level expression of deep shear zones of the order of several tens of kilometres in width. Evidence for the nature of major displacements at deep **crustal** levels comes from the study of the large shear zones in **Precambrian** terrains. These have been studied particularly in high-grade **Archean** and Early **Proterozoic** terrains of Greenland, north-west Scotland and Canada, where they form important boundary zones between **mobile belts**. For example, a major thrust-sense shear zone, 50 km wide, separates the Nagssugtoqidian **mobile belt** from the **Archean craton** in west Greenland. There are also a number of major **strike-slip** shear zones cutting the south Greenland craton. These form a conjugate set of NE–SW **sinistral** and NW–SE **dextral** zones which have been used to indicate the Early **Proterozoic** compression direction (Watterson, 1978). Many major shear zones exhibit very high **strains** ($\gamma \approx 10$) which have resulted in the rotation of all planar structures into near-parallelism with the **shear** plane, and of linear structures with the **shear** direction. **Sheath folds** are characteristic. [RGP]

Escher, A., Sorensen, K. & Zeck, H.P. (1976) Nagssugtoqidian mobile belt in West Greenland. *In*: Escher, A. & Watt, W.S. (eds) *Geology of Greenland*, pp. 76–96. Geological Survey of Greenland.

Ramsay, J.G. (1980) Shear zone geometry: a review. *Journal of Structural Geology* **2**: 83–99.

Watterson, J. (1978) Proterozoic intraplate deformation in the light of south-east Asian neotectonics. *Nature* **273**: 636–40.

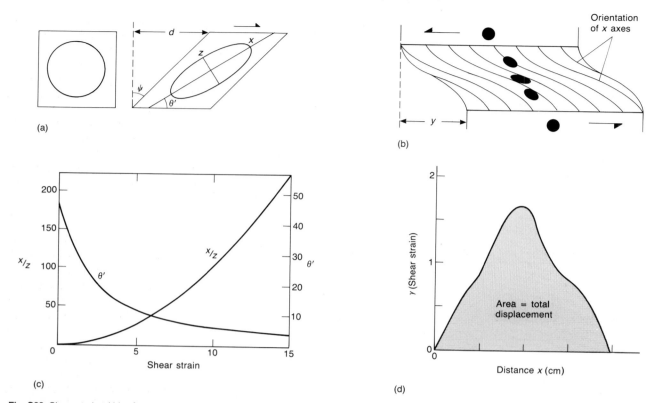

Fig. S28 Shear strain within **shear zones**. (a) Relationship between shear strain and strain axes. (b) Continuous variation in shear strain across a shear zone shown by a change in the orientation of the *x* strain axis and in the strain ratio *x/z*. The angle θ' made by *x* with the shear direction is 45° at the margin of the shear zone and decreases to a minimum in the centre. (c) Plot showing the variation of the angle θ' and the strain ratio *x/z* with increase in the shear strain, γ. (d) Plot of γ against *x* (distance across the shear zone) for the shear zone of (b). The shear displacement *y* is equal to the area under the curve. (After Ramsay, 1980.)

sheath fold A highly *non-cylindrical fold*, with *hinge lines* strongly curved within *fold axial* surfaces. The name arises from the resemblance of the folded surface to a sheath. Such folds represent high **strain, ductile deformation**. They are mostly found in the internal, deeper parts of mountain belts. In **shear zones** where there is a strongly defined **mineral lineation** indicating the *shear direction*, sheath folds are elongate parallel to this direction. Folded layers form closed, often elliptical traces in sections normal or oblique to the **shear** direction. Sheath folds are commonly identified in approximately **simple shear strain** regimes, but can also form under other **strain** geometries, such as **pure shear**. Although formation of such extremely curvilinear *fold hinges* requires considerable **strain** variations across each layer, the shape change on a larger scale can be homogeneous. Cobbold & Quinquis (1980) modelled formation of sheath folds both theoretically and using silicone putty, and found that these geometries develop passively from irregularities in layering at **shear strains** of about 10 to 15. [SB]

Cobbold, P.R. & Quinquis, H. (1980) Development of sheath folds in shear regimes. *Journal of Structural Geology* **2**: 119–26.

sheet erosion The removal of fine-grained, surficial materials by sheets of flowing water rather than by channelized streams of water which tend to erode **rills** and *gullies*; also termed *sheetflood erosion*. Sheet erosion occurs in desert environments where sporadic rainfall produces **runoff** by **sheetflow** on *fans* and **pediments** which do not have *gullies* or stream channels incised into their surfaces. However, because **sheetflow** is not channelized, it is not as highly **turbulent** as **streamflow** so sheet erosion is less effective at **eroding** coarse particles. [AW]

sheetflood An unconfined flow of water over a terrestrial surface resulting in the deposition of mud, silt, **sand** or **gravel**. Sheetfloods are distinct from **crevasse splays** as the former generally originate from channel bank over-topping rather than breaching and the resulting flow is less channelized. Sheetflood deposits are sharp based, though not always **erosional** and contain sedimentary structures indicative of deposition from moving water. These deposits frequently show fining upwards profiles but may coarsen up if the sediment load peak lags behind the flood discharge peak. [JA]

sheetflow Though defined by R.E. Horton (1945) as **overland flow** which is not concentrated in channels larger than **rills**, modern usage restricts the term to flow over smooth *soil* surfaces, and treats **rill** wash separately. The **flow** is usually very shallow near the divide but may become deeper and **turbulent** downslope. Sheetflow, or *sheetwash*, may be an effective **erosional** agent even without the concentration of water into **rills**. The interaction of *rainsplash* and *sheetwash* may be particularly **erosive**. [TPB]

Horton, R.E. (1945) Erosional development of streams and their drainage basins: hydrophysical approach to quantitative morphology. *Geological Society of America Bulletin* **56**: 275–370.

sheeting The process whereby shells or sheets of rock are split off along **joints** that run roughly parallel to the rock surface. This process results from the **denudation** of material overlying a rock, thereby leading to **pressure release**. [NJM]

shell pavement Accumulation of shell valves where selective **erosional** winnowing of mixtures of **sand** and shell has left a superficial *lag* deposit of the coarse shell material. Pavements of this type are common in coastal situations, where, for instance, fine sediment can be moved by **tidal** currents or by aeolian processes above the high-water mark. [ASG]

Carter, R.W.G. (1976) Formation, maintenance, and geomorphological significance of an aeolian shell pavement. *Journal of Sedimentary Petrology* **46**: 418–29.

shield A large area of exposed basement rock, which has remained unaffected by mountain-building for a lengthy period of geological time. It commonly has a very gently convex surface and is surrounded by sediment-covered platforms. Shields are generally composed of *Precambrian* rocks, and constitute the central position of each continent, such as the Baltic shield of northern Europe, the Laurentian shield of North America, the continent of Africa and the Brazilian Plateau. [NJM]

Windley, B.F. (1984) *The Evolving Continents* (2nd edn). John Wiley & Sons, Chichester.

shield volcano A type of volcano, consisting of many **lava flows**, with subordinate **pyroclastic rocks**, having an approximately circular or elliptical shape in plan view, and having gently dipping flanks (2–8°). There are many examples in Iceland, where they were named from their resemblance in shape to Viking shields laid face up on the ground. Shield volcanoes range from about 10 to 100 km in diameter. They are commonly surmounted by a **crater**, or a cluster or line of **craters**, and/or a **caldera**. Some shields are circular in plan, with a central vent area (see **volcanic vent**), but most are elongate along a main fissure (see **fissure eruption**). Nearly all shield volcanoes are overwhelmingly **basaltic** in composition, but small volumes of intermediate or silicic lavas and **pyroclastic rocks** may be present.

Shield volcanoes typically occur in overlapping groups. The island of Hawaii consists of the large, overlapping and interfingered shield volcanoes of Mauna Kea, Mauna Loa, Hualalai and Kilauea. These shields have diameters up to about 100 km (for the parts of the shields above sea-level). The Snake River Plain, Idaho, consists of multiple, interfingering very low-angle shield volcanoes, each with a diameter of about 15 km. Greeley (1982) called this coalescence of numerous, closely-spaced, shield volcanoes **basaltic** plains volcanism. Nevertheless, some shield volcanoes occur as individual volcanoes (Macdonald, 1972).

Rarely, shield volcanoes consist of **lavas** of intermediate, rather than **basaltic** composition. Webb & Weaver (1975) described **trachyte** shield volcanoes up to 50 km in diameter from the Kenya **rift** valley. These shields comprise a gently-sloping flank zone which consists of **trachyte** lavas with some mafic lavas, and interbedded **pyroclastic rocks**, and a complex, central source zone which consists of plugs, **scoria** cones and **dykes**. Shield volcanoes of **andesitic**, **dacitic** or **rhyolitic** composition do not seem to occur, because of the tendency of **magmas** of these compositions to erupt explosively, forming thick aprons of **pyroclastic rock** around the volcanoes. (See **plateau lava, volcanic eruption**.) [PTL]

Greeley, R. (1982) The Snake River Plain, Idaho: representative of a new category of volcanism. *Journal of Geophysical Research* **87**: 2705–12.

Macdonald, G.A. (1972) *Volcanoes*. Prentice-Hall, Englewood Cliffs, NJ.

Webb, P.K. & Weaver, S.D. (1975) Trachyte shield volcanoes: a new volcanic form from south Turkana, Kenya. *Bulletin Volcanologique* **39**: 1–19.

Shields beta A dimensionless measure of **shear stress** which represents the ratio of the fluid forces encouraging sediment movement to the submerged weight of a single layer of grains, or the **gravity** forces, that are opposing motion. Shields beta, β, is given by:

$$\beta = \tau/(\rho_s - \rho_f)g.d$$

where τ is the mean **shear stress** acting upon a unit area of bed (which is related to the average **drag force** and, at constant **Reynolds numbers**, to the **lift force**), ρ_s and ρ_f are the sediment and fluid densities respectively, g is acceleration due to **gravity** and d is the grain size. (See **Shields diagram**.) [JLB]

Shields diagram A graphical relationship proposed by Shields in 1936 depicting the threshold of sediment entrainment on a plot of dimensionless **shear stress** (**Shields beta**) against particle **Reynolds number** (see Fig. S29). Although there is much scatter about the curve representing this movement threshold, this diagram can

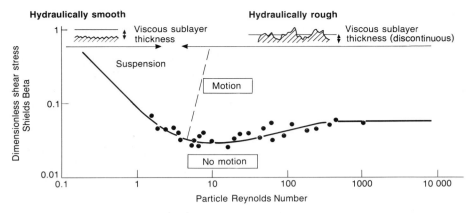

Fig. S29 Shields diagram. The threshold of sediment transport as represented by the Shields diagram of dimensionless shear stress (Shields beta) plotted against particle Reynolds number. The approximate field of sediment suspension is marked together with an indication of the hydraulic roughness of the bed. (After Richards, 1982.)

be used to interpret the forces influential in the entrainment of sediment irrespective of the fluid and sediment densities. At low particle **Reynolds numbers** a high **Shields beta** at the entrainment threshold reflects both the influence of sediment cohesitivity and the submergence of grains within the **viscous sublayer**. After declining to a minimum dimensionless **shear stress** at a particle **Reynolds number** of approximately 10, where cohesionless grains protrude through the **viscous sublayer**, **Shields beta** increases slightly with increasing **Reynolds number** as greater particle mass and mutual particle interference become influential upon initial movement. Large departures from the Shields curve are to be expected in poorly sorted sediments or when the bed is not essentially flat due to bedform generation. [JLB]

Richards, K.S. (1982) *Rivers: Form and Process in Alluvial Channels.* Methuen, London.

shipborne geophysical survey Geophysical survey that can be performed from ships and can be **gravity**, **magnetic** or **seismic**.

Gravity surveys are performed either by lowering a remote instrument to the sea-bed or, more efficiently, by using a shipborne **gravimeter**. The latter is mounted on a stable table and damped so that the horizontal and vertical accelerations associated with **wave** motions are minimized. Because of these necessities, the accuracy is decreased to 10 g.u. at best when using **gravimeters** based on a suspended beam. Greater accuracy of only a few gravity units can be attained with modern axisymmetric meters, such as the *Bell gravimeter*.

Magnetic surveys are performed by mounting the sensor of a **magnetometer** in a 'fish' which is towed at least two and a half shiplengths behind the vessel to minimize its magnetic effects.

Continuous **seismic reflection** surveys are performed by using a source of seismic energy which can be fired every few seconds, such as an airgun or sparker. **Seismic reflections** are received at a towed array of **hydrophones** mounted in a streamer which can be several kilometres in length. **Seismic refraction** experiments can be performed using two ships, which reverse roles as shooter and recorder. The second ship can be replaced by a **sonobuoy**, a disposable **hydrophone** and transmitter. This decreases the cost of the survey, but makes reversal of the refraction line rather more difficult. [PK]

Bell, R.E. & Watts, A.B. (1986) Evaluation of the BGM-3 sea gravity meter system onboard R/V *Conrad*. *Geophysics* **51**: 1480–93.
Dehlinger, P. (1978) '*Marine Gravity*'. Elsevier, Amsterdam.
Kearey, P. & Brooks, M. (1991) *An Introduction to Geophysical Exploration* (2nd edn). Blackwell Scientific Publications, Oxford.

shock remanent magnetization The **natural remanent magnetization** acquired as a result of rapidly applied **stress**, such as a *meteorite* impact. [DHT]

shore platform Low gradient intertidal rock surface created by the recession of **sea cliffs**. The form has been variously called a *wave-cut bench* or *platform* or a *marine abrasion platform*, but because processes other than **wave** quarrying and **abrasion** are involved, the term shore platform is now preferred. There are three broad groups of platforms and of platform processes. Platforms produced by quarrying and **abrasion** are most common in *storm* **wave** environments and they develop intertidal gradients that steepen as **tidal** range increases. Lithological variations produce marked internal relief on such platforms (Davies, 1980). The alternate wetting and drying of rocks produces platforms close to high **tide** mark as this is the highest level of permanent saturation. Such platforms display very slight slopes or are absolutely horizontal and are most common in low latitudes where lower **wave** energy encourages their preservation. The same low latitude bias is shown by platforms produced by solution and biological **erosion**. These are formed at or slightly above low water mark and are generally associated with calcareous rocks. In high latitudes, extensive shore platforms may be formed by intertidal frost action amongst other processes, e.g. **strandflat**. In common with **sea cliffs**, many shore platforms at present sea-level may in fact be re-exhumed surfaces related to former sea-levels. [JDH]

Davies, J.L. (1980) *Geographical Variation in Coastal Development.* Longman, London.

Trenhaile, A.S. (1987) *The Geomorphology of Rock Coasts.* Clarendon Press, Oxford.

shortening/shortening strain A linear measure of shape change (**strain**) using the decrease in length of a line or lines in any direction, defined thus:

$$\frac{\text{undeformed length} - \text{deformed length}}{\text{undeformed length}}$$

It is widely used as a descriptive term, as in **axially symmetric shortening** which describes a decrease in length in one **principal strain** direction (the axis of symmetry), and equal length changes in the two perpendicular principal strain directions. Since shortening refers to the undeformed length of a line, which is in general unknown, its use in analysis of **penetrative strain** is restricted. It is widely used, however, in studies of large-scale **deformation** in **thrust belts**, and in gently folded areas, where **deformation** does not involve large internal **strain** in the beds, but **displacement** on *thrust faults* and **interlayer slip** during **folding** are common. The arc length of **folded** and **faulted** beds represents their undeformed lengths, and the deformed length is given by the length they now occupy.

Shortening is often quoted for **balanced cross-sections**, in which the separations of *pin lines* or reference points on the balanced and restored sections form the deformed and undeformed lengths respectively. As shown in Fig. S30, the shortening measured depends on the distance over which it is measured. To facilitate comparison between different areas and examples, many workers in **thrust belts** quote the total **fault displacement**, and bed lengths in initial and final states. (See **elongation**, **stretch**, **quadratic elongation**.) [SB]

Butler, R.W.H. (1987) Thrust systems. *Journal of the Geological Society, London* **144**: 619–34.

shot The detonation of an explosive, often to create an impulsive **seismic source**. More generally the term is used to describe any artificial **seismic source**. [GS]

side-scan sonar A sideways scanning, marine, acoustic surveying method in which the sea-bed to either side of a ship is insonified by beams of high frequency sound (30–110 kHz) emitted by hull-mounted or towed transceiving transducers. The method provides information on bathymetric features in a broad swath and so allows wide areal coverage of a region in a very time-effective fashion. Sea-bed features, such as rock outcrops or sedimentary structures, which face the ship reflect energy back to the transducers, while features facing away from the ship reflect energy away, as does a flat sea-bed (Fig. S31a). A display of the sonar returns is known as a *sonograph*. The distortion arising from the variably oblique travel path

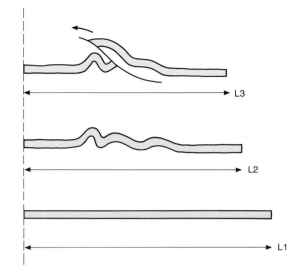

Fig. S30 Shortening strain. In this sequential development of a model fold/thrust structure:

$$\text{the total shortening} = \frac{L1 - L3}{L1}, \quad \text{shortening by folding} = \frac{L1 - L2}{L1},$$

$$\text{shortening by thrusting} = \frac{L2 - L3}{L2}. \text{ (After Butler, 1987.)}$$

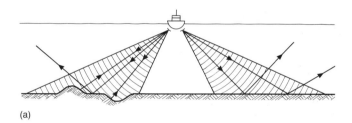

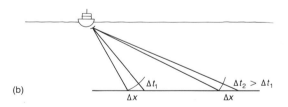

Fig. S31 Side-scan sonar. (a) Selected raypaths. (b) Scale distortion resulting from oblique incidence. (After Kearey & Brooks, 1991.)

(Fig. S31b) can be corrected to provide an isometric plan of the sea-floor.

Side-scan sonar has been extensively used to study the morphology of *ocean ridge* crests, and is useful in locating artefacts on the sea-floor such as wrecks. [PK]

Kearey, P. & Brooks, M. (1991) *An Introduction to Geophysical Exploration* (2nd edn). Blackwell Scientific Publications, Oxford.

siderite (FeCO₃) A **carbonate mineral**, occasionally important as an **ore** of **iron** (in clay *ironstones* and black-band **ore**). [DJV]

siemen (S) SI unit of electrical conductance, conductance being the reciprocal of resistance (see **Ohm's law**), 1 siemen is $1\,\Omega^{-1}$ (defined in terms of basic units as $kg^{-1}\,m^{-2}\,s^3\,A^2$). [ATB]

sieve deposit Sediment deposited as a result of rapid discharge reduction of surface flow due to percolation of water into previously deposited coarse sediments. Sieve deposit generally refers to the fine sediment that infiltrates recently deposited *conglomerate* as **flow** wanes and *flow competence* is lost (Hooke, 1967). The term sieve deposit has also been applied to coarse sediment lobes on semi-arid **alluvial fans** that result from rapid discharge reduction due to water **infiltration**. [JA]

Hooke, R. Le B. (1967) Processes on arid-region alluvial fans. *Journal of Geology* 75: 438–60.

signal to noise ratio The ratio of the amplitude or energy of a particular signal to the surrounding unwanted background **noise** or undesired signal. [GS]

silcrete Surface and near surface terrestrial material composed of **silica**. It occurs as nodular or highly indurated and massive forms and results from the **cementation** and **replacive** introduction of **silica** into *soils*, indurated or **weathered** rock and sediments. It can form in arid or more humid climates and is particularly widespread in Australia and Africa. It is a striking cap-rock layer on extensive **erosion** surfaces in these areas. Because of its resistant nature it often occurs as extensive sheets of eroded debris to form 'gibber' plains. [VPW]

Langford-Smith, T. (ed.) (1978) *Silcrete in Australia*. Department of Geography, University of New England, Armidale, NSW.
Summerfield, M.A. (1983) Silcrete. In: Goudie, A.S. & Pye, K. (eds) *Chemical Sediments and Geomorphology*, pp. 59–91. Academic Press, London.

silica minerals A group of minerals based on the formula SiO₂. **Quartz** is by far the most common form, but there are a large number of **polymorphs**, and also a number of impure, often poorly crystalline varieties which are semiprecious, such as **agate** and **opal**. **Quartz** is an industrially important **mineral**, in glass making, as an abrasive, and as the central component in **quartz** oscillators and clocks.

A phase diagram for the silica minerals is given in Fig. S32. **Quartz** itself has a tightly packed structure and for this reason is usually nearly pure SiO₂, but the high temperature forms **tridymite** and **cristobalite** have much more open structures which permit more extensive substitution, including that of **aluminium** for silicon. Low (or α-) quartz

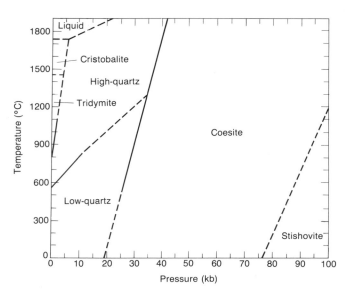

Fig. S32 Pressure – temperature phase diagram for the **silica minerals**. (After Hurlbut & Klein, 1985.)

is trigonal in symmetry, but at 573°C it inverts rapidly and non-quenchably to hexagonal high (or β-) **quartz**, a displacive structural change which involves twisting of linked (SiO₄) tetrahedra (Fig. S33), but no fundamental reconstruction. **Tridymite** (stable above 870°C) is hexagonal above 140°C but has a quite different structure and its transformation to **quartz** is reconstructive and sluggish. At 140°C it exhibits a metastable displacive transformation to monoclinic or orthorhombic low **tridymite**. Similarly **cristobalite**, stable from 1470°C to the melting point, also has a different, cubic, structure and inverts sluggishly to **tridymite**. High **cristobalite** has a metastable displacive transformation to tetragonal low **cristobalite** at 268°C.

Coesite and **stishovite** are very dense (low **quartz** 2.65, **coesite** 3.01, **stishovite** $4.35\,Mg\,m^{-3}$) high pressure **polymorphs** with monoclinic and tetragonal symmetry respectively. In **stishovite** silicon is octahedrally co-ordinated with oxygen. High **tridymite** and high **cristobalite** have low densities (2.22 and $2.20\,Mg\,m^{-3}$).

The structures of high and low **quartz** are illustrated in Fig. S33. Both are characterized by screw symmetry axes, parallel to which the SiO₄ tetrahedra are arranged in a spiral. The sense of screw may be clockwise or anticlockwise, so that **quartz** occurs in enantiomorphous right- and left-handed forms, and exhibits optical activity, the ability to rotate the plane of polarization of light travelling parallel to the Z crystallographic axis. **Quartz** develops an electric **potential** across the *x y u* axes when deformed in the appropriate direction and this piezoelectric property forms the basis of its use in **quartz** oscillators and clocks. When an alternating **potential** is applied to a suitably cut plate the crystal deforms and relaxes. Because of the very perfect elastic properties, these plates can then be used to stabilize an alternating frequency to extremely high precision.

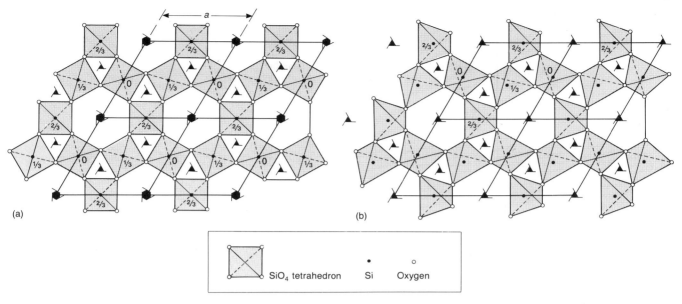

Fig. S33 Silica minerals. Crystal structure, projected on (0001), of (a) high or α-quartz, (b) low or β-quartz. Note how the tetrahedra twist in going from (a) to (b) but no linkages are broken. (After Deer *et al.*, 1963.)

Quartz makes up about 30% of **granitic** rocks, is the main detrital constituent of clastic sediments and an important cement in **diagenesis**. It is usually colourless and easily distinguished from **feldspar** by its lack of **cleavage** and absence of turbidity. **Quartz** often contains minute cavities containing fluid, representing in some cases the fluid trapped at the time the crystals grew. Coloured varieties include **citrine** (yellow), **amethyst** (purple) and **smoky quartz** or *Cairngorm* (brown to black). **Tridymite** and **cristobalite** are much less common and are found in siliceous volcanic rocks. **Coesite** is extremely rare but has been discovered in high pressure metamorphic rocks from the Alps and in **kimberlite**. It and **stishovite** are otherwise known only from *meteorite* impact **craters**.

Cryptocrystalline or near amorphous varieties of silica include **chalcedony**, a fibrous, waxy-looking form of fine-grained **quartz**. **Agate** and **onyx** are concentrically banded variants. Granular fine-grained varieties include **chert**, **flint** and red **jasper**. **Opal** is a hydrated variety in which poorly crystalline silica occurs as minute (300 nm) spheres. *Opalescence* is caused by thin-film interference of light along planes of voids between the packed spheres. All these varieties form from low temperature solutions, in environments ranging from cavities in **basalts** (**agate**, **jasper**) to pore spaces or replacive bodies in sediments (**flint, opal**). [IP]

Deer, W.A., Howie, R.A. & Zussman, J. (1963) *Rock Forming Minerals* Vol. 4: *Framework Silicates*. Longman, London.

Hurbut, C.S., Jr & Klein, C. (1985) *Manual of Mineralogy* (20th edn). John Wiley & Sons, New York.

Sosman, R.B. (1965) *The Phases of Silica*. Rutgers University Press, New Brunswick.

silica solubility The concentration of **silica** (SiO_2) that can be dissolved in simple, dilute aqueous solutions before saturation is reached is well known from a large base of experimental data that encompasses a wide range of temperatures and pressures. At 25°C in such aqueous solutions **silica** dissolves at **pH** values of <9 to produce the undissociated monomeric tetrahydrate molecule $Si_4(OH)_4°$ according to the reaction

$$SiO_{2(s)} + H_2O_{(1)} = Si_4(OH)_4°$$

The resulting solubility of **silica** is directly a function of which **silica polymorph** is in contact with the solution and is being dissolved. Amorphous **silica** is by far the most soluble of the **polymorphs**, **cristobalite**, **tridymite** and **chalcedony** successively less so, and **quartz** the least so. This solubility series essentially reflects the degree of crystallinity and structural order in the **polymorphs**. Additionally, however, the solubility of silica is strongly dependent upon the particle size, surface area and shape of the particular **polymorph** concerned because the excess free Gibbs energy associated with all solid surfaces significantly influences the solubility of solids which have large surface to volume ratios. As a consequence, poorly ordered, low crystallinity **polymorphs** with small particle size, large surface area and highly irregular shape — such as amorphous **silica** — are the most soluble, whilst well ordered, highly crystalline **polymorphs** forming euhedral crystals — such as **quartz** — are the least soluble.

Silica solubility is also significantly influenced by a variety of external physico-chemical parameters. As **pH** increases above 9.5 the tetrahedral hydrate dissociates to the monovalent species

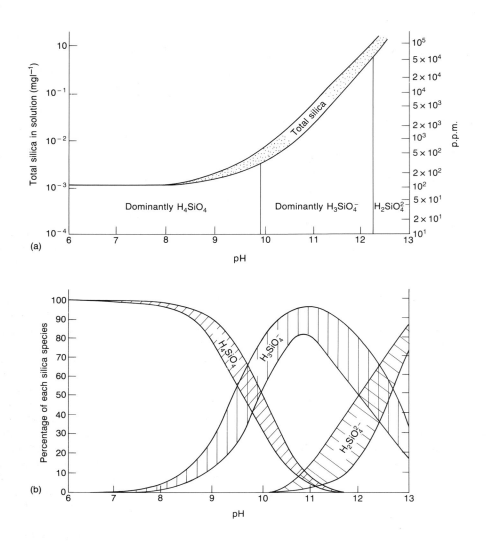

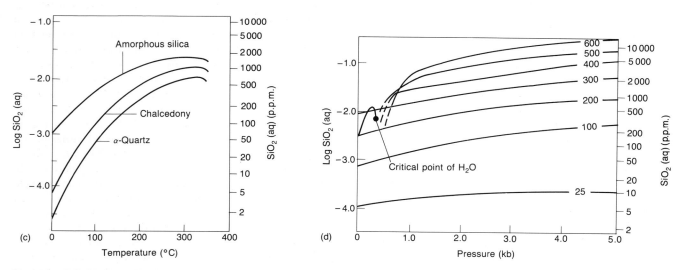

Fig. S34 (a) The increase in **silica solubility** in water with increasing pH at 25°C. (b) Relative amounts of dissolved silica species in solution in water at 25°C as a function of pH. (c) The effect of temperature on the solubility of several silica polymorphs. (d) The effect of pressure at varying temperatures on quartz solubility.

$Si_4(OH)_4^\circ = H_3SiO_4^- + H^+$

and, ultimately, at **pH** values above 12, ionizes to the divalent species

$H_3SiO_4 = H_2SiO_4^- + H^+$

Thus as solutions become more basic the solubility of silica increases markedly (Fig. S34a and b).

The solubility of all silica **polymorphs** increases exponentially with increasing temperature (Fig. S34c). Similarly, the dissociation of silicon tetrahydrate is strongly temperature dependent such that ionization takes place at progressively lower **pH** values with increasing temperature. Increasing pressure also produces an increase in silica solubility although the amount and rate of change is small compared with that produced through fluctuation in temperature (Fig. S34d).

The behaviour of **silica** in complex, high salinity aqueous solutions is less well known. In highly silica-supersaturated solutions **silica** has been documented to polymerize, forming both simple oligmers and larger molecular species with spherical diameters up to 5 nm and molecular weights in the order of 10^4 according to the general equation:

$4Si_4(OH)_4^\circ = Si_4O_6(OH)_6^{2-} + 4H_2O + 2H^+$

Larger particles that form may remain in suspension either as colloidal sols in high **pH**, low salinity solutions or as gels in low **pH**, high salinity solutions.

Other dissolved species have varying effects on the solubility of silica. Monovalent metallic cations do not apparently affect the solubility of the silicon tetrahydrate in dilute aqueous systems. Multivalent cations, however, drastically lower solubility. By contrast, some organic hydroxy acids slightly increase silica solubility whilst silicon forms soluble complexes with benzine derivatives typical of many humic substances. The presence of simple salts in solution reduces silica sol formation whilst colloid sized silica polymers can be *flocculated* in the presence of cations, hydroxyl ions or metal hydroxides.

Our present understanding of silica solubility explains well the **opal**-A → **opal** CT → **quartz** transformation during early **diagenesis**. However, it is clear that we need to learn much more regarding the solubility of silica in the deep burial environment. The source of silica and **precipitation** of large **quartz** overgrowths in **sandstones** during deep burial remains problematical. Massive supersaturation with respect to silica can be achieved in high silica solutions through any one or a combination of a drop in **pH**, temperature or pressure. Known levels of silica solubility, however, require these processes to be repeated many times over with a consequential throughput of vast numbers of supersaturated pore volumes of formation waters completely to cement pores with **quartz** overgrowths. Most concepts of **basin** evolution concede a finite volume of **pore-water** available in a sedimentary sequence orders of magnitude too small for such a mechanism to be viable. (See **burial diagenesis**, **mass solute transfer, precipitation, pressure dissolution, subsurface fluid migration**.) [SDB]

Iler, R.K. (1979) *Chemistry of Silica*. Wiley-Interscience, New York.
Williams, L.A. & Crerar, D.A. (1985) Silica diagenesis, II: General mechanisms. *Journal of Sedimentary Petrology* **55**: 312–21.

silication The process by which a rock is converted into, or replaced by, silicates; especially the conversion of carbonate rocks to **skarns**. [AME]

siliceous ooze A **pelagic** biogenic sediment composed of the skeletal remains of siliceous microfossils such as **radiolaria** or *diatoms*, e.g. **radiolarian** ooze. Siliceous ooze commonly occurs below the **calcite compensation depth**, or above it in regions of high productivity. [AESK]

silicification The **alteration** by **cementation** or **replacement** of precursor sediment by amorphous, cryptocrystalline, microcrystalline or macrocrystalline **silica** (i.e. **cherty** or **opaline silica**).

Diagenetic silica is widespread in many sedimentary rocks, either as a cement phase, or replacing a precursor material. The presence of this secondary **silica** indicates that **silica** is, to some extent, soluble in natural waters. Its mobility is dependent on the presence of soluble **silica polymorphs** — amorphous biogenic **silica** (**opal**-A) dissolves to give solutions of around 120 p.p.m. at room temperature. Siliceous **sponge** spicules and the tests of **radiolaria** and *diatoms*, which consist of **opaline silica** containing a few per cent of water, are relatively easily dissolved, particularly in the subsurface where solubility increases rapidly with the rise in temperature and pressure. The solubility of crystalline **silica** (**quartz**) is around 5% of that of **opal**, but again increases in regions of elevated temperature or **stress**. 'Pressure dissolution' of **quartz** in siliceous **sands** results in the transfer of **silica** away from highly stressed grain contacts into areas of lower **stress**. **Silica** is typically present in solution in the form of orthosilicic acid, H_4SiO_4, which dissociates into $H_3SiO_4^-$ and $H_2SiO_4^{2-}$ with increasing alkalinity.

Oversaturation results in the **precipitation** of silica, the form of the precipitate being dependent on the conditions. At low temperatures, the difficulty of precipitating an ordered phase (micro- or macrocrystalline **quartz**) results in amorphous **silica** precipitates, but at higher temperatures crystalline **quartz** is produced. Consequently, most **silica**-cemented **sandstones** that have been buried to moderate depths contain **quartz** cements.

The transformation of relatively unstable **opal**-A to **quartz** proceeds through a series of **dissolution–precipitation** reactions. An intermediate product, **opal**-CT, is formed during the transformations, and may be preserved as spherical microcrystalline aggregates of bladed crystals known as *lepispheres*. The initial transformation of **opal**-A to **opal**-CT is aided by the presence of Mg^{2+} ions.

Not surprisingly, silicification is often associated with **dolomitization**.

One of the most important examples of silicification is the formation of **cherts** in **limestones**, a process involving the **replacement** of **calcite** or **aragonite** by **silica**. The reaction can be simply represented by the following equation:

$$CaCO_3 + H_2O + CO_2 + H_4SiO_4 =$$
$$SiO_2 + Ca^{2+} + 2HCO_3^- + 2H_2O$$

and is therefore favoured by relatively acidic environments. **Chert** formed by this process generally consists of cryptocrystalline or microcrystalline **quartz**. Distinctive radiating textures of fibrous microcrystalline **quartz** (**chalcedony**) are also common. A little extra-crystalline water is often present (<1%), while crystalline impurities (clays, **calcite**, **hematite**) generally total less than 10%. Crystal size varies from less than one micron to a few tens of microns, and a wide variety of textures is visible using the **electron microscope**.

The silicification mechanism involves the production of **silica**-rich solutions by the **dissolution** or calcitization of siliceous tests. These solutions can cause the wholesale replacement of **carbonate minerals**, resulting in **chert nodules** and beds. Very finely detailed precursor **fabrics** may also be preserved, suggesting that *chertification* occurred by a contemporaneous **dissolution–precipitation** mechanism. Non-replacive **cherts** are also found — burrows may be filled with **chert**, or whole beds, composed originally of siliceous skeletal material, may form in the absence of terrigenous or calcareous sediments.

Cherts are common in sequences of basinal or **pelagic** carbonates. The **Cretaceous** Chalk of north-west Europe contains a variety of **cherts** of primary or replacive origin. Calcareous *turbidites* are also commonly silicified. The presence of relatively unstable **carbonate minerals** (e.g. **aragonite** and high-**magnesium calcite**) derived from shelf areas, and the rapid sedimentation rates, which prevent the flux of **silica**-rich solutions to the sea, combine to produce **chert** beds and **nodules**.

The heavy isotopic composition of many **cherts** indicates that they form at near-surface temperatures in marine waters. Lighter values in some examples have been used to suggest that final stabilization took place in **meteoric groundwaters**. [DMP]

The term silicification is also applied to a form of wall rock **alteration** involving an increase in the proportion of quartz or crypto-crystalline silica (i.e. cherty or opaline silica) in the altered rock. [AME]

sill A concordant tabular or sheetlike **igneous body**. Sills range from a few centimetres to hundreds of metres in thickness, although the commonest sills are *c.* 1–20 m in thickness, and are commonly intruded into sedimentary sequences. Although largely concordant, sills may show local transgressions across individual stratigraphic horizons. Sills are distinguished from **lava flows** by the absence of lava surface features and by the occurrence of fine-grained chilled margins at the upper and lower contacts. Well-known examples of sills with the features described above are the **Triassic** Palisade Sill of New York and the late **Carboniferous** Whin Sill of northern England. [RST]

sillar Informal name for weakly-consolidated, rather fine-grained **ash-flow tuff**. The consolidation may be a result of incipient welding, or of vapour-phase crystallization. [PTL]

sillimanite (Al_2SiO_5) A *nesosilicate* mineral found in metamorphosed aluminous rocks. [DJV]

sillimanite minerals The anhydrous **aluminium** silicate group **andalusite**, **kyanite** and **sillimanite** are known in industry either under this title or as the *kyanite group*. They are highly refractory minerals. The theoretical alumina content of 62.93% is never found in nature, but commercial **grades** must contain a minimum of 56% alumina and 42% **silica** with less than 1% acid soluble Fe_2O_3, 1.2% TiO_2 and 0.1% each of CaO and MgO. This material is calcined to produce a **mullite–silica** mixture which is extremely refractory, has a small coefficient of expansion and can withstand abrasion and slag **erosion**. Only a few countries produce sillimanite minerals and the most important is the RSA. [AME]

Harben, P.W. & Bates, R.L. (1984) *Geology of the Nonmetallics.* Metal Bulletin Inc., New York.

Silurian The third system of the Lower **Paleozoic**. It is believed to span a period of about 23 Ma from 435 to 412 Ma (Fig. S35). It was named by R.I. Murchison after the early British tribe, the 'Silures', who inhabited the type area, and was based on rocks in eastern Wales and the Welsh Borderland. Murchison first published the term 'Silurian' in 1835. In the same year he read a joint paper with Adam Sedgwick on the **Cambrian** and Silurian systems which was published in 1836. As initially defined, the Silurian included rocks which Sedgwick regarded as **Cambrian** and which Lapworth in 1879 included in the **Ordovician** System. Throughout the later part of the 19th century 'Silurian' was used by some geologists, including the British Geological Survey, to include **Ordovician** rocks, but was restricted by others to post-**Ordovician** sequences. Finally at the 1960 meeting of the International Geological Congress the term 'Silurian' was formally accepted for the upper system of the Lower **Paleozoic**, taking precedence over the term *Gotlandian* which had been widely used in continental Europe.

As defined now, the Silurian has its base at the bottom of the *Parakidograptus acuminatus Biozone* in the Dob's Linn section near Moffat in the Southern Uplands of Scotland.

							Europe			N.E. Siberia Mirnyy Creek	North America

Silurian system

Period	Series		British Stages		British Biozones	Ma	Europe — Wenlock Edge and Ludlow, England	Bohemia	Gotland	N.E. Siberia Mirnyy Creek	North America
	Epoch		Age								
D	Early Devonian		Gedinnian		Monograptus uniformis	412	Dittonian	Lochkovium		Mirnyy	
	Pridoli				Monograptus ultimus	414	Red Downtonian / Temeside Shales / Downton Castle	Pridoli-schichten	Sundre / Hamra / Burgsvik / Eke	Bizon	Cayugan
	Ludlow		Ludfordian		Bohemograptus		Whitcliffe				
					Saetograptus leintwardinensis		Leintwardine	Kopanina-schichten			
			Gorstian		Pristiograptus tumescens/ Saetograptus incipiens		Bringewood		Hemse		
					Lobograptus scanicus						
					Neodiversograptus nilssoni	421	Elton		Klinteberg		
Silurian (S)	Wenlock		Homerian		Monograptus ludensis		Wenlock / Tickwood			Upper Sandugan	Lockportian
					Gothograptus nassa						
					Cyrtograptus lundgreni		Coalbrookdale		Mulde / Halla		
			Sheinwoodian		Cyrtograptus ellesae						
					Cyrtograptus linnarssoni				Slite / Tofta		
					Cyrtograptus rigidus				Högklint		
					Monograptus riccartonensis				Upper Visby		
					Cyrtograptus murchisoni		Buildwas		Lower Visby		
					Cyrtograptus centrifugus	428					
	Llandovery		Telychian		Monoclimacis crenulata		Woolhope				Tonawandan
					Monoclimacis griestoniensis		Wych			Anika	
					Monograptus crispus			Liten-schichten			Ontarian
					Monograptus turriculatus						
			Aeronian		Monograptus sedgwickii		Cowleigh Park				
					Monograptus convolutus						
					Coronograptus gregarius: argentus / magnus / triangulatus					Chalmak.	Alexandrian
			Rhuddanian		Coronograptus cyphus: cyphus / acinaces						
					Cystograptus vesiculosus = atavus						
					Parakidograptus acuminatus						
					Glyptograptus persculptus	435				Tirekhtyakh.	
O	Ashgill		Hirnantian								

Fig. S35 Stratigraphic table showing the division of the **Silurian** system into series, stages and biozones in Britain and elsewhere.

The top of the System is taken at the base of the *Monograptus uniformis Biozone* in the boundary stratotype section in the Barrandian area of the Prague Basin in Czechoslovakia.

Murchison recognized three major subdivisions of his Upper Silurian: the *Llandovery, Wenlock* and *Ludlow*. These divisions now have series status and are formally related to stratotypes in Wales and the Welsh Borderlands. A fourth series, the *Downtonian*, came into use in the 1930s for fluviatile and marginal marine rocks above the *Ludlow* in the type area. The *Downtonian* rocks have restricted marine or freshwater faunas including **fish**, which are not easy to use for correlation. Consequently the new *Pridoli Series* based on a section in the Barrandian area of Czechoslovakia was approved by the Subcommission on Silurian Stratigraphy in 1984 (Fig. S35).

Silurian rocks are most effectively correlated using **graptolites** which allow the system to be divided into about 32 *biozones*. **Conodonts** have been widely used to correlate rocks of carbonate facies and **acritarchs** and

Chitinozoa can be useful in mudstone sequences. **Brachiopods** and to a lesser extent *trilobites* have been effectively used to correlate within regions and even between continents because **brachiopods** are unusually cosmopolitan in the Silurian until the *Pridoli* when there was an increasing provinciality of faunas.

The associations of suspension-feeding animals, such as **brachiopods**, **Bryozoa**, **corals** and *crinoids*, and pelagic faunas such as **graptolites**, suffered a major phase of **extinction** immediately prior to the Silurian, but most of the general faunal associations were re-established during the early Silurian.

Silurian shallow marine benthic biotas commonly include **brachiopods**, **Bryozoa**, **corals**, *crinoids, trilobites, ostracods, bivalves* and **gastropods**. **Stromatoporoids** and **calcareous algae** are common, together with rugose and tabulate **corals** in *reefs*. In more offshore sediments, *trilobites, nautiloids, bivalves, hyolithids* and **graptolites** are typical elements of the fauna. **Fish** started to diversify in the late Silurian. *Cephalaspids, anaspids* and heterostracan

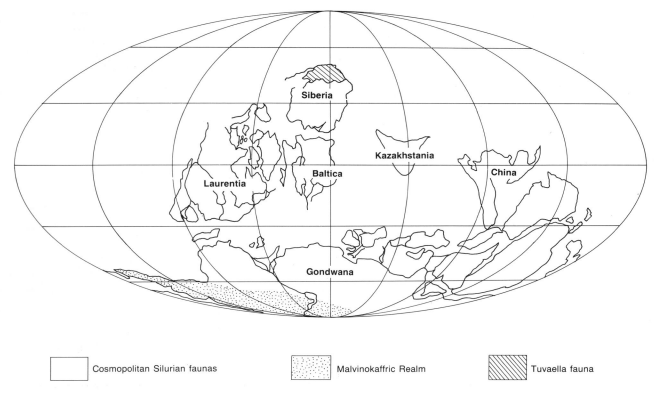

Cosmopolitan Silurian faunas   Malvinokaffric Realm   Tuvaella fauna

Fig. S36 Possible reconstruction of the continents in the middle **Silurian** (Wenlock). (After Scotese *et al.*, 1979.)

ostracoderms are known from the lower or middle parts of the Silurian, but become more diverse from the *Ludlow* onwards. Thelodont scales are known from marine and brackish Silurian sediments.

The presence of **land plants** in the earlier part of the Silurian is suggested by the occurrence of *spores*. Vascular plant megafossils are known first from the *Wenlock* and the relatively diverse trilete *spore* assemblages of the *Wenlock* and post-Wenlock probably have a vascular plant source.

The work of Ziegler (1965) established the presence of depth-related 'communities' in clastic sequences formed on marine shelves in the upper *Llandovery* (the *Lingula, Eocoelia, Pentamerus, Costistricklandia* and *Clorinda* communities in offshore sequence) and similar types of benthic assemblages have been recognized in other series of the Silurian. Benthic 'communities' in the carbonate sequence of the Wenlock Limestone were established by Hurst (1975).

A possible distribution of the continents in the middle Silurian is shown in Fig. S36. **Laurentia**, *Baltica*, part of Siberia, *Khazakhstania* and China all lay within the tropics, while **Gondwanaland** lay across the southern pole but probably extended to the southern tropics where Australia might have been positioned. A large ocean occupied the northern arctic region.

Sea-level was generally high during the Silurian and most continents were covered by shallow epeiric seas. The generally cosmopolitan benthic faunas of the Silurian appear to reflect the widespread distribution of animals in the tropical or warm temperate epeiric seas. There are, however, two distinctly different types of Silurian fauna. One, the *Clarkeia* fauna, is found in the *Malvinokaffric Province* which embraces South America and formerly adjacent parts of South and North Africa and the second, *Tuvaella* fauna, occurs in Mongolia and adjacent parts of Siberia. Both are low diversity faunas and are thought to have occupied cool temperate climatic regions, with the *Clarkeia* fauna in the southern hemisphere and the *Tuvaella* fauna in the northern one.

Most of the continents which lay in tropical regions have Silurian sequences dominated by carbonates, often with associated *reefs* from later *Llandovery* times onwards. *Glacial* deposits of probable Silurian age are known from Brazil, reflecting *ice caps* on **Gondwanaland**, but these appear to have been smaller than in the Upper **Ordovician**.

Major clastic sequences associated with **subduction zones** are found in Mongolia, Scotland, Kazakhstan and Australia.

Sea-level rose sharply immediately prior to the Silurian as the *ice caps* of **Gondwanaland** were reduced in size. There was a general rise in sea-level throughout the *Llandovery* but with a sharper rise followed by a fall in the late *Llandovery*. In addition there appear to have been many smaller changes of relative sea-level on a time-scale of 1–2 Ma throughout the Silurian, though it is not clear

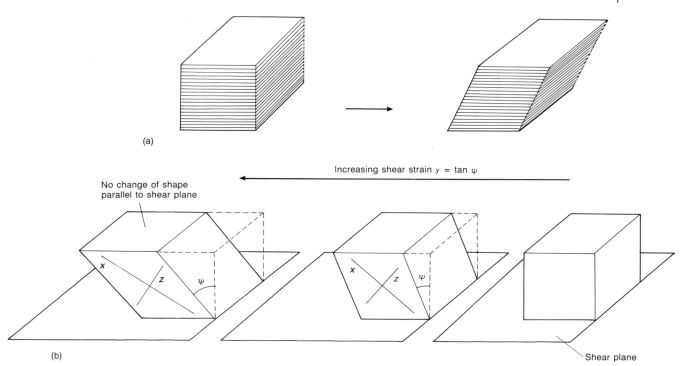

Fig. S37 (a) **Simple shear** on a deck of cards by sliding each card past its neighbour in a constant direction. There is no deformation in the plane of an individual card, but any lines or angles drawn on the sides of the card block are distorted. The shear plane lies parallel to the plane of individual cards in this case. (b) Simple shear in three dimensions acting on a unit cube showing the angular shear strain with increasing finite strain. Note the increasing ratio of principal strain axes *x* and *z*, and the decreasing angle between *x* and the shear plane.

which of these have a local **tectonic** control and which are **eustatic**. There was lowering of sea-level in Britain, but not generally in the *Ludlow*, but possibly a more widespread sea-level fall in the *Pridoli*. [PJB]

Boucot, A. (1975) *Evolution and Extinction Rate Controls*. Developments in Palaeontology and Stratigraphy. Elsevier, Amsterdam.

Gray, J. (1985) Ordovician–Silurian land plants: the interdependence of ecology and evolution. In: Bassett, M.G. & Lawson, J.D. (eds) *Autecology of Silurian Organisms*, pp. 281–95. *Special Paper on Palaeontology* **32**.

Holland, C.H. (1984) Series and stages of the Silurian System. *Episodes* **8**: 101–3.

Hurst, J.M. (1975) Wenlock carbonate, level bottom, brachiopod-dominated communities from Wales and the Welsh Borderland. *Palaeogeography, Palaeoclimatology, Palaeoecology* **17**: 227–55.

Scotese, C.R., Bambach, R.K., Barton, C., Van der Voo, R. & Ziegler, A.M. Paleozoic base maps. *Journal of Geology* **87**: 217–77.

Ziegler, A.M. (1965) Silurian marine communities and their environmental significance. *Nature* **207**: 270–2.

Ziegler, A.M., Hanson, K.S., Johnson, M.E. *et al.* (1977) Silurian continental distributions, paleogeography, climatology and biogeography. *Tectonophysics* **40**: 13–51.

silver (Ag) A rare native metal mineral. (See **native elements**.) Though too soft for functional tools, silver has been valued throughout history for its appearance, and the rarely found native silver was collected prior to smelting of **ores**. Some excellent examples of silverwork are seen in Celtic art, including the famous Gunderstrup Cauldron (Denmark) with figures in relief. Later, in the classical world of the Mediterranean, silver coins were an important part of the economy, while the major source of silver was the mines of Laurion near Athens, Greece. [DJV/OWT]

simple shear A shape change (*distortional* **strain**) in which all particles in a body move in parallel lines, and the amount of movement depends on the distance of each particle from a given plane, in which there are no **displacements**. The relationship of **displacement** to distance from this plane can be linear, or continuously varying, giving *homogeneous simple shear*, or abruptly changing, resulting in *heterogeneous simple shear*. A plane parallel to the particle displacements is known as the **shear plane**. This **deformation** can be simply modelled using shapes drawn on the sides of a pack of cards; the cards, parallel to the **shear plane**, slide past each other. The shape of the sides of the pack of cards changes, but each card is not itself deformed. So, in simple shear, there is no **strain** parallel to the **shear plane**. As the **displacement** and so the **strain** increases, the **principal strain** axes rotate so that *x*, the maximum **principal strain** axis, lies closer to the **shear plane**. Simple shear is thus a non-**coaxial deformation**. Many **shear zones** show **deformation** approximating to simple shear, or to combinations of simple shear with other ideal **strain** distributions such as **pure shear** (Fig. S37). (See **shear strain**.) [SB]

Fig. S38 Site investigation. Part of the core from a borehole.

singing sand (*booming dune*) When in motion certain **dune sands** generate clearly audible sounds that have variously been reported as booming, roaring, squeaking, singing and musical (Haff, 1986). A unique combination of grain properties appears to be responsible (Van Rooyen & Verster, 1983), including amongst others a high *sorting* value, uniform grain size and a high degree of *roundness* of grains. [ASG]

Haff, P.K. (1986) Booming dunes. *American Scientist* **74**: 376–81.
Van Rooyen, T.H. & Verster, E. (1983) Granulometric properties of the roaring sands in the south-eastern Kalahari. *Journal of Arid Environments* **6**: 215–22.

sinhalite (Mg(Al, Fe)BO$_4$) A rare **gem** mineral. [DJV]

sinistral Sense of relative movement across a boundary or zone, in which the opposite side to the observer moves to the left. The term is used for the sense of **strike-slip** displacement on **faults** and **shear zones**. [RGP]

sinkhole An American term for a simple closed depression found in **karst** terrain. These features are also referred to as **dolines** or **swallow holes**. [HAV]

sinuosity A measure of the degree to which a channel winds around the shortest distance path down a valley, and is thus usually defined as the quotient of channel length to valley length. Where the valley itself meanders, more complex indices may be needed. Adjustment of sinuosity is one method by which a river system can respond to changes in the external controls of **discharge** and sediment load. [TPB]

site investigation In order to ensure that engineering design and construction work are carried out in a safe and economic manner and that engineering structures will adequately serve their designed function, it is necessary to investigate the site to be affected by the proposals.

Site investigation provides the information upon which decisions regarding the suitability of a proposed site for engineering works can be based and it is also the primary means by which the ground conditions and properties of geomaterials are taken into account during civil engineering design and construction work. The work includes a consideration of the general character of the site as well as the nature of both the *in situ* ground and any rocks and *soils* utilized as construction materials; for example as decorative or structural stone, **aggregate**, fill or other purpose. Failure to take full account of the ground conditions beneath a foundation, for example, could lead to the structure's suffering cracking due to excessive settlement. More serious ground movement could result in structural failure, rendering the structure unfit for its intended purpose. Thus, overestimating the support capabilities of the ground may lead to increased costs through the need for remedial works and it could even cause loss of life. On the other hand, an over-conservative design could give rise to increased construction costs. Site investigations are also carried out to provide data about the safety of existing structures, for the design of extensions and the analysis of failures. Other applications include data acquisition relating to land use planning, the disposal by tipping or landfill of solid and liquid wastes (see **sanitary landfill**), particularly hazardous ones, and the exploitation of **groundwater**, **hydrothermal** and *hydrocarbon* **resources**. Increased awareness of the need for the care of the environment by the use of engineering solutions which harmonize with their surroundings should heighten the need for such studies.

Clayton *et al.* (1982) give a comprehensive account of the methods and procedures of site investigation. They also review the historical development of the activity pointing out that, whereas in early times the nature of the ground may have been determined, more recent developments have enabled the behaviour of earth materials to be quantified. Significant developments were achieved in *soil* sampling and testing techniques in the USA and Europe during the 1920s and 1930s. The first draft *Civil Engineering Code of Practice for Site Investigations* was produced in 1949. This was followed by an actual code of practice in 1950 and the *British Standard Code of Practice CP 2001* in 1957. The latter document has now been replaced by *British Standard BS 5930* (Anon, 1981a) which forms the basis for almost all site investigation work in the UK. Similar codes exist for other countries (see *American Society for Testing Materials Annual Year Book*, Vol. 04.08, Anon, 1975; Anon, 1981b).

British Standard BS 5930 (Anon, 1981a) identifies the following main objectives for site investigation:
1 to assess the general suitability of the site and its environs,
2 to enable a safe and economic design to be prepared for both temporary and permanent works,
3 to plan the best method of construction, to foresee any diffulties or delays due to the ground or other local conditions, to explore sources of indigenous construction materials and to plan for the disposal of surplus materials,
4 to determine and assess the effect on the works, adjacent works and the environment in general of any changes which may arise in the ground and environmental conditions either naturally or as a result of the works and
5 where alternatives exist, to advise on the relative merit of different sites or different parts of the same site.

The data to be acquired and the methods employed are a function of the situation and type of ground, the complexity of the area, particularly with respect to its geology, and the engineering proposals. Relatively minor engineering works do not warrant a large-scale investigation but, on the other hand, a very extensive exploration is usually required for the redevelopment of sites such as derelict industrial areas. Although most studies relate specifically to the site of proposed engineering works, some are more general in nature. The latter investigations often form part of feasibility or planning studies and thus may result, for example, in maps showing areas of ground liable to become unstable or to be threatened by a particular hazard. In the case of specific engineering proposals, the information obtained must enable the design engineer to produce a safe and economic design for both the construction process and the structure itself. Care needs to be exercised to determine whether additional expenditure on investigation could result in large savings in construction costs. Most site investigations are divided into at least three stages in which data from early stages influence the work done during later ones.

1 Preliminary or desk study. In this all relevant existing information is collected together and reviewed. In practice, data are obtained from a great variety of sources, including all available published and unpublished topographic, land classification, *soil* survey and **geological maps**, memoirs and reports. Information regarding past land uses, particularly if this involves any mining activity, is germane, and unpublished data from previous site investigations and other sources, *aerial photographs*, terrain evaluation and various forms of **remote sensing** survey might also be considered. Walk-over surveys or more detailed **geomorphological** mapping of some areas might also be carried out. At this stage only a general outline of the ground conditions is required to provide for the assessment of the feasibility of proposals, the elimination of alternative construction and siting options and the planning of later stages. However, depending on the amount and type of information forthcoming, some preliminary ground exploration involving, for instance, **geological mapping**, the drilling of boreholes and geotechnical testing may be commissioned.
2 Main or design investigation. The objective at this stage of the investigation is to provide an appropriate type and amount of information for the design and execution of the engineering proposals. It involves verifying and expanding the data obtained during earlier stages. The investigation may involve obtaining sufficient data such

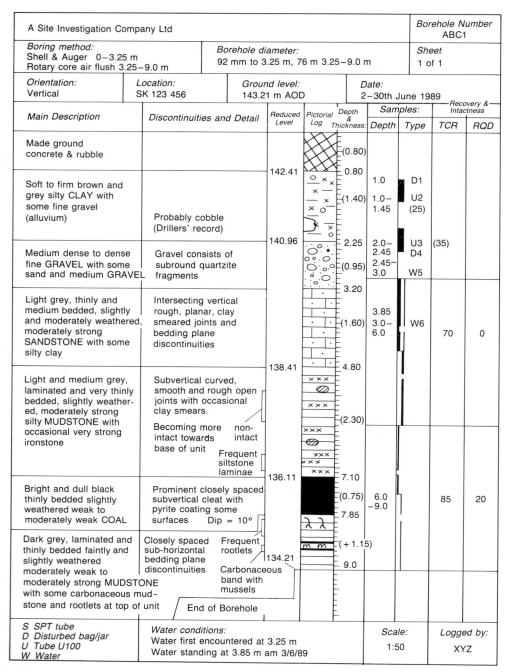

Fig. S39 Site investigation. A typical boitehole log.

that the spatial distribution and engineering behaviour of all the materials to be affected by the engineering proposals can be determined. In addition, the **groundwater** conditions are investigated. The distribution of the different rock and *soil* units present is determined by the use of standard **geological** correlation and **mapping** techniques using natural and artificial exposures, including trial pits and cored boreholes. **Geophysical** studies aimed at determining the distribution and/or engineering properties of

particular units within the area might also be carried out. Additionally, information about the location and character of particular disadvantageous conditions such as **faults**, **karst** features or abandoned mine workings is sought.

The need for effective communication of geoscience data between geologists and engineers and its application in engineering work has been greatly assisted by the standardization and quantification of geological descriptions. Accordingly, in addition to colour, grain size and lithology, *BS 5930* (Anon, 1981a) also provides definitions for the description of **strength**, **weathering** grade and the spacing and other features of discontinuities. Part of the **core** for a borehole together with a typical **log** are shown in Figs S38 and S39. The data recorded include the quantity of core recovered and a quantitative assessment of the intactness of the material (*Rock Quality Designation — RQD*).

Decisions regarding the load-carrying capacity, settlement characteristics and other engineering properties of *soils* and rocks are usually based on the results of *in situ* or laboratory geotechnical testing of the material (see Head, 1982). In the case of *soils*, particularly frequently performed tests include the measurement of *Atterberg limits* (liquid and plastic limits), particle size distribution, compaction characteristics, **shear strength**, compressibility and **permeability**. Depending on the application, the latter three tests as well as the determination of their resistance to **weathering** action may be performed on rocks. These tests provide numerical values for various rock and *soil* property parameters that are used in engineering design work.

3 Investigation during construction. This is carried out in order to confirm that the geological conditions are in accordance with those assumed for design purposes. Where significant differences do occur, observations during construction will be used to arrive at modified designs or changes to the construction procedures. In some cases designs may be finalized only when the actual ground conditions can be ascertained during construction. An engineering geologist or geotechnical engineer might also advise on the use of materials in construction and the stability and design of any temporary structures, slopes and excavations at this stage.

The results of both site-specific and general investigations are usually presented in the form of a report. All such reports include the factual data obtained, while some also provide an interpretation of the data in terms of the engineering proposals. The factual data would include plans indicating the location of boreholes together with **logs** describing the geological and **engineering geological** conditions. An interpretation of these data could consist of a series of maps and cross-sections indicating the surface and subsurface conditions between the boreholes. It might also predict the performance of the materials in terms of the engineering proposals and make design and construction recommendations. (See **engineering geology**.) [JCC]

Anon (1975) *Recommendations on Site Investigation Techniques*. Commission on recommendations on site investigation techniques. International Society for Rock Mechanics.

Anon (1981a) British Standard BS 5930:1981 *Code of Practice for Site Investigations*. British Standards Institution, London.

Anon (1981b) Report of the International Association of Engineering Geology Commission on Site Investigation. *Bulletin of International Association of Engineering Geologists* 24: 185–226.

Clayton, C.R.I., Symonds, N.E. & Matthews, M.C. (1982) *Site Investigation — A Handbook for Engineers*. Granada, London.

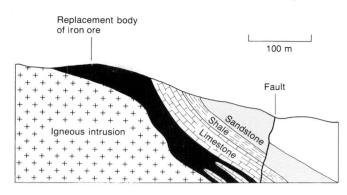

Fig. S40 Skarn deposit at Iron Springs, Utah. (After Evans, 1993.)

Head, K.H. (1982) *Manual of Soil Laboratory Testing* Vols 1, 2 and 3. Pentech Press, London.

skarn/pyrometasomatic deposit Replacement **ore** deposit generally formed at high temperatures in metamorphosed carbonate-rich sediments at contacts with medium to large igneous intrusions. The **orebodies** are characterized by the development of calc-silicate minerals such as **diopside**, **wollastonite**, **andradite garnet** and **actinolite**. These deposits are extremely irregular in shape (Fig. S40); tongues of **ore** may project along any available planar structure — **bedding**, **joints**, **faults**, etc., and the distribution within the *contact aureole* is often apparently capricious. Structural changes may cause abrupt termination of the **orebodies**. The principal materials produced from pyrometasomatic deposits are: **iron**, **copper**, **tungsten**, **graphite**, **zinc**, **lead**, **molybdenum**, **tin** and **uranium**. **Orebodies** developed in metasediments are often referred to as *exoskarns* and those where **replacement** of the intrusion has occurred are termed *endoskarns*. [AME]

Einudi, M.T., Meinert, L.D. & Newberry, R.J. (1981) Skarns deposits. *Economic Geology 75th Anniversary Volume*: 317–91.

Evans, A.M. (1993) *Ore Geology and Industrial Minerals: An Introduction* (3rd edn). Blackwell Scientific Publications, Oxford.

skutterudite ($(Co,Ni)As_3$) An **ore** mineral of **cobalt** and **nickel**. [DJV]

slab failure A common form of **weathering** on hard-rock cliffs. Cliff **erosion** releases the lateral **confining pressure** with the effect that tension **joints** open up. Over time the **joint** gradually opens up until the **tensile strength** of the rock at the edges of the slab is exceeded and the slab falls. Slab failures often occur on the lower part of a cliff leaving an overhang; this may well become the site for further slab failure. Slab failure may be an extremely slow process in the most massive of rocks, so that many rock walls in deglaciated valleys still show no sign of slab failure. Rarely, the gradual process of **joint** opening and **failure** has been documented, e.g. the fall of 'Threatening Rock' on to an abandoned village site in southern USA (Schumm

& Chorley, 1964). On unbuttressed slopes (where the rock **dips** towards the cliff) natural **joints** may provide sites of weakness for slab failure to occur. [TPB]

Schumm, S.A. & Chorley, R.J. (1964) The fall of Threatening Rock. *American Journal of Science* **262**: 1041–54.

slab pull Mechanism whereby the gravitational pull of a downgoing **lithosphere** slab exerts a lateral **extensional** force on the *plate* descending in a **subduction zone**. The force arises from the negative buoyancy of the denser oceanic **lithosphere** of the slab, compared with the surrounding **asthenosphere**. Bott & Kusznir (1984) estimate a magnitude of 0–50 MPa for the **extensional stress** resulting from slab pull. The slab pull force is potentially the largest of the renewable *plate* boundary forces, but is opposed by variable resistance forces that may apparently be large enough to overcome the slab pull **extension** and place the slab in a state of down-dip compression throughout (e.g. see Isacks & Molnar, 1969). This effect may be typical of long slabs of old **lithosphere** that have encountered increasing resistance on descending through the **asthenosphere** into the **mesosphere**.

The subduction process may also create an extensional **stress** in the upper plate through the **subduction suction** force. (See also **ridge push, forces acting on plates**.) [RGP]

Bott, M.P. & Kusznir, N.J. (1984) Origins of tectonic stress in the lithosphere. *Tectonophysics* **105**: 1–14.
Isacks, B. & Molnar, P. (1969) Mantle earthquake mechanisms and the sinking of the lithosphere. *Nature* **223**: 1121.

slaking The disintegration of a loosely consolidated material on the introduction of water or exposure to the atmosphere. Clays and **shales** are especially prone to this form of **failure**, especially in the presence of saline waters. [ASG]

Badger, C.W., Cummings, A.D. & Whitmore, R.L. (1956) The disintegration of shales in water. *Journal of the Institute of Fuel* **29**: 417–23.
Franklin, J.A. & Chandra, R. (1972) The slake-durability test. *International Journal of Rock Mechanics and Mining Science* **9**: 325–41.

slate A fine-grained low-grade or regionally-metamorphosed **mudrock** characterized by a well-developed true penetrative or *slaty cleavage*. Such **cleavage** is a type of **foliation** involving well-developed parallel alignment of submicroscopic *phyllosilicate* **minerals** such that the rock may be split into platy sheets which may be used as building (particularly roofing) materials.

A *slate belt* is a linear zone within an **orogenic belt** in which the pelitic rocks are dominantly in the form of slates. Such belts are characterized by the lowest grades (*prehnite–pumpellyite* and *greenschist*) of regional metamorphism, but exhibit moderate to high **strains**. The rocks possess a steeply dipping **cleavage** subparallel to the belt axis. For example, in the British Caledonides, the Southern

Uplands of Scotland, Wales, and the Lake District all represent *slate belts*.

Slate was used in antiquity mainly as a minor **building stone** and was used for roofing from the Roman period onwards. Smaller objects of slate include Christian monuments (crosses, graves) from Scotland, occasional British stone **axes**, and 3rd millenium BC Egyptian statues and carvings. [RGP/RST/OWT]

slickenside A smooth or polished **fault** surface. Both natural and experimental **fault** surfaces often have a shiny appearance which may be caused by a number of processes including polishing during **fault** movement, and growth of crystal fibres or cement; the exact mechanisms are uncertain in many cases. These surfaces are slickensides following the original usage; they commonly have a **lineation** referred to as a *slickenside lineation*, or *slickenline*, as a separate feature; this has also incorrectly been referred to as a slickenside. The **lineations** can be shown to be parallel to the direction of **displacement** and occur on many scales from millimetres to metres in length. They often terminate abruptly at one end, in a *slickenside step*, which is a short, narrow offset of a slickensided **fault** surface, usually approximately perpendicular to a **lineation** on the **fault** surface. The step therefore gives an asymmetry to the **lineation**, in profile along its length. The overriding side of the step is congruous if it **faces** in the sense of **displacement**, or incongruous if opposed to the motion of the overriding *fault block*. The recognition that some **lineations** may be incongruous means that they may not be generally used to determine the **shear** sense of the **fault**, but only the **displacement** direction, unless the exact mechanism is known. It has also been proposed that the length of **lineations** may be equal to the **displacement** during which they were formed, but this again depends on a knowledge of the mechanism of formation. There are at least six different mechanisms:

1 *Asperity ploughing*. The **abrasion** of the **fault** surface by a hard asperity leaves an incongrous **lineation**, which has been referred to as a *scratch, groove, tool track* or *gutter*.
2 *Debris streaking*. Abraded products of an asperity accumulate on either side in the direction of displacement; they may be congruous or incongruous.
3 *Erosional sheltering*. Debris is deposited in the direction of slip behind an asperity, creating an incongruous 'tail'.
4 *Fibre growth*. Elongate crystals grow on a **fault** surface in the movement direction, which is congruous.
5 *Slickolite formation*. Solution of material oblique to the *fault plane* leaves oblique, incongruous **stylolites** in the direction of compression; these are also referred to as *spikes*.
6 *Nesting lineations*. Some **lineations** are perfectly matched across the *fault plane*; they do not usually have steps. The reason for their formation is unknown, but it is not one of the five above. Larger linear features on **fault** planes have been described as *corrugations*. [TGB]

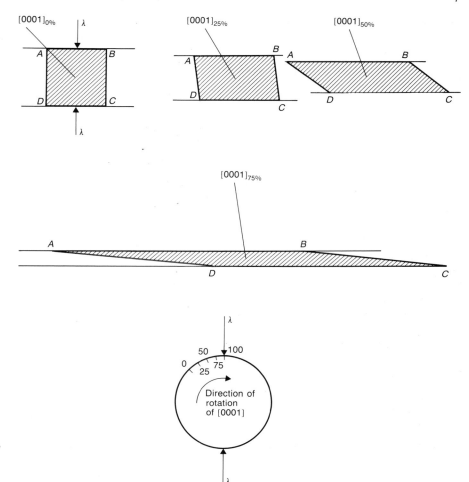

Fig. S41 The development of preferred orientation by crystallographic **slip**: progressive shear parallel to *AB* with shortening normal to λ, causes clockwise rotation of quartz c-axes [0001] towards λ by means of slip along the 0001 planes. (After Hobbs *et al.*, 1976.)

Fleuty, M.J. (1975) Slickensides and slickenlines. *Geological Magazine* **112**: 319–22.

Hancock, P.L. & Barka, A.A. (1987) Kinematic indicators on active normal faults. *Journal of Structural Geology* **9**: 573–84.

Means, W.D. (1987) A newly recognised type of slickenside striation. *Journal of Structural Geology* **9**: 585–90.

slide (1) Any downslope **mass movement** involving the translation of sediment where the motive force includes **gravity**. **Displacements** occur along recognized **shear** surfaces where the ruptured mass moves essentially as discrete units. Any lithology may be affected, and laminae and beds may be folded, rolled, contorted and/or rotated during translations. *Slide folds* are believed to develop preferentially with axial surfaces dipping upslope and **vergence** directions downslope; however, considerable movement may lead to isolated **fold** units rotating such that *fold axes* are aligned at high angles to the **strike** of a slope. Sediment slides result from many different processes that cause slope instability and **failure**. (See also **olistolith, wet-sediment deformation, slump**.)

(2) Term originally used by Bailey (1910) for a **fault** believed to have developed during major *recumbent folding*. Slides were divided into *thrust* and *lag* types depending on whether stratigraphy was repeated or omitted. In practice, it has proved difficult in metamorphic terrains to distinguish between **faults** developed prior to **folding** and those developed synchronously, particularly since later movements frequently localize on earlier **displacement** surfaces. The term is not widely used except in the *Caledonian* terrains of Britain and Ireland.[RGP/KTP]

Bailey, E.B. (1910) Recumbent folds in the schists of the Scottish Highlands. *Quarterly Journal of the Geological Society, London* **69**: 586–620.

slip Process of **displacement** within individual crystals during **plastic deformation** (Fig. S41). The slip process results in a crystallographic **preferred orientation** that is geometrically related to the *strain axes*. **Displacement** takes place parallel to a '*slip direction*' within a '*slip plane*', both of which represent crystallographic planes of weakness. For example, in **quartz**, the slip direction may be parallel to the *c*, *a*, or (*c* + *a*) axes. (See **deformation mechanism**.) [RGP]

Hobbs, B.E., Means, W.D. & Williams, P.F. (1976) *An Outline of Structural Geology*. John Wiley & Sons, New York.

slip clay A **kaolin**-rich clay used for glazes. [AME]

slip system The **shear plane** and *shear direction* within the plane of **simple shear**, including all symmetrically equivalent combinations. A slip system is independent if it produces a **strain** that cannot be the result of a combination of **strains** on other slip systems. A general **strain** at constant volume (excluding rotation) requires the simultaneous operation of five independent slip systems; this is known as *Von Mises' criterion*, and is a special case of the more general rule that the number of independent slip systems required to accommodate a **strain** is equal to the number of independent components of the *deformation tensor* specified.

This theory was first developed for *glide* of dislocations in *crystal plastic deformation* in which slip systems are specified by crystallographic planes and directions. For example, **quartz** at temperatures of a few hundred degrees commonly deforms by slip on the basal plane (0001) in one of the three *a* directions ($< a >$): the slip system is given as $a > (0001)$. The slip systems that can operate in a crystal depend primarily on the temperature. In **quartz**, for example, the *prism planes* and *rhomb planes* can also act as slip systems. Different directions of slip take place on the same planes, e.g. slip parallel to the *a* axes or the *c* axis can occur on **quartz** prism planes. Only a limited number of the many possible slip systems are used during **deformation** of a **mineral**. The activity of each slip direction on each slip plane depends on the orientation, and on the characteristic *critical resolved shear stress (Schmid Factor)* for each system. Thus under different **stress** fields, different slip systems are active. Understanding of **mineral** slip systems is important for modelling and interpreting crystallographic **preferred orientations**.

Recently the theory has been extended to encompass **deformation** on multiple sets of **faults**. Each **fault** can be considered as two independent slip systems, so that the number of **faults** necessary for accommodating specified states of **strain** can be evaluated; for example a general, constant volume, **strain** can be accommodated by four **faults**. [SB/TGB]

Nicholas, A. & Poirier, J.P. (1976) *Crystalline Plasticity and Solid State Flow in Metamorphic Rocks*. Wiley–Interscience, London.
Poirier, J.P. (1986) *Creep of Crystals*. Cambridge University Press, Cambridge.
Reches, Z. (1978) Analysis of faulting in a three-dimensional strain field. *Tectonophysics* **47**: 109–29.

slip-face Common feature on the **lee slope** of **dunes** where **sand** accumulates at the *angle of repose* of **sand**, normally 30° to 34°. [RDS]

slip-off slope A low-gradient slope on the inside of a meander bend. It forms part of the characteristic asymmetrical shape of the channel cross-section at the apex of the meander bend. Deposition of sediment on the slip-off slope forms part of the mechanism for meander migration. [TPB]

slope apron The region between the margins and floors of **basins**. In ocean basins, the slope apron comprises the continental slope and rise around *continental margins*, but slope aprons also flank **seamounts**, *oceanic ridges* and plateaus. Slope aprons range from less than 1 km to more than 200 km in width, with gentle gradients that seldom exceed about 10°, and commonly vary between 2° and 5°. These slopes may be **erosional** or entirely depositional, smooth to rugged, laterally homogeneous or extremely heterogeneous, with canyons and *gullies* eroded into the slope. The slopes around the margins of ocean basins are major centres for sediment accumulation as well as areas of considerable sediment instability, from which *mass wastage* may initiate various **sediment gravity flows**. Slope modification by contour currents to produce **contourite** drifts may be important along ocean margin slope aprons. Faulted slope aprons tend to be the site of submarine **fault** scarps, along which rubble, debris and other relatively coarse-grained sediments may preferentially accumulate. Carbonate slope aprons, either faulted or smooth, typically develop on the basinward side of major shelf carbonate/ *reef* accumulations. [KTP]

slough channel A channel which may develop in a **braided stream** in which there is relatively little flow under most conditions and deposition is dominated by settling out of fine suspended material. This channel may be one which is in the process of being abandoned or may develop into a more active channel. Slough channels may become courses of flow during periods of high channel **discharge**. [JA]

slump A type of sediment **slide** where the downward slipping mass of rock or unconsolidated material of any size moves as a unit or several related units, usually with backward rotation on an approximately horizontal axis parallel to the slope from which it descends. Also referred to as a *rotational slide*. Slumps are generally internally highly deformed, whereas **slides** contain less internal **strain**. (See **slide**, **wet-sediment deformation**.) [KTP]

small circle Any circle on the Earth's surface that does not correspond to a circumference of the Earth, e.g. all latitudes (except the Equator) are small circles. (See also **great circle**.) [DHT]

smaltite A mineral name synonymous with **skutterudite**. [DJV]

smectite clays and attapulgite A group of clays made up mainly of **montmorillonite** or attapulgite (*palygorskite*). There are numerous variants and relatives and the term most generally applied to the whole group is **bentonite**. Sodium **bentonites** have the important property of swelling in water. Finely ground swelling **bentonite** will increase the **viscosity**, suspending power and *thixotropy* of water and the major use is in oil drilling muds. Attapulgite and **sepiolite** can also be used but are inferior to swelling **bentonite**. **Bentonite** is also used as a bonding agent in the pelletizing of *iron ores*, in moulding sands, as an absorbent (**Fuller's earth**) and has many other industrial uses. Many countries produce these industrial clays but the USA is pre-eminent. [AME]

Harben, P.W. & Bates, R.L. (1984) *Geology of the Nonmetallics*. Metal Bulletin Inc., New York.

smithsonite ($ZnCO_3$) A **zinc ore** mineral of supergene origin. (See **carbonate minerals**.) [DJV]

smoky quartz A brown to black variety of **quartz**. (See **silica minerals**.) [DJV]

soapstone A name for compact, massive **talc**. [DJV]

sodalite ($Na_8(AlSiO_4)_6Cl_2$) A comparatively rare rock-forming mineral of the feldspathoid group. (See **feldspathoid minerals**.) [DJV]

soil conservation The protection of the *soil* from chemical and physical degradation, and preservation of its potential productivity. Soil conservation is practised mainly in agricultural areas where **accelerated erosion** and *soil* deterioration have resulted from disruption of the natural vegetation cover. The methods of conservation generally involve modification of agricultural practices such as cultivation techniques and cropping procedures, but broad restrictions on the type of land brought into production may also be applied.

Soil erosion usually occurs through the action of wind on bare *soil*, or *rainsplash* and **runoff** on the *soil* surface, and by *soil creep*. The deleterious effects of these processes can be minimized in different ways. **Deflation** of topsoil can be reduced by planting tree-belts and hedgerows which increase surface roughness over the landscape thereby decreasing the potential for airflow to mobilize *soil* particles. However, wind **erosion** may be reduced more effectively by minimizing the length of time that the bare *soil* is exposed — especially at the times of year which are especially dry and windy. For example, in parts of the Midwest of the United States, spring ploughing has been shown to cause an increase in **soil erosion** by wind even in years when the amount of annual rainfall is close to the long-term mean (Chagnon, 1983). Similarly, the amount of **soil erosion** by *rainsplash* and surface **runoff** increases when the *soil* surface is devoid of vegetation. In areas of markedly seasonal rainfall, it is sometimes advisable to defer ploughing and sowing of crops so that the erosive effect of the first rains on the dry *soil* surface is less pronounced.

In areas where the amount of *soil* production by natural processes equals the amount of *soil* lost through **erosion**, serious problems may not occur. Elsewhere, however, there is a gradual loss in fertility and agricultural productivity which may be irreversible in the short term. In many parts of Africa, **soil erosion** by surface **runoff** is tackled by employing strip cultivation on hillslopes. By leaving narrow, unploughed strips and constructing low embankments parallel to the contours, **runoff** from the intervening cultivated plots does not generate sufficient energy to carry significant quantities of *soil* from the hillside. When such techniques are adopted in conjunction with multiple cropping procedures, **soil erosion** can be markedly reduced. Crops such as maize provide little ground cover, even when mature, so detachment of *soil* particles by *rainsplash* occurs whenever there is heavy rainfall. This detachment increases the likelihood of **erosion** by surface wash. However, such crops can be planted along with legumes which provide a denser ground cover thereby reducing the potential for **soil erosion**. Soil conservation projects in the Third World aim to increase farmers' awareness of these problems and ways of averting them by adopting different cropping techniques.

In developed countries with technologically advanced farming methods, the threat of **soil erosion** and *soil* deterioration is perhaps as great as in the Third World where farming practices are comparatively unsophisticated. Decreasing *soil* fertility owing to overcropping or inadequate crop rotation is common because the application of fertilizers is insufficient to replenish the nutrients taken up by the crops. The only remedy is to modify farming practices. Similar gradual deterioration in productivity may result from poor drainage of farmland or mismanagement of irrigation schemes. Waterlogging and *soil* salinization may be slow to make cultivation of an area uneconomical, so it may be difficult to convince farmers — or even government organizations — to adopt soil conservation measures while the land is still productive. Nevertheless, this is essential, since once *soils* are depleted of nutrients, or have become excessively saline, they may be impossible to reclaim and will remain barren. [AW]

Bennett, H.H. (1947) *Elements of Soil Conservation*. McGraw-Hill, New York.
Chagnon, S.A. (1983) Record dust storms in Illinois: causes and implications. *Journal of Soil and Water Conservation* **38**: 58–63.
Hudson, N. (1971) *Soil Conservation*. Batsford, London.

soil erosion The removal of material from the *soil* zone — especially the topsoil — by natural processes. While the leaching of water-soluble salts and **erosion** of the subsoil by **piping** and **tunnel erosion** are also forms of soil erosion, usually the phenomenon is associated with the depletion or removal of the surface of the *soil* zone by wind, water, and *soil creep*. If the rate of **erosion** is increased by human activities, particularly clearance of natural vegetation for agriculture, **accelerated erosion** results. However, even under natural conditions beneath a dense vegetation cover, a portion of the *soil* zone is lost through **erosion**.

In semi-arid and arid environments, *soils* are often only poorly developed owing to the sparsity of vegetation. This limits the supply of nutrients to the *soil*, and also permits the **deflation** of fine particulate material. Many desert surfaces are protected from wind **erosion** by a *lag* of coarse **gravel**, or by saline crusts. However, if these *armoured surfaces* are disrupted, the underlying, unconsolidated material can be subject to rapid **erosion**. Dry, sandy *soils* which have a low clay content are most susceptible to **deflation**. The presence of moisture, clays, and salts generally provides some **cohesion** and resistance to erosion (Gillette *et al.*, 1982; Nickling, 1984).

The rates of erosion by surface **runoff** are greatest in areas where rainfall is markedly seasonal — for example, in *monsoonal*, mediterranean and some semi-arid environments. Surface **runoff** as channelled flow on hillslopes results in the formation of **rills** and *gullies*. On a broader scale, **sheet erosion** may accomplish the removal of large amounts of *soil*. This process is especially effective when *soil* particles are detached from the surface by the impact of raindrops. *Rainsplash* itself does not transport material in a net downslope direction, but once *soil* particles have been loosened from the surface, they are more readily eroded by any surface **runoff**. Such erosional processes are most effective on dry, exposed *soil* surfaces. On cultivated land, **soil conservation** techniques aim to reduce the rate of removal of the topsoil.

In environments where the *soil* zone is mantled by a protective mat of vegetation — whether woodland or grassland — soil erosion by wind and *sheetwash* is not pronounced. In such areas, deep and well-developed *soils* can form. On hillslopes, however, *mass wasting* processes can maintain a rate of erosion equal to the rate of *soil* formation. *Soil creep, solifluction*, **slumping**, *avalanching*, and **mudflows** may occur when seasonal or periodic fluctuations in the moisture content of the *soils* on slopes results in a loss of **cohesion**. In cold environments where there is *permafrost*, freeze–thaw processes can bring about soil erosion by *gelifluction*. [AW]

Finlayson, B.L. (1984) Soil creep: a formidable fossil of misconception. In: Richards, K.S., Arnett, R.R. & Ellis, S. (eds) *Geomorphology and Soils*, pp. 141–58. Allen & Unwin, London.

Gillette, D.A., Adams, J., Muhs, D. & Kihl, R. (1982) Threshold friction velocities and rupture moduli for crusted desert soils for the input of soil particles into the air. *Journal of Geophysical Research* **87**: 9003–15.

Kirkby, M.J. & Morgan, R.P.C. (eds) (1980) *Soil Erosion*. John Wiley & Sons, London.

Nickling, W.G. (1984) The stabilizing role of bonding agents on the entrainment of sediment by wind. *Sedimentology* **31**: 111–7.

sole mark Cast of structures on the base of **beds**, best observed where **sandstones**, *conglomerates* and **limestones** overlie relatively finer-grained cohesive **mudrocks** and siltstones. Selective **erosion** of the relatively fine-grained substrate beneath a flow, either by vortices in the current or suspended, **saltating** and rolling tools (**tool marks**), followed by the rapid infilling of the scour favours the preservation of sole marks. Post-depositional differential loading of a sediment into the underlying material, to form *load casts*, is a particular type of sole mark. Sole marks generally are associated with *turbidite*, shallow-marine storm-surge, fluviatile crevasse **sandstone** and semi-arid region **sheetflood** environments. Sole marks occur in many geometric forms, with a considerable range of descriptive terminology, but the most common types include *flute, groove, chevron*, **prod-and-bounce** and *frondescent marks, gutter casts, load casts* and **trace fossils**. [KTP]

solfatara A **fumarole** in a volcanically active area in which the escaping **volcanic gases** contain a high proportion of **sulphur**, commonly forming bright yellow deposits. [PTL]

solid Earth This term usually applies to those parts of the Earth which exclude the atmosphere, oceans and biosphere, i.e. the **crust** of the Earth and its interior, although this is not all strictly solid. There are also definition problems in some instances. For example, paleontology is both a biological and solid Earth science, and geomagnetism is solid Earth in terms of the origin of the field inside the Earth, but its study extends far into outer space. Similarly, the study of the surfaces and interiors of

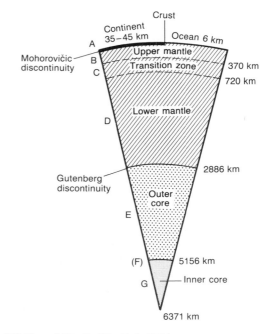

Fig. S42 The **solid Earth**. (After Bott, 1982.)

other planets can be related to understanding terrestrial properties. (See Fig. S42.) [DHT]

Bott, M.H.P. (1982) *The Interior of the Earth.* Arnold, London.

solution An important chemical **weathering** process involving the dissociation of a mineral in contact with a solvent into its component ions, forming a solution. This process is also known as **dissolution**. A distinction may be made between true solution, which involves the complete dissociation of a mineral in a solvent (as occurs with carbonates and sulphates), and the dissociation of silicates by *hydrolysis* which leaves a residual clay. Solution of calcium carbonate has been well studied by **karst** geomorphologists and has always been regarded as an important process in the development of **karst** terrain. More recently, solution has been recognized as an important process in many other areas. The rate of solution of a mineral in any environment is determined by the **aggressivity** of the solvent towards the mineral, the movement of the solvent, and by the solubility of the mineral. Mineral solubility is itself influenced by the **pH** of the solvent, although a general series of mineral solubilities can be identified. [HAV]

Gunn, J. (1986) Solute processes and karst landforms. In: Trudgill, S.T. (ed.) *Solute Processes*, pp. 363–437. John Wiley & Sons, London.

sonde Any device lowered down a borehole for measuring or testing purposes. (See **geophysical borehole logging**.) [PK]

sonic log An **acoustic log** that measures the travel-time of an acoustic wave over a fixed distance down a borehole. It is usual for this measurement to be stated in microseconds per foot, which is the inverse of the velocity in the formation. Information from the sonic log can be integrated down the borehole to give a travel-time with depth curve and is also used for determining the **porosity** of the rocks. The device used to collect the travel-time information is usually of the 'borehole compensated' variety, which consists of two sources at either end of the instrument which transmit alternately. The travel-time is measured over a fixed distance from both directions and averaged such that errors due to tilting of the instrument and variations in the borehole width are minimized. (See **geophysical borehole logging**.) [GS]

sonobuoy A buoy used mainly in marine **seismic refraction** work which is attached to or contains instrumentation for detecting seismic information from a distant **seismic source** and transmitting it back to a recording boat. A sonobuoy is usually attached to a number of **hydrophones** and contains its own power source and means of transmitting information over distances. A **seismic shooting** boat with a **seismic source** in tow trans-mitting **seismic waves** then moves away from a group of sonobuoys so that a **time–distance curve** can be produced. A sonobuoy can refer to a buoy used in positioning that emits a radio signal. [GS]

sövite A coarse-grained, light-coloured calcium carbonate **carbonatite**. [RST]

specific gravity Specific gravity has been used in archeological work to determine the composition of binary alloy (e.g. **gold/silver**) coins. [OWT]

spectral analysis (*power spectrum analysis*) Determination of the relationship between frequency (or **wavenumber**) and energy for a waveform. For a **potential** field the log-power spectrum above the noise level should have a linear gradient whose magnitude provides an estimate of the depth to the source. The presence of multiple bodies at different depths should provide a series of linear segments from which the different depths could be calculated. [PK]

Spector, A. & Grant, F.S. (1970) Statistical models for interpreting aeromagnetic data. *Geophysics* **35**: 293–302.

specularite A platy metallic variety of **hematite**. [DJV]

speleology The scientific study of **caves**. Speleology began with the earliest explorations of **caves** and cave systems. During the 19th century interest in **caves** grew as their importance as archeological sites was discovered. Much of the early work in speleology was carried out by gifted and enthusiastic amateurs, who still play a role in the subject today. The science of speleology involves many branches of knowledge pertinent to underground environments. Exploration and surveying of cave systems are vital parts of speleology. The geology, **geomorphology, hydrology**, climatology, ecology and archeology of **caves** are also studied by cave scientists. Many important cave systems are found in **karst** areas, and the development of underground drainage systems is an important component of **karst** landscapes. Speleology therefore contributes much to **karst geomorphology**, and vice versa. [HAV]

Ford, T.D. & Cullingford, C.H.D. (1976) *The Science of Speleology*. Academic Press, London.

speleothem The generic term for chemical precipitates in **caves**, which include *stalactites, stalagmites, helictites, flowstone, curtains, rimstone dams* and *moonmilk*. **Calcite** is the most important mineral **precipitated** in **caves** as carbon dioxide degasses from percolation water entering the **cave** atmosphere causing calcium carbonate to be **precipitated**. Other minerals forming speleothems in some circumstances include **gypsum** and **halite**. Speleothems have recently become an important focus of study in **karst**

geomorphology, as they can be dated and used to provide information upon the history of landscape evolution [HAV]

sperrylite ($PtAs_2$) A rare **ore** mineral of **platinum**. [DJV]

spessartine ($Mn_3Al_2Si_3O_{12}$) A mineral of the garnet group. (See **garnet minerals**.) [DJV]

spessartite A calc-alkaline lamprophyre containing **amphibole** (**hornblende**), **pyroxene** (*clinopyroxene*) ± **olivine** with **plagioclase** in excess of **alkali feldspar**. [RST]

sphalerite (ZnS) The major **ore** mineral of **zinc**. [DJV]

sphenochasm The triangular gap of **oceanic crust** separating two **cratonic** blocks with **fault** margins converging to a point, and interpreted as having originated by the rotation of one of the blocks with respect to the other (e.g. the Bay of Biscay). A *sphenopiezm*, by contrast, is a wedge-shaped squeezing together (e.g. the Pyrenees). [ASG]

spherical harmonic analysis The equivalent of *Fourier analysis* in spherical coordinates. The **International Geomagnetic Reference Field** formula is a spherical harmonic representation of the **geomagnetic field**. It is derived by making successive approximations to the field which become progressively more accurate as the order of the harmonics increases. [PK]

spheroid A body approximately spherical in shape. Usually a mathematical description of the **figure of the Earth**. [DHT]

spheroidal weathering The process of rock disintegration characterized by mechanical spalling and surficial chemical weathering of boulders; also known as *desquammation* and *onion-skin weathering*. Spheroidal weathering is often associated with **deep weathering** of crystalline rocks such as **basalts, dolerites**, and **granites**. As water moves through the rocks, chemical **weathering** occurs along the **joint** networks which developed when the igneous rock masses cooled and contracted. The angular blocks of rock become rounded **corestones** set within a matrix of clayey or silty **saprolite**. Spheroidal weathering proceeds as the surface of the **corestone** is attacked by chemical processes, or as the **weathering rind** on exposed boulders is spalled off by *desquammation* [AW]

Ollier, C.D. (1967) Spheroidal weathering, exfoliation and constant volume alteration. *Zeitschrift für Geomorphologie* NF **11**: 103–8.

spherulite A near-spherical structure in a silicic volcanic rock consisting wholly or partly of radially-disposed needle-shaped (acicular) crystals. Spherulites normally range from 0.1 to 2 cm in diameter. Spherulites normally consist predominantly of **alkali feldspar**. They are regarded as a product of devitrification of hydrated silicic glass (see **perlite**). If many spherulites are present in a rock, it is said to have a *spherulitic texture*, often clearly seen in hand-specimens in which the light-coloured spherulites contrast with the black and dark grey hydrated glass. Sometimes spherulites are arranged in trains or layers within the rock. *Spherulitic texture* can commonly be observed petrographically, even when the rock has been entirely devitrified. In such cases, the presence of the texture strongly suggests that the rock was originally **glassy rock**. Spherulites occur in volcanic **glassy rocks** of **rhyolitic**, rhyodacitic and **trachytic** composition, and in **lava flows**, **dykes**, unconsolidated **tephra** and in **welded tuffs**. [PTL]

spilite A mafic volcanic rock, commonly **pillow lava**, dominantly consisting of **albite** and **chlorite**. Spilites are regarded as strongly **altered basalts**, and have high abundances of Na, H_2O and CO_2, and low Si relative to fresh **basalt**. The spilite suite comprises spilite, *keratophyre*, and *quartz keratophyre*, thought to represent altered mafic, intermediate and silicic volcanic rocks respectively. The *keratophyres* consist of **albite**, **chlorite**, **epidote** and Fe–Ti oxides, with **quartz** appearing in the *quartz keratophyres*. The nomenclature of the spilite suite was once used extensively to describe volcanic rocks in Ancient rocks, but has tended to fall into disuse as a result of modern trends to describe altered volcanic rocks by comparison with geologically recent, fresh volcanic suites. [PTL]

spinel The name given to a group of isostructural **oxide minerals** of general formula M_3O_4 and specifically to the member of the group with the composition $MgAl_2O_4$. [DJV]

spinner magnetometer **Magnetometer** in which the sample is rotated within a detecting system, such as a **fluxgate magnetometer**, so that a signal is generated in the detector that has an amplitude proportional to the **intensity of magnetization** of the sample, and the phase of which is directly related to its direction. [DHT]

spit A long narrow accumulation of **beach** sediment with one end attached to the mainland and the other projecting out into a large water body. Spits occur as a result of **longshore drift** carrying sediment beyond an abrupt change in coastal orientation and are thus common at river exits and **estuaries** and on indented coasts. Spits depend on sediment delivery from updrift and they tend to have narrow updrift or proximal sections and broader downdrift or distal sections. **Wave refraction** around the end carries sediment into fingerlike recurves. In the sheltered lee deposition of fine sediments promotes the development of **salt marsh**. [JDH]

Schwartz, M.L. (ed.) (1972) *Spits and Bars*. Dowden, Hutchinson & Ross, Stroudsburg, PA.

spodumene (LiAlSi$_2$O$_6$) A **pyroxene mineral** found almost exclusively in **lithium**-rich *pegmatites* and exploited as a source of **lithium**. [DJV]

sponges/Porifera The phylum now considered to be intermediate between *protozoans* and *metazoans*, with a restricted variety of poorly organized cell types. Although the gross morphology is quite variable and particular species may themselves exhibit variation in form in response to local environmental criteria, a typical sponge comprises a sack-like shape attached to the substratum at the base and with a central cavity opening upwards. Pores on the external surface of the sponge lead to canals within the body wall, and hence to chambers. The chambers are lined by specialized (collar) cells, each equipped with a sticky collar and a long flagellum. Spiral action of the flagellae, acting in concert, pull gentle currents through the system and expel them into the central cavity, thence upwards and out of the aperture. The sticky collars trap particulate matter, and thus the sponge is strictly a filter feeder.

The skeleton or body wall is a jelly colloid, spongin, calcareous or siliceous spicules or some combination of these. The spicules are often fossilized, certainly as far back as the Lower **Cambrian**; they are very common, for example, in **black shales** of **Ordovician** or **Carboniferous** age in the United Kingdom. Based partly upon the nature of the skeleton, sponges are grouped into three classes: one has siliceous spicules with 60° or 120° rays and/or spongin (*Demospongea*); a second has calcareous spicules (*Calcarea*); the third has **opaline silica** spicules with rays at 90° (*Hexactinellida*) (fossils once thought to be **corals**, the *Chaetetida*, are now placed in the first class). All three classes range from **Cambrian** to Recent, but show few major changes in this time and are perhaps best regarded as an evolutionary offshoot, a blind alley, from the *protozoans*. Some *protozoans* closely resemble the sponge collar cells.

About 80% of sponges are marine, the remainder being freshwater. The *hexactinellids* occupy relatively deeper water to the other classes which prefer a shallow shelf environment, sometimes as partial *reef* builders, albeit with a low physical profile. [RBR]

Fry, W.G. (ed.) (1970) *The Biology of the Porifera*. Symposium of the Zoological Society of London No. 25, pp. i–xxviii, 1–512. Academic Press, London.

spontaneous magnetization The magnetization that arises naturally in a **ferromagnetic** (*s.l*) substance due to its quantum-mechanical exchange or super-exchange forces, even in the absence of an external **magnetic field**. [DHT]

spreading rate The rate at which an ocean is opening as determined from *oceanic magnetic anomalies* or other dating methods that can be applied to the **oceanic crust**. (See **half spreading rate**.) [DHT]

spring A localized flow of water from an **unconfined aquifer** which occurs where the **water table** intersects the surface, e.g. along a **fault** or fissure. [AME]

stable isotopes of carbon and oxygen in sediments The stable isotopic compositions of sedimentary rocks and their mineral and fossil components are used to determine *paleotemperatures* and to fingerprint changes in fluid compositions in depositional and **diagenetic** environments.

Isotope determinations are expressed in 'delta' values which record the 'permil' difference in ratio of the abundance of two stable isotopes in a sample compared to the equivalent ratio in an international standard. For example:

$$\delta = \left(\frac{R_{\text{sample}}}{R_{\text{standard}}} - 1 \right) 1000$$

where R is the ratio of ^{13}C/^{12}C or ^{18}O/^{16}O. Carbon isotope values are usually expressed as difference in the ratio ^{13}C/^{12}C between the sample and the PDB standard (a Jurassic *belemnite* carbonate from the PeeDee Formation to which secondary standards have been calibrated).

Oxygen values are expressed in terms of the ratio of ^{18}O/^{16}O compared either to PDB or to SMOW — a Standard Mean Ocean Water. The two scales are simply related (Friedman & O'Neil, 1977).

Sedimentary carbonate values are generally reported with respect to PDB for both **carbon** and oxygen. Silicate oxygen values are generally reported on the SMOW scale.

To obtain analytical precision capable of resolving the very small changes in stable isotope ratios in sedimentary systems, measurements are generally made on gases extracted from the geological materials and analysed by direct comparison with reference gases in special dual-inlet isotope ratio **mass spectrometers**. This analytical technique necessitates careful sample preparation methods, often involving the painstaking mechanical separation of discrete mineral or fossil components prior to analysis. Gas extraction usually involves the preparation of carbon dioxide; for carbonate phases this involves simple phosphoric acid digestion under vacuum; for silicates, oxygen is evolved by reaction with BrF$_5$ or ClF$_3$ and then converted to carbon dioxide for analysis. Development of laser and ion probe techniques offers very attractive possibilities for direct sampling of individual phases.

The **precipitation** of a **mineral** from solution generally involves *isotopic fractionation*, i.e. the isotopic composition of the **mineral** is different from that of the fluid. Fractionation effects can be subdivided into two types: equilibrium and kinetic. The effects of temperature on equilibrium fractionation is systematic but different for different minerals and for different isotopes. In the formation of carbonates, for example, oxygen compositions of mineral phases are about 30% heavier (more positive) than the fluid from which they formed and the difference decreases systematically with increasing temperature; carbon equilibrium *isotope fractionation* effects are small and temperature has only a minor effect. Kinetic effects, where the *isotopic fractionation* is affected by the rate of crystal growth, are much more poorly understood and are commonly ignored in geological interpretation.

In sedimentological work it is commonly assumed that minerals form in isotopic equilibrium with depositional or **diagenetic** fluids and thus that isotopic ratios in solid phases are determined solely by the isotopic composition of the fluid and the temperature of formation.

Equilibrium formation is probably valid for mineral cements which form relatively slowly but cannot be assumed for more rapidly **precipitated** skeletal components or for minerals such as *tufas* which are formed rapidly in near-surface environments. Non-equilibrium fractionations in the **precipitation** of biogenic carbonate are collectively known as *'vital effects'*. These are more pronounced in some groups of organisms (**corals**, *algae*) than others (**brachiopods**, most **molluscs**). *Vital effects* have been shown to be more pronounced in early growth stages where calcification is most rapid.

TEMPERATURE AND ENVIRONMENT. Oxygen isotope values are commonly employed in the calculation of depositional and **diagenetic** temperatures and can be used to assess *paleosalinities*.

Equations used to calculate formation temperatures are specific for a particular mineral phase and assume equilibrium **precipitation** and that values have not been altered by later exchange, **cementation** or recrystallization. *Paleotemperature* work on fossil material can only be carried out on unaltered material (assessed by mineralogical and **fabric** criteria, backed up by **trace element** geochemistry and, increasingly, by **cathodoluminescence** investigation).

Temperature equations are of two types. Those derived from high-temperature experimental data are expressed in terms of the *fractionation factor* alpha (α). For example, for calcite

$$10^3 \ln\alpha = 2.78(10^6 T^{-2}) - 2.89$$

where T is the temperature in kelvin and $10^3 \ln\alpha$ is approximately the difference between the oxygen isotopic composition of the mineral and that of the fluid from which it formed.

Equations derived from low-temperature experimental growth of carbonate-producing organisms have the following form; for example, for **calcite**

$$T°C = 16.0 - 4.14(\delta_c - \delta_w) + 0.13(\delta_c - \delta_w)^2$$

where δ_c is the oxygen isotopic composition of the **calcite** (w.r.t. PDB) and δ_w is the oxygen isotopic composition of the water (w.r.t. SMOW). Both types of equation relate isotopic composition to temperature and to the isotopic composition of the fluid.

The isotopic composition of surface and subsurface fluids can be very variable and can exert the dominant control on the isotopic value. In sediments with two or more minerals which evidently formed at the same time (carbonate/phosphate or **quartz/kaolinite** pairs for example), it is possible directly to calculate both temperature and water composition. In most sedimentological work, however, only a single phase is available and there are two 'unknowns'; we can measure the isotopic composition of the mineral but must estimate either the fluid composition in order to derive a temperature or a temperature to derive the water value.

Where water values cannot be estimated, isotopic values can only indicate changes between samples. Absolute temperature estimates depend on an accurate estimate of the ambient fluid composition. Marine temperatures are relatively simple; modern open marine waters have essentially uniform isotopic composition (SMOW of zero) and estimates for preglacial oceans have been made (values of around −1‰ have been assumed for the **Jurassic**, **Cretaceous** and Early *Tertiary*). Where **meteoric** (light) or evaporated (heavy) or deep basinal fluids may have been involved, meaningful temperatures are very difficult to determine. Isotopic variation within a suite of well-preserved fossils may be interpreted in terms of variation in temperature and salinity.

OCEANIC CHANGES. Stable isotopic compositions of carbon and oxygen in carbonate sediments and fossils have been used to monitor long-term changes in oceanographic conditions.

In samples from **Deep Sea Drilling Project** cores, oxygen isotopic variation in planktonic and benthonic **Foraminifera** give a very sensitive record of changes in temperature and oceanic water composition during the Late **Cenozoic** and *Tertiary*. The isotopic changes during *Pleistocene glacial* advances have been correlated to well-dated patterns of **magnetic reversals** and give rise to a detailed *isotopic stratigraphy* for the **Quaternary**.

The recognition of marked peaks in carbon isotope composition in **limestones** near certain stratigraphic boundaries has helped in the recognition of *Oceanic Anoxic Events* in the **Mesozoic**. The 'events' which may have a duration of around a million years, reflect times of **black shale** deposition and enhanced removal of isotopically 'light' organic carbon from the world's oceans.

Oxygen isotopic compositions of **Paleozoic** and **Precambrian** carbonates and their unaltered components (fossils and early marine cements) provide evidence for changes in seawater isotopic chemistry which can be linked to other global events.

DIAGENETIC ENVIRONMENTS. Stable isotope compositions of **authigenic** minerals are used to constrain environments of **diagenesis**. Marine, **meteoric**, evaporitic

[588]

and burial fluids may have very different oxygen isotopic compositions. Oxygen values in mineral **precipitates** are used to interpret formation temperatures and identify fluids responsible for **diagenetic** reactions. Differentiation between temperature and fluid composition as the cause of isotopic change is difficult for single phases but may be facilitated by evidence from petrography, **trace element** geochemistry and the analysis of *fluid inclusions*. Regional changes in the oxygen composition of a cement may be interpreted in terms of **paleohydrological** or *paleoheat-flow* patterns.

Marine bicarbonate has a narrow range of isotopic composition but large variations in carbon isotope ratios are associated with fractionation reactions involving organic matter. Microbially mediated reactions (in oxic, sub-oxic and **anoxic** conditions) in near surface environments give rise to distinct zones with characteristic isotopic signatures in organic-rich sediments. Concretionary products of **diagenetic** reactions involving sulphate *reduction*, *methanogenesis* and thermal **decarboxylation** reactions are common in organic-rich sediments: isotopic signatures in the carbonate and sulphide phases are powerful aids in the recognition of the responsible processes.

Shallow meteoric **diagenesis** is associated with light carbon signatures where *soil*-derived CO_2 is incorporated into cements. Negative carbon signatures are associated with bicarbonate derived from the abiogenic thermal breakdown of organic matter during deeper burial.

Where shell material has recrystallized or undergone mineralogical stabilization, isotopic values can help to constrain the degree of water–rock interaction. **Diagenetic** reactions involving carbonate stabilization or **clay mineral neoformation** in systems with a relatively low water to rock ratio can cause changes in the composition of basinal fluids.

Isotopic studies of **dolomites** are constrained by uncertainties in **dolomite** fractionation factors; in several instances, however, carbon and oxygen values have helped distinguish **evaporite**-related **dolomites** from those formed in more dilute **brines**. [JDM]

Coleman, M.L. (1985) Geochemistry of diagenetic non-silicate minerals: kinetic considerations. *Philosophical Transactions of the Royal Society, London* **A315**: 35–56.
Friedman, I. & O'Neil, J.R. (1977) *Compilation of Stable Isotope Fractionation Factors of Geochemical Interest.* Professional Paper of the United States Geological Survey **440**.
Hudson, J.D. (1977) Stable isotopes and limestone lithification. *Journal of the Geological Society, London* **133**: 637–60.

stable sliding Displacement on a **fault** at constant or slowly varying **strain rate**. This mode of **fault** movement contrasts with rapidly oscillating *strain rates* known as **stick-slip faulting** behaviour. It is known on both natural and experimental **faults**. [TGB]

Paterson, M.S. (1978) *Experimental Rock Deformation — The Brittle Field,* Springer-Verlag, Berlin.

stable tectonic zone Applied to part of the **continental crust**, implies a lack of **tectonic** activity. The term is applied to *plate* interiors, in contrast to **plate boundaries**, to undeformed blocks within *orogenic belts*, or to **cratons**, in contrast to **mobile belts**. The concept is, however, purely relative, since all parts of the **crust** undergo continuous activity on a geological time-scale, particularly in terms of vertical movements. What distinguishes stable from unstable zones is, essentially, the slow rate at which the movements take place. [RGP]

stack A **seismic record** that has been produced by summing the traces from a number of different records. [GS]

stadial A short cold period of the *Pleistocene* with smaller ice volumes than the full *glacial* stages. The warmer intervals between them are called **interstadials**. [ASG]

staining for carbonate minerals Simple chemical techniques used to distinguish the various **carbonate minerals** in thin section. Identifying **carbonate minerals** by their optical properties is notoriously difficult. Consequently, a large number of organic and inorganic reagents have been used to impart stains to the minerals in uncovered thin sections and hand specimens. Most widely used in **limestone** petrology is the combination of acidified Alizarin Red S and Potassium Ferricyanide to distinguish **calcite** (which stains pink to red-brown), **dolomite** (colourless), ferroan **calcite** (mauve to blue) and *ferroan dolomite* (pale blue). The intensity of the colour induced by the stain is generally related to the degree of etching caused by the acidic solution. Fine-grained **fabrics** etch more deeply, and hence display a stronger stain. Stains for identifying **aragonite** ('Feigl's solution') and magnesian **calcites** (e.g. titan yellow) have also been used. [DMP]

Dickson, J.A.D. (1965) A modified staining technique for carbonates in thin section. *Nature* **205**: 587.
Friedman, G.M. (1959) Identification of carbonate minerals by staining methods. *Journal of Sedimentary Petrology* **29**: 87–97.

standing wave Water surface wave that is stationary. Standing waves occur in many situations such as in association with upper **flow regime** conditions and **antidunes**, at **hydraulic jumps** or where wave reflection is created by an obstacle, for instance a **beach** or cliff line. [JLB]

stannite (Cu_2FeSnS_4) An **ore** mineral of **tin**. [DJV]

star dune A pyramidal shaped **dune** with three arms radiating from a high central dome. Although **sand** moves from one part of the **dune** to another, often in response to seasonal prevailing winds from different directions, there is rarely a net movement. [RDS].

starlite The name given to **gem zircon** heated in a *reducing* environment to give a blue colour. [DJV]

static correction A constant time correction applied to a **seismic record** so that the **arrival times** are what would have been observed if the **shot** and detector were on the same, usually horizontal, surface. This correction aims to eliminate the effect of variation in topography, weathered layer thickness and weathered layer velocity on the **arrival times**. Field static corrections are applied to a **seismic reflection record** from computations on data obtained by direct measurement of near-surface travel times (*uphole shooting*) or refraction surveys (low velocity layer (LVL) surveys). Residual or automatic statics are a later stage process and are applied to **moveout**-corrected traces to account for short wavelength static correction variations. [GS]

staurolite $((Fe,Mg)_4Al_{17}(Si,Al)_8O_{44}(OH)_4)$ A *nesosilicate* mineral found in regionally metamorphosed **aluminium-rich** rocks. [DJV]

steady-state creep **Creep** takes place by a range of mechanisms, many of which occur without any change in the physical state of the material. These are termed steady-state creep and characterize **mantle** convective motions. Non steady-state creep involves changes between solids, liquids and gases. (See **creep**.) [DHT]

steatite An alternative name for **soapstone**, a compact and massive form of **talc**. The commercial product is essentially talc with less than 1.5% each of CaO and Fe_2O_3 and less than 4% alumina. It is suitable for certain ceramic bodies.

Steatite is a frequent find on archeological sites in many parts of the world. It was used mainly for small objects (bowls, lamps, seals, jewellery, carvings) and occasionally, as in southern Africa, for larger sculptures. It was used in pre-Dynastic and later Egypt, the Near East, India and the Americas and is found from the Neolithic onwards in the Mediterranean and north-west Europe. The Vikings and Eskimos made extensive use of steatite as an alternative to pottery. Because of its resistance to high temperature, steatite was also used in early **copper** working, for moulds and crucibles. [AME/OWT/DJV]

step faults Set of parallel *normal faults* dipping in the same direction and with the same sense of movement. Step faults typically separate tilted *fault blocks* in **extensional fault zones**. [RGP]

stephanite (Ag_5SbS_4) A rare **ore mineral** of **silver**. [DJV]

stereographic projection Graphical method of portraying three-dimensional geometrical data in two dimensions, and of solving three-dimensional geometrical problems. In geology, the method is used mainly for solving problems involving the orientations of lines and planes in crystallography and structural geology. Such problems involve the angular relationship between lines or planes rather than their spatial relationships. Planes plot as lines (usually curved) and lines as points on such a projection.

The orientation of a plane (e.g. **bedding**, **foliation** or crystal face) is represented by imagining the plane to pass through the centre of a sphere of radius R, the projection sphere. The plane intersects the sphere in a circle with radius R called a **great circle**. The two-dimensional projection of the plane is produced in a horizontal plane through the centre O of the projection sphere (see Fig. S43a). In structural geology, the lower hemisphere only is used for projection, whereas in crystallography, the upper hemisphere is used. In the following account, the lower hemisphere projection is described. Upper hemisphere projections will be mirror-images across a vertical plane of symmetry. For lower hemisphere projection, each point on the lower half of the **great circle** is connected to the point T where the vertical line through O cuts the top of the projection sphere, producing a set of lines TA, TB, etc., that intersect the projection plane in an arc known as the *cyclographic trace* of the **great circle**. The *cyclographic trace* of the horizontal plane (i.e. the projection plane) is called the *primitive circle*. The *cyclographic trace* of a vertical plane is a straight line through O. If the *primitive circle* is given geographic co-ordinates, then the orientation of any specified plane may be represented. Thus Fig. S43(b) shows the *cyclographic traces* of a set of **great circles** with a N–S **strike dipping** at 10° intervals from 10° W through the vertical to 10° E.

In practice, stereographic projection is carried out by means of a protractor termed a *stereographic net* or **Wulff net** (Fig. S43c) which gives the *cyclographic traces* of the complete set of great circles at 2° intervals about a common axis XX'. The net also gives the *cyclographic traces* of a family of circular cones about the same axis XX', again at 2° intervals. These, termed **small circles**, are used to graduate angular distances along the **great circles**.

Fig. S43 (*Opposite*) (a) The principle of **stereographic projection**. (b) Stereographic projections of planes dipping due east or due west at the various indicated angles. (c) Wulff or equi-angular net. (d) Stereographic projection of a plane with a strike of 067° dipping at 56° towards 157°. A lineation, *L*, pitches 60° north-east in this plane. The plunge of the lineation is 45° toward 111°. (e) Schmidt net. (f) The principle of the equal-area projection. *O* is the centre of the projection sphere, radius *R*. *OB* is the trace of a plane dipping at angle ϕ. *X* is a point on the equal-area projection of the plane, *PQ*. (After Hobbs *et al.*, 1973.)

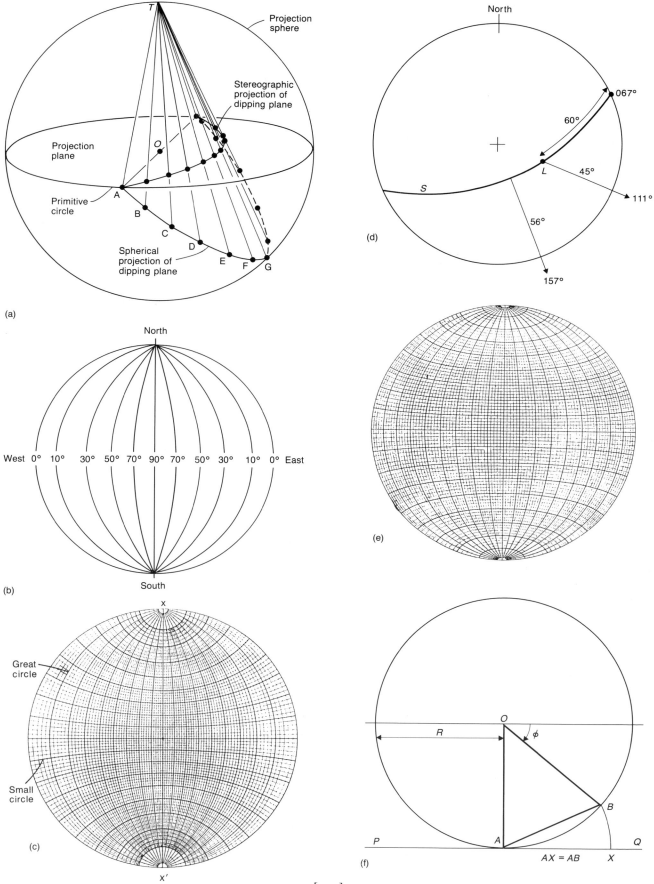

Projection sphere

Stereographic projection of dipping plane

Projection plane

O

Primitive circle

A

B

C

D

E

F

G

Spherical projection of dipping plane

(a)

North

West 0° 10° 30° 50° 70° 90° 70° 50° 30° 10° 0° East

South

(b)

x

Great circle

Small circle

x′

(c)

North

067°

60°

45°

111°

L

56°

S

157°

(d)

(e)

O

R

φ

P

A

B

Q

AX = AB

X

(f)

[591]

The procedure for plotting geometric data is as follows: a piece of tracing paper is laid over the stereographic net and fixed with a pin through the centre O. The *primitive circle* is marked, and given geographical coordinates. The plot produced on the tracing paper is termed a *stereogram*. To plot the position of a plane with **strike** bearing (azimuth) 067°, the position of 067° is marked on the *primitive circle*, by counting clockwise from north, and the *stereogram* rotated over the net until 067° corresponds with the axis XX' of the net. If the plane has a **dip** of 56° to the south-east, the appropriate **great circle** trace is found by counting 56° inwards from the perimeter (*primitive circle*) towards the centre (Fig. S43d). Alternatively, if the **dip** azimuth is given (56° to 157°), the position corresponding to 157° is marked, the net rotated until the point 90° clockwise of X corresponds with 157°, and the appropriate *great circle* trace found as before. It should be noted that planes **dipping** to the east plot in the right side of the *stereogram* and vice versa.

To plot the position of a line with a **plunge** of 45° on a bearing of 111° say, a similar procedure is followed. The bearing 111° is found as before. Since a **plunge** is measured in a vertical plane, one of the two vertical planes (represented by straight lines on the net) must be rotated to correspond with the position of 111° on the *stereogram*. The **plunge** angle is counted inwards from the perimeter to a point situated 45° along the line towards the centre (Fig. S43d).

Angles between any two lines may be found by placing their projected points on a common **great circle** trace. Angles between any two planes may be found by plotting the *poles* (normals) to the two planes and obtaining the angle between the *poles*. The *pole* to a plane is the line (point) situated 90° from the centre of the **great circle** trace, measured along the vertical trace. Many geometrical problems in structural geology can be solved using various combinations of these simple procedures. For example the **plunge** of a *fold axis* may be obtained by plotting the intersection of **great circles** representing two planar *fold limbs*, or representing various positions on the **fold** surface (see **beta (β) diagram**). *Poles* to a *cylindroidal fold* surface plot on a **great circle** trace, the pole of which represents the *fold axis* (see **pi (π) diagram**).

In the statistical analysis of the distribution of structural elements, it is more important to be able to compare the density of lines (points) in different areas of the projection sphere than to compare angles between lines. For the former purpose, a different type of projection is used, in which equal areas on the projection sphere plot as equal areas on the *stereogram*. The net used for this purpose is termed an *equal-area* or **Schmidt net** (Fig. S43e). The method of projection is illustrated in Fig. S43(f). Equal angles on the projection sphere are not projected as equal angles on the *stereogram*, and the cyclographic traces of the **great circles** are fourth-order quadric curves. The **Schmidt net** may be used to plot planes and lines, and measure their angular relationship in exactly the same way as the **Wulff net**, and is often employed by structural geologists in preference to the latter.

For details of procedures in plotting and analysing density distributions of structural or **fabric** data, see Turner & Weiss (1963) and Phillips (1971). (See **pole figure**, **preferred orientation**.) [RGP]

Hobbs, B.E., Means, W.D. & Williams, P.F. (1973) *An Outline of Structural Geology*. John Wiley & Sons, New York.
Phillips, F.C. (1971) *The Use of Stereographic Projection in Structural Geology*. Arnold, London.
Turner, F.C. & Weiss, L.E. (1963) *Structural Analysis of Metamorphic Tectonites*. McGraw-Hill, New York.

stibnite (Sb_2S_3) The major **ore** mineral of **antimony**. [DJV]

stick-slip **Fault** behaviour consisting of alternations of a period of slow **strain rate** followed by a rapid **slip** event. This type of faulting is described from frictional sliding experiments and also from alternations of slow and rapid **creep** events on natural **faults**; it contrasts with the **stable sliding** mode of **fault** displacement. A very simple mechanical explanation can be given by suggesting that the 'stick' period of the oscillation occurs while an applied *elastic* **stress** builds up to a critical value necessary to overcome a static *coefficient of friction*: the 'slip event' releases this **stress** with a smaller dynamic *coefficient of frictional sliding*. It has been suggested that this is an explanation of **earthquake** faulting with the seismic event representing the slip part of the cycle. (See **elastic rebound theory**.)

The potential applicability of the stick-slip mechanism to **earthquake prediction** has created many mathematical models in which the *coefficent of friction* varies according to a complex function and a single **fault** may be represented by multiple, interconnected individual systems. Stick-slip behaviour may follow **stable sliding** in experiments; the factors promoting stick-slip have been investigated thoroughly, and include lithology, surface roughness, development of *gouge*, *normal stress*, **strain rate**, temperature and **fabric**. [TGB]

Paterson, M.S. (1978) *Experimental Rock Deformation — The Brittle Field*. Springer-Verlag, Berlin.

stilbite ($CaAl_2Si_7O_{18} . 7H_2O$) A **zeolite mineral**. [DJV]

stilpnomelane (*c.* $K_{0.6}(Mg,Fe^{2+},Fe^{3+})_6Si_8Al(O,OH)_{27} . 2–4H_2O$) A *phyllosilicate* mineral. [DJV]

stishovite (SiO_2) A high pressure form of **silica** found in areas such as those of *meteorite* impact. (See **silica minerals**.) [DJV]

stockwork A closely-spaced, interlocking network of veinlets (Fig. S44). Stockworks most commonly occur in

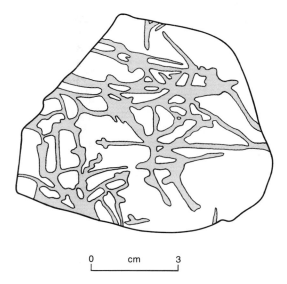

Fig. S44 **Stockwork** of molybdenite – bearing quartz veinlets in granite which has undergone phyllic alteration. Run of the mill ore, Climax, Colorado. (After Evans, 1993.)

acid to intermediate **plutonic** igneous intrusions but they may extend across the contact into the *country rocks* and a few are developed wholly in sedimentary or metamorphic rocks. They are important controls of mineralization in some **disseminated gold** and **molybdenum deposits** and in **porphyry copper** and other **deposits**. (See **crackle brecciation**.) [AME]

stolzite ($PbWO_4$) A tungstate mineral. [DJV]

stone pavement Area of flat desert covered with a surface layer of rounded pebbles or **gravels**. Stone pavements are *lag* deposits resulting form the action of wind and wash **erosion** removing the finer grains from the surface. [RDS]

stone-line A horizon of angular **gravel** and cobbles found within the *soil* zone of fine-grained sediments. Stone-lines often lie parallel to the surface topography. They are probably *lags* of coarse particles which originally formed at the landsurface as a result of **erosion** of finer material by wind or water. Subsequent sedimentation has then buried the gravelly accumulation. [AW]

Stocking, M.A. (1978) Interpretation of stone-lines. *South African Geographical Journal* **60**: 517–22.

stope An excavation within a mine made to extract **ore**. [AME]

stoping The process of mechanical incorporation of blocks of **wall rock** into a **magma** body during uprise. Stoping of **xenoliths** of **wall rock** may therefore be an important mechanism for the emplacement of **plutons**. [RST]

stored deformation energy Potential energy contained within a deformed material. The energy may be stored by **elastic** distortion of a *crystal lattice*, or by imperfections in the *crystal lattice*, including **dislocations** and *point defects*. The stored energy is the driving force for the processes of **recovery** and primary recrystallization. [TGB]

storm deposit The body of sediment deposited as a result of the passage of one *storm* event (Fig. S45). Storm deposits are particularly common on continental shelves of low tidal range and exposed to strong prevailing winds. Storm

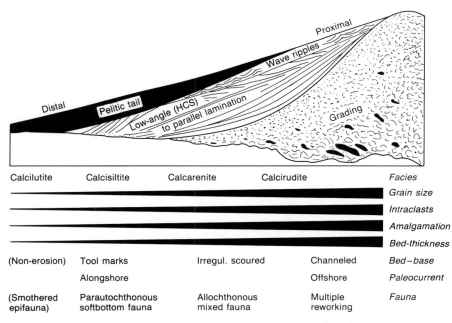

				Facies
Calcilutite	Calcisiltite	Calcarenite	Calcirudite	
				Grain size
				Intraclasts
				Amalgamation
				Bed-thickness
(Non-erosion)	Tool marks	Irregul. scoured	Channeled	Bed–base
	Alongshore		Offshore	Paleocurrent
(Smothered epifauna)	Parautochthonous softbottom fauna	Allochthonous mixed fauna	Multiple reworking	Fauna

Fig. S45 **Storm deposit**. Generalized trends in vertical and lateral variation of 'ideal' tempestite: proximality as a guide for environmental interpretation. (After Aigner, 1985.)

[593]

deposits, or *tempestites*, commonly show proximal–distal trends along storm-flow transport paths. In 'proximal' settings storm layers generally have erosive bases, locally with *tool marks* and *gutter casts*, basal *lag* deposits of *intraclasts* (mud *clasts*, shell, plant debris, rock fragments), horizontal to low-angle **cross-stratification** (including **hummocky cross-stratification**), wave-rippled upper units and a burrowed top. 'Distal' storm **sands** are finer-grained and thinner, have erosive or non-erosive bases, internal flat lamination and rare grading. Transported shell layers are generally absent in the more distal storm **sands**. The German Bight of the North Sea is an excellent example where these proximal–distal trends occur. However, this is a region where rivers enter a relatively confined embayment, and such trends may not be entirely characteristic of storm-dominated shelves where **geostrophic flows** move sediment for large distances essentially alongshore, as on the eastern seaboard of the USA.

The processes causing widespread deposition of storm **sands** have received considerable scrutiny by physical oceanographers and historical geologists. Early views were very much influenced by the study of Hurricane Carla in the Gulf Coast region. It was originally thought that sediment transport and deposition followed the build up of waters in the coastal region when there was a vigorous seaward return flow, or *storm surge ebb*. It has been shown, however, that the maximum currents accompanied Hurricane Carla rather than followed it. The most widely held view is now that pile-up of water against the coast (*storm set-up*) causes a pressure field which drives an offshore flow which is deflected by **Coriolis** effects into a roughly alongshore orientation. This **geostrophic** unidirectional flow is superimposed by the oscillatory motion of storm waves, producing a combined flow regime. The along-shelf passage of the **geostrophic** flow is retarded by losses of energy due to friction on the bed, causing sediment to be progressively deposited and thereby giving a systematic along-shelf grading of storm deposits. [PAA]

Aigner, T. (1985) *Storm Depositional Systems*. Lecture Notes in Earth Sciences 3. Springer-Verlag, Berlin.

Aigner, T. & Reineck, H.E. (1982) Proximity trends in modern storm sands from the Helgoland Bight (North Sea) and their implications for basin analysis. *Senckenbergiana maritima*. **14**: 183–215.

Swift, D.J.P., Han, G. & Vincent, C.E. (1986) Fluid processes and sea floor response on a modern storm-dominated shelf: middle Atlantic shelf of North America. Part 1: the storm-current regime. In: Knight, R. John & McLean, J. Ross (eds) *Shelf Sands and Sandstones*. Canadian Society of Petroleum Geologists Memoir **11**: 99–120.

storm surge Elevation of sea-level caused by extreme meteorological events. There are two interrelated causes of storm surges. Onshore winds tend to pile up water in the nearshore zone and extreme low pressure allows the sea surface to rise. These effects enhance the height of the normal tidal **wave** creating especially severe increases in enclosed seas and regions of extensive shallow water. Coincidence of a storm surge and the high water of normal spring **tides** may lead to catastrophic floods in low-lying coastal areas. In the tropics, severe storm surges are generated by cyclones, hurricanes and typhoons. [JDH]

Steers, J.A. (1953) The East Coast Floods, 31 January–1 February, 1953. *Geographical Journal* **3**: 280–99.

stoss side/slope The upstream side/slope of a body or obstruction that commonly experiences increasing fluid velocities and **shear stresses** as the flow moves over the object. Stoss side may be applied at many different physical scales: the windward side of a hill, the shallow upstream-dipping slope of a **ripple** or the upstream face of a particle exposed to the flow. [JLB]

strain A change in shape or internal arrangement of rock caused by **tectonic** activity. **Deformation** is used qualitatively to refer to the same thing. Strain or straining also refers to the progressive shape change during **deformation**. A particular strain can be visualized in terms of the **strain ellipsoid** (or ellipse in two dimensions), the deformed shape of a unit sphere. The **principal strains**, i.e. the **stretches** of the axes of the **strain ellipsoid**, and their orientations, define the **finite strain**. A unit cube can also be used, and is helpful when modelling *strain sequences*, but less clear in general than the sphere because of the initial variation in dimensions. The *strain parallelepiped* is a geometrical construction for analysis of *infinitesimal strain*. It is the shape taken up by a unit cube after a vanishingly small *strain increment* and is used for mathematical analysis of strain sequences, as discussed by Means (1979).

Strain in general refers to the net change in position of every point in a body; this can always be factorized (see Fig. S46) into *translation*, or change of position, *rigid body rotation*, or change of orientation, affecting each point identically, and *distortional strain*, or shape change, i.e. **deformation** that can affect each point differently (a '*pure strain*', as discussed by Means). Each of these components of a general strain can be represented by a matrix (or more properly a tensor, as discussed by Means, 1979), which transforms the initial co-ordinates of each point to those of their strained positions. Each small change in shape of a body within a more extensive **deformation** or **deformation** sequence is termed a *strain increment*. Vanishingly small *strain increments* are known as *infinitesimal strains*. *Strain superposition* is the application of successive *strain increments* to a body of rock to produce the observed **finite strain**. It is thus a method of modelling or describing a *deformation history*. The increments of strain may be related to several separate **deformation** events, or may all be part of a larger-scale complex strain episode. *Strain superposition* can be represented by multiplication of matrices representing each *strain increment* (Coward & Kim, 1981). Both these processes are non-commutative, i.e. the order of matrix multiplication and hence of *strain superposition* affects the final result, as shown in Fig. S47. The *strain*

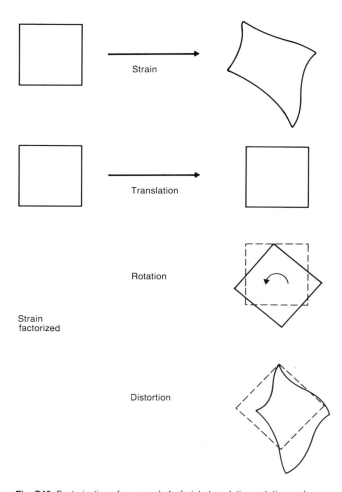

Fig. S46 Factorization of a general **strain** into translation, rotation and distortion.

increments are the same, but the order of operation and the **finite strain** differ. It is not generally possible to deduce a unique *strain history*, or any actual *strain increments*, given only the **finite strain**.

The state of strain in rock is in general spatially variable, i.e. heterogeneous. Homogeneous strains, in which lines initially straight remain straight, are rare. As shown in Fig. S48, strain can be continuous or discontinuous. **Faults, decollements** or high strain zones such as **mylonite** zones are the most common examples of strain discontinuities, which can isolate differing strain domains. Strain distributed over volumes of rock can result in the formation of **foliations, lineations**, and mineral **preferred orientations**. These are volumes of rock that show strain differing abruptly from that in its surroundings. Examples include **deformation bands** in crystals and **fault rock** along *fault zones*.

Strain is always measured as a relative quantity; **elongation, stretch** and **quadratic elongation** measure length changes relative to the final strained line length. **Shortening** uses the initial line length, and so is only of use in certain geological situations (see **balanced cross-section, thrust belt, fold**). Changes in angular relationships between lines define the **shear strain**. Measurement of strain depends on quantifying changes in lengths and orientations of lines, using *strain markers* where present. A *strain marker* is an object of known initial shape, geometry or shape distribution, present in a deformed rock and allowing measurement of strain. In general, many such objects are needed so that accurate, statistical analysis of the **deformation** can be carried out. Different *strain markers* are needed for measuring strain in different ways. Fossils such as **brachiopods** and **trilobites** are used as *strain markers* because of their initial bilateral symmetry; how-

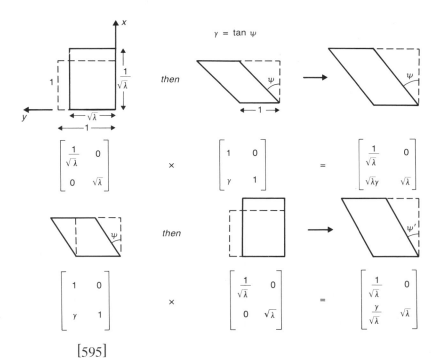

Fig. S47 The effect of superimposition of **strains** in different orders on a square. The two strains are the same but the finite strains differ.

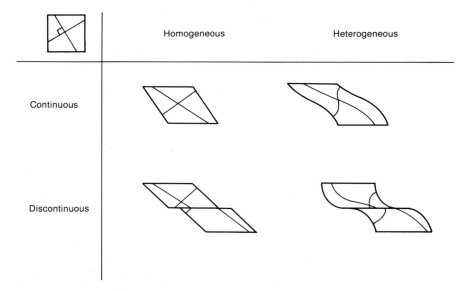

	Homogeneous	Heterogeneous
Continuous		
Discontinuous		

Fig. S48 The effects of different types of **strain** distribution on a square.

ever, they are rarely found in sufficient numbers for reliable results. Structures that are initially statistically normal to **bedding**, such as certain worm burrow **trace fossils**, or randomly orientated parallel to **bedding**, such as mud cracks, are also used. Assemblages of objects with close to elliptical or ellipsoidal shapes, e.g. pebbles in *conglomerates*, **ooliths** or **sand** grains, are used with the assumption that they initially had no or simple **preferred orientations** (discussed by Wheeler, 1984, 1986). In all uses of *strain markers*, it is important to assess to what extent the measured objects deform homogeneously with the matrix, and whether strain is also taken up by other processes, as these factors control how much the results represent the strain of the whole rock. Most *strain markers* relate initial shape to final shape, and so give the *finite strain*. Information on incremental strain is useful, but only rarely available. Fibrous mineral growth in **pressure shadows** or in **veins** can indicate the order of superposition of strain increments for some of the strain history; these methods are fully discussed by Ramsay & Huber (1983). Strain measured on a small scale, such as the outcrop scale, can represent local **deformation**. Extrapolation to larger scales is possible if *strain* is *homogeneous*, or systematically variable. Localization of **deformation** onto high strain zones, or formation of **decollements**, and the formation of strain domains on any scale, make a complete description of strain from isolated local measurements difficult. For example, in **thrust belts**, large values of **shortening** occur without high penetrative strains in most of the **allochthonous** rock. Measuring strain within the *thrust sheets* while neglecting **deformation** localized onto **faults** would give very misleading results.

Strain in rocks results from **tectonic** activity. Widespread regional metamorphism and **deformation** leads to distributed strain, often with high and low strain domains. Anastomosing networks of **shear zones**, with varying foliation and **lineation** development are a feature of much deep **crustal deformation**. **Deformation** mechanisms such as *crystal plasticity*, and other forms of **creep** dominate under these conditions. At higher levels in the **crust**, **cataclasis**, **pressure solution**, and *crystal plasticity* are important. **Deformation mechanism** maps show the interactions between physical factors such as temperature, **confining pressure**, **differential stress** and **strain rate**, controlling the dominant **deformation** process operating. The various **deformation mechanisms** can occur in localized zones, or distributed over wide areas, i.e. as continuous or discontinuous strain (discussed by Rutter, 1986). The net **strain rates** resulting from a particular mechanism will vary depending on the volume of the straining rock. This is an important factor when assessing the significance of rock **fabrics** in a wider **tectonic** setting. (See also **Poisson's ratio, elastic constants**.) [SB]

Coward, M.P. & Kim, J.H. (1981) Strain in thrust sheets. In: McClay, K. & Price, N.J. (eds) *Thrust and Nappe Tectonics*. Geological Society Special Publication No. 9. Blackwell Scientific Publications, Oxford.
Means, W.D. (1979) *Stress and Strain*. Wiley International, New York.
Ramsay, J.G. & Huber, M.I. (1983) *The Techniques of Modern Structural Geology* Vol. 1. Academic Press, London.
Rutter, E.H. (1986) On the nomenclature of mode of failure transitions in rocks. *Tectonophysics* **122**: 381–7.
Wheeler, J. (1984) A new plot to display the strain of elliptical markers. *Journal of Structural Geology* **6**: 417–25.
Wheeler, J. (1986) Strain analysis in rocks with pretectonic fabrics. *Journal of Structural Geology* **8**: 887–96.

strain band Planar zone of high **strain** in a set of asymmetrically folded layers, corresponding to the superimposed short *limbs* of the asymmetric **folds**. The strain band boundaries correspond to the *axial surfaces* of the **folds**. **Kink bands** are a type of strain band produced by *kink folds*. A set of strain bands alternating with less strained zones forms a *crenulation cleavage*. [RGP]

strain ellipse/ellipsoid The shape taken up by a unit circle or sphere after a given **deformation**; it represents and makes easy to visualize the shape change associated with that **strain** (Fig. S49). It can be used to represent **finite strain**, **strain** increments or *infinitesimal strains*, as well as in **strain** analysis. The three axes of symmetry of the ellipsoid form the **principal strain** *axes*; their **stretches** form the **principal strains**, usually expressed as a ratio. x, y and z, are used to describe respectively the maximum, intermediate and minimum **principal strains**. The *principal planes* of the strain ellipsoid are the three planes normal to each *strain axis* in turn; they are described using the *strain axes* that they contain, i.e. the xz plane is normal to the y axis. For two-dimensional **strains**, a strain ellipse, and two **principal strains** are used.

The **principal strains**, as vector quantities, describe both the nature of the shape change and the orientation of directions of **extension** and contraction, with respect to an external reference frame. The sizes of the **principal strains** allow classification of strains into uniaxial, biaxial and triaxial classes, thus:

uniaxial two of x, y, z are equal, i.e. $x = 1$, $y = z$; or $z = 1$, $x = y$;
biaxial $y = 1$, $x = 1/z$;
triaxial x not equal to y not equal to z.

The variation in ellipsoid shapes is most commonly expressed on a **Flinn diagram** (and on the **Hsu diagram**; see also Ramsay, 1967). Ellipsoids are divided into two groups; prolate ellipsoids are 'cigar-shaped', with $x \gg y > z$, and oblate ellipsoids are pancake shaped, with $x > y \gg z$. In the absence of volume change (**dilation**) these ellipsoid shapes represent *constrictional* and *flattening strains* respectively. Ellipsoids in which $yy = xz$ represent **plane strain deformation**.

The strain ellipsoid can be used to visualize the effect of **deformation** on lines of differing orientation, by considering these lines as joining the centre of the unit sphere to the edge. The transformation to the ellipsoid affects the line length and its intrinsic angular relationships with other material lines. An ellipsoid in general has two circular sections (through the origin) which divide it into fields in which lines increase in length and fields in which they decrease in length during **deformation**. These are known as the extensional and contractional fields, and change position with respect to material lines during non-**coaxial deformation** such as **simple shear**.

Measurements of **strain** using *strain markers* can represent the strain ellipsoid, if the markers have no initial **preferred orientation**. Where an initial **fabric**, such as a sedimentary grain alignment, does exist, it is necessary to define the initial **fabric** ellipsoid, and the final **fabric** ellipsoid, to calculate the strain ellipsoid representing the transformation that links them. (See **strain**.) [SB]

Ramsay, J.G. (1967) *Folding and Fracturing of Rocks*. McGraw-Hill, New York.

strain rate The rate of change of a measure of **strain**, i.e. shape change, with respect to time. Strain rate can be quoted for measurements of longitudinal **strain** such as **stretch**, **shortening**, **elongation**, or **quadratic elongation**; **shear strain** rate is quoted where changes in angular relationships between lines are measured. As **strain** is always measured as a relative quantity, strain rate has dimensions of 'per unit time'. Seconds, years and million years are all quoted.

In order to link strain rates deduced from *microstructures* and their associated **deformation mechanisms** (see below) to the rates observed in other geological processes, it is necessary to know the volume of rock involved in the **strain** process. Consider a small cube of rock undergoing **simple shear** at a steady rate, as in Fig. S50, which shows the **deformation** resulting after one time increment. If this volume of rock is the full thickness of the **shear zone**, then the rock on each side of the **shear zone** can only be displaced a small amount in each increment of time. If the **shear zone** is much wider, and still deforming at the same rate, a much greater **displacement** is achieved in the same increment of time. Both the strain rate deduced from the **deformation** of the rock, and the extent of deforming rock determine the overall **strain**.

Geological strain rates are thought to be in the range $>10^{-6}\,\text{s}^{-1}$ to $10^{-16}\,\text{s}^{-1}$. 'Fast' strain rates are observed in **earthquakes** and some long duration rock **deformation**

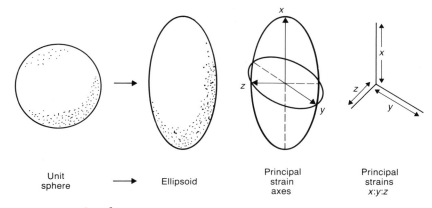

Fig. S49 The **strain ellipsoid**. A unit sphere becomes an ellipsoid under strain, and its axes form the principal strain axes x, y, z.

Unit sphere → Ellipsoid → Principal strain axes → Principal strains $x{:}y{:}z$

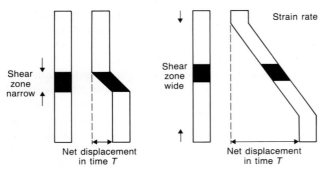

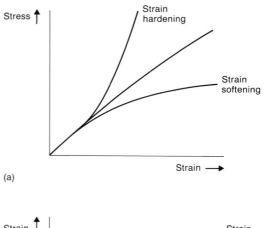

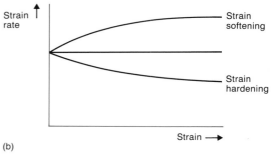

Fig. S50 The dark square represents a unit of rock deformed at constant **strain rate** in broad and narrow model shear zones. Notice that small and large offsets on these model shear zones occur in the same time (T) so that different bulk displacement rates result from constant strain rate deformation over small and large volumes.

Fig. S51 Strain softening. Plots to show characteristics of rock deformation, (a) at constant strain rate, (b) at constant stress. Based on experimental data.

experiments. Behaviour at lower strain rates is inferred from observation of *microstructures* resulting from natural examples, experimental samples at high temperatures and pressures, and analogue materials such as metals and ceramics. Identification of **deformation mechanisms** such as **pressure solution**, or *crystal plasticity*, leads to understanding of the rate-determining step, i.e. the slowest process in the chain of events that make up the mechanism. Thus geological strain rates can depend on, for example, the rate of atomic diffusion in the crystal lattice (*Nabarro–Herring creep*) which has in some cases been measured experimentally. In general, quantitative estimates of strain rate are difficult to achieve from isolated microstructural observations. Qualitative observations of change in strain rate can usefully be made and incorporated into models of *deformation history*. On a large scale, rates of **shortening** and **extension** based on syntectonic sediment distribution around **faults**, and rates of **lithosphere** *plate* movements provide a framework for a more detailed understanding of strain rate histories in rock **deformation**. [SB]

strain softening The phenomenon of **deformation** occurring at an increased rate with increasing **strain**, or as a decrease in the applied **stress** needed for **deformation** at a constant rate, as **strain** increases. It gives a characteristic form to plots of **stress** against **strain**, or **strain rate** against **strain**, as shown in Fig. S51. This behaviour is well known in materials science, where it is more commonly known as *work softening*. The converse behaviour, an increase in **strength**, or a decrease in **strain rate** at constant **stress**, or an increase in **stress** at constant **strain rate**, as **strain** increases, is called *strain hardening*, or *work hardening*.

Strain softening is significant in **deformation** of rocks because of its role in **strain** localization. Where **strain** is heterogeneous and softening occurs, higher **strain domains** can deform faster than surrounding areas. Thus **strain** increases in and around these zones, which tend to enlarge, and often accommodate all the **deformation**.

Conversely, where hardening occurs with increasing **strain**, **deformation** spreads over a wider volume of rock as **strain** increases. Softening can occur by a variety of processes, discussed more fully by White *et al.* (1980). Grain size reduction during **deformation**, allowing operation of a different, and faster, **deformation mechanism**, is a common geological process promoting **strain** localization. (See **deformation mechanism**.) [SB]

White, S.H., Burrows, S.E., Carreras, J., Shaw, N.D. & Humphries, F.J. (1980) On mylonites in ductile shear zones. *Journal of Structural Geology* **2**: 175–87.

strandflat A partly submerged, undulating rocky lowland close to sea-level in western Norway. Similar features occur in Svalbard, Iceland, Greenland, Baffin Island, Antarctica and possibly in western Ireland (Guilcher *et al.*, 1986). Innumerable **stacks** and skerries rise above the surface which is up to 50 km wide. The origin is problematic but may be summarized by two views. The first of these requires extensive frost action on the shoreline whereas the second regards strandflats as lowlands **eroded** by the sea but subsequently modified by **glacier** ice. Whatever view is adopted, it seems likely that strandflats in different parts of the world have developed in different ways over a wide range of time spans. [JDH]

Gjessing, J. (1966) Norway's paleic surface. *Norsk Geografisk Tidsskrift* **21**: 69–132.

Guilcher, A., Bodere, J.-C., Coude, A. *et al.* (1986) Le problème des strandflats en cinq pays de hautes latitudes. *Revue de Géologie Dynamique et Géographie Physique* **27**(1): 47–79.

Trenhaile, A.S. (1987) *The Geomorphology of Rock Coasts.* Clarendon Press, Oxford.

strata-bound mineral deposit Deposit, irrespective of its morphology or genesis, that is restricted to a fairly limited stratigraphical range within the strata of a particular region. It may, or may not, be **stratiform**. [AME]

stratiform mineral deposit A deposit having considerable development in two dimensions parallel to the principal planar structure in the *host rock*, e.g. **bedding** or igneous **lamination**, and a limited development perpendicular to it. The term should not be confused with **strata-bound**. [AME]

streak The colour of a **mineral** in a finely powdered form. It is useful in **mineral** identification because although the colour in a bulk hand specimen may be variable, the streak is usually constant. The streak is determined by rubbing the **mineral** on a piece of unglazed white procelain known as a *streak plate*. Because the *streak plate* has a Mohs' Hardness of about 7, it cannot be used with minerals of greater hardness. The streak is actually the qualitative expression of what is measured quantitatively by *diffuse reflectance spectroscopy*. [DJV]

stream flow The volume of water passing a given point in a given unit of time. The term **discharge** is synonymous. In large rivers, stream flow is usually measured in $m^3 s^{-1}$ ('cumecs') or in $ft^3 s^{-1}$ ('cusecs'). In small streams, $1 s^{-1}$ is a more convenient unit. Stream flow can be measured using a variety of techniques; many rely on the product of channel cross-section area and average flow velocity (which can be measured in many ways, e.g. current meter). In other cases, permanent structures (weirs and flumes) may be built which provide a set relationship between river level (*stage*) and discharge. *Stage* is measured continuously in most applications using water-level recorders which have pen and chart or, more recently, pressure transducer and data logger. Direct estimation of stream flow can be achieved using *dilution gauging* techniques, although these are only feasible in small streams. [TPB]

stream order A topological classification of links within the stream network. Exterior links, or headwater tributaries, have an order and magnitude of unity. Under the *Strahler system* a stream of order $n+1$ is initiated at the confluence of two streams of order n. By this method the entry of a stream of lower order does not increase the order of the main stream. *Shreve's magnitude* approach adds together the order values at each confluence to yield the magnitude of the downstream link. This produces a measure of order which is more equivalent to increases in **discharge** within the network. One feature of the *Strahler* (1964) *system* is that the number of streams of each order in a large **basin** tends to approximate an inverse geometric series. The varying character of this series is expressed by the **bifurcation ratio**. [TPB]

Strahler, A.N. (1964) Quantitative geomorphology of drainage basins and channel networks. In: Chow, V.T. (ed.) *Handbook of Applied Hydrology*, Section 4–11. McGraw-Hill, New York.

streamlines Lines drawn to depict the motion of a fluid which are tangential to the flow direction at any point along the line. Streamlines are often curved and never cross as the intersection would demand the impossibility of a fluid particle having two directions of motion, one parallel to each streamline. A streamline has no width and no volume and essentially can be thought of as representing the path of a neutrally buoyant particle in the flow. [JLB]

strength The maximum *principal stress* that a material can withstand before **failure**. This depends strongly on the conditions, so that there are several ways of quoting strength. The *uniaxial compressive* or *tensile strength*, under a single *compressive* or *tensile stress*, is one of the most commonly used values; the former is also referred to as the *cohesive strength*. In *triaxial stress*, the strength is given by the *failure criterion*, or relationship between *principal stresses* at **failure**. When **failure** occurs by **fracture**, the strength is known as the *brittle strength*; if it occurs after *ductility*, it is referred to as the *ultimate strength*. Since **strain rate** may also affect strength, it may be necessary to distinguish long- and short-term strengths, or specify the **strain rate**. The long-term strength is sometimes referred to as the *creep strength* (see **creep**). The short-term strength is known as the *instantaneous strength*.

Unfortunately laboratory determinations of rock strength in conventional apparatus produce widely discrepant results, because it seems that experimental conditions are also an extremely important variable. The more recent attempts to characterize strength attempt to measure parameters which should be independent of experimental procedure such as *critical crack extension force* or *fracture toughness*. [TGB]

Jaeger, J.C. & Cook, N.G.W. (1979) *Fundamentals of Rock Mechanics* (3rd edn), Chapter 4. Chapman and Hall, London.

stress The force acting on unit area of a surface of a solid together with the equal and opposite force representing the reaction of the material. The SI unit of stress is the **pascal** (Pa) ($1 Pa = 1 N m^{-2}$); other commonly used units are given in Table S4 with conversion factors. Stresses acting perpendicular to a surface are *normal stresses*, those parallel to the surface are **shear stresses**. The state of stress

Table S4 Stress conversion factors

Unit	Pascal (Pa)	Megapascal (MPa)	Gigapascal (GPa)	Bar (b)	Kilobar (Kb)	Pounds in^{-2} (p.s.i.)	Dynes cm^{-2}	Atmosphere
1 Pascal	1	10^{-6}	10^{-9}	10^{-5}	10^{-8}	1.45×10^{-4}	10	9.9×10^{-6}
1 Megapascal	10^{6}	1	10^{-3}	10	10^{-2}	1.45×10^{2}	10^{7}	9.9
1 Gigapascal	10^{9}	10^{3}	1	10^{4}	10^{1}	1.45×10^{5}	10^{10}	9.9×10^{3}
1 Bar	10^{5}	10^{-1}	10^{-4}	1	10^{-3}	1.45×10^{1}	10^{6}	9.9×10^{-1}
1 Kilobar	10^{8}	10^{2}	10^{-1}	10^{3}	1	1.45×10^{4}	10^{9}	9.9×10^{2}
1 Pound in^{-2}	6.9×10^{3}	6.9×10^{-3}	6.9×10^{-6}	6.9×10^{-2}	6.9×10^{-5}	1	6.9×10^{4}	6.8×10^{-2}
1 Dyne cm^{-2}	10^{-1}	10^{-7}	10^{-10}	10^{-6}	10^{-9}	1.45×10^{-5}	1	9.9×10^{-7}
1 Atmosphere	1.01×10^{5}	1.1×10^{-1}	1.01×10^{-4}	1.01	1.01×10^{-3}	1.47×10^{1}	1.01×10^{6}	1

at a point can be described by nine quantities, the *stress components*, which are the normal and **shear stresses** on the faces of an infinitesimal cube, in the coordinate axes of Fig. S52. These are conveniently rewritten in a matrix with components σ_{ij} where i is the direction of the normal to the face, and j is the direction of the stress. This is a particularly useful notation for manipulating stress algebraically, known as tensor notation; it is also possible to use a conventional cartesian notation with reference to axes x, y and z, in which case the **shear stress** components are often represented by the symbol τ.

σ_{11} σ_{12} σ_{13}
σ_{21} σ_{22} σ_{23}
σ_{31} σ_{32} σ_{33}

It can be seen that the normal components are given by σ_{ij} with $i = j$ and the **shear stresses** by σ_{ij} with $i \neq j$. There are three *normal stress* components along the three axes, and six **shear stress** components; however, it can be shown that $\sigma_{ij} = \sigma_{ji}$ so that the number of shear components reduces to three, requiring only six independent quantities to specify the stress. Stress is a type of quantity known as a second-order tensor.

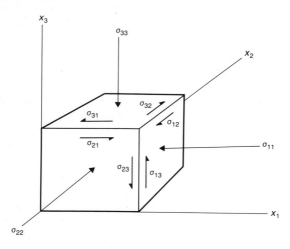

Fig. S52 Description of the state of **stress** at a point represented by an infinitesimally small cube. The normal stresses are σ_{ij} where $i = j$, the shear stresses are σ_{ij} where $i \neq j$, i is the direction of the normal to the face and j is the direction of the stress.

One of its properties is that there are three mutually perpendicular planes on which there are no **shear stresses**. These are known as the *principal planes of stress*, and their normals are the *principal axes of stress* (stress axial cross). The values of the stresses on these axes are the *maximum* (σ_1), *intermediate* (σ_2) and *minimum* (σ_3) *principal stresses*, which are taken as positive for compression in geological literature. The state of stress at a 'point' can be completely described by the magnitudes of the *principal stresses* and their directions alone.

Three quantities exist which are useful for calculating the orientations and magnitudes of the *principal stresses*; these are the *stress invariants*, labelled I_1, I_2 and I_3:

$$I_1 = \sigma_1 + \sigma_2 + \sigma_3$$
$$I_2 = -(\sigma_1\sigma_2 + \sigma_3\sigma_1 + \sigma_2\sigma_3)$$
$$I_3 = \sigma_1\,\sigma_2\,\sigma_3$$

It is useful to describe certain states of stress by special names:
uniaxial stress has a finite value of σ_1, $\sigma_2 = \sigma_3 = 0$;
biaxial or *plane stress* has $\sigma_1 > \sigma_2$, $\sigma_3 = 0$;
triaxial stress is the general state, $\sigma_1 > \sigma_2 > \sigma_3$;
hydrostatic stress, $\sigma_1 = \sigma_2 = \sigma_3$. No **shear stresses** exist in a state of *hydrostatic stress*: it causes volume change only.

A general *triaxial stress* can always be separated into two parts, the mean stress, σ_m (also known as the hydrostatic, octahedral normal, σ_{oct}, or non-deviatoric part) and the part remaining when the mean stress is subtracted: this is the *deviatoric stress*, σ'_{ij}, and its components are the *stress deviators*.

mean stress $\sigma_m = (\sigma_{11} + \sigma_{22} + \sigma_{33})/3 = I_3/3 = \sigma_{oct}$

$$\begin{bmatrix} \sigma_{11} & \sigma_{12} & \sigma_{13} \\ \sigma_{21} & \sigma_{22} & \sigma_{23} \\ \sigma_{31} & \sigma_{32} & \sigma_{33} \end{bmatrix} = \begin{bmatrix} \sigma_m & 0 & 0 \\ 0 & \sigma_m & 0 \\ 0 & 0 & \sigma_m \end{bmatrix}$$

General stress   =   Mean stress

$$= \begin{bmatrix} (\sigma_{11} - \sigma_m) & \sigma_{12} & \sigma_{13} \\ \sigma_{12} & (\sigma_{22} - \sigma_m) & \sigma_{23} \\ \sigma_{13} & \sigma_{23} & (\sigma_{33} - \sigma_m) \end{bmatrix}$$

=   Deviatoric stress

Stress invariants can be defined for the deviatoric part of the stress tensor just as for the complete tensor. A final useful quantity is the octahedral **shear stress**, τ_{oct}, where:

$$\tau_{oct} = \frac{((\sigma_1 - \sigma_2)^2 + (\sigma_2 - \sigma_3)^2 + (\sigma_3 - \sigma_1)^2)^{1/2}}{3}$$

$$= \frac{\sqrt{2}}{3}(I_1^2 + 3I_2)^{1/2}$$

There are several ways of representing the state of stress and its variation. The **Mohr diagram** is a two-dimensional plot of the *normal* and **shear stress** variation with direction at a point; it can be constructed to show both two- and three-dimensional states of stress. The complete variation of stress with direction can be imagined as the surface of an ellipsoid, with the length of any radius giving the magnitude of the normal stress on the conjugate plane to the radius. This is known as the *stress ellipsoid*, which has axes parallel and proportional to the directions and magnitudes of the three *principal stresses*.

Orientations of the *principal axes of stress* in three dimensions are usually plotted in **stereographic projection**. Inhomogeneous stress distribution is most simply shown by *isobars*, which are lines of equal *principal stress* magnitude. These are distinct from *stress trajectories*, which are lines giving the orientations of the *principal stresses*. Where the *principal stresses* are equal, *isotropic points* are defined; if they are zero, such *isotropic points* are also *singular points*. *Isochromatics* are lines of equal maximum **shear stress** magnitude, obtained directly by photoelastic methods. *Isopachs* are lines of equal mean stress magnitude. *Isoclinics* are lines showing where the *principal stress* orientations are similar. Finally *slip lines* show the direction of maximum **shear stress**: they bisect the *stress trajectories*.

The state of stress in the Earth can be considered as the sum of the stress due to **gravity** and that due to a **tectonic** element, which is directly or indirectly related to the Earth's thermal **convective** system driving **lithospheric** *plate* motions.

Because **gravity** necessarily contributes a proportion of the stress field, a common assumption is that one *principal stress* in the **crust** is vertical, and the other two are therefore horizontal. If the maximum *principal stress* is due to **gravity**, extensional structures, e.g. *normal faults*, may be anticipated; if **gravity** is the intermediate *principal stress*, conditions are appropriate for **strike-slip faulting**, and for *thrust faulting* if **gravity** is the least *principal stress* (Anderson, 1953).

Sources of **lithospheric** stresses are conveniently divided into *plate boundary forces* (**ridge push**, **slab pull** and **subduction suction**) and *compensated load forces* (*continental margin* and **plateau uplift**). (See **forces acting on plates**.) Since these cannot be measured at depths greater than a few kilometres, knowledge of stress at deeper levels depends largely on modelling the behaviour of Earth materials based on laboratory experiments of frictional sliding or triaxial testing to **failure** or yield. Modelling is also useful to determine stress distribution within structures on a much smaller scale such as **folds** and **faults**.

However, near-surface stresses can be measured by a number of *in situ* techniques. *Overcoring* (also known as the *doorstopper technique*) involves measuring small **strains** produced following the release of stress around a cylinder at the end of a borehole; alternatively borehole stress is measured directly using an *inclusion stress meter*, or by measuring the fluid pressure necessary to fracture or dilate the borehole: this is the *hydrofrac technique*. Elongation of the circular borehole, recorded by *borehole breakout* logs, has also been used to determine the orientation and magnitude of horizontal stresses. A thin metal membrane known as a *flatjack* can be inserted in rock at surface or in underground workings to determine stress directly.

The orientation of stresses to depths of a few tens of kilometres can also be deduced from sufficiently detailed records of **earthquakes** by *fault plane solutions*.

Some inferences about past states of stress can be made from certain structures, such as **fault** sets (see Anderson, above), or microscopic **fractures** such as microcracks or *deformation lamellae*. *Piezometry* is the measurement of pressure or stress and the study of previous states of stress in rocks is known as *paleopiezometry*. A number of studies have attempted to deduce the magnitudes of differential stress required for **deformation** by comparing natural and experimental features including grain size, subgrain size, **dislocation** density and **twinning**. [TGB]

Anderson, E.M. (1951) *The Dynamics of Faulting*. Oliver & Boyd, Edinburgh.
Jaeger, J.C. & Cook, N.G.W. (1979) *Fundamentals of Rock Mechanics* (3rd edn), Chapters 2, 14, 15. Chapman and Hall, London.
Jaeger, J.C. (1962) *Elasticity, Fracture and Flow* (2nd edn). Methuen, London.
Kusznir, N.J. & Park, R.G. (1984) Intraplate lithosphere deformation and the strength of the lithosphere. *Geophysical Journal of the Royal Astronomical Society* 79: 513–38.
Means, W.D. (1979) *Stress and Strain* (2nd edn), Chapters 5–13. Springer-Verlag, New York.

stretch A linear measure of the shape change (**strain**) of a body, defined as the ratio of the deformed length of a line to its undeformed length:

$$\text{stretch } S = \frac{\text{deformed length}}{\text{undeformed length}} = 1 + \textbf{elongation}$$

The stretch is greater than one for **extension**, and less than one for **contraction**. (See **quadratic elongation**.) [SB]

stretching direction/lineation Direction of maximum **extension** in a given plane, typically a **cleavage** plane. The stretching direction is often marked by a **lineation**, known as the 'stretching lineation', produced by the **elongation** of minute grains or grain aggregates or even fossils on the **cleavage** surface. Stretching lineations are useful in determining the **principal strain** axes where the **cleavage** corresponds to a **principal strain** plane, as does true *slaty cleavage*. In such a case, the stretching lineation is the *maximum strain axis*. [RGP]

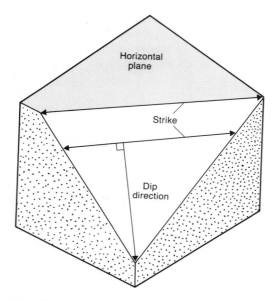

Fig. S53 A plane shown cutting across the corner of a block with a horizontal upper surface, with lines showing the **strike**.

striation Marking on the surfaces of pebbles and bedrock exposures in *glacial* environments which is produced by interference between particles and between particles and the bed. Studies in **glaciers** have shown that only a small proportion of *clasts* become striated, depending on such factors as debris concentration, relative velocities, *clast/bed* properties (e.g. hardness), etc. [ASG]

strike The unique direction of a horizontal straight line on an inclined planar surface. Strike is measured with a compass as an azimuth, and is used routinely to define the orientation of **bedding**, **faults**, **foliations** and other geological surfaces. On any surface, the **true dip** direction is 90° from the strike (Fig. S53).

A *strike line* is a line on a **geological map** parallel to the strike of a rock unit, **fault**, etc. Several *strike lines* are usually drawn, at uniform height intervals. For planar layers such as **bedding**, of constant **dip**, *strike lines* are straight and parallel, with uniform spacing on the map. An increase in angle of **dip** causes *strike line* spacing to decrease; *strike lines* aid fast interpretation of geological structure. A bed curved in space, i.e. its strike and **dip** both vary, has curved strike lines of non-uniform spacing, properly described as **structure contours**. (See also **form line**.) [SB]

strike-slip Relating to **displacement** parallel to the **strike** of the *displacement plane*; thus 'strike-slip **fault**', 'strike-slip **shear zone**'. *Oblique-slip* **displacements** may be divided into strike-slip and **dip-slip** components. (See **fault**.) [RGP]

strike-slip fault system Set of related **faults** on which the individual **displacements** produce a net **strike-slip displacement** in the system as a whole. Such fault systems are characteristic of *strike-slip tectonic regimes*, which are associated with major *conservative plate boundaries* (**transform fault** zones) but also occur within continental *plates* such as, for example, the numerous active *strike-slip fault* zones within the Asian *plate* north of the current **plate boundary** with the Indian *plate*. The main features of a continental **strike-slip** regime were analysed by Reading (1980) who pointed out the importance of differential vertical movements, which created uplifted blocks and depressed **basins** (see **pull-apart basin**).

Major **strike-slip** zones such as the San Andreas *fault zone* are of the order of 100 km across, within which quite complex tectonic effects take place. If the zone is considered as a deformable sheet, the **deformation** may be visualized as a combination of **pure shear** and **simple shear**, the **pure shear** component arising from net **extension** or compression across the zone, and the **simple shear** component from the **strike-slip displacement** of the boundaries of the zone. Considering only the **simple shear** component, the possible structures are summarized in Fig. S54, which shows the orientations of *normal faults*, *reverse faults*, and **conjugate** sets of *strike-slip faults*, as well as **folds** within the zone. The orientation of these subsidiary structures within a **strike-slip** zone thus conveys important information as to the sense of **strike-slip** motion. The addition of an **extensional** component across the zone produces **transtension** and a compressional component produces **transpression** (Sanderson & Marchini, 1984). **Transtension** results in clockwise rotation of the **extension** axis of Fig. S54 (i.e. for **dextral shear**) and **transpression** in an anticlockwise rotation. The reverse would hold for **sinistral shear**.

Other important geometrical effects result from the irregularity of the **fault** geometry. **Fault** offsets produce localized **strains** around the terminations that are either compressional or **extensional**, depending on the sense of movement (Fig. S55). Local changes of direction in *strike-*

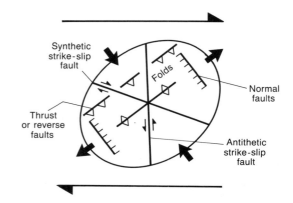

Fig. S54 Strike-slip fault system. Diagrammatic representation (in plan view) of the structural pattern produced by dextral simple shear. (After Reading, 1980.)

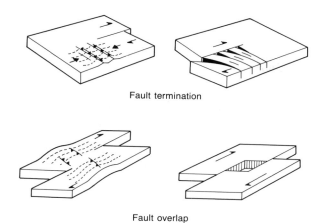

Fault termination

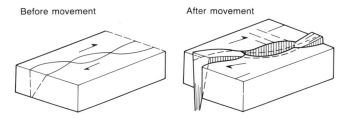

Fault overlap

Fig. S55 Strike-slip fault system. Local compressional and extensional structures produced by fault terminations and fault overlaps in strike-slip faulting. (After Reading, 1980.)

Before movement   After movement

Fig. S56 Formation of raised and depressed wedge-shaped blocks by local transpression and transtension on a branching **strike-slip fault system.**

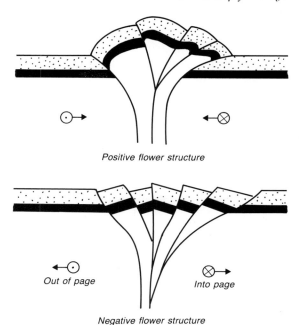

Positive flower structure

Out of page   *Into page*

Negative flower structure

Fig. S57 Strike-slip fault system. Positive and negative flower structures produced by convergence and divergence respectively in strike-slip motion. Dot and cross symbols within circles indicate out-of-page and into-page components of motion, respectively. (After Park, 1988.)

slip faults create local zones of **transtension** or **transpression**. In complex **fault** networks where branching **faults** have opposed **dips**, this process leads to alternate zones of raised and depressed blocks resulting from the convergence and divergence of wedge-shaped blocks (Fig. S56). The

combination of **folds** and **faults** produced in these local zones are termed *flower structures*. These are positive for uplifted blocks and negative for depressed blocks (Fig. S57). *Strike-slip duplexes* (Fig. S58) may be formed at bends in a *strike-slip fault* as a result of the progressive migration of the active **fault** into one wall. The *duplexes* may be either **transtensional** or **transpressional**, and are morphologically equivalent to extensional and thrust *duplexes* respectively. Large pieces of **crust**, or **terranes**, may become detached

Fig. S58 Strike-slip fault system. Diagrams illustrating the formation of strike-slip duplex structure in transpression (a) and transtension (b). Note that the structures are analogous morphologically to compressional and extensional dip-slip fault duplexes, respectively. (After Park, 1988)

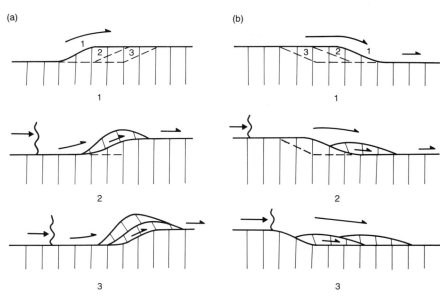

and isolated from one block, transferred to the opposing block, and transported far from their sites of origin, becoming a *displaced terrane*.

Major **strike-slip** zones, particularly those that constitute **plate boundaries**, are presumed to continue at depth as **strike−slip shear zones**. Studies of deeply eroded **Precambrian shield** regions demonstrate the importance of major **strike-slip shear zones** at deep crustal levels. Other **strike-slip** zones may detach on subhorizontal **décollement** zones within or at the base of the **crust**. [RGP]

Park, R.G. (1988) *Geological Structures and Moving Plates*. Blackie, Glasgow.
Reading, H.G. (1980) *Characteristics and Recognition of Strike-slip Fault Systems*. Special Publication of the International Association of Sedimentologists **4**: 7–26.
Sanderson, D.J. & Marchini, W.R.D. (1984) Transpression. *Journal of Structural Geology* **6**: 449–58.

strip mining Form of **open pit mining** in which strips of country are worked and the excavation is backfilled as mining proceeds. [AME]

stripping ratio The ratio of **waste** to **ore** removed during **open pit mining**. [AME]

stromatactis A type of cavity structure characterized by a smooth floor of internal sediment and an irregular roof, the infilling cement usually comprising fibrous and/or drusy **calcite**. The origin of stromatactis is unclear. Local seafloor **cementation** and scouring, and sediment **dewatering** may be responsible for these structures which are characteristic of mud-mound **limestones**. (See **geopetal structure**.) [DMP]

Pratt, B.R. (1982) Stromatolitic framework of carbonate mud-mounds. *Journal of Sedimentary Petrology* **52**: 1203–27.

stromatolite Laminated, microbial deposit, among the oldest fossils known. Stromatolites are typically calcareous and *cyanobacteria* are probably the principal organisms involved, together with *bacteria*, and *algae* such as diatoms and chlorophytes. Most stromatolites occur in shallow aqueous environments at normal temperatures, although metalliferous deep-sea and siliceous *hot spring* examples have been reported. Stromatolites are often referred to as *algal mats* because *cyanobacteria* are also known as *blue-green algae*. They exhibit major changes during their long history, the most important of which relate to the grazing, disturbing and displacive effects of animals. *Metazoan* evolution resulted in a rapid decline in the Late *Precambrian* from which stromatolites never recovered.

DEFINITION. Genetically, stromatolites can be defined as organo-sedimentary structures produced by sediment trapping, binding and/or **precipitation** as a result of the growth and metabolic activity of micro-organisms, prin-

cipally cyanophytes. Descriptively, they can be regarded as an attached, laminated lithified sedimentary growth structure, accretionary away from a point or limited surface of initiation. However, the precise biological origins of fossil stromatolites can rarely be established with certainty and their physical appearance is liable to be confused with other laminated calcareous deposits such as **calcrete** and *sinter*.

CLASSIFICATION. Three processes contribute to stromatolite formation: (i) trapping of sedimentary particles; (ii) calcification (*biomineralization*) of organic tissue; (iii) surficial **precipitation** of minerals on organisms and/ or sediment. These result in the following types of stromatolites:

1 Agglutinated stromatolites, produced by trapping/ binding of sediment;
   (a) fine-grained, well-laminated;
     (i) smooth lamination, subaqueous origin;
     (ii) crinkly lamination, with **fenestrae**, due to exposure and desiccation;
   (b) coarse-grained, crudely-laminated.
2 Skeletal stromatolites, produced by *biomineralization* of organisms during life.
3 Tufa stromatolites, produced by **precipitation** of minerals on organic substrates.

At the present day, trapping is far more important than calcification in marine stromatolites whereas *biomineralization* and **precipitation** are usually the dominant processes in freshwater stromatolites. Large grains and calcification disrupt lamination so that coarse-grained and skeletal stromatolites tend to be crudely laminated. Rapid **precipitation** in *tufa* may result in thick laminae. In contrast, fine-grained agglutinated stromatolites, especially those in subaqueous environments, may develop delicate regular lamination, which in some cases has been interpreted as diurnal. Exposure of stromatolites, as on tidal flats, results in irregular laminae due to desiccation. Most other stromatolites are essentially subaqueous in origin.

Tufa stromatolites are characteristic of fluviatile environments and commonly occur as *nodules* (**oncoids** or *oncolites*). Skeletal stromatolites contain the organisms responsible for their formation preserved as calcified fossils. They resemble tufa stromatolites but lack the precipitated crusts which form in tufa during or after the death of the organisms. Skeletal stromatolites are marine or lacustrine but are not known from Recent marine environments although they were locally common in shallow seas during the **Paleozoic** and **Mesozoic**. Agglutinated stromatolites dominate the geological record, and the vast majority are fine-grained. However, well-known modern stromatolites at **Shark Bay**, for example, are coarse-grained.

Despite these important variations in internal **fabric**, most description of stromatolites has concentrated on their external morphology. Attached stromatolites may be flat (stratiform), domical, columnar, or branching in side view. These morphologies do not necessarily reflect the degree of original relief (*synoptic profile*), but may merely be products of repeated growth of essentially flat mats on the same spot. However, if the laminae are high and enveloping then this does indicate relief. Stromatolites can form large *reef*-like masses. **Oncoids** (nodular stromatolites) develop around a loose nucleus, rather than being attached to the substrate. Flat (stratiform) stromatolites are typical of tidal flats, whereas columnar forms and **oncoids** are characteristically subaqueous.

Initially, classification of morphotypes followed a binomial Linnean system, e.g. *Cryptozoon proliferum* Hall. However, the terms 'groups' and 'forms' have been used in preference to 'genera' and 'species' for stromatolites because they are organo-sedimentary structures. Groups are based on column morphology and lamina shape; forms are based on finer details of morphology, together with microstructure. In 1964, Logan *et al.* proposed a geometric classification with the following major categories: laterally linked hemispheroids (LLH), vertically stacked hemispheroids (SH), and spheroidal structures (SS). In practice, geometric terms have been applied to **Phanerozoic** stromatolites whereas the binomial system has been retained for the more complicated **Precambrian** forms. The latter are generally divided into branched (e.g. *Baicalia*, *Gymnosolen*, *Tungussia*), falsely branched (e.g. *Kussiella*), unbranched (e.g. *Conophyton*), and stratiform (e.g. *Stratifera*) types.

Thrombolites have a macroscopic clotted **fabric**. They are not laminated, and cannot therefore be strictly regarded as stromatolites, although they have a similar external morphology, are believed to be microbial in origin, and sometimes show intergradation to stromatolites. They are most common in the **Cambrian** and **Ordovician**, and form domes or simple columns up to 1 m in height and width which may be amalgamated into **biohermal** masses. The origin of *thrombolites* is currently uncertain. They may be disturbed (? burrowed stromatolites), or a type of calcified *cyanobacterial* deposit, or both. Their concentration in the early **Paleozoic** suggests that they may represent a microbial sedimentary response to *metazoan* evolution following the crisis affecting stromatolites in the late *Precambrian*.

HISTORY OF RESEARCH. The problems presented by stromatolites are gradually being resolved, mainly through study of Recent analogues. Key developments are as follows:

1908 E. Kalkowsky, working with the **Triassic** of northern Germany, introduced the term stromatolith (stroma, layer; lithos, rock).

1914 C.D. Walcott compared *Riphean* stromatolites of the Belt Supergroup in Montana with Recent freshwater *tufa* deposits. This was the first clear connection to be made between stromatolites and *cyanobacteria*.

1933 M. Black discovered modern intertidal stromatolites on Andros Island (Bahamas) and recognized that they formed by trapping particulate sediment.

1961 B.W. Logan compared columnar stromatolites at **Shark Bay**, Western Australia, with *Precambrian* forms.

1970 P. Garrett explained the general limitation of modern marine stromatolites to tidal flats as being due to grazing by invertebrates, particularly *gastropods*.

1982 B.R. Pratt suggested that coarse bioclastic sediment also contributed to the **Phanerozoic** decline of stromatolites.

1983 J.J. Dravis recognized Recent columnar stromatolites in normal seawater at Eleuthera, Bahamas, and attributed their development to tidal currents sweeping **sand** over them and thereby inhibiting grazing invertebrates.

1986 R.F. Dill, E.A. Shinn, A.T. Kelly and R.P. Steinen discover similar, but larger, 'giant' stromatolites from near Lee Stocking Island, Bahamas.

1988 S.M. Awramik and R. Riding reported a significant diatom component in **Shark Bay** subtidal columnar stromatolites, and suggested that *algae* (as opposed to *cyanobacteria*) are important in trapping coarse sediment. They contrasted fine-grained, well-laminated *cyanobacterial* stromatolites with coarse-grained, crudely laminated, *algal–cyanobacterial* stromatolites and doubted whether the essentially coarse-grained **Shark Bay** and Bahamian columnar stromalites are good analogues for most **Precambrian** stromatolites.

Major changes in stromatolite formation through time correspond to physical and biological alterations in the environment. Six stages can be recognized:

1 Archean. The earliest (3.5 Ga Warrawoona Group, Western Australia) stromatolites are probably *bacterial*. They occurred in a nearshore environment and are currently the earliest confidently recognized fossils.

2 Proterozoic, to end-*Riphean*. The golden age of stromatolites, initiated by the appearance of *cyanobacteria* near the end of the **Archean**. Presumably these were formed mainly by trapping of fine sediment. Greatest diversity and abundance occurred during the *Riphean* (800–1650 Ma) and workers in the former USSR have pioneered their use as guide-fossils.

3 *Vendian*. The appearance of *metazoans* initiated a major decline in marine stromatolites.

4 Cambrian. Stromatolites formed by sediment trapping continued to decline due to *metazoan* disturbance and due to the coarseness of much bioclastic sediment. However, *biomineralization* of *cyanobacteria* led to the formation of skeletal stromatolites in marine environments. *Thrombolites* appeared near the *Precambrian–Cambrian boundary* and were common in the early **Paleozoic**.

5 Ordovician to **Cretaceous**. Fine-grained agglutinating stromatolites progressively became restricted to tidal flats, whereas skeletal stromatolites retained moderate success in marine environments.

6 *Cenozoic.* Skeletal stromatolites disappeared from marine environments, possibly due to changes in seawater chemistry lowering CaCO₃ **precipitation** rates. Formation of coarse-grained stromatolites was initiated by association of *diatoms*, chlorophytes and *cyanobacteria* in an *algal–cyanobacterial* consortium. At the present day, fine-grained agglutinating stromatolites are most widespread on tidal flats; coarse-grained stromatolites occur very rarely in subtidal marine environments where salinity or sediment movement inhibit grazers; and skeletal and tufa stromatolites are associated with calcareous lakes and streams.

Stromatolites continue to present a challenge to research. They are more varied than is often imagined and reflect important changes in life and environments at the Earth's surface throughout their long history. [RR]

Flügel, E. (ed.) (1977) *Fossil Algae, Recent Results and Developments.* Springer-Verlag, Berlin.

Logan, B.W., Rezak, R. & Ginsburg, R.N. (1964) Classification and environmental significance of algal stromatolites. *Journal of Geology* **72**: 68–83.

Walter, M.R. (ed.) (1976) *Stromatolites.* Developments in Sedimentology No. 20. Elsevier, Amsterdam.

stromatoporoids Problematic extinct organisms with mesh-like skeletons, common in **Paleozoic** (especially **Ordovician** to **Devonian**) carbonate sediments. They occur as sheet-like, domed, discoidal or dendroid masses, often forming *reef* structures in association with **corals**. The skeleton, or coenosteum, is built of vertical elements such as pillars and of horizontal laminae and tabulae (Fig. S59). Some forms contain horizontal systems of stellate branching canals linked by discontinuous vertical tubes; these astrorhizae appear to have permitted the removal of excurrent waters from galleries in the coenosteum and may be represented by stellate grooves on the upper surface. The skeletal surface also commonly displays polygonal markings and may have small swellings (mamelons) at intervals. Although always preserved as **calcite**, the original skeleton may have been **aragonitic** in composition.

The classificatory status of the stromatoporoids is controversial. Stearn (1982) regarded them as multicellular animals with a grade of organization somewhere between the **Porifera** (**sponges**) and *Cnidaria* (**corals**), while other authors have allied them to specific **sponge** or *coelenterate* groups. They have been linked at various times with *sclerosponges*, *scleractinian corals*, *tabulate corals* and *hydrozoans*, or have been assigned to their own separate phylum. The discovery of filamentous and coccoid *cyanobacteria*-like microfossils in some well-preserved **Devonian** specimens led Kazmierczak (1976) to suggest a *cyanobacterian* (blue-green) affinity, but most specialists are sceptical, considering the microfossils to represent symbionts or contaminants.

An assignment to the **Porifera** found increased favour with the discovery of living *sclerosponges* in Jamaican waters. These encrusting forms possess a skeleton of

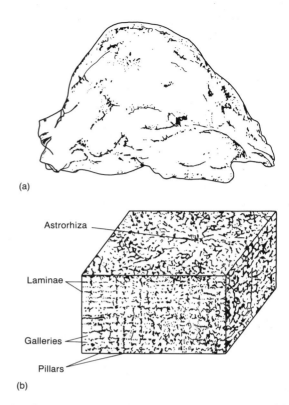

Fig. S59 Stromatoporoids. (a) Domal stromatoporoid coenosteum. (b) Diagram of coenosteum structure. (After Murray, 1985.)

calcareous or siliceous spicules, which compare with spicule pseudomorphs recognized in the calcified structures of **Mesozoic** genera that have been referred to the stromatoporoids. Spicules have not been found in **Paleozoic** stromatoporoids, which may belong to a separate group exhibiting only convergent features to those of the **Mesozoic**. In the absence of a continuous stratigraphic record, it is arguable that conclusions based on **Mesozoic** taxa should not be extrapolated to the **Paleozoic**.

The earliest unequivocal stromatoporoids are from Lower **Ordovician** rocks, although some early **Cambrian** stromatoporoid-like fossils of uncertain relationships have been reported from the former USSR. The group diversified during the mid-**Ordovician**, but important **extinctions** at the end of the period resulted in the disappearance of large, specialized, columnar forms. In the **Silurian** and for much of the **Devonian**, stromatoporoids associated with **corals** as important *reef* and **bioherm** builders. *Reef* masses composed dominantly of stromatoporoids reach diameters of 200 m in the **Silurian** strata of Gotland, Sweden. In both *reef* and non-reef carbonate environments the shape of stromatoporoid skeletons is strongly affected by sedimentation and environmental energy. The late **Devonian** (*Frasnian/Famennian*) **extinction** event saw the collapse of the **coral**–stromatoporoid *reef* community and the disappearance of many genera, although stromatoporoids continued to be represented on a diminished scale to the end of the **Paleozoic**. [RJA]

Kazmierczak, J. (1976) Cyanophycean nature of stromatoporoids. *Nature* **264**: 49–51.
Murray, J.W. (ed.) (1985) *Atlas of Invertebrate Macrofossils*. Longman, Harlow.
Stearn, C.W. (1982) The unity of the Stromatoporoidea. *Proceedings of the 3rd North American Paleontological Convention* **2**: 511–6.
Webby, B.D. (1986) Early stromatoporoids. In: Hoffman, A. & M.H. Nitecki (eds) *Problematic Fossil Taxa*, pp. 148–66. Oxford University Press, New York.

stromeyerite $((Ag,Cu)_2S)$ A rare **ore** mineral of **silver** and **copper**. [DJV]

strontianite $(SrCO_3)$ A **carbonate mineral** used as a source of **strontium**. [DJV]

strontium (Sr) At. no. 38, at. wt. 87.62. An alkaline earth metal. Strontium compounds are used in pyrotechnics and flares, glasses and ceramics, permanent magnets, etc. Obtained commercially from **evaporite deposits**. [AME]

strontium isotope analysis Strontium isotope ratios, determined by **mass spectrometry** can be used to distinguish between sources of **obsidian**, **alabaster** and **marble** used in antiquity, and to relate artefacts of these rocks to their quarries. [OWT]

structural relations of rivers *Anteconsequent streams* are those which flow as consequents on an early **uplift** but antecedent to later stages of the same **tectonic uplift**. *Consequent streams* are those which flow in the direction of the original slope of the land surface. *Inconsequent streams* are those not related to land surface features or major geological controls, but follow minor surface features without being developed into an organized pattern overall. *Insequent streams* are those which develop as a result of features which are not determinable. *Obsequents* are those *streams* which are the tributaries of a *subsequent stream* and flow in a direction opposite to the regional **dip** of the land surface. *Resequent streams*, while they follow the original direction of drainage, developed at a later stage. This might apply to streams on the back slope of a **cuesta** of resistant rock which did not outcrop on the originally exposed land surface which guided initial stream development. *Subsequent streams* are those which follow a course determined by the structure of the local bedrock. [ASG]

structural terrace Local flattening in an area of more steeply inclined *strata*. The converse structure is termed a **monocline**. [RGP]

structure Any geological feature defined geometrically. A fundamental subdivision of structures may be made into sedimentary, igneous and **tectonic**. A **tectonic** structure is produced by **deformation**, and *structural geology* is that branch of the Earth sciences dealing with this type of structure. Deformational structures are geometrical arrangements of planes, lines and surfaces whose orientation and shape are the result of interaction of pre-existing rock bodies with deformational forces. **Folds**, **faults**, and **fabrics** are all important types of deformational structure. [RGP]

structure contour A line, usually curved, drawn on a geological map to follow constant height on a geological surface, i.e. parallel to **strike** at each point. If the surface is planar, the line is straight, and called a *strike line*. Curved and **folded** surfaces such as **beds** have curvilinear structure contours which, drawn at uniform intervals of height, represent their shape in the same way as contours on a topographic map indicate the land surface shape. [SB]

stylolite A surface within a rock along which **dissolution** has taken place. In carbonate rocks they are common, and often marked by seams of insoluble **clay minerals**, remaining when dissolved **calcite** has diffused away. Stylolites may be planar, wavy or jagged, and can result from both **diagenesis** and **deformation**; they form by **pressure solution**. Examples of stylolites are shown in **pressure solution**. [SB]

subcrop (1) Subsurface *outcrop*. A stratigraphical formation may intersect a subsurface plane, e.g. an **unconformity** or **fault**, in a subcrop, which represents the area of the plane lying between the lines of intersection (**feather edges**) of the boundaries of the formation.
(2) In mining geology: any near surface development of a rock or **orebody**, usually under superficial deposits. [AME/RGP]

subduction suction Mechanism whereby a subducting slab may exert an **extensional** force on the upper *plate* of a **subduction zone**. The mechanism was originally proposed by Elsasser (1971) as *trench suction* and renamed 'subduction suction' by Bott & Kusznir (1984). The corresponding force on the downgoing *plate* is termed *slab pull*. The force is a gravitational effect created by the negative buoyancy of the descending slab. Because the upper and lower *plates* are in contact, force may be transmitted from the descending *plate* into the upper *plate*. (See **forces acting on plates**.) [RGP]

Bott, M.P. & Kusznir, N.J. (1984) Origins of tectonic stress in the lithosphere. *Tectonophysics* **105**: 1–14.
Elsasser, W.M. (1971) Sea floor spreading as thermal convection. *Geophysical Journal of the Royal Astronomical Society* **70**: 295–321.

subduction zone The region in which oceanic **lithosphere** is carried down into the **mantle** (Fig. S60). It is mainly

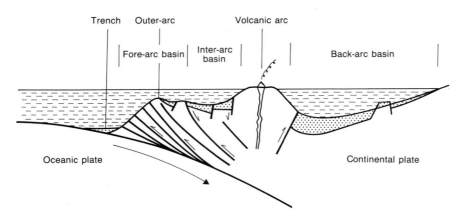

Trench   Outer-arc   Volcanic arc

Fore-arc basin | Inter-arc
                 basin   Back-arc basin

Oceanic plate   Continental plate

Fig. S60 Subduction zone. Generalized classification of tectonic elements and basins in a zone of crustal destruction. In an oceanic realm, the back-arc basin will be one-sided and may be expanded through secondary crustal spreading. (After Stoneley, 1981.)

marked by a series of **geophysical** and topographic **anomalies**, combined with **island arc** sequences of volcanic activity, and may or may not include a *marginal basin* between the **island arc** and continent. The main topographic feature is the *oceanic trench*, which is normally convex towards the **lithosphere** being subducted, but some are linear, i.e. lie on **great circles** (e.g. the Tonga–Kermadec Trench). The trench may be almost empty of sediments or almost entirely filled with sediment that is commonly transparent to **seismic waves** passing through it, i.e. appears to have no systematic structure. The topography of the trench is paralleled on the oceanic side by a slight rise in the ocean floor and, over the subducted **lithosphere**, a complex topography comprising a **fore-arc basin**, overlying an **accretionary prism**, the volcanic **island arc** itself, which may form part of the continent (as in the Andes), relatively small continental blocks (as in Japan), or entirely **oceanic crust** (as in Tonga–Kermadec). The main **geophysical anomaly** is the strong **seismicity**, particularly along the **Benioff–Wadati zone(s)**, associated with the subducted **lithosphere**, and also **earthquake** activity resulting from the upward migration of volcanic fluids in the **mantle** overlying the **Benioff–Wadati zone**. The **gravity** field, as observed at the Earth's surface, is also strongly distorted, with a positive **gravity anomaly** over the topographic high on the oceanward side, a very strong negative anomaly which is very slightly offset from the greatest depth of the trench towards the **Benioff–Wadati zone**, and a large positive **gravity anomaly** associated with the volcanic arc itself. However, the **gravity** field as determined from satellites shows little disturbance over these zones, suggesting that the sources of the anomalies are within the upper 100 km or so. The strong negative **gravity anomaly** is now thought to be mainly associated with very low density sediments within the **accretionary prism** close to the trench itself. There are no clear **magnetic anomalies**, other than a higher intensity associated with the volcanic rocks of the **island arc**. **Heat flow** is very low within the *oceanic trench*, but locally very high along the volcanic arc, and this is reflected in the degree of metamorphism found in both present and ancient sub-

duction zones. The distributions of **heat flow**, pressure and hydration (from the subducting plate) result in a high pressure–low temperature metamorphic zone (such as *prehenite–pumpellyite*) close to the trench, and a higher grade, hydrous (**amphibolite**) metamorphic zone close to the **island arc**, thus forming a **paired metamorphic belt**. The systematic variation in chemical composition of the volcanics overlying the **Benioff–Wadati zone** is attributed to fluids (mostly from **serpentines** when they reach temperatures of *c.* 600°C, and from **amphiboles** when the ambient temperatures reach *c.* 800°C), driven from the subducting oceanic **lithosphere**, which rise into the **mantle** rocks overlying it. The ascent of these hydrous fluids carries heat upwards, but the main cause of *partial melting* and the formation of **magmas**, is probably due to the effect of these fluids in reducing the melting temperature of the minerals within the **mantle**, even without an increase in actual temperature, and hence the formation of **magmas** the composition of which varies systematically as a function of the ambient pressure, temperature and water pressures at different depths along the **Benioff–Wadati zone**. The actual mechanisms causing the oceanic *plate* to be subducted are not established. As newly created oceanic **lithosphere** moves from an *oceanic ridge*, it cools and becomes thicker. When sufficiently thick and cold, the **basalt** could convert to *eclogite*, with a density of some $3.39 \, \mathrm{Mg \, m^{-3}}$, causing a negative buoyancy. However, there are subduction zones in the south-west Pacific (e.g. Solomons and the New Hebrides) where thin, young **lithosphere** is being subducted in preference to older Pacific **lithosphere**. It is also possible that **convection** currents actually transport the oceanic **lithosphere** downwards, aided by any increase in density by such phase changes, but not dependent on them. However, when the oceanic **lithosphere** reaches a depth of some 250 km, major increases in density occur, as in the **upper mantle**, but these occur at somewhat higher levels in the colder subducting *plate* than in the surrounding **mantle**. There is thus a major gravitational force causing sinking of the oceanic **lithosphere**, which is further increased as even higher pressure phase changes occur, again at somewhat higher levels than in the surrounding, warmer **mantle**. Such forces result in a **subduction pull** that is commonly considered to be the driving force of *plate tectonics*. However, while such forces are real, there is evidence that they may be only locally important as the subducted **lithosphere** often appears to have become decoupled from the shallower **lithosphere**. [DHT]

Barazangi, M. & Isaacks, B. (1971) Lateral variations of seismic wave attenuation in the upper mantle above the inclined earthquake zone for the Tonga Island Arc: deep anomaly in the upper mantle. *Journal of Geophysical Research* **76**: 8493–516.
Gill, J. (1981) *Orogenic Andesites and Plate Tectonics.* Springer-Verlag, Berlin.
Green, D.H. (1980) Island-arc magmatism and continental-building magmatism — a review of petrogenic models based on experimental petrology and geochemistry. *Tectonophysics* **63**: 367–85.
Karig, D.E. & Sharman III, G.F. (1975) Subduction and accretion in trenches. *Geological Society of America Bulletin* **86**: 377–89.
Stoneley, R. (1981) In: Tarling, D.H. (ed.) *Economic Geology and Geotectonics.* Blackwell Scientific Publications, Oxford.
Toksöz, M.N., Uyeda, S. & Francheteau, J. (eds) (1980) *Oceanic Ridges and Arcs.* Elsevier, Amsterdam.

subgrain A region of a *crystal lattice* differing in orientation from the surrounding mineral but without internal distortion. Subgrains are bounded by three-dimensional **crystal defects**, called subgrain walls or boundaries. These comprise ordered arrays of **dislocations**, which accommodate the lattice misorientations. Increased ordering of **dislocations** indicates **recovery**. Arrays of subgrains give rise to undulose extinction in minerals such as **quartz** in thin section. Subgrain shapes can be equant or elongate in section. Their shape appears to be controlled by planes within the crystal lattice, called *slip planes*, along which **dislocations** move. [SB]

submarine canyon A steep sided or canyon-like trench cut into the continental shelf and sometimes crossing the continental slope into deep water. Close to the coastline the smooth walls of the trench are often overhanging and canyon-like but further seaward the trench may be only slightly incised into the surface of broad submarine *fans*. The canyon gradient is usually steeper than a river valley but similarities include knick points and tributary valleys. Globally, 81% of submarine canyons have some relationship with river valleys, past or present. For example, Cap Breton canyon in south-west France lies off the 15th century mouth of the Adour River which now exits to the south. The relationship between canyons and river valleys has led to the view that fluvial incision during periods of low sea-level was responsible for canyon development. The great abundance of submarine canyons in the Mediterranean is held as evidence of this process, the canyons being later flooded by subsidence and sea-level rise. An alternative view sees canyons as products of **erosion** by submarine **turbidity currents**; fast (1.3 m s^{-1}), high density flows that have been recorded in many canyons and capable of moving large volumes of sediment. Whatever their origin it is likely that, once cut, the

canyons may be kept open by **turbidity currents**. Many canyons show two or more stages of incision with filling in the intervals and dating has shown that they existed as canyons as far back as the **Pliocene**. [JDH]

Shepard, F.P. (1978) *Geological Oceanography.* Heinemann, London.

submarine fan (*deep-sea fan, deep-water fan*) A broadly convex-up deep-water siliciclastic or carbonate system with positive relief above the adjacent **basin** floor, and developed from a point-source sediment supply, or several closely-spaced point-sources, such as **submarine canyons**. *Fans* generally show a broadly radial geometry unless severely constrained by topographic highs. Dimensions range from 10–30 km in the **basins** offshore from the Californian Borderland, to about 3000 km in length for the Bengal Fan. Many *continental margin* fans are located on the lower continental slope, rise and even **oceanic crust** near the ocean–continent boundary. Submarine fans tend to **aggrade** and prograde most rapidly during periods of relative low-stands in sea-level. Fans typically show a threefold division of environments from the feeder canyon(s) to the **basin plain**: (i) upper or inner fan, with one or several large channel-**levée** complexes; (ii) middle fan, with a complex system of anastomosing, meandering and/or braided channel-**levée** systems, and (iii) lower or outer fan, with small, ephemeral, discontinuous channels and lobate sheet-like sediments called 'lobes'. [KTP]

submerged forest An area of forest vegetation, generally a layer of **peat** with eroded tree stumps in the growth position, that has become inundated by the sea. Exposed either intertidally or below low **tide** mark, it is indicative of a rise in relative sea-level since the period of forest growth. Accurate heighting and dating of the organic material gives an indication of the amount of **eustatic** sea-level rise, or of **isostatic** land submergence, since the forest grew. Most of the **peat** beds and tree stumps exposed on the foreshores of Britain are post-glacial in age but some are older, e.g. in East Anglia, middle *Pleistocene* forest beds occur close to high **tide** mark. [JDH]

subsurface fluid migration **Pore-fluids** within a sedimentary **basin** are rarely static, but tend to move along permeable *carrier beds* from regions of relative high pressure to regions of relative low pressure. This fluid flow within a sedimentary **basin** encompasses subsurface fluid migration.

Most sedimentary formations have the ability to transmit fluids, but the rate of flow varies greatly depending upon a number of variables, including the **permeability** of the sediment, type of fluid involved (water, oil, gas) relative balance of opposing flows, gross sedimentary geometry, *hydraulic conductivity*, **aquifer** gradient and overall hydrodynamic regime of the **basin**. In the simplest terms,

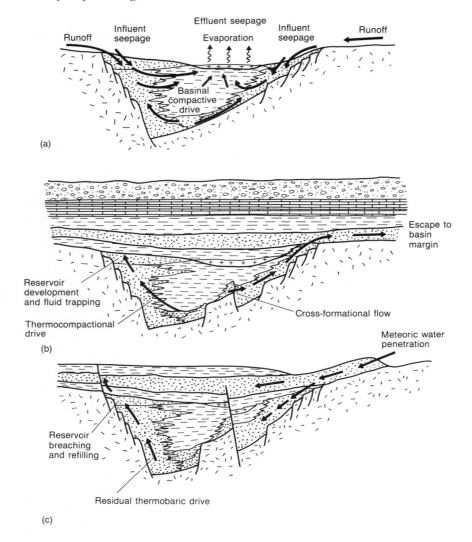

Fig. S61 Subsurface fluid migration.
Schematic development and evolution of
hydrodynamic regimes in an idealized
sedimentary basin. (a) Active subsidence in an
immature basin with opposing compactive and
eogenetic pore-water flows. (b) Mature basin
dominated by thermocompactional flow. (c)
Supermature basin following structural inversion
with topographic head-driven meteoric water
ingress.

three types of sedimentary **basins** can be defined in the
framework of hydrodynamic regimes (Fig. S61):

1 Immature **basins**, undergoing active subsidence which
results in gradual compaction of the component sediments
as burial progresses. **Meteoric groundwater** circulation
may be established on the margins of continental **basins**
but penetration into the **basin** will be opposed by overburden pressures from the deeper parts of the **basin**
where **pore-waters** are expelled from compacting **mudrock**
lithologies. The expelled waters will thus include **connate
water** deposited with the sediment and **meteoric water**
buried below the zone of active **meteoric water** circulation.

2 Mature **basins** which have attained sufficient temperatures and pressures in their deeper parts to initiate dehydration reactions of **clay** and other hydrous **minerals**
which may release large volumes of water. Organic matter
undergoes **decarboxylation** releasing water and *hydrocarbons*. The resulting thermocompactional flow is driven
by temperature and pressure gradients generated by either
phase changes (generation of *hydrocarbons*, release of

mineral-bound water, etc.) or lithostatic loading. The
overall fluid movement is upwards and out towards the
basin margins although the exact pathway may be highly
tortuous.

3 Supermature **basins**, characterized by either fully compacted sediments or the condition where **compaction** has
been arrested through structural **inversion** and **uplift**.
Meteoric water ingress is the main fluid movement, and
depending upon the topographic head available, can be
driven for considerable distances into the **basin**. Water
flow is recharged by **infiltration** of atmospheric precipitation and surface **runoff** that moves down the topographic gradient in the direction of decreasing gravitational
energy.

As **basins** evolve with time, so will the hydrodynamic
regimes, and hence the style and type of fluid movement.
Fluid flow within compacting immature or early mature
basins will tend to move from deep basinal, low **permeability** media to high **permeability basin** margin zones.
The exact flow paths will be a function of the geometry

and interconnectivity of the various permeable conduits and, therefore, be largely constrained by the local geology. However, on a **basin**-wide scale, the overall tendency will be for fluid migration up-dip, along pressure, salinity or density gradients. Theoretically, flow will continue until prevented by stratigraphic, structural or **diagenetic** barriers. Once the mature stage has been attained in the **basin** development, copious quantities of water are produced as a result of dehydration reactions and cross-formational flow is widespread due to the high degree of inhomogeneity in **basin**-fill sequences and lithological restrictions to vertical flow. During the supermature stage, **meteoric waters** that penetrate to considerable depths are warmed under the ambient **geothermal gradient** and, as a result, become less dense and rise back towards the base **hydrological** level establishing major **groundwater** circulation. Once the base level is reached there is usually no head advantage to be attained by further movement, so there is little potential for lateral flow beneath the **basin** centre. In old, supermature **basins** characterized by long-term **tectonic** stability and **peneplanation** of the flanking topographic relief, the movement of **meteoric water** gradually slows and the hydrodynamic regime tends towards (although rarely achieves) effective stagnation. (See also **burial diagenesis, compaction, mass solute transfer, overpressuring, pore-waters and their evolution, water table**.) [SDB]

Bogomolov, Y.G., Kudelsky, A.V. & Lapshin, N.N. (1978) Hydrogeology of large sedimentary basins. In: *Hydrogeology of Great Sedimentary Basins*. International Association of Scientific Hydrology, Special Publication **120**: 117–22.
Magara, K. (1978) *Compaction and Fluid Migration — Practical Petroleum Geology*. Developments in Petroleum Science No. 9. Elsevier, Amsterdam.

subsurface mapping The production of subsurface maps is almost fully dependent on borehole data and the work is essentially of an applied nature in relation to a commercial project. Oil companies are foremost in the production and use of a wide variety of different types of subsurface plans, but they are also used in *mineral exploration* (particularly **open-pit**) and in **engineering geology**. A good treatment of this work in regard to oil is given by Bishop (1960), Low (1977) and Moore (1963).

All subsurface maps are prepared with a number of reference points, often widely spaced with no information between. All these reference points should be marked on the map with the relevant information recorded against them. From borehole data it is possible to establish the depth (and hence elevation relative to a base datum) of geological features such as sedimentary formations, the thickness, lithologies, etc. Analyses of samples from boreholes provide information, etc., on the rock types, chemical and **assay** data, all of which can be incorporated on the map. Only with cores is it possible to determine the amount of **dip** of the rocks, *fault planes*, etc., but because of rotation within the core barrel the direction cannot be determined accurately from a single core.

Problems are encountered when the borehole deviates from the line intended, since this can lead to false thickness recordings and incorrect depths. With vertical drilling the collar position of the hole and the position where it encounters the horizon of interest would have the same grid coordinates, but inclined holes will not, and in this case it is the position where the rock is encountered and not the collar position that should be recorded on the map.

The production of reliable and meaningful maps under these circumstances requires the application of a wide range of geological knowledge and experience in an area. However, the maps produced will be no better than the quality of the basic borehole data.

STRUCTURAL MAPS. These are contour maps of a chosen datum and illustrate the present subsurface attitude and disposition of the datum **bed** or structure. The elevations at the data points are contoured at an interval selected to depict the structure. Contouring should be done carefully with accurate interpolation and even spacing between control points, but, more importantly, with geological thought, since mechanical contouring can sometimes fail to show features which can be identified by a skilled and experienced interpreter. The contours should be curved and 'flow' evenly across the map, rather than be angular and abrupt as is the case when straight lines join control points.

Widely spaced data points will only bring out generalities in the geology, and whilst this may suffice for most cases of simple geology such as gently inclined, uniform sediments, in most cases it is inadequate. Wide spacing may be able to demonstrate that a **coal seam** is present for some kilometres ahead of present colliery workings, but it would not detect those small **faults** (2–3 m throw) which cause substantial mining problems. Structural maps should show *axial plane* traces of **folds** (or *fold hinge* lines); the position and amount of *want* on **extensional faults** (*fault plane* should also be contoured if possible), which would be detected in the drilling by an abnormal thinning of some beds, absence of beds, or an abrupt change in the amount of **dip** (strike **fault**) or **strike** (dip or oblique **faults**); and *thrusts* which would be identified through repetition of strata and abnormalities in the contour direction or spacing. Sections drawn through boreholes would aid in detecting these structures.

In **oilfield** exploration it would be necessary to produce **structure contour** plans at several depths (shallow, intermediate and deep) particularly if **unconformities** are present, the base of which should also be contoured. In **coalfield** exploration all seams of interest should be contoured, and in the case of **opencast coal** mining where there is a lot of shallow drilling, there should be enough control to show almost all structures. In metalliferous mining this technique is used most widely for **ores** of the sedimentary association where there is stratigraphic control over the distribution.

THICKNESS PLANS. These include **isopachyte** and *isochore* maps. An **isopachyte** (*isopach*) is a contour line representing equal stratigraphic (true) thickness and an **isopachyte** map is one where **isopachytes** show the variation in the true thickness of a formation. Low (1977) includes in this category a group of formations but Moore (1963) prefers to use the term *isochore* for these, regarding *isochore* maps as equal interval maps. A better use of the term *isochore* is for a map that shows contours of vertical (drilled) thickness of formations irrespective of their true thickness. It is vital that *isopach* and *isochore* values are not combined on the same map. However, subsurface maps representing drilled thicknesses are commonly called **isopachyte** maps, not *isochore*, and this can lead to difficulties in cases where the **dips** of the strata are greater than about 10°. In very gently inclined strata (as is the case of much **oilfield** drilling) it makes little difference to the final map but in areas of steep **dip** (e.g. some **Proterozoic stratiform copper deposits**), the distinction must be made. Where inclined drilling is used for exploratory purposes the *isochore* map has little meaning and all thicknesses represented on maps should be true thicknesses and hence **isopachyte** maps produced. Also, whilst it may be easier not to have to calculate true thicknesses from steeply dipping strata where the **true dip** is difficult to determine, or even more so in the case of inclined drilling oblique to inclined strata, for all aspects of **basin** analysis the maps should be **isopachyte** maps in the correct sense.

As for structural maps, thickness maps should also be contoured with care to show the variations in thickness based on geological reasoning and not just numerical values. A combined structural and **isopachyte** map is a very useful exploration tool.

FACIES MAPS. These maps illustrate the lateral changes in lithology of a formation, group or system within a sedimentary **basin** and in order for them to have any meaning that stratigraphic unit must be clearly defined. These maps enable complex stratigraphic data to be presented in a simple and readily appreciated form. There are many types of lithofacies maps (Bishop, 1960; Low, 1977).

A *ratio contour map* shows the ratio of the aggregate thickness of one lithological class to that of the remaining classes constituting the total stratigraphic unit section.

A *percentage contour map* shows the percentage value of one lithology in the total thickness of the stratigraphic unit. An *isolith map* shows the aggregate thickness of beds of one lithology in a stratigraphic succession composed of several lithologies. Where many lithologies are present, particularly if intertonguing occurs, it is necessary to produce several *isolith plans* (one for each lithology) for the stratigraphic unit. The sum of the thicknesses for any point would be the same as that for the **isopachyte** map.

Isofacies maps are produced to denote the limits of facies which may include several rock types, e.g. a sequence of alternating **shales** and **sandstones** or a carbonate facies

which is composed of several different types of **limestone** and **dolomite**. Lithofacies maps are best coloured as well as contoured since it is possible to show gradual changes from one lithology to another by alternating bands of colours giving a striped effect. *Alternation frequency maps* are produced to enhance the information on lithofacies maps which do not distinguish between a sequence of rapidly alternating thin units and one which has the same lithologies, but grouped into a few thick units. On these maps the number of alternations within the defined unit are determined at the data points and these values contoured.

In combination, **structure contour** maps, **isopachyte** maps and *facies maps* (Fig. S62) enable **basin** studies to be carried out to greatest effect and in skilled hands the amount of information that can be gained from them is surprisingly great when one considers how few and widely spaced are the actual data points. The range of contour maps that can be produced after detailed laboratory analysis is wide in the oil industry alone and could include **porosity** or **permeability** maps.

ASSAY AND OTHER MINING RELATED PLANS. *Assay plans* show the variation in **grade** and the distribution of metals throughout the **mineral** deposit. Only a very small sample is actually assayed and consequently care has to be taken to ensure that this is representative of the area influenced by the data points. **Grades** and metal content vary to a much greater extent than lithofacies and a different range of skills is required to contour and interpret the data. *Assay plans* can show, in addition to the contoured values of a metal (one plan for each metal in a polymetallic body), the ratio of one metal to another, e.g. lead:zinc ratio in a *Mississippi Valley type deposit*.

For **industrial minerals** quality parameters are of great importance, since a product has to meet the strict requirements of the consumers and consequently maps showing the distribution and amount of deleterious compounds need to be produced. Many paralic **coal seams** are fairly consistent in quality over the take of a colliery or **opencast mine** but *limnic coals* are much more variable. In order to plan mining efficiently, plans of these seams are produced to show such factors as the percentage distribution of **ash**, **sulphur** and volatile matter. **Coal seam** plans should also show areas of **washout** or other absences of the **coal** as well as seam-splits. In **opencast** coal **mining** in the UK, many areas explored as potential sites are those which have undergone previous underground working to varying degrees and consequently the seam plans show an impression of the extent to which the seams have been worked. This will be a statistical representation since even with close drilling it would not be possible to define individual pillars and cavities.

PALEOGEOLOGICAL MAPS. These are maps which show the distribution of rock types below an **unconformity**. It could be argued that maps which show the incrop of **beds**

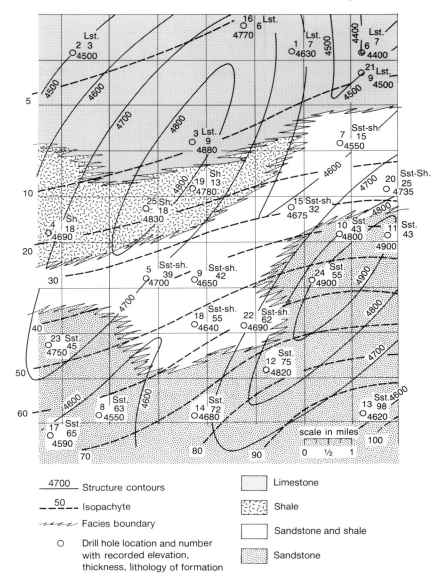

Fig. S62 Subsurface mapping. Combined structural, isopach and facies map of a formation. (After Low, 1977.)

4700 — Structure contours

---- 50 ---- Isopachyte

~~~~ Facies boundary

○ Drill hole location and number with recorded elevation, thickness, lithology of formation

Limestone

Shale

Sandstone and shale

Sandstone

into the base of the *Pleistocene* drift deposits are of this type and plans for **open pit mining** in drift-covered areas certainly need to show the limit of the deposit against the base of the drift. In the oil industry, reconstructions of pre-**unconformity** maps are produced from borehole data as an aid to detecting **unconformity** traps and it must always be borne in mind that **unconformities** need not be flat. The British Geological Survey have produced several maps of this type, such as the Pre-**Permian** Geology of South Britain. [DER]

Bishop, M.S. (1960) *Subsurface Mapping*. John Wiley & Sons, New York.

Low, J.W. (1977) Subsurface maps and illustrations. In: Le Roy, L.W. & Le Roy, D.O. (eds) *Subsurface Geology*. Colorado School of Mines, Golden, CO.

Moore, C.A. (1973) *Handbook of Subsurface Geology*. Harper & Row, New York.

sudoite $(Mg_2(Al,Fe^{3+})_3Si_3AlO_{10}(OH)_8)$ A member of the **chlorite** group of *phyllosilicate* minerals. [DJV]

suffosion The **erosion** of incoherent surface sediments by slumping into underground cavities produced by bedrock **dissolution**. In areas of **limestone** bedrock, suffosion may produce **dolines**. In Cheshire, north-west England, suffosion of *glacial* **sands** into solution cavities in underlying **halite** beds has formed **flashes**. [AW]

sulphide minerals These metal–sulphur compounds constitute the single most important group of **ore** minerals. They may be classified on the basis of their **crystal structures** into nine major groups as shown in Table S5. (Craig & Vaughan, 1990). Although several hundred sulphide minerals are now known, many are relatively rare phases.

Table S5 Structural groups of **sulphide minerals**

THE DISULPHIDE GROUP

Pyrite structure	*Marcasite structure*	*Arsenopyrite structure*	*Loellingite structure*
FeS_2 pyrite	FeS_2 marcasite	FeAsS arsenopyrite	$FeAs_2$ loellingite
CoS_2 cattierite		FeSbS gudmundite	$CoAs_2$ safflorite
			$NiAs_2$ rammelsbergite

derived by As/S ordered substitution
(Co,Fe)AsS cobaltite
(Ni,Co,Fe)AsS gersdorffite (I)

THE GALENA GROUP
PbS galena
α-MnS alabandite

THE SPHALERITE GROUP

Sphalerite structure	→ *derived by ordered substitution*	→ *stuffed derivatives*
β-ZnS sphalerite	$CuFeS_2$ chalcopyrite	$Cu_9Fe_8S_{16}$ talnakhite
CdS hawleyite	Cu_2FeSnS_4 stannite	$Cu_9Fe_9S_{16}$ mooihoekite
Hg(S,Se) metacinnabar	Cu_2ZnSnS_4 kesterite	$Cu_4Fe_5S_8$ haycockite

THE WURTZITE GROUP

Wurtzite structure	→ *composite structure derivatives*	→ *?further derivatives*
α-ZnS wurtzite	$CuFe_2S_3$ cubanite	$Cu_2Fe_2SnS_6$ hexastannite
CdS greenockite	$?AgFe_2S_3$ argentopyrite	

derived by ordered substitution
Cu_3AsS_4 enargite

THE NICKEL ARSENIDE GROUP

NiAs structure	→ *distorted derivatives*	→ *ordered ommission derivatives*
NiAs niccolite	FeS troilite	Fe_7S_8 monoclinic pyrrhotite
NiSb breithauptite	CoAs modderite	Fe_9S_{10}, $Fe_{11}S_{12}$ hexagonal pyrrhotite, etc.?

THE THIOSPINEL GROUP
Co_3S_4 linnaeite
$FeNi_2S_4$ violarite
$CuCo_2S_4$ carrollite

THE LAYER SULPHIDES GROUP

Molybdenite structure	*Tetragonal PbO structure*	*Covellite structure*
MoS_2 molybdenite	$(Fe,Co,Ni,Cr,Cu)_{1+x}S$	CuS covellite
WS_2 tungstenite	mackinawite	$c.Cu_3FeS_4$ idaite

METAL EXCESS GROUP

Pentlandite structure	*Argentite structure*	*Chalcocite structure*
$(Ni,Fe)_9S_8$ pentlandite	Ag_2S argentite	Cu_2S chalcocite
Co_9S_8 cobalt pentlandite		↘ *?derivative*
		$Cu_{1.96}S$ djurleite

Digenite structure	→ *derived by ordered substitution*	*Nickel sulphide structures*
Cu_9S_5 digenite	Cu_7S_4 anilite	NiS millerite
		Ni_3S_2 heazlewoodite

RING OR CHAIN STRUCTURE GROUP

Stibnite structure	*Realgar structure*	*Cinnabar structure*
Sb_2S_3 stibnite	As_4S_4 realgar	HgS cinnabar
Bi_2S_3 bismuthinite		

The most abundant members of this group are the minerals **pyrite**, **pyrrhotite**, **chalcopyrite**, **galena**, **sphalerite** and the group of **copper** sulphide minerals which includes **chalcocite** and **covellite**. [DJV]

Craig, J.R. & Vaughan, D.J. (1990) Compositional and textural variations of the major iron and base-metal sulphide minerals. In: Gray, P.M.J., Bowyer, G.J., Castle, J.F., Vaughan D.J. & Warner N.A. (eds) *Sulphide Deposits — their Origin and Processing*. Institute of Mineralogy and Metallurgy, London.

Vaughan, D.J. & Craig, J.R. (1978) *Mineral Chemistry of Metal Sulphides*. Cambridge University Press, Cambridge.

sulphur (S) A mineral occurring as a **native element**. Sulphur is recorded as a medicine for the treatment of skin disease in ancient Egypt. [OWT/DJV]

supergene enrichment An increase in the **grade** of mineralization brought about by secondary processes.

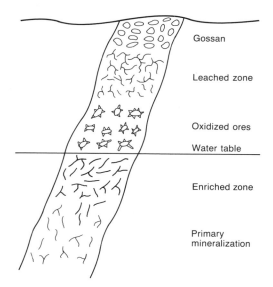

Fig. S63 Generalized section through a sulphide-bearing vein showing **supergene enrichment**. (After Evans, 1987.)

Labels on figure:
Gossan
Leached zone
Oxidized ores
Water table
Enriched zone
Primary mineralization

More commonly applied to the enrichment of sulphide deposits, the term supergene enrichment has been extended to include similar processes affecting oxide or carbonate **ores** and rocks such as those of **iron** and **manganese**. In supergene sulphide enrichment the metals of economic interest are carried down into *hypogene* (*primary*) *ore* where they are **precipitated** with a resultant increase in metal content. The upper portion of the *primary ore* from which the metals have been leached is termed **gossan**, see Fig. S63. In the case of **iron** and **manganese** ores it is chiefly the **gangue** material that is mobilized and carried away to leave behind a purer metal deposit. [AME]

Evans, A.M. (1993) *Ore Geology and Industrial Minerals: An Introduction* (3rd edn). Blackwell Scientific Publications, Oxford.

superimposed drainage A drainage pattern which bears little relation to the existing underlying rocks because the stream network was developed on a rock cover that has now been removed by **denudation**. This explanation is believed to account for some areas of discordant drainage such as in the Lake District of Britain. [NJM]

superimposition of structures **Deformation** of a pre-existing structure may be referred to as 'superimposition'. The term is used for **folds**, **foliations** and **fabrics**, but not for **faults**. Superimposed **folds** produce **fold interference structures**. (See also **overprinting, strain**.) [RGP]

Superior-type iron formation A type of **banded iron formation** (BIF) developed as thinly banded rocks belonging to the oxide, carbonate and silicate facies. They are usually free of clastic material. The rhythmical banding of **iron**-rich and **iron**-poor **cherty** layers, which normally range in thickness from a centimetre or so up to a metre, is a prominent feature and this distinctive feature allows correlation of BIF over considerable distances. Superior-type BIF may extend for hundreds of kilometres along **strike** and thicken from a few tens of metres to several hundred metres. It is stratigraphically closely associated with **quartzite** and **black** carbonaceous **shale** and usually also with *conglomerate*, **dolomite**, massive **chert**, **chert** breccia and argillite. Volcanic rocks are not always directly associated with this BIF, but they are nearly always present somewhere in the stratigraphical column.

The development of Superior BIF reached its acme during the early **Proterozoic**, and Ronov has calculated that BIF accounts for 15% of the total thickness of sedimentary rocks of this age. Stratigraphical studies show that BIF frequently extended right around early **Proterozoic** sedimentary **basins** and Gross (1980) suggested that BIF was once present around the entire shoreline of the Ungava Craton for a distance of more than 3200 km. [AME]

Gross, G.A. (1980) A classification of iron formations based on depositional environments. *Canadian Mineralogist* **18**: 215–22.
McConchie, D. (1984) A depositional environment for the Hamersley Group: paleogeography and geochemistry. In: Muhling, J.R., Groves, D.I. & Blake, T.S. (eds) *Archean and Proterozoic Basins of the Pilbara, Western Australia: Evolution and Mineralization Potential*, pp. 144–77. Publication No. 9, Geology Department & University Extension, University of Western Australia, Nedlands.

superparamagnetism **Ferromagnetic** particles that are very fine, usually less than 0.5 μm, are incapable of retaining a remanence for more than a few minutes, but for that period of time, behave as **ferromagnetic** particles. This gives them a stronger magnetization than that of even smaller particles while their magnetization is being measured within an applied field, but this remanence decays rapidly after removal of the field. [DHT]

superposition of strata The principle, first enunciated by William Smith, stating that each **bed** in a stratified sequence is younger than the underlying **bed** and older than the overlying **bed**. This is one of the basic laws of stratigraphy. The principle does not apply to strata that have been inverted in *recumbent folds* and **nappes**. [RGP]

suprafan Term used by some workers to describe the most rapidly **aggrading** part of **sand**-rich **submarine fans**, which shows strong positive relief as a convex-up mound above the surrounding *fan* surface. The term is generally synonymous with the middle *fan* and proximal parts of the outer *fan*, where many channels are developed. [KTP]

surf The mass of broken foaming water that forms as **waves** break on the coast. The *surf zone* is the area between the breaker zone and the **swash** zone and is characterized by bores of shoreward moving water formed from spilling or plunging breakers. On gently sloping *beaches* the *surf zone* may be very wide but on steep pebble **beaches** where the breakers impact directly on the **beach** face the *surf zone* is narrow. Expenditure of **wave** energy at the *breaker zone* and across the *surf zone* causes currents to develop both normal to the shore and alongshore. (See **longshore drift**.) [JDH]

Davies, J.L. (1980) *Geographical Variation in Coastal Development* (2nd edn). Longman, London.

suspended load Solid material transported in streams which is not in contact with the channel bed. The material is supported by the vertical velocity component of turbulent fluid eddies. It may be derived from wash **erosion** of hillslopes or from the channel bed. Suspended sediment transport increases dramatically with increasing **discharge** such that in most rivers most of the suspended sediment load is moved in just a few days each year. Suspended sediment is non-uniformly distributed with depth and across the channel so that in all but the smallest **turbulent** streams a sophisticated sampling procedure is needed to establish the suspended sediment concentration. Often this is achieved by using water samplers and filtration, though more recently *turbidity meters*, which incorporate a light source, have been used to provide continuous measurement. [TPB]

suspension The transport of sediment kept aloft within the body of a flow that does not come into contact with the bed (i.e. not **bedload**). Fully uninterrupted suspension demands that the upward vertical fluid velocities created by **turbulence** that are needed to keep grains in suspension must exceed the **settling velocity** of the particles themselves. Some grains may be kept in partial suspension by impacts from other grains that have been lifted from the bed in **saltation**. The suspended load includes both coarser suspended bed material and finer-grained sediment that can be suspended at the lowest **discharges**, this latter sediment often being termed the *wash load*. The finer sediment is usually more uniformly distributed throughout the flow depth whereas the coarser fraction of the suspended load is concentrated near the bed where the **turbulence** intensities are greatest. Consequently, suspended sediment concentration profiles characteristically display a negative exponential trend with distance away from the bed. [JLB]

Bagnold, R.A. (1966) *An Approach to the Sediment Transport Problem from General Physics*. Professional Paper of United States Geological Survey **422I**.
Vanoni, V.A. (1946) Transportation of suspended sediment by water. *Transactions of the American Society of Civil Engineers* **11**: 67–133.
Yalin, M.S. (1977) *Mechanics of Sediment Transport*. Pergamon, Oxford.

suture Surface, passing through the **crust**, separating the continental parts of two opposing *plates* in zones of *continent–continent collision*. *Orogenic belts* typically contain one or several sutures representing the sites of former collisions. Sutures may be identified from **paleomagnetic** or faunal mismatches across them. In **Precambrian shield** regions, where sutures might be expected to be frequent, identification is often more tentative and depends on criteria such as the identification of mafic complexes as **obducted ophiolites**. [RGP]

swale (1) An area of low-lying, often *marshy* land. Shallow depressions in undulating landscapes of *glacial moraine*, and low-lying areas enclosed by **sand dunes**, are termed swales.
(2) A shallow trough between storm ridges on a **beach**. [AW]

swallet/swallow hole A vertical or steeply sloping shaft where surface water disappears underground in **limestone** areas. The equivalent Slav term is **ponor**. A large and spectacular example is Gaping Gill Hole in the Craven district, northern England. [HAV]

swamp A waterlogged area with characteristic vegetation. Swamps form in several environments, and in both coastal and inland situations. **Mangrove swamps** are one important type of swamp. [HAV]

swash The shoreward moving uprush of **turbulent** water formed as a **wave** breaks on a **beach**. As the uprush derives its energy from the **wave**, the velocities reached are generally of higher magnitude but of shorter duration than the return flow or **backwash**, which is gravity-fed. **Beach** sediment moves onshore as a result of the swash velocities and although sediment is returned downbeach by the **backwash**, the coarser sediment may remain on the upper **beach** unmoved by the lower velocity **backwash**. The *swash zone* is the portion of the **beach** face alternately covered by the swash and exposed by the **backwash**. [JDH]

Komar, P.D. (1976) *Beach Processes and Sedimentation*. Prentice-Hall, Englewood Cliffs, NJ.

swash bar (*longshore bar*) **Sand** ridge that occurs on many **beaches**. Swash bars are thought to be produced by the steepening of a low gradient foreshore within the *swash zone* by constructional **waves**. They generally occupy positions on the foreshore where the **tide** remains for the maximum time, thus giving the **swash** the greatest time to build a steeper **swash** slope. There may be only one swash bar on a **beach** (if the tidal range is micro- to low-mesotidal) or multiple swash bars (if the **tidal** range is high meso-tidal or macro-tidal) forming a **ridge and runnel** system. It is likely that a combination of **swash** slope steepening and partial **wave** breaking and reformation, at higher stages of the **tide**, combine in promoting bar formation.

Swash bars are only found on **fetch**-limited coastlines

where shorter period, steeper **waves** are critical in their formation. They are annually destroyed by the first winter *storm*, only to be rebuilt during the fairweather period when they move slightly landward. [RK]

King, C.A.M. (1972) *Beaches and Coastlines* (2nd edn). Arnold, London.

sweep Inrush of relatively high velocity fluid that penetrates downwards into the **viscous sublayer** of the **turbulent boundary layer**. Sweeps are characterized by downstream velocities higher than the time mean average and a vertical velocity component towards the bed. These high velocity fluid inrushes are separated by lower velocity streak regions in the spanwise plane which display a spanwise spacing, λ_s, that is a function of the fluid **shear velocity**, $U_\star$, and **kinematic viscosity**, ν, such that

$$\lambda_s \cdot U_\star / \nu \simeq 100$$

The influence of fluid sweep events upon a cohesionless sediment bed is instrumental in the initiation of sediment movement and propagation of bedforms. (See **burst**, **primary current lineation**.) [JLB]

Allen, J.R.L. (1985) *Principles of Physical Sedimentology*. Allen & Unwin, London.
Smith, C.R. & Metzler, S.P. (1983) The characteristics of low-speed streaks in the near-wall region of a turbulent boundary layer. *Journal of Fluid Mechanics* **129**: 27–54.

syenite A phaneritic **alkaline igneous rock** composed essentially of **alkali feldspar** (exceeding 67% of the total **feldspar**) with **accessory quartz** or **nepheline** (5–20%) and termed respectively **quartz** and **nepheline syenites**. Syenites are hence compositionally equivalent to extrusive **trachyte** (**nepheline** syenite is equivalent to **phonolite**). Syenites occur in alkaline intrusive complexes, particularly those emplaced in continental environments. [RST]

sylvanite ($(Au,Ag)Te_2$) A rare **ore** mineral of **gold** and **silver**. [DJV]

sylvite (KCl) An **evaporite** mineral. [DJV]

symplectite A textural feature of certain **gabbroic** rocks in which bulbous **myrmekite**-like extensions of **plagioclase** crystals contain vermicular inclusions of *orthopyroxene*. Such intergrowths represent a late stage in crystallization or recrystallization of the rock. [RST]

syncline Fold containing younger rocks in its core. The term is also used synonymously, but incorrectly, with **synform**. A synclinal **fold** may close in any direction, downwards and sideways as well as upwards, and its attitude must be further defined by the use of terms such as **synformal**, or by a description of *limb* orientation. (See **fold**, **fold closure**.) [RGP]

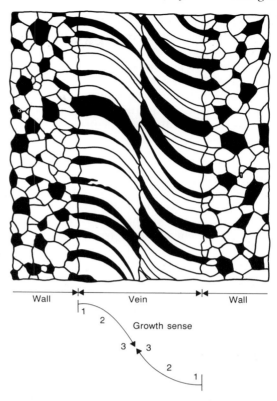

Fig. S64 Characteristics of **syntaxial veins**. The black and white areas are crystals of the same species but with differing crystallographic orientations.

synform Fold that closes downwards, i.e. in which the *limbs* converge downwards. [RGP]

syngenetic deposit Deposit which formed at the same time as the rocks in which it occurs. It is sometimes part of a stratigraphical succession like an **iron-** or **manganese**-rich sedimentary **bed**. Stratiform **chromite** deposits as in the **Bushveld Complex** (see also **orthomagmatic segregation deposit**) are igneous examples of syngenetic deposits. Many mineral deposits appear to have formed during the early or late **diagenesis** of sediments and are then referred to as *syndiagenetic*. [AME]

syntaxial vein/growth Vein filling grown from the walls towards the centre. Syntaxial growth is characterized by host and veinfill of different compositions, lack of crystallographic continuity across the **vein** wall, oblique contact of veinfill and **vein** wall, single crystallographic veinfills and inclusions of **wall rock** fragments along the median line and in subparallel bands (Fig. S64). The latter point is diagnostic of veinfill by the **crack-seal mechanism**. This mode of **vein** growth contrasts with *antitaxial growth*. [TGB]

Ramsay, J.G. & Huber, M.I. (1983) *The Techniques of Modern Structural Geology* Vol. 1. Academic Press, London.

syntaxy The cement on a crystal which grows in crystallographic (and hence optical) continuity with its substrate, resulting in a single enlarged crystal. This phenomenon is commonly observed in **limestones** containing **echinoderm** plates; these are often overgrown with a continuous rim of **calcite** in optical continuity with the skeletal element. Many **silica** cements on **quartz** grains in **sandstones** also form in this way. (See **diagenesis**, **epitaxy**.) [DMP]

synthetic Structures having the prevailing orientation or sense of asymmetry (**vergence**). The term is usually applied to **faults**, **shear zones** and particularly to **thrust belts**. For example, in a synthetic **thrust belt**, the orientation of the dominant *thrusts* is parallel to the plane of overthrusting in *continent–continent collision* belts, or of **underthrusting** in **subduction zones**. Within such a **thrust belt**, **antithetic** *thrusts* would have the opposite sense of movement (**backthrusts**). An **antithetic thrust belt** has the opposite sense of **overthrusting** to that of the nearest **plate boundary**. [RGP]

synthetic seismogram Computer-generated **seismogram**, derived by using mathematical expressions describing the propagation of **seismic waves** in a particular geological structure. They are a valuable aid to interpreting seismic data. The synthetic seismograms are compared to real data and the seismological and/or geological model is varied until an acceptable match is found. In some cases this procedure can be automated. [RAC]

T

tachylite A **basaltic** glass. Tachylite is a component of rapidly-quenched **basaltic magma** erupted as **lava flows** and **pyroclastic** material and also occurs in the chilled margins of **basaltic sills** and on the rims of **basaltic pillow lavas**. (Tachylite should not be confused with frictionally melted glass formed at fault zones and termed **pseudotachylite**.) (See **glassy rock**.) [RST]

taconite **Banded iron formation** suitable for concentration of its **magnetite** and **hematite** content by fine grinding and magnetic or other **benefication** methods. [AME]

taenite (FeNi) A naturally occurring alloy found in *meteorites*. (See **native elements**.) [DJV]

tafoni Pits and hollows on the surface of rock outcrops and boulders; cavernous **weathering** features. Tafoni are common on crystalline rocks (especially **granites**) in desert environments. However, the term was originally given to the features on rocks in Corsica; they also occur in Antarctica. Tafoni may develop when salt **weathering** and wind scour enlarge small depressions on the surface of exposed rocks. Once formed, the accumulation of moisture in the hollow may enhance the **weathering** process. Some tafoni may form when **case-hardened** surfaces are breached and the softer rock beneath is **weathered** and **eroded**. [AW]

Gill, E.D. (1981) Rapid honeycomb weathering (tafoni formation) in greywacke, S.E. Australia. *Earth Surface Processes and Landforms* **6**: 81–3.
Mustoe, G.E. (1982) The origin of honeycomb weathering. *Geological Society of America Bulletin* **93**: 108–15.
Mustoe, G.E. (1983) Cavernous weathering in the Capitol Reef Desert, Utah. *Earth Surface Processes and Landforms* **8**: 517–26.

talc ($Mg_3Si_4O_{10}(OH)_2$) A *phyllosilicate* mineral formed by the **alteration** of **magnesium** silicates such as **olivine**, **pyroxene** and **amphiboles**.

Carvings of talc are found amongst the Gandharan sculptures of the 1–6th centuries AD in north-west Pakistan. [OWT/DJV]

tangent-arc method Method of constructing a vertical **cross-section** through a set of **folds** assuming them to be *parallel*. Normals are drawn to the **fold** surface at a number of points along the **cross-section** where **dip** readings are available. The **dips** are regarded as tangents to an arc of the **fold** surface. Each successive pair of normals intersects

at a point which is the centre of curvature for that particular **fold** arc. The method is illustrated in Fig. T1, and is usually termed the *Busk method*. [RGP]

Busk, H.G. (1929) *Earth Flexures*. Cambridge University Press, Cambridge.
Ragan, D.M. (1973) *Structural Geology: An Introduction to Geometrical Techniques* (2nd edn). John Wiley & Sons, New York.

tantalite ($(Fe,Mn)Ta_2O_6$) An **ore** mineral of **tantalum**. [DJV]

tantalum (Ta) At. no. 73, at. wt. 180.9470. d 16.6. A group V element. Used in resistant alloys and in surgical appliances. It is won from **carbonatite**, **pegmatite** and **placer deposits**. [AME]

tanzanite A name given to a blue, **gem** quality **zoisite**. [DJV]

tar sand Normally bituminous **sandstone** impregnated with oil that is too viscous and heavy to be extracted by conventional drilling techniques. Other rock types may be similarly oil-saturated such as the large Hit 'tar mat' in Iraq which is a **limestone**. When underground deposits are exploitable using techniques such as cyclic steam injection under pressure, these deposits are called *heavy oil sands*. At the present time the only tar sands that might be exploited economically are those that can be mined by **open pit** methods. Of the twenty or so largest tar sand deposits there are three gigantic accumulations accessible to surface exploitation: those of the Orinoco Oil Belt in Venezuela, the Athabaska Tar Sands of Alberta and the Olenek Tar Sands in northern Siberia. [AME]

North, F.K. (1985) *Petroleum Geology*. Allen & Unwin, Boston.

Taylor number (Ta) A dimensionless parameter that depends on the scale of a **convective** cell, its **kinematic viscosity** and the rate of rotation. If Ta $\geq$ 1 then rotational effects are significant. [DHT]

Tritton, D.J. (1988) *Physical Fluid Dynamics*. Clarendon Press, Oxford.

tectonic Relating (i) to **structures** produced by **deformation** (in which sense the term is synonymous with 'structural'); (ii) to major Earth structure and its formation. Thus 'tectonic activity' in *orogenic belts* covers igneous and metamorphic as well as structural phenomena contribut-

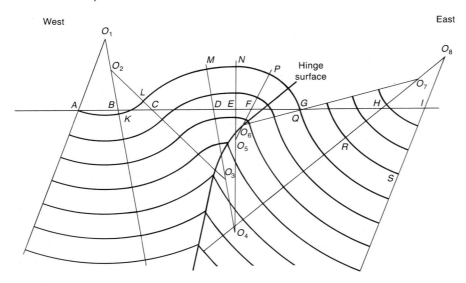

West

East

Hinge surface

Fig. T1 Tangent-arc method. Reconstruction of parallel folds, showing the trace of the hinge surface. (After Ragan, 1973.)

ing to the formation of the belt. *Tectonics* is the branch of geology dealing with the study of major Earth structure and its formation (see also **geotectonics**). **Plate tectonics** is the application of *plate* theory to tectonics.

A *tectonic province* is part of the **continental crust** distinguished from adjoining parts by differences in tectonic history. The term is used particularly in **Precambrian shield** regions to distinguish **mobile belts** and **cratons** of different age. Thus the Superior and Slave provinces of the Canadian **shield** are mobile zones in the **Archean** but **cratons** during the Early **Proterozoic**, when the Churchill province was a mobile zone. The term is also used for subdivision of *orogenic belts* and **cratons**; thus in the USA, the **Basin-and-Range** province and the Colorado plateau are examples of *tectonic provinces* within the Cordilleran *orogenic belt*. The term *tectonic province* is analogous to **domain** but on a larger scale.

Tectonic attenuation is the thinning of strata due to **extensional deformation**, associated with **slides**, for example, rather than to original stratigraphic variation. *Tectonic erosion* is the process whereby **erosion** results in the *detachment* of a sheet that subsequently slides under **gravity**. This may occur when **erosion** breaches a **competent** layer which then slides over more incompetent material below (see **gravity gliding**). The term *tectonic style* covers aspects of **fold profile** geometry used to distinguish **folds** of different type and origin. *Tectonic transport* is a term much used in older literature (e.g. see Turner & Weiss, 1963) to refer to the direction of relative movement of rock masses. Certain **lineations** were ascribed to movement in the direction of *tectonic transport*. The term **displacement** is preferable. [RGP]

Turner, F.J. & Weiss, L.E. (1963) *Structural Analysis of Metamorphic Tectonites*. McGraw Hill, New York.

tectonic earthquake The vast majority of **earthquakes** (about 95% by number) take place around the margins of *plates*. **Stresses** caused by the motion of the *plates* cause an accumulation of **shear strain** in the rocks locked together at the **plate boundaries**. When the rocks fail, an **earthquake** occurs. This mechanism also leads to earthquakes within the tectonic *plates*, because the **stresses** are transmitted across the whole *plate*. They can cause minor **earthquakes** on pre-existing **faults** or in regions where the shape of the *plate* causes **stresses** to concentrate. (See **elastic rebound theory, focal mechanism solution**.) [RAC]

tektite Glassy object probably formed from terrestrial material melted and displaced by the impact of an extraterrestial body, such as a *meteorite* or comet material. Tektites were occasionally collected by prehistoric man, probably mainly as curios but perhaps sometimes in mistake for **obsidian**. They have been reported at Paleolithic and Neolithic sites in central Europe, including the settlement of Willendorf in Austria. [OWT]

teleseism Because of the complicating effect of **crustal** and uppermost **mantle** structures on **seismic waves** (reflections, refractions, etc.), **seismograms** recorded within a few hundreds of kilometres ('local' events) or up to some 2000–3000 km ('regional distance') of their source are relatively complex. Beyond this, *body wave* paths are mostly in the relatively uniform **mantle**. The range in *epicentral angle* of 30–100° is termed teleseismic range, and the **seismograms** are teleseisms. (See **shadow zone**.) [RAC]

telethermal deposit Very low temperature deposit believed to be formed at great distances from the source of the **hydrothermal solutions** that gave rise to it. [AME]

telluric Currents induced in the Earth by **geomagnetic field** variations, usually of **diurnal** period and its harmonics. The **potential** differences due to telluric currents are of the order of millivolts over hundreds of metres. Long-term records of telluric current data have been used to derive the Earth's **electrical conductivity** structure to several kilometres depth (see **magnetotellurics**). Mostly, however, tellurics are merely **noise** in electrical methods, their very low frequency making filtering difficult. In 'd.c.' methods it is usual to work with a low frequency (a few Hz) alternating current input, averaging the results over a number of full cycles to cancel out the bias introduced by the telluric currents. [ATB]

tellurium (Te) At. no. 52, at. wt. 127.60. A **sulphur**-group element. Used in alloys, particularly with **lead**, **copper** and stainless steel. Recovered from flue dust produced during the smelting of sulphide **ores**. [AME]

temporary strain (*recoverable strain*) A transient shape change (**strain**) after which the material concerned reverts to its original shape. *Permanent strain*, in contrast, is a non-recoverable shape change. In **elastic** materials, the original shape is regained instantaneously after **stress** decrease, but more slowly in the case of **viscoelastic** behaviour. Rocks show **elastic** behaviour at low **strains**. (See **elastic strain**.) [SB]

tenacity The name given to the resistance that a **mineral** offers to mechanical **deformation** (bending or stretching) or to disintegration (tearing or crushing). The following terms are used to describe tenacity in minerals.
Brittle minerals are those that break and powder easily.
Ductile minerals can be drawn into a wire.
Malleable minerals are those that can be hammered out into thin sheets, such as native **copper**, **silver** and **gold**.
Sectile minerals can be cut by a knife although they still powder under a hammer. An example is **cerargyrite** or **horn silver** (AgCl).
Elastic minerals will bend and then resume their original shape upon release of pressure. This property is shown by **cleavage** flakes of **mica**.
Flexible minerals bend but do not resume their original shape upon release of pressure. Examples include **cleavage** sheets of **chlorite** and **talc**.
These are, of course, qualitative and descriptive terms although the **deformation** properties of **minerals** (and rocks) can be measured in various ways. It is also true to say that the property described here as tenacity depends ultimately on the nature of chemical bonding and **crystal structure** and, in certain cases, of the **microstructure** of the material being considered. However, the relationships between tenacity, bonding, structure and **microstructure** are by no means simple. [DJV]

tennantite ($Cu_{12}As_4S_{13}$) An **ore** mineral of **copper**. [DJV]

tenorite (CuO) A black supergene **copper** mineral. (See **oxide minerals**.) [DJV]

tensile Refers to a **force**, **stress** or **strain** acting in opposite directions away from a reference point or plane. A tension is a **stress** state produced by two equal **forces** acting in opposite directions away from a reference point or plane. Examples of tensile structures include *tension*, *tensile-* or *extension fractures*, *cracks*, **joints** and **faults**, *tension gashes*, **boudins**, **pinch-and-swell structure**, **necks** and **veins**. Tensile **displacement** on **fractures** is referred to as *mode I* or *opening mode*. [TGB]

tepee An **overthrust** sheet of **limestone** which appears as an inverted V in a two-dimensional exposure, and is so named because of its resemblance to the shape of the hide dwellings of early American Indians. Tepees (Warren, 1983) are found in tidal areas, in **calcrete**, and around salt lakes (**playas**) as a result of **deformation** or desiccation and contraction processes related either to fluctuations in water levels or to changes in the nature of chemical **precipitation**. [ASG]

Warren, J.K. (1983) Tepees, modern (Southern Australia) and ancient (Permian — Texas and New Mexico) — a comparison. *Sedimentary Geology* 34: 1–19.

tephra **Pyroclastic** material explosively fragmented during the course of a **volcanic eruption**. The term is used for all **magma** compositions. Tephra may be deposited by any mechanism (e.g. **pyroclastic flow** or air-fall), but does not include redeposited (*epiclastic*) material. Tephra consists of various proportions of **ash**-sized glass shards, **pumice** and/or **scoria** *clasts*, crystals derived from *phenocrysts* in the **magma**, and lithic *clasts*. Tephra may or may not be welded. Tephra is commonly formed when **magma** experiences decompression upon eruption, so that **volcanic gases**, dissolved in the **magma**, rapidly **exsolve** and explosively tear the **magma** apart. The style of explosive eruption will vary according to a range of parameters, including **magma** composition, gas content, vent size and magmatic mass eruption rate. The texture of the tephra produced will likewise vary. Mafic **magma** with moderate gas content and at moderate mass eruption rates will erupt in **fire fountain** or *strombolian* activity to produce scoriaceous tephra (Wilson & Head, 1981). **Silicic magmas** commonly have high gas contents, and their eruption can be very explosive, producing **pumice** and abundant **ash** deposited as air-fall, **ash-flow** or **pyroclastic** surge deposits in various proportions according to eruption parameters such as **vent** size, mass eruption rate, and gas content.

Tephra can also be formed when **magma** of any composition comes into contact with water. In this case, explosive hyrdovolcanic interactions may occur. Such eruptions commonly generate **tuff rings** composed of air-fall tephra, **pyroclastic** surge deposits, and **mudflow** deposits. Hydrovolcanic tephra may be deposited dry and hot, or wet with condensed water, and relatively cool.

Tephrachronology is the study of widespread tephra deposits to establish age relations. (See **pyroclastic rock.**) [PTL]

Fisher, R.V. & Schminke, H.-U. (1984) *Pyroclastic Rocks.* Springer-Verlag, Berlin.
Heiken, G. & Wohletz, K. *Volcanic Ash.* University of California Press, Berkeley.
Kokelaar, P. (1986) Magma–water interactions in subaqueous and emergent basaltic volcanism. *Bulletin of Volcanology* **48**: 275–89.
Self, S. & Sparks, R.S.J. (eds). (1981) *Tephra Studies.* Reidel, Dordrecht.
Wilson, L. & Head, J.W. (1981) Ascent and eruption of basaltic magma on the Earth and Moon. *Journal of Geophysical Research* **86**: 2971–3001.

tephrite An **olivine**-free alkaline volcanic rock composed essentially of calcic **plagioclase**, *clinopyroxene* and **feldspathoid minerals** (**feldspathoid** generally exceeding 10% of the felsic minerals). Tephrites grade into **basanites** with the occurrence of **olivine**, and into **phonolites** (phonolitic tephrites) by the occurrence of **alkali feldspar**. Tephrites occur on *oceanic islands* and in varied continental settings (cf. **alkaline igneous rock**). [RST]

tephroite (Mn_2SiO_4) The **manganese** end-member of the **olivine mineral** group. [DJV]

terminal fan A specific type of **alluvial fan** where the amount of surface **runoff** decreases down the *fan* due to **infiltration** and evaporation such that under most conditions little or no water leaves the system as surface flow. Terminal fans are characterized by a reduction in the proportion and size of channelized deposits down the *fan* which is frequently associated with an increase in **paleosol** maturity. [JA]

termite activity Termites, of which there are several thousand species, are members of the order Isoptera. Though 'fierce, sinister and often repulsive' (Maeterlinck, 1927), they are remarkable for having lived in highly organized communities for as much as 150–200 Ma, and much of their success is due to their development of elaborate architectural, behavioural, morphological and chemical strategies for colony defence. Maeterlinck regarded them as 'the most tenacious, the most deeply rooted, the most formidable, of all the occupants and conquerors of this globe'. The numbers of termite species decrease rapidly with increasing latitude and altitude; they are very much a tropical phenomenon. In the tropics their **geomorphological** role is considerable, contributing to the translocation of bare *soil* to the ground surface, to the formation of **laterite**, calcareous *concretions*, **stone-lines** and duplex *soils*. Above all, they produce sometimes imposing termite mounds (though not all species do so). Maeterlinck describes such forms 'as wrinkled hillocks, battered sugar-loafs, gigantic stalagmites, portentous sponges, ricks of storm-tossed hay or corn'. [ASG]

Maeterlinck, M. (1927) *The Life of the White Ant.* Allen & Unwin, London.
Lee, K.E. & Wood, T.G. (1971) *Termites and Soils.* Academic Press, London.

terra rossa Red clay *soil* developed on **limestones** and associated with **karstic** features. Such *soils* are well developed in areas of Mediterranean climates with their strong seasonal contrasts in precipitation. **Dissolution** of the **limestone** occurs during the wet season and clay and **iron** compounds are released. The strong red colour is the result of the conversion of hydrated ferric oxides to **hematite** during the dry season. The clays are the residual products of the **dissolution** but may be relict in old landscapes, having formed during earlier wetter periods. The term is widely used by geologists to refer to any residual **red clays** on **limestones** but it should strictly be used only for *soils* of the Red Mediterranean type (classifiable as *rhodustalfs* or *chromic luvisols*). (See **karst**.) [VPW]

Duchaufour, P. (1982) *Pedology. Pedogenesis and Classification.* Allen & Unwin, London.

terracette Miniature terrace or ridge extending across a slope, usually normal to the direction of maximum slope. Terracettes are rarely more than 0.5 m wide and deep. Their origin is still a matter for debate. Some may be animal tracks, but as others occur in areas where animals are very rare it would seem that some other mechanism is involved. They are probably a consequence of *soil* mantle instability on steep slopes. [ASG]

terrane The term 'terrane' is the American equivalent of the English word 'terrain'. However, it has acquired in addition a specific geological meaning as a piece of **crust**, defined by clear boundaries, which differs significantly in its **tectonic** evolution from neighbouring regions. Terranes are described as *displaced* or *exotic* if they are provably **allochthonous**, and *suspect* if there are grounds for believing them to be displaced.

The distribution and nature of more than 50 *suspect terranes* in the western North American Cordillera are summarized by Coney *et al.* (1980) who discuss the principles involved in their recognition. A terrane exhibits internal homogeneity and continuity of stratigraphy, and of **tectonic** style and sequence, but is distinguishable from adjoining terranes by discontinuities of structure or stratigraphy that cannot be explained on the basis of normal facies or **tectonic** changes. *Displaced terranes* ideally should contain **paleomagnetic** or faunal records that support their **allochthonous** origin. [RGP]

Coney, P.J., Jones, D.L. & Monger, J.W.H. (1980) Cordilleran suspect terranes. *Nature* **270**: 329–33.

teschenite An **andesine**-bearing **gabbro** or **dolerite**, composed essentially of calcium-rich **plagioclase** (near **labradorite**), *clinopyroxene* (usually *titanaugite*), and *analcite*, with accessory Fe–Ti oxide, **amphibole**, **biotite**, with or without **olivine**. Teschenites may occur as minor intrusions, as coarse *pegmatitic* **schlieren**, or as finer **veins** within differentiated minor intrusions such as **sills**. [RST]

tesla (T) SI unit of **magnetic field** strength, with dimensions Vsm^{-2}. [PK]

Tethys The hypothetical ocean that lay between **Gondwanaland** and **Laurasia** and was often considered to be the ancestor of the present Mediterranean. It is now recognized that an older ocean, Paleo-Tethys, existed between parts of Asia, and then gave way to a Neo-Tethys in the same region. It is unclear to what extent either of these Asian oceans extended through Europe into the Caribbean region, although shallow seas may have provided links for marine organisms. [DHT]

Audley-Charles, M.G. & Hallam, A. (1988) *Gondwana and Tethys.* Geological Society Special Publication No. 37. Oxford University Press, Oxford.

tetrahedrite ($Cu_{12}Sb_4S_{13}$) An **ore** mineral of **copper**. [DJV]

thalassostatic Related to a period of static sea-level. Generally used to describe river terraces caused by fluctuations in sea-level. Low sea-levels under *glacial* conditions cause river incision, and high sea-levels in **interglacial** conditions cause **aggradation**. Such **aggradation** occurred during the *Pleistocene* **interglacials** and thalassostatic terraces were formed as low *glacial* sea-levels induced incision. Close to the coastline such terraces may be related to **raised beaches**. [JDH]

thalweg A line connecting the points of deepest flow within successive downstream channel cross-sections and representing the planform pattern of the greatest channel depth. [JLB]

thermal conductivity A measure of the ability of a material to conduct heat, defined as the normal flow of heat across unit area in the material in unit time when the temperature gradient normal to the area is unity. The unit of thermal conductivity is W m^{-1}K^{-1} and is a tensor which, in combination with the temperature gradient, yields the conductive **heat flow**. In isotropic solids, the heat flow in any direction is the product of the thermal conductivity and the thermal gradient in that direction.

Thermal conductivities of rocks vary with composition, temperature and pressure. There are empirical formulae available for the determination of bulk conductivity in terms of the component mineral content, but other factors such as texture, **porosity**, **foliation**, and **veining** also affect it, especially by creating **anisotropy**. The effect of increasing temperature and pressure is to decrease the conductivity. The variation of conductivity with temperature, while not accurately known for most Earth materials, is of significance even at a few kilometres depth, and affects the reliability of subsurface temperature predictions and other **geothermal** calculations. [ATB]

Carslaw, H.S. & Jaeger, J.C. (1959) *Conduction of Heat in Solids.* Oxford University Press, Oxford.
Kappelmeyer, O. & Haenel, R. (1974) *Geothermics with Special Reference to Application.* Geoexploration Monographs Series 1, No. 4. Gebrüder Borntraeger, Berlin.

thermal diffusivity The parameter which controls the rate at which heat propagates through a substance. Considerations of energy conservation give rise to the diffusion equation

$$\nabla . (\nabla T) = \kappa \frac{\partial T}{\partial t}$$

where ∇ is the first order vector partial differential operator, T the instaneous temperature, t the time; κ, the thermal diffusivity, is the ratio of the **thermal conductivity** of the medium to the product of its density and specific heat. [ATB]

thermocline The depth in the ocean at which the rate of decrease of temperature with increasing depth is greatest. A *permanent* thermocline persists throughout the seasons at low and middle latitudes. The top of the permanent thermocline varies in depth from about 25 to 200 m and extends to depths of between 200 and 1000 m, being thickest and deepest in intermediate latitudes. The permanent thermocline separates the surface mixed layer of uniform temperature from intermediate and bottom waters. Its presence is related to the sinking and equatorward flow of more uniformly cold water from high latitudes. A temporary seasonal thermocline develops throughout the oceans within the surface 50 m in response to surface temperature rise during the summer. As a significant density interface (*pycnocline*) the thermocline acts as a barrier to the transfer of water to the **photic zone**, hence limiting primary productivity. The thermocline is sedimentologically significant in that plumes of suspended sediment can be transported oceanward along it. In some areas a diurnal surface thermocline a few metres thick forms in response to daily solar heating. [AESK]

Pickard, G.L. & Emery, W.J. (1982) *Descriptive Physical Oceanography: An Introduction* (4th edn). Pergamon, Oxford.

thermohaline current A current which is driven by density differences due to variation in temperature and/or salinity. The currents responsible for the deep circulation of the oceans are largely thermohaline in origin. Winter cooling of water in high latitudes of the North Atlantic causes it to sink and spread out at a depth appropriate to its density, forming North Atlantic deep water (NADW). In Antarctic high latitude areas such as the Weddel Sea and Ross Sea, freezing to produce sea ice enriches surface waters in salt which then sink and traverse the ocean floor as Antarctic bottom water (AABW). Although substantial areas of the seafloor are affected by thermohaline currents, they are intensified along the western margins of continents due to the Earth's rotation. The resultant currents, such as the Western Boundary Undercurrent which flows south-westward near the base of the eastern North American continental slope, have typical speeds of between 0.1 and 0.3 m s^{-1}, are associated with **nepheloid layers** and are capable of transporting and depositing significant quantities of sediment as **contourite** drifts. [AESK]

Warren, B.A. (1982) Deep circulation of the World Ocean. In: Warren, B.A. & Wunsch, C. (eds) *Evolution of Physical Oceanography*. MIT Press, Cambridge, MA.

Killworth, P.D. (1983) Deep convection in the World Ocean. *Reviews of Geophysics and Space Physics* 21: 1–26.

thermoluminescence dating A dating method suitable for rocks and inorganic artefacts containing small amounts of natural *radioactivity*. Some minerals often present in rocks and pottery, such as **zircon**, contain naturally *radioactive* elements (e.g. **thorium**, **uranium**) which decay, causing ionization (removing electrons from atoms) within the mineral. Some electrons may become trapped at defects in the *crystal lattice*, and remain there until released by heating to *c*. 500 °C in the form of visible light (thermoluminescence) which is in addition to the normal electromagnetic spectrum expected from heating to that temperature. Since any thermoluminescence present in a rock is released at this temperature of 500 °C or over, thermoluminescence begins re-accumulating, in a rock, at the time of its last melting and solidification, and, in a heated artefact such as pottery, at the time of its firing. Careful measurement of the amount of thermoluminescence present, the concentration of *radioactive* elements in or near the sample, the sensitivity of the sample (i.e. the amount of thermoluminescence produced per rad of ionizing radiation) will give a determination of the date of the rock or artefact. There are complications to the method (for example, abnormal leaching of thermoluminescence, non-linear accumulation of thermoluminescence with radiation dose), but it has been successfully used to date **calcite** from *stalagmites* and *stalagtites*, archeological pottery, to determine whether **flint** and similar material has been heated in antiquity, and to test the authenticity of artefacts. [OWT]

thermoremanent magnetization When a heated **ferromagnetic** material cools in a **magnetic field** from a temperature above its **Curie temperature**, it acquires a **remanent magnetization** (TRM) that is dependent on the quantity and **susceptibility** of the **ferromagnetic** material, the strength of the ambient **magnetic field**, and the rate of cooling. If single **magnetic domain** particles are present, then this **remanent magnetization** can be preserved for many billions of years provided that the magnetic minerals are not chemically altered or reheated. All igneous and most metamorphic rocks may thus possess a thermal remanence from the time that they last cooled. If heated to a temperature below their **Curie temperature**, any pre-existing thermal remanence acquired below the **Curie temperature** and above the reheating temperature will remain unaffected, but all magnetizations originally acquired at and below the reheating temperature will be replaced by the new partial thermal remanence acquired as it cools from the reheating temperature. [DHT]

Tarling, D.H. (1983) *Palaeomagnetism*. Chapman & Hall, London.

tholeiite A silica-oversaturated **basalt** (a '**quartz** tholeiite' or an '**olivine** tholeiite'; see **basalt**) composed of calcium-rich **plagioclase**, **pyroxene** (abundance of *orthopyroxene* or **pigeonite** exceeding calcium-rich **pyroxene**) and Fe–Ti oxide possibly with groundmass *clinopyroxene* (**augite**) and ± **olivine**, in a siliceous (glassy or quartzo-feldspathic) groundmass. The **olivine** may be associated with *orthopyroxene* as a result of reaction with siliceous residual melt. Tholeiitic **basalts** form the most abundant **basalt** group, and form the **basaltic** (**olivine** tholeiite) layers of the **oceanic crust** (Layer 2). [RST]

thomsonite ($NaCa_2(Al_5Si_5O_{20}) . 6H_2O$) A **zeolite mineral**. [DJV]

thorianite (ThO_2) An **ore** mineral of **thorium**. [DJV]

thorite ($ThSiO_4$) A **thorium** mineral isostructural with **zircon**. [DJV]

thorium (Th) At. no. 90, at. wt. 232.04, d 11.7. A *radioactive* metal used in the manufacture of gas mantles, as a catalyst, in strengthening **nickel** and as an oxygen remover (getter) in the electronics industry. Obtained as a by-product in the treatment of **monazite** from **beach placers**. [AME]

threshold slope A graded or equilibrium slope produced by **denudational** processes. Above this angle, rapid **erosion** reduces the slope to the *threshold angle*. Slopes are not produced at lower angles since there is a minimum angle below which **erosional** processes cannot operate. W.M. Davis recognized that the graded slope would decline in angle over time as the **regolith weathers** progressively over time. Following from the work of Carson & Petley (1970), modern interpretation of the term relates the threshold slope to the straight segment commonly found on many hillslopes. In the engineering sense, Young (1961) has defined an upper limiting angle as the maximum angle at which a **regolith** can stand with respect to its frictional **strength**, and a lower limiting angle below which the slope is stable with respect to rapid mass movement even when the *soil* is fully saturated. Characteristic maximum slope angles, often associated with the midslope straight segment, relate to the limiting *threshold angle* of stability at various stages in the **weathering** of the **regolith**. [TPB]

Carson, M.A. & Petley, D.J. (1970) The existence of threshold hillslopes in the denudation of the landscape. *Transactions of the Institute of British Geographers* **49**: 71–95.

Young, A. (1961) Characteristic and limiting slope angles. *Zeitschrift für Geomorphologie* Supp. **5**: 126–31.

throughflow Flow of water through the *soil* in a downslope direction, approximately parallel to the ground surface. The term *interflow* may also be used; *subsurface flow* is preferred by some hillslope hydrologists as a generic alternative to throughflow. Such lateral flow is generated where there is a decrease in the **hydraulic conductivity** of the *soil* with depth; this decrease may occur gradually, but is often abrupt, being associated with a *soil* horizon boundary or with the *soil*-bedrock interface. If **infiltration** rates exceed the rate of percolation into the lower horizon, *soil* moisture will accumulate within the upper horizon. This may lead to *soil saturation* — a feature often referred to as the *saturated wedge* because of its extension upslope. Throughflow will provide all the *baseflow* in drainage **basins** where there is no true **groundwater**. In addition, throughflow may occur rapidly enough after precipitation to yield a significant amount of *storm* **runoff**. In **basins** where permeable *soils* overlie impermeable bedrock, and where steep slopes feed directly into the stream, throughflow may provide most or even all of the flood **runoff** response. Where slopes are shallower or where there is a more extensive **floodplain**, throughflow will accumulate at the slope base so that the *soil* becomes completely saturated. **Runoff** from these saturated areas — termed *saturation overland flow* — will then provide the main storm **runoff** response. This **runoff** is a mixture of *direct runoff* (precipitation falling directly on to the saturated *soil* and unable to infiltrate) and *return flow* (throughflow which *exfiltrates* from the *soil* profile at the slope base). The extent of such saturated zones may vary spatially seasonally, and within individual *storm* events. This notion forms the basis of the *variable source area model* of Hewlett which describes the production of storm **runoff** from limited areas of a drainage **basin**. [TPB]

Kirkby, M.J. (ed.) (1978) *Hillslope Hydrology*, Chapters 1, 4 and 7. John Wiley & Sons, London.

thrust belt A **tectonic** zone in which most of the **shortening** is taken up on *thrust faults*. Such regions generally comprise the external parts of *orogenic belts*, and form at *destructive plate margins*. The Appalachians, the Himalaya, and the Canadian Rockies are good examples of this type of **tectonic** regime.

Thrust belts form at relatively shallow crustal levels in *continent–continent collision* zones following the closure of an ocean by subduction. *Foreland thrust belts* are zones in which the *thrust faults* are generally discrete, and affect rocks that can be recognized in the adjacent **foreland**. They are often seen as components of major thrust belts that include **penetratively** strained and metamorphosed rocks, as, for example, the Salt Ranges are part of the Himalayan chain as a whole. Small-scale thrust belts are also associated with **subduction zones** where no collision has occurred. Present-day submarine thrusting at the continental margin above **subduction zones** has been documented from seismic, sonar and borehole information, e.g. in the Lesser Antilles Arc. The repeated thrusting that occurs to scrape ocean floor sediments off the downgoing oceanic slab tilts and folds them to form a wedge-shaped prism known as an **accretionary prism**. Uplifted examples such as the Makran show a combination of **tectonic** and soft-sediment **deformation** affecting the sediments.

During *continent–continent collision*, **shortening** is transferred from the **subduction zone** to shallower levels in the **crust**, and **strain** affects both oceanic and continental material. Thus a thrust belt can contain lithologies originating from all parts of a continental margin. One important aspect of structural analysis of thrust belts is reconstruction of the pre-existing margin, to enable examination of its shape, structure and sedimentary facies variations. **Palinspastic reconstructions** showing the relative positions of **paleogeographical** domains before shortening have been used for many years. These methods have been developed into the techniques of producing **balanced cross-sections**, in which the *thrust* **displacements** and order of development are analysed to give a viable model for the development of the structures seen (Butler, 1987). This method relies on deducing the order of development of the observed structures, using structural analysis as outlined below.

Thrust belts show certain common features that can be recognized and used to analyse their **strain** histories in this way. Thrusts often form **imbricate structures**, in which several closely spaced **faults** converge at depth onto a *sole thrust* or **décollement** surface, which generally **dips** gently (less than 5°) from the **foreland** parts of thrust belts where the **faults** can be emergent, towards the hinterland (Fig. T2). On a regional scale, the *thrust faults* are concave upwards (*listric*), but in detail they tend to follow a staircase trajectory, composed of long **bedding** plane parallel or subhorizontal surfaces, known as *flats*, and shorter segments called **ramps**, which cross-cut bedding or other datum surfaces. Branching of *thrust faults* to form several adjacent **ramps** results in the formation of **imbricate structures**. In most cases, the *thrust faults* develop closer to the **foreland** than existing **faults** and related **folding**, i.e.

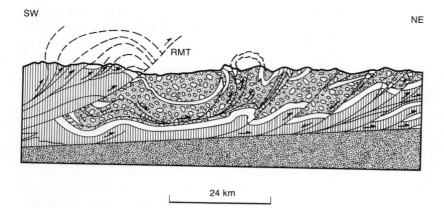

SW

NE

RMT

24 km

Fig. T2 Thrust belt. A cross-section through the Canadian Cordillera showing thrust faults and related structures. (After Price, 1986.)

a *piggyback propagation sequence* dominates. The *thrust* sequence can be deduced when an existing *thrust fault* or related **strain** zone is cut by a later-formed **fault**. When the later-formed **fault** propagates at a lower level than an existing **fault**, and cuts up through it, it is said to be a *breaching thrust*. If such a **fault** occurs within an existing imbricate system, rather than in the **foreland** to such a structure, it is out of sequence, and is called a *break-back thrust*. Within a thrust belt, thrusts with **vergence** towards the **foreland** are the normal case. **Backthrusts**, with a transport direction towards the hinterland, can occur in similar imbricate systems, as well as with **backfolds** of the same **vergence**. Single **backthrusts** are also found as the roof thrust to *duplex* systems at the mountain front.

The relationships between **faulting** and **folding** of different ages can be deduced most clearly where there existed a recognizable stratigraphy before **deformation**. Simple and laterally extensive sequences such as continental shelf clastics and carbonates result in relatively simple thrust belt structures, such as are seen in the Moine thrust belt. In regions where the thrust belt dismembers a more complex *continental margin*, features such as pre-existing **extensional fault systems**, with their associated sediment distribution patterns, may affect the compressional structures. Where *normal faults* are synsedimentary, abrupt changes in sediment thickness may be a factor in the subsequent **deformation** pattern. The distribution of salt and other **evaporite** minerals, related to the early stages of **rifting** and ocean development, will strongly affect the positions of *thrust faults* because they are considerably weaker than other rocks. The nature of the pre-existing **extensional fault** pattern is important because of the possibility of re-use of the earlier **fault** surfaces by *thrust faults*. Re-use, or *reactivation*, of such **faults**, and cases where younger *thrusts* cut across earlier extensional **faults**, are both seen in thrust belts such as the Alps.

Application of these methods of structural analysis to some more recent thrust belts has allowed the resulting **displacement** estimates to be correlated with **lithosphere** *plate* movements recorded in *oceanic magnetic anomalies*. In ancient examples, this information is lacking. Relatively large amounts of **displacement** and **shortening** are seen in short distances in most thrust belts; in the Moine thrust belt, part of the Caledonides, restored sediment widths of 54 km and 64 km are telescoped into 12 km and 8 km. **Seismic reflection** profiling by COCORP in the Appalachians shows similarly large amounts of **shortening**, as sedimentary rocks are seen for many tens of kilometres beneath major crystalline *thrust sheets* (Cook *et al.*, 1979). Emplacement of **oceanic crust** and ocean-bottom sediments onto continental material, as **obducted ophiolite** sheets, also implies considerable *thrust* transport.

Thrust belts also include deformed and metamorphosed rock from the more internal parts of the *orogenic belt*, where **shortening** associated with *continent–continent collision* has occurred at deeper levels in the **crust**. In these areas, **deformation** can be localized in **shear zones**, or can be distributed over wider areas. Unravelling of the metamorphic areas of mountain belts requires understanding of metamorphic facies variations, three-dimensional **strain** analysis as at higher levels, and of the large-scale **shortening** distribution from area-balancing methods (Butler, 1987) (see **balanced section**).

When considering thrust belts as a whole, information from all levels of **deformation** must be integrated. The frontal parts of thrust belts are especially useful in this when the *thrust faults* break the surface, and become emergent. **Erosion** of the uplifted *thrust sheet* leads to the formation of *syntectonic sediments*, often in the **foreland basin**, which may be overridden by *thrust sheets* as **deformation** continues. Thus *flysch* and *molasse* sediments can be used to date *thrust* movements, depending on whether they are transported and/or folded by *thrust* movements or unconformably overlie **fault** segments. Analysis of thrust belts requires the combination of large-scale and detailed observations on many aspects of structure, and of metamorphic and sedimentary facies variations, to give as complete a picture as possible of their complex evolutionary history. [SB]

Butler, R.W.H. (1986) Structural evolution in the Moine of northwest Scotland: A Caledonian linked thrust system? *Geological Magazine* **123**: 1–11.

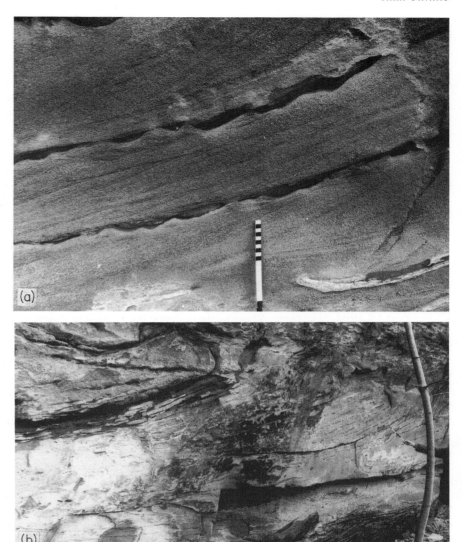

Fig. T3 Tidal bundle. Tidal bundles from a Miocene subtidal sandwave in the molasse of Switzerland (Bois du Devin, Marly near Fribourg). (a) Close-up view showing inclined depositional units of sand separated by fine-sediment drapes. (b) General view showing the right-to-left dipping foresets composed of tidal bundles. Note that the bundles change from being thick in the centre of the photograph, to being very thin in the centre-left, indicating a change from spring tide deposition to neap tide deposition (Allen & Homewood, 1984).

Butler, R.W.H. (1987) Thrust sequences. *Journal of the Geological Society, London* **144**: 619–34.

Cook, F.A., Albaugh, D.S., Brown, L.D., *et al.* (1979) Thin-skinned tectonics in the crystalline southern Appalachians; COCORP seismic reflection profiling of the Blue Ridge and Piedmont. *Geology* **7**: 563–7.

Price, R.A. (1986) Tectonic wedging in the Canadian Cordillera. *Journal of Structural Geology* **8**: 239–54.

thufur A small earth hummock found in *periglacial* areas. Thufurs are usually 100–500 mm high and 0.3–1 m in diameter, and are found north of the treeline in Iceland and similar environments. They generally consist of mineral *soil*, sometimes with a turf cover. The formation of thufurs is a matter of some debate, but they are known to be caused by differential freezing of the ground. [HAV]

tidal bundle A unit of sediment deposited during a tidal cycle consisting of **sand** units (transported by the main ebb and/or flood **tides**) separated by **non-conformities** or **pause planes** which are commonly draped by fine-grained sediment, reflecting the fall-out of material during slack water periods. Boersma first defined a bundle in 1969 as representing the sediment deposited on the **slip-face** of a bedform during the dominant tidal current. It should therefore be bounded below by the slack water material deposited after the subordinate current, and bounded above by the slack water material deposited after the dominant current. Since the slack water drape following the dominant **tide** is commonly partly or entirely **eroded** by the subordinate **tide**, this makes field definition of bundles difficult. Some authors therefore use the term tidal bundle to refer to both the deposits of the dominant and subordinate **tides**, bounded below and above by prominent slack water drapes.

Because tidal bundles are deposited during a known period of time (the tidal cycle), it is possible to calculate the mass transport of sediment that has taken place and also the tidal paleovelocities. Tidal bundles are therefore an invaluable aid in **paleohydrological** reconstructions. (See Fig. T3.) [PAA]

Allen, P.A. & Homewood, P. (1984) Evolution and mechanics of a Miocene tidal sandwave. *Sedimentology* **31**: 63–82.

Boersma, J.R. (1969) Internal structures of some tidal megaripples on a shoal in the Westerschelde estuary, the Netherlands. *Geologie en Mijnbouw* **48**: 409–14.

tidal correction The correction applied to **gravity** measurements during **reduction** for the variation caused by solid *Earth tides*. They cause a change in **gravity** with a maximum amplitude of 3 g.u. and a period of about 12 h. If base observations are made at a greater periodicity than that of the *Earth tides*, compensation is automatically made during the **drift correction**. If, however, base readings are only taken at the beginning and end of the day, a separate correction for *Earth tides* must be made. [PK]

tidal current Horizontal water movements generated by the gravitational attraction of the Moon and Sun on the Earth and hydrodynamically linked to vertical water movements or **tides**. They are usually more powerful where the tidal range is great although strong tidal currents are also found in coastal constrictions like straits where high or low **tide** occurs at different times at either end. Tidal currents are reversing, yet inequalities between the flood and ebb velocities gives rise to a residual tidal current that may result in net sediment transport in one direction. The geometry and size of coastal inlets also affect the velocity and location of tidal currents. [JDH]

Howarth, M.J. (1982) Tidal currents of the continental shelf. In: Stride, A.H. (ed.) *Offshore Tidal Sands*. Chapman & Hall, London.

tidal delta A fan-shaped accumulation of sediment formed at the mouth of a **tidal inlet** caused by the deposition of sediment as tidal flows expand and decelerate after emergence from the adjacent **tidal inlet**. Two types of tidal deltas are distinguished by their position relative to the tidal inlet. These are flood tidal deltas, located on the landward side of the **tidal inlet** and formed by the flood **tide**, and ebb tidal deltas produced by the ebb **tide** and situated on the seaward side of the **tidal inlet**.

The standard model for flood tidal deltas associated with meso-tidal barrier inlets is shown in Fig. T4(a). The ebb tidal flow is weakly developed in micro-tidal barrier systems, the flood tidal deltas therefore do not have the ebb-produced features of Fig. T4(a) and assume a multi-lobate **delta** form. Flood delta morphology also varies with the amount of **wave** influence which is dependent on the size of the back barrier **lagoon** and the concomitant **wave fetch**.

The standard model depicting the major components of ebb tidal deltas is shown in Fig. T4(b). They are best developed in meso-tidal barrier systems in association with tide-dominated **tidal inlets**. Their morphology is strongly dependent on the ratio of the open ocean **wave** power to tidal current energy. The dominant **wave** approach directions also play an important role in the symmetry of the ebb tidal delta about the tidal channel. [RK]

Boothroyd, J.C. (1985) Tidal inlets and tidal deltas. In: Davies, R.A. (ed.) *Coastal Sedimentary Environments*. Springer-Verlag, Berlin.

Hayes, M.O. (1980) General morphology and sediment patterns in tidal inlets. *Sedimentary Geology* **26**: 139–56.

tidal friction The Moon exerts a gravitational pull on the Earth which causes oceanic **tides**, and also much smaller distortions of the solid Earth known as solid *Earth tides*. Oceanic **tides** are in phase with the lunar motion, but, because of the Earth's internal friction, the high *Earth tide* lags behind the Moon's position, a phenomenon known as tidal friction. Tidal friction results in a continuous loss of rotational energy from the Earth–Moon system. In order to conserve momentum over geological time, this has caused the rotation of the Moon to slow to the point at which it

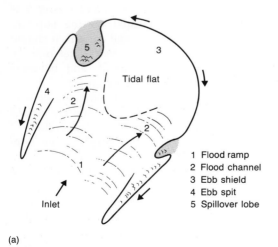

(a)

1 Flood ramp
2 Flood channel
3 Ebb shield
4 Ebb spit
5 Spillover lobe

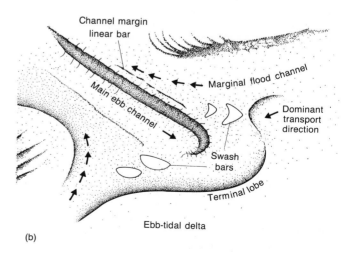

(b)

Fig. T4 Tidal delta. Morphologic models of (a) flood-tidal deltas, (b) ebb-tidal deltas. The arrows indicate the dominant direction of tidal-current flow. (After Hayes, 1980.)

always presents the same face to the Earth, and also the Moon to progressively recede from the Earth. [PK]

Bott, M.H.P. (1982) *The Interior of the Earth*. Arnold, London.

tidal inlet The intervening subaqueous channel between **barrier island** segments, through which water flows alternately landward with the flood **tide** and seaward with the ebb **tide**. The occurrence of tidal inlets is limited to non macro-tidal shorelines. On macro-tidal coastlines no **barrier islands** form; on such coastlines large tidal inlets are termed **estuaries**.

Morphological differences between tidal inlets are controlled by the relative magnitudes of **wave** and tidal energy. Three major types of tidal inlet are distinguished on this basis. These are **tide**-dominated, **wave**-dominated and transitional tidal inlets. **Tide**-dominated tidal inlets are associated with meso-tidal barrier systems and have deep, ebb-oriented narrow tidal channels. **Wave**-dominated tidal inlets characterize micro-tidal **barrier islands** and have generally flood-dominated tidal channels. Transitional tidal inlets vary widely in morphology between the tidal and **wave**-dominated forms. [RK]

Boothroyd, J.C. (1985) Tidal inlets and tidal deltas. In: Davies, R.A. (ed.) *Coastal Sedimentary Environments*. Springer-Verlag, Berlin.

tidal paleomorph Oversized, often meandering valley thought to be a former tidal channel created when sea-level was higher that it is now. Such valleys have **underfit streams** in them (Geyl, 1985). [ASG]

Geyl, W.F. (1985) *Tidal Paleomorphs in Eastern Virginia*. Research Paper in Geography No. 29. University of Newcastle, New South Wales.

tide Vertical water movement generated by the gravitational attraction of the Moon and Sun on the Earth. Tides are important both on the coast, as they create a marked rise and fall in water level, and at sea because of the **tidal currents** that are produced. The main tide-raising force is that of the Moon with the attraction of the Sun about half as great. These attractions affect the entire Earth but the oceans respond most clearly with water bulging out in the direction of pull on the side closest the Sun or Moon. The response of the solid Earth is known as an *Earth tide*. Due to the so-called centrifugal force, a tidal bulge also appears on the far side of the Earth and so two tidal bulges are produced each day. Every 14 days the lunar and solar attractions combine to give large tidal ranges, or spring tides, at new and full Moon. Smaller tidal ranges, or neap tides, occur between these maxima when the lunar and solar attractions tend to cancel each other. In coastal regions tides suffer modification by shallow water, blocking and partial reflection from coastlines and deflection by the Earth's rotation (see **Coriolis effect**). A reflected tidal **wave** may interfere with an incoming tidal **wave** to produce an **amphidromic point** with zero tidal

range around which the tide rotates. Due to the irregularity of both coastlines and bathymetry several **amphidromic points** may exist, e.g. three occur in the North Sea. [JDH]

Defant, A. (1958) *Ebb and Flow*. University of Michigan Press, Ann Arbor.

tiger-eye A fibrous yellow **quartz** *pseudomorphous* after **crocidolite** and exhibiting **chatoyancy**.

Tiger-eye is an ornamental stone from South Africa formed when *crocidolite* (blue asbestiform **riebeckite**) in metamorphosed **banded iron formation** is wholly or partially replaced by **quartz**. Some tiger-eye retains the blue colour of the *crocidolite*, but more commonly the **iron** has oxidized to a golden brown. [AME/DJV]

tight sand/tight gas sand Reservoir rock whose **permeability** is so low that it is incapable of producing more than five barrels of oil per day or equivalent without costly stimulation of the **wells**. [AME]

tile clay Clay similar to **brick clay** and suitable for tile making. [AME]

tiltmeter An instrument designed to measure the ultra-long period (days and longer), angular displacement of the Earth's surface. The main purposes of such observations are to detect ground distortion as part of **earthquake prediction** campaigns and to measure *tides* of the *solid Earth*. The instrument designs are usually centred on either suspended pendulums or measurement of height differences in the fluid-filling pairs of connected reservoirs. A practical difficulty is separating the effects of horizontal accelerations from those due to tilt itself. [RAC]

time correction A correction applied to a time base. (See also **static correction** and **dynamic correction**.) [GS]

time–distance curve A plot of the **arrival time** against the **shot** to **geophone** distance (range) of a particular seismic signal. The slopes of the segments of the curves give the reciprocal of the apparent velocities.

Reduced travel-time curves are sometimes plotted. These plot the difference between the observed refraction travel-time and the time that would have been observed if only one velocity material had been present. The result yields the total **delay time** if the assumed velocity is that of the refractor. [GS]

time-term method A method of **seismic refraction** interpretation used commonly in **crustal** surveys. The time-term represents the difference in time taken for a **head wave** to travel to a detector on the surface from a refractor and the time for a wave to travel to a point on the refractor perpendicularly below the detector point. Thus the time-term relates to the depth of the refractor and the method has advantages over other similar techniques as it does not require **shots** and detectors to be in a line. [GS]

Willmore, P.L. & Bancroft, A.M. (1960) The time-term approach to refraction seismology. *Geophysical Journal of the Royal Astronomical Society* **3**: 419–32.

tin (Sn) A very rare native metal mineral. (See **native element**.)

Tin is of importance archeologically because of its use in tin–**copper bronze** from the mid-3rd millenium BC onwards. The main sources of tin in the Old World were Cornwall (England), north-west Spain and Bohemia. In the New World tin was used for vessels in Mexico and alloyed in rare tin **bronzes** produced in parts of middle and south America in pre-Hispanic times. [OWT/DJV]

tincalconite ($Na_2B_4O_5(OH) . 3H_2O$) A **mineral** formed by the **alteration** of **borax**. [DJV]

titanite ($CaTiO(SiO_4)$) A common **accessory mineral** in a variety of igneous and metamorphic rocks, also sometimes exploited as an **ore** of **titanium** (formerly known as *sphene*). [DJV]

titanium (Ti) At. no. 20, at. wt. 47. A common metal. It is used extensively as a lightweight construction material and in titanium oxide pigments. Production is from the **rutile** of **beach placer deposits** and a few **orthomagmatic ilmenite deposits**. [AME]

todorokite ($(Mn,Ca,Mg)Mn_3O_7 . H_2O$)) An **ore** mineral of **manganese**. [DJV]

tombolo A **bar** or **spit** of **sand** or shingle linking an island either to the mainland or to another island. There are three main methods of formation but the first is the stricter, more limited, usage. Deposition occurring in the sheltered lee of an island close to the mainland results in **spits** forming which eventually extend to the mainland. The **spits** are cuspate, shaped by **wave refraction** round the island. Islands also become linked to the mainland by **longshore drift** extending the coastal orientation and by a rising sea-level driving *barrier beaches* landwards. Tombolos are common in flooded *drumlin* topography, e.g. Clew Bay in western Ireland. [JDH]

Zenkovitch, V.P. (1967) *Processes of Coastal development*. Oliver and Boyd, Edinburgh.

tonalite A phaneritic rock composed essentially of **plagioclase** (over 90% of the **feldspar**) and **quartz** (20–60%) with **alkali feldspar** either absent or present in low proportion (below 10%). Tonalites are associated with the **diorite–granodiorite–granite** associations forming **plutons/batholiths**. [RST]

topaz ($Al_2SiO_4(F,OH)_2$) A *nesosilicate* mineral found in **granites** and *pegmatites* and valued as a **gemstone**. [DJV]

topographic correction (*terrain correction*) Correction applied to a **gravity** measurement during **reduction** for the attraction of topography within about 100 km of the observation. The correction is calculated by using a **Hammer chart** on topographic maps of suitable scale. This procedure cannot fully be accomplished by computer, and accounts for much labour in the **reduction** process. [PK]

tor Exposure of bare bedrock, isolated by free faces on all sides, and the result of differential **weathering** followed by mass wasting and stripping. **Joint** control is important in controlling the location of **weathering** and in determining tor architecture. Tors occur in a wide range of climatic environments (arctic, temperate, desert and tropical) and on a wide range of rock types (**granite**, **sandstone**, **dolomite**, **quartzite**, **dolerite,** etc). [ASG]

Caine, N. (1967) The tors of Ben Lomond, Tasmania. *Zeitschrift für Geomorphologie* NF **11**: 418–29.

torbernite ($Cu(UO_2)_2(PO_4)_2 . 8–12H_2O$) A green secondary mineral found in the zone of **oxidation** and **weathering** of **uranium ore** deposits. [DJV]

toroidal field A **magnetic field** that is confined within a sphere as it has no radial components. The Earth's **core** is thought to contain a toroidal field that may be some 100 times greater than the **poloidal field** observed at the Earth's surface. [DHT]

Levy, E.H. (1979) Dynamo magnetic field generation. *Reviews of Geophysics and Space Physics* **17**: 277–81.
Parkinson, W.D. (1983) *Introduction to Geomagnetism*. Scottish Academic Press, Edinburgh.

torsion balance An early device for the relative measurement of **gravity** based on the oscillation of a beam suspended from a torsion fibre. Now superseded by the **gravimeter**. [PK]

touchstone A piece of stone held in the hand and scratched across a **precious metal** or alloy in order to assess its purity. The metal would leave a **streak** on the touchstone, which was examined for colour and size to estimate the amount of, typically, **gold** in the sample. Touchstones are recorded in Greek and Roman literature, and were the standard method of assaying **gold** in medieval Europe. Recent studies of medieval British and modern Czech touchstones show that they are made of a variety of rock types, including **tuff**, silicified siltstone, **chert**, *greywacke* and **quartz**-bearing **slates**. [OWT]

tourmaline   ((Na,K,Ca)(Li,Mg,Fe,Mn,Al)$_3$(Al,Fe,Cr,V)$_6$(BO$_3$)$_3$(Si$_6$O$_{18}$)(O,OH,F)$_4$) A complex *cyclosilicate* mineral characteristically found in **granite** *pegmatites*, but also as an **accessory mineral** in igneous and metamorphic rocks. Varieties include the **iron**-bearing **schorl**, **magnesium**-bearing **dravite**, **lithium**-bearing **elbaite** and **liddicoatite**. [DJV]

tourmalinization A form of **wall rock alteration** in which significant amounts of new **tourmaline** are developed. It is usually associated with medium and high temperature mineralization. [AME]

tower karst A **karst** landscape characterized by steep, flat-topped **limestone** towers rising from a flat plain, usually found in humid tropical areas. Notable examples of tower karst are found in southern China, Perak in Malaysia, and Belize. Brook & Ford (1978) discovered a type of tower karst forming in the subarctic climate of the Canadian Northwest Territories and challenged the old idea that tower karst only formed under humid tropical conditions. Structural conditions are now thought to be important to the development of tower karst, and there is still debate over the processes involved. [HAV]

Brook, G.A. & Ford, D.C. (1978) The origin of labyrinth and tower karst and the climatic conditions necessary for their development. *Nature* **275**: 493–6.
McDonald, R.C. (1979) Tower karst geomorphology in Belize. *Zeitschrift für Geomorphologie* Supp. **32**: 35–45.

trace element Element present in such low concentrations within a rock (usually in the part per million range) that it does not control the appearance of minerals such as **feldspar** and **pyroxene**, but occurs within such minerals as a result of ionic substitution for major elements. Such trace elements are important because their relative concentrations in a given suite of associated rocks (and their contained minerals) may be used to place constraints on the nature of the source (i.e. the mineralogy and composition) and the degrees of *partial melting* and/or *fractional crystallization* involved (see **magmatic diversification**). Trace elements may show much greater degrees of variation (1–3 orders of magnitude) in comparison with major elements (generally below 1 order of magnitude) as a result of such processes.

Petrogenetic models based upon trace elements may be quantified by use of the *partition*, or *distribution, coefficient* (K_D) which is the ratio of the concentration of an element between a mineral and coexisting melt such that:

$$K_D = \frac{\text{concentration of element in mineral}}{\text{concentration of element in melt}}$$

Such *partition coefficients* depend upon the pressure, temperature and chemical composition of the system. Trace elements are termed incompatible if $K_D \ll 1$ and compatible if $K_D > 1$.

For trace elements that are concentrated in **minerals** in surface equilibrium with a melt, the concentration of the element in the melt (C_l) in comparison with the solid (C_0) is given by:

$$\frac{C_l}{C_0} = F^{D-1} \tag{1}$$

where C_l = concentration of element in the residual melt, C_0 = concentration of element in original melt, F = fraction of melt remaining (e.g. 10% melt; $F = 0.10$).

For the behaviour of trace elements the simplest *partial melting* model is one in which the melt remains at the site of melting, in equilibrium with the solid residue, before escaping as a single batch of **magma**. Such a model is hence termed a *batch melting model*. For such a model the concentration of the element in the melt (liquid, C_l), is related to that in the original source (C_0) by:

$$\frac{C_l}{C_0} = \frac{1}{F + D - FD} \tag{2}$$

Equations such as (1) and (2) may be used together with *partition coefficient* data to evaluate the extent of *partial fusion* and/or *fractional crystallization* involved responsible for the formation of a genetically-related association of volcanic rocks. [RST]

Henderson, P. (1982) *Inorganic Geochemistry*. Pergamon, Oxford.

trace elements in sedimentary carbonates Carbonate **minerals** can incorporate a variety of trace metals within their *crystal lattice*. Cation concentrations in depositional and **diagenetic** fluids may be very different. Accordingly where element substitution is systematic (i.e. governed by known *distribution coefficients*) elemental abundances in sedimentary carbonates can be used as tracers in determining the type and relative abundance of fluids responsible for mineral **precipitation** or stabilization. [JDM]

Veizer, J. (1983) Trace elements and isotopes in sedimentary carbonates. In: Reeder, R.J. (ed.) *Carbonates: Mineralogy and Chemistry*. Reviews in Mineralogy 11, pp. 265–99. Mineralogical Society of America, Washington, DC.

trace fossil Structure in sediment produced by the activity of an ancient organism. The terms 'trace', '*Lebensspur*' (German literature) and '*trace d'activité animale*' (French literature) refer to such things in unconsolidated sediment, while sedimentologists often prefer the synonym *biogenic sedimentary structure*. Trace fossils resemble 'body fossils' (remains or representation of all or part of the body of a once living organism) in the sense that they may reflect some aspect of an organism's morphology, but they differ in that trace fossils represent

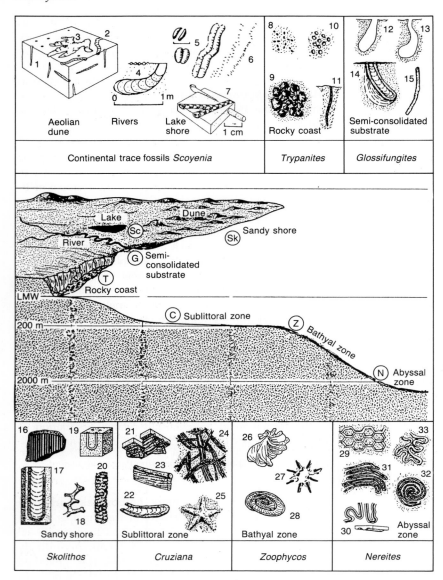

Fig. T5 Schematic representation of the common ichnofacies with examples of characteristic **trace fossils**. 1, ? *Skolithos*; 2, ? *Muensteria*; 3, Vertebrate tracks; 4, *Beaconites*; 5, *Isopodichnus*; 6, *Acripes*; 7, *Scoyenia*; 8, borings of *Polydora*; 9, *Entobia*; 10, echinoid borings; 11, *Trypanites*; 12, 13, pholadid burrows; 14, *Diplocraterion*; 15, unlined crab burrow; 16, *Skolithos*; 17, *Diplocraterion*; 18, *Thalassinoides*; 19, *Arenicolites*; 20, *Ophiomorpha*; 21, *Phycodes*; 22, *Rhizocorallium*; 23, *Teichichnus*; 24, *Cruziana*; 25, *Asteriacites*; 26, *Zoophycos*; 27, *Lorenzinia*; 28, *Zoophycos*; 29, *Paleodictyon*; 30, *Taphrhelminthopsis*; 31, *Helminthoida*; 32, *Spirorhaphe*; 33, cosmorhape. (After Ekdale *et al.*, 1984.)

the behaviour of the organism as well. *Ichnology* is the branch of science studying trace fossils or their recent counterparts.

Some common trace fossils are illustrated in Fig. T5 and include the following: (a) tracks (individual footprints, 3), (b) trackways (sets of multiple tracks indicating movement, 6), (c) trails (continuous ribbon-like locomotion traces, 30), (d) burrows (structures made in unlithified sediment 16–25), (e) borings (structures excavated in hard substrates, 8–13), (f) *bioerosion* traces (drillings or raspings on hard substrates), (g) fecal material or structures (*coprolites* (fossil feces), fecal pellets, fecal castings and regurgitates). As they are both fossils and sedimentary structures, trace fossils often are superior to body fossils as indicators of ancient environments. They are commonly restricted to narrow facies ranges and often occur in otherwise unfossiliferous **clastic rocks**. Virtually

always they are found *in situ*. Many preservable traces were made in sediment by soft bodied organisms responding to environmental factors, providing information which allows a more complete picture of the trophic structure and total diversity of ancient communities to be reconstructed.

Although the study of trace fossils had its roots in 19th century descriptive and systematic paleontology of *ichnites* (footprints) and *fucoids* (burrows, erroneously considered to be fossil seaweeds) the pragmatic value of these structures has been demonstrated only in the past two decades (Seilacher, 1964; Frey, 1975; Ekdale *et al.*, 1984). Trace fossils are nowadays described and classified on the basis of both observational and interpretational criteria. Observational criteria refer to the mode of preservation (*toponomy*) and shape and structure (*morphology*). Toponomic terminology refers to the relationship of the trace fossil to the preserving bed; on the top (*epichnia*), on the base (*hypichnia*), within (*endichnia*) or infilled within another substrate (*exichnia*). Full morphology includes attitude in the sediment, outline shape and internal structure and is the basis for the ichnogeneric name, usually derived from Latin or Greek roots (Häntzschel, 1975). The discrete morphology of each ichnogenus reflects the distinct and recurrent behaviour pattern of the producer (ethological interpretation), which often is environmentally related (*ichnofacies* interpretation). Seven intergrading ethological (behavioural) groups are recognized among trace fossils, namely: locomotion traces (*repichnia*), grazing traces (*pascichnia*), farming traces (*agrichnia*), feeding structures (*fodinichnia*), dwelling structures (*domichnia*), resting structures (*cubichnia*) and escape structures (*fugichnia*) (Häntzschel, 1975; Ekdale *et al.*, 1984). *Ichnofacies*, although each named after one characteristic ichnogenus, are based on recurring associations of trace fossils, lithologies and inorganic sedimentary structures, which reflect discrete sedimentary environments (see Fig. T5 and below).

The value of trace fossils in both the academic and applied aspects of paleontology, **paleoecology**, stratigraphy and sedimentology is demonstrated by the wealth of recent literature. More papers have been published in this field since 1970 than in the previous 170 years (Ekdale *et al.*, 1984). Paleontological studies have focused not only on fossil behaviour of **arthropods**, vertebrates and deep water marine sediment feeders, but also on the evolution and ecology of infaunal animal communities (Frey, 1975; Ekdale *et al.*, 1984). Trace fossils have provided valuable evidence to fix the position of the *Precambrian–Cambrian boundary* in rock sequences in many parts of the world. Workable **biostratigraphic** schemes for correlation of **Paleozoic sandstones** based on *Cruziana* (*trilobite* trails) and **Permian** and **Triassic** *red beds* using **reptilian** footprints, have been devised (Frey, 1975). However, it is in the fields of sedimentology and **sedimentary facies** analysis that the unique value of trace fossils is most apparent.

As the behaviour and ecology of most infaunal animals is controlled by the nature of the substrate and environmental factors, trace fossils are often indicators of oxygen level and firmness of the substrate (e.g. *Chondrites* produced by a low O_2-tolerant burrower or the change from burrowers to borers in the development of a **hardground**). Escape traces frequently indicate upward producer movement in pace with sedimentation and so provide an indicator of relative rate of sedimentation. Resting traces of bilobed or ovoid outlined organisms sometimes show current response orientation and so may be used as *paleocurrent indicators* (e.g. *bivalve* or *trilobite* resting traces). *Bioturbation* patterns produced by sediment feeders frequently reflect such phenomena as *storm*-generated bedforms in near-shore sequences.

The most fundamental advance in the use of trace fossils for *paleoenvironmental analysis* is the *ichnofacies* concept, first formulated by Seilacher (1964). He recognized that some trace fossils show facies restriction and constantly recur in similar sedimentary associations throughout the geological column, while other traces are facies crossing irrespective of geological age. The *ichnofacies* model he proposed has been extended to a seven facies model based also on substrate firmness, energy level and *paleosalinity*. (See Fig. T5, *ichnofacies* Sc–N and Ekdale *et al.*, 1984.) Many ichnological studies of the past decade have shown the basic validity of this model but also that trace fossil–sediment relationships can be further refined and applied to problems of environmental reconstruction and **basin** analysis, especially in cored sequences.

Trace fossils of the *Nereites ichnofacies* (Fig. T5, *ichnofacies* N, Nos 29–33), complex meandering, spiral and network horizontal grazing and farming traces, occur widely in fossil *flysch* sequences and modern deep ocean *carbonate oozes*. Complex feeding burrows of *Zoophycos* type (*ichnofaces* Z 26, 28), while occurring widely in deep ocean sediments (**Deep Sea Drilling Project** cores) are also common in fossil carbonates formed in quiet conditions offshore, but often at no great water depth. The greatest behavioural diversity and abundance of trace fossils occurs in offshore clastic sediments of the *Cruziana ichnofacies* (C). Here locally subichnofacies can often be recognized dominated by such feeding burrows as *Teichichnus* (23), *Rhizocorallium* (22) or *Thalassinoides* (18). The prime importance of substrate type and energy level control of trace fossil distribution is demonstrated in the shoreface to intertidal environments represented by *Skolithos* (**sand**, Sk), *Glossifungites* (firm mud, G) and *Trypanites* (**hard-ground**, T) *ichnofacies*. Both diversity and abundance of trace fossils can be low in these high stress situations. Continental aquatic environments tend to be ephemeral or highly intergrading, so discrete trace fossil associations comparable to those in marine facies are not yet recognized. **Deltaic** sequences contain both marginal marine (traces 16–20) and fluviolacustrine (traces 4–6), mainly burrows and trackways. The *Scoyenia* association (Sc) (traces 5–7) occurs in ephemeral fluvial or lake margin situations often associated with *red beds* and vertebrate footprints. In fully terrestrial sediments such as aeolian **sands** and **paleosols**, **plant** root traces and **insect** dwelling burrows (traces 1, 2) form the predominant *ichnofauna* in both ancient and recent examples. [JEP]

Ekdale, A.A., Bromley, R.G. & Pemberton, S.G. (1984) *Ichnology — Trace Fossils in Sedimentology and Stratigraphy*. Short Course No. 15. Society of Economic Paleontologists and Mineralogists, Tulsa, OK.
Frey, R.W. (ed.) (1975) *The Study of Trace Fossils*. Springer-Verlag, Berlin.
Häntzschel, W. (1975) Trace fossils and problematica (2nd edn). In: Teichert, C. (ed.) *Treatise of Invertebrate Paleontology Part W. Miscellanea Supp. 1*. Geological Society of America and University of Kansas Press, Lawrence, KS.
Seilacher, A. (1964) Biogenic sedimentary structures. In: Imbrie, J. & Newall, N.D. (eds) *Approaches to Paleoecology*, pp. 96–316. John Wiley & Sons, New York.

trachyte A **silica**-saturated, or **silica**-oversaturated alkaline volcanic rock of intermediate composition.

Trachytes contain **alkali feldspar** (**sanidine** or **anorthoclase**) as the main *phenocryst* and groundmass phase. They are normally weakly **porphyritic**, containing about 10% modal *phenocrysts* of the assemblage **alkali feldspar**, *clinopyroxene* close to *ferrohedenbergite* in composition, Fe–Ti oxides, **amphibole**, **fayalite**, **aenigmatite** and **biotite**. Trachytes normally have a well-crystallized groundmass, which commonly has a *trachytic texture* — a strong flow-alignment of small, acicular **alkali feldspar** crystals. Trachytic **tephra** is commonly glassy (see **glassy rock**), but glass is rather rare in trachyte **lava flows**.

Trachytes are distinguished from *trachydacites* and **dacites** by their higher $Na_2O + K_2O$ contents, and absence of **plagioclase** *phenocrysts*. Trachytes grade into weakly silica-undersaturated alkaline intermediate rocks (**phonolitic** trachytes), which can be distinguished from trachytes by containing normative **nepheline**, and the presence of **nepheline** and/or *analcite* in the groundmass. Trachytes may or may not be *peralkaline* (defined as a molecular excess of $(Na_2O + K_2O)$ over Al_2O_3). *Peralkaline* trachytes are divided into *comenditic* trachytes and more iron-rich **pantelleritic** trachytes on the basis of Al_2O_3 and total iron contents (Macdonald, 1974). These *peralkaline* trachytes grade into the peralkaline **rhyolites**, the *comendites* and **pantellerites**.

The names *trachybasalt* and *trachyandesite* are used broadly to classify volcanic rocks intermediate in composition between **basalt** and trachyte. The nomenclature of these rocks is chaotic, and classification inevitably arbitrary in view of the spectrum of volcanic compositions. Recommendations for the classification of these rocks by abundances of total alkalis ($Na_2O + K_2O$) and SiO_2 were presented by Cox *et al.* (1979) and Le Bas *et al.* (1986). *Trachybasalts* have similar abundances of SiO_2 to **basalts** (45–52 wt%), but contain relatively sodic **plagioclase** (near **andesine**). They can be chemically separated from **basalts** by their high alkali contents ($Na_2O + K_2O > 5\%$). They can be divided into sodic types and potassic types (**hawaiites** and potassic *trachybasalts* respectively). *Trachyandesites* have higher abundances of total alkalis and SiO_2 than *trachybasalts*, and normally contain **oligoclase**. They can be divided into sodic and potassic types. The sodic types can be further divided into *mugearites*, and more SiO_2-rich **benmoreites**. The potassic types can be further divided into *shoshonites*, and more SiO_2-rich **latites**. The division between *trachyandesites* and trachytes is also arbitrary, and can be defined in several ways, all approximately corresponding to a SiO_2 content of 59%. [PTL]

Cox, K.G., Bell, J.D. & Pankhurst, R.J. (1979) *The Interpretation of Igneous Rocks.* Allen & Unwin, London.
Le Bas, M.J., Le Maitre, R.W., Streckeisen, A. & Zanettin, B. (1986) *Journal of Petrology* **27**: 745–50.
Macdonald, R. (1974) Nomenclature and petrochemistry of the peralkaline oversaturated extrusive rocks. *Bulletin Volcanologique* **38**: 498–516.

transform fault *Transcurrent faults* are faults in which the relative motion on opposite sides are parallel to the **strike** of the fault line, i.e. the motion is **strike-slip**. A special form of this type of fault, the transform fault, was distinguished by Wilson (1965) as *transcurrent faults* linking other forms of **plate margins** (**subduction zones** and *ocean ridges*) of which the commonest is the oceanic **fracture zone** that is essentially quiescent on the flanks of an *oceanic ridge*, but seismically active where it links between two offset ridge crests. [DHT]

Wilson, J.T. (1965) A new class of faults and their bearing on continental drift. *Nature* **207**: 343–7.

transient creep The slow **deformation** which takes place in rocks when subjected to loads at environmental temperatures and pressures at or very close to the surface of the Earth. (See **creep**.) [DHT]

transition zone Although this term can apply to any boundary zone, it is commonly used to mean that part of the Earth's **mantle** between the base of the **upper mantle**, at *c.* 350 km depth and the top of the **lower mantle** at *c.* 700 km depth. It is sometimes included within the **upper mantle**. [DHT]

translation The **deformation** of a rock body may be regarded as a combination of translation, distortion and rotation. Distortion and rotation are aspects of **strain**, whereas translation is the **displacement** of the centre of the deformed body to a new position. [RGP]

transmission coefficient The ratio of the amplitude of the transmitted wave to that of the wave incident on an **acoustic impedance** contrast. For a wave impinging the contrast at right angles, i.e. the transmitted wave travels straight through the contact unrefracted, the transmission coefficient can be calculated as

$$\frac{2A_1}{A_1 + A_2}$$

Where A_1 and A_2 are the **acoustic impedances** of the media through which the arriving and transmitted waves are travelling. For a wave that has passed through the interface going down from the surface and during its upward journey back to the surface, the two-way transmission coefficient is given by $1 - R^2$, where R is the **reflection coefficient**. In the more general case of a wave incident at any angle to a contact, both shear and compressional waves may be transmitted and the amplitudes of the various waves are described by **Zoeppritz' equations**. [GS]

transmissivity The product of *hydraulic conductivity* (see **permeability**) and the thickness of a rock layer that is acting as an **aquifer**. [AME]

transportation slope A slope which undergoes neither **erosion** nor deposition because at each point the material brought in from upslope is equal to that carried away downslope. Usually these features occur at the base of a slope and are concave in form since the slope is either formed by *soil wash* **erosion** or because the **regolith** becomes finer downslope. The concavity represents the minimum slope across which material delivered from upslope can be moved downslope. Over the longer term the transportational slope may undergo a gradual decrease in slope angle so that the assumption of mass balance is effectively one of short-term equilibrium. [TPB]

transposition Creation of a new **fabric** by **deformation** of an earlier one. The term is most commonly applied to the production of a transposed **foliation**, in which a new **foliation** is created, often by very tight **folding** and disruption of folded layers, from an original set of surfaces in highly **strained** rocks. The early **foliation** may have been **bedding** or a previous transposed **foliation**. The process can be recognized where both primary and derived **foliations** can be seen together, or from features such as preserved isolated *fold hinges*. [TGB]

transpression Compressional effects associated directly with *transcurrent* motion. Mainly dependent on the interaction between pre-existing structural features and the **tectonic** forces operating on the *transcurrent* or **transform fault** systems. (See also **transtension**.) [DHT]

transtension Tensional effects associated directly with *transcurrent* motions. Mainly dependent on the interaction between pre-existing structural features and the **tectonic** forces operating on the *transcurrent* or **transform fault** systems. (See also **transpression**.) [DHT]

transverse dune Dune which has its long axis perpendicular to the predominant **sand**-moving wind direction. The most common forms are **barchans**, barchanoid ridges and transverse ridges. [RDS]

travertine Accumulation of calcareous material **precipitated** from flowing freshwater. Freshwater is usually undersaturated with respect to $CaCO_3$, but passage through calcareous rocks or sediments may raise the $CaCO_3$ content of the waters. Expulsion of CO_2 from the waters, either at hot springs or **waterfalls**, results in **precipitation** of $CaCO_3$, a process perhaps aided by biochemical activity. These deposits are known as travertine. **Tufa** (or calcareous **tufa**) refers to the spongy and porous varieties of travertine associated with *algal* colonies. [DMP]

tremolite ($Ca_2Mg_5Si_8O_{22}(OH)_2$) An *inosilicate* of the amphibole group. (See **amphibole minerals**.) [DJV]

Table T1 Triassic

Major subdivisions	Traditional German classification	Stages	
Upper	Keuper	Rhetian	
		Norian	
		Carnian	
Middle	Muschelkalk	Ladinian	
		Anisian	Spathian
Lower	Bunter	Scythian	Smithian
			Dienerian
			Griesbachian

Triassic Stratigraphic system taking its name from the traditional threefold German classification, but the subdivision of stages also presented in Table T1 is based on *ammonites* and was originally established in the Austrian Alps. More complete and tectonically less disturbed successions are present, however, in North America, which in consequence is a better place to establish a global standard. One important proposed change is to subdivide the *Scythian* into four stages. Some biostratigraphers would either relegate the *Rhetian* to a substage of the *Norian* or abandon the term altogether.

During the period there were only two main *continental margins*, the circum-Pacific, which could be regarded as an external margin, and the **Tethys**, which appears as an internal margin, the two being separated by the huge land areas of **Laurasia** and **Gondwanaland**. The available **paleomagnetic** data suggest that there was no significant movement of **Pangea**, but it is likely that some small landmasses, such as south-east Asia, accreted onto eastern **Laurasia**. Tectonically the Triassic appears to have been relatively quiescent, with no major mountain-building episode, but there were substantial eruptions of *flood basalts*. The so-called Siberian Traps straddle the *Permian–Triassic boundary* and the *Karoo* **igneous rocks** of southern Africa, and **basalts** in eastern North America and North Africa, the *Triassic–Jurassic boundary*. All these rocks are associated with tensional **tectonic** activity; both the *Karoo* and Atlantic rocks mark an early stretching phase that anticipates the later opening of the North Atlantic and western Indian Oceans.

Sea-level was very low at the start of the Triassic. A more or less progressive rise led to a major marine transgression in Middle Triassic times, the so-called *Muschelkalk* transgression, when the sea extended from **Tethys** over the lands flanking it to the north and south, from Europe and North Africa to China. In other places an early Triassic transgression is very marked (e.g. eastern Greenland, western Australia, China). Upper Triassic rocks are also transgressive in some places, when maximum sea-level was probably attained late in the period, in *Norian* times, but the end of the period was marked by a pronounced regression in many parts of the world.

Climatically the Triassic was an equable period, with no indication of polar *ice caps*. Vast areas of the continents were subjected to an arid climate, as evidenced by extensive **evaporite deposits**, but **coals** were formed in *monsoonal* climates of seasonal rainfall in mid latitudes of the eastern parts of **Laurasia** and **Gondwanaland** (former USSR, China and Australia). The **Tethyan** zone extending from the Alps to the Himalaya and Indonesia is characterized by a distinctive suite of rocks markedly different from the continental *red beds* so common elsewhere, dominated by thick **limestones** and **dolomites** of biogenic, sometimes *reefal* origin. These give rise to spectacular mountain complexes such as the Northern Calcareous Alps and the Dolomites.

Triassic faunas are substantially different from those of the late **Paleozoic**, a direct consequence of the end-**Permian** *mass extinction* phase, the greatest of the **Phanerozoic**, which affected a high percentage of the world's biota, both marine and terrestrial. With regard to terrestrial vertebrates, throughout the first two-thirds of the period the most important **reptiles** were advanced members of the so-called mammal-like **reptiles**, the *therapsids*. By early in the late Triassic (*Carnian*) these had been largely replaced by *archosaurs*, which underwent an evolutionary radiation to give rise to *dinosaurs*, *crocodiles* and *pterosaurs*. By the end of the period true **mammals** had appeared. Other less conspicuous but notable **reptiles** include *lizards*, *rhynchosaurs* and *phytosaurs*, while the most common and widely distributed of all Triassic continental vertebrates are the *labyrinthodonts*, which almost completely died out at the end of the period.

Among the marine faunas, stratigraphically most useful are *ammonites* and **conodonts**. New groups of **corals**, **bryozoans** and **echinoderms** emerged, and *bivalves* are the most important macrofossils in a wide variety of facies; **brachiopods** never again reached their **Paleozoic** dominance. The major *decapod* order of the *crustaceans* began in the Triassic, and the unrelated king crabs survive little-changed to the present day. Major Tethyan *reef* builders include **corals**, **stromatoporoids**, **sponges**, *hydrozoans* and *algae*. Terrestrial floras were dominated by *gymnosperms*, *horsetails* and *ferns*. None of the fauna or flora exhibit any significant provinciality, apart from tropical groups being confined to **Tethys**, because of global equability and coherence of major continental masses. (See **land plants**.) [AH]

Haq, B.U., Hardenbol, J. & Vail, P.R. (1987) Chronology of fluctuating sea levels since the Triassic. *Science* **235**: 1156–67. This article contains biostratigraphic zonal schemes as well as a sea-level curve based on seismic stratigraphy.

Kent, P.E., Audley-Charles, M.G., Balchin, D.A. *et al.* (1970) Triassic rocks of the British Isles. *Quarterly Journal of the Geological Society, London* **126**: 1–291.

Tozer, E.T. (1967) A standard for Triassic time. *Geological Survey of Canada Bulletin* **156**: 1–103.

Zapfe, H. (ed.) (1974) *The Stratigraphy of the Alpine-Mediterranean Triassic*. Springer-Verlag, Vienna.

tridymite (SiO_2) A high-temperature form of **silica**. (See **silica minerals**.) [DJV]

triphylite ($Li(Fe,Mn)PO_4$) A mineral found in *pegmatites*. [DJV]

triple junction A point where three **plate margins** intersect. These may comprise any combination of **subduction zone**, **transform fault** or *oceanic ridge* **plate boundaries**, but certain configurations are unstable and impersistent. Also an important term in mineral textures. [DHT]

McKenzie, D.P. & Morgan, W.J. (1969) Evolution of triple junctions. *Nature* **224**: 125–33.

tripoli Microcrystalline, soft, friable, porous siliceous material used as a fine **abrasive** in toothpaste, industrial soaps, as a buffing and polishing compound in lacquer finishing, in the automobile industry and as a filler in paints, plastics, rubber and various enamels. The USA is the only country with significant production. [AME]

troilite (FeS) An **iron sulphide mineral** found chiefly in *meteorites*. [DJV]

trona ($Na_3H(CO_3)_2 . 2H_2O$) A **mineral** found in saline lake deposits. [DJV]

trondhjemite A phaneritic **quartz–albite**-rich **plagioclase** rock with low **colour index** (a leucocratic **tonalite**). Trondhjemites, as so defined, include *plagiogranites* within **ophiolite** complexes, **tonalites** within **calc-alkaline batholiths** and form part of a distinctive **Archean** rock association. Such occurrences are bimodal trondhjemite (*grey gneiss*) and metamorphosed **basaltic** rocks without any rocks of intermediate composition. Such associations appear to have been important in the formation of early **crustal** nuclei. The absence of intermediate rocks (e.g. **andesites**) suggests that the petrogenesis of trondhjemites may be different from that of **calc-alkaline magmas** and may result from *partial melting* of the hydrated basic igneous *amphibolite*. [RST]

Barker, F. (1979) *Trondhjemites, Dacites and Related Rocks*. Elsevier, Amsterdam.

troostite A **manganiferous** variety of the mineral **willemite**. [DJV]

trottoir A narrow, intertidal zone, organic *reef*. Trottoirs are found commonly on Mediterranean coastlines and also in the tropics and are formed by vermetids, coralline *algae* and other organisms. They act as protective features on **limestone** shorelines. [HAV]

Focke, J.W. (1978) Limestone cliff morphology on Curacao (Netherlands Antilles), with special attention to the origin of notches and vermetid/coralline algal surf benches ('cornices', 'trottoirs'). *Zeitschrift für Geomorphologie* **22**: 329–49.

trough line Line joining the topographically lowest points on a folded surface. The trough line is normally parallel to the *hinge line* of a **fold** but is only coincident with it in the case of *upright folds*. [RGP]

true dip The angle between the surface of a layer (typically **bedding**) and the horizontal, measured in a vertical plane perpendicular to the **strike** of the layer. The true dip is the maximum angle from the horizontal that can be measured for a given plane. Lines in any other orientation on the surface are at a smaller inclination to the horizontal; these angles represent **apparent dips**. When measuring the orientation of surfaces such as **bedding** or **faults**, it is usual to quote the direction of **dip**, i.e. the direction in which the surface becomes lower, as well as the amount. [SB]

true north The direction towards the Earth's axis of rotation (the geographic pole) as distinct from magnetic north which is the direction towards the geomagnetic north pole. [DHT]

true polar wander The agreement in *paleolatitude* between *paleoclimatic* and **paleomagnetic** observations over geological time suggests that the Earth's axis of rotation has not changed significantly, relative to the plane of the ecliptic, during geological time. As such, true polar wander probably does not occur. However, if the Earth's surface, as a whole, moved relative to this pole, then this would, in effect, result in effective polar wander. This was originally thought to occur to explain the apparently changing position of the Earth's climatic pole, but this concept was largely subsumed within the **plate tectonic** hypothesis. Nonetheless, a summation of the total plate motions on the Earth today shows a net motion towards the northern Pacific, due to the large number of **subduction zones** within this region. (See **apparent polar wander**.) [DHT]

Goldreich, P. & Toomre, A. (1969) Some remarks on polar wandering. *Journal of Geophysical Research* **74**: 2555–67.

Jurdy, D.M. & Van der Voo, R. (1974) A method for the separation of true polar wander and continental drift, including results from the last 55 m.y. *Journal of Geophysical Research* **50**: 35–54.

Lambeck, K. (1980) *The Earth's Variable Rotation*. Cambridge University Press, Cambridge.

tschermakite ($Ca_2Mg_3(Al,Fe)_2Al_2Si_6O_{22}(OH)_2$) An **amphibole mineral**. [DJV]

tsunami Often inappropriately called *tidal waves*, tsunamis are long-wavelength water **waves** caused by sudden sea-floor movements such as sediment slumps, **earthquakes** and volcanic eruptions. Rarely noticed at sea because of their long period (several minutes) and small amplitude (less than a metre or so), tsunami amplitudes grow when they reach shallow coastal water, reaching 10–20 m or more. Travelling at speeds of tens of m s^{-1}, they can be highly destructive, even at great distances from their source. A Seismic Sea Wave Warning System has been established in Honolulu to monitor circum-Pacific **earthquakes** and tide guages and issue tsunami alerts. (See also **seiche**.) [RAC]

Bolt, B.A. (1978) *Earthquakes — Primer*, Chapter 5, W.H. Freeman, San Francisco.

tuff A volcanic sediment. Tuffs have been employed as **building stones** in many European countries including Italy, where they were used at Pompeii and Herculaneum. In Polynesia statues were produced; several hundred Easter Island heads, dated to *c.* AD 700, are of tuff. In Britain one of the most important **axe** groups of the Neolithic period was the epidotized tuff of the Lake District where extensive **axe factories** have been located. [OWT]

tuff ring A kind of monogenetic volcano produced from **phreatomagmatic eruptions**, having a low rim and a broad, flattish **crater**. **Magma** may be of any composition, but **basalt** is most common. Tuff rings consist of phreatic air-fall breccias and finely-bedded air-fall and **pyroclastic** surge beds. Maximum thickness of the deposits are 50 m or less, and the maximum depositional **dips** of the **tephra** are 3–12°. *Tuff cones* are similar to tuff rings, also being hydrovolcanic, but they form generally larger, more steep-sided structures having maximum thicknesses of 100–350 m and higher maximum **dips** of deposition (24–30°). *Tuff cones* contain a higher proportion of **mudflow** deposits than tuff rings.

At many tuff rings, hydrovolcanic eruptions were accompanied by sagging or collapse of the crater, so that the whole resulting structure may be termed a **maar**. [PTL]

Wohletz, K.H. & Sheridan, M.F. (1983) Hydrovolcanic explosions II. Evolution of basaltic tuff rings and tuff cones. *American Journal of Science* **283**: 385–413.

tungstates and molybdates Two groups of minerals having related structures. Tungstates contain **tungsten** in the form of the $(WO_4)^{2-}$ anion unit (hence as W^{6+}); molybdates contain **molybdenum** in the form of the $(MoO_4)^{2-}$ anion unit (hence as Mo^{6+}). Both anion units have the central cation coordinated to four oxygens at the corners of a flattened tetrahedron. Being of similar ionic radius, **molybdenum** and **tungsten** may freely substitute for each other in this distorted tetrahedral site, although in nature **molybdenum** and **tungsten** commonly follow separate geochemical paths and the primary tungstates and molybdates may be relatively free of the other species.

The minerals in the tungstate and molybdate families fall mainly into two isostructural groups:

1 Wolframite group. A complete *solid solution* exists between **ferberite** ($FeWO_4$) and **huebnerite** ($MnWO_4$) with the intermediate members being most common (($Fe,Mn)WO_4$ with 20–80 atomic % Fe) and being termed **wolframite**. Between the tetrahedral WO_4 groups, Fe^{2+} and Mn^{2+} occur in octahedral coordination in this monoclinic structure. **Wolframite** is the major **ore** mineral of **tungsten**.

2 Scheelite group. Consists of compounds with larger ions (Ca^{2+}, Pb^{2+}) in eightfold coordination with the oxygens of WO_4^{2-} and of MoO_4^{2-} groups. The members include **scheelite** ($CaWO_4$) **powellite** ($CaMoO_4$), **stolzite** ($PbWO_4$) and **wulfenite** ($PbMoO_4$). Substitution of **molybdenum** for **tungsten** occurs forming partial *solid solution* series between **scheelite** and **powellite**, and between **stolzite** and **wulfenite**. **Scheelite** is also significant as an **ore** of **tungsten**. [DJV]

tungsten (W) At. no. 74, at. wt. 183.8. A metal used in steel alloys, lamp and heating filaments. It is recovered principally from **skarn** and **vein** deposits. [AME]

tunnel erosion A form of *pipe erosion* initiated by an abnormally high **hydraulic gradient** possibly in combination with a dispersible *soil* layer. Only under these circumstances may particulate **erosion** develop as a result of subsurface flow, in the absence of pre-existing cracks, since normally *soil* particles are larger than the void spaces and so cannot be carried through the *soil*. This type of piping is rare under natural conditions but has been reported for earth dams where the *hydraulic gradient* may be very large. Tunnels may develop naturally as the result of the **erosion** of cracks in the *soil*; the term **pipe** is more usually applied to this latter case. [TPB]

turam EM method *Mineral exploration* method based on **electromagnetic induction**, employing a large fixed source. In a typical turam survey, the source could be a cable several kilometres long, parallel to the expected **strike** of the body, and the receivers are a pair of coils kept at a fixed separation of about 10 m and moved along profiles perpendicular to the source cable. Vertical field components are measured at one or two frequencies, of the order of hundreds of Hz, on each survey. The instrument allows the measurement of ratios of the signal amplitudes at the two coils as well as their phase difference. Any phase difference indicates the presence of anomalous **electrical conductivities**, as does any departure of the amplitude ratio from that expected theoretically for the primary field (due to the different distances of the coils from the linear source). The theoretical ratio is used to 'normalize' the field data, and standard curves for normalized ratios for certain types of bodies are available

as an aid to interpretation; turam is ideally suited for sheet-like bodies, but as a general electromagnetic method it is no longer widely used. [ATB]

Telford, W.M., Geldart, L.P. & Sheriff, R.E. (1990) *Applied Geophysics* (2nd edn). Cambridge University Press, Cambridge.

turbidity current Subaqueous density current of suspended sediment with sufficient potential energy for **gravity** to provide the motive force for flow. Turbidity currents may flow on slopes of less than 1°. These currents occur in non-marine to marine, shallow to deep water. They can be generated by a wide range of sediment suspension processes, including shallow-marine *storms*, river plumes entering the sea or a lake, shelf currents transporting sediment over the shelf-break, and various forms of slope **failure**. Sediment **slides**, **debris flows** and other sediment flows (see **sediment gravity flow**) may transform into turbidity currents under appropriate hydrodynamic conditions. These currents typically possess a bulbous 'head' and a relatively elongated region behind called the 'body', the rear being the 'tail' of the flow. To maintain equilibrium flow conditions, sediment must be continuously supplied from the body to head region to compensate for energy losses due to *turbulence* and frictional losses from the upper and lower flow boundaries. Experiments suggest that for **Froude numbers** >0.75, the head is thicker than the body, whereas for lower values the body is thicker. Turbidity currents may flow with low to high velocities, up to metres per second, and comprise low to high density suspensions. Flow thickness is greatest in very low density (a few $mg\,l^{-1}$), low velocity ($0.1\,m\,s^{-1}$) currents where the thickness may reach hundreds of metres to about 1 km, as a **nepheloid layer**. Flow length ranges up to thousands of kilometres, but typically varies between tens and hundreds of kilometres. Increased gradients and flow constriction may lead to flow acceleration and net **erosion**, with enhanced sediment entrainment. **Erosion** at the base of a flow can produce various *scour* or *sole marks*. Deposition of sediment from a turbidity current occurs when there is a net loss of energy due to, for example, flow expansion, rapid deceleration and/or a decrease or reversal in bottom gradient. Deposition from a waning turbidity current can produce a vertical sequence of sedimentary structures, or *turbidite*, showing the upper to lower flow regime **Bouma sequence**. [KTP]

turbulent flow Flow characterized by a three-dimensional motion in which velocity at a point fluctuates in time and in which the **streamlines** need not be parallel. Turbulent flow is present at **Reynolds numbers** in excess of 2000 and can be spatially characterized by velocities of flow along three mutually perpendicular axes — downstream (u), vertical (v) and spanwise (w). Mean velocity in each direction has superimposed upon it a fluctuating component which is attributable to *turbulence* within the flow.

Turbulence consists of a series of quasi-random motions operating at different physical scales within a fluid. As a result of mixing within a turbulent flow the turbulent **boundary layer** has a steep velocity gradient near the bed and may be divided into three regions: the **viscous sublayer** adjacent to the boundary; a region of maximum *turbulence* generation which occupies approximately 10% of the flow depth and above this a zone of *turbulence* dissipation where eddies generated near the bed grow and then decay.

Turbulence is normally expressed as the root mean square of the instantaneous fluctuating component, x', or $\sqrt{\overline{x'^2}}$, or as a *turbulence intensity* which is equivalent to the root mean square divided by the mean flow velocity, $\bar{x}$, or $\sqrt{\overline{x'^2}}/\bar{x}$. [JLB]

Bradshaw, P. (1971) *An Introduction to Turbulence and its Measurement.* Pergamon, Oxford.

Cantwell, B.J. (1981) Organised motion in turbulent flow. *Annual Review of Fluid Mechanics* **13**: 457–515.

turlough A seasonal lake, as found in the **glacially** influenced **karst** area of western Ireland. Turloughs are shallow hollows up to $5\,km^2$ in area which become lakes for 5 to 11 months of the year, and are dry for the remainder of the time. They fill and empty through **springs** and **sinkholes**, and have similar **hydrological** behaviour to **poljes**. [HAV]

Coxon, C. & Drew, D.P. (1986) Groundwater flow in the lowland limestone aquifer of eastern Co. Galway and eastern Co. Mayo, western Ireland. In: Paterson, K. & Sweeting, M.M. (eds) *New Directions in Karst*, pp. 259–80. Geo Books, Norwich.

turmkarst German term for **tower karst**, i.e. a **karst** landscape with steep, flat-topped towers rising from a plain and usually found in humid tropical areas. [HAV]

turquoise $(CuAl_6(PO_4)_4(OH)_8 \cdot 4H_2O)$ A blue-green secondary **copper** mineral valued as a **gemstone**.

Turquoise was highly prized in the ancient world for making beads, jewellery and small decorative objects. Turquoise was obtained from the Middle East, in ancient Persia and from Sinai, and was widely used in India. It was also used extensively in the New World by the Aztecs and Toltecs, and by the North American Indians of the 1st millenium AD, especially for beads and amulets. [OWT/DJV]

twinning The formation of composite crystals in which the two individuals have different orientations which bear a simple crystallographic relationship to each other. The orientation relationship frequently gives rise to *re-entrant angles* by which twins can at once be recognized (Fig. T6).

Although such composite crystals have usually grown as a twin, with the two orientations present from the start, the orientation relationship is described as the symmetry operation which would convert one orientation into the other. The twin individuals may be related by 180° rota-

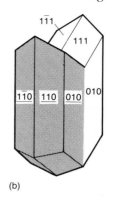
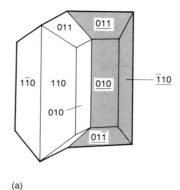

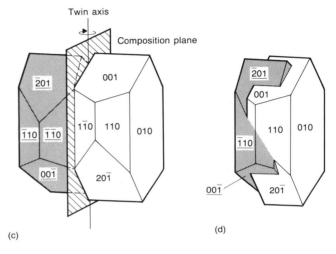

Fig. T6 Twinning. Examples of twinned crystals. (a) Aragonite twinned on (110). The twins are related by reflection in a plane (110) and in contact on that plane. An alternative description of the twin relationship is a 180° rotation about an axis normal to the (110) twin plane. (b) Gypsum twinned on (100). (c) A Carlsbad twin in orthoclase feldspar drawn to show the parallel twin relationship with the twin axis [001] lying in the (010) composition plane. (d) An interpenetrant Carlsbad twin in orthoclase. (After Bishop, 1967.)

tion about a *twin axis* or by reflection in a *twin plane*, these being equivalent if the crystal possesses a centre of symmetry.

The plane along which the two individuals are joined is called the *composition plane* and there is usually no discontinuity across it, the two orientations of the crystal fitting together almost exactly across this plane. The *composition plane* is often, but not always, the same as the *twin plane*.

Two common types of rotation twin are:

1 *Normal twins*, in which the *twin axis* is normal to the *composition* plane. In centrosymmetric crystals this is equivalent to, and looks like, reflection in the *composition plane* (Fig. T6a and b).

2 *Parallel twins*, in which the *twin axis* lies in the *composition plane* (Fig. T6c).

tyuyamunite

The twins with portions on either side of a *composition plane* so far described are *contact twins*. *Interpenetrant twins* are also found in which the two individuals appear to have grown through each other (Fig. T6d).

A twinned crystal may appear to have higher symmetry than it actually possesses: twinning which closely imitates a higher symmetry is referred to as *mimetic twinning*. Twinning often originates from *pseudo-symmetry* in the **crystal structure**, when the regular arrangement of the atoms is nearly but not exactly symmetrical about the *twin axis* or over the *twin plane*.

Macroscopically well-formed twin crystals have usually grown as two individuals from the beginning but twinning may be brought about in other ways too:
Glide twinning occurs when an applied **stress** forces part of a crystal to move into the twin-related orientation.
Inversion twinning results from a phase transformation in which the crystal symmetry is lowered. The two twin orientations are equivalent in the higher symmetry. When the symmetry is lowered the two orientations are equally likely to be adopted and may be found in equal numbers, often as closely spaced lamellae (*polysynthetic twinning*).

The observation of twinning within a crystal may thus provide information on the **stress** to which it has been subjected or on the phase changes, and hence pressure and temperature changes, it has undergone. [JEC]

Berry, L.G. & Mason, B. (1983) *Mineralogy: Concepts, Descriptions, Determinations* (2nd edn), Chapter 2. W.H. Freeman, San Francisco.
Bishop, A.C. (1967) *An Outline of Crystal Morphology*, Chapter 11. Hutchinson, London.
Phillips, F.C. (1971) *An Introduction to Crystallography* (4th edn), Chapter VII, Longman, London.

tyuyamunite ($Ca(UO_2)_2(VO_4)_2 . 5-8\frac{1}{2}H_2O$) A greenish secondary **ore** mineral of **uranium** and **vanadium**. [DJV]

U

ulexite. $(NaCaB_5O_6(OH)_6 \cdot 5H_2O)$ A borate mineral found in association with **borax** and deposited from **brines** formed in enclosed **basins** in arid regions. [DJV]

ultramafic rock A rock (generally ultrabasic in chemical composition) that is dominantly composed (over 90%) of Fe–Mg minerals. Such rocks are composed dominantly of **olivine**, *orthopyroxene* and *clinopyroxene* ($\pm$ **amphibole**) and may be classified in terms of these minerals (see **dunite, lherzolite, peridotite, websterite**). Ultramafic rocks occur in continental settings in association with **gabbroic** intrusions and within **ophiolite** complexes where they may show evidence of igneous lamination or layering. [RST]

ulvöspinel (Fe_2TiO_4) A **spinel** structure mineral found terrestrially only as fine **exsolution** bodies within **magnetite** but as coarser grains in lunar rocks. (See **oxide minerals**.) [DJV]

umber A naturally occurring brown pigment. Hydrated **iron** and **manganese** oxides are the principal components, with minor quantities of **silica**, alumina and **lime**. Deep continental **weathering** of rocks rich in ferromagnesian minerals can give rise to pigments such as umber. [DMP]

unconfined aquifer An **aquifer** lacking an upper **aquiclude** or confining bed so that its upper saturation limit is the **water table**. (See **artesian aquifer**.) [AME]

unconformity Break in the stratigraphic record, representing a period of geological time not marked by sediment deposition. Allied terms are: *diastem, non-sequence, paraconformity* and *disconformity* (Fig. U1).

Diastems represent pauses in sedimentation, marked by abrupt changes in sediment type, producing surfaces of discontinuity (**bedding** planes) but no other evidence of a time gap. *Non-sequences*, or *paraconformities*, are similar to *diastems* but exhibit faunal or other evidence of a time gap. *Disconformities* are marked by evidence of **erosion** during the sedimentary break, but the **bedding** below the **erosion surface** is parallel to that above, i.e. there has been no **deformation** of the lower series of **beds** prior to **erosion**. This is also known as a *hiatus*.

Unconformities are distinguished from other stratigraphic breaks by angular discordance between the older **beds** below the unconformity surface and the younger **beds** above. Hence an unconformity represents the following sequence of events: deposition of lower *strata*; tilting or other **deformation** of lower strata; **erosion**; deposition of upper strata. The structure produced by the discordance of younger upon older strata is termed **overstep** or *overlap*, and the basal beds of the younger series are said to 'overstep' the various strata of the older series truncated by the **erosion surface** (Fig. U1). (See **onlap, offlap**.) [RGP]

Roberts, J.L. (1982) *Introduction to Geological Maps and Structures.* Pergamon, Oxford.

unconformity-associated uranium deposit Epigenetic **deposit** of **uranium** where most of the mineralization occurs at or just below an **unconformity**. Deposits of this nature have only come into prominence during the last twenty years but are now known to include the largest (Jabiluka, Northern Territory, 200 000 t contained U_3O_8), highest **grade** (Cigar Lake, Sask., 9.04% U) and the most valuable **uranium** deposits (Key Lake, 200 Mlb [million pounds (weight)] contained 8% U) in the world (apart from Olympic Dam, South Australia). The Athabaska Basin in Saskatchewan and the Northern Territory of Australia host the main deposits so far found and these contain about half the western world's low cost **uranium** reserves.

Orebodies of this deposit type range from very small up

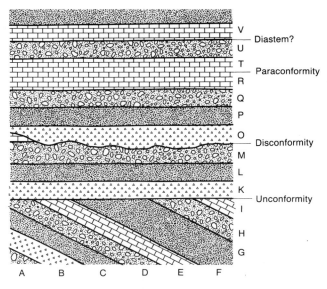

Fig. U1 Unconformity. Schematic cross-section illustrating the various types of stratigraphic break defined above. (After Roberts, 1982.)

to more than 50 Mt and grade from 0.3 to over 5%. They may also have many metal by-products, e.g. **gold** and **nickel**. The geological environment is that of middle **Proterozoic**, **sandstone**-dominated sequences unconformably overlying older metamorphosed **Proterozoic** basement rocks. Most of the mineralization occurs at or just below the **unconformity**, where it is intersected by **faults** passing through carbonaceous **schists** in the basement (Fig. U2). The **orebodies** tend to be tubular to flattened cigar-shaped and they grade outwards into stratiform disseminations and **fracture** fillings. [AME]

Clark, R.J.McH., Homeniuk, L.A. & Bonnar, R. (1982) Uranium geology in the Athabaska and a comparison with other Canadian Proterozoic basins. *Canadian Institute of Mining and Metallurgy Bulletin* 75 April: 91–8.
Evans, A.M. (1993) *Ore Geology and Industrial Minerals: An Introduction* (3rd edn). Blackwell Scientific Publications, Oxford.

underclay A term applied to any fine-grained sediment lying immediately beneath a **coal seam** whatever its state of induration. The term **seat-earth** implies an underclay in which the plants that decayed to form the **coal** were rooted. When root impressions are not found in the underclay, it is probable that the vegetation was transported to form the **coal** above it. Plant growth may be responsible for the general absence of **bedding** in underclays and for the strong *leaching* that has led to the high **kaolinite** content of most underclays. The thickness of underclays may show no direct relation to the thickness of any **coal seam** that may occur above them. This is probably because once a substantial thickness of **peat** had accumulated the plants were able to root in the organic debris itself. (See **fireclay**.) [AME]

underfit stream Large valley channels, often with a regular meandering pattern, which contain within them smaller alluvial channels on the valley floor. Thus, the current channel is considered to be too small to account for the valley channels, and is termed *underfit*. Dury (1965) has argued that the *underfit* channel reflects a climatic change from **discharge** conditions which formed the valley channels. Hack (1965) and others have reasoned that **valley meanders** may be equilibrium forms in cases where the channel is cut into bedrock. [TPB]

Dury, G.H. (1965) *Theoretical Implications of Underfit Streams*. Professional Paper of the United States Geological Survey **452C**.
Hack, J.T. (1965) *Postglacial Drainage Evolution and Stream Geometry in the Ontonagon Area*, Michigan. Professional Paper of the United States Geological Survey **504B**.

underground mapping Mapping carried out in tunnel drivage, working mines and abandoned mines. In the working environment there is continuous fresh exposure after each blast which should be examined immediately despite the difficulties of having drillers and explosives in the way. As a result of mining activity, walls are often

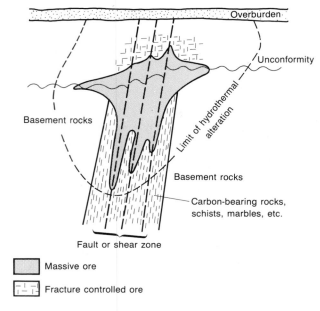

Fig. U2 Generalized diagram of an **unconformity-associated uranium deposit**. (After Evans, 1993.)

quickly covered in dust or mud together with an atomized mixture of oil and water from drilling equipment and should be washed before a geological examination is made. In abandoned mines the walls are rarely clean and often have iron staining or sulphur efflorescence on them neccessitating chipping to expose fresh rock.

Mapping within a mine is carried out at a large scale, often 1:100 or 1:250 (alternatively 1 inch to 10 feet or 1 inch to 20 feet) with areas of great importance meriting a scale of 1:10. Many old maps are on a scale of 2 chains to the inch and depths recorded in fathoms. Major mining companies have survey departments which accurately survey all workings and these plans serve as a useful base for geological work. These are unlikely to be available for abandoned mines (which even if available are probably incomplete and on an obsolete scale with an unconventional local grid), for mines developed by small companies and even for the most recent developments in major mines. The extent of workings shown varies, and many maps of old **coal** mines show a blanket area of old workings which may have been pillar and stall with possibly less than 20% extraction.

Under these circumstances the geologist has to prepare his own plans, normally by a tape and compass survey, and this, like all underground mapping, is best executed by at least two people. The key position at the base of the shaft or mine portal should be accurately surveyed and the tape laid out in a straight line to the next station. A series of primary stations is set up in this way and offsets to the walls are made. A clear and comprehensive account of these survey methods, including those where it is not feasible to use a compass because of metallic machinery or a nearby magnetic **orebody**, is given by Peters (1987, Chapter 12).

A plan is produced of each level within the mine. It is a practice to use a datum at waist height, 1.0 m height, chest height or eye height depending on the company procedure, and whichever is used it should be consistent throughout. This is normally a horizontal datum, but in some gently inclined deposits the datum itself will be inclined. The geological data are projected to the datum and plotted onto the plan. In some cases the vertical sections of both side walls (ribs) are also mapped. This gives a complete three-dimensional impression of the geology. In gently inclined or horizontal deposits, one side wall would be mapped rather than the back (top surface of cavity; referred to as roof in **coal** mining). The geology is not plotted in its true position as observed in the back since this is normally variable in height and often inaccessible; the floor is not mapped because it is covered with mud, water, waste, rails, concrete, etc. The geology is recorded in the normal way but more rock-mechanic data has to be collected than in conventional surface mapping.

Shafts, winzes and raises are mapped as wall sections because they give excellent information on steeply dipping bodies such as **veins**. **Stopes** are shown on level plans by dashed lines above or below datum level and in working mines are mapped after each round is blasted, the map growing with the development of the **stope**. Abandoned **stopes** are more difficult to map since they may be inaccessible and consequently projections and distance viewing are often necessary. A series of datum levels or cross-sections are used to display the geology of the stope.

Mine maps are usually presented level by level. A composite map of all levels is practical only if the structure is moderately to gently inclined and it is possible to plot each level on one sheet without there being any overlap (Fig. U3). An alternative presentation is to produce plans on transparent overlays which can be superimposed on each other, or to make a model of the mine to show the geology of each level, the position of shafts, tunnels, raises, etc., as well as data derived from underground and surface exploratory drilling. These models are an effective tool of communication. Mine sections are also produced through the various levels (side wall mapping is a great help in producing these). The surface and boreholes are also shown, enabling a good interpretation of the unworked deposit to be depicted.

Apart from the maps of the basic geology, other mine data are frequently produced in map form. These include: *structure contour plans* of the **orebody** (either in horizontal projection or vertical plane projection for steeply dipping deposits such as **veins**); *assay plans* where the metal content and or metal ratios are contoured; quality parameters (e.g. percentage $MgCO_3$ or $CaCO_3$ in **limestones**); thickness contour plans, which are a valuable tool in conjunction with *structure contour plans*; *thickness-grade plans* where the multiple of the **grade** and thickness is recorded ($8 \, \mathrm{g \, t^{-1}}$

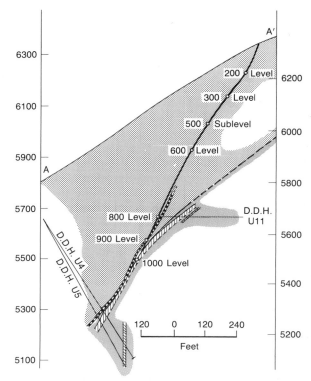

Fig. U3 Underground mapping. Red Rose tungsten mine, Canada. Plan showing vertical section AA′, normal to vein. (After Guilbert & Park, 1986.)

gold over an **ore** width of 3 m is recorded as $24 \, \mathrm{mg \, t^{-1}}$ **gold**). Plans showing geotechnic properties are increasing in importance. [DER]

Guibert, J.M. & Park, C.F. (1986) *The Geology of Ore Deposits.* W.H. Freeman, San Francisco.

Peters, W.C. (1987) *Exploration and Mining Geology* (2nd edn). John Wiley & Sons, New York.

undersaturated oil pool An oil pool that is undersaturated with gas. This may arise when a saturated oil pool becomes more deeply buried. [AME]

underthrusting Movement of a *thrust sheet* or series of *thrust sheets* under the leading edge of a *plate*, rather than over (**overthrusting**). Underthrusting is typical of **subduction zones** where *thrust sheets* forming part of the **accretionary prism** are driven underneath the edge of the upper *plate* as a result of the subduction of the lower *plate*. In *continent–continent collision*, the **crust** may become detached from the **mantle lithosphere** of the upper *plate*, allowing the **crust** of the lower *plate* to underthrust it. This process is termed A- (Ampferer) subduction. Since all **displacements** are relative, underthrusting of rock mass A below mass B implies that B is also **overthrusting** A.

However, the terms are usually applied to the movement of material from the margin of a *plate* towards the *plate* interior. [RGP]

unequal slopes A theory proposed by G.K. Gilbert (1877) which considered the **erosional** development of a **badland** divide separating two slopes of unequal *declivity*. In homogeneous material and with equal quantities of water, the rate of **erosion** of two slopes depends upon their gradient (or *declivity*, as Gilbert termed it). The steeper is degraded faster. When the two slopes are on opposite sides of a divide, the more rapid **erosion** of the steeper carries the divide towards the side of the gentler. The divide becomes stationary only when it has been rendered symmetrical. Gilbert argued that slopes in **badland** areas acquire their smooth curves according to his *Law of Divides*, which states that the nearer the divide the steeper the slope, with all points on a single slope being interdependent. By extension of his argument concerning unequal slopes, Gilbert formulated a *Law of Equal Declivities* which views slopes on both sides of a divide as being interdependent. One result of the *Law of Equal Declivities* is that a **badland** ridge separating two streams at the same elevation will stand midway between them. [TPB]

Gilbert, G.K. (1877) *The Geology of the Henry Mountains*. United States Geographical and Geological Survey, Washington, DC.

uniclinal shifting The migration of a river in the direction of the **dip** in an area of inclined strata, resulting from the tendency of a river to flow along the **strike** of less resistant rocks. [NJM]

universal soil loss equation (*USLE*) This is the most widely used method of *soil loss prediction*. USLE was developed in the USA as a result of a co-operative research programme which involved more than 8000 plot-years of **erosion** data being collected from 36 locations in 21 states, using standard plots 22.13 m long on a uniform slope of 9%. Ease of application was a prime consideration in the development of USLE, which resulted largely through the work of W.H. Wischmeier & D.D. Smith:

$$A = RKLSCP$$

where: *A = soil loss, R = rainfall erosivity factor, K = soil erodibility factor, S = slope gradient factor, C = crop management factor*, and, *P = erosion control practice factor*.

The equation was designated 'universal' because it was free of some of the geographical and climatic restrictions of earlier methods of *soil loss prediction*. However, it was developed largely in the eastern two-thirds of the USA and so is not necessarily applicable outside that area. [TPB]

Wischmeier, W.H. & Smith, D.D. (1978) Predicting rainfall losses — a guide to conservation planning. *United States Department of Agriculture, Agriculture Handbook* No. 537.

unloading The removal of **stress**. There are two common contexts: (i) Removal of gravitational **stress**, for example by **erosion**, melting of ice or **gliding** of a *thrust sheet*, followed by the effects of **pressure release**. Unloading leads to the relief of **lithostatic pressure** and may be followed by tectonic **uplift**. This may initiate characteristic **weathering** processes such as **exfoliation**. (ii) Removal of **stress** in experiments. This may also cause **fracturing** of rock specimens. [TGB/AW]

unsaturated zone The zone above the **water table**, but below the surface, that is not saturated with water. [AME]

uplift Used as a verb, the upward movement of part of the Earth's surface; used as a noun, a structure caused by such a process in a **stable tectonic zone** or **craton**. Uplifts and **basins** are the main types of structure found in stable *plate* interiors. Uplifts are typically large, of the order of 500–1000 km across, and may exist for long periods of geological time, undergoing slow upward movements, and through **erosion** providing sediment to fill adjoining **basins**. The Ukrainian uplift and the Baltic **shield** are examples of long-lived uplifts on the East European **craton**. Two important causes of uplift appear to be **isostatic** gravitational forces and **flexure of the lithosphere**. [RGP]

Park, R.G. (1988) *Geological Structures and Moving Plates*. Blackie, Glasgow.

upper continental crust The upper 10–12 km of the continental crust, often known as the 'granitic' layer as it has seismic velocities similar to those of a **granite** and an average composition of a **granodiorite**. (See **continental crust**.) [DHT]

upper mantle That part of the **mantle** above some 700 km depth and characterized by a range of seismic discontinuities, mostly associated with the collapse of *crystal lattices* to higher density forms, without change in the total composition. Some authorities do not include the **mantle transition zone** in the upper mantle, i.e. they restrict it to the zone between the seismic **Mohorovičić discontinuity** and a depth of 350 km, but this boundary is so diffuse that the upper mantle is more simply defined as that part of the **mantle** above the **lower mantle**. [DHT]

upper-stage plane bed Flat bed present in all **sand** grade sediment produced at **Froude numbers** approaching unity and over which there is intense sediment transport. Two sedimentary features characterize these beds: extensive

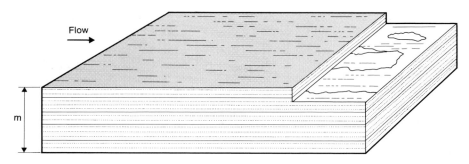

Fig. U4 Planar lamination produced by aggradation during **upper-stage plane-bed** transport. The transport surface shows current lineation, and the internal surfaces show parting lineation.

parallel laminae which have been linked to the migration of very low amplitude sediment waves over the **plane bed** and **primary current lineation**, attributable to the impact of fluid **sweeps** upon the bed surface (Fig. U4). (See **bedform theory**.) [JLB]

Allen, J.R.L. (1985) *Principles of Physical Sedimentology*. Allen & Unwin, London.

upthrust A *reverse fault* that flattens upwards from a steep attitude to become a *thrust* at shallow depths. Structures of this type have been attributed to differential vertical movements, and are also an important component of positive *flower structures* in **strike-slip fault zones**. [RGP]

Harding, T.P. & Lowell, J.D. (1979) Structural styles, their plate tectonic habitats, and hydrocarbon traps in petroleum provinces. *American Association of Petroleum Geologists Bulletin* **63**: 1016–58.

upwelling The transport of subsurface waters to the surface. This process occurs on a wide variety of scales in oceans, **estuaries** and lakes. Upwelling is generally a response to wind **stress** acting on the surface. The most significant and geologically important upwelling occurs on the western margins of continents. The location of this upwelling is governed by the effects of the Earth's rotation (**Coriolis effect**). In the northern hemisphere, winds generate a current at 90° to the right of the wind direction (**Ekman spiral**). Thus a southward wind off western North America generates an offshore flow. This oceanward displacement of surface waters results in the transport of colder waters to the sea surface and in the disruption of the **thermocline**. These subsurface waters are relatively rich in nutrients and the introduction of these to the **photic zone** causes high levels of primary productivity. The sediments which accumulate below upwelling systems are commonly rich in both organic **carbon** and also phosphatic material. The major zones of modern coastal upwelling occur off western North America, Peru, northwest and south-west Africa and off Oman. Upwelling also occurs due to the divergence of oceanic currents, for example along the equatorial zone in the Pacific Ocean (divergence upwelling). The location of upwelling zones has varied with differing continental configurations through time and such fossil upwelling zones coincide with the development of organic-rich and **petroleum** *source rocks* (Parrish, 1982; Parrish & Curtis, 1982). [AESK]

Parrish, J.T. (1982) Upwelling and petroleum source beds with reference to the Palaeozoic. *American Association of Petroleum Geologists Bulletin* **66**: 750–74.

Parrish, J.T. & Curtis, R.L. (1982) Atmospheric circulation, upwelling and organic-rich rocks in the Mesozoic and Cenozoic Eras. *Palaeogeography, Palaeoclimatology, Palaeoecology* **40**: 31–66.

uraniferous calcrete Calcrete mineralized with **uranium**. **Carnotite** is developed in the calcreted drainage systems of some arid and semi-arid areas, sometimes in sufficient amount to form **orebodies**. The Yeelirrie deposit in Western Australia is the best known. **Calcrete** extends for about 15 km upstream from the **orebody**. The valley narrows downstream and this is thought to have been important in concentrating the flow of **groundwater** to form the **orebody** which is 6 km long, 0.5 km wide and 8 m thick. Similar deposits occur in Namibia and Somalia. [AME]

Hutchison, C.S. (1983) *Economic Deposits and Their Tectonic Setting*. Macmillan, London.

uraninite (UO_2) The major **ore** mineral of **uranium**. (See **oxide minerals**.) [DJV]

uranium (U) At. no. 92, at. wt. 238.029, d 19.07. A heavy, *radioactive* metal used in nuclear reactors and weapons, as an additive in steel and catalytically, as the carbide, in the formation of ammonia. It is recovered economically from many types of deposit, e.g. fossil **placer**, **pegmatite**, **unconformity-associated uranium** and **vein deposits**. [AME]

uvala A term of Slav origin referring to complex closed depressions in **karst** areas. These complex depressions commonly have uneven floors and are often caused by the coalescence of **dolines**. [HAV]

uvarovite ($Ca_3Cr_2Si_3O_{12}$) A rare green-coloured mineral of the **garnet** group. (See **garnet minerals**.) [DJV]

V

vadose zone The shallow subsurface zone of **infiltration** and percolation of rainfall, **runoff** or melt water between the ground surface and the **water table** up to tens of metres thick in which the intergranular pores and fissures are unsaturated with water and contain air at atmospheric pressure. Also known as the *zone of aeration* or *unsaturated zone* (see **phreatic zone**). The vadose zone is characterized by episodic saturation and seasonal capillarity and by evaporation and transpiration. Intense mineral *leaching* is often characteristic of the upper part of the vadose zone, whilst in the capillary zone above the **water table**, particularly in tropical and semi-arid climates, there tends to be reprecipitation of carbonates, sulphates, **quartz** or **iron** oxyhydroxides which form **calcretes**, *gypcretes*, **silcretes** or *hard pans*, respectively. (See also **meteoric water diagenesis**.) [SDB/VPW]

Goudie, A.S. (1973) *Duricrusts in Tropical and Subtropical Landscapes.* Oxford University Press, Oxford.

valence bond theory A view of chemical bonding based on pairs of electrons. The pairs may be thought of as shared between atoms (covalent), localized on atoms (often ionic) or on separate atoms but with opposed spins (non-bonding). The valence electrons are arranged in pairs amongst the atoms of a molecule to form 'structures'. In each structure the pairs are disposed in a different way and each structure is characterized by a specific energy. The structures interact with each other to form a representation for the molecule more stable than any one structure and in which the most stable structures predominate. This interaction is termed *resonance*.

The excessive ionic character associated with **molecular orbitals** can be resolved, for the hydrogen molecule, by calculating the most stable mix of ionic structures ($H_A^+H_B^-$ and $H_A^-H_B^+$) and covalent structures ($H_A:H_B$). In order that structures can be constructed for more complicated molecules involving polyvalent atoms it is necessary that the valence shell **atomic orbitals** (AOs) be rearranged so as to point in specific spatial directions. This process is termed *hybridization* and requires prior knowledge of the shape of the molecule. It also requires that AOs of different energies be allowed to participate in the construction of suitable linear orbital combinations. If the valence AOs of silicon are to be *hybridized* to form four orbitals orientated tetrahedrally, as would be suitable for $[SiO_4]^{4-}$, then both 3s and 3p orbitals must participate. A typical 'sp^3' hybrid has the form,

$$\psi \text{ (hybrid)} = 0.5 \, (s + P_x + P_y + P_z)$$

Four such hybrid orbitals can then make four covalent electron pair bonds with the four O^- ligands to generate structure (a) for silicate. This and some other possible structures are:

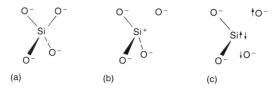

In (b) Si^+ makes three covalent bonds with three O^- anions and an ionic bond with O^{2-}. Structure (c) has only two covalent bonds, the other electrons being non-bonding, either paired on the silicon or on the two oxygen anions. [DSU]

Pauling, L. (1940) *The Nature of the Chemical Bond.* Oxford University Press, Oxford.

valley bulge (*valley-bottom bulge*) Sedimentary strata that have been bulged up in the floor of a valley as a result of **erosive** processes acting upon rocks of different character. The mechanism of formation invoked for those of Gloucestershire, England (Ackermann & Cave, 1967) is that during the *Pleistocene* severe **erosion** and valley incision occurred at a time when *permafrost* conditions pertained. At the end of the cold period the rocks thawed out and susceptible clays, silts and **sands**, highly charged with water, became **plastic**, and under the weight of the more competent **limestones** above were extruded through the weakest points of the recently developed valley floors. **Cambering** of strata would occur on the valley sides. [ASG]

Ackermann, K.J. & Cave, R. (1977) Superficial deposits and structures, including landslip, in the Stroud District, Gloucestershire. *Proceedings of the Geologists' Association of London* 78: 567–86.

valley meander Meander which usually has a greater wavelength than those of the currently functioning river system. Generally they are cut into bedrock and have been produced by formerly higher **runoff** conditions. [ASG]

vallon de gélivation A small valley formed by the widening of bedrock **joints** by frost action rather than by normal fluvial processes. [ASG]

vanadinite ($Pb_5(VO_4)_3Cl$) A rare secondary mineral found in the oxidized portions of **lead veins**. [DJV]

[646]

vanadium (V) At. no. 23, at. wt. 50.9415. A Group V transition element metal used mainly in steel and **aluminium** alloys. The chief sources are **sandstone–uranium–vanadium-base metal** and **orthomagmatic magnetite deposits**. [AME]

variometer An early form of **magnetometer** based on a small magnet suspended in the **geomagnetic field**. Measurements were confined to the land surface. Now superseded by more efficient instruments. [PK]

variscite (Al(PO₄).2H₂O) A blue-green massive mineral somewhat resembling **turquoise** and used as a **gem**. [DJV]

varve dating Measurement and counting of annual varves in lake beds may allow a chronology to be established for these deposits. Such chronologies for varves in Sweden and the USA (Minnesota) have been linked to **radiocarbon dates** (by comparison of associated *pollen* remains and direct **radiocarbon dating** of material in the lake deposit respectively) to show comparison of the dates from the two methods. This is particularly useful for the period of time before *c.* 5000 BC, when comparisons by **dendrochronology** are not available. Like **dendrochronology** some varve dating indicates that **radiocarbon dates** before *c.* 1000 years BC may be too low because of fluctuations in the carbon-14 reservoir. [OWT]

vasques Wide (up to several decimetres), shallow pools with flat bottoms, which form a network consisting of a tiered, terrace-like series of steps on **limestone** coastal platforms, especially in **aeolianite**. The pools are separated from each other by winding, narrow, lobed ridges, 10–200 mm in height, and running continuously for tens of metres. They develop between high and low **tide** levels, especially in inter-tropical and Mediterranean climatic regions. [ASG]

Battistini, R. (1983) La morphogénèse des plateformes de corrosion littorale dans les grès calcaires (plateforme supériere et plateforme à vasques) et le problème des vasques, d'après des observations faites à Madagascar. *Revue de Géomorphologie Dynamique* **30**: 81–94.

vector Strictly, a line with a specific orientation representing both magnitude and direction of motion. Vectors are used in the study of both **relative** and **absolute plate motion**. The term is also used less precisely for the relative **displacement** direction of two rock masses that have undergone relative motion, e.g. of a *thrust sheet* in relation to the **autochthon**. (See **plate tectonics**.) [RGP]

vein/vein system Sheet-like or tabular, discordant, mineralized body formed by partial or complete filling of **fractures** within igneous, metamorphic and sedimentary rocks. (See Figs P5 and P6, p. 467.) They are epigenetic in origin and usually have a high **dip**: They may have parallel walls but more commonly they branch and may display **pinch-and-swell structure**. They may also show regularities in their orientation to form vein systems (see Fig. C9, p. 90). Such veins comprise minerals deposited from **hydrothermal solutions** at submagmatic temperatures (commonly 100–600°C), so are termed *hydrothermal veins*. The most common vein minerals are **quartz** and **calcite** together with other low-grade hydrothermal-metamorphic minerals, and veins may show textural and mineralogical zoning, commonly across the width of a vein, along the horizontal extent, with depth and according to the *country rock*.

Complex veins may contain suites of different minerals arranged in bands in sequence from the walls to the centre of the vein (see Fig. C42, p. 142). These variations reflect the changing mineral species that were being **precipitated**, as time progressed, from hot solutions passing through the vein. Successive layers, usually of lower temperature minerals, are produced (e.g. **quartz** followed by **calcite**) and at the centre there may be an open cavity. Metallic sulphides may form minor constituents of the vein in which case the vein may form an economic mineral deposit. However, since veins are narrow in the short dimensions (by definition) they are only profitable to mine if they are high **grade**. Consequently they may form important sources for **gold, silver, tin** and **uranium** and **industrial minerals** such as **fluorite** and **barite**. [AME/RST]

Edwards, R. & Atkinson, K. (1986) *Ore Deposit Geology*. Chapman & Hall, London.
Evans, A.M. (1993) *Ore Geology and Industrial Minerals: An Introduction* (3rd edn). Blackwell Scientific Publications, Oxford.

velocity defect law A relationship which describes the shape of the outer region of a *turbulent* **boundary layer** where the **Karman–Prandtl velocity law** is inapplicable. This law relates the velocity difference between the free-stream velocity, U_m, and the mean velocity at a point, $\bar{U}$, (the velocity defect $U_m - \bar{U}$) at a distance y from the wall to the **shear velocity**, U_*, and the **boundary layer** thickness, δ. In wide, open channels the relationship

$$\frac{U_m - \bar{U}}{U_*} = -5.75 \log_{10}\left(\frac{y}{\delta}\right)$$

(where δ can be taken as the flow depth) produces a logarithmic distribution which fits data very closely for both **hydraulically smooth and rough** boundaries and permits prediction of velocity at any point within the flow. [JLB]

Middleton, G.V. & Southard, J.B. (1984) *Mechanics of Sediment Transport* (2nd edn). Short Course No. 3 Lecture Notes. Society of Economic Paleontologists and Mineralogists, Tulsa, OK.

ventifact An object which has been modified by means of wind action. The sculpting of objects is primarily achieved by means of **sand abrasion**. Good examples most commonly occur in desert areas where a moderate **sand**

vergence

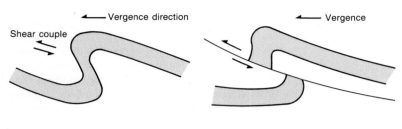

West

East

Fig. V1 Vergence direction for an asymmetrical fold is that of the upper component of the shear couple acting on the short limb in this example. The vergence of a thrust fault is defined as the direction of transport of the hangingwall, or thrust sheet. The overall vergence of a thrust belt is towards the foreland. Both the structures shown verge to the West.

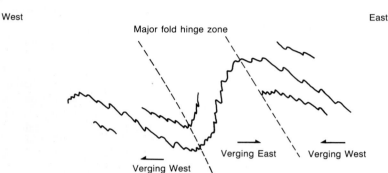

Fig. V2 In regions of parasitic folding, changes in the **vergence** direction of minor folds indicate the positions of major fold hinge zones.

supply is moved by the wind across an immobile surface. This condition is met in areas on the periphery of *sand seas*. (See **dreikanter, yardang**.) [RDS]

vergence The sense of **shear** deduced from asymmetry of minor structures, e.g. **folds**, **faults**, in *orogenic belts*. In the case of **fold systems**, minor asymmetrical **folds** show a sense of **shear** deduced from the rotation of their shorter *limbs*, described as **sinistral**/anticlockwise (*s folds*), or **dextral**/clockwise (*z folds*). A change in minor **fold** vergence indicates the position of a major *fold hinge*. Direction of vergence can be quoted geographically, and is defined by Roberts (1974) as the horizontal direction, in the **fold profile** plane, of the upper component of the **shear**; when examining minor *parasitic folds*, this is the direction in which the adjacent **macroscopic antiformal** *hinge zone* is found. Figures V1 and V2 shows this use of vergence for **major** and **minor folds** (and *thrust faults*, see below). In regions of complex folding, account must be taken of **younging**, and **facing** of **folds**. In all such analysis, **fold** asymmetry must be assessed in a consistent orientation with respect to **macroscopic fold plunge**, otherwise spurious vergence changes result.

In *orogenic belts* and **thrust belts**, vergence is defined as the direction of overall **shear** movement. In a **thrust belt** this is from the hinterland to **foreland**. Individual *thrust faults* verge in the direction of movement of the **hanging wall** (see Fig. V1). *Thrust faults* and **folds** with vergence opposed to that of the **thrust belt** as a whole are known as **backthrusts** and **backfolds**. [SB]

Roberts, J.L. (1974) The structure of the Dalradian rocks in the North Ballachulish district of Scotland. *Journal of the Geological Society, London* **132**: 139–54.

vermiculite $((Mg,Ca)_{0.3}(Mg,Fe,Al)_{3.0}(Al,Si)_4O_{10}-(OH)_2 . nH_2O)$ A *phyllosilicate* mineral formed chiefly by **alteration** of **biotite**. [DJV]

vertical electrical sounding **Resistivity method** in which the variation of **electrical resistivity** with depth is investigated. The method, which is specially suited for a horizontally stratified, homogeneous subsurface, uses electrode configurations which are symmetrically disposed about the point of investigation. The principle involved is that the **depth of penetration** of the current increases as the current electrode separation is increased, thereby allowing a deeper *geoelectric section* to influence the potential distribution, and hence the measured **apparent resistivity**, ρ_a. Comparing log–log plots of observed ρ_a versus electrode separation with theoretical curves, the resistivities of the layers and their thicknesses can be estimated (Fig. V3). [ATB]

very long baseline interferometry Certain cosmic sources, quasars, emit very regular radio signals which can be received at different stations. Determination of the phase differences between the signal at different stations enables the distance between the stations to be determined with an accuracy of a few centimetres per 1000 km. This enables **tectonic** movement between different sites to be measured if their distances are resurveyed every few years. [DHT]

Niell, A.E., Ong, K.M., MacDoran, P.F. *et al.* (1979) Comparison of a radiointerferometric differential baseline measurement with conventional geodesy. *Tectonophysics* **52**: 49–58.

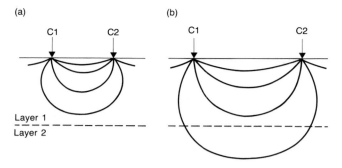

(a) (b)

C1 C2   C1 C2

Layer 1
Layer 2

Fig. V3 Vertical electrical sounding. Effect of current electrode separation (C1–C2) on depth of current flow. (a) Separation too small for layer 2 to influence the flow pattern; ρ_a will be much the same as first layer resistivity. (b) Significant flow of current in layer 2, ρ_a now depending upon resistivities of both layers.

vesicle A gas-filled cavity in **magma** or volcanic rock. Growth of vesicles in **magma** is called *vesiculation*, and is caused by **exsolution** of **volcanic gases** initially dissolved in the **magma**. *Vesiculation* of **magma** is an important volcanological process, influencing the way **magma** erupts, and hence the kind of volcanic rocks formed (Sparks, 1978; Williams & McBirney, 1979; Fisher & Schmincke, 1984). After initial *nucleation* of vesicles, their further growth can result from several processes: diffusion of gas from the **magma** into the vesicles; coalescence of vesicles; and fall of confining pressure as the **magma**, or vesicles, or both, rise. **Magma** rising slowly through the vent and/or with a low content of initially dissolved volcanic gases will normally be erupted as a vesicular **lava flow** or **dome**. Many **magmas** contain sufficient dissolved gas to undergo intense *vesiculation* as they experience decompression upon uprise in volcanic conduits, such that the **magma** becomes volumetrically dominated by vesicles. Further expansion of the vesicles in such **magmas** in the **volcanic vent** can explosively tear apart the **magma** to form a mixture of pyroclasts suspended in up-rushing gas. The level in the **volcanic vent** or conduit that this occurs is called the *fragmentation level*. The mixture of gas and suspended pyroclasts (or **tephra**) forms an upward-directed jet above the vent. The **tephra** are normally deposited as a widely dispersed air-fall layer. (See **amygdale, pyroclastic rock, volcanic eruption, volcanic gas**.)

Lavas with vesicular structure have been used preferentially for **millstones** both in the ancient world and in modern times, as the vesicles provide good cutting edges for the grain or other product being milled. (See **Mayen lava, millstone**.) [PTL/OWT]

Fisher, R.V. & Schmincke, H.-U. (1984) *Pyroclastic Rocks*. Springer-Verlag, Berlin.
Sparks, R.S.J. (1978) The dynamics of bubble formation and growth in magmas: a review and analysis. *Journal of Volcanology and Geothermal Research* **3**: 1–37.
Williams, H. & McBirney, A.R. (1979) *Volcanology*. Freeman, Cooper, San Francisco.

vesuvianite ($Ca_{10}(Mg,Fe)_2Al_4(SiO_4)_5(Si_2O_7)_2(OH)_4$) A *cyclosilicate* mineral (formerly known as **idocrase**) and formed as a result of **contact metamorphism** of impure **limestones**. [DJV]

vibration magnetometer A **magnetometer** in which the sample is vibrated in order to produce an oscillating voltage in the pick-up system of the same frequency as the vibration. [DHT]

Vibroseis® A **seismic source** in which a vibratory mechanism is used to generate a controlled wavetrain whose frequencies vary with time. Both land and marine Vibroseis sources exist, though truck mounted Vibroseis sources are most common. Vibroseis is a low energy source that is enviromentally friendly and transmits energy into the ground over tens of seconds, to be collapsed back as if from an impulsive shot during a later processing stage. A conventional **seismic record** is produced by the **correlation** of the Vibroseis wavetrain with the detected signal, which consists of the overlapping superposition of a series of long input wavetrains produced by **refraction** and **reflection** in the subsurface. Vibroseis is a trade mark of Continental Oil Company. [GS]

Vine–Matthews hypothesis A model to explain the origin of linear *oceanic magnetic anomalies* associated with the **ocean crust** by means of **thermal remanent magnetization** acquired as newly injected oceanic **lithosphere** cooled in the Earth's **magnetic field** at different times when the **geomagnetic field** was of opposing polarities, thus giving rise to normal and **reverse magnetizations** in the rocks. The model in which the newly injected oceanic rocks were emplaced within a very narrow zone at the precise crest of the *oceanic ridges* was developed very shortly afterwards. (See also **seafloor spreading**.) [DHT]

Vine, F.J. & Matthews, D.H. (1963) Magnetic anomalies over oceanic ridges. *Nature* **199**: 947–9.
Vine, F.J. (1966) Spreading of the ocean floor: new evidence. *Science* **154**: 1405–15.

viscoelasticity Type of **deformation** characterized by **elastic** and **viscous** components acting in parallel. A viscoelastic response to applied **stress** consists of an initial rapid **strain** exponentially decreasing with time, tending towards a finite value given by **stress** divided by **Young's modulus**. The time taken to reach half of this value is known as the *retardation time*, r, which is equal to the ratio of the **viscosity** to **Young's modulus**. On removal of the **stress**, **strain** is **recovered** exponentially towards a limit represented by the original value. This behaviour models some aspects of **creep** experiments in rocks, and is also known as the *Kelvin, Voigt*, or *Kelvin–Voigt model*.

When combined with a component of *elasticity*, visco-

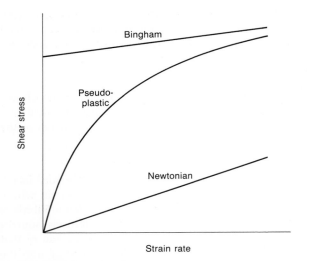

Fig. V4 Viscosity. Covariance of strain rate with shear stress for substances with Newtonian, pseudo-plastic, and Bingham rheologies.

elastic behaviour describes **deformation** of many solids closely and corresponds to the *standard linear solid*. [TGB]

Ramsay, J.G. (1967) *Folding and Fracturing of Rocks*. McGraw-Hill, New York.

viscosity The **rheological** physical property of a material that deforms such that the ratio of **shear stress** (τ) to **shear strain rate** ($\dot{\gamma}$) is a constant, as in a fluid; '*viscous*' describes a material exhibiting such a property.

Viscosity $\eta = \dot{\gamma}/\tau$

Viscosity is a measure of the resistance of a fluid to **flow**. The SI unit of viscosity is the Ns m^{-2} or Pa s. The cgs unit is the poise, P, where $1\,\mathrm{Ns\,m^{-2}} = 10\,\mathrm{P}$. Viscosities of rocks vary from 10^{-15} to 10^{21} Pa s and decrease with temperature.

For a pure **Newtonian fluid** (Fig. V4), the **stress** is a linear function of **strain rate**, which can be expressed in tensor notation as

$$\sigma'_{ij} = C_{ijkl}\dot{\varepsilon}_{kl}$$

where σ'_{ij} are the components of the *deviatoric stress* tensor and $\dot{\varepsilon}_{kl}$ are the components of the **strain rate** tensor.

The coefficients of proportionality (C_{ijkl}) are the *viscosity coefficients*. The reciprocal of viscosity ($1/\eta$) is the *fluidity*. The concept of viscosity does not necessarily imply a pure *Newtonian* (linear) relationship between **shear stress** and **shear strain rates** or a fluid rather than **plastic rheology**. Any positive relationship between **shear stress** and **strain rate** can thus be called '*viscous*' behaviour, *viscous flow*, or *viscous strain*. To emphasize that viscosity is simply an observed quantity, it is sometimes referred to as *equivalent viscosity*.

Viscosities of **magmas** are strongly dependent on the chemical composition of **magma**; **silica**-rich **magmas** have viscosities several orders of magnitude greater than mafic ones. There has been much recent discussion of the structures of silicate melts (e.g. Mysen *et al.*, 1982). They are thought to be polymerized to various degrees, the polymers being formed by chains of silicon and **aluminium** cations (the network-forming cations) and bridging oxygen ions. Most metal cations (e.g. **magnesium**, calcium, sodium — the network-modifying cations) form simple bonds with oxygen ions (the non-bridging oxygens, NBO); the latter are then not available for incorporation into the network. The degree of polymerization of a **magma** is sometimes expressed as the inverse to the ratios NBO/silicon, or NBO/T (where T is the number of tetrahedrally co-ordinated cations). Mafic **magmas** are weakly polymerized, have high NBO/silicon ratios and therefore have low viscosities; silicic **magmas** are highly polymerized, have low NBO/silicon ratios, and have much higher viscosities. Dissolved volatile species (e.g. H_2O, F) act as network-modifiers in **magmas**, and viscosity of **magma** decreases with increasing dissolved volatile content. As **volcanic gases** exsolve from **magmas** to form **vesicles**, **magma** viscosity initially decreases with increasing vesicularity. However, when **vesicles** form a significant volume of the **magma**, so that adjacent **vesicles** touch, and have common bubble walls, this trend is reversed. The viscosities of forming **magma** from which *pumice* is derived are thought to be very high. Viscosities of **magmas** also increase with increasing crystal content, and with decreasing pressure and temperature.

Viscous (*fluid*) *flow* is an important **deformation** process in unlithified sediments and **evaporites** at shallow levels of the **crust**, and in all rocks as the temperature increases with depth in the **lower crust**, **mantle** and **core**. [TGB/PTL]

McBirney, A.R. & Murase, T. (1984) Rheological properties of magmas. *Annual Review of Earth and Planetary Sciences* **12**: 337–57.
Means, W.D. (1979) *Stress and Strain*, Chapter 26. Springer-Verlag, New York.
Mysen, B.O., Virgo, D. & Seifert, F.A. (1982) The structure of silicate melts: implications for chemical and physical properties of natural magma. *Reviews of Geophysics and Space Physics* **20**: 353–83.
Pinkerton, H. & Sparks, R.S.J. (1978) Field measurements on the rheology of lava. *Nature* **276**: 383–4.

viscous remanent magnetization The **remanent magnetization** acquired by a sample, usually rock, as a function of the time in which it is held in a **magnetic field**, such as that of the Earth. Usually of relatively low magnetic stability and readily removed by partial *demagnetization*. [DHT]

viscous sublayer The region of flow closest to the bed in a *turbulent* **boundary layer** that is dominated by viscous fluid forces and in which velocity displays a linear increase with height above the bed. The thickness of the sublayer, δ, is usually only a maximum of 1% of the total boundary layer thickness and is dependent upon both the **shear**

velocity of the fluid, U_*, and its **kinematic viscosity**, ν, in the general relation

$$\delta = 11.5 \, \nu/U_*$$

(See **turbulent flow**, **hydraulically smooth and rough**.) [JLB]

vitrified fort Celtic forts of the later 1st millenium BC and early 1st millenium AD in France, Germany and especially in Scotland, which have been burned so intensively that the stone of which they are built is fused together. The burning may have been deliberate, and the forts were built of **sandstones**, *conglomerates*, **gneisses** and **igneous rocks**. [OWT]

vivianite ($Fe_3(PO_4)_2 . 8H_2O$) A rare mineral formed by **weathering** of primary phosphates in **pegmatite deposits**. [DJV]

VLF method (*very low (radio) frequency method*) **Electromagnetic induction method** in which the frequencies used (15–25 kHz) are far higher than in other electromagnetic methods and are the pre-fixed carrier frequencies of military communications signals transmitted continuously by a dozen or so stations around the world. The magnetic vector of the primary electromagnetic waves — plane waves because the sources are distant — will be horizontal over non-anomalous ground, but tilted and elliptically polarized near conductors; both tilt and phase are measured in a VLF receiver. The receiver has two orthogonal pick-up coils tuned to the transmitter frequency — one measuring the horizontal 'primary' **magnetic field**, and the other measuring the *tilt-angle* of the polarization vector — and provides readings of percentage **real** and **imaginary components** of the secondary field.

VLF elicits a stronger ground response than other electromagnetic methods because of the higher frequencies used, but the **depth of penetration** is correspondingly smaller. The receiver is highly portable, and easily and quickly operated by one person, which accounts for its popularity as a rapid reconnaissance method. It is ideal for the detection of **dykes**, **faults**, etc., provided that a transmitter approximately along **strike** is available as the source. [ATB]

Keller, G.V. & Frischknecht, F.C. (1966) *Electrical Methods in Geophysical Prospecting*. Pergamon, Oxford.
Telford, W.M., Geldart, L.P., Sheriff, R.E. & Keys, D.A. (1990) *Applied Geophysics* (2nd edn). Cambridge University Press, Cambridge.

vogesite A **calc-alkaline lamprophyre** with *phenocrysts* of **amphibole** (**hornblende**), **pyroxene** and possibly **olivine** in a matrix containing **alkali feldspar** in excess of **plagioclase**. [RST]

volcanic earthquake Volcanically active areas experience large numbers of **earthquakes**. Many seem to have a different origin from **tectonic earthquakes**. Based on their seismic signatures, they can be grouped loosely into three kinds.

1 'High-frequency' **earthquakes**. The most common. Shallow (<5–10 km), small (*body wave* **magnitude** <5) **tectonic earthquakes** in the general vicinity of the volcanic area. Originate from regional **stresses**, perhaps modified by volumes of **magma** emplaced in the Earth's **crust**.

2 'Low-frequency' **earthquakes**. Have an emergent onset to the *P wave* and a weak or absent *S wave*. Frequency content is limited to a maximum of about 5 Hz. Empirically they are associated with eruptions and outgassing events. **Hypocentres** are hard to determine because of the imprecise **arrival times**.

3 'Harmonic tremors'. Unusual **seismic events**, near-monochromatic (single-frequency) and constant-amplitude signals of 1–5 Hz, which can grow gradually from background **noise**. Can continue for hours or even days. Only detected within a few kilometres of volcanic centres. Suggested causes include oscillations of liquid and/or gas in *magma chambers* and **magma** flow in conduits being compressed by *hydrostatic pressure*.

Understanding them may allow better prediction of eruptions. Other **seismic events** are directly associated with injection and eruption blasts. [RAC]

Minakami, T. (1961) Study of eruptions and earthquakes originating from volcanoes. *International Geology Review* **5**: 712–19.

volcanic eruption Arrival of **magma** at the Earth's surface and/or expulsion of broken rock from a **volcanic vent**. Volcanic eruptions do not include emissions purely of **volcanic gases** from **fumaroles**, or ejection of hot water from **geysers**. Volcanic eruptions vary in character from quiet extrusion of **lava flows** to very explosive eruption of **tephra**, and in size from single eruptions of **lava**, **scoria** or **ash** with volumes of less than $0.1 \, km^3$, and *phreatic* blasts in which no **magma** is erupted, to eruptions of lava as *flood basalts* and **tephra** as **ignimbrites** which can exceed $1000 \, km^3$ in volume. In duration, volcanic eruptions range from individual, isolated blasts lasting minutes, to essentially continuous activity lasting many months. Bursts of activity separated by quiet periods are considered as phases of one eruption, unless the quiet period exceeds about three months (Simkin *et al.*, 1981).

Classification of volcanic eruptions is made primarily by comparison of historical observations of eruptions. *Flood basalt* and **ignimbrite** eruptions of very large volume (and also **komatiite** eruptions which only occurred in the **Archean**) have never been witnessed, and their character must be inferred from the geologic record.

Hawaiian eruptions are characterized by **fire fountaining** behaviour and quiet extrusion of lava. The **magma** is always mafic in composition, and the eruptions typically occur in *ocean island* settings (e.g. Hawaii, Iceland). In *strombolian eruptions*, **scoria** and **ash** are thrown out of the **volcanic vent** in a large number of distinct explosions. The **scoria** accumulates close to the vent to form a *cinder cone*, while the **ash** is usually carried away in a plume. The **scoria** *clasts* in the *cinder cone* may weld to form *agglutinate*. In larger *cinder cones*, the **scoria** are redeposited as **talus screes** on the flanks of the cones. Strombolian activity is commonly associated with extrusion of a **lava flow**. Most strombolian activity is **basaltic**, but can be **lamproitic** or intermediate (andesitic or trachytic) in composition. Strombolian activity is common in all volcanic settings on land.

Eruption of mafic **magma** in deep water is characterized by non-explosive formation of **pillow lava**. At progressively shallower water depths of less than 1–2 km, several processes contribute to increasing explosivity of **magma**–water interactions (Kokelaar, 1986), and eruption of **magma** in shallow water is commonly very explosive and called *Surtseyan activity*. It is characterized by an almost continuous series of blasts producing 'cypressoid' or 'cock's tail' clouds. The **tephra** is fine-grained and deposited as finely-bedded **base surge** and air-fall beds, sometimes interbedded with massive deposits that resemble **mudflows**. The characteristic feature found is a *tuff cone* or a **tuff ring**. Similar **phreatomagmatic** activity can occur on land where rising **magma** encounters an **aquifer**.

When volatile-rich **magma** is erupted continuously at a high rate, the **magma** is torn apart by *vesiculation* (see **vesicle**), and the resulting mixture of **pumice** or **scoria**, **ash** and **volcanic gas** generates a column rising rapidly above the vent. This mixture can mix with air which, when heated by the **pumice** and **ash**, imparts a buoyancy to the column relative to the atmosphere. The rapidly-rising buoyant column can rise to great heights (up to 50 km), and the **tephra** is supported in the column by the upwelling hot air and **volcanic gas**. Such eruptions are called *plinian*, and generate widely-dispersed sheet-like air-fall deposits, characterized by good *sorting* (at individual localities). Most *plinian eruptions* are **rhyolitic**, but some are intermediate or even mafic in composition. *Sub-plinian eruptions* are similar to *plinian eruptions*, but are less violent, producing less widely-dispersed **tephra** (Wright *et al.*, 1980).

Vulcanian eruptions are discrete, powerful, cannon-like blasts which expel mixtures of lithic *clasts*, **scoria** *clasts* (usually of intermediate composition), **ash** and gas at high velocity from vents. The eruptions generate a short-lived plume of **ash**. The **tephra** is deposited as an air-fall layer, consisting of **ash** and large lithic and **scoria** *clasts*. The mechanism of vulcanic eruptions is poorly understood. Some workers (e.g. Fisher & Schminke, 1984) regard *vulcanian eruptions* as **phreatomagmatic**, others (e.g. Walker, 1982) regard them as a consequence of build-up of pressure of **volcanic gases** in a vent trapped below a cap of viscous **magma**.

In *peléean eruptions*, small-volume (<1 km³) **ash flows** are initiated near the summit of a steep-sided volcano, and sweep down the flanks to deposit block and **ash flows** as radiating fans. These **ash flows** may be initiated by several mechanisms, including fragmentation of unstable viscous **magma** forming extrusive **domes**, and explosive emission of **tephra** from the vent (Fisher & Heiken, 1982). The active **ash flows** are accompanied by a dilute, **ash**-rich cloud called a **nuée ardente**.

Ignimbrite eruptions involve the explosive eruption of large volumes of intermediate or silicic **magma** as **pumice**-rich **ash flow tuffs**. The flows are thought to be generated by gravitational collapse of a *plinian eruption* column, or by direct extrusion of **vesiculated** fragmented **pumice** and **ash** as **ash flows**. (See **ash flow**, **base surge**, **hyaloclastite**, **ignimbrite**, **lahar**, **lava flow**, **mudflow**, **phreatic eruption**, **tephra**, **viscosity**.) [PTL]

Cas, R.A.F. & Wright, J.V. (1987) *Volcanic Successions Modern and Ancient*. Allen & Unwin, London.

Fisher, R.V. & Heiken, G. (1982) Mt Pelée, Martinique: May 8 and 20, 1902, pyroclastic flows and surges. *Journal of Volcanology and Geothermal Research* **13**: 339–71.

Fisher, R.V. & Schmincke, H.-U. (1984) *Pyroclastic Rocks*. Springer-Verlag, Berlin.

Kokelaar, P. (1986) Magma–water interactions in subaqueous and emergent basaltic magmatism. *Bulletin of Volcanology* **48**: 275–89.

Simkin, T., Siebert, L., McLelland, L. *et al.* (1981) *Volcanoes of the World*. Hutchinson Ross, Stroudsburg, PA.

Walker, G.P.L. (1981) Plinian eruptions and their products. *Bulletin Volcanologique* **44**: 223–40.

Walker, G.P.L. (1982) In: Thorpe, R.S. (ed.) *Andesites*. John Wiley & Sons, Chichester.

Wilson, L., Pinkerton, H. & Macdonald, R. (1987) Physical processes in volcanic eruptions. *Annual Review of Earth and Planetary Science* **15**: 73–95.

Wright, J.V., Smith, A.L. & Self, S. (1980) A working terminology of pyroclastic deposits. *Journal of Volcanology and Geothermal Research* **8**, 315–36.

volcanic gas Gas discharged to the atmosphere as a consequence of **magmatic** activity. Some gases are **exsolved** from **magma** by decompression during eruption, causing **vesicles** to grow in the **magma**. Such *vesiculation* is a major driving force of explosive **volcanic eruptions**. Other volcanic gases are emitted, independent of **volcanic eruption**, through *fumaroles*. The common volcanic gases are H_2O, CO_2, CO, sulphur gases, H_2, N_2, HCl, HF, CH_4 and NH_4. Solubilities of such volatile species in **magma** are variable, so that the composition of **exsolved** volcanic gas will not represent the volatile composition of the **magma** (e.g. CO_2 is less soluble in most **magmas** than H_2O, F and Cl). Some volcanic gases emitted from **fumaroles** may never have been dissolved in the underlying **magma**, but may have been derived from meteoric sources, or from **alteration** of rock. During **phreatomagmatic eruptions**, large quantities of steam are co-erupted with **magma** which is derived from seawater, lakes, or **aquifers**.

Mafic **magmas** derived directly by *partial melting* of the **mantle** contain dissolved volcanic gases also derived from the **mantle**. Their presence is regarded as an outgassing phenomenon of the Earth. Bailey (1980) argued that there are differences in the composition of volcanic gases between different tectonic settings of magmatism. In **sub-duction zones** volcanic gases are rich in H_2O, whereas in **rift**-related settings, volcanic gases are rich in CO_2, F and Cl. Some volcanic gases initially dissolved in **magma** may have been derived from the **crust** by **assimilation** of **crust** by **magma** during its uprise. Most of the volatile constituents of certain silicic **magmas** thought to have been generated by *partial melting* of **crust** must have a crustal provenance.

Some indication of the composition of volcanic gases dissolved in **magma** can be gained by studies of *fluid inclusions* trapped in *phenocrysts*. [PTL]

Bailey, D.K. (1980) Volcanism, Earth degassing and replenished lithosphere mantle. *Philosophical Transactions of the Royal Society, London* **A297**: 309–22.
Tazieff, H. & Sabroux, J.C. (1983) *Forecasting Volcanic Events*. Elsevier, Amsterdam.

volcanic hazard A risk to life or property from volcanic activity. Volcanic hazards can be assessed by analysis of historical records of **volcanic eruptions**, although these are usually too short to provide a meaningful hazards dossier. More emphasis is put on detailed mapping of prehistoric volcanic deposits, and comparison with styles of activity at volcanoes with similar evolutions. There has been a recent trend towards the assembly of volcanic hazard zoning maps which indicate zones around a volcano with different degrees of probability and intensity for each type of volcanic event. In some cases, such maps have met with some success in forecasting volcanic events, as in the case of Mount St Helen's (Crandell *et al.* in Sheets & Grayson, 1979, cf. Lipman & Mullineaux, 1981). Volcanic hazard zoning maps are important tools for the rationalization of land use in volcanic territory in order to minimize risk of damage to residential, industrial and agricultural districts. Two recent volumes (Sheets & Grayson, 1979; Tazieff & Sabroux, 1983) contain many volcanic hazard assessment case studies.

Lava flows tend to move slowly, and rarely pose a serious risk to life, but are extremely damaging to property in their path. **Lava flows** can, in some cases, be diverted by cooling with water, or by artificial barricades. Very large flood lava eruptions (see **plateau lava**) have not occurred in geologically recent time. Air-fall deposits tend to accumulate slowly, and present no serious immediate risk to life. Thick air-fall deposits may cause buildings to collapse, and deposits even a few centimetres thick, spread over very large areas, have disastrous effects on crops and livestock (Thorarinsson in Sheets & Grayson, 1979). **Base surges** are very destructive, because of their fast lateral movement, but rarely travel more than about 4 km from a **volcanic vent. Ash flows** of small volume ($<1\,km^3$) are very destructive, but tend to be strongly controlled by topography and restricted to valley bottoms. Associated clouds (see **nuée ardente**) may flow over a wider area and can be extremely destructive, as in the type example from Mt Pelée (Francis, 1976; Fisher & Heiken,

1982). Large-volume **ash flow** (**ignimbrite**) eruptions, associated with **caldera** formation, are probably the most destructive kind of volcanic event. There are examples in the geological record with volumes of over $1000\,km^2$ of erupted **magma**, and which covered areas of about $20\,000\,km^2$ with a blanket of **ash** about 20 m thick. During the only well-documented historic **ignimbrite** eruption in a remote part of Alaska in 1912, only $15\,km^3$ of **magma** was erupted. Most **ignimbrites** are thought to travel with velocities in the range $30-200\,m\,s^{-1}$ (Wilson & Walker, 1982), and while they tend to pond into valleys, they can also surmount hills over 500 m high which lie in their path. **Ignimbrites** are thought to generate large volumes of widely-dispersed air-fall **tephra**. Because no **ignimbrite** eruption has ever been scientifically monitored, it is not known how they can be forecast. They are very rare events; many volcanoes never experience **ignimbrite** eruption, and few experience more than three. Nevertheless, there are probably thousands of **ignimbrites** recorded in the geologic record.

Lahars are extremely destructive, but tend to flow only along the bottoms of valleys. **Lahars** are very mobile and have been known to travel over 100 km. They commonly occur on volcanoes covered by snow, or which have **crater** lakes. They can occur in any climatic zone, however, and are commonly caused by the heavy rains which often fall during, or after, eruptions, and which can mobilize unconsolidated, freshly-fallen **tephra**.

Floods may be caused indirectly by **volcanic eruptions**, by flowage of **lahars** or **ash flows** into lakes in dammed river valleys, by release of water from **crater** lakes, or by melting of ice. **Volcanic eruptions** below *ice sheets* in Iceland melt large volumes of ice which can be released catastrophically as floods called *jökulhlaups*. **Tsunamis** ('tidal-waves') can be generated by several volcanic events; **caldera** collapse and entry of **ash flows** or *debris avalanches* into the sea. **Tsunamis** are major volcanic hazards in volcanic areas with long coastlines, such as Indonesia and Japan. The **tsunamis** generated by the 1883 eruption of Krakatau, Indonesia, killed about 36 000 people in coastal towns and villages (Francis, 1976). *Debris avalanches* can exceed $1\,km^3$ in volume, can travel up to about 100 km from source and are commonly initiated by slope **failure** on steep-sided volcanoes. Siebert *et al.* (1987) reviewed the volcanic hazards associated with volcanic *debris avalanches*.

Hazards from the common **volcanic gases** were documented by Faivre-Pierret and Le Guern (in Tazieff & Sabroux, 1983). Carbon dioxide is denser than air and may pond in topographic depressions, or flow down valleys, displacing air, with potentially fatal results (e.g. Thorarinsson in Sheets & Grayson, 1979). The toxic volcanic gases CO, COS, CS_2, SO_2, H_2S and HCl are not expected to be present in sufficiently high concentrations to be fatal, except very close to vents (Faivre-Pierret and Le Guern in Sheets & Grayson, 1979). Fluorine is strongly toxic, and is soluble in volcanic glass, so that it is deposited with air-fall **tephra**. Livestock which have eaten grass polluted by **tephra** have been reported to have developed the potentially fatal condition of *fluorosis* (Thorarinsson in Sheets & Grayson, 1979). [PTL]

Fisher, R.V. & Heikin, G. (1982) Mt Pelee, Martinique: May 8 and 20, 1902, Pyroclastic flows and surges. *Journal of Volcanology and Geothermal Research* **13**: 339–71.
Francis, P. (1976) *Volcanoes.* Penguin, Harmondsworth, Middlesex.
Lipman, P.W. & Mullineaux, D.R. (eds) (1981) *The 1980 Eruptions of Mount St Helen's, Washington.* Professional Paper of the United States Geological Survey **1250**.
Sheets, P.D. & Grayson, D.K. (1979) *Volcanic Activity and Human Ecology.* Academic Press, London.
Siebert, L., Clicken, H. & Ui, T. (1987) Volcanic hazards from Bezymianny- and Bandai-type eruptions. *Bulletin of Volcanology* **49**: 435–59.
Tazieff, H. & Sabroux, J.C. (1983) *Forecasting Volcanic Events.* Elsevier, Amsterdam.
Wilson, C.J.N. & Walker, G.P.L. (1982) Ignimbrite depositional facies: the anatomy of a pyroclastic flow. *Journal of the Geological Society, London* **139**: 581–92.

volcanic vent An orifice through which a **volcanic eruption** has occurred. Volcanic vents may be approximately circular in plan, representing a cross-section of a *volcanic pipe*, or they may be elongate, as in the case of a **fissure eruption**, and underlain by a **dyke**. *Volcanic pipes* are approximately cylindrical conduits immediately underlying vents. They are usually between 50 m and 1 km in diameter and may flare towards the surface (see **diatreme**). A *volcanic plug* is a solid mass of material filling a *volcanic pipe*. *Volcanic plugs* are commonly exposed by **erosion** of extinct volcanoes, and are seen to consist either of a rather homogeneous mass of solidified **magma**, or a chaotic breccia. Such breccias consist of fragments of chilled magmatic material, fragments of *country rock*, and commonly separate grains of sediment derived from the disaggregation of the surrounding rock. The breccias are commonly intruded by small **dykes**. A fine-grained matrix is commonly present in the breccias, which consists of finely-divided magmatic fragments and grains of sediment. This matrix usually shows signs of having been **fluidized**, having been intruded along tiny cracks in blocks in the breccia. Such fine-grained matrix is called *tuffisite*. *Volcanic plugs* are commonly more resistant to **erosion** than the sedimentary and poorly-consolidated **pyroclastic** material they intrude, so they often form positive topographic features called *volcanic necks*.

When volcanic vents on active volcanoes are filled by a solid mass of chilled, or very viscous, **magma**, the solid mas — the *volcanic plug* — may be explosively fragmented and expelled from the vent as **tephra** consisting of large blocks and **ash**. Such explosive expulsion of *volcanic plugs* may occur as an initial event of a renewed phase of activity at a volcano. [PTL]

Macdonald, G.A. (1972) *Volcanoes.* Prentice-Hall, Englewood Cliffs, NJ.

volcanic-associated massive oxide deposit Stratiform oxide **ore** similar to a **massive sulphide deposit**. It is frequently characterized by the paragenesis **magnetite–hematite–apatite** and occurs in volcanic or volcanic–sedimentary terrains. The best known example is the famous Kiruna deposit of northern Sweden and this deposit type is often called *Kiruna-type*. There may be a whole spectrum of deposits from massive sulphide to massive oxide through **magnetite–pyrite ores** with minor **chalcopyrite** such as Savage River, Tasmania. Some, like **massive sulphide deposits**, are known to be underlain by **stockworks** and they too may be **exhalites**, but others, like the **magnetite–hematite–apatite lava flows** in Chile, may be of extrusive origin. [AME]

Evans, A.M. (1993) *Ore Geology and Industrial Minerals: An Introduction* (3rd edn). Blackwell Scientific Publications, Oxford.

Von Karman's constant A universal constant relating the mean size, or *mixing length*, of eddies within a flow, l, to the depth of the flow, y, such that

$$l = \kappa \cdot y$$

where κ is von Karman's constant which is commonly taken as 0.4184. This constant provides an important assumption underlying the derivation of **shear stress** in the **Karman–Prandtl velocity law**. Past work has suggested the value of von Karman's κ is decreased by the presence of suspended sediment, although recent work has thrown doubt upon this. [JLB]

Francis, J.R.D. (1975) *Fluid Mechanics for Engineering Students.* Arnold, London.
Coleman, N.L. (1981) Velocity profiles with suspended sediment. *Journal of Hydraulic Research* **19**: 211–29.

vorticity The average **angular velocity** of material lines (i.e. lines between known points) in a rock undergoing **strain**. It is much used in modelling ideal **strain** geometries. Vorticity has the symbol w, which is positive or negative depending on the sense of rotation. It is used to describe *rotational strains*, e.g. those resulting from a non-**coaxial strain** path such as that of **simple shear**. A zero vorticity number does not, however, imply a **coaxial deformation** history; a non-**coaxial strain** ($w \neq 0$) combined with a **rigid body rotation** ($w \neq 0$) could result in an irrotational bulk **strain** ($w = 0$). [SB]

Lister, G.S. & Williams, P.F. (1983) The partitioning of deformation in flowing rock masses. *Tectonophysics* **92**: 1–33.

W

wad A name given to a fine-grained mixture of **manganese** oxide and hydrated **oxide minerals**. [DJV]

wadi (also *wady* and, in French, *ouady* or *oued*) An Arabic term for an **arroyo**; a steep-sided watercourse found in arid regions. Wadis (correctly *widyan*) may flow only occasionally and sporadically after heavy rainfall. The fluvial sediments which accumulate as desert *fans* at the mouths of wadis are often **gravelly** and are typically poorly sorted. [AW]

wall rock The rock or rocks adjacent to a mineral deposit or igneous intrusion. [AME]

Walther's law Concept (rather than a law *sensu stricto*) derived by Walther in 1894 suggesting that the conformable vertical succession of **sedimentary facies** is a reflection of their former lateral juxtaposition. This concept only applies to successions without major breaks, since an **erosional** hiatus represented by an **unconformity** may mark the passage of a number of environments with distinctive facies which have not been preserved. Walther's law has been particularly successfully applied to the progradation of a **delta** causing a vertical succession of prodelta muds to **delta** slope, front and finally coastal plain. [PAA]

Walther, J. (1894) *Einleitung in die Geologie als Historische Wissenschaft, Bd. 3: Lithogenesis der Gegenwart*, pp. 535–1055. Fischer Verlag, Jena.

waning slope Weathered **scree** is **eroded** by **overland flow** to form a concave *footslope*. This was the lowest of the four slope elements defined by Wood (1942). The waning slope develops once stream incision has ceased to be the dominant process controlling slope form; as lateral **erosion** ensues, the waning slope gradually extends upslope so consuming the constant slope above. (See also **waxing slope**.) [TPB]

Wood, A. (1942) The development of hillside slopes. *Proceedings of the Geologists' Association* **53**: 128–40.

washout A lenticular body of clastic sediment, usually **sandstone**, that projects downwards from the roof of a **coal seam** replacing all or a part of it. Washouts are usually elongate or sinuous in plan as they represent scour-and-fill structures cut by stream or river activity. [AME]

washover (*fan*) Sediment deposited by the action of overwash of **barrier islands**. They are characteristic of microtidal **barrier islands** where **storm surge**-enhanced seas

are unable to pass through poorly developed **tidal inlets**. The sea thus breaks through topographic lows in the **barrier island**. Sediment eroded from the **beach** and shoreface is transported landward through the breaches in the barrier, forming washover fan deposits on the landward side of the barrier. [RK]

waste Unwanted material disposed of during mining and quarrying. (See **gangue**, **stripping ratio**.) [AME]

water drive The **buoyancy** effect of underlying water that drives oil through permeable rocks during its secondary **migration**. [AME]

water table The surface of **groundwater**, defined by the level of free-standing water within intergranular pores or fissures at the top of the **phreatic zone**, below which the pores of the host sediment are saturated with water, i.e. the upper limit of the **saturated zone**, see Fig. A8, p. 31. This is an equilibrium surface, at which the *hydrostatic pressure* in the pores is equal to atmospheric pressure in the pores of the overlying **vadose zone**. The water table is uneven, varying with topography, and is variable in its height, fluctuating with rainfall and internal drainage in the host sediment. During periods of high rainfall the water table rises whilst during periods of drought it falls. Deserts characteristically have low water tables; river valleys and coastal systems have characteristically high water tables. The lowest level to which the water table falls in any given locality is known as the *permanent water table*. Where the water table coincides with the ground surface, **seepages** and **springs** form. When below ground level, its depth can be measured in a **well**. [SDB/AME]

Freeze, R.A. & Cherry, J.A. (1979) *Groundwater*. Prentice-Hall, Englewood Cliffs, NJ.

waterfall A stream falling over a precipice. It is often believed that the great majority of waterfalls occur where a soft rock is eroded from beneath a harder rock (the *caprock model*), but many waterfalls exist without such undercutting and may have distinct buttressing at their bases (Young, 1985). [ASG]

Young, R.W. (1985) Waterfalls: form and process. *Zeitschrift für Geomorphologie* Supp. NF **55**: 81–95.

watten An area of tidal marshland and sand flats lying between the mainland and an offshore **bar** or island. Such

wave ripple

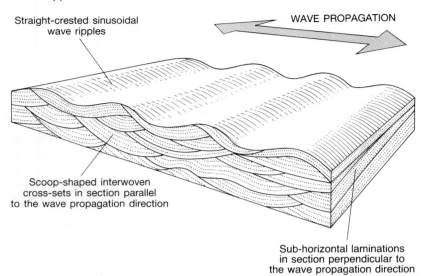

Straight-crested sinusoidal
wave ripples

WAVE PROPAGATION

Scoop-shaped interwoven
cross-sets in section parallel
to the wave propagation direction

Sub-horizontal laminations
in section perpendicular to
the wave propagation direction

Fig. W1 Wave ripple. The three-dimensional structure of stratification produced by wave oscillation. (After Boersma, 1970 in Reineck & Singh, 1975.)

land is exposed only at low tide. The Waddenzee — part of the former Zuider Zee — is an area of watten and deep tidal channels lying between the Dutch mainland and the Frisian Islands. [AW]

wave ripple The characteristic periodic bedform generated by progressive *gravity waves* (see **oscillatory flow**). Wave ripples are commonly symmetrical to slightly asymmetrical and trochoidal to sinusoidal in cross-section, and occur at wavelengths of less than 10 mm, when they are termed *miniripples*, to over 1 m.

A wide range of wave ripple morphologies has been described from experimental studies, ranging from very flat varieties which have variously been called *rolling grain ripples*, *decaying ripples* and *post-vortex ripples*, to steep varieties termed *vortex ripples*, *growing ripples* and *orbital ripples*. For steep *vortex ripples* there is a relationship between the wavelength of the ripple-mark and the orbital diameter of the near-bed water motion:

wavelength = 0.65 orbital diameter

However, above a critical **wave Reynolds number**, this relationship breaks down. The steep *vortex ripples* begin to flatten until eventually **plane bed** or **sheetflow** conditions occur.

The variability of wave-generated ripples is even greater in the natural environment. Wave ripples with regularly spaced crestlines (*vortex ripples*) are common, but less regular bifurcated or chaotic types are found as disequilibrium bedforms. Such ripple types change in form in response to the variable wave climate.

The equilibrium forms of *vortex ripples* are distinct from those produced under unidirectional flows in being more symmetrical, steeper, and having more regular crestlines in plan view. Trochoidal crests are particularly diagnostic of **wave** oscillation. Internally, wave ripples typically con-

tain unidirectional *cross-laminations* in isolated lenses or highly scoop shaped laminations in interwoven *cross-sets* in the section parallel to the direction of **wave** propagation, and commonly subhorizontal parallel laminations in the section perpendicular to this (Fig. W1). Because of the mass transports caused by progressive *gravity waves* and the likelihood of additional superimposed unidirectional currents, wave ripples can migrate on- or offshore and take on geometries intermediate between true unidirectional current ripples and steep *vortex ripples*. These intermediate types are called combined flow ripples. (See **ripple**, **oscillatory flow**.) [PAA]

Reineck, H.-E. & Singh, I.B. (1975) *Depositional Sedimentary Environments*. Springer-Verlag, Berlin.

wavefront The surface over which the phase (i.e. peak or trough) of a travelling wave disturbance is the same. Wave propagation can be thought of as the motion of a wavefront through a medium and ray paths are perpendicular to wavefronts. Wavefronts propagate according to *Huygen's principle*, which states that each point on an advancing wavefront can be regarded as the centre of a fresh disturbance. The later wavefront is the envelope of the resultant of the superposition of the waves from the secondary sources. [GS]

wavellite $(Al_3(PO_4)_2(OH)_3 \cdot 5H_2O)$ A secondary mineral found in low grade metamorphic rocks. [DJV]

wavenumber The equivalent of frequency for a spatial waveform, such as a magnetic profile. Expressed in cycles per unit distance. [PK]

waves Regular oscillations of the water surface of oceans and lakes created by wind-generated pressure differences.

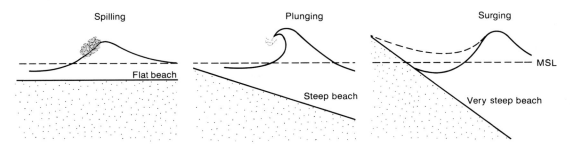

Fig. W2 The three principal types of breaking **waves** and beaches upon which they break. In reality there is a continuum between these waves and beach gradients.

The magnitude of oscillation depends on the strength and duration of the generating wind and on the **fetch** length over which it blows. Waves are also produced as a result of **earthquakes** or **volcanic eruptions** but the **tsunamis** so produced are both irregular and unpredictable (see also **seiche**). Within the generation area of wind-waves a wide range of wave sizes is present but the principal characteristic of these 'sea' waves is the dominance of short wavelengths. As waves move out of the generating area their velocity is governed by wavelength and so the longer waves travel faster and become size-sorted to arrive on a distant shore as regular 'swell' waves.

All waves undergo transformations as they approach the shore. **Refraction** of the **wave** occurs and the wave ultimately breaks. Refraction of water waves obeys the same law as light refraction (*Snell's law*) in which the change in angle across an interface is determined by the accompanying change in velocity. As velocity is depth-controlled the wave crest bends by regular amounts over a shelving bottom. This serves to focus wave energy on certain sections of the coast and to reduce it on others, with important implications for **erosion**, transportation and deposition of coastal sediment. Waves entering shallow water undergo increases in wave height and steepness that eventually induce breaking. In general, three main *breaker* types are recognized (Fig. W2): spilling *breakers* where the face foams and breaks section by section over a wide zone, plunging *breakers* where the face steepens until vertical then curls over and plunges downwards over a narrower zone and surging *breakers* where the base of the wave rushes up the **beach** without fully breaking over a very steep **beach** face. Surging *breakers* are associated with flat, low waves and steep **beaches** whereas spilling *breakers* are associated with high, short waves and flat **beaches**.

The importance of wave refraction and breaking lies in their capacity to perform **geomorphological** work. The type of breaking wave influences the asymmetry of onshore/offshore water velocities which in turn controls the amount and type of sediment movement normal to the coast. The **refraction** associated with those **waves** controls the distribution alongshore of wave energy and the magnitude of the longshore currents which move sediment. (See **longshore drift**.) [JDH]

Komar, P.D. (1976) *Beach Processes and Sedimentation*. Prentice-Hall, Englewood Cliffs, NJ.
Pethick, J.S. (1984) *An Introduction to Coastal Geomorphology*. Arnold, London.

waxing slope A convex slope formed at the crest of a cliff (Wood, 1942). Wood argued that cliff **weathering** attacked the initial angular intersection between the cliff and the **interfluve plateau** to form the convexity. The processes of **creep** and *rainsplash* may also contribute to the development of the convexity (Gilbert, 1909). (See **waning slope**.) [TPB]

Gilbert, G.K. (1909) The convexity of hilltops. *Journal of Geology* **17**: 344–51.
Wood, A. (1942) The development of hillside slopes. *Proceedings of the Geologists' Association* **53**: 128–40.

weathering The disintegration and decomposition of rocks and sediments by mechanical and chemical processes acting at or near the surface of the Earth. Weathering of rock is often a prerequisite for its **erosion** since the action of wind, water, and ice may have only a small effect on exposed, unweathered rock. In effect, the understanding of weathering mechanisms is a critical step in our interpretation of landforms and the evolution of the landscape. The study of weathering processes is an integral part of **geomorphology**.

Mechanical weathering involves the fracturing and granular disintegration of rock. On a regional scale, processes of **faulting** and compression of the Earth's **crust** may initiate fissuring of solid rock. On a smaller scale, however, such influences may be masked by other forms of weathering. **Pressure release** is a process which creates characteristic weathering features; it causes **exfoliation** and spalling of huge sheets of rock on exposed masses of bedrock — especially crystalline rocks. This process is usually associated with the release of **lithostatic pressures** as deeply buried rocks are exhumed by **erosion**. However, it has been suggested that the forces generated by **diastrophism** also play a role in this form of weathering (Twidale, 1973).

Other processes of mechanical weathering occur at a smaller scale and are often environmentally controlled. **Insolation weathering**, for example, has been said to occur in deserts where large diurnal temperature fluctuations can cause rapid expansion and contraction of rocks leading to **fracturing**. The likelihood that such processes occur under natural conditions has been disputed but, in some environments, fire may effect the shattering of rock. In cold climates, *frost shattering* is a significant weathering process. Either as a result of the volumetric expansion of water as it freezes, or through the growth of ice crystals, rock is fractured as water freezes in small **joints** and fissures. Typically, frost weathering produces **screes** of angular rock fragments at the base of rocky outcrops.

In arid environments, **salt weathering** is an important mechanical weathering process. The evaporation of saline water which has infiltrated rock along cracks and **joints**, and through pore spaces, results in salt crystal *nucleation* and crystal growth. The forces generated by unidirectional crystal growth of common salts such as sodium chloride, sodium sulphate, and **gypsum** are sufficient to shatter the rock (Cooke, 1981; Goudie, 1985; Sperling & Cooke, 1985; Winkler & Singer, 1972). The presence of salts in fissures within the rock may also stimulate their disintegration as a result of volumetric changes during hydration and dehydration — for example, *hemihydrate* ($CaSO_4 . \frac{1}{2}H_2O$) to **gypsum** ($CaSO_4 . 2H_2O$) and *thenardite* (Na_2SO_4) to *mirabilite* ($Na_2SO_4 . 10H_2O$) (Winkler & Wilhelm, 1970). Simple thermal expansion may also accomplish this (Mortensen, 1933). **Salt weathering** is most common in arid regions where evaporation rates are high; however, it is not restricted to the Earth's warm deserts, it is also an important weathering process in parts of Antarctica.

There are a number of other mechanical weathering processes which are related to specific rock types or environments. These include rock disintegration caused by volumetric changes resulting from wetting and drying — this occurs in **shales** which contain expansive **clay minerals** such as **smectites**. In some environments, certain plants are said to be able to split rocks as their root systems penetrate **joints** and fissures. In reality, trees such as the 'rock-splitter fig' (*Ficus* sp.) probably only widen existing cracks in the rock. In deserts, exposed rock surfaces may be weathered through sand-blasting by airborne particles or by vortices in 'clean' airflow which detach grains from the rock surface. Because the weathered material is immediately removed by the wind, this process is usually regarded as an **erosional** process rather than weathering.

Chemical weathering is the decomposition of rocks and rock minerals by direct **dissolution** or chemical **alteration** and **eluviation**. All rock-forming minerals — even **quartz** — are soluble to some degree in pure water. Some, such as **halite** and **gypsum** are highly soluble and are readily weathered in humid environments. Others, such as **limestones**, are dissolved through a combined process of **dissolution** and chemical reaction with ions in solution (see **carbonation**). Certain minerals such as **orthoclase** are significantly more soluble in water containing other salts in solution. The **dissolution** of rocks by **meteoric water** and **groundwater** can produce characteristic landforms

such as **karstic** terrain. It is also an important process in the mobilization of secondary minerals formed by other weathering processes. **Eluviation** and *cheluviation* in weathering profiles leads to the formation of **laterites** and some **duricrusts; illuviation** can result in the **accretion** of specific minerals in the lower horizons of **deep weathering** profiles — some **silcretes** develop in this way.

The **dissolution** of minerals in rocks is an important first step in many weathering processes since it allows freer access of water through enlarged **joints**, fissures, and pores. Subsequent chemical **alteration** can be accomplished thorough decomposition and rotting of the bedrock. The main processes are hydration and *hydrolysis*, **carbonation**, *redox reactions* (see **oxidation**), and **chelation**. The addition of water (hydration) and hydroxyl (OH^- ions) (*hydrolysis*) not only causes chemical **alteration** of minerals, but often also physical changes which further enhance the weathering process. For example, the hydration of *anhydrite* ($CaSO_4$) to **gypsum** ($CaSO_4 . 2H_2O$) causes volumetric expansion and heaving which further disrupt the rock mass. **Orthoclase** — a common constituent of acid crystalline rocks — is *hydrolysed* to clay (usually **kaolin**) in the presence of carbonated water. Indeed, *hydrolysis* and **carbonation** by water containing dissolved carbon dioxide can reduce most common rock-forming minerals — with the main exception of **quartz** — to clays such as **chlorite**, **kaolinite**, and **gibbsite**. Subsequent **oxidation** of **iron** compounds by water containing dissolved oxygen often stains these clayey weathering products red. The rates of chemical decomposition of different minerals vary but factors such as the chemistry of **groundwater** and *soil* moisture, and the structural characteristics of the rock can result in markedly different weathering rates on similar rock types. Near the land surface, the influence of organic compounds and the action of *bacteria* also aid the weathering process. The desilicification of **clay minerals** and the **chelation** of **iron** and **aluminium** ions which occurs in **lateritic** weathering profiles appear to be related to the presence of complex organic substances and microorganisms.

In broad terms, the character of the weathering processes in any environment is closely linked to climatic conditions. In cold climates and in arid environments, mechanical weathering often predominates over chemical processes. In humid regions, chemical weathering is often most important. However, rock weathering is often accomplished through the combined action of a variety of agencies. Often, chemical weathering is enhanced once a rock is fractured by mechanical processes such as **pressure release**, frost action, or **salt weathering**. Conversely, the partial rotting of rock surfaces by chemical processes may allow the penetration of water into **weathering rinds** which may then be spalled off the less weathered rock mass beneath by mechanical processes such as salt crystallization (see **desquammation**). In practice, a wide variety of environmental factors including rock type and structure, climate, vegetation and *soils*, **hydrology**, and human influences will determine the nature and rate of weathering. [AW]

Cooke, R.U. (1981) Salt weathering in deserts. *Proceedings of the Geologists' Association of London* **92**: 1–16.
Goudie, A.S. (1981) Weathering. In: A.S. Goudie (ed.) *Geomorphological Techniques*, pp. 139–55. Allen & Unwin, London.
Goudie, A.S. (1985) *Salt Weathering*. Research Paper 32. School of Geography, Oxford University.
Mortensen, H. (1933) Die 'Salzsprengung' und ihre Bedeutung für die regionklimatische Gliederung der Wüsten. *Petermanns Mitteilungen aus Justus Perthes geographischer Anstalt* **79**: 130–5.
Ollier, C.D. (1969) *Weathering*. Oliver & Boyd, Edinburgh.
Sperling, C.H.B. & Cooke, R.U. (1985) Laboratory simulation of rock weathering by salt crystallization and hydration processes in hot, arid environments. *Earth Surface Processes and Landforms* **10**: 541–55.
Trudgill, S.T. (ed.) (1986) *Solute Processes*. John Wiley & Sons, Chichester.
Twidale, C.R. (1973) On the origin of sheet jointing. *Rock Mechanics* **5**: 163–87.
Winkler, E.M. & Singer, P.C. (1972) Crystallization pressure of salts in stone and concrete. *Geological Society of America Bulletin* **83**: 3509–13.
Winkler, E.M. & Wilhelm, E.J. (1970) Salt burst by hydration pressures in architectural stone in urban atmosphere. *Geological Society of America Bulletin* **81**: 567–72.

weathering front The transition zone or interface between unweathered bedrock and the overlying **saprolite** of a **deep weathering** profile. At the weathering front, chemical processes cause the **alteration** and disintegration of minerals in the sound rock; this is the first step in the decomposition of the rock. In areas of crystalline bedrock (such as **basalts**, **dolerites**, **granites**, and **gneisses**) in the humid tropics and subtropics, the depth of the weathering front can reach 100 m or more if the land surface is stable. However, the depth of the weathering front can vary locally owing to differences in the bedrock's susceptibilty to **weathering**. When such landscapes are **denuded**, the zones where the weathering front was shallowest may be exhumed to create *bornhardts* and **inselbergs**. [AW]

Ollier, C.D. (1969) *Weathering*. Oliver & Boyd, Edinburgh.

weathering rind A layer of partially **weathered** material forming the surface of a boulder or rock outcrop. Weathering rinds are often yellow, orange, or red in colour as a result of **oxidation** of **iron** minerals. Though they may resemble **desert varnish**, the rinds usually penetrate the rock for as much as a few centimetres, and a transition from heavily weathered material to sounder rock is evident. In contrast, **desert varnish** comprises a thin (less than 1 mm) coating on the rock surface. The induration of weathering rinds through mineral **cementation** can result in **case-hardening** of the rocks. The spalling of weathering rinds from rocks and boulders is called **desquammation**. [AW]

Ollier, C.D. (1969) *Weathering*. Oliver & Boyd, Edinburgh.

websterite A *pyroxenite* composed mainly (over 95%) of *orthopyroxene* and *clinopyroxene* (each exceeding 10%).

Websterites occur as small parts of large ultramafic intrusions and as ultramafic **xenoliths** in **basalt**. [RST]

wehrlite A **peridotite** composed dominantly (over 95%) of *clinopyroxene* and **olivine** (>40%) with minor (below 5%) *orthopyroxene*. Wehrlites occur as small parts of large ultramafic intrusions and as ultramafic **xenoliths** in **basalt**. [RST]

welded tuff A volcanic rock composed of explosively fragmented **tephra** which, upon deposition, retained enough heat for the glass shards and **pumice** *clasts* to deform and stick together, to form a compact, lava-like rock. In geologically young rocks, the degree of welding of **tephra** can be estimated by inspection, or by measuring **porosity**. In rocks affected by burial and lithification, the degree of welding can be determined petrographically, using criteria outlined by Ross & Smith (1961). The term welded tuff does not imply any composition or mode of deposition, but most are intermediate or silicic in composition. Welded spatter deposited ballistically in *strombolian* or **fire fountain** eruptions (commonly of **basaltic** composition) is called *agglutinate*.

Silicic welded tuffs can always be traced laterally (at least before **erosion**) into non-welded portions of the same deposit. Most silicic welded tuffs were deposited as **ash-flow tuffs** (Smith, 1960). In these, the degree of welding increases towards the vent, and with increasing thickness of the deposit. Silicic welded air-fall tuffs are common only on some volcanoes, notably where **magma** composition is strongly alkaline or where the air-fall **tuffs** are derived from mafic–silicic mixed **magma**. [PTL]

Cas, R.A.F. & Wright, J.V. (1987) *Volcanic Successions Ancient and Modern*. Allen & Unwin, London.
Fisher, R.V. & Schmincke, H.-U. (1984) *Pyroclastic Rocks*. Springer-Verlag, Berlin.
Ross, C.S. & Smith, R.L. (1961) *Ash-flow tuffs: their origin, geologic relations, and identification*. Professional Paper of the United States Geological Survey **366**: 1–77.
Smith, R.L. (1960) Ash flows. *Geological Society of America Bulletin* **71**: 795–842.

well A hole that yields any fluid is a well no matter whether it is dug or drilled or whether it produces **brine**, gas, oil or water. [AME]

Wenner configuration One of the popular arrangements of electrodes used in **resistivity methods**. In this array the distance between adjacent electrodes is a constant, a, with the two outer electrodes usually carrying the current, I, and the inner ones measuring the potential difference, V; then the expression for the **apparent resistivity** ρ_a is

$$\rho_a = 2\pi a \frac{V}{I}$$

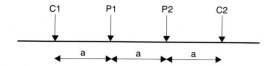

Fig. W3 Wenner electrode array. P1 and P2 are the potential electrodes and C1 and C2 the current electrodes.

The Wenner array is the most efficient in horizontal profiling (see **constant separation traversing**) because each new measurement need only involve one extra electrode to be inserted if the array is moved on by *a* each time. It is also widely used in **vertical electrical sounding** along with the **Schlumberger configuration** (see Fig. W3). [ATB]

westward drift The westward motion by about 0.2° longitude a^{-1} of most features of the **geomagnetic non-dipole field** during at least this century. Once attributed to differential rotation between the Earth's **core** and **mantle**, it is now generally ascribed to the passage of magnetic lines of force through the Earth's outer **core**. [DHT]

Bullard, E.C., Freedman, C., Gellman, H. & Nixon, J. (1950) The westward drift of the Earth's magnetic field. *Philosophical Transactions of the Royal Society* **A234**: 67–92.

wet-sediment deformation Also called *soft-sediment deformation*, the term 'wet' being preferred because it describes an inherent physical feature of sediment that may liquefy or be fluidized whereas 'soft' only implies malleable or subject to **deformation**; for example **strain** under high-grade dry metamorphic conditions may affect soft sediment. Wet-sediment deformation describes any post- or syn-depositional **liquefaction** and/or **fluidization** of sediment and is identified where **strain** has altered the geometry of primary sedimentary lamination, **bedding**, **fabric** and texture. **Deformation** may occur at or near the sediment surface, or at considerable depths where **overpressuring** occurs. Structures may show brittle to ductile features, and occur at low to high **strain rates**. Typical wet-sediment deformation includes **slides**, **slumps**, **convolute-lamination**, oversteepened and recumbent **cross-stratification**, **dish-structure**, mud or *sand* injection **dykes**, **sills** and *volcanoes*, *load structure*, *ball-and-pillow structure*, and sedimentary **faults** with perhaps up to tens of metres throw as in *growth faults*. Any process causing the applied normal or **shear** forces to overcome the *yield strength* of a sediment may produce wet-sediment deformation, for example **earthquakes**, sedimentary oversteepening of slopes, undercutting of slopes by currents or *bioturbation*, cyclic loading due to **wave** activity, and current **shear** above unconsolidated sediment. Wet-sediment deformation is generated by **tectonic** as well as sedimentary processes, as in **accretionary prisms** where *thrusts* affect *underconsolidated sediments*. [KTP]

whetstone One of the several types of sharpening stone used since the beginnings of metallurgy for sharpening metal blades (other sharpening stones include honestones and scythestones). A whetstone is typically rectangular in shape and may be pierced at one end for hanging or carrying. A large variety of rocks were used for whetstone production, but classical references from Pliny (1st century AD), tell of specialized quarries in Greece and Turkey for whetstones, so clearly the rocks were carefully selected. **Schists** and **sandstones** (in which the **quartz** grains provided the **abrasive**) were popular for whetstones in Britain in the Bronze Age and medieval periods, while other rocks used include **basalt**, *greywacke* and some **limestones**. [OWT]

white smoker Plume of *hydrothermal water* issuing from a part of the ocean floor like a **black smoker** but producing clouds of white sulphate minerals rather than black sulphides. [AME]

willemite (Zn_2SiO_4) A *nesosilicate* mineral that occurs as an **ore** in the mines at Franklin, New Jersey, USA. [DJV]

Wilson cycle The hypothesis, propounded by J.T. Wilson, that **continental splitting** occurs, followed by ocean formation and eventual closure with associated **orogeny**. The new continental block then tends to split along approximately the same lines as before to produce yet another ocean in approximately the same location as the preceding one. This model was based on the formation of the Atlantic Ocean which appears to have formed in approximately the same places as the original **Iapetus** Ocean, which preceded the collision between North America-Greenland and Europe-Africa, forming the Caledonian Mountains. [DHT]

Burke, K., Dewey, J.F. & Kidd, W.S.F. (1976) Precambrian paleomagnetic results compatible with contemporary operation of the Wilson Cycle. *Tectonophysics* **33**: 287–99.

wind ripple A small-scale aeolian bedform, moulded by wind from dry **sand** or granules (see **adhesion structure** for damp surface aeolian bedforms). Wavelengths and heights are controlled in part by grain size such that they reach 20 m and 1 m respectively in the coarsest wind ripples, which therefore overlap in size with fine-grained **dunes**. However, for any grade of sediment and wind velocity, there is a distinct gap between the size ranges of ripples and **dunes**. Ripples are frequently superimposed upon **dunes**, both on **stoss slopes** and on **lee slopes** approaching the *angle of repose* of dry **sand**, but individual ripples do not develop into **dunes**.

Impact ripples (*ballistic ripples*) are the most common type of wind ripple. There is a continuous spectrum of morphologies which change with grain size. Both grain size and *sorting* control the maximum size to which *impact ripples* may grow. In the former case the relationship is direct, in the latter inverse.

In fine to medium **sand**, *impact ripple* wavelengths vary from less than a centimetre to over a metre and heights from a few millimetres to several centimetres, but are typically less than a centimetre. Such ripples have low relief and high *ripple form indices* (ratios of wavelength to height), ranging from 10 to 70, with typical values of 15–20. Sharp (1963) found that the *ripple form index* varies directly with wind velocity and inversely with grain size. Ripple crests run transverse to the wind and are usually continuous and only slightly sinuous, occasionally *bifurcating*, though rare sinuous and linguoid forms have been recorded. Profiles are slightly asymmetrical, the lee face (downwind side) being steeper than the stoss, which tends to be slightly convex-up. According to Sharp (1963), the degree of asymmetry varies directly with grain size and inversely with wind velocity. They are dissimilar to aqueous current **ripples** (i) in plan form, (ii) in having **lee slope** gradients generally far less than the *angle of repose* of loose **sand**, (iii) in the frequent absence of internal *cross-laminae* and (iv) in being **inverse graded** as a result of the concentration of finer grains in the ripple troughs. Characteristics (ii) to (iv) reflect the fact that ripple migration due to **lee side** deposition is not the result of avalanching, but rather of the preferential trapping of grains moving by **saltation**; **sand** grains impacting at a high angle of incidence on the **stoss side** tend to bounce and/or set others in motion, whereas the lower angle of incidence on the **lee side** favours trapping of grains. **Inverse grading** results from the **saltation** population's domination by finer **sand**, whereas grains of all sizes move over ripple surfaces by **reptation**. Migrating fine-medium **sand** impact ripples produce **inverse graded** laminae (see **aeolian stratification**). *Cross-laminae* are seen only occasionally, taking the form of fine layers deposited on ripple lees during periods of reduced wind velocity. Ripple wavelength does not change with time and is observed to be proportional to wind velocity. This has long been thought to be the result of the equivalence of wavelength to the length of the average **saltation** trajectory, which is directly proportional to wind velocity, but Anderson (1987) has shown that the equivalence is in fact between ripple wavelength and six times the mean **reptation** length, which also is directly proportional to wind velocity.

At the coarser end of the grain size-controlled spectrum of *impact ripples* (in coarse **sand**, very coarse **sand** and granules) are bedforms called 'aeolian sand', 'pebble' 'residue', or 'granule ridges', 'grit waves', or *'granule ripples'*. Because of their grain size, the coarsest of these are restricted to **interdune** and **deflationary** settings. They are similar to their finer counterparts in being **inverse graded** and asymmetrical in profile, but they differ in that they can grow much larger (wavelengths and heights can reach 20 m and 1 m respectively), in having **stoss sides** which may be slightly concave-up, in being more irregular in plan, often showing a cuspate form similar to **barchan**

dunes, and in sometimes showing *cross-bedding*, at least in their upper parts. The third of these characteristics reflects their common bimodal composition, with a fine **sand** mode moved by **saltation** and a coarse mode moved by *creep* (**reptation** of grains too large to be lifted by wind). The relative immobility of the coarse mode causes *granule ripples*, unlike fine-medium **sand** *impact ripples*, to migrate only slowly and to grow vertically with time, probably because they have not reached the size limit imposed by their grain size and sorting.

Aerodynamic ripples have a similar size range to *impact ripples* and, like them, have low relief and high *ripple indices*, but they are much more sinuous in plan and elongate parallel to the wind. They are thought to be moulded by vortices close to the bed. They often occur superimposed on *impact ripples*.

Wind tunnel experiments show that at high velocities, with very fine **sand**, *impact ripples* of the typical form described above are replaced by steeper, sinuous crested bedforms similar to subaqueous three-dimensional *current ripples*. For all grades of **sand**, very high wind velocities produce an *aeolian* **plane bed** (see **aeolian stratification**). [DAR]

Anderson, R.S. (1987) A theoretical model for aeolian impact ripples. *Sedimentology* **34**: 943–56.
Sharp, R.A. (1963) Wind ripples. *Journal of Geology* **71**: 617–36.

witherite ($BaCO_3$) A relatively rare mineral found in **veins** associated with **galena**. [DJV]

wolframite (($Fe,Mn)WO_4$) The major **ore** mineral of **tungsten**. [DJV]

wollastonite ($CaSiO_3$) A **pyroxenoid mineral** chiefly formed by **contact metamorphism** of **limestones**. [DJV]

wood opal Fossil wood with **opal** as the petrifying material. [DJV]

wood tin A variety of **cassiterite** (SnO_2) with a fibrous appearance. [DJV]

wrinkle ridge Linear to sinuous topographic high having an asymmetric profile and considerable morphologic complexity. Wrinkle ridges are commonly found on the lunar maria and the plains units of Mars and Mercury, but analogues are known from Earth (Plescia & Golombek, 1986). They may result from the **deformation** of surface rocks over *thrust faults*. [ASG]

Plescia, J.B. & Golombek, M.P. (1986) Origin of planetary wrinkle ridges based on the study of terrestrial analogs. *Geological Society of America Bulletin* **97**: 1289–99.

wulfenite (PbMoO$_4$) An orange-red molybdate mineral found in the oxidized portions of **lead veins**. [DJV]

Wulff net Type of *stereographic net* in which equal angles on the surface of a sphere project as equal distances on the net. Such a net is also known as an *equal-angle net*. The Wulff net is used primarily for calculating angular relationships between planes and lines, in crystallography and in structural geology. It does not conserve area, and the **Schmidt net** (*equal-area net*) is used where comparison of areas is important. (See **stereographic projection**.) [RGP]

wurtzite (ZnS) A **polymorph** of **sphalerite**. [DJV]

X

X-ray diffraction The **diffraction** of X-rays by crystalline materials. The wavelengths of X-rays are the same order of magnitude as the distances between atoms in crystals (c.0.1 nm), so that the regular arrangement of atoms in a crystal or an individual grain of a polycrystalline aggregate can act like a grating and give a **diffraction** pattern. This possibility was first envisaged by von Laue in 1912. Wavelengths now in common use are CuKα = 0.15418 nm and FeKα = 0.19373 nm.

X-rays (which are electromagnetic waves) are scattered by the electronic charges surrounding the atomic nuclei. It was shown by the early work of von Laue and Bragg that the net effect of this scattering by a crystal could be interpreted as equivalent to the reflection of X-rays by a set of lattice planes (hkl). In the case of mirror reflection only, the path difference between waves scattered by two points in a plane is zero, and constructive interference (reinforcement) occurs, see Fig. X1. From the geometry, *Bragg's law* is obtained:

$$n\lambda = 2d_{hkl} \sin \theta$$

where n is an integer, λ is the X-ray wavelength, d_{hkl} is the spacing of the (hkl) planes and 2θ is the angle between the incident and diffracted beams. The interplanar spacing depends upon the **crystal structure** and since θ varies with d_{hkl}, individual reflections can be located by setting a detector at appropriate angles from the incident beam.

The scattering power of atoms varies with their electronic configurations. The intensities of X-ray beams diffracted by a crystal therefore depend upon the nature of the constituent atoms and their arrangements on the *crystal lattice*. The diffracted intensity from a crystal is proportional to the square of the structure factor, which gives the amplitude of radiation resulting from waves of various amplitudes and phase, scattered from the individual atoms within a single unit cell of the structure. For particular (hkl) planes of the possible crystal space groups, the structure factor can be zero, i.e. there is no diffracted intensity and the Bragg reflection is said to be 'forbidden'. Particular sequences of forbidden reflections occur for given space groups and are known as systematic absences. They are one of the fundamental signatures by which X-rays are used to elucidate the structure of minerals. X-ray diffraction measurements are able to give information about both the *crystal lattice* and the atom or group of atoms (the basis) associated with each lattice point. The important science known as *X-ray crystallography* has thus grown from the first experiments to determine **crystal structure** by X-ray diffraction.

Numerous different techniques, based upon X-ray diffraction, now exist for the study of crystalline materials. These usually combine some form of *goniometer* and *diffractometer*. The former enables a specimen to be positioned and oriented at a chosen angle with respect to fixed axes, while the latter allows the distribution and intensity of diffracted radiation to be measured. Such instruments now display a wide range of sophistication and accuracy, as indicated below, for specific systems. But many important X-ray diffraction methods still employ photographic film for recording diffracted intensities, and there may be both reflection and transmission versions of basically the same method of investigation. Systems also exist that are designed to subject specimens to variations or extremes of temperature or pressure during investigation, in order to follow changes in *lattice parameter* or to effect *phase transitions*. However, the most important and commonplace techniques are described below.

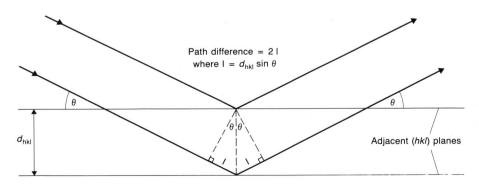

Path difference = 2 l
where l = $d_{hkl} \sin \theta$

d_{hkl}

Adjacent (hkl) planes

Fig. X1 X-ray diffraction.

POWDER DIFFRACTION TECHNIQUES. These are used to identify powdered crystalline materials or to measure the *lattice parameters* of a mineral. The essential features are a narrow beam of monochromatic X-rays directed onto a small volume of finely powdered, randomly oriented particles. Measurements of the resulting spatial distribution of diffracted X-rays can be carried out with various degrees of sophistication, but can be considered under two headings.

1 *The Debye–Scherrer Method.* A filtered beam of Kα X-rays strikes a small cylindrical powder sample, often within a thin-walled glass capillary, with axis normal to the beam. Cones of diffracted intensity are intercepted by a strip of photographic film cylindrically disposed around the sample and contained within the Debye–Scherrer camera. After short exposures (*c.*30 min), development of the film produces a series of lines (actually arcs) that correspond to the various interplanar spacings of the powdered sample. Measurements of the positions and intensities of the lines are potentially diagnostic when compared with a file of published powder diffraction data (ASTM index). It is necessary to prepare a powder of a suitable particle size without causing **plastic** distortions (which affect X-ray linewidth). The brittle nature of many minerals helps in this task. Various X-ray focusing arrangements enable greater resolution of close X-ray lines to be obtained without longer exposures. Amongst these, Guinier-type cameras are particularly popular and widely used.

2 *The powder diffractometer.* This instrument uses a focusing geometry, with a counter moving on a circle, centred on a small powder sample. As the sample is driven to rotate through an angle θ, the counter rotates through 2θ about a common axis. The counter (usually either proportional or *scintillation* type) produces measurements of the 2θ positions and intensity profiles of X-rays diffracted from the sample. More sophisticated *goniometer* arrangements also exist (e.g. spectrometers to increase peak-to-background ratios). In the simplest systems the data are plotted via a strip chart recorder, but on-line computer controlled systems are available that select lines of interest and process the resulting data.

POLE FIGURE METHODS. These are techniques for the generation of data from polycrystalline specimens that possess **fabrics** with **preferred orientations**. The specimens, in the form of slabs (*c.*2 mm thick for reflection and *c.*100 μm thick for transmission) are mounted in a *goniometer* and rotated in their own planes. A specimen with random orientation will have an equal number of crystallites diffracting at all angular settings and give uniform recorded intensity. If there is **preferred orientation**, the intensity for a particular 2θ varies with the orientation of the sample with respect to the beam and these data constitute the X-ray pole figure, or **fabric** diagram.

SINGLE CRYSTAL METHODS. Numerous types of apparatus exist for the study of single crystal specimens. Such work may have several purposes, e.g. to orientate a crystal in order to obtain oriented specimens by cutting, to study **twinning**, to refine the values of structural parameters, etc. All the methods mount the crystal on a *goniometer*. The following are the most frequently encountered techniques.

1 *Laue back-reflection method.* This simple method directs the incident beam at the crystal via a collimating tube through the centre of a flat piece of photographic film. Back-diffracted radiation produces a pattern of spots on the exposed, developed film, which is interpreted with the aid of a net, and gives the orientation of the crystal.

2 *The Weissenberg camera.* The X-ray beam is incident normal to the axis of rotation of the crystal and *goniometer*. A cylindrical film is concentric with the rotation axis and moves back and forth along the axis in a linked motion as the crystal oscillates through a large rotation angle. Metal screens limit the diffracted beams recorded to those from the zero layer of the reciprocal lattice. Effectively the layer data are projected over the whole film so that no overlap of diffraction spots occurs. A different incident beam inclination is used for other lattice layers. Interpretation of photographs is facilitated with grids.

3 *Lang topographic method.* This is a diffraction method for the direct imaging of microstructural **defects** in sections of relatively perfect crystal. A high intensity X-ray source, collimating slits and double crystal spectrometer are used to produce a monochromatic beam through which the crystal is translated. The specimen is carefully oriented for a strong Bragg reflection and a film recording the transmitted radiation is moved in synchronism with the crystal. Intensity is transmitted only where the elastic **strain** fields of a **dislocation**, etc. (which distorts the lattice planes) produces deviation from the Bragg condition. The technique gives no direct magnification and has low resolution compared with **defect** imaging by transmission **electron microscopy**. [DJB]

Cullity, B.D. (1978) *Elements of X-ray Diffraction*. Addison Wesley, Reading, MA.

Klug, H.P. & Alexander, L.E. (1974) *X-ray Diffraction Procedures*. John Wiley & Sons, London.

Peiser, H.S., Rooksby, H.P. & Wilson. A.J.C. (1960) *X-ray Diffraction by Polycrystalline Materials*. Chapman & Hall, London.

X-ray fluorescence analysis *(XRF)* X-ray fluorescence analysis has been widely applied in archeology to characterize artefacts and their natural sources. Applications include pottery studies, **obsidian**, pigments used in, for example, wall-paintings, ancient glasses, and **igneous rock millstones**. In each case the aim of the analysis is to gain information about manufacturing techniques, and/or to determine the natural source of the material used in the artefact. [OWT]

xanthophyllite A synonym for the *brittle mica* clintonite (see **mica minerals**.) [DJV]

xenolith/xenocryst A foreign rock/crystal inclusion within an **igneous rock**. Xenoliths are distinguished from inclusions that might represent concentrations of early-formed crystals **precipitated** from the **magma**. Such gene- tically related inclusions are termed *autoliths* (see **pluton**). Xenoliths represent blocks of genetically unrelated ('foreign') **wall rock** that have been mechanically incor- porated into the **magma**, by the process of **stoping**. [RST]

Y

yardang Elongated landform resembling the up-turned hull of a boat sculptured by wind action from weakly consolidated rocks. The term was introduced to the literature by Sven Hedin in 1903 following an expedition though the Taklimakan Desert of China, where it is used locally to describe these features. Yardangs have widespread occurrence in desert regions with poor *soil* development and restricted **sand** supply, commonly in the form of clusters of parallel ridges. They vary greatly in scale, ranging from 1 to 200 m high and 10 to more than 1000 m long. [RDS]

Greeley, R. & Iversen, J.D. (1985) *Wind as a Geological Process.* Cambridge University Press, Cambridge.

McCauley, J.F., Grolier, M.J., & Breed, C.S. (1977) Yardangs of Peru and other desert regions. United States Geological Survey, Interagency Report. *Astrogeology* **81**: 1–177.

yazoo A tributary stream that runs parallel to the main river for some distance. [ASG]

yield depression curve A graph drawn by plotting the rate of extraction of water from a **well** against the **drawdown** corresponding to different pumping rates. This technique is used to assess the behaviour of a **well** because the greater the **drawdown**, the more energy that is required to overcome the **head** loss and bring water to the surface. Thus to keep pumping costs down the **drawdown** should be kept to a minimum. [AME]

Price, M. (1985) *Introducing Groundwater.* Allen & Unwin, London.

Young's modulus One of several measures of the elastic properties of a material. It is defined as the ratio of applied stress p to the resultant extension e along the axis of a plane-ended cylinder. Together with **Poisson's ratio** (the ratio of p to the lateral **contraction**) it can also be used to define the **elastic moduli** and hence P and S *wave* velocities. (See **elastic moduli**.) [RAC]

younging The property, in strata, of becoming stratigraphically younger in a particular direction, also called *way-up*. The younging direction is defined as the direction in which stratigraphically younger rocks are found. For undisturbed sedimentary rocks, the younging is upwards; overturned **beds** young downwards. This information is useful in folded sediments, where changes in younging direction indicate **folding**. In an **anticline**, **beds** young away from the *fold axial surface*, and in a **syncline**, towards the *axial surface*. Younging is usually determined from sedimentary structures such as **sole marks**, **cross-stratification** and **graded bedding**, known as *way-up criteria*. On a larger scale, a known stratigraphic succession indicates the local geographical younging direction. (See also **facing**, **fold**, **fold system**.) [SB]

ytterbium (Yb) At. no. 70, at. wt. 173.04. A lanthanide metal used in steel alloys. It is recovered commercially from the **monazite** concentrates resulting from the exploitation of **beach placers**. [AME]

yttrium (Y) At. no. 39, at. wt. 88.91. A Group III element. Y_2O_3 containing *europium* is used as a red phosphor in colour television tubes and yttrium **iron garnets** are used as microwave filters. It is also used in lasers and as a catalyst. It occurs with **ytterbium** in **monazite**. [AME]

Z

zeolite minerals Hydrated aluminosilicates with a framework structure enclosing cavities occupied by large ions and water molecules, both of which have considerable freedom of movement permitting ion exchange and reversible dehydration. Chemically, they are related to **feldspars** but have much more open structures; they may be transformed to **feldspars** when subjected to thermal metamorphism. They are generally colourless or white when pure, but many specimens are coloured due to the presence of finely divided **iron oxides** or other impurities. Their refractive indices are low (generally between 1.47 and 1.52) and they have a relatively low birefringence (0–0.015). Zeolites have long been recognized as occurring in **amygdales** and fissures in basic volcanic rocks where they often form well-developed crystals. More recently they have been recognized as being important and often abundant **authigenic** constituents in sedimentary rocks. They have been found as major constituents of volcanic **tuffs** in ancient saline lake deposits and in thick marine **tuffs**. In low grade metamorphic rocks they occur as a result of **hydrothermal alteration** and burial metamorphism. Although zeolitic **tuffs** have been used as building materials for at least 4000 years, it is only within the last 40 years that the zeolite contents of these materials have been recognized, and in the same period their industrial application as cation-exchangers, adsorbents, *molecular sieves*, dessicants, etc., has been increasingly realized.

There are some 45 naturally occurring zeolites with more than 25 different structures. The synthetic production of zeolites is largely aimed at tailoring their properties to specific industrial needs. **Laumontite, heulandite–***clinoptilolite* and *analcite* are the most abundant naturally occurring zeolites, although *analcite* is notable for its somewhat higher temperature paragenesis, being found as an apparent primary mineral of late formation in some basic **igneous rocks** as well as in zeolitic **vesicles** and as an **authigenic** mineral in **tuffs** and **sandstones**.

The structure of the minerals of the zeolite group is basically a three-dimensional framework, but fibrous and lamellar habits are known, as for example in **natrolite** and *clinoptilolite* respectively. Classifications based on the framework topology have been proposed by Breck (1974), Flanigen *et al.* (1971) and Meier (1988). Each structure is described in terms of fundamental building units of TO_4 tetrahedra, with a central ion (T) of Si^{4+} or Al^{3+}, secondary building units (SBU) consisting of single rings of 4, 5, 6, 8, 10 and 12 tetrahedra and double rings of 4, 6 and 8, and larger symmetrical polyhedra such as a truncated cuboctahedron, truncated octahedron, or the 11-hedron (or **cancrinite** unit). Breck's classification (Table Z1) divides zeolites into seven groups according to the type of SBU contained in the structure (Fig. Z1).

Thus Group 6 forms complex chains of six-tetrahedra which are linked to each other in various ways (Fig. Z2) giving channels with different sizes of aperture, whereas the 6-ring zeolites form structural cages, e.g. **chabazite** and *faujasite* and the synthetic zeolites X and Y which are in widespread use in the catalytic cracking of **petroleum**. The channels present in the structure are formed by various combinations of the linked rings of tetrahedra and vary in diameter from 2.2 to 7.4 Å; the wider the channels at their narrowest parts, the larger the cation that can be introduced into the structure. Those with 8- and 12-membered rings have channels large enough for the admission of

Table Z1 Breck's classification of **zeolites**

Group	Secondary building unit
1	Single 4-ring (S4R)
2	Single 6-ring (S6R)
3	Double 4-ring (D4R)
4	Double 6-ring (D6R)
5	Complex 4–1, T_5O_{10} unit
6	Complex 5–1, T_8O_{16} unit
7	Complex 4–4–1, $T_{10}O_{20}$ unit

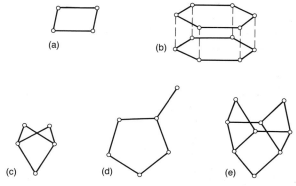

Fig. Z1 The secondary building units (SBU) corresponding with the structural classification of **zeolites**. (a) The single four-ring (S4R). (b) The six-ring, single or double (S6R and D6R). (c) The natrolite unit (4–1). (d) The mordenite unit (5–1). (e) The stilbite unit (4–4–1). Every corner or vertex (T position) is a four-connected nodal point in the three-dimensional network. Oxygen atoms occupy approximately the mid-points of the branches. (After Breck, 1974.)

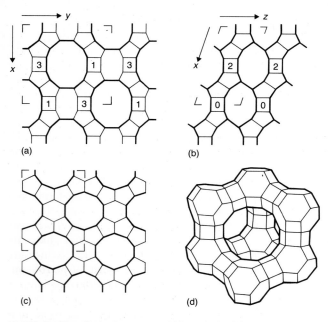

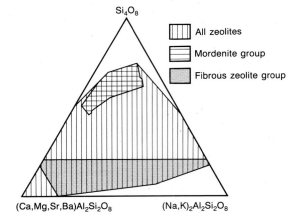

Fig. Z2 Zeolites. Skeletal structures of (a) mordenite, (b) dachiardite, (c) ferrierite, projected along the axes of the main channels. The numbers give the heights of the 4-membered rings as multiples of one quarter of the respective cell constant. (d) The cage-like unit of the structure of faujasite. The T atoms are at the intersections of the lines.

Fig. Z3 The triangular diagram $[Si_4O_8]-D[Al_2Si_2O_8]-M_2[Al_2Si_2O_8]$ showing the compositional area of all **zeolites**. (After Gottardi, 1978.)

organic molecules as well as cations; thus zeolites can act as ion or *molecular sieves*, each zeolite having its characteristic upper limit for the size of ion or molecule to which it is permeable. Channel width, however, is not the only criterion for **permeability**, as the presence of many cations may block the channels, and ionic or molecular diffusion is also affected by water content. Cation-exchange capacity in general diminishes with loss of water; cations are most mobile in zeolites with low cation content.

A structural classification of zeolites proposed by Gottardi & Galli (1985) is based on typical features of the framework structures rather than secondary building units, but is not applicable to a number of synthetic zeolites. In an entirely different approach, the various three-dimensional frameworks of zeolites can be placed in numerical order based on topological density values (Meier, 1988).

Chemically the zeolites form a well-defined group of hydrated silicates of **aluminium** and the alkalis or alkaline earths, and are characterized also by having the molecular ratio $(Ca, Sr, Ba, Na_2, K_2)O:Al_2O_3$ equal to unity and the ratio $(Al + Si):O$ always $1:2$. A general formula can be written $M_xD_y[Al_{x+2y}Si_zO_{2x+4y+2z}] . nH_2O$, where $M =$ Na, K or other monovalent cations and $D =$ Mg, Ca, Sr, Ba and other divalent cations (Fig. Z3). Thus the common natural zeolite **heulandite** can be written as $CaO . Al_2O_3 . 7SiO_2 . 6H_2O$ or as $Ca_4[Al_8Si_{28}O_{72}] . 24H_2O$ on a *unit cell* basis, the cations in the first group being the exchangeable ions and those in the square brackets representing the structural cations which with oxygen

make up the tetrahedral framework of the structure. The ratio $SiO_2:Al_2O_3$ is always $\geqslant 2:1$. The chemistry of many zeolites is rather variable and sometimes deviates considerably from the schematic formulae of Table Z2, which are at best only an approximation of what the composition of a particular zeolite may be.

The synthesis of *molecular sieves* not known in nature was first accomplished in 1948 and their widespread industrial use as adsorbents began in 1954. The production of synthetic zeolites tailored according to specific industrial needs is now possible and this has led also to the production of a new set of compounds with a very high **silica** content (Si:Al ratios ranging $10-100$) with *silicalite* as the pure **silica** end-member (Flanigen *et al.*, 1978). In the absence of tetrahedral Al, *silicalite* does not have ion-exchange properties and has a strong hydrophobic character; it cannot be considered strictly as a zeolite but certainly qualifies to be called a *molecular sieve*. In many syntheses large organic cations have been used in the production of structures with 'cages' occupied by them; e.g. *tetrapropylammonium* (TPA) and *tetramethylammonium* (TMA) ions have been used to produce large-pore 'zeolites' such as N-A, ZK-4, ZSM-5, etc., and in the production of *silicalite* a precursor phase with composition $4(TPA)OH . 96SiO_2$ was used.

The identification of individual zeolite species is often difficult due to the similarity of their optical and other properties. Recourse may be had to their **X-ray diffraction** patterns, *differential thermal analysis* behaviour, and infrared spectra.

The occurrence and paragenesis of natural zeolites have been extensively studied in the last two decades. Being hydrous aluminosilicates, they are relatively sensitive to changes in temperature and pressure. The lowest temperatures at which zeolites form (4°C) are in deep-sea sediments, where **phillipsite** crystallizes at the sediment–water interface and continues to grow in the sediment. **Phillipsite** is commonest in **Miocene** and younger sedi-

Table Z2 Approximate structural formulae of the more common natural and synthetic **zeolites**

Name	Typical unit cell contents	Space group
Group 1 (S4R)		
Analcite	$Na_{16}[Al_{16}Si_{32}O_{96}] . 16H_2O$	Ia_3d
Wairakite	$Ca_8[Al_{16}Si_{32}O_{96}] . 16H_2O$	Pc
Phillipsite	$(Ca_{0.5},Na,K)_6[Al_6Si_{10}O_{32}] . 12H_2O$	$P2_1/m$
Harmotome	$Ba_2[Al_4Si_{12}O_{32}] . 12H_2O$	$P2_1/m$
Gismondine	$Ca_4[Al_8Si_8O_{32}] . 16H_2O$	$P2/c$
Laumontite	$Ca_4[Al_8Si_{16}O_{48}] . 16H_2O$	Am
Yugawaralite	$Ca_2[Al_4Si_{12}O_{32}] . 8H_2O$	Pc
Group 2 (S6R)		
Erionite	$K_2NaCa_{1.5}Mg[Al_9Si_{27}O_{72}] . 28H_2O$	$P6_3/mmc$
Levyne	$(Ca,Na_2,K_2)_3[Al_6Si_{12}O_{36}] . 18H_2O$	$R\bar{3}m$
Group 3 (D4R)		
A	$Na_{12}[Al_{12}Si_{12}O_{48}] . 27H_2O$	$Pm\bar{3}m$
Group 4 (D6R)		
Faujasite	$(Na_2,Ca,Mg)_{32}[Al_{64}Si_{128}O_{384}] . 256H_2O$	Fd_3m
X	$Na_{86}[Al_{86}Si_{106}O_{384}] . 264H_2O$	Fd_3m
Y	$Na_{56}[Al_{56}Si_{136}O_{384}] . 250H_2O$	Fd_3m
Chabazite	$Ca_2[Al_4Si_8O_{24}] . 12H_2O$	$R\bar{3}m$
Gmelinite	$(Na_2Ca)_4[Al_8Si_{16}O_{48}] . 24H_2O$	$P6_3/mmc$
Group 5 (T$_5$O$_{10}$)		
Natrolite	$Na_{16}[Al_{16}Si_{24}O_{80}] . 16H_2O$	$Fdd2$
Scolecite	$Ca_8[Al_{16}Si_{24}O_{80}] . 24H_2O$	Cc
Mesolite	$Na_{16}Ca_{16}[Al_{48}Si_{72}O_{240}] . 64H_2O$	$Fdd2$
Thomsonite	$Na_4Ca_8[Al_{20}Si_{20}O_{80}] . 24H_2O$	$Pnna$
Group 6 (T$_8$O$_{16}$)		
Mordenite	$Na_3KCa_2[Al_8Si_{40}O_{96}] . 28H_2O$	$Cmcm$
Dachiardite	$(Na,K,Ca_{0.5})_5[Al_5Si_{19}O_{48}] . 12H_2O$	C_2/m
Ferrierite	$(Na,K)_{1.5}Mg_2[Al_{5.5}Si_{30.5}O_{72}] . 18H_2O$	$Immm$
Epistilbite	$Ca_3[Al_6Si_{18}O_{48}] . 16H_2O$	$C2/m$
Group 7 (T$_{10}$O$_{20}$)		
Stilbite	$Na_2Ca_4[Al_{10}Si_{26}O_{72}] . 28H_2O$	$C2/m$
Heulandite	$(Ca,Na_2)_4[Al_8Si_{28}O_{72}] . 24H_2O$	Cm
Clinoptilolite	$(Na,K,Ca_{0.5})_6[Al_6Si_{30}O_{72}] . 24H_2O$	Cm

zonation of zeolites as **meteoric water** moves through the system: thus **clay minerals** (commonly **smectites**) are formed by *hydrolysis* of vitric **ash**, altering to zeolites such as *clinoptilolite* at depth and then *analcite* at still greater depth.

In areas of **hydrothermal alteration**, **zeolitization** may also demonstrate a zonal pattern, with *clinoptilolite* or *mordenite* in the shallowest and coolest zones while progressively deeper zones may contain *analcite*, **heulandite**, **laumontite** and *wairakite*. Various zeolites were recorded in the masonry of Roman hot water baths (Daubrée, 1879) and **laumontite** is found in radiators fed by natural *hot spring* water in Kamchatka.

The regional occurrence of zeolites in low-grade metamorphic rocks was first recognized by Coombs (1954) in southern New Zealand. A zeolite facies was later defined as 'that set of mineral assemblages that is characterized by the association calcium zeolite–**chlorite**–**quartz** in rocks of favourable bulk composition'. Such an **alteration** sequence typically involving the breakdown of calcic **plagioclase** to **laumontite** or **heulandite** is due chiefly to the increasing temperature associated with progressive burial. Such zeolite occurrences are relatively common in the circum-Pacific area such as Japan and the western coast of North America. Zeolites in any type of occurrence change ultimately to thermodynamically stable **alkali feldspar** or other non-zeolite minerals, through intermediate zeolites such as *analcite* and **laumontite** due to the increase in pressure and temperature, to the increase in concentration of **pore water**, or to ageing. The time taken for the *analcite*–**albite** transformation during **burial diagenesis** has been estimated to be 400 000 years. In areas where intercalated **lavas** and intrusive rocks are present, **zeolitization** may have been influenced by emplacement of the igneous intrusives to some distance from the contact, e.g. in Mull (McLintock, 1915).

The best-known occurrences of zeolites are probably those which form well-developed crystals in the **vesicles** and **amygdales** of **basalts**. These zeolites are generally considered to be the result of late-stage fluids that permeated the **basalts** after their extrusion. Typically such **amygdales** contain more than one species of zeolite, e.g. **chabazite** and *analcite*, often associated also with **calcite**. Certain zeolite assemblages are often observed to be typical of individual **basalt** flows, though in the *Tertiary* lavas of Northern Ireland and of eastern Iceland it has been reported that the zeolite zones cut discordantly across the lava stratigraphy. Other famous occurrences of zeolites in **basaltic** rocks include those from the Deccan Traps of India, the Watchung **basalt** of New Jersey and the Columbia River **basalts** of the western USA.

COMMERCIAL APPLICATIONS. A combination of ion-exchange, catalytic and *molecular sieve* properties has led to the extensive use of zeolites, both natural and synthetic, in a great range of applications. Such industrial uses include their use as fillers in high-brightness paper, the

ments whereas *clinoptilolite* is abundant in **Cretaceous** and early *Tertiary* sediments, though it has been reported that **phillipsite** is derived from **basaltic** glass and *clinoptilolite* from acidic glass. However, deep-sea zeolites may be found in the **fabric** of **radiolarian** tests, suggesting a non-volcanic origin, the **silica** perhaps being supplied from biogenic **opal**. At the land surface, zeolites are formed readily from suitable materials (volcanic glass, **tuff, nepheline, feldspar**, etc.) in conditions of high **pH** resulting from the concentration of sodium carbonate–bicarbonate by evaporation in an arid or semi-arid environment. In saline, alkali lakes zeolites are common and widespread; in the western USA *clinoptilolite*, **phillipsite** and *erionite* are found replacing **rhyolitic tuffs** in **Plio-Pleistocene** lake deposits. Reactions are relatively rapid, with vitric **tuff** altering to zeolites in about 1000 years. In water of moderately high salinity these zeolites can alter to *analcite*; in highly saline water, all zeolites can alter to K-**feldspar**. In open hydrological systems, **tephra** may show a vertical upgrading of natural gas with low heat values, the removal of *radioactive* isotopes from nuclear wastes, their addition as dietary supplements for pigs and poultry, their use in purification of **petroleum** products, as carriers of pesticides, as traps for heavy metals in the food chain, in reducing various offensive odours from decaying organic material and even as a polishing agent in fluoride-containing toothpaste (Mumpton, 1977). *Erionite*, a fibrous zeolite which occurs in altered volcanic glass in the Cappadocia area of Turkey, has been implicated in the endemic occurrence of malignant mesotheliomas among residents of the area. [RAH]

Breck, D.W. (1974) *Zeolite Molecular Sieves*. Wiley–Interscience, New York.
Coombs, D.S. (1954) The nature and alteration of some Triassic sediments from Southland, New Zealand. *Transactions of the Royal Society of New Zealand* **82**: 65–109.
Daubrée, A. (1879) *Etudes Synthetiques de Géologie Experimentale*. Paris.
Dyer, A. (1988) *An Introduction to Zeolite Molecular Sieves*. John Wiley & Sons, Chichester.
Flanigen, E.M., Khatemi, H. & Szymanski, H.A. (1971) Infrared structural studies of zeolite frameworks. *Advances in Chemistry Series* **101**: 201–29.
Flanigen, E.M., Bennett, J.M., Grose, R.W. *et al.* (1978) Silicalite, a new hydrophobic crystalline silica molecular sieve. *Nature* **271**: 512–16.
Gottardi, G. (1978) Mineralogy and crystal chemistry of zeolites. In: Sand, L.B. & Mumpton, F.H. (eds) *Natural Zeolites: Occurrence, Properties, Use*, pp. 31–44. Pergamon, Oxford.
Gottardi, G. & Galli, E. (1985) *Natural Zeolites*. Springer-Verlag, Berlin.
McLintock, W.F.P. (1915) On the zeolites and associated minerals from the Tertiary lavas around Ben More, Mull. *Transactions of the Royal Society of Edinburgh* **51**: 1–32.
Meier, W.M. (1988) The structural classification of zeolites. In: Kalló, D. & Sherry, H.S. (eds) *Occurrence, Properties and Utilization of Natural Zeolites*, pp. 217–21. Akadémiai Kiadó, Budapest.
Mumpton, F.A. (ed.) (1977) *Mineralogy and Geology of Natural Zeolites*. Short Course Notes, Vol. 4. Mineralogical Society of America, Washington, DC.
Sand, L.B. & Mumpton, F.A. (1978) *Natural Zeolites: Occurrence, Properties, Use*. Pergamon, Oxford.

zeolitization A form of **wall rock alteration** marked by the development of **stilbite**, **natrolite**, **heulandite** and other **zeolites**. This **alteration** often accompanies native **copper** mineralization in **amygdaloidal basalts** and **calcite**, **prehnite**, **apophyllite** and **datolite** are generally also present. [AME]

zeuge A tabular mass of rock perched on a pinnacle of softer rock as a result of differential **erosion**, possibly by wind, of the underlying materials. [ASG]

zibar Coarse-grained, slipfaceless, low relief, aeolian depositional feature. Zibars have regular spacings of up to 400 m and a maximum relief of less than 10 m. They form undulatory surfaces on **sand sheets** and **interdune** corridors (Nielson & Kocurek, 1986). [ASG]

Nielson, J. & Kocurek, G. (1986) Climbing zibars of the Algodones. *Sedimentary Geology* **48**: 1–15.

zinc (Zn) At. no. 30, at. wt. 65.38. A transition element. Used in alloys, e.g. brass, solder; for die-casting and in galvanizing steel. ZnO is used in paints, plastics, textiles, etc., and as an inert filler in electrical components. ZnS is used in cathode-ray tubes (TV, X-ray), fluorescent lights and luminous paint. The principal **ore** mineral is **sphalerite** (*blende*) which occurs in many deposit types, e.g. **carbonate-hosted base metal deposits**, **massive sulphide deposits**. [AME]

zincblende An alternative name for **sphalerite**. [DJV]

zincite (ZnO) An **ore** mineral of **zinc**. [DJV]

zircon ($ZrSiO_4$) A *nesosilicate* mineral that is a common accessory phase in all types of **igneous rocks**. When transparent, it is valued as a **gemstone**. [DJV]

zirconium (Zr) At. no. 40, at. wt. 91.22. A transition element of Group IV. It is a metal that is steel-like in appearance, very resistant to corrosion and suitable for making laboratory ware. It has low neutron absorption and alloys are used in nuclear reactor construction. Nb–Zr alloys are used in superconducting magnets. The principal **ore** mineral is **zircon**, which is won from **beach placers**, and *baddeleyite* which is a by-product in the working of some **carbonatite** deposits. [AME]

Zoeppritz' equations A lengthy suite of formulae describing how **seismic wave** energy is partitioned between **reflected**, transmitted, and 'mode-converted' waves when a *P* or *S wave* is incident upon a planar discontinuity between two solid rock units. For example, when a *P wave* meets a boundary at some oblique angle of incidence, it will not only be transmitted through the boundary but will also generate a **reflected** *P wave*. Some of its energy will be converted to *S waves*. So there will exist both transmitted and **reflected** *S waves* also. Zoeppritz' equations give the transmitted and **reflected** *P* and *S wave amplitudes* as a fraction of the incident wave amplitude. The closely-related *Knott's equations* define how the wave energy is partitioned. In both, the result depends on the *P wave* and *S wave* velocities and densities in the rocks on each side of the boundary, the incident wave type (P or S, and the polarization of S) and its angle of incidence relative to the boundary. At near-vertical incidence, most energy is transmitted; **reflected** *P* amplitudes are only some 5–10% of the incident value. At angles shallower than the **critical angles** for *P* and *S* critical refraction, all energy is reflected.

The angles of reflection and refraction depend on velocity ratios, as expressed by *Snell's law*. *Knott's equations* were completed in 1899, Zoeppritz' in 1907. Some special cases (e.g. waves incident from below and reflected at the Earth's surface) can be described more simply. (See **attenuation, seismic discontinuity.**) [RAC]

Aki, K. & Richards, P.G. (1980) *Quantitative Seismology: Theory and Methods* (2 Vols). W.H. Freeman, San Francisco.
Young, G.B. & Braile, L.W. (1976) A computer program for the application of Zoeppritz' Amplitude Equations and Knott's Energy Equations. *Bulletin of the Seismological Society of America* 66: 1881–5.

zoisite ($Ca_2Al_3O(SiO_4)(Si_2O_7OH)$) A *sorosilicate* mineral found in metamorphic rocks, an orthorhombic **polymorph** of the more common **clinozoisite**. [DJV]

zone of fluctuation The zone lying between the highest levels reached by the **water table** in an area and the lowest. In northern Europe the highest levels occur in Spring and the lowest in the Autumn. [AME]

zone refining A process for **magma diversification** during *partial melting* in which a succession of 'zones' of molten or partially molten **magmas** advance through a solid by causing melting at the front of each zone followed by crystallization and deposition behind. This process would strongly affect the distribution of incompatible elements (with low *partition coefficients*) since these will be preferentially partitioned into the melt and carried forward by each advancing zone. Zone refining may be important in **magma** genesis but may be difficult to distinguish from other processes of **magma diversification**. [RST]

zoning of ore deposits Changes in the mineralogy of **ore**, **industrial** or **gangue** minerals, or of their chemistry from place to place in a mineral deposit or a mineralized district. Zoning was first described from **epigenetic vein deposits** but it is also present in other types of deposit. For example, **syngenetic deposits** may show zoning parallel to a former shoreline as is the case with the *iron ores* of the Mesabi Range, Minnesota; **alluvial deposits** may show zoning along the course of a river leading from the source area; some **exhalative syngenetic** sulphide **deposits** show a marked zonation of their metals and *pyrometasomatic* (**skarn**) deposits often show a zoning running parallel to the igneous–sedimentary contact. In this discussion, attention will be focused on the zoning of **epigenetic hydrothermal deposits**.

Epigenetic hydrothermal zoning can be divided into three intergradational classes; regional, district and **orebody** zoning. Regional zoning occurs on a very large scale, often corresponding to large sections of *orogenic belts* and their **foreland** (Fig. M8, p. 391). Zoning on this scale occurs in the circum-Pacific *orogenic belts*. Some of this regional zoning, e.g. in the Andes, appears to be related to the depth of the underlying **Benioff Zone** which suggests a deep level origin for the metals as well as the associated **magmas**. District zoning is the zoning seen in individual orefields such as Cornwall, England (Fig. Z4).

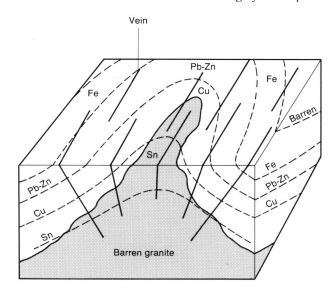

Fig. Z4 Zoning of ore deposits. Diagrammatic illustration of district zoning in the orefield of south-west England showing the relationship of the zonal boundaries to the granite–metasediment contact.

Zoning of this type is most clearly displayed in **hypothermal deposits** where the mineralization is of considerable vertical extent and was formed at depth where changes in the pressure and temperature gradients were very gradual. If deposition took place near to the surface, then steep temperature gradients may have caused superimposition of what would, at deeper levels, be distinct zones, thus giving rise to the effect known as *telescoping*. Some **geothermal systems** also show metal zoning, e.g. Broadlands, New Zealand where there is **antimony, gold** and thallium enrichment near the surface and **lead, zinc, silver, copper**, etc., at depth. **Orebody** zoning takes the form of changes in the mineralization within a single **orebody**. A good example occurs in the Emperor Gold Mine, Fiji, where vertical zoning of **gold–silver** tellurides in one of the main *ore shoots* gives rise to an increase in the Ag:Au ratio with depth.

There are two principal schools of thought concerning the transportation of metal ions in **hydrothermal solutions**. The first favours transportation as bisulphide complexes, and the second as complex chloride ions in chloride-rich, sulphide-poor **brines**. The first hypothesis can account for the zoning seen in **epigenetic** and **syngenetic hydrothermal deposits**. Clearly, the relative stabilities of com-

Table Z3 Zoning of ore deposits. Predicted sequence of stabilities of bisulphide complexes (in kilocalories)

			Least soluble . . . . Most soluble				
Fe	Ni	Sn	Zn	Cu	Pb	Ag	Hg
79	84	126	132	135	153	157	226

plex metal bisulphide ions will control their relative times of **precipitation** and hence both the resulting **paragenetic sequence** and any zoning which may be developed. These stabilities are shown in Table Z3. The data in this table suggest that **iron** and **tin** would be **precipitated** early in the **paragenetic sequence** and would be present in the lowest zone of a zoned deposit, whilst **silver** and **mercury** would be late **precipitates** which would travel furthest from the source of the mineralizing solutions. This is exactly what is found in Cornwall and other zoned orefields. [AME]

Barnes, H.L. (1975) Zoning of ore deposits: types and causes. *Transactions of the Royal Society, Edinburgh* **69**: 295–311.

Index

Note: Keywords appear in SMALL CAPITALS.
The principal page reference for any particular entry appears in **bold** type.
An asterisk (*) denotes a commonly occurring word;
in these cases only the principal page reference is given.